**$62.50**

# Craftsman 2011

**59th Edition**

# NATIONAL CONSTRUCTION ESTIMATOR

## Edited by Dave Ogershok and Richard Pray

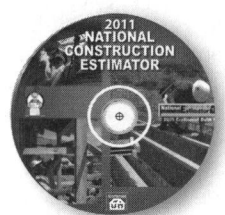

Craftsman

VMS, Inc.
ALLEGAN MI 49010
www.VMS-online.com
1-800-343-6430

6058 Cor[...]d, CA 92018

# Contents
## A complete index begins on page 646

Copyright 2010 Craftsman Book Company ISBN 978-1-57218-242-4 1st printing November 2010 for the year 2011

# This Book Is an Encyclopedia of 2011 Building Costs

**The 2011 National Construction Estimator** lists estimated construction costs to general contractors performing the work with their own crews, as of mid-2011. Overhead & profit are not included.

**This Manual Has Two Parts**; the Residential Construction Division begins on page 17. Use the figures in this division when estimating the cost of homes and apartments with a wood, steel or masonry frame. The Industrial and Commercial Division begins on page 309 and can be used to estimate costs for nearly all construction not covered by the Residential Division.

The Residential Construction Division is arranged in alphabetical order by construction trade and type of material. The Industrial and Commercial Division follows MasterFormat™ 2004. A complete index begins on page 646.

**National Estimator** (inside the back cover) is an electronic version of this book. To watch a short video guide to National Estimator, click "Help" on the menu bar. Then click "Launch The Tutorial."

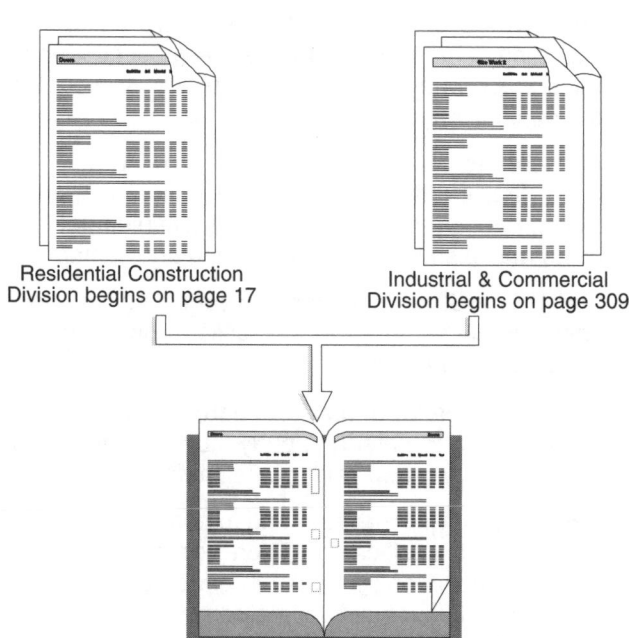

Residential Construction Division begins on page 17

Industrial & Commercial Division begins on page 309

2011 National Construction Estimator

2011 National Estimator is an electronic version of this book

## Material Costs

**Material Costs** for each item are listed in the column headed "Material." These are neither retail nor wholesale prices. They are estimates of what most contractors who buy in moderate volume will pay suppliers as of mid-2011. Discounts may be available for purchases in larger volume.

**Quarterly prices updates on the Web are free** and automatic all during 2011. You'll be prompted when it's time to collect the next update. A connection to the Web is required.

**Add Delivery Expense** to the material cost for other than local delivery of reasonably large quantities. Cost of delivery varies with the distance from source of supply, method of transportation, and quantity to be delivered. But most material dealers absorb the delivery cost on local delivery (5 to 15 miles) of larger quantities to good customers. Add the expense of job site delivery when it is a significant part of the material cost.

**Add Sales Tax** when sales tax will be charged to the contractor buying the materials.

**Waste and Coverage** loss is included in the installed material cost. The cost of many materials per unit after installation is greater than the purchase price for the same unit because of waste, shrinkage or coverage loss during installation. For example, about 120 square feet of nominal 1" x 4" square edge boards will be needed to cover 100 square feet of floor or wall. There is no coverage loss with plywood sheathing, but waste due to cutting and fitting will average about 6%.

Costs in the "Material" column of this book assume normal waste and coverage loss. Small and irregular jobs may require a greater waste allowance. Materials priced without installation (with no labor cost) do not include an allowance for waste and coverage except as noted.

## Labor Costs

**Labor Costs** for installing the material or doing the work described are listed in the column headed "Labor." The labor cost per unit is the labor cost per hour multiplied by the manhours per unit shown after the @ sign in the "Craft@Hours" column. Labor cost includes the basic wage, the employer's contribution to welfare, pension, vacation and apprentice funds and all tax and insurance charges based on wages. Hourly labor costs for the various crafts are listed on page 10 (for the Residential Division) and page 308 (for the Industrial and Commercial Division).

Hourly labor costs used in the Industrial and Commercial Division are higher than those used in the

Residential Division, reflecting the fact that craftsmen on industrial and commercial jobs are often paid more than craftsmen on residential jobs.

**Supervision Expense** to the general contractor is not included in the labor cost. The cost of supervision and non-productive labor varies widely from job to job. Calculate the cost of supervision and non-productive labor and add this to the estimate.

**Payroll Taxes and Insurance** included in the labor cost are itemized in the sections beginning on pages 181 and 284.

**Manhours per Unit** and the Craft performing the work are listed in the "Craft@Hrs" column. Pages 7 through 9 explain the "Craft@Hrs" column. To find the units of work done per man in an 8-hour day, divide 8 by the manhours per unit. To find the units done by a crew in an 8-hour day, multiply the units per man per 8-hour day by the number of crew members.

**Manhours Include** all productive labor normally associated with installing the materials described. This will usually include tasks such as:

■ Unloading and storing construction materials, tools and equipment on site.

■ Moving tools and equipment from a storage area or truck on site at the beginning of the day.

■ Returning tools and equipment to a storage area or truck on site at the end of the day.

■ Normal time lost for work breaks.

■ Planning and discussing the work to be performed.

■ Normal handling, measuring, cutting and fitting.

■ Keeping a record of the time spent and work done.

■ Regular cleanup of construction debris.

■ Infrequent correction or repairs required because of faulty installation.

**Adjust the Labor Cost** to the job you are figuring when your actual hourly labor cost is known or can be estimated. The labor costs listed on pages 10 and 308 will apply within a few percent on many jobs. But labor costs may be much higher or much lower on the job you are estimating.

If the hourly wage rates listed on page 10 or page 308 are not accurate, divide your known or estimated cost per hour by the listed cost per hour. The result is your adjustment for any figure in the "Labor" column for that craft. See page 11 for more information on adjusting labor costs.

**Adjust for Unusual Labor Productivity.** Costs in the labor column are for normal conditions: experienced craftsmen working on reasonably well planned and managed new construction with fair to good productivity. Labor estimates assume that materials are standard grade, appropriate tools are on hand, work done by other crafts is adequate, layout and installation are relatively uncomplicated, and working conditions don't slow progress.

Working conditions at the job site have a major effect on labor cost. Estimating experience and careful analysis can help you predict the effect of most changes in working conditions. Obviously, no single adjustment will apply on all jobs. But the adjustments that follow should help you produce more accurate labor estimates. More than one condition may apply on a job.

■ Add 10% to 15% when working temperatures are below 40 degrees or above 95 degrees.

■ Add 15% to 25% for work on a ladder or a scaffold, in a crawl space, in a congested area or remote from the material storage point.

■ Deduct 10% when the work is in a large open area with excellent access and good light.

■ Add 1% for each 10 feet that materials must be lifted above ground level.

■ Add 5% to 50% for tradesmen with below average skills. Deduct 5% to 25% for highly motivated, highly skilled tradesmen.

■ Deduct 10% to 20% when an identical task is repeated many times for several days at the same site.

■ Add 30% to 50% on small jobs where fitting and matching of materials is required, adjacent surfaces have to be protected and the job site is occupied during construction.

■ Add 25% to 50% for work done following a major flood, fire, earthquake, hurricane or tornado while skilled tradesmen are not readily available. Material costs may also be higher after a major disaster.

■ Add 10% to 35% for demanding specs, rigid inspections, unreliable suppliers, a difficult owner or an inexperienced architect.

**Use an Area Modification Factor** from pages 12 through 15 if your material, hourly labor or equipment costs are unknown and can't be estimated.

Here's how: Use the labor and material costs in this manual without modification. Then add or deduct the percentage shown on pages 12 through 15 to estimated costs to find your local estimated cost.

## Equipment Costs

**Equipment Costs** for major equipment (such as cranes and tractors) are listed in the column headed "Equipment." Costs for small tools and expendable supplies (such as saws and tape) are usually considered overhead expense and do not appear in the Equipment cost column.

Equipment costs are based on rental rates listed in the section beginning on page 319 and assume that the equipment can be used productively for an entire 8-hour day. Add the cost of moving equipment on and off the site. Allow for unproductive time when equipment can't be used for the full rental period. For example, the equipment costs per unit of work completed will be higher when a tractor is used for 4 hours during a day and sits idle for the remaining 4 hours. Generally, an 8-hour day is the minimum rental period for most heavy equipment. Many sections describe the equipment being used, the cost per hour and a suggested minimum job charge.

## Subcontracted Work

**Subcontractors** do most of the work on construction projects. That's because specialty contractors can often get the work done at competitive cost, even after adding overhead and profit.

Many sections of this book cover work usually done by subcontractors. If you see the word "subcontract" in a section description, assume that costs are based on quotes by subcontractors and include typical subcontractor markup (about 30% on labor and 15% on material). Usually no material or labor costs will appear in these sections. The only costs shown will be in the "Total" column and will include all material, labor and equipment expense.

If you don't see the word "subcontract" in a section description, assume that costs are based on work done by a general contractor's crew. No markup is included in these costs. If the work is done by a subcontractor, the specialty contractor may be able to perform the work for the cost shown, even after adding overhead and profit.

## Markup

**The General Contractor's Markup** is not included in any costs in this book. On page 207 we suggest a 20% markup on the contract price for general contractors handling residential construction. Apply this markup or some figure you select to all costs, including both subcontract items and work done by your own crews.

To realize a gross profit of 20% on the contract price, you'll have to mark up costs by 25%. See page 207 for an example of how markup is calculated. Markup includes overhead and profit and may be the most difficult item to estimate.

## Keep In Mind

**Labor and Material Costs Change.** These costs were compiled in the fall of 2010 and projected to mid-2011 by adding 3% to 6%. This estimate will be accurate for some materials but inaccurate for others. No one can predict material price changes accurately.

**How Accurate Are These Figures?** As accurate as possible considering that the estimators who wrote this book don't know your subcontractors or material suppliers, haven't seen the plans or specifications, don't know what building code applies or where the job is, had to project material costs at least 6 months into the future, and had no record of how much work the crew that will be assigned to the job can handle.

You wouldn't bid a job under those conditions. And we don't claim that all construction is done at these prices.

**Estimating Is an Art**, not a science. On many jobs the range between high and low bid will be 20% or more. There's room for legitimate disagreement on what the correct costs are, even when complete plans and specifications are available, the date and site are established, and labor and material costs are identical for all bidders.

No cost fits all jobs. Good estimates are custom made for a particular project and a single contractor through judgment, analysis and experience.

This book is not a substitute for judgment, analysis and sound estimating practice. It's an aid in developing an informed opinion of cost. If you're using this book as your sole cost authority for contract bids, you're reading more into these pages than the editors intend.

**Use These Figures** to compile preliminary estimates, to check your costs and subcontract bids and when no actual costs are available. This book will reduce the chance of error or omission on bid estimates, speed "ball park" estimates, and be a good guide when there's no time to get a quote.

**Where Do We Get These Figures?** From the same sources all professional estimators use: contractors and subcontractors, architectural and engineering firms, material suppliers, material price services, analysis of plans, specifications, estimates and completed project costs, and both published and unpublished cost studies. In addition, we conduct nationwide mail and phone surveys and have the use of several major national estimating databases.

 **We'll Answer Your Questions** about any part of this book and explain how to apply these costs. Free telephone assistance is available from 8 a.m. until 5 p.m. California time Monday through Friday except holidays. Phone 760-438-7828 x 2. We don't accept collect calls and won't estimate the job for you. But if you need clarification on something in this manual, we can help.

# Abbreviations

| | | | | | | |
|---|---|---|---|---|---|
| **AASHO** | American Assn. of State Highway Officials | **FAA** | Federal Aviation Administration | **OC** | spacing from center to center |
| **ABS** | acrylonitrile butadiene styrene | **FICA** | Federal Insurance Contributions Act (Social Security, Medicare tax) | **OD** | outside diameter |
| **AC** | alternating current | | | **OS & Y** | outside screw & yoke |
| **AISC** | American Institute of Steel Construction Inc. | | | **oz** | ounce |
| **APP** | attactic polypropylene | **FOB** | freight on board | **perf** | perforated |
| **ASHRAE** | American Society of Heating, Refrigerating and Air Conditioning Engineers | **FPM** | feet per minute | **Pr** | pair |
| | | **FRP** | fiberglass reinforced plastic | **PSF** | pounds per square foot |
| | | **FS** | Federal Specification | **PSI** | pounds per square inch |
| | | **ft-lbs** | foot pounds | **PV** | photovoltaic |
| **ASME** | American Society of Mechanical Engineers | **FUTA** | Federal Unemployment Compensation Act Tax | **PVC** | polyvinyl chloride |
| **ASTM** | American Society for Testing Materials | **Gal** | gallon | **Qt** | quart |
| | | **GFCI** | ground fault circuit interruptor | **R** | thermal resistance |
| **AWPA** | American Wood Products Association | | | **R/L** | random length(s) |
| **AWWA** | American Water Works Association | **GPH** | gallon(s) per hour | **R/W/L** | random widths and lengths |
| | | **GPM** | gallon(s) per minute | | |
| **Ba** | bay | **H** | height | **RPM** | revolutions per minute |
| **Bdle** | bundle | **HP** | horsepower | **RSC** | rigid steel conduit |
| **BF** | board foot | **Hr(s)** | hour(s) | **S1S2E** | surfaced 1 side, 2 edges |
| **BHP** | boiler horsepower | **IMC** | intermediate metal conduit | **S2S** | surfaced 2 sides |
| **Btr** | better | **ID** | Inside diameter | **S4S** | surfaced 4 sides |
| **Btu** | British thermal unit | **KD** | kiln dried or knocked down | **Sa** | sack |
| **B & W** | black & white | | | **SBS** | styrene butyl styrene |
| **C** | thermal conductance | **KSI** | kips per square inch | **SDR** | size to diameter ratio |
| **C** | one hundred | **KV** | kilovolt(s) | **SF** | square foot |
| **CF** | cubic foot | **KVA** | 1,000 volt amps | **SFCA** | square feet of form in contact with concrete |
| **CFM** | cubic feet per minute | **kw** | kilowatt(s) | | |
| **CLF** | 100 linear feet | **kwh** | kilowatt hour | **Sq** | 100 square feet |
| **cm** | centimeter | **L** | length | **SSB** | single strength B quality glass |
| **CPE** | chlorinated polyethylene | **Lb(s)** | pound(s) | | |
| **CPM** | cycles per minute | **LF** | linear foot | **STC** | sound transmission class |
| **CPVC** | chlorinated polyvinyl chloride | **LP** | liquified propane | **Std** | standard |
| | | **LS** | lump sum | **SY** | square yard |
| **CSPE** | chloro sulphinated polyethylene | **M** | one thousand | **T** | thick |
| | | **Mb** | million bytes (characters) | **T&G** | tongue & groove edge |
| **CSF** | 100 square feet | **MBF** | 1,000 board feet | **TV** | television |
| **CSY** | 100 square yards | **MBtu** | 1,000 British thermal units | **UBC** | Uniform Building Code |
| **CY** | cubic yard | **MCM** | 1,000 circular mils | **UL** | Underwriter's Laboratory |
| **d** | penny | **MDO** | medium density overlaid | **USDA** | United States Dept. of Agriculture |
| **D** | depth | **MH** | manhour | | |
| **DC** | direct current | **Mi** | mile | | |
| **dia** | diameter | **MLF** | 1,000 linear feet | **VLF** | vertical linear foot |
| **DSB** | double strength B quality glass | **MPH** | miles per hour | **W** | width |
| | | **mm** | millimeter(s) | **Wk** | week |
| **DWV** | drain, waste, vent piping | **Mo** | month | **W/** | with |
| **Ea** | each | **MSF** | 1,000 square feet | **x** | by or times |
| **EMT** | electric metallic tube | **NEMA** | National Electrical Manufacturer's Association | | |
| **EPDM** | ethylene propylene diene monomer | | | | |
| **equip.** | equipment | **NFPA** | National Fire Protection Association | | |
| **exp.** | exposure | **No.** | number | | |
| **F** | Fahrenheit | **NRC** | noise reduction coefficient | | |

## Symbols

| | |
|---|---|
| **/** | per |
| **—** | through or to |
| **@** | at |
| **%** | per 100 or percent |
| **$** | U.S. dollars |
| **'** | feet |
| **"** | inches |
| **#** | pound or number |

# Craft Codes, Hourly Costs and Crew Compositions

Both the Residential Division and Commercial and Industrial Division of this book include a column titled *Craft@Hrs*. Letters and numbers in this column show our estimates of:

■ Who will do the work (the craft code)

■ An @ symbol which means at

■ How long the work will take (manhours).

For example, on page 51 you'll find estimates for installing BC plywood wall sheathing by the square foot. The *Craft@Hrs* column opposite ½" plywood wall sheathing shows:

B1@.016

That means we estimate the installation rate for crew B1 at .016 manhours per square foot. That's the same as 16 manhours per 1,000 square feet.

The table that follows defines each of the craft codes used in this book. Notice that crew B1 is composed of two craftsmen: one laborer and one carpenter.

To install 1,000 square feet of ½" BC wall sheathing at .016 manhours per square foot, that crew would need 16 manhours (one 8-hour day for a crew of two).

Notice also in the table below that the cost per manhour for crew B1 is listed as $33.31. That's the average for a residential laborer (listed at $29.67 per hour on page 10) and a Residential carpenter (listed at $36.94 per hour): $29.67 plus $36.94 is $66.61. Divide by 2 to get $33.31, the average cost per manhour for crew B1.

In the table below, the cost per manhour is the sum of hourly costs of all crew members divided by the number of crew members. That's the average cost per manhour.

Costs in the Labor column in this book are the installation time (in manhours) multiplied by the cost per manhour. For example, on page 51 the labor cost listed for ½" BC wall sheathing is $.53 per square foot. That's the installation time (.016 manhours per square foot) multiplied by $33.31, the average cost per manhour for crew B1.

## Residential Division

| Craft Code | Cost Per Manhour | Crew Composition | Craft Code | Cost Per Manhour | Crew Composition |
|---|---|---|---|---|---|
| B1 | $33.31 | 1 laborer and 1 carpenter | BR | $35.88 | 1 lather |
| B2 | $34.52 | 1 laborer, 2 carpenters | BS | $32.09 | 1 marble setter |
| B3 | $32.09 | 2 laborers, 1 carpenter | CF | $35.02 | 1 cement mason |
| B4 | $35.09 | 1 laborer<br>1 operating engineer<br>1 reinforcing iron worker | CT | $33.98 | 1 mosaic & terrazzo worker |
| | | | D1 | $35.66 | 1 drywall installer<br>1 drywall taper |
| B5 | $35.45 | 1 laborer, 1 carpenter<br>1 cement mason<br>1 operating engineer<br>1 reinforcing iron worker | DI | $35.58 | 1 drywall installer |
| | | | DT | $35.74 | 1 drywall taper |
| | | | HC | $28.30 | 1 plasterer helper |
| B6 | $32.35 | 1 laborer, 1 cement mason | OE | $40.04 | 1 operating engineer |
| B7 | $30.10 | 1 laborer, 1 truck driver | P1 | $35.76 | 1 laborer, 1 plumber |
| B8 | $34.86 | 1 laborer<br>1 operating engineer | PM | $41.85 | 1 plumber |
| | | | PP | $33.47 | 1 painter, 1 laborer |
| B9 | $32.63 | 1 bricklayer<br>1 bricklayer's helper | PR | $37.45 | 1 plasterer |
| | | | PT | $37.26 | 1 painter |
| BB | $37.47 | 1 bricklayer | R1 | $35.42 | 1 roofer, 1 laborer |
| BC | $36.94 | 1 carpenter | RI | $35.57 | 1 reinforcing iron worker |
| BE | $39.59 | 1 electrician | RR | $41.17 | 1 roofer |
| BF | $34.00 | 1 floor layer | SW | $41.04 | 1 sheet metal worker |
| BG | $34.26 | 1 glazier | T1 | $32.56 | 1 tile layer, 1 laborer |
| BH | $27.78 | 1 bricklayer's helper | TL | $35.44 | 1 tile layer |
| BL | $29.67 | 1 laborer | TR | $30.52 | 1 truck driver |

# Commercial and Industrial Division

| Craft Code | Cost Per Manhour | Crew Composition |
|---|---|---|
| A1 | $49.43 | 1 asbestos worker<br>1 laborer |
| AT | $42.34 | 1 air tool operator |
| AW | $58.98 | 1 asbestos worker |
| BM | $60.48 | 1 boilermaker |
| BT | $40.38 | 1 bricklayer tender |
| C1 | $40.82 | 4 laborers, 1 truck driver |
| C2 | $47.80 | 1 laborer, 2 truck drivers<br>2 tractor operators |
| C3 | $46.48 | 1 laborer, 1 truck driver<br>1 tractor operator |
| C4 | $41.45 | 2 laborers, 1 truck driver |
| C5 | $44.83 | 2 laborers, 1 truck driver<br>1 tractor operator |
| C6 | $43.84 | 6 laborers, 2 truck drivers<br>2 tractor operators |
| C7 | $46.24 | 2 laborers , 3 truck drivers<br>1 crane operator<br>1 tractor operator |
| C8 | $45.83 | 1 laborer, 1 carpenter |
| C9 | $47.55 | 1 laborer, 1 crane operator |
| CB | $52.86 | 1 bricklayer |
| CC | $51.78 | 1 carpenter |
| CD | $50.60 | 1 drywall Installer |
| CE | $58.37 | 1 electrician |
| CG | $52.10 | 1 glazier |
| CL | $39.88 | 1 laborer |
| CM | $50.16 | 1 cement mason |
| CO | $55.21 | 1 crane operator |
| CV | $53.55 | 1 elevator constructor |
| D2 | $45.24 | 1 drywall installer<br>1 laborer |
| D3 | $52.02 | 1 laborer, 1 iron worker<br>(structural), 1 millwright |
| D4 | $46.25 | 1 laborer, 1 millwright |
| D5 | $50.18 | 1 boilermaker, 1 laborer |
| D6 | $53.41 | 2 millwrights<br>1 tractor operator |
| D7 | $45.66 | 1 painter, 1 laborer |
| D9 | $48.37 | 2 millwrights, 1 laborer |
| E1 | $50.30 | 2 electricians, 2 laborers<br>1 tractor operator |
| E2 | $49.13 | 2 electricians, 2 laborers |
| E3 | $50.01 | 2 electricians, 2 laborers<br>2 carpenters |
| E4 | $49.13 | 1 electrician, 1 laborer |
| F5 | $47.02 | 3 carpenters, 2 laborers |

| Craft Code | Cost Per Manhour | Crew Composition |
|---|---|---|
| F6 | $47.66 | 2 carpenters, 2 laborers<br>1 tractor operator |
| F7 | $49.61 | 2 carpenters, 1 laborer<br>1 tractor operator |
| F8 | $50.61 | 2 plasterers<br>1 plasterer's helper |
| F9 | $45.19 | 1 laborer, 1 floor layer |
| FL | $50.49 | 1 floor layer |
| G1 | $45.99 | 1 glazier, 1 laborer |
| H1 | $52.55 | 1 carpenter, 1 laborer<br>1 iron worker (structural)<br>1 tractor operator |
| H2 | $49.90 | 1 crane operator<br>1 truck driver |
| H3 | $45.20 | 1 carpenter, 3 laborers<br>1 crane operator<br>1 truck driver |
| H4 | $60.14 | 1 crane operator<br>6 iron workers (structural)<br>1 truck driver |
| H5 | $52.42 | 1 crane operator<br>2 iron workers (structural)<br>2 laborers |
| H6 | $51.72 | 1 iron worker (structural)<br>1 laborer |
| H7 | $60.78 | 1 crane operator<br>2 iron workers (structural) |
| H8 | $59.01 | 1 crane operator<br>4 iron workers (structural)<br>1 truck driver |
| H9 | $58.42 | 1 electrician<br>1 sheet metal worker |
| IW | $63.56 | 1 iron worker (structural) |
| LA | $47.43 | 1 lather |
| M1 | $46.62 | 1 bricklayer<br>1 bricklayer's tender |
| M2 | $43.85 | 1 carpenter, 2 laborers |
| M3 | $48.76 | 1 plasterer<br>1 plasterer's helper |
| M4 | $45.86 | 1 laborer, 1 marble setter |
| M5 | $49.71 | 1 pipefitter, 1 laborer, |
| M6 | $52.80 | 1 asbestos worker<br>1 laborer, 1 pipefitter |
| M8 | $54.63 | 3 pipefitters, 1 laborer |
| M9 | $58.96 | 1 electrician, 1 pipefitter |
| MI | $52.99 | 2 pipefitters, 1 laborer |
| MS | $51.83 | marble setter |
| MT | $48.72 | mosaic & terrazzo worker |
| MW | $52.62 | millwright |

# Commercial and Industrial Division

| Craft Code | Cost Per Manhour | Crew Composition |
|---|---|---|
| P5 | $43.84 | 3 laborers<br>1 tractor operator<br>1 truck driver |
| P6 | $50.22 | 1 laborer, 1 plumber |
| P8 | $45.02 | 1 laborer, 1 cement mason |
| P9 | $47.27 | 1 carpenter, 1 laborer<br>1 cement mason |
| PA | $51.43 | 1 painter |
| PD | $53.93 | 1 pile driver |
| PF | $59.54 | 1 pipefitter |
| PH | $43.18 | 1 plasterer's helper |
| PL | $60.55 | 1 plumber |
| PS | $54.33 | 1 plasterer |
| R3 | $45.42 | 2 roofers, 1 laborer |
| RB | $58.31 | 1 reinforcing iron worker |
| RF | $48.19 | 1 roofer |
| S1 | $47.44 | 1 laborer<br>1 tractor operator |
| S3 | $49.79 | 1 truck driver<br>1 tractor operator |
| S4 | $39.88 | 3 laborers |
| S5 | $42.74 | 5 laborers<br>1 crane operator<br>1 truck driver |
| S6 | $44.92 | 2 laborers<br>1 tractor operator |
| S7 | $47.44 | 3 laborers<br>3 tractor operators |

| Craft Code | Cost Per Manhour | Crew Composition |
|---|---|---|
| S8 | $48.91 | 2 pile drivers, 2 laborers<br>1 truck driver<br>1 crane operator<br>1 tractor operator |
| S9 | $46.65 | 1 pile driver, 2 laborers<br>1 tractor operator<br>1 truck driver |
| SM | $58.47 | 1 sheet metal worker |
| SP | $61.78 | 1 sprinkler fitter |
| SS | $49.95 | 1 laborer<br>2 tractor operators |
| T2 | $51.40 | 3 laborers, 3 carpenters<br>3 iron workers (structural)<br>1 crane operator<br>1 truck driver |
| T3 | $49.10 | 1 laborer<br>1 reinforcing iron worker |
| T4 | $44.30 | 1 laborer, 1 mosaic worker |
| T5 | $49.18 | 1 sheet metal worker<br>1 laborer |
| T6 | $52.27 | 2 sheet metal workers<br>1 laborer |
| TD | $44.58 | 1 truck driver |
| TO | $54.99 | 1 tractor operator |
| U1 | $48.83 | 1 plumber, 2 laborers<br>1 tractor operator |
| U2 | $46.77 | 1 plumber, 2 laborers |

# Residential Division

| Craft | 1<br>Base wage per hour | 2<br>Taxable fringe benefits (@5.15% of base wage) | 3<br>Insurance and employer taxes (%) | 4<br>Insurance and employer taxes ($) | 5<br>Non-taxable fringe benefits (@4.55% of base wage) | 6<br>Total hourly cost used in this book |
|---|---|---|---|---|---|---|
| Bricklayer | $27.44 | $1.41 | 25.54% | $7.37 | $1.25 | $37.47 |
| Bricklayer's Helper | 20.34 | 1.05 | 25.54 | 5.46 | 0.93 | 27.78 |
| Building Laborer | 20.55 | 1.06 | 32.93 | 7.12 | 0.94 | 29.67 |
| Carpenter | 25.80 | 1.33 | 31.83 | 8.64 | 1.17 | 36.94 |
| Cement Mason | 26.09 | 1.34 | 23.35 | 6.40 | 1.19 | 35.02 |
| Drywall installer | 26.42 | 1.36 | 23.77 | 6.60 | 1.20 | 35.58 |
| Drywall Taper | 26.53 | 1.37 | 23.77 | 6.63 | 1.21 | 35.74 |
| Electrician | 30.27 | 1.56 | 20.04 | 6.38 | 1.38 | 39.59 |
| Floor Layer | 25.19 | 1.30 | 24.02 | 6.36 | 1.15 | 34.00 |
| Glazier | 25.00 | 1.29 | 25.98 | 6.83 | 1.14 | 34.26 |
| Lather | 27.12 | 1.40 | 21.48 | 6.13 | 1.23 | 35.88 |
| Marble Setter | 24.24 | 1.25 | 21.56 | 5.50 | 1.10 | 32.09 |
| Millwright | 26.24 | 1.35 | 21.44 | 5.92 | 1.19 | 34.70 |
| Mosiac & Terrazzo Worker | 25.67 | 1.32 | 21.56 | 5.82 | 1.17 | 33.98 |
| Operating Engineer | 29.35 | 1.51 | 25.42 | 7.84 | 1.34 | 40.04 |
| Painter | 27.38 | 1.41 | 25.08 | 7.22 | 1.25 | 37.26 |
| Plasterer | 26.75 | 1.38 | 28.78 | 8.10 | 1.22 | 37.45 |
| Plasterer Helper | 20.22 | 1.04 | 28.78 | 6.12 | 0.92 | 28.30 |
| Plumber | 30.90 | 1.59 | 24.47 | 7.95 | 1.41 | 41.85 |
| Reinforcing Ironworker | 25.40 | 1.31 | 28.81 | 7.70 | 1.16 | 35.57 |
| Roofer | 26.33 | 1.36 | 44.34 | 12.28 | 1.20 | 41.17 |
| Sheet Metal Worker | 29.90 | 1.54 | 26.21 | 8.24 | 1.36 | 41.04 |
| Sprinkler Fitter | 30.35 | 1.56 | 25.28 | 8.07 | 1.38 | 41.36 |
| Tile Layer | 26.77 | 1.38 | 21.56 | 6.07 | 1.22 | 35.44 |
| Truck Driver | 22.20 | 1.14 | 26.42 | 6.17 | 1.01 | 30.52 |

## Hourly Labor Cost

The labor costs shown in Column 6 were used to compute the manhour costs for crews on page 7 and the figures in the "Labor" column of the Residential Division of this manual. Figures in the "Labor" column of the Industrial and Commercial Division of this book were computed using the hourly costs shown on page 308. All labor costs are in U.S. dollars per manhour.

It's important that you understand what's included in the figures in each of the six columns above. Here's an explanation:

**Column 1**, the base wage per hour, is the craftsman's hourly wage. These figures are representative of what many contractors will be paying craftsmen working on residential construction in 2011.

**Column 2**, taxable fringe benefits, includes vacation pay, sick leave and other taxable benefits. These fringe benefits average 5.15% of the base wage for many construction contractors. This benefit is in addition to the base wage.

**Column 3**, insurance and employer taxes in percent, shows the insurance and tax rate for construction trades. The cost of insurance in this column includes workers' compensation and contractor's casualty and liability coverage. Insurance rates vary widely from state to state and depend on a contractor's loss experience. Typical rates are shown in the Insurance section of this manual beginning on page 181. Taxes are itemized in the section on page 284. Note that taxes and insurance increase the hourly labor cost by 30 to 35% for most trades. There is no legal way to avoid these costs.

Column 4, insurance and employer taxes in dollars, shows the hourly cost of taxes and insurance for each construction trade. Insurance and taxes are paid on the costs in both columns 1 and 2.

Column 5, non-taxable fringe benefits, includes employer paid non-taxable benefits such as medical coverage and tax-deferred pension and profit sharing plans. These fringe benefits average 4.55% of the base wage for many construction contractors. The employer pays no taxes or insurance on these benefits.

Column 6, the total hourly cost in dollars, is the sum of columns 1, 2, 4, and 5.

These hourly labor costs will apply within a few percent on many jobs. But wage rates may be much higher or lower in some areas. If the hourly costs shown in column 6 are not accurate for your work, develop modification factors that you can apply to the labor costs in this book. The following paragraphs explain the procedure.

## Adjusting Labor Costs

Here's how to customize the labor costs in this book if your wage rates are different from the wage rates shown on page 10 or 308.

Start with the taxable benefits you offer. Assume craftsmen on your payroll get one week of vacation each year and one week of sick leave each year. Convert these benefits into hours. Your computation might look like this:

$$\begin{array}{r} 40 \text{ vacation hours} \\ + \ 40 \text{ sick leave hours} \\ \hline 80 \text{ taxable leave hours} \end{array}$$

Then add the regular work hours for the year:

$$\begin{array}{r} 2{,}000 \text{ regular hours} \\ + \ 80 \text{ taxable benefit hours} \\ \hline 2{,}080 \text{ total hours} \end{array}$$

Multiply these hours by the base wage per hour. If you pay carpenters $10.00 per hour, the calculation would be:

$$\begin{array}{r} 2{,}080 \text{ hours} \\ \times \ \$10.00 \text{ per hour} \\ \hline \$20{,}800 \text{ per year} \end{array}$$

Next determine the tax and insurance rate for each trade. If you know the rates that apply to your jobs, use those rates. If not, use the rates in column 3 on page

10. Continuing with our example, we'll use 31.83%, the rate for carpenters in column 3 on page 10. To increase the annual taxable wage by 31.83%, we'll multiply by 1.3183:

$$\begin{array}{r} \$20{,}800 \text{ per year} \\ \times \ 1.3183 \text{ tax \& insurance rate} \\ \hline \$27{,}421 \text{ annual cost} \end{array}$$

Then add the cost of non-taxable benefits. Suppose your company has no pension or profit sharing plan but does provide medical insurance for employees. Assume that the cost for your carpenter is $343.67 per month or $4,124 per year.

$$\begin{array}{r} \$4{,}124 \text{ medical plan} \\ + \ 27{,}421 \text{ annual cost} \\ \hline \$31{,}545 \text{ total annual cost} \end{array}$$

Divide this total annual cost by the actual hours worked in a year. This gives the contractor's total hourly labor cost including all benefits, taxes and insurance. Assume your carpenter will work 2,000 hours a year:

$$\frac{\$31{,}545}{2{,}000} = \$15.77 \text{ per hour}$$

Finally, find your modification factor for the labor costs in this book. Divide your total hourly labor cost by the total hourly labor cost shown on page 10. For the carpenter in our example, the figure in column 6 is $36.94.

$$\frac{\$15.77}{\$36.94} = .43$$

Your modification factor is 43%. Multiply any building carpenter (Craft Code BC) labor costs in the Residential Division of this book by .43 to find your estimated cost. For example, on page 24 the labor cost for installing an 18" long towel bar is $10.30 per each bar. If installed by your carpenter working at $10.00 per hour, your estimated cost would be 43% of $10.30 or $4.43. The manhours would remain the same @ .280, assuming normal productivity.

## If the Labor Rate Is Unknown

On some estimates you may not know what labor rates will apply. In that case, use both labor and material figures in this book without making any adjustment. When all labor, equipment and material costs have been compiled, add or deduct the percentage shown in the area modification table on pages 12 through 15.

Adjusting the labor costs in this book will make your estimates much more accurate.

# Area Modification Factors

Construction costs are higher in some areas than in other areas. Add or deduct the percentages shown on the following pages to adapt the costs in this book to your job site. Adjust your cost estimate by the appropriate percentages in this table to find the estimated cost for the site selected. Where 0% is shown, it means no modification is required.

Modification factors are listed alphabetically by state and province. Areas within each state are listed alphabetically. For convenience, one representative city is identified in each three-digit zip or range of zips. Percentages are based on the average of all data points in the table. Factors listed for each state and province are the average of all data points in that state or province. Figures for three-digit zips are the average of all five-digit zips in that area. Figures in the Total column are the weighted average of factors for Labor, Material and Equipment.

The National Estimator program will apply an area modification factor for any five-digit zip you select. Click Utilities. Click Options. Then select the Area Modification Factors tab.

These percentages are composites of many costs and will not necessarily be accurate when estimating the cost of any particular part of a building. But when used to modify costs for an entire structure, they should improve the accuracy of your estimates.

| Location | Zip | Mat. | Lab. | Equip. | Total Wtd. Avg. |
|---|---|---|---|---|---|
| **Alabama Average** | | -1 | -15 | 0 | -7% |
| Anniston | 362 | -3 | -22 | -1 | -12% |
| Auburn | 368 | 0 | -19 | 0 | -9% |
| Bellamy | 369 | -2 | -19 | -1 | -10% |
| Birmingham | 350-352 | -3 | 5 | -1 | 1% |
| Dothan | 363 | -1 | -17 | 0 | -8% |
| Evergreen | 364 | -2 | -27 | -1 | -13% |
| Gadsden | 359 | -4 | -17 | -1 | -10% |
| Huntsville | 358 | 1 | -11 | 0 | -4% |
| Jasper | 355 | -2 | -30 | -1 | -15% |
| Mobile | 365-366 | 4 | 5 | 1 | 4% |
| Montgomery | 360-361 | -1 | -11 | 0 | -5% |
| Scottsboro | 357 | 0 | -17 | 0 | -8% |
| Selma | 367 | -1 | -23 | 0 | -11% |
| Sheffield | 356 | -1 | -9 | 0 | -4% |
| Tuscaloosa | 354 | 0 | -16 | 0 | -7% |
| **Alaska Average** | | 14 | 58 | 4 | 34% |
| Anchorage | 995 | 16 | 68 | 5 | 40% |
| Fairbanks | 997 | 16 | 54 | 5 | 33% |
| Juneau | 998 | 17 | 50 | 6 | 32% |
| Ketchikan | 999 | 3 | 59 | 1 | 28% |
| King Salmon | 996 | 16 | 57 | 5 | 35% |
| **Arizona Average** | | 1 | -11 | 0 | -5% |
| Chambers | 865 | 1 | 4 | 0 | 3% |
| Douglas | 855 | 0 | -26 | 0 | -12% |
| Flagstaff | 860 | 2 | -21 | 1 | -9% |
| Kingman | 864 | 1 | 5 | 0 | 3% |
| Mesa | 852-853 | 1 | 1 | 0 | 1% |
| Phoenix | 850 | 1 | 1 | 0 | 1% |
| Prescott | 863 | 3 | -22 | 1 | -8% |
| Show Low | 859 | 1 | -21 | 0 | -9% |
| Tucson | 856-857 | 0 | -11 | 0 | -5% |
| Yuma | 853 | -2 | -24 | -1 | -13% |
| **Arkansas Average** | | -2 | -15 | -1 | -8% |
| Batesville | 725 | 0 | -33 | 0 | -15% |
| Camden | 717 | -4 | 29 | -1 | 11% |
| Fayetteville | 727 | 0 | -15 | 0 | -7% |
| Fort Smith | 729 | -2 | -18 | -1 | -9% |
| Harrison | 726 | -1 | -31 | 0 | -15% |
| Hope | 718 | -3 | -1 | -1 | -2% |
| Hot Springs | 719 | -2 | -33 | -1 | -16% |
| Jonesboro | 724 | -1 | -23 | 0 | -11% |
| Little Rock | 720-722 | -1 | -9 | 0 | -5% |
| Pine Bluff | 716 | -4 | -11 | -1 | -7% |
| Russellville | 728 | 0 | -16 | 0 | -7% |
| West Memphis | 723 | -3 | -17 | -1 | -10% |
| **California Average** | | 2 | 20 | 1 | 10% |
| Alhambra | 917-918 | 3 | 21 | 1 | 11% |
| Bakersfield | 932-933 | 0 | 6 | 0 | 3% |
| El Centro | 922 | 1 | 3 | 0 | 2% |
| Eureka | 955 | 1 | 0 | 0 | 1% |
| Fresno | 936-938 | 0 | -1 | 0 | -1% |

| Location | Zip | Mat. | Lab. | Equip. | Total Wtd. Avg. |
|---|---|---|---|---|---|
| Herlong | 961 | 2 | 3 | 1 | 2% |
| Inglewood | 902-905 | 3 | 22 | 1 | 12% |
| Irvine | 926-927 | 3 | 30 | 1 | 15% |
| Lompoc | 934 | 3 | 8 | 1 | 5% |
| Long Beach | 907-908 | 3 | 23 | 1 | 12% |
| Los Angeles | 900-901 | 3 | 19 | 1 | 10% |
| Marysville | 959 | 1 | 2 | 0 | 2% |
| Modesto | 953 | 1 | 4 | 0 | 3% |
| Mojave | 935 | 0 | 17 | 0 | 8% |
| Novato | 949 | 3 | 31 | 1 | 16% |
| Oakland | 945-947 | 3 | 44 | 1 | 22% |
| Orange | 928 | 3 | 28 | 1 | 14% |
| Oxnard | 930 | 3 | 13 | 1 | 8% |
| Pasadena | 910-912 | 3 | 21 | 1 | 12% |
| Rancho Cordova | 956-957 | 2 | 20 | 1 | 10% |
| Redding | 960 | 1 | 3 | 0 | 2% |
| Richmond | 948 | 2 | 43 | 1 | 21% |
| Riverside | 925 | 1 | 9 | 0 | 5% |
| Sacramento | 958 | 1 | 21 | 0 | 10% |
| Salinas | 939 | 3 | 12 | 1 | 7% |
| San Bernardino | 923-924 | 0 | 15 | 0 | 7% |
| San Diego | 919-921 | 3 | 20 | 1 | 11% |
| San Francisco | 941 | 3 | 58 | 1 | 28% |
| San Jose | 950-951 | 3 | 40 | 1 | 20% |
| San Mateo | 943-944 | 4 | 44 | 1 | 22% |
| Santa Barbara | 931 | 3 | 17 | 1 | 10% |
| Santa Rosa | 954 | 3 | 20 | 1 | 11% |
| Stockton | 952 | 1 | 11 | 0 | 6% |
| Sunnyvale | 940 | 3 | 45 | 1 | 23% |
| Van Nuys | 913-916 | 3 | 20 | 1 | 10% |
| Whittier | 906 | 3 | 21 | 1 | 11% |
| **Colorado Average** | | 1 | -1 | 1 | 1% |
| Aurora | 800-801 | 2 | 10 | 1 | 6% |
| Boulder | 803-804 | 3 | 1 | 1 | 2% |
| Colorado Springs | 808-809 | 2 | -4 | 1 | -1% |
| Denver | 802 | 2 | 10 | 1 | 6% |
| Durango | 813 | 1 | -7 | 0 | -3% |
| Fort Morgan | 807 | 1 | -11 | 0 | -5% |
| Glenwood Springs | 816 | 2 | 17 | 1 | 9% |
| Grand Junction | 814-815 | 1 | 1 | 0 | 1% |
| Greeley | 806 | 2 | 5 | 1 | 4% |
| Longmont | 805 | 3 | 0 | 1 | 2% |
| Pagosa Springs | 811 | 0 | -15 | 0 | -7% |
| Pueblo | 810 | -1 | 0 | 0 | 0% |
| Salida | 812 | 1 | -17 | 0 | -7% |
| **Connecticut Average** | | 1 | 24 | 0 | 12% |
| Bridgeport | 066 | 1 | 23 | 0 | 11% |
| Bristol | 060 | 1 | 30 | 0 | 14% |
| Fairfield | 064 | 2 | 28 | 1 | 14% |
| Hartford | 061 | 0 | 28 | 0 | 13% |
| New Haven | 065 | 1 | 26 | 0 | 12% |
| Norwich | 063 | 0 | 15 | 0 | 7% |
| Stamford | 068-069 | 4 | 30 | 1 | 16% |
| Waterbury | 067 | 1 | 23 | 0 | 11% |
| West Hartford | 062 | 1 | 17 | 0 | 8% |

| Location | Zip | Mat. | Lab. | Equip. | Total Wtd. Avg. |
|---|---|---|---|---|---|
| **Delaware Average** | | | 6 | 0 | 3% |
| Dover | 199 | 1 | -14 | 0 | -6% |
| Newark | 197 | 2 | 17 | 1 | 8% |
| Wilmington | 198 | 0 | 16 | 0 | 7% |
| **District of Columbia Average** | | 2 | 23 | 1 | 12% |
| Washington | 200-205 | 2 | 23 | 1 | 12% |
| **Florida Average** | | 0 | -13 | 0 | -6% |
| Altamonte Springs | 327 | -1 | -12 | 0 | -6% |
| Bradenton | 342 | 0 | -15 | 0 | -6% |
| Brooksville | 346 | 0 | -18 | 0 | -8% |
| Daytona Beach | 321 | -2 | -24 | -1 | -12% |
| Fort Lauderdale | 333 | 2 | 2 | 1 | 2% |
| Fort Myers | 339 | 0 | -17 | 0 | -8% |
| Fort Pierce | 349 | -2 | -21 | -1 | -11% |
| Gainesville | 326 | -1 | -23 | 0 | -11% |
| Jacksonville | 322 | -1 | -3 | 0 | -2% |
| Lakeland | 338 | -3 | -17 | -1 | -9% |
| Melbourne | 329 | -2 | -15 | -1 | -8% |
| Miami | 330-332 | 2 | 0 | 1 | 1% |
| Naples | 341 | 3 | -7 | 1 | -1% |
| Ocala | 344 | -2 | -29 | -1 | -14% |
| Orlando | 328 | 0 | -2 | 0 | -1% |
| Panama City | 324 | -2 | -22 | -1 | -11% |
| Pensacola | 325 | 0 | -20 | 0 | -9% |
| Saint Augustine | 320 | -1 | -14 | 0 | -7% |
| Saint Cloud | 347 | -1 | -11 | 0 | -5% |
| St. Petersburg | 337 | 0 | -9 | 0 | -4% |
| Tallahassee | 323 | 0 | -16 | 0 | -7% |
| Tampa | 335-336 | -1 | 7 | 0 | 3% |
| West Palm Beach | 334 | 1 | -4 | 0 | -2% |
| **Georgia Average** | | -1 | -13 | -1 | -6% |
| Albany | 317 | -2 | -23 | -1 | -12% |
| Athens | 306 | 0 | -15 | 0 | -7% |
| Atlanta | 303 | 3 | 22 | 1 | 12% |
| Augusta | 308-309 | -2 | -14 | -1 | -7% |
| Buford | 305 | 0 | -9 | 0 | -4% |
| Calhoun | 307 | -1 | -28 | 0 | -13% |
| Columbus | 318-319 | -1 | -19 | 0 | -9% |
| Dublin/Fort Valley | 310 | -3 | -19 | -1 | -10% |
| Hinesville | 313 | -2 | -12 | -1 | -7% |
| Kings Bay | 315 | -2 | -24 | -1 | -12% |
| Macon | 312 | -2 | -10 | -1 | -6% |
| Marietta | 300-302 | 1 | 5 | 0 | 3% |
| Savannah | 314 | -2 | -6 | -1 | -3% |
| Statesboro | 304 | -2 | -19 | -1 | -10% |
| Valdosta | 316 | -2 | -21 | -1 | -11% |
| **Hawaii Average** | | 17 | 38 | 6 | 27% |
| Aliamanu | 968 | 17 | 41 | 6 | 28% |
| Ewa | 967 | 17 | 37 | 6 | 26% |
| Halawa Heights | 967 | 17 | 37 | 6 | 26% |
| Hilo | 967 | 17 | 37 | 6 | 26% |

12

# Area Modification Factors

| Location | Zip | Mat. | Lab. | Equip. | Total Wtd. Avg. |
|---|---|---|---|---|---|
| Honolulu | 968 | 17 | 41 | 6 | 28% |
| Kailua | 968 | 17 | 41 | 6 | 28% |
| Lualualei | 967 | 17 | 37 | 6 | 26% |
| Mililani Town | 967 | 17 | 37 | 6 | 26% |
| Pearl City | 967 | 17 | 37 | 6 | 26% |
| Wahiawa | 967 | 17 | 37 | 6 | 26% |
| Waianae | 967 | 17 | 37 | 6 | 26% |
| Wailuku (Maui) | 967 | 17 | 37 | 6 | 26% |
| **Idaho Average** | | **0** | **-17** | **0** | **-8%** |
| Boise | 837 | 1 | -2 | 0 | 0% |
| Coeur d'Alene | 838 | 0 | -18 | 0 | -8% |
| Idaho Falls | 834 | -1 | -17 | 0 | -8% |
| Lewiston | 835 | 0 | -22 | 0 | -10% |
| Meridian | 836 | 0 | -16 | 0 | -7% |
| Pocatello | 832 | -1 | -22 | 0 | -11% |
| Sun Valley | 833 | 0 | -24 | 0 | -11% |
| **Illinois Average** | | **-1** | **15** | **0** | **7%** |
| Arlington Heights | 600 | 2 | 44 | 1 | 21% |
| Aurora | 605 | 2 | 40 | 1 | 20% |
| Belleville | 622 | -2 | 7 | -1 | 2% |
| Bloomington | 617 | 1 | 6 | 0 | 3% |
| Carbondale | 629 | -3 | -10 | -1 | -6% |
| Carol Stream | 601 | 2 | 41 | 1 | 20% |
| Centralia | 628 | -3 | -12 | -1 | -7% |
| Champaign | 618 | -1 | 3 | 0 | 1% |
| Chicago | 606-608 | 2 | 44 | 1 | 21% |
| Decatur | 623 | -2 | -14 | -1 | -7% |
| Galesburg | 614 | -2 | -9 | -1 | -6% |
| Granite City | 620 | -3 | 27 | -1 | 11% |
| Green River | 612 | -1 | 6 | 0 | 2% |
| Joliet | 604 | 0 | 41 | 0 | 19% |
| Kankakee | 609 | -2 | 8 | -1 | 3% |
| Lawrenceville | 624 | -4 | -4 | -1 | -4% |
| Oak Park | 603 | 3 | 48 | 1 | 23% |
| Peoria | 615-616 | -1 | 14 | 0 | 6% |
| Peru | 613 | 0 | 12 | 0 | 6% |
| Quincy | 602 | 3 | 44 | 1 | 22% |
| Rockford | 610-611 | -2 | 22 | -1 | 9% |
| Springfield | 625-627 | -2 | 5 | -1 | 1% |
| Urbana | 619 | -3 | -7 | -1 | -4% |
| **Indiana Average** | | **-2** | **-3** | **-1** | **-2%** |
| Aurora | 470 | -1 | -4 | 0 | -2% |
| Bloomington | 474 | 1 | -8 | 0 | -3% |
| Columbus | 472 | 0 | -10 | 0 | -4% |
| Elkhart | 465 | -2 | 0 | -1 | -1% |
| Evansville | 476-477 | -2 | 11 | -1 | 4% |
| Fort Wayne | 467-468 | -3 | -5 | -1 | -4% |
| Gary | 463-464 | -4 | 26 | -1 | 10% |
| Indianapolis | 460-462 | -1 | 13 | 0 | 5% |
| Jasper | 475 | -1 | -14 | 0 | -7% |
| Jeffersonville | 471 | 0 | -7 | 0 | -3% |
| Kokomo | 469 | -2 | -12 | -1 | -7% |
| Lafayette | 479 | -1 | -10 | 0 | -5% |
| Muncie | 473 | -4 | -18 | -1 | -10% |
| South Bend | 466 | -4 | 2 | -1 | -1% |
| Terre Haute | 478 | -4 | -2 | -1 | -3% |
| **Iowa Average** | | **-2** | **-11** | **-1** | **-6%** |
| Burlington | 526 | 0 | -11 | 0 | -5% |
| Carroll | 514 | -3 | -29 | -1 | -15% |
| Cedar Falls | 506 | -1 | -10 | 0 | -5% |
| Cedar Rapids | 522-524 | 0 | 3 | 0 | 1% |
| Cherokee | 510 | -2 | -12 | -1 | -7% |
| Council Bluffs | 515 | -2 | -7 | -1 | -4% |
| Creston | 508 | -3 | -8 | -1 | -5% |
| Davenport | 527-528 | -1 | 0 | 0 | -1% |
| Decorah | 521 | -2 | -8 | -1 | -5% |
| Des Moines | 500-503 | -2 | 6 | -1 | 2% |
| Dubuque | 520 | -1 | 0 | 0 | -1% |
| Fort Dodge | 505 | -2 | -12 | -1 | -7% |
| Mason City | 504 | 0 | -8 | 0 | -4% |
| Ottumwa | 525 | 0 | -16 | 0 | -7% |
| Sheldon | 512 | 0 | -24 | 0 | -11% |

| Location | Zip | Mat. | Lab. | Equip. | Total Wtd. Avg. |
|---|---|---|---|---|---|
| Shenandoah | 516 | -3 | -28 | -1 | -14% |
| Sioux City | 511 | -2 | -3 | -1 | -3% |
| Spencer | 513 | -1 | -21 | 0 | -10% |
| Waterloo | 507 | -4 | -12 | -1 | -7% |
| **Kansas Average** | | **-2** | **-12** | **-1** | **-7%** |
| Colby | 677 | -1 | 23 | 0 | 10% |
| Concordia | 669 | -1 | -34 | 0 | -16% |
| Dodge City | 678 | -2 | -13 | -1 | -7% |
| Emporia | 668 | -3 | -16 | -1 | -9% |
| Fort Scott | 667 | -2 | -20 | -1 | -10% |
| Hays | 676 | -2 | -31 | -1 | -15% |
| Hutchinson | 675 | -3 | -19 | -1 | -10% |
| Independence | 673 | -3 | -11 | -1 | -7% |
| Kansas City | 660-662 | 0 | 14 | 0 | 6% |
| Liberal | 679 | -2 | -20 | -1 | -11% |
| Salina | 674 | -3 | -17 | -1 | -9% |
| Topeka | 664-666 | -3 | -6 | -1 | -4% |
| Wichita | 670-672 | -2 | -10 | -1 | -6% |
| **Kentucky Average** | | **-1** | **-10** | **0** | **-5%** |
| Ashland | 411-412 | -3 | -9 | -1 | -6% |
| Bowling Green | 421 | 0 | -10 | 0 | -4% |
| Campton | 413-414 | -1 | -23 | 0 | -11% |
| Covington | 410 | -1 | 1 | 0 | 0% |
| Elizabethtown | 427 | -1 | -18 | 0 | -9% |
| Frankfort | 406 | 1 | -13 | 0 | -6% |
| Hazard | 417-418 | -1 | -19 | 0 | -9% |
| Hopkinsville | 422 | -2 | -16 | -1 | -9% |
| Lexington | 403-405 | 1 | -1 | 0 | 0% |
| London | 407-409 | -1 | -16 | 0 | -8% |
| Louisville | 400-402 | -1 | 3 | 0 | 1% |
| Owensboro | 423 | -2 | -14 | -1 | -8% |
| Paducah | 420 | -2 | -8 | -1 | -5% |
| Pikeville | 415-416 | -3 | -8 | -1 | -5% |
| Somerset | 425-426 | 0 | -20 | 0 | -9% |
| White Plains | 424 | -3 | 8 | -1 | 2% |
| **Louisiana Average** | | **1** | **16** | **0** | **8%** |
| Alexandria | 713-714 | -3 | -5 | -1 | -4% |
| Baton Rouge | 707-708 | 5 | 35 | 2 | 19% |
| Houma | 703 | 3 | 48 | 1 | 24% |
| Lafayette | 705 | 0 | 19 | 0 | 9% |
| Lake Charles | 706 | -3 | 26 | -1 | 11% |
| Mandeville | 704 | 4 | 23 | 1 | 13% |
| Minden | 710 | -2 | -10 | -1 | -6% |
| Monroe | 712 | -2 | -12 | -1 | -6% |
| New Orleans | 700-701 | 5 | 41 | 2 | 21% |
| Shreveport | 711 | -2 | -7 | -1 | -4% |
| **Maine Average** | | **-1** | **-16** | **0** | **-8%** |
| Auburn | 042 | -1 | -14 | 0 | -7% |
| Augusta | 043 | -1 | -19 | 0 | -9% |
| Bangor | 044 | -2 | -15 | -1 | -8% |
| Bath | 045 | 0 | -19 | 0 | -8% |
| Brunswick | 039-040 | 1 | -7 | 0 | -3% |
| Camden | 048 | -1 | -24 | 0 | -12% |
| Cutler | 046 | -1 | -22 | 0 | -11% |
| Dexter | 049 | -2 | -15 | -1 | -8% |
| Northern Area | 047 | -2 | -26 | -1 | -13% |
| Portland | 041 | 2 | -1 | 1 | 1% |
| **Maryland Average** | | **1** | **7** | **0** | **4%** |
| Annapolis | 214 | 3 | 19 | 1 | 10% |
| Baltimore | 210-212 | -1 | 18 | 0 | 8% |
| Bethesda | 208-209 | 3 | 27 | 1 | 14% |
| Church Hill | 216 | 2 | -12 | 1 | -4% |
| Cumberland | 215 | -4 | -13 | -1 | -8% |
| Elkton | 219 | 2 | 6 | 1 | 4% |
| Frederick | 217 | 1 | 11 | 0 | 5% |
| Laurel | 206-207 | 2 | 21 | 1 | 11% |
| Salisbury | 218 | 1 | -12 | 0 | -5% |

| Location | Zip | Mat. | Lab. | Equip. | Total Wtd. Avg. |
|---|---|---|---|---|---|
| **Massachusetts Average** | | **2** | **28** | **1** | **14%** |
| Ayer | 015-016 | 1 | 19 | 0 | 9% |
| Bedford | 017 | 3 | 40 | 1 | 20% |
| Boston | 021-022 | 3 | 72 | 1 | 35% |
| Brockton | 023-024 | 2 | 46 | 1 | 22% |
| Cape Cod | 026 | 2 | 8 | 1 | 5% |
| Chicopee | 010 | 1 | 17 | 0 | 8% |
| Dedham | 019 | 2 | 36 | 1 | 18% |
| Fitchburg | 014 | 2 | 29 | 1 | 14% |
| Hingham | 020 | 3 | 48 | 1 | 23% |
| Lawrence | 018 | 2 | 37 | 1 | 18% |
| Nantucket | 025 | 3 | 23 | 1 | 12% |
| New Bedford | 027 | 2 | 15 | 1 | 8% |
| Northfield | 013 | 1 | 0 | 0 | 1% |
| Pittsfield | 012 | 1 | 4 | 0 | 2% |
| Springfield | 011 | -1 | 21 | 0 | 9% |
| **Michigan Average** | | **-2** | **2** | **-1** | **0%** |
| Battle Creek | 490-491 | -3 | 3 | -1 | 0% |
| Detroit | 481-482 | 0 | 18 | 0 | 8% |
| Flint | 484-485 | -3 | 0 | -1 | -2% |
| Grand Rapids | 493-495 | -1 | 0 | 0 | -1% |
| Grayling | 497 | 1 | -21 | 0 | -9% |
| Jackson | 492 | -3 | 2 | -1 | -1% |
| Lansing | 488-489 | 0 | 1 | 0 | 0% |
| Marquette | 498-499 | -1 | -9 | 0 | -5% |
| Pontiac | 483 | -3 | 27 | -1 | 11% |
| Royal Oak | 480 | -2 | 19 | -1 | 8% |
| Saginaw | 486-487 | -2 | -8 | -1 | -5% |
| Traverse City | 496 | -1 | -7 | 0 | -4% |
| **Minnesota Average** | | **0** | **1** | **0** | **0%** |
| Bemidji | 566 | -1 | 5 | 0 | 2% |
| Brainerd | 564 | 0 | -3 | 0 | -2% |
| Duluth | 556-558 | -2 | 9 | -1 | 3% |
| Fergus Falls | 565 | -1 | -17 | 0 | -8% |
| Magnolia | 561 | 0 | -20 | 0 | -9% |
| Mankato | 560 | 0 | -4 | 0 | -2% |
| Minneapolis | 553-555 | 1 | 27 | 0 | 13% |
| Rochester | 559 | 0 | -4 | 0 | -1% |
| St. Cloud | 563 | -1 | 3 | 0 | 1% |
| St. Paul | 550-551 | 1 | 25 | 0 | 12% |
| Thief River Falls | 567 | 0 | -6 | 0 | -3% |
| Willmar | 562 | -1 | -4 | 0 | -2% |
| **Mississippi Average** | | **-2** | **-16** | **-1** | **-9%** |
| Clarksdale | 386 | -3 | -23 | -1 | -12% |
| Columbus | 397 | -1 | -22 | 0 | -10% |
| Greenville | 387 | -4 | -23 | -1 | -13% |
| Greenwood | 389 | -3 | -24 | -1 | -13% |
| Gulfport | 395 | 3 | 2 | 1 | 3% |
| Jackson | 390-392 | -3 | -9 | -1 | -6% |
| Laurel | 394 | -3 | -3 | -1 | -3% |
| McComb | 396 | -2 | -22 | -1 | -11% |
| Meridian | 393 | -2 | -14 | -1 | -8% |
| Tupelo | 388 | -1 | -24 | 0 | -12% |
| **Missouri Average** | | **-1** | **-3** | **0** | **-2%** |
| Cape Girardeau | 637 | -2 | 17 | -1 | 7% |
| Caruthersville | 638 | -2 | -22 | -1 | -11% |
| Chillicothe | 646 | -2 | -12 | -1 | -7% |
| Columbia | 652 | 1 | -12 | 0 | -5% |
| East Lynne | 647 | -1 | -8 | 0 | -4% |
| Farmington | 636 | -3 | -9 | -1 | -6% |
| Hannibal | 634 | 0 | 5 | 0 | 2% |
| Independence | 640 | -1 | 19 | 0 | 8% |
| Jefferson City | 650-651 | 1 | -11 | 0 | -4% |
| Joplin | 648 | -2 | -22 | -1 | -11% |
| Kansas City | 641 | -2 | 22 | -1 | 9% |
| Kirksville | 635 | 0 | -21 | 0 | -10% |
| Knob Noster | 653 | 0 | -18 | 0 | -8% |
| Lebanon | 654-655 | -1 | -26 | 0 | -13% |
| Poplar Bluff | 639 | -1 | -14 | 0 | -7% |
| Saint Charles | 633 | 1 | 7 | 0 | 3% |
| Saint Joseph | 644-645 | -3 | 53 | -1 | 23% |

13

# Area Modification Factors

| Location | Zip | Mat. | Lab. | Equip. | Total Wtd. Avg. |
|---|---|---|---|---|---|
| Springfield | 656-658 | -2 | -15 | -1 | -8% |
| St. Louis | 630-631 | -2 | 19 | -1 | 7% |
| **Montana Average** | | **0** | **-13** | **0** | **-6%** |
| Billings | 590-591 | 0 | -4 | 0 | -2% |
| Butte | 597 | 1 | -9 | 0 | -4% |
| Fairview | 592 | -1 | -11 | 0 | -5% |
| Great Falls | 594 | -1 | -12 | 0 | -6% |
| Havre | 595 | -1 | -25 | 0 | -12% |
| Helena | 596 | 0 | -12 | 0 | -5% |
| Kalispell | 599 | 2 | -19 | 1 | -8% |
| Miles City | 593 | 0 | -13 | 0 | -6% |
| Missoula | 598 | 1 | -13 | 0 | -5% |
| **Nebraska Average** | | **-1** | **-22** | **0** | **-10%** |
| Alliance | 693 | -1 | -30 | 0 | -14% |
| Columbus | 686 | 0 | -17 | 0 | -8% |
| Grand Island | 688 | 0 | -28 | 0 | -13% |
| Hastings | 689 | 0 | -17 | 0 | -8% |
| Lincoln | 683-685 | 0 | -11 | 0 | -5% |
| McCook | 690 | 0 | -31 | 0 | -14% |
| Norfolk | 687 | -3 | -18 | -1 | -10% |
| North Platte | 691 | 0 | -28 | 0 | -13% |
| Omaha | 680-681 | -1 | -1 | 0 | -1% |
| Valentine | 692 | -2 | -37 | -1 | -18% |
| **Nevada Average** | | **2** | **4** | **1** | **3%** |
| Carson City | 897 | 2 | -11 | 1 | -4% |
| Elko | 898 | 1 | 15 | 0 | 7% |
| Ely | 893 | 2 | -32 | 1 | -14% |
| Fallon | 894 | 2 | 10 | 1 | 6% |
| Las Vegas | 889-891 | 2 | 31 | 1 | 15% |
| Reno | 895 | 2 | 10 | 1 | 6% |
| **New Hampshire Average** | | **1** | **-4** | **0** | **-2%** |
| Charlestown | 036 | 1 | -14 | 0 | -6% |
| Concord | 034 | 1 | -6 | 0 | -2% |
| Dover | 038 | 0 | 4 | 0 | 2% |
| Lebanon | 037 | 2 | -13 | 1 | -5% |
| Littleton | 035 | -1 | -12 | 0 | -6% |
| Manchester | 032-033 | 0 | 0 | 0 | 0% |
| New Boston | 030-031 | 1 | 11 | 0 | 6% |
| **New Jersey Average** | | **1** | **34** | **0** | **16%** |
| Atlantic City | 080-084 | -2 | 26 | -1 | 11% |
| Brick | 087 | 2 | 18 | 1 | 9% |
| Dover | 078 | 1 | 34 | 0 | 16% |
| Edison | 088-089 | 1 | 38 | 0 | 18% |
| Hackensack | 076 | 3 | 30 | 1 | 15% |
| Monmouth | 077 | 3 | 38 | 1 | 19% |
| Newark | 071-073 | 1 | 33 | 0 | 16% |
| Passaic | 070 | 2 | 35 | 1 | 17% |
| Paterson | 074-075 | 2 | 27 | 1 | 13% |
| Princeton | 085 | -2 | 43 | -1 | 19% |
| Summit | 079 | 3 | 45 | 1 | 22% |
| Trenton | 086 | -3 | 39 | -1 | 16% |
| **New Mexico Average** | | **0** | **-21** | **0** | **-10%** |
| Alamogordo | 883 | -1 | -21 | 0 | -10% |
| Albuquerque | 870-871 | 2 | -3 | 1 | 0% |
| Clovis | 881 | -2 | -24 | -1 | -12% |
| Farmington | 874 | 2 | -7 | 1 | -2% |
| Fort Sumner | 882 | -3 | 2 | -1 | -1% |
| Gallup | 873 | 1 | -19 | 0 | -8% |
| Holman | 877 | 2 | -29 | 1 | -12% |
| Las Cruces | 880 | -1 | -26 | 0 | -13% |
| Santa Fe | 875 | 3 | -23 | 1 | -9% |
| Socorro | 878 | 1 | -47 | 0 | -21% |
| Truth or Consequences | 879 | -1 | -30 | 0 | -15% |
| Tucumcari | 884 | -1 | -24 | 0 | -11% |

| Location | Zip | Mat. | Lab. | Equip. | Total Wtd. Avg. |
|---|---|---|---|---|---|
| **New York Average** | | **0** | **21** | **0** | **9%** |
| Albany | 120-123 | 0 | 20 | 0 | 9% |
| Amityville | 117 | 2 | 31 | 1 | 16% |
| Batavia | 140 | -3 | 6 | -1 | 1% |
| Binghamton | 137-139 | -4 | 4 | -1 | 0% |
| Bronx | 104 | 2 | 39 | 1 | 19% |
| Brooklyn | 112 | 3 | 24 | 1 | 12% |
| Buffalo | 142 | -4 | 8 | -1 | 2% |
| Elmira | 149 | -4 | 5 | -1 | 0% |
| Flushing | 113 | 3 | 50 | 1 | 24% |
| Garden City | 115 | 3 | 42 | 1 | 21% |
| Hicksville | 118 | 3 | 38 | 1 | 19% |
| Ithaca | 148 | -4 | 3 | -1 | -1% |
| Jamaica | 114 | 3 | 47 | 1 | 23% |
| Jamestown | 147 | -4 | -8 | -1 | -5% |
| Kingston | 124 | 0 | -5 | 0 | -2% |
| Long Island | 111 | 3 | 76 | 1 | 37% |
| Montauk | 119 | 1 | 29 | 0 | 14% |
| New York (Manhattan) | 100-102 | 3 | 77 | 1 | 37% |
| New York City | 100-102 | 3 | 77 | 1 | 37% |
| Newcomb | 128 | -1 | 0 | 0 | -1% |
| Niagara Falls | 143 | -4 | -12 | -1 | -8% |
| Plattsburgh | 129 | 1 | -5 | 0 | -2% |
| Poughkeepsie | 125-126 | 0 | 5 | 0 | 2% |
| Queens | 110 | 4 | 49 | 1 | 25% |
| Rochester | 144-146 | -3 | 7 | -1 | 2% |
| Rockaway | 116 | 3 | 35 | 1 | 18% |
| Rome | 133-134 | -3 | -2 | -1 | -3% |
| Staten Island | 103 | 3 | 31 | 1 | 16% |
| Stewart | 127 | -1 | -11 | 0 | -6% |
| Syracuse | 130-132 | -3 | 5 | -1 | 1% |
| Tonawanda | 141 | -4 | 3 | -1 | -1% |
| Utica | 135 | -4 | -4 | -1 | -4% |
| Watertown | 136 | -2 | 0 | -1 | -1% |
| West Point | 109 | 1 | 21 | 0 | 10% |
| White Plains | 105-108 | 3 | 40 | 1 | 20% |
| **North Carolina Average** | | **0** | **-12** | **0** | **-5%** |
| Asheville | 287-289 | 1 | -18 | 0 | -8% |
| Charlotte | 280-282 | 1 | 4 | 0 | 2% |
| Durham | 277 | 2 | 7 | 1 | 4% |
| Elizabeth City | 279 | 1 | -21 | 0 | -9% |
| Fayetteville | 283 | -1 | -22 | 0 | -11% |
| Goldsboro | 275 | 1 | -8 | 0 | -3% |
| Greensboro | 274 | 0 | -9 | 0 | -4% |
| Hickory | 286 | -1 | -24 | 0 | -11% |
| Kinston | 285 | -1 | -27 | 0 | -13% |
| Raleigh | 276 | 3 | -3 | 1 | 0% |
| Rocky Mount | 278 | -1 | -18 | 0 | -9% |
| Wilmington | 284 | 1 | -10 | 0 | -4% |
| Winston-Salem | 270-273 | 0 | -12 | 0 | -5% |
| **North Dakota Average** | | **-1** | **5** | **0** | **2%** |
| Bismarck | 585 | 0 | 9 | 0 | 4% |
| Dickinson | 586 | 1 | 10 | 0 | 4% |
| Fargo | 580-581 | 0 | 2 | 0 | 1% |
| Grand Forks | 582 | 0 | 8 | 0 | 4% |
| Jamestown | 584 | -1 | 4 | 0 | 1% |
| Minot | 587 | -1 | 7 | 0 | 3% |
| Nekoma | 583 | -1 | -11 | 0 | -5% |
| Williston | 588 | -1 | 11 | 0 | 5% |
| **Ohio Average** | | **-2** | **4** | **-1** | **1%** |
| Akron | 442-443 | -2 | 0 | -1 | -1% |
| Canton | 446-447 | -3 | -7 | -1 | -4% |
| Chillicothe | 456 | -2 | -2 | -1 | -2% |
| Cincinnati | 450-452 | -1 | 10 | 0 | 4% |
| Cleveland | 440-441 | -3 | 11 | -1 | 4% |
| Columbus | 432 | 0 | 12 | 0 | 5% |
| Dayton | 453-455 | -3 | 1 | -1 | -1% |
| Lima | 458 | -3 | -9 | -1 | -6% |
| Marietta | 457 | -3 | 16 | -1 | 6% |
| Marion | 433 | -3 | -16 | -1 | -9% |
| Newark | 430-431 | -1 | 13 | 0 | 6% |

| Location | Zip | Mat. | Lab. | Equip. | Total Wtd. Avg. |
|---|---|---|---|---|---|
| Sandusky | 448-449 | -1 | 0 | 0 | -1% |
| Steubenville | 439 | -3 | 3 | -1 | 0% |
| Toledo | 434-436 | -1 | 18 | 0 | 8% |
| Warren | 444 | -4 | -6 | -1 | -5% |
| Youngstown | 445 | -5 | 6 | -2 | 0% |
| Zanesville | 437-438 | -2 | 19 | -1 | 8% |
| **Oklahoma Average** | | **-3** | **-15** | **-1** | **-8%** |
| Adams | 739 | -2 | -19 | -1 | -10% |
| Ardmore | 734 | -3 | -16 | -1 | -9% |
| Clinton | 736 | -3 | -2 | -1 | -3% |
| Durant | 747 | -4 | -11 | -1 | -7% |
| Enid | 737 | -4 | -19 | -1 | -10% |
| Lawton | 735 | -3 | -22 | -1 | -12% |
| McAlester | 745 | -4 | -18 | -1 | -10% |
| Muskogee | 744 | -2 | -24 | -1 | -12% |
| Norman | 730 | -2 | -11 | -1 | -6% |
| Oklahoma City | 731 | -2 | -9 | -1 | -5% |
| Ponca City | 746 | -3 | -3 | -1 | -3% |
| Poteau | 749 | -2 | -22 | -1 | -11% |
| Pryor | 743 | -2 | -27 | -1 | -14% |
| Shawnee | 748 | -4 | -28 | -1 | -15% |
| Tulsa | 740-741 | -1 | -3 | 0 | -2% |
| Woodward | 738 | -4 | -1 | -1 | -2% |
| **Oregon Average** | | **1** | **-8** | **1** | **-3%** |
| Adrian | 979 | -1 | -35 | 0 | -16% |
| Bend | 977 | 1 | -8 | 0 | -3% |
| Eugene | 974 | 2 | -4 | 1 | -1% |
| Grants Pass | 975 | 2 | -9 | 1 | -3% |
| Klamath Falls | 976 | 2 | -17 | 1 | -7% |
| Pendleton | 978 | 0 | -8 | 0 | -4% |
| Portland | 970-972 | 2 | 18 | 1 | 9% |
| Salem | 973 | 2 | 0 | 1 | 1% |
| **Pennsylvania Average** | | **-3** | **1** | **-1** | **-1%** |
| Allentown | 181 | -2 | 18 | -1 | 7% |
| Altoona | 166 | -3 | -11 | -1 | -7% |
| Beaver Springs | 178 | -3 | -7 | -1 | -5% |
| Bethlehem | 180 | -1 | 18 | 0 | 8% |
| Bradford | 167 | -4 | -17 | -1 | -10% |
| Butler | 160 | -4 | 4 | -1 | 0% |
| Chambersburg | 172 | -1 | -9 | 0 | -5% |
| Clearfield | 168 | 2 | -7 | 1 | -3% |
| DuBois | 158 | -2 | -27 | -1 | -13% |
| East Stroudsburg | 183 | 0 | -9 | 0 | -4% |
| Erie | 164-165 | -3 | -10 | -1 | -6% |
| Genesee | 169 | -4 | -14 | -1 | -8% |
| Greensburg | 156 | -4 | -2 | -1 | -3% |
| Harrisburg | 170-171 | -2 | 12 | -1 | 4% |
| Hazleton | 182 | -3 | -6 | -1 | -5% |
| Johnstown | 159 | -4 | -13 | -1 | -8% |
| Kittanning | 162 | -4 | -9 | -1 | -6% |
| Lancaster | 175-176 | -3 | 12 | -1 | 4% |
| Meadville | 163 | -4 | -27 | -1 | -15% |
| Montrose | 188 | -3 | -14 | -1 | -8% |
| New Castle | 161 | -4 | 1 | -1 | -2% |
| Philadelphia | 190-191 | -3 | 38 | -1 | 16% |
| Pittsburgh | 152 | -4 | 17 | -1 | 6% |
| Pottsville | 179 | -4 | -10 | -1 | -7% |
| Punxsutawney | 157 | -4 | -8 | -1 | -6% |
| Reading | 195-196 | -4 | 15 | -1 | 5% |
| Scranton | 184-185 | -2 | -3 | -1 | -2% |
| Somerset | 155 | -4 | -8 | -1 | -6% |
| Southeastern | 193 | 0 | 26 | 0 | 12% |
| Uniontown | 154 | -4 | -6 | -1 | -5% |
| Valley Forge | 194 | -3 | 36 | -1 | 15% |
| Warminster | 189 | 1 | 32 | 0 | 14% |
| Warrendale | 150-151 | -4 | 17 | -1 | 6% |
| Washington | 153 | -4 | 17 | -1 | 6% |
| Wilkes Barre | 186-187 | -3 | -3 | -1 | -3% |
| Williamsport | 177 | -3 | -7 | -1 | -5% |
| York | 173-174 | -3 | 9 | -1 | 3% |

# Area Modification Factors

| Location | Zip | Mat. | Lab. | Equip. | Total Wtd. Avg. |
|---|---|---|---|---|---|
| **Rhode Island Average** | | 1 | 14 | 0 | 8% |
| Bristol | 028 | 1 | 13 | 0 | 7% |
| Coventry | 028 | 1 | 13 | 0 | 7% |
| Cranston | 029 | 1 | 18 | 0 | 9% |
| Davisville | 028 | 1 | 13 | 0 | 7% |
| Narragansett | 028 | 1 | 13 | 0 | 7% |
| Newport | 028 | 1 | 13 | 0 | 7% |
| Providence | 029 | 1 | 18 | 0 | 9% |
| Warwick | 028 | 1 | 13 | 0 | 7% |
| **South Carolina Average** | | 0 | -10 | 0 | -5% |
| Aiken | 298 | 0 | 8 | 0 | 3% |
| Beaufort | 299 | -1 | -14 | 0 | -7% |
| Charleston | 294 | -1 | -5 | 0 | -3% |
| Columbia | 290-292 | 0 | -9 | 0 | -4% |
| Greenville | 296 | 0 | -10 | 0 | -5% |
| Myrtle Beach | 295 | 0 | -20 | 0 | -9% |
| Rock Hill | 297 | 0 | -18 | 0 | -9% |
| Spartanburg | 293 | -1 | -11 | 0 | -6% |
| **South Dakota Average** | | -1 | -20 | 0 | -9% |
| Aberdeen | 574 | -1 | -13 | 0 | -6% |
| Mitchell | 573 | 0 | -20 | 0 | -10% |
| Mobridge | 576 | -1 | -36 | 0 | -17% |
| Pierre | 575 | -1 | -29 | 0 | -14% |
| Rapid City | 577 | -1 | -13 | 0 | -7% |
| Sioux Falls | 570-571 | 0 | -1 | 0 | 0% |
| Watertown | 572 | -1 | -25 | 0 | -12% |
| **Tennessee Average** | | -1 | -9 | 0 | -4% |
| Chattanooga | 374 | -1 | -5 | 0 | -3% |
| Clarksville | 370 | 1 | -2 | 0 | 0% |
| Cleveland | 373 | -1 | -7 | 0 | -4% |
| Columbia | 384 | -1 | -21 | 0 | -10% |
| Cookeville | 385 | 0 | -21 | 0 | -10% |
| Jackson | 383 | -1 | -11 | 0 | -6% |
| Kingsport | 376 | 0 | -11 | 0 | -5% |
| Knoxville | 377-379 | -1 | -6 | 0 | -3% |
| McKenzie | 382 | -2 | -24 | -1 | -12% |
| Memphis | 380-381 | -2 | 4 | -1 | 1% |
| Nashville | 371-372 | 1 | 10 | 0 | 5% |
| **Texas Average** | | -2 | -6 | -1 | -4% |
| Abilene | 795-796 | -4 | 1 | -1 | -2% |
| Amarillo | 790-791 | -2 | -5 | -1 | -4% |
| Arlington | 760 | -1 | 0 | 0 | -1% |
| Austin | 786-787 | 1 | 4 | 0 | 2% |
| Bay City | 774 | -1 | 46 | -1 | 20% |
| Beaumont | 776-777 | -3 | 22 | -1 | 8% |
| Brownwood | 768 | -3 | -21 | -1 | -11% |
| Bryan | 778 | 0 | -19 | 0 | -9% |
| Childress | 792 | -3 | -33 | -1 | -17% |
| Corpus Christi | 783-784 | -3 | 6 | -1 | 1% |
| Dallas | 751-753 | -1 | 9 | 0 | 3% |
| Del Rio | 788 | -3 | -37 | -1 | -19% |
| El Paso | 798-799 | -3 | -23 | -1 | -12% |
| Fort Worth | 761-762 | -2 | 0 | -1 | -1% |
| Galveston | 775 | -3 | 20 | -1 | 8% |
| Giddings | 789 | 0 | -18 | 0 | -8% |
| Greenville | 754 | -3 | -1 | -1 | -2% |
| Houston | 770-772 | -1 | 21 | 0 | 9% |
| Huntsville | 773 | -2 | 16 | -1 | 6% |
| Longview | 756 | -3 | -5 | -1 | -4% |
| Lubbock | 793-794 | -3 | -17 | -1 | -10% |
| Lufkin | 759 | -3 | -16 | -1 | -9% |
| McAllen | 785 | -4 | -32 | -1 | -17% |
| Midland | 797 | -4 | 10 | -1 | 3% |
| Palestine | 758 | -2 | -14 | -1 | -8% |
| Plano | 750 | -1 | 8 | 0 | 3% |
| San Angelo | 769 | -3 | -21 | -1 | -11% |
| San Antonio | 780-782 | -3 | -2 | -1 | -2% |
| Texarkana | 755 | -3 | -22 | -1 | -12% |
| Tyler | 757 | -1 | -22 | 0 | -11% |
| Victoria | 779 | -3 | -7 | -1 | -5% |
| Waco | 765-767 | -3 | -13 | -1 | -8% |
| Wichita Falls | 763 | -3 | -20 | -1 | -11% |
| Woodson | 764 | -3 | -20 | -1 | -11% |

| Location | Zip | Mat. | Lab. | Equip. | Total Wtd. Avg. |
|---|---|---|---|---|---|
| **Utah Average** | | 1 | -10 | 1 | -4% |
| Clearfield | 840 | 2 | -3 | 1 | 0% |
| Green River | 845 | 1 | -4 | 0 | -1% |
| Ogden | 843-844 | 0 | -22 | 0 | -10% |
| Provo | 846-847 | 2 | -23 | 1 | -9% |
| Salt Lake City | 841 | 2 | 1 | 1 | 1% |
| **Vermont Average** | | 0 | -13 | 0 | -5% |
| Albany | 058 | 1 | -18 | 0 | -8% |
| Battleboro | 053 | 0 | -8 | 0 | -3% |
| Beecher Falls | 059 | 1 | -23 | 0 | -10% |
| Bennington | 052 | -1 | -17 | 0 | -8% |
| Burlington | 054 | 2 | 2 | 1 | 2% |
| Montpelier | 056 | 2 | -13 | 1 | -5% |
| Rutland | 057 | -1 | -14 | 0 | -7% |
| Springfield | 051 | -1 | -11 | 0 | -5% |
| White River Junction | 050 | 1 | -11 | 0 | -5% |
| **Virginia Average** | | 0 | -11 | 0 | -5% |
| Abingdon | 242 | -2 | -18 | -1 | -9% |
| Alexandria | 220-223 | 3 | 22 | 1 | 12% |
| Charlottesville | 229 | 1 | -17 | 0 | -7% |
| Chesapeake | 233 | -1 | -8 | 0 | -4% |
| Culpeper | 227 | 2 | -9 | 1 | -3% |
| Farmville | 239 | -2 | -28 | -1 | -14% |
| Fredericksburg | 224-225 | 1 | -14 | 0 | -6% |
| Galax | 243 | -2 | -18 | -1 | -9% |
| Harrisonburg | 228 | 1 | -19 | 0 | -8% |
| Lynchburg | 245 | -2 | -25 | -1 | -12% |
| Norfolk | 235-237 | 0 | -5 | 0 | -3% |
| Petersburg | 238 | -2 | -11 | -1 | -6% |
| Radford | 241 | -2 | -25 | -1 | -12% |
| Reston | 201 | 3 | 22 | 1 | 12% |
| Richmond | 232 | -1 | 5 | 0 | 2% |
| Roanoke | 240 | -2 | -18 | -1 | -9% |
| Staunton | 244 | 0 | -25 | 0 | -11% |
| Tazewell | 246 | -3 | -14 | -1 | -8% |
| Virginia Beach | 234 | 0 | -8 | 0 | -4% |
| Williamsburg | 230-231 | 0 | -6 | 0 | -3% |
| Winchester | 226 | 0 | -14 | 0 | -7% |
| **Washington Average** | | 1 | 2 | 0 | 2% |
| Clarkston | 994 | 0 | -12 | 0 | -5% |
| Everett | 982 | 2 | 9 | 1 | 5% |
| Olympia | 985 | 2 | -1 | 1 | 1% |
| Pasco | 993 | 0 | 6 | 0 | 3% |
| Seattle | 980-981 | 3 | 29 | 1 | 15% |
| Spokane | 990-992 | 0 | -2 | 0 | -1% |
| Tacoma | 983-984 | 1 | 10 | 0 | 5% |
| Vancouver | 986 | 2 | 5 | 1 | 3% |
| Wenatchee | 988 | 1 | -13 | 0 | -6% |
| Yakima | 989 | 0 | -10 | 0 | -5% |
| **West Virginia Average** | | -2 | -10 | -1 | -5% |
| Beckley | 258-259 | 1 | -17 | 0 | -8% |
| Bluefield | 247-248 | -1 | 0 | 0 | -1% |
| Charleston | 250-253 | 1 | 7 | 0 | 3% |
| Clarksburg | 263-264 | -3 | -9 | -1 | -6% |
| Fairmont | 266 | 0 | -25 | 0 | -12% |
| Huntington | 255-257 | -2 | -1 | -1 | -1% |
| Lewisburg | 249 | -2 | -30 | -1 | -14% |
| Martinsburg | 254 | -1 | -8 | 0 | -4% |
| Morgantown | 265 | -3 | -3 | -1 | -3% |
| New Martinsville | 262 | -3 | -25 | -1 | -13% |
| Parkersburg | 261 | -3 | 33 | -1 | 13% |
| Romney | 267 | -4 | -15 | -1 | -9% |
| Sugar Grove | 268 | -3 | -26 | -1 | -13% |
| Wheeling | 260 | -3 | -15 | -1 | -8% |
| **Wisconsin Average** | | -1 | 3 | 0 | 1% |
| Amery | 540 | 0 | -2 | 0 | -1% |
| Beloit | 535 | 0 | 11 | 0 | 5% |
| Clam Lake | 545 | -1 | -5 | 0 | -3% |
| Eau Claire | 547 | -1 | -3 | 0 | -1% |
| Green Bay | 541-543 | 0 | 2 | 0 | 1% |

| Location | Zip | Mat. | Lab. | Equip. | Total Wtd. Avg. |
|---|---|---|---|---|---|
| La Crosse | 546 | -2 | -3 | -1 | -2% |
| Ladysmith | 548 | -2 | -7 | -1 | -5% |
| Madison | 537 | 2 | 16 | 1 | 8% |
| Milwaukee | 530-534 | 0 | 27 | 0 | 12% |
| Oshkosh | 549 | -1 | 5 | 0 | 2% |
| Portage | 539 | 0 | 4 | 0 | 2% |
| Prairie du Chien | 538 | -2 | -8 | -1 | -5% |
| Wausau | 544 | -1 | 4 | 0 | 1% |
| **Wyoming Average** | | 0 | -5 | 0 | -3% |
| Casper | 826 | -2 | 6 | -1 | 2% |
| Cheyenne/ Laramie | 820 | 1 | -9 | 0 | -4% |
| Gillette | 827 | -1 | 16 | 0 | 7% |
| Powell | 824 | 0 | -21 | 0 | -10% |
| Rawlins | 823 | 0 | 2 | 0 | 1% |
| Riverton | 825 | -2 | -12 | -1 | -6% |
| Rock Springs | 829-831 | 0 | 5 | 0 | 2% |
| Sheridan | 828 | 0 | -10 | 0 | -4% |
| Wheatland | 822 | 0 | -23 | 0 | -11% |

**UNITED STATES TERRITORIES**

| Location | Zip | Mat. | Lab. | Equip. | Total Wtd. Avg. |
|---|---|---|---|---|---|
| Guam | 53 | | -21 | -5 | 18% |
| Puerto Rico | 2 | | -47 | -5 | -21% |

**VIRGIN ISLANDS (U.S.)**

| Location | | Mat. | Lab. | Equip. | Total Wtd. Avg. |
|---|---|---|---|---|---|
| St. Croix | 18 | | -15 | -4 | 2% |
| St. John | 52 | | -15 | -4 | 20% |
| St. Thomas | 23 | | -15 | -4 | 5% |

**CANADIAN AREA MODIFIERS**

These figures assume an exchange rate of $1.07 Canadian to $1.00 U.S.

| Location | Mat. | Lab. | Equip. | Total Wtd. Avg. |
|---|---|---|---|---|
| **Alberta Average** | 16.4 | 14.1 | 13.1 | 15.1% |
| Calgary | 15.2 | 9.8 | 10.2 | 12.4% |
| Edmonton | 16.7 | 15.4 | 14.5 | 16.0% |
| Ft. McMurray | 17.3 | 17.0 | 16.7 | 17.1% |
| **British Columbia Avg** | 15.2 | 15.0 | 22.5 | 15.1% |
| Fraser Valley | 15.2 | 11.9 | 19.4 | 13.4% |
| Okanagan | 11.0 | 8.6 | 16.1 | 9.7% |
| Vancouver | 19.6 | 24.7 | 32.1 | 22.1% |
| **Manitoba Average** | 17.9 | 19.2 | 12.8 | 18.4% |
| North Manitoba | 21.8 | 31.2 | 18.9 | 26.5% |
| South Manitoba | 18.0 | 18.9 | 10.2 | 18.3% |
| Selkirk | 15.2 | 10.6 | 11.6 | 12.6% |
| Winnipeg | 16.7 | 16.4 | 10.2 | 16.5% |
| **New Brunswick Average** | 12.1 | 15.2 | 15.2 | 13.7% |
| Moncton | 12.1 | 15.2 | 15.2 | 13.7% |
| **Nova Scotia Average** | 17.7 | 20.2 | 16.0 | 18.9% |
| Amherst | 17.4 | 19.0 | 16.0 | 18.1% |
| Nova Scotia | 16.0 | 13.8 | 16.0 | 14.8% |
| Sydney | 19.6 | 27.8 | 16.0 | 23.7% |
| **Newfoundland/Labrador Average** | 15.2 | 11.6 | 17.4 | 13.2% |
| **Ontario Average** | 17.2 | 17.9 | 10.7 | 17.5% |
| London | 18.2 | 22.0 | 10.4 | 20.0% |
| Thunder Bay | 13.8 | 9.9 | 10.2 | 11.6% |
| Toronto | 19.6 | 22.0 | 11.6 | 20.1% |
| **Quebec Average** | 17.1 | 24.9 | 16.0 | 21.0% |
| Montreal | 18.2 | 30.9 | 16.0 | 24.7% |
| Quebec City | 16.0 | 18.9 | 16.0 | 17.4% |
| **Saskatchewan Average** | 17.0 | 23.8 | 12.1 | 20.4% |
| La Ronge | 18.2 | 28.7 | 16.0 | 23.6% |
| Prince Albert | 16.7 | 22.6 | 10.2 | 19.7% |
| Saskatoon | 16.0 | 20.0 | 10.2 | 17.9% |

# Credits and Acknowledgments

This book has over 30,000 cost estimates for 2011. To develop these estimates, the editors relied on information supplied by hundreds of construction cost authorities. We offer our sincere thanks to the contractors, engineers, design professionals, construction estimators, material suppliers and manufacturers who, in the spirit of cooperation, have assisted in the preparation of this Fifty Ninth Edition of the National Construction Estimator. Many of the cost authorities who supplied information for this volume are listed below.

**A.A. Drafting Service.,** Andrew Gutt
**ABC Construction Co.,** Wayne Czubernat
**ABTCO, Inc.,** Tom Roe
**Advance Lifts,** Jim Oriowski
**Advanced Stainless & Aluminum,** Louie Sprague
**AFCO Roofing Supply,** John Ording
**AGRA Earth and Environmental,** Jack Eagan
**Allied Barricade,** Bob Johnson
**Alply Incorporated.,** Alan E. Sleppy
**Amerec Products,** John Gunderson
**Applied Waterproofing Technology, Inc.,** Mike Weinert
**Architectural Illust.,** Bruce Wickern
**Area Home Inspection,** Anthony Galeota
**Atascadero Glass,** Jim Stevens
**Atlas Mineral & Chemical Co.,** Glenn Kresge
**Bell Blueprint,** Ed Moore
**Bessler Stairways,** Morris Hudspeth
**Beveled Glass Design,** Lorna Koss
**Brooks Products,** Lynn Diaz
**Brower Pease,** Mark Brossert
**Buy-Rite Decorating House,** Hank Jordan
**California Pneumatic Systems,** Clayton Taylor
**Cement Cutting, Inc.,** Greg Becker
**Classic Ceilings,** Donna Morgan
**Cold Spring Granite,** Paul Dierkhising
**Columbia Equipment Co.,** Arthur Cohen
**Creative Ceilings,** Bob Warren
**C. R. Laurence, Inc.,** Michael Sedano
**David Backer Limited,** David D. Backer
**Delta Pacific Builders,** Jeff Andrews
**Dial One,** Charlie Gindele
**Dryvit System, Inc.,** Roger Peevier
**Eco-Block, LLC,** Jennifer Hansell
**Eliason Corporation,** John Duncan
**Envel Design Corporation,** Quinn Mayer
**Escondido Roof Truss Co.,** David Truax
**Finlandia Sauna Rooms,** Reino J. Tarkianen
**Fiorentine Company,** Tom Forsyth
**Flexi-Wall Systems,** Nancy West
**Fortifiber Corporation,** Dave Olson

**GAF Premium Products,** Sandra Stern
**Georgia Pacific,** Travis Hemborg
**Giles and Kendall, Inc.,** Todd M. Green
**Glen-Gery Brick,** David M. Reitz
**Heintschel Plastering,** Marvin J. Heintschel, Jr.
**H&H Specialties,** Reid Nesiage
**I. C. C.,** Mike McInnis
**Inclinator Company,** Boc Crede
**Int'l Conf. Of Bldg. Officials,** Dennis McCreary
**Iron-A-Way Company,** Reg Smidt
**J.A. Crawford Company,** Jim Crawford
**Jewett Refrigerator Co.,** Thomas D. Reider
**J. H. Baxter Inc.,** Dean Rogers
**Kentucky Wood Floors,** John Stern
**Kirsch Company,** Dennis Kirby
**Koffler Sales Corporation,** Mary Holman
**Landscape Structures, Inc.,** Mary Allen Grossman
**Lasco Panels,** Mike Bunke, Sales Manager
**Lombardi, Diamond Core Drilling,** Kit Sanders
**Maccaferri Gabions, Inc.,** Ahmad Kasem
**Marbleworks of San Diego,** Charlene Butler
**Mashburn Waste & Recycling,** Bill Dean
**M.C. Nottingham Company**
**Mel-Northey Co.,** Mel Northey
**Merli Concrete Pumping,** John Caole
**Microphor,** Ross Beck
**Morrow Equipment Co.,** John Morrow
**Multi Water Systems,** John Kratz
**National Floor Products,** Edsel Holden
**National Schedule Masters,** Steve Scholl
**Nicolon Mirafi Group,** Ted Pope
**North County Interiors,** Rod Gould
**Oregon Lumber Company,** John Couch
**Penhall Construction Services,** Bruce Varney
**Perma Grain Products,** Thomas W. Gunuskey
**Photocom,** Ron Kenedi
**Pittsburgh Corning Corp.,** Joe Zavada
**Plastmo Raingutters,** Ed Gunderson
**Prof. Photographic Service,** Larry Hoagland
**Rally Rack,** Shirlee Turnipseed

**RCP Block & Brick Co.,** Gina Adams
**Rheem Manufacturing,** Rick Miller
**Ryan Roulette Photography,** Ryan Roulette
**R. W. Little Company, Inc.,** Ron Little
**Safeway Steel Products,** Patrick Whelan
**Sauder Manufacturing Co.,** John Gascho
**SDGE,** Don Altevers
**Sico, Inc.,** Charles Williamson
**Simpler Solar Systems,** Al Simpler
**Slater Waterproofing,** Merle Slater, Jr.
**Southern Bleacher Co.,** Glen McNatt
**Specialty Wood Company,** Bob DiNello
**Stark Ceramics, Inc.,** Paul Tauer
**Stebbins Company,** Mrs. Pat Stebbins
**Stone Products Corp.,** Robert Heath
**Superior Fireplace Co.,** Mike Hammer
**Superior Rock Products,** John Knieff
**Superior Stakes,** George Kellett
**Supradur Manufacturing,** Marilyn Mueller
**The Parking Block Store,** Jim Headley
**Towill INC.,** Jim Smith
**U. S. Gypsum Company,** Roger Merchet
**Ultraflo Corporation,** Doug Didion
**Unique Building Products,** Charlie Gindele
**United Panel Partnership,** Donna Korp
**U. S. Tile Company,** Brian Austin
**Velux-America, Inc.,** John Carigan
**Vintage Lumber & Const. Co.,** Mark Harbold
**Vital Services Co.,** Myrna Humphrey
**Wasco Products,** Estelle Gore
**West Coast Lumber Inspect. Bur.,** Jim Kneaper
**Western Colloid Products,** Tim Griffin
**Western Exterminators,** Bill Sharp
**Western Wood Products Assn.,** Frank Stewart
**Westingconsul, Inc.,** Verne T. Kelling
**Weyerhaeuser Company,** Doug Barkee
**York Spiral Stair,** Jim A. Guerette
**Zenith Aerial Photo,** Lew Roberts
**Zeus Lightning Protection,** Dave Silvester

A special thanks to the following people. Special assistants to the editors: Ray Hicks, James Thomson. Layout & Images: Joan Hamilton. Software production: Bilandra Chase.

Cover design: Bill Grote
Cover photography by *Jim Karnik Films*

**Adhesives** See also, Caulking, Flooring, Roofing and Tile in the Residential Division

**Panel adhesives** Better quality, gun applied in continuous bead to wood or metal framing or furring members, material only, add labor below. Per 100 SF of wall, floor, or ceiling including 6% waste. Based on 10.5 ounce tubes.

| | | | Bead diameter | | |
|---|---|---|---|---|---|
| | Unit | 1/8" | 1/4" | 3/8" | 1/2" |
| Subfloor adhesive, on floors | | | | | |
| 12" OC members | CSF | 1.08 | 4.32 | 9.71 | 17.30 |
| 16" OC members | CSF | .81 | 3.24 | 7.29 | 13.00 |
| 20" OC members | CSF | .65 | 2.59 | 5.83 | 10.40 |
| Wall sheathing or shear panel adhesive, on walls | | | | | |
| 16" OC members | CSF | 2.16 | 8.63 | 19.40 | 34.50 |
| 20" OC members | CSF | 1.73 | 6.91 | 15.50 | 27.60 |
| 24" OC members | CSF | 1.44 | 5.76 | 13.00 | 23.00 |
| Polystyrene or polyurethane foam panel adhesive, on walls | | | | | |
| 12" OC members | CSF | 2.70 | 10.80 | 24.30 | 43.20 |
| 16" OC members | CSF | 2.03 | 8.10 | 18.20 | 32.40 |
| 20" OC members | CSF | 1.62 | 6.48 | 14.60 | 25.90 |
| 24" OC members | CSF | 1.35 | 5.40 | 12.20 | 21.60 |
| Gypsum drywall adhesive | | | | | |
| 12" OC members | CSF | 1.19 | 4.77 | 10.70 | 19.10 |
| 16" OC members | CSF | .89 | 3.57 | 8.04 | 14.30 |
| 20" OC members | CSF | .71 | 2.86 | 6.43 | 11.40 |
| 24" OC members | CSF | .60 | 2.38 | 5.36 | 9.53 |
| Hardboard or plastic panel adhesive, on walls | | | | | |
| 12" OC members | CSF | 2.77 | 11.10 | 24.90 | 44.30 |
| 16" OC members | CSF | 2.07 | 8.30 | 18.70 | 33.20 |
| 20" OC members | CSF | 1.66 | 6.64 | 14.90 | 26.60 |
| 24" OC members | CSF | 1.38 | 5.53 | 12.50 | 22.10 |

| | Craft@Hrs | Unit | Material | Labor | Total |
|---|---|---|---|---|---|
| Labor to apply adhesive to framing members, 1/8" to 1/2" bead diameter, no material included | | | | | |
| Floor or ceiling joists | | | | | |
| 12" OC members | BC@.075 | CSF | — | 2.77 | 2.77 |
| 16" OC members | BC@.056 | CSF | — | 2.07 | 2.07 |
| 20" OC members | BC@.052 | CSF | — | 1.92 | 1.92 |
| 24" OC members | BC@.042 | CSF | — | 1.55 | 1.55 |
| Interior and exterior wall members | | | | | |
| 12" OC members | BC@.100 | CSF | — | 3.69 | 3.69 |
| 16" OC members | BC@.090 | CSF | — | 3.32 | 3.32 |
| 20" OC members | BC@.084 | CSF | — | 3.10 | 3.10 |
| 24" OC members | BC@.084 | CSF | — | 3.10 | 3.10 |

**Special purpose adhesives**

| | Craft@Hrs | Unit | Material | Labor | Total |
|---|---|---|---|---|---|
| Construction adhesive, for counters, cabinets, paneling, brick veneer, ceramic fixtures, shelving, sets in 10 minutes, 10.5 oz cartridge | — | Ea | 2.70 | — | 2.70 |
| Epoxy-tie adhesive, bonds concrete, fills cracks to 3/4", seals doors and windows, anchor grouting, non-shrink, 22 oz. cartridge | — | Ea | 28.90 | — | 28.90 |
| Gypsum drywall adhesive, waterproof, bonds to wood metal, masonry, concrete, 29 oz. cartridge | — | Ea | 3.99 | — | 3.99 |

# Adhesives

| | Craft@Hrs | Unit | Material | Labor | Total |
|---|---|---|---|---|---|
| Latex FRP adhesive, indoor, bonds wood, concrete, drywall, foamboard, trowel grade, gallon | — | Ea | 21.00 | — | 21.00 |
| Panel insulation adhesive, waterproof, for polyurethane & polystyrene panels, bonds to wood, metal, masonry, concrete, 16 oz | — | Ea | 27.70 | — | 27.70 |
| Premium floor tile adhesive, solvent base, quart | — | Ea | 9.90 | — | 9.90 |
| Marine adhesive/sealant, 10 oz. | — | Ea | 12.99 | — | 12.99 |
| Project adhesive, rubber-based, bonds plywood, hardboard fiberglass, drywall, foam board, shelving, ceramic fixtures, 10 minute work time, 10.5 oz cartridge | — | Ea | 2.69 | — | 2.69 |
| Tile cement, solvent base, for installation of ceramic, quarry, slate and marble, gallon | — | Ea | 24.99 | — | 24.99 |

| | Unit | 1 quart can | 1 gallon can |
|---|---|---|---|
| **General purpose adhesives** | | | |
| Acoustic tile adhesive, solvent base, waterproof, sound deadening type | Ea | — | 19.50 |
| Aliphatic resin woodworking glue | Ea | 9.50 | 22.00 |
| Carpet cement, outdoor | Ea | 10.00 | — |
| Professional carpenter's glue, multi-purpose | Ea | — | 20.00 |
| Contact cement, rubber based, waterproof, bonds veneers to plywood, particleboard, wallboard | Ea | 15.00 | 30.00 |
| Gel contact cement | Ea | 15.00 | — |
| Resilient flooring adhesive, latex base, adheres to concrete, plywood, felt, sheet flooring | Ea | 6.50 | 16.00 |

| | Craft@Hrs | Unit | Material | Labor | Total |
|---|---|---|---|---|---|
| **Aggregate** Typical prices, 5 mile haul, 24 ton minimum. See also Roofing, Built-up | | | | | |
| Crushed stone (1.4 tons per CY) | | | | | |
| 3/8" stone | — | Ton | 21.50 | — | 21.50 |
| 3/4" (Number 3) | — | Ton | 20.50 | — | 20.50 |
| 1-1/2" (Number 2) | — | Ton | 21.50 | — | 21.50 |
| Crushed slag, typical prices where available | | | | | |
| 3/4" slag | — | Ton | 13.80 | — | 13.80 |
| 1-1/2" | — | Ton | 14.10 | — | 14.10 |
| Washed gravel (1.4 tons per CY) | | | | | |
| 3/4" gravel | — | Ton | 21.50 | — | 21.50 |
| 1-1/2" | — | Ton | 21.50 | — | 21.50 |
| Fill sand (1.35 tons per CY) | — | Ton | 10.00 | — | 10.00 |
| Add per ton less than 24 tons | — | Ton | 6.60 | — | 6.60 |
| Add for delivery over 5 miles, one way | — | Mile | 5.30 | — | 5.30 |

**Appraisal Fees** Costs for determining the value of existing buildings, land, and equipment. Actual fee charged is based on the level of difficulty and the time spent on appraisal plus travel to location and cost of support services, if any. Costs include research and report by a professional state licensed appraiser. Client may request an appraisal on a "fee not to exceed" basis. Fees shown are averages and are not quoted as a percentage of value or contingent on value. The fee for cancelling an appraisal after the inspection is equal to the original appraisal fee.

| | Craft@Hrs | Unit | Material | Labor | Total |
|---|---|---|---|---|---|

Single family residences, condominiums, planned unit developments (PUDs). Fees for complex, expensive (over $1 million) or atypical properties or that require higher licensure than a state license are usually negotiated. Based on square feet of gross living area (excluding the garage).

| | Craft@Hrs | Unit | Material | Labor | Total |
|---|---|---|---|---|---|
| To 2,500 square feet | — | LS | — | — | 350.00 |
| Over 2,500 SF to 3,500 SF | — | LS | — | — | 400.00 |
| Over 3,500 SF to 5,000 SF | — | LS | — | — | 500.00 |
| Over 5,000 SF | — | LS | — | — | 750.00 |

Small residential income properties (duplex, triplex, quadriplex)

| | Craft@Hrs | Unit | Material | Labor | Total |
|---|---|---|---|---|---|
| 2 to 4 units | — | LS | — | — | 650.00 |

Apartment houses, commercial and industrial buildings

| | Craft@Hrs | Unit | Material | Labor | Total |
|---|---|---|---|---|---|
| To $300,000 valuation | — | LS | — | — | 3,000.00 |
| Over $300,000 to $1,000,000 valuation | — | LS | — | — | 4,000.00 |
| Over $1,000,000 to $3,000,000 valuation | — | LS | — | — | 4,500.00 |
| Over $3,000,000 to $5,000,000 valuation | — | LS | — | — | 5,500.00 |

Other services

| | Craft@Hrs | Unit | Material | Labor | Total |
|---|---|---|---|---|---|
| Additional photograph (each) | — | LS | — | — | 25.00 |
| Outlying area fee | — | LS | — | — | 50.00 |
| Satisfactory Completion (old Form 442 / new Form 1004d) | — | LS | — | — | 125.00 |
| Appraisal Update same comps (Form 1004d) | — | LS | — | — | 150.00 |
| Operating Income Statement (FNMA 216) | — | LS | — | — | 75.00 |
| SFR Rental Survey (FNMA 1007) | — | LS | — | — | 75.00 |
| Appraisal review (desk) | — | LS | — | — | 150.00 |
| Appraisal review (field) | — | LS | — | — | 350.00 |
| Drive-by appraisal (Form 2055) starting at | — | LS | — | — | 250.00 |
| Drive-by appraisal (Form 2070/2075) | — | LS | — | — | 200.00 |

Machinery (Fee is based on total value of equipment appraised) Additional charges for travel and lodging may be required

| | Craft@Hrs | Unit | Material | Labor | Total |
|---|---|---|---|---|---|
| To $30,000 valuation | — | LS | — | — | 570.00 |
| Over $30,000 to $100,000 valuation | — | LS | — | — | 770.00 |
| Over $100,000 to $500,000 valuation | — | LS | — | — | 1,300.00 |
| Over $500,000 to $1,000,000 valuation | — | LS | — | — | 1,800.00 |
| Over $1,000,000 to $5,000,000 valuation | — | LS | — | — | 3,100.00 |
| Court testimony (excluding preparation) | — | Day | — | — | 1,400.00 |
| Consulting fees are billed at an hourly rate | — | Hour | — | — | 175.00 |

Outside of office meetings or court appearance are billed including travel time and expenses round trip, minimum 4 hours.

**Arbitration and Mediation Fees** These are administrative fees paid to the American Arbitration Association (AAA). Rules and fees are subject to change. The AAA's web site at www.adr.org has current rules and fee information. Arbitrators are chosen from the National Roster of Construction Arbitrators and are paid a fee by the parties. Legal representation, if desired (although not necessary), is at the expense of each party. These fees do not include rental of a hearing room. An initial filing fee is payable in full by a filing party when a claim, counterclaim or additional claim is filed. A case service fee is payable at the time the first hearing is scheduled. This fee will be refunded at the conclusion of the case if no hearings have occurred so long as the Association was notified of cancellation at least 24 hours before the first scheduled hearing. The minimum filing fee for any case having three or more arbitrators is $2,750 plus a $1,000 case service fee.

# Arbitration and Mediation Fees

| | Unit | Initial Filing Fee | Case Service Fee |
|---|---|---|---|
| Claims to $10,000 | LS | 750.00 | 200.00 |
| Claims over $10,000 to $75,000 | LS | 975.00 | 300.00 |
| Claims over $75,000 to $150,000 | LS | 1,850.00 | 750.00 |
| Claims over $150,000 to $300,000 | LS | 2,800.00 | 1,250.00 |
| Claims over $300,000 to $500,000 | LS | 4,350.00 | 1,750.00 |
| Claims over $500,000 to $1,000,000 | LS | 6,200.00 | 2,500.00 |
| Claims over $1,000,000 to $5,000,000 | LS | 8,200.00 | 3,250.00 |
| Claims over $5,000,000 to $10,000,000 | LS | 10,200.00 | 4,000.00 |
| Claims over $10,000,000 * | LS | * | 6,000.00 |
| No Amount Stated ** | LS | 3,350.00 | 1,250.00 |

* $10 million and above – Base fee is $12,800 plus .01% of the amount of claim above $10 million. Filing fees are capped at $65,000.

**This fee is applicable when a claim or counterclaim is not for a monetary amount. Where a monetary claim is not known, parties will be required to state a range of claims or be subject to the highest possible filing fee.

| | Craft@Hrs | Unit | Material | Labor | Total |
|---|---|---|---|---|---|
| **Mediation, mobile** Generally will meet at the offices of the attorneys involved in the dispute. Does not replace legal counsel but is used to maintain equity between the mediating parties. Typical fees. | | | | | |
| Nonrefundable retainer (half from each party) | — | Ea | — | — | 280.00 |
| Weekday sessions after the initial consultation | — | Hr | — | — | 140.00 |
| Nonrefundable retainer for weekend mediation | — | Ea | — | — | 336.00 |
| Weekend mediation sessions | — | Hr | — | — | 167.00 |
| **Architectural Illustrations** Full color painting on watercolor board with matted frame with title and credit on matte. Glass and elaborate framing are extra. Costs for pen and ink illustrations with color mylar overlay are similar to cost for watercolor illustrations. Typical fees. | | | | | |
| Custom home, eye level view | | | | | |
| Simple rendering | — | LS | — | — | 891.00 |
| Complex rendering | — | LS | — | — | 1,310.00 |
| Custom home, bird's eye view | | | | | |
| Simple rendering | — | LS | — | — | 1,070.00 |
| Complex rendering | — | LS | — | — | 1,520.00 |
| Tract homes in groups of five or more (single floor plans, multiple elevations), eye level view | | | | | |
| Simple rendering | — | LS | — | — | 526.00 |
| Complex rendering | — | LS | — | — | 729.00 |
| Tract homes in groups of five or more (single floor plans, multiple elevations), bird's eye view | | | | | |
| Simple rendering | — | LS | — | — | 678.00 |
| Complex rendering | — | LS | — | — | 1,080.00 |
| Tract homes or condominium project, overall bird's eye view | | | | | |
| 10-25 homes or living units | — | LS | — | — | 3,350.00 |
| Typical commercial structure | | | | | |
| Eye level view | — | LS | — | — | 1,470.00 |
| Bird's eye view | — | LS | — | — | 1,650.00 |
| Complex commercial structure | | | | | |
| Eye level view | — | LS | — | — | 2,170.00 |
| Bird's eye view | — | LS | — | — | 2,810.00 |

| | Craft@Hrs | Unit | Material | Labor | Total |
|---|---|---|---|---|---|
| Deduct for pen and ink drawings (no color) | — | % | — | — | -33.0 |
| Computer generated perspective drawings using CAD system for design studies | | | | | |
| Custom home | — | LS | — | — | 567.00 |
| Large condo or apartment projects | — | LS | — | — | 1,420.00 |
| Tract homes | — | LS | — | — | 354.00 |
| Commercial structure, line drawing | — | LS | — | — | 1,070.00 |

**Awnings and Canopies for Doors and Windows** Costs for awnings include all hardware. All have adjustable support arms to control angle height and preferred amount of window coverage. For larger size aluminum awnings, price two awnings and add for splice kit below. For commercial grade awnings, see the Industrial and Commercial Division section 10, Specialties.

| | Craft@Hrs | Unit | Material | Labor | Total |
|---|---|---|---|---|---|
| Natural aluminum ribbed awning with clear, weather-resistant finish and 26" arms | | | | | |
| 36" wide x 30" long | SW@1.45 | Ea | 59.50 | 59.50 | 119.00 |
| 48" wide x 30" long | SW@1.86 | Ea | 76.80 | 76.30 | 153.10 |
| 60" wide x 30" long | SW@2.07 | Ea | 95.50 | 85.00 | 180.50 |
| 72" wide x 30" long | SW@2.27 | Ea | 108.00 | 93.20 | 201.20 |
| Add for door canopy with 17" drop sides | — | % | 50.0 | — | — |
| Custom colored window awnings in stripes or solids, with baked enamel finish and ventilation panels | | | | | |
| 30" wide x 24" high | SW@1.45 | Ea | 96.30 | 59.50 | 155.80 |
| 36" wide x 36" high | SW@1.45 | Ea | 145.00 | 59.50 | 204.50 |
| 48" wide x 48" high | SW@1.87 | Ea | 245.00 | 76.70 | 321.70 |
| 48" wide x 60" high | SW@1.87 | Ea | 346.00 | 76.70 | 422.70 |
| 60" wide x 72" high | SW@2.07 | Ea | 380.00 | 85.00 | 465.00 |
| 72" wide x 72" high | SW@2.26 | Ea | 520.00 | 92.80 | 612.80 |
| Add for splice kit with overlap slats | SW@.218 | Ea | 20.10 | 8.95 | 29.05 |

Security roll-up awning with pull cord assembly and folding arms, clear weather-resistant finish. Awning rolls down to cover whole window. 48" long, 24" arms

| | Craft@Hrs | Unit | Material | Labor | Total |
|---|---|---|---|---|---|
| 36" wide | SW@1.52 | Ea | 184.00 | 62.40 | 246.40 |
| 48" wide | SW@1.94 | Ea | 220.00 | 79.60 | 299.60 |

Plastic awning with baked-on acrylic finish, ventilated side panels, and reinforced metal frame, hardware included. 24" drop, 24" projection

| | Craft@Hrs | Unit | Material | Labor | Total |
|---|---|---|---|---|---|
| 36" wide | BC@1.58 | Ea | 123.00 | 58.40 | 181.40 |
| 42" wide | BC@1.75 | Ea | 137.00 | 64.60 | 201.60 |
| 48" wide | BC@2.02 | Ea | 158.00 | 74.60 | 232.60 |
| 60" wide | BC@2.26 | Ea | 177.00 | 83.50 | 260.50 |
| 72" wide | BC@2.47 | Ea | 190.00 | 91.20 | 281.20 |
| 96" wide | BC@2.79 | Ea | 241.00 | 103.00 | 344.00 |

Plastic door canopy with 36" projection

| | Craft@Hrs | Unit | Material | Labor | Total |
|---|---|---|---|---|---|
| 42" wide | BC@1.80 | Ea | 260.00 | 66.50 | 326.50 |

Traditional fabric awning, with waterproof, acrylic duck, colorfast fabric, double stitched seams, and tubular metal framing and pull cord assembly. 24" drop, 24" projection

| | Craft@Hrs | Unit | Material | Labor | Total |
|---|---|---|---|---|---|
| 30" wide | BC@1.35 | Ea | 48.90 | 49.90 | 98.80 |
| 36" wide | BC@1.58 | Ea | 64.20 | 58.40 | 122.60 |
| 42" wide | BC@1.80 | Ea | 59.90 | 66.50 | 126.40 |
| 48" wide | BC@2.02 | Ea | 70.40 | 74.60 | 145.00 |
| Add for 30" drop, 30" projection | — | % | 10.0 | 20.0 | — |

Cloth canopy patio cover, with front bar and tension support rafters, 9" valance and 8' projection

| | Craft@Hrs | Unit | Material | Labor | Total |
|---|---|---|---|---|---|
| 8' x 10' | BC@2.03 | Ea | 297.00 | 75.00 | 372.00 |
| 8' x 15' | BC@2.03 | Ea | 409.00 | 75.00 | 484.00 |

# Barricades, Construction Safety

| | Craft@Hrs | Unit | Material | Labor | Total |
|---|---|---|---|---|---|

**Barricades, Construction Safety** Purchase prices except as noted. See also Equipment Rental in the index.

Heavy Duty Plastic Type I barricade, Engineer Grade Reflectivity, 8" height Top Panel

| | | | | | |
|---|---|---|---|---|---|
| Quantities from 1 - 15 units | — | Ea | 51.60 | — | 51.60 |
| Quantities from 16 - 31 units | — | Ea | 43.00 | — | 43.00 |
| Pallet quantities, multiples of 32 units | — | Ea | 39.00 | — | 39.00 |

Reflectorized plastic, injected barricade, NCHRP-350 certified meets MUTCD specifications 8" to 12" wide rail, 4" to 6" wide stripes, 40" legs, no light

Type I, 2' wide, 3' high, Top panel 12" height, 1 reflectorized rail each side

| | | | | | |
|---|---|---|---|---|---|
| Quantities from 1 - 15 units | — | Ea | 56.20 | — | 56.20 |
| Quantities from 16 - 31 units | — | Ea | 46.90 | — | 46.90 |
| Pallet quantities, multiples of 32 units | — | Ea | 40.00 | — | 40.00 |

Type II, 2' wide, 3' high, Top panel 8" height, 1 reflectorized rail each side

| | | | | | |
|---|---|---|---|---|---|
| Quantities from 1 - 15 units | — | Ea | 59.80 | — | 59.80 |
| Quantities from 16 - 31 units | — | Ea | 52.80 | — | 52.80 |
| Pallet quantities, multiples of 32 units | — | Ea | 47.80 | — | 47.80 |

Type III, 4' wide, 5' high, 3 reflectorized rails each side, wood & steel legs

| | | | | | |
|---|---|---|---|---|---|
| Quantities from 1 - 15 units | — | Ea | 58.70 | — | 58.70 |
| Quantities from 16 - 31 units | — | Ea | 53.90 | — | 53.90 |
| Pallet quantities, multiples of 32 units | — | Ea | 46.50 | — | 46.50 |

Add for lighted units without batteries (batteries last 2 months)

| | | | | | |
|---|---|---|---|---|---|
| To 100 units | — | Ea | 14.00 | — | 14.00 |
| Over 100 units | — | Ea | 14.00 | — | 14.00 |
| Batteries, 6 volt (2 needed) | — | Ea | 2.75 | — | 2.75 |

Utility barricade solar powered lights, compliant with M.U.T.C.D. Specification 6E-5 and ITE Specifications for Flashing / Steady Burn warning light, 180 degree swivel base axis.

| | | | | | |
|---|---|---|---|---|---|
| | — | Ea | 34.50 | — | 34.50 |
| Lighted units, rental, per day | — | Ea | .35 | — | .35 |
| Unlighted units, rental per day | — | Ea | .25 | — | .25 |
| Add for pickup and delivery, per trip | — | Ea | 35.00 | — | 35.00 |
| "Road Closed", reflectorized, 30" x 48", Engineer Grade | — | Ea | 82.20 | — | 82.20 |
| "Construction Zone", 4' x 4' high intensity grade | — | Ea | 223.00 | — | 223.00 |
| High-rise tripod with 3 orange flags | — | Ea | 173.00 | — | 173.00 |
| Flags | — | Ea | 2.95 | — | 2.95 |

Traffic cones, PVC

Non-reflectorized type

| | | | | | |
|---|---|---|---|---|---|
| 18" high | — | Ea | 9.20 | — | 9.20 |
| 28" high | — | Ea | 13.30 | — | 13.30 |

Reflectorized type

| | | | | | |
|---|---|---|---|---|---|
| 18" high | — | Ea | 10.00 | — | 10.00 |
| 28" high | — | Ea | 17.80 | — | 17.80 |

Lane delineator, 42" orange plastic cylinder with 2 reflectors on a 12 pound

| | | | | | |
|---|---|---|---|---|---|
| rubber base | — | Ea | 19.00 | — | 19.00 |

Mesh signs, orange, 48" x 48", includes brace and

| | | | | | |
|---|---|---|---|---|---|
| clamp | — | Ea | 62.50 | — | 62.50 |
| Hand-held traffic paddles, "Stop" and "Slow" | — | Ea | 23.90 | — | 23.90 |

Typical labor cost, place and remove any barricade

| | | | | | |
|---|---|---|---|---|---|
| Per use | BL@.160 | Ea | — | 4.75 | 4.75 |

| | Craft@Hrs | Unit | Material | Labor | Total |
|---|---|---|---|---|---|
| Orange plastic safety fencing | | | | | |
| 4' x 50' roll | — | Ea | 54.30 | — | 54.30 |
| 5' x 50' roll | — | Ea | 108.00 | — | 108.00 |
| 6' x 50' roll | — | Ea | 111.00 | — | 111.00 |
| 4' x 100', light duty | — | Ea | 32.20 | — | 32.20 |
| Barricade tape for marking construction sites, landscaping, wet paint | | | | | |
| Yellow plastic 3" x 1000', "Caution" | — | Ea | 9.60 | — | 9.60 |
| Various colors, 3" x 200', "Danger" | — | Ea | 5.90 | — | 5.90 |

**Basement Doors** Good quality 12 gauge primed steel, center opening basement doors. Costs include assembly and installation hardware. No concrete, masonry, anchor placement or finish painting included. Bilco Company www.bilco.com

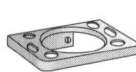

| | Craft@Hrs | Unit | Material | Labor | Total |
|---|---|---|---|---|---|
| Doors (overall dimensions) | | | | | |
| 30"H, 47"W, 58"L | BC@3.41 | Ea | 358.00 | 126.00 | 484.00 |
| 22"H, 51"W, 64"L | BC@3.41 | Ea | 365.00 | 126.00 | 491.00 |
| 19-1/2"H, 55"W, 72"L | BC@3.41 | Ea | 399.00 | 126.00 | 525.00 |
| 52"H, 51"W, 43-1/4"L | BC@3.41 | Ea | 428.00 | 126.00 | 554.00 |
| Door extensions (available for 19-1/2"H, 55"W, 72"L door only) | | | | | |
| 6" deep | BC@1.71 | Ea | 99.10 | 63.20 | 162.30 |
| 12" deep | BC@1.71 | Ea | 118.00 | 63.20 | 181.20 |
| 18" deep | BC@1.71 | Ea | 147.00 | 63.20 | 210.20 |
| 24" deep | BC@1.71 | Ea | 176.00 | 63.20 | 239.20 |
| Stair stringers, steel, pre-cut for 2" x 10" wood treads (without treads) | | | | | |
| 32" to 39" stair height | BC@1.71 | Ea | 83.70 | 63.20 | 146.90 |
| 48" to 55" stair height | BC@1.71 | Ea | 105.00 | 63.20 | 168.20 |
| 56" to 64" stair height | BC@1.71 | Ea | 118.00 | 63.20 | 181.20 |
| 65" to 72" stair height | BC@1.71 | Ea | 132.00 | 63.20 | 195.20 |
| 73" to 78" stair height | BC@1.71 | Ea | 182.00 | 63.20 | 245.20 |
| 81" to 88" stair height | BC@1.71 | Ea | 197.00 | 63.20 | 260.20 |
| 89" to 97" stair height | BC@1.71 | Ea | 210.00 | 63.20 | 273.20 |

**Bathroom Accessories** Average quality. Better quality brass accessories cost 75% to 100% more. See also Medicine Cabinets and Vanities

| | Craft@Hrs | Unit | Material | Labor | Total |
|---|---|---|---|---|---|
| Cup and toothbrush holder, chrome | BC@.258 | Ea | 10.70 | 9.53 | 20.23 |
| Cup holder, porcelain, surface mounted | BC@.258 | Ea | 7.55 | 9.53 | 17.08 |
| Cup, toothbrush & soap holder, recessed | BC@.258 | Ea | 28.50 | 9.53 | 38.03 |
| Cup, toothbrush holder, polished brass | BC@.258 | Ea | 15.20 | 9.53 | 24.73 |
| Electrical plates, chrome plated | | | | | |
| Switch plate, single | BE@.154 | Ea | 4.01 | 6.10 | 10.11 |
| Switch plate, double | BE@.154 | Ea | 5.28 | 6.10 | 11.38 |
| Duplex receptacle plate | BE@.154 | Ea | 4.01 | 6.10 | 10.11 |
| Duplex receptacle and switch | BE@.154 | Ea | 5.28 | 6.10 | 11.38 |
| Grab bars | | | | | |
| Tubular chrome plated, with anchor plates | | | | | |
| Straight bar, 16" | BC@.414 | Ea | 23.50 | 15.30 | 38.80 |
| Straight bar, 24" | BC@.414 | Ea | 28.50 | 15.30 | 43.80 |
| Straight bar, 32" | BC@.414 | Ea | 31.10 | 15.30 | 46.40 |
| "L"- shaped bar, 16" x 32" | BC@.620 | Ea | 71.90 | 22.90 | 94.80 |

# Bathroom Accessories

| | Craft@Hrs | Unit | Material | Labor | Total |
|---|---|---|---|---|---|
| Stainless steel, with anchor plates | | | | | |
| Straight bar, 16" | BC@.414 | Ea | 35.50 | 15.30 | 50.80 |
| Straight bar, 24" | BC@.414 | Ea | 42.80 | 15.30 | 58.10 |
| Straight bar, 32" | BC@.414 | Ea | 46.50 | 15.30 | 61.80 |
| "L"- shaped bar, 16" x 32" | BC@.620 | Ea | 100.00 | 22.90 | 122.90 |
| Mirrors, stainless steel framed, surface mount, no light or cabinet | | | | | |
| 16" high x 20" wide | BG@.420 | Ea | 56.70 | 14.40 | 71.10 |
| 18" high x 24" wide | BG@.420 | Ea | 65.90 | 14.40 | 80.30 |
| 18" high x 36" wide | BG@.420 | Ea | 101.00 | 14.40 | 115.40 |
| 24" high x 36" wide | BG@.420 | Ea | 118.00 | 14.40 | 132.40 |
| 48" high x 24" wide | BG@.420 | Ea | 145.00 | 14.40 | 159.40 |
| Mirrors, wood framed, surface mount, better quality | | | | | |
| 18" x 29" rectangular | BG@.420 | Ea | 70.40 | 14.40 | 84.80 |
| 20" x 27" oval, oak | BG@.420 | Ea | 100.00 | 14.40 | 114.40 |
| Robe hook | | | | | |
| Chrome | BC@.258 | Ea | 21.10 | 9.53 | 30.63 |
| Double, solid brass | BC@.258 | Ea | 20.30 | 9.53 | 29.83 |

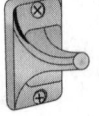

| | Craft@Hrs | Unit | Material | Labor | Total |
|---|---|---|---|---|---|
| Shower curtain rods, chrome plated | | | | | |
| 60", recessed | BC@.730 | Ea | 24.70 | 27.00 | 51.70 |
| 66", recessed | BC@.730 | Ea | 27.20 | 27.00 | 54.20 |
| Soap holder, surface mounted, with drain holes | | | | | |
| Williamsburg, satin chrome | BC@.258 | Ea | 50.20 | 9.53 | 59.73 |
| Polished brass | BC@.258 | Ea | 16.50 | 9.53 | 26.03 |
| Facial tissue holder, stainless steel, recessed | BC@.258 | Ea | 33.80 | 9.53 | 43.33 |
| Toilet tissue roll holder, chrome, recessed | BC@.258 | Ea | 15.80 | 9.53 | 25.33 |
| Toothbrush holder, chrome, surface mount | BC@.258 | Ea | 5.83 | 9.53 | 15.36 |

| | Craft@Hrs | Unit | Material | Labor | Total |
|---|---|---|---|---|---|
| Towel bars, 3/4" round bar | | | | | |
| 18" long, chrome | BC@.280 | Ea | 12.90 | 10.30 | 23.20 |
| 24" long, chrome | BC@.280 | Ea | 15.30 | 10.30 | 25.60 |
| 30" long, chrome | BC@.280 | Ea | 18.10 | 10.30 | 28.40 |
| 36" long, chrome | BC@.280 | Ea | 20.70 | 10.30 | 31.00 |
| 18" long, solid brass | BC@.280 | Ea | 42.10 | 10.30 | 52.40 |
| 24" long, solid brass | BC@.280 | Ea | 44.20 | 10.30 | 54.50 |
| Towel rack, swing-arm, chrome, 3 bars, 12" L | BC@.280 | Ea | 14.60 | 10.30 | 24.90 |

| | Craft@Hrs | Unit | Material | Labor | Total |
|---|---|---|---|---|---|
| Towel rings | | | | | |
| Williamsburg chrome and brass | BC@.280 | Ea | 39.90 | 10.30 | 50.20 |
| Williamsburg chrome and porcelain | BC@.280 | Ea | 38.00 | 10.30 | 48.30 |
| Towel shelf, chrome, 24" L with bar below | BC@.280 | Ea | 37.20 | 10.30 | 47.50 |

Heated towel racks, 16" width, mounted on wall with brackets. Wall plug in or direct connection, 700 watt output. Add for electrical work

| | Craft@Hrs | Unit | Material | Labor | Total |
|---|---|---|---|---|---|
| 26" high, standard colors | BE@.850 | Ea | 646.00 | 33.70 | 679.70 |
| Add for control panel | BE@.850 | Ea | 169.00 | 33.70 | 202.70 |
| 26" high, chrome finish | BE@.850 | Ea | 1,040.00 | 33.70 | 1,073.70 |
| Add for control panel | BE@.850 | Ea | 194.00 | 33.70 | 227.70 |
| 35" high, standard colors | BE@1.00 | Ea | 719.00 | 39.60 | 758.60 |
| Add for control panel | BE@1.00 | Ea | 172.00 | 39.60 | 211.60 |
| 35" high, chrome finish | BE@1.00 | Ea | 1,130.00 | 39.60 | 1,169.60 |
| Add for control panel | BE@1.00 | Ea | 198.00 | 39.60 | 237.60 |
| 44" high, standard colors | BE@1.10 | Ea | 808.00 | 43.50 | 851.50 |
| Add for control panel | BE@1.10 | Ea | 187.00 | 43.50 | 230.50 |
| 44" high, chrome finish | BE@1.10 | Ea | 1,230.00 | 43.50 | 1,273.50 |
| Add for control panel | BE@1.10 | Ea | 215.00 | 43.50 | 258.50 |
| Add for 24" width rack | BE@1.10 | Ea | 40.40 | 43.50 | 83.90 |

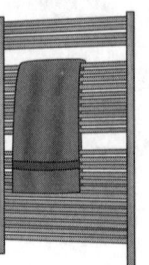

| | Craft@Hrs | Unit | Material | Labor | Total |
|---|---|---|---|---|---|
| Heated towel rack accessories | | | | | |
|   Programmable control panel | BE@.250 | Ea | 240.00 | 9.90 | 249.90 |
|   Pull out drying racks | BE@.250 | Ea | 65.00 | 9.90 | 74.90 |
|   Robe knobs | BE@.150 | Ea | 42.00 | 5.94 | 47.94 |
|   Towel bar accents with standard colors | BE@.150 | Ea | 85.00 | 5.94 | 90.94 |
|   Towel bar accents with chrome finish | BE@.150 | Ea | 115.00 | 5.94 | 120.94 |
|   Robe knob accents (all styles) | BE@.150 | Ea | 199.00 | 5.94 | 204.94 |
| Tub or shower chairs | | | | | |
|   Wall-hung elevated fixed seat | BE@.750 | Ea | 29.10 | 29.70 | 58.80 |
|   White adjustable seat | BE@.750 | Ea | 56.00 | 29.70 | 85.70 |

**Beds, Folding** Concealed-in-wall type. Steel framed, folding wall bed system. Bed requires 18-5/8" or 22" deep recess. Includes frame, lift mechanism, all hardware. Installed in framed opening. Padded vinyl headboard. Bed face panel accepts paint, wallpaper, vinyl or laminate up to 1/4" thick. Add for box spring and mattress. Sico Incorporated www.sico-wallbeds.com

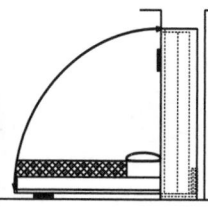

| | Craft@Hrs | Unit | Material | Labor | Total |
|---|---|---|---|---|---|
| Twin, 41"W x 83-1/2"H x 18-5/8"D | B1@5.41 | Ea | 1,070.00 | 180.00 | 1,250.00 |
|   Box spring and mattress | — | Ea | 369.00 | — | 369.00 |
| Twin-long, 41"W x 88-5/8"H x 18-5/8"D | B1@5.41 | Ea | 1,090.00 | 180.00 | 1,270.00 |
|   Box spring and mattress | — | Ea | 386.00 | — | 386.00 |
| Double, 56"W x 83-1/2"H x 18-5/8"D | B1@5.41 | Ea | 1,070.00 | 180.00 | 1,250.00 |
|   Box spring and mattress | — | Ea | 461.00 | — | 461.00 |
| Double-long, 56"W x 88-5/8-1/2"H x 18-5/8"D | B1@5.41 | Ea | 1,090.00 | 180.00 | 1,270.00 |
|   Box spring and mattress | — | Ea | 488.00 | — | 488.00 |
| Queen, 62-1/2"W x 88-5/8-1/2"H x 18-5/8"D | B1@5.41 | Ea | 1,160.00 | 180.00 | 1,340.00 |
|   Box spring and mattress | — | Ea | 542.00 | — | 542.00 |
| King, 79"W x 88-5/8"H x 18-5/8"D | B1@5.41 | Ea | 1,930.00 | 180.00 | 2,110.00 |
|   Box spring and mattress | — | Ea | 779.00 | — | 779.00 |
| Add for chrome frame, any bed | — | LS | 233.00 | — | 233.00 |
| Add for brass frame, any bed | — | LS | 250.00 | — | 250.00 |

**Blueprinting** (Reproduction only) Assumes original is on semi-transparent drafting paper or film. See also Architectural Illustration and Drafting. Cost per square foot reproduced except as noted. Stapled edge and binder included. Diazo

| | Craft@Hrs | Unit | Material | Labor | Total |
|---|---|---|---|---|---|
| Blueline or blackline prints | | | | | |
|   1-100 SF | — | SF | — | — | .20 |
|   101-1,000 SF | — | SF | — | — | .18 |
|   1,001-2,000 SF | — | SF | — | — | .17 |
|   2,001-to 3,000 SF | — | SF | — | — | .16 |
|   3,001-4,000 SF | — | SF | — | — | .09 |
|   4,001 SF and higher | — | SF | — | — | .08 |
|   Presentation blackline (heavy paper) | — | SF | — | — | .50 |
|   Sepia | — | SF | — | — | .75 |
|   Mylar | — | SF | — | — | 2.50 |
| Xerographic prints | | | | | |
|   Vellum | — | SF | — | — | 1.00 |
|   Erasable vellum | — | SF | — | — | 1.25 |
|   Mylar | — | SF | — | — | 2.50 |
|   Enlargements (bond) | — | SF | — | — | .60 |
|   Reductions (bond, per sheet) | — | Ea | — | — | 2.50 |

# Building Inspection Service

| | Craft@Hrs | Unit | Material | Labor | Total |
|---|---|---|---|---|---|
| Plotting prints | | | | | |
| Translucent bond | — | SF | — | — | 1.50 |
| Vellum | — | SF | — | — | 1.75 |
| Erasable vellum | — | SF | — | — | 2.50 |
| Mylar | — | SF | — | — | 2.50 |
| Photo prints | | | | | |
| Mylar | — | SF | — | — | 5.65 |
| Add for local pickup and delivery, round trip | — | LS | — | — | 7.50 |

**Building Inspection Service** (Home inspection service) Inspection of all parts of building by qualified engineer or certified building inspection technician. Includes written report covering all doors and windows, electrical system, foundation, heating and cooling system, insulation, interior and exterior surface conditions, landscaping, plumbing system, roofing, and structural integrity.

| | Craft@Hrs | Unit | Material | Labor | Total |
|---|---|---|---|---|---|
| Single-family residence | | | | | |
| Base fee (up to 2,500 SF) | — | LS | — | — | 325.00 |
| Add for additional 1,000 SF or fraction | — | LS | — | — | 100.00 |
| Add for out buildings (each) | — | LS | — | — | 50.00 |
| Add for houses over 50 years old | — | LS | — | — | 75.00 |
| Add per room for houses with over 10 rooms | — | Ea | — | — | 60.00 |
| Add per room for houses with over 15 rooms | — | Ea | — | — | 65.00 |
| Add for swimming pool, spa or sauna | — | LS | — | — | 200.00 |
| Add for soil testing (expansive soil only) | — | LS | — | — | 200.00 |
| Add for water testing (coliform only) | — | LS | — | — | 70.00 |
| Add for warranty protection | | | | | |
| Houses to 10 rooms & 50 years old | — | LS | — | — | 250.00 |
| Houses over 50 years old | — | LS | — | — | 270.00 |
| Houses over 10 rooms | — | LS | — | — | 270.00 |
| Multi-family structures | | | | | |
| Two family residence base fee | — | LS | — | — | 450.00 |
| Apartment or condominium base fee | — | LS | — | — | 250.00 |
| Warranty protection (base cost) | — | LS | — | — | 250.00 |
| Add for each additional unit | — | LS | — | — | 50.00 |
| Add for each family living unit | | | | | |
| Standard inspection | — | LS | — | — | 50.00 |
| Detailed inspection | — | LS | — | — | 75.00 |
| Add for swimming pool, spa, sauna | — | LS | — | — | 75.00 |
| Add for potable water quality testing | — | LS | — | — | 225.00 |
| Add for water quantity test, per well | — | LS | — | — | 150.00 |
| Add for soil testing (EPA toxic) | — | LS | — | — | 1,500.00 |
| Add for soil testing (lead) | — | LS | — | — | 45.00 |
| Add for lead paint testing, full analysis, per room | — | LS | — | — | 40.00 |
| Hazards testing for single and multi-family dwellings | | | | | |
| Urea-formaldehyde insulation testing | — | LS | — | — | 175.00 |
| Asbestos testing | — | LS | — | — | 175.00 |
| Radon gas testing | — | LS | — | — | 125.00 |
| Geotechnical site examination, typical price | — | LS | — | — | 400.00 |

# Building Paper

| | Craft@Hrs | Unit | Material | Labor | Total |
|---|---|---|---|---|---|

**Building Paper**  See also Roofing for roof applications and Polyethylene Film. Costs include 7% coverage allowance for 2" lap and 5% waste allowance. See installation costs at the end of this section.

Asphalt felt, 36" wide

| | | | | | |
|---|---|---|---|---|---|
| 15 lb., ASTM F45 (432 SF roll) | — | SF | .04 | — | .04 |
| 15 lb., F40 (432 SF roll) | — | SF | .06 | — | .06 |
| 30 lb. (216 SF roll) | — | SF | .13 | — | .13 |

Asphalt shake felt, 18" x 120'

| | | | | | |
|---|---|---|---|---|---|
| 30 lb. (180 SF roll) | — | SF | .16 | — | .16 |

Building paper, 40" wide

| | | | | | |
|---|---|---|---|---|---|
| Single ply, black (1,078 SF roll) | — | SF | .02 | — | .02 |
| Two ply (539 SF roll) | — | SF | .03 | — | .03 |

Aquabar™, two layer, laminated with asphalt, Fortifiber™ products

Class A, 36" wide, 30-50-30

| | | | | | |
|---|---|---|---|---|---|
| (1,000 SF roll) | — | SF | .04 | — | .04 |

Class B, 36" wide, 30-30-30

| | | | | | |
|---|---|---|---|---|---|
| (500 SF roll) | — | SF | .05 | — | .05 |

Ice and water shield, self-adhesive rubberized asphalt and poly

| | | | | | |
|---|---|---|---|---|---|
| 225 SF roll | — | SF | 1.00 | — | 1.00 |
| 201 SF roll | — | SF | .79 | — | .79 |

"Jumbo Tex" gun grade sheathing paper, 40" wide, asphalt saturated

| | | | | | |
|---|---|---|---|---|---|
| (324 SF roll) | — | SF | .05 | — | .05 |

"Jumbo Tex" black building paper, 36", 40" wide, asphalt saturated

| | | | | | |
|---|---|---|---|---|---|
| (500 SF roll) | — | SF | .03 | — | .03 |

"Super Jumbo Tex" two-ply, 60 minute, asphalt saturated Kraft

| | | | | | |
|---|---|---|---|---|---|
| (162 SF roll) | — | SF | .11 | — | .11 |

Red rosin sized sheathing (duplex sheathing) 36" wide

| | | | | | |
|---|---|---|---|---|---|
| (484 SF roll at $13.90) | — | SF | .03 | — | .03 |

Bruce rosin paper, floor underlay

| | | | | | |
|---|---|---|---|---|---|
| (500 SF roll at $13.20) | — | SF | .04 | — | .04 |

Moistop flashing paper, Fortifiber™, 12" wide x 300' long

| | | | | | |
|---|---|---|---|---|---|
| (300 SF roll) | — | SF | .13 | — | .13 |

Vycor window and door flashing, Grace, 6" x 75' roll

| | | | | | |
|---|---|---|---|---|---|
| 40 mil thickness | — | SF | .83 | — | .83 |

Vycor deck protector, self adhesive joist flashing, 4" x 75' roll

| | | | | | |
|---|---|---|---|---|---|
| Roll covers 66 SF of deck | — | SF | .30 | — | .30 |

Plasterkraft grade D weather-resistive sheathing paper, 40" wide

| | | | | | |
|---|---|---|---|---|---|
| Super 60 minute (500 SF roll) | — | SF | .04 | — | .04 |
| Ten minute 2-ply (500 SF roll) | — | SF | .03 | — | .03 |

Roof flashing paper, seals around skylights, dormers, vents, valleys and eaves, rubberized, fiberglass reinforced, self-adhesive, GAF StormGuard™

| | | | | | |
|---|---|---|---|---|---|
| Roll covers 200 SF | — | SF | .64 | — | .64 |

Roof underlay, for use under shingles, granular surface, Weatherlock®

| | | | | | |
|---|---|---|---|---|---|
| 200 SF roll | — | SF | .33 | — | .33 |

Tri-Flex 30 roof underlay, for use under shingles, tile, slate or metal roofing, 41.5" x 116'

| | | | | | |
|---|---|---|---|---|---|
| 400 SF roll | — | SF | .21 | — | .21 |

Above grade vapor barrier, no waste or coverage allowance included. Fortifiber™ products

Vaporstop 298, (fiberglass reinforcing and asphaltic adhesive between 2 layers of Kraft)

| | | | | | |
|---|---|---|---|---|---|
| 32" x 405' roll (1,080 SF roll) | — | SF | .11 | — | .11 |

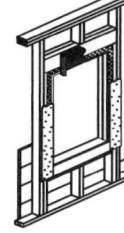

# Building Paper

| | Craft@Hrs | Unit | Material | Labor | Total |
|---|---|---|---|---|---|
| Pyro-Kure™ 600, (2 layers of heavy Kraft with adhesive fire retardant edge reinforced with fiberglass) | | | | | |
| 32" x 405' roll (1,080 SF roll) | — | SF | .18 | — | .18 |
| Foil Barrier 718 (fiberglass reinforced aluminum foil and adhesive between 2 layers of Kraft) | | | | | |
| 52" x 231' roll (1,000 SF roll) | — | SF | .32 | — | .32 |
| Below grade vapor barrier, Fortifiber™ Moistop (fiberglass reinforced Kraft between 2 layers of polyethylene) | | | | | |
| 8' x 250' roll (2,000 SF roll) | — | SF | .15 | — | .15 |
| Concrete curing papers, Fortifiber™ | | | | | |
| Orange Label Sisalkraft (fiberglass and adhesive between 2 layers of Kraft), 4.8 lbs. per CSF | | | | | |
| 48" x 125' roll, (500 SF roll) | — | SF | .23 | — | .23 |
| Sisalkraft SK-10, economy papers, fiberglass and adhesive between 2 layers of Kraft, 4.2 lbs. per CSF, 48" x 300' roll (1,200 SF roll) | — | SF | .07 | — | .07 |
| Protective paper, Fortifiber™ Seekure (fiberglass reinforcing strands and nonstaining adhesive between 2 layers of Kraft) | | | | | |
| 48" x 300' roll, (1,200 SF) | — | SF | .11 | — | .11 |
| Tyvek™ house wrap by DuPont | | | | | |
| Air infiltration barrier (high-density polyethylene fibers in sheet form) | | | | | |
| 3' x 100' rolls or | | | | | |
| 9' x 100' rolls | — | SF | .20 | — | .20 |
| House wrap tape, 2" x 165' | — | LF | .11 | — | .11 |
| Labor to install building papers | | | | | |
| Felts, vapor barriers, infiltration barriers, building papers on walls | | | | | |
| Tack stapled, typical | BC@.002 | SF | — | .07 | .07 |
| Heavy stapled, typical | BC@.003 | SF | — | .11 | .11 |
| Felts, vapor barriers, infiltration barriers, building papers on ceilings and roofs | | | | | |
| Tack stapled, typical | BC@.004 | SF | — | .15 | .15 |
| Heavy stapled, typical | BC@.006 | SF | — | .22 | .22 |
| Self-adhesive, typical | BC@.006 | SF | — | .22 | .22 |
| Curing papers, protective papers and vapor barriers, minimal fasteners | BC@.001 | SF | — | .04 | .04 |
| Flashing papers, 6" to 8" wide | BC@.010 | LF | — | .37 | .37 |

**Building Permit Fees** Fees are set by each jurisdiction and can vary widely. Building departments publish current fee schedules. The permit fee will usually be doubled when work is started without securing a permit. When the valuation of the proposed construction exceeds $1,000, plans are usually required. Estimate the plan check fee at 65% of the permit fee for residences and 100% of the permit fee for non-residential buildings. Estimate the fee for reinspection at $100 per hour. Inspections outside normal business hours are about $100 per hour with a two hour minimum. Estimate the fee for additional plan review required by changes, additions or revisions to approved plans at $100 per hour with a one-half hour minimum. Plumbing, electrical and mechanical work will usually require a separate permit based on a similar fee schedule. Valuations are based on a table published by the I.C.C. at: http://www.iccsafe.org/cs/Documents/BVD.pdf

The minimum fee for construction values to $500 is $70.00

| | Craft@Hrs | Unit | Material | Labor | Total |
|---|---|---|---|---|---|
| $500 to $2,000, for the first $500 | — | LS | — | — | 70.00 |
| each extra $100 or fraction, to $2,000 | — | LS | — | — | 3.70 |
| $2,000 to $25,000, for the first $2,000 | — | LS | — | — | 125.65 |
| each extra $1,000 or fraction, to $25,000 | — | LS | — | — | 24.50 |
| $25,000 to $50,000, for the first $25,000 | — | LS | — | — | 689.00 |
| each extra $1,000 or fraction to $50,000 | — | LS | — | — | 17.65 |

| | Craft@Hrs | Unit | Material | Labor | Total |
|---|---|---|---|---|---|
| $50,000 to $100,000, for the first $50,000 | — | LS | — | — | 1,130.00 |
| each extra $1,000 or fraction, to $100,000 | — | LS | — | — | 12.20 |
| $100,000 to $500,000, for the first $100,000 | — | LS | — | — | 1,740.00 |
| each extra $1,000 or fraction, to $500,000 | — | LS | — | — | 9.79 |
| $500,000 to $1,000,000, for the first $500,000 | — | LS | — | — | 6,560.00 |
| each extra $1,000 or fraction, to $1,000,000 | — | LS | — | — | 8.30 |
| For $1,000,000 | — | LS | — | — | 9,810.00 |
| each extra $1,000 or fraction thereof | — | LS | — | — | 5.84 |

**Cabinets, Kitchen** See also Vanities. Good quality mill-made modular units with solid hardwood face frames, hardwood door frames and drawer fronts, hardwood veneer on raised door panels (front and back), glued mortise, dowel, and dado joint construction, full backs (1/8" vinyl laminated plywood), vinyl laminated cabinet interiors, vinyl laminated composition drawer bodies with nylon and metal guides. Includes self-closing hinges, door and drawer pulls, mounting hardware and adjustable shelves. See illustrations for unit types. See the price adjustments below for pricing of other units. No countertops included. See Countertops.

Kitchen cabinet costs vary widely. The prices listed in this section are for standard grade cabinets. Add 65% to material costs for premium grade cabinets with solid hardwood fronts and frames, mitered corners and solid wood drawer bodies with steel guides and ball bearings. Deduct 45% from material costs for economy grade cabinets, melamine laminated to particleboard.

**Cabinets, Rule of Thumb** Cabinet cost per running foot of cabinet installed. These figures are based on a set of mill-fabricated and assembled kitchen cabinets including a sink base cabinet, one 3-drawer base cabinet and six door base cabinets. Per linear foot of front or back edge, whichever is longer.

| | Craft@Hrs | Unit | Material | Labor | Total |
|---|---|---|---|---|---|
| Base cabinets, 34-1/2" high, 24" deep | BC@.521 | LF | 144.00 | 19.20 | 163.20 |
| Wall cabinets, 30" high, 12" deep | BC@.340 | LF | 75.20 | 12.60 | 87.80 |

**Cabinets, Kitchen** (See the note above concerning cabinet costs.)

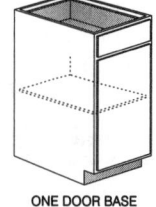

ONE DOOR BASE

DRAWER BASE

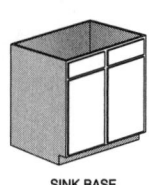

SINK BASE

| | Craft@Hrs | Unit | Material | Labor | Total |
|---|---|---|---|---|---|
| One door base cabinets, 34-1/2" high, 24" deep | | | | | |
| 9" wide, tray divider | BC@.461 | Ea | 134.00 | 17.00 | 151.00 |
| 12" wide, 1 door, 1 drawer | BC@.461 | Ea | 182.00 | 17.00 | 199.00 |
| 15" wide, 1 door, 1 drawer | BC@.638 | Ea | 193.00 | 23.60 | 216.60 |
| 18" wide, 1 door, 1 drawer | BC@.766 | Ea | 203.00 | 28.30 | 231.30 |
| 21" wide, 1 door, 1 drawer | BC@.766 | Ea | 222.00 | 28.30 | 250.30 |
| 24" wide, 1 door, 1 drawer | BC@.911 | Ea | 228.00 | 33.70 | 261.70 |
| Drawer base cabinets, 34-1/2" high, 24" deep | | | | | |
| 15" wide, 4 drawers | BC@.638 | Ea | 198.00 | 23.60 | 221.60 |
| 18" wide, 4 drawers | BC@.766 | Ea | 211.00 | 28.30 | 239.30 |
| 24" wide, 4 drawers | BC@.911 | Ea | 244.00 | 33.70 | 277.70 |
| Sink base cabinets, 34-1/2" high, 24" deep | | | | | |
| 24" wide, 1 door, 1 drawer front | BC@.740 | Ea | 193.00 | 27.30 | 220.30 |
| 30" wide, 2 doors, 2 drawer fronts | BC@.766 | Ea | 234.00 | 28.30 | 262.30 |
| 33" wide, 2 doors, 2 drawer fronts | BC@.766 | Ea | 244.00 | 28.30 | 272.30 |
| 36" wide, 2 doors, 2 drawer fronts | BC@.766 | Ea | 250.00 | 28.30 | 278.30 |
| 42" wide, 2 doors, 2 drawer fronts | BC@.911 | Ea | 277.00 | 33.70 | 310.70 |
| 48" wide, 2 doors, 2 drawer fronts | BC@.911 | Ea | 304.00 | 33.70 | 337.70 |

# Cabinets, Kitchen

| | Craft@Hrs | Unit | Material | Labor | Total |
|---|---|---|---|---|---|
| **Two door base cabinets, 34-1/2" high, 24" deep** | | | | | |
| 27" wide, 2 doors, 2 drawers | BC@1.25 | Ea | 304.00 | 46.20 | 350.20 |
| 30" wide, 2 doors, 2 drawers | BC@1.25 | Ea | 321.00 | 46.20 | 367.20 |
| 33" wide, 2 doors, 2 drawers | BC@1.25 | Ea | 335.00 | 46.20 | 381.20 |
| 36" wide, 2 doors, 2 drawers | BC@1.35 | Ea | 353.00 | 49.90 | 402.90 |
| 42" wide, 2 doors, 2 drawers | BC@1.50 | Ea | 368.00 | 55.40 | 423.40 |
| 48" wide, 2 doors, 2 drawers | BC@1.71 | Ea | 408.00 | 63.20 | 471.20 |

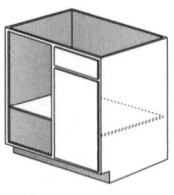

BLIND CORNER BASE

| | Craft@Hrs | Unit | Material | Labor | Total |
|---|---|---|---|---|---|
| **Blind corner base cabinets, 34-1/2" high** | | | | | |
| Minimum 36", maximum 39" at wall | BC@1.39 | Ea | 222.00 | 51.30 | 273.30 |
| Minimum 39", maximum 42" at wall | BC@1.50 | Ea | 239.00 | 55.40 | 294.40 |
| **45-degree corner base, revolving, 34-1/2" high** | | | | | |
| 36" wide at each wall | BC@2.12 | Ea | 344.00 | 78.30 | 422.30 |

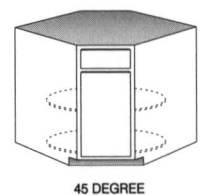

| | Craft@Hrs | Unit | Material | Labor | Total |
|---|---|---|---|---|---|
| **Corner sink front, 34-1/2" high** | | | | | |
| 40" wide at walls | BC@2.63 | Ea | 193.00 | 97.20 | 290.20 |

45 DEGREE
CORNER BASE CABINET

| | Craft@Hrs | Unit | Material | Labor | Total |
|---|---|---|---|---|---|
| **Wall cabinets, adjustable shelves, 30" high, 12" deep** | | | | | |
| 9" wide, 1 door | BC@.461 | Ea | 122.00 | 17.00 | 139.00 |
| 12" wide or 15" wide, 1 door | BC@.461 | Ea | 131.00 | 17.00 | 148.00 |
| 18" wide, 1 door | BC@.638 | Ea | 157.00 | 23.60 | 180.60 |
| 21" wide, 1 door | BC@.638 | Ea | 163.00 | 23.60 | 186.60 |
| 24" wide, 1 door | BC@.766 | Ea | 175.00 | 28.30 | 203.30 |
| 27" wide, 2 doors | BC@.766 | Ea | 203.00 | 28.30 | 231.30 |
| 30" wide, 2 doors | BC@.911 | Ea | 203.00 | 33.70 | 236.70 |
| 33" wide, 2 doors | BC@.911 | Ea | 222.00 | 33.70 | 255.70 |
| 36" wide, 2 doors | BC@1.03 | Ea | 235.00 | 38.00 | 273.00 |
| 42" wide, 2 doors | BC@1.03 | Ea | 250.00 | 38.00 | 288.00 |
| 48" wide, 2 doors | BC@1.16 | Ea | 270.00 | 42.90 | 312.90 |

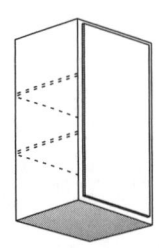

1 DOOR WALL CABINET

| | Craft@Hrs | Unit | Material | Labor | Total |
|---|---|---|---|---|---|
| **Above-appliance wall cabinets, 12" deep** | | | | | |
| 12" high, 30" wide, 2 doors | BC@.461 | Ea | 128.00 | 17.00 | 145.00 |
| 15" high, 30" wide, 2 doors | BC@.461 | Ea | 147.00 | 17.00 | 164.00 |
| 15" high, 33" wide, 2 doors | BC@.537 | Ea | 158.00 | 19.80 | 177.80 |
| 15" high, 36" wide, 2 doors | BC@.638 | Ea | 163.00 | 23.60 | 186.60 |
| 18" high, 18" wide, 2 doors | BC@.537 | Ea | 122.00 | 19.80 | 141.80 |
| 18" high, 30" wide, 2 doors | BC@.766 | Ea | 163.00 | 28.30 | 191.30 |
| 18" high, 36" wide, 2 doors | BC@.911 | Ea | 180.00 | 33.70 | 213.70 |

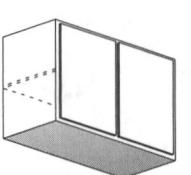

ABOVE-APPLIANCE
WALL CABINET

| | Craft@Hrs | Unit | Material | Labor | Total |
|---|---|---|---|---|---|
| **Corner wall cabinets, 30" high, 12" deep** | | | | | |
| 24" at each wall, fixed shelves | BC@1.03 | Ea | 222.00 | 38.00 | 260.00 |
| 24" at each wall, revolving shelves | BC@1.03 | Ea | 298.00 | 38.00 | 336.00 |
| **Blind corner wall cabinets, 30" high** | | | | | |
| 24" minimum, 1 door | BC@1.03 | Ea | 168.00 | 38.00 | 206.00 |
| 36" minimum, 1 door | BC@1.32 | Ea | 203.00 | 48.80 | 251.80 |
| 42" minimum, 2 doors | BC@1.20 | Ea | 256.00 | 44.30 | 300.30 |

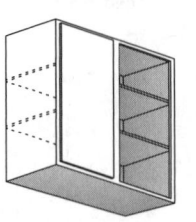

BLIND CORNER WALL CABINET

| | Craft@Hrs | Unit | Material | Labor | Total |
|---|---|---|---|---|---|
| **Utility cabinets, 66" high, 12" deep, no shelves** | | | | | |
| 18" wide | BC@1.32 | Ea | 256.00 | 48.80 | 304.80 |
| 24" wide | BC@1.71 | Ea | 292.00 | 63.20 | 355.20 |
| **Utility cabinets, 66" high, 24" deep, add shelf cost below** | | | | | |
| 18" wide | BC@1.24 | Ea | 270.00 | 45.80 | 315.80 |
| 24" wide | BC@1.71 | Ea | 321.00 | 63.20 | 384.20 |

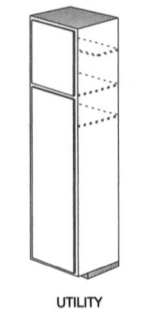

UTILITY

|  | Craft@Hrs | Unit | Material | Labor | Total |
|---|---|---|---|---|---|
| Add for utility cabinet revolving shelves, includes mounting hardware | | | | | |
| 18" wide x 24" deep | BC@.360 | Ea | 234.00 | 13.30 | 247.30 |
| 24" wide x 24" deep | BC@.360 | Ea | 272.00 | 13.30 | 285.30 |
| Add for utility cabinet plain shelves | | | | | |
| 18" wide x 24" deep | BC@.541 | Ea | 78.80 | 20.00 | 98.80 |
| 24" wide x 24" deep | BC@.541 | Ea | 82.40 | 20.00 | 102.40 |
| Oven cabinets, 66" high, 24" deep | | | | | |
| 27" wide, single oven | BC@2.19 | Ea | 325.00 | 80.90 | 405.90 |
| 27" wide, double oven | BC@2.19 | Ea | 245.00 | 80.90 | 325.90 |
| Microwave cabinet, with trim, | | | | | |
| 21" high, 20" deep, 30" wide | BC@.986 | Ea | 195.00 | 36.40 | 231.40 |
| Additional labor costs for cabinets | | | | | |
| Tall utility, pantry, or oven cabinets | BC@2.00 | Ea | — | 73.90 | 73.90 |
| Tall wall cabinet to counter level | BC@1.50 | Ea | — | 55.40 | 55.40 |
| Hood cabinet over a range with vent cutout | BC@1.50 | Ea | — | 55.40 | 55.40 |
| 3/4" raised end panels applied to cabinet ends | BC@.250 | Ea | — | 9.24 | 9.24 |
| Refrigerator end panels, cut, fit, install | BC@.500 | Ea | — | 18.50 | 18.50 |
| Cabinet end panels, most sizes, per panel | | | | | |
| Refrigerator end panels | BC@.557 | Ea | — | 20.60 | 20.60 |
| Refrigerator end panels with return | BC@1.00 | Ea | — | 36.90 | 36.90 |
| Applied decorative end panels | BC@.334 | Ea | — | 12.30 | 12.30 |
| Mitered island end or back panels | BC@.667 | Ea | — | 24.60 | 24.60 |
| Dishwasher return panels | BC@.667 | Ea | — | 24.60 | 24.60 |
| Precut filler panels installed between cabinets, per panel | | | | | |
| Most base or wall fillers | BC@.200 | Ea | — | 7.39 | 7.39 |
| Most base or wall fillers with overlays | BC@.400 | Ea | — | 14.80 | 14.80 |
| Tall filler panels over 36" | BC@.268 | Ea | — | 9.90 | 9.90 |
| Tall filler panels over 36" with overlays | BC@.535 | Ea | — | 19.80 | 19.80 |
| Corner filler panels | BC@.224 | Ea | — | 8.27 | 8.27 |
| Corner fillers panels with overlays | BC@.448 | Ea | — | 16.50 | 16.50 |
| Angled corner filler panels | BC@.268 | Ea | — | 9.90 | 9.90 |
| Angled corner filler panels with overlays | BC@.536 | Ea | — | 19.80 | 19.80 |
| Moldings and trim for cabinet work | | | | | |
| Scribe molding, per 8' length | BC@.224 | Ea | — | 8.27 | 8.27 |
| Scribe molding, per linear foot | BC@.028 | LF | — | 1.03 | 1.03 |
| Crown molding, per 8' length, miter cut | BC@.448 | Ea | — | 16.50 | 16.50 |
| Crown molding, per 8' length, butt joint | BC@.180 | Ea | — | 6.65 | 6.65 |
| Crown plate and riser, per miter cut | BC@.180 | Ea | — | 6.65 | 6.65 |
| Crown plate and riser, per straight cut | BC@.112 | Ea | — | 4.14 | 4.14 |
| Applied molding, per miter cut | BC@.112 | Ea | — | 4.14 | 4.14 |
| Applied molding, per straight cut | BC@.067 | Ea | — | 2.47 | 2.47 |
| Light rail molding, per miter cut | BC@.224 | Ea | — | 8.27 | 8.27 |
| Light rail molding, per straight cut | BC@.067 | Ea | — | 2.47 | 2.47 |
| Furniture toe kick, per 8' length | BC@.334 | Ea | — | 12.30 | 12.30 |
| Furniture toe kick, per linear foot | BC@.042 | LF | — | 1.55 | 1.55 |
| Valance, straight, per 8' length | BC@.334 | Ea | — | 12.30 | 12.30 |
| Valance, with side returns or cap, 8' length | BC@.667 | Ea | — | 24.60 | 24.60 |
| Mantle or hood molding, length to 8' | BC@.800 | Ea | — | 29.60 | 29.60 |
| Corbels, plain | BC@.224 | Ea | — | 8.27 | 8.27 |
| Corbels, on uprights | BC@.334 | Ea | — | 12.30 | 12.30 |

SINGLE
OVEN CABINET

DOUBLE
OVEN CABINET

# Carpentry, Rule of Thumb

|  | Craft@Hrs | Unit | Material | Labor | Total |
|---|---|---|---|---|---|
| Enkeboll molding, miter cut, 8' length | BC@.667 | Ea | — | 24.60 | 24.60 |
| Enkeboll molding, straight cut, 8' length | BC@.180 | Ea | — | 6.65 | 6.65 |
| Posts, to 8' length | BC@.500 | Ea | — | 18.50 | 18.50 |
| Pilasters | BC@.334 | Ea | — | 12.30 | 12.30 |
| Cabinet feet, per cabinet | BC@.334 | Ea | — | 12.30 | 12.30 |
| Cabinet extras |  |  |  |  |  |
| Scribing in wood tops | BC@.133 | LF | — | 4.91 | 4.91 |
| Job built 2 x 4 support wall to 42" high | BC@.334 | LF | — | 12.30 | 12.30 |
| Framing for range support or sink cutout | BC@.667 | Ea | — | 24.60 | 24.60 |

**Carpentry** See also carpentry items in other sections: Carpentry Steel Framing, Cabinets, Ceilings, Closet Door Systems, Countertops, Cupolas, Doors, Entrances, Flooring, Framing Connectors, Hardboard, Hardware, Lumber, Moulding, Paneling, Shutters, Siding, Skylights, Soffits, Stairs, Thresholds, and Weatherstripping.

**Carpentry Rule of Thumb** Typical rough carpentry (framing) cost per square foot of floor in living area. These figures will apply on many residential jobs where joists, studs and rafters are 16" on center. Unconventional designs and complex framing plans will cost more. See detailed cost breakdowns in the next section.

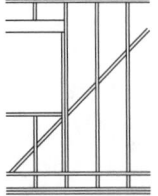

| | Craft@Hrs | Unit | Material | Labor | Total |
|---|---|---|---|---|---|
| Using framing lumber at | — | MBF | 479.00 | — | 479.00 |
| Using 7/16" OSB sheathing at | — | MSF | 265.00 | — | 265.00 |
| Single story, conventional foundation | B1@.237 | SF | 3.61 | 7.89 | 11.50 |
| Single story, concrete slab foundation | B1@.180 | SF | 2.53 | 6.00 | 8.53 |
| First of two floors, conventional foundation | B1@.203 | SF | 2.80 | 6.76 | 9.56 |
| First of two floors, concrete slab foundation | B1@.146 | SF | 1.72 | 4.86 | 6.58 |
| Second floor of a two-story residence | B1@.191 | SF | 2.83 | 6.36 | 9.19 |

### Framing a single story residence, conventional crawl-space foundation

| | Craft@Hrs | Unit | Material | Labor | Total |
|---|---|---|---|---|---|
| Sills, pier blocks, floor beams (145 BF per 1,000 SF) | B1@.018 | SF | .07 | .60 | .67 |
| Floor joists, doublers, blocking, bridging (1,480 BF per 1,000 SF) | B1@.028 | SF | .71 | .93 | 1.64 |
| Subflooring, 7/16" OSB (1,150 SF per 1,000 SF) | B1@.011 | SF | .30 | .37 | .67 |
| Layout, studs, sole plates, top plates, header and end joists, backing, blocking, bracing and framing for openings (2,250 BF per 1,000 SF) | B1@.093 | SF | 1.08 | 3.10 | 4.18 |
| Ceiling joists, header and end joists, backing, blocking and bracing (1,060 BF per 1,000 SF) | B1@.045 | SF | .51 | 1.50 | 2.01 |
| Rafters, braces, collar beams, ridge boards, 2" x 8" rafters 16" OC, (1,340 BF per 1000 SF) | B1@.032 | SF | .64 | 1.07 | 1.71 |
| Roof sheathing, 7/16" OSB (1,150 SF per 1,000 SF) | B1@.010 | SF | .30 | .33 | .63 |
| **Total framing, single story, conventional foundation** | B1@.237 | SF | 3.61 | 7.90 | 11.51 |

### Framing a single story residence, concrete slab foundation

| | Craft@Hrs | Unit | Material | Labor | Total |
|---|---|---|---|---|---|
| Layout, sole plates, anchors, studs, top plates, header and end joists, backing, blocking, bracing and framing for openings (2,250 BF per 1,000 SF) | B1@.093 | SF | 1.08 | 3.10 | 4.18 |
| Ceiling joists, header and end joists, backing, blocking and bracing (1,060 BF per 1,000 SF) | B1@.045 | SF | .51 | 1.50 | 2.01 |
| Rafters, braces, collar beams, ridge boards, 2" x 8" rafters 16" OC, (1,340 BF per 1,000 SF) | B1@.032 | SF | .64 | 1.07 | 1.71 |
| Roof sheathing, 7/16" OSB (1,150 SF per 1,000 SF) | B1@.010 | SF | .30 | .33 | .63 |
| **Total framing, single story, concrete slab foundation** | B1@.180 | SF | 2.53 | 6.00 | 8.53 |

| | Craft@Hrs | Unit | Material | Labor | Total |
|---|---|---|---|---|---|
| **Framing the first of two floors, conventional crawl-space foundation** | | | | | |
| Sills, pier blocks, floor beams | | | | | |
| (145 BF per 1,000 SF) | B1@.018 | SF | .07 | .60 | .67 |
| Floor joists, doublers, blocking, bridging | | | | | |
| (1,480 BF per 1,000 SF) | B1@.028 | SF | .71 | .93 | 1.64 |
| Subflooring, 7/16" OSB | | | | | |
| (1,150 SF per 1,000 SF) | B1@.011 | SF | .30 | .37 | .67 |
| Layout, studs, sole plates, top plates, header and | | | | | |
| end joists, backing, blocking, bracing and framing for openings | | | | | |
| (2,250 BF per 1,000 SF) | B1@.093 | SF | 1.08 | 3.10 | 4.18 |
| Rough stairway, 15 risers and landing (96 BF of dimension lumber | | | | | |
| and 128 SF of plywood per 1,000 SF) | B1@.008 | SF | .13 | .27 | .40 |
| Ceiling joists, header and end joists, backing, blocking and bracing | | | | | |
| (1,060 BF per 1,000 SF) | B1@.045 | SF | .51 | 1.50 | 2.01 |
| Total framing, 1st of 2 floors, crawl-space foundation | | | | | |
| | B1@.203 | SF | 2.80 | 6.77 | 9.57 |
| **Framing the first of two floors, concrete slab foundation** | | | | | |
| Layout, sole plates, anchors, studs, top plates, header and | | | | | |
| end joists, backing, blocking, bracing and framing for openings | | | | | |
| (2,250 BF per 1,000 SF) | B1@.093 | SF | 1.08 | 3.10 | 4.18 |
| Rough stairway, 15 risers and landing (96 BF of dimension lumber | | | | | |
| and 128 SF of plywood per 1,000 SF) | B1@.008 | SF | .13 | .27 | .40 |
| Ceiling joists, header and end joists, backing, blocking and bracing | | | | | |
| (1,060 BF per 1,000 SF) | B1@.045 | SF | .51 | 1.50 | 2.01 |
| Total framing, first of two floors, concrete foundation | | | | | |
| | B1@.146 | SF | 1.72 | 4.87 | 6.59 |
| **Framing the second story of a residence** | | | | | |
| Subflooring, 7/16" OSB | | | | | |
| (1,150 SF per 1,000 SF) | B1@.011 | SF | .30 | .37 | .67 |
| Layout, studs, sole plates, top plates, header and | | | | | |
| end joists, backing, blocking, bracing and framing for openings | | | | | |
| (2,250 BF per 1,000 SF) | B1@.093 | SF | 1.08 | 3.10 | 4.18 |
| Ceiling joists, header and end joists, backing, blocking and bracing | | | | | |
| (1,060 BF per 1,000 SF) | B1@.045 | SF | .51 | 1.50 | 2.01 |
| Rafters, braces, collar beams, ridge boards, 2" x 8" rafters 16" OC | | | | | |
| (1,340 BF per 1,000 SF) | B1@.032 | SF | .64 | 1.07 | 1.71 |
| Roof sheathing, 7/16" OSB | | | | | |
| (1,150 SF per 1,000 SF) | B1@.010 | SF | .30 | .33 | .63 |
| Total framing, second floor of a two-story residence | | | | | |
| | B1@.191 | SF | 2.83 | 6.37 | 9.20 |

**Floor Assemblies** Costs for wood framed floor joists with subflooring and R-19 insulation. These costs include the floor joists, subflooring as described, blocking, nails and 6-1/4" thick R-1.9 fiberglass insulation between the floor joists. Figures include box or band joists and typical double joists. No beams included. Planked subflooring is based on 1.24 BF per square foot of floor. Costs shown are per square foot of area covered and include normal waste. Deduct for openings over 25 SF.

| | Craft@Hrs | Unit | Material | Labor | Total |
|---|---|---|---|---|---|
| Using framing lumber at | — | MBF | 479.00 | — | 479.00 |
| Using 5/8" CDX plywood subfloor at | — | MSF | 622.00 | — | 622.00 |
| Using 3/4" CDX plywood subfloor at | — | MSF | 746.00 | — | 746.00 |
| Using 7/16" OSB subfloor at | — | MSF | 265.00 | — | 265.00 |
| Using 5/8" OSB subfloor at | — | MSF | 420.00 | — | 420.00 |
| Using 3/4" OSB T&G subfloor at | — | MSF | 543.00 | — | 543.00 |

# Carpentry, Assemblies

| | Craft@Hrs | Unit | Material | Labor | Total |
|---|---|---|---|---|---|
| **Floor joists 16" OC, R-19 insulation and OSB subflooring** | | | | | |
| 7/16" OSB subfloor | | | | | |
| 2" x 6" joists | B1@.040 | SF | 1.37 | 1.33 | 2.70 |
| 2" x 8" joists | B1@.041 | SF | 1.60 | 1.37 | 2.97 |
| 2" x 10" joists | B1@.043 | SF | 1.89 | 1.43 | 3.32 |
| 2" x 12" joists | B1@.044 | SF | 2.37 | 1.47 | 3.84 |
| **Floor joists 16" OC, R-19 insulation and OSB subflooring** | | | | | |
| 5/8" OSB subfloor | | | | | |
| 2" x 6" joists | B1@.040 | SF | 1.52 | 1.33 | 2.85 |
| 2" x 8" joists | B1@.041 | SF | 1.76 | 1.37 | 3.13 |
| 2" x 10" joists | B1@.043 | SF | 2.04 | 1.43 | 3.47 |
| 2" x 12" joists | B1@.044 | SF | 2.52 | 1.47 | 3.99 |
| 3/4" OSB subfloor | | | | | |
| 2" x 6" joists | B1@.042 | SF | 1.64 | 1.40 | 3.04 |
| 2" x 8" joists | B1@.043 | SF | 1.88 | 1.43 | 3.31 |
| 2" x 10" joists | B1@.045 | SF | 2.17 | 1.50 | 3.67 |
| 2" x 12" joists | B1@.046 | SF | 2.65 | 1.53 | 4.18 |
| **Floor joists 16" OC, R-19 insulation and plywood subflooring** | | | | | |
| 5/8" plywood subfloor | | | | | |
| 2" x 6" joists | B1@.040 | SF | 1.72 | 1.33 | 3.05 |
| 2" x 8" joists | B1@.041 | SF | 1.95 | 1.37 | 3.32 |
| 2" x 10" joists | B1@.043 | SF | 2.24 | 1.43 | 3.67 |
| 2" x 12" joists | B1@.044 | SF | 2.72 | 1.47 | 4.19 |
| 3/4" plywood subfloor | | | | | |
| 2" x 6" joists | B1@.042 | SF | 1.85 | 1.40 | 3.25 |
| 2" x 8" joists | B1@.043 | SF | 2.08 | 1.43 | 3.51 |
| 2" x 10" joists | B1@.045 | SF | 2.37 | 1.50 | 3.87 |
| 2" x 12" joists | B1@.046 | SF | 2.85 | 1.53 | 4.38 |
| For different type insulation | | | | | |
| Fiberglass batts | | | | | |
| 10" thick R-30, add | — | SF | .27 | — | .27 |
| 12" thick R-38, add | — | SF | .36 | — | .36 |

**Wall Assemblies** Costs for wood framed stud walls with wall finish treatment on both sides. These costs include wall studs at 16" center to center, double top plates, single bottom plates, fire blocking, nails and wall finish treatment as described. No headers or posts included. All lumber is Std & Btr. 2" x 4" walls have 1.12 BF per SF of wall and 2" x 6" walls have 1.68 BF per SF of wall. Costs shown are per SF or LF of wall measured on one face and include normal waste.

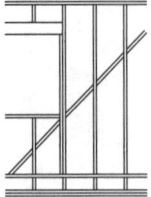

| | Craft@Hrs | Unit | Material | Labor | Total |
|---|---|---|---|---|---|
| Using 2 x 4 lumber at | — | MBF | 473.00 | — | 473.00 |
| Using 2 x 6 lumber at | — | MBF | 465.00 | — | 465.00 |
| Using 1/2" gypsum wallboard at | — | SF | .25 | — | .25 |

### Interior wall assemblies

| | Craft@Hrs | Unit | Material | Labor | Total |
|---|---|---|---|---|---|
| 2" x 4" stud walls with 1/2" gypsum drywall both sides, ready for painting | | | | | |
| Cost per square foot of wall | B1@.064 | SF | .89 | 2.13 | 3.02 |
| Cost per running foot, for 8' high walls | B1@.512 | LF | 7.08 | 17.10 | 24.18 |
| 2" x 4" stud walls with 1/2" gypsum drywall one side, ready for painting | | | | | |
| Cost per square foot of wall | B1@.046 | SF | .64 | 1.53 | 2.17 |
| Cost per running foot, for 8' high walls | B1@.368 | LF | 5.08 | 12.30 | 17.38 |

| | Craft@Hrs | Unit | Material | Labor | Total |
|---|---|---|---|---|---|
| 2" x 4" stud walls with 5/8" gypsum fire rated drywall both sides, ready for painting | | | | | |
| Cost per square foot of wall | B1@.068 | SF | .99 | 2.27 | 3.26 |
| Cost per running foot, for 8' high walls | B1@.544 | LF | 7.88 | 18.10 | 25.98 |
| 2" x 4" stud walls with 5/8" gypsum fire rated drywall one side, ready for painting | | | | | |
| Cost per square foot of wall | B1@.048 | SF | .69 | 1.60 | 2.29 |
| Cost per running foot, for 8' high walls | B1@.384 | LF | 5.48 | 12.80 | 18.28 |
| 2" x 6" stud walls with 1/2" gypsum drywall both sides, ready for painting | | | | | |
| Cost per square foot of wall | B1@.072 | SF | 1.08 | 2.40 | 3.48 |
| Cost per running foot, for 8' high walls | B1@.576 | LF | 8.62 | 19.20 | 27.82 |
| 2" x 6" stud walls with 1/2" gypsum drywall one side, ready for painting | | | | | |
| Cost per square foot of wall | B1@.054 | SF | .83 | 1.80 | 2.63 |
| Cost per running foot, for 8' high walls | B1@.432 | LF | 6.62 | 14.40 | 21.02 |
| 2" x 6" stud walls with 5/8" gypsum fire rated drywall both sides, ready for painting | | | | | |
| Cost per square foot of wall | B1@.076 | SF | 1.18 | 2.53 | 3.71 |
| Cost per running foot, for 8' high walls | B1@.608 | LF | 9.42 | 20.30 | 29.72 |
| 2" x 6" stud walls with 5/8" gypsum fire rated drywall one side, ready for painting | | | | | |
| Cost per square foot of wall | B1@.056 | SF | .88 | 1.87 | 2.75 |
| Cost per running foot, for 8' high walls | B1@.448 | LF | 7.02 | 14.90 | 21.92 |

## Exterior wall assemblies

2" x 4" stud walls with drywall interior, wood siding exterior, 1/2" gypsum drywall inside face ready for painting, over 3-1/2" R-11 insulation with 5/8" thick rough sawn T-1-11 exterior grade plywood siding on the outside face.

| | Craft@Hrs | Unit | Material | Labor | Total |
|---|---|---|---|---|---|
| Using 5/8" rough sawn T-1-11 siding at | — | MSF | 1,300.00 | — | 1,300.00 |
| Cost per square foot of wall | B1@.068 | SF | 2.20 | 2.27 | 4.47 |
| Cost per running foot, for 8' high walls | B1@.544 | LF | 17.60 | 18.10 | 35.70 |

2" x 6" stud walls with drywall interior, wood siding exterior, same construction as above, except with 6-1/4" R-19 insulation

| | Craft@Hrs | Unit | Material | Labor | Total |
|---|---|---|---|---|---|
| Cost per square foot of wall | B1@.077 | SF | 2.53 | 2.56 | 5.09 |
| Cost per running foot, for 8' high walls | B1@.616 | LF | 20.20 | 20.50 | 40.70 |

2" x 4" stud walls with drywall interior, 1/2" gypsum drywall on inside face ready for painting, over 3-1/2" R-11 insulation with 1" x 6" southern yellow pine drop siding, D grade, 1.19 BF per SF at 5-1/4" exposure on the outside face.

| | Craft@Hrs | Unit | Material | Labor | Total |
|---|---|---|---|---|---|
| Using D grade yellow pine drop siding at | — | MSF | 2,450.00 | — | 2,450.00 |
| Cost per square foot of wall | B1@.074 | SF | 3.35 | 2.46 | 5.81 |
| Cost per running foot, for 8' high wall | B1@.592 | LF | 26.80 | 19.70 | 46.50 |

2" x 6" stud walls with drywall interior, 1" x 6" drop siding exterior, same construction as above, except with 6-1/4" R-19 insulation

| | Craft@Hrs | Unit | Material | Labor | Total |
|---|---|---|---|---|---|
| Cost per square foot of wall | B1@.083 | SF | 3.67 | 2.76 | 6.43 |
| Cost per running foot, for 8' high wall | B1@.664 | LF | 29.40 | 22.10 | 51.50 |

2" x 4" stud walls with drywall interior, stucco exterior, 1/2" gypsum drywall on inside face ready for painting, over 3-1/2" R-11 insulation and a three-coat exterior plaster (stucco) finish with integral color on the outside face

| | Craft@Hrs | Unit | Material | Labor | Total |
|---|---|---|---|---|---|
| Cost per square foot of wall | B1@.050 | SF | 3.18 | 1.67 | 4.85 |
| Cost per running foot, for 8' high wall | B1@.400 | LF | 25.50 | 13.30 | 38.80 |

2" x 6" stud walls with drywall interior, stucco exterior, same construction as above, except with 6-1/4" R-19 insulation

| | Craft@Hrs | Unit | Material | Labor | Total |
|---|---|---|---|---|---|
| Cost per square foot of wall | B1@.059 | SF | 3.51 | 1.97 | 5.48 |
| Cost per running foot, for 8' high wall | B1@.472 | LF | 28.00 | 15.70 | 43.70 |

# Carpentry, Assemblies

| | Craft@Hrs | Unit | Material | Labor | Total |
|---|---|---|---|---|---|
| Add for different type gypsum board | | | | | |
| 1/2" or 5/8" moisture resistant greenboard | | | | | |
|   Cost per SF, greenboard per side, add | — | SF | .08 | — | .08 |
| 1/2" or 5/8" moisture resistant greenboard | | | | | |
|   Cost per running foot per side 8' high | — | LF | .70 | — | .70 |
| 5/8" thick fire rated type X gypsum drywall | | | | | |
|   Cost per SF, per side, add | — | SF | .11 | — | .11 |
| 5/8" thick fire rated type X gypsum drywall | | | | | |
|   Cost per running foot per side 8' high | — | LF | .84 | — | .84 |

**Ceiling Assemblies** Costs for wood framed ceiling joists with ceiling finish and fiberglass insulation, based on performing the work at the construction site. These costs include the ceiling joists, ceiling finish as described, blocking, nails and 3-1/2" thick R-11 fiberglass insulation batts between the ceiling joists. Figures in parentheses indicate board feet per square foot of ceiling framing including end joists and typical header joists. No beams included. Ceiling joists and blocking are based on standard and better grade lumber. Costs shown are per square foot of area covered and include normal waste. Deduct for openings over 25 SF.

| | Craft@Hrs | Unit | Material | Labor | Total |
|---|---|---|---|---|---|
| Ceiling joists with regular gypsum drywall taped and sanded smooth finish, ready for paint | | | | | |
|   Using 2" x 4" at | — | MBF | 473.00 | — | 473.00 |
|   Using 2" x 6" at | — | MBF | 465.00 | — | 465.00 |
|   Using 2" x 8" at | — | MBF | 465.00 | — | 465.00 |
| 2" x 4" ceiling joists at 16" on center (.59 BF per SF), with insulation and 1/2" gypsum drywall | B1@.053 | SF | .95 | 1.77 | 2.72 |
| 2" x 6" ceiling joists at 16" on center (.88 BF per SF), with insulation and 1/2" gypsum drywall | B1@.055 | SF | 1.16 | 1.83 | 2.99 |
| 2" x 8" ceiling joists at 16" on center (1.17 BF per SF), with insulation and 1/2" gypsum drywall | B1@.057 | SF | 1.37 | 1.90 | 3.27 |
| For spray applied plaster finish (sometimes called "popcorn" or "cottage cheese" texture) | | | | | |
|   Add for ceiling texture | DT@.003 | SF | .11 | .11 | .22 |
| For different type gypsum drywall with taped and sanded smooth finish, ready for paint | | | | | |
|   1/2" moisture resistant greenboard, add | — | SF | .09 | — | .09 |
|   5/8" fire rated board, add | — | SF | .11 | — | .11 |
| For different ceiling joist center to center dimensions | | | | | |
|   2" x 4" ceiling joists | | | | | |
|     12" on center, add | B1@.004 | SF | .09 | .13 | .22 |
|     20" on center, deduct | — | SF | -.05 | -.06 | -.11 |
|     24" on center, deduct | — | SF | -.08 | -.12 | -.20 |
|   2" x 6" ceiling joists | | | | | |
|     12" on center, add | B1@.006 | SF | .12 | .20 | .32 |
|     20" on center, deduct | — | SF | -.07 | -.06 | -.13 |
|     24" on center, deduct | — | SF | -.11 | -.11 | -.22 |
|   2" x 8" ceiling joists | | | | | |
|     12" on center, add | B1@.006 | SF | .16 | .20 | .36 |
|     20" on center, deduct | — | SF | -.09 | -.08 | -.17 |
|     24" on center, deduct | — | SF | -.15 | -.13 | -.28 |
| For different type insulation | | | | | |
|   Fiberglass batts | | | | | |
|     6-1/4" thick R-19, add | — | SF | .12 | — | .12 |
|     10" thick R-30, add | — | SF | .28 | — | .28 |
|   Blown-in fiberglass | | | | | |
|     8" thick R-19, add | — | SF | .04 | — | .04 |

# Carpentry, Assemblies

| | Craft@Hrs | Unit | Material | Labor | Total |
|---|---|---|---|---|---|

**Roofing Assemblies** Costs for wood framed roof assemblies with roof finish material as shown based on performing the work at the construction site. Costs shown include all material and labor required above the top plate or ledger on the supporting walls. These costs assume the supporting wall structure is in-place and suitable for the assembly described.

**Flat roof assembly** Based on using 2" x 12" joists Std & Btr grade at 16" on center including blocking, ripped strips and normal waste. The roof surface is built-up 3 ply asphalt consisting of 2 plies of 15 lb. felt hot mopped with a 90 lb. mineral-coated cap sheet.

| | Craft@Hrs | Unit | Material | Labor | Total |
|---|---|---|---|---|---|
| Using 2" x 12" joists at | — | MBF | 563.00 | — | 563.00 |
| Using 2" x 8" joists at | — | MBF | 465.00 | — | 465.00 |
| Using 1/2" CDX plywood at | — | MSF | 507.00 | — | 507.00 |
| Using 7/16" OSB at | — | MSF | 265.00 | — | 265.00 |
| Flat roof assembly as described above | | | | | |
| Framing, (w/ plywood sheathing), per 100 SF | B1@5.00 | Sq | 192.00 | 167.00 | 359.00 |
| Framing, (w/ OSB sheathing), per 100 SF | B1@5.00 | Sq | 166.00 | 167.00 | 333.00 |
| Built-up roofing, per 100 square feet | R1@1.25 | Sq | 107.00 | 44.30 | 151.30 |

**Conventionally framed roof assemblies** Based on straight gable type roof (no hips, valleys, or dormers) with 6" in 12" rise or less. Cost per 100 square feet of plan area under the roof, not actual roof surface area. Framing includes 2" x 8" common rafters Std & Btr grade at 24" on center. Cost includes blocking, ridge and normal bracing. Roof sheathing is either 7/16" OSB (oriented strand board) or 1/2" CDX plywood.

Conventionally framed roof assembly with built-up rock-finish roofing. The finished roof treatment is built-up 3 ply asphalt consisting of 1 ply 30 lb. felt, 2 plies of 15 lb. felt and 3 hot mop coats of asphalt with decorative crush rock spread at 180 lbs. per 100 square feet.

| | Craft@Hrs | Unit | Material | Labor | Total |
|---|---|---|---|---|---|
| Framing, (w/ plywood sheathing), per 100 SF | B1@4.10 | Sq | 107.00 | 137.00 | 244.00 |
| Framing, (w/ OSB sheathing), per 100 SF | B1@4.10 | Sq | 81.50 | 137.00 | 218.50 |
| Built-up rock finish roofing, per 100 square feet | R1@2.15 | Sq | 112.00 | 76.20 | 188.20 |

Conventionally framed roof assembly with 25 year fiberglass shingle roofing

| | Craft@Hrs | Unit | Material | Labor | Total |
|---|---|---|---|---|---|
| Framing, (w/ plywood sheathing), per 100 SF | B1@4.10 | Sq | 107.00 | 137.00 | 244.00 |
| Framing, (w/ OSB sheathing), per 100 SF | B1@4.10 | Sq | 81.50 | 137.00 | 218.50 |
| Fiberglass shingle roofing, per 100 square feet | B1@1.83 | Sq | 67.40 | 61.00 | 128.40 |

Conventionally framed roof assembly with 25 year asphalt shingle roofing

| | Craft@Hrs | Unit | Material | Labor | Total |
|---|---|---|---|---|---|
| Framing, (w/ plywood sheathing), per 100 SF | B1@4.10 | Sq | 107.00 | 137.00 | 244.00 |
| Framing, (w/ OSB sheathing), per 100 SF | B1@4.10 | Sq | 81.50 | 137.00 | 218.50 |
| Asphalt shingle roofing, per 100 square feet | B1@2.07 | Sq | 60.40 | 69.00 | 129.40 |

Conventionally framed roof assembly, 28 gauge galvanized corrugated steel roofing, no sheathing

| | Craft@Hrs | Unit | Material | Labor | Total |
|---|---|---|---|---|---|
| Framing, (w/ plywood sheathing), per 100 SF | B1@4.10 | Sq | 107.00 | 137.00 | 244.00 |
| Framing, (w/ OSB sheathing), per 100 SF | B1@4.10 | Sq | 81.50 | 137.00 | 218.50 |
| 28 gauge steel roofing, per 100 square feet | B1@2.70 | Sq | 220.00 | 89.90 | 309.90 |

Add for gable studs

| | Craft@Hrs | Unit | Material | Labor | Total |
|---|---|---|---|---|---|
| 2" x 4" spaced 16" OC (.54 BF per SF) | B1@.023 | SF | .26 | .77 | 1.03 |

Add for purlins (purling), Std & Btr, installed below roof rafters. Figures in parentheses indicate board feet per LF including 5% waste

| | Craft@Hrs | Unit | Material | Labor | Total |
|---|---|---|---|---|---|
| Using 2" x 8" joists at | — | MBF | 465.00 | — | 465.00 |
| 2" x 8" (1.40 BF per LF) | B1@.023 | LF | .65 | .77 | 1.42 |

# Carpentry, Piecework

| | Craft@Hrs | Unit | Material | Labor | Total |
|---|---|---|---|---|---|

**Piecework Rough Carpentry** Rough carpentry on residential tracts is usually done by framing subcontractors who bid at piecework rates (such as per square foot of floor). The figures below list typical piecework rates for repetitive framing work and assume all materials are supplied to the framing subcontractor. No figures appear in the Craft@Hrs column because the work is done for a fixed price per square foot and the labor productivity can be expected to vary widely.

**Layout and plating**  Piecework rates
Lay out wall plates according to the plans (snap chalk lines for wall plates, mark location for studs, windows, doors and framing details), cut top and bottom plates and install bottom plates. Costs per square foot of floor (excluding garage).

| | | | | | |
|---|---|---|---|---|---|
| Custom or more complex jobs | — | SF | — | .27 | .27 |
| Larger job, longer runs | — | SF | — | .18 | .18 |

**Wall framing**  Piecework rates
Measure, cut, fit, assemble, and tip up walls, including studs, plates, cripples, let-in braces, trimmers and blocking. Costs per square foot of floor.

| | | | | | |
|---|---|---|---|---|---|
| Complex job, radius walls, rake walls | — | SF | — | .74 | .74 |
| Larger job, 8' high walls, fewer partitions | — | SF | — | .27 | .27 |

**Plumb and align framed walls**  Piecework rates
Force walls into alignment, adjust walls to vertical and install temporary wood braces as needed, fasten nuts on anchor bolts or shoot power driven fasteners through wall plates into the slab. Based on accuracy to 3/16". Costs per square foot of floor are shown below.

| | | | | | |
|---|---|---|---|---|---|
| Small or complex job, many braces | — | SF | — | .30 | .30 |
| Larger job, fewer braces | — | SF | — | .18 | .18 |

**Floor joists or ceiling joists**  Piecework rates
Lay out, cut and install floor or ceiling joists, including rim joists, doubled joists, straps, joist hangers, blocking at 8' OC and ceiling backing for drywall. Based on larger jobs with simple joist layouts set 16" OC and pre-cut blocking supplied by the general contractor. Add the cost of floor beams, if required. Costs per square foot of horizontal joist area. More complex jobs with shorter runs may cost 50% more.

| | | | | | |
|---|---|---|---|---|---|
| 2" x  8" ceiling or floor joists | — | SF | — | .21 | .21 |
| 2" x 10" ceiling or floor joists | — | SF | — | .24 | .24 |
| 2" x 12" ceiling or floor joists | — | SF | — | .26 | .26 |
| 2" x 14" ceiling or floor joists | — | SF | — | .30 | .30 |
| Add for 12" OC spacing | — | SF | — | .09 | .09 |
| Deduct for 20" OC spacing | — | SF | — | -.03 | -.03 |
| Deduct for 24" OC spacing | — | SF | — | -.06 | -.06 |

**Floor sheathing**  Piecework rates
Lay out, cut, fit and install 5/8" or 3/4" tongue and groove plywood floor sheathing, including blocking as required. Based on nailing done with a pneumatic nailer and nails supplied by the general contractor. Costs per square foot of sheathing installed

| | | | | | |
|---|---|---|---|---|---|
| Smaller, cut-up job | — | SF | — | .24 | .24 |
| Larger job, longer runs | — | SF | — | .18 | .18 |
| Add for 1-1/8" sheathing | — | SF | — | .08 | .08 |

| | Craft@Hrs | Unit | Material | Labor | Total |
|---|---|---|---|---|---|

**Stair framing** Piecework rates

Lay out, cut, fit and install straight, "U"- or "L"-shaped 30" to 36" wide stairs made from plywood and 2" x 12" stringers set 16" OC. These costs include blocking in the adjacent stud wall and a 1" skirt board. Costs per 7-1/2" riser. Framing more complex stairs may cost up to $500 per flight.

| | | | | | |
|---|---|---|---|---|---|
| Small job, short runs | — | Ea | — | 17.20 | 17.20 |
| Larger job, longer runs | — | Ea | — | 12.90 | 12.90 |
| Add per 3' x 3' landing, including supports | — | Ea | — | 44.30 | 44.30 |
| Add for larger landings, including supports | — | Ea | — | 109.90 | 109.90 |

**Shear panels** Piecework rates

Lay out, cut, fit and install structural 3/8" or 1/2" OSB or plywood wall panels. These figures assume shear studs were set correctly by others and that panel nails are driven at 4" OC with a pneumatic nailer. Not including hold-down straps, posts, shear blocking or extra studs. Costs per 4' x 9' panel installed.

| | | | | | |
|---|---|---|---|---|---|
| Small job, many openings, 2nd floor | — | Ea | — | 11.90 | 11.90 |
| Larger job, few openings, 1st floor | — | Ea | — | 7.34 | 7.34 |

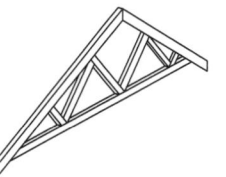

**Roof trusses** Piecework rates

Setting and nailing engineered gable and hip roof trusses 24" OC on prepared wall plates. These figures assume that lifting equipment is provided by the general contractor and that the truss supplier provides a fill package, spreader blocks for each plate and the ridge and jack rafters (if required). Includes installation of ceiling backing where required and catwalks at the bottom chord. Costs per square foot of plan area under the truss.

Small job assumes multiple California fill between roof surfaces and understacking

| | | | | | |
|---|---|---|---|---|---|
| Small job, rake fill above a partition wall | — | SF | — | .65 | .65 |
| Larger job, little or no fill or understacking | — | SF | — | .36 | .36 |

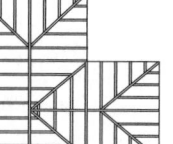

**Conventional roof framing** Piecework rates

Calculate lengths, lay out, cut and install 2" x 10" or 2" x 12" common, hip, valley and jack rafters on parallel and horizontal plates. Costs per square foot of plan area under the roof.

Small job, cut-up roof, few common rafters

| | | | | | |
|---|---|---|---|---|---|
| Rafters 12" OC | — | SF | — | 1.32 | 1.32 |
| Rafters 16" OC | — | SF | — | 1.12 | 1.12 |
| Rafters 20" OC | — | SF | — | .88 | .88 |
| Rafters 24" OC | — | SF | — | .65 | .65 |

Larger job, longer runs, nearly all common rafters

| | | | | | |
|---|---|---|---|---|---|
| Rafters 12" OC | — | SF | — | .65 | .65 |
| Rafters 16" OC | — | SF | — | .58 | .58 |
| Rafters 20" OC | — | SF | — | .45 | .45 |
| Rafters 24" OC | — | SF | — | .37 | .37 |
| Add for slope over 6 in 12 | — | SF | — | .22 | .22 |
| Deduct for 2" x 6" or 2" x 8" rafters | — | SF | — | -.08 | -.08 |

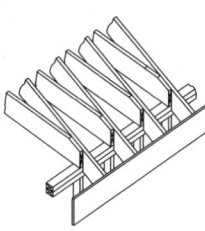

**Fascia** Piecework rates

Applied to rafter tails and as a barge rafter on gable ends. Includes trimming the rafter tails to the correct length and installing outlookers at gable ends. Costs per linear foot of 2" x 8" fascia installed.

| | | | | | |
|---|---|---|---|---|---|
| Small job, short runs, with moulding | — | LF | — | 2.86 | 2.86 |
| Larger job, longer runs | — | LF | — | 1.64 | 1.64 |

# Carpentry, Detailed Breakdown

| | Craft@Hrs | Unit | Material | Labor | Total |
|---|---|---|---|---|---|

**Roof sheathing** Piecework rates
Lay out, cut, fit and install 1/2" or 5/8" OSB or plywood roof sheathing, including blocking and 1" x 8" starter board on overhangs as required. Based on nailing done with a pneumatic nailer and nails supplied by the general contractor. Costs per square foot of sheathing installed.

| | Craft@Hrs | Unit | Material | Labor | Total |
|---|---|---|---|---|---|
| Smaller, cut-up hip and valley job | — | SF | — | .25 | .25 |
| Larger job, longer runs | — | SF | — | .19 | .19 |
| Add for slope over 6 in 12 | — | SF | — | .12 | .12 |

**Carpentry Cost, Detailed Breakdown** This section is arranged in the order of construction. Material costs shown here can be adjusted to reflect your actual lumber cost: divide your actual lumber cost (per MBF) by the cost listed (per MBF). Then multiply the cost in the material column by this adjustment factor. No waste included..

**Lally columns** (Residential adjustable basement column) 7' 9" to 8' 1", steel tube

| | Craft@Hrs | Unit | Material | Labor | Total |
|---|---|---|---|---|---|
| Material only | — | Ea | 55.40 | — | 55.40 |
| Add for installation, per column to 12' high | B1@.458 | Ea | — | 15.30 | 15.30 |

**Precast pier blocks** Posts set on precast concrete pier block, including pier block with anchor placed on existing grade, temporary reuseable 1" x 6" bracing (8 LF) and stakes (2). Cost is for each post set. Add for excavation if required

| | Craft@Hrs | Unit | Material | Labor | Total |
|---|---|---|---|---|---|
| Heights to 8', cost of post not included | BL@.166 | Ea | 9.18 | 4.93 | 14.11 |

**Pier pads** 2" x 6", treated, #2 & Btr

| | Craft@Hrs | Unit | Material | Labor | Total |
|---|---|---|---|---|---|
| Using #2 & Btr treated lumber at | — | MBF | 4,760.00 | — | 4,760.00 |
| 1.10 BF per LF | B1@.034 | LF | 5.24 | 1.13 | 6.37 |

**Posts** 4" x 4", material costs include 10% waste (1.47 BF per LF). See also Lally columns above and Posts in the Lumber section

| | Craft@Hrs | Unit | Material | Labor | Total |
|---|---|---|---|---|---|
| Fir, rough Std & Btr, K.D. | — | MBF | 598.00 | — | 598.00 |
| Fir, rough Std & Btr, K.D. 4" x 4" | — | LF | .88 | — | .88 |
| Red cedar, rough green constr. | — | MBF | 1,600.00 | — | 1,600.00 |
| Red cedar, rough green constr. | — | LF | 2.35 | — | 2.35 |
| Redwood, S4S construction heart | — | MBF | 1,870.00 | — | 1,870.00 |
| Redwood, S4S construction heart | — | LF | 2.74 | — | 2.74 |
| Southern yellow pine #2, treated | — | MBF | 1040.00 | — | 1040.00 |
| Southern yellow pine #2, treated | — | LF | 1.54 | — | 1.54 |

**Girders** Standard and better lumber, first floor work. Figures in parentheses show board feet per linear foot of girder, including 7% waste

| | Craft@Hrs | Unit | Material | Labor | Total |
|---|---|---|---|---|---|
| 4" x 6", per MBF | B1@15.8 | MBF | 635.00 | 526.00 | 1,161.00 |
| 4" x 8", 10", 12", per MBF | B1@15.8 | MBF | 635.00 | 526.00 | 1,161.00 |
| 6" x 6", per MBF | B1@15.8 | MBF | 942.00 | 526.00 | 1,468.00 |
| 6" x 8", 10", 12", 8" x 8", per MBF | B1@15.8 | MBF | 942.00 | 526.00 | 1,468.00 |
| 4" x 6" (2.15 BF per LF) | B1@.034 | LF | 1.37 | 1.13 | 2.50 |
| 4" x 8" (2.85 BF per LF) | B1@.045 | LF | 1.81 | 1.50 | 3.31 |
| 4" x 10" (3.58 BF per LF) | B1@.057 | LF | 2.27 | 1.90 | 4.17 |
| 4" x 12" (4.28 BF per LF) | B1@.067 | LF | 2.72 | 2.23 | 4.95 |
| 6" x 6" (3.21 BF per LF) | B1@.051 | LF | 3.02 | 1.70 | 4.72 |
| 6" x 8" (4.28 BF per LF) | B1@.067 | LF | 4.03 | 2.23 | 6.26 |
| 6" x 10" (5.35 BF per LF) | B1@.083 | LF | 5.04 | 2.76 | 7.80 |
| 6" x 12" (6.42 BF per LF) | B1@.098 | LF | 6.05 | 3.26 | 9.31 |
| 8" x 8" (5.71 BF per LF) | B1@.088 | LF | 5.38 | 2.93 | 8.31 |

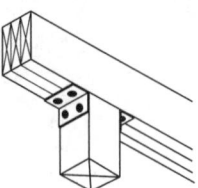

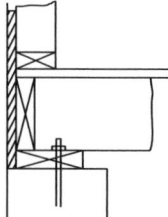

| | Craft@Hrs | Unit | Material | Labor | Total |
|---|---|---|---|---|---|

**Sill plates** (At foundation) SYP #2 pressure treated lumber, drilled and installed with foundation bolts at 48" OC, no bolts, nuts or washers included. See also plates in this section. Figures in parentheses indicate board feet per LF of foundation, including 5% waste and wolmanized treatment.

| | Craft@Hrs | Unit | Material | Labor | Total |
|---|---|---|---|---|---|
| Sill plates, per MBF | — | MBF | 858.00 | — | 858.00 |
| 2" x 3" (.53 BF per LF) | B1@.020 | LF | .32 | .67 | .99 |
| 2" x 4" (.70 BF per LF) | B1@.023 | LF | .59 | .77 | 1.36 |
| 2" x 6" (1.05 BF per LF) | B1@.024 | LF | .86 | .80 | 1.66 |
| 2" x 8" (1.40 BF per LF) | B1@.031 | LF | 1.28 | 1.03 | 2.31 |

**Floor joists** Per SF of area covered. Figures in parentheses indicate board feet per square foot of floor including box or band joist, typical double joists, and 6% waste. No beams, blocking or bridging included. Deduct for openings over 25 SF. Costs shown are based on a job with 1,000 SF of area covered. For scheduling purposes, estimate that a two-man crew can complete 750 SF of area per 8-hour day for 12" center to center framing; 925 SF for 16" OC; 1,100 SF for 20" OC; or 1,250 SF for 24" OC.

| | Craft@Hrs | Unit | Material | Labor | Total |
|---|---|---|---|---|---|
| **2" x 6" Std & Btr** | | | | | |
| 2" x 6" floor joists, per MBF | — | MBF | 465.00 | — | 465.00 |
| 12" centers (1.28 BF per SF) | B1@.021 | SF | .59 | .70 | 1.29 |
| 16" centers (1.02 BF per SF) | B1@.017 | SF | .47 | .57 | 1.04 |
| 20" centers (.88 BF per SF) | B1@.014 | SF | .41 | .47 | .88 |
| 24" centers (.73 BF per SF) | B1@.012 | SF | .34 | .40 | .74 |
| **2" x 8" Std & Btr** | | | | | |
| 2" x 8" floor joists, per MBF | — | MBF | 465.00 | — | 465.00 |
| 12" centers (1.71 BF per SF) | B1@.023 | SF | .80 | .77 | 1.57 |
| 16" centers (1.36 BF per SF) | B1@.018 | SF | .63 | .60 | 1.23 |
| 20" centers (1.17 BF per SF) | B1@.015 | SF | .54 | .50 | 1.04 |
| 24" centers (1.03 BF per SF) | B1@.013 | SF | .48 | .43 | .91 |
| **2" x 10" Std & Btr** | | | | | |
| 2" x 10" floor joists, per MBF | — | MBF | 484.00 | — | 484.00 |
| 12" centers (2.14 BF per SF) | B1@.025 | SF | 1.04 | .83 | 1.87 |
| 16" centers (1.71 BF per SF) | B1@.020 | SF | .83 | .67 | 1.50 |
| 20" centers (1.48 BF per SF) | B1@.016 | SF | .72 | .53 | 1.25 |
| 24" centers (1.30 BF per SF) | B1@.014 | SF | .63 | .47 | 1.10 |
| **2" x 12" Std & Btr** | | | | | |
| 2" x 12" floor joists, per MBF | — | MBF | 563.00 | — | 563.00 |
| 12" centers (2.56 BF per SF) | B1@.026 | SF | 1.44 | .87 | 2.31 |
| 16" centers (2.05 BF per SF) | B1@.021 | SF | 1.15 | .70 | 1.85 |
| 20" centers (1.77 BF per SF) | B1@.017 | SF | 1.00 | .57 | 1.57 |
| 24" centers (1.56 BF per SF) | B1@.015 | SF | .88 | .50 | 1.38 |

**Floor joist wood, TJI truss type** Suitable for residential use, 50 PSF floor load design. Costs shown are per square foot (SF) of floor area, based on joists at 16" OC, for a job with 1,000 SF of floor area. Figure 1.22 SF of floor area for each LF of joist. Add the cost of beams, supports and blocking. For scheduling purposes, estimate that a two-man crew can install 900 to 950 SF of joists in an 8-hour day.

| | Craft@Hrs | Unit | Material | Labor | Total |
|---|---|---|---|---|---|
| 9-1/2" TJI/15 | B1@.017 | SF | 1.93 | .57 | 2.50 |
| 11-7/8" TJI/15 | B1@.017 | SF | 2.11 | .57 | 2.68 |
| 14" TJI/35 | B1@.018 | SF | 3.05 | .60 | 3.65 |
| 16" TJI/35 | B1@.018 | SF | 3.34 | .60 | 3.94 |

# Carpentry, Detailed Breakdown

|  | Craft@Hrs | Unit | Material | Labor | Total |
|---|---|---|---|---|---|

**Bridging or blocking** Installed between 2" x 6" thru 2" x 12" joists. Costs shown are per each set of cross bridges or per each block for solid bridging, and include normal waste. The spacing between the bridging or blocking, sometimes called a "bay," depends on job requirements. Labor costs assume bridging is cut to size on site.

| | Craft@Hrs | Unit | Material | Labor | Total |
|---|---|---|---|---|---|
| 1" x 4" cross | | | | | |
|   Joist bridging, per MBF | — | MBF | 1,150.00 | — | 1,150.00 |
|   Joists on 12" centers | B1@.034 | Ea | .57 | 1.13 | 1.70 |
|   Joists on 16" centers | B1@.034 | Ea | .78 | 1.13 | 1.91 |
|   Joists on 20" centers | B1@.034 | Ea | .99 | 1.13 | 2.12 |
|   Joists on 24" centers | B1@.034 | Ea | 1.21 | 1.13 | 2.34 |
| 2" x 6" solid, Std & Btr | | | | | |
|   2" x 6" blocking, per MBF | — | MBF | 465.00 | — | 465.00 |
|   Joists on 12" centers | B1@.042 | Ea | .51 | 1.40 | 1.91 |
|   Joists on 16" centers | B1@.042 | Ea | .68 | 1.40 | 2.08 |
|   Joists on 20" centers | B1@.042 | Ea | .85 | 1.40 | 2.25 |
|   Joists on 24" centers | B1@.042 | Ea | 1.02 | 1.40 | 2.42 |
| 2" x 8" solid, Std & Btr | | | | | |
|   2" x 8" blocking, per MBF | — | MBF | 465.00 | — | 465.00 |
|   Joists on 12" centers | B1@.042 | Ea | .68 | 1.40 | 2.08 |
|   Joists on 16" centers | B1@.042 | Ea | .91 | 1.40 | 2.31 |
|   Joists on 20" centers | B1@.042 | Ea | 1.14 | 1.40 | 2.54 |
|   Joists on 24" centers | B1@.042 | Ea | 1.36 | 1.40 | 2.76 |
| 2" x 10" solid, Std & Btr | | | | | |
|   2" x 10" blocking, per MBF | — | MBF | 484.00 | — | 484.00 |
|   Joists on 12" centers | B1@.057 | Ea | .89 | 1.90 | 2.79 |
|   Joists on 16" centers | B1@.057 | Ea | 1.18 | 1.90 | 3.08 |
|   Joists on 20" centers | B1@.057 | Ea | 1.48 | 1.90 | 3.38 |
|   Joists on 24" centers | B1@.057 | Ea | 1.77 | 1.90 | 3.67 |
| 2" x 12" solid, Std & Btr | | | | | |
|   2" x 12" blocking, per MBF | — | MBF | 563.00 | — | 563.00 |
|   Joists on 12" centers | B1@.057 | Ea | 1.24 | 1.90 | 3.14 |
|   Joists on 16" centers | B1@.057 | Ea | 1.65 | 1.90 | 3.55 |
|   Joists on 20" centers | B1@.057 | Ea | 2.06 | 1.90 | 3.96 |
|   Joists on 24" centers | B1@.057 | Ea | 2.48 | 1.90 | 4.38 |
| Steel, no nail type, cross | | | | | |
|   Snap-on joist bridging | B1@.020 | Ea | 1.69 | .67 | 2.36 |

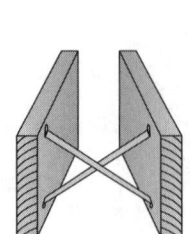

## Subflooring

Board sheathing, 1" x 6" #3 & Btr (1.18 BF per SF)

| | Craft@Hrs | Unit | Material | Labor | Total |
|---|---|---|---|---|---|
|   Board sheathing, per MBF | — | MBF | 1,820.00 | — | 1,820.00 |
|   With 12% shrinkage, 5% waste & nails | B1@.020 | SF | 2.15 | .67 | 2.82 |
|   Add for diagonal patterns | B1@.001 | SF | .22 | .03 | .25 |

Plywood sheathing, CD standard exterior grade. Material costs shown include 5% for waste & fasteners

| | Craft@Hrs | Unit | Material | Labor | Total |
|---|---|---|---|---|---|
|   3/8" | B1@.011 | SF | .46 | .37 | .83 |
|   1/2" | B1@.012 | SF | .53 | .40 | .93 |
|   5/8" | B1@.012 | SF | .65 | .40 | 1.05 |
|   3/4" | B1@.013 | SF | .78 | .43 | 1.21 |

OSB sheathing, Material costs shown include 5% for waste & fasteners

| | Craft@Hrs | Unit | Material | Labor | Total |
|---|---|---|---|---|---|
|   3/8" | B1@.011 | SF | .29 | .37 | .66 |
|   7/16" | B1@.011 | SF | .28 | .37 | .65 |
|   1/2" | B1@.012 | SF | .32 | .40 | .72 |

| | Craft@Hrs | Unit | Material | Labor | Total |
|---|---|---|---|---|---|
| 5/8" | B1@.012 | SF | .44 | .40 | .84 |
| 3/4" | B1@.014 | SF | .57 | .47 | 1.04 |
| 7/8" | B1@.014 | SF | .96 | .47 | 1.43 |
| 1" | B1@.015 | SF | 1.40 | .50 | 1.90 |

Engineered floating subfloor, 2' x 2' panels, trade name DRIcore, resilient engineered wood core with a water-resistant polyethylene membrane on the underside. www.dricore.com

| | Craft@Hrs | Unit | Material | Labor | Total |
|---|---|---|---|---|---|
| Per SF costs including 10% for waste | B1@.020 | SF | 1.75 | .67 | 2.42 |

**Plates** (Wall plates) Std & Btr, untreated. For pressure treated plates, see also Sill Plates in this section. Figures in parentheses indicate board feet per LF. Costs shown include 10% for waste and nails

| | Craft@Hrs | Unit | Material | Labor | Total |
|---|---|---|---|---|---|
| 2" x 3" wall plates, per MBF | — | MBF | 423.00 | — | 423.00 |
| 2" x 3" (.55 BF per LF) | B1@.010 | LF | .23 | .33 | .56 |
| 2" x 4" wall plates, per MBF | — | MBF | 473.00 | — | 473.00 |
| 2" x 4" (.73 BF per LF) | B1@.012 | LF | .35 | .40 | .75 |
| 2" x 6" wall plates, per MBF | — | MBF | 465.00 | — | 465.00 |
| 2" x 6" (1.10 BF per LF) | B1@.018 | LF | .51 | .60 | 1.11 |
| 2" x 8" wall plates, per MBF | — | MBF | 465.00 | — | 465.00 |
| 2" x 8" (1.40 BF per LF) | B1@.020 | LF | .65 | .67 | 1.32 |

**Studding** Per square foot of wall area. Do not subtract for openings less than 16' wide. Figures in parentheses indicate typical board feet per SF of wall area, measured on one side, and include normal waste. Add for each corner and partition from below. Costs include studding, nails. Add for plates from the section above and also, door and window opening framing, backing, let-in bracing, fire blocking, and sheathing for shear walls from the sections that follow. Labor includes layout, plumb and align.

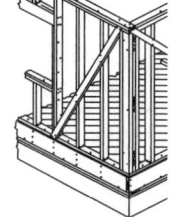

| | Craft@Hrs | Unit | Material | Labor | Total |
|---|---|---|---|---|---|
| 2" x 3", Std & Btr | | | | | |
| 2" x 3" studs, 16" centers, per MBF | — | MBF | 423.00 | — | 423.00 |
| 12" centers (.55 BF per SF) | B1@.024 | SF | .23 | .80 | 1.03 |
| 16" centers (.41 BF per SF) | B1@.018 | SF | .17 | .60 | .77 |
| 20" centers (.33 BF per SF) | B1@.015 | SF | .14 | .50 | .64 |
| 24" centers (.28 BF per SF) | B1@.012 | SF | .12 | .40 | .52 |
| Add for each corner or partition | B1@.083 | Ea | 5.49 | 2.76 | 8.25 |
| 2" x 4", Std & Btr | | | | | |
| 2" x 4" studs, 16" centers, per MBF | — | MBF | 474.00 | — | 474.00 |
| 12" centers (.73 BF per SF) | B1@.031 | SF | .35 | 1.03 | 1.38 |
| 16" centers (.54 BF per SF) | B1@.023 | SF | .26 | .77 | 1.03 |
| 20" centers (.45 BF per SF) | B1@.020 | SF | .21 | .67 | .88 |
| 24" centers (.37 BF per SF) | B1@.016 | SF | .18 | .53 | .71 |
| Add for each corner or partition | B1@.083 | Ea | 8.04 | 2.76 | 10.80 |
| 2" x 6", Std & Btr | | | | | |
| 2" x 6" studs, 24" centers, per MBF | — | MBF | 465.00 | — | 465.00 |
| 12" centers (1.10 BF per SF) | B1@.044 | SF | .51 | 1.47 | 1.98 |
| 16" centers (.83 BF per SF) | B1@.033 | SF | .39 | 1.10 | 1.49 |
| 20" centers (.67 BF per SF) | B1@.027 | SF | .31 | .90 | 1.21 |
| 24" centers (.55 BF per SF) | B1@.022 | SF | .26 | .73 | .99 |
| Add for each corner or partition | B1@.083 | Ea | 11.10 | 2.76 | 13.86 |
| 4" x 4", Std & Btr | | | | | |
| Installed in wall framing, with 10% waste | B1@.064 | LF | .88 | 2.13 | 3.01 |

# Carpentry, Detailed Breakdown

| | Craft@Hrs | Unit | Material | Labor | Total |
|---|---|---|---|---|---|

**Door opening framing** In wall studs, based on walls 8' in height. Costs shown are per each door opening and include header of appropriate size, double vertical studs each side of the opening less than 8' wide (triple vertical studs each side of openings 8' wide or wider), cripples, blocking, nails and normal waste. Width shown is size of finished opening. Figures in parentheses indicate header size

2" x 4" wall studs, Std & Btr, opening size as shown

| | Craft@Hrs | Unit | Material | Labor | Total |
|---|---|---|---|---|---|
| 2" x 4" wall studs, per MBF | — | MBF | 473.00 | — | 473.00 |
| 4" x 4" and 4" x 6" headers, per MBF | — | MBF | 598.00 | — | 598.00 |
| 4" x 8", 10", 12" and 14" headers, per MBF | — | MBF | 635.00 | — | 635.00 |
| To 3' wide (4" x 4" header) | B1@.830 | Ea | 14.20 | 27.60 | 41.80 |
| Over 3' to 4' wide (4" x 6" header) | B1@1.11 | Ea | 17.30 | 37.00 | 54.30 |
| Over 4' to 5' wide (4" x 6" header) | B1@1.39 | Ea | 19.00 | 46.30 | 65.30 |
| Over 5' to 6' wide (4" x 8" header) | B1@1.66 | Ea | 23.40 | 55.30 | 78.70 |
| Over 6' to 8' wide (4" x 10" header) | B1@1.94 | Ea | 36.60 | 64.60 | 101.20 |
| Over 8' to 10' wide (4" x 12" header) | B1@1.94 | Ea | 45.80 | 64.60 | 110.40 |
| Over 10' to 12' wide (4" x 14" header) | B1@2.22 | Ea | 56.40 | 73.90 | 130.30 |
| Add per foot of height for 4" walls | | | | | |
| over 8' in height | — | LF | 1.89 | — | 1.89 |

2" x 6" wall studs, Std & Btr, opening size as shown

| | Craft@Hrs | Unit | Material | Labor | Total |
|---|---|---|---|---|---|
| 2" x 6" wall studs, per MBF | — | MBF | 465.00 | — | 465.00 |
| 6" x 4" headers, per MBF | — | MBF | 635.00 | — | 635.00 |
| 6" x 8", 10", 12" and 14" headers, per MBF | — | MBF | 942.00 | — | 942.00 |
| To 3' wide (6" x 4" header) | B1@1.11 | Ea | 22.80 | 37.00 | 59.80 |
| Over 3' to 4' wide (6" x 6" header) | B1@1.39 | Ea | 27.30 | 46.30 | 73.60 |
| Over 4' to 5' wide (6" x 6" header) | B1@1.66 | Ea | 30.60 | 55.30 | 85.90 |
| Over 5' to 6' wide (6" x 8" header) | B1@1.94 | Ea | 39.50 | 64.60 | 104.10 |
| Over 6' to 8' wide (6" x 10" header) | B1@2.22 | Ea | 62.50 | 73.90 | 136.40 |
| Over 8' to 10' wide (6" x 12" header) | B1@2.22 | Ea | 82.60 | 73.90 | 156.50 |
| Over 10' to 12' wide (6" x 14" header) | B1@2.50 | Ea | 106.00 | 83.30 | 189.30 |
| Add per foot of height for 6" walls | | | | | |
| over 8' in height | — | LF | 2.79 | — | 2.79 |

**Window opening framing** In wall studs, based on walls 8' in height. Costs shown are per window opening and include header of appropriate size, sub-sill plate (double sub-sill if opening is 8' wide or wider), double vertical studs each side of openings less than 8' wide (triple vertical studs each side of openings 8' wide or wider), top and bottom cripples, blocking, nails and normal waste. Figures in parentheses indicate header size.

2" x 4" wall studs, Std & Btr, opening size shown is width of finished opening

| | Craft@Hrs | Unit | Material | Labor | Total |
|---|---|---|---|---|---|
| 2" x 4" wall studs, per MBF | — | MBF | 473.00 | — | 473.00 |
| 4" x 4" and 4" x 6" headers, per MBF | — | MBF | 635.00 | — | 635.00 |
| 4" x 8", 10", 12" and 14" headers, per MBF | — | MBF | 635.00 | — | 635.00 |
| To 2' wide (4" x 4" header) | B1@1.00 | Ea | 13.10 | 33.30 | 46.40 |
| Over 2' to 3' wide (4" x 4" header) | B1@1.17 | Ea | 16.30 | 39.00 | 55.30 |
| Over 3' to 4' wide (4" x 6" header) | B1@1.45 | Ea | 20.10 | 48.30 | 68.40 |
| Over 4' to 5' wide (4" x 6" header) | B1@1.73 | Ea | 22.50 | 57.60 | 80.10 |
| Over 5' to 6' wide (4" x 8" header) | B1@2.01 | Ea | 27.20 | 67.00 | 94.20 |
| Over 6' to 7' wide (4" x 8" header) | B1@2.29 | Ea | 34.60 | 76.30 | 110.90 |
| Over 7' to 8' wide (4" x 10" header) | B1@2.57 | Ea | 41.60 | 85.60 | 127.20 |
| Over 8' to 10' wide (4" x 12" header) | B1@2.57 | Ea | 52.10 | 85.60 | 137.70 |
| Over 10' to 12' wide (4" x 14" header) | B1@2.85 | Ea | 67.80 | 94.90 | 162.70 |
| Add per foot of height for walls | | | | | |
| over 8' high | — | LF | 1.89 | — | 1.89 |

# Carpentry, Detailed Breakdown

|  | Craft@Hrs | Unit | Material | Labor | Total |
|---|---|---|---|---|---|
| 2" x 6" wall studs, Std & Btr, opening size as shown | | | | | |
| 2" x 6" wall studs, per MBF | — | MBF | 465.00 | — | 465.00 |
| 4" x 4" and 6" x 4" headers, per MBF | — | MBF | 635.00 | — | 635.00 |
| 6" x 6", 8", 10", 12" and 14" headers, per MBF | — | MBF | 942.00 | — | 942.00 |
| To 2' wide (4 x 4) | B1@1.17 | Ea | 16.20 | 39.00 | 55.20 |
| Over 2' to 3' wide (6" x 4" header) | B1@1.45 | Ea | 20.80 | 48.30 | 69.10 |
| Over 3' to 4' wide (6" x 6" header) | B1@1.73 | Ea | 29.90 | 57.60 | 87.50 |
| Over 4' to 5' wide (6" x 6" header) | B1@2.01 | Ea | 35.40 | 67.00 | 102.40 |
| Over 5' to 6' wide (6" x 8" header) | B1@2.29 | Ea | 43.20 | 76.30 | 119.50 |
| Over 6' to 8' wide (6" x 10" header) | B1@2.85 | Ea | 67.40 | 94.90 | 162.30 |
| Over 8' to 10' wide (6" x 12" header) | B1@2.85 | Ea | 91.80 | 94.90 | 186.70 |
| Over 8' to 12' wide (6" x 14" header) | B1@3.13 | Ea | 117.00 | 104.00 | 221.00 |
| Add per foot of height for 6" walls over 8' high | — | LF | 2.79 | — | 2.79 |

**Bracing**  See also sheathing for plywood bracing and shear panels

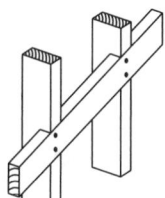

| | Craft@Hrs | Unit | Material | Labor | Total |
|---|---|---|---|---|---|
| Let-in wall bracing, using standard and better lumber | | | | | |
| 1" x 4" | B1@.021 | LF | .64 | .70 | 1.34 |
| 1" x 6" | B1@.027 | LF | .94 | .90 | 1.84 |
| 2" x 4" | B1@.035 | LF | .32 | 1.17 | 1.49 |
| Steel strap bracing, 1-1/4" wide | | | | | |
| 9'6" or 11'6" lengths | B1@.010 | LF | .49 | .33 | .82 |
| Steel "V" bracing, 3/4" x 3/4" | | | | | |
| 9'6" or 11'6" lengths | B1@.010 | LF | .78 | .33 | 1.11 |
| Temporary wood frame wall bracing, assumes salvage at 50% and 3 uses | | | | | |
| 1" x 4" Std & Btr | B1@.006 | LF | .16 | .20 | .36 |
| 1" x 6" Std & Btr | B1@.010 | LF | .23 | .33 | .56 |
| 2" x 4" utility | B1@.012 | LF | .08 | .40 | .48 |
| 2" x 6" utility | B1@.018 | LF | .12 | .60 | .72 |

**Fireblocks**  Installed in wood frame walls, per LF of wall to be blocked. Figures in parentheses indicate board feet of fire blocking per linear foot of wall including 10% waste. See also Bridging and Backing and Nailers in this section

| | Craft@Hrs | Unit | Material | Labor | Total |
|---|---|---|---|---|---|
| 2" x 3" blocking, Std & Btr | | | | | |
| 2" x 3" fireblocks, 16" centers, per MBF | — | MBF | 423.00 | — | 423.00 |
| 12" OC members (.48 BF per LF) | B1@.016 | LF | .20 | .53 | .73 |
| 16" OC members (.50 BF per LF) | B1@.015 | LF | .21 | .50 | .71 |
| 20" OC members (.51 BF per LF) | B1@.012 | LF | .22 | .40 | .62 |
| 24" OC members (.52 BF per LF) | B1@.010 | LF | .22 | .33 | .55 |
| 2" x 4" blocking, Std & Btr | | | | | |
| 2" x 4" fireblocks, 16" centers, per MBF | — | MBF | 473.00 | — | 473.00 |
| 12" OC members (.64 BF per LF) | B1@.020 | LF | .30 | .67 | .97 |
| 16" OC members (.67 BF per LF) | B1@.015 | LF | .32 | .50 | .82 |
| 20" OC members (.68 BF per LF) | B1@.012 | LF | .32 | .40 | .72 |
| 24" OC members (.69 BF per LF) | B1@.010 | LF | .33 | .33 | .66 |
| 2" x 6" blocking, Std & Btr | | | | | |
| 2" x 6" fireblocks, 24" centers, per MBF | — | MBF | 465.00 | — | 465.00 |
| 12" OC members (.96 BF per LF) | B1@.021 | LF | .45 | .70 | 1.15 |
| 16" OC members (1.00 BF per LF) | B1@.016 | LF | .46 | .53 | .99 |
| 20" OC members (1.02 BF per LF) | B1@.012 | LF | .47 | .40 | .87 |
| 24" OC members (1.03 BF per LF) | B1@.010 | LF | .48 | .33 | .81 |

# Carpentry, Detailed Breakdown

| | Craft@Hrs | Unit | Material | Labor | Total |
|---|---|---|---|---|---|

**Beams** Std & Btr. Installed over wall openings and around floor, ceiling and roof openings or where a flush beam is called out in the plans, including 10% waste. Do not use these beams for door or window headers in framed walls, use door or window opening framing assemblies.

| | Craft@Hrs | Unit | Material | Labor | Total |
|---|---|---|---|---|---|
| 2" x 6" beams, per MBF | — | MBF | 465.00 | — | 465.00 |
| 2" x 8" beams, per MBF | — | MBF | 465.00 | — | 465.00 |
| 2" x 10" beams, per MBF | — | MBF | 484.00 | — | 484.00 |
| 2" x 12" beams, per MBF | — | MBF | 563.00 | — | 563.00 |
| 4" x 6" beams, per MBF | — | MBF | 635.00 | — | 635.00 |
| 4" x 8", 10", 12", 14" beams, per MBF | — | MBF | 635.00 | — | 635.00 |
| 6" x 6", 8", 10", 12", 14" beams, per MBF | — | MBF | 942.00 | — | 942.00 |
| 2" x 6" (1.10 BF per LF) | B1@.028 | LF | .51 | .93 | 1.44 |
| 2" x 8" (1.47 BF per LF) | B1@.037 | LF | .68 | 1.23 | 1.91 |
| 2" x 10" (1.83 BF per LF) | B1@.046 | LF | .89 | 1.53 | 2.42 |
| 2" x 12" (2.20 BF per LF) | B1@.057 | LF | 1.24 | 1.90 | 3.14 |
| 4" x 6" (2.20 BF per LF) | B1@.057 | LF | 1.40 | 1.90 | 3.30 |
| 4" x 8" (2.93 BF per LF) | B1@.073 | LF | 1.86 | 2.43 | 4.29 |
| 4" x 10" (3.67 BF per LF) | B1@.094 | LF | 2.33 | 3.13 | 5.46 |
| 4" x 12" (4.40 BF per LF) | B1@.112 | LF | 2.79 | 3.73 | 6.52 |
| 4" x 14" (5.13 BF per LF) | B1@.115 | LF | 3.26 | 3.83 | 7.09 |
| 6" x 6" (3.30 BF per LF) | B1@.060 | LF | 3.11 | 2.00 | 5.11 |
| 6" x 8" (4.40 BF per LF) | B1@.080 | LF | 4.14 | 2.66 | 6.80 |
| 6" x 10" (5.50 BF per LF) | B1@.105 | LF | 5.18 | 3.50 | 8.68 |
| 6" x 12" (6.60 BF per LF) | B1@.115 | LF | 6.22 | 3.83 | 10.05 |
| 6" x 14" (7.70 BF per LF) | B1@.120 | LF | 7.25 | 4.00 | 11.25 |

**Posts** S4S, green. Posts not in wall framing. Material costs include 10% waste. For scheduling purposes, estimate that a two-man crew can complete 100 to 125 LF per 8-hour day.

| | Craft@Hrs | Unit | Material | Labor | Total |
|---|---|---|---|---|---|
| 4" x 4", 4" x 6" posts, per MBF | — | MBF | 598.00 | — | 598.00 |
| 4" x 8" to 4" x 12" posts, per MBF | — | MBF | 635.00 | — | 635.00 |
| 6" x 6" to 8" x 12" posts, per MBF | — | MBF | 942.00 | — | 942.00 |
| 4" x 4" per LF | B1@.110 | LF | .87 | 3.66 | 4.53 |
| 4" x 6" per LF | B1@.120 | LF | 1.31 | 4.00 | 5.31 |
| 4" x 8" per LF | B1@.140 | LF | 1.86 | 4.66 | 6.52 |
| 4" x 10" per LF | B1@.143 | LF | 2.33 | 4.76 | 7.09 |
| 4" x 12" per LF | B1@.145 | LF | 2.79 | 4.83 | 7.62 |
| 6" x 6" per LF | B1@.145 | LF | 3.11 | 4.83 | 7.94 |
| 6" x 8" per LF | B1@.145 | LF | 3.43 | 4.83 | 8.26 |
| 6" x 10" per LF | B1@.147 | LF | 5.18 | 4.90 | 10.08 |
| 6" x 12" per LF | B1@.150 | LF | 6.22 | 5.00 | 11.22 |
| 8" x 8" per LF | B1@.150 | LF | 5.53 | 5.00 | 10.53 |
| 8" x 10" per LF | B1@.155 | LF | 6.91 | 5.16 | 12.07 |
| 8" x 12" per LF | B1@.166 | LF | 8.29 | 5.53 | 13.82 |

**Ceiling joists and soffits** Per SF of area covered. Figures in parentheses indicate board feet per square foot of ceiling including end joists, header joists, and 5% waste. No beams, bridging, blocking, or ledger strips included. Deduct for openings over 25 SF. Costs shown are based on a job with 1,000 SF of area covered. For scheduling purposes, estimate that a two-man crew can complete 650 SF of area per 8-hour day for 12" center to center framing; 800 SF for 16" OC; 950 SF for 20" OC; or 1,100 SF for 24" OC.

| | Craft@Hrs | Unit | Material | Labor | Total |
|---|---|---|---|---|---|
| 2" x 4", Std & Btr grade | | | | | |
| 2" x 4" ceiling joists, per MBF | — | MBF | 473.00 | — | 473.00 |
| 12" centers (.78 BF per SF) | B1@.026 | SF | .37 | .87 | 1.24 |
| 16" centers (.59 BF per SF) | B1@.020 | SF | .28 | .67 | .95 |
| 20" centers (.48 BF per SF) | B1@.016 | SF | .23 | .53 | .76 |
| 24" centers (.42 BF per SF) | B1@.014 | SF | .20 | .47 | .67 |

| | Craft@Hrs | Unit | Material | Labor | Total |
|---|---|---|---|---|---|
| **2" x 6", Std & Btr grade** | | | | | |
| 2" x 6" ceiling joists, per MBF | — | MBF | 465.00 | — | 465.00 |
| 12" centers (1.15 BF per SF) | B1@.026 | SF | .53 | .87 | 1.40 |
| 16" centers ( .88 BF per SF) | B1@.020 | SF | .41 | .67 | 1.08 |
| 20" centers ( .72 BF per SF) | B1@.016 | SF | .33 | .53 | .86 |
| 24" centers ( .63 BF per SF) | B1@.014 | SF | .29 | .47 | .76 |
| **2" x 8", Std & Btr grade** | | | | | |
| 2" x 8" ceiling joists, per MBF | — | MBF | 465.00 | — | 465.00 |
| 12" centers (1.53 BF per SF) | B1@.028 | SF | .71 | .93 | 1.64 |
| 16" centers (1.17 BF per SF) | B1@.022 | SF | .54 | .73 | 1.27 |
| 20" centers ( .96 BF per SF) | B1@.018 | SF | .45 | .60 | 1.05 |
| 24" centers ( .84 BF per SF) | B1@.016 | SF | .39 | .53 | .92 |
| **2" x 10", Std & Btr grade** | | | | | |
| 2" x 10" ceiling joists, per MBF | — | MBF | 484.00 | — | 484.00 |
| 12" centers (1.94 BF per SF) | B1@.030 | SF | .94 | 1.00 | 1.94 |
| 16" centers (1.47 BF per SF) | B1@.023 | SF | .71 | .77 | 1.48 |
| 20" centers (1.21 BF per SF) | B1@.019 | SF | .59 | .63 | 1.22 |
| 24" centers (1.04 BF per SF) | B1@.016 | SF | .50 | .53 | 1.03 |
| **2" x 12", Std & Btr grade** | | | | | |
| 2" x 12" ceiling joists, per MBF | — | MBF | 563.00 | — | 563.00 |
| 12" centers (2.30 BF per SF) | B1@.033 | SF | 1.29 | 1.10 | 2.39 |
| 16" centers (1.76 BF per SF) | B1@.025 | SF | .99 | .83 | 1.82 |
| 20" centers (1.44 BF per SF) | B1@.020 | SF | .81 | .67 | 1.48 |
| 24" centers (1.26 BF per SF) | B1@.018 | SF | .71 | .60 | 1.31 |

**Catwalks** 1" x 6" solid planking walkway, #2 & Btr grade

| | Craft@Hrs | Unit | Material | Labor | Total |
|---|---|---|---|---|---|
| ( 0.50 BF per SF) | B1@.017 | SF | .91 | .57 | 1.48 |

**Backing and nailers** Std & Btr, for wall finishes, "floating" backing for drywall ceilings, trim, z-bar, appliances and fixtures, etc. See also Bridging and Fireblocking in this section. Figures in parentheses show board feet per LF including 10% waste.

| | Craft@Hrs | Unit | Material | Labor | Total |
|---|---|---|---|---|---|
| 1" x  4" ( .37 BF per LF) | B1@.011 | LF | .43 | .37 | .80 |
| 1" x  6" ( .55 BF per LF) | B1@.017 | LF | .68 | .57 | 1.25 |
| 1" x  8" ( .73 BF per LF) | B1@.022 | LF | .90 | .73 | 1.63 |
| 2" x  4" ( .73 BF per LF) | B1@.023 | LF | .35 | .77 | 1.12 |
| 2" x  6" (1.10 BF per LF) | B1@.034 | LF | .51 | 1.13 | 1.64 |
| 2" x  8" (1.47 BF per LF) | B1@.045 | LF | .68 | 1.50 | 2.18 |
| 2" x 10" (1.83 BF per LF) | B1@.057 | LF | .89 | 1.90 | 2.79 |

**Ledger strips** Std & Btr, nailed to faces of studs, beams, girders, joists, etc. See also Ribbons in this section for let-in type. Figures in parentheses indicate board feet per LF including 10% waste.

| | Craft@Hrs | Unit | Material | Labor | Total |
|---|---|---|---|---|---|
| 1" x 2", 3" or 4" ledger, per MBF | — | MBF | 1,150.00 | — | 1,150.00 |
| 1" x 2" ( .18 BF per LF) | B1@.010 | LF | .21 | .33 | .54 |
| 1" x 3" ( .28 BF per LF) | B1@.010 | LF | .32 | .33 | .65 |
| 1" x 4" ( .37 BF per LF) | B1@.010 | LF | .43 | .33 | .76 |
| 2" x 2", 3" or 4" ledger, per MBF | — | MBF | 473.00 | — | 473.00 |
| 2" x 2" ( .37 BF per LF) | B1@.010 | LF | .18 | .33 | .51 |
| 2" x 3" ( .55 BF per LF) | B1@.010 | LF | .26 | .33 | .59 |
| 2" x 4" ( .73 BF per LF) | B1@.010 | LF | .35 | .33 | .68 |

# Carpentry, Detailed Breakdown

| | Craft@Hrs | Unit | Material | Labor | Total |
|---|---|---|---|---|---|

**Ribbons** (Ribbands) Let in to wall framing. See also Ledgers in this section. Figures in parentheses indicate board feet per LF including 10% waste.

| | Craft@Hrs | Unit | Material | Labor | Total |
|---|---|---|---|---|---|
| 1" x 3", 4" or 6" ribbon, per MBF | — | MBF | 1,150.00 | — | 1,150.00 |
| 1" x 3", Std & Btr ( .28 BF per LF) | B1@.020 | LF | .32 | .67 | .99 |
| 1" x 4", Std & Btr ( .37 BF per LF) | B1@.020 | LF | .43 | .67 | 1.10 |
| 1" x 6", Std & Btr ( .55 BF per LF) | B1@.030 | LF | .63 | 1.00 | 1.63 |
| 2" x 3", 4" ribbon, per MBF | — | MBF | 473.00 | — | 473.00 |
| 2" x 3", Std & Btr ( .55 BF per LF) | B1@.041 | LF | .26 | 1.37 | 1.63 |
| 2" x 4", Std & Btr ( .73 BF per LF) | B1@.041 | LF | .35 | 1.37 | 1.72 |
| 2" x 6" ribbon, per MBF | — | MBF | 465.00 | — | 465.00 |
| 2" x 6", Std & Btr (1.10 BF per LF) | B1@.045 | LF | .51 | 1.50 | 2.01 |
| 2" x 8" ribbon, per MBF | — | MBF | 465.00 | — | 465.00 |
| 2" x 8", Std & Btr (1.47 BF per LF) | B1@.045 | LF | .68 | 1.50 | 2.18 |

**Rafters** Flat, shed, or gable roofs, up to 5 in 12 slope (5/24 pitch), maximum 25' span. Figures in parentheses indicate board feet per SF of actual roof surface area (not roof plan area), including rafters, ridge boards, collar beams and normal waste, but no blocking, bracing, purlins, curbs, or gable walls.

| | Craft@Hrs | Unit | Material | Labor | Total |
|---|---|---|---|---|---|
| 2" x 4", Std & Btr | | | | | |
| 2" x 4" rafters, per MBF | — | MBF | 473.00 | — | 473.00 |
| 12" centers (.89 BF per SF) | B1@.021 | SF | .42 | .70 | 1.12 |
| 16" centers (.71 BF per SF) | B1@.017 | SF | .34 | .57 | .91 |
| 24" centers (.53 BF per SF) | B1@.013 | SF | .25 | .43 | .68 |
| 2" x 6", Std & Btr | | | | | |
| 2" x 6" rafters, per MBF | — | MBF | 465.00 | — | 465.00 |
| 12" centers (1.29 BF per SF) | B1@.029 | SF | .60 | .97 | 1.57 |
| 16" centers (1.02 BF per SF) | B1@.023 | SF | .47 | .77 | 1.24 |
| 24" centers ( .75 BF per SF) | B1@.017 | SF | .35 | .57 | .92 |
| 2" x 8", Std & Btr | | | | | |
| 2" x 8" rafters, per MBF | — | MBF | 465.00 | — | 465.00 |
| 12" centers (1.71 BF per SF) | B1@.036 | SF | .80 | 1.20 | 2.00 |
| 16" centers (1.34 BF per SF) | B1@.028 | SF | .62 | .93 | 1.55 |
| 24" centers (1.12 BF per SF) | B1@.024 | SF | .52 | .80 | 1.32 |
| 2" x 10", Std & Btr | | | | | |
| 2" x 10" rafters, per MBF | — | MBF | 484.00 | — | 484.00 |
| 12" centers (2.12 BF per SF) | B1@.039 | SF | 1.03 | 1.30 | 2.33 |
| 16" centers (1.97 BF per SF) | B1@.036 | SF | .95 | 1.20 | 2.15 |
| 24" centers (1.21 BF per SF) | B1@.022 | SF | .59 | .73 | 1.32 |
| 2" x 12", Std & Btr | | | | | |
| 2" x 12" rafters, per MBF | — | MBF | 563.00 | — | 563.00 |
| 12" centers (2.52 BF per SF) | B1@.045 | SF | 1.42 | 1.50 | 2.92 |
| 16" centers (1.97 BF per SF) | B1@.035 | SF | 1.11 | 1.17 | 2.28 |
| 24" centers (1.43 BF per SF) | B1@.026 | SF | .80 | .87 | 1.67 |
| Add for hip roof | B1@.007 | SF | — | .23 | .23 |
| Add for slope over 5 in 12 | B1@.015 | SF | — | .50 | .50 |

**Trimmers and curbs** At stairwells, skylights, dormers, etc. Figures in parentheses show board feet per LF, Std & Btr grade, including 10% waste.

| | Craft@Hrs | Unit | Material | Labor | Total |
|---|---|---|---|---|---|
| 2" x 4" ( .73 BF per LF) | B1@.018 | LF | .35 | .60 | .95 |
| 2" x 6" (1.10 BF per LF) | B1@.028 | LF | .51 | .93 | 1.44 |
| 2" x 8" (1.47 BF per LF) | B1@.038 | LF | .68 | 1.27 | 1.95 |
| 2" x 10" (1.83 BF per LF) | B1@.047 | LF | .89 | 1.57 | 2.46 |
| 2" x 12" (2.20 BF per LF) | B1@.057 | LF | 1.24 | 1.90 | 3.14 |

|  | Craft@Hrs | Unit | Material | Labor | Total |
|---|---|---|---|---|---|
| **Collar beams & collar ties**  Std & Btr grade, including 10% waste | | | | | |
| Collar beams, 2" x 6" | B1@.013 | LF | .51 | .43 | .94 |
| Collar ties, 1" x 6" | B1@.006 | LF | .68 | .20 | .88 |

**Purlins** (Purling)  Std & Btr, installed below roof rafters. Figures in parentheses indicate board feet per LF including 5% waste.

|  | Craft@Hrs | Unit | Material | Labor | Total |
|---|---|---|---|---|---|
| 2" x 4" ( .70 BF per LF) | B1@.012 | LF | .33 | .40 | .73 |
| 2" x 6" (1.05 BF per LF) | B1@.017 | LF | .49 | .57 | 1.06 |
| 2" x 8" (1.40 BF per LF) | B1@.023 | LF | .65 | .77 | 1.42 |
| 2" x 10" (1.75 BF per LF) | B1@.027 | LF | .85 | .90 | 1.75 |
| 2" x 12" (2.10 BF per LF) | B1@.030 | LF | 1.18 | 1.00 | 2.18 |
| 4" x 6" (2.10 BF per LF) | B1@.034 | LF | 1.33 | 1.13 | 2.46 |
| 4" x 8" (2.80 BF per LF) | B1@.045 | LF | 1.67 | 1.50 | 3.17 |

|  | Craft@Hrs | Unit | Material | Labor | Total |
|---|---|---|---|---|---|
| **Dormer studs**  Std & Btr, per square foot of wall area, including 10% waste | | | | | |
| 2" x 4" ( .84 BF per SF) | B1@.033 | SF | .40 | 1.10 | 1.50 |

**Roof trusses**  24" OC, any slope from 3 in 12 to 12 in 12, total height not to exceed 12' high from bottom chord to highest point on truss. Prices for trusses over 12' high will be up to 100% higher. Square foot (SF) costs, where shown, are per square foot of area to be covered.

Scissor truss

|  | Craft@Hrs | Unit | Material | Labor | Total |
|---|---|---|---|---|---|
| 2" x 4" top and bottom chords | | | | | |
| Up to 38' span | B1@.022 | SF | 2.25 | .73 | 2.98 |
| 40' to 50' span | B1@.028 | SF | 2.81 | .93 | 3.74 |
| Fink truss "W" (conventional roof truss) | | | | | |
| 2" x 4" top and bottom chords | | | | | |
| Up to 38' span | B1@.017 | SF | 1.95 | .57 | 2.52 |
| 40' to 50' span | B1@.022 | SF | 2.35 | .73 | 3.08 |
| 2" x 6" top and bottom chords | | | | | |
| Up to 38' span | B1@.020 | SF | 2.41 | .67 | 3.08 |
| 40' to 50' span | B1@.026 | SF | 2.89 | .87 | 3.76 |
| Truss with gable fill at 16" OC | | | | | |
| 28' span, 5 in 12 slope | B1@.958 | Ea | 162.00 | 31.90 | 193.90 |
| 32' span, 5 in 12 slope | B1@1.26 | Ea | 200.00 | 42.00 | 242.00 |
| 40' span, 5 in 12 slope | B1@1.73 | Ea | 281.00 | 57.60 | 338.60 |

**Fascia**  Material column includes 10% waste

|  | Craft@Hrs | Unit | Material | Labor | Total |
|---|---|---|---|---|---|
| White Wood, #3 & Btr | | | | | |
| 1" x 4", per MBF | — | MBF | 1,780.00 | — | 1,780.00 |
| 1" x 6", per MBF | — | MBF | 1,910.00 | — | 1,910.00 |
| 1" x 8", per MBF | — | MBF | 1,950.00 | — | 1,950.00 |
| 1" x 4", per LF | B1@.044 | LF | .65 | 1.47 | 2.12 |
| 1" x 6", per LF | B1@.044 | LF | 1.05 | 1.47 | 2.52 |
| 1" x 8", per LF | B1@.051 | LF | 1.43 | 1.70 | 3.13 |
| Hem-fir, S4S, dry, Std and Btr | | | | | |
| 2" x 6", per MBF | — | MBF | 471.00 | — | 471.00 |
| 2" x 8", per MBF | — | MBF | 493.00 | — | 493.00 |
| 2" x 10", per MBF | — | MBF | 473.00 | — | 473.00 |
| 2" x 6", per LF | B1@.044 | LF | .52 | 1.47 | 1.99 |
| 2" x 8", per LF | B1@.051 | LF | .72 | 1.70 | 2.42 |
| 2" x 10", per LF | B1@.051 | LF | .87 | 1.70 | 2.57 |

# Carpentry, Detailed Breakdown

| | Craft@Hrs | Unit | Material | Labor | Total |
|---|---|---|---|---|---|
| **Redwood, S4S kiln dried, clear** | | | | | |
| 1" x 4", per MBF | — | MBF | 2,450.00 | — | 2,450.00 |
| 1" x 6", per MBF | — | MBF | 3,140.00 | — | 3,140.00 |
| 1" x 8", per MBF | — | MBF | 2,160.00 | — | 2,160.00 |
| 2" x 4", per MBF | — | MBF | 1,420.00 | — | 1,420.00 |
| 2" x 6", per MBF | — | MBF | 1,350.00 | — | 1,350.00 |
| 2" x 8", per MBF | — | MBF | 1,450.00 | — | 1,450.00 |
| 1" x 4", per LF | B1@.044 | LF | .90 | 1.47 | 2.37 |
| 1" x 6", per LF | B1@.044 | LF | 1.73 | 1.47 | 3.20 |
| 1" x 8", per LF | B1@.051 | LF | 1.58 | 1.70 | 3.28 |
| 2" x 4", per LF | B1@.044 | LF | 1.04 | 1.47 | 2.51 |
| 2" x 6", per LF | B1@.044 | LF | 1.49 | 1.47 | 2.96 |
| 2" x 8", per LF | B1@.051 | LF | 2.13 | 1.70 | 3.83 |
| **Sheathing, roof**  Per SF of roof surface including 10% waste | | | | | |
| CDX plywood sheathing, rough | | | | | |
| 1/2" | B1@.013 | SF | .54 | .43 | .97 |
| 1/2", 4-ply | B1@.013 | SF | .56 | .43 | .99 |
| 1/2", 5-ply | B1@.013 | SF | .60 | .43 | 1.03 |
| 5/8", 4-ply | B1@.013 | SF | .68 | .43 | 1.11 |
| 5/8", 5-ply | B1@.013 | SF | .69 | .43 | 1.12 |
| 3/4" | B1@.013 | SF | .82 | .43 | 1.25 |
| OSB sheathing | | | | | |
| 1/2" | B1@.013 | SF | .34 | .43 | .77 |
| 5/8" | B1@.013 | SF | .46 | .43 | .89 |
| 3/4" | B1@.014 | SF | .60 | .47 | 1.07 |
| 7/8" | B1@.014 | SF | 1.01 | .47 | 1.48 |
| 1-1/8" | B1@.015 | SF | 1.46 | .50 | 1.96 |
| Add for hip roof | B1@.007 | SF | — | .23 | .23 |
| Add for steep pitch or cut-up roof | B1@.015 | SF | — | .50 | .50 |
| Board sheathing, 1" x 6" or 1" x 8" utility T&G laid diagonal | | | | | |
| 1" utility T&G lumber, per MBF | — | MBF | 1,350.00 | — | 1,350.00 |
| (1.13 BF per SF) | B1@.026 | SF | 1.68 | .87 | 2.55 |
| Add for hip roof | B1@.007 | SF | — | .23 | .23 |
| Add for steep pitch or cut-up roof | B1@.015 | SF | — | .50 | .50 |

**Roof decking**  Std & Btr grade. See also Sheathing in this section. Flat, shed, or gable roofs to 5 in 12 slope (5/24 pitch). Figures in parentheses indicate board feet per SF of actual roof area (not roof plan area), including 5% waste.

| | Craft@Hrs | Unit | Material | Labor | Total |
|---|---|---|---|---|---|
| 2" x 6", per MBF | — | MBF | 465.00 | — | 465.00 |
| 2" x 8", per MBF | — | MBF | 465.00 | — | 465.00 |
| 3" x 6", per MBF | — | MBF | 512.00 | — | 512.00 |
| 2" x 6" (2.28 BF per SF) | B1@.043 | SF | 1.06 | 1.43 | 2.49 |
| 2" x 8" (2.25 BF per SF) | B1@.043 | SF | 1.05 | 1.43 | 2.48 |
| 3" x 6" (3.43 BF per SF) | B1@.047 | SF | 1.75 | 1.57 | 3.32 |
| Add for steep pitch or cut-up roof | B1@.019 | SF | — | .63 | .63 |

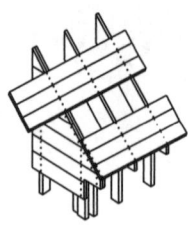

**Cant strips**  Std & Btr grade lumber

| | Craft@Hrs | Unit | Material | Labor | Total |
|---|---|---|---|---|---|
| 2" x 4" beveled | B1@.020 | LF | .52 | .67 | 1.19 |
| For short or cut-up runs | B1@.030 | LF | .52 | 1.00 | 1.52 |

| | Craft@Hrs | Unit | Material | Labor | Total |
|---|---|---|---|---|---|

**Furring** Per SF of surface area to be covered. Figures in parentheses show coverage including 7% waste, typical job.

Over masonry, Utility grade

| | Craft@Hrs | Unit | Material | Labor | Total |
|---|---|---|---|---|---|
| 1" furring, per MBF | — | MBF | 1,150.00 | — | 1,150.00 |
| 12" OC, 1" x 2" (.24 BF per SF) | B1@.025 | SF | .28 | .83 | 1.11 |
| 16" OC, 1" x 2" (.20 BF per SF) | B1@.020 | SF | .23 | .67 | .90 |
| 20" OC, 1" x 2" (.17 BF per SF) | B1@.018 | SF | .20 | .60 | .80 |
| 24" OC, 1" x 2" (.15 BF per SF) | B1@.016 | SF | .17 | .53 | .70 |
| 12" OC, 1" x 3" (.36 BF per SF) | B1@.025 | SF | .41 | .83 | 1.24 |
| 16" OC, 1" x 3" (.29 BF per SF) | B1@.020 | SF | .33 | .67 | 1.00 |
| 20" OC, 1" x 3" (.25 BF per SF) | B1@.018 | SF | .29 | .60 | .89 |
| 24" OC, 1" x 3" (.22 BF per SF) | B1@.016 | SF | .25 | .53 | .78 |

Over wood frame, Utility grade

| | Craft@Hrs | Unit | Material | Labor | Total |
|---|---|---|---|---|---|
| 12" OC, 1" x 2" (.24 BF per SF) | B1@.016 | SF | .28 | .53 | .81 |
| 16" OC, 1" x 2" (.20 BF per SF) | B1@.013 | SF | .23 | .43 | .66 |
| 20" OC, 1" x 2" (.17 BF per SF) | B1@.011 | SF | .20 | .37 | .57 |
| 24" OC, 1" x 2" (.15 BF per SF) | B1@.010 | SF | .17 | .33 | .50 |
| 12" OC, 1" x 3" (.36 BF per SF) | B1@.016 | SF | .41 | .53 | .94 |
| 16" OC, 1" x 3" (.29 BF per SF) | B1@.013 | SF | .33 | .43 | .76 |
| 20" OC, 1" x 3" (.25 BF per SF) | B1@.011 | SF | .29 | .37 | .66 |
| 24" OC, 1" x 3" (.22 BF per SF) | B1@.010 | SF | .25 | .33 | .58 |

Over wood subfloor, Utility grade

| | Craft@Hrs | Unit | Material | Labor | Total |
|---|---|---|---|---|---|
| 12" OC, 1" x 2" (.24 BF per SF) | B1@.033 | SF | .28 | 1.10 | 1.38 |
| 16" OC, 1" x 2" (.20 BF per SF) | B1@.028 | SF | .23 | .93 | 1.16 |
| 20" OC, 1" x 2" (.17 BF per SF) | B1@.024 | SF | .20 | .80 | 1.00 |
| 24" OC, 1" x 2" (.15 BF per SF) | B1@.021 | SF | .17 | .70 | .87 |
| 2" furring, per MBF | — | MBF | 473.00 | — | 473.00 |
| 12" OC, 2" x 2" Std & Btr (.48 BF per SF) | B1@.033 | SF | .23 | 1.10 | 1.33 |
| 16" OC, 2" x 2" Std & Btr (.39 BF per SF) | B1@.028 | SF | .18 | .93 | 1.11 |
| 20" OC, 2" x 2" Std & Btr (.34 BF per SF) | B1@.024 | SF | .16 | .80 | .96 |
| 24" OC, 2" x 2" Std & Btr (.30 BF per SF) | B1@.021 | SF | .14 | .70 | .84 |

**Grounds** Screeds for plaster or stucco, #3, 1" x 2" lumber. Based on .18 BF per LF including 10% waste.

| | Craft@Hrs | Unit | Material | Labor | Total |
|---|---|---|---|---|---|
| 1" grounds, per MBF | — | MBF | 1,150.00 | — | 1,150.00 |
| Over masonry | B1@.047 | LF | .21 | 1.57 | 1.78 |
| Over wood | B1@.038 | LF | .21 | 1.27 | 1.48 |

**Sheathing, wall** Per SF of wall surface, including 5% waste

BC plywood sheathing, plugged and touch sanded, interior grade

| | Craft@Hrs | Unit | Material | Labor | Total |
|---|---|---|---|---|---|
| 1/4" | B1@.013 | SF | .83 | .43 | 1.26 |
| 3/8" | B1@.015 | SF | .87 | .50 | 1.37 |
| 1/2" | B1@.016 | SF | 1.02 | .53 | 1.55 |
| 5/8" | B1@.018 | SF | 1.28 | .60 | 1.88 |
| 3/4" | B1@.020 | SF | 1.49 | .67 | 2.16 |

OSB sheathing

| | Craft@Hrs | Unit | Material | Labor | Total |
|---|---|---|---|---|---|
| 1/4" | B1@.013 | SF | .27 | .43 | .70 |
| 3/8" | B1@.015 | SF | .29 | .50 | .79 |
| 7/16" | B1@.015 | SF | .28 | .50 | .78 |
| 1/2" | B1@.016 | SF | .32 | .53 | .85 |
| 5/8" | B1@.018 | SF | .44 | .60 | 1.04 |
| 3/4" | B1@.018 | SF | .57 | .60 | 1.17 |
| 7/8" | B1@.020 | SF | .96 | .67 | 1.63 |
| 1-1/8" | B1@.020 | SF | 1.40 | .67 | 2.07 |

# Carpentry, Finish

| | Craft@Hrs | Unit | Material | Labor | Total |
|---|---|---|---|---|---|
| Board sheathing 1" x 6" or 1" x 8" utility T&G | | | | | |
| 1" x 6" or 8" utility T&G, per MBF | — | MBF | 1,320.00 | — | 1,320.00 |
| (1.13 BF per SF) | B1@.020 | SF | 1.39 | .67 | 2.06 |
| Add for diagonal patterns | B1@.006 | SF | — | .20 | .20 |

**Sleepers** Std & Btr pressure treated lumber, including 5% waste

| | Craft@Hrs | Unit | Material | Labor | Total |
|---|---|---|---|---|---|
| 2" x 6" sleepers, per MBF | — | MBF | 820.00 | — | 820.00 |
| 2" x 6" sleepers, per LF | B1@.017 | LF | .86 | .57 | 1.43 |
| Add for taper cuts on sleepers, per cut | B1@.009 | Ea | — | .30 | .30 |

## Finish Carpentry
These figures assume all materials to be installed are supplied by others. Labor costs assume good quality stain-grade work up to 9 feet above floor level. Paint-grade work will cost from 20% to 33% less. Add the cost of scaffolding, staining or painting, if required.

**Blocking and backing** Installation before drywall is hung. Cost per piece of blocking or backing set

| | Craft@Hrs | Unit | Material | Labor | Total |
|---|---|---|---|---|---|
| 2" x 4" backing or blocking | BC@.040 | Ea | — | 1.48 | 1.48 |
| 2" x 4" strongback 10' to 16' long | | | | | |
| screwed to ceiling joists | BC@.350 | Ea | — | 12.90 | 12.90 |

**Furring** Nailed over studs and adjusted with shims to bring the wall surface flush. Costs will be higher if the wall plane is very uneven or if the furred wall has to be exceptionally even.
Cost per linear foot of furring installed

| | Craft@Hrs | Unit | Material | Labor | Total |
|---|---|---|---|---|---|
| 1" x 4" furring, few shims required | BC@.010 | LF | — | .37 | .37 |

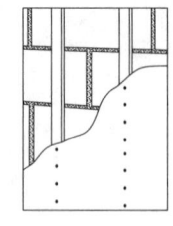

**Straightening wall studs** Costs per stud straightened
Cut saw kerf and drive shim or screw into the kerf

| | Craft@Hrs | Unit | Material | Labor | Total |
|---|---|---|---|---|---|
| reinforce with a plywood scab | BC@.300 | Ea | — | 11.10 | 11.10 |
| Remove and replace a badly warped stud | BC@.300 | Ea | — | 11.10 | 11.10 |

**Cased openings** Install jambs only. See costs for casing below. Costs per opening cased

| | Craft@Hrs | Unit | Material | Labor | Total |
|---|---|---|---|---|---|
| Opening to 4' wide | BC@.800 | Ea | — | 29.60 | 29.60 |
| Openings over 4' wide | BC@1.05 | Ea | — | 38.80 | 38.80 |

**Exterior doors** Based on hanging doors in prepared openings. No lockset, or casing included. See the sections that follow. Costs will be higher if the framing has to be adjusted before the door can be set.
Hang a new door in an existing jamb

| | Craft@Hrs | Unit | Material | Labor | Total |
|---|---|---|---|---|---|
| Cut door, set butt hinges, install door | BC@1.46 | Ea | — | 53.90 | 53.90 |

Prehung flush or panel 1-3/4" thick solid core single doors, with felt insulation, based on an adjustable metal sill

| | Craft@Hrs | Unit | Material | Labor | Total |
|---|---|---|---|---|---|
| Up to 3' wide | BC@.750 | Ea | — | 27.70 | 27.70 |
| Over 3' wide | BC@1.25 | Ea | — | 46.20 | 46.20 |
| Add for door with sidelights | BC@.750 | Ea | — | 27.70 | 27.70 |
| Sliding glass doors, opening to 8' wide | BC@1.75 | Ea | — | 64.60 | 64.60 |
| French doors, 1-3/4" thick, to 6' wide | BC@1.75 | Ea | — | 64.60 | 64.60 |

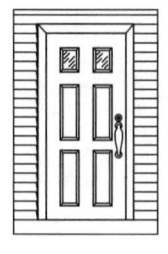

**Interior doors** Based on hanging 1-3/8" hollow core doors in prepared openings. No locksets or casing included. See the sections that follow.

| | Craft@Hrs | Unit | Material | Labor | Total |
|---|---|---|---|---|---|
| Prehung interior doors | BC@.500 | Ea | — | 18.50 | 18.50 |
| Add for assembling a factory-cut | | | | | |
| mortised and rabbeted jamb | BC@.200 | Ea | — | 7.39 | 7.39 |
| Pocket doors | | | | | |
| Assemble and install the door frame | BC@.981 | Ea | — | 36.20 | 36.20 |
| Install door and valances in existing frame | BC@.500 | Ea | — | 18.50 | 18.50 |

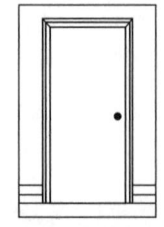

| | Craft@Hrs | Unit | Material | Labor | Total |
|---|---|---|---|---|---|
| Bifold doors installed in cased openings. See costs for cased openings above | | | | | |
|     Single bifold door to 4' wide | BC@.700 | Ea | — | 25.90 | 25.90 |
|     Double bifold door to 8' wide, pair of doors | BC@1.20 | Pr | — | 44.30 | 44.30 |
| Bypass closet doors installed on a track in a cased opening. See cased opening manhours above | | | | | |
|     Double doors, per opening | BC@.660 | Ea | — | 24.40 | 24.40 |
|     Triple doors, per opening | BC@1.00 | Ea | — | 36.90 | 36.90 |
| Rehang most hollow core doors in an existing opening, | | | | | |
|     re-shim jamb and plane door, per door | BC@1.50 | Ea | — | 55.40 | 55.40 |
| Rehang large or heavy solid core doors in an existing opening, | | | | | |
|     re-shim jamb and plane door, per door | BC@1.75 | Ea | — | 64.60 | 64.60 |
| Correcting a sagging door hinge. Remove door from the jamb, remove hinge leaf, | | | | | |
|     deepen the mortise and rehang the door | BC@.600 | Ea | — | 22.20 | 22.20 |
| Hang a new door in an existing jamb. Mark the door to fit opening, cut the door, set butt hinges, install door in the jamb, install door hardware | | | | | |
|     Per door hung | BC@1.75 | Ea | — | 64.60 | 64.60 |
| Prehanging doors off site. Cost per door hung, excluding door cost | | | | | |
|     Make and assemble frame, cost per frame | BC@.750 | Ea | — | 27.70 | 27.70 |
| Set hinge on door and frame, panel or flush door, per hinge | | | | | |
|     Using jigs and router | BC@.160 | Ea | — | 5.91 | 5.91 |
|     Using hand tools | BC@.500 | Ea | — | 18.50 | 18.50 |
|     Add for French doors, per door | BC@.250 | Ea | — | 9.24 | 9.24 |

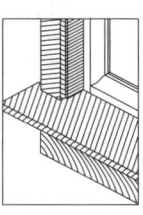

**Lockset installation**  Set in either a solid core or hollow core door, includes setting the strike plate

| | Craft@Hrs | Unit | Material | Labor | Total |
|---|---|---|---|---|---|
| Costs per lockset | | | | | |
|     Standard 2-1/8" bore, factory bored | BC@.160 | Ea | — | 5.91 | 5.91 |
|     Standard 2-1/8" bore, site bored | BC@1.00 | Ea | — | 36.90 | 36.90 |
|     Mortise type lockset, using a boring machine | BC@1.50 | Ea | — | 55.40 | 55.40 |
|     Mortise type lockset, using hand tools | BC@3.00 | Ea | — | 111.00 | 111.00 |
| Pocket door hardware | | | | | |
|     Privacy latch | BC@.500 | Ea | — | 18.50 | 18.50 |
|     Passage latch | BC@.300 | Ea | — | 11.10 | 11.10 |
|     Edge pull | BC@.300 | Ea | — | 11.10 | 11.10 |

**Setting door stop moulding**  Set at the head and on two sides. Costs per opening

| | Craft@Hrs | Unit | Material | Labor | Total |
|---|---|---|---|---|---|
| Panel or flush door | BC@.250 | Ea | — | 9.24 | 9.24 |
| French door | BC@.330 | Ea | — | 12.20 | 12.20 |
| Pocket door | BC@.250 | Ea | — | 9.24 | 9.24 |

**Set casing around a framed opening**  Mitered or butted moulding corners. Costs per opening, per side, using a pneumatic nail gun

| | Craft@Hrs | Unit | Material | Labor | Total |
|---|---|---|---|---|---|
| Opening to 4' wide using casing to 2-1/4" wide | BC@.250 | Ea | — | 9.24 | 9.24 |
| Opening over 4' wide to 8' wide | | | | | |
|     using casing up to 2-1/4" wide | BC@.300 | Ea | — | 11.10 | 11.10 |
| Add for casing over 2-1/4" wide to 4" wide | BC@.160 | Ea | — | 5.91 | 5.91 |
| Add for block and plinth door casing | BC@.300 | Ea | — | 11.10 | 11.10 |
| Add for hand nailing | BC@.275 | Ea | — | 10.20 | 10.20 |

**Window trim**  Based on trim set around prepared openings. Costs per window opening on square windows. Extra tall or extra wide windows will cost more.

| | Craft@Hrs | Unit | Material | Labor | Total |
|---|---|---|---|---|---|
| Extension jambs cut, assembled, shimmed and installed | | | | | |
|     Up to 3/4" thick, 3 or 4 sides, per window | BC@.330 | Ea | — | 12.20 | 12.20 |
|     Over 3/4" thick, per LF of extension jamb | BC@.100 | LF | — | 3.69 | 3.69 |

| | Craft@Hrs | Unit | Material | Labor | Total |
|---|---|---|---|---|---|
| **Window stool** | | | | | |
| Set to sheetrock wraps | | | | | |
| nailed directly on framing | BC@.500 | Ea | — | 18.50 | 18.50 |
| When stool has to be shimmed up from framing | | | | | |
| Add per LF of stool | BC@.050 | LF | — | 1.85 | 1.85 |
| Stool set to wooden jambs, per opening | BC@.400 | Ea | — | 14.80 | 14.80 |
| When stool has to be shimmed up from jambs | | | | | |
| Add per LF of stool | BC@.050 | LF | — | 1.85 | 1.85 |
| Add for mitered returns on horns | | | | | |
| Per opening | BC@.500 | Ea | — | 18.50 | 18.50 |
| Window apron moulding without mitered returns | | | | | |
| Per opening | BC@.160 | Ea | — | 5.91 | 5.91 |
| Window apron moulding with mitered returns | | | | | |
| Per opening | BC@.320 | Ea | — | 11.80 | 11.80 |
| Window casing, straight casing with square corners, per opening | | | | | |
| 3 sides (stool and apron), up to 5' on any side | BC@.330 | Ea | — | 12.20 | 12.20 |
| 4 sides (picture frame), up to 5' on any side | BC@.500 | Ea | — | 18.50 | 18.50 |
| Add for any side over 5' long, per side | BC@.090 | Ea | — | 3.32 | 3.32 |
| Add for angle corners (trapezoid or triangle) | | | | | |
| Per corner | BC@.250 | Ea | — | 9.24 | 9.24 |
| Installing a pre-manufactured arched top casing only, per opening | | | | | |
| To 4' wide set against rosettes | BC@.330 | Ea | — | 12.20 | 12.20 |
| To 4' wide, mitered corners | BC@.500 | Ea | — | 18.50 | 18.50 |
| Over 4' wide set against rosettes | BC@.660 | Ea | — | 24.40 | 24.40 |
| Over 4' wide, mitered corners | BC@1.00 | Ea | — | 36.90 | 36.90 |

**Ceiling treatments** All figures assume a wall height not over 9' with work done from scaffolding. Add the cost of scaffolding

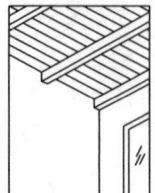

| | Craft@Hrs | Unit | Material | Labor | Total |
|---|---|---|---|---|---|
| Set up and dismantle scaffolding, based on a typical 10' x 16' room | | | | | |
| Per room | BC@1.75 | Ea | — | 64.60 | 64.60 |
| Wood strip ceiling, T&G or square edged, applied wall to wall with no miters | | | | | |
| Per square foot of ceiling | BC@.050 | SF | — | 1.85 | 1.85 |
| Plywood panel ceiling | | | | | |
| Screwed in place, moulding strips nailed with a pneumatic nailer, including normal layout | | | | | |
| Per square foot of ceiling covered | BC@.160 | SF | — | 5.91 | 5.91 |
| Applied beams, including building up the beam, layout and installation on a flat ceiling | | | | | |
| Per LF of beam applied | BC@.310 | LF | — | 11.50 | 11.50 |
| Additional costs for ceiling treatments | | | | | |
| Add for each joint or beam scribed to the wall | BC@.660 | Ea | — | 24.40 | 24.40 |
| Add for each corbel | BC@.400 | Ea | — | 14.80 | 14.80 |
| Add for each knee brace | BC@.600 | Ea | — | 22.20 | 22.20 |

**Cornice or crown moulding** Including layout and coped joints, per LF of moulding

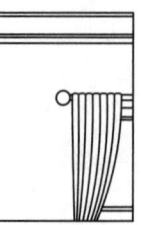

| | Craft@Hrs | Unit | Material | Labor | Total |
|---|---|---|---|---|---|
| Up to 2-1/4" wide | BC@.050 | LF | — | 1.85 | 1.85 |
| Over 2-1/4" wide to 3-1/2" wide | BC@.060 | LF | — | 2.22 | 2.22 |
| Over 3-1/2" wide | BC@.070 | LF | — | 2.59 | 2.59 |

**Wall paneling** No moulding included. See moulding below. Costs per square foot of wall covered

| | Craft@Hrs | Unit | Material | Labor | Total |
|---|---|---|---|---|---|
| Board wainscot or paneling, 3-1/2" to 5-1/2" wide tongue and groove | | | | | |
| Per square foot of wall covered | BC@.080 | SF | — | 2.96 | 2.96 |

| | Craft@Hrs | Unit | Material | Labor | Total |
|---|---|---|---|---|---|

Frame and panel wainscot. Installation only. Add for fabricating the frames and panels

| | Craft@Hrs | Unit | Material | Labor | Total |
|---|---|---|---|---|---|
| Per square foot of wall covered | BC@.090 | SF | — | 3.32 | 3.32 |
| Add for frame attached with screws and covered with wood plugs | BC@.010 | SF | — | .37 | .37 |

Hardwood sheet paneling installed on wood frame or masonry walls

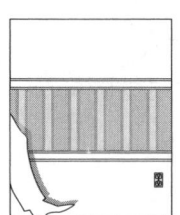

| | Craft@Hrs | Unit | Material | Labor | Total |
|---|---|---|---|---|---|
| Per square foot of wall covered | BC@.040 | SF | — | 1.48 | 1.48 |
| Add for running moulding if required | BC@.040 | SF | — | 1.48 | 1.48 |
| Add if vapor barrier must be installed with mastic over a masonry wall | BC@.010 | SF | — | .37 | .37 |
| Add when 1" x 4" furring is nailed on a masonry wall with a powder-actuated nail gun | | | | | |
| No shimming or straightening included | BC@.020 | SF | — | .74 | .74 |
| Add if sheet paneling is scribed around stone veneer | | | | | |
| Per linear foot scribed | BC@.150 | LF | — | 5.54 | 5.54 |
| Minimum time for scribing sheet paneling | BC@.500 | Ea | — | 18.47 | 18.47 |

**Running mouldings** Single piece mitered installations nailed by hand. Add 100% for two-piece moulding. Costs per LF of moulding

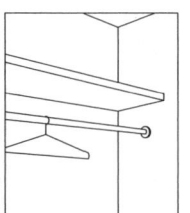

| | Craft@Hrs | Unit | Material | Labor | Total |
|---|---|---|---|---|---|
| Picture rail | BC@.050 | LF | — | 1.85 | 1.85 |
| Chair rail | BC@.040 | LF | — | 1.48 | 1.48 |
| Wainscot cap or bed moulding | BC@.040 | LF | — | 1.48 | 1.48 |
| Baseboard | BC@.040 | LF | — | 1.48 | 1.48 |

Kneewall cap up to 8" wide and 1-1/2" thick. Complex job layout and bed moulding will increase costs

| | Craft@Hrs | Unit | Material | Labor | Total |
|---|---|---|---|---|---|
| Kneewall cap | BC@.160 | LF | — | 5.91 | 5.91 |

**Closet poles and shelving** Including layout and installation

1-1/4" closet pole supported with a metal bracket every 32"

| | Craft@Hrs | Unit | Material | Labor | Total |
|---|---|---|---|---|---|
| Per LF of pole | BC@.040 | LF | — | 1.48 | 1.48 |
| Add for mitered corner joints, per corner | BC@.300 | Ea | — | 11.10 | 11.10 |

3/4" x 12" hat shelf supported with a metal shelf bracket every 32"

| | Craft@Hrs | Unit | Material | Labor | Total |
|---|---|---|---|---|---|
| Per LF of shelf | BC@.100 | LF | — | 3.69 | 3.69 |
| Add for mitered corner joint, per corner | BC@.300 | Ea | — | 11.10 | 11.10 |

Site-built closet storage unit with drawers using plate biscuit joiners and 3/4" board or plywood stock, including fabrication, layout, assembly and installation

| | Craft@Hrs | Unit | Material | Labor | Total |
|---|---|---|---|---|---|
| Per square foot of storage unit face | BC@.680 | SF | — | 25.10 | 25.10 |
| Add for each drawer | BC@.850 | Ea | — | 31.40 | 31.40 |

**Cabinet installation** Including layout, leveling and installation. No fabrication, assembly or countertops included. Costs per cabinet

Base cabinets

| | Craft@Hrs | Unit | Material | Labor | Total |
|---|---|---|---|---|---|
| Cabinets to 36" wide | BC@.766 | Ea | — | 28.30 | 28.30 |
| Cabinets over 36" wide | BC@.911 | Ea | — | 33.70 | 33.70 |

Wall cabinets

| | Craft@Hrs | Unit | Material | Labor | Total |
|---|---|---|---|---|---|
| Cabinets to 18" wide | BC@.638 | Ea | — | 23.60 | 23.60 |
| Cabinets over 18" to 36" wide | BC@.766 | Ea | — | 28.30 | 28.30 |
| Cabinets over 36" wide | BC@1.03 | Ea | — | 38.00 | 38.00 |
| Oven or utility cabinets, 70" high, to 30" wide | BC@1.71 | Ea | — | 63.20 | 63.20 |

Filler strip between cabinets, measured, cut, fit and installed

| | Craft@Hrs | Unit | Material | Labor | Total |
|---|---|---|---|---|---|
| Per strip installed | BC@.200 | Ea | — | 7.39 | 7.39 |

# Carpentry, Finish

| | Craft@Hrs | Unit | Material | Labor | Total |
|---|---|---|---|---|---|

**Fireplace surrounds and mantels** Costs will be higher when a mantel or moulding must be scribed around irregular masonry surfaces

Layout, cutting, fitting, assembling and installing a vertical pilaster on two sides and a horizontal frieze board above the pilasters. Additional mouldings will take more time

| | Craft@Hrs | Unit | Material | Labor | Total |
|---|---|---|---|---|---|
| Costs per linear foot of pilaster and frieze | BC@.350 | LF | — | 12.90 | 12.90 |

Mantel shelf attached to masonry, including cutting, fitting and installation

| | | | | | |
|---|---|---|---|---|---|
| Per mantel | BC@.750 | Ea | — | 27.70 | 27.70 |

Installing a prefabricated mantel and fireplace surround

| | | | | | |
|---|---|---|---|---|---|
| Per mantel and surround installed | BC@1.50 | Ea | — | 55.40 | 55.40 |

**Wood flooring** Including layout, vapor barrier and installation with a power nailer. Add for underlayment, sanding and finishing. Borders, feature strips and mitered corners will increase costs. Costs per square foot of floor covered

| | | | | | |
|---|---|---|---|---|---|
| Tongue & groove strip flooring | | | | | |
| 3/4" x 2-1/4" strips | BF@.036 | SF | — | 1.22 | 1.22 |
| Straight edge plank flooring, 3/4" x 6" to 12" wide | | | | | |
| Set with biscuit joiner plates | BF@.055 | SF | — | 1.87 | 1.87 |
| Add for plank flooring | | | | | |
| set with screws and butterfly plugs | BF@.010 | SF | — | .34 | .34 |
| Filling and sanding a wood floor with drum and disc sanders | | | | | |
| 3 passes (60, 80, 100 grit), per square foot | BF@.023 | SF | — | .78 | .78 |
| Finishing a wood floor, including filler, shellac, | | | | | |
| polyurethane or waterborne urethane | BF@.024 | SF | — | .82 | .82 |

**Skirt boards for stairways** Including layout, fitting, assembly and installation. Based on a 9' floor-to-floor rise where the stair carriage has already been installed by others. Costs will be higher if re-cutting, shimming, or adjusting of an existing stair carriage is required.

Cutting and installing a skirt board on the closed side of a stairway. Costs per skirt board

| | | | | | |
|---|---|---|---|---|---|
| Treads and risers butted against skirt boards, | | | | | |
| no dadoes | BC@2.00 | Ea | — | 73.90 | 73.90 |
| Treads and risers set in dadoes | BC@3.75 | Ea | — | 139.00 | 139.00 |

Cutting and installing a skirt board on the open side of a stairway

| | | | | | |
|---|---|---|---|---|---|
| Risers miter-jointed into the board | BC@5.50 | Ea | — | 203.00 | 203.00 |

**Treads and risers for stairs** Including layout, cutting, fitting, and installation of treads, risers and typical moulding under the tread nosing. Includes a miter return on one side of each tread. Cost per 9' rise of 14 treads and 15 risers

| | | | | | |
|---|---|---|---|---|---|
| 9' rise stairway | BC@2.50 | Ea | — | 92.40 | 92.40 |

**Handrail for stairs** Including layout, cutting, fitting and installation. Based on milled hardwood rail, posts and fittings

| | | | | | |
|---|---|---|---|---|---|
| Setting newel posts, costs per post | BC@2.50 | Ea | — | 92.40 | 92.40 |
| Setting post-to-post handrail | | | | | |
| Costs per 10 linear feet of rail | BC@1.50 | Ea | — | 55.40 | 55.40 |
| Setting over-the-post handrail | | | | | |
| Costs per 10 linear feet of rail | BC@2.50 | Ea | — | 92.40 | 92.40 |
| Add for each transition fitting | | | | | |
| Turn, easing or gooseneck | BC@1.75 | Ea | — | 64.60 | 64.60 |

**Balusters for stair railing** Including layout, cutting, fitting and installation. Based on mill-made hardwood balusters. Cost per baluster

| | | | | | |
|---|---|---|---|---|---|
| Balusters with round top section | BC@.330 | Ea | — | 12.20 | 12.20 |
| Balusters with square top section | BC@.750 | Ea | — | 27.70 | 27.70 |

| | Craft@Hrs | Unit | Material | Labor | Total |
|---|---|---|---|---|---|

**Steel framing rule of thumb** Light gauge steel framing cost per square foot of living area. These figures will apply on most crawl space and basement homes. Labor and material cost will be about 30% less for framing over a slab on grade. Complex framing jobs will cost more. The seven lines of cost data under "Steel framing rule of thumb cost breakdown" below show the cost breakdown for framing a medium cost house. Add $1.50 to $3.00 per SF of floor if engineer approval is required (plans not based on code tables). Material cost assumes precut materials fastened with screws. Labor cost includes layout, assembly, plumb and align.

| | Craft@Hrs | Unit | Material | Labor | Total |
|---|---|---|---|---|---|
| Low cost house | B1@.223 | SF | 5.62 | 7.43 | 13.05 |
| Medium cost house | B1@.268 | SF | 6.39 | 8.93 | 15.32 |
| Better quality house | B1@.355 | SF | 8.75 | 11.80 | 20.55 |

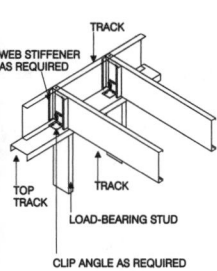

**Steel framing rule of thumb cost breakdown** Costs per square foot of living area. These figures were used to compile the framing cost for a medium cost home in the section immediately above.

| | Craft@Hrs | Unit | Material | Labor | Total |
|---|---|---|---|---|---|
| Using light gauge steel framing materials at | — | Lb | .62 | — | .62 |
| Using 1/2" OSB subfloor at | — | MSF | 306.00 | — | 306.00 |
| Using 7/16" OSB roof sheathing at | — | MSF | 265.00 | — | 265.00 |
| Using 5/8" CDX plywood subfloor at | — | MSF | 622.00 | — | 622.00 |
| Using 1/2" CDX plywood roof sheathing at | — | MSF | 507.00 | — | 507.00 |
| Sills, pier blocks, floor beams (270 Lb per 1,000 SF of floor) | B1@.020 | SF | .17 | .67 | .84 |
| Floor joists, doublers, blocking, bridging (2,200 Lb per 1,000 SF of floor) | B1@.031 | SF | 1.36 | 1.03 | 2.39 |
| Subflooring, 1/2" OSB sheathing (1,150 SF per 1,000 SF) | B1@.011 | SF | .35 | .37 | .72 |
| Subfloor, 5/8" CDX plywood (1,150 SF per 1,000 SF of floor) | B1@.012 | SF | .72 | .40 | 1.12 |
| Layout, studs, top and bottom track, headers, backing, blocking, bracing and framing for openings (2,230 Lb per 1,000 SF of floor) | B1@.102 | SF | 1.38 | 3.40 | 4.78 |
| Ceiling joists, header and end joists, backing, blocking and bracing (1,810 Lb per 1,000 SF of floor) | B1@.050 | SF | 1.12 | 1.67 | 2.79 |
| Rafters, braces, collar beams, ridge members, 8" rafters, 24" OC (2,300 Lb per 1000 SF of roof) | B1@.042 | SF | 1.43 | 1.40 | 2.83 |
| Roof sheathing, 1/2" CDX plywood (1,150 SF per 1,000 SF of roof) | B1@.011 | SF | .58 | .37 | .95 |
| Roof sheathing, 7/16" OSB sheathing (1,150 SF per 1,000 SF) | B1@.010 | SF | .30 | .33 | .63 |

**Steel lally columns** Residential basement columns, 3-1/2" diameter, steel tube

| | Craft@Hrs | Unit | Material | Labor | Total |
|---|---|---|---|---|---|
| Material only | — | LF | 8.30 | — | 8.30 |
| Add for base or cap plate | — | Ea | 3.40 | — | 3.40 |
| Add for installation, per column to 12' high | B1@.458 | Ea | — | 15.30 | 15.30 |

**Wood sill plates for steel framing** (At foundation) SYP #2 pressure treated lumber, drilled and installed with foundation bolts at 48" OC, no bolts, nuts or washers included. See also wood plates in this section. Figures in parentheses indicate board feet per LF of foundation, including 5% waste and wolmanized treatment.

| | Craft@Hrs | Unit | Material | Labor | Total |
|---|---|---|---|---|---|
| 2" x 3" sill plates, per MBF | — | MBF | 601.00 | — | 601.00 |
| 2" x 3" (.53 BF per LF) | B1@.020 | LF | .32 | .67 | .99 |
| 2" x 4" sill plates, per MBF | — | MBF | 843.00 | — | 843.00 |
| 2" x 4" (.70 BF per LF) | B1@.023 | LF | .59 | .77 | 1.36 |
| 2" x 6" sill plates, per MBF | — | MBF | 820.00 | — | 820.00 |
| 2" x 6" (1.05 BF per LF) | B1@.024 | LF | .86 | .80 | 1.66 |
| 2" x 8" sill plates, per MBF | — | MBF | 911.00 | — | 911.00 |
| 2" x 8" (1.40 BF per LF) | B1@.031 | LF | 1.20 | 1.03 | 2.23 |

# Carpentry, Steel Framing

| | Craft@Hrs | Unit | Material | Labor | Total |
|---|---|---|---|---|---|

**Foundation anchors** Set and aligned in forms before concrete is placed. Equivalent Simpson Strong-Tie Co. model numbers are shown in parentheses

Mud sill anchors, embedded in concrete, folded over the track and screw fastened

| | Craft@Hrs | Unit | Material | Labor | Total |
|---|---|---|---|---|---|
| (MAS) | B1@.051 | Ea | .79 | 1.70 | 2.49 |

Anchor bolts, including bolts, nuts and washers as required

| | Craft@Hrs | Unit | Material | Labor | Total |
|---|---|---|---|---|---|
| 5/8" x 17" (SSTB16) | B1@.157 | Ea | 3.67 | 5.23 | 8.90 |
| 5/8" x 21" (SSTB20) | B1@.158 | Ea | 4.47 | 5.26 | 9.73 |
| 5/8" x 25" (SSTB24) | B1@.159 | Ea | 4.51 | 5.30 | 9.81 |
| 7/8" x 29" (SSTB28) | B1@.163 | Ea | 11.50 | 5.43 | 16.93 |
| 7/8" x 34" (SSTB34) | B1@.164 | Ea | 12.30 | 5.46 | 17.76 |
| 7/8" x 36" (SSTB36) | B1@.167 | Ea | 13.10 | 5.56 | 18.66 |

Hold down brackets (one per anchor bolt to distribute pull-out forces over a greater area)

| | Craft@Hrs | Unit | Material | Labor | Total |
|---|---|---|---|---|---|
| 24-1/8" x 10" x 6-1/4" (HPAHD22-2P) | B1@.165 | Ea | 10.40 | 5.50 | 15.90 |
| 24-1/8" x 14" x 6-1/4" (S/HPAHD22) | B1@.169 | Ea | 8.48 | 5.63 | 14.11 |
| 13-7/8" x 2-1/2" x 1-1/2" (S/HD8) | B1@.650 | Ea | 24.90 | 21.70 | 46.60 |
| 16-1/8" x 2-1/2" x 1-1/2" (S/HD10) | B1@.655 | Ea | 25.90 | 21.80 | 47.70 |

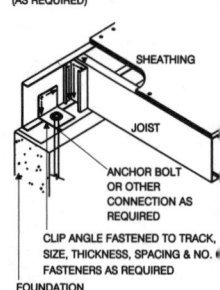

WEB STIFFENER (AS REQUIRED)
SHEATHING
JOIST
ANCHOR BOLT OR OTHER CONNECTION AS REQUIRED
CLIP ANGLE FASTENED TO TRACK, SIZE, THICKNESS, SPACING & NO. FASTENERS AS REQUIRED
FOUNDATION

**Steel first floor joists** Per SF of area covered. Includes joists and rim joists, typical screws, stiffeners, splices, blocking and strapping. Add the cost of support beams, bridging, trimmer joists and sheathing. See the beam, bridging, trimmer joist and sheathing costs below. Deduct for openings over 25 SF. Figures in parentheses show weight of steel per square foot of floor. Assumes minimal waste.

| | Craft@Hrs | Unit | Material | Labor | Total |
|---|---|---|---|---|---|
| Using light gauge steel framing materials at | — | Lb | .62 | — | .62 |
| Deduct for multiple spans (supported at midpoint) | — | % | -6.0 | — | — |

6" steel floor joists 16" OC, 2" flange

| | Craft@Hrs | Unit | Material | Labor | Total |
|---|---|---|---|---|---|
| 600S200-33 (20 gauge, 1.22 Lbs./SF) | B1@.019 | SF | .76 | .63 | 1.39 |
| 600S200-43 (18 gauge, 1.57 Lbs./SF) | B1@.019 | SF | .97 | .63 | 1.60 |
| 600S200-54 (16 gauge, 1.96 Lbs./SF) | B1@.021 | SF | 1.22 | .70 | 1.92 |
| 600S200-68 (14 gauge, 2.42 Lbs./SF) | B1@.023 | SF | 1.50 | .77 | 2.27 |
| 600S200-97 (12 gauge, 3.37 Lbs./SF) | B1@.025 | SF | 2.09 | .83 | 2.92 |

6" steel floor joists 24" OC, 1-5/8" flange

| | Craft@Hrs | Unit | Material | Labor | Total |
|---|---|---|---|---|---|
| 600S162-33 (20 gauge, .78 Lbs./SF) | B1@.013 | SF | .48 | .43 | .91 |
| 600S162-43 (18 gauge, 1.01 Lbs./SF) | B1@.013 | SF | .63 | .43 | 1.06 |
| 600S162-54 (16 gauge, 1.25 Lbs./SF) | B1@.014 | SF | .78 | .47 | 1.25 |
| 600S162-68 (14 gauge, 1.55 Lbs./SF) | B1@.016 | SF | .96 | .53 | 1.49 |
| 600S162-97 (12 gauge, 2.15 Lbs./SF) | B1@.018 | SF | 1.33 | .60 | 1.93 |

6" steel floor joists 24" OC, 2" flange

| | Craft@Hrs | Unit | Material | Labor | Total |
|---|---|---|---|---|---|
| 600S200-33 (20 gauge, .86 Lbs./SF) | B1@.013 | SF | .53 | .43 | .96 |
| 600S200-43 (18 gauge, 1.11 Lbs./SF) | B1@.013 | SF | .69 | .43 | 1.12 |
| 600S200-54 (16 gauge, 1.38 Lbs./SF) | B1@.014 | SF | .86 | .47 | 1.33 |
| 600S200-68 (14 gauge, 1.71 Lbs./SF) | B1@.016 | SF | 1.06 | .53 | 1.59 |
| 600S200-97 (12 gauge, 2.37 Lbs./SF) | B1@.018 | SF | 1.47 | .60 | 2.07 |

8" steel floor joists 12" OC, 1-5/8" flange

| | Craft@Hrs | Unit | Material | Labor | Total |
|---|---|---|---|---|---|
| 800S162-33 (20 gauge, 1.47 Lbs./SF) | B1@.025 | SF | .91 | .83 | 1.74 |
| 800S162-43 (18 gauge, 2.24 Lbs./SF) | B1@.025 | SF | 1.39 | .83 | 2.22 |
| 800S162-54 (16 gauge, 2.78 Lbs./SF) | B1@.028 | SF | 1.72 | .93 | 2.65 |
| 800S162-68 (14 gauge, 3.44 Lbs./SF) | B1@.031 | SF | 2.13 | 1.03 | 3.16 |
| 800S162-97 (12 gauge, 4.79 Lbs./SF) | B1@.034 | SF | 2.97 | 1.13 | 4.10 |

| | Craft@Hrs | Unit | Material | Labor | Total |
|---|---|---|---|---|---|
| **8" steel floor joists 12" OC, 2" flange** | | | | | |
| 800S200-33 (20 gauge, 1.87 Lbs./SF) | B1@.025 | SF | 1.16 | .83 | 1.99 |
| 800S200-43 (18 gauge, 2.42 Lbs./SF) | B1@.025 | SF | 1.50 | .83 | 2.33 |
| 800S200-54 (16 gauge, 3.00 Lbs./SF) | B1@.028 | SF | 1.86 | .93 | 2.79 |
| 800S200-68 (14 gauge, 3.74 Lbs./SF) | B1@.031 | SF | 2.32 | 1.03 | 3.35 |
| 800S200-97 (12 gauge, 5.20 Lbs./SF) | B1@.034 | SF | 3.22 | 1.13 | 4.35 |
| **8" steel floor joists 16" OC, 1-5/8" flange** | | | | | |
| 800S162-33 (20 gauge, 1.35 Lbs./SF) | B1@.020 | SF | .84 | .67 | 1.51 |
| 800S162-43 (18 gauge, 1.73 Lbs./SF) | B1@.020 | SF | 1.07 | .67 | 1.74 |
| 800S162-54 (16 gauge, 2.14 Lbs./SF) | B1@.022 | SF | 1.33 | .73 | 2.06 |
| 800S162-68 (14 gauge, 2.66 Lbs./SF) | B1@.024 | SF | 1.65 | .80 | 2.45 |
| 800S162-97 (12 gauge, 3.70 Lbs./SF) | B1@.026 | SF | 2.29 | .87 | 3.16 |
| **8" steel floor joists 16" OC, 2" flange** | | | | | |
| 800S200-33 (20 gauge, 1.45 Lbs./SF) | B1@.020 | SF | .90 | .67 | 1.57 |
| 800S200-43 (18 gauge, 1.87 Lbs./SF) | B1@.020 | SF | 1.16 | .67 | 1.83 |
| 800S200-54 (16 gauge, 2.32 Lbs./SF) | B1@.022 | SF | 1.44 | .73 | 2.17 |
| 800S200-68 (14 gauge, 2.89 Lbs./SF) | B1@.024 | SF | 1.79 | .80 | 2.59 |
| 800S200-97 (12 gauge, 4.01 Lbs./SF) | B1@.026 | SF | 2.49 | .87 | 3.36 |
| **8" steel floor joists 24" OC, 1-5/8" flange** | | | | | |
| 800S162-33 (20 gauge, .95 Lbs./SF) | B1@.014 | SF | .59 | .47 | 1.06 |
| 800S162-43 (18 gauge, 1.22 Lbs./SF) | B1@.014 | SF | .76 | .47 | 1.23 |
| 800S162-54 (16 gauge, 1.51 Lbs./SF) | B1@.016 | SF | .94 | .53 | 1.47 |
| 800S162-68 (14 gauge, 1.87 Lbs./SF) | B1@.017 | SF | 1.16 | .57 | 1.73 |
| 800S162-97 (12 gauge, 2.61 Lbs./SF) | B1@.019 | SF | 1.62 | .63 | 2.25 |
| **8" steel floor joists 24" OC, 2" flange** | | | | | |
| 800S200-33 (20 gauge, 1.02 Lbs./SF) | B1@.014 | SF | .63 | .47 | 1.10 |
| 800S200-43 (18 gauge, 1.32 Lbs./SF) | B1@.014 | SF | .82 | .47 | 1.29 |
| 800S200-54 (16 gauge, 1.63 Lbs./SF) | B1@.016 | SF | 1.01 | .53 | 1.54 |
| 800S200-68 (14 gauge, 2.03 Lbs./SF) | B1@.017 | SF | 1.26 | .57 | 1.83 |
| 800S200-97 (12 gauge, 2.83 Lbs./SF) | B1@.019 | SF | 1.75 | .63 | 2.38 |
| **10" steel floor joists 12" OC, 1-5/8" flange** | | | | | |
| 1000S162-43 (18 gauge, 2.62 Lbs./SF) | B1@.028 | SF | 1.62 | .93 | 2.55 |
| 1000S162-54 (16 gauge, 3.24 Lbs./SF) | B1@.030 | SF | 2.01 | 1.00 | 3.01 |
| 1000S162-68 (14 gauge, 4.04 Lbs./SF) | B1@.033 | SF | 2.50 | 1.10 | 3.60 |
| 1000S162-97 (12 gauge, 5.63 Lbs./SF) | B1@.037 | SF | 3.49 | 1.23 | 4.72 |
| **10" steel floor joists 12" OC, 2" flange** | | | | | |
| 1000S200-43 (18 gauge, 2.81 Lbs./SF) | B1@.028 | SF | 1.74 | .93 | 2.67 |
| 1000S200-54 (16 gauge, 3.48 Lbs./SF) | B1@.030 | SF | 2.16 | 1.00 | 3.16 |
| 1000S200-68 (14 gauge, 4.32 Lbs./SF) | B1@.033 | SF | 2.68 | 1.10 | 3.78 |
| 1000S200-97 (12 gauge, 6.05 Lbs./SF) | B1@.037 | SF | 3.75 | 1.23 | 4.98 |
| **10" steel floor joists 16" OC, 1-5/8" flange** | | | | | |
| 1000S162-43 (18 gauge, 2.02 Lbs./SF) | B1@.022 | SF | 1.25 | .73 | 1.98 |
| 1000S162-54 (16 gauge, 2.50 Lbs./SF) | B1@.024 | SF | 1.55 | .80 | 2.35 |
| 1000S162-68 (14 gauge, 3.12 Lbs./SF) | B1@.027 | SF | 1.93 | .90 | 2.83 |
| 1000S162-97 (12 gauge, 4.35 Lbs./SF) | B1@.029 | SF | 2.70 | .97 | 3.67 |
| **10" steel floor joists 16" OC, 2" flange** | | | | | |
| 1000S200-43 (18 gauge, 2.17 Lbs./SF) | B1@.022 | SF | 1.35 | .73 | 2.08 |
| 1000S200-54 (16 gauge, 2.69 Lbs./SF) | B1@.024 | SF | 1.67 | .80 | 2.47 |
| 1000S200-68 (14 gauge, 3.34 Lbs./SF) | B1@.027 | SF | 2.07 | .90 | 2.97 |
| 1000S200-97 (12 gauge, 4.67 Lbs./SF) | B1@.029 | SF | 2.90 | .97 | 3.87 |

| | Craft@Hrs | Unit | Material | Labor | Total |
|---|---|---|---|---|---|
| **10" steel floor joists 24" OC, 1-5/8" flange** | | | | | |
| 1000S162-43 (18 gauge, 1.42 Lbs./SF) | B1@.015 | SF | .88 | .50 | 1.38 |
| 1000S162-54 (16 gauge, 1.76 Lbs./SF) | B1@.017 | SF | 1.09 | .57 | 1.66 |
| 1000S162-68 (14 gauge, 2.20 Lbs./SF) | B1@.019 | SF | 1.36 | .63 | 1.99 |
| 1000S162-97 (12 gauge, 3.06 Lbs./SF) | B1@.020 | SF | 1.90 | .67 | 2.57 |
| **10" steel floor joists 24" OC, 2" flange** | | | | | |
| 1000S200-43 (18 gauge, 1.52 Lbs./SF) | B1@.015 | SF | .94 | .50 | 1.44 |
| 1000S200-54 (16 gauge, 1.89 Lbs./SF) | B1@.017 | SF | 1.17 | .57 | 1.74 |
| 1000S200-68 (14 gauge, 2.35 Lbs./SF) | B1@.019 | SF | 1.46 | .63 | 2.09 |
| 1000S200-97 (12 gauge, 3.29 Lbs./SF) | B1@.020 | SF | 2.04 | .67 | 2.71 |
| **12" steel floor joists 12" OC, 1-5/8" flange** | | | | | |
| 1200S162-54 (16 gauge, 3.72 Lbs./SF) | B1@.029 | SF | 2.31 | .97 | 3.28 |
| 1200S162-68 (14 gauge, 4.63 Lbs./SF) | B1@.032 | SF | 2.87 | 1.07 | 3.94 |
| 1200S162-97 (12 gauge, 6.47 Lbs./SF) | B1@.035 | SF | 4.01 | 1.17 | 5.18 |
| **12" steel floor joists 12" OC, 2" flange** | | | | | |
| 1200S200-54 (16 gauge, 3.95 Lbs./SF) | B1@.029 | SF | 2.45 | .97 | 3.42 |
| 1200S200-68 (14 gauge, 4.92 Lbs./SF) | B1@.032 | SF | 3.05 | 1.07 | 4.12 |
| 1200S200-97 (12 gauge, 6.88 Lbs./SF) | B1@.035 | SF | 4.27 | 1.17 | 5.44 |
| **12" steel floor joists 16" OC, 1-5/8" flange** | | | | | |
| 1200S162-54 (16 gauge, 2.88 Lbs./SF) | B1@.023 | SF | 1.79 | .77 | 2.56 |
| 1200S162-68 (14 gauge, 3.57 Lbs./SF) | B1@.025 | SF | 2.21 | .83 | 3.04 |
| 1200S162-97 (12 gauge, 4.99 Lbs./SF) | B1@.028 | SF | 3.09 | .93 | 4.02 |
| **12" steel floor joists 16" OC, 2" flange** | | | | | |
| 1200S200-54 (16 gauge, 3.05 Lbs./SF) | B1@.023 | SF | 1.89 | .77 | 2.66 |
| 1200S200-68 (14 gauge, 3.80 Lbs./SF) | B1@.025 | SF | 2.36 | .83 | 3.19 |
| 1200S200-97 (12 gauge, 5.31 Lbs./SF) | B1@.028 | SF | 3.29 | .93 | 4.22 |
| **12" steel floor joists 24" OC, 1-5/8" flange** | | | | | |
| 1200S162-54 (16 gauge, 2.03 Lbs./SF) | B1@.017 | SF | 1.26 | .57 | 1.83 |
| 1200S162-68 (14 gauge, 2.52 Lbs./SF) | B1@.018 | SF | 1.56 | .60 | 2.16 |
| 1200S162-97 (12 gauge, 3.52 Lbs./SF) | B1@.020 | SF | 2.18 | .67 | 2.85 |
| **12" steel floor joists 24" OC, 2" flange** | | | | | |
| 1200S200-54 (16 gauge, 2.15 Lbs./SF) | B1@.017 | SF | 1.33 | .57 | 1.90 |
| 1200S200-68 (14 gauge, 2.68 Lbs./SF) | B1@.018 | SF | 1.66 | .60 | 2.26 |
| 1200S200-97 (12 gauge, 3.75 Lbs./SF) | B1@.020 | SF | 2.33 | .67 | 3.00 |

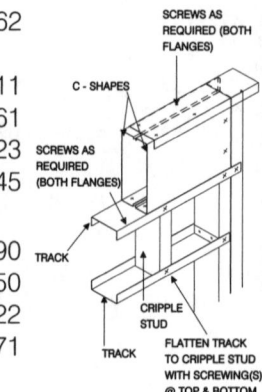

TRACK OR C-SHAPE DEPTH AS REQUIRED

JOIST

SCREWS THROUGH EACH LEG OF CUP ANGLE, BOTH ENDS LENGTH AS REQUIRED

FLAT STRAP

**Built-up steel support beams and girders** Box or back-to-back beams made from 1-5/8" flange C-shaped sections. Includes typical fasteners and minimal waste.

| | Craft@Hrs | Unit | Material | Labor | Total |
|---|---|---|---|---|---|
| Using built-up steel sections at | — | Lb | .62 | — | .62 |
| **6" beams and girders built-up from two members** | | | | | |
| 2-600S162-43 (18 gauge, 3.04 Lbs./LF) | B1@.037 | LF | 1.88 | 1.23 | 3.11 |
| 2-600S162-54 (16 gauge, 3.78 Lbs./LF) | B1@.038 | LF | 2.34 | 1.27 | 3.61 |
| 2-600S162-68 (14 gauge, 4.72 Lbs./LF) | B1@.039 | LF | 2.93 | 1.30 | 4.23 |
| 2-600S162-97 (12 gauge, 6.58 Lbs./LF) | B1@.041 | LF | 4.08 | 1.37 | 5.45 |
| **8" beams and girders built-up from two members** | | | | | |
| 2-800S162-43 (18 gauge, 3.66 Lbs./LF) | B1@.049 | LF | 2.27 | 1.63 | 3.90 |
| 2-800S162-54 (16 gauge, 4.56 Lbs./LF) | B1@.050 | LF | 2.83 | 1.67 | 4.50 |
| 2-800S162-68 (14 gauge, 5.68 Lbs./LF) | B1@.051 | LF | 3.52 | 1.70 | 5.22 |
| 2-800S162-97 (12 gauge, 7.96 Lbs./LF) | B1@.053 | LF | 4.94 | 1.77 | 6.71 |

SCREWS AS REQUIRED (BOTH FLANGES)

C - SHAPES

SCREWS AS REQUIRED (BOTH FLANGES)

TRACK

CRIPPLE STUD

TRACK

FLATTEN TRACK TO CRIPPLE STUD WITH SCREWING(S) @ TOP & BOTTOM (BOTH SIDES)

| | Craft@Hrs | Unit | Material | Labor | Total |
|---|---|---|---|---|---|
| **10" beams and girders built-up from two members** | | | | | |
| 2-1000S162-43 (18 gauge, 4.26 Lbs./LF) | B1@.063 | LF | 2.64 | 2.10 | 4.74 |
| 2-1000S162-54 (16 gauge, 5.23 Lbs./LF) | B1@.064 | LF | 3.24 | 2.13 | 5.37 |
| 2-1000S162-68 (14 gauge, 6.66 Lbs./LF) | B1@.065 | LF | 4.13 | 2.17 | 6.30 |
| 2-1000S162-97 (12 gauge, 9.34 Lbs./LF) | B1@.069 | LF | 5.79 | 2.30 | 8.09 |
| **12" beams and girders built-up from two members** | | | | | |
| 2-1200S162-54 (16 gauge, 6.10 Lbs./LF) | B1@.073 | LF | 3.78 | 2.43 | 6.21 |
| 2-1200S162-68 (14 gauge, 7.62 Lbs./LF) | B1@.075 | LF | 4.72 | 2.50 | 7.22 |
| 2-1200S162-97 (12 gauge, 10.72 Lbs./LF) | B1@.079 | LF | 6.65 | 2.63 | 9.28 |

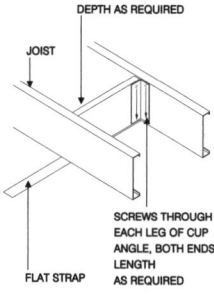

TRACK OR C-SHAPE DEPTH AS REQUIRED

JOIST

SCREWS THROUGH EACH LEG OF CUP ANGLE, BOTH ENDS LENGTH AS REQUIRED

FLAT STRAP

**Steel joist solid blocking** Installed between 6" thru 12" joists. 18 gauge, C-shape, with angle. Including typical screws, stiffeners and minimal waste.

| | Craft@Hrs | Unit | Material | Labor | Total |
|---|---|---|---|---|---|
| Using light gauge steel framing materials at | — | Lb | .62 | — | .62 |
| **Blocking for 6" steel joists** | | | | | |
| Joists on 12" centers | B1@.042 | Ea | 1.02 | 1.40 | 2.42 |
| Joists on 16" centers | B1@.042 | Ea | 1.36 | 1.40 | 2.76 |
| Joists on 20" centers | B1@.042 | Ea | 1.71 | 1.40 | 3.11 |
| Joists on 24" centers | B1@.042 | Ea | 2.04 | 1.40 | 3.44 |
| **Blocking for 8" steel joists** | | | | | |
| Joists on 12" centers | B1@.050 | Ea | 1.53 | 1.67 | 3.20 |
| Joists on 16" centers | B1@.050 | Ea | 2.04 | 1.67 | 3.71 |
| Joists on 20" centers | B1@.050 | Ea | 2.54 | 1.67 | 4.21 |
| Joists on 24" centers | B1@.050 | Ea | 3.06 | 1.67 | 4.73 |
| **Blocking for 10" steel joists** | | | | | |
| Joists on 12" centers | B1@.057 | Ea | 1.78 | 1.90 | 3.68 |
| Joists on 16" centers | B1@.057 | Ea | 2.38 | 1.90 | 4.28 |
| Joists on 20" centers | B1@.057 | Ea | 2.98 | 1.90 | 4.88 |
| Joists on 24" centers | B1@.057 | Ea | 3.54 | 1.90 | 5.44 |
| **Blocking for 12" steel joists** | | | | | |
| Joists on 12" centers | B1@.064 | Ea | 2.06 | 2.13 | 4.19 |
| Joists on 16" centers | B1@.064 | Ea | 2.74 | 2.13 | 4.87 |
| Joists on 20" centers | B1@.064 | Ea | 3.41 | 2.13 | 5.54 |
| Joists on 24" centers | B1@.064 | Ea | 4.09 | 2.13 | 6.22 |

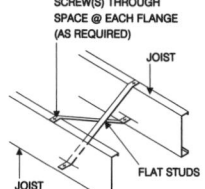

SCREW(S) THROUGH SPACE @ EACH FLANGE (AS REQUIRED)

JOIST

FLAT STUDS

JOIST

**Steel cross bridging for steel floor joists** Using strap material or manufactured straps installed between 6" thru 12" joists, cost per pair of straps including typical fasteners.

| | Craft@Hrs | Unit | Material | Labor | Total |
|---|---|---|---|---|---|
| Joists on 12" centers | B1@.025 | Pr | 1.49 | .83 | 2.32 |
| Joists on 16" centers | B1@.025 | Pr | 1.55 | .83 | 2.38 |
| Joists on 20" centers | B1@.025 | Pr | 1.70 | .83 | 2.53 |
| Joists on 24" centers | B1@.025 | Pr | 1.98 | .83 | 2.81 |

**Board sheathing subfloor** 1" x 6" #3 & Btr (1.28 BF per SF), screwed or pinned to steel floor joists, including fasteners.

| | Craft@Hrs | Unit | Material | Labor | Total |
|---|---|---|---|---|---|
| Board sheathing, per MBF | — | MBF | 1,820.00 | — | 1,820.00 |
| With 12% shrinkage, 5% waste & screws | B1@.022 | SF | 2.15 | .73 | 2.88 |
| Add for diagonal patterns | B1@.001 | SF | .21 | .03 | .24 |

**Plywood subfloor** CD standard exterior grade. Material costs shown include 5% for waste, screwed or pinned to steel floor joists, including fasteners

| | Craft@Hrs | Unit | Material | Labor | Total |
|---|---|---|---|---|---|
| 3/8" | B1@.012 | SF | .46 | .40 | .86 |
| 1/2" | B1@.013 | SF | .53 | .43 | .96 |
| 5/8" | B1@.013 | SF | .65 | .43 | 1.08 |
| 3/4" | B1@.014 | SF | .78 | .47 | 1.25 |
| 1-1/8" | B1@.015 | SF | 1.44 | .50 | 1.94 |

# Carpentry, Steel Framing

| | Craft@Hrs | Unit | Material | Labor | Total |
|---|---|---|---|---|---|
| OSB sheathing, Material costs shown include 5% for waste & fasteners | | | | | |
| 7/16" | B1@.011 | SF | .28 | .37 | .65 |
| 1/2" | B1@.012 | SF | .32 | .40 | .72 |
| 5/8" | B1@.012 | SF | .44 | .40 | .84 |
| 3/4" tongue and groove | B1@.014 | SF | .57 | .47 | 1.04 |

**Steel interior wall studs** Per square foot of partition wall area measured one side. Do not subtract for openings less than 16' wide. Costs include studs, top and bottom track, screws and splices. Add for each corner or wall junction as indicated below. Add the cost of wood plates (if used), support beams, door opening framing and corners from below. All studs with 1-1/4" flange are assumed to have a 3/16" stiffening lip. Labor includes layout, assembly, plumb and align. Figures in parentheses show weight of steel per square foot of wall measured one side. Assumes minimal waste. See the following section for exterior wall estimates.

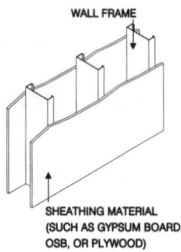

WALL FRAME

SHEATHING MATERIAL
(SUCH AS GYPSUM BOARD,
OSB, OR PLYWOOD)

| | Craft@Hrs | Unit | Material | Labor | Total |
|---|---|---|---|---|---|
| Using light gauge steel framing materials at | — | Lb | .62 | — | .62 |
| 2-1/2" steel interior wall framing, 16" OC, 1-1/4" flange | | | | | |
| 250S125-18 (25 gauge, .33 Lbs./SF) | B1@.018 | SF | .20 | .60 | .80 |
| 250S125-27 (22 gauge, .49 Lbs./ SF) | B1@.018 | SF | .30 | .60 | .90 |
| 250S125-30 (20DW gauge, .54 Lbs./SF) | B1@.019 | SF | .33 | .63 | .96 |
| 250S125-33 (20 gauge, .60 Lbs./SF) | B1@.020 | SF | .37 | .67 | 1.04 |
| 2-1/2" steel interior wall framing, 24" OC, 1-1/4" flange | | | | | |
| 250S125-18 (25 gauge, .25 Lbs./SF) | B1@.017 | SF | .16 | .57 | .73 |
| 250S125-27 (22 gauge, .37 Lbs./SF) | B1@.017 | SF | .23 | .57 | .80 |
| 250S125-30 (20DW gauge, .41 Lbs./SF) | B1@.018 | SF | .25 | .60 | .85 |
| 250S125-33 (20 gauge, .45 Lbs./SF) | B1@.019 | SF | .28 | .63 | .91 |
| Add for each 2-1/2" corner or wall junction | B1@.091 | Ea | 2.43 | 3.03 | 5.46 |
| 3-1/2" steel interior wall framing, 16" OC, 1-1/4" flange | | | | | |
| 350S125-18 (25 gauge, .39 Lbs./SF) | B1@.023 | SF | .24 | .77 | 1.01 |
| 350S125-27 (22 gauge, .59 Lbs./SF) | B1@.023 | SF | .37 | .77 | 1.14 |
| 350S125-30 (20DW gauge, .65 Lbs./SF) | B1@.024 | SF | .40 | .80 | 1.20 |
| 350S125-33 (20 gauge, .72 Lbs./SF) | B1@.025 | SF | .45 | .83 | 1.28 |
| 3-1/2" steel interior wall framing, 24" OC, 1-1/4" flange | | | | | |
| 350S125-18 (25 gauge, .29 Lbs./SF) | B1@.020 | SF | .18 | .67 | .85 |
| 350S125-27 (22 gauge, .44 Lbs./SF) | B1@.020 | SF | .27 | .67 | .94 |
| 350S125-30 (20DW gauge, .49 Lbs./SF) | B1@.022 | SF | .30 | .73 | 1.03 |
| 350S125-33 (20 gauge, .54 Lbs./SF) | B1@.023 | SF | .33 | .77 | 1.10 |
| Add for each 3-1/2" corner or wall junction | B1@.091 | Ea | 2.59 | 3.03 | 5.62 |
| 3-5/8" steel interior wall framing, 16" OC, 1-1/4" flange | | | | | |
| 362S125-18 (25 gauge, .40 Lbs./SF) | B1@.023 | SF | .25 | .77 | 1.02 |
| 362S125-27 (22 gauge, .60 Lbs./SF) | B1@.023 | SF | .37 | .77 | 1.14 |
| 362S125-30 (20DW gauge, .66 Lbs./SF) | B1@.024 | SF | .41 | .80 | 1.21 |
| 362S125-33 (20 gauge, .73 Lbs./SF) | B1@.025 | SF | .45 | .83 | 1.28 |
| 3-5/8" steel interior wall framing, 24" OC, 1-1/4" flange | | | | | |
| 362S125-18 (25 gauge, .30 Lbs./SF) | B1@.021 | SF | .19 | .70 | .89 |
| 362S125-27 (22 gauge, .45 Lbs./SF) | B1@.021 | SF | .28 | .70 | .98 |
| 362S125-30 (20DW gauge, .50 Lbs./SF) | B1@.022 | SF | .31 | .73 | 1.04 |
| 362S125-33 (20 gauge, .55 Lbs./SF) | B1@.023 | SF | .34 | .77 | 1.11 |
| Add for each 3-5/8" corner or wall junction | B1@.091 | Ea | 2.59 | 3.03 | 5.62 |

| | Craft@Hrs | Unit | Material | Labor | Total |
|---|---|---|---|---|---|
| **5-1/2" steel interior wall framing, 16" OC, 1-1/4" flange** | | | | | |
| 550S125-18 (25 gauge, .52 Lbs./SF) | B1@.029 | SF | .32 | .97 | 1.29 |
| 550S125-27 (22 gauge, .78 Lbs./SF) | B1@.029 | SF | .48 | .97 | 1.45 |
| 550S125-30 (20DW gauge, .86 Lbs./SF) | B1@.030 | SF | .53 | 1.00 | 1.53 |
| 550S125-33 (20 gauge, .95 Lbs./SF) | B1@.032 | SF | .60 | 1.07 | 1.67 |
| **5-1/2" steel interior wall framing, 24" OC, 1-1/4" flange** | | | | | |
| 550S125-18 (25 gauge, .39 Lbs./SF) | B1@.028 | SF | .24 | .93 | 1.17 |
| 550S125-27 (22 gauge, .59 Lbs./SF) | B1@.028 | SF | .37 | .93 | 1.30 |
| 550S125-30 (20DW gauge, .65 Lbs./SF) | B1@.029 | SF | .40 | .97 | 1.37 |
| 550S125-33 (20 gauge, .71 Lbs./SF) | B1@.030 | SF | .44 | 1.00 | 1.44 |
| Add for each 5-1/2" corner or wall junction | B1@.091 | Ea | 3.41 | 3.03 | 6.44 |
| **6" steel interior wall framing, 16" OC, 1-1/4" flange** | | | | | |
| 600S125-18 (25 gauge, .55 Lbs./SF) | B1@.028 | SF | .34 | .93 | 1.27 |
| 600S125-27 (22 gauge, .83 Lbs./SF) | B1@.029 | SF | .51 | .97 | 1.48 |
| 600S125-30 (20DW gauge), .91 Lbs./SF | B1@.030 | SF | .56 | 1.00 | 1.56 |
| 600S125-33 (20 gauge, 1.01 Lbs./SF) | B1@.032 | SF | .63 | 1.07 | 1.70 |
| **6" steel interior wall framing, 24" OC, 1-1/4" flange** | | | | | |
| 600S125-18 (25 gauge, .41 Lbs./SF) | B1@.027 | SF | .25 | .90 | 1.15 |
| 600S125-27 (22 gauge, .62 Lbs./SF) | B1@.027 | SF | .38 | .90 | 1.28 |
| 600S125-30 (20DW gauge, .68 Lbs./SF) | B1@.028 | SF | .42 | .93 | 1.35 |
| 600S125-33 (20 gauge, .76 Lbs./SF) | B1@.030 | SF | .47 | 1.00 | 1.47 |
| Add for each 6" corner or wall junction | B1@.091 | Ea | 3.61 | 3.03 | 6.64 |

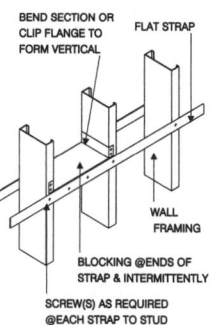

BEND SECTION OR CLIP FLANGE TO FORM VERTICAL

FLAT STRAP

WALL FRAMING

BLOCKING @ENDS OF STRAP & INTERMITTENTLY

SCREW(S) AS REQUIRED @EACH STRAP TO STUD

**Steel exterior wall studs** Per square foot of wall area measures one side. Do not subtract for openings less than 16' wide. Add for each corner and partition from below. Costs include studs, top and bottom track, screws and splices. Add for door and window opening framing, backing, cross-bracing and sheathing for shear walls from the sections that follow. Add the cost of wood plates (if used), support beams, door opening framing, corners and wall junctions from below. All studs with 1-5/8" flange are assumed to have a 1/2" stiffening lip. Labor includes layout, assembly, plumb and align. Figures in parentheses show weight of steel per square foot of wall measured one side. Assumes minimal waste. See the previous section for interior (partition) wall estimates.

| | Craft@Hrs | Unit | Material | Labor | Total |
|---|---|---|---|---|---|
| Using light gauge steel framing materials at | — | Lb | .62 | — | .62 |
| **3-1/2" steel exterior wall framing, 12" OC, 1-5/8" flange** | | | | | |
| 350S162-33 (20 gauge, 1.10 Lbs./SF) | B1@.026 | SF | .68 | .87 | 1.55 |
| 350S162-43 (18 gauge, 1.43 Lbs./SF) | B1@.026 | SF | .87 | .87 | 1.74 |
| 350S162-54 (16 gauge, 1.76 Lbs./SF) | B1@.028 | SF | 1.09 | .93 | 2.02 |
| 350S162-68 (14 gauge, 2.19 Lbs./SF) | B1@.029 | SF | 1.36 | .97 | 2.33 |
| **3-1/2" steel exterior wall framing, 16" OC, 1-5/8" flange** | | | | | |
| 350S162-33 (20 gauge, .88 Lbs./SF) | B1@.025 | SF | .55 | .83 | 1.38 |
| 350S162-43 (18 gauge, 1.14 Lbs./SF) | B1@.025 | SF | .71 | .83 | 1.54 |
| 350S162-54 (16 gauge, 1.41 Lbs./SF) | B1@.028 | SF | .87 | .93 | 1.80 |
| 350S162-68 (14 gauge, 1.75 Lbs./SF) | B1@.029 | SF | 1.09 | .97 | 2.06 |
| **3-1/2" steel exterior wall framing, 24" OC, 1-5/8" flange** | | | | | |
| 350S162-33 (20 gauge, .66 Lbs./SF) | B1@.023 | SF | .41 | .77 | 1.18 |
| 350S162-43 (18 gauge, .86 Lbs./SF) | B1@.023 | SF | .53 | .77 | 1.30 |
| 350S162-54 (16 gauge, 1.06 Lbs./SF) | B1@.025 | SF | .66 | .83 | 1.49 |
| 350S162-68 (14 gauge, 1.31 Lbs./SF) | B1@.026 | SF | .81 | .87 | 1.68 |
| Add for each 3-1/2" corner or partition (33 mil) | B1@.091 | Ea | 3.66 | 3.03 | 6.69 |

# Carpentry, Steel Framing

| | Craft@Hrs | Unit | Material | Labor | Total |
|---|---|---|---|---|---|
| **3-5/8" steel exterior wall framing, 12" OC, 1-5/8" flange** | | | | | |
| 362S162-33 (20 gauge, 1.11 Lbs./SF) | B1@.026 | SF | .69 | .87 | 1.56 |
| 362S162-43 (18 gauge, 1.45 Lbs./SF) | B1@.026 | SF | .90 | .87 | 1.77 |
| 362S162-54 (16 gauge, 1.80 Lbs./SF) | B1@.028 | SF | 1.12 | .93 | 2.05 |
| 362S162-68 (14 gauge, 2.23 Lbs./SF) | B1@.029 | SF | 1.38 | .97 | 2.35 |
| **3-5/8" steel exterior wall framing, 16" OC, 1-5/8" flange** | | | | | |
| 362S162-33 (20 gauge, .89 Lbs./SF) | B1@.025 | SF | .55 | .83 | 1.38 |
| 362S162-43 (18 gauge, 1.16 Lbs./SF) | B1@.025 | SF | .72 | .83 | 1.55 |
| 362S162-54 (16 gauge, 1.44 Lbs./SF) | B1@.028 | SF | .89 | .93 | 1.82 |
| 362S162-68 (14 gauge, 1.78 Lbs./SF) | B1@.029 | SF | 1.10 | .97 | 2.07 |
| **3-5/8" steel exterior wall framing, 24" OC, 1-5/8" flange** | | | | | |
| 362S162-33 (20 gauge, .67 Lbs./SF) | B1@.023 | SF | .42 | .77 | 1.19 |
| 362S162-43 (18 gauge, .87 Lbs./SF) | B1@.023 | SF | .54 | .77 | 1.31 |
| 362S162-54 (16 gauge, 1.08 Lbs./SF) | B1@.025 | SF | .67 | .83 | 1.50 |
| 362S162-68 (14 gauge, 1.34 Lbs./SF) | B1@.026 | SF | .83 | .87 | 1.70 |
| Add for each 3-5/8" corner or partition (33 mil) | B1@.091 | Ea | 3.74 | 3.03 | 6.77 |
| **5-1/2" steel exterior wall framing, 12" OC, 1-5/8" flange** | | | | | |
| 550S162-33 (20 gauge, 1.39 Lbs./SF) | B1@.035 | SF | .86 | 1.17 | 2.03 |
| 550S162-43 (18 gauge, 1.80 Lbs./SF) | B1@.036 | SF | 1.12 | 1.20 | 2.32 |
| 550S162-54 (16 gauge, 2.25 Lbs./SF) | B1@.039 | SF | 1.40 | 1.30 | 2.70 |
| 550S162-68 (14 gauge, 2.80 Lbs./SF) | B1@.041 | SF | 1.74 | 1.37 | 3.11 |
| **5-1/2" steel exterior wall framing, 16" OC, 1-5/8" flange** | | | | | |
| 550S162-33 (20 gauge, 1.11 Lbs./SF) | B1@.032 | SF | .69 | 1.07 | 1.76 |
| 550S162-43 (18 gauge, 1.44 Lbs./SF) | B1@.033 | SF | .89 | 1.10 | 1.99 |
| 550S162-54 (16 gauge, 1.80 Lbs./SF) | B1@.035 | SF | 1.12 | 1.17 | 2.29 |
| 550S162-68 (14 gauge, 2.24 Lbs./SF) | B1@.037 | SF | 1.39 | 1.23 | 2.62 |
| **5-1/2" steel exterior wall framing, 20" OC, 1-5/8" flange** | | | | | |
| 550S162-33 (20 gauge, .94 Lbs./SF) | B1@.031 | SF | .58 | 1.03 | 1.61 |
| 550S162-43 (18 gauge, 1.22 Lbs./SF) | B1@.032 | SF | .76 | 1.07 | 1.83 |
| 550S162-54 (16 gauge, 1.53 Lbs./SF) | B1@.034 | SF | .95 | 1.13 | 2.08 |
| 550S162-68 (14 gauge, 1.90 Lbs./SF) | B1@.036 | SF | 1.18 | 1.20 | 2.38 |
| **5-1/2" steel exterior wall framing, 24" OC, 1-5/8" flange** | | | | | |
| 550S162-33 (20 gauge, .83 Lbs./SF) | B1@.030 | SF | .51 | 1.00 | 1.51 |
| 550S162-43 (18 gauge, 1.08 Lbs./SF) | B1@.031 | SF | .67 | 1.03 | 1.70 |
| 550S162-54 (16 gauge, 1.35 Lbs./SF) | B1@.033 | SF | .84 | 1.10 | 1.94 |
| 550S162-68 (14 gauge, 1.68 Lbs./SF) | B1@.035 | SF | 1.04 | 1.17 | 2.21 |
| Add for each 5-1/2" corner or partition (43 mil) | B1@.091 | Ea | 4.65 | 3.03 | 7.68 |
| **6" steel exterior wall framing, 12" OC, 1-5/8" flange** | | | | | |
| 600S162-33 (20 gauge, 1.46 Lbs./SF) | B1@.035 | SF | .91 | 1.17 | 2.08 |
| 600S162-43 (18 gauge, 1.90 Lbs./SF) | B1@.036 | SF | 1.18 | 1.20 | 2.38 |
| 600S162-54 (16 gauge, 2.36 Lbs./SF) | B1@.039 | SF | 1.46 | 1.30 | 2.76 |
| 600S162-68 (14 gauge, 2.95 Lbs./SF) | B1@.041 | SF | 1.83 | 1.37 | 3.20 |
| 600S162-97 (12 gauge, 4.11 Lbs./SF) | B1@.043 | SF | 2.55 | 1.43 | 3.98 |

| | Craft@Hrs | Unit | Material | Labor | Total |
|---|---|---|---|---|---|
| **6" steel exterior wall framing, 16" OC, 1-5/8" flange** | | | | | |
| 600S162-33 (20 gauge, 1.17 Lbs./SF) | B1@.032 | SF | .73 | 1.07 | 1.80 |
| 600S162-43 (18 gauge, 1.52 Lbs./SF) | B1@.033 | SF | .94 | 1.10 | 2.04 |
| 600S162-54 (16 gauge, 1.89 Lbs./SF) | B1@.035 | SF | 1.17 | 1.17 | 2.34 |
| 600S162-68 (14 gauge, 2.36 Lbs./SF) | B1@.037 | SF | 1.46 | 1.23 | 2.69 |
| 600S162-97 (12 gauge, 3.29 Lbs./SF) | B1@.039 | SF | 2.04 | 1.30 | 3.34 |
| **6" steel exterior wall framing, 20" OC, 1-5/8" flange** | | | | | |
| 600S162-33 (20 gauge, .99 Lbs./SF) | B1@.031 | SF | .61 | 1.03 | 1.64 |
| 600S162-43 (18 gauge, 1.29 Lbs./SF) | B1@.032 | SF | .80 | 1.07 | 1.87 |
| 600S162-54 (16 gauge, 1.60 Lbs./SF) | B1@.034 | SF | .99 | 1.13 | 2.12 |
| 600S162-68 (14 gauge, 2.00 Lbs./SF) | B1@.036 | SF | 1.24 | 1.20 | 2.44 |
| 600S162-97 (12 gauge), 2.79 Lbs./SF | B1@.038 | SF | 1.73 | 1.27 | 3.00 |
| **6" steel exterior wall framing, 24" OC, 1-5/8" flange** | | | | | |
| 600S162-33 (20 gauge, .88 Lbs./SF) | B1@.030 | SF | .55 | 1.00 | 1.55 |
| 600S162-43 (18 gauge, 1.14 Lbs./SF) | B1@.031 | SF | .71 | 1.03 | 1.74 |
| 600S162-54 (16 gauge, 1.42 Lbs./SF) | B1@.033 | SF | .88 | 1.10 | 1.98 |
| 600S162-68 (14 gauge, 1.77 Lbs./SF) | B1@.035 | SF | 1.10 | 1.17 | 2.27 |
| 600S162-97 (12 gauge, 2.47 Lbs./SF) | B1@.037 | SF | 1.53 | 1.23 | 2.76 |
| Add for each 6" corner or partition (43 mil) | B1@.091 | Ea | 4.87 | 3.03 | 7.90 |

**Door opening steel framing** In wall studs, based on walls 8' high. Costs shown are per each door opening and include header of appropriate size, double king studs each side of the opening for doors greater than 3' wide, double trimmer studs each side of the opening for doors greater than 8' wide, cripples, blocking, screws and normal waste. Width shown is size of finished opening. Figures in parentheses indicate header size.

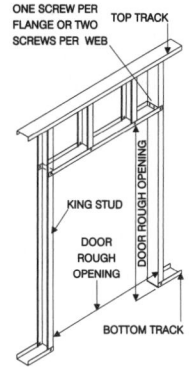

| | Craft@Hrs | Unit | Material | Labor | Total |
|---|---|---|---|---|---|
| Using wall studs and headers, per Lb | — | Lb | .62 | — | .62 |
| **Door opening steel framing in 3-1/2" wall, 20 gauge (33 mil) stud wall, opening size as shown** | | | | | |
| To 3' wide, 6" header, 18 gauge, two 600S162-43, 40.05 Lbs | B1@1.00 | Ea | 24.80 | 33.30 | 58.10 |
| Over 3' to 4' wide, 6" header, 16 gauge, two 600S162-54, 56.48 Lbs | B1@1.33 | Ea | 35.00 | 44.30 | 79.30 |
| Over 4' to 5' wide, 8" header, 16 gauge, two 800S162-54, 63.85 Lbs | B1@1.67 | Ea | 39.60 | 55.60 | 95.20 |
| Over 5' to 6' wide, 8" header, 16 gauge, two 800S162-54, 68.43 Lbs | B1@1.99 | Ea | 42.40 | 66.30 | 108.70 |
| Over 6' to 8' wide, 10" header, 14 gauge, two 1000S162-68, 107.25 Lbs | B1@2.33 | Ea | 66.50 | 77.60 | 144.10 |
| Over 8' to 10' wide, 12" header, 14 gauge, two 1200S162-68, 130.15 Lbs | B1@2.33 | Ea | 104.00 | 77.60 | 181.60 |
| Over 10' to 12' wide, 12" header, 12 gauge, two 1200S162-97, 182.60 Lbs | B1@2.67 | Ea | 113.00 | 88.90 | 201.90 |
| Add per foot of height for 3-1/2" walls over 8' in height | — | LF | 3.57 | — | 3.57 |

| | Craft@Hrs | Unit | Material | Labor | Total |
|---|---|---|---|---|---|
| Door opening steel framing in 5-1/2" wall, 18 gauge (43 mil) stud wall, opening size as shown | | | | | |
| To 3' wide, 6" header, 18 gauge, | | | | | |
| two 600S162-43, 53.75 Lbs | B1@1.00 | Ea | 33.30 | 33.30 | 66.60 |
| Over 3' to 4' wide, 6" header, 16 gauge, | | | | | |
| two 600S162-54, 82.80 Lbs | B1@1.33 | Ea | 51.30 | 44.30 | 95.60 |
| Over 4' to 5' wide, 8" header, 16 gauge, | | | | | |
| two 800S162-54, 90.00 Lbs | B1@1.67 | Ea | 55.80 | 55.60 | 111.40 |
| Over 5' to 6' wide, 8" header, 16 gauge, | | | | | |
| two 800S162-54, 94.55 Lbs | B1@1.99 | Ea | 58.60 | 66.30 | 124.90 |
| Over 6' to 8' wide, 10" header, 14 gauge, | | | | | |
| two 1000S162-68, 120.00 Lbs | B1@2.33 | Ea | 74.40 | 77.60 | 152.00 |
| Over 8' to 10' wide, 12" header, 14 gauge, | | | | | |
| two 1200S162-68, 162.60 Lbs | B1@2.33 | Ea | 101.00 | 77.60 | 178.60 |
| Over 10' to 12' wide, 12" header, 12 gauge, | | | | | |
| two 1200S162-97, 215.05 Lbs | B1@2.67 | Ea | 133.00 | 88.90 | 221.90 |
| Add per foot of height for 5-1/2" | | | | | |
| walls over 8' in height | — | LF | 5.81 | — | 5.81 |

**Window opening steel framing** In wall studs, based on walls 8' high. Costs shown are per window opening and include header of appropriate size, sill track, double king studs each side of the opening for windows greater than 3' wide, double trimmer studs each side of the opening for windows greater than 8' wide, top and bottom cripples, blocking, screws and normal waste. Width shown is size of finished opening.

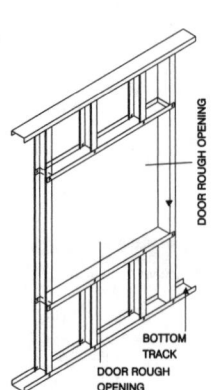

| | Craft@Hrs | Unit | Material | Labor | Total |
|---|---|---|---|---|---|
| Using steel framing materials at | — | Lb | .62 | — | .62 |
| Window opening steel framing in 3-1/2" wall, 20 gauge (33 mil) stud wall, opening size shown | | | | | |
| To 2' wide, 4" header, 18 gauge, | | | | | |
| two 400S162-43, 35.18 Lbs | B1@1.20 | Ea | 21.80 | 40.00 | 61.80 |
| Over 2' to 3' wide, 6" header, 18 gauge, | | | | | |
| two 600S162-43, 40.48 Lbs | B1@1.40 | Ea | 25.10 | 46.60 | 71.70 |
| Over 3' to 4' wide, 6" header, 16 gauge, | | | | | |
| two 600S162-54, 61.85 Lbs | B1@1.74 | Ea | 38.30 | 58.00 | 96.30 |
| Over 4' to 5' wide, 8" header, 16 gauge, | | | | | |
| two 800S162-54, 71.45 Lbs | B1@2.08 | Ea | 44.30 | 69.30 | 113.60 |
| Over 5' to 6' wide, 8" header, 16 gauge, | | | | | |
| two 800S162-54, 77.55 Lbs | B1@2.41 | Ea | 48.10 | 80.30 | 128.40 |
| Over 6' to 7' wide, 8" header, 14 gauge, | | | | | |
| two 800S162-68, 91.45 Lbs | B1@2.75 | Ea | 56.70 | 91.60 | 148.30 |
| Over 7' to 8' wide, 10" header, 14 gauge, | | | | | |
| two 1000S162-68, 106.23 Lbs | B1@3.08 | Ea | 65.90 | 103.00 | 168.90 |
| Over 8' to 10' wide, 12" header, 14 gauge, | | | | | |
| two 1200S162-68, 144.20 Lbs | B1@3.08 | Ea | 89.40 | 103.00 | 192.40 |
| Over 10' to 12' wide, 12" header, 12 gauge, | | | | | |
| two 1200S162-97, 199.68 Lbs | B1@3.42 | Ea | 124.00 | 114.00 | 238.00 |
| Add per foot of height for 3-1/2" | | | | | |
| walls over 8' high | — | LF | 3.57 | — | 3.57 |
| Window opening steel framing in 5-1/2" wall, 18 gauge, 43 mil stud wall, opening size as shown | | | | | |
| To 2' wide, 4" header, 18 gauge, | | | | | |
| two 400S162-43, 55.00 Lbs | B1@1.20 | Ea | 34.10 | 40.00 | 74.10 |
| Over 2' to 3' wide, 6" header, 18 gauge, | | | | | |
| two 600S162-43, 61.18 Lbs | B1@1.40 | Ea | 37.90 | 46.60 | 84.50 |

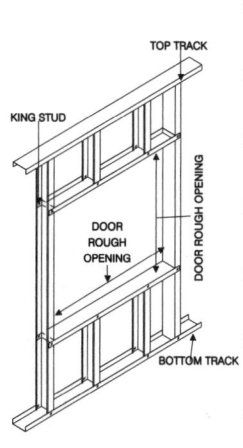

| | Craft@Hrs | Unit | Material | Labor | Total |
|---|---|---|---|---|---|
| Over 3' to 4' wide, 6" header, 16 gauge, two 600S162-54, 92.62 Lbs | B1@1.74 | Ea | 57.40 | 58.00 | 115.40 |
| Over 4' to 5' wide, 8" header, 16 gauge, two 800S162-54, 103.15 Lbs | B1@2.08 | Ea | 64.00 | 69.30 | 133.30 |
| Over 5' to 6' wide, 8" header, 16 gauge, two 800S162-54, 110.30 Lbs | B1@2.41 | Ea | 68.40 | 80.30 | 148.70 |
| Over 6' to 8' wide, 10" header, 14 gauge, two 1000S162-68, 140.95 Lbs | B1@3.08 | Ea | 87.40 | 103.00 | 190.40 |
| Over 8' to 10' wide, 12" header, 14 gauge, two 1200S162-68, 188.78 Lbs | B1@3.08 | Ea | 117.00 | 103.00 | 220.00 |
| Over 8' to 12' wide, 12" header, 12 gauge, two 1200S162-97, 246.43 Lbs | B1@3.42 | Ea | 153.00 | 114.00 | 267.00 |
| Add per foot of height for 5-1/2" walls over 8' high | — | LF | 5.80 | — | 5.80 |

**Steel wall bracing** See also sheathing for plywood bracing and shear panels.

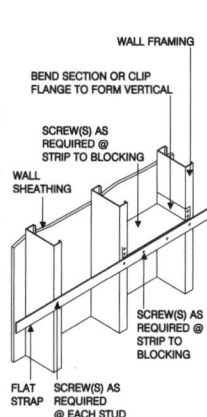

Steel strap bracing

| | Craft@Hrs | Unit | Material | Labor | Total |
|---|---|---|---|---|---|
| 4" wide 33 mil with gusset plates | B1@.015 | LF | .20 | .50 | .70 |
| 4" wide 43 mil with gusset plates | B1@.015 | LF | .26 | .50 | .76 |
| 8.8" wide 33 mil | B1@.015 | LF | .40 | .50 | .90 |
| 8.8" wide 43 mil | B1@.015 | LF | .53 | .50 | 1.03 |
| 8.8" wide 54 mil | B1@.015 | LF | .66 | .50 | 1.16 |

Intermediate bracing  3-1/2", 33 mil studs or track

| | Craft@Hrs | Unit | Material | Labor | Total |
|---|---|---|---|---|---|
| 350S162-33 ( 0.80 Lbs./LF) | B1@.019 | LF | .50 | .63 | 1.13 |

Temporary wood bracing for walls, assumes salvage at 50% and 3 uses

| | Craft@Hrs | Unit | Material | Labor | Total |
|---|---|---|---|---|---|
| 1" x 4" Std & Btr | B1@.006 | LF | .16 | .20 | .36 |
| 1" x 6" Std & Btr | B1@.010 | LF | .23 | .33 | .56 |
| 2" x 4" utility | B1@.012 | LF | .08 | .40 | .48 |
| 2" x 6" utility | B1@.018 | LF | .12 | .60 | .72 |

**Wood wall plates** Std & Btr, untreated. For pressure treated plates, see Wood sill plates for steel framing previously in this section. Figures in parentheses indicate board feet per LF. Costs shown include 10% waste.

| | Craft@Hrs | Unit | Material | Labor | Total |
|---|---|---|---|---|---|
| 2" x 3" wall plates, per MBF | — | MBF | 423.00 | — | 423.00 |
| 2" x 3" (.55 BF per LF) | B1@.010 | LF | .23 | .33 | .56 |
| 2" x 4" wall plates, per MBF | — | MBF | 473.00 | — | 473.00 |
| 2" x 4" (.73 BF per LF) | B1@.012 | LF | .35 | .40 | .75 |
| 2" x 6" wall plates, per MBF | — | MBF | 465.00 | — | 465.00 |
| 2" x 6" (1.10 BF per LF) | B1@.018 | LF | .51 | .60 | 1.11 |
| 2" x 8" wall plates, per MBF | — | MBF | 465.00 | — | 465.00 |
| 2" x 8" (1.10 BF per LF) | B1@.018 | LF | .65 | .60 | 1.25 |

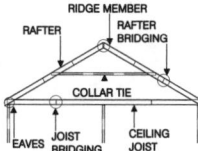

**Steel ceiling joists and soffits** Per SF of area covered. Including typical stiffeners, splices, bracing, rim joists and screws. Add the cost of beams, bridging, blocking, ledger strips and trimmers as required. See costs for beams, bridging, backing, blocking and trimmers elsewhere in this section. Deduct for openings over 25 SF. Labor includes layout, assembly and alignment. Figures in parentheses show weight of steel per square foot of ceiling. Assumes minimal waste.

| | Craft@Hrs | Unit | Material | Labor | Total |
|---|---|---|---|---|---|
| Using light gauge steel framing materials at | — | Lb | .62 | — | .62 |

3-1/2" steel ceiling joists, 16" OC, 1-5/8" flange

| | Craft@Hrs | Unit | Material | Labor | Total |
|---|---|---|---|---|---|
| 350S162-33 (20 gauge, .88 Lbs./SF) | B1@.022 | SF | .55 | .73 | 1.28 |
| 350S162-43 (18 gauge), 1.11 Lbs./SF | B1@.024 | SF | .69 | .80 | 1.49 |
| 350S162-54 (16 gauge, 1.35 Lbs./SF) | B1@.027 | SF | .84 | .90 | 1.74 |
| 350S162-68 (14 gauge, 1.65 Lbs./SF) | B1@.029 | SF | 1.02 | .97 | 1.99 |

# Carpentry, Steel Framing

|  | Craft@Hrs | Unit | Material | Labor | Total |
|---|---|---|---|---|---|
| **3-1/2" steel ceiling joists, 24" OC, 1-5/8" flange** | | | | | |
| 350S162-33 (20 gauge, .63 Lbs./SF) | B1@.015 | SF | .39 | .50 | .89 |
| 350S162-43 (18 gauge, .79 Lbs./SF) | B1@.017 | SF | .49 | .57 | 1.06 |
| 350S162-54 (16 gauge, .96 Lbs./SF) | B1@.019 | SF | .60 | .63 | 1.23 |
| 350S162-68 (14 gauge, 1.17 Lbs./SF) | B1@.020 | SF | .73 | .67 | 1.40 |
| **4" steel ceiling joists, 16" OC, 1-5/8" flange** | | | | | |
| 400S162-33 (20 gauge, .93 Lbs./SF) | B1@.022 | SF | .58 | .73 | 1.31 |
| 400S162-43 (18 gauge), 1.17 Lbs./SF | B1@.024 | SF | .73 | .80 | 1.53 |
| 400S162-54 (16 gauge, 1.44 Lbs./SF) | B1@.027 | SF | .89 | .90 | 1.79 |
| 400S162-68 (14 gauge, 1.76 Lbs./SF) | B1@.029 | SF | 1.09 | .97 | 2.06 |
| **4" steel ceiling joists, 24" OC, 1-5/8" flange** | | | | | |
| 400S162-33 (20 gauge, .67 Lbs./SF) | B1@.015 | SF | .42 | .50 | .92 |
| 400S162-43 (18 gauge, .84 Lbs./SF) | B1@.017 | SF | .52 | .57 | 1.09 |
| 400S162-54 (16 gauge, 1.02 Lbs./SF) | B1@.019 | SF | .63 | .63 | 1.26 |
| 400S162-68 (14 gauge, 1.25 Lbs./SF) | B1@.020 | SF | .78 | .67 | 1.45 |
| **5-1/2" steel ceiling joists, 12" OC, 1-5/8" flange** | | | | | |
| 550S162-33 (20 gauge, 1.41 Lbs./SF) | B1@.029 | SF | .87 | .97 | 1.84 |
| 550S162-43 (18 gauge, 1.79 Lbs./SF) | B1@.032 | SF | 1.11 | 1.07 | 2.18 |
| 550S162-54 (16 gauge, 2.21 Lbs./SF) | B1@.035 | SF | 1.37 | 1.17 | 2.54 |
| 550S162-68 (14 gauge, 2.72 Lbs./SF) | B1@.038 | SF | 1.69 | 1.27 | 2.96 |
| **5-1/2" steel ceiling joists, 16" OC, 1-5/8" flange** | | | | | |
| 550S162-33 (20 gauge, 1.09 Lbs./SF) | B1@.022 | SF | .68 | .73 | 1.41 |
| 550S162-43 (18 gauge, 1.39 Lbs./SF) | B1@.024 | SF | .86 | .80 | 1.66 |
| 550S162-54 (16 gauge, 1.71 Lbs./SF) | B1@.027 | SF | 1.06 | .90 | 1.96 |
| 550S162-68 (14 gauge, 2.10 Lbs./SF) | B1@.029 | SF | 1.30 | .97 | 2.27 |
| **5-1/2" steel ceiling joists, 24" OC, 1-5/8" flange** | | | | | |
| 550S162-33 (20 gauge, .78 Lbs./SF) | B1@.016 | SF | .48 | .53 | 1.01 |
| 550S162-43 (18 gauge, .99 Lbs./SF) | B1@.018 | SF | .61 | .60 | 1.21 |
| 550S162-54 (16 gauge, 1.21 Lbs./SF) | B1@.020 | SF | .75 | .67 | 1.42 |
| 550S162-68 (14 gauge, 1.48 Lbs./SF) | B1@.022 | SF | .92 | .73 | 1.65 |
| **6" steel ceiling joists, 12" OC, 1-5/8" flange** | | | | | |
| 600S162-33 (20 gauge, 1.48 Lbs./SF) | B1@.029 | SF | .92 | .97 | 1.89 |
| 600S162-43 (18 gauge, 1.89 Lbs./SF) | B1@.032 | SF | 1.17 | 1.07 | 2.24 |
| 600S162-54 (16 gauge, 2.32 Lbs./SF) | B1@.035 | SF | 1.44 | 1.17 | 2.61 |
| 600S162-68 (14 gauge, 2.86 Lbs./SF) | B1@.038 | SF | 1.77 | 1.27 | 3.04 |
| **6" steel ceiling joists, 16" OC, 1-5/8" flange** | | | | | |
| 600S162-33 (20 gauge, 1.15 Lbs./SF) | B1@.022 | SF | .71 | .73 | 1.44 |
| 600S162-43 (18 gauge, 1.46 Lbs./SF) | B1@.024 | SF | .91 | .80 | 1.71 |
| 600S162-54 (16 gauge, 1.79 Lbs./SF) | B1@.027 | SF | 1.11 | .90 | 2.01 |
| 600S162-68 (14 gauge, 2.21 Lbs./SF) | B1@.029 | SF | 1.37 | .97 | 2.34 |
| **6" steel ceiling joists, 24" OC, 1-5/8" flange** | | | | | |
| 600S162-33 (20 gauge, .82 Lbs./SF) | B1@.016 | SF | .51 | .53 | 1.04 |
| 600S162-43 (18 gauge, 1.04 Lbs./SF) | B1@.018 | SF | .64 | .60 | 1.24 |
| 600S162-54 (16 gauge, 1.27 Lbs./SF) | B1@.020 | SF | .79 | .67 | 1.46 |
| 600S162-68 (14 gauge, 1.56 Lbs./SF) | B1@.022 | SF | .97 | .73 | 1.70 |
| **8" steel ceiling joists, 12" OC, 1-5/8" flange** | | | | | |
| 800S162-33 (20 gauge, 1.78 Lbs./SF) | B1@.031 | SF | 1.10 | 1.03 | 2.13 |
| 800S162-43 (18 gauge, 2.26 Lbs./SF) | B1@.034 | SF | 1.40 | 1.13 | 2.53 |
| 800S162-54 (16 gauge, 2.79 Lbs./SF) | B1@.037 | SF | 1.73 | 1.23 | 2.96 |
| 800S162-68 (14 gauge, 3.44 Lbs./SF) | B1@.041 | SF | 2.13 | 1.37 | 3.50 |
| 800S162-97 (12 gauge, 4.76 Lbs./SF) | B1@.049 | SF | 2.95 | 1.63 | 4.58 |

| | Craft@Hrs | Unit | Material | Labor | Total |
|---|---|---|---|---|---|
| **8" steel ceiling joists, 16" OC, 1-5/8" flange** | | | | | |
| 800S162-33 (20 gauge, 1.38 Lbs./SF) | B1@.024 | SF | .86 | .80 | 1.66 |
| 800S162-43 (18 gauge, 1.75 Lbs./SF) | B1@.027 | SF | 1.09 | .90 | 1.99 |
| 800S162-54 (16 gauge, 2.15 Lbs./SF) | B1@.029 | SF | 1.33 | .97 | 2.30 |
| 800S162-68 (14 gauge, 2.65 Lbs./SF) | B1@.032 | SF | 1.64 | 1.07 | 2.71 |
| 800S162-97 (12 gauge, 3.67 Lbs./SF) | B1@.039 | SF | 2.28 | 1.30 | 3.58 |
| **8" steel ceiling joists, 20" OC, 1-5/8" flange** | | | | | |
| 800S162-33 (20 gauge, 1.14 Lbs./SF) | B1@.020 | SF | .71 | .67 | 1.38 |
| 800S162-43 (18 gauge, 1.44 Lbs./SF) | B1@.022 | SF | .89 | .73 | 1.62 |
| 800S162-54 (16 gauge, 1.77 Lbs./SF) | B1@.024 | SF | 1.10 | .80 | 1.90 |
| 800S162-68 (14 gauge, 2.18 Lbs./SF) | B1@.026 | SF | 1.35 | .87 | 2.22 |
| 800S162-97 (12 gauge, 3.01 Lbs./SF) | B1@.032 | SF | 1.87 | 1.07 | 2.94 |
| **8" steel ceiling joists, 24" OC, 1-5/8" flange** | | | | | |
| 800S162-33 (20 gauge, .98 Lbs./SF) | B1@.018 | SF | .61 | .60 | 1.21 |
| 800S162-43 (18 gauge, 1.24 Lbs./SF) | B1@.019 | SF | .77 | .63 | 1.40 |
| 800S162-54 (16 gauge, 1.52 Lbs./SF) | B1@.021 | SF | .94 | .70 | 1.64 |
| 800S162-68 (14 gauge, 1.87 Lbs./SF) | B1@.023 | SF | 1.16 | .77 | 1.93 |
| 800S162-97 (12 gauge, 2.57 Lbs./SF) | B1@.028 | SF | 1.59 | .93 | 2.52 |
| **10" steel ceiling joists, 12" OC, 1-5/8" flange** | | | | | |
| 1000S162-43 (18 gauge, 2.63 Lbs./SF) | B1@.033 | SF | 1.63 | 1.10 | 2.73 |
| 1000S162-54 (16 gauge, 3.24 Lbs./SF) | B1@.036 | SF | 2.59 | 1.20 | 3.79 |
| 1000S162-68 (14 gauge, 4.02 Lbs./SF) | B1@.040 | SF | 2.01 | 1.33 | 3.34 |
| 1000S162-97 (12 gauge, 5.58 Lbs./SF) | B1@.048 | SF | 3.46 | 1.60 | 5.06 |
| **10" steel ceiling joists, 16" OC, 1-5/8" flange** | | | | | |
| 1000S162-43 (18 gauge, 2.03 Lbs./SF) | B1@.025 | SF | 1.26 | .83 | 2.09 |
| 1000S162-54 (16 gauge, 2.50 Lbs./SF) | B1@.028 | SF | 1.55 | .93 | 2.48 |
| 1000S162-68 (14 gauge, 3.10 Lbs./SF) | B1@.031 | SF | 1.92 | 1.03 | 2.95 |
| 1000S162-97 (12 gauge, 4.29 Lbs./SF) | B1@.037 | SF | 2.66 | 1.23 | 3.89 |
| **10" steel ceiling joists, 20" OC, 1-5/8" flange** | | | | | |
| 1000S162-43 (18 gauge, 1.67 Lbs./SF) | B1@.022 | SF | 1.04 | .73 | 1.77 |
| 1000S162-54 (16 gauge, 2.06 Lbs./SF) | B1@.024 | SF | 1.28 | .80 | 2.08 |
| 1000S162-68 (14 gauge, 2.54 Lbs./SF) | B1@.027 | SF | 1.57 | .90 | 2.47 |
| 1000S162-97 (12 gauge, 3.52 Lbs./SF) | B1@.032 | SF | 2.18 | 1.07 | 3.25 |
| **10" steel ceiling joists, 24" OC, 1-5/8" flange** | | | | | |
| 1000S162-43 (18 gauge, 1.44 Lbs./SF) | B1@.019 | SF | .89 | .63 | 1.52 |
| 1000S162-54 (16 gauge, 1.76 Lbs./SF) | B1@.021 | SF | 1.09 | .70 | 1.79 |
| 1000S162-68 (14 gauge, 2.18 Lbs./SF) | B1@.023 | SF | 1.35 | .77 | 2.12 |
| 1000S162-97 (12 gauge, 3.01 Lbs./SF) | B1@.027 | SF | 1.87 | .90 | 2.77 |
| **12" steel ceiling joists, 12" OC, 1-5/8" flange** | | | | | |
| 1200S162-54 (16 gauge, 3.71 Lbs./SF) | B1@.034 | SF | 2.30 | 1.13 | 3.43 |
| 1200S162-68 (14 gauge, 4.60 Lbs./SF) | B1@.038 | SF | 2.85 | 1.27 | 4.12 |
| 1200S162-97 (12 gauge, 6.39 Lbs./SF) | B1@.045 | SF | 3.96 | 1.50 | 5.46 |
| **12" steel ceiling joists, 16" OC, 1-5/8" flange** | | | | | |
| 1200S162-54 (16 gauge, 2.86 Lbs./SF) | B1@.028 | SF | 1.77 | .93 | 2.70 |
| 1200S162-68 (14 gauge, 3.54 Lbs./SF) | B1@.030 | SF | 2.19 | 1.00 | 3.19 |
| 1200S162-97 (12 gauge, 4.92 Lbs./SF) | B1@.036 | SF | 3.05 | 1.20 | 4.25 |
| **12" steel ceiling joists, 20" OC, 1-5/8" flange** | | | | | |
| 1200S162-54 (16 gauge, 2.35 Lbs./SF) | B1@.022 | SF | 1.46 | .73 | 2.19 |
| 1200S162-68 (14 gauge, 2.90 Lbs./SF) | B1@.024 | SF | 1.80 | .80 | 2.60 |
| 1200S162-97 (12 gauge, 4.03 Lbs./SF) | B1@.029 | SF | 2.50 | .97 | 3.47 |

# Carpentry, Steel Framing

| | Craft@Hrs | Unit | Material | Labor | Total |
|---|---|---|---|---|---|
| 12" steel ceiling joists, 24" OC, 1-5/8" flange | | | | | |
|     1200S162-54 (16 gauge, 2.01 Lbs./SF) | B1@.020 | SF | 1.25 | .67 | 1.92 |
|     1200S162-68 (14 gauge, 2.49 Lbs./SF) | B1@.022 | SF | 1.54 | .73 | 2.27 |
|     1200S162-97 (12 gauge, 3.45 Lbs./SF) | B1@.026 | SF | 2.14 | .87 | 3.01 |

**Backing for drywall ceiling** Floating backing for drywall ceilings, trim, appliances, fixtures, etc

| | Craft@Hrs | Unit | Material | Labor | Total |
|---|---|---|---|---|---|
|     1-1/2" x 1-1/2" (.54 Lbs./LF) | B1@.012 | LF | .33 | .40 | .73 |

**Steel rafters** Shed or gable roof with spans from 7'0" to 38'0" and pitch of 3 in 12 to 12 in 12 as published in light gauge steel framing span tables. Costs per square foot of roof surface. Includes rafters, bracing, screws and minimal waste but no blocking, purlins, ridge boards, collar beams, or gable walls. See costs for blocking, purlins, ridgeboards and collar beams elsewhere in this section. Labor includes layout, assembly, and alignment. Figures in parentheses show weight of steel per square foot of roof.

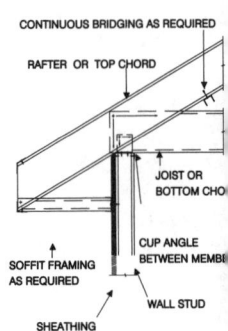

| | Craft@Hrs | Unit | Material | Labor | Total |
|---|---|---|---|---|---|
| Using light gauge steel rafters at | — | Lb | .62 | — | .62 |
| Add for hip roof | B1@.009 | SF | — | .30 | .30 |
| 5-1/2" steel rafters, 12" OC, 1-5/8" flange | | | | | |
|     550S162-33 (20 gauge, 1.18 Lbs./SF) | B1@.037 | SF | .73 | 1.23 | 1.96 |
|     550S162-43 (18 gauge, 1.51 Lbs./SF) | B1@.041 | SF | .94 | 1.37 | 2.31 |
|     550S162-54 (16 gauge, 1.87 Lbs./SF) | B1@.045 | SF | 1.16 | 1.50 | 2.66 |
|     550S162-68 (14 gauge, 2.31 Lbs./SF) | B1@.050 | SF | 1.43 | 1.67 | 3.10 |
| 5-1/2" steel rafters, 16" OC, 1-5/8" flange | | | | | |
|     550S162-33 (20 gauge, .90 Lbs./SF) | B1@.031 | SF | .56 | 1.03 | 1.59 |
|     550S162-43 (18 gauge, 1.15 Lbs./SF) | B1@.034 | SF | .71 | 1.13 | 1.84 |
|     550S162-54 (16 gauge, 1.42 Lbs./SF) | B1@.037 | SF | .88 | 1.23 | 2.11 |
|     550S162-68 (14 gauge, 1.75 Lbs./SF) | B1@.041 | SF | 1.09 | 1.37 | 2.46 |
| 5-1/2" steel rafters, 24" OC, 1-5/8" flange | | | | | |
|     500S162-33 (20 gauge, .63 Lbs./SF) | B1@.024 | SF | .39 | .80 | 1.19 |
|     500S162-43 (18 gauge, .79 Lbs./SF) | B1@.026 | SF | .49 | .87 | 1.36 |
|     500S162-54 (16 gauge, .97 Lbs./SF) | B1@.029 | SF | .60 | .97 | 1.57 |
|     500S162-68 (14 gauge, 1.19 Lbs./SF) | B1@.032 | SF | .74 | 1.07 | 1.81 |
| 8" steel rafters, 12" OC, 1-5/8" flange | | | | | |
|     800S162-33 (20 gauge, 1.48 Lbs./SF) | B1@.048 | SF | .92 | 1.60 | 2.52 |
|     800S162-43 (18 gauge, 1.90 Lbs./SF) | B1@.053 | SF | 1.18 | 1.77 | 2.95 |
|     800S162-54 (16 gauge, 2.35 Lbs./SF) | B1@.058 | SF | 1.46 | 1.93 | 3.39 |
|     800S162-68 (14 gauge, 2.91 Lbs./SF) | B1@.064 | SF | 1.80 | 2.13 | 3.93 |
|     800S162-97 (12 gauge, 4.05 Lbs./SF) | B1@.070 | SF | 2.51 | 2.33 | 4.84 |
| 8" steel rafters, 16" OC, 1-5/8" flange | | | | | |
|     800S162-33 (20 gauge, 1.13 Lbs./SF) | B1@.040 | SF | .70 | 1.33 | 2.03 |
|     800S162-43 (18 gauge, 1.44 Lbs./SF) | B1@.044 | SF | .89 | 1.47 | 2.36 |
|     800S162-54 (16 gauge, 1.78 Lbs./SF) | B1@.048 | SF | 1.10 | 1.60 | 2.70 |
|     800S162-68 (14 gauge, 2.20 Lbs./SF) | B1@.053 | SF | 1.36 | 1.77 | 3.13 |
|     800S162-97 (12 gauge, 3.06 Lbs./SF) | B1@.058 | SF | 1.90 | 1.93 | 3.83 |
| 8" steel rafters, 24" OC, 1-5/8" flange | | | | | |
|     800S162-33 (20 gauge, .78 Lbs./SF) | B1@.032 | SF | .48 | 1.07 | 1.55 |
|     800S162-43 (18 gauge, .99 Lbs./SF) | B1@.035 | SF | .61 | 1.17 | 1.78 |
|     800S162-54 (16 gauge, 1.21 Lbs./SF) | B1@.039 | SF | .75 | 1.30 | 2.05 |
|     800S162-68 (14 gauge, 1.49 Lbs./SF) | B1@.043 | SF | .92 | 1.43 | 2.35 |
|     800S162-97 (12 gauge, 2.06 Lbs./SF) | B1@.047 | SF | 1.28 | 1.57 | 2.85 |

| | Craft@Hrs | Unit | Material | Labor | Total |
|---|---|---|---|---|---|
| **10" steel rafters, 12" OC, 1-5/8" flange** | | | | | |
| 1000S162-43 (18 gauge, 2.20 Lbs./SF) | B1@.060 | SF | 1.36 | 2.00 | 3.36 |
| 1000S162-54 (16 gauge, 2.73 Lbs./SF) | B1@.066 | SF | 1.69 | 2.20 | 3.89 |
| 1000S162-68 (14 gauge, 3.40 Lbs./SF) | B1@.072 | SF | 2.11 | 2.40 | 4.51 |
| 1000S162-97 (12 gauge, 4.74 Lbs./SF) | B1@.080 | SF | 2.91 | 2.67 | 5.58 |
| **10" steel rafters, 16" OC, 1-5/8" flange** | | | | | |
| 1000S162-43 (18 gauge, 1.67 Lbs./SF) | B1@.048 | SF | 1.04 | 1.60 | 2.64 |
| 1000S162-54 (16 gauge, 2.07 Lbs./SF) | B1@.053 | SF | 1.28 | 1.77 | 3.05 |
| 1000S162-68 (14 gauge, 2.57 Lbs./SF) | B1@.058 | SF | 1.59 | 1.93 | 3.52 |
| 1000S162-97 (12 gauge, 3.57 Lbs./SF) | B1@.064 | SF | 2.21 | 2.13 | 4.34 |
| **10" steel rafters, 24" OC, 1-5/8" flange** | | | | | |
| 1000S162-43 (18 gauge, 1.14 Lbs./SF) | B1@.039 | SF | .71 | 1.30 | 2.01 |
| 1000S162-54 (16 gauge, 1.40 Lbs./SF) | B1@.042 | SF | .87 | 1.40 | 2.27 |
| 1000S162-68 (14 gauge, 1.74 Lbs./SF) | B1@.047 | SF | 1.08 | 1.57 | 2.65 |
| 1000S162-97 (12 gauge, 2.41 Lbs./SF) | B1@.051 | SF | 1.49 | 1.70 | 3.19 |
| **12" steel rafters, 12" OC, 1-5/8" flange** | | | | | |
| 1200S162-54 (16 gauge, 3.12 Lbs./SF) | B1@.066 | SF | 1.93 | 2.20 | 4.13 |
| 1200S162-68 (14 gauge, 3.88 Lbs./SF) | B1@.073 | SF | 2.41 | 2.43 | 4.84 |
| 1200S162-97 (12 gauge, 5.43 Lbs./SF) | B1@.080 | SF | 3.37 | 2.66 | 6.03 |
| **12" steel rafters, 16" OC, 1-5/8" flange** | | | | | |
| 1200S162-54 (16 gauge, 2.36 Lbs./SF) | B1@.053 | SF | 1.46 | 1.77 | 3.23 |
| 1200S162-68 (14 gauge, 2.93 Lbs./SF) | B1@.059 | SF | 1.82 | 1.97 | 3.79 |
| 1200S162-97 (12 gauge, 4.09 Lbs./SF) | B1@.064 | SF | 2.54 | 2.13 | 4.67 |
| **12" steel rafters, 24" OC, 1-5/8" flange** | | | | | |
| 1200S162-54 (16 gauge, 1.60 Lbs./SF) | B1@.043 | SF | .99 | 1.43 | 2.42 |
| 1200S162-68 (14 gauge, 1.98 Lbs./SF) | B1@.047 | SF | 1.23 | 1.57 | 2.80 |
| 1200S162-97 (12 gauge, 2.75 Lbs./SF) | B1@.052 | SF | 1.71 | 1.73 | 3.44 |

HEADER JOIST
JOIST
TRIMMER JOIST
CLIP ANGLE
JOIST
JOIST
TRIMMER JOIST

**Steel trimmers** At floor openings, stairwells, skylights, dormers, etc. Consists of one C-shaped member nested in same size track. Labor includes layout, assembly and alignment. Figures in parentheses show weight of steel per linear foot of trimmer. Includes screws and minimal waste.

| | Craft@Hrs | Unit | Material | Labor | Total |
|---|---|---|---|---|---|
| Using light gauge steel framing materials | — | Lb | .62 | — | .62 |
| **3-1/2" steel trimmers, 1-5/8" flange** | | | | | |
| 350S162-33 (20 gauge, 1.76 Lbs./LF) | B1@.020 | LF | 1.09 | .67 | 1.76 |
| 350S162-43 (18 gauge, 2.29 Lbs./LF) | B1@.020 | LF | 1.42 | .67 | 2.09 |
| 350S162-54 (16 gauge, 2.85 Lbs./LF) | B1@.022 | LF | 1.77 | .73 | 2.50 |
| 350S162-68 (14 gauge, 3.60 Lbs./LF) | B1@.023 | LF | 2.23 | .77 | 3.00 |
| **5-1/2" steel trimmers, 1-5/8" flange** | | | | | |
| 550S162-33 (20 gauge, 2.23 Lbs./LF) | B1@.031 | LF | 1.38 | 1.03 | 2.41 |
| 550S162-43 (18 gauge, 2.90 Lbs./LF) | B1@.031 | LF | 1.80 | 1.03 | 2.83 |
| 550S162-54 (16 gauge, 3.63 Lbs./LF) | B1@.034 | LF | 2.25 | 1.13 | 3.38 |
| 550S162-68 (14 gauge, 4.54 Lbs./LF) | B1@.035 | LF | 2.81 | 1.17 | 3.98 |
| **8" steel trimmers, 1-5/8" flange** | | | | | |
| 800S162-33 (20 gauge, 2.82 Lbs./LF) | B1@.042 | LF | 1.75 | 1.40 | 3.15 |
| 800S162-43 (18 gauge, 3.67 Lbs./LF) | B1@.042 | LF | 2.28 | 1.40 | 3.68 |
| 800S162-54 (16 gauge, 4.59 Lbs./LF) | B1@.046 | LF | 2.85 | 1.53 | 4.38 |
| 800S162-68 (14 gauge, 5.75 Lbs./LF) | B1@.047 | LF | 3.57 | 1.57 | 5.14 |
| 800S162-97 (12 gauge, 8.13 Lbs./LF) | B1@.048 | LF | 5.04 | 1.60 | 6.64 |
| **10" steel trimmers, 1-5/8" flange** | | | | | |
| 1000S162-43 (18 gauge, 4.28 Lbs./LF) | B1@.052 | LF | 2.65 | 1.73 | 4.38 |
| 1000S162-54 (16 gauge, 5.35 Lbs./LF) | B1@.052 | LF | 3.32 | 1.73 | 5.05 |
| 1000S162-68 (14 gauge, 6.72 Lbs./LF) | B1@.057 | LF | 4.17 | 1.90 | 6.07 |
| 1000S162-97 (12 gauge, 9.51 Lbs./LF) | B1@.058 | LF | 5.90 | 1.93 | 7.83 |

| | Craft@Hrs | Unit | Material | Labor | Total |
|---|---|---|---|---|---|
| 12" steel trimmers, 1-5/8" flange | | | | | |
|     1200S162-54 (16 gauge, 6.13 Lbs./LF) | B1@.063 | LF | 3.80 | 2.10 | 5.90 |
|     1200S162-68 (14 gauge, 7.69 Lbs./LF) | B1@.066 | LF | 4.77 | 2.20 | 6.97 |
|     1200S162-97 (12 gauge, 10.89 Lbs./LF) | B1@.069 | LF | 6.75 | 2.30 | 9.05 |

**Steel collar beams** Cost per linear foot of collar beam or rafter brace, including screws.

| | Craft@Hrs | Unit | Material | Labor | Total |
|---|---|---|---|---|---|
|     350S162-33 (3-1/2", 33 mil, 20 gauge) | B1@.007 | LF | .41 | .23 | .64 |
|     550S162-43 (5-1/2", 43 mil, 18 gauge) | B1@.014 | LF | .71 | .47 | 1.18 |

**Steel purlins** Purlins or "perling" installed below roof rafters.

| | Craft@Hrs | Unit | Material | Labor | Total |
|---|---|---|---|---|---|
|     350S162-33 (3-1/2", 33 mil, 20 gauge) | B1@.013 | LF | .54 | .43 | .97 |
|     350S162-43 (3-1/2", 43 mil, 18 gauge) | B1@.019 | LF | .41 | .63 | 1.04 |

**Steel dormer studs** Per square foot of wall area

| | Craft@Hrs | Unit | Material | Labor | Total |
|---|---|---|---|---|---|
|     350S162-33 (3-1/2", 33 mil, 20 gauge) | B1@.036 | SF | .53 | 1.20 | 1.73 |

**Steel ridgeboard** Horizontal support at top of rafter ends. Consists of one C-shaped member nested in same size track. Labor includes layout, assembly and alignment. Figures in parentheses show weight of steel per linear foot of ridgeboard. Includes screws and minimal waste.

| | Craft@Hrs | Unit | Material | Labor | Total |
|---|---|---|---|---|---|
| Using light gauge steel framing materials | — | Lb | .62 | — | .62 |
| 5-1/2" steel ridgeboard, 1-5/8" flange | | | | | |
|     550S162-33 (20 gauge, 2.23 Lbs./LF) | B1@.031 | LF | 1.38 | 1.03 | 2.41 |
|     550S162-43 (18 gauge, 2.90 Lbs./LF) | B1@.031 | LF | 1.80 | 1.03 | 2.83 |
|     550S162-54 (16 gauge, 3.63 Lbs./LF) | B1@.034 | LF | 2.25 | 1.13 | 3.38 |
|     550S162-68 (14 gauge, 4.54 Lbs./LF) | B1@.035 | LF | 2.82 | 1.17 | 3.99 |
| 8" steel ridgeboard, 1-5/8" flange | | | | | |
|     800S162-33 (20 gauge, 2.82 Lbs./LF) | B1@.042 | LF | 1.75 | 1.40 | 3.15 |
|     800S162-43 (18 gauge, 3.67 Lbs./LF) | B1@.042 | LF | 2.28 | 1.40 | 3.68 |
|     800S162-54 (16 gauge, 4.59 Lbs./LF) | B1@.046 | LF | 2.85 | 1.53 | 4.38 |
|     800S162-68 (14 gauge, 5.75 Lbs./LF) | B1@.047 | LF | 3.57 | 1.57 | 5.14 |
|     800S162-97 (12 gauge, 8.13 Lbs./LF) | B1@.048 | LF | 5.04 | 1.60 | 6.64 |
| 10" steel ridgeboard, 1-5/8" flange | | | | | |
|     1000S162-43 (18 gauge, 4.28 Lbs./LF) | B1@.052 | LF | 2.65 | 1.73 | 4.38 |
|     1000S162-54 (16 gauge, 5.35 Lbs./LF) | B1@.052 | LF | 3.32 | 1.73 | 5.05 |
|     1000S162-68 (14 gauge, 6.72 Lbs./LF) | B1@.057 | LF | 4.17 | 1.90 | 6.07 |
|     1000S162-97 (12 gauge, 9.51 Lbs./LF) | B1@.058 | LF | 5.90 | 1.93 | 7.83 |
| 12" steel ridgeboard, 1-5/8" flange | | | | | |
|     1200S162-54 (16 gauge, 6.13 Lbs./LF) | B1@.063 | LF | 3.80 | 2.10 | 5.90 |
|     1200S162-68 (14 gauge, 7.69 Lbs./LF) | B1@.066 | LF | 4.77 | 2.20 | 6.97 |
|     1200S162-97 (12 gauge, 10.89 Lbs./LF) | B1@.069 | LF | 6.75 | 2.30 | 9.05 |

**Steel roof trusses** Site-fabricated and set by hand 24" OC for any roof slope from 3 in 12 to 12 in 12 so long as the total truss height does not exceed 12' from bottom chord to highest point on truss. Trusses over 12' high may cost up to 100% more. Costs are per square foot (SF) of horizontal area covered.

| | Craft@Hrs | Unit | Material | Labor | Total |
|---|---|---|---|---|---|
| Using light gauge steel framing materials at | — | Lb | .62 | — | .62 |
| 4" C-shaped top and bottom chords, members unpunched | | | | | |
|     Up to 38' span | B1@.029 | SF | 2.85 | .97 | 3.82 |
|     40' to 50' span | B1@.037 | SF | 3.57 | 1.23 | 4.80 |
| Fink truss "W" (conventional roof truss). 4" C-shaped top and bottom chords | | | | | |
|     Up to 38' span | B1@.023 | SF | 2.50 | .77 | 3.27 |
|     40' to 50' span | B1@.029 | SF | 2.95 | .97 | 3.92 |

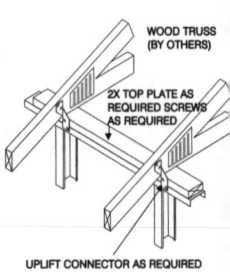

WOOD TRUSS
(BY OTHERS)

2X TOP PLATE AS
REQUIRED SCREWS
AS REQUIRED

UPLIFT CONNECTOR AS REQUIRED

| | Craft@Hrs | Unit | Material | Labor | Total |
|---|---|---|---|---|---|
| 6" C-shaped top and bottom chords, members unpunched | | | | | |
|     Up to 38' span | B1@.027 | SF | 3.00 | .90 | 3.90 |
|     40' to 50' span | B1@.035 | SF | 3.62 | 1.17 | 4.79 |
| Truss with gable fill at 16" OC | | | | | |
|     28' span, 5 in 12 slope | B1@1.27 | Ea | 203.00 | 42.30 | 245.30 |
|     32' span, 5 in 12 slope | B1@1.68 | Ea | 251.00 | 56.00 | 307.00 |
|     40' span, 5 in 12 slope | B1@2.30 | Ea | 350.00 | 76.60 | 426.60 |

**Piecework Steel Framing** Steel framing on residential tracts is sometimes done by sub-contractors who bid at piecework rates (such as per square foot of floor). The figures below list typical piecework rates for repetitive framing and assume all materials are supplied to the framing subcontractor. No figures appear in the Craft@Hrs column because the work is done for a fixed price per square foot and the labor productivity can be expected to vary widely.

**Steel frame layout and bottom track** Piecework rates
Lay out wall track according to the plans (snap chalk lines for wall track, mark location for studs, windows, doors and framing details), cut top and bottom track and install bottom track. Costs per square foot of floor (excluding garage).

| | | Unit | | Labor | Total |
|---|---|---|---|---|---|
|     Custom or more complex jobs | — | SF | — | .29 | .29 |
|     Larger job, longer runs | — | SF | — | .19 | .19 |

**Steel exterior wall framing** Piecework rates
Measure, cut, fit, assemble, and tip up walls, including studs, track, cripples, x-bracing, trimmers and blocking. Costs per square foot of floor.

| | | Unit | | Labor | Total |
|---|---|---|---|---|---|
|     Complex job, radius walls, rake walls | — | SF | — | .83 | .83 |
|     Larger job, 8' high walls, fewer partitions | — | SF | — | .29 | .29 |

**Plumb and align steel framed walls** Piecework rates
Align walls, adjust walls to vertical and install temporary braces as needed, fasten nuts on anchor bolts or shoot power driven fasteners through wall track into the slab. Based on accuracy to 3/16". Costs per square foot of floor are shown below.

| | | Unit | | Labor | Total |
|---|---|---|---|---|---|
|     Small or complex job, many braces | — | SF | — | .28 | .28 |
|     Larger job, fewer braces | — | SF | — | .17 | .17 |

**Install steel floor or ceiling joists** Piecework rates
Lay out, cut and install floor or ceiling joists, including rim track, doubled joists, straps, joist hangers, blocking at 12' OC and ceiling backing for drywall. Based on larger jobs with simple joist layouts set 24" OC and pre-cut blocking supplied by the general contractor. Add the cost of floor beams, if required. Costs per square foot of horizontal joist area. More complex jobs with shorter runs may cost 50% more.

| | | Unit | | Labor | Total |
|---|---|---|---|---|---|
| Install 6" ceiling or floor joists 24" OC | | | | | |
|     600S200-33 (2" flange, 20 gauge) | — | SF | — | .20 | .20 |
|     600S200-43 (2" flange, 18 gauge) | — | SF | — | .20 | .20 |
|     600S200-54 (2" flange, 16 gauge) | — | SF | — | .21 | .21 |
|     600S200-68 (2" flange, 14 gauge) | — | SF | — | .22 | .22 |
| Install 8" ceiling or floor joists 24" OC | | | | | |
|     800S200-43 (2" flange, 18 gauge) | — | SF | — | .20 | .20 |
|     800S200-54 (2" flange, 16 gauge) | — | SF | — | .21 | .21 |
|     800S200-68 (2" flange, 14 gauge) | — | SF | — | .22 | .22 |
|     800S200-97 (2" flange, 12 gauge) | — | SF | — | .24 | .24 |

| | Craft@Hrs | Unit | Material | Labor | Total |
|---|---|---|---|---|---|
| Install 10" ceiling or floor joists 24" OC | | | | | |
| 1000S200-43 (2" flange, 18 gauge) | — | SF | — | .21 | .21 |
| 1000S200-54 (2" flange, 16 gauge) | — | SF | — | .22 | .22 |
| 1000S200-68 (2" flange, 14 gauge) | — | SF | — | .23 | .23 |
| 1000S200-97 (2" flange, 12 gauge) | — | SF | — | .25 | .25 |
| Install 12" ceiling or floor joists 24" OC | | | | | |
| 1200S200-54 (2" flange, 16 gauge) | — | SF | — | .23 | .23 |
| 1200S200-68 (2" flange, 14 gauge) | — | SF | — | .24 | .24 |
| 1200S200-97 (2" flange, 12 gauge) | — | SF | — | .26 | .26 |
| Add for other than 24" joist spacing | | | | | |
| Add for 12" OC spacing | — | SF | — | .16 | .16 |
| Add for 16" OC spacing | — | SF | — | .06 | .06 |
| Add for 20" OC spacing | — | SF | — | .03 | .03 |

**Install plywood floor sheathing on steel joists** Piecework rates
Lay out, cut, fit and install 5/8" or 3/4" tongue and groove plywood floor sheathing, including blocking as required. Based on nailing done with a pneumatic nailer and pins supplied by the general contractor, or screwing done with a collated screw gun and screws supplied by the general contractor. Costs per square foot of sheathing installed.

| | Craft@Hrs | Unit | Material | Labor | Total |
|---|---|---|---|---|---|
| Smaller, cut-up job | — | SF | — | .26 | .26 |
| Larger job, longer runs | — | SF | — | .21 | .21 |
| Add for 1-1/8" sheathing | — | SF | — | .06 | .06 |

**Install structural shear panels on steel studs** Piecework rates
Lay out, cut, fit and install structural 3/8" or 1/2" plywood wall panels. These figures assume steel shear studs were set correctly by others and that panel nails or screws are driven at 4" OC with a pneumatic nailer or collated screwgun. Not including hold-down straps, posts, shear blocking or extra studs. Costs per 4' x 9' panel installed.

| | Craft@Hrs | Unit | Material | Labor | Total |
|---|---|---|---|---|---|
| Small job, many openings, 2nd floor | — | Ea | — | 13.20 | 13.20 |
| Larger job, few openings, 1st floor | — | Ea | — | 8.19 | 8.19 |

**Steel roof trusses** Piecework rates
Setting and anchoring engineered gable and hip steel roof trusses 24" OC on prepared wall top track. These figures assume that lifting equipment is provided by the general contractor (if required) and that the truss supplier provides a fill package, spreader blocks for each plate and the ridge and jack rafters (if required). Includes installation of ceiling backing where required and rat runs at the bottom chord. Costs per square foot of plan area under the truss. Small job assumes multiple California fill between roof surfaces and understacking.

| | Craft@Hrs | Unit | Material | Labor | Total |
|---|---|---|---|---|---|
| Small job, rake fill above a partition wall | — | SF | — | .88 | .88 |
| Larger job, little or no fill or understacking | — | SF | — | .51 | .51 |

**Conventional roof framing** Piecework rates
Calculate lengths, lay out, cut and install steel 8", 10" or 12" common, hip, valley and jack rafters on parallel and horizontal top track. Costs per square foot of plan area under the roof. Slope to 6 in 12.

| | Craft@Hrs | Unit | Material | Labor | Total |
|---|---|---|---|---|---|
| Add for slope over 6 in 12 | — | SF | — | .29 | .29 |
| Steel rafters 12" OC, small job, cut-up roof, few common rafters | | | | | |
| 800S162-33 (8", 1-5/8" flange, 20 gauge) | — | SF | — | 1.78 | 1.78 |
| 800S162-43 (8", 1-5/8" flange, 18 gauge) | — | SF | — | 1.80 | 1.80 |
| 800S162-54 (8", 1-5/8" flange, 16 gauge) | — | SF | — | 1.83 | 1.83 |
| 800S162-68 (8", 1-5/8" flange, 14 gauge) | — | SF | — | 1.91 | 1.91 |
| 800S162-97 (8", 1-5/8" flange, 12 gauge) | — | SF | — | 2.01 | 2.01 |
| 1000S162-43 (10", 1-5/8" flange, 18 gauge) | — | SF | — | 1.85 | 1.85 |

| | Craft@Hrs | Unit | Material | Labor | Total |
|---|---|---|---|---|---|
| 1000S162-54 (10", 1-5/8" flange, 16 gauge) | — | SF | — | 1.87 | 1.87 |
| 1000S162-68 (10", 1-5/8" flange, 14 gauge) | — | SF | — | 1.95 | 1.95 |
| 1000S162-97 (10", 1-5/8" flange, 12 gauge) | — | SF | — | 2.07 | 2.07 |
| 1200S162-54 (12", 1-5/8" flange, 16 gauge) | — | SF | — | 1.93 | 1.93 |
| 1200S162-68 (12", 1-5/8" flange, 14 gauge) | — | SF | — | 2.00 | 2.00 |
| 1200S162-97 (12", 1-5/8" flange, 12 gauge) | — | SF | — | 2.13 | 2.13 |
| **Steel rafters 16" OC, small job, cut-up roof, few common rafters** | | | | | |
| 800S162-33 (8", 1-5/8" flange, 20 gauge) | — | SF | — | 1.50 | 1.50 |
| 800S162-43 (8", 1-5/8" flange, 18 gauge) | — | SF | — | 1.51 | 1.51 |
| 800S162-54 (8", 1-5/8" flange, 16 gauge) | — | SF | — | 1.53 | 1.53 |
| 800S162-68 (8", 1-5/8" flange, 14 gauge) | — | SF | — | 1.61 | 1.61 |
| 800S162-97 (8", 1-5/8" flange, 12 gauge) | — | SF | — | 1.71 | 1.71 |
| 1000S162-43 (10", 1-5/8" flange, 18 gauge) | — | SF | — | 1.56 | 1.56 |
| 1000S162-54 (10", 1-5/8" flange, 16 gauge) | — | SF | — | 1.57 | 1.57 |
| 1000S162-68 (10", 1-5/8" flange, 14 gauge) | — | SF | — | 1.64 | 1.64 |
| 1000S162-97 (10", 1-5/8" flange, 12 gauge) | — | SF | — | 1.76 | 1.76 |
| 1200S162-54 (12", 1-5/8" flange, 16 gauge) | — | SF | — | 1.63 | 1.63 |
| 1200S162-68 (12", 1-5/8" flange, 14 gauge) | — | SF | — | 1.71 | 1.71 |
| 1200S162-97 (12", 1-5/8" flange, 12 gauge) | — | SF | — | 1.79 | 1.79 |
| **Steel rafters 24" OC, small job, cut-up roof, few common rafters** | | | | | |
| 800S162-33 (8", 1-5/8" flange, 20 gauge) | — | SF | — | .89 | .89 |
| 800S162-43 (8", 1-5/8" flange, 18 gauge) | — | SF | — | .90 | .90 |
| 800S162-54 (8", 1-5/8" flange, 16 gauge) | — | SF | — | .91 | .91 |
| 800S162-68 (8", 1-5/8" flange, 14 gauge) | — | SF | — | .95 | .95 |
| 800S162-97 (8", 1-5/8" flange, 12 gauge) | — | SF | — | 1.01 | 1.01 |
| 1000S162-43 (10", 1-5/8" flange, 18 gauge) | — | SF | — | .92 | .92 |
| 1000S162-54 (10", 1-5/8" flange, 16 gauge) | — | SF | — | .93 | .93 |
| 1000S162-68 (10", 1-5/8" flange, 14 gauge) | — | SF | — | .97 | .97 |
| 1000S162-97 (10", 1-5/8" flange, 12 gauge) | — | SF | — | 1.04 | 1.04 |
| 1200S162-54 (12", 1-5/8" flange, 16 gauge) | — | SF | — | .96 | .96 |
| 1200S162-68 (12", 1-5/8" flange, 14 gauge) | — | SF | — | 1.01 | 1.01 |
| 1200S162-97 (12", 1-5/8" flange, 12 gauge) | — | SF | — | 1.08 | 1.08 |
| **Steel rafters 12" OC, larger job, longer runs, nearly all common rafters** | | | | | |
| 800S162-33 (8", 1-5/8" flange, 20 gauge) | — | SF | — | .89 | .89 |
| 800S162-43 (8", 1-5/8" flange, 18 gauge) | — | SF | — | .90 | .90 |
| 800S162-54 (8", 1-5/8" flange, 16 gauge) | — | SF | — | .91 | .91 |
| 800S162-68 (8", 1-5/8" flange, 16 gauge) | — | SF | — | .95 | .95 |
| 800S162-97 (8", 1-5/8" flange, 12 gauge) | — | SF | — | 1.01 | 1.01 |
| 1000S162-43 (10", 1-5/8" flange, 18 gauge) | — | SF | — | .92 | .92 |
| 1000S162-54 (10", 1-5/8" flange, 16 gauge) | — | SF | — | .93 | .93 |
| 1000S162-68 (10", 1-5/8" flange, 14 gauge) | — | SF | — | .97 | .97 |
| 1000S162-97 (10", 1-5/8" flange, 12 gauge) | — | SF | — | 1.04 | 1.04 |
| 1200S162-54 (12", 1-5/8" flange, 16 gauge) | — | SF | — | .96 | .96 |
| 1200S162-68 (12", 1-5/8" flange, 14 gauge) | — | SF | — | 1.01 | 1.01 |
| 1200S162-97 (12", 1-5/8" flange, 12 gauge) | — | SF | — | 1.08 | 1.08 |
| **Steel rafters 16" OC, larger job, longer runs, nearly all common rafters** | | | | | |
| 800S162-33 (8", 1-5/8" flange, 20 gauge) | — | SF | — | .81 | .81 |
| 800S162-43 (8", 1-5/8" flange, 18 gauge) | — | SF | — | .81 | .81 |
| 800S162-54 (8", 1-5/8" flange, 16 gauge) | — | SF | — | .82 | .82 |
| 800S162-68 (8", 1-5/8" flange, 16 gauge) | — | SF | — | .85 | .85 |
| 800S162-97 (8", 1-5/8" flange, 12 gauge) | — | SF | — | .88 | .88 |

| | Craft@Hrs | Unit | Material | Labor | Total |
|---|---|---|---|---|---|
| 1000S162-43 (10", 1-5/8" flange, 18 gauge) | — | SF | — | .83 | .83 |
| 1000S162-54 (10", 1-5/8" flange, 16 gauge) | — | SF | — | .84 | .84 |
| 1000S162-68 (10", 1-5/8" flange, 14 gauge) | — | SF | — | .86 | .86 |
| 1000S162-97 (10", 1-5/8" flange, 12 gauge) | — | SF | — | .90 | .90 |
| 1200S162-54 (12", 1-5/8" flange, 16 gauge) | — | SF | — | .86 | .86 |
| 1200S162-68 (12", 1-5/8" flange, 14 gauge) | — | SF | — | .88 | .88 |
| 1200S162-97 (12", 1-5/8" flange, 12 gauge) | — | SF | — | .92 | .92 |

Steel rafters 24" OC, larger job, longer runs, nearly all common rafters

| | Craft@Hrs | Unit | Material | Labor | Total |
|---|---|---|---|---|---|
| 800S162-33 (8", 1-5/8" flange, 20 gauge) | — | SF | — | .52 | .52 |
| 800S162-43 (8", 1-5/8" flange, 18 gauge) | — | SF | — | .53 | .53 |
| 800S162-54 (8", 1-5/8" flange, 16 gauge) | — | SF | — | .53 | .53 |
| 800S162-68 (8", 1-5/8" flange, 16 gauge) | — | SF | — | .55 | .55 |
| 800S162-97 (8", 1-5/8" flange, 12 gauge) | — | SF | — | .58 | .58 |
| 1000S162-43 (10", 1-5/8" flange, 18 gauge) | — | SF | — | .54 | .54 |
| 1000S162-54 (10", 1-5/8" flange, 16 gauge) | — | SF | — | .54 | .54 |
| 1000S162-68 (10", 1-5/8" flange, 14 gauge) | — | SF | — | .57 | .57 |
| 1000S162-97 (10", 1-5/8" flange, 12 gauge) | — | SF | — | .59 | .59 |
| 1200S162-54 (12", 1-5/8" flange, 16 gauge) | — | SF | — | .55 | .55 |
| 1200S162-68 (12", 1-5/8" flange, 14 gauge) | — | SF | — | .58 | .58 |
| 1200S162-97 (12", 1-5/8" flange, 12 gauge) | — | SF | — | .60 | .60 |

**Fascia, wood** Piecework rates
Applied to rafter tails and as a barge rafter on gable ends. Includes trimming the rafter tails to the correct length and installing outlookers at gable ends. Costs per linear foot of 2" x 8" fascia installed.

| | Craft@Hrs | Unit | Material | Labor | Total |
|---|---|---|---|---|---|
| Small job, short runs, with moulding | — | LF | — | 2.96 | 2.96 |
| Larger job, longer runs | — | LF | — | 1.61 | 1.61 |

**Install roof sheathing on steel rafters** Piecework rates
Lay out, cut, fit and install 1/2" or 5/8" plywood roof sheathing, including blocking and 1" x 8" starter board on overhangs as required. Based on nailing done with a pneumatic nailer and nails or collated screwgun and screws supplied by the general contractor. Slope to 6 in 12. Costs per square foot of sheathing installed.

| | Craft@Hrs | Unit | Material | Labor | Total |
|---|---|---|---|---|---|
| Smaller, cut-up hip and valley job | — | SF | — | .27 | .27 |
| Larger job, longer runs | — | SF | — | .21 | .21 |
| Add for slope over 6 in 12 | — | SF | — | .11 | .11 |

**Carpeting, Subcontract** Costs listed are average prices for complete residential jobs and include consultation, measurement, pad, carpet, and professional installation of pad and carpet using tack strips and hot melt tape on seams. Prices can be expected to vary, up or down, by 50% depending on quality and quantity of materials required.

| | Craft@Hrs | Unit | Material | Labor | Total |
|---|---|---|---|---|---|
| Minimum quality, 25 to 35 oz. face weight nylon carpet | | | | | |
| Including a 1/2" (4 lb density) rebond pad | — | SY | — | — | 27.60 |
| Medium quality, 35 to 50 oz. face weight nylon carpet | | | | | |
| Including a 1/2" (6 lb density) rebond pad | — | SY | — | — | 39.00 |
| Better quality, 50 (plus) oz. face weight nylon carpet | | | | | |
| Including a 1/2" (6 lb density) rebond pad | — | SY | — | — | 54.10 |
| Wool Berber carpet (large loop) installed | — | SY | — | — | 71.80 |
| Installation cost alone (residential) | — | SY | — | — | 6.76 |
| Installation cost alone (commercial) | — | SY | — | — | 5.72 |
| Add for typical furniture moving | — | SY | — | — | 2.08 |
| Add for waterfall (box steps) stairways | — | Riser | — | — | 7.80 |

| | Craft@Hrs | Unit | Material | Labor | Total |
|---|---|---|---|---|---|
| Add for wrapped steps (open riser), sewn | — | Step | — | — | 10.40 |
| Add for sewn edge treatment on one side | — | Riser | — | — | 12.50 |
| Add for circular stair steps, costs vary depending on frame | | | | | |
| Add for circular stairs, typical cost | — | Step | — | — | 16.70 |

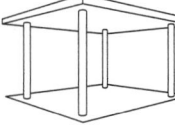

**Carports** Material costs include all hardware. Labor is based on bolting posts to an existing concrete slab. Steel frame, 4" x 4" galvanized 14 gauge posts, 14 gauge purlins, 26 gauge rib metal roofing, 90 MPH wind load.

| | Craft@Hrs | Unit | Material | Labor | Total |
|---|---|---|---|---|---|
| 18' wide x 18' long x 8' high, four posts, 20 PSF snow load | | | | | |
|   Natural zinc finish | B1@8.00 | Ea | 1,470.00 | 266.00 | 1,736.00 |
|   White enamel finish | B1@8.00 | Ea | 1,680.00 | 266.00 | 1,946.00 |
| 18' wide x 18' long x 8' high, six posts, 40 PSF snow load | | | | | |
|   Natural zinc finish | B1@9.00 | Ea | 1,780.00 | 300.00 | 2,080.00 |
|   White enamel finish | B1@9.00 | Ea | 2,020.00 | 300.00 | 2,320.00 |
| 36' wide x 18' long x 8' high, six posts, 20 PSF snow load | | | | | |
|   Natural zinc finish | B1@10.0 | Ea | 2,120.00 | 333.00 | 2,453.00 |
|   White enamel finish | B1@10.0 | Ea | 2,420.00 | 333.00 | 2,753.00 |
| 36' wide x 18' long x 8' high, nine posts, 40 PSF snow load | | | | | |
|   Natural zinc finish | B1@11.5 | Ea | 2,400.00 | 383.00 | 2,783.00 |
|   White enamel finish | B1@11.5 | Ea | 2,750.00 | 383.00 | 3,133.00 |
| 15' wide x 35' long x 15' high, ten posts, 40 PSF snow load | | | | | |
|   Natural zinc finish | B1@13.0 | Ea | 2,800.00 | 433.00 | 3,233.00 |
|   White enamel finish | B1@13.0 | Ea | 3,150.00 | 433.00 | 3,583.00 |

**Caulking** Material costs are typical costs for bead diameter listed. Figures in parentheses indicate approximate coverage including 5% waste. Labor costs are for good quality application on smooth to slightly irregular surfaces. Per LF of bead length.

| | Craft@Hrs | Unit | Material | Labor | Total |
|---|---|---|---|---|---|
| Concrete sealant with polyurethane | | | | | |
|   Concrete sealant, per ounce | — | Oz | .59 | — | .59 |
|   1/4" (2.91 LF per fluid oz.) | BC@.025 | LF | .20 | .92 | 1.12 |
|   3/8" (1.29 LF per fluid oz.) | BC@.030 | LF | .46 | 1.11 | 1.57 |
|   1/2" (.728 LF per fluid oz.) | BC@.033 | LF | .81 | 1.22 | 2.03 |
| Acoustical caulk, flexible, sound deadening | | | | | |
|   Acoustical caulk, per ounce | — | Oz | 2.49 | — | 2.49 |
|   1/8" (11.6 LF per fluid oz.) | BC@.018 | LF | .21 | .66 | .87 |
|   1/4" (2.91 LF per fluid oz.) | BC@.025 | LF | .85 | .92 | 1.77 |
|   3/8" (1.29 LF per fluid oz.) | BC@.030 | LF | 1.93 | 1.11 | 3.04 |
|   1/2" (.728 LF per fluid oz.) | BC@.033 | LF | 3.42 | 1.22 | 4.64 |
| Acrylic latex caulk, premium quality, 35 year life expectancy | | | | | |
|   Acrylic latex caulk, per ounce | — | Oz | .21 | — | .21 |
|   1/8" (11.6 LF per fluid oz.) | BC@.018 | LF | .02 | .66 | .68 |
|   1/4" (2.91 LF per fluid oz.) | BC@.025 | LF | .07 | .92 | .99 |
|   3/8" (1.29 LF per fluid oz.) | BC@.030 | LF | .17 | 1.11 | 1.28 |
|   1/2" (.728 LF per fluid oz.) | BC@.033 | LF | .29 | 1.22 | 1.51 |
| Painters acrylic latex caulk, paintable in 30 minutes | | | | | |
|   Painters acrylic latex caulk, per ounce | — | Oz | .13 | — | .13 |
|   1/8" (11.6 LF per fluid oz.) | BC@.018 | LF | .01 | .66 | .67 |
|   1/4" (2.91 LF per fluid oz.) | BC@.025 | LF | .04 | .92 | .96 |
|   3/8" (1.29 LF per fluid oz.) | BC@.030 | LF | .10 | 1.11 | 1.21 |
|   1/2" (.728 LF per fluid oz.) | BC@.033 | LF | .18 | 1.22 | 1.40 |

# Caulking

| | Craft@Hrs | Unit | Material | Labor | Total |
|---|---|---|---|---|---|
| **Elastomeric latex sealant. Clearer than silicone** | | | | | |
| Elastomeric latex sealant, per ounce | — | Oz | .60 | — | .60 |
| 1/8" (11.6 LF per fluid oz.) | BC@.018 | LF | .05 | .66 | .71 |
| 1/4" (2.91 LF per fluid oz.) | BC@.025 | LF | .20 | .92 | 1.12 |
| 3/8" (1.29 LF per fluid oz.) | BC@.030 | LF | .46 | 1.11 | 1.57 |
| 1/2" (.728 LF per fluid oz.) | BC@.033 | LF | .82 | 1.22 | 2.04 |
| **Tub & tile caulk, for tubs, showers and sinks** | | | | | |
| Tub & tile caulk, per ounce | — | Oz | .47 | — | .47 |
| 1/8" (11.6 LF per fluid oz.) | BC@.018 | LF | .04 | .66 | .70 |
| 1/4" (2.91 LF per fluid oz.) | BC@.025 | LF | .16 | .92 | 1.08 |
| 3/8" (1.29 LF per fluid oz.) | BC@.030 | LF | .36 | 1.11 | 1.47 |
| 1/2" (.728 LF per fluid oz.) | BC@.033 | LF | .64 | 1.22 | 1.86 |
| **Fire barrier latex caulk, four hour fire rating for pipes up to 30 inches** | | | | | |
| Fire barrier latex caulk, per ounce | — | Oz | 1.40 | — | 1.40 |
| 1/8" (11.6 LF per fluid oz.) | BC@.018 | LF | .12 | .66 | .78 |
| 1/4" (2.91 LF per fluid oz.) | BC@.022 | LF | .48 | .81 | 1.29 |
| 3/8" (1.29 LF per fluid oz.) | BC@.027 | LF | 1.09 | 1.00 | 2.09 |
| 1/2" (.728 LF per fluid oz.) | BC@.030 | LF | 1.92 | 1.11 | 3.03 |
| **Silicone caulk** | | | | | |
| Siliconized acrylic caulk, per ounce | — | Oz | .64 | — | .64 |
| 1/8" (11.6 LF per fluid oz.) | BC@.018 | LF | .06 | .66 | .72 |
| 1/4" (2.91 LF per fluid oz.) | BC@.025 | LF | .22 | .92 | 1.14 |
| 3/8" (1.29 LF per fluid oz.) | BC@.030 | LF | .50 | 1.11 | 1.61 |
| 1/2" (.728 LF per fluid oz.) | BC@.033 | LF | .88 | 1.22 | 2.10 |
| **Marine adhesive/sealant, premium quality** | | | | | |
| Silicon rubber sealant, per ounce | — | Oz | 1.27 | — | 1.27 |
| 1/8" (11.6 LF per fluid oz.) | BC@.043 | LF | .11 | 1.59 | 1.70 |
| 1/4" (2.91 LF per fluid oz.) | BC@.048 | LF | .44 | 1.77 | 2.21 |
| 3/8" (1.29 LF per fluid oz.) | BC@.056 | LF | .99 | 2.07 | 3.06 |
| 1/2" (.728 LF per fluid oz.) | BC@.060 | LF | 1.75 | 2.22 | 3.97 |
| **Anti-algae and mildew-resistant tub caulk, premium quality white or clear silicone** | | | | | |
| Anti-algae tub caulk, per ounce | — | Oz | .41 | — | .41 |
| 1/8" (11.6 LF per fluid oz.) | BC@.018 | LF | .04 | .66 | .70 |
| 1/4" (2.91 LF per fluid oz.) | BC@.022 | LF | .14 | .81 | .95 |
| 3/8" (1.29 LF per fluid oz.) | BC@.027 | LF | .32 | 1.00 | 1.32 |
| 1/2" (.728 LF per fluid oz.) | BC@.030 | LF | .57 | 1.11 | 1.68 |
| **Vinyl window and siding sealant, permanently flexible, paintable** | | | | | |
| Vinyl window and siding sealant, per ounce | — | Oz | .56 | — | .56 |
| 1/8" (11.6 LF per fluid oz.) | BC@.018 | LF | .05 | .66 | .71 |
| 1/4" (2.91 LF per fluid oz.) | BC@.022 | LF | .19 | .81 | 1.00 |
| 3/8" (1.29 LF per fluid oz.) | BC@.027 | LF | .43 | 1.00 | 1.43 |
| 1/2" (.728 LF per fluid oz.) | BC@.030 | LF | .77 | 1.11 | 1.88 |
| **Gutter and flashing sealant for gutters, downspouts, lap joints, flashing, vents, aluminum siding, and other metal joints. Meets ASTM C-1085** | | | | | |
| Gutter and flashing sealant, per ounce | — | Oz | .64 | — | .64 |
| 1/8" (11.6 LF per fluid oz.) | BC@.018 | LF | .06 | .66 | .72 |
| 1/4" (2.91 LF per fluid oz.) | BC@.022 | LF | .22 | .81 | 1.03 |
| 3/8" (1.29 LF per fluid oz.) | BC@.027 | LF | .50 | 1.00 | 1.50 |
| 1/2" (.728 LF per fluid oz.) | BC@.030 | LF | .88 | 1.11 | 1.99 |
| **Polyurethane door, window & siding sealant, for installing or repairing door and window frames** | | | | | |
| Gutter and flashing sealant, per ounce | — | Oz | .56 | — | .56 |
| 1/8" (11.6 LF per fluid oz.) | BC@.018 | LF | .05 | .66 | .71 |
| 1/4" (2.91 LF per fluid oz.) | BC@.022 | LF | .19 | .81 | 1.00 |

| | Craft@Hrs | Unit | Material | Labor | Total |
|---|---|---|---|---|---|
| 3/8" (1.29 LF per fluid oz.) | BC@.027 | LF | .43 | 1.00 | 1.43 |
| 1/2" (.728 LF per fluid oz.) | BC@.030 | LF | .77 | 1.11 | 1.88 |
| Sanded caulk, comes in a variety of colors to match colored grout | | | | | |
| Sanded caulk, per ounce | — | Oz | .65 | — | .65 |
| 1/8" (11.6 LF per fluid oz.) | BC@.018 | LF | .06 | .66 | .72 |
| 1/4" (2.91 LF per fluid oz.) | BC@.022 | LF | .22 | .81 | 1.03 |
| 3/8" (1.29 LF per fluid oz.) | BC@.027 | LF | .51 | 1.00 | 1.51 |
| 1/2" (.728 LF per fluid oz.) | BC@.030 | LF | .90 | 1.11 | 2.01 |
| Tub and wall caulk strip, white, 1-5/8" x 11' | | | | | |
| Tub and wall sealant, per LF | BC@.025 | LF | .57 | .92 | 1.49 |
| Caulking gun, professional type, heavy duty | | | | | |
| Cordless caulk gun | — | Ea | 278.00 | — | 278.00 |
| "Caulk Master" caulking gun | — | Ea | 42.00 | — | 42.00 |
| Caulking gun, economy grade | | | | | |
| 1/10 gallon tube, ratchet rod | — | Ea | 6.58 | — | 6.58 |
| Quart capacity, smooth rod | — | Ea | 9.31 | — | 9.31 |

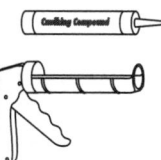

**Cedar Closet Lining** Material costs are per SF covered including 7% waste. Labor is for good quality installation in typical residential closet over studs or gypsum wallboard.

| | Craft@Hrs | Unit | Material | Labor | Total |
|---|---|---|---|---|---|
| Solid cedar planks | | | | | |
| 18 SF package | BC@.060 | SF | 1.30 | 2.22 | 3.52 |
| 37.75 SF package | BC@.060 | SF | 1.30 | 2.22 | 3.52 |

**Ceiling Domes** Complete ceiling dome kits, includes grid, hardware, panels and crating, from flat to 30 degree pitch in 5 degree increments. Prices shown are for best quality. Prices may be as much as 40% lower depending on the requirements. Envel Design Corp.

| | Craft@Hrs | Unit | Material | Labor | Total |
|---|---|---|---|---|---|
| 9' diameter | B1@10.7 | Ea | 3,750.00 | 356.00 | 4,106.00 |
| 10' diameter | B1@10.7 | Ea | 6,200.00 | 356.00 | 6,556.00 |
| 11' diameter | B1@17.2 | Ea | 9,700.00 | 573.00 | 10,273.00 |
| UV stabilized commercial grade acrylic with decorative patterns (Costs vary from $3.90 to $6.50 per SF) | | | | | |
| Typical cost per SF | B1@.100 | SF | 5.58 | 3.33 | 8.91 |
| Custom, dimensional Envelex™ Designs | | | | | |
| Standard sizes and patterns (2' x 2', 2' x 4') | B1@.006 | SF | 34.90 | .20 | 35.10 |
| Irregular shapes | | | | | |
| Oversize and custom designs | B1@.006 | SF | 40.90 | .20 | 41.10 |
| Add for cutting panels | B1@.033 | Ea | — | 1.10 | 1.10 |

**Ceiling Panel Suspension Grids** Typical costs, plain white. Add panel costs below.

| | Craft@Hrs | Unit | Material | Labor | Total |
|---|---|---|---|---|---|
| Main runner, 12' long | — | Ea | 6.04 | — | 6.04 |
| Main runner, 12' long | — | LF | .50 | — | .50 |
| Cross tee, 2' long | — | Ea | 1.43 | — | 1.43 |
| Cross tee, 2' long | — | LF | .71 | — | .71 |
| Cross tee, 4' long | — | Ea | 2.58 | — | 2.58 |
| Cross tee, 4' long | — | LF | .65 | — | .65 |
| Wall mould, 12' long | — | Ea | 3.74 | — | 3.74 |
| Wall mould, 12' long | — | LF | .31 | — | .31 |
| Add for colored | — | % | 10.0 | — | — |
| Add for fire-rated white | — | % | 100.0 | — | — |
| Add for chrome or gold color grid | — | % | 40.0 | — | — |
| Suspension grid system including runners, tees, wall mould, hooks and hanging wire (but no tile or lighting fixtures) | | | | | |
| 2' x 2' grid | BC@.017 | SF | .73 | .63 | 1.36 |
| 2' x 4' grid | BC@.015 | SF | .56 | .55 | 1.11 |
| Add for under 400 SF job | BC@.007 | SF | — | .26 | .26 |

# Ceiling Panels, Suspension Grids

| | Craft@Hrs | Unit | Material | Labor | Total |
|---|---|---|---|---|---|
| Troffer lighting fixtures for ceiling suspension grid, includes installation, connection, lens and tubes | | | | | |
| 2 tube 2' x 2' fixture | BE@1.00 | Ea | 41.80 | 39.60 | 81.40 |
| 2 tube 2' x 4' fixture | BE@1.00 | Ea | 72.60 | 39.60 | 112.20 |
| 3 tube 2' x 4' fixture | BE@1.00 | Ea | 81.60 | 39.60 | 121.20 |
| 4 tube 2' x 4' fixture | BE@1.00 | Ea | 53.30 | 39.60 | 92.90 |
| Light-diffusing panels for installation in 2' x 4' suspension system under fluorescent lighting. | | | | | |
| Styrene | | | | | |
| .065" thick, cracked ice, 2' x 2', white | BC@.020 | Ea | 3.21 | .74 | 3.95 |
| .065" thick, cracked ice, 2' x 4', white | BC@.029 | Ea | 9.06 | 1.07 | 10.13 |
| .095" thick, prismatic, 2' x 2', clear | BC@.020 | Ea | 3.06 | .74 | 3.80 |
| .095" thick, prismatic, 2' x 2', white | BC@.020 | Ea | 3.06 | .74 | 3.80 |
| .095" thick, prismatic, 2' x 4', clear | BC@.029 | Ea | 5.09 | 1.07 | 6.16 |
| .095" thick, prismatic, 2' x 4', white | BC@.029 | Ea | 5.09 | 1.07 | 6.16 |
| .095" thick, Victorian pattern 2' x 4', | BC@.029 | Ea | 14.80 | 1.07 | 15.87 |
| Egg crate louvers, 1/2" x 1/2" x 1/2" cell size | | | | | |
| Frost | BC@.188 | Ea | 8.38 | 6.94 | 15.32 |
| White styrene | BC@.188 | Ea | 11.70 | 6.94 | 18.64 |
| Silver finish | BC@.188 | Ea | 20.70 | 6.94 | 27.64 |
| Metallic styrene | BC@.188 | Ea | 25.80 | 6.94 | 32.74 |
| Anodized aluminum | BC@.188 | Ea | 50.20 | 6.94 | 57.14 |
| Parabolic louvers, 5/8" x 5/8" x 7/16" cell size | | | | | |
| Silver styrene | BC@.188 | Ea | 31.60 | 6.94 | 38.54 |
| Gold styrene | BC@.188 | Ea | 33.30 | 6.94 | 40.24 |
| Silver acrylic | BC@.188 | Ea | 53.90 | 6.94 | 60.84 |
| Gold acrylic | BC@.188 | Ea | 53.90 | 6.94 | 60.84 |
| Stained glass look, one dimension, 2' x 2' | | | | | |
| Standard colors | BC@.033 | Ea | 37.40 | 1.22 | 38.62 |
| Antiqued colors | BC@.035 | Ea | 49.90 | 1.29 | 51.19 |

**Ceiling Panels, Lay-in Type** Costs per SF of ceiling area covered. No waste included. Add suspension grid system cost from preceding pages.

| | Craft@Hrs | Unit | Material | Labor | Total |
|---|---|---|---|---|---|
| Acoustical rated suspended ceiling panels | | | | | |
| 2' x 2', random texture, "A" fire rated | BC@.012 | SF | .59 | .44 | 1.03 |
| 2' x 2' x 5/8", USG "5th Avenue" | BC@.012 | SF | .50 | .44 | .94 |
| 2' x 2', random texture, "A" fire rated | BC@.012 | SF | .95 | .44 | 1.39 |
| 2' x 2', random textured, Armstrong | BC@.012 | SF | .81 | .44 | 1.25 |
| 2' x 2' x 5/8", slant edge, USG "Alpine" | BC@.012 | SF | .91 | .44 | 1.35 |
| 2' x 2' "A" fire rated, with Bioguard™ | BC@.012 | SF | 1.29 | .44 | 1.73 |
| 2' x 2', "A" fire rated, Armstrong "Sahara" | BC@.012 | SF | 1.35 | .44 | 1.79 |
| 2' x 2' x 5/8", square edge, USG "Astro" | BC@.012 | SF | 1.09 | .44 | 1.53 |
| 2' x 2', Armstrong "Bravada II" | BC@.012 | SF | 1.83 | .44 | 2.27 |
| 2' x 2', Amstrong "Classic" | BC@.012 | SF | 1.83 | .44 | 2.27 |
| 2' x 2' x 3/4", foil back, USG "Cheyenne" | BC@.012 | SF | 1.64 | .44 | 2.08 |
| 2' x 2' x 3/4", USG "Luna", fine textured | BC@.012 | SF | 1.66 | .44 | 2.10 |
| 2' x 4', textured | BC@.007 | SF | .50 | .26 | .76 |
| 2' x 4', textured, class "A" fire rated | BC@.007 | SF | .52 | .26 | .78 |
| 2' x 4' x 5/8", USG "5th Avenue" | BC@.007 | SF | .48 | .26 | .74 |
| 2' x 4' x 5/8", fissured fire code | BC@.007 | SF | .74 | .26 | 1.00 |
| 2' x 4', square edge, scored | BC@.007 | SF | .63 | .26 | .89 |
| 2' x 4', fiberglass class "A" fire rated | BC@.007 | SF | .75 | .26 | 1.01 |
| 2' x 4' x 5/8", "A" fire rated, USG "5th Ave" | BC@.007 | SF | .62 | .26 | .88 |
| 2' x 4' x 5/8", perforated, USG "Tabaret" | BC@.007 | SF | .76 | .26 | 1.02 |

| | Craft@Hrs | Unit | Material | Labor | Total |
|---|---|---|---|---|---|
| **Non-acoustical suspended ceiling panels** | | | | | |
| 2' x 2' x 5/8", USG "ClimaPlus" | BC@.012 | SF | .39 | .44 | .83 |
| 2' x 2' x 5/8", square edge, USG "Radar" | BC@.012 | SF | .55 | .44 | .99 |
| 2' x 2', tapered edge, USG "Radar" | BC@.012 | SF | .73 | .44 | 1.17 |
| 2' x 2', Armstrong "Brighton" | BC@.012 | SF | .79 | .44 | 1.23 |
| 2' x 2', Armstrong "Royal Oak" | BC@.012 | SF | 1.19 | .44 | 1.63 |
| 2' x 2' x 3/4", fissured, re-panel | BC@.012 | SF | 1.02 | .44 | 1.46 |
| 2' x 2' x 5/8", USG "Ceramic ClimaPlus" | BC@.012 | SF | 1.21 | .44 | 1.65 |
| 2' x 2' x 3/4", USG "Saville Row" | BC@.012 | SF | 1.46 | .44 | 1.90 |
| 2' x 2', grid pattern, Armstrong "Prestige" | BC@.012 | SF | .35 | .44 | .79 |
| 2' x 4', mineral fiber, USG "Plateau" | BC@.007 | SF | .36 | .26 | .62 |
| 2' x 4' x 5/8", square edge, USG "Radar" | BC@.007 | SF | .47 | .26 | .73 |
| 2' x 4' x 9/16", mineral, USG "Stonehurst" | BC@.007 | SF | .48 | .26 | .74 |
| 2' x 4', plain white, square edge | BC@.007 | SF | .66 | .26 | .92 |
| 2' x 4', Armstrong "Grenoble" | BC@.007 | SF | .56 | .26 | .82 |
| 2' x 4' x 5/8", fissured fiberglass | BC@.007 | SF | .69 | .26 | .95 |
| 2' x 4' x 5/8", Armstrong "Esprit" | BC@.007 | SF | .69 | .26 | .95 |
| 2' x 4' x 3/4", slant edge, USG "Illusions" | BC@.007 | SF | .73 | .26 | .99 |
| 2' x 4' random texture, "A" fire rated | BC@.007 | SF | .71 | .26 | .97 |
| 2' x 4' clean room, non-perforated | BC@.007 | SF | 2.04 | .26 | 2.30 |

Fiberglass reinforced 24" x 24" gypsum panels for T-bar systems. Weight is 3 to 4 pounds each, depending on pattern

| | Craft@Hrs | Unit | Material | Labor | Total |
|---|---|---|---|---|---|
| White painted finish | BC@.015 | Ea | 6.35 | .55 | 6.90 |
| Wood grain finish | BC@.020 | Ea | 12.70 | .74 | 13.44 |

**Ceiling Pans, Metal** Flat, 2' x 2' pans
Aluminum with concealed grid

| | Craft@Hrs | Unit | Material | Labor | Total |
|---|---|---|---|---|---|
| 2' x 2' painted | BC@.083 | SF | 7.20 | 3.07 | 10.27 |
| 2' x 2' polished | BC@.083 | SF | 10.50 | 3.07 | 13.57 |
| Stainless steel with concealed grid | | | | | |
| 2' x 2', polished clear mirror | BC@.097 | SF | 15.70 | 3.58 | 19.28 |
| Stainless steel with exposed grid | | | | | |
| 2' x 2', polished clear mirror | BC@.074 | SF | 8.96 | 2.73 | 11.69 |

**Ceiling Tile** Class "C" panels for non-acoustical and non-fire rated applications. Tongue and groove 12" x 12" tile nailed or stapled to ceilings. Costs per SF of ceiling area covered including fasteners, trim and typical waste.

| | Craft@Hrs | Unit | Material | Labor | Total |
|---|---|---|---|---|---|
| 12" x 12", Washable, White | BC@.035 | SF | .72 | 1.29 | 2.01 |
| 12" x 12", Armstrong "Grenoble" | BC@.036 | SF | 1.03 | 1.33 | 2.36 |
| 12" x 12", USG lace wood fiber | BC@.035 | SF | .84 | 1.29 | 2.13 |
| 12" x 12", USG white wood fiber | BC@.035 | SF | .95 | 1.29 | 2.24 |
| 12' x 12", USG "Tivoli" | BC@.035 | SF | .94 | 1.29 | 2.23 |
| 12" x 12", Armstrong "Glenwood" | BC@.035 | SF | .82 | 1.29 | 2.11 |
| 12" x 12", Armstrong tin ceiling design | BC@.035 | SF | 1.69 | 1.29 | 2.98 |
| 6" x 48", Armstrong classic plank | BC@.054 | SF | 1.35 | 1.99 | 3.34 |

**Ceilings, Tin look** Various patterns
Embossed panels

| | Craft@Hrs | Unit | Material | Labor | Total |
|---|---|---|---|---|---|
| 2' x 2' square panels, cost each | B1@.147 | Ea | 13.80 | 4.90 | 18.70 |
| Embossed plates | | | | | |
| 12" x 12" plates, cost per SF | B1@.042 | SF | 2.04 | 1.40 | 3.44 |
| 12" x 24" plates, cost per SF | B1@.040 | SF | 2.33 | 1.33 | 3.66 |
| 24" x 24" plates, cost per SF | B1@.037 | SF | 3.25 | 1.23 | 4.48 |
| 24" x 48" plates, cost per SF | B1@.035 | SF | 3.05 | 1.17 | 4.22 |

# Cement

| | Craft@Hrs | Unit | Material | Labor | Total |
|---|---|---|---|---|---|

**Cement** See also Adhesives, Aggregates, and Concrete

| | Craft@Hrs | Unit | Material | Labor | Total |
|---|---|---|---|---|---|
| Portland cement, type I and II, 94 lb sack | — | Sa | 10.10 | — | 10.10 |
| Plastic cement, 94 lb sack | — | Sa | 9.98 | — | 9.98 |
| High early strength cement, 60 lb sack | — | Sa | 4.93 | — | 4.93 |
| White cement, 94 lb sack | — | Sa | 19.10 | — | 19.10 |
| Concrete mix, fast setting, 50 lb sack | — | Sa | 4.98 | — | 4.98 |
| Light-weight concrete, 2/3 C.F sack | — | Sa | 3.42 | — | 3.42 |
| Masonry cement, type N, 70 lb sack | — | Sa | 8.62 | — | 8.62 |
| Mortar mix, 60 lb sack | — | Sa | 4.34 | — | 4.34 |
| Non-shrink construction grout, 50 lb sack | — | Sa | 11.40 | — | 11.40 |
| Fiber-reinforced stucco mix, 90 lb sack | — | Sa | 15.10 | — | 15.10 |
| Plaster patch, 25 lb sack | — | Sa | 15.10 | — | 15.10 |
| Blacktop cold patch mix, 50 lb sack | — | Sa | 6.52 | — | 6.52 |
| Sand, 60 lb sack | — | Sa | 5.17 | — | 5.17 |
| Deduct for pallet quantities | — | % | -15.0 | — | — |

**Cleaning, Final** Interior surfaces, plumbing fixtures and windows. Includes removing construction debris and grime. Add for removing paint splatter or mastic. Work performed using hand tools and commercial non-toxic cleaning products or solvent. Use $80.00 as a minimum job charge.

| | Craft@Hrs | Unit | Material | Labor | Total |
|---|---|---|---|---|---|
| Cabinets, per linear foot of cabinet face. Wall or base cabinet, (bathroom, kitchen or linen) | | | | | |
|   Cleaned inside and outside | BL@.048 | LF | .08 | 1.42 | 1.50 |
|   Add for polish, exposed surface only | BL@.033 | LF | .06 | .98 | 1.04 |
| Countertops, per linear foot of front, bathroom, family room or kitchen | | | | | |
|   Marble or granite | BL@.040 | LF | .06 | 1.19 | 1.25 |
|   Tile, clean (including grout joints) | BL@.046 | LF | .08 | 1.36 | 1.44 |
|   Vinyl or plastic | BL@.035 | LF | .07 | 1.04 | 1.11 |
| Painted surfaces, wipe down by hand | | | | | |
|   Base board, trim or moulding | BL@.003 | LF | .06 | .09 | .15 |
|   Cased openings, per side cleaned | BL@.060 | Ea | .66 | 1.78 | 2.44 |
|   Ceilings, painted | BL@.003 | SF | .06 | .09 | .15 |
|   Walls, painted or papered | BL@.002 | SF | .06 | .06 | .12 |
|   Window sills | BL@.004 | LF | .06 | .12 | .18 |
| Doors, painted or stained | | | | | |
|   Cleaned two sides, including casing | BL@.150 | Ea | .08 | 4.45 | 4.53 |
|   Add for polishing two sides | BL@.077 | Ea | .06 | 2.28 | 2.34 |
| Floors | | | | | |
|   Hardwood floor, clean, wax and polish | BL@.025 | SF | .11 | .74 | .85 |
|   Mop floor | BL@.002 | SF | .03 | .06 | .09 |
|   Vacuum carpet or sweep floor | BL@.002 | SF | .03 | .06 | .09 |
| Plumbing fixtures | | | | | |
|   Bar sink, including faucet | BL@.380 | Ea | .16 | 11.30 | 11.46 |
|   Bathtubs, including faucets & shower heads | BL@.750 | Ea | .16 | 22.30 | 22.46 |
|   Kitchen sink, including faucets | BL@.500 | Ea | .16 | 14.80 | 14.96 |
|   Lavatories, including faucets | BL@.500 | Ea | .16 | 14.80 | 14.96 |
|   Toilets, including seat and tank | BL@.600 | Ea | .16 | 17.80 | 17.96 |
| Shower stalls, including walls, floor, shower rod and shower head | | | | | |
|   Three-sided | BL@1.80 | Ea | .26 | 53.40 | 53.66 |
|   Add for glass doors | BL@.500 | Ea | .08 | 14.80 | 14.88 |
| Sliding glass doors, clean and polish both sides including cleaning of track and screen | | | | | |
|   Up to 8' wide | BL@.650 | Ea | .08 | 19.30 | 19.38 |
|   Over 8' wide | BL@.750 | Ea | .13 | 22.30 | 22.43 |

| | Craft@Hrs | Unit | Material | Labor | Total |
|---|---|---|---|---|---|
| **Windows, clean and polish, inside and outside, first or second floors** | | | | | |
| Bay windows | BL@.750 | Ea | .13 | 22.30 | 22.43 |
| Sliding, casement or picture windows | BL@.300 | Ea | .08 | 8.90 | 8.98 |
| Jalousie type | BL@.350 | Ea | .13 | 10.40 | 10.53 |
| Remove window screen, hose wash, replace | BL@.500 | Ea | — | 14.80 | 14.80 |
| **Typical final cleanup for new residential construction** | | | | | |
| Per SF of floor (no glass cleaning) | BL@.004 | SF | .12 | .12 | .24 |

## Clothesline Units

| | Craft@Hrs | Unit | Material | Labor | Total |
|---|---|---|---|---|---|
| **Retractable type, 5-line, indoor-outdoor, one end is wall-mounted, lines retract into aluminum case** | | | | | |
| Cost per unit as described | B1@.860 | Ea | 64.00 | 28.60 | 92.60 |
| **Two "T"-shaped galvanized steel pole, set in concrete** | | | | | |
| 4 prong (not including cost of concrete) | BL@1.91 | Ea | 225.00 | 56.70 | 281.70 |
| Concrete mix, 60 lb sack | — | Sa | 3.15 | — | 3.15 |
| **Umbrella-type, 8' galvanized steel center post, requires 10' x 10' area, folds for storage** | | | | | |
| One post, set in concrete | BL@.885 | Ea | 54.00 | 26.30 | 80.30 |
| Concrete mix, 60 lb sack | — | Sa | 3.15 | — | 3.15 |

## Columns and Porch Posts  Colonial style round wood, standard cap and base, hemlock/pine

| | Craft@Hrs | Unit | Material | Labor | Total |
|---|---|---|---|---|---|
| 8" x 8' | B1@2.00 | Ea | 143.00 | 66.60 | 209.60 |
| 10" x 10' | B1@2.00 | Ea | 221.00 | 66.60 | 287.60 |
| Add for fluted wood, either of above | — | Ea | 32.00 | — | 32.00 |
| **Round aluminum, standard cap and base** | | | | | |
| 8" x 9' | B1@2.00 | Ea | 166.00 | 66.60 | 232.60 |
| 10" x 10' | B1@2.00 | Ea | 258.00 | 66.60 | 324.60 |
| 12" x 12' | B1@3.00 | Ea | 433.00 | 99.90 | 532.90 |
| Add for ornamental wood cap, typical price | — | Ea | 61.90 | — | 61.90 |
| Add for ornamental aluminum cap, typical | — | Ea | 60.80 | — | 60.80 |
| **Turncraft round columns** | | | | | |
| 8" x 8' fluted | B1@2.00 | Ea | 181.00 | 66.60 | 247.60 |
| 8" x 8' plain | B1@2.00 | Ea | 119.00 | 66.60 | 185.60 |
| 10" x 10' plain | B1@2.00 | Ea | 203.00 | 66.60 | 269.60 |
| 10" x 12' plain | B1@3.00 | Ea | 275.00 | 99.90 | 374.90 |
| Add for cap | — | Ea | 16.10 | — | 16.10 |
| Add for base | — | Ea | 20.40 | — | 20.40 |
| **Porch posts, clear laminated west coast hemlock, solid turned** | | | | | |
| 3-1/4" x 8' | B1@1.04 | Ea | 106.00 | 34.60 | 140.60 |
| 4-1/4" x 8' | B1@1.18 | Ea | 89.00 | 39.30 | 128.30 |
| 5-1/4" x 8' | B1@1.28 | Ea | 146.00 | 42.60 | 188.60 |

## Concrete  Ready-mix delivered by truck. Typical prices for most cities. Includes delivery up to 20 miles for 10 CY or more, 3" to 4" slump. Material cost only, no placing or pumping included. See forming and finishing costs on the following pages.

| | Craft@Hrs | Unit | Material | Labor | Total |
|---|---|---|---|---|---|
| **Footing and foundation concrete, 1-1/2" aggregate** | | | | | |
| 2,000 PSI, 4.8 sack mix | — | CY | 94.40 | — | 94.40 |
| 2,500 PSI, 5.2 sack mix | — | CY | 96.90 | — | 96.90 |
| 3,000 PSI, 5.7 sack mix | — | CY | 98.60 | — | 98.60 |
| 3,500 PSI, 6.3 sack mix | — | CY | 102.00 | — | 102.00 |
| 4,000 PSI, 6.9 sack mix | — | CY | 105.00 | — | 105.00 |
| **Slab, sidewalk, and driveway concrete, 1" aggregate** | | | | | |
| 2,000 PSI, 5.0 sack mix | — | CY | 96.00 | — | 96.00 |
| 2,500 PSI, 5.5 sack mix | — | CY | 98.60 | — | 98.60 |

# Concrete

| | Craft@Hrs | Unit | Material | Labor | Total |
|---|---|---|---|---|---|
| 3,000 PSI, 6.0 sack mix | — | CY | 100.00 | — | 100.00 |
| 3,500 PSI, 6.6 sack mix | — | CY | 102.00 | — | 102.00 |
| 4,000 PSI, 7.1 sack mix | — | CY | 104.00 | — | 104.00 |
| Pea gravel pump mix / grout mix, 3/8" aggregate | | | | | |
| 2,000 PSI, 6.0 sack mix | — | CY | 104.00 | — | 104.00 |
| 2,500 PSI, 6.5 sack mix | — | CY | 105.00 | — | 105.00 |
| 3,000 PSI, 7.2 sack mix | — | CY | 109.00 | — | 109.00 |
| 3,500 PSI, 7.9 sack mix | — | CY | 112.00 | — | 112.00 |
| 5,000 PSI, 8.5 sack mix | — | CY | 119.00 | — | 119.00 |
| Fiber mesh integral concrete reinforcing, polypropylene fiber, add per cubic yard of concrete | | | | | |
| Short fiber polypropylene (slabs) | — | CY | 5.00 | — | 5.00 |
| Long fiber polypropylene (structural) | — | CY | 7.00 | — | 7.00 |
| Extra delivery costs for ready-mix concrete | | | | | |
| Add for delivery over 20 miles | — | Mile | 8.00 | — | 8.00 |
| Add for standby charge in excess of 5 minutes per CY delivered | | | | | |
| per minute of extra time | — | Ea | 1.50 | — | 1.50 |
| Add for less than 10 CY per load delivered | — | CY | 25.00 | — | 25.00 |
| Add for Saturday delivery, per CY | — | CY | 4.00 | — | 4.00 |
| Extra costs for non-standard aggregates | | | | | |
| Add for lightweight aggregate, typical | — | CY | 44.00 | — | 44.00 |
| Add for lightweight aggregate, pump mix | — | CY | 46.00 | — | 46.00 |
| Add for granite aggregate, typical | — | CY | 6.90 | — | 6.90 |
| Extra costs for non-standard mix additives | | | | | |
| High early strength 5 sack mix | — | CY | 12.50 | — | 12.50 |
| High early strength 6 sack mix | — | CY | 16.50 | — | 16.50 |
| Add for white cement (architectural) | — | CY | 57.00 | — | 57.00 |
| Add for 1% calcium chloride | — | CY | 1.60 | — | 1.60 |
| Add for chemical compensated shrinkage | — | CY | 21.00 | — | 21.00 |
| Add for super plasticized mix, 7" - 8" slump | — | % | 8.0 | — | — |

Add for colored concrete, typical prices. The ready-mix supplier charges for cleanup of the ready-mix truck used for delivery of colored concrete. The usual practice is to provide one truck per day for delivery of all colored concrete for a particular job. Add the cost below to the cost per cubic yard for the design mix required. Also add the cost for truck cleanup.

| | Craft@Hrs | Unit | Material | Labor | Total |
|---|---|---|---|---|---|
| Adobe | — | CY | 20.00 | — | 20.00 |
| Black | — | CY | 31.50 | — | 31.50 |
| Blended red | — | CY | 17.90 | — | 17.90 |
| Brown | — | CY | 23.10 | — | 23.10 |
| Green | — | CY | 35.70 | — | 35.70 |
| Yellow | — | CY | 24.20 | — | 24.20 |
| Colored concrete, truck cleanup, per day | — | LS | 78.80 | — | 78.80 |

**Concrete Expansion Joints** Fiber or asphaltic-felt sided. Per linear foot

| Depth | Wall (thickness) | | |
|---|---|---|---|
| | 1/4" | 3/8" | 1/2" |
| 3" deep | .18 | .23 | .30 |
| 3-1/2" deep | .22 | .30 | .37 |
| 4" deep | .43 | .57 | .72 |
| 6" deep | .59 | .79 | .99 |

| | Craft@Hrs | Unit | Material | Labor | Total |
|---|---|---|---|---|---|
| Labor to install concrete expansion joints | | | | | |
| 3" to 6" width, 1/4" to 1/2" thick | B1@.030 | LF | — | 1.00 | 1.00 |

| | Craft@Hrs | Unit | Material | Labor | Total |
|---|---|---|---|---|---|

**Concrete Snap Ties**  Short end ties, price each based on cartons of 200

| | Craft@Hrs | Unit | Material | Labor | Total |
|---|---|---|---|---|---|
| 6" form | — | Ea | .99 | — | .99 |
| 8" form | — | Ea | 1.03 | — | 1.03 |
| 10" form | — | Ea | 1.07 | — | 1.07 |
| 12" form | — | Ea | 1.06 | — | 1.06 |
| 14" form | — | Ea | 1.16 | — | 1.16 |
| 16" form | — | Ea | 1.30 | — | 1.30 |
| 18" form | — | Ea | 1.36 | — | 1.36 |
| 24" form | — | Ea | 1.37 | — | 1.37 |
| 6" wedge form tie | — | Ea | .45 | — | .45 |
| 8" wedge form tie | — | Ea | .47 | — | .47 |
| 12" wedge form tie | — | Ea | .80 | — | .80 |
| 14" wedge form tie | — | Ea | .84 | — | .84 |
| 16" wedge form tie | — | Ea | 1.16 | — | 1.16 |
| 24" wedge form tie | — | Ea | 2.19 | — | 2.19 |
| Add for long end ties | — | % | 12.0 | — | — |

**Concrete Form Excavation**  By hand using a pick and shovel, to 3'-0" deep. For wall footings, grade beams and column footings. Digging and one throw only, no disposal or back filling included. Per CY. See also the Excavation section.

| | Craft@Hrs | Unit | Material | Labor | Total |
|---|---|---|---|---|---|
| Light soil | BL@1.10 | CY | — | 32.60 | 32.60 |
| Average soil | BL@1.65 | CY | — | 49.00 | 49.00 |
| Heavy soil or loose rock | BL@2.14 | CY | — | 63.50 | 63.50 |

**Concrete Form Stakes**  Steel, 3/8 thick x 1-1/2" wide. Costs per stake.

| | Craft@Hrs | Unit | Material | Labor | Total |
|---|---|---|---|---|---|
| 18" flat | — | Ea | 3.82 | — | 3.82 |
| 24" flat | — | Ea | 4.52 | — | 4.52 |
| 36" flat | — | Ea | 6.49 | — | 6.49 |
| 48" flat | — | Ea | 7.23 | — | 7.23 |
| 18" round | — | Ea | 2.91 | — | 2.91 |
| 24" round | — | Ea | 3.92 | — | 3.92 |
| 36" round | — | Ea | 5.33 | — | 5.33 |

**Insulated Concrete Forms (ICFs)**  Expanded polystyrene (EPS) forms for residential and commercial reinforced concrete walls 4", 6" or 8" thick. Based on standard sized block connected with plastic clips and reinforced with steel bars before cores are filled with concrete. Costs assume forms are left in place as wall insulation and that no concrete finishing is required. See www.eco-block.com. The crew is a carpenter and a laborer. Material costs include EPS forms and connectors as described but no concrete or rebar.

| | Craft@Hrs | Unit | Material | Labor | Total |
|---|---|---|---|---|---|
| Standard block, 5.3 SF per block | | | | | |
| 16" H, 48" L, with 12 panel connectors | — | Ea | 15.80 | — | 15.80 |
| Corner blocks, each includes 8 panel connectors | | | | | |
| 4" cavity, 16" H, 32" L | — | Ea | 15.70 | — | 15.70 |
| 6" cavity, 16" H, 32" L | — | Ea | 15.70 | — | 15.70 |
| 8" cavity, 16" H, 32" L | — | Ea | 15.70 | — | 15.70 |
| Connectors | | | | | |
| 4" connectors, per box of 1000 | — | Ea | 163.00 | — | 163.00 |
| 6" connectors, per box of 500 | — | Ea | 81.20 | — | 81.20 |
| 8" connectors, per box of 500 | — | Ea | 81.20 | — | 81.20 |
| 10" connectors, per box of 500 | — | Ea | 81.20 | — | 81.20 |
| Splice connectors, per box of 1000 | — | Ea | 73.10 | — | 73.10 |
| Brick ledge rails, per box of 50 rails | — | Ea | 36.70 | — | 36.70 |
| 1" extension "Texas T", per box of 500 | — | Ea | 33.50 | — | 33.50 |

# Concrete Formwork

| | Craft@Hrs | Unit | Material | Labor | Total |
|---|---|---|---|---|---|
| 16" 90 degree panel connectors, per box of 40 | — | Ea | 108.00 | — | 108.00 |
| 24" 90 degree panel connectors, per box of 40 | — | Ea | 135.20 | — | 135.20 |
| 16" 45 degree panel connectors, per box of 40 | — | Ea | 108.00 | — | 108.00 |
| 24" 45 degree panel connectors, per box of 40 | — | Ea | 135.20 | — | 135.20 |

Material and labor for EPS forms. These figures assume footings or concrete slab are in place and ready to accept the first course of forms. Material includes forms, connectors, rebar and concrete for a 6" wall. Labor includes erecting forms, setting rebar and pouring concrete in walls to 8' high. Assume that each square foot of wall requires 1 linear foot of #4 rebar.

| | Craft@Hrs | Unit | Material | Labor | Total |
|---|---|---|---|---|---|
| Per square foot of wall, below grade | B1@.060 | SF | 4.42 | 2.00 | 6.42 |
| Per standard block (5.3 SF), below grade | B1@.318 | Ea | 23.40 | 10.60 | 34.00 |
| Per square foot of wall, above grade | B1@.080 | SF | 4.42 | 2.66 | 7.08 |
| Per standard block (5.3 SF), above grade | B1@.424 | Ea | 23.40 | 14.10 | 37.50 |
| Add for 8" walls | — | % | 12.0 | 5.0 | — |
| Deduct for 4" walls | — | % | 8.0 | 5.0 | — |

**Cement and expanded polystyrene forms**  Similar to EPS forms above but with cement added to the mix: Recycled polystyrene (85%) and concrete (15%) building blocks with a grid of internal chambers designed to be placed either horizontally or vertically. Available brands are Rastra Block and Grid-WALL. Once the panels are in place they are filled with concrete providing a high strength, fireproof structural grid. Includes cement filling and grouting. When cut for odd sizes use per sq ft pricing.
Material unit costs for Grid-WALL systems
Standard wall block, 5.3 SF per block

| | Craft@Hrs | Unit | Material | Labor | Total |
|---|---|---|---|---|---|
| 48" L, 16" H, | — | Ea | 23.40 | — | 23.40 |

Labor and material for wall construction
Material includes cost of Grid-WALL but not concrete or rebar. Labor includes stacking, leveling, adding rebar, bracing and prepping the wall but not the concrete pour.

| | Craft@Hrs | Unit | Material | Labor | Total |
|---|---|---|---|---|---|
| Per square foot of wall | B1@.058 | SF | 4.42 | 1.93 | 6.35 |
| Per standard block (5.3 SF) | B1@.313 | Ea | 23.40 | 10.40 | 33.80 |
| Add to labor if concrete pour included | — | % | — | 20.0 | — |

Use to calculate concrete and rebar costs:
1.1 cubic yards of concrete are required to fill 100 square feet of Grid-WALL.
1 Grid-WALL block will take 8 ft. of rebar, unless engineered otherwise.

**Concrete Formwork**  Multiple use of forms assumes that forms can be removed, partially disassembled and cleaned and that 75% of the form material can be reused. Costs include assembly, oiling, setting, stripping and cleaning. Material costs in this section can be adjusted to reflect your actual cost of lumber. Here's how: Divide your actual lumber cost per MBF by the assumed cost (listed in parentheses). Then multiply the cost in the material column by this adjustment factor. For expanded coverage of concrete and formwork, see the *National Concrete & Masonry Estimator* at http://CraftsmanSiteLicense.com.

**Board forming and stripping**  For wall footings, grade beams, column footings, site curbs and steps. Includes 5% waste. Per SF of contact area. When forms are required on both sides of the concrete, include the contact surface for each side.
2" thick forms and bracing, using 2.85 BF of form lumber per SF. Includes nails, ties, and form oil

| | Craft@Hrs | Unit | Material | Labor | Total | |
|---|---|---|---|---|---|---|
| Using 2" lumber, per MBF | — | MBF | 469.00 | — | 469.00 | |
| Using nails, ties and form oil, per SF | — | SF | .22 | — | .22 | 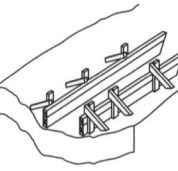 |
| 1 use | B2@.115 | SF | 1.60 | 3.97 | 5.57 | |
| 3 use | B2@.115 | SF | .80 | 3.97 | 4.77 | |
| 5 use | B2@.115 | SF | .64 | 3.97 | 4.61 | |

Add for keyway beveled on two edges, one use. No stripping included

| | Craft@Hrs | Unit | Material | Labor | Total | |
|---|---|---|---|---|---|---|
| 2" x 4" | B2@.027 | LF | .33 | .93 | 1.26 | 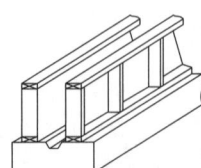 |
| 2" x 6" | B2@.027 | LF | .49 | .93 | 1.42 | |

|  | Craft@Hrs | Unit | Material | Labor | Total |
|---|---|---|---|---|---|

**Driveway and walkway edge forms** Material costs include stakes, nails, and form oil and 5% waste. Per LF of edge form. When forms are required on both sides of the concrete, include the length of each side plus the end widths.

| 2" x 4" edge form | | | | | |
|---|---|---|---|---|---|
| Using 2" x 4" lumber, .7 BF per LF | — | MBF | 473.00 | — | 473.00 |
| Using nails, ties and form oil, per LF | — | LF | .22 | — | .22 |
| 1 use | B2@.050 | LF | .49 | 1.73 | 2.22 |
| 3 use | B2@.050 | LF | .31 | 1.73 | 2.04 |
| 5 use | B2@.050 | LF | .27 | 1.73 | 2.00 |

| 2" x 6" edge form | | | | | |
|---|---|---|---|---|---|
| Using 2" x 6" lumber, 1.05 BF per LF | — | MBF | 465.00 | — | 465.00 |
| 1 use | B2@.050 | LF | .65 | 1.73 | 2.38 |
| 3 use | B2@.050 | LF | .38 | 1.73 | 2.11 |
| 5 use | B2@.050 | LF | .32 | 1.73 | 2.05 |

| 2" x 8" edge form | | | | | |
|---|---|---|---|---|---|
| Using 2" x 8" lumber, 1.4 BF per LF | — | MBF | 465.00 | — | 465.00 |
| 1 use | B2@.055 | LF | .85 | 1.90 | 2.75 |
| 3 use | B2@.055 | LF | .47 | 1.90 | 2.37 |
| 5 use | B2@.055 | LF | .39 | 1.90 | 2.29 |

| 2" x 10" edge form | | | | | |
|---|---|---|---|---|---|
| Using 2" x 10" lumber, 1.75 BF per LF | — | MBF | 484.00 | — | 484.00 |
| 1 use | B2@.055 | LF | 1.01 | 1.90 | 2.91 |
| 3 use | B2@.055 | LF | .54 | 1.90 | 2.44 |
| 5 use | B2@.055 | LF | .44 | 1.90 | 2.34 |

| 2" x 12" edge form | | | | | |
|---|---|---|---|---|---|
| Using 2" x 12" lumber, 2.1 BF per LF | — | MBF | 563.00 | — | 563.00 |
| 1 use | B2@.055 | LF | 1.34 | 1.90 | 3.24 |
| 3 use | B2@.055 | LF | .69 | 1.90 | 2.59 |
| 5 use | B2@.055 | LF | .55 | 1.90 | 2.45 |

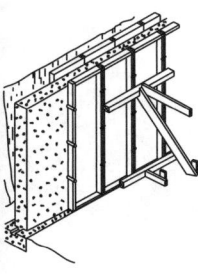

**Plywood forming** For foundation walls, building walls and retaining walls, using 3/4" plyform with 10% waste and 2" bracing with 20% waste. All material costs include nails, ties, clamps, form oil and stripping. Costs shown are per SF of contact area. When forms are required on both sides of the concrete, include the contact surface area for each side.

| | | | | | |
|---|---|---|---|---|---|
| Using 3/4" plyform, per MSF | — | MSF | 2,890.00 | — | 2,890.00 |
| Using 2" bracing, per MBF | — | MBF | 469.00 | — | 469.00 |
| Using nails, ties, clamps & form oil, per SF | — | SFCA | .22 | — | .22 |

Walls up to 4' high (1.10 SF of plywood and .42 BF of bracing per SF of form area)

| | | | | | |
|---|---|---|---|---|---|
| 1 use | B2@.051 | SFCA | 3.60 | 1.76 | 5.36 |
| 3 use | B2@.051 | SFCA | 1.74 | 1.76 | 3.50 |
| 5 use | B2@.051 | SFCA | 1.33 | 1.76 | 3.09 |

Walls 4' to 8' high (1.10 SF of plywood and .60 BF of bracing per SF of form area)

| | | | | | |
|---|---|---|---|---|---|
| 1 use | B2@.060 | SFCA | 3.68 | 2.07 | 5.75 |
| 3 use | B2@.060 | SFCA | 1.78 | 2.07 | 3.85 |
| 5 use | B2@.060 | SFCA | 1.36 | 2.07 | 3.43 |

Walls 8' to 12' high (1.10 SF of plywood and .90 BF of bracing per SF of form area)

| | | | | | |
|---|---|---|---|---|---|
| 1 use | B2@.095 | SFCA | 3.82 | 3.28 | 7.10 |
| 3 use | B2@.095 | SFCA | 1.84 | 3.28 | 5.12 |
| 5 use | B2@.095 | SFCA | 1.41 | 3.28 | 4.69 |

# Concrete Reinforcing Steel

| | Craft@Hrs | Unit | Material | Labor | Total |
|---|---|---|---|---|---|
| **Walls 12' to 16' high (1.10 SF of plywood and 1.05 BF of bracing per SF of form area)** | | | | | |
| 1 use | B2@.128 | SFCA | 3.89 | 4.42 | 8.31 |
| 3 use | B2@.128 | SFCA | 1.87 | 4.42 | 6.29 |
| 5 use | B2@.128 | SFCA | 1.43 | 4.42 | 5.85 |
| **Walls over 16' high (1.10 SF of plywood and 1.20 BF of bracing per SF of form area)** | | | | | |
| 1 use | B2@.153 | SFCA | 3.96 | 5.28 | 9.24 |
| 3 use | B2@.153 | SFCA | 1.90 | 5.28 | 7.18 |
| 5 use | B2@.153 | SFCA | 1.46 | 5.28 | 6.74 |

**Form stripping** Labor to remove forms and bracing, clean, and stack on the job site.
Board forms at wall footings, grade beams and column footings. Per SF of contact area

| | Craft@Hrs | Unit | Material | Labor | Total |
|---|---|---|---|---|---|
| 1" thick lumber | BL@.010 | SFCA | — | .30 | .30 |
| 2" thick lumber | BL@.012 | SFCA | — | .36 | .36 |
| Keyways, 2" x 4" and 2" x 6" | BL@.006 | LF | — | .18 | .18 |
| **Slab edge forms. Per LF of edge form** | | | | | |
| 2" x 4" to 2" x 6" | BL@.012 | LF | — | .36 | .36 |
| 2" x 8" to 2" x 12" | BL@.013 | LF | — | .39 | .39 |
| **Walls, plywood forms. Per SF of contact area stripped** | | | | | |
| To 4' high | BL@.017 | SFCA | — | .50 | .50 |
| Over 4' to 8' high | BL@.018 | SFCA | — | .53 | .53 |
| Over 8' to 12' high | BL@.020 | SFCA | — | .59 | .59 |
| Over 12' to 16' high | BL@.027 | SFCA | — | .80 | .80 |
| Over 16' high | BL@.040 | SFCA | — | 1.19 | 1.19 |

**Concrete Reinforcing Steel** Steel reinforcing bars (rebar), ASTM A615 Grade 60. Material costs are for deformed steel reinforcing rebars, including 10% lap allowance, cutting and bending. These costs also include detailed shop drawings and delivery to jobsite with identity tags per shop drawings. Add for epoxy or galvanized coating of rebars, chairs, splicing, spiral caissons and round column reinforcing, if required, from the sections following the data below. Costs per pound (Lb) and per linear foot (LF) including tie wire and tying.

| | Craft@Hrs | Unit | Material | Labor | Total |
|---|---|---|---|---|---|
| **Reinforcing steel placed and tied in footings and grade beams** | | | | | |
| 1/4" diameter, #2 rebar | RI@.015 | Lb | .95 | .53 | 1.48 |
| 1/4" diameter, #2 rebar (.17 Lb per LF) | RI@.003 | LF | .16 | .11 | .27 |
| 3/8" diameter, #3 rebar | RI@.011 | Lb | .60 | .39 | .99 |
| 3/8" diameter, #3 rebar (.38 Lb per LF) | RI@.004 | LF | .23 | .14 | .37 |
| 1/2" diameter, #4 rebar | RI@.010 | Lb | .58 | .36 | .94 |
| 1/2" diameter, #4 rebar (.67 Lb per LF) | RI@.007 | LF | .39 | .25 | .64 |
| 5/8" diameter, #5 rebar | RI@.009 | Lb | .52 | .32 | .84 |
| 5/8" diameter, #5 rebar (1.04 Lb per LF) | RI@.009 | LF | .54 | .32 | .86 |
| 3/4" diameter, #6 rebar | RI@.008 | Lb | .50 | .28 | .78 |
| 3/4" diameter, #6 rebar (1.50 Lb per LF) | RI@.012 | LF | .75 | .43 | 1.18 |
| 7/8" diameter, #7 rebar | RI@.008 | Lb | .60 | .28 | .88 |
| 7/8" diameter, #7 rebar (2.04 Lb per LF) | RI@.016 | LF | 1.22 | .57 | 1.79 |
| 1" diameter, #8 rebar | RI@.008 | Lb | .52 | .28 | .80 |
| 1" diameter, #8 rebar (2.67 Lb per LF) | RI@.021 | LF | 1.39 | .75 | 2.14 |
| **Reinforcing steel placed and tied in structural slabs** | | | | | |
| 1/4" diameter, #2 rebar | RI@.014 | Lb | .95 | .50 | 1.45 |
| 1/4" diameter, #2 rebar (.17 Lb per LF) | RI@.002 | LF | .16 | .07 | .23 |
| 3/8" diameter, #3 rebar | RI@.010 | Lb | .60 | .36 | .96 |
| 3/8" diameter, #3 rebar (.38 Lb per LF) | RI@.004 | LF | .23 | .14 | .37 |
| 1/2" diameter, #4 rebar | RI@.009 | Lb | .58 | .32 | .90 |
| 1/2" diameter, #4 rebar (.67 Lb per LF) | RI@.006 | LF | .39 | .21 | .60 |

| | Craft@Hrs | Unit | Material | Labor | Total |
|---|---|---|---|---|---|
| 5/8" diameter, #5 rebar | RI@.008 | Lb | .52 | .28 | .80 |
| 5/8" diameter, #5 rebar (1.04 Lb per LF) | RI@.008 | LF | .54 | .28 | .82 |
| 3/4" diameter, #6 rebar | RI@.007 | Lb | .50 | .25 | .75 |
| 3/4" diameter, #6 rebar (1.50 Lb per LF) | RI@.011 | LF | .75 | .39 | 1.14 |
| 7/8" diameter, #7 rebar | RI@.007 | Lb | .60 | .25 | .85 |
| 7/8" diameter, #7 rebar (2.04 Lb per LF) | RI@.014 | LF | 1.22 | .50 | 1.72 |
| 1" diameter, #8 rebar | RI@.007 | Lb | .52 | .25 | .77 |
| 1" diameter, #8 rebar (2.67 Lb per LF) | RI@.019 | LF | 1.39 | .68 | 2.07 |

Reinforcing steel placed and tied in walls

| | Craft@Hrs | Unit | Material | Labor | Total |
|---|---|---|---|---|---|
| 1/4" diameter, #2 rebar | RI@.017 | Lb | .95 | .60 | 1.55 |
| 1/4" diameter, #2 rebar (.17 Lb per LF) | RI@.003 | LF | .16 | .11 | .27 |
| 3/8" diameter, #3 rebar | RI@.012 | Lb | .60 | .43 | 1.03 |
| 3/8" diameter, #3 rebar (.38 Lb per LF) | RI@.005 | LF | .23 | .18 | .41 |
| 1/2" diameter, #4 rebar | RI@.011 | Lb | .58 | .39 | .97 |
| 1/2" diameter, #4 rebar (.67 Lb per LF) | RI@.007 | LF | .39 | .25 | .64 |
| 5/8" diameter, #5 rebar | RI@.010 | Lb | .52 | .36 | .88 |
| 5/8" diameter, #5 rebar (1.04 Lb per LF) | RI@.010 | LF | .54 | .36 | .90 |
| 3/4" diameter, #6 rebar | RI@.009 | Lb | .50 | .32 | .82 |
| 3/4" diameter, #6 rebar (1.50 Lb per LF) | RI@.014 | LF | .75 | .50 | 1.25 |
| 7/8" diameter, #7 rebar | RI@.009 | Lb | .60 | .32 | .92 |
| 7/8" diameter, #7 rebar (2.04 Lb per LF) | RI@.018 | LF | 1.22 | .64 | 1.86 |
| 1" diameter, #8 rebar | RI@.009 | Lb | .52 | .32 | .84 |
| 1" diameter, #8 rebar (2.67 Lb per LF) | RI@.024 | LF | 1.39 | .85 | 2.24 |

Add for reinforcing steel coating. Coating applied prior to shipment to the job site. Labor for placing coated rebar is the same as that shown for plain rebar above. Cost per pound of reinforcing steel.

| | | Unit | Material | Labor | Total |
|---|---|---|---|---|---|
| Epoxy coating | | | | | |
| Any size rebar, add | — | Lb | .30 | — | .30 |
| Galvanized coating | | | | | |
| Any #2 rebar, #3 rebar or #4 rebar, add | — | Lb | .39 | — | .39 |
| Any #5 rebar or #6 rebar, add | — | Lb | .37 | — | .37 |
| Any #7 rebar or #8 rebar, add | — | Lb | .35 | — | .35 |

Add for reinforcing steel chairs. Labor for placing rebar shown above includes time for setting chairs.

Chairs, individual type, cost each

| | | Unit | Material | Labor | Total |
|---|---|---|---|---|---|
| 3" high | | | | | |
| Steel | — | Ea | .55 | — | .55 |
| Plastic | — | Ea | .23 | — | .23 |
| Galvanized steel | — | Ea | 1.00 | — | 1.00 |
| 5" high | | | | | |
| Steel | — | Ea | .85 | — | .85 |
| Plastic | — | Ea | .58 | — | .58 |
| Galvanized steel | — | Ea | 1.30 | — | 1.30 |
| 8" high | | | | | |
| Steel | — | Ea | 1.79 | — | 1.79 |
| Plastic | — | Ea | 1.20 | — | 1.20 |
| Galvanized steel | — | Ea | 2.82 | — | 2.82 |

Chairs, continuous type, with legs 8" OC, cost per linear foot

| | | Unit | Material | Labor | Total |
|---|---|---|---|---|---|
| 3" high | | | | | |
| Steel | — | LF | .77 | — | .77 |
| Plastic | — | LF | .48 | — | .48 |
| Galvanized steel | — | LF | 1.16 | — | 1.16 |

# Concrete Reinforcing Steel

| | Craft@Hrs | Unit | Material | Labor | Total |
|---|---|---|---|---|---|
| **6" high** | | | | | |
| Steel | — | LF | 1.05 | — | 1.05 |
| Plastic | — | LF | .68 | — | .68 |
| Galvanized steel | — | LF | 1.63 | — | 1.63 |
| **8" high** | | | | | |
| Steel | — | LF | 1.62 | — | 1.62 |
| Plastic | — | LF | 1.25 | — | 1.25 |
| Galvanized steel | — | LF | 2.30 | — | 2.30 |
| **Waterstop, 3/8"** | | | | | |
| Rubber, 6" | B2@.059 | LF | 3.32 | 2.04 | 5.36 |
| Rubber, 9" | B2@.059 | LF | 4.72 | 2.04 | 6.76 |

**Spiral caisson and round column reinforcing** 3/8" diameter hot rolled steel spirals with main vertical bars as shown. Costs include typical engineering and shop drawings. Based on shop fabricated spirals delivered, tagged, ready to install. Cost per vertical linear foot (VLF)

| | Craft@Hrs | Unit | Material | Labor | Total |
|---|---|---|---|---|---|
| 16" diameter with 6 #6 bars, 16.8 Lbs per VLF | RI@.041 | VLF | 32.70 | 1.46 | 34.16 |
| 24" diameter with 6 #6 bars, 28.5 Lbs per VLF | RI@.085 | VLF | 55.90 | 3.02 | 58.92 |
| 36" diameter with 8 #10 bars, 51.2 Lbs per VLF | RI@.176 | VLF | 102.00 | 6.26 | 108.26 |

Welded wire mesh, steel, electric weld, including 15% waste and overlap

| | Craft@Hrs | Unit | Material | Labor | Total |
|---|---|---|---|---|---|
| 2" x 2" W.9 x W.9 (#12 x #12), slabs | RI@.004 | SF | .95 | .14 | 1.09 |
| 2" x 2" W.9 x W.9 (#12 x #12), beams and columns | RI@.020 | SF | .93 | .71 | 1.64 |
| 4" x 4" W1.4 x W1.4 (#10 x #10), slabs | RI@.003 | SF | .44 | .11 | .55 |
| 4" x 4" W2.0 x W2.0 (#8 x #8), slabs | RI@.004 | SF | .51 | .14 | .65 |
| 4" x 4" W2.9 x W2.9 (#6 x #6), slabs | RI@.005 | SF | .58 | .18 | .76 |
| 4" x 4" W4.0 x W4.0 (#4 x #4), slabs | RI@.006 | SF | .76 | .21 | .97 |
| 6" x 6" W1.4 x W1.4 (#10 x #10), slabs | RI@.003 | SF | .20 | .11 | .31 |
| 6" x 6" W2.0 x W2.0 (#8 x #8), slabs | RI@.004 | SF | .34 | .14 | .48 |
| 6" x 6" W2.9 x W2.9 (#6 x #6), slabs | RI@.004 | SF | .39 | .14 | .53 |
| 6" x 6" W4.0 x W4.0 (#4 x #4), slabs | RI@.005 | SF | .54 | .18 | .72 |
| Add for lengthwise cut, LF of cut | RI@.002 | LF | — | .07 | .07 |

Welded wire mesh, galvanized steel, electric weld, including 15% waste and overlap

| | Craft@Hrs | Unit | Material | Labor | Total |
|---|---|---|---|---|---|
| 2" x 2" W.9 x W.9 (#12 x #12), slabs | RI@.004 | SF | 1.85 | .14 | 1.99 |
| 4" x 4" W1.4 x W1.4 (#10 x #10), slabs | RI@.003 | SF | .87 | .11 | .98 |
| 4" x 4" W2.0 x W2.0 (#8 x #8), slabs | RI@.004 | SF | 1.03 | .14 | 1.17 |
| 4" x 4" W2.9 x W2.9 (#6 x #6), slabs | RI@.005 | SF | 1.16 | .18 | 1.34 |
| 4" x 4" W4.0 x W4.0 (#4 x #4), slabs | RI@.006 | SF | 1.50 | .21 | 1.71 |
| 6" x 6" W1.4 x W1.4 (#10 x #10), slabs | RI@.003 | SF | .48 | .11 | .59 |
| 6" x 6" W2.0 x W2.0 (#8 x #8), slabs | RI@.004 | SF | .61 | .14 | .75 |
| 6" x 6" W2.9 x W2.9 (#6 x #6), slabs | RI@.004 | SF | .76 | .14 | .90 |
| 6" x 6" W4.0 x W4.0 (#4 x #4), slabs | RI@.005 | SF | 1.09 | .18 | 1.27 |
| Add for lengthwise cut, LF of cut | RI@.002 | LF | — | .07 | .07 |

Welded wire mesh, epoxy coated steel, electric weld, including 15% waste and overlap

| | Craft@Hrs | Unit | Material | Labor | Total |
|---|---|---|---|---|---|
| 2" x 2" W.9 x W.9 (#12 x #12), slabs | RI@.004 | SF | 3.81 | .14 | 3.95 |
| 4" x 4" W1.4 x W1.4 (#10 x #10), slabs | RI@.003 | SF | 3.39 | .11 | 3.50 |
| 4" x 4" W2.0 x W2.0 (#8 x #8), slabs | RI@.004 | SF | 3.37 | .14 | 3.51 |
| 4" x 4" W2.9 x W2.9 (#6 x #6), slabs | RI@.005 | SF | 3.43 | .18 | 3.61 |
| 4" x 4" W4.0 x W4.0 (#4 x #4), slabs | RI@.006 | SF | 3.60 | .21 | 3.81 |
| 6" x 6" W1.4 x W1.4 (#10 x #10), slabs | RI@.003 | SF | 3.09 | .11 | 3.20 |
| 6" x 6" W2.0 x W2.0 (#8 x #8), slabs | RI@.004 | SF | 3.17 | .14 | 3.31 |
| 6" x 6" W2.9 x W2.9 (#6 x #6), slabs | RI@.004 | SF | 3.22 | .14 | 3.36 |
| 6" x 6" W4.0 x W4.0 (#4 x #4), slabs | RI@.005 | SF | 3.39 | .18 | 3.57 |
| Add for lengthwise cut, LF of cut | RI@.002 | LF | — | .07 | .07 |

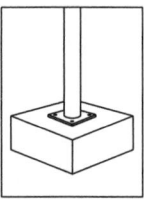

| | Craft@Hrs | Unit | Material | Labor | Total |
|---|---|---|---|---|---|

**Column Footings** Concrete material and placing only. Material costs are for 3,000 PSI, 5.7 sack mix, with 1-1/2" aggregate with 5% waste included. Labor cost is for placing concrete in an existing form. Add for excavation, board forming and steel reinforcing.

| | Craft@Hrs | Unit | Material | Labor | Total |
|---|---|---|---|---|---|
| Using concrete (before waste allowance) | — | CY | 98.60 | — | 98.60 |
| Typical cost per CY | B1@.875 | CY | 104.00 | 29.10 | 133.10 |

**Concrete Walls** Cast-in-place concrete walls for buildings or retaining walls. Material costs for concrete placed direct from chute are based on 2,000 PSI, 5.0 sack mix, with 1" aggregate, including 5% waste. Pump mix costs include an additional $15.00 per CY for the pump. Labor costs are for placing only. Add the cost of excavation, formwork, steel reinforcing, finishes and curing. Square foot costs are based on SF of wall measured on one face only. Costs do not include engineering, design or foundations. For expanded coverage of concrete and formwork, see the *National Concrete & Masonry Estimator* at http://CraftsmanSiteLicense.com

| | Craft@Hrs | Unit | Material | Labor | Total |
|---|---|---|---|---|---|
| Using concrete (before 5% waste allowance) | — | CY | 94.40 | — | 94.40 |
| 4" thick walls (1.23 CY per CSF) | | | | | |
| To 4' high, direct from chute | B1@.013 | SF | 1.22 | .43 | 1.65 |
| 4' to 8' high, pumped | B3@.015 | SF | 1.40 | .48 | 1.88 |
| 8' to 12' high, pumped | B3@.017 | SF | 1.40 | .55 | 1.95 |
| 12' to 16' high, pumped | B3@.018 | SF | 1.40 | .58 | 1.98 |
| 16' high, pumped | B3@.020 | SF | 1.40 | .64 | 2.04 |
| 6" thick walls (1.85 CY per CSF) | | | | | |
| To 4' high, direct from chute | B1@.020 | SF | 1.83 | .67 | 2.50 |
| 4' to 8' high, pumped | B3@.022 | SF | 2.11 | .71 | 2.82 |
| 8' to 12' high, pumped | B3@.025 | SF | 2.11 | .80 | 2.91 |
| 12' to 16' high, pumped | B3@.027 | SF | 2.11 | .87 | 2.98 |
| 16' high, pumped | B3@.030 | SF | 2.11 | .96 | 3.07 |
| 8" thick walls (2.47 CY per CSF) | | | | | |
| To 4' high, direct from chute | B1@.026 | SF | 2.88 | .87 | 3.75 |
| 4' to 8' high, pumped | B3@.030 | SF | 2.82 | .96 | 3.78 |
| 8' to 12' high, pumped | B3@.033 | SF | 2.82 | 1.06 | 3.88 |
| 12' to 16' high, pumped | B3@.036 | SF | 2.82 | 1.16 | 3.98 |
| 16' high, pumped | B3@.040 | SF | 2.82 | 1.28 | 4.10 |
| 10" thick walls (3.09 CY per CSF) | | | | | |
| To 4' high, direct from chute | B1@.032 | SF | 3.06 | 1.07 | 4.13 |
| 4' to 8' high, pumped | B3@.037 | SF | 3.53 | 1.19 | 4.72 |
| 8' to 12' high, pumped | B3@.041 | SF | 3.53 | 1.32 | 4.85 |
| 12' to 16' high, pumped | B3@.046 | SF | 3.53 | 1.48 | 5.01 |
| 16' high, pumped | B3@.050 | SF | 3.53 | 1.60 | 5.13 |
| 12" thick walls (3.70 CY per CSF) | | | | | |
| To 4' high, placed direct from chute | B1@.040 | SF | 3.67 | 1.33 | 5.00 |
| 4' to 8' high, pumped | B3@.045 | SF | 4.22 | 1.44 | 5.66 |
| 8' to 12' high, pumped | B3@.050 | SF | 4.22 | 1.60 | 5.82 |
| 12' to 16' high, pumped | B3@.055 | SF | 4.22 | 1.76 | 5.98 |
| 16' high, pumped | B3@.060 | SF | 4.22 | 1.93 | 6.15 |

**Concrete Footings, Grade Beams and Stem Walls** Use the figures below for preliminary estimates. Concrete costs are based on 2,000 PSI, 5.0 sack mix with 1" aggregate placed directly from the chute of a ready-mix truck. Figures in parentheses show the cubic yards of concrete per linear foot of foundation (including 5% waste). Costs shown include concrete, 60 pounds of reinforcing per CY of concrete, and typical excavation using a 3/4 CY backhoe with excess backfill spread on site.

# Concrete Footings

**Footings and grade beams** Cast directly against the earth, no forming or finishing required. These costs assume the top of the foundation will be at finished grade. For scheduling purposes estimate that a crew of 3 can lay out, excavate, place and tie the reinforcing steel and place 13 CY of concrete in an 8-hour day. Use $900.00 as a minimum job charge.

| | Craft@Hrs | Unit | Material | Labor | Equipment | Total |
|---|---|---|---|---|---|---|
| Using 2,000 PSI concrete, per CY | — | CY | 96.00 | — | — | 96.00 |
| Using #4 reinforcing bars, per pound | — | Lb | .58 | — | — | .58 |
| Typical cost per CY | B4@1.80 | CY | 131.00 | 63.20 | 21.30 | 215.50 |
| 12" W x  6" D (0.019 CY per LF) | B4@.034 | LF | 2.49 | 1.19 | .40 | 4.08 |
| 12" W x  8" D (0.025 CY per LF) | B4@.045 | LF | 3.28 | 1.58 | .53 | 5.39 |
| 12" W x 10" D (0.032 CY per LF) | B4@.058 | LF | 4.19 | 2.04 | .69 | 6.92 |
| 12" W x 12" D (0.039 CY per LF) | B4@.070 | LF | 5.11 | 2.46 | .83 | 8.40 |
| 12" W x 18" D (0.058 CY per LF) | B4@.104 | LF | 7.60 | 3.65 | 1.23 | 12.48 |
| 16" W x  8" D (0.035 CY per LF) | B4@.063 | LF | 4.59 | 2.21 | .75 | 7.55 |
| 16" W x 10" D (0.043 CY per LF) | B4@.077 | LF | 5.63 | 2.70 | .91 | 9.24 |
| 16" W x 12" D (0.052 CY per LF) | B4@.094 | LF | 6.81 | 3.30 | 1.11 | 11.22 |
| 18" W x  8" D (0.037 CY per LF) | B4@.067 | LF | 4.85 | 2.35 | .79 | 7.99 |
| 18" W x 10" D (0.049 CY per LF) | B4@.088 | LF | 6.42 | 3.09 | 1.04 | 10.55 |
| 18" W x 24" D (0.117 CY per LF) | B4@.216 | LF | 15.30 | 7.58 | 2.56 | 25.44 |
| 20" W x 12" D (0.065 CY per LF) | B4@.117 | LF | 8.52 | 4.11 | 1.39 | 14.02 |
| 24" W x 12" D (0.078 CY per LF) | B4@.140 | LF | 10.20 | 4.91 | 1.66 | 16.77 |
| 24" W x 24" D (0.156 CY per LF) | B4@.281 | LF | 20.40 | 9.86 | 3.33 | 33.59 |
| 24" W x 30" D (0.195 CY per LF) | B4@.351 | LF | 25.50 | 12.30 | 4.16 | 41.96 |
| 24" W x 36" D (0.234 CY per LF) | B4@.421 | LF | 30.70 | 14.80 | 4.99 | 50.49 |
| 30" W x 36" D (0.292 CY per LF) | B4@.526 | LF | 38.30 | 18.50 | 6.24 | 63.04 |
| 30" W x 42" D (0.341 CY per LF) | B4@.614 | LF | 44.70 | 21.50 | 7.28 | 73.48 |
| 36" W x 48" D (0.467 CY per LF) | B4@.841 | LF | 61.20 | 29.50 | 9.97 | 100.67 |

**Tricks of the trade:** To estimate the cost of footing or grade beam sizes not shown, multiply the width in inches by the depth in inches and divide the result by 3700. This is the CY of concrete per LF of footing including 5% waste. Multiply by the "Typical cost per CY" in the prior table to find the cost per LF.

**Continuous concrete footing with foundation stem wall** These figures assume the foundation stem wall projects 24" above the finished grade and extends into the soil 18" to the top of the footing. Costs shown include typical excavation using a 3/4 CY backhoe with excess backfill spread on site, forming both sides of the foundation wall and the footing, based on three uses of the forms and 2 #4 rebar. Use $1,200.00 as a minimum cost for this type work.

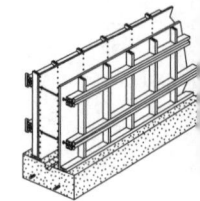

| | Craft@Hrs | Unit | Material | Labor | Equipment | Total |
|---|---|---|---|---|---|---|
| Typical cost per CY | B5@7.16 | CY | 159.00 | 254.00 | 50.90 | 463.90 |
| Typical single story structure, footing 12" W x 8" D, wall 6" T x 42" D (.10 CY per LF) | B5@.716 | LF | 15.90 | 25.40 | 5.09 | 46.39 |
| Typical two story structure, footing 18" W x 10" D, wall 8" T x 42" D (.14 CY per LF) | B5@1.00 | LF | 22.30 | 35.50 | 7.11 | 64.91 |
| Typical three story structure, footing 24" W x 12" D, wall 10" T x 42" D (.19 CY per LF) | B5@1.36 | LF | 30.20 | 48.20 | 9.67 | 88.07 |

**Slabs, Walks and Driveways** Typical costs for reinforced concrete slabs-on-grade including fine grading, slab base, forms, vapor barrier, wire mesh, 3,000 PSI concrete, finishing and curing. For expanded coverage of concrete and formwork, see the *National Concrete & Masonry Estimator* at http://CraftsmanSiteLicense.com. For thickened edge slabs, add the area of the thickened edge. Use 500 square feet as a minimum.

| | Craft@Hrs | Unit | Material | Labor | Total |
|---|---|---|---|---|---|
| 2" thick | B5@.067 | SF | 1.68 | 2.38 | 4.06 |
| 3" thick | B5@.068 | SF | 1.99 | 2.41 | 4.40 |
| 4" thick | B5@.069 | SF | 2.30 | 2.45 | 4.75 |
| 5" thick | B5@.070 | SF | 2.61 | 2.48 | 5.09 |
| 6" thick | B5@.071 | SF | 2.92 | 2.52 | 5.44 |

**Slab base** Aggregate base for slabs. No waste included. Labor costs are for spreading aggregate from piles only. Add for fine grading, using hand tools.

| Crushed stone base 1.4 tons equal 1 cubic yard | | | | | |
|---|---|---|---|---|---|
| Using crushed stone, per CY | — | CY | 30.10 | — | 30.10 |
| 1" base (.309 CY per CSF) | BL@.001 | SF | .09 | .03 | .12 |
| 2" base (.617 CY per CSF) | BL@.003 | SF | .19 | .09 | .28 |
| 3" base (.926 CY per CSF) | BL@.004 | SF | .28 | .12 | .40 |
| 4" base (1.23 CY per CSF) | BL@.006 | SF | .37 | .18 | .55 |
| 5" base (1.54 CY per CSF) | BL@.007 | SF | .46 | .21 | .67 |
| 6" base (1.85 CY per CSF) | BL@.008 | SF | .56 | .24 | .80 |
| Sand fill base 1.40 tons equal 1 cubic yard | | | | | |
| Using sand, per CY | — | CY | 13.50 | — | 13.50 |
| 1" fill (.309 CY per CSF) | BL@.001 | SF | .04 | .03 | .07 |
| 2" fill (.617 CY per CSF) | BL@.002 | SF | .08 | .06 | .14 |
| 3" fill (.926 CY per CSF) | BL@.003 | SF | .13 | .09 | .22 |
| 4" fill (1.23 CY per CSF) | BL@.004 | SF | .17 | .12 | .29 |
| 5" fill (1.54 CY per CSF) | BL@.006 | SF | .21 | .18 | .39 |
| 6" fill (1.85 CY per CSF) | BL@.007 | SF | .25 | .21 | .46 |
| Add for fine grading for slab on grade | BL@.008 | SF | — | .24 | .24 |

**Slab membrane** Material costs are for 10' wide membrane with 6" laps. Labor costs are to lay membrane over a level slab base.

| | | | | | |
|---|---|---|---|---|---|
| 4 mil polyethylene | BL@.002 | SF | .07 | .06 | .13 |
| 6 mil polyethylene | BL@.002 | SF | .08 | .06 | .14 |
| Detaching membrane, Ditra, polyethylene laminate, SF costs include 5% waste | | | | | |
| Laid on a level slab | BL@.002 | SF | 2.06 | .06 | 2.12 |
| Add for adhesive cemented seams | BL@.001 | SF | .06 | .03 | .09 |

**Curbs and gutters** Small quantities. Material costs are for 2,000 PSI, 5.0 sack mix with 1" aggregate. No waste included. Labor cost is for placing and finishing concrete only. Add for excavation, wood forming, steel reinforcing.

| | | | | | |
|---|---|---|---|---|---|
| Per CY of concrete | B6@1.61 | CY | 96.00 | 52.10 | 148.10 |

**Steps on grade** Costs include layout, fabrication, and placing forms, setting and tying steel reinforcing, installation of steel nosing, placing 2,000 PSI concrete directly from the chute of a ready-mix truck, finishing, stripping forms, and curing. Costs assume excavation, back filling, and compacting have been completed. For scheduling purposes, estimate that a crew of 3 can set forms, place steel, and pour 4 to 5 CY of concrete cast-in-place steps in an 8 hour day. See also Cast-in-place Concrete Steps on Grade in the Industrial Section 3. Costs shown are for concrete steps with a 6" rise height and 12" tread depth supported by a 6" thick monolithic slab.

| | | | | | |
|---|---|---|---|---|---|
| Total cost of steps per CY of concrete | B1@5.43 | CY | 274.00 | 181.00 | 455.00 |

**Concrete Finishing**
Slab finishes

| | | | | | |
|---|---|---|---|---|---|
| Broom finish | B6@.012 | SF | — | .39 | .39 |
| Float finish | B6@.010 | SF | — | .32 | .32 |

# Concrete Finishing

|  | Craft@Hrs | Unit | Material | Labor | Total |
|---|---|---|---|---|---|
| Trowel finishing | | | | | |
| Steel, machine work | B6@.015 | SF | — | .49 | .49 |
| Steel, hand work | B6@.018 | SF | — | .58 | .58 |
| Finish treads and risers | | | | | |
| No abrasives, no plastering, per LF of tread | B6@.042 | LF | — | 1.36 | 1.36 |
| With abrasives, plastered, per LF of tread | B6@.063 | LF | .60 | 2.04 | 2.64 |
| Scoring concrete surface, hand work | B6@.005 | LF | — | .16 | .16 |
| Sweep, scrub and wash down | B6@.006 | SF | .02 | .19 | .21 |
| Liquid curing and sealing compound, Acrylic Concrete Cure and Seal, | | | | | |
| spray-on, 400 SF at $57.00 per 5 gallon | B6@.003 | SF | .05 | .10 | .15 |
| Exposed aggregate (washed, including finishing), | | | | | |
| no disposal of slurry | B6@.017 | SF | .30 | .55 | .85 |
| Non-metallic color and hardener, troweled on, 2 applications, red, gray or black | | | | | |
| 60 pounds per 100 SF | B6@.021 | SF | .44 | .68 | 1.12 |
| 100 pounds per 100 SF | B6@.025 | SF | .72 | .81 | 1.53 |
| Add for other standard colors | — | % | 50.0 | — | — |
| Wall finishes | | | | | |
| Snap ties and patch | B6@.011 | SF | .13 | .36 | .49 |
| Remove fins | B6@.008 | LF | .05 | .26 | .31 |
| Grind smooth, typical | B6@.021 | SF | .10 | .68 | .78 |
| Burlap rub with cement grout | B6@.022 | SF | .06 | .71 | .77 |
| Brush hammer green concrete | B6@.061 | SF | .05 | 1.97 | 2.02 |
| Bush hammer cured concrete | B6@.091 | SF | .11 | 2.94 | 3.05 |
| Wash with acid and rinse | B6@.004 | SF | .20 | .13 | .33 |
| Break fins, patch voids, Carborundum dry rub | B6@.035 | SF | .07 | 1.13 | 1.20 |
| Carborundum wet rub | B6@.057 | SF | .13 | 1.84 | 1.97 |
| Specialty finishes | | | | | |
| Monolithic natural aggregate topping | | | | | |
| 3/16" | B6@.020 | SF | .26 | .65 | .91 |
| 1/2" | B6@.022 | SF | .30 | .71 | 1.01 |
| Integral colors | | | | | |
| Various colors, 10 oz. bottle | — | Ea | 5.10 | — | 5.10 |
| Colored release agent | — | Lb | 2.00 | — | 2.00 |
| Dry shake colored hardener | — | Lb | .45 | — | .45 |
| Stamped finish (embossed concrete) of joints, if required | | | | | |
| Diamond, square, octagonal patterns | B6@.047 | SF | — | 1.52 | 1.52 |
| Spanish paver pattern | B6@.053 | SF | — | 1.71 | 1.71 |
| Add for grouting | B6@.023 | SF | .30 | .74 | 1.04 |

**Countertops** Custom-fabricated straight, "U"- or "L"-shaped tops. See also Tile for ceramic tile countertops.

Laminated plastic countertops (such as Formica, Textolite or Wilsonart) on a particleboard base. 22" to 25" wide. Cost per LF of longest edge (either back or front) in solid colors with 3-1/2" to 4" backsplash.

|  | Craft@Hrs | Unit | Material | Labor | Total |
|---|---|---|---|---|---|
| Square edge, separate backsplash | B1@.181 | LF | 15.20 | 6.03 | 21.23 |
| Rolled drip edge (post formed) | B1@.181 | LF | 21.60 | 6.03 | 27.63 |
| Full wrap (180 degree) front edge | B1@.181 | LF | 23.20 | 6.03 | 29.23 |
| Accent color front edge | B1@.181 | LF | 29.50 | 6.03 | 35.53 |
| Inlaid front and back edge | B1@.181 | LF | 34.80 | 6.03 | 40.83 |
| Additional costs for laminated countertops | | | | | |
| Add for textures, patterns | — | % | 20.0 | — | — |
| Add for square end splash | — | Ea | 18.40 | — | 18.40 |

| | Craft@Hrs | Unit | Material | Labor | Total |
|---|---|---|---|---|---|
| Add for contour end splash | — | Ea | 26.40 | — | 26.40 |
| Add for mitered corners | — | Ea | 21.80 | — | 21.80 |
| Add for seamless tops | — | Ea | 31.60 | — | 31.60 |
| Add for sink, range or vanity cutout | — | Ea | 8.53 | — | 8.53 |
| Add for drilling 3 plumbing fixture holes | — | LS | 10.90 | — | 10.90 |
| Add for quarter round corner | — | Ea | 21.90 | — | 21.90 |
| Add for half round corner | — | Ea | 31.40 | — | 31.40 |

Granite countertops, custom made to order from a template. 22" to 25" wide. Per square foot of top and backsplash, not including cutting, polishing, sink cut out, range cut out and a flat polish edge. Add the cost of edge polishing for undermount sink, faucet holes, angle cuts, radius ends, edge detail and jobsite delivery as itemized below. 7/8" granite weighs about 30 pounds per square foot. 1-1/4" granite weighs about 45 pounds per square foot. 7/8" tops should be set on 3/4" plywood and set in silicon adhesive every 2'. 1-1/4" granite tops can be set directly on base cabinets with silicon adhesive.

| | Craft@Hrs | Unit | Material | Labor | Total |
|---|---|---|---|---|---|
| Most 3/4" (2cm) granite tops | B1@.128 | SF | 39.80 | 4.26 | 44.06 |
| Most 1-1/4" (3cm) granite tops | B1@.128 | SF | 52.00 | 4.26 | 56.26 |
| Add for radius or beveled edge | — | LF | 7.14 | — | 7.14 |
| Add for half bullnose edge | — | LF | 10.20 | — | 10.20 |
| Add for full bullnose edge | — | LF | 20.40 | — | 20.40 |
| Add for 1/2" beveled edge | — | LF | 7.14 | — | 7.14 |
| Add for backsplash seaming and edging | — | LF | 13.10 | — | 13.10 |
| Add for cooktop overmount cutout | — | Ea | 132.00 | — | 132.00 |
| Add for delivery (typical) | — | Ea | 154.00 | — | 154.00 |
| Add for eased edge | — | LF | 5.10 | — | 5.10 |
| Add for electrical outlet cutout | — | Ea | 35.90 | — | 35.90 |
| Add for faucet holes | — | Ea | 26.80 | — | 26.80 |
| Add for rounded corners | — | Ea | 30.80 | — | 30.80 |
| Add for undermount sink, polished | — | Ea | 306.00 | — | 306.00 |

Engineered stone countertops, quartz particles with acrylic or epoxy binder. Trade names include Crystalite, Silestone and Cambria. Custom made from a template. Per square foot of top and backsplash surface measured one side.

| | Craft@Hrs | Unit | Material | Labor | Total |
|---|---|---|---|---|---|
| Small light chips, 3/4" square edge | B1@.128 | SF | 50.00 | 4.26 | 54.26 |
| Large dark chips, 3/4" square edge | B1@.128 | SF | 56.20 | 4.26 | 60.46 |
| Add for 3/4" bullnose edge | — | LF | 25.60 | — | 25.60 |
| Add for 3/4" ogee edge | — | LF | 33.60 | — | 33.60 |
| Add for 1-1/2" bullnose edge | — | LF | 49.00 | — | 49.00 |
| Add for 1-1/2" ogee edge | — | LF | 85.60 | — | 85.60 |
| Add for sink cutout | — | Ea | 256.00 | — | 256.00 |
| Add for electrical outlet cutout | — | Ea | 30.60 | — | 30.60 |
| Add for radius corner or end | — | Ea | 189.00 | — | 189.00 |
| Add for delivery | — | Ea | 154.00 | — | 154.00 |

Cultured stone vanity tops with integral back splash and sink, 22" deep

| | Craft@Hrs | Unit | Material | Labor | Total |
|---|---|---|---|---|---|
| Cultured marble | | | | | |
| 24" long | B1@2.60 | Ea | 318.00 | 86.60 | 404.60 |
| 36" long | B1@2.60 | Ea | 409.00 | 86.60 | 495.60 |
| 48" long | B1@2.60 | Ea | 500.00 | 86.60 | 586.60 |
| Cultured onyx | | | | | |
| 24" long | B1@3.42 | Ea | 428.00 | 114.00 | 542.00 |
| 36" long | B1@3.42 | Ea | 548.00 | 114.00 | 662.00 |
| 48" long | B1@3.42 | Ea | 667.00 | 114.00 | 781.00 |

# Cupolas

| | Craft@Hrs | Unit | Material | Labor | Total |
|---|---|---|---|---|---|

Corian vanity tops and integral sink, 1/2" thick, 22" deep. Series 100 includes matching integrated bowls. Series 300 has a solid bowl with different deck colors

| | Craft@Hrs | Unit | Material | Labor | Total |
|---|---|---|---|---|---|
| Series 100 | | | | | |
| 25" long | B1@.654 | Ea | 374.00 | 21.80 | 395.80 |
| 31" long | B1@.742 | Ea | 432.00 | 24.70 | 456.70 |
| 37" long | B1@.830 | Ea | 490.00 | 27.60 | 517.60 |
| 43" long | B1@.924 | Ea | 549.00 | 30.80 | 579.80 |
| 49" long | B1@1.02 | Ea | 619.00 | 34.00 | 653.00 |
| Series 300 | | | | | |
| 25" long | B1@.654 | Ea | 502.00 | 21.80 | 523.80 |
| 31" long | B1@.742 | Ea | 525.00 | 24.70 | 549.70 |
| 37" long | B1@.830 | Ea | 619.00 | 27.60 | 646.60 |
| 43" long | B1@.924 | Ea | 677.00 | 30.80 | 707.80 |
| 49" long | B1@1.02 | Ea | 747.00 | 34.00 | 781.00 |
| All side splashes, 22" L | B1@.314 | Ea | 30.00 | 10.50 | 40.50 |

**Cupolas** Add the cost of weather vanes below. Estimate the delivery cost at $75 to $100 per unit.

| | Craft@Hrs | Unit | Material | Labor | Total |
|---|---|---|---|---|---|
| White vinyl with copper roof covering | | | | | |
| 18" square base x 24" high | B1@.948 | Ea | 599.00 | 31.60 | 630.60 |
| 22" square base x 29" high | B1@.948 | Ea | 800.00 | 31.60 | 831.60 |
| 26" square base x 35" high | B1@1.43 | Ea | 1,030.00 | 47.60 | 1,077.60 |
| 30" square base x 44" high | B1@2.02 | Ea | 1,370.00 | 67.30 | 1,437.30 |
| White pine with copper roof covering | | | | | |
| 18" square base x 24" high | B1@.948 | Ea | 465.00 | 31.60 | 496.60 |
| 22" square base x 29" high | B1@.948 | Ea | 600.00 | 31.60 | 631.60 |
| 26" square base x 35" high | B1@1.43 | Ea | 770.00 | 47.60 | 817.60 |
| 30" square base x 44" high | B1@2.02 | Ea | 975.00 | 67.30 | 1,042.30 |
| Weather vanes for cupolas | | | | | |
| Standard aluminum, baked black finish | | | | | |
| 30" high | B1@.275 | Ea | 32.00 | 9.16 | 41.16 |
| Deluxe model, gold-bronze finish aluminum | | | | | |
| 46" high, ornamental | B1@.368 | Ea | 180.00 | 12.30 | 192.30 |

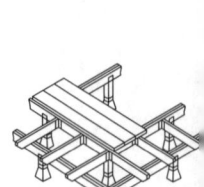

**Decks** These costs assume 5/4" x 6" southern yellow pine decking, 2" x 8" redwood decking, 5/4" x 6" recycled plastic lumber decking or 2" x 8" recycled composite lumber decking supported by #2 & better 2" x 8" pressure treated joists spaced 24" on center over 4" x 8" pressure treated beams. Beams are supported by 4" x 4" pressure treated posts set 6' on center in concrete. Post length is assumed to be 2' above grade. Fasteners are 16d galvanized nails, galvanized steel joist hangers and post anchors. Decking is spaced at 1/8" to permit drainage. Costs are based on a rectangular deck, with one side fastened to an existing structure. Larger decks will reduce labor costs. Where better support is required, add the cost of round concrete forms (Sonotubes) and additional framing fasteners. Deck 3' above ground level with joists, beams, posts and hardware as described above and with decking material as shown.

| | Craft@Hrs | Unit | Material | Labor | Total |
|---|---|---|---|---|---|
| 10' x 45' | | | | | |
| Pine, pressure treated decking | B1@.170 | SF | 5.47 | 5.66 | 11.13 |
| Redwood, select heart decking | B1@.170 | SF | 7.18 | 5.66 | 12.84 |
| 5/4" x 6" recycled composite lumber decking | B1@.225 | SF | 13.50 | 7.49 | 20.99 |
| 2" x 8" recycled composite lumber decking | B1@.225 | SF | 13.80 | 7.49 | 21.29 |
| 15' x 45' | | | | | |
| Pine, pressure treated decking | B1@.170 | SF | 4.83 | 5.66 | 10.49 |
| Redwood, select heart decking | B1@.170 | SF | 6.53 | 5.66 | 12.19 |
| 5/4" x 6" recycled composite lumber decking | B1@.225 | SF | 12.80 | 7.49 | 20.29 |
| 2" x 8" recycled composite lumber decking | B1@.225 | SF | 11.10 | 7.49 | 18.59 |

| | Craft@Hrs | Unit | Material | Labor | Total |
|---|---|---|---|---|---|
| **15' x 65'** | | | | | |
| Pine, pressure treated decking | B1@.170 | SF | 4.83 | 5.66 | 10.49 |
| Redwood, select heart decking | B1@.170 | SF | 6.53 | 5.66 | 12.19 |
| 5/4" x 6" recycled composite lumber decking | B1@.225 | SF | 12.80 | 7.49 | 20.29 |
| 2" x 8" recycled composite lumber decking | B1@.225 | SF | 11.10 | 7.49 | 18.59 |
| **20' x 35'** | | | | | |
| Pine, pressure treated decking | B1@.170 | SF | 4.35 | 5.66 | 10.01 |
| Redwood, select heart decking | B1@.170 | SF | 6.06 | 5.66 | 11.72 |
| 5/4" x 6" recycled composite lumber decking | B1@.225 | SF | 12.30 | 7.49 | 19.79 |
| 2" x 8" recycled composite lumber decking | B1@.225 | SF | 10.60 | 7.49 | 18.09 |
| **20' x 55'** | | | | | |
| Pine, pressure treated decking | B1@.170 | SF | 4.12 | 5.66 | 9.78 |
| Redwood, select heart decking | B1@.170 | SF | 5.82 | 5.66 | 11.48 |
| 5/4" x 6" recycled composite lumber decking | B1@.225 | SF | 12.10 | 7.49 | 19.59 |
| 2" x 8" recycled composite lumber decking | B1@.225 | SF | 10.40 | 7.49 | 17.89 |
| **Add for diagonal pattern decking** | | | | | |
| Pine decking, pressure treated | B1@.007 | SF | .16 | .23 | .39 |
| Redwood decking | B1@.007 | SF | .40 | .23 | .63 |
| 5/4" x 6" recycled composite lumber decking | B1@.007 | SF | 1.28 | .23 | 1.51 |
| 2" x 8" recycled composite lumber decking | B1@.007 | SF | 1.03 | .23 | 1.26 |
| Add for stain and sealer finish | B1@.006 | SF | .10 | .20 | .30 |

**Deck cover framing** Can also be referred to as a pergola. Conventionally framed wood cover joists. Per SF of area covered. Figures in parentheses indicate board feet per square foot of ceiling including end joists, header joists, and 5% waste. Costs shown are based on a job with 200 SF of area covered and include normal waste. Add the cost of posts and finish roof.

| | Craft@Hrs | Unit | Material | Labor | Total |
|---|---|---|---|---|---|
| **2" x 4", Std & Btr grade** | | | | | |
| 16" centers (.59 BF per SF) | B1@.020 | SF | .28 | .67 | .95 |
| 24" centers (.42 BF per SF) | B1@.014 | SF | .20 | .47 | .67 |
| **2" x 6", Std & Btr grade** | | | | | |
| 16" centers (.88 BF per SF) | B1@.020 | SF | .41 | .67 | 1.08 |
| 24" centers (.63 BF per SF) | B1@.015 | SF | .29 | .50 | .79 |

Add for posts to support conventionally framed wood cover joists. Posts are usually required at each corner. Includes one sack of concrete but no excavation.

| | Craft@Hrs | Unit | Material | Labor | Total |
|---|---|---|---|---|---|
| 4" x 4" pressure treated, 10' long | B1@.750 | Ea | 11.50 | 25.00 | 36.50 |
| 6" x 6" pressure treated, 10' long | B1@.750 | Ea | 27.70 | 25.00 | 52.70 |

**Deck roof cover** Cover over conventionally framed wood ceiling joists.
Flat deck built-up roofing. Using 1/2" CDX roof sheathing with 3 ply asphalt surface consisting of 2 plies of 15 lb felt with and 3 hot mop coats of asphalt and 90 lb cap sheet. Add the deck cover framing cost from the previous section.

| | Craft@Hrs | Unit | Material | Labor | Total |
|---|---|---|---|---|---|
| Flat roof cover cost per SF | B1@.029 | SF | 1.41 | .97 | 2.38 |

**Deck railing** 42" high handrail. 2" x 2" paling 5" OC run vertically, lag bolted to the edge of the deck, one 2" x 4" horizontal top rail and one 2" x 6" placed on edge directly below the top rail.

| | Craft@Hrs | Unit | Material | Labor | Total |
|---|---|---|---|---|---|
| Pine, pressure treated, per running foot | B1@.333 | LF | 2.41 | 11.10 | 13.51 |
| Redwood, select heart, per running foot | B1@.333 | LF | 7.10 | 11.10 | 18.20 |
| Recycled composite lumber, per running foot | B1@.333 | LF | 17.30 | 11.10 | 28.40 |

# Deck Stairs

| | Craft@Hrs | Unit | Material | Labor | Total |
|---|---|---|---|---|---|

**Deck stairs** All stairs assume 12" tread (step) composed of two pieces of 2" x 6" framing lumber, a riser and supporting stringers.

One or two steps high, box construction. Cost per linear foot of step. A 4' wide stairway with 2 steps has 8 linear feet of step.

| | Craft@Hrs | Unit | Material | Labor | Total |
|---|---|---|---|---|---|
| Pine treads, open risers | B1@.270 | LF | 4.02 | 8.99 | 13.01 |
| Pine treads, with pine risers | B1@.350 | LF | 7.30 | 11.70 | 19.00 |
| Redwood treads, open risers | B1@.270 | LF | 3.87 | 8.99 | 12.86 |
| Redwood treads, with redwood risers | B1@.350 | LF | 6.86 | 11.70 | 18.56 |
| 5/4" recycled composite lumber treads, open risers | B1@.270 | LF | 8.54 | 8.99 | 17.53 |
| 5/4" recycled composite lumber treads, with recycled composite lumber risers | B1@.350 | LF | 18.10 | 11.70 | 29.80 |
| 2" x 6" recycled composite lumber treads, open risers | B1@.270 | LF | 8.78 | 8.99 | 17.77 |
| 2" x 6" recycled composite lumber treads, with recycled composite lumber risers | B1@.350 | LF | 18.80 | 11.70 | 30.50 |

Three or more steps high, 36" wide. 3 stringers cut from 2" x 12" pressure treated lumber. Cost per riser

| | Craft@Hrs | Unit | Material | Labor | Total |
|---|---|---|---|---|---|
| Pine treads, open risers | B1@.530 | Ea | 14.40 | 17.70 | 32.10 |
| Pine treads, with pine risers | B1@.620 | Ea | 24.30 | 20.70 | 45.00 |
| Redwood treads, open risers | B1@.530 | Ea | 14.00 | 17.70 | 31.70 |
| Redwood treads, with redwood risers | B1@.620 | Ea | 23.00 | 20.70 | 43.70 |
| 5/4" recycled composite lumber treads, open risers | B1@.270 | LF | 29.40 | 8.99 | 38.39 |
| 5/4" recycled composite lumber treads, with recycled composite lumber risers | B1@.350 | LF | 58.10 | 11.70 | 69.80 |
| 2" x 6" recycled composite lumber treads, open risers | B1@.270 | LF | 30.10 | 8.99 | 39.09 |
| 2" x 6" recycled composite lumber treads, with recycled composite lumber risers | B1@.350 | LF | 60.20 | 11.70 | 71.90 |

Concrete footing to support stairway. Cost includes brackets and fasteners to attach stairway to the footing. Add the cost of excavation.

| | Craft@Hrs | Unit | Material | Labor | Total |
|---|---|---|---|---|---|
| 18" x 18" x 12" footing | B1@.530 | Ea | 21.00 | 17.70 | 38.70 |

**Demolition** Itemized costs for demolition of building components when building is being remodeled, repaired or rehabilitated, not completely demolished. Costs include protecting adjacent areas and normal clean-up. Costs are to break out the items listed and pile debris on site only. No hauling, dump fees or salvage value included. Equipment cost includes one air compressor and a pneumatic breaker with jackhammer bits. (Figures in parentheses show the volume before and after demolition and weight of the materials after demolition.) Use $250.00 as a minimum charge.

| | Craft@Hrs | Unit | Material | Labor | Equipment | Total |
|---|---|---|---|---|---|---|

Brick walls and walks, cost per SF removed, measured on one face, using pneumatic breaker

| | Craft@Hrs | Unit | Material | Labor | Equipment | Total |
|---|---|---|---|---|---|---|
| 4" thick walls (60 SF per CY and 36 lbs. per SF) | BL@.061 | SF | — | 1.81 | .82 | 2.63 |
| 8" thick walls (30 SF per CY and 72 lbs. per SF) | BL@.110 | SF | — | 3.26 | 1.47 | 4.73 |
| 12" thick walls (20 SF per CY and 108 lbs. per SF) | BL@.133 | SF | — | 3.95 | 1.78 | 5.73 |

| | Craft@Hrs | Unit | Material | Labor | Equipment | Total |
|---|---|---|---|---|---|---|

Brick sidewalks, cost per SF removed, 5 brick (3-3/4" x 2-1/2" x 8") per SF, salvage condition
2-1/2" thick bricks on sand base, no mortar bed, removed using hand tools

| | | | | | | |
|---|---|---|---|---|---|---|
| (100 SF per CY and 28 lbs. per SF) BL@.020 | SF | — | .59 | — | .59 |

Brick pavers up to 4-1/2" thick with mortar bed, removed using a pneumatic breaker

| | | | | | | |
|---|---|---|---|---|---|---|
| (55 SF per CY and 50 lbs. per SF) BL@.050 | SF | — | 1.48 | .67 | 2.15 |

Ceiling tile, cost per in-place SF removed using hand tools
Stapled or glued 12" x 12"

| | | | | | | |
|---|---|---|---|---|---|---|
| (250 SF per CY and .25 lbs. per SF) BL@.015 | SF | — | .45 | — | .45 |

Suspended panels 2' x 4'

| | | | | | | |
|---|---|---|---|---|---|---|
| (250 SF per CY and .25 lbs. per SF) BL@.010 | SF | — | .30 | — | .30 |

Concrete masonry walls, cost per in-place SF removed, using a pneumatic breaker
4" thick walls

| | | | | | | |
|---|---|---|---|---|---|---|
| (60 SF per CY and 19 lbs. per SF) BL@.066 | SF | — | 1.96 | .88 | 2.84 |

6" thick walls

| | | | | | | |
|---|---|---|---|---|---|---|
| (40 SF per CY and 28 lbs. per SF) BL@.075 | SF | — | 2.23 | 1.00 | 3.23 |

8" thick walls

| | | | | | | |
|---|---|---|---|---|---|---|
| (30 SF per CY and 34 lbs. per SF) BL@.098 | SF | — | 2.91 | 1.31 | 4.22 |

12" thick walls

| | | | | | | |
|---|---|---|---|---|---|---|
| (20 SF per CY and 46 lbs. per SF) BL@.140 | SF | — | 4.15 | 1.88 | 6.03 |
| Reinforced or grouted walls add | — | % | — | 50.0 | 50.0 | — |

Concrete foundations (footings), steel reinforced, removed using a pneumatic breaker. Concrete at 3,900 pounds per in place cubic yard.

| | | | | | | |
|---|---|---|---|---|---|---|
| Cost per CY (.75 CY per CY) | BL@3.96 | CY | — | 117.00 | 53.10 | 170.10 |

Cost per LF with width and depth as shown

| | | | | | | |
|---|---|---|---|---|---|---|
| 6" W x 12" D (35 LF per CY) | BL@.075 | LF | — | 2.23 | 1.00 | 3.23 |
| 8" W x 12" D (30 LF per CY) | BL@.098 | LF | — | 2.91 | 1.31 | 4.22 |
| 8" W x 16" D (20 LF per CY) | BL@.133 | LF | — | 3.95 | 1.78 | 5.73 |
| 8" W x 18" D (18 LF per CY) | BL@.147 | LF | — | 4.36 | 1.97 | 6.33 |
| 10" W x 12" D (21 LF per CY) | BL@.121 | LF | — | 3.59 | 1.62 | 5.21 |
| 10" W x 16" D (16 LF per CY) | BL@.165 | LF | — | 4.90 | 2.21 | 7.11 |
| 10" W x 18" D (14 LF per CY) | BL@.185 | LF | — | 5.49 | 2.48 | 7.97 |
| 12" W x 12" D (20 LF per CY) | BL@.147 | LF | — | 4.36 | 1.97 | 6.33 |
| 12" W x 16" D (13 LF per CY) | BL@.196 | LF | — | 5.82 | 2.63 | 8.45 |
| 12" W x 20" D (11 LF per CY) | BL@.245 | LF | — | 7.27 | 3.28 | 10.55 |
| 12" W x 24" D ( 9 LF per CY) | BL@.294 | LF | — | 8.72 | 3.94 | 12.66 |

Concrete sidewalks, to 4" thick, cost per SF removed. Concrete at 3,900 pounds per in place cubic yard.
Non-reinforced, removed by hand

| | | | | | | |
|---|---|---|---|---|---|---|
| (60 SF per CY) | BL@.050 | SF | — | 1.48 | — | 1.48 |

Reinforced, removed using pneumatic breaker

| | | | | | | |
|---|---|---|---|---|---|---|
| (55 SF per CY) | BL@.033 | SF | — | .98 | .44 | 1.42 |

Concrete slabs, non-reinforced, removed using pneumatic breaker. Concrete at 3,900 pounds per in-place cubic yard.

| | | | | | | |
|---|---|---|---|---|---|---|
| Cost per CY (.80 CY per CY) | BL@3.17 | CY | — | 94.10 | 42.50 | 136.60 |

Cost per SF with thickness as shown

| | | | | | | |
|---|---|---|---|---|---|---|
| 3" slab thickness (90 SF per CY ) | BL@.030 | SF | — | .89 | .40 | 1.29 |
| 4" slab thickness (60 SF per CY ) | BL@.040 | SF | — | 1.19 | .54 | 1.73 |
| 6" slab thickness (45 SF per CY ) | BL@.056 | SF | — | 1.66 | .75 | 2.41 |
| 8" slab thickness (30 SF per CY ) | BL@.092 | SF | — | 2.73 | 1.23 | 3.96 |

# Demolition

| | Craft@Hrs | Unit | Material | Labor | Equipment | Total |
|---|---|---|---|---|---|---|
| Concrete slabs, steel reinforced, removed using pneumatic breaker. Concrete at 3,900 pounds per in-place cubic yard. | | | | | | |
| Cost per CY (1.33 CY per CY) | BL@3.96 | CY | — | 117.00 | 53.10 | 170.10 |
| Cost per SF with thickness as shown | | | | | | |
| 3" slab thickness (80 SF per CY) | BL@.036 | SF | — | 1.07 | .48 | 1.55 |
| 4" slab thickness (55 SF per CY) | BL@.050 | SF | — | 1.48 | .67 | 2.15 |
| 6" slab thickness (40 SF per CY) | BL@.075 | SF | — | 2.23 | 1.00 | 3.23 |
| 8" slab thickness (30 SF per CY) | BL@.098 | SF | — | 2.91 | 1.31 | 4.22 |
| Concrete walls, steel reinforced, removed using pneumatic breaker. Concrete at 3,900 pounds per in-place cubic yard. | | | | | | |
| Cost per CY (1.33 CY per CY) | BL@4.76 | CY | — | 141.00 | 63.80 | 204.80 |
| Cost per SF with thickness as shown (deduct openings over 25 SF) | | | | | | |
| 3" wall thickness (80 SF per CY) | BL@.045 | SF | — | 1.34 | .60 | 1.94 |
| 4" wall thickness (55 SF per CY) | BL@.058 | SF | — | 1.72 | .78 | 2.50 |
| 5" wall thickness (45 SF per CY) | BL@.075 | SF | — | 2.23 | 1.00 | 3.23 |
| 6" wall thickness (40 SF per CY) | BL@.090 | SF | — | 2.67 | 1.21 | 3.88 |
| 8" wall thickness (30 SF per CY) | BL@.120 | SF | — | 3.56 | 1.61 | 5.17 |
| 10" wall thickness (25 SF per CY) | BL@.147 | SF | — | 4.36 | 1.97 | 6.33 |
| 12" wall thickness (20 SF per CY) | BL@.176 | SF | — | 5.22 | 2.36 | 7.58 |
| Curbs and 18" gutter, concrete, removed using pneumatic breaker. Concrete at 3,900 pounds per in-place cubic yard. | | | | | | |
| Cost per LF (8 LF per CY) | BL@.422 | LF | — | 12.50 | 5.65 | 18.15 |
| Doors, typical cost to remove door, frame and hardware | | | | | | |
| 3' x 7' metal (2 doors per CY) | BL@1.00 | Ea | — | 29.70 | — | 29.70 |
| 3' x 7' wood (2 doors per CY) | BL@.497 | Ea | — | 14.70 | — | 14.70 |
| Flooring, cost per in-place SY of floor area removed (1 square yard = 9 square feet) | | | | | | |
| Ceramic tile | | | | | | |
| (25 SY per CY and 34 lbs. per SY) | BL@.263 | SY | — | 7.80 | — | 7.80 |
| Hardwood, nailed | | | | | | |
| (25 SY per CY and 18 lbs. per SY) | BL@.290 | SY | — | 8.60 | — | 8.60 |
| Hardwood, glued | | | | | | |
| (25 SY per CY and 18 lbs. per SY) | BL@.503 | SY | — | 14.90 | — | 14.90 |
| Laminated wood, floating, engineered, "Pergo" | | | | | | |
| (30 SY per CY and 5 lbs. per SY) | BL@.013 | SY | — | .39 | — | .39 |
| Linoleum, sheet | | | | | | |
| (30 SY per CY and 3 lbs. per SY) | BL@.056 | SY | — | 1.66 | — | 1.66 |
| Carpet | | | | | | |
| (40 SY per CY and 5 lbs. per SY) | BL@.028 | SY | — | .83 | — | .83 |
| Carpet pad | | | | | | |
| (35 SY per CY and 1.7 lbs. per SY) | BL@.014 | SY | — | .42 | — | .42 |
| Resilient tile | | | | | | |
| (30 SY per CY and 3 lbs. per SY) | BL@.070 | SY | — | 2.08 | — | 2.08 |
| Terrazzo | | | | | | |
| (25 SY per CY and 34 lbs. per SY) | BL@.286 | SY | — | 8.49 | — | 8.49 |
| Remove mastic, using power sander | BL@.125 | SY | — | 3.71 | — | 3.71 |
| Insulation, fiberglass batts, cost per in-place SF removed | | | | | | |
| R-11 to R-19 | | | | | | |
| (500 SF per CY and .3 lb per SF) | BL@.003 | SF | — | .09 | — | .09 |
| Pavement, asphaltic concrete (bituminous) to 6" thick, removed using pneumatic breaker | | | | | | |
| Cost per SY | | | | | | |
| (4 SY per CY and 660 lbs. per SY) | BL@.250 | SY | — | 7.42 | 3.35 | 10.77 |

| | Craft@Hrs | Unit | Material | Labor | Equipment | Total |
|---|---|---|---|---|---|---|

Plaster walls or ceilings, interior, cost per in-place SF removed by hand
| (150 SF per CY and 8 lbs. per SF) | BL@.015 | SF | — | .45 | — | .45 |

Roofing, cost per in-place square (Sq), removed using hand tools (1 square = 100 square feet)
Built-up, 5 ply
| (2.5 Sq per CY and 250 lbs per Sq) | BL@1.50 | Sq | — | 44.50 | — | 44.50 |

Shingles, asphalt, single layer
| (2.5 Sq per CY and 240 lbs per Sq) | BL@1.33 | Sq | — | 39.50 | — | 39.50 |

Shingles, asphalt, double layer
| (1.2 Sq per CY and 480 lbs per Sq) | BL@1.58 | Sq | — | 46.90 | — | 46.90 |

Shingles, slate. Weight ranges from 600 lbs. to 1,200 lbs. per Sq
| (1 Sq per CY and 900 lbs per Sq). | BL@1.79 | Sq | — | 53.10 | — | 53.10 |

Shingles, wood
| (1.7 Sq per CY and 400 lbs per Sq) | BL@2.02 | Sq | — | 59.90 | — | 59.90 |

Sheathing, up to 1" thick, cost per in-place SF, removed using hand tools
Gypsum
| (250 SF per CY and 2.3 lbs. per SF) | BL@.017 | SF | — | .50 | — | .50 |

Plywood
| (200 SF per CY and 2 lbs. per SF) | BL@.017 | SF | — | .50 | — | .50 |

Siding, up to 1" thick, cost per in-place SF, removed using hand tools
Metal
| (200 SF per CY and 2 lbs. per SF) | BL@.027 | SF | — | .80 | — | .80 |

Plywood
| (200 SF per CY and 2 lbs. per SF) | BL@.017 | SF | — | .50 | — | .50 |

Wood, boards
| (250 SF per CY and 2 lbs. per SF) | BL@.030 | SF | — | .89 | — | .89 |

Stucco walls or soffits, cost per in-place SF, removed using hand tools
| (150 SF per CY and 8 lbs. per SF) | BL@.036 | SF | — | 1.07 | — | 1.07 |

Trees. See Excavation

Wallboard up to 1" thick, cost per in-place SF, removed using hand tools
Gypsum, walls or ceilings
| (250 SF per CY and 2.3 lbs. per SF) | BL@.018 | SF | — | .53 | — | .53 |

Plywood or insulation board
| (200 SF per CY and 2 lbs. per SF) | BL@.018 | SF | — | .53 | — | .53 |

Windows, typical cost to remove window, frame and hardware, cost per SF of window
| Metal (36 SF per CY) | BL@.058 | SF | — | 1.72 | — | 1.72 |
| Wood (36 SF per CY) | BL@.063 | SF | — | 1.87 | — | 1.87 |

Wood stairs, cost per in-place SF or LF
| Risers (with tread) (25 SF per CY) | BL@.116 | SF | — | 3.44 | — | 3.44 |
| Landings (50 SF per CY) | BL@.021 | SF | — | .62 | — | .62 |
| Handrails (100 LF per CY) | BL@.044 | LF | — | 1.31 | — | 1.31 |
| Posts (200 LF per CY) | BL@.075 | LF | — | 2.23 | — | 2.23 |

| | Craft@Hrs | Unit | Material | Labor | Total |
|---|---|---|---|---|---|

**Wood Framing Demolition** Typical costs for demolition of wood frame structural components using hand tools. Normal center to center spacing (12" thru 24" OC) is assumed. Costs include removal of type NM non-metallic (Romex) cable as necessary, protecting adjacent areas, and normal clean-up. Costs do not include temporary shoring to support structural elements above or adjacent to the work area. (Add for demolition of flooring, roofing, sheathing, etc.) Figures in parentheses give the approximate "loose" volume and weight of the materials after being demolished.
Ceiling joists, per in-place SF of ceiling area removed
| 2" x 4" (720 SF per CY and 1.18 lbs. per SF) | BL@.009 | SF | — | .27 | .27 |
| 2" x 6" (410 SF per CY and 1.76 lbs. per SF) | BL@.013 | SF | — | .39 | .39 |

# Demolition

| | Craft@Hrs | Unit | Material | Labor | Total |
|---|---|---|---|---|---|
| 2" x 8" (290 SF per CY and 2.34 lbs. per SF) | BL@.018 | SF | — | .53 | .53 |
| 2" x 10" (220 SF per CY and 2.94 lbs. per SF) | BL@.023 | SF | — | .68 | .68 |
| 2" x 12" (190 SF per CY and 3.52 lbs. per SF) | BL@.027 | SF | — | .80 | .80 |
| **Floor joists, per in-place SF of floor area removed** | | | | | |
| 2" x 6" (290 SF per CY and 2.04 lbs. per SF) | BL@.016 | SF | — | .47 | .47 |
| 2" x 8" (190 SF per CY and 2.72 lbs. per SF) | BL@.021 | SF | — | .62 | .62 |
| 2" x 10" (160 SF per CY and 3.42 lbs. per SF) | BL@.026 | SF | — | .77 | .77 |
| 2" x 12" (120 SF per CY and 4.10 lbs. per SF) | BL@.031 | SF | — | .92 | .92 |
| **Rafters, per in-place SF of actual roof area removed** | | | | | |
| 2" x 4" (610 SF per CY and 1.42 lbs. per SF) | BL@.011 | SF | — | .33 | .33 |
| 2" x 6" (360 SF per CY and 2.04 lbs. per SF) | BL@.016 | SF | — | .47 | .47 |
| 2" x 8" (270 SF per CY and 2.68 lbs. per SF) | BL@.020 | SF | — | .59 | .59 |
| 2" x 10" (210 SF per CY and 3.36 lbs. per SF) | BL@.026 | SF | — | .77 | .77 |
| 2" x 12" (180 SF per CY and 3.94 lbs. per SF) | BL@.030 | SF | — | .89 | .89 |
| **Stud walls, interior or exterior, includes allowance for plates and blocking, per in-place SF of wall area removed, measured on one face** | | | | | |
| 2" x 3" (430 SF per CY and 1.92 lbs. per SF) | BL@.013 | SF | — | .39 | .39 |
| 2" x 4" (310 SF per CY and 2.58 lbs. per SF) | BL@.017 | SF | — | .50 | .50 |
| 2" x 6" (190 SF per CY and 3.74 lbs. per SF) | BL@.025 | SF | — | .74 | .74 |

**Wood Framed Assemblies Demolition** Typical costs for demolition of wood framed structural assemblies using hand tools when a building is being remodeled, repaired or rehabilitated but not completely demolished. Costs include removal of type NM non-metallic (Romex) cable as necessary, protecting adjacent areas and normal clean-up. Costs do not include temporary shoring to support structural elements above or adjacent to the work area, loading, hauling, dump fees or salvage value. Costs shown are per in-place square foot (SF) of area being demolished. Normal center to center spacing (12" thru 24" OC) is assumed. Figures in parentheses give the approximate "loose" volume and weight of the materials (volume after being demolished and weight.)

| | Craft@Hrs | Unit | Material | Labor | Total |
|---|---|---|---|---|---|
| **Ceiling assemblies. Remove ceiling joists, fiberglass insulation and gypsum board up to 5/8" thick** | | | | | |
| 2" x 4" (135 SF per CY and 3.78 lbs. per SF) | BL@.030 | SF | — | .89 | .89 |
| 2" x 6" (118 SF per CY and 4.36 lbs. per SF) | BL@.034 | SF | — | 1.01 | 1.01 |
| 2" x 8" (106 SF per CY and 4.94 lbs. per SF) | BL@.039 | SF | — | 1.16 | 1.16 |
| 2" x 10" (95 SF per CY and 5.54 lbs. per SF) | BL@.043 | SF | — | 1.28 | 1.28 |
| 2" x 12" (89 SF per CY and 6.12 lbs. per SF) | BL@.048 | SF | — | 1.42 | 1.42 |
| **Floor assemblies. Remove floor joists and plywood subfloor up to 3/4" thick** | | | | | |
| 2" x 8" (97 SF per CY and 4.72 lbs. per SF) | BL@.035 | SF | — | 1.04 | 1.04 |
| 2" x 10" (88 SF per CY and 5.42 lbs. per SF) | BL@.040 | SF | — | 1.19 | 1.19 |
| 2" x 12" (75 SF per CY and 6.10 lbs. per SF) | BL@.043 | SF | — | 1.28 | 1.28 |
| For gypsum board ceiling removed below floor joists, add to any of above (250 SF per CY and 2.3 lbs. per SF) | BL@.018 | SF | — | .53 | .53 |
| **Wall assemblies, interior. Remove studs, 2 top plates, 1 bottom plate and gypsum drywall up to 5/8" thick on both sides** | | | | | |
| 2" x 3" (97 SF per CY and 6.52 lbs. per SF) | BL@.048 | SF | — | 1.42 | 1.42 |
| 2" x 4" (89 SF per CY and 7.18 lbs. per SF) | BL@.052 | SF | — | 1.54 | 1.54 |
| 2" x 6" (75 SF per CY and 8.34 lbs. per SF) | BL@.059 | SF | — | 1.75 | 1.75 |
| **Wall assemblies, exterior. Remove studs, 2 top plates, 1 bottom plate, insulation, gypsum drywall up to 5/8" thick on one side, plywood siding up to 3/4" thick on other side** | | | | | |
| 2" x 4" (70 SF per CY and 7.18 lbs. per SF) | BL@.067 | SF | — | 1.99 | 1.99 |
| 2" x 6" (75 SF per CY and 8.34 lbs. per SF) | BL@.074 | SF | — | 2.20 | 2.20 |
| For stucco in lieu of plywood, add to any of above (SF per CY is the same and add 6 lbs. per SF) | BL@.006 | SF | — | .18 | .18 |

| | Craft@Hrs | Unit | Material | Labor | Total |
|---|---|---|---|---|---|

**Demolition, Subcontract** Costs for demolishing and removing an entire building using a dozer and front-end loader. Costs shown include loading and hauling up to 5 miles, but no dump fees. Dump charges vary widely; see Disposal Fees in the Site Work Section of the Industrial and Commercial division of this book. Costs shown are for above ground structures only, per in-place SF of total floor area with ceilings up to 10' high. No salvage value assumed. Add the cost of foundation demolition. Figures in parentheses give the approximate "loose" volume of the materials (volume after being demolished).
Light wood-frame structures, up to three stories in height. Based on 2,500 SF job. No basements included. Estimate each story separately. Cost per in-place SF floor area

| | | | | | |
|---|---|---|---|---|---|
| First story (8 SF per CY) | — | SF | — | — | 2.45 |
| Second story (8 SF per CY) | — | SF | — | — | 3.21 |
| Third story (8 SF per CY) | — | SF | — | — | 4.15 |

Masonry building, cost per in-place SF floor area

| based on 5,000 SF job (50 SF per CY) | — | SF | — | — | 4.04 |
|---|---|---|---|---|---|

Concrete building, cost per in-place SF floor area

| based on 5,000 SF job (30 SF per CY) | — | SF | — | — | 5.26 |
|---|---|---|---|---|---|

Reinforced concrete building, cost per in-place SF floor area

| based on 5,000 SF job (20 SF per CY) | — | SF | — | — | 5.96 |
|---|---|---|---|---|---|

**Door Chimes** Electric, typical installation, add transformer and wire below.
Surface mounted, 2 notes

| | | | | | |
|---|---|---|---|---|---|
| Simple white plastic chime, 4" x 7" | BE@3.50 | Ea | 22.80 | 139.00 | 161.80 |
| Modern oak grained finish, plastic, 8" x 5" | BE@3.50 | Ea | 46.50 | 139.00 | 185.50 |
| White, with repeating tones | BE@3.50 | Ea | 68.00 | 139.00 | 207.00 |
| Maple finish, plastic, 8" x 5" | BE@3.50 | Ea | 46.70 | 139.00 | 185.70 |

Designer, wood-look plastic strips across a cloth grille

| 9" x 7" | BE@3.50 | Ea | 46.00 | 139.00 | 185.00 |
|---|---|---|---|---|---|

Modern real oak wood, with brass plated chimes

| 9" x 7" | BE@3.50 | Ea | 76.40 | 139.00 | 215.40 |
|---|---|---|---|---|---|

Surface mounted, 4 to 8 notes, solid wood door chime

| Three 12" long tubes | BE@3.64 | Ea | 210.00 | 144.00 | 354.00 |
|---|---|---|---|---|---|
| Four 55" long tubes | BE@3.64 | Ea | 270.00 | 144.00 | 414.00 |

Any of above

| Add for chime transformer | — | Ea | 12.00 | — | 12.00 |
|---|---|---|---|---|---|
| Add for bell wire, 65' per coil | — | Ea | 5.00 | — | 5.00 |
| Add for lighted push buttons | — | Ea | 9.00 | — | 9.00 |

Non-electric, push button type, door mounted

| Modern design, 2-1/2" x 2-1/2" | BE@.497 | Ea | 27.00 | 19.70 | 46.70 |
|---|---|---|---|---|---|
| With peep sight, 7" x 3-1/2" | BE@.497 | Ea | 16.20 | 19.70 | 35.90 |

Wireless door chimes, with push button transmitter, 3-year battery, 150' range, 2" speaker, 128 codes

| Oak housing with satin chime tubes | BE@.358 | Ea | 47.50 | 14.20 | 61.70 |
|---|---|---|---|---|---|

**Door Closers**
Residential and light commercial door closers, doors to 140 lbs and 38" wide. 150 degree maximum swing, non-hold open, adjustable closing speed, ANSI A156.4

| Brown or white finish | BC@.750 | Ea | 15.50 | 27.70 | 43.20 |
|---|---|---|---|---|---|

Heavy duty door closers, doors to 170 lbs. and 36" wide. Schlage

| Aluminum | BC@1.04 | Ea | 59.30 | 38.40 | 97.70 |
|---|---|---|---|---|---|
| Aluminum with cover | BC@1.04 | Ea | 75.40 | 38.40 | 113.80 |

Screen door closers, brass finish

| Pneumatic closer, light duty | BC@.546 | Ea | 12.50 | 20.20 | 32.70 |
|---|---|---|---|---|---|

Sliding door closers, pneumatic, aluminum finish

| Screen | BC@.250 | Ea | 18.50 | 9.24 | 27.74 |
|---|---|---|---|---|---|
| Glass | BC@.250 | Ea | 39.20 | 9.24 | 48.44 |

# Doors, Wood Exterior Slab

| | Craft@Hrs | Unit | Material | Labor | Total |
|---|---|---|---|---|---|

**Wood Exterior Slab Doors** Exterior doors are usually 1-3/4" thick and solid core – a thin veneer laid over a frame but with foam or composition board filling the door cavity. Flush doors are smooth and flat, with no decorative treatment. Slab doors are the door alone, as distinguished from prehung doors which are sold with jamb, hinges, trim and threshold attached. Installation times for slab doors will be up to 50% faster when an experienced carpenter has the use of a router, door jig and pneumatic nailer.

Flush hardboard exterior doors. Solid core. 1-3/4" thick. Primed and ready to paint. Hardboard doors are made of composition wood veneer, usually the least expensive material available. Labor includes setting three hinges on the door and jamb and hanging the door in an existing cased opening.

| | | | | | |
|---|---|---|---|---|---|
| 30" x 80" | BC@1.45 | Ea | 57.10 | 53.60 | 110.70 |
| 32" x 80" | BC@1.45 | Ea | 57.00 | 53.60 | 110.60 |
| 34" x 80" | BC@1.45 | Ea | 58.40 | 53.60 | 112.00 |
| 36" x 80" | BC@1.45 | Ea | 57.30 | 53.60 | 110.90 |

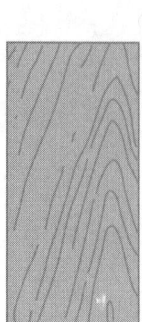

Flush hardwood exterior doors. Solid core. 1-3/4" thick. Labor includes setting three hinges on the door and jamb and hanging the door in an existing cased opening.

| | | | | | |
|---|---|---|---|---|---|
| 28" x 80" | BC@1.45 | Ea | 57.50 | 53.60 | 111.10 |
| 30" x 80" | BC@1.45 | Ea | 57.40 | 53.60 | 111.00 |
| 32" x 80" | BC@1.45 | Ea | 57.50 | 53.60 | 111.10 |
| 36" x 80" | BC@1.45 | Ea | 57.70 | 53.60 | 111.30 |
| 36" x 84" | BC@1.45 | Ea | 58.40 | 53.60 | 112.00 |
| 30" x 80", diamond light | BC@1.45 | Ea | 68.90 | 53.60 | 122.50 |
| 30" x 80", square light | BC@1.45 | Ea | 68.90 | 53.60 | 122.50 |
| 32" x 80", square light | BC@1.45 | Ea | 71.30 | 53.60 | 124.90 |
| 36" x 80", diamond light | BC@1.45 | Ea | 74.70 | 53.60 | 128.30 |
| 36" x 80", square light | BC@1.45 | Ea | 74.70 | 53.60 | 128.30 |

Flush birch exterior doors. Solid core.1-3/4" thick. Stainable. Labor includes setting three hinges on the door and jamb and hanging the door in an existing cased opening.

| | | | | | |
|---|---|---|---|---|---|
| 32" x 80" | BC@1.45 | Ea | 57.40 | 53.60 | 111.00 |
| 34" x 80" | BC@1.45 | Ea | 62.20 | 53.60 | 115.80 |
| 36" x 80" | BC@1.45 | Ea | 57.60 | 53.60 | 111.20 |

Flush lauan exterior doors. Solid core. 1-3/4" thick. Stainable. Lauan is a tropical hardwood like mahogany. Labor includes setting three hinges on the door and jamb and hanging the door in an existing cased opening.

| | | | | | |
|---|---|---|---|---|---|
| 30" x 80" | BC@1.45 | Ea | 55.20 | 53.60 | 108.80 |
| 30" x 84" | BC@1.45 | Ea | 57.30 | 53.60 | 110.90 |
| 32" x 80" | BC@1.45 | Ea | 54.80 | 53.60 | 108.40 |
| 36" x 80" | BC@1.45 | Ea | 55.00 | 53.60 | 108.60 |

Red oak exterior doors. Wood stave core. 1-3/4" thick. Labor includes setting three hinges on the door and jamb and hanging the door in an existing cased opening.

| | | | | | |
|---|---|---|---|---|---|
| 30" x 80" | BC@1.45 | Ea | 65.00 | 53.60 | 118.60 |
| 32" x 80" | BC@1.45 | Ea | 66.60 | 53.60 | 120.20 |
| 34" x 80" | BC@1.45 | Ea | 70.40 | 53.60 | 124.00 |
| 36" x 80" | BC@1.45 | Ea | 72.60 | 53.60 | 126.20 |

6-panel fir exterior doors. 1-3/4" thick. Beveled, hip-raised panels. Doweled stile and rail joints. 4-1/2" wide stiles. Unfinished clear fir. Ready to paint or stain. Labor includes setting three hinges on the door and jamb and hanging the door in an existing cased opening.

| | | | | | |
|---|---|---|---|---|---|
| 30" x 80" | BC@1.45 | Ea | 166.00 | 53.60 | 219.60 |
| 32" x 80" | BC@1.45 | Ea | 168.00 | 53.60 | 221.60 |
| 36" x 80" | BC@1.45 | Ea | 173.00 | 53.60 | 226.60 |

4-panel 4-lite fir exterior doors. Vertical grain Douglas fir. 1-3/4" thick. 3/4" thick panels. Unfinished. Veneered and doweled construction. 1/8" tempered safety glass. Labor includes setting three hinges on the door and jamb and hanging the door in an existing cased opening.

| | | | | | |
|---|---|---|---|---|---|
| 36" x 80" x 1-3/4" | BC@1.70 | Ea | 202.00 | 62.80 | 264.80 |
| 36" x 80" x 1-3/4" | BC@1.70 | Ea | 165.00 | 62.80 | 227.80 |

| | Craft@Hrs | Unit | Material | Labor | Total |
|---|---|---|---|---|---|

**Entry Slab Doors** Any exterior door can be use at the front entry of a home. But entry doors usually include decorative elements, such as glass panels set in a border of brass caming. Slab doors are the door alone, as distinguished from prehung doors which are sold with jamb, hinges, trim and threshold attached. Installation times for entry doors will be up to 50% faster when an experienced carpenter has the use of a router, door jig and pneumatic nailer.

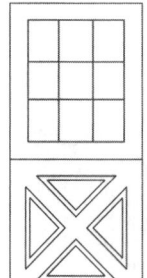

Fir crossbuck doors. 1-3/4" thick. Solid core. Labor includes setting three hinges on the door and jamb and hanging the door in an existing cased opening. Crossbuck doors have a raised "X" design in the bottom half of the door.

| | Craft@Hrs | Unit | Material | Labor | Total |
|---|---|---|---|---|---|
| 32" x 80" x 1-3/4" | BC@1.45 | Ea | 203.00 | 53.60 | 256.60 |
| 36" x 80" | BC@1.45 | Ea | 207.00 | 53.60 | 260.60 |

Mahogany entry doors. 1-3/4" thick. Beveled tempered glass lite. Unfinished solid wood. Ready to paint or stain. Labor includes setting three hinges on the door and jamb and hanging the door in an existing cased opening.

| | Craft@Hrs | Unit | Material | Labor | Total |
|---|---|---|---|---|---|
| 30" x 80", arch lite | BC@1.45 | Ea | 432.00 | 53.60 | 485.60 |
| 30" x 80", moon lite | BC@1.45 | Ea | 432.00 | 53.60 | 485.60 |
| 30" x 80", oval lite | BC@1.45 | Ea | 454.00 | 53.60 | 507.60 |

2-panel 9-lite entry doors. 1-3/4" thick. Single-pane glass. Solid core with doweled joints. Unfinished. Suitable for stain or paint. Labor includes setting three hinges on the door and jamb and hanging the door in an existing cased opening.

| | Craft@Hrs | Unit | Material | Labor | Total |
|---|---|---|---|---|---|
| 30" x 80" | BC@1.70 | Ea | 194.00 | 62.80 | 256.80 |
| 32" x 80" | BC@1.70 | Ea | 194.00 | 62.80 | 256.80 |
| 36" x 80" | BC@1.70 | Ea | 194.00 | 62.80 | 256.80 |

1-lite clear fir entry doors. 1-3/4" thick. Clear fir surface ready to stain or paint. Insulated glass. Labor includes setting three hinges on the door and jamb and hanging the door in an existing cased opening.

| | Craft@Hrs | Unit | Material | Labor | Total |
|---|---|---|---|---|---|
| 30" x 80" | BC@1.70 | Ea | 187.00 | 62.80 | 249.80 |
| 32" x 80" | BC@1.70 | Ea | 191.00 | 62.80 | 253.80 |
| 36" x 80" | BC@1.70 | Ea | 193.00 | 62.80 | 255.80 |

Fir 15-lite entry doors. Single-pane glass. Ready for stain or paint. 1-3/4" thick. Doweled stile and rail joints. Labor includes setting three hinges on the door and jamb and hanging the door in an existing cased opening.

| | Craft@Hrs | Unit | Material | Labor | Total |
|---|---|---|---|---|---|
| 30" x 80" x 1-3/4" | BC@1.70 | Ea | 194.00 | 62.80 | 256.80 |
| 32" x 80" x 1-3/4" | BC@1.70 | Ea | 194.00 | 62.80 | 256.80 |
| 36" x 80" x 1-3/4" | BC@1.70 | Ea | 194.00 | 62.80 | 256.80 |

Fir fan lite entry doors. Vertical grain Douglas fir. 1-3/4" thick. 3/4" thick panels. Unfinished. Veneer and doweled construction. 1/8" tempered safety glass. Labor includes setting three hinges on the door and jamb and hanging the door in an existing cased opening.

| | Craft@Hrs | Unit | Material | Labor | Total |
|---|---|---|---|---|---|
| 32" x 80" | BC@1.45 | Ea | 218.00 | 53.60 | 271.60 |
| 30" x 80" | BC@1.45 | Ea | 208.00 | 53.60 | 261.60 |

Deco fan lite entry doors. 1-3/4" thick. 3/4" thick panels. Unfinished solid wood. Ready to paint or stain. Beveled insulated glass. Labor includes setting three hinges on the door and jamb and hanging the door in an existing cased opening

| | Craft@Hrs | Unit | Material | Labor | Total |
|---|---|---|---|---|---|
| 36" x 80" | BC@1.45 | Ea | 235.00 | 53.60 | 288.60 |

Flush oak entry doors. 1-3/4" thick. Prefinished. Labor includes setting three hinges on the door and jamb and hanging the door in an existing cased opening.

| | Craft@Hrs | Unit | Material | Labor | Total |
|---|---|---|---|---|---|
| 36" x 80" | BC@1.45 | Ea | 941.00 | 53.60 | 994.60 |

# Doors, Beveled Glass

|  | Craft@Hrs | Unit | Material | Labor | Total |
|---|---|---|---|---|---|

**Beveled glass entry doors** Meranti mahogany with beveled glass insert. Matching sidelites. Transom designed to fit over one 36" door and two 14" sidelites combination. Door dimensions are 36" x 80" x 1-3/4". Labor cost includes hanging door only. Add the cost of hinges and lockset.

| | Craft@Hrs | Unit | Material | Labor | Total |
|---|---|---|---|---|---|
| Bolero, oval insert | | | | | |
|     Door | BC@1.45 | Ea | 1,130.00 | 53.60 | 1,183.60 |
|     Sidelight | BC@1.00 | Ea | 764.00 | 36.90 | 800.90 |
|     Matching glass transom | BC@1.00 | Ea | 641.00 | 36.90 | 677.90 |
| Minuet, oval insert | | | | | |
|     Door | BC@1.45 | Ea | 1,180.00 | 53.60 | 1,233.60 |
|     Sidelight | BC@1.00 | Ea | 617.00 | 36.90 | 653.90 |
|     Matching glass transom | BC@1.00 | Ea | 701.00 | 36.90 | 737.90 |
| Concerto, arched insert | | | | | |
|     Door | BC@1.45 | Ea | 982.00 | 53.60 | 1,035.60 |
|     Sidelight | BC@1.00 | Ea | 739.00 | 36.90 | 775.90 |
|     Matching glass transom | BC@1.00 | Ea | 744.00 | 36.90 | 780.90 |
| Sonata, narrow arched insert | | | | | |
|     Door | BC@1.45 | Ea | 1,030.00 | 53.60 | 1,083.60 |
|     Sidelight | BC@1.00 | Ea | 789.00 | 36.90 | 825.90 |
|     Matching glass transom | BC@1.00 | Ea | 766.00 | 36.90 | 802.90 |
| Tremolo, full rectangular insert | | | | | |
|     Door | BC@1.45 | Ea | 1,290.00 | 53.60 | 1,343.60 |
|     Sidelight | BC@1.00 | Ea | 743.00 | 36.90 | 779.90 |
|     Matching glass transom | BC@1.00 | Ea | 736.00 | 36.90 | 772.90 |

**Trim for Exterior Slab Doors** Trim for an exterior slab door includes two side jambs, head jamb, threshold, stop molding, interior casing and exterior molding. Installation time for exterior door trim will be up to 33% faster when an experienced carpenter has the use of a pneumatic nailer.

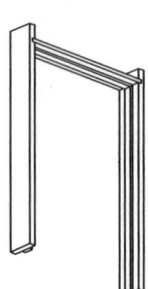

Primed exterior door jamb. Primed finger joint pine. 1-1/4" thick. Set for a 3' door consists of two 6'8" legs and one 3' head.

| | Craft@Hrs | Unit | Material | Labor | Total |
|---|---|---|---|---|---|
|     4-5/8" wide, per 6'8" leg | BC@.204 | Ea | 27.60 | 7.54 | 35.14 |
|     4-5/8" wide, per set | BC@.500 | Ea | 55.20 | 18.50 | 73.70 |
|     5-1/4" wide, per 6'8" leg | BC@.204 | Ea | 32.60 | 7.54 | 40.14 |
|     5-1/4" wide, per set | BC@.500 | Ea | 65.20 | 18.50 | 83.70 |
|     6-9/16" wide, per 6'8" leg | BC@.204 | Ea | 23.30 | 7.54 | 30.84 |
|     6-9/16" wide, per set | BC@.500 | Ea | 46.60 | 18.50 | 65.10 |

Primed kerfed exterior door jamb. 1-1/4" thick. Set for a 3' door consists of two 6'8" legs and one 3' head.

| | Craft@Hrs | Unit | Material | Labor | Total |
|---|---|---|---|---|---|
|     4-5/8" x 7' leg | BC@.204 | Ea | 25.90 | 7.54 | 33.44 |
|     4-5/8" x 3' head | BC@.092 | Ea | 9.70 | 3.40 | 13.10 |
|     6-5/8" x 7' leg | BC@.204 | Ea | 31.10 | 7.54 | 38.64 |
|     6-5/8" x 3' head | BC@.092 | Ea | 10.40 | 3.40 | 13.80 |

Oak high boy threshold. Fits standard door sizes.

| | Craft@Hrs | Unit | Material | Labor | Total |
|---|---|---|---|---|---|
|     5/8" x 3-1/2" x 36" | BC@.518 | Ea | 13.60 | 19.10 | 32.70 |

Stop molding. Nails to face of the door jamb to prevent door from swinging through the opening. Solid pine.

| | Craft@Hrs | Unit | Material | Labor | Total |
|---|---|---|---|---|---|
|     3/8" x 1-1/4" | BC@.015 | LF | .72 | .55 | 1.27 |
|     7/16" x 3/4" | BC@.015 | LF | .53 | .55 | 1.08 |
|     7/16" x 1-3/8" | BC@.015 | LF | .81 | .55 | 1.36 |
|     7/16" x 1-5/8" | BC@.015 | LF | .80 | .55 | 1.35 |
|     7/16" x 2-1/8" | BC@.015 | LF | 1.25 | .55 | 1.80 |

Solid pine ranch casing. Casing is the trim around each side and head of the interior side of an exterior door. Casing covers the joint between jamb and the wall.

| | Craft@Hrs | Unit | Material | Labor | Total |
|---|---|---|---|---|---|
|     9/16" x 2-1/4" | BC@.015 | LF | .73 | .55 | 1.28 |
|     9/16" x 2-1/4" x 7' | BC@.105 | Ea | 5.14 | 3.88 | 9.02 |

| | Craft@Hrs | Unit | Material | Labor | Total |
|---|---|---|---|---|---|
| 11/16" x 3-1/2" | BC@.015 | LF | 1.79 | .55 | 2.34 |
| 11/16" x 2-1/2" x 7' | BC@.015 | Ea | 9.18 | .55 | 9.73 |

Primed finger joint ranch door casing.

| | Craft@Hrs | Unit | Material | Labor | Total |
|---|---|---|---|---|---|
| 9/16" x 3-1/4" | BC@.015 | LF | 1.34 | .55 | 1.89 |
| 11/16" x 2-1/4" | BC@.015 | LF | .95 | .55 | 1.50 |
| 11/16" x 2-1/4" x 7' | BC@.105 | Ea | 6.62 | 3.88 | 10.50 |
| 11/16" x 2-1/2" | BC@.015 | LF | 1.10 | .55 | 1.65 |
| 11/16" x 2-1/2" x 7' | BC@.105 | Ea | 7.53 | 3.88 | 11.41 |

Fluted pine casing and rosette set. Includes 1 header piece, 2 side pieces, 2 rosettes and 2 base blocks.

| | Craft@Hrs | Unit | Material | Labor | Total |
|---|---|---|---|---|---|
| Set | BC@.250 | Ea | 37.60 | 9.24 | 46.84 |

Brick mold set. Trim for the exterior side of an exterior door. Two legs and one head. Primed finger joint pine. 36" x 80" door. 1-1/4" x 2".

| | Craft@Hrs | Unit | Material | Labor | Total |
|---|---|---|---|---|---|
| Finger joint pine | BC@.250 | Ea | 37.80 | 9.24 | 47.04 |
| Pressure treated pine | BC@.250 | Ea | 22.70 | 9.24 | 31.94 |

Brick mold. Brick mold and stucco mold are the casing around each side and head on the exterior side of an exterior door.

| | Craft@Hrs | Unit | Material | Labor | Total |
|---|---|---|---|---|---|
| 11/16" x 1-5/8", finger joint pine | BC@.015 | LF | 1.59 | .55 | 2.14 |
| 1-3/16" x 2", hemlock | BC@.015 | LF | 1.87 | .55 | 2.42 |
| 1-1/4" x 2", fir | BC@.015 | LF | 1.26 | .55 | 1.81 |
| 1-1/4" x 2", solid pine | BC@.015 | LF | 2.45 | .55 | 3.00 |

Stucco molding. Redwood.

| | Craft@Hrs | Unit | Material | Labor | Total |
|---|---|---|---|---|---|
| 13/16" x 1-1/2" | BC@.015 | LF | 1.03 | .55 | 1.58 |
| 1" x 1-3/8" | BC@.015 | LF | 1.03 | .55 | 1.58 |

T-astragal molding. Covers the vertical joint where a pair of doors meet.

| | Craft@Hrs | Unit | Material | Labor | Total |
|---|---|---|---|---|---|
| 1-1/4" x 2" x 7', pine | BC@.529 | Ea | 22.10 | 19.50 | 41.60 |

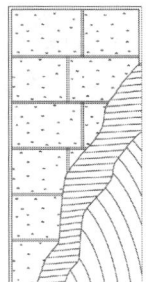

**Fire Rated Doors** Fire rated doors are designed to meet building code requirements for specialized occupancies, such as hotel rooms and dormitories. Most building codes require that fire doors be used between a house and an attached garage.

6-panel prehung steel fire-rated exterior doors. 24-gauge galvanized steel. Foam core. Caming matches hinge and sill finishes. Jamb guard security plate. No brick mold. Fixed sill. 4-9/16" primed jamb. Triple sweep compression weatherstripping. Factory primed. Thermal-break threshold. 20-minute fire rating. 10-year limited warranty.

| | Craft@Hrs | Unit | Material | Labor | Total |
|---|---|---|---|---|---|
| 30" x 80" | BC@1.00 | Ea | 156.00 | 36.90 | 192.90 |
| 32" x 80" | BC@1.00 | Ea | 156.00 | 36.90 | 192.90 |
| 36" x 80" | BC@1.00 | Ea | 156.00 | 36.90 | 192.90 |

Prehung hardboard fire-rated doors. 20-minute fire rating. 80" high. Labor includes hanging the door in an existing framed opening.

| | Craft@Hrs | Unit | Material | Labor | Total |
|---|---|---|---|---|---|
| 30" x 80" | BC@1.00 | Ea | 156.00 | 36.90 | 192.90 |
| 32" x 80" | BC@1.00 | Ea | 161.00 | 36.90 | 197.90 |
| 36" x 80" | BC@1.00 | Ea | 165.00 | 36.90 | 201.90 |

6-panel prehung fire-rated interior doors. 4-9/16" jamb. 20-minute fire rating.

| | Craft@Hrs | Unit | Material | Labor | Total |
|---|---|---|---|---|---|
| 32" x 80" | BC@.750 | Ea | 216.00 | 27.70 | 243.70 |
| 36" x 80" | BC@.750 | Ea | 220.00 | 27.70 | 247.70 |

Flush fire-rated doors. 1-3/4" thick. 20-minute fire rating. Labor includes setting three hinges on the door and jamb and hanging the door in an existing cased opening.

| | Craft@Hrs | Unit | Material | Labor | Total |
|---|---|---|---|---|---|
| 32" x 80", hardboard veneer | BC@1.45 | Ea | 62.40 | 53.60 | 116.00 |
| 36" x 80", hardboard veneer | BC@1.45 | Ea | 62.40 | 53.60 | 116.00 |
| 32" x 80", birch veneer | BC@1.45 | Ea | 72.20 | 53.60 | 125.80 |
| 36" x 80", birch veneer | BC@1.45 | Ea | 72.20 | 53.60 | 125.80 |

Reversible hollow metal fire doors.

| | Craft@Hrs | Unit | Material | Labor | Total |
|---|---|---|---|---|---|
| 34" x 80" | BC@.721 | Ea | 156.00 | 26.60 | 182.60 |
| 36" x 80" | BC@.721 | Ea | 184.00 | 26.60 | 210.60 |

## Doors, Prehung Wood Exterior

| | Craft@Hrs | Unit | Material | Labor | Total |
|---|---|---|---|---|---|

Knock down steel door frames. Bored for (3) 4-1/2" hinges, 4-7/8" strike plate and 2-3/4" backset. 90-minute fire label manufactured to N.Y.C. M.E.A. procedures. Hollow core. For drywall and masonry applications. Galvanized steel.

| | Craft@Hrs | Unit | Material | Labor | Total |
|---|---|---|---|---|---|
| 30" x 80" | BC@.944 | Ea | 80.00 | 34.90 | 114.90 |
| 32" x 80" | BC@.944 | Ea | 78.50 | 34.90 | 113.40 |
| 34" x 80" | BC@.944 | Ea | 78.50 | 34.90 | 113.40 |
| 36" x 80" | BC@.944 | Ea | 80.20 | 34.90 | 115.10 |
| 36" x 84" | BC@.944 | Ea | 80.90 | 34.90 | 115.80 |

Knock down trimless door frames.

| | Craft@Hrs | Unit | Material | Labor | Total |
|---|---|---|---|---|---|
| 36" x 80" | BC@.944 | Ea | 67.60 | 34.90 | 102.50 |

90-minute fire-rated door frame parts.

| | Craft@Hrs | Unit | Material | Labor | Total |
|---|---|---|---|---|---|
| 4-7/8" x 80" legs, pack of 2 | BC@.778 | Ea | 71.90 | 28.70 | 100.60 |
| 4-7/8" x 36" head | BC@.166 | Ea | 19.10 | 6.13 | 25.23 |

Steel cellar doors. Durable heavy gauge steel.

| | Craft@Hrs | Unit | Material | Labor | Total |
|---|---|---|---|---|---|
| 57" x 45" x 24-1/2" | BC@1.00 | Ea | 347.00 | 36.90 | 383.90 |
| 63" x 49" x 22" | BC@1.00 | Ea | 354.00 | 36.90 | 390.90 |
| 71" x 53" x 19-1/2" | BC@1.00 | Ea | 369.00 | 36.90 | 405.90 |

**Prehung Wood Exterior Doors**  Prehung doors are sold as a package with the door, jamb, hinges, weatherstripping, threshold and trim already assembled.

Flush hardwood prehung exterior doors. 1-3/4" door. Solid particleboard core. Lauan veneer. Weatherstripped. Includes oak sill and threshold with vinyl insert. Finger joint 4-9/16" jamb and finger joint brick mold. 2-1/8" bore for lockset with 2-3/8" backset. Flush doors are smooth and flat with no decorative treatment. Labor includes hanging the door in an existing framed opening.

| | Craft@Hrs | Unit | Material | Labor | Total |
|---|---|---|---|---|---|
| 32" x 80" | BC@1.00 | Ea | 111.00 | 36.90 | 147.90 |
| 36" x 80" | BC@1.00 | Ea | 111.00 | 36.90 | 147.90 |

Flush birch prehung exterior doors. 1-3/4" door. Solid particleboard core. Weatherstripped. Includes oak sill and threshold. Finger joint 4-9/16" jamb and finger joint brick mold. 2-1/8" bore for lockset with 2-3/8" backset. Labor includes hanging the door in an existing framed opening.

| | Craft@Hrs | Unit | Material | Labor | Total |
|---|---|---|---|---|---|
| 36" x 80" | BC@1.00 | Ea | 110.00 | 36.90 | 146.90 |

Flush hardboard prehung exterior doors. Primed. Solid core. Weatherstripped. Includes sill and threshold. Finger joint 4-9/16" jamb and finger joint brick mold. 2-1/8" bore for lockset with 2-3/8" backset. Labor includes hanging the door in an existing framed opening.

| | Craft@Hrs | Unit | Material | Labor | Total |
|---|---|---|---|---|---|
| 32" x 80" | BC@1.00 | Ea | 111.00 | 36.90 | 147.90 |
| 36" x 80" | BC@1.00 | Ea | 111.00 | 36.90 | 147.90 |

Fan lite prehung hemlock exterior doors. Weatherstripped. Includes sill and threshold. Finger joint 4-9/16" jamb and finger joint brick mold. 2-1/8" bore for lockset with 2-3/8" backset. Labor includes hanging the door in an existing framed opening.

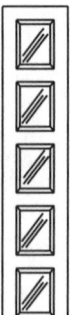

| | Craft@Hrs | Unit | Material | Labor | Total |
|---|---|---|---|---|---|
| 36" x 80" | BC@1.00 | Ea | 322.00 | 36.90 | 358.90 |

9-lite fir prehung exterior doors. Inward swing. No brick mold. Solid core. Weatherstripped. Includes sill and threshold. Finger joint 4-9/16" jamb and finger joint brick mold. 2-1/8" bore for lockset with 2-3/8" backset. Brick mold (also called stucco mold) is the casing around each side and head on the exterior side of an exterior door. Finger joint trim is made from two or more lengths of wood joined together with a finger-like joint that yields nearly as much strength as trim made of a single, solid piece. Labor includes hanging the door in an existing framed opening.

| | Craft@Hrs | Unit | Material | Labor | Total |
|---|---|---|---|---|---|
| 36" x 80" | BC@1.00 | Ea | 260.00 | 36.90 | 296.90 |

Sidelites for wood prehung exterior doors, 1-3/4" thick, with insulated glass, assembled.

| | Craft@Hrs | Unit | Material | Labor | Total |
|---|---|---|---|---|---|
| 6'8" x 1'2", fir or hemlock | BC@1.36 | Ea | 437.00 | 50.20 | 487.20 |

| | Craft@Hrs | Unit | Material | Labor | Total |
|---|---|---|---|---|---|

**Prehung Steel Exterior Doors** Prehung doors are sold as a package with the door, jamb, hinges, weatherstripping, threshold and trim already assembled. Flush doors have a smooth, flat face with no decorative treatment.

Utility flush prehung steel exterior doors. No brick mold. 26-gauge galvanized steel. Foam core. Compression weatherstrip. Triple fin sweep. Non-thermal threshold. Single bore

| | Craft@Hrs | Unit | Material | Labor | Total |
|---|---|---|---|---|---|
| 32" x 80" | BC@1.00 | Ea | 130.00 | 36.90 | 166.90 |
| 36" x 80" | BC@1.00 | Ea | 130.00 | 36.90 | 166.90 |

Premium flush prehung steel exterior doors. No brick mold. 24-gauge galvanized steel. Foam core. Compression weatherstrip. Triple fin sweep. Thermal threshold.

| | Craft@Hrs | Unit | Material | Labor | Total |
|---|---|---|---|---|---|
| 30" x 80" | BC@1.00 | Ea | 156.00 | 36.90 | 192.90 |
| 32" x 80" | BC@1.00 | Ea | 156.00 | 36.90 | 192.90 |
| 36" x 80" | BC@1.00 | Ea | 156.00 | 36.90 | 192.90 |

6-panel prehung steel exterior doors. With brick mold. Inward swing. Non-thermal threshold. Bored for lockset with 2-3/4" backset. Includes compression weatherstrip and thermal-break threshold. Panel doors have decorative raised panels in the colonial style. Six or nine panels are most common.

| | Craft@Hrs | Unit | Material | Labor | Total |
|---|---|---|---|---|---|
| 32" x 80" | BC@1.00 | Ea | 130.00 | 36.90 | 166.90 |
| 36" x 80" | BC@1.00 | Ea | 130.00 | 36.90 | 166.90 |

6-panel premium prehung exterior steel doors. 4-9/16" jamb. Adjustable thermal-break sill. Inward swing. 24-gauge galvanized steel. Bored for 2-3/8" lockset. 12" lock block. Impact-resistant laminated glass. Polyurethane core. Triple sweep and compression weatherstripping. Factory primed.

| | Craft@Hrs | Unit | Material | Labor | Total |
|---|---|---|---|---|---|
| 32" x 80", with brick mold | BC@1.00 | Ea | 156.00 | 36.90 | 192.90 |
| 32" x 80", no brick mold | BC@1.00 | Ea | 156.00 | 36.90 | 192.90 |
| 36" x 80", no brick mold | BC@1.00 | Ea | 156.00 | 36.90 | 192.90 |
| 36" x 80", with brick mold | BC@1.00 | Ea | 156.00 | 36.90 | 192.90 |
| 72" x 80", no brick mold (double door) | BC@1.50 | Ea | 312.00 | 55.40 | 367.40 |

6-panel prehung steel entry door with sidelites. Two 10"-wide sidelites. Sidelites have 5 divided lites.

| | Craft@Hrs | Unit | Material | Labor | Total |
|---|---|---|---|---|---|
| 36" x 80" | BC@2.00 | Ea | 482.00 | 73.90 | 555.90 |

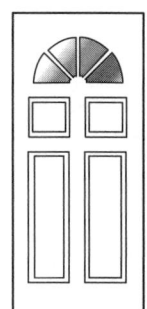

Fan lite premium prehung steel exterior doors. With brick mold. 24-gauge galvanized steel. Foam core. Fixed sill. Compression weatherstripping. Factory primed. Thermal-break threshold.

| | Craft@Hrs | Unit | Material | Labor | Total |
|---|---|---|---|---|---|
| 32" x 80" | BC@1.00 | Ea | 207.00 | 36.90 | 243.90 |
| 36" x 80" | BC@1.00 | Ea | 207.00 | 36.90 | 243.90 |

Half lite prehung steel exterior doors with mini-blind. 22" x 36" light. Adjustable sill. 24-gauge galvanized steel. Non-yellowing, white, warp-resistant, paintable, high-performance light frames. Brick mold applied.

| | Craft@Hrs | Unit | Material | Labor | Total |
|---|---|---|---|---|---|
| 32" x 80", 4-9/16" jamb | BC@1.00 | Ea | 302.00 | 36.90 | 338.90 |
| 32" x 80", 6-9/16" jamb | BC@1.00 | Ea | 322.00 | 36.90 | 358.90 |
| 36" x 80", 4-9/16" jamb | BC@1.00 | Ea | 302.00 | 36.90 | 338.90 |
| 36" x 80", 6-9/16" jamb | BC@1.00 | Ea | 322.00 | 36.90 | 358.90 |

Full lite prehung steel exterior doors. 22" x 64" light. 4-9/16" jamb. Adjustable sill. No brick mold. 24-gauge galvanized steel.

| | Craft@Hrs | Unit | Material | Labor | Total |
|---|---|---|---|---|---|
| 32" x 80" | BC@1.00 | Ea | 257.00 | 36.90 | 293.90 |
| 36" x 80" | BC@1.00 | Ea | 257.00 | 36.90 | 293.90 |

2-lite, 4-panel prehung steel doors. 4-9/16" jamb. Adjustable sill. Ready-to-install door and jamb system. 24-gauge galvanized steel. Non-yellowing, white, warp-resistant, paintable, high-performance light frames. Brick mold applied.

| | Craft@Hrs | Unit | Material | Labor | Total |
|---|---|---|---|---|---|
| 30" x 80" | BC@1.00 | Ea | 196.00 | 36.90 | 232.90 |
| 34" x 80" | BC@1.00 | Ea | 196.00 | 36.90 | 232.90 |
| 36" x 80" | BC@1.00 | Ea | 196.00 | 36.90 | 232.90 |

# Doors, Flush Ventilating Prehung Steel Exterior

|  | Craft@Hrs | Unit | Material | Labor | Total |
|---|---|---|---|---|---|

9-lite prehung steel exterior doors. Insulating glass with internal 9-lite grille. Adjustable sill. Ready-to-install door and jamb system. 24-gauge galvanized steel. Non-yellowing, white, warp-resistant, paintable, high-performance light frames. Brick mold applied.

| | Craft@Hrs | Unit | Material | Labor | Total |
|---|---|---|---|---|---|
| 32" x 80", 4-9/16" jamb | BC@1.00 | Ea | 224.00 | 36.90 | 260.90 |
| 32" x 80", 6-9/16" jamb | BC@1.00 | Ea | 224.00 | 36.90 | 260.90 |

10-lite prehung steel exterior doors. 4-9/16" jamb. 22" x 36" light. Adjustable sill. With brick mold.

| | Craft@Hrs | Unit | Material | Labor | Total |
|---|---|---|---|---|---|
| 32" x 80" | BC@1.00 | Ea | 257.00 | 36.90 | 293.90 |
| 36" x 80" | BC@1.00 | Ea | 257.00 | 36.90 | 293.90 |

15-lite prehung steel exterior doors. Multi-paned insulated glass. 24-gauge hot-dipped galvanized steel. Polyurethane foam core. Compression weatherstrip. Thermal-break construction. No brick mold. Adjustable seal. Aluminum sill with composite adjustable threshold.

| | Craft@Hrs | Unit | Material | Labor | Total |
|---|---|---|---|---|---|
| 32" x 80" | BC@1.00 | Ea | 257.00 | 36.90 | 293.90 |
| 36" x 80" | BC@1.00 | Ea | 257.00 | 36.90 | 293.90 |

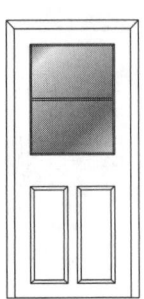

**Flush Ventilating Prehung Steel Exterior Doors** No brick mold. 4-9/16" primed jamb. 24-gauge galvanized steel. Foam core. Magnetic weatherstrip. Triple seal sweep. Thermal-break threshold. Bored for lockset and deadbolt. 20-minute fire rating. Ventilating doors include a window that can be opened.

| | Craft@Hrs | Unit | Material | Labor | Total |
|---|---|---|---|---|---|
| 30" x 80" | BC@1.00 | Ea | 257.00 | 36.90 | 293.90 |
| 32" x 80" | BC@1.00 | Ea | 257.00 | 36.90 | 293.90 |
| 36" x 80" | BC@1.00 | Ea | 257.00 | 36.90 | 293.90 |

**Decorative Prehung Steel Entry Doors** Entry doors are decorative exterior doors, often with glass panels set in a border of brass caming. Prehung doors are sold as a package with the door, jamb, hinges, threshold and trim already assembled.

Center arch lite prehung steel entry doors. 24-gauge galvanized steel. High-performance light frames. Decorative impact-resistant glass. Insulated core. 4-9/16" jamb with adjustable sill. Includes brick mold.

| | Craft@Hrs | Unit | Material | Labor | Total |
|---|---|---|---|---|---|
| 36" x 80" | BC@1.00 | Ea | 336.00 | 36.90 | 372.90 |

Camber top prehung steel entry doors. 4-9/16" jamb.

| | Craft@Hrs | Unit | Material | Labor | Total |
|---|---|---|---|---|---|
| 36" x 80", with brick mold | BC@1.00 | Ea | 294.00 | 36.90 | 330.90 |
| 36" x 80", no brick mold | BC@1.00 | Ea | 294.00 | 36.90 | 330.90 |

Camber top 6" jamb prehung steel entry doors. 6-9/16" jamb. With brick mold.

| | Craft@Hrs | Unit | Material | Labor | Total |
|---|---|---|---|---|---|
| 36" x 80" | BC@1.00 | Ea | 312.00 | 36.90 | 348.90 |

Decorative oval lite prehung steel entry door. 40" x 18" decorative oval laminated glass with brass caming to match hinge and sill finish. 24-gauge galvanized steel. Polyurethane core. Adjustable thermal-break sill. Triple sweep compression weatherstripping. Non-yellowing and warp-resistant light frame. Factory primed.

| | Craft@Hrs | Unit | Material | Labor | Total |
|---|---|---|---|---|---|
| 36" x 80", with brick mold, right hinge 4-9/16" jamb | BC@1.00 | Ea | 369.00 | 36.90 | 405.90 |

Decorative fan lite prehung steel entry doors. Polyurethane core. Adjustable thermal-break sill with composite saddle. 24-gauge galvanized steel. Laminated glass. Brass caming matches hinge and sill. Triple sweep and compression weatherstrip. Non-yellowing and warp-resistant light frame. Factory primed. Includes brick mold.

| | Craft@Hrs | Unit | Material | Labor | Total |
|---|---|---|---|---|---|
| 36" x 80", no brick mold | BC@1.00 | Ea | 257.00 | 36.90 | 293.90 |
| 36" x 80", with brick mold | BC@1.00 | Ea | 257.00 | 36.90 | 293.90 |

Decorative half lite prehung steel entry doors. 22" x 36" laminated safety glass. Brass caming matches sill and hinges. 24-gauge galvanized insulated steel.

| | Craft@Hrs | Unit | Material | Labor | Total |
|---|---|---|---|---|---|
| 36" x 80", no brick mold | BC@1.00 | Ea | 313.00 | 36.90 | 349.90 |
| 36" x 80", with brick mold | BC@1.00 | Ea | 313.00 | 36.90 | 349.90 |

**Fiberglass Prehung Entry Doors** Fiberglass doors require very little maintenance and are available in simulated wood finish. Prehung doors include the door, jamb, hinges, threshold and trim already assembled.

|  | Craft@Hrs | Unit | Material | Labor | Total |
|---|---|---|---|---|---|

6-panel fiberglass prehung exterior doors. Polyurethane core. PVC stiles and rails. Double bore. 4-9/16" prefinished jambs. Adjustable thermal-break brass threshold. Weatherstripped. With brick mold.

| | | | | | |
|---|---|---|---|---|---|
| 32" x 80" | BC@1.00 | Ea | 188.00 | 36.90 | 224.90 |
| 36" x 80" | BC@1.00 | Ea | 188.00 | 36.90 | 224.90 |

Fan lite prehung light oak fiberglass entry doors. With brick mold. Factory prefinished. Triple pane insulated glass with brass caming. Adjustable brass thermal-break threshold and fully weatherstripped. 4-9/16" jamb.

| | | | | | |
|---|---|---|---|---|---|
| 36" x 80" | BC@1.00 | Ea | 415.00 | 36.90 | 451.90 |

3/4 oval lite prehung fiberglass entry doors. With brick mold. Ready to finish.

| | | | | | |
|---|---|---|---|---|---|
| 36" x 80", 4-9/16" jamb | BC@1.00 | Ea | 369.00 | 36.90 | 405.90 |
| 36" x 80", 6-9/16" jamb | BC@1.00 | Ea | 392.00 | 36.90 | 428.90 |

3/4 oval lite prehung light oak fiberglass doors. With brick mold. Prefinished. Polyurethane core. Triple pane insulated glass. Glue chip and clear 30" x 18" beveled glass surrounded by brass caming. Double bore. Extended wood lock block. 4-9/16" primed jamb. Thermal-break brass threshold with adjustable oak cap.

| | | | | | |
|---|---|---|---|---|---|
| 36" x 80", 4-9/16" jamb | BC@1.00 | Ea | 527.00 | 36.90 | 563.90 |
| 36" x 80", 6-9/16" jamb | BC@1.00 | Ea | 566.00 | 36.90 | 602.90 |

Full height oval lite prehung medium oak fiberglass entry doors.

| | | | | | |
|---|---|---|---|---|---|
| 36" x 80", 4-9/16" jamb | BC@1.00 | Ea | 561.00 | 36.90 | 597.90 |
| 36" x 80", 6-9/16" jamb | BC@1.00 | Ea | 640.00 | 36.90 | 676.90 |

Center arch lite medium oak prehung fiberglass doors. Factory prefinished. Triple pane insulated glass with brass caming. Adjustable brass thermal-break threshold and fully weatherstripped. 4-9/16" jamb except as noted. PVC stiles and rails.

| | | | | | |
|---|---|---|---|---|---|
| 36" x 80", no brick mold | BC@1.00 | Ea | 442.00 | 36.90 | 478.90 |
| 36" x 80", with brick mold | BC@1.00 | Ea | 470.00 | 36.90 | 506.90 |
| 36" x 80", with brick mold, 6-9/16" jamb | BC@1.00 | Ea | 504.00 | 36.90 | 540.90 |

Fan lite prehung fiberglass entry doors. With brick mold. Ready to finish. Insulated glass. Adjustable thermal-break brass threshold. 4-9/16" jamb except as noted.

| | | | | | |
|---|---|---|---|---|---|
| 32" x 80" | BC@1.00 | Ea | 302.00 | 36.90 | 338.90 |
| 36" x 80" | BC@1.00 | Ea | 302.00 | 36.90 | 338.90 |
| 36" x 80", 6-9/16" jamb | BC@1.00 | Ea | 324.00 | 36.90 | 360.90 |

9-lite fiberglass prehung exterior doors. Inward swing. With brick mold. Smooth face.

| | | | | | |
|---|---|---|---|---|---|
| 32" x 80" | BC@1.00 | Ea | 268.00 | 36.90 | 304.90 |
| 36" x 80" | BC@1.00 | Ea | 268.00 | 36.90 | 304.90 |

15-lite smooth prehung fiberglass entry doors. No brick mold. Factory prefinished. Left or right hinge. Triple pane, insulated glass with brass caming. Adjustable brass thermal-break threshold and fully weatherstripped. 4-9/16" jamb.

| | | | | | |
|---|---|---|---|---|---|
| 32" x 80" | BC@1.00 | Ea | 280.00 | 36.90 | 316.90 |
| 36" x 80" | BC@1.00 | Ea | 280.00 | 36.90 | 316.90 |

Full lite prehung smooth fiberglass doors. With brick mold. Inward swing.

| | | | | | |
|---|---|---|---|---|---|
| 36" x 80", 4-9/16" jamb | BC@1.00 | Ea | 582.00 | 36.90 | 618.90 |
| 36" x 80", 6-9/16" jamb | BC@1.00 | Ea | 604.00 | 36.90 | 640.90 |

Smooth prehung fiberglass ventlite exterior doors. Ready to finish. No brick mold.

| | | | | | |
|---|---|---|---|---|---|
| 32" x 80" | BC@1.00 | Ea | 280.00 | 36.90 | 316.90 |
| 36" x 80" | BC@1.00 | Ea | 280.00 | 36.90 | 316.90 |

**Screen Doors, Wood** 3-3/4"-wide stile and rail with 7-1/4" bottom rail. Removable screen. Mortise and tenon solid construction. Wide center push bar. Ready to finish.

| | | | | | |
|---|---|---|---|---|---|
| 32" x 80" | BC@2.17 | Ea | 79.90 | 80.20 | 160.10 |
| 36" x 80" | BC@2.17 | Ea | 79.90 | 80.20 | 160.10 |

| | Craft@Hrs | Unit | Material | Labor | Total |
|---|---|---|---|---|---|

T-style wood screen doors. 3-3/4"-wide stile and rail with 7-1/4" bottom rail. Removable screen. Mortise and tenon solid construction. All joints sealed with water-resistant adhesive. Wide center push bar.

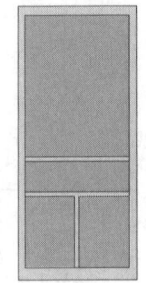

| | | | | | |
|---|---|---|---|---|---|
| 30" x 80" | BC@2.17 | Ea | 21.60 | 80.20 | 101.80 |
| 32" x 80" | BC@2.17 | Ea | 21.60 | 80.20 | 101.80 |
| 36" x 80" | BC@2.17 | Ea | 21.60 | 80.20 | 101.80 |

Metal screen doors. Includes hardware and fiberglass screening. 1-piece steel frame. Heavy-duty pneumatic closer. 5" kickplate.

| | | | | | |
|---|---|---|---|---|---|
| 32" x 80" | BC@1.35 | Ea | 46.80 | 49.90 | 96.70 |
| 36" x 80" | BC@1.35 | Ea | 46.80 | 49.90 | 96.70 |

Five-bar solid vinyl screen doors. Accepts standard screen door hardware.

| | | | | | |
|---|---|---|---|---|---|
| 32" x 80" | BC@1.35 | Ea | 90.70 | 49.90 | 140.60 |
| 36" x 80" | BC@1.35 | Ea | 74.50 | 49.90 | 124.40 |

Vinyl screen doors. Right or left side installation. 32" to 36" wide. Hardware included.

| | | | | | |
|---|---|---|---|---|---|
| 80" high, T-bar grid | BC@1.35 | Ea | 63.00 | 49.90 | 112.90 |
| 80" high, full panel grid | BC@1.35 | Ea | 84.10 | 49.90 | 134.00 |
| 80" high, decorative | BC@1.35 | Ea | 107.00 | 49.90 | 156.90 |

Steel security screen doors. Class I rated to withstand 300 pounds. All welded steel construction for strength and durability. Double lockbox with extension plate. Baked-on powder coating.

| | | | | | |
|---|---|---|---|---|---|
| 36" x 80" | BC@1.35 | Ea | 74.50 | 49.90 | 124.40 |

**Storm Doors** Self-storing aluminum storm and screen doors. Pneumatic closer and sweep. Tempered safety glass. Maintenance-free finish. Push-button hardware. 1" x 2-1/8" frame size. Bronze or white. Includes screen.

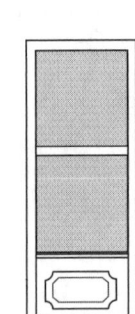

| | | | | | |
|---|---|---|---|---|---|
| 30" x 80" x 1-1/4" | BC@1.45 | Ea | 99.00 | 53.60 | 152.60 |
| 32" x 80" x 1-1/4" | BC@1.45 | Ea | 99.00 | 53.60 | 152.60 |
| 36" x 80" x 1-1/4" | BC@1.45 | Ea | 99.00 | 53.60 | 152.60 |

Self-storing vinyl-covered wood storm doors. White vinyl-covered wood core. Push-button handle set. Self-storing window and screen. Single black closer.

| | | | | | |
|---|---|---|---|---|---|
| 30" x 80" | BC@1.45 | Ea | 94.00 | 53.60 | 147.60 |
| 32" x 80" | BC@1.45 | Ea | 94.00 | 53.60 | 147.60 |
| 34" x 80" | BC@1.45 | Ea | 94.00 | 53.60 | 147.60 |
| 36" x 80" | BC@1.45 | Ea | 94.00 | 53.60 | 147.60 |

Store-in-Door™ storm doors. 1-1/2"-thick polypropylene frame. Reversible hinge. Slide window or screen, completely concealed when not in use. Brass-finished keylock hardware system with separate color-matched deadbolt. Triple-fin door sweep. 2 color-matched closers. Full-length piano hinge. Flexible weather seal. White.

| | | | | | |
|---|---|---|---|---|---|
| 30" x 80", crossbuck style | BC@1.45 | Ea | 280.00 | 53.60 | 333.60 |
| 32" x 80", crossbuck style | BC@1.45 | Ea | 280.00 | 53.60 | 333.60 |
| 36" x 80", crossbuck style | BC@1.45 | Ea | 280.00 | 53.60 | 333.60 |
| 30" x 80", traditional style | BC@1.45 | Ea | 280.00 | 53.60 | 333.60 |
| 36" x 80", traditional style | BC@1.45 | Ea | 291.00 | 53.60 | 344.60 |

Aluminum and wood storm doors. Aluminum over wood core. Self-storing window and screen. Black lever handle set. Separate deadbolt. Single black closer. White finish. Traditional panel style.

| | | | | | |
|---|---|---|---|---|---|
| 30" x 80" | BC@1.45 | Ea | 155.00 | 53.60 | 208.60 |
| 32" x 80" | BC@1.45 | Ea | 155.00 | 53.60 | 208.60 |
| 36" x 80" | BC@1.45 | Ea | 155.00 | 53.60 | 208.60 |

Triple-track storm doors. Low-maintenance aluminum over solid wood core. Ventilates from top, bottom or both. Solid brass exterior handle. Color-matched interior handle. Separate deadbolt. 80" high. White, almond or bronze.

| | | | | | |
|---|---|---|---|---|---|
| 30" x 80" | BC@1.45 | Ea | 183.00 | 53.60 | 236.60 |
| 32" x 80" | BC@1.45 | Ea | 215.00 | 53.60 | 268.60 |
| 34" x 80" | BC@1.45 | Ea | 183.00 | 53.60 | 236.60 |
| 36" x 80" | BC@1.45 | Ea | 183.00 | 53.60 | 236.60 |

|  | Craft@Hrs | Unit | Material | Labor | Total |
|---|---|---|---|---|---|

Colonial triple-track storm doors. 1-piece composite frame. 12-lite grille. Vents from top, bottom, or both. Solid-brass handle set with deadbolt security. Brass-finished sweep ensures tight seal across the entire threshold. Wood-grain finish. White, almond or sandstone.

| 32" x 80" | BC@1.45 | Ea | 291.00 | 53.60 | 344.60 |

Full-view brass all-season storm doors. Water-diverting rain cap. Brass-finished, color-matched sweep. One closer. Window and insect screen snap in and out. Reversible. Opens 180 degrees when needed. Separate deadbolt for added security. Tempered glass. Heavy-gauge one-piece 1" aluminum construction with reinforced corners and foam insulation.

| 36" x 80" | BC@1.45 | Ea | 322.00 | 53.60 | 375.60 |

Full-view aluminum storm doors. 1-1/2"-thick heavy-gauge aluminum frame with foam insulation. Extra perimeter weatherstripping and water-diverting rain cap. Solid brass handle set and double-throw deadbolt. Brass-finished sweep. Full-length, piano-style hinge. Reversible hinge for left- or right-side entry door handle. Two heavy-duty, color-matched closers. Solid brass perimeter locks. Insect screen. White, brown, green or almond.

| 30" x 80" | BC@1.45 | Ea | 294.00 | 53.60 | 347.60 |
| 32" x 80" | BC@1.45 | Ea | 294.00 | 53.60 | 347.60 |
| 36" x 80" | BC@1.45 | Ea | 291.00 | 53.60 | 344.60 |

Full-view woodcore storm doors. Aluminum over solid wood core. Extra perimeter weatherstripping and water-diverting rain cap. Black push-button handle with night latch. Color-matched sweep. One heavy-duty closer. Snap-in, snap-out window and insect screen retainer system. White or bronze.

| 32" x 84" | BC@1.45 | Ea | 133.00 | 53.60 | 186.60 |
| 36" x 84" | BC@1.45 | Ea | 135.00 | 53.60 | 188.60 |

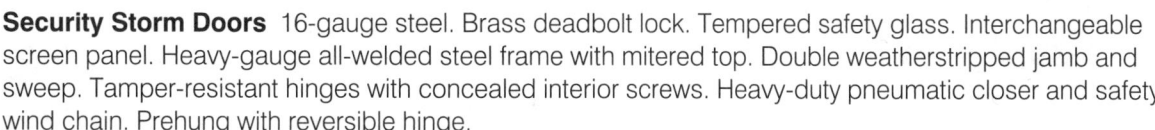

**Security Storm Doors** 16-gauge steel. Brass deadbolt lock. Tempered safety glass. Interchangeable screen panel. Heavy-gauge all-welded steel frame with mitered top. Double weatherstripped jamb and sweep. Tamper-resistant hinges with concealed interior screws. Heavy-duty pneumatic closer and safety wind chain. Prehung with reversible hinge.

| 30" x 80" | BC@1.45 | Ea | 355.00 | 53.60 | 408.60 |
| 32" x 80" | BC@1.45 | Ea | 355.00 | 53.60 | 408.60 |
| 36" x 80" | BC@1.45 | Ea | 355.00 | 53.60 | 408.60 |

Class II security doors. Rated to withstand 400 pounds. Double lockbox with extension plate. Decorative geometric ornaments. All welded steel construction. Expanded, galvanized metal screen. 3/4" square frame design.

| 32" x 80" | BC@1.45 | Ea | 100.00 | 53.60 | 153.60 |
| 36" x 80" | BC@1.45 | Ea | 102.00 | 53.60 | 155.60 |

Heavy-duty security door with screen. Right- or left-hand exterior mounting. All welded steel construction. Baked-on powder coating. Double lockbox with tamper-resistant extension plate. Uses standard 2-3/8" backset lock (not included). Galvanized perforated metal security screen. 1" x 1" door frame. 1" x 1-1/2" jambs, prehung on hinge side.

| 32" x 80" | BC@1.45 | Ea | 138.00 | 53.60 | 191.60 |
| 36" x 80" | BC@1.45 | Ea | 140.00 | 53.60 | 193.60 |

Paradise style security door. Five semi-concealed hinges. Heavy-duty deadbolt. All steel frame.

| 30" x 80" | BC@1.45 | Ea | 406.00 | 53.60 | 459.60 |
| 32" x 80" | BC@1.45 | Ea | 406.00 | 53.60 | 459.60 |
| 36" x 80" | BC@1.45 | Ea | 406.00 | 53.60 | 459.60 |

Folding security gates. Riveted 3/4" top and bottom channel. Gate pivots and folds flat to one side. Provides air circulation and positive visibility.

| 48" max width, 79" high | BC@1.25 | Ea | 118.00 | 46.20 | 164.20 |

# Doors, Interior Slab

| | Craft@Hrs | Unit | Material | Labor | Total |
|---|---|---|---|---|---|

**Interior Slab Doors** Slab doors are the door alone, as distinguished from prehung doors which are sold with jamb, hinges and trim attached and bored for the door latch. Interior doors are usually 1-3/8" thick and have a "hollow" core. A wood frame made from stiles and rails is filled with a matrix of corrugated paper and covered with a thin wood veneer. Stiles are vertical and every door has two, a lock stile and the hinge stile. Rails run horizontally at the top and bottom of the door. Flush doors are smooth and flat, with no decorative treatment.

Flush hardboard interior doors. Hollow core. Bored for lockset. Ready to paint or stain. 1-3/8" thick. Hardboard doors are made of composition wood veneer, usually the least expensive material available.

| | Craft@Hrs | Unit | Material | Labor | Total |
|---|---|---|---|---|---|
| 24" x 80" | BC@1.15 | Ea | 23.30 | 42.50 | 65.80 |
| 28" x 80" | BC@1.15 | Ea | 25.00 | 42.50 | 67.50 |
| 30" x 80" | BC@1.15 | Ea | 26.10 | 42.50 | 68.60 |
| 32" x 80" | BC@1.15 | Ea | 27.20 | 42.50 | 69.70 |
| 36" x 80" | BC@1.15 | Ea | 28.70 | 42.50 | 71.20 |

Flush birch interior doors. Hollow core. 1-3/8" thick. Wood veneer. Stainable and paintable. All wood stile and rail.

| | Craft@Hrs | Unit | Material | Labor | Total |
|---|---|---|---|---|---|
| 18" x 80" | BC@1.15 | Ea | 26.40 | 42.50 | 68.90 |
| 24" x 80" | BC@1.15 | Ea | 31.00 | 42.50 | 73.50 |
| 28" x 80" | BC@1.15 | Ea | 33.60 | 42.50 | 76.10 |
| 30" x 80" | BC@1.15 | Ea | 35.20 | 42.50 | 77.70 |
| 32" x 80" | BC@1.15 | Ea | 36.90 | 42.50 | 79.40 |
| 36" x 80" | BC@1.15 | Ea | 39.00 | 42.50 | 81.50 |

Flush lauan interior doors. Hollow core. 1-3/8" thick. Wood veneer. Stainable and paintable. Lauan is a tropical hardwood like mahogany. All wood stile and rail.

| | Craft@Hrs | Unit | Material | Labor | Total |
|---|---|---|---|---|---|
| 18" x 80" | BC@1.15 | Ea | 21.50 | 42.50 | 64.00 |
| 24" x 80" | BC@1.15 | Ea | 23.50 | 42.50 | 66.00 |
| 28" x 80" | BC@1.15 | Ea | 25.30 | 42.50 | 67.80 |
| 30" x 80" | BC@1.15 | Ea | 26.30 | 42.50 | 68.80 |
| 32" x 80" | BC@1.15 | Ea | 28.30 | 42.50 | 70.80 |
| 36" x 80" | BC@1.15 | Ea | 29.90 | 42.50 | 72.40 |

Flush oak interior doors. Hollow core. 1-3/8" thick. Wood veneer. Stainable and paintable. All wood stile and rail

| | Craft@Hrs | Unit | Material | Labor | Total |
|---|---|---|---|---|---|
| 18" x 80" | BC@1.15 | Ea | 29.30 | 42.50 | 71.80 |
| 24" x 80" | BC@1.15 | Ea | 30.70 | 42.50 | 73.20 |
| 28" x 80" | BC@1.15 | Ea | 34.20 | 42.50 | 76.70 |
| 30" x 80" | BC@1.15 | Ea | 35.70 | 42.50 | 78.20 |
| 32" x 80" | BC@1.15 | Ea | 40.50 | 42.50 | 83.00 |
| 36" x 80" | BC@1.15 | Ea | 40.50 | 42.50 | 83.00 |

Flush hardboard solid core interior doors. Primed. 1-3/8" thick. Hardboard doors are made of composition wood veneer and are usually the least expensive.

| | Craft@Hrs | Unit | Material | Labor | Total |
|---|---|---|---|---|---|
| 30" x 80" | BC@1.15 | Ea | 48.00 | 42.50 | 90.50 |
| 32" x 80" | BC@1.15 | Ea | 49.60 | 42.50 | 92.10 |
| 36" x 80" | BC@1.15 | Ea | 53.50 | 42.50 | 96.00 |

Flush lauan solid core interior doors. 1-3/8" thick. Stainable. Lauan doors are finished with a veneer of Philippine hardwood from the mahogany family.

| | Craft@Hrs | Unit | Material | Labor | Total |
|---|---|---|---|---|---|
| 28" x 80" | BC@1.15 | Ea | 49.70 | 42.50 | 92.20 |
| 32" x 80" | BC@1.15 | Ea | 54.20 | 42.50 | 96.70 |
| 36" x 80" | BC@1.15 | Ea | 53.60 | 42.50 | 96.10 |
| 30" x 84" | BC@1.15 | Ea | 52.90 | 42.50 | 95.40 |
| 32" x 84" | BC@1.15 | Ea | 55.10 | 42.50 | 97.60 |
| 34" x 84" | BC@1.15 | Ea | 59.40 | 42.50 | 101.90 |
| 36" x 84" | BC@1.15 | Ea | 59.40 | 42.50 | 101.90 |

Flush birch solid core interior doors. 1-3/8" thick.

| | Craft@Hrs | Unit | Material | Labor | Total |
|---|---|---|---|---|---|
| 24" x 80" (water heater closet door) | BC@1.15 | Ea | 76.30 | 42.50 | 118.80 |

| | Craft@Hrs | Unit | Material | Labor | Total |
|---|---|---|---|---|---|
| 32" x 80" | BC@1.15 | Ea | 57.30 | 42.50 | 99.80 |
| 36" x 80" | BC@1.15 | Ea | 62.40 | 42.50 | 104.90 |

Flush hardboard interior doors with lite. 1-3/8" thick.

| | | | | | |
|---|---|---|---|---|---|
| 30" x 80" x 1-3/8" | BC@1.15 | Ea | 110.00 | 42.50 | 152.50 |
| 32" x 80" x 1-3/8" | BC@1.15 | Ea | 116.00 | 42.50 | 158.50 |

Flush hardboard walnut finish stile and rail hollow core doors. 1-3/8" thick.

| | | | | | |
|---|---|---|---|---|---|
| 24" x 80" | BC@1.15 | Ea | 24.80 | 42.50 | 67.30 |
| 28" x 80" | BC@1.15 | Ea | 28.10 | 42.50 | 70.60 |
| 30" x 80" | BC@1.15 | Ea | 28.10 | 42.50 | 70.60 |
| 32" x 80" | BC@1.15 | Ea | 30.20 | 42.50 | 72.70 |
| 36" x 80" | BC@1.15 | Ea | 31.30 | 42.50 | 73.80 |

**Interior Panel Doors**  Panel doors have decorative molding applied to the face.

Hardboard 6-panel molded face interior doors. 1-3/8" thick. Molded and primed hardboard. Embossed simulated woodgrain. Hollow core. Ready to paint. Bored for lockset.

| | | | | | |
|---|---|---|---|---|---|
| 24" x 80" | BC@1.15 | Ea | 36.50 | 42.50 | 79.00 |
| 28" x 80" | BC@1.15 | Ea | 38.30 | 42.50 | 80.80 |
| 30" x 80" | BC@1.15 | Ea | 39.10 | 42.50 | 81.60 |
| 32" x 80" | BC@1.15 | Ea | 41.10 | 42.50 | 83.60 |
| 36" x 80" | BC@1.15 | Ea | 43.10 | 42.50 | 85.60 |

Lauan 6-panel molded face interior doors. Bored for latchset. 1-3/8" thick.

| | | | | | |
|---|---|---|---|---|---|
| 24" x 80" | BC@1.15 | Ea | 38.60 | 42.50 | 81.10 |
| 28" x 80" | BC@1.15 | Ea | 42.20 | 42.50 | 84.70 |
| 30" x 80" | BC@1.15 | Ea | 43.70 | 42.50 | 86.20 |
| 32" x 80" | BC@1.15 | Ea | 44.30 | 42.50 | 86.80 |
| 36" x 80" | BC@1.15 | Ea | 46.00 | 42.50 | 88.50 |

Hardwood colonial 6-panel molded face interior doors. Primed. Solid core. 1-3/8" thick.

| | | | | | |
|---|---|---|---|---|---|
| 24" x 80" | BC@1.15 | Ea | 67.00 | 42.50 | 109.50 |
| 28" x 80" | BC@1.15 | Ea | 68.00 | 42.50 | 110.50 |
| 30" x 80" | BC@1.15 | Ea | 70.00 | 42.50 | 112.50 |
| 32" x 80" | BC@1.15 | Ea | 71.30 | 42.50 | 113.80 |
| 36" x 80" | BC@1.15 | Ea | 74.50 | 42.50 | 117.00 |

Pine 6-panel stile and rail interior doors. Ready to paint or stain. 1-3/8" thick.

| | | | | | |
|---|---|---|---|---|---|
| 24" x 80" | BC@1.15 | Ea | 84.30 | 42.50 | 126.80 |
| 28" x 80" | BC@1.15 | Ea | 89.70 | 42.50 | 132.20 |
| 30" x 80" | BC@1.15 | Ea | 93.00 | 42.50 | 135.50 |
| 32" x 80" | BC@1.15 | Ea | 98.50 | 42.50 | 141.00 |
| 36" x 80" | BC@1.15 | Ea | 102.00 | 42.50 | 144.50 |

Fir 1-panel stile and rail interior doors. 1-3/8" thick.

| | | | | | |
|---|---|---|---|---|---|
| 28" x 80" | BC@1.15 | Ea | 175.00 | 42.50 | 217.50 |
| 30" x 80" | BC@1.15 | Ea | 177.00 | 42.50 | 219.50 |
| 32" x 80" | BC@1.15 | Ea | 178.00 | 42.50 | 220.50 |

Fir 3-panel stile and rail interior doors. 1-3/8" thick.

| | | | | | |
|---|---|---|---|---|---|
| 28" x 80" | BC@1.15 | Ea | 214.00 | 42.50 | 256.50 |
| 30" x 80" | BC@1.15 | Ea | 214.00 | 42.50 | 256.50 |
| 32" x 80" | BC@1.15 | Ea | 215.00 | 42.50 | 257.50 |

Oak 6-panel stile and rail interior doors. Red oak. Double bevel hip raised panels. 1-3/8" thick.

| | | | | | |
|---|---|---|---|---|---|
| 24" x 80" | BC@1.15 | Ea | 74.50 | 42.50 | 117.00 |
| 28" x 80" | BC@1.15 | Ea | 80.00 | 42.50 | 122.50 |
| 30" x 80" | BC@1.15 | Ea | 84.74 | 42.50 | 127.24 |
| 32" x 80" | BC@1.15 | Ea | 97.00 | 42.50 | 139.50 |
| 36" x 80" | BC@1.15 | Ea | 103.00 | 42.50 | 145.50 |

# Door Jambs, Interior

| | Craft@Hrs | Unit | Material | Labor | Total |
|---|---|---|---|---|---|

French stile and rail interior doors. Clear pine. Ready to paint or stain. 1-3/8" thick. Pre-masked tempered glass. French doors are mostly glass, with one, to as many as fifteen panes set in wood sash.

| | Craft@Hrs | Unit | Material | Labor | Total |
|---|---|---|---|---|---|
| 24" x 80", 10 lite | BC@1.40 | Ea | 116.00 | 51.70 | 167.70 |
| 28" x 80", 10 lite | BC@1.40 | Ea | 118.00 | 51.70 | 169.70 |
| 30" x 80", 15 lite | BC@1.40 | Ea | 129.00 | 51.70 | 180.70 |
| 32" x 80", 15 lite | BC@1.40 | Ea | 137.00 | 51.70 | 188.70 |
| 36" x 80", 15 lite | BC@1.40 | Ea | 145.00 | 51.70 | 196.70 |

Half louver stile and rail interior doors. 1-3/8" thick. Louver doors include wood louvers that allow air to circulate but obscure vision, an advantage for enclosing a closet, pantry or water heater.

| | | | | | |
|---|---|---|---|---|---|
| 24" x 80" | BC@1.15 | Ea | 166.00 | 42.50 | 208.50 |
| 28" x 80" | BC@1.15 | Ea | 173.00 | 42.50 | 215.50 |
| 30" x 80" | BC@1.15 | Ea | 178.00 | 42.50 | 220.50 |
| 32" x 80" | BC@1.15 | Ea | 185.00 | 42.50 | 227.50 |
| 36" x 80" | BC@1.15 | Ea | 189.00 | 42.50 | 231.50 |

Full louver café (bar) doors. Per pair of doors. With hardware. Bar (café) doors are familiar to everyone who has seen a Western movie.

| | | | | | |
|---|---|---|---|---|---|
| 24" wide, 42" high | BC@1.15 | Ea | 80.70 | 42.50 | 123.20 |
| 30" wide, 42" high | BC@1.15 | Ea | 88.40 | 42.50 | 130.90 |
| 32" wide, 42" high | BC@1.15 | Ea | 97.00 | 42.50 | 139.50 |
| 36" wide, 42" high | BC@1.15 | Ea | 99.00 | 42.50 | 141.50 |

Stile and rail café doors. Per pair of doors. With hardware.

| | | | | | |
|---|---|---|---|---|---|
| 30" x 42" | BC@1.15 | Ea | 50.40 | 42.50 | 92.90 |
| 32" x 42" | BC@1.15 | Ea | 52.00 | 42.50 | 94.50 |
| 36" x 42" | BC@1.15 | Ea | 53.90 | 42.50 | 96.40 |

**Interior Door Jambs**  Finger joint pine interior jamb set. Unfinished. Two 6'-8" legs and one 3' head.

| | | | | | |
|---|---|---|---|---|---|
| 4-9/16" x 11/16" | BC@.500 | Ea | 31.10 | 18.50 | 49.60 |
| 5-1/4" x 11/16" | BC@.500 | Ea | 65.20 | 18.50 | 83.70 |

Solid clear pine interior jamb. Varnish or stain. Jamb set includes two 6'-8" legs and one 3' head.

| | | | | | |
|---|---|---|---|---|---|
| 11/16" x 4-9/16" x 7' leg | BC@.210 | Ea | 9.66 | 7.76 | 17.42 |
| 11/16" x 4-9/16" x 3' head | BC@.090 | Ea | 11.10 | 3.32 | 14.42 |
| 11/16" x 4-9/16" jamb set | BC@.500 | Ea | 43.00 | 18.50 | 61.50 |

Door jamb set with hinges.

| | | | | | |
|---|---|---|---|---|---|
| 80" x 4-9/16", per set | BC@.500 | Ea | 50.30 | 18.50 | 68.80 |
| 80" x 6-9/16", per set | BC@.500 | Ea | 54.50 | 18.50 | 73.00 |

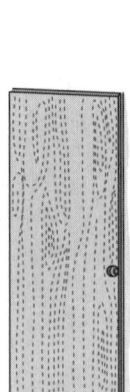

**Prehung Interior Doors**  Prehung doors are sold as a package with the door, jamb, hinges, and trim already assembled and the door bored for a latchset. Flush doors are smooth and flat, with no decorative treatment.

Flush prehung hardboard interior doors. 1-3/8" thick. Hollow core. Primed. Ready to paint.

| | | | | | |
|---|---|---|---|---|---|
| 24" x 80" | BC@.750 | Ea | 53.90 | 27.70 | 81.60 |
| 28" x 80" | BC@.750 | Ea | 60.60 | 27.70 | 88.30 |
| 30" x 80" | BC@.750 | Ea | 59.80 | 27.70 | 87.50 |
| 32" x 80" | BC@.750 | Ea | 61.90 | 27.70 | 89.60 |
| 36" x 80" | BC@.750 | Ea | 64.10 | 27.70 | 91.80 |

Flush prehung lauan interior doors. 1-3/8" thick. 4-9/16" jamb. Bored for lockset.

| | | | | | |
|---|---|---|---|---|---|
| 18" x 78" | BC@.750 | Ea | 54.00 | 27.70 | 81.70 |
| 24" x 78" | BC@.750 | Ea | 58.80 | 27.70 | 86.50 |
| 28" x 78" | BC@.750 | Ea | 59.90 | 27.70 | 87.60 |
| 30" x 78" | BC@.750 | Ea | 61.60 | 27.70 | 89.30 |
| 32" x 78" | BC@.750 | Ea | 62.90 | 27.70 | 90.60 |
| 36" x 78" | BC@.750 | Ea | 65.10 | 27.70 | 92.80 |

| | Craft@Hrs | Unit | Material | Labor | Total |
|---|---|---|---|---|---|
| 18" x 80" | BC@.750 | Ea | 56.90 | 27.70 | 84.60 |
| 24" x 80" | BC@.750 | Ea | 57.20 | 27.70 | 84.90 |
| 28" x 80" | BC@.750 | Ea | 58.30 | 27.70 | 86.00 |
| 30" x 80" | BC@.750 | Ea | 61.60 | 27.70 | 89.30 |
| 32" x 80" | BC@.750 | Ea | 63.70 | 27.70 | 91.40 |
| 36" x 80" | BC@.750 | Ea | 65.90 | 27.70 | 93.60 |

Flush prehung lauan interior doors. 1-3/8" thick. Hollow core. Bored for 2-1/8" lockset with 2-3/8" backset. Casing and lockset sold separately. Wood stile construction. Can be stained or finished naturally. Two brass hinges. Paint grade pine split jambs and 2-1/4" casings. 4" to 4-5/8" adjustable jamb.

| | Craft@Hrs | Unit | Material | Labor | Total |
|---|---|---|---|---|---|
| 24" x 80" | BC@.750 | Ea | 73.70 | 27.70 | 101.40 |
| 28" x 80" | BC@.750 | Ea | 76.60 | 27.70 | 104.30 |
| 30" x 80" | BC@.750 | Ea | 77.10 | 27.70 | 104.80 |
| 32" x 80" | BC@.750 | Ea | 79.00 | 27.70 | 106.70 |
| 36" x 80" | BC@.750 | Ea | 81.20 | 27.70 | 108.90 |

Flush prehung birch interior doors. Stain grade. 1-3/8" thick. Wood stiles. Bored for lockset. Multi-finger jointed door jamb 11/16" x 4-9/16". 3 brass finish hinges 3-1/2" x 3-1/2".

| | Craft@Hrs | Unit | Material | Labor | Total |
|---|---|---|---|---|---|
| 24" x 80" | BC@.750 | Ea | 71.10 | 27.70 | 98.80 |
| 28" x 80" | BC@.750 | Ea | 74.50 | 27.70 | 102.20 |
| 30" x 80" | BC@.750 | Ea | 76.90 | 27.70 | 104.60 |
| 32" x 80" | BC@.750 | Ea | 79.10 | 27.70 | 106.80 |
| 36" x 80" | BC@.750 | Ea | 80.90 | 27.70 | 108.60 |

Flush prehung oak interior doors. Stain grade. 1-3/8" thick. Wood stiles. Bored for lockset. Multi-finger jointed door jamb 11/16" x 4-9/16". 3 brass finish hinges 3-1/2" x 3-1/2".

| | Craft@Hrs | Unit | Material | Labor | Total |
|---|---|---|---|---|---|
| 24" x 80" | BC@.750 | Ea | 63.50 | 27.70 | 91.20 |
| 28" x 80" | BC@.750 | Ea | 66.50 | 27.70 | 94.20 |
| 30" x 80" | BC@.750 | Ea | 67.90 | 27.70 | 95.60 |
| 32" x 80" | BC@.750 | Ea | 71.00 | 27.70 | 98.70 |
| 36" x 80" | BC@.750 | Ea | 72.60 | 27.70 | 100.30 |

Interior prehung lauan heater closet doors.

| | Craft@Hrs | Unit | Material | Labor | Total |
|---|---|---|---|---|---|
| 24" x 60" | BC@.750 | Ea | 75.60 | 27.70 | 103.30 |

6-panel colonial prehung interior doors. 11/16" x 4-9/16" jamb. Bored for lockset. Casing and hardware sold separately. Hollow core. 1-3/8" thick. Panel doors have raised decorative wood panels that resemble doors popular in colonial America. Six or nine panels are most common.

| | Craft@Hrs | Unit | Material | Labor | Total |
|---|---|---|---|---|---|
| 18" x 80", 3-panel | BC@.750 | Ea | 48.40 | 27.70 | 76.10 |
| 24" x 80" | BC@.750 | Ea | 47.50 | 27.70 | 75.20 |
| 28" x 80" | BC@.750 | Ea | 47.50 | 27.70 | 75.20 |
| 30" x 80" | BC@.750 | Ea | 47.50 | 27.70 | 75.20 |
| 32" x 80" | BC@.750 | Ea | 47.50 | 27.70 | 75.20 |
| 36" x 80" | BC@.750 | Ea | 47.50 | 27.70 | 75.20 |

6-panel colonist prehung interior doors. 6-9/16" jamb. Hollow core.

| | Craft@Hrs | Unit | Material | Labor | Total |
|---|---|---|---|---|---|
| 24" x 80" | BC@.750 | Ea | 111.00 | 27.70 | 138.70 |
| 30" x 80" | BC@.750 | Ea | 113.00 | 27.70 | 140.70 |
| 32" x 80" | BC@.750 | Ea | 119.00 | 27.70 | 146.70 |
| 36" x 80" | BC@.750 | Ea | 126.00 | 27.70 | 153.70 |

6-panel hardboard prehung interior doors. 4-9/16" adjustable split jambs. Hollow core. Paint grade casing applied. Pre-drilled for lockset. Ready to paint or gel stain.

| | Craft@Hrs | Unit | Material | Labor | Total |
|---|---|---|---|---|---|
| 24" x 80" | BC@.750 | Ea | 58.60 | 27.70 | 86.30 |
| 28" x 80" | BC@.750 | Ea | 58.70 | 27.70 | 86.40 |
| 30" x 80" | BC@.750 | Ea | 54.70 | 27.70 | 82.40 |
| 32" x 80" | BC@.750 | Ea | 58.80 | 27.70 | 86.50 |
| 36" x 80" | BC@.750 | Ea | 58.90 | 27.70 | 86.60 |

9 PANEL

## Doors, Prehung Interior

|  | Craft@Hrs | Unit | Material | Labor | Total |
|---|---|---|---|---|---|
| **6-panel hardboard prehung double interior doors. 1-3/8" thick. Flat jamb.** | | | | | |
| 48" x 80" | BC@1.25 | Ea | 149.00 | 46.20 | 195.20 |
| 60" x 80" | BC@1.25 | Ea | 159.00 | 46.20 | 205.20 |
| **6-panel oak prehung interior doors. Hollow core. 1-3/8" thick. 11/16" x 4-9/16" clear pine jamb.** | | | | | |
| 24" x 80" | BC@.750 | Ea | 124.00 | 27.70 | 151.70 |
| 28" x 80" | BC@.750 | Ea | 129.00 | 27.70 | 156.70 |
| 30" x 80" | BC@.750 | Ea | 131.00 | 27.70 | 158.70 |
| 32" x 80" | BC@.750 | Ea | 136.00 | 27.70 | 163.70 |
| 36" x 80" | BC@.750 | Ea | 139.00 | 27.70 | 166.70 |
| **6-panel prehung interior double doors. Bored for lockset. Ready to paint or stain. 1-3/8" thick. 4-9/16" jambs.** | | | | | |
| 4'0" x 6'8" | BC@1.50 | Ea | 181.00 | 55.40 | 236.40 |
| 6'0" x 6'8" | BC@1.50 | Ea | 186.00 | 55.40 | 241.40 |
| **6-panel hemlock prehung interior doors. Bored for lockset. Ready to paint or stain. 1-3/8" thick. 4-9/16" jambs.** | | | | | |
| 24" x 80" | BC@.750 | Ea | 166.00 | 27.70 | 193.70 |
| 28" x 80" | BC@.750 | Ea | 171.00 | 27.70 | 198.70 |
| 30" x 80" | BC@.750 | Ea | 174.00 | 27.70 | 201.70 |
| 32" x 80" | BC@.750 | Ea | 175.00 | 27.70 | 202.70 |
| 36" x 80" | BC@.750 | Ea | 178.00 | 27.70 | 205.70 |
| **2-panel knotty alder prehung interior doors. Bored for lockset. 1-3/8" thick.** | | | | | |
| 24" x 80" | BC@.750 | Ea | 291.00 | 27.70 | 318.70 |
| 28" x 80" | BC@.750 | Ea | 296.00 | 27.70 | 323.70 |
| 30" x 80" | BC@.750 | Ea | 301.00 | 27.70 | 328.70 |
| 32" x 80" | BC@.750 | Ea | 308.00 | 27.70 | 335.70 |
| 36" x 80" | BC@.750 | Ea | 312.00 | 27.70 | 339.70 |
| **Pine full louver prehung interior doors. Ready to install. Provides ventilation while maintaining privacy. 11/16" x 4-9/16" clear face pine door jamb. 1-3/8" thick. Includes (3) 3-1/2" x 3-1/2" hinges with brass finish. Bored for lockset.** | | | | | |
| 24" x 80" | BC@.750 | Ea | 125.00 | 27.70 | 152.70 |
| 28" x 80" | BC@.750 | Ea | 150.00 | 27.70 | 177.70 |
| 30" x 80" | BC@.750 | Ea | 119.00 | 27.70 | 146.70 |
| 32" x 80" | BC@.750 | Ea | 130.00 | 27.70 | 157.70 |
| 36" x 80" | BC@.750 | Ea | 127.00 | 27.70 | 154.70 |
| **10-lite wood French prehung double interior doors. Pre-masked tempered glass for easy finishing. True divided light design. Ponderosa pine. 1-3/8" thick. French doors are mostly glass, with one, to as many as fifteen panes set in wood sash.** | | | | | |
| 48" x 80", unfinished | BC@1.25 | Ea | 328.00 | 46.20 | 374.20 |
| 60" x 80", unfinished | BC@1.25 | Ea | 351.00 | 46.20 | 397.20 |
| 48" x 80", primed | BC@1.25 | Ea | 264.00 | 46.20 | 310.20 |
| 60" x 80", primed | BC@1.25 | Ea | 275.00 | 46.20 | 321.20 |
| **Oak French prehung double interior doors. Bored for lockset. 1-3/8" thick.** | | | | | |
| 48" x 80" | BC@1.25 | Ea | 474.00 | 46.20 | 520.20 |
| 60" x 80" | BC@1.25 | Ea | 507.00 | 46.20 | 553.20 |
| 72" x 80" | BC@1.25 | Ea | 379.00 | 46.20 | 425.20 |

## Patio Doors

10-lite swinging patio double doors. White finish. Dual-glazed with clear tempered glass. Fully weatherstripped. Pre-drilled lock-bore.

|  | Craft@Hrs | Unit | Material | Labor | Total |
|---|---|---|---|---|---|
| 6'-0" wide x 6'-8" high | B1@3.58 | Ea | 599.00 | 119.00 | 718.00 |

6 PANEL

| | Craft@Hrs | Unit | Material | Labor | Total |
|---|---|---|---|---|---|

Vinyl-clad hinged patio double doors, Frenchwood®, Andersen. White exterior. Perma-Shield® low-maintenance finish. Clear pine interior. 1-lite. 3-point locking system. Adjustable hinges with ball-bearing pivots. Low-E2 tempered insulating glass. Mortise and tenon joints.

| | Craft@Hrs | Unit | Material | Labor | Total |
|---|---|---|---|---|---|
| 6'-0" x 6'-8" | B1@3.58 | Ea | 1,300.00 | 119.00 | 1,419.00 |

Center hinge 15-lite aluminum patio double doors. Sized for replacement. Unfinished clear pine interior. White extruded aluminum-clad exterior. Weatherstripped frame. Self-draining sill. Bored for lockset. 3/4" Low-E insulated safety glass.

| 6'-0" x 6'-8" | B1@3.58 | Ea | 855.00 | 119.00 | 974.00 |
|---|---|---|---|---|---|

Aluminum inward swing 1-lite patio double doors. Low-E insulated safety glass. Clear wood interior ready to paint or stain.

| 6'-0" x 6'-8" | B1@3.58 | Ea | 787.00 | 119.00 | 906.00 |
|---|---|---|---|---|---|

Aluminum inward swing 10-lite patio double doors. Unfinished interior. Low-E2 argon-filled insulated glass. Thermal-break sill. Bored for lockset. With head and foot bolts.

| 6'-0" x 6'-8" | B1@3.58 | Ea | 829.00 | 119.00 | 948.00 |
|---|---|---|---|---|---|

200 series Narroline™ sliding wood patio doors, Andersen. Dual-pane Low-E glass. Anodized aluminum track with stainless steel cap. With hardware and screen. White.

| 6'-0" x 6'-8" | B1@3.58 | Ea | 1,030.00 | 119.00 | 1,149.00 |
|---|---|---|---|---|---|

Vinyl-clad gliding patio doors, Andersen. White. Tempered 1" Low-E insulating glass. Anodized aluminum track.

| 6'-0" x 6'-8" | B1@3.58 | Ea | 728.00 | 119.00 | 847.00 |
|---|---|---|---|---|---|

10-lite sliding patio doors. Solid pine core. Weatherstripped on 4 sides of operating panels. Butcher block 4-1/2"-wide stiles and 9-1/2" bottom rail. 4-9/16" jamb width. White. Actual dimensions: 71-1/4" wide x 79-1/2" high. Pre-drilled lock-bore.

| 6'-0" wide x 6'-8" high | B1@3.58 | Ea | 610.00 | 119.00 | 729.00 |
|---|---|---|---|---|---|

Aluminum sliding patio doors. Unfinished interior. White aluminum-clad exterior. With lockset. Insulating Low-E2 glass.

| 6'-0" x 6'-8", 1 lite | B1@3.58 | Ea | 755.00 | 119.00 | 874.00 |
|---|---|---|---|---|---|
| 6'-0" x 6'-8", 15 lites | B1@3.58 | Ea | 755.00 | 119.00 | 874.00 |

Aluminum sliding patio doors. Insulating glass. Reversible, self-aligning ball bearing rollers. Zinc finish lock. With screen. By nominal width x height. Actual height and width are 1/2" less.

| 5' x 6'-8", bronze finish | B1@3.58 | Ea | 275.00 | 119.00 | 394.00 |
|---|---|---|---|---|---|
| 6' x 6'-8", bronze finish | B1@3.58 | Ea | 290.00 | 119.00 | 409.00 |
| 6' x 6'-8", mill finish | B1@3.58 | Ea | 214.00 | 119.00 | 333.00 |

Aluminum sliding patio doors. With screen. Insulated glass except as noted. Reversible, self-aligning ball bearing rollers. Zinc finish lock. By nominal width x height. Actual height and width are 1/2" less.

| 5'-0" x 6'-8", mill finish, single glazed | B1@3.58 | Ea | 190.00 | 119.00 | 309.00 |
|---|---|---|---|---|---|
| 5'-0" x 6'-8", white finish | B1@3.58 | Ea | 269.00 | 119.00 | 388.00 |
| 5'-0" x 8'-0", white finish, with grids | B1@3.58 | Ea | 306.00 | 119.00 | 425.00 |
| 6'-0" x 6'-8", bronze finish, single glazed | B1@3.58 | Ea | 317.00 | 119.00 | 436.00 |
| 6'-0" x 6'-8", bronze finish, with grids | B1@3.58 | Ea | 306.00 | 119.00 | 425.00 |
| 6'-0" x 6'-8", white finish, single glazed | B1@3.58 | Ea | 317.00 | 119.00 | 436.00 |
| 8'0" x 8'-0", white finish | B1@4.60 | Ea | 339.00 | 153.00 | 492.00 |

Screen for sliding or swinging patio doors. 72-1/2" wide x 6'8" high.

| White | B1@.250 | Ea | 139.00 | 8.33 | 147.33 |
|---|---|---|---|---|---|

Vinyl sliding patio doors. Includes screen. Low-E insulating glass.

| 5'-0" x 6'-8" | B1@3.58 | Ea | 583.00 | 119.00 | 702.00 |
|---|---|---|---|---|---|
| 6'-0" x 6'-8" | B1@3.58 | Ea | 620.00 | 119.00 | 739.00 |

Vinyl sliding patio doors. Single glazed. By opening size. Actual width and height are 1/2" less.

| 5'-0" x 6'-8", with grille | B1@3.58 | Ea | 420.00 | 119.00 | 539.00 |
|---|---|---|---|---|---|
| 6'-0" x 6'-8" | B1@3.58 | Ea | 420.00 | 119.00 | 539.00 |
| 6'-0" x 6'-8", with grille | B1@3.58 | Ea | 453.00 | 119.00 | 572.00 |
| 8'-0" x 6'-8" | B1@3.58 | Ea | 528.00 | 119.00 | 647.00 |
| 8'-0" x 6'-8", with grille | B1@3.58 | Ea | 561.00 | 119.00 | 680.00 |

# Doors, Accordion Folding

| | Craft@Hrs | Unit | Material | Labor | Total |
|---|---|---|---|---|---|
| 1-lite swinging prehung steel patio double doors. Polyurethane core. High-performance weatherstrip. Adjustable thermal-break sill. Non-yellowing frame and grille. Inward swing. | | | | | |
| 6'-0" x 6'-8" | B1@3.58 | Ea | 323.00 | 119.00 | 442.00 |
| 1-lite swinging steel patio double door with internal mini-blinds. Prehung. Double-pane insulated safety glass. | | | | | |
| 6'-0" x 6'-8", outward swing | B1@3.58 | Ea | 527.00 | 119.00 | 646.00 |
| 6'-0" x 6'-8", inward swing, with brick mold | B1@3.58 | Ea | 593.00 | 119.00 | 712.00 |
| 6'-0" x 6'-8", outward swing, Dade County-approved (high wind) | B1@3.58 | Ea | 669.00 | 119.00 | 788.00 |
| 10-lite swinging steel patio double doors. Inward swing. Insulated Low-E glass. | | | | | |
| 5'-0" x 6'-8" | B1@3.58 | Ea | 465.00 | 119.00 | 584.00 |
| 6'-0" x 6'-8" | B1@3.58 | Ea | 462.00 | 119.00 | 581.00 |
| 10-lite venting steel patio double doors. Inward swing. Insulated glass. | | | | | |
| 6'-0" x 6'-8" | B1@3.58 | Ea | 604.00 | 119.00 | 723.00 |
| 8'-0" x 6'-8" | B1@3.58 | Ea | 729.00 | 119.00 | 848.00 |
| 15-lite swinging steel patio double doors. Inward swing | | | | | |
| 5'-0" x 6'-8", insulated glass | B1@3.58 | Ea | 323.00 | 119.00 | 442.00 |
| 6'-0" x 6'-8", Low-E insulated glass | B1@3.58 | Ea | 430.00 | 119.00 | 549.00 |
| 6'-0" x 6'-8", single glazed | B1@3.58 | Ea | 323.00 | 119.00 | 442.00 |
| 15-lite prehung swinging fiberglass patio double doors. Outward swing. Insulated glass. Adjustable mill finish thermal-break threshold | | | | | |
| 6'-0" x 6'-8" | B1@3.58 | Ea | 636.00 | 119.00 | 755.00 |
| **Accordion Folding Doors** Solid PVC slats with flexible hinges. Vertical embossed surface. Includes hardware. Fully assembled. 32" x 80". Hung in a cased opening. | | | | | |
| Light weight, white | BC@.750 | Ea | 18.40 | 27.70 | 46.10 |
| Standard weight, gray | BC@.750 | Ea | 37.50 | 27.70 | 65.20 |
| Standard weight, white | BC@.750 | Ea | 36.80 | 27.70 | 64.50 |
| Premium accordion folding doors. Vertical embossed surface. Preassembled. Prefinished. Adjustable width and height. Flexible vinyl hinges. Hung in a cased opening. | | | | | |
| 8" x 80" section | BC@.750 | Ea | 19.40 | 27.70 | 47.10 |
| 32" x 80" | BC@.750 | Ea | 57.30 | 27.70 | 85.00 |
| Folding door lock. | | | | | |
| Door lock | BC@.250 | Ea | 4.20 | 9.24 | 13.44 |
| **Pet Doors** Entrance doors, lockable, swinging, aluminum frame with security panel | | | | | |
| For small dogs, 8-1/2" x 12-1/2" | BC@.800 | Ea | 27.00 | 29.60 | 56.60 |
| For medium dogs, 11-1/2" x 16-1/2" | BC@.800 | Ea | 36.90 | 29.60 | 66.50 |
| For large dogs, 15" x 20" | BC@.850 | Ea | 44.90 | 31.40 | 76.30 |
| Sliding screen or patio doors, adjustable full length 1/2" panel, 80" high, "Lexan" plastic above lockable, swinging PVC door, aluminum frame | | | | | |
| For cats and miniature dogs, panel is 11-1/2" wide with 5" x 7-1/2" door | BC@.500 | Ea | 270.00 | 18.50 | 288.50 |
| For small dogs, panel is 15" wide with 8-1/2" x 12-1/2" door | BC@.500 | Ea | 309.00 | 18.50 | 327.50 |
| For medium dogs, panel is 18" wide with 11-1/2" x 16-1/2" door | BC@.500 | Ea | 358.00 | 18.50 | 376.50 |
| Pet screen, vinyl coated 36" x 84" Resists tears and punctures | BC@.500 | Ea | 15.30 | 18.50 | 33.80 |

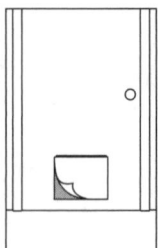

| | Craft@Hrs | Unit | Material | Labor | Total |
|---|---|---|---|---|---|

**Bypass Closet Doors** Bypass closet doors slide left or right on an overhead track, never revealing more than one-half of the interior.

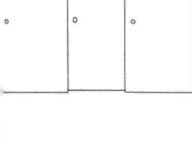

White bypass closet doors. Prefinished white vinyl-covered hardboard panels instead of mirrors. White steel frame. Jump-proof bottom track. 1-1/2" dual race ball bearing wheels. Injection-molded top guides. Dimensions are width x height. Per pair of doors.

| | Craft@Hrs | Unit | Material | Labor | Total |
|---|---|---|---|---|---|
| 48" x 80" | BC@.750 | Ea | 59.40 | 27.70 | 87.10 |
| 60" x 80" | BC@1.00 | Ea | 69.10 | 36.90 | 106.00 |
| 72" x 80" | BC@1.00 | Ea | 82.00 | 36.90 | 118.90 |
| 48" x 96" | BC@.750 | Ea | 68.00 | 27.70 | 95.70 |
| 60" x 96" | BC@1.00 | Ea | 84.20 | 36.90 | 121.10 |
| 72" x 96" | BC@1.00 | Ea | 99.20 | 36.90 | 136.10 |

Economy mirrored bypass closet doors. Gold finished steel. Plate mirror with double strength glass. Bottom rail with jump-proof track. Dimensions are width x height. Per pair of doors.

| | Craft@Hrs | Unit | Material | Labor | Total |
|---|---|---|---|---|---|
| 48" x 80" | BC@.750 | Ea | 80.00 | 27.70 | 107.70 |
| 60" x 80" | BC@1.00 | Ea | 99.40 | 36.90 | 136.30 |
| 72" x 80" | BC@1.00 | Ea | 110.00 | 36.90 | 146.90 |
| 96" x 80" | BC@1.25 | Ea | 184.00 | 46.20 | 230.20 |

Good quality mirrored bypass closet doors. White frame. Safety-backed mirror. Dimensions are width x height. Per pair of doors.

| | Craft@Hrs | Unit | Material | Labor | Total |
|---|---|---|---|---|---|
| 48" x 80", silver nickel mirror | BC@.750 | Ea | 107.00 | 27.70 | 134.70 |
| 60" x 80", silver nickel mirror | BC@1.00 | Ea | 131.00 | 36.90 | 167.90 |
| 72" x 80", silver nickel mirror | BC@1.00 | Ea | 153.00 | 36.90 | 189.90 |
| 48" x 80", trimline bronze mirror door | BC@.750 | Ea | 210.00 | 27.70 | 237.70 |
| 60" x 80", trimline bronze mirror door | BC@1.00 | Ea | 232.00 | 36.90 | 268.90 |
| 72" x 80", trimline bronze mirror door | BC@1.00 | Ea | 256.00 | 36.90 | 292.90 |
| 48" x 80", satin mirror door | BC@.750 | Ea | 172.00 | 27.70 | 199.70 |
| 60" x 80", satin mirror door | BC@1.00 | Ea | 195.00 | 36.90 | 231.90 |
| 72" x 80", satin mirror door | BC@1.00 | Ea | 218.00 | 36.90 | 254.90 |

Beveled mirror bypass closet doors. 1/2" bevel on both sides of glass. Safety-backed mirror. Includes hardware. Dimensions are width x height. Per pair of doors.

| | Craft@Hrs | Unit | Material | Labor | Total |
|---|---|---|---|---|---|
| 48" x 80" | BC@.750 | Ea | 135.00 | 27.70 | 162.70 |
| 60" x 80" | BC@1.00 | Ea | 158.00 | 36.90 | 194.90 |
| 72" x 80" | BC@1.00 | Ea | 179.00 | 36.90 | 215.90 |
| 96" x 80" | BC@1.25 | Ea | 225.00 | 46.20 | 271.20 |

Premium mirrored bypass closet doors. Brushed nickel. Anodized aluminum finish with color-coordinated glazing vinyl. Color-matched mirror handles. "Select" quality 3mm plate mirror. Jump-proof bottom track and 1-1/2" dual race ball bearing wheels. Mitered frame corners. Dimensions are width x height. Per pair of doors.

| | Craft@Hrs | Unit | Material | Labor | Total |
|---|---|---|---|---|---|
| 48" x 81" | BC@.750 | Ea | 184.00 | 27.70 | 211.70 |
| 60" x 81" | BC@1.00 | Ea | 209.00 | 36.90 | 245.90 |
| 72" x 81" | BC@1.00 | Ea | 237.00 | 36.90 | 273.90 |
| 96" x 81" | BC@1.25 | Ea | 288.00 | 46.20 | 334.20 |
| Bypass door accessories | | | | | |
| Bumper | BC@.175 | Ea | 2.18 | 6.46 | 8.64 |
| Carpet riser | BC@.175 | Ea | 1.89 | 6.46 | 8.35 |
| Guide | BC@.175 | Ea | 1.84 | 6.46 | 8.30 |

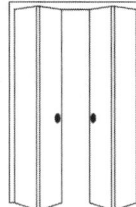

**Bi-Fold Closet Doors** Bi-fold closet doors fold in half when open, revealing nearly the full interior.

Lauan bi-fold flush closet doors. 1-3/8" thick. 2 prehinged panels. Ready to paint, stain, or varnish.

| | Craft@Hrs | Unit | Material | Labor | Total |
|---|---|---|---|---|---|
| 24" x 80" | BC@.700 | Ea | 30.40 | 25.90 | 56.30 |
| 30" x 80" | BC@.700 | Ea | 34.20 | 25.90 | 60.10 |
| 32" x 80" | BC@.700 | Ea | 37.30 | 25.90 | 63.20 |
| 36" x 80" | BC@.700 | Ea | 38.00 | 25.90 | 63.90 |

## Doors, Bi-Fold Closet

| | Craft@Hrs | Unit | Material | Labor | Total |
|---|---|---|---|---|---|
| **Primed hardboard bi-fold flush closet doors. 1-3/8" thick. 2 prehinged panels.** | | | | | |
| 24" x 80" | BC@.700 | Ea | 32.10 | 25.90 | 58.00 |
| 30" x 80" | BC@.700 | Ea | 35.30 | 25.90 | 61.20 |
| 32" x 80" | BC@.700 | Ea | 38.00 | 25.90 | 63.90 |
| 36" x 80" | BC@.700 | Ea | 40.10 | 25.90 | 66.00 |
| **4-panel birch bi-fold flush closet doors. Wood stile and rail construction. 1-3/8" thick. Open left or right. Includes track and hardware.** | | | | | |
| 24" x 80" | BC@.700 | Ea | 36.80 | 25.90 | 62.70 |
| 30" x 80" | BC@.700 | Ea | 39.00 | 25.90 | 64.90 |
| 36" x 80" | BC@.700 | Ea | 42.10 | 25.90 | 68.00 |
| 48" x 80" | BC@.800 | Ea | 66.90 | 29.60 | 96.50 |
| **4-panel lauan flush bi-fold closet doors. Wood stile and rail construction. 1-3/8" thick. Open left or right. Includes track and hardware.** | | | | | |
| 48" x 80" | BC@.800 | Ea | 48.70 | 29.60 | 78.30 |
| **4-panel red oak flush bi-fold closet doors. Wood stile and rail construction. 1-3/8" thick. Open left or right. Includes track and hardware.** | | | | | |
| 24" x 80" | BC@.700 | Ea | 36.70 | 25.90 | 62.60 |
| 30" x 80" | BC@.700 | Ea | 41.20 | 25.90 | 67.10 |
| 36" x 80" | BC@.700 | Ea | 45.50 | 25.90 | 71.40 |
| 48" x 80" | BC@.800 | Ea | 69.00 | 29.60 | 98.60 |
| **Pine colonial panel bi-fold doors. Includes hardware and track. Can be painted or stained. Two hinged panels.** | | | | | |
| 24" x 80" | BC@.700 | Ea | 56.03 | 25.90 | 81.93 |
| 30" x 80" | BC@.700 | Ea | 60.90 | 25.90 | 86.80 |
| 36" x 80" | BC@.700 | Ea | 65.30 | 25.90 | 91.20 |
| **Clear pine 2-panel colonial bi-fold doors. Clear pine. Recessed rails. Unfinished. Open left or right. Includes all hardware and track. Two hinged panels.** | | | | | |
| 24" x 80" | BC@.700 | Ea | 98.50 | 25.90 | 124.40 |
| 30" x 80" | BC@.700 | Ea | 102.00 | 25.90 | 127.90 |
| 32" x 80" | BC@.700 | Ea | 106.00 | 25.90 | 131.90 |
| 36" x 80" | BC@.700 | Ea | 111.00 | 25.90 | 136.90 |
| **Oak panel bi-fold doors. Includes track and hardware. 1-3/8" thick. Ready to finish. Classic. Two hinged panels.** | | | | | |
| 24" x 80" | BC@.700 | Ea | 127.00 | 25.90 | 152.90 |
| 30" x 80" | BC@.700 | Ea | 137.00 | 25.90 | 162.90 |
| 36" x 80" | BC@.700 | Ea | 148.00 | 25.90 | 173.90 |
| **Pine louver over louver bi-fold doors. 1-1/8" thick ponderosa pine. 3-11/16" wide rails. 1-1/4" wide stiles. Two prehinged panels. Unfinished.** | | | | | |
| 24" x 80" | BC@.700 | Ea | 43.30 | 25.90 | 69.20 |
| 30" x 80" | BC@.700 | Ea | 47.30 | 25.90 | 73.20 |
| 32" x 80" | BC@.700 | Ea | 53.60 | 25.90 | 79.50 |
| 36" x 80" | BC@.700 | Ea | 56.70 | 25.90 | 82.60 |
| **Pine louver over panel bi-fold doors. Recessed rails. Open left or right. 1/2" swing space. Two prehinged panels. Includes track and hardware. Unfinished stain grade.** | | | | | |
| 24" x 80" | BC@.700 | Ea | 43.00 | 25.90 | 68.90 |
| 30" x 80" | BC@.700 | Ea | 67.00 | 25.90 | 92.90 |
| 36" x 80" | BC@.700 | Ea | 76.10 | 25.90 | 102.00 |
| **Bi-fold door hardware** | | | | | |
| Bi-fold hinge | BC@.175 | Ea | 3.19 | 6.46 | 9.65 |
| Carpet riser | BC@.175 | Ea | 1.95 | 6.46 | 8.41 |
| Hardware set | BC@.175 | Ea | 7.38 | 6.46 | 13.84 |
| 48" track | — | Ea | 23.60 | — | 23.60 |

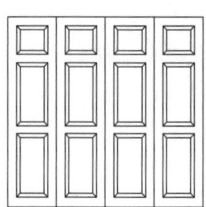

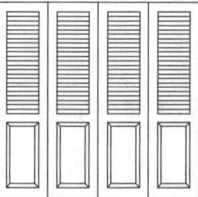

| | Craft@Hrs | Unit | Material | Labor | Total |
|---|---|---|---|---|---|
| 60" track | — | Ea | 24.70 | — | 24.70 |
| 72" track | — | Ea | 27.20 | — | 27.20 |

**Mirrored Bi-Fold Closet Doors** Framed bi-fold mirror doors. White frame. Top-hung with single wheel top hanger. Snap-in bottom guide and fascia. Low rise bottom track. Non-binding door operation. Reversible top and bottom track. 3 mm safety-backed mirror. Two hinged panels.

| | Craft@Hrs | Unit | Material | Labor | Total |
|---|---|---|---|---|---|
| 24" x 80" | BC@.700 | Ea | 89.60 | 25.90 | 115.50 |
| 30" x 80" | BC@.700 | Ea | 101.00 | 25.90 | 126.90 |
| 36" x 80" | BC@.700 | Ea | 112.00 | 25.90 | 137.90 |

Bevel edge bi-fold mirror doors. 1/2" vertical bevel with bearing hinges. White finished steel frame with safety-backed mirror. Includes hardware. Full access to closet opening. Reversible high- and low-profile top and bottom tracks. Two hinged panels.

| | Craft@Hrs | Unit | Material | Labor | Total |
|---|---|---|---|---|---|
| 24" x 80" | BC@.700 | Ea | 103.00 | 25.90 | 128.90 |
| 30" x 80" | BC@.700 | Ea | 113.00 | 25.90 | 138.90 |

Frameless bi-fold mirror doors. Frameless style maximizes the mirror surface. Beveled edge glass. All preassembled hinges. Safety-reinforced mirror backing. Two hinged panels. Includes hardware.

| | Craft@Hrs | Unit | Material | Labor | Total |
|---|---|---|---|---|---|
| 36" x 80" | BC@.700 | Ea | 92.40 | 25.90 | 118.30 |
| 30" x 80" | BC@.700 | Ea | 112.00 | 25.90 | 137.90 |
| 36" x 80" | BC@.700 | Ea | 124.00 | 25.90 | 149.90 |

Chrome bi-fold mirror doors. Two hinged panels. Includes hardware.

| | Craft@Hrs | Unit | Material | Labor | Total |
|---|---|---|---|---|---|
| 24" x 80" | BC@.700 | Ea | 161.00 | 25.90 | 186.90 |
| 30" x 80" | BC@.700 | Ea | 171.00 | 25.90 | 196.90 |
| 36" x 80" | BC@.700 | Ea | 183.00 | 25.90 | 208.90 |

**Drafting** Per SF of floor area, architectural only, not including engineering fees. See also Architectural Illustrations and Blueprinting. Typical prices.

| | Craft@Hrs | Unit | Material | Labor | Total |
|---|---|---|---|---|---|
| Apartments | — | SF | — | — | 2.04 |
| Warehouses and storage buildings | — | SF | — | — | 1.81 |
| Office buildings | — | SF | — | — | 3.53 |
| Residences | | | | | |
| Minimum quality tract work | — | SF | — | — | 2.75 |
| Typical work | — | SF | — | — | 3.40 |
| Detailed jobs, exposed woods, hillsides | — | SF | — | — | 4.24 |

**Drainage Piping** Corrugated plastic drainage pipe, plain or perforated and snap-on fittings. Installed in trenches or at foundation footings. No excavation, gravel or backfill included.

| | Craft@Hrs | Unit | Material | Labor | Total |
|---|---|---|---|---|---|
| Polyethylene pipe | | | | | |
| 3" pipe perforated | BL@.010 | LF | .68 | .30 | .98 |
| 4" pipe solid | BL@.010 | LF | .58 | .30 | .88 |
| 6" pipe slotted | BL@.012 | LF | 1.29 | .36 | 1.65 |
| Corrugated drain snap pipe fittings | | | | | |
| 3" internal coupling | P1@.100 | Ea | 2.33 | 3.58 | 5.91 |
| 3" perforated end cap | P1@.055 | Ea | 2.82 | 1.97 | 4.79 |
| 3" snap adapter | P1@.100 | Ea | 3.55 | 3.58 | 7.13 |
| 3" corrugated x smooth adapter | P1@.100 | Ea | 2.00 | 3.58 | 5.58 |
| 4" internal coupling | P1@.100 | Ea | 2.13 | 3.58 | 5.71 |
| 4" universal adapter | P1@.100 | Ea | 2.94 | 3.58 | 6.52 |
| 4" x 3" reducer | P1@.100 | Ea | 3.51 | 3.58 | 7.09 |
| 4" snap adapter | P1@.100 | Ea | 2.40 | 3.58 | 5.98 |
| 6" snap adapter | P1@.110 | Ea | 4.83 | 3.93 | 8.76 |
| 6" x 4" reducer | P1@.110 | Ea | 4.83 | 3.93 | 8.76 |
| 6" snap coupling | P1@.110 | Ea | 4.20 | 3.93 | 8.13 |
| 6" wye | P1@.150 | Ea | 12.10 | 5.36 | 17.46 |
| 6" tee | P1@.150 | Ea | 10.80 | 5.36 | 16.16 |

# Drain Systems

| | Craft@Hrs | Unit | Material | Labor | Total |
|---|---|---|---|---|---|
| 6" blind tee | P1@.150 | Ea | 9.02 | 5.36 | 14.38 |
| 6" split end cap | P1@.060 | Ea | 3.74 | 2.15 | 5.89 |

Rock or sand fill for drainage systems. Labor cost is for spreading base and covering pipe. 3 mile haul, dumped on site and placed by wheelbarrow

| | Craft@Hrs | Unit | Material | Labor | Total |
|---|---|---|---|---|---|
| Crushed stone (1.4 tons per CY) | | | | | |
|    3/4" (Number 3) | BL@.700 | CY | 28.70 | 20.80 | 49.50 |
|    1-1/2" (Number 2) | BL@.700 | CY | 30.10 | 20.80 | 50.90 |
| Crushed slag (1.86 tons per CY) | | | | | |
|    3/4" | BL@.700 | CY | 25.70 | 20.80 | 46.50 |
| Sand (1.35 tons per CY) | BL@.360 | CY | 13.50 | 10.70 | 24.20 |

**Precast Residential Rain Drain System** System includes 5 interlocking tongue and groove connection channels with galvanized steel slotted grate, one meter long each, 4" outlet and closing end caps, catch basin, and 4" PVC pipe outlet. Channel has 4" internal width. System withstands imposed loadings up to 3,500 lbs. Set in concrete, not including concrete or forming costs.

| System components | Craft@Hrs | Unit | Material | Labor | Total |
|---|---|---|---|---|---|
| 1 meter channel with grate | — | Ea | 72.00 | — | 72.00 |
| 4" outlet cap | — | Ea | 6.36 | — | 6.36 |
| 4" closed end cap | — | Ea | 6.36 | — | 6.36 |
| Catch basin with PVC trash bucket | | | | | |
|    and steel grating | — | Ea | 84.70 | — | 84.70 |
| 4" PVC pipe outlet | — | Ea | 19.80 | — | 19.80 |
| Typical 5 meter rain drain system | BL@4.0 | Ea | 477.00 | 119.00 | 596.00 |

**Draperies, Custom, Subcontract** Custom draperies include 4" double hems and headers, 1-1/2" side hems, and weighted corners. Finished sizes allow 4" above, 4" below (except for sliding glass openings), and 5" on each side of wall openings. Prices are figured at one width, 48" of fabric, pleated to 16". Costs listed include fabric, full liner, manufacturing, all hardware, and installation. Add 50% per LF of width for jobs outside working range of metropolitan centers. Use $300 as a minimum job charge.

Minimum quality, 200% fullness, average is 6 panels of fabric, limited choice of fabric styles, colors, and textures

| | Craft@Hrs | Unit | Material | Labor | Total |
|---|---|---|---|---|---|
| To 95" high | — | LF | — | — | 45.19 |
| To 68" high | — | LF | — | — | 42.43 |
| To 54" high | — | LF | — | — | 41.00 |

Good quality, fully lined, 250% fullness, average is 5 panels of fabric, better selection of fabric styles, colors, and textures, weighted seams

| | Craft@Hrs | Unit | Material | Labor | Total |
|---|---|---|---|---|---|
| To 95" high | — | LF | — | — | 107.10 |
| To 84" high | — | LF | — | — | 101.59 |
| To 68" high | — | LF | — | — | 96.08 |
| To 54" high | — | LF | — | — | 90.47 |
| To 44" high | — | LF | — | — | 87.21 |
| Deduct for multi-unit jobs | — | % | — | — | -25.0 |
| Add for insulated fabrics or liners | | | | | |
|    Pleated shade liner | — | SY | — | — | 29.68 |
|    Thermal liner | — | SY | — | — | 33.95 |

Better quality, fully lined, choice of finest fabric styles, colors, and textures, does not include special treatments such as elaborate swags, or custom fabrics

| | Craft@Hrs | Unit | Material | Labor | Total |
|---|---|---|---|---|---|
| To 95" high | — | LF | — | — | 133.62 |
| To 84" high | — | LF | — | — | 128.52 |
| To 68" high | — | LF | — | — | 114.24 |
| To 54" high | — | LF | — | — | 105.06 |
| To 44" high | — | LF | — | — | 98.33 |

| | Craft@Hrs | Unit | Material | Labor | Total |
|---|---|---|---|---|---|

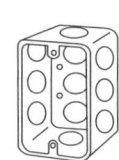

**Electrical Work, Subcontract** Costs listed below are for wiring new residential and light commercial buildings with Romex cable and assume circuit lengths averaging 40 feet. If flex cable is required, add $15.00 for 15 amp circuits and $35 for 20 amp circuits. No fixtures or appliances included except as noted. Work on second and higher floors may cost 25% more. Work performed by a qualified subcontractor.

Rule of thumb: Total cost for electrical work, performed by a qualified subcontractor, per SF of floor area. Includes service entrance, outlets, switches, basic lighting fixtures and connecting appliances only.

| | Craft@Hrs | Unit | Material | Labor | Total |
|---|---|---|---|---|---|
| All wiring and fixtures | — | SF | — | — | 4.83 |
| Lighting fixtures only (no wiring) | — | SF | — | — | 1.39 |
| Air conditioners (with thermostat) | | | | | |
| Central, 2 ton (220 volt) | — | LS | — | — | 663.00 |
| First floor room (115 volt) | — | LS | — | — | 198.00 |
| Second floor room (115 volt) | — | LS | — | — | 399.00 |
| Add for thermostat on second floor | — | LS | — | — | 134.00 |
| Alarms. See also Security Alarms | | | | | |
| Fire or smoke detector, wiring and outlet box only | — | LS | — | — | 150.00 |
| Add for detector unit | — | LS | — | — | 76.00 |
| Bathroom fixtures, wiring only, no fixtures or equipment included | | | | | |
| Mirror lighting (valance or side lighted mirrors) | — | LS | — | — | 134.00 |
| Sauna heater (40 amp branch circuit) | — | LS | — | — | 400.00 |
| Steam bath generator | — | LS | — | — | 400.00 |
| Heat lamp with timer wall switch | — | LS | — | — | 198.00 |
| Whirlpool bath system and wall switch | — | LS | — | — | 333.00 |
| Clock outlets (recessed) | — | LS | — | — | 100.00 |
| Closet lighting | | | | | |
| Ceramic "pull chain" ceiling fixture | — | LS | — | — | 100.00 |
| Switch-operated fixture with wall switch | — | LS | — | — | 134.00 |

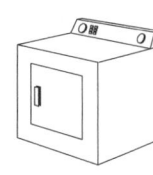

| | Craft@Hrs | Unit | Material | Labor | Total |
|---|---|---|---|---|---|
| Clothes dryers | | | | | |
| Gas dryer, receptacle only | — | LS | — | — | 134.00 |
| Direct connection for up to 5,760 watt (30 amp) electric dryer, 220 volt | — | LS | — | — | 198.00 |
| Direct connection for over 5,760 watt (40 amp) electric dryer, 220 volt | — | LS | — | — | 232.00 |
| Clothes washers (115 volt) | — | LS | — | — | 100.00 |
| Dishwashers | — | LS | — | — | 134.00 |
| Door bells, rough wiring, front and rear with transformer | | | | | |
| Add the cost of two push-buttons and the chime | — | LS | — | — | 198.00 |
| Fans (exhaust) | | | | | |
| Attic fans (wiring only) | — | LS | — | — | 134.00 |
| Bathroom fans (includes 70 CFM fan) | — | LS | — | — | 198.00 |
| Garage fans (wiring only) | — | LS | — | — | 134.00 |
| Kitchen fans (includes 225 CFM fan) | — | LS | — | — | 266.00 |
| Furnace wiring and blower hookup only | — | LS | — | — | 165.00 |
| Garage door opener, wiring and hookup only | — | LS | — | — | 134.00 |
| Garbage disposers, wiring, switch and connection only | | | | | |
| No disposer included | — | LS | — | — | 165.00 |
| Grounding devices (see also Lightning protection systems below) | | | | | |
| Grounding entire electrical system | — | LS | — | — | 198.00 |
| Grounding single appliance | — | LS | — | — | 82.00 |

# Electrical Work, Subcontract

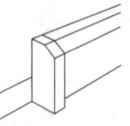

| | Craft@Hrs | Unit | Material | Labor | Total |
|---|---|---|---|---|---|
| Ground fault circuit interrupter (GFCI), rough and finish wiring | | | | | |
| Includes outlet box and GFCI | — | LS | — | — | 165.00 |
| Heaters | | | | | |
|   Baseboard (115 volt) per branch circuit | — | LS | — | — | 134.00 |
|   Bathroom (ceiling type) wiring, switch | | | | | |
|     connection (with GFCI) only | — | LS | — | — | 158.00 |
|   Ceiling heat system (radiant-resistance type) 1,000 watt, 120 volt, | | | | | |
|     per branch circuit, including thermostat | — | LS | — | — | 305.00 |
|   Space heating (flush in-wall type) up to 2,000 | | | | | |
|     watt, 220 volt | — | LS | — | — | 278.00 |
| Humidifiers, central | — | LS | — | — | 134.00 |
| Exterior lamp posts, wiring and hookup only, to 25' | | | | | |
|   Using buried wire and conduit | — | LS | — | — | 794.00 |
| Lighting fixture outlets (rough wiring and box only) | | | | | |
|   Ceiling | — | Ea | — | — | 66.00 |
|   Floor | — | Ea | — | — | 100.00 |
|   Set in concrete | — | Ea | — | — | 132.00 |
|   Set in masonry | — | Ea | — | — | 132.00 |
|   Underground | — | Ea | — | — | 173.00 |
|   Valance | — | Ea | — | — | 79.00 |
|   Wall outlet | — | Ea | — | — | 66.00 |
| Lightning protection systems (static electric grounding system) Residential, ridge protection including | | | | | |
| one chimney and connected garage (houses with cut-up roofs will cost more) | | | | | |
|   Typical home system | — | LS | — | — | 2,832.00 |
|   Barns and light commercial buildings | — | LS | — | — | 5,512.00 |
| Ovens, wall type (up to 10' of wiring and hookup only), 220 volt | | | | | |
|   To 4,800 watts (20 amp) | — | LS | — | — | 162.00 |
|   4,800 to 7,200 watts (30 amp) | — | LS | — | — | 189.00 |
|   7,200 to 9,600 watts (40 amp) | — | LS | — | — | 246.00 |
| Ranges, (up to 10' of wiring and hookup only), 220 volt | | | | | |
|   Countertop type | | | | | |
|     To 4,800 watts (20 amp) | — | LS | — | — | 131.00 |
|     4,800 to 7,200 watts (30 amp) | — | LS | — | — | 189.00 |
|     7,200 to 9,600 watts (40 amp) | — | LS | — | — | 246.00 |
|   Freestanding type (50 amp) | — | LS | — | — | 270.00 |
| Receptacle outlets (rough wiring, box, receptacle and plate) | | | | | |
|   Ceiling | — | Ea | — | — | 54.00 |
|   Countertop wall | — | Ea | — | — | 59.00 |
|   Floor outlet | — | Ea | — | — | 81.00 |
|   Split-wired | — | Ea | — | — | 102.00 |
|   Standard indoor, duplex wall outlet | — | Ea | — | — | 54.00 |
|   Waterproof, with ground fault circuit | — | Ea | — | — | 87.00 |
| Refrigerator or freezer wall outlet | — | Ea | — | — | 65.00 |
| Service entrance connections, complete (panel box hookup but no wiring) | | | | | |
|   100 amp service including meter socket, main switch, GFCI and 5 single pole breakers | | | | | |
|     in 20 breaker space exterior panel box | — | LS | — | — | 1,031.00 |
|   200 amp service including meter socket, main switch, 2 GFCI and 15 single pole breakers | | | | | |
|     in 40 breaker space exterior panel box | — | LS | — | — | 1,685.00 |
| Sub panel connections (panel box hookup but no wiring) | | | | | |
|   40 amp circuit panel including 12 single pole breakers | | | | | |
|     in 12 breaker indoor panel box | — | LS | — | — | 381.00 |
|   50 amp circuit panel including 16 single pole breakers | | | | | |
|     in 16 breaker indoor panel box | — | LS | — | — | 381.00 |

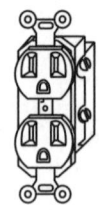

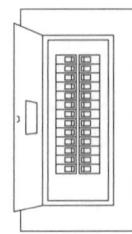

# Electrical Work, Digital Networking, Residential

| | Craft@Hrs | Unit | Material | Labor | Total |
|---|---|---|---|---|---|
| Sump pump connection | | | | | |
|     including 15' of underground conduit | — | Ea | — | — | 194.00 |
| Switches (includes rough wiring, box, switch and plate) | | | | | |
|   Dimmer | — | Ea | — | — | 57.00 |
|   Lighted | — | Ea | — | — | 57.00 |
|   Quiet | — | Ea | — | — | 46.00 |
|   Mercury | — | Ea | — | — | 57.00 |
|   Standard indoor wall switch | — | Ea | — | — | 45.00 |
|   Stair-wired (3 way) | — | Ea | — | — | 120.00 |
|   Waterproof | — | Ea | — | — | 97.00 |
| Television outlet wiring (300 ohm wire) | — | Ea | — | — | 46.00 |
| Trash compactor, 15" (includes compactor installed under counter and wiring) | | | | | |
|   Standard | — | LS | — | — | 510.00 |
|   Deluxe | — | LS | — | — | 596.00 |
| Vacuuming system, central | | | | | |
|   Central unit hookup | — | LS | — | — | 115.00 |
|   Remote vacuum outlets | | | | | |
|     Receptacle wiring only | — | Ea | — | — | 54.00 |
| Water heaters | | | | | |
|   (up to 10' of wiring and hookup only), 220 volt | — | LS | — | — | 169.00 |
| Water pumps, domestic potable water | | | | | |
|   Connection only | — | LS | — | — | 120.00 |

**Residential computer wiring and network setup** These costs are based on work done with wall cavities exposed. Labor costs in existing dwellings will be up to four times higher if cable has to be fished through enclosed walls. Single story homes use stranded wire. Multi story homes use solid wire on vertical runs. Includes normal (5%) waste. Costs assume cable runs terminate at the point of phone or cable service entry and that connections are made at a wall-mounted punch down block with RJ45 connectors. The local building code may require wire runs in conduit or use of plenum type cable. Network cable, twisted pair category 5e, rated for 350 Mbps, material price based on 1,000' rolls. Add the cost of conduit, if required. Costs per linear foot of cable run.

| | Craft@Hrs | Unit | Material | Labor | Total |
|---|---|---|---|---|---|
|   4 pair UTP PVC stranded wire | BE@.008 | LF | .13 | .32 | .45 |
|   4 pair UTP PVC solid wire | BE@.008 | LF | .13 | .32 | .45 |
|   Add for plenum grade wire | — | % | 60.0 | — | — |

Termination to structured cable panel. Includes structured cable panel with telephone, data and video connectors for 6, 12 or 18 stations. Also includes RJ45 keystone jack and wall plate at the room/station end.

| | Craft@Hrs | Unit | Material | Labor | Total |
|---|---|---|---|---|---|
|   12 station termination | BE@1.40 | Ea | 162.00 | 55.40 | 217.40 |
|   18 station termination | BE@2.10 | Ea | 183.00 | 83.10 | 266.10 |
|   24 station termination | BE@2.80 | Ea | 211.00 | 111.00 | 322.00 |

Cable testing using a dedicated cable tester

| | Craft@Hrs | Unit | Material | Labor | Total |
|---|---|---|---|---|---|
|   Shielded/Unshielded twisted pair, per cable | E4@.125 | Ea | — | 6.14 | 6.14 |

Distribution of Internet connection to network switch or router. Includes 10/100 network switch/wireless router and patch cables

| | Craft@Hrs | Unit | Material | Labor | Total |
|---|---|---|---|---|---|
|   6 station connection | BE@.166 | Ea | 82.20 | 6.57 | 88.77 |
|   12 station connection | BE@.333 | Ea | 85.50 | 13.20 | 98.70 |
|   18 station connection | BE@.500 | Ea | 126.00 | 19.80 | 145.80 |

Wireless access points. Preferred location is one foot below the ceiling on a small shelf. Each access point counts as one station termination and connection. Add the cost of network cable wiring to the access point and an electrical outlet at the access point.

| | Craft@Hrs | Unit | Material | Labor | Total |
|---|---|---|---|---|---|
|   Wireless access point | BE@.100 | Ea | 65.80 | 3.96 | 69.76 |

# Electrical Work, Digital Networking, Subcontract

| | Craft@Hrs | Unit | Material | Labor | Total |
|---|---|---|---|---|---|

Remote high gain antenna for a wireless network access point. (Extends the range of wireless connections beyond the usual 50 to 100 feet.) Add the cost of cable connection to the wireless router/access point.

| | Craft@Hrs | Unit | Material | Labor | Total |
|---|---|---|---|---|---|
| 8 Dbi high gain panel antenna | BE@.000 | Ea | 70.00 | 0.00 | 70.00 |

Cable for high gain antennae, thin ethernet, 50 ohm, (RG58). Used to connect a high gain antenna to the wireless access points.

| | | | | | |
|---|---|---|---|---|---|
| PVC cable | BE@.008 | LF | .14 | .32 | .46 |
| Plenum cable | BE@.008 | LF | .39 | .32 | .71 |

Connection of live internet feed to each computer. Includes physical connection of DSL or cable internet feed, internet setup per carriers' instruction, setup and testing of live feed at each outlet and wireless access point. Testing assumes the computers are installed and running.

| | | | | | |
|---|---|---|---|---|---|
| Connection and setup of Internet feed | BE@2.00 | Ea | — | 79.20 | 79.20 |
| Testing of live feed at each computer | BE@1.00 | Ea | — | 39.60 | 39.60 |

Post installation of room/station computer setup. Assumes a peer to peer network with resources shared per the customers' preference, including browser and email setup.

| | | | | | |
|---|---|---|---|---|---|
| Cost per computer configuration | — | LS | — | — | 50.00 |

**Digital Networking Systems, Subcontract**  Modular and expandable electronic gateway for the in-home distribution of external services such as CATV, satellite TV, broadcast TV, telephone and high speed internet access. System also processes and redistributes in-home generated signals such as those from VCRs, DVD players, audio, security cameras and computers. Provides (6) computer local area network (LAN) outlets, (6) telephone jacks and (6) cable hookups where required. Costs include stand alone console, 1,000' of each type of cable (phone, network, video), installation and connection.

Basic networking system

| | | | | | |
|---|---|---|---|---|---|
| Six TV and eight voice locations | — | LS | — | — | 603.00 |

Apartment type networking system

| | | | | | |
|---|---|---|---|---|---|
| Six video, six data and six voice locations | — | LS | — | — | 658.00 |

Amplified networking system. Video and voice surge protection

| | | | | | |
|---|---|---|---|---|---|
| Six video, six data and 15 voice locations | — | LS | — | — | 767.00 |

Large amplified networking system. Video and voice surge protection

| | | | | | |
|---|---|---|---|---|---|
| Six video, 8 data and 18 voice locations | — | LS | — | — | 855.00 |

**Electrical generators, residential**  Suitable for standby or extended primary power generation. Dual fuel (liquid propane or natural gas) two-stroke, V-twin cylinder, 60 Hz, single phase, UL and CSA listed. Emissions certified to EPA Tier 3 rating. 65 decibels sound output rating at 10 feet. Control package includes digital engine and generator control panel, ground fault interruption, lightning shield and combination voltage regulator and automatic transfer switch with automatic dual utility connection lockout. Includes fused main circuit breaker, electronic load-sharing module for generator ganging, electric starter motor, battery and trickle charger. Add the cost of connecting electrical loads and electrical controls. Add $247.50 per day for a 4,000 lb. capacity rough-terrain fork lift, if required. Includes set in place only.

| | | | | | |
|---|---|---|---|---|---|
| 7 KW, 120 volt | B8@1.00 | Ea | 3,500.00 | 34.90 | 3,534.90 |
| 10 KW, 120 volt | B8@1.00 | Ea | 4,300.00 | 34.90 | 4,334.90 |
| 12 KW, 120 volt | B8@1.00 | Ea | 7,000.00 | 34.90 | 7,034.90 |
| 15 KW, 120 volt | B8@1.00 | Ea | 9,000.00 | 34.90 | 9,034.90 |
| 20 KW, 120 volt | B8@1.00 | Ea | 14,000.00 | 34.90 | 14,034.90 |
| 65 KW, 400/480 volt | B8@1.00 | Ea | 58,500.00 | 34.90 | 58,534.90 |
| 100 KW, 400/480 volt | B8@1.00 | Ea | 90,000.00 | 34.90 | 90,034.90 |
| 150 KW, 400/480 volt | B8@1.00 | Ea | 13,5000.00 | 34.90 | 135,034.90 |
| 200 KW, 400/480 volt | B8@1.00 | Ea | 180,000.00 | 34.90 | 180,034.90 |
| 250 KW, 400/480 volt | B8@1.00 | Ea | 225,000.00 | 34.90 | 225,034.90 |

| | Craft@Hrs | Unit | Material | Labor | Total |
|---|---|---|---|---|---|
| Residential motor generator installation | | | | | |
| Form and pour slab foundation | B5@.069 | SF | 2.57 | 2.45 | 5.02 |
| Tie-down and leveling | B1@1.00 | Ea | 25.00 | 33.30 | 58.30 |
| Piping for twin fuel trains, lube oil cooling, and dedicated fire protection | B9@0.04 | LF | 2.25 | 1.31 | 3.56 |
| Mount and wire automatic transfer switch to household main breaker box | BE@.184 | LF | .35 | 7.28 | 7.63 |
| Check and test wiring | BE@.250 | LF | .35 | 9.90 | 10.25 |
| RFI, lightning and ground fault wiring | BE@.184 | LF | .35 | 7.28 | 7.63 |
| 350 gallon propane tank | B9@1.00 | Ea | 575.00 | 32.60 | 607.60 |
| Fire protection CO2 extinguisher and piping | B9@2.00 | Ea | 150.00 | 65.30 | 215.30 |

Soundproof "doghouse" enclosure for reduction of noise level to 10 decibels at 10 feet distance. Includes vertical support studs, spacers, fasteners and trim. Complies with American Institute of Sound Engineering specifications. Per square foot of wall or ceiling covered.

| | Craft@Hrs | Unit | Material | Labor | Total |
|---|---|---|---|---|---|
| Sidewall coverage | D1@.040 | SF | 1.46 | 1.43 | 2.89 |
| Ceiling coverage | D1@.060 | SF | 1.46 | 2.14 | 3.60 |

Utility shed generator enclosure, 5/8" plywood with 2 x 4 frame, prefabricated kit, weatherized, with mansard roof, double door, lock assembly, and dual galvanized metal screened lintel air inlet and exhaust vents

| | Craft@Hrs | Unit | Material | Labor | Total |
|---|---|---|---|---|---|
| Generator enclosure | D1@.020 | SF | 5.50 | .71 | 6.21 |

Combination digital load/voltage controller, usage meter with main breaker box and controls tie-in

| | Craft@Hrs | Unit | Material | Labor | Total |
|---|---|---|---|---|---|
| Controller | BE@.184 | Ea | 1,500.00 | 7.28 | 1,507.28 |

Zone control fused panel box to shield sensitive electronics and home medical electronics for handicapped

| | Craft@Hrs | Unit | Material | Labor | Total |
|---|---|---|---|---|---|
| Fused panel box | BE@.184 | Ea | 650.00 | 7.28 | 657.28 |

Uninterruptible power supply and trickle charger for sensitive electronic and medical appliances

| | Craft@Hrs | Unit | Material | Labor | Total |
|---|---|---|---|---|---|
| UPS | BE@.184 | Ea | 575.00 | 7.28 | 582.28 |

Automatic supervisory control and data acquisition (SCADA) notification circuit to local emergency services to prevent electrical shock of electrical utility repair technicians or fire department personnel

| | Craft@Hrs | Unit | Material | Labor | Total |
|---|---|---|---|---|---|
| Notification circuit | BE@.184 | Ea | 250.00 | 7.28 | 257.28 |

Generator-activated power failure lighting system. Self-diagnostic emergency lighting fixtures, microprocessor based, continuously monitored, with battery and lamps.

| | Craft@Hrs | Unit | Material | Labor | Total |
|---|---|---|---|---|---|
| Two-lamp emergency fixture | BE@.825 | Ea | 173.63 | 32.70 | 206.33 |

**Wind turbine generators** Vertical dual-use ventilator or conventional horizontal type, rated and certified to California Construction Code and American Wind Energy Association standards. Technospin, Rooftop Wind Power LLC, or equal. Rated wind speed 11 m/s (24.6 mph) start-up wind: 2.5 m/s (5.6 mph) survival wind: 60 m/s (134 mph) max. Permanent magnet generator. Generator voltage for battery charging: 12 to 48 V DC. Voltage for grid connection is adjusted to requirements of inverter. Overspeed protection, mechanical and electrical system. Maximum axis load: 150 Kg force (330 lb). Installation specifications are for rooftop mounting. For tax incentives, see http://energytaxincentives.org/. Costs include weather-proof metal enclosure, wind rotor, generator, main wiring and breaker, overspeed protection, control modules and basic power meter. Add the cost of design work, regulatory fees and insurance inspection, electrical interconnection to the power grid and battery backup, if required.

| Wind turbine electrical generator | Craft@Hrs | Unit | Material | Labor | Total |
|---|---|---|---|---|---|
| 100 watts | R1@2.50 | Ea | 400.00 | 88.60 | 488.60 |
| 500 watts | R1@3.00 | Ea | 1,750.00 | 106.00 | 1,856.00 |
| 750 watts | R1@3.00 | Ea | 3,000.00 | 106.00 | 3,106.00 |
| 1.0 Kw | R1@3.50 | Ea | 4,000.00 | 124.00 | 4,124.00 |
| 1.5 Kw | R1@3.50 | Ea | 6,000.00 | 124.00 | 6,124.00 |
| 2.0 Kw | R1@4.00 | Ea | 8,000.00 | 142.00 | 8,142.00 |

## Elevators and Lifts, Subcontract

| | Craft@Hrs | Unit | Material | Labor | Total |
|---|---|---|---|---|---|
| Base plate for rooftop mount | BE @1.00 | Ea | 125.00 | 39.59 | 164.59 |
| Tie-down for generator | R1 @.250 | Ea | 300.00 | 8.06 | 308.86 |
| Electric and control modules | BE @.350 | Ea | 25.00 | 13.86 | 38.86 |
| Install electrical wiring (typical) | BE @.150 | LF | 1.75 | 5.94 | 7.69 |
| Install controls | BE @.500 | Ea | — | 19.80 | 19.80 |
| Commission and test | BE @4.00 | Ea | — | 158.40 | 158.40 |

**Elevators and Lifts, Subcontract** Elevators for apartments or commercial buildings, hydraulic vertical cab, meets code requirements for public buildings, 2,500 pound capacity, to 13 passengers. Includes side opening sliding door, illuminated controls, emergency light, alarm, code-approved fire service operation, motor control unit, wiring in hoistway, cab and door design options.

| | | | | | |
|---|---|---|---|---|---|
| Basic 2 stop, typical costs | | | | | |
| 100 feet per minute | — | LS | — | — | 42,700.00 |
| 125 feet per minute | — | LS | — | — | 46,800.00 |
| 150 feet per minute | — | LS | — | — | 50,200.00 |
| Add per stop to 5 stops | — | LS | — | — | 4,600.00 |
| Add for infrared door protection | — | Ea | — | — | 1,730.00 |
| Add for hall position indicator | — | Ea | — | — | 424.00 |
| Add for car position indicator | — | Ea | — | — | 424.00 |
| Add for car direction light & tone | — | Ea | — | — | 422.00 |
| Add for hall lantern with audible tone | — | Ea | — | — | 424.00 |

Elevators for private residences, electric cable, motor driven, 500 pound capacity, includes complete installation of elevator, controls and safety equipment. Meets code requirements for residences

| | | | | | |
|---|---|---|---|---|---|
| 2 stops (up to 10' rise) | — | LS | — | — | 22,200.00 |
| 3 stops (up to 20' rise) | — | LS | — | — | 25,300.00 |
| 4 stops (up to 30' rise) | — | LS | — | — | 28,300.00 |
| 5 stops (up to 40' rise) | — | LS | — | — | 31,300.00 |
| Add for additional gate | — | Ea | — | — | 969.00 |

| | | | | | |
|---|---|---|---|---|---|
| Dumbwaiters, electric | | | | | |
| 75 pound, 2 stop, 24" x 24" x 30" | — | LS | — | — | 8,590.00 |
| 75 pound, add for each extra stop | — | Ea | — | — | 1,550.00 |
| 100 pound, 2 stop, 24" x 30" x 36" | — | LS | — | — | 9,600.00 |
| 100 pound, add for each extra stop | — | Ea | — | — | 1,750.00 |
| 100 pound, 2 stop, 24" x 24" x 36", tray type | — | LS | — | — | 13,400.00 |
| 100 pound tray type, add for each extra stop | — | Ea | — | — | 2,060.00 |
| 300 pound, 2 stop, 30" x 30" x 36", tray type | — | LS | — | — | 19,400.00 |
| 300 pound, add for each extra stop | — | Ea | — | — | 3,090.00 |
| 500 pound, 2 stop, 23" x 56" x 48" | — | LS | — | — | 20,600.00 |
| 500 pound, add for each extra stop | — | Ea | — | — | 3,090.00 |

Hydraulic elevators, private residence, 750 pound capacity. Includes controls and safety equipment with standard unfinished birch plywood cab and one gate.

| | | | | | |
|---|---|---|---|---|---|
| 2 stops (up to 15' rise) | — | LS | — | — | 24,200.00 |
| 3 stops (up to 22' rise) | — | LS | — | — | 27,300.00 |
| 4 stops (up to 28' rise) | — | LS | — | — | 30,300.00 |
| 5 stops (up to 34' rise) | — | LS | — | — | 35,400.00 |
| Add for adjacent opening | — | Ea | — | — | 567.00 |
| Add for additional gate | — | Ea | — | — | 979.00 |
| Add for plastic laminate cab | — | Ea | — | — | 772.00 |
| Add for oak raised panel cab | — | Ea | — | — | 3,290.00 |

Stairlifts (incline lifts), single passenger, indoor. Lift to 18' measured diagonally. 300 pound capacity. Includes top and bottom call button, safety chair, track, power unit, installation by a licensed elevator contractor. Single family residential use

| | | | | | |
|---|---|---|---|---|---|
| Straight stairs | — | LS | — | — | 3,920.00 |

| | Craft@Hrs | Unit | Material | Labor | Total |
|---|---|---|---|---|---|

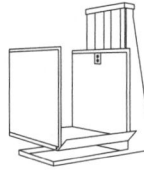

Wheelchair lift (porch lift), screw driven, fully enclosed sides.
Typical electrical connections and control included. Up to 5' rise

| | | | | | |
|---|---|---|---|---|---|
| Residential wheelchair lift | — | LS | — | — | 8,860.00 |

**Engineering Fees** Typical cost for consulting work.
Acoustical engineering (Based on an hourly wage of $70.00 to $135.00)

Environmental noise survey of a lot; includes technician with measuring equipment, data reduction and analysis, and written report with recommendations

| | | | | | |
|---|---|---|---|---|---|
| Written report with recommendations | — | LS | — | — | 1,760.00 |
| Exterior to interior noise analysis | — | LS | — | — | 576.00 |

Measure "Impact Insulation Class" and "Sound Transmission Class" in existing buildings, (walls, floors, and ceilings)

| | | | | | |
|---|---|---|---|---|---|
| Minimum cost | — | LS | — | — | 1,150.00 |

Priced by wall, floor, or ceiling sections

| | | | | | |
|---|---|---|---|---|---|
| Per section, approx. 100 SF | — | Ea | — | — | 576.00 |

Analyze office, conference room, or examining room to assure acoustical privacy

| | | | | | |
|---|---|---|---|---|---|
| Per room analyzed | — | LS | — | — | 505.00 |

Prepare acoustical design for church, auditorium, or lecture hall, (fees vary greatly with complexity of acoustics desired)

| | | | | | |
|---|---|---|---|---|---|
| Typical cost | — | LS | — | — | 2,150.00 |

Evaluate noise problems from proposed building

| | | | | | |
|---|---|---|---|---|---|
| Impact on surrounding environment | — | LS | — | — | 1,150.00 |

Evaluate noise problems from surrounding environment

| | | | | | |
|---|---|---|---|---|---|
| Impact on a proposed building | — | LS | — | — | 936.00 |

Front end scheduling. Coordinate architectural, engineering and other consultant services, plan check times and other approval times

| | | | | | |
|---|---|---|---|---|---|
| For residential and commercial projects | — | LS | — | — | 6,700.00 |

Project scheduling using CPM (Critical Path Method). Includes consultation, review of construction documents, development of construction logic, and graphic schedule. Comprehensive schedules to meet government or owner specifications will cost more.

| | | | | | |
|---|---|---|---|---|---|
| Wood frame buildings, 1 or 2 stories | — | LS | — | — | 4,460.00 |

Structural engineering plan check, typical 2-story, 8 to 10 unit apartment building

| | | | | | |
|---|---|---|---|---|---|
| Typical price per apartment building | — | LS | — | — | 1,180.00 |
| Add for underground parking | — | LS | — | — | 960.00 |

**Entrances, Colonial** An entrance consists of two pilasters (one at each side of the exterior opening) and a crosshead at the top of the door. A decorative pediment may rest on the crosshead. No crosshead is required if the pediment includes a breastboard. No door, frame or sill costs are included in the estimates below. Unpainted urethane.

Molded one-piece pilasters with plinth blocks, per set of two pilasters

| | Craft@Hrs | Unit | Material | Labor | Total |
|---|---|---|---|---|---|
| 1-1/2" wide x 86" high | BC@.800 | Ea | 48.50 | 29.60 | 78.10 |
| 3-1/2" wide x 81" high | BC@.800 | Ea | 70.50 | 29.60 | 100.10 |
| 3-1/2" wide x 108" high | BC@.800 | Ea | 90.00 | 29.60 | 119.60 |
| 5" wide x 108" high | BC@.800 | Ea | 108.00 | 29.60 | 137.60 |
| 7" wide x 90" high | BC@.800 | Ea | 118.00 | 29.60 | 147.60 |
| 9" wide x 90" high | BC@.800 | Ea | 140.00 | 29.60 | 169.60 |
| 9" wide x 108" high | BC@.800 | Ea | 220.00 | 29.60 | 249.60 |
| 11" wide x 144" high | BC@.800 | Ea | 300.00 | 29.60 | 329.60 |

Molded entrance crosshead, without pediment

| | Craft@Hrs | Unit | Material | Labor | Total |
|---|---|---|---|---|---|
| 40" wide by 6" high | BC@.400 | Ea | 47.50 | 14.80 | 62.30 |
| 48" wide by 6" high | BC@.400 | Ea | 54.50 | 14.80 | 69.30 |
| 78" wide by 6" high | BC@.400 | Ea | 86.00 | 14.80 | 100.80 |
| 40" wide by 10" high | BC@.400 | Ea | 86.00 | 14.80 | 100.80 |

| | Craft@Hrs | Unit | Material | Labor | Total |
|---|---|---|---|---|---|
| 48" wide by 10" high | BC@.400 | Ea | 95.00 | 14.80 | 109.80 |
| 78" wide by 10" high | BC@.400 | Ea | 165.00 | 14.80 | 179.80 |
| 40" wide by 14" high | BC@.400 | Ea | 138.00 | 14.80 | 152.80 |
| 48" wide by 14" high | BC@.400 | Ea | 170.00 | 14.80 | 184.80 |
| 78" wide by 14" high | BC@.400 | Ea | 188.00 | 14.80 | 202.80 |
| 96" wide by 14" high | BC@.400 | Ea | 255.00 | 14.80 | 269.80 |
| **Molded pediment without breastboard (crosshead required)** | | | | | |
| 38" wide, 10" high peaked cap | BC@.400 | Ea | 90.00 | 14.80 | 104.80 |
| 48" wide, 13" high peaked cap | BC@.400 | Ea | 95.00 | 14.80 | 109.80 |
| 78" wide, 21" high peaked cap | BC@.400 | Ea | 166.00 | 14.80 | 180.80 |
| 42" wide, 12" high decorative acorn | BC@.400 | Ea | 78.00 | 14.80 | 92.80 |
| 56" wide, 13" high decorative acorn | BC@.400 | Ea | 89.00 | 14.80 | 103.80 |
| 78" wide, 16" high decorative acorn | BC@.400 | Ea | 180.00 | 14.80 | 194.80 |
| 49" wide, 18" high ram's head | BC@.400 | Ea | 190.00 | 14.80 | 204.80 |
| 61" wide, 21" high ram's head | BC@.400 | Ea | 190.00 | 14.80 | 204.80 |
| 91" wide, 29" high ram's head | BC@.400 | Ea | 200.00 | 14.80 | 214.80 |
| **Molded pediment with breastboard (no crosshead required)** | | | | | |
| 40" wide, 17" high peaked cap | BC@.400 | Ea | 110.00 | 14.80 | 124.80 |
| 48" wide, 19" high peaked cap | BC@.400 | Ea | 129.00 | 14.80 | 143.80 |
| 80" wide, 29" high peaked cap | BC@.400 | Ea | 215.00 | 14.80 | 229.80 |
| 50" wide, 21" high decorative acorn | BC@.400 | Ea | 110.00 | 14.80 | 124.80 |
| 58" wide, 22" high decorative acorn | BC@.400 | Ea | 168.00 | 14.80 | 182.80 |
| 80" wide, 26" high decorative acorn | BC@.400 | Ea | 215.00 | 14.80 | 229.80 |
| 50" wide, 24" high ram's head | BC@.400 | Ea | 170.00 | 14.80 | 184.80 |
| 58" wide, 25" high ram's head | BC@.400 | Ea | 215.00 | 14.80 | 229.80 |
| 80" wide, 36" high ram's head | BC@.400 | Ea | 260.00 | 14.80 | 274.80 |

**Excavation and Backfill by Hand** Using hand tools.

General excavation, using a pick and shovel (loosening and one throw)

| | Craft@Hrs | Unit | Material | Labor | Total |
|---|---|---|---|---|---|
| Light soil | BL@1.10 | CY | — | 32.60 | 32.60 |
| Average soil | BL@1.70 | CY | — | 50.40 | 50.40 |
| Heavy soil or loose rock | BL@2.25 | CY | — | 66.80 | 66.80 |
| **Backfilling (one shovel throw from stockpile)** | | | | | |
| Sand | BL@.367 | CY | — | 10.90 | 10.90 |
| Average soil | BL@.467 | CY | — | 13.90 | 13.90 |
| Rock or clay | BL@.625 | CY | — | 18.50 | 18.50 |
| Add for compaction, average soil or sand | BL@.400 | CY | — | 11.90 | 11.90 |
| Fine grading | BL@.008 | SF | — | .24 | .24 |
| **Footings, light soil, using a pick and shovel** | | | | | |
| 6" deep x 12" wide (1.85 CY per CLF) | BL@.034 | LF | — | 1.01 | 1.01 |
| 8" deep x 12" wide (2.47 CY per CLF) | BL@.050 | LF | — | 1.48 | 1.48 |
| 8" deep x 16" wide (3.29 CY per CLF) | BL@.055 | LF | — | 1.63 | 1.63 |
| 8" deep x 18" wide (3.70 CY per CLF) | BL@.060 | LF | — | 1.78 | 1.78 |
| 10" deep x 12" wide (3.09 CY per CLF) | BL@.050 | LF | — | 1.48 | 1.48 |
| 10" deep x 16" wide (4.12 CY per CLF) | BL@.067 | LF | — | 1.99 | 1.99 |
| 10" deep x 18" wide (4.63 CY per CLF) | BL@.075 | LF | — | 2.23 | 2.23 |
| 12" deep x 12" wide (3.70 CY per CLF) | BL@.060 | LF | — | 1.78 | 1.78 |
| 12" deep x 16" wide (4.94 CY per CLF) | BL@.081 | LF | — | 2.40 | 2.40 |
| 12" deep x 20" wide (6.17 CY per CLF) | BL@.100 | LF | — | 2.97 | 2.97 |
| 12" deep x 24" wide (7.41 CY per CLF) | BL@.125 | LF | — | 3.71 | 3.71 |
| 16" deep x 16" wide (6.59 CY per CLF) | BL@.110 | LF | — | 3.26 | 3.26 |
| Add for average soil - loose rock or roots | — | % | — | 50.0 | — |

# Excavation – Trenching and Backfill with Heavy Equipment

| | Craft@Hrs | Unit | Material | Labor | Total |
|---|---|---|---|---|---|
| Loading trucks (shoveling, one throw) | | | | | |
| Average soil | BL@1.45 | CY | — | 43.00 | 43.00 |
| Rock or clay | BL@2.68 | CY | — | 79.50 | 79.50 |
| Pits to 5' (pits over 5' require shoring) | | | | | |
| Light soil | BL@1.34 | CY | — | 39.80 | 39.80 |
| Average soil | BL@2.00 | CY | — | 59.30 | 59.30 |
| Heavy soil | BL@2.75 | CY | — | 81.60 | 81.60 |
| Shaping trench bottom for pipe | | | | | |
| To 10" pipe | BL@.024 | LF | — | .71 | .71 |
| 12" to 20" pipe | BL@.076 | LF | — | 2.25 | 2.25 |
| Shaping embankment slopes | | | | | |
| Up to 1 in 4 slope | BL@.060 | SY | — | 1.78 | 1.78 |
| Over 1 in 4 slope | BL@.075 | SY | — | 2.23 | 2.23 |
| Add for top crown or toe | — | % | — | 50.0 | — |
| Add for swales | — | % | — | 90.0 | — |
| Spreading material piled on site | | | | | |
| Average soil | BL@.367 | CY | — | 10.90 | 10.90 |
| Stone or clay | BL@.468 | CY | — | 13.90 | 13.90 |
| Strip and pile top soil, depths to 6" | BL@.024 | SF | — | .71 | .71 |
| Tamping, hand tamp only | BL@.612 | CY | — | 18.20 | 18.20 |
| Trenches to 5', soil piled beside trench | | | | | |
| Light soil | BL@1.13 | CY | — | 33.50 | 33.50 |
| Average soil | BL@1.84 | CY | — | 54.60 | 54.60 |
| Heavy soil or loose rock | BL@2.86 | CY | — | 84.90 | 84.90 |

| | Craft@Hrs | Unit | Material | Labor | Equipment | Total |
|---|---|---|---|---|---|---|

**Trenching and Backfill with Heavy Equipment** These costs and productivity are based on utility line trenches and continuous footings where the spoil is piled adjacent to the trench. Linear feet (LF) and cubic yards (CY) per hour shown are based on a crew of two. Reduce productivity by 10% to 25% when spoil is loaded in trucks. Shoring, dewatering or unusual conditions are not included.

| | Craft@Hrs | Unit | Material | Labor | Equipment | Total |
|---|---|---|---|---|---|---|
| Wheel loader 55 HP, with integral backhoe | | | | | | |
| 12" wide bucket, for 12" wide trench. Depths 3' to 5' | | | | | | |
| Light soil (60 LF per hour) | B8@.033 | LF | — | 1.15 | .59 | 1.74 |
| Medium soil (55 LF per hour) | B8@.036 | LF | — | 1.25 | .64 | 1.89 |
| Heavy or wet soil (35 LF per hour) | B8@.057 | LF | — | 1.99 | 1.01 | 3.00 |
| 18" wide bucket, for 18" wide trench. Depths 3' to 5' | | | | | | |
| Light soil (55 LF per hour) | B8@.036 | LF | — | 1.25 | .64 | 1.89 |
| Medium soil (50 LF per hour) | B8@.040 | LF | — | 1.39 | .71 | 2.10 |
| Heavy or wet soil (30 LF per hour) | B8@.067 | LF | — | 2.34 | 1.19 | 3.53 |
| 24" wide bucket, for 24" wide trench. Depths 3' to 5' | | | | | | |
| Light soil (50 LF per hour) | B8@.040 | LF | — | 1.39 | .71 | 2.10 |
| Medium soil (45 LF per hour) | B8@.044 | LF | — | 1.53 | .78 | 2.31 |
| Heavy or wet soil (25 LF per hour) | B8@.080 | LF | — | 2.79 | 1.42 | 4.21 |
| Backfill trenches from loose material piled adjacent to trench. No compaction included. | | | | | | |
| Soil, previously excavated | | | | | | |
| Front-end loader 60 HP | | | | | | |
| (50 CY per hour) | B8@.040 | CY | — | 1.39 | .58 | 1.97 |
| D-3 crawler dozer | | | | | | |
| (25 CY per hour) | B8@.080 | CY | — | 2.79 | 1.67 | 4.46 |
| 3/4 CY crawler loader | | | | | | |
| (33 CY per hour) | B8@.061 | CY | — | 2.13 | 1.24 | 3.37 |

# Excavation with Heavy Equipment

|  | Craft@Hrs | Unit | Material | Labor | Equipment | Total |
|---|---|---|---|---|---|---|
| D-7 crawler dozer | | | | | | |
| (130 CY per hour) | B8@.015 | CY | — | .52 | .77 | 1.29 |
| Sand or gravel bedding | | | | | | |
| 3/4 CY wheel loader | | | | | | |
| (80 CY per hour) | B8@.025 | CY | — | .87 | .39 | 1.26 |
| Compaction of soil in trenches in 8" layers | | | | | | |
| Pneumatic tampers | | | | | | |
| (40 CY per hour) | BL@.050 | CY | — | 1.48 | .88 | 2.36 |
| Vibrating rammers, gasoline powered "Jumping Jack" | | | | | | |
| (20 CY per hour) | BL@.100 | CY | — | 2.97 | .93 | 3.90 |

|  | Craft@Hrs | Unit | Material | Labor | Total |
|---|---|---|---|---|---|
| **Excavation with Heavy Equipment** These figures assume a crew of one operator unless noted otherwise. Only labor costs are included here. See equipment rental costs at the end of this section. Use the productivity rates listed here to determine the number of hours or days that the equipment will be needed. For larger jobs and commercial work, see excavation costs under Site Work in the Commercial and Industrial division of this book. | | | | | |
| Excavation rule of thumb for small jobs | — | CY | — | 2.85 | 2.85 |
| Backhoe, operator and one laborer, 3/4 CY bucket | | | | | |
| Light soil (13.2 CY per hour) | B8@.152 | CY | — | 5.30 | 5.30 |
| Average soil (12.5 CY per hour) | B8@.160 | CY | — | 5.58 | 5.58 |
| Heavy soil (10.3 CY per hour) | B8@.194 | CY | — | 6.76 | 6.76 |
| Sand (16 CY per hour) | B8@.125 | CY | — | 4.36 | 4.36 |
| Add when using 1/2 CY bucket | — | % | — | 25.0 | — |
| Bulldozer, 65 HP unit | | | | | |
| Backfill (36 CY per hour) | 0E@.027 | CY | — | 1.08 | 1.08 |
| Clearing brush (900 SF per hour) | 0E@.001 | SF | — | .04 | .04 |
| Add for thick brush | — | % | — | 300.0 | — |
| Spread dumped soil (42 CY per hour) | 0E@.024 | CY | — | .96 | .96 |
| Strip topsoil (17 CY per hour) | 0E@.059 | CY | — | 2.36 | 2.36 |
| Bulldozer, 90 HP unit | | | | | |
| Backfill (40 CY per hour) | 0E@.025 | CY | — | 1.00 | 1.00 |
| Clearing brush (1,000 SF per hour) | 0E@.001 | SF | — | .04 | .04 |
| Add for thick brush | — | % | — | 300.0 | — |
| Spread dumped soil (45 CY per hour) | 0E@.023 | CY | — | .92 | .92 |
| Strip top soil (20 CY per hour) | 0E@.050 | CY | — | 2.00 | 2.00 |
| Bulldozer, 140 HP unit | | | | | |
| Backfill (53 CY per hour) | 0E@.019 | CY | — | .76 | .76 |
| Clearing brush (1,250 SF per hour) | 0E@.001 | SF | — | .04 | .04 |
| Add for thick brush | — | % | — | 200.0 | — |
| Spread dumped soil (63 CY per hour) | 0E@.016 | CY | — | .64 | .64 |
| Strip top soil (25 CY per hour) | 0E@.040 | CY | — | 1.60 | 1.60 |
| Dump truck, spot, load, unload, travel | | | | | |
| 3 CY truck | | | | | |
| Short haul (9 CY per hour) | B7@.222 | CY | — | 6.68 | 6.68 |
| 2-3 mile haul (6 CY per hour) | B7@.333 | CY | — | 10.00 | 10.00 |
| 4 mile haul (4 CY per hour) | B7@.500 | CY | — | 15.10 | 15.10 |
| 5 mile haul (3 CY per hour) | B7@.666 | CY | — | 20.00 | 20.00 |
| 4 CY truck | | | | | |
| Short haul (11 CY per hour) | B7@.182 | CY | — | 5.48 | 5.48 |
| 2-3 mile haul (7 CY per hour) | B7@.286 | CY | — | 8.61 | 8.61 |

| | Craft@Hrs | Unit | Material | Labor | Total |
|---|---|---|---|---|---|
| 4 mile haul (5.5 CY per hour) | B7@.364 | CY | — | 11.00 | 11.00 |
| 5 mile haul (4.5 CY per hour) | B7@.444 | CY | — | 13.40 | 13.40 |
| 5 CY truck | | | | | |
|   Short haul (14 CY per hour) | B7@.143 | CY | — | 4.30 | 4.30 |
|   2-3 mile haul (9.5 CY per hour) | B7@.211 | CY | — | 6.35 | 6.35 |
|   4 mile haul (6.5 CY per hour) | B7@.308 | CY | — | 9.27 | 9.27 |
|   5 mile haul (5.5 CY per hour) | B7@.364 | CY | — | 11.00 | 11.00 |
| Jackhammer, one laborer (per CY of unloosened soil) | | | | | |
|   Average soil (2-1/2 CY per hour) | BL@.400 | CY | — | 11.90 | 11.90 |
|   Heavy soil (2 CY per hour) | BL@.500 | CY | — | 14.80 | 14.80 |
|   Igneous or dense rock (.7 CY per hour) | BL@1.43 | CY | — | 42.40 | 42.40 |
|   Most weathered rock (1.2 CY per hour) | BL@.833 | CY | — | 24.70 | 24.70 |
|   Soft sedimentary rock (2 CY per hour) | BL@.500 | CY | — | 14.80 | 14.80 |
|   Air tamp (3 CY per hour) | BL@.333 | CY | — | 9.88 | 9.88 |
| Loader (tractor shovel) | | | | | |
|   Piling earth on premises, 1 CY bucket | | | | | |
|     Light soil (43 CY per hour) | OE@.023 | CY | — | .92 | .92 |
|     Heavy soil (35 CY per hour) | OE@.029 | CY | — | 1.16 | 1.16 |
|   Loading trucks, 1 CY bucket | | | | | |
|     Light soil (53 CY per hour) | OE@.019 | CY | — | .76 | .76 |
|     Average soil (43 CY per hour) | OE@.023 | CY | — | .92 | .92 |
|     Heavy soil (40 CY per hour) | OE@.025 | CY | — | 1.00 | 1.00 |
|   Deduct for 2-1/4 CY bucket | — | % | — | -43.0 | — |
| Sheepsfoot roller (18.8 CSF per hour) | OE@.001 | SF | — | .04 | .04 |
| Sprinkling with truck, (62.3 CSF per hour) | TR@.001 | SF | — | .03 | .03 |
| Tree and brush removal, labor only (clear and grub, one operator and one laborer) | | | | | |
|   Light brush | B8@10.0 | Acre | — | 349.00 | 349.00 |
|   Heavy brush | B8@12.0 | Acre | — | 418.00 | 418.00 |
|   Wooded | B8@64.0 | Acre | — | 2,230.00 | 2,230.00 |
| Tree removal, cutting trees, removing branches, cutting into short lengths with chain saws and axes, by tree diameter, labor only | | | | | |
|   8" to 12" (2.5 manhours per tree) | BL@2.50 | Ea | — | 74.20 | 74.20 |
|   13" to 18" (3.5 manhours per tree) | BL@3.50 | Ea | — | 104.00 | 104.00 |
|   19" to 24" (5.5 manhours per tree) | BL@5.50 | Ea | — | 163.00 | 163.00 |
|   25" to 36" (7.0 manhours per tree) | BL@7.00 | Ea | — | 208.00 | 208.00 |
| Tree stump removal, using a small dozer, by tree diameter, operator only | | | | | |
|   6" to 10" (1.6 manhours per stump) | OE@1.60 | Ea | — | 64.10 | 64.10 |
|   11" to 14" (2.1 manhours per stump) | OE@2.10 | Ea | — | 84.10 | 84.10 |
|   15" to 18" (2.6 manhours per stump) | OE@2.60 | Ea | — | 104.00 | 104.00 |
|   19" to 24" (3.1 manhours per stump) | OE@3.10 | Ea | — | 124.00 | 124.00 |
|   25" to 30" (3.3 manhours per stump) | OE@3.30 | Ea | — | 132.00 | 132.00 |
| Trenching machine, crawler-mounted | | | | | |
|   Light soil (27 CY per hour) | OE@.037 | CY | — | 1.48 | 1.48 |
|   Heavy soil (21 CY per hour) | OE@.048 | CY | — | 1.92 | 1.92 |
|   Add for pneumatic-tired machine | — | % | — | 15.0 | — |

**Excavation Equipment Rental Costs** Typical cost not including fuel, delivery or pickup charges. Add labor costs from the previous section. Half day minimums.

| | 1/2 day | Day | Week |
|---|---|---|---|
| Backhoe, wheeled mounted | | | |
|   55 HP unit, 1/8 to 3/4 CY bucket | 186.00 | 265.00 | 825.00 |
|   65 HP unit, 1/4 to 1 CY bucket | 224.00 | 320.00 | 960.00 |
|   Add for each delivery or pickup | 100.00 | 100.00 | 100.00 |

# Excavation Equipment Rental Costs

|  | 1/2 day | Day | Week |
|---|---|---|---|
| Bulldozer, crawler tractor | | | |
| 65 HP | 238.00 | 340.00 | 1,060.00 |
| 90 to 105 HP unit, D-4 or D-5 | 326.00 | 465.00 | 1,320.00 |
| 140 HP unit, D-6 | 497.00 | 710.00 | 2,150.00 |
| Add for each delivery or pickup | 100.00 | 100.00 | 100.00 |
| Dump truck, on-highway type | | | |
| 3 CY | 147.00 | 210.00 | 550.00 |
| 5 CY | 161.00 | 230.00 | 820.00 |
| 10 CY | 196.00 | 280.00 | 1,070.00 |
| Wheel loaders, front-end load and dump, diesel | | | |
| 3/4 CY bucket, 4WD, articulated | 165.00 | 235.00 | 670.00 |
| 1 CY bucket, 4WD, articulated | 203.00 | 290.00 | 845.00 |
| 2 CY bucket, 4WD, articulated | 238.00 | 340.00 | 1,050.00 |
| 3-1/4 CY bucket, 4WD, articulated | 490.00 | 700.00 | 2,150.00 |
| 5 CY bucket, 4WD, articulated | 711.00 | 1,020.00 | 3,030.00 |
| Add for each delivery or pickup | 100.00 | 100.00 | 100.00 |
| Sheepsfoot roller, towed type, 40" diameter, | | | |
| 48" wide, double drum | 119.00 | 170.00 | 520.00 |
| Trenching machine, to 5'6" depth, 6" to 16" width | | | |
| capacity, crawler-mounted type | 193.00 | 275.00 | 810.00 |
| Add for each delivery or pickup | 100.00 | 100.00 | 100.00 |
| Compactor, 32" plate width, manually guided, | | | |
| gasoline engine, vibratory plate type | 98.00 | 140.00 | 460.00 |
| Chain saw, 18" to 23" bar, gasoline engine | 39.00 | 55.00 | 200.00 |
| Breakers, pavement, medium duty, with blade | | | |
| Including 50' of 1" hose | 39.00 | 55.00 | 163.00 |
| Air compressors, trailer-mounted, gas powered, silenced | | | |
| 100 CFM to 150 CFM | 70.00 | 100.00 | 300.00 |
| 175 CFM | 112.00 | 160.00 | 510.00 |

## Fans

|  | Craft@Hrs | Unit | Material | Labor | Total |
|---|---|---|---|---|---|

Economy ceiling and wall exhaust fan. For baths, utility, and recreation rooms. For baths up to 45 square feet, other rooms up to 60 square feet. Installs in ceiling or in wall. Plastic duct collar. White polymeric grille. UL listed for use in tub and shower enclosure with GFI branch circuit wiring. Housing dimensions: 8-1/16-inch length, 7-3/16-inch width, 3-7/8-inch depth. Grille size, 8-11/16-inch x 9-1/2-inch. 3-inch duct. 0.75 amp. CFM (cubic feet of air per minute). Labor includes setting and connecting only. Add the cost of wiring and ducting.

| | Craft@Hrs | Unit | Material | Labor | Total |
|---|---|---|---|---|---|
| 50 CFM, 2.5 sones | BE@1.00 | Ea | 13.80 | 39.60 | 53.40 |
| 70 CFM, 4.0 sones | BE@1.00 | Ea | 31.30 | 39.60 | 70.90 |

Vertical discharge bath fan. Galvanized steel housing, with built-in damper and spin-on white polymeric grille. Polymeric fan blade and duct connectors. Built-in double steel mounting ears with keyhole slots. Fits 8-inch ducts. UL listed. CFM (cubic feet of air per minute). Labor includes setting and connecting only. Add the cost of wiring and ducting.

| | Craft@Hrs | Unit | Material | Labor | Total |
|---|---|---|---|---|---|
| 80 CFM, 1.0 sones | BE@1.00 | Ea | 127.00 | 39.60 | 166.60 |

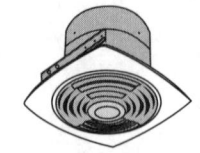

Vertical discharge utility fan. Mounts in ceiling. Discharges through duct to roof or wall. Pre-wired motor with plug-in receptacle. Adjustable hanger bars for 16- or 24-inch on-center joists. Silver anodized aluminum grille. Fits 7-inch round ducts. Housing dimensions: 11-inch diameter x 5-5/8-inch deep. UL listed. CFM (cubic feet of air per minute). Labor includes setting and connecting only. Add the cost of wiring and ducting.

| | Craft@Hrs | Unit | Material | Labor | Total |
|---|---|---|---|---|---|
| 210 CFM, 6.5 sones | BE@1.00 | Ea | 81.20 | 39.60 | 120.80 |

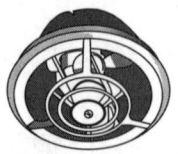

| | Craft@Hrs | Unit | Material | Labor | Total |
|---|---|---|---|---|---|

Ceiling or wall bath exhaust fan. White polymeric grille. Torsion spring grille mounting requires no tools. Plug-in, permanently lubricated motor. Centrifugal blower wheel. Rugged, 26 gauge galvanized steel housing. Sturdy key holes mounting brackets for quick, accurate installation. Tapered, polymeric duct fitting with built-in backdraft damper. U.L. listed for use over bathtubs and showers when connected to a GFCI circuit. 120 volts, 0.5 amps, 2.5 sones, 80 CFM (HVI-2100 certified) and 4-inch round duct. Labor includes setting and connecting only. Add the cost of wiring and ducting.

| 80 CFM, 2.5 sones | BE@1.00 | Ea | 72.70 | 39.60 | 112.30 |

Bath exhaust fan. Pre-wired outlet box for plug-in receptacle, torsion spring-mounted low-profile grille. Steel housing, damper to eliminate back drafts. Labor includes setting and connecting only. Add the cost of wiring and ducting.

| 80 CFM, 2.5 sones, to 80 SF bath | BE@1.00 | Ea | 101.00 | 39.60 | 140.60 |

Thru-the-wall utility fan. Steel housing with built-in damper and white polymeric grille. Permanently lubricated motor. Rotary on and off switch. Housing dimensions: 14-1/4-inch length, 14-1/4-inch width, 4-1/2-inch depth. UL listed. CFM (cubic feet of air per minute). Labor includes setting and connecting only. Add the cost of wiring and exterior finish.

| 180 CFM, 5.0 sones | BE@1.00 | Ea | 74.80 | 39.60 | 114.40 |

QuieTTest® low-sound bath fan. For 105 square foot bath. Pre-wired outlet box with plug-in receptacle. Adjustable hanger brackets. Fits 4-inch round ducts. Low-profile white polymeric grille, torsion-spring mounted. Housing dimensions: 9-3/8-inch length, 11-1/4-inch width, 7-7/8-inch depth. Grille size, 14-1/4-inch x 12-1/16-inch. UL listed for use in tub and shower when used with GFI branch circuit wiring. Not recommended for kitchen use. CFM (cubic feet of air per minute). Labor includes setting and connecting only. Add the cost of wiring and ducting.

| 110 CFM, 2.0 sones | BE@1.00 | Ea | 128.00 | 39.60 | 167.60 |

QuieTTest® low-sound ceiling blower. For 375 square foot room. Pre-wired outlet box with plug-in receptacle. Rounded, low-profile white polymeric grille with silver anodized trim at each end. Fits 3-1/4-inch x 10-inch ducts. Housing dimensions: 14-1/4-inch length, 10-inch width, 9-inch depth. Grille size, 16-1/2-inch x 12-3/32 inch. CFM (cubic feet of air per minute). Labor includes setting and connecting only. Add the cost of wiring and ducting.

| 300 CFM, 4.5 sones | BE@1.00 | Ea | 153.00 | 39.60 | 192.60 |

## Lighted Bath Exhaust Fans

ValueTest™ economy bath fan and light, NuTone. Fan and light operate separately or together. Fits 4-inch ducts. White polymeric grille with break-resistant lens. Uses 100-watt lamp (not included). Housing dimensions: 9-inch length, 9-inch width, 5-1/2-inch depth. Grille size, 10-3/4-inch x 12-1/8-inch. UL listed for use in tub and shower when used with GFI branch circuit wiring. CFM (cubic feet of air per minute). Labor includes setting and connecting only. Add the cost of wiring and ducting.

| 50 CFM, 2.5 sones, 45 SF bath | BE@1.00 | Ea | 40.70 | 39.60 | 80.30 |
| 70 CFM, 4.5 sones, 65 SF bath | BE@1.00 | Ea | 66.60 | 39.60 | 106.20 |

Decorative bath exhaust fan with light, Broan Manufacturing. Corrosion-resistant finish. Frosted melon glass globe. 60 watt max. CFM (cubic feet of air per minute). Labor includes setting and connecting only. Add the cost of wiring and ducting.

| 70 CFM, 3.5 sones | BE@1.00 | Ea | 94.70 | 39.60 | 134.30 |

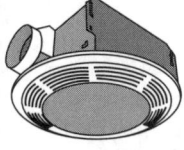

Exhaust Air deluxe bath fan with light, NuTone. Ventilation for baths up to 95 square feet, other rooms up to 125 square feet. 100 and 7 watt ceiling and night-light or energy-saving fluorescent light (lamps not included). Snap-on grille assembly. Housing dimensions: 9-inch length, 9-inch width, 6-inch depth. Polymeric white grille, 15-inch diameter. UL listed for use in tub and shower when used with GFI branch circuit wiring. CFM (cubic feet of air per minute). Labor includes setting and connecting only. Add the cost of wiring and ducting.

| 100 CFM, 3.5 sones | BE@1.00 | Ea | 128.00 | 39.60 | 167.60 |

# Fans

| | Craft@Hrs | Unit | Material | Labor | Total |
|---|---|---|---|---|---|

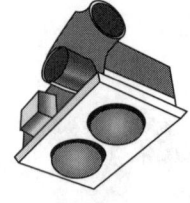

## Bath Exhaust Fan with Heater

Infrared bulb heater and fan. 4-point adjustable mounting brackets span up to 24-inches. Uses 250 watt, 120 volt R-40 infrared bulbs (not included). Heater and fan units include 70 CFM, 3.5 sones ventilator fan. Damper and duct connector included. Plastic matte-white molded grille and compact housings. CFM (cubic feet of air per minute). Labor includes setting and connecting only. Add the cost of wiring and ducting.

| | | | | | |
|---|---|---|---|---|---|
| 1-bulb, 70 CFM, 3.5 sones | BE@1.00 | Ea | 54.50 | 39.60 | 94.10 |
| 2-bulb heater, 7" duct | BE@1.10 | Ea | 68.70 | 43.50 | 112.20 |

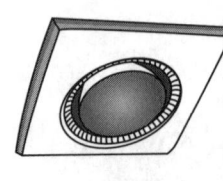

Heat-A-Lamp® bulb heater and fan, NuTone. Non-IC. Swivel-mount. Torsion spring holds plates firmly to ceiling. 250-watt radiant heat from one R-40 infrared heat lamp. Uses 4-inch duct. Adjustable socket for ceilings up to 1-inch thick. Adjustable hanger bars. Automatic reset for thermal protection. White polymeric finish. UL listed. Btu (British Thermal Unit). 6-7/8" high. Labor includes setting and connecting only. Add the cost of wiring and ducting.

| | | | | | |
|---|---|---|---|---|---|
| 2.6 amp, 1 lamp, 12-1/2" x 10" | BE@1.50 | Ea | 56.50 | 59.40 | 115.90 |
| 5.0 amp, 2 lamp, 15-3/8" x 11" | BE@1.50 | Ea | 72.70 | 59.40 | 132.10 |

Surface-mount ceiling bath resistance heater and fan. 1250 watt. 4266 Btu (British Thermal Unit). 120 volt. Chrome alloy wire element for instant heat. Built-in fan. Automatic overheat protection. Low-profile housing. Permanently lubricated motor. Mounts to standard 3-1/4-inch round or 4-inch octagonal ceiling electrical box. Satin-finish aluminum grille extends 2-3/4-inch from ceiling. 11" diameter x 2-3/4" deep. Labor includes setting and connecting only. Add the cost of wiring and ducting.

| | | | | | |
|---|---|---|---|---|---|
| 10.7 amp | BE@.750 | Ea | 57.50 | 29.70 | 87.20 |

Designer Series bath heater and exhaust fan, Broan Manufacturing. 1500-watt fan-forced heater. 120-watt light capacity, 7-watt night-light (bulbs not included). Permanently lubricated motor. Polymeric damper prevents cold back drafts. 4-point adjustable mounting brackets with keyhole slots. Torsion-spring grille mounting, no tools needed. Non-glare light diffusing glass lenses. Fits single gang opening. Suitable for use with insulation. Includes 4-function control unit. CFM (cubic feet of air per minute). 100 CFM, 1.2 sones. Labor includes setting and connecting only. Add the cost of wiring and ducting.

| | | | | | |
|---|---|---|---|---|---|
| 4" duct, 100 watt lamp | BE@1.00 | Ea | 232.00 | 39.60 | 271.60 |

## Exhaust Fan Accessories

Timer switch

| | | | | | |
|---|---|---|---|---|---|
| 60-minute timer | BE@.250 | Ea | 21.20 | 9.90 | 31.10 |

Exhaust fan switch

| | | | | | |
|---|---|---|---|---|---|
| 60 minute timer, white | BE@.300 | Ea | 27.10 | 11.90 | 39.00 |

SensAire® 4-function wall control switch. Fits single gang opening. Top switch for on-auto-off. Other switches for light and night-light. For use with SensAire® fans and lights.

| | | | | | |
|---|---|---|---|---|---|
| White | BE@.250 | Ea | 17.50 | 9.90 | 27.40 |

Bath vent kit, Deflect-O. 4" x 8' UL listed Supurr-Flex® duct, aluminum roof vent, 3" to 4" increaser and 2 clamps. Fire resistant.

| | | | | | |
|---|---|---|---|---|---|
| Flexible duct | SW@.650 | Ea | 21.20 | 26.70 | 47.90 |

Roof vent kit. For venting kitchen or bath exhaust fan through slanted roof. Works with both 3-inch and 4-inch ducted units.

| | | | | | |
|---|---|---|---|---|---|
| 8-foot length | SW@.650 | Ea | 20.70 | 26.70 | 47.40 |

## Ventilators

Roof turbine vent with base

| | | | | | |
|---|---|---|---|---|---|
| 12", galvanized | SW@1.00 | Ea | 37.50 | 41.00 | 78.50 |
| 12", weathered wood | SW@1.00 | Ea | 39.50 | 41.00 | 80.50 |
| 12", white aluminum | SW@1.00 | Ea | 34.10 | 41.00 | 75.10 |

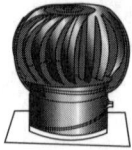

| | Craft@Hrs | Unit | Material | Labor | Total |
|---|---|---|---|---|---|
| Add for steep pitch base | SW@.440 | Ea | 7.98 | 18.10 | 26.08 |
| Add for weather cap | — | Ea | 9.26 | — | 9.26 |

**Belt drive attic exhaust fan, 1/3 HP, with aluminum shutter and plenum boards**

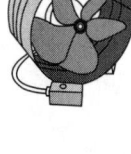

| | Craft@Hrs | Unit | Material | Labor | Total |
|---|---|---|---|---|---|
| 30" diameter | SW@3.00 | Ea | 303.00 | 123.00 | 426.00 |
| 36" diameter | SW@3.00 | Ea | 308.00 | 123.00 | 431.00 |
| 12-hour timer switch | BE@.950 | Ea | 31.50 | 37.60 | 69.10 |

**Power attic gable vent**

| | Craft@Hrs | Unit | Material | Labor | Total |
|---|---|---|---|---|---|
| 1,280 CFM | SW@1.65 | Ea | 49.90 | 67.70 | 117.60 |
| 1,540 CFM | SW@1.65 | Ea | 75.80 | 67.70 | 143.50 |
| 1,600 CFM | SW@1.65 | Ea | 102.00 | 67.70 | 169.70 |
| Shutter for gable vent | SW@.450 | Ea | 43.20 | 18.50 | 61.70 |
| Humidistat control | BE@.550 | Ea | 31.70 | 21.80 | 53.50 |
| Thermostat control | BE@.550 | Ea | 25.30 | 21.80 | 47.10 |

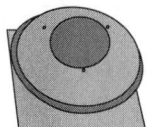

**Roof-mount power ventilator**

| | Craft@Hrs | Unit | Material | Labor | Total |
|---|---|---|---|---|---|
| 1,250 CFM | SW@1.65 | Ea | 89.80 | 67.70 | 157.50 |
| 1,600 CFM | SW@1.65 | Ea | 125.00 | 67.70 | 192.70 |

## Ceiling Fans

Ceiling fans, decorative, 3-speed reversible motors, lighting kits optional (see below). Includes 10' of type NM cable, installation and connection.

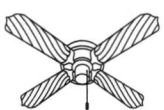

| | Craft@Hrs | Unit | Material | Labor | Total |
|---|---|---|---|---|---|
| 42" four-blade fan, white finish | BE@1.50 | Ea | 72.70 | 59.40 | 132.10 |
| 44" five blade fan with 3 spotlights | BE@1.50 | Ea | 79.40 | 59.40 | 138.80 |
| 44" five-blade reversible fan | BE@1.50 | Ea | 97.00 | 59.40 | 156.40 |
| 48" five rattan blade fan | BE@1.50 | Ea | 144.00 | 59.40 | 203.40 |
| 56" five-blade leaf design premium fan | BE@1.50 | Ea | 224.00 | 59.40 | 283.40 |

Optional lighting fixtures for ceiling fans

| | Craft@Hrs | Unit | Material | Labor | Total |
|---|---|---|---|---|---|
| vaulted ceiling mount | BE@.350 | Ea | 19.50 | 13.90 | 33.40 |
| Schoolhouse or round globe, brass trim | BE@.283 | Ea | 17.50 | 11.20 | 28.70 |
| Add for ball pull chain | BE@.017 | Ea | 5.00 | .67 | 5.67 |
| Add for pack of four ceiling fan light bulbs | BE@.017 | Pack | 8.59 | .67 | 9.26 |

**Garage exhaust fans**   For enclosed automotive garage space. U.S. Green Building Council, Underwriters Laboratories and CSA approved. Includes tamper-proof metal enclosure, sensor, fan actuator, weatherproof exhaust grille and cover, outlet duct and relay switch. Add for electrical connection.

Home garage exhaust fan, roof or upper sidewall mount, 110 volts.

| | Craft@Hrs | Unit | Material | Labor | Total |
|---|---|---|---|---|---|
| 118 CFM | SW@2.00 | Ea | 470.00 | 82.10 | 552.10 |
| 235 CFM | SW@2.00 | Ea | 525.00 | 82.10 | 607.10 |
| 416 CFM | SW@2.50 | Ea | 552.00 | 103.00 | 655.00 |

Parking garage exhaust fan, 208 volts, remote sensor controlled

| | Craft@Hrs | Unit | Material | Labor | Total |
|---|---|---|---|---|---|
| 400 CFM | SW@2.00 | Ea | 470.00 | 82.10 | 552.10 |
| 600 CFM | SW@2.00 | Ea | 525.00 | 82.10 | 607.10 |
| 800 CFM | SW@2.50 | Ea | 552.00 | 103.00 | 655.00 |
| 1,000 CFM | SW@2.75 | Ea | 598.00 | 113.00 | 711.00 |
| 1,500 CFM | SW@3.00 | Ea | 871.00 | 123.00 | 994.00 |
| 2,000 CFM | SW@3.50 | Ea | 1,100.00 | 144.00 | 1,244.00 |
| Mount sensors | SW@.250 | Ea | 450.00 | 10.30 | 460.30 |
| Apply fan frame sealant | SW@.150 | LF | 2.40 | 6.16 | 8.56 |
| Install electrical wiring | BE@.150 | LF | 1.75 | 5.94 | 7.69 |
| Install controls | BE@.500 | Ea | — | 19.80 | 19.80 |
| Commission and test | P1@4.00 | Ea | — | 143.00 | 143.00 |

# Fencing, Chain Link

| | Craft@Hrs | Unit | Material | Labor | Total |
|---|---|---|---|---|---|

## Fencing, Chain Link

Galvanized steel, 11.5 gauge 2" x 2" fence fabric, 2-3/8" terminal post with post caps, 1-5/8" line posts at 10' with loop line post caps, 1-3/8" top rail with rail ends, tension bands, tension bars and required bolts with nuts. Material includes 2/3 CF sack of concrete per post. Use 50 LF of fence as a minimum job cost.

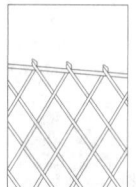

| | Craft@Hrs | Unit | Material | Labor | Total |
|---|---|---|---|---|---|
| 36" high | BL@.125 | LF | 9.14 | 3.71 | 12.85 |
| 42" high | BL@.125 | LF | 9.74 | 3.71 | 13.45 |
| 48" high | BL@.125 | LF | 10.30 | 3.71 | 14.01 |
| 60" high | BL@.125 | LF | 11.90 | 3.71 | 15.61 |
| 72" high | BL@.125 | LF | 13.40 | 3.71 | 17.11 |
| Add for redwood filler strips | BL@.024 | LF | 6.89 | .71 | 7.60 |
| Add for aluminum filler strips | BL@.018 | LF | 11.00 | .53 | 11.53 |
| Add for PVC filler strips | BL@.018 | LF | 5.50 | .53 | 6.03 |

Gates, driveway or walkway. Same construction as fencing shown above, assembled units ready to install. Add for gate hardware from below.

| Driveway gates, costs per linear foot (LF) | Craft@Hrs | Unit | Material | Labor | Total |
|---|---|---|---|---|---|
| 36" high | BL@.025 | LF | 13.90 | .74 | 14.64 |
| 42" high | BL@.025 | LF | 14.60 | .74 | 15.34 |
| 48" high | BL@.025 | LF | 15.20 | .74 | 15.94 |
| 60" high | BL@.033 | LF | 16.40 | .98 | 17.38 |
| 72" high | BL@.033 | LF | 17.70 | .98 | 18.68 |

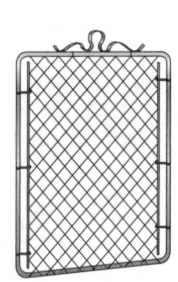

| 3'6" wide walkway gate, costs per gate | Craft@Hrs | Unit | Material | Labor | Total |
|---|---|---|---|---|---|
| 36" high | BL@.500 | Ea | 31.30 | 14.80 | 46.10 |
| 42" high | BL@.500 | Ea | 35.20 | 14.80 | 50.00 |
| 48" high | BL@.500 | Ea | 39.80 | 14.80 | 54.60 |
| 60" high | BL@.666 | Ea | 56.10 | 19.80 | 75.90 |
| 72" high | BL@.666 | Ea | 58.00 | 19.80 | 77.80 |

Add for gate hardware, per gate. Labor is included with labor shown for gate above. Includes latch, hanger brackets, bolts and hinge.

| Driveway gate hardware | Craft@Hrs | Unit | Material | Labor | Total |
|---|---|---|---|---|---|
| 36" or 42" high | — | Ea | 74.40 | — | 74.40 |
| 48" or 60" high | — | Ea | 94.30 | — | 94.30 |
| 72" high | — | Ea | 99.90 | — | 99.90 |
| Walkway gate hardware | | | | | |
| 36" or 42" high | — | Ea | 29.90 | — | 29.90 |
| 48" or 60" high | — | Ea | 36.40 | — | 36.40 |
| 72" high | — | Ea | 39.40 | — | 39.40 |

Pet enclosure fencing. Chain link fence, galvanized steel, 11.5 gauge, 2" x 2" mesh fence fabric, 1-3/8" tubular steel frame, 72" high, lockable gate. Posts set in concrete. Add for concrete slab, if required.

| | Craft@Hrs | Unit | Material | Labor | Total |
|---|---|---|---|---|---|
| 4' x 10' | BL@4.00 | Ea | 566.00 | 119.00 | 685.00 |
| 4' x 10', with chain link cover | BL@5.00 | Ea | 778.00 | 148.00 | 926.00 |
| 8' x 10' | BL@4.75 | Ea | 665.00 | 141.00 | 806.00 |
| 10' x 15' | BL@5.00 | Ea | 827.00 | 148.00 | 975.00 |

## Fence, Galvanized Wire Mesh
With 1-3/8" galvanized steel posts, set without concrete 10' OC.

12-1/2 gauge steel, 12" vertical wire spacing

| | Craft@Hrs | Unit | Material | Labor | Total |
|---|---|---|---|---|---|
| 36" high, 6 horizontal wires | BL@.127 | LF | .64 | 3.77 | 4.41 |
| 48" high, 8 horizontal wires | BL@.127 | LF | .80 | 3.77 | 4.57 |
| 48" high, 8 horizontal wires (11 gauge) | BL@.127 | LF | 1.17 | 3.77 | 4.94 |

11 gauge fence fabric, 9" vertical wire spacing

| | Craft@Hrs | Unit | Material | Labor | Total |
|---|---|---|---|---|---|
| 42" high, 8 horizontal wires | BL@.127 | LF | 2.11 | 3.77 | 5.88 |
| 48" high, 9 horizontal wires | BL@.127 | LF | 2.32 | 3.77 | 6.09 |

|  | Craft@Hrs | Unit | Material | Labor | Total |
|---|---|---|---|---|---|
| 14 gauge fence fabric, 6" vertical wire spacing | | | | | |
| 36" high, 8 horizontal wires | BL@.127 | LF | 1.43 | 3.77 | 5.20 |
| 48" high, 9 horizontal wires | BL@.127 | LF | 2.13 | 3.77 | 5.90 |
| 60" high, 10 horizontal wires | BL@.127 | LF | 2.73 | 3.77 | 6.50 |
| Add for components for electrified galvanized wire mesh fence | | | | | |
| 15 gauge aluminum wire, single strand | BL@.050 | LF | .10 | 1.48 | 1.58 |
| 12-1/2 gauge barbed wire, 4 barbs per LF | BL@.050 | LF | 2.04 | 1.48 | 3.52 |
| Charging unit for electric fence, Solar powered, | | | | | |
| 10 mile range, built in 6V battery | BL@.500 | Ea | 180.00 | 14.80 | 194.80 |
| Charging unit, 110V, | | | | | |
| 30 mile range, indoors or out | BL@.500 | Ea | 106.00 | 14.80 | 120.80 |
| Charging unit, 110V, | | | | | |
| 2 mile range, indoors or out | BL@.500 | Ea | 40.80 | 14.80 | 55.60 |
| Fence stays for barbed wire, 48" high | BL@.250 | Ea | .87 | 7.42 | 8.29 |
| Add for electric fence warning sign | | | | | |
| Yellow with black lettering | BL@.125 | Ea | .82 | 3.71 | 4.53 |
| Add for polyethylene insulators | BL@.250 | Ea | .21 | 7.42 | 7.63 |
| Add for non-conductive gate fastener | BL@.500 | Ea | 4.02 | 14.80 | 18.82 |
| Add for lightning arrestor kit | BL@.500 | Ea | 9.70 | 14.80 | 24.50 |

## Fence Post Holes

Based on holes 1' diameter and 2' deep with easy access on level terrain. Adjustments for other conditions are below. Per hole dug.

|  | Craft@Hrs | Unit | Material | Labor | Total |
|---|---|---|---|---|---|
| Excavate post hole by hand | | | | | |
| Light soil, sand or loam | BL@.400 | Ea | — | 11.90 | 11.90 |
| Medium soil, gravel or light clay | BL@.600 | Ea | — | 17.80 | 17.80 |
| Heavy soil, shale, sandstone, rocky | BL@1.00 | Ea | — | 29.70 | 29.70 |
| Excavate post hole with power auger. Add the auger cost. | | | | | |
| Light soil, sand or loam | BL@.200 | Ea | — | 5.93 | 5.93 |
| Medium soil, gravel or light clay | BL@.250 | Ea | — | 7.42 | 7.42 |
| Heavy soil, shale, sandstone, rocky | BL@.300 | Ea | — | 8.90 | 8.90 |

## Setting Fence Posts

Posts set with concrete, including temporary 1" x 6" bracing and stakes. Heights to 6'. Concrete assumes post hole is 1' x 1' x 2' deep. Per post set, using either wood or metal post.

|  | Craft@Hrs | Unit | Material | Labor | Total |
|---|---|---|---|---|---|
| Set and brace wood or metal fence post | | | | | |
| Lay in gravel for drainage | BL@.010 | Ea | .02 | .30 | .32 |
| Set 4" x 4" x 8' treated wood post | BL@.015 | Ea | 8.71 | .45 | 9.16 |
| Set 18 gauge 1-7/8" metal tube post with cap | BL@.015 | Ea | 15.80 | .45 | 16.25 |
| Check and adjust post alignment | BL@.010 | Ea | — | .30 | .30 |
| Plumb and brace post (5 uses) | BL@.100 | Ea | .50 | 2.97 | 3.47 |
| Backfill with concrete or soil | BL@.030 | Ea | — | .89 | .89 |
| Tamp soil or rod concrete | BL@.010 | Ea | — | .30 | .30 |
| Trowel drainage crown | BL@.010 | Ea | — | .30 | .30 |
| Strip bracing and inspect | BL@.020 | Ea | — | .59 | .59 |
| Concrete for post base, 60 lb. sack, mixed | BL@.125 | Ea | 5.00 | 3.71 | 8.71 |
| Total per fence post (wood) | BL@.330 | Ea | 14.23 | 9.79 | 24.02 |
| Wood posts 10' OC, per LF of fence | BL@.033 | LF | 1.42 | .98 | 2.40 |
| Total per fence post (metal) | BL@.330 | Ea | 21.32 | 9.79 | 31.11 |
| Metal posts 10' OC, per LF of fence | BL@.033 | LF | 2.13 | .98 | 3.11 |

# Fencing, Wood

|  | Craft@Hrs | Unit | Material | Labor | Total |
|---|---|---|---|---|---|

## Setting Fence Rail on Wood Posts

Stringing 2" x 4" wood rail between fence posts, including 2" x 4" pressure treated rail at $.68 per linear foot.

|  | Craft@Hrs | Unit | Material | Labor | Total |
|---|---|---|---|---|---|
| Measure opening between posts | BL@.010 | Ea | — | .30 | .30 |
| Measure and cut 10' pressure treated rail | BL@.015 | Ea | 9.59 | .45 | 10.04 |
| Measure rail position on post | BL@.010 | Ea | — | .30 | .30 |
| Nail one end in place | BL@.010 | Ea | .02 | .30 | .32 |
| Position other end and check for level | BL@.020 | Ea | — | .59 | .59 |
| Nail second end in place | BL@.010 | Ea | .02 | .30 | .32 |
| Recheck level and adjust as needed | BL@.015 | Ea | — | .45 | .45 |
| Paint cut ends, post connections, rail side | BL@.100 | Ea | — | 2.97 | 2.97 |
| Set rail nailed in place, total | BL@.190 | Ea | 9.63 | 5.64 | 15.27 |
| Set two rails, per LF of fence, 10' rails | BL@.038 | LF | 1.93 | 1.13 | 3.06 |

## Setting Fence Rail on Metal Posts

Stringing 2" x 4" wood fence rail between metal posts, including 2" x 4" pressure treated rail at $.68 per linear foot and one 5" lag bolt connecting adjacent rail at each post.

|  | Craft@Hrs | Unit | Material | Labor | Total |
|---|---|---|---|---|---|
| Measure opening between posts | BL@.010 | Ea | — | .30 | .30 |
| Measure and cut two 10' rail | BL@.030 | Ea | 19.20 | .89 | 20.09 |
| Measure rail position on post | BL@.010 | Ea | — | .30 | .30 |
| Drill 2 rail and metal tube post for lag bolt | BL@.167 | Ea | — | 4.95 | 4.95 |
| Set two rail on tube post with lag bolt | BL@.133 | Ea | .31 | 3.95 | 4.26 |
| Recheck level and adjust as needed | BL@.015 | Ea | — | .45 | .45 |
| Paint cut ends, post connections, rail side | BL@.100 | Ea | — | 2.97 | 2.97 |
| Set two rails, bolted in place, total | BL@.465 | Ea | 19.51 | 13.81 | 33.32 |
| Set two rails, bolted in place, per LF of fence | BL@.047 | LF | 1.95 | 1.38 | 3.33 |

## Installing Fence Face Board

Hand nailing or stapling face board (pickets) to existing rail. Deduct 50% from labor costs for power nailing or power stapling. Includes measuring and cutting each picket, nailing or stapling to the top rail, plumbing with a level and nailing or stapling to the bottom rail. Two nails or staples at each rail. Cost per LF of fence. Based on 1" x 4" x 6', 1" x 6" x 6' and 1" x 8" x 6' face boards. Face board costs vary widely.

| Nailing 4" wide pressure treated southern pine fence face board (3.4 per LF of fence) | Craft@Hrs | Unit | Material | Labor | Total |
|---|---|---|---|---|---|
| French gothic, 1" x 4" x 3-1/2' ($1.10 ea) | BL@.102 | LF | 3.29 | 3.03 | 6.32 |
| French gothic, 5/8" x 4" x 4' ($1.90 ea) | BL@.102 | LF | 5.71 | 3.03 | 8.74 |
| Dog-eared, 7/16" x 4" x 6' ($1.76 ea) | BL@.136 | LF | 5.28 | 4.04 | 9.32 |
| Dog-eared, 1" x 4", 6' ($2.32 ea) | BL@.136 | LF | 6.96 | 4.04 | 11.00 |
| Nailing 5-1/2" wide pressure treated southern pine fence face board (2.4 per LF of fence) |  |  |  |  |  |
| Dog-eared, 5/8" x 5-1/2" x 6' ($1.80 ea) | BL@.120 | LF | 3.92 | 3.56 | 7.48 |
| Nailing 6" wide pressure treated southern pine fence face board (2.2 per LF of fence) |  |  |  |  |  |
| Dog-eared, 1" x 6" x 6' ($2.71 ea) | BL@.110 | LF | 5.41 | 3.26 | 8.67 |
| Dog-eared Prem S4S, 1" x 6" x 6' ($3.23 ea) | BL@.110 | LF | 6.45 | 3.26 | 9.71 |
| Full cut premium, 1" x 6" x 6' ($2.94 ea) | BL@.110 | LF | 5.87 | 3.26 | 9.13 |
| Dog-eared, 5/8" x 5-1/2" x 8' ($3.64 ea) | BL@.110 | LF | 7.28 | 3.26 | 10.54 |
| Nailing 4" wide white wood fence face board (3.4 per LF of fence) |  |  |  |  |  |
| Gothic, 3/4" x 4" x 6' ($1.72 ea) | BL@.136 | LF | 5.15 | 4.04 | 9.19 |
| Flat top, 1" x 4" x 6' ($1.89 ea) | BL@.136 | LF | 5.68 | 4.04 | 9.72 |
| Dog-eared, 1" x 4" x 6' ($2.02 ea) | BL@.136 | LF | 6.07 | 4.04 | 10.11 |
| Dog-eared, 1" x 4" x 8' ($2.18 ea) | BL@.136 | LF | 6.53 | 4.04 | 10.57 |

| | Craft@Hrs | Unit | Material | Labor | Total |
|---|---|---|---|---|---|
| Nailing 4" wide cedar fence face board (3.4 per LF of fence) | | | | | |
| Dog-eared, No. 2, 3/4" x 4" x 8' ($3.67 ea) | BL@.136 | LF | 11.00 | 4.04 | 15.04 |
| Gothic, 3/4" x 4" x 6' ($2.02 ea) | BL@.136 | LF | 6.06 | 4.04 | 10.10 |
| Dog-eared, 9/16" x 4" x 6' ($1.66 ea) | BL@.136 | LF | 4.97 | 4.04 | 9.01 |
| Flat top, full-cut, 1" x 4" x 6' ($3.15 ea) | BL@.136 | LF | 9.46 | 4.04 | 13.50 |
| No. 1 flat top, full-cut, 1" x 4" x 8' ($3.80 ea) | BL@.136 | LF | 11.40 | 4.04 | 15.44 |
| Nailing 6" wide cedar fence face board (2.2 per LF of fence) | | | | | |
| Dog-eared, No. 2, 9/16" x 6" x 6' ($2.85 ea) | BL@.110 | LF | 5.70 | 3.26 | 8.96 |
| Dog-eared, No. 2, 9/16" x 6" x 8' ($5.30 ea) | BL@.110 | LF | 10.60 | 3.26 | 13.86 |
| Dog-eared, clear, 1" x 6" x 6' ($3.69 ea) | BL@.110 | LF | 7.37 | 3.26 | 10.63 |
| Flat top, full-cut, 1" x 6" x 6' ($4.23 ea) | BL@.110 | LF | 8.45 | 3.26 | 11.71 |
| Nailing 4" wide dog-eared common redwood fence face board (3.4 per LF of fence) | | | | | |
| 3/4" x 4" x 6' ($2.53 ea) | BL@.136 | LF | 7.59 | 4.04 | 11.63 |
| Nailing 6" wide dog-eared common redwood fence face board (2.2 per LF of fence) | | | | | |
| 3/4" x 6" x 5' ($2.46 ea) | BL@.110 | LF | 4.91 | 3.26 | 8.17 |
| 3/4" x 6" x 6' ($2.70 ea) | BL@.110 | LF | 5.39 | 3.26 | 8.65 |
| 3/4" x 6" x 6' ($4.74 ea) | BL@.110 | LF | 9.47 | 3.26 | 12.73 |
| Nailing 8" wide dog-eared common redwood fence face board (1.6 per LF of fence) | | | | | |
| 3/4" x 8" x 6' ($6.14 ea) | BL@.088 | LF | 9.21 | 2.61 | 11.82 |
| 3/4" x 8" x 6' Premium Const. Heart ($9.00 ea) | BL@.088 | LF | 13.50 | 2.61 | 16.11 |
| 1" x 8" x 8' ($12.10 ea) | BL@.088 | LF | 18.20 | 2.61 | 20.81 |
| Nailing 10" wide dog-eared common redwood fence face board (1.3 per LF of fence) | | | | | |
| 3/4" x 10" x 5' ($7.34 ea) | BL@.072 | LF | 8.81 | 2.14 | 10.95 |
| 3/4" x 10" x 6' ($8.58 ea) | BL@.072 | LF | 10.30 | 2.14 | 12.44 |
| Nailing 12" wide dog-eared common redwood fence face board (1.05 per LF of fence) | | | | | |
| 3/4" x 12" x 5' Rough ($11.00 ea) | BL@.063 | LF | 11.00 | 1.87 | 12.87 |
| 3/4" x 12" x 6' ($11.90 ea) | BL@.063 | LF | 11.90 | 1.87 | 13.77 |
| Nailing 6" wide flat top redwood fence face board (2.2 per LF of fence) | | | | | |
| 3/4" x 6" x 6' Const. Heart ($4.96 ea) | BL@.110 | LF | 9.91 | 3.26 | 13.17 |
| 3/4" x 6" x 6' Merch. and Better ($4.96 ea) | BL@.110 | LF | 9.91 | 3.26 | 13.17 |
| 3/4" x 6" x 6' Ridge Valley Redwood ($7.45 ea) | BL@.110 | LF | 14.90 | 3.26 | 18.16 |
| 3/4" x 6" x 8' Ridge Valley Redwood ($12.50 ea) | BL@.110 | LF | 25.00 | 3.26 | 28.26 |
| Nailing 8" wide flat top redwood fence face board (1.6 per LF of fence) | | | | | |
| 3/4" x 8" x 5' ($4.80 ea) | BL@.088 | LF | 7.20 | 2.61 | 9.81 |
| 3/4" x 8" x 5' T&G ($8.67 ea) | BL@.088 | LF | 13.00 | 2.61 | 15.61 |
| 3/4" x 8" x 6' ($5.78 ea) | BL@.088 | LF | 8.67 | 2.61 | 11.28 |
| 3/4" x 8" x 6' Construction Heart ($9.93 ea) | BL@.088 | LF | 14.90 | 2.61 | 17.51 |
| 1" x 8" x 8' Construction Heart ($12.10 ea) | BL@.088 | LF | 18.20 | 2.61 | 20.81 |
| Nailing 10" and 12" wide flat top redwood fence face board (1.3 per LF of fence for 10", 1.0 for 12") | | | | | |
| 3/4" x 10" x 8' Redwood ($11.00 ea) | BL@.072 | LF | 13.20 | 2.14 | 15.34 |
| 3/4" x 12" x 6' Dog-eared ($12.30 ea) | BL@.072 | LF | 14.70 | 2.14 | 16.84 |
| Deduct for power stapling or power nailing | — | % | — | -50.0 | — |

## Building Fence on a Hillside

Post hole excavation, post setting and fence erection will take longer when the site is obstructed or on a hillside. Adjust manhours as follows for sloping terrain.

| | | | | | |
|---|---|---|---|---|---|
| Add for slope to 4 in 12 | — | % | — | 20.0 | — |
| Add for slope over 4 in 12 to 8 in 12 | — | % | — | 50.0 | — |
| Add for slope over 8 in 12 to 12 in 12 | — | % | — | 100.0 | — |

# Fencing, Wood

| | Craft@Hrs | Unit | Material | Labor | Total |
|---|---|---|---|---|---|

**Build and Hang a Wood Fence Gate** 5' high by 3' wide wood gate, 2" x 4" frame using 1" x 6" face boards. Add the cost of fence posts. Based on treated 2" x 4" at $.81 per linear foot and treated 6" x 5' face boards at $1.75 each. Units are for each gate.

| | Craft@Hrs | Unit | Material | Labor | Total |
|---|---|---|---|---|---|
| Measure, design and lay out gate | BL@.200 | Ea | — | 5.93 | 5.93 |
| Measure and cut frame (11' x 2" x 4") | BL@.050 | Ea | 9.41 | 1.48 | 10.89 |
| Nail frame together | BL@.100 | Ea | .10 | 2.97 | 3.07 |
| Square up and measure 4' diagonal brace | BL@.050 | Ea | 3.42 | 1.48 | 4.90 |
| Cut cross brace and nail in place | BL@.083 | Ea | .08 | 2.46 | 2.54 |
| Square up and add metal corner brackets | BL@.333 | Ea | 1.00 | 9.88 | 10.88 |
| Cut and nail seven face boards($3.38 ea) | BL@.397 | Ea | 23.60 | 11.80 | 35.40 |
| Mark top pattern on face boards | BL@.133 | Ea | — | 3.95 | 3.95 |
| Cut top pattern on face boards | BL@.089 | Ea | — | 2.64 | 2.64 |
| Set hinges on gate and fence post | BL@.250 | Ea | 8.60 | 7.42 | 16.02 |
| Set latch on gate and strike on fence post | BL@.100 | Ea | 4.50 | 2.97 | 7.47 |
| Check and adjust as needed | BL@.125 | Ea | — | 3.71 | 3.71 |
| Build and hang wood fence gate, total | BL@1.91 | Ea | 50.71 | 56.67 | 107.38 |

**Fence Assemblies, Wood** Costs shown include excavation of post holes, concrete and labor to set posts, attaching rails with nails or lag bolts, and hand nailing of face boards where applicable. See additional costs at the end of this section.

**Basketweave fence**, redwood, "B" grade, 1" x 6" boards, 2" x 4" stringers or spreaders, 4" x 4" posts

| | Craft@Hrs | Unit | Material | Labor | Total |
|---|---|---|---|---|---|
| Tight weave, 4' high, posts at 4' OC | B1@.332 | LF | 14.60 | 11.10 | 25.70 |
| Tight weave, 6' high, posts at 4' OC | B1@.522 | LF | 11.30 | 17.39 | 28.69 |
| Wide span, 4' high, posts at 8' OC | B1@.291 | LF | 6.91 | 9.69 | 16.60 |
| Wide span, 8' high, posts at 8' OC | B1@.481 | LF | 9.40 | 16.00 | 25.40 |
| 4' high gate, tight weave | B1@.906 | Ea | 64.30 | 30.20 | 94.50 |
| 4' high gate, wide span | B1@.906 | Ea | 49.80 | 30.20 | 80.00 |
| 6' high gate, tight weave | B1@1.11 | Ea | 30.50 | 37.00 | 67.50 |
| 8' high gate, wide span | B1@1.26 | Ea | 41.40 | 42.00 | 83.40 |

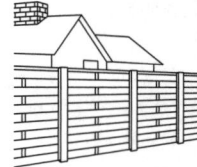

**Board fence, using 4" wide boards,** hand nailed, posts at 10' OC
Pressure treated Southern Pine, 1" x 4" boards

| | Craft@Hrs | Unit | Material | Labor | Total |
|---|---|---|---|---|---|
| 6' high, 2 rail, wood posts | B1@.267 | LF | 7.73 | 8.89 | 16.62 |
| 6' high, 2 rail, metal posts | B1@.276 | LF | 8.27 | 9.19 | 17.46 |
| Deduct for power stapling or power nailing | — | % | — | -22.0 | — |

Cedar, 1" x 4" boards

| | Craft@Hrs | Unit | Material | Labor | Total |
|---|---|---|---|---|---|
| 6' high, 2 rail, wood posts | B1@.267 | LF | 12.10 | 8.89 | 20.99 |
| 8' high, 2 rail, wood posts | B1@.267 | LF | 13.20 | 8.89 | 22.09 |
| 6' high, 2 rail, metal posts | B1@.276 | LF | 12.60 | 9.19 | 21.79 |
| 8' high, 2 rail, metal posts | B1@.276 | LF | 13.80 | 9.19 | 22.99 |
| Deduct for power stapling or power nailing | — | % | — | -22.0 | — |

Redwood, 3/4" x 4" boards

| | Craft@Hrs | Unit | Material | Labor | Total |
|---|---|---|---|---|---|
| 6' high, 2 rail, wood posts | B1@.267 | LF | 9.78 | 8.89 | 18.67 |
| 6' high, 2 rail, metal posts | B1@.276 | LF | 10.30 | 9.19 | 19.49 |
| Deduct for power stapling or power nailing | — | % | — | -22.0 | — |

Whitewoods, 1" x 4" boards

| | Craft@Hrs | Unit | Material | Labor | Total |
|---|---|---|---|---|---|
| 6' high, 2 rail, wood posts | B1@.267 | LF | 6.94 | 8.89 | 15.83 |
| 8' high, 2 rail, wood posts | B1@.267 | LF | 8.83 | 8.89 | 17.72 |
| 6' high, 2 rail, metal posts | B1@.276 | LF | 7.49 | 9.19 | 16.68 |
| 8' high, 2 rail, metal posts | B1@.276 | LF | 9.38 | 9.19 | 18.57 |
| Deduct for power stapling or power nailing | — | % | — | -22.0 | — |

**Board fence, using 6" wide boards,** hand nailed, posts at 10' OC
Pressure treated Southern Pine, 1" x 6" boards

| | Craft@Hrs | Unit | Material | Labor | Total |
|---|---|---|---|---|---|
| 6' high, 2 rail, wood posts | B1@.241 | LF | 8.40 | 8.03 | 16.43 |
| 8' high, 2 rail, wood posts | B1@.241 | LF | 9.68 | 8.03 | 17.71 |
| 6' high, 2 rail, metal posts | B1@.250 | LF | 8.95 | 8.33 | 17.28 |
| 8' high, 2 rail, metal posts | B1@.250 | LF | 10.20 | 8.33 | 18.53 |
| Deduct for power stapling or power nailing | — | % | — | -22.0 | — |

| | Craft@Hrs | Unit | Material | Labor | Total |
|---|---|---|---|---|---|
| **Cedar, 9/16" x 6" boards** | | | | | |
| 6' high, 2 rail, wood posts | B1@.241 | LF | 8.25 | 8.03 | 16.28 |
| 8' high, 2 rail, wood posts | B1@.241 | LF | 12.70 | 8.03 | 20.73 |
| 6' high, 2 rail, metal posts | B1@.250 | LF | 8.80 | 8.33 | 17.13 |
| 8' high, 2 rail, metal posts | B1@.250 | LF | 13.20 | 8.33 | 21.53 |
| Deduct for power stapling or power nailing | — | % | — | -22.0 | — |
| **Redwood, 3/4" x 6" boards** | | | | | |
| 6' high, 2 rail, wood posts | B1@.241 | LF | 16.60 | 8.03 | 24.63 |
| 8' high, 2 rail, wood posts | B1@.241 | LF | 16.50 | 8.03 | 24.53 |
| 6' high, 2 rail, metal posts | B1@.250 | LF | 17.10 | 8.33 | 25.43 |
| 8' high, 2 rail, metal posts | B1@.250 | LF | 24.60 | 8.33 | 32.93 |
| Deduct for power stapling or power nailing | — | % | — | -22.0 | — |
| | | | | | |
| **Split rail fence,** red cedar, 4" posts at 8' OC | | | | | |
| Split 2" x 4" rails, 3' high, 2 rail | B1@.095 | LF | 3.99 | 3.16 | 7.15 |
| Split 2" x 4" rails, 4' high, 3 rail | B1@.102 | LF | 4.67 | 3.40 | 8.07 |
| Rustic 4" round, 3' high, 2 rail | B1@.095 | LF | 4.79 | 3.16 | 7.95 |
| Rustic 4" round, 4' high, 3 rail | B1@.102 | LF | 5.62 | 3.40 | 9.02 |
| | | | | | |
| **Picket fence,** 1" x 2" pickets, 2" x 4" rails, 4" x 4" posts at 8' OC | | | | | |
| Pressure treated Southern Pine | | | | | |
| 3' high, 2 rail | B1@.270 | LF | 2.42 | 8.99 | 11.41 |
| 5' high, 3 rail | B1@.385 | LF | 3.30 | 12.80 | 16.10 |
| 3' high gate | B1@.770 | Ea | 10.60 | 25.60 | 36.20 |
| 5' high gate | B1@.887 | Ea | 14.60 | 29.50 | 44.10 |
| Red cedar, #3 & Better | | | | | |
| 3' high, 2 rail | B1@.270 | LF | 4.28 | 8.99 | 13.27 |
| 5' high, 3 rail | B1@.385 | LF | 5.88 | 12.80 | 18.68 |
| 3' high gate | B1@.770 | Ea | 18.90 | 25.60 | 44.50 |
| 5' high gate | B1@.887 | Ea | 25.80 | 29.50 | 55.30 |
| Redwood, B grade | | | | | |
| 3' high, 2 rail | B1@.270 | LF | 5.57 | 8.99 | 14.56 |
| 5' high, 3 rail | B1@.385 | LF | 7.25 | 12.80 | 20.05 |
| 3' high gate | B1@.770 | Ea | 24.50 | 25.60 | 50.10 |
| 5' high gate | B1@.887 | Ea | 31.90 | 29.50 | 61.40 |
| Picket fence gate hardware | | | | | |
| Latch, standard duty | B1@.281 | Ea | 4.33 | 9.36 | 13.69 |
| Latch, heavy duty | B1@.281 | Ea | 15.60 | 9.36 | 24.96 |
| Hinges, standard, per pair | B1@.374 | Pr | 6.97 | 12.50 | 19.47 |
| Hinges, heavy duty, per pair | B1@.374 | Pr | 18.10 | 12.50 | 30.60 |
| No-sag cable kit | B1@.281 | Ea | 10.90 | 9.36 | 20.26 |
| Self closing spring | B1@.281 | Ea | 7.44 | 9.36 | 16.80 |

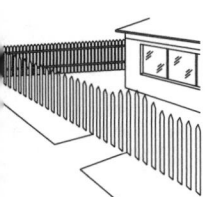

## Fencing, Aluminum

Decorative aluminum, 72-1/2" panel width, 2" x 2" posts, .060" wall thickness, 3-13/16" picket spacing, 1" x .055" top rail with 1-1/2" x .082" sides. Pre-routed holes in posts. Gates include hardware. Pool code compliant model has 48" space between horizontal rails

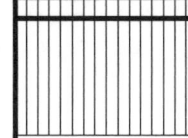

| | Craft@Hrs | Unit | Material | Labor | Total |
|---|---|---|---|---|---|
| 48" high panel | BL@.170 | Ea | 42.00 | 5.04 | 47.04 |
| 54" pool code compliant panel | BL@.170 | Ea | 51.00 | 5.04 | 56.04 |
| 60" high panel | BL@.170 | Ea | 55.00 | 5.04 | 60.04 |
| 72" high panel | BL@.170 | Ea | 69.00 | 5.04 | 74.04 |

## Fencing, Vinyl

| | Craft@Hrs | Unit | Material | Labor | Total |
|---|---|---|---|---|---|
| Fence posts | | | | | |
| 48" high | BL@.830 | Ea | 19.50 | 24.60 | 44.10 |
| 60" high | BL@.830 | Ea | 21.50 | 24.60 | 46.10 |
| 72" high | BL@.830 | Ea | 27.10 | 24.60 | 51.70 |
| Single swing gates, 5' wide | | | | | |
| 48" high | BL@.175 | Ea | 255.00 | 5.19 | 260.19 |
| 60" high | BL@.175 | Ea | 280.00 | 5.19 | 285.19 |
| 72" high | BL@.175 | Ea | 318.00 | 5.19 | 323.19 |
| Gate posts | | | | | |
| 48" high | BL@.830 | Ea | 36.10 | 24.60 | 60.70 |
| 60" high | BL@.830 | Ea | 44.00 | 24.60 | 68.60 |
| 72" high | BL@.830 | Ea | 53.40 | 24.60 | 78.00 |
| Add for setting post in concrete | BL@.330 | Ea | 3.02 | 9.79 | 12.81 |

### Fencing, Vinyl

Costs shown include concrete and labor to set posts.

Vinyl picket fence, 1-1/2" pickets with 2" spacing, 2 rail, 2" x 4" rails, 4" x 4" posts at 8' OC, pre-assembled in 8' sections. Prices include post mounts, post top, screws and one post.

| | Craft@Hrs | Unit | Material | Labor | Total |
|---|---|---|---|---|---|
| 4' high | B1@.110 | LF | 4.25 | 3.66 | 7.91 |
| Gate with 4" posts | B1@1.00 | Ea | 275.00 | 33.30 | 308.30 |
| Add for additional fence post with top | B1@.500 | Ea | 23.20 | 16.70 | 39.90 |

Vinyl picket fence, 3-1/2" pickets with 3" spacing, 2 rail, 2" x 4" rails, 4" x 4" posts at 6' OC, pre-assembled in 6' sections. Prices include post mounts, post top, screws and one post.

| | Craft@Hrs | Unit | Material | Labor | Total |
|---|---|---|---|---|---|
| 4' high | B1@.110 | LF | 7.32 | 3.66 | 10.98 |
| Gate with 4" posts | B1@1.00 | Ea | 68.40 | 33.30 | 101.70 |
| Add for additional fence post with top | B1@.500 | Ea | 23.40 | 16.70 | 40.10 |

Vinyl shadowbox fence, 5-1/2" offset pickets, 3 rail, with galvanized steel support, posts at 71" OC. Prices include post mounts, post top, screws and one post.

| | Craft@Hrs | Unit | Material | Labor | Total |
|---|---|---|---|---|---|
| 4' high | B1@.130 | LF | 14.80 | 4.33 | 19.13 |
| 5' high | B1@.140 | LF | 21.20 | 4.66 | 25.86 |
| 6' high | B1@.140 | LF | 18.90 | 4.66 | 23.56 |
| Gate with 4" posts | B1@1.20 | Ea | 76.30 | 40.00 | 116.30 |
| Add for additional fence post with top | B1@.500 | Ea | 20.60 | 16.70 | 37.30 |

Vinyl privacy fence, 6" boards, 3 rail, posts at 6' OC. Prices include post mounts, post top, screws and one post.

| | Craft@Hrs | Unit | Material | Labor | Total |
|---|---|---|---|---|---|
| 6' high | B1@.140 | LF | 14.80 | 4.66 | 19.46 |
| 6' high gate | B1@1.20 | Ea | 112.00 | 40.00 | 152.00 |
| Add for additional fence post with top | B1@.500 | Ea | 29.20 | 16.70 | 45.90 |

Vinyl post & rail fence, 2" x 6" x 96" notched rail, posts 8' OC. Includes post mounts, post top, and one post.

| | Craft@Hrs | Unit | Material | Labor | Total |
|---|---|---|---|---|---|
| 40" high, 2 rail, 4" routed post | B1@.110 | LF | 6.82 | 3.66 | 10.48 |
| 57" high, 3 rail, 4" routed post | B1@.120 | LF | 8.42 | 4.00 | 12.42 |
| 57" high, 3 rail, 5" routed post | B1@.120 | LF | 9.50 | 4.00 | 13.50 |
| 54" high, 4 rail, 5" routed post | B1@.120 | LF | 11.30 | 4.00 | 15.30 |
| Gate with 4" posts | B1@1.00 | Ea | 111.00 | 33.30 | 144.30 |
| Gate with 5" posts | B1@1.00 | Ea | 134.00 | 33.30 | 167.30 |
| Add for additional 4" fence post with top | B1@.500 | Ea | 17.20 | 16.70 | 33.90 |
| Add for additional 5" fence post with top | B1@.500 | Ea | 26.20 | 16.70 | 42.90 |

**Silt fence**, black 6 mil poly with stakes 8' OC, staked and buried 9" deep

| | Craft@Hrs | Unit | Material | Labor | Total |
|---|---|---|---|---|---|
| 3' high, 100 LF roll at $28.20 | BL@.030 | LF | .28 | .89 | 1.17 |

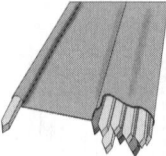

| | Craft@Hrs | Unit | Material | Labor | Total |
|---|---|---|---|---|---|

**Fiberglass Panels** Nailed or screwed onto wood frame.

Corrugated, 8', 10', 12' panels, 2-1/2" corrugations, standard colors. Costs include 10% loss for coverage and waste.

| | Craft@Hrs | Unit | Material | Labor | Total |
|---|---|---|---|---|---|
| Polycarbonate, 26" wide, gray or white | BC@.012 | SF | 1.33 | .44 | 1.77 |
| Polycarbonate, 26" wide, clear | BC@.012 | SF | 1.33 | .44 | 1.77 |
| PVC, 26" wide panels | BC@.012 | SF | .87 | .44 | 1.31 |

Flat panels, clear, green or white

| | Craft@Hrs | Unit | Material | Labor | Total |
|---|---|---|---|---|---|
| .06" flat sheets, 4' x 8', 10', 12' | BC@.012 | SF | 1.18 | .44 | 1.62 |
| .03" rolls, 24", 36", 48" x 50' | BC@.012 | SF | 1.26 | .44 | 1.70 |
| .037" rolls, 24", 36", 48" x 50' | BC@.012 | SF | 1.45 | .44 | 1.89 |

Accessories

Nails, ring shank with rubber washer, 1-3/4" (labor included with panels)

| | Craft@Hrs | Unit | Material | Labor | Total |
|---|---|---|---|---|---|
| 100 per box (covers 130 SF) | — | SF | .08 | — | .08 |

Self-tapping screws, 1-1/4", rust-proof (labor included with panels)

| | Craft@Hrs | Unit | Material | Labor | Total |
|---|---|---|---|---|---|
| 100 per box (covers 130 SF) | — | SF | .29 | — | .29 |

Horizontal closure strips, corrugated (labor included with panels)

| | Craft@Hrs | Unit | Material | Labor | Total |
|---|---|---|---|---|---|
| Redwood, 2-1/2" x 1-1/2" | — | LF | .32 | — | .32 |
| Poly-foam, 1" x 1" | — | LF | .52 | — | .52 |
| Rubber, 1" x 1" | — | LF | .43 | — | .43 |

Vertical crown moulding

| | Craft@Hrs | Unit | Material | Labor | Total |
|---|---|---|---|---|---|
| Redwood, 1-1/2" x 1" or poly-foam, 1" x 1" | — | LF | .32 | — | .32 |
| Rubber, 1" x 1" | — | LF | .81 | — | .81 |

Ridge cap, covers the joint where panels meet at the roof peak

| | Craft@Hrs | Unit | Material | Labor | Total |
|---|---|---|---|---|---|
| Polycarbonate, 50" long | — | LF | 4.05 | — | 4.05 |

**Ornamental Iron, Railings and Fence, Subcontract** Typical prices, fabricated, delivered and installed.

Handrail, 1-1/2" x 1" rectangular top rail tubing, 1" square bottom rail tubing,

1/2" square pickets with 5-1/2" space between

| | Craft@Hrs | Unit | Material | Labor | Total |
|---|---|---|---|---|---|
| 36" high railing for single family dwelling | — | LF | — | — | 35.70 |
| 42" high apartment style railing | — | LF | — | — | 28.50 |
| Add for inclined runs | — | LF | — | — | 6.55 |

Pool fence

| | Craft@Hrs | Unit | Material | Labor | Total |
|---|---|---|---|---|---|
| 5' high, 4-1/2" centers | — | LF | — | — | 35.70 |
| Add for gate, with self-closer and latch | — | LS | — | — | 167.00 |

**Fire Sprinkler Systems** Typical subcontract prices for systems installed in a single family residence not over two stories high in accord with NFPA Section 13D. Includes connection to domestic water line inside garage (1" pipe size minimum), exposed copper riser with shut-off and drain valve, CPVC concealed distribution piping and residential type sprinkler heads below finished ceilings. These costs include hydraulic design calculations and inspection of completed system for compliance with design requirements. Costs shown are per SF of protected area. When estimating the square footage of protected area, include non-inhabited areas such as bathrooms, closets and attached garages. For scheduling purposes, estimate that a crew of 2 men can install the rough-in piping for 1,500 to 1,600 SF of protected area in an 8-hour day and about the same amount of finish work in another 8-hour day.

Fabricate, install, and test system

| | Craft@Hrs | Unit | Material | Labor | Total |
|---|---|---|---|---|---|
| Single residence | — | SF | — | — | 3.70 |
| Tract work | — | SF | — | — | 3.17 |
| Condominiums or apartments | — | SF | — | — | 5.00 |

# Fireplaces

| | Craft@Hrs | Unit | Material | Labor | Total |
|---|---|---|---|---|---|

## Masonry Fireplaces

Fire box lined with refractory firebrick, backed with common brick and enclosed in a common brick wall. Includes fireplace throat and smoke chamber. Overall height of the fireplace and throat is 5'. Chimney (flue) is made from common brick and a flue liner. Including lintels, damper and 12" thick reinforced concrete foundation. For more on masonry fireplaces, see *National Concrete & Masonry Estimator,* http://CraftsmanSiteLicense.com.

| | Craft@Hrs | Unit | Material | Labor | Total |
|---|---|---|---|---|---|
| 30" wide x 29" high x 16" deep fire box | B9@39.7 | LS | 463.00 | 1,300.00 | 1,763.00 |
| 36" wide x 29" high x 16" deep fire box | B9@43.5 | LS | 488.00 | 1,420.00 | 1,908.00 |
| 42" wide x 32" high x 16" deep fire box | B9@48.3 | LS | 541.00 | 1,580.00 | 2,121.00 |
| 48" wide x 32" high x 18" deep fire box | B9@52.8 | LS | 593.00 | 1,720.00 | 2,313.00 |
| 12" x 12" chimney to 12' high | B9@9.50 | LS | 274.00 | 310.00 | 584.00 |
| Add for higher chimney, per LF | B9@1.30 | LF | 37.40 | 42.40 | 79.80 |
| Add for 8" x 10" ash cleanout door | B9@.500 | SF | 17.00 | 16.30 | 33.30 |
| Add for combustion air inlet kit | B9@.750 | SF | 25.50 | 24.50 | 50.00 |
| Add for common brick veneer interior face | B9@.451 | SF | 3.91 | 14.70 | 18.61 |
| Add for rubble or cobble stone veneer face | B9@.480 | SF | 10.70 | 15.70 | 26.40 |
| Add for limestone veneer face or hearth | B9@.451 | SF | 20.30 | 14.70 | 35.00 |
| Add for 1-1/4" marble face or hearth | B9@.451 | SF | 30.50 | 14.70 | 45.20 |
| Add for modern design 11" x 77" pine mantel | B1@3.81 | Ea | 600.00 | 127.00 | 727.00 |
| Add for gas valve, log lighter and key | PM@.750 | Ea | 69.00 | 31.40 | 100.40 |
| Angle iron (lintel support), lengths 25" to 96" | | | | | |
| 3" x 3" x 3/16" | B9@.167 | LF | 5.10 | 5.45 | 10.55 |
| Ash drops, cast iron top, galvanized container | | | | | |
| 16-3/4" x 12-1/2" x 8" | B9@.750 | Ea | 94.90 | 24.50 | 119.40 |
| Ash dumps, cast iron | | | | | |
| 5-3/8" x 9-1/2" overall | B9@.500 | Ea | 14.50 | 16.30 | 30.80 |
| 5-3/8" x 9-1/2" overall, vented | B9@.500 | Ea | 27.80 | 16.30 | 44.10 |
| Chimney anchors, per pair | | | | | |
| 48" x 1-1/2" x 3/16" | — | Pr | 13.80 | — | 13.80 |
| 72" x 1-1/2" x 3/16" | — | Pr | 18.20 | — | 18.20 |
| Cleanout doors, cast iron, overall dimensions | | | | | |
| 10-1/2" wide x 10-1/2" high | B9@.500 | Ea | 34.50 | 16.30 | 50.80 |
| 11" wide x 17-1/2" high | B9@.500 | Ea | 82.80 | 16.30 | 99.10 |
| 15-1/2" wide x 15-1/2" high | B9@.500 | Ea | 89.00 | 16.30 | 105.30 |
| Dampers for single opening firebox, heavy steel with rockwool insulation | | | | | |
| 14" high, 25" to 30" wide, 10" deep, 25 lbs | B9@1.33 | Ea | 62.10 | 43.40 | 105.50 |
| 14" high, 36" wide, 10" deep, 31 lbs | B9@1.33 | Ea | 67.40 | 43.40 | 110.80 |
| 14" high, 42" wide, 10" deep, 38 lbs | B9@1.33 | Ea | 82.40 | 43.40 | 125.80 |
| 14" high, 48" wide, 10" deep, 43 lbs | B9@1.75 | Ea | 107.00 | 57.10 | 164.10 |
| 16" high, 54" wide, 13" deep, 66 lbs | B9@2.50 | Ea | 199.00 | 81.60 | 280.60 |
| 16" high, 60" wide, 13" deep, 74 lbs | B9@2.50 | Ea | 219.00 | 81.60 | 300.60 |
| 16" high, 72" wide, 13" deep, 87 lbs | B9@2.50 | Ea | 264.00 | 81.60 | 345.60 |
| Dampers for multiple opening or corner opening firebox, high capacity, steel downdraft shelf support | | | | | |
| 35" wide | B9@1.33 | Ea | 255.00 | 43.40 | 298.40 |
| 41" wide | B9@1.33 | Ea | 291.00 | 43.40 | 334.40 |
| 47" wide | B9@1.75 | Ea | 373.00 | 57.10 | 430.10 |
| Chimney-top dampers, with cable to the firebox and handle mount, aluminum, by flue size | | | | | |
| 8" x 8" | B9@1.33 | Ea | 165.00 | 43.40 | 208.40 |
| 8" x 13" | B9@1.33 | Ea | 168.00 | 43.40 | 211.40 |
| 8" x 17" | B9@1.33 | Ea | 181.00 | 43.40 | 224.40 |

| | Craft@Hrs | Unit | Material | Labor | Total |
|---|---|---|---|---|---|
| 13" x 13" | B9@1.75 | Ea | 173.00 | 57.10 | 230.10 |
| 13" x 17" | B9@2.50 | Ea | 208.00 | 81.60 | 289.60 |
| 17" x 17" | B9@2.50 | Ea | 249.00 | 81.60 | 330.60 |

| Fuel grates | | | | | |
|---|---|---|---|---|---|
| 21" x 16" | — | Ea | 60.80 | — | 60.80 |
| 29" x 16" | — | Ea | 78.00 | — | 78.00 |
| 40" x 30" | — | Ea | 120.00 | — | 120.00 |

**Factory-built wood-burning fireplaces** Including refractory-lined firebox, fireplace front, combustion air kit and damper with control. By hearth opening width. 21" depth overall.

| | Craft@Hrs | Unit | Material | Labor | Total |
|---|---|---|---|---|---|
| 36" wide | B9@3.50 | Ea | 425.00 | 114.00 | 539.00 |
| 42" wide | B9@3.50 | Ea | 625.00 | 114.00 | 739.00 |
| 50" wide | B9@4.00 | Ea | 1,870.00 | 131.00 | 2,001.00 |
| 36" wide, 1 open end | B9@3.50 | Ea | 925.00 | 114.00 | 1,039.00 |
| 36" wide, 3 open sides | B9@4.00 | Ea | 1,500.00 | 131.00 | 1,631.00 |
| 36" wide, see-through | B9@4.00 | Ea | 1,400.00 | 131.00 | 1,531.00 |
| 25" wide, EPA Phase II compliant | B9@4.00 | Ea | 3,550.00 | 131.00 | 3,681.00 |
| Add for 8" Class A vent, per vertical foot | B9@.250 | LF | 50.00 | 8.16 | 58.16 |

**Gas-burning direct vent zero clearance fireplaces** Vents directly through an exterior wall or the roof using sealed double chamber vent pipe. See vent pipe costs below. Waste gases exit through the vent inner chamber. Outside air enters the firebox through the vent outside chamber. Framing may be applied directly against the unit. Circulates heated air into the room. Includes brick firebox, fireplace front, bi-fold glass doors, temperature control, masonry logs and fuel grate. Electric ignition. Overall dimensions.

| | Craft@Hrs | Unit | Material | Labor | Total |
|---|---|---|---|---|---|
| 36" high x 33" wide x 19" deep | B9@3.50 | Ea | 1,490.00 | 114.00 | 1,604.00 |
| 43" high x 39" wide x 19" deep | B9@3.50 | Ea | 1,490.00 | 114.00 | 1,604.00 |
| 42" high x 36" wide x 23" deep | B9@3.50 | Ea | 2,050.00 | 114.00 | 2,164.00 |
| 41" high x 36" wide x 19" deep, 1 open end | B9@3.50 | Ea | 2,720.00 | 114.00 | 2,834.00 |
| 40" high x 36" wide x 19" deep, 2 open ends | B9@4.00 | Ea | 2,820.00 | 131.00 | 2,951.00 |
| 37" high x 44" wide x 24" deep, see-through | B9@4.00 | Ea | 3,260.00 | 131.00 | 3,391.00 |
| Deduct for conventional B-vent units | — | Ea | -140.00 | — | -140.00 |

Direct vent fireplace flue (chimney), double wall, typical cost for straight vertical installation

| | Craft@Hrs | Unit | Material | Labor | Total |
|---|---|---|---|---|---|
| 4" inside diameter 6-5/8" outside diameter | B9@.167 | LF | 12.50 | 5.45 | 17.95 |
| 5" inside diameter, 8" outside diameter | B9@.167 | LF | 15.50 | 5.45 | 20.95 |
| Flue spacers, for passing through ceiling | | | | | |
| 1" or 2" clearance | B9@.167 | Ea | 13.60 | 5.45 | 19.05 |
| Typical flue offset | B9@.250 | Ea | 85.70 | 8.16 | 93.86 |
| Flashing and storm collar, one required per roof penetration | | | | | |
| Flat to 6/12 pitch roofs | B9@3.25 | Ea | 54.50 | 106.00 | 160.50 |
| 6/12 to 12/12 pitch roofs | B9@3.25 | Ea | 79.20 | 106.00 | 185.20 |
| Spark arrestor top | B9@.500 | Ea | 111.00 | 16.30 | 127.30 |

Blower kit for heat-circulating fireplace, requires 110 volt electrical receptacle outlet (electrical wiring not included)

| | Craft@Hrs | Unit | Material | Labor | Total |
|---|---|---|---|---|---|
| Blower unit for connection to existing outlet | B9@.500 | Ea | 110.00 | 16.30 | 126.30 |
| Add variable speed and thermostatic control | B9@.100 | Ea | 80.00 | 3.26 | 83.26 |

**Electric fireplaces** Prefabricated, 120 volt, including grate, ember bed, interior face, thermostat, brightness control and logs.

| | Craft@Hrs | Unit | Material | Labor | Total |
|---|---|---|---|---|---|
| 23" wide opening, 120 volt, 1,500 watts | B9@2.50 | Ea | 530.00 | 81.60 | 611.60 |
| 26" wide opening, 120 volt, 1,500 watts | B9@2.50 | Ea | 630.00 | 81.60 | 711.60 |

# Flooring, Sheet Vinyl

| | Craft@Hrs | Unit | Material | Labor | Total |
|---|---|---|---|---|---|

## Sheet Vinyl Flooring

Armstrong Solarian sheet vinyl flooring, no wax. Bonded with adhesive at perimeter and at seams. Ten year limited warranty. Performance Appearance Rating 4.1 Includes 10% waste and adhesive.

| | Craft@Hrs | Unit | Material | Labor | Total |
|---|---|---|---|---|---|
| Bellissimo Solarian | BF@.250 | SY | 35.40 | 8.50 | 43.90 |
| Rhythms Solarian | BF@.250 | SY | 33.30 | 8.50 | 41.80 |
| Starstep Solarian | BF@.250 | SY | 29.10 | 8.50 | 37.60 |
| Traditions Solarian | BF@.250 | SY | 29.10 | 8.50 | 37.60 |
| Successor Fundamentals | BF@.250 | SY | 16.00 | 8.50 | 24.50 |

Armstrong sheet vinyl flooring. By Performance Appearance Rating (PAR). Bonded with adhesive at perimeter and at seams. Including 10% waste and adhesive.

| | Craft@Hrs | Unit | Material | Labor | Total |
|---|---|---|---|---|---|
| Medley, urethane wear layer, 15 year | BF@.250 | SY | 18.30 | 8.50 | 26.80 |
| Caspian, PAR 3.8, urethane wear layer, 15 year | BF@.250 | SY | 14.90 | 8.50 | 23.40 |
| Themes, PAR 2.9, urethane wear layer, 10 year | BF@.250 | SY | 11.50 | 8.50 | 20.00 |
| Sundial, urethane wear layer, .098", 10 year | BF@.250 | SY | 12.60 | 8.50 | 21.10 |
| Metro, PAR 2.3, vinyl wear layer, 5 year | BF@.250 | SY | 6.57 | 8.50 | 15.07 |
| Cambray FHA, PAR 1.7, vinyl wear layer, 5 year | BF@.250 | SY | 6.57 | 8.50 | 15.07 |
| Memories, 12 year | BF@.250 | SY | 21.70 | 8.50 | 30.20 |
| Successor, 10 year | BF@.250 | SY | 16.30 | 8.50 | 24.80 |
| Enhancements, 6 year | BF@.250 | SY | 16.60 | 8.50 | 25.10 |
| Initiator, 5 year | BF@.250 | SY | 11.30 | 8.50 | 19.80 |
| Sentinel, 5 year | BF@.250 | SY | 9.21 | 8.50 | 17.71 |
| Apartment One, 1 year | BF@.250 | SY | 9.00 | 8.50 | 17.50 |
| Royelle, PAR 1.3, vinyl wear layer, 1 year | BF@.250 | SY | 4.55 | 8.50 | 13.05 |

Tarkett sheet vinyl, bonded with adhesive at perimeter and at seams. Including 10% waste and adhesive.

| | Craft@Hrs | Unit | Material | Labor | Total |
|---|---|---|---|---|---|
| Simply Brite | BF@.250 | SY | 26.90 | 8.50 | 35.40 |
| Signature | BF@.250 | SY | 23.30 | 8.50 | 31.80 |
| Naturelle | BF@.250 | SY | 22.70 | 8.50 | 31.20 |
| Presence Felt | BF@.250 | SY | 17.90 | 8.50 | 26.40 |
| Style Brite Flex | BF@.250 | SY | 15.40 | 8.50 | 23.90 |
| Style Brite Felt | BF@.250 | SY | 15.30 | 8.50 | 23.80 |
| Simply Brite Felt | BF@.250 | SY | 15.10 | 8.50 | 23.60 |
| Fiber Floor | BF@.250 | SY | 9.06 | 8.50 | 17.56 |
| Sierrra | BF@.250 | SY | 8.10 | 8.50 | 16.60 |
| Yosemite | BF@.250 | SY | 7.28 | 8.50 | 15.78 |
| Preference | BF@.250 | SY | 6.38 | 8.50 | 14.88 |
| Sandran | BF@.250 | SY | 4.24 | 8.50 | 12.74 |

Sheet vinyl, no wax, Mannington. Bonded with adhesive at perimeter and at seams. Including 10% waste and adhesive.

| | Craft@Hrs | Unit | Material | Labor | Total |
|---|---|---|---|---|---|
| Gold Series | BF@.250 | SY | 40.00 | 8.50 | 48.50 |
| Silver Series | BF@.250 | SY | 28.60 | 8.50 | 37.10 |
| Bronze Series | BF@.250 | SY | 28.60 | 8.50 | 37.10 |
| Villa Series | BF@.250 | SY | 20.50 | 8.50 | 29.00 |
| Ceramica | BF@.250 | SY | 25.60 | 8.50 | 34.10 |
| Silverado | BF@.250 | SY | 23.10 | 8.50 | 31.60 |
| Aurora | BF@.250 | SY | 20.70 | 8.50 | 29.20 |
| Lumina | BF@.250 | SY | 17.10 | 8.50 | 25.60 |
| Vega II, Omni | BF@.250 | SY | 13.30 | 8.50 | 21.80 |

## Vinyl 12" x 12" Tile Flooring

Armstrong vinyl tile, 12" x 12", including 10% waste and adhesive

| | Craft@Hrs | Unit | Material | Labor | Total |
|---|---|---|---|---|---|
| Solarian Stonegate, .094" | BF@.023 | SF | 2.91 | .78 | 3.69 |
| Natural Images, .094" | BF@.023 | SF | 2.91 | .78 | 3.69 |

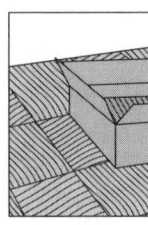

| | Craft@Hrs | Unit | Material | Labor | Total |
|---|---|---|---|---|---|
| Chelsea (urethane), .08" | BF@.023 | SF | 2.31 | .78 | 3.09 |
| Chesapeake (urethane), .08" | BF@.023 | SF | 1.87 | .78 | 2.65 |
| Harbour, .08" | BF@.023 | SF | 1.79 | .78 | 2.57 |
| StoneTex, 1/8", | BF@.021 | SF | 1.62 | .71 | 2.33 |
| Stylistik, .065", self-adhesive | BF@.023 | SF | 1.27 | .78 | 2.05 |
| Themes .08" (urethane) | BF@.023 | SF | 1.30 | .78 | 2.08 |
| Vernay, Metro, .045" | BF@.023 | SF | 1.09 | .78 | 1.87 |
| MultiColor, 1/8" | BF@.021 | SF | .93 | .71 | 1.64 |
| Excelon Civic Square, 1/8" | BF@.021 | SF | .69 | .71 | 1.40 |
| Afton, .075" | BF@.023 | SF | 1.19 | .78 | 1.97 |
| Builder urethane, .080", self-adhesive | BF@.023 | SF | 1.98 | .78 | 2.76 |
| Builder urethane, .080", dry back | BF@.023 | SF | 1.87 | .78 | 2.65 |

Elegant Images™ Series vinyl floor tile, Armstrong. 12" x 12". ToughGuard® durability. Resists stains, scratches, scuffs and indentations. 25-year warranty. Per square foot, including adhesive and 10% waste.

| | Craft@Hrs | Unit | Material | Labor | Total |
|---|---|---|---|---|---|
| Cornwall, Burnt Almond | BF@.023 | SF | 2.43 | .78 | 3.21 |
| Modern, Rose/Green Inset | BF@.023 | SF | 1.64 | .78 | 2.42 |

Rubber studded tile. 12" x 12". Highly resilient to reduce fatigue. 10-year wear warranty. Slip-resistant. Sound absorbent. High abrasion resistance. For kitchens and utility rooms.

| | Craft@Hrs | Unit | Material | Labor | Total |
|---|---|---|---|---|---|
| All colors | BF@.025 | SF | 7.15 | .85 | 8.00 |

## Vinyl Flooring Accessories

Resilient flooring adhesive

| | Craft@Hrs | Unit | Material | Labor | Total |
|---|---|---|---|---|---|
| Sheet vinyl adhesive, per gallon | — | Gal | 14.20 | — | 14.20 |
| Sheet vinyl adhesive, 350 SF per gallon | — | SF | .04 | — | .04 |
| Vinyl tile adhesive, per gallon | — | Gal | 13.50 | — | 13.50 |
| Vinyl tile adhesive, 350 per gallon | — | SF | .04 | — | .04 |
| Vinyl seam sealer kit (100 LF) | — | LF | .13 | — | .13 |

Resilient flooring accessories

| | Craft@Hrs | Unit | Material | Labor | Total |
|---|---|---|---|---|---|
| Patching and leveling compound, 25 pound bag at $17.80, | | | | | |
|   1 pound per SF at .25" depth | BF@.016 | SF | .72 | .54 | 1.26 |
| Latex primer (used over dusting, porous or below grade concrete) | | | | | |
|   300 SF per gallon at $8.47 | BF@.008 | SF | .03 | .27 | .30 |
| Liquid underlay (application over existing vinyl floor) | | | | | |
|   Gallon at $27.80 covers 75 SF | BF@.010 | SF | .37 | .34 | .71 |
| Lining felt, 15 Lb. asphalt felt (used over board subfloor) | | | | | |
|   432 SF roll at $25.50 covers 350 SF | BF@.004 | SF | .06 | .14 | .20 |
| Tempered hardboard underlayment (levels uneven subfloor) | | | | | |
|   1/8" x 4' x 8' sheet at $11.00 | BF@.016 | SF | .36 | .54 | .90 |

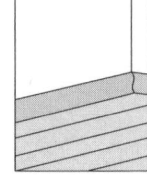

Cove base, colors,

| | Craft@Hrs | Unit | Material | Labor | Total |
|---|---|---|---|---|---|
| 2-1/2" high, .08" thick | BF@.025 | LF | .44 | .85 | 1.29 |
| 4" high, .08" thick | BF@.025 | LF | .59 | .85 | 1.44 |
| 4" high, self-adhesive | BF@.016 | LF | 1.10 | .54 | 1.64 |

Cove base cement

| | Craft@Hrs | Unit | Material | Labor | Total |
|---|---|---|---|---|---|
| 200 LF of 4" base per gallon at $16.00 | — | LF | .08 | — | .08 |

Accent strip, 1/8" x 24" long

| | Craft@Hrs | Unit | Material | Labor | Total |
|---|---|---|---|---|---|
| 1/2" wide | BF@.025 | LF | .34 | .85 | 1.19 |
| 1" wide | BF@.025 | LF | .43 | .85 | 1.28 |
| 2" wide | BF@.025 | LF | .80 | .85 | 1.65 |

Vinyl edging strip, 1" wide

| | Craft@Hrs | Unit | Material | Labor | Total |
|---|---|---|---|---|---|
| 1/16" thick | — | LF | .56 | — | .56 |
| 1/8" thick | — | LF | .70 | — | .70 |

# Flooring, Wood Strip

|  | Craft@Hrs | Unit | Material | Labor | Total |
|---|---|---|---|---|---|

**Wood Strip Flooring** Unfinished tongue and groove 3/4" x 2-1/4" strip flooring. Bundle covers 19.5 square feet. See installation, sanding and finishing cost below.

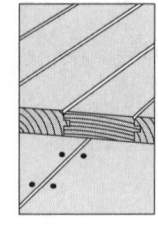

|  | Craft@Hrs | Unit | Material | Labor | Total |
|---|---|---|---|---|---|
| Select red oak, $64.80 per bundle | — | SF | 3.32 | — | 3.32 |
| Select white oak, $53.90 per bundle | — | SF | 2.76 | — | 2.76 |
| No. 1 common red oak, $53.10 per bundle | — | SF | 2.72 | — | 2.72 |
| No. 1 common white oak, $50.10 per bundle | — | SF | 2.57 | — | 2.57 |
| No. 2 common red oak, $47.20 per bundle | — | SF | 2.42 | — | 2.42 |
| Select ash, $59.10 per bundle | — | SF | 3.03 | — | 3.03 |
| Red oak stair nosing, 3/4" | — | LF | 4.34 | — | 4.34 |
| Red oak reducer strip, 3/4" x 2" | — | LF | 3.45 | — | 3.45 |
| Red oak quarter round, 3/4" x 3/4" | — | LF | 2.18 | — | 2.18 |

Labor installing, sanding and finishing hardwood strip flooring on a prepared subfloor. Nailed in place.

|  | Craft@Hrs | Unit | Material | Labor | Total |
|---|---|---|---|---|---|
| Install 1-1/2" width strips | BF@.066 | SF | — | 2.24 | 2.24 |
| Install 2" or 2-1/4" width strips | BF@.062 | SF | — | 2.11 | 2.11 |
| Install 3-1/4" width strips | BF@.060 | SF | — | 2.04 | 2.04 |
| Install stair nose or reducer strip | BF@.040 | LF | — | 1.36 | 1.36 |
| Sanding, 3 passes (60/80/100 grit) | BF@.023 | SF | .07 | .78 | .85 |
| 2 coats of stain/sealer | BF@.010 | SF | .11 | .34 | .45 |
| 2 coats of urethane | BF@.012 | SF | .14 | .41 | .55 |
| 2 coats of lacquer | BF@.024 | SF | .19 | .82 | 1.01 |

**Unfinished antique tongue and groove plank flooring** 25/32" thick, random widths and lengths. Includes typical cutting waste. Installed over a prepared subfloor.

|  | Craft@Hrs | Unit | Material | Labor | Total |
|---|---|---|---|---|---|
| Ash, select, 3" to 7" wide | BF@.057 | SF | 4.55 | 1.94 | 6.49 |
| Chestnut, American, 3" to 7" wide | BF@.057 | SF | 9.45 | 1.94 | 11.39 |
| Chestnut, distressed, 3" to 7" wide | BF@.057 | SF | 11.10 | 1.94 | 13.04 |
| Chestnut, distressed, over 7" wide | BF@.057 | SF | 14.60 | 1.94 | 16.54 |
| Cherry, #1, 3" to 7" wide | BF@.057 | SF | 10.60 | 1.94 | 12.54 |
| Cherry, #1, 6" wide and up | BF@.057 | SF | 12.90 | 1.94 | 14.84 |
| Hickory, #1, 3" to 7" wide | BF@.057 | SF | 4.42 | 1.94 | 6.36 |
| Hemlock, 3" to 7" wide | BF@.057 | SF | 7.60 | 1.94 | 9.54 |
| Hemlock, over 7" wide | BF@.057 | SF | 9.84 | 1.94 | 11.78 |
| Oak, #1, red or white, 3" to 7" wide | BF@.057 | SF | 4.05 | 1.94 | 5.99 |
| Oak, #1, red or white, over 7" wide | BF@.057 | SF | 6.39 | 1.94 | 8.33 |
| Oak, 3" to 7" wide | BF@.057 | SF | 8.11 | 1.94 | 10.05 |
| Oak, over 7" wide | BF@.057 | SF | 9.83 | 1.94 | 11.77 |
| Oak, distressed, 3" to 7" wide | BF@.057 | SF | 9.22 | 1.94 | 11.16 |
| Oak, distressed, over 7" wide | BF@.057 | SF | 11.70 | 1.94 | 13.64 |
| Maple, FAS, 3" to 7" wide | BF@.057 | SF | 5.59 | 1.94 | 7.53 |
| Poplar, 3" to 8" wide | BF@.057 | SF | 3.94 | 1.94 | 5.88 |
| Poplar, over 8" wide | BF@.057 | SF | 5.30 | 1.94 | 7.24 |
| Walnut, #3, 3" to 7" wide | BF@.057 | SF | 6.39 | 1.94 | 8.33 |
| White pine, 3" to 7" | BF@.057 | SF | 5.97 | 1.94 | 7.91 |
| White pine, distressed, 3" to 7" | BF@.057 | SF | 6.51 | 1.94 | 8.45 |
| White pine, distressed, over 7" wide | BF@.057 | SF | 9.31 | 1.94 | 11.25 |
| White pine, knotty grade, 12" and up wide | BF@.057 | SF | 9.56 | 1.94 | 11.50 |
| Yellow pine, 3" to 5" wide | BF@.057 | SF | 6.96 | 1.94 | 8.90 |
| Yellow pine, 6" to 10" wide | BF@.057 | SF | 9.43 | 1.94 | 11.37 |

| | Craft@Hrs | Unit | Material | Labor | Total |
|---|---|---|---|---|---|
| **Unfinished northern hard maple strip flooring** | | | | | |
| 25/32" x 2-1/4" (138.3 BF per 100 SF) | | | | | |
| 1st grade | BF@.036 | SF | 5.74 | 1.22 | 6.96 |
| 2nd grade | BF@.036 | SF | 5.14 | 1.22 | 6.36 |
| 3rd grade | BF@.036 | SF | 4.73 | 1.22 | 5.95 |
| 25/32" x 1-1/2" (155 BF per 100 SF) | | | | | |
| 1st grade | BF@.040 | SF | 6.85 | 1.36 | 8.21 |
| 2nd grade | BF@.040 | SF | 6.15 | 1.36 | 7.51 |
| 3rd grade | BF@.040 | SF | 5.68 | 1.36 | 7.04 |
| 33/32" x 1-1/2" (170 BF per 100 SF) | | | | | |
| 1st grade | BF@.036 | SF | 6.38 | 1.22 | 7.60 |
| 2nd grade | BF@.036 | SF | 5.48 | 1.22 | 6.70 |
| 3rd grade | BF@.036 | SF | 5.19 | 1.22 | 6.41 |

**Unfinished long plank flooring**  Random 3" to 5" widths, 1' to 16' lengths, 3/4" thick.

| | Craft@Hrs | Unit | Material | Labor | Total |
|---|---|---|---|---|---|
| Select grade, ash | BF@.062 | SF | 5.83 | 2.11 | 7.94 |
| Common grade, cherry | BF@.062 | SF | 7.69 | 2.11 | 9.80 |
| Select grade, cherry | BF@.062 | SF | 9.61 | 2.11 | 11.72 |
| Select grade, rock maple | BF@.062 | SF | 7.22 | 2.11 | 9.33 |
| Select grade, figured maple | BF@.062 | SF | 12.50 | 2.11 | 14.61 |
| Select grade, red oak | BF@.062 | SF | 8.10 | 2.11 | 10.21 |
| Select grade, white oak | BF@.062 | SF | 6.68 | 2.11 | 8.79 |

**Unfinished wide plank flooring**  Random 6" to 8" widths, 1' to 16' lengths, 3/4" thick, including 8% waste, Specialty Wood Floors.

| | Craft@Hrs | Unit | Material | Labor | Total |
|---|---|---|---|---|---|
| Knotty grade, pine | BF@.074 | SF | 4.54 | 2.52 | 7.06 |
| Select grade, cherry | BF@.074 | SF | 10.70 | 2.52 | 13.22 |
| Select grade, maple | BF@.074 | SF | 9.70 | 2.52 | 12.22 |
| Select grade, red oak | BF@.074 | SF | 9.11 | 2.52 | 11.63 |
| Select grade, white oak | BF@.074 | SF | 8.18 | 2.52 | 10.70 |

**Softwood flooring**  Material costs include typical waste and coverage loss. Estimates for sanding and finishing appear at the end of this section.

Unfinished Douglas fir, C Grade and better, dry, vertical grain flooring. 1.25 board feet cover 1 square foot including end cutting and typical waste.

| | Craft@Hrs | Unit | Material | Labor | Total |
|---|---|---|---|---|---|
| 1" x 2", square edge | BF@.037 | SF | 3.17 | 1.26 | 4.43 |
| 1" x 3", square edge | BF@.037 | SF | 3.19 | 1.26 | 4.45 |
| 1" x 4", tongue and groove | BF@.037 | SF | 2.60 | 1.26 | 3.86 |
| Deduct for D Grade | — | SF | -.77 | — | -.77 |
| Add for B Grade and better | — | SF | .77 | — | .77 |
| Add for specific lengths | — | SF | .15 | — | .15 |
| Add for diagonal patterns | BF@.004 | SF | — | .14 | .14 |

Unfinished select kiln dried heart pine flooring, 97% dense heartwood, random lengths, including 5% waste. 1" to 1-1/4" thick.

| | Craft@Hrs | Unit | Material | Labor | Total |
|---|---|---|---|---|---|
| 4" wide | BF@.042 | SF | 7.80 | 1.43 | 9.23 |
| 5" wide | BF@.042 | SF | 8.04 | 1.43 | 9.47 |
| 6" wide | BF@.042 | SF | 8.12 | 1.43 | 9.55 |
| 8" wide | BF@.046 | SF | 8.73 | 1.56 | 10.29 |
| 10" wide | BF@.046 | SF | 10.10 | 1.56 | 11.66 |
| Add for diagonal patterns | BF@.004 | SF | — | .14 | .14 |

# Flooring, Wood Strip, Prefinished

| | Craft@Hrs | Unit | Material | Labor | Total |
|---|---|---|---|---|---|
| Unfinished plank flooring, 3/4" thick, 3" to 8" widths, 2' to 12' lengths. | | | | | |
| Antique white pine | BF@.100 | SF | 8.33 | 3.40 | 11.73 |
| Antique yellow pine | BF@.100 | SF | 7.69 | 3.40 | 11.09 |
| Milled heart pine | BF@.100 | SF | 9.11 | 3.40 | 12.51 |
| Distressed oak | BF@.100 | SF | 9.86 | 3.40 | 13.26 |
| Milled oak | BF@.100 | SF | 8.95 | 3.40 | 12.35 |
| Distressed American chestnut | BF@.100 | SF | 9.16 | 3.40 | 12.56 |
| Milled American chestnut | BF@.100 | SF | 9.86 | 3.40 | 13.26 |
| Aged hemlock/fir | BF@.100 | SF | 9.13 | 3.40 | 12.53 |
| Antique poplar | BF@.100 | SF | 4.20 | 3.40 | 7.60 |
| Aged cypress | BF@.100 | SF | 4.99 | 3.40 | 8.39 |
| Add for 9" to 15" widths | — | % | 15.0 | — | — |
| Add for sanding and finishing. | | | | | |
| Sanding, 3 passes (60/80/100 grit) | BF@.023 | SF | .06 | .78 | .84 |
| 2 coats of stain/sealer | BF@.010 | SF | .11 | .34 | .45 |
| 2 coats of urethane | BF@.012 | SF | .15 | .41 | .56 |
| 2 coats of lacquer | BF@.024 | SF | .20 | .82 | 1.02 |

## Prefinished strip flooring

Tongue and groove, 25/32" thick, 2-1/4" wide. Includes typical cutting waste. Installed over a prepared subfloor

| | Craft@Hrs | Unit | Material | Labor | Total |
|---|---|---|---|---|---|
| Marsh, case at $69 covers 20 SF | BF@.062 | SF | 3.44 | 2.11 | 5.55 |
| Rustic oak, case at $69 covers 20 SF | BF@.062 | SF | 3.44 | 2.11 | 5.55 |
| Butterscotch, case at $92 covers 20 SF | BF@.062 | SF | 4.58 | 2.11 | 6.69 |
| Natural, case at $92 covers 20 SF | BF@.062 | SF | 4.58 | 2.11 | 6.69 |
| Gunstock, case at $101 covers 20 SF | BF@.062 | SF | 5.05 | 2.11 | 7.16 |
| Toffee, case at $96 covers 24 SF | BF@.062 | SF | 3.99 | 2.11 | 6.10 |
| Red oak, case at $96 covers 24 SF | BF@.062 | SF | 3.99 | 2.11 | 6.10 |
| Prefinished oak strip flooring trim | | | | | |
| 7' 10" quarter round | BF@.320 | Ea | 17.20 | 10.90 | 28.10 |
| 3' 11" reducer | BF@.320 | Ea | 23.00 | 10.90 | 33.90 |
| 3' 11" stair nose | BF@.320 | Ea | 23.00 | 10.90 | 33.90 |
| 3' 11" T-moulding | BF@.320 | Ea | 23.00 | 10.90 | 33.90 |

**Floating hardwood flooring** Tongue and groove. Install over concrete or other subfloor using an underfloor membrane. Edges and ends glued. Based on 3/8" gap at walls covered with 1/2" base or quarter round molding. The labor estimates below include laying the membrane, hardwood flooring and gluing.

| | Craft@Hrs | Unit | Material | Labor | Total |
|---|---|---|---|---|---|
| Comfort Guard membrane | — | SF | .20 | — | .20 |
| Everseal adhesive | — | SF | .08 | — | .08 |
| 1/2" tongue and groove hardwood flooring, | | | | | |
| various colors | FL@.051 | SF | 3.71 | 2.57 | 6.28 |

**Laminate wood floor, Pergo** Plastic laminate on pressed wood base, 7-7/8" x 47-1/4" x 3/8" (20 cm x 120 cm x 8 mm), loose lay with glued tongue and groove joints. Many wood finishes available. 20.67 SF (1.9 square meters) per pack. Includes 10% waste and joint glue (bottle covers 150 SF).

| | Craft@Hrs | Unit | Material | Labor | Total |
|---|---|---|---|---|---|
| Pergo Presto, on floors | | | | | |
| Most finishes, $45.00 per case | BF@.051 | SF | 2.13 | 1.73 | 3.86 |
| Glued on stairs (treads and risers) | BF@.115 | SF | 5.16 | 3.91 | 9.07 |
| Add for Silent Step underlay | | | | | |
| (102 SF per pack) | BF@.010 | SF | .60 | .34 | .94 |

| | Craft@Hrs | Unit | Material | Labor | Total |
|---|---|---|---|---|---|
| Pergo sound block foam | BF@.002 | SF | .18 | .07 | .25 |
| Add for 8 mil poly vapor barrier (over concrete) | BF@.002 | SF | .08 | .07 | .15 |
| T mould strip (joins same level surface) | BF@.040 | LF | 5.37 | 1.36 | 6.73 |
| Reducer strip (joins higher surface) | BF@.040 | LF | 4.82 | 1.36 | 6.18 |
| End molding (joins lower surface) | BF@.040 | LF | 3.32 | 1.36 | 4.68 |
| Wall base, 4" high | BF@.040 | LF | 2.35 | 1.36 | 3.71 |
| Stair nosing | BF@.050 | LF | 5.11 | 1.70 | 6.81 |

**Laminate wood floor, Dupont**  Plastic laminate on pressed wood base, 11-1/2" x 46-9/16" x .0462", foam back, tongue and groove joints. Pack covers 17 SF with 8% waste.

| | Craft@Hrs | Unit | Material | Labor | Total |
|---|---|---|---|---|---|
| Dupont Real Touch Elite on floors | | | | | |
| Most finishes, $66.00 per pack | BF@.051 | SF | 3.91 | 1.73 | 5.64 |
| Glued on stairs (treads and risers) | BF@.115 | SF | 6.31 | 3.91 | 10.22 |
| 6 mil vapor barrier (over concrete) | BF@.002 | SF | .08 | .07 | .15 |

**Bamboo flooring**
Laminated plank bamboo flooring, tongue and groove, 72" long panels. Treated and kiln dried. Installs over wood subfloor. Use 1/2" nail gun. Space fasteners 8" to 10" apart, with a minimum of two nails per board. Assorted shades and textures.

| | Craft@Hrs | Unit | Material | Labor | Total |
|---|---|---|---|---|---|
| 5/8" x 3-3/4" prefinished | BF@.062 | SF | 4.41 | 2.11 | 6.52 |
| 1/2" x 3" unfinished | BF@.062 | SF | 4.41 | 2.11 | 6.52 |
| Laminated bamboo baseboard | | | | | |
| 3" beveled edge prefinished | BC@.040 | LF | 4.30 | 1.48 | 5.78 |
| Laminated prefinished bamboo stair nose | | | | | |
| 1/2" x 3-1/4" | BC@.040 | LF | 6.61 | 1.48 | 8.09 |
| Laminated prefinished bamboo threshold | | | | | |
| 9/16" x 3-5/8" flat grain | BC@.174 | LF | 5.06 | 6.43 | 11.49 |
| 9/16" x 3-5/8" edge grain | BC@.174 | LF | 5.06 | 6.43 | 11.49 |
| 5/8" x 2" | BC@.174 | LF | 6.61 | 6.43 | 13.04 |
| Add for sanding and finishing | | | | | |
| Sanding, 3 passes (60/80/100 grit) | BF@.023 | SF | .06 | .78 | .84 |
| 2 coats of stain/sealer | BF@.010 | SF | .11 | .34 | .45 |
| 2 coats of urethane | BF@.012 | SF | .15 | .41 | .56 |
| 2 coats of lacquer | BF@.024 | SF | .20 | .82 | 1.02 |

**Parquet block flooring**  Includes typical waste and coverage loss, installed over a prepared subfloor.
Unfinished parquet block flooring, 3/4" thick, 10" x 10" to 39" x 39" blocks, includes 5% waste. Cost varies by pattern with larger blocks costing more. These are material costs only. Add labor cost below. Kentucky Wood Floors.

| | Craft@Hrs | Unit | Material | Labor | Total |
|---|---|---|---|---|---|
| Ash or quartered oak, select | — | SF | 17.60 | — | 17.60 |
| Ash or quartered oak, natural | — | SF | 12.80 | — | 12.80 |
| Cherry or walnut select grade | — | SF | 20.10 | — | 20.10 |
| Cherry or walnut natural grade | — | SF | 15.10 | — | 15.10 |
| Oak, select plain | — | SF | 14.30 | — | 14.30 |
| Oak, natural plain | — | SF | 12.30 | — | 12.30 |

Prefinished parquet block flooring, 5/16" thick, select grade, specified sizes, includes 5% waste. Cost varies by pattern and size, with larger blocks costing more. These are material costs only. Add labor cost below. Kentucky Wood Floors.

| | Craft@Hrs | Unit | Material | Labor | Total |
|---|---|---|---|---|---|
| Cherry or walnut | | | | | |
| 9" x 9" | — | SF | 2.74 | — | 2.74 |
| 13" x 13" | — | SF | 3.26 | — | 3.26 |

# Flooring, Wood, Parquet

| | Craft@Hrs | Unit | Material | Labor | Total |
|---|---|---|---|---|---|
| Oak | | | | | |
| 12" x 12" chestnut | — | SF | 1.90 | — | 1.90 |
| 12" x 12" desert | — | SF | 1.90 | — | 1.90 |
| Teak | | | | | |
| 12" x 12" | — | SF | 2.44 | — | 2.44 |
| Prefinished natural hardwood, commodity grade | | | | | |
| 12" x 12" | — | SF | 1.57 | — | 1.57 |
| Labor and mastic for installing parquet block flooring | | | | | |
| Parquet laid in mastic | BF@.055 | SF | .29 | 1.87 | 2.16 |
| Edge-glued over high density foam | BF@.052 | SF | .49 | 1.77 | 2.26 |
| Custom or irregular-shaped tiles, with adhesive | BF@.074 | SF | .49 | 2.52 | 3.01 |
| Add for installations under 1,000 SF | BF@.025 | SF | .05 | .85 | .90 |
| Add for sanding and finishing for unfinished parquet block flooring | | | | | |
| Sanding, 3 passes (60/80/100 grit) | BF@.023 | SF | .06 | .78 | .84 |
| 2 coats of stain/sealer | BF@.010 | SF | .11 | .34 | .45 |
| 2 coats of urethane | BF@.012 | SF | .15 | .41 | .56 |
| 2 coats of lacquer | BF@.024 | SF | .20 | .82 | 1.02 |

Prefinished cross-band 3-ply birch parquet flooring, 6" to 12" wide by 12" to 48" blocks, Oregon Lumber.
Costs include cutting, fitting and 5% waste, but no subfloor preparation.

| | Craft@Hrs | Unit | Material | Labor | Total |
|---|---|---|---|---|---|
| "Saima", 9/16" thick, typical price | BF@.050 | SF | 4.40 | 1.70 | 6.10 |
| Bonded blocks, 12" x 12", 1/8" thick, laid in mastic | | | | | |
| Vinyl bonded cork | BF@.050 | SF | 3.60 | 1.70 | 5.30 |
| Vinyl bonded domestic woods | BF@.062 | SF | 7.19 | 2.11 | 9.30 |
| Vinyl bonded exotic woods | BF@.062 | SF | 14.60 | 2.11 | 16.71 |
| Acrylic (Permagrain) impregnated wood flooring. 12" x 12" x 5/16" thick block laid in mastic | | | | | |
| Oak block | BF@.057 | SF | 9.90 | 1.94 | 11.84 |
| Ash block | BF@.057 | SF | 10.50 | 1.94 | 12.44 |
| Cherry block | BF@.057 | SF | 11.00 | 1.94 | 12.94 |
| Maple block | BF@.057 | SF | 11.10 | 1.94 | 13.04 |
| Acrylic (Permagrain) impregnated 2-7/8" x random length plank laid in mastic | | | | | |
| Oak plank | BF@.057 | SF | 8.18 | 1.94 | 10.12 |
| Maple plank | BF@.057 | SF | 10.90 | 1.94 | 12.84 |
| Tupelo plank | BF@.057 | SF | 10.10 | 1.94 | 12.04 |
| Lindenwood plank | BF@.057 | SF | 11.20 | 1.94 | 13.14 |
| Cherry plank | BF@.057 | SF | 11.00 | 1.94 | 12.94 |

Wood flooring adhesive. For laminated wood plank and wood parquet flooring. Covers 40 to 50 square feet per gallon when applied with a 1/8" x 1/8" x 1/8" square-notch trowel for parquet or a 1/4" x 1/4" x 1/4" V-notch trowel for laminated plank.

| | Craft@Hrs | Unit | Material | Labor | Total |
|---|---|---|---|---|---|
| Quart | — | Ea | 5.54 | — | 5.54 |
| 4 gallons | — | Ea | 37.20 | — | 37.20 |

**Foundation Bolts** Galvanized, with nut and washer attached. Costs per bolt or bag.

| | Craft@Hrs | Unit | Material | Labor | Total |
|---|---|---|---|---|---|
| 1/2" x 6", pack of 1 | B1@.107 | Ea | .92 | 3.56 | 4.48 |
| 1/2" x 6", pack of 50 | B1@5.35 | Ea | 43.30 | 178.00 | 221.30 |
| 1/2" x 8", pack of 1 | B1@.107 | Ea | .97 | 3.56 | 4.53 |
| 1/2" x 8", pack of 50 | B1@5.35 | Ea | 45.30 | 178.00 | 223.30 |
| 1/2" x 10", pack of 1 | B1@.107 | Ea | 1.23 | 3.56 | 4.79 |
| 1/2" x 10", pack of 50 | B1@5.35 | Ea | 49.40 | 178.00 | 227.40 |
| 1/2" x 12", pack of 1 | B1@.107 | Ea | 1.33 | 3.56 | 4.89 |
| 1/2" x 12", pack of 50 | B1@5.35 | Ea | 51.60 | 178.00 | 229.60 |
| 1/2" x 18", pack of 1 | B1@.107 | Ea | 1.80 | 3.56 | 5.36 |
| 1/2" x 18", pack of 50 | B1@5.35 | Ea | 89.00 | 178.00 | 267.00 |
| 5/8" x 8", pack of 1 | B1@.107 | Ea | 2.41 | 3.56 | 5.97 |

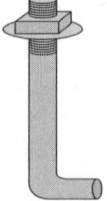

| | Craft@Hrs | Unit | Material | Labor | Total |
|---|---|---|---|---|---|
| 5/8" x 8", pack of 25 | B1@2.68 | Ea | 55.40 | 89.30 | 144.70 |
| 5/8" x 10", pack of 1 | B1@.107 | Ea | 2.41 | 3.56 | 5.97 |
| 5/8" x 10", pack of 25 | B1@2.68 | Ea | 58.50 | 89.30 | 147.80 |

**Framing Connectors** All bolted connectors include nuts, bolts, and washers as needed. Costs are based on bulk purchases. Add 20% for smaller quantities.
Equivalent Simpson Strong-Tie Co. model numbers are shown in parentheses.
Angle clips (A34 and A35), 18 gauge, galvanized, no nails included

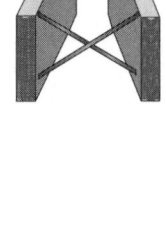

| | Craft@Hrs | Unit | Material | Labor | Total |
|---|---|---|---|---|---|
| 3" x 5" connector (A35 or A35F) | BC@.030 | Ea | .36 | 1.11 | 1.47 |
| 3" x 4" connector (A34) | BC@.030 | Ea | .28 | 1.11 | 1.39 |

Angle clips ("L") for general utility use, 16 gauge, galvanized, no nails included

| | | | | | |
|---|---|---|---|---|---|
| 3" (L30) | BC@.027 | Ea | .57 | 1.00 | 1.57 |
| 5" (L50) | BC@.027 | Ea | .95 | 1.00 | 1.95 |
| 7" (L70) | BC@.027 | Ea | 2.38 | 1.00 | 3.38 |
| 9" (L90) | BC@.027 | Ea | 3.05 | 1.00 | 4.05 |

Brick wall ties (BTR), 22 gauge, per 1,000

| | | | | | |
|---|---|---|---|---|---|
| 7/8" x 6", corrugated | — | M | 49.00 | — | 49.00 |

Bridging. See also Carpentry.

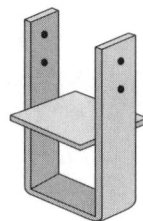

Cross bridging (NC), steel, per pair, no nails required
12" joist spacing on center

| | | | | | |
|---|---|---|---|---|---|
| 2" x  8" to 2" x 16" joists (NC2X8-16) | BC@.022 | Pr | 3.11 | .81 | 3.92 |

16" joist spacing on center

| | | | | | |
|---|---|---|---|---|---|
| 2" x  8" to 2" x 12" joists (NC2X8-12) | BC@.022 | Pr | 3.11 | .81 | 3.92 |
| 2" x 14" to 2" x 16" joists (NC2X14-16) | BC@.022 | Pr | 3.42 | .81 | 4.23 |

24" joist spacing on center

| | | | | | |
|---|---|---|---|---|---|
| 2" x 10" to 2" x 12" joists (NC2X10-12) | BC@.022 | Pr | 3.19 | .81 | 4.00 |

Column bases, heavy duty (CB). Includes nuts and bolts. Labor shown is for placing column base before pouring concrete

| | | | | | |
|---|---|---|---|---|---|
| 4" x  4" (CB44) | BC@.121 | Ea | 12.90 | 4.47 | 17.37 |
| 4" x  6" (CB46) | BC@.121 | Ea | 13.60 | 4.47 | 18.07 |
| 4" x  8" (CB48) | BC@.121 | Ea | 15.30 | 4.47 | 19.77 |
| 6" x  6" (CB66) | BC@.121 | Ea | 17.70 | 4.47 | 22.17 |
| 6" x  8" (CB68) | BC@.121 | Ea | 18.60 | 4.47 | 23.07 |
| 6" x 10" (CB610) | BC@.131 | Ea | 20.00 | 4.84 | 24.84 |
| 6" x 12" (CB612) | BC@.131 | Ea | 21.70 | 4.84 | 26.54 |
| 8" x  8" (CB88) | BC@.131 | Ea | 36.60 | 4.84 | 41.44 |
| 8" x 10" (CB810) | BC@.131 | Ea | 43.80 | 4.84 | 48.64 |
| 8" x 12" (CB812) | BC@.150 | Ea | 47.50 | 5.54 | 53.04 |
| 10" x 10" (CB1010) | BC@.150 | Ea | 48.50 | 5.54 | 54.04 |
| 10" x 12" (CB1012) | BC@.150 | Ea | 55.80 | 5.54 | 61.34 |
| 12" x 12" (CB1212) | BC@.150 | Ea | 60.50 | 5.54 | 66.04 |

Column caps or end column caps, heavy duty (CC) or (ECC). Includes nuts and bolts
Column caps for solid lumber
Beam size x post size

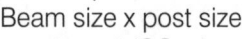

| | | | | | |
|---|---|---|---|---|---|
| 4" x  4" (CC44) | BC@.260 | Ea | 29.30 | 9.60 | 38.90 |
| 4" x  6" (CC46) | BC@.300 | Ea | 41.40 | 11.10 | 52.50 |
| 6" x  4" (CC64) | BC@.300 | Ea | 62.10 | 11.10 | 73.20 |
| 6" x  6" (CC66) | BC@.400 | Ea | 62.10 | 14.80 | 76.90 |
| 6" x  8" (CC68) | BC@.600 | Ea | 62.10 | 22.20 | 84.30 |
| 8" x  6" (CC86) | BC@.600 | Ea | 68.50 | 22.20 | 90.70 |
| 8" x  8" (CC88) | BC@.750 | Ea | 68.50 | 27.70 | 96.20 |
| 10" x  6" (CC10X) | BC@.750 | Ea | 102.00 | 27.70 | 129.70 |

# Framing Connectors

| | Craft@Hrs | Unit | Material | Labor | Total |
|---|---|---|---|---|---|
| Column caps for glu-lam beams | | | | | |
| Beam size x post size | | | | | |
| 3-1/4" x 4" (CC3-14-4) | BC@.260 | Ea | 53.20 | 9.60 | 62.80 |
| 3-1/4" x 6" (CC3-1/4-6) | BC@.300 | Ea | 53.20 | 11.10 | 64.30 |
| 5-1/4" x 6" (CC5-1/4-6) | BC@.400 | Ea | 64.70 | 14.80 | 79.50 |
| 5-1/4" x 8" (CC5-1/4-8) | BC@.600 | Ea | 64.70 | 22.20 | 86.90 |
| 7" x 6" (CC76) | BC@.600 | Ea | 64.80 | 22.20 | 87.00 |
| 7" x 7" (CC77) | BC@.750 | Ea | 64.80 | 27.70 | 92.50 |
| 7" x 8" (CC78) | BC@.750 | Ea | 64.80 | 27.70 | 92.50 |
| 9" x 6" (CC96) | BC@.750 | Ea | 72.90 | 27.70 | 100.60 |
| 9" x 8" (CC98) | BC@.750 | Ea | 72.90 | 27.70 | 100.60 |

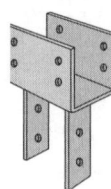

| | Craft@Hrs | Unit | Material | Labor | Total |
|---|---|---|---|---|---|
| Floor tie anchors (FTA) | | | | | |
| 2" x 38" (FTA2) | BC@.200 | Ea | 29.50 | 7.39 | 36.89 |
| 3-1/2" x 45" (FTA5) | BC@.200 | Ea | 30.90 | 7.39 | 38.29 |
| 3-1/2" x 56" (FTA7) | BC@.320 | Ea | 54.40 | 11.80 | 66.20 |
| Header hangers (HH), 16 gauge, galvanized | | | | | |
| 4" header (HH4) | BC@.046 | Ea | 4.36 | 1.70 | 6.06 |
| 6" header (HH6) | BC@.046 | Ea | 8.62 | 1.70 | 10.32 |

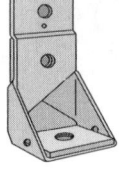

| | Craft@Hrs | Unit | Material | Labor | Total |
|---|---|---|---|---|---|
| Hold downs (HD), including typical nuts, bolts and washers for studs. Add for foundation bolts | | | | | |
| HD2A | BC@.310 | Ea | 6.08 | 11.50 | 17.58 |
| HD5A | BC@.310 | Ea | 14.70 | 11.50 | 26.20 |
| HD8A | BC@.420 | Ea | 20.40 | 15.50 | 35.90 |
| HD10A | BC@.500 | Ea | 22.00 | 18.50 | 40.50 |
| HD20A | BC@.500 | Ea | 42.80 | 18.50 | 61.30 |
| Threaded rod (ATR) 24" long. Labor includes drilling through 6" of wood | | | | | |
| 1/2" rod | BC@.185 | Ea | 4.84 | 6.83 | 11.67 |
| 5/8" rod | BC@.185 | Ea | 6.20 | 6.83 | 13.03 |
| 3/4" rod | BC@.200 | Ea | 13.20 | 7.39 | 20.59 |
| 1" rod | BC@.220 | Ea | 22.40 | 8.13 | 30.53 |
| Hurricane and seismic ties (H) | | | | | |
| H1 tie | BC@.035 | Ea | .65 | 1.29 | 1.94 |
| H2 tie | BC@.027 | Ea | .86 | 1.00 | 1.86 |
| H2.5 tie | BC@.027 | Ea | .29 | 1.00 | 1.29 |
| H3 tie | BC@.027 | Ea | .29 | 1.00 | 1.29 |
| H4 tie | BC@.027 | Ea | .23 | 1.00 | 1.23 |
| H5 tie | BC@.027 | Ea | .28 | 1.00 | 1.28 |
| H6 tie | BC@.045 | Ea | 3.42 | 1.66 | 5.08 |
| H7 tie | BC@.055 | Ea | 5.36 | 2.03 | 7.39 |

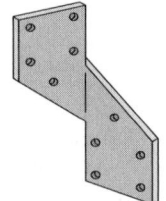

| | Craft@Hrs | Unit | Material | Labor | Total |
|---|---|---|---|---|---|
| Mud sill anchors (MA) | | | | | |
| 2" x 4" (MA4) | BC@.046 | Ea | 1.32 | 1.70 | 3.02 |
| 2" x 6" (MA6) | BC@.046 | Ea | 1.53 | 1.70 | 3.23 |
| Panelized plywood roof hangers (F24N, F26N), for panelized plywood roof construction | | | | | |
| 2" x 4" (F24N) | — | Ea | .53 | — | .53 |
| 2" x 6" (F26N) | — | Ea | .89 | — | .89 |
| Purlin hangers (HU), without nails, 7 gauge | | | | | |
| 4" x 4" (HU44) | BC@.052 | Ea | 3.68 | 1.92 | 5.60 |
| 4" x 6" (HU46) | BC@.052 | Ea | 4.14 | 1.92 | 6.06 |
| 4" x 8" (HU48) | BC@.052 | Ea | 5.01 | 1.92 | 6.93 |
| 4" x 10" (HU410) | BC@.070 | Ea | 6.19 | 2.59 | 8.78 |
| 4" x 12" (HU412) | BC@.070 | Ea | 7.13 | 2.59 | 9.72 |
| 4" x 14" (HU414) | BC@.087 | Ea | 9.07 | 3.21 | 12.28 |
| 4" x 16" (HU416) | BC@.087 | Ea | 10.50 | 3.21 | 13.71 |

| | Craft@Hrs | Unit | Material | Labor | Total |
|---|---|---|---|---|---|
| 6" x 6" (HU66) | BC@.052 | Ea | 7.27 | 1.92 | 9.19 |
| 6" x 8" (HU68) | BC@.070 | Ea | 7.41 | 2.59 | 10.00 |
| 6" x 10" (HU610) | BC@.087 | Ea | 10.50 | 3.21 | 13.71 |
| 6" x 12" (HU612) | BC@.087 | Ea | 15.50 | 3.21 | 18.71 |
| 6" x 16" (HU616) | BC@.117 | Ea | 20.50 | 4.32 | 24.82 |
| 8" x 10" (HU810) | BC@.087 | Ea | 11.60 | 3.21 | 14.81 |
| 8" x 12" (HU812) | BC@.117 | Ea | 14.70 | 4.32 | 19.02 |
| 8" x 14" (HU814) | BC@.117 | Ea | 17.60 | 4.32 | 21.92 |
| 8" x 16" (HU816) | BC@.117 | Ea | 21.40 | 4.32 | 25.72 |
| **Top flange hangers (LB)** | | | | | |
| 2" x 6" (LB26) | BC@.046 | Ea | 2.23 | 1.70 | 3.93 |
| 2" x 8" (LB28) | BC@.046 | Ea | 2.96 | 1.70 | 4.66 |
| 2" x 10" (LB210) | BC@.046 | Ea | 3.68 | 1.70 | 5.38 |
| 2" x 12" (LB212) | BC@.046 | Ea | 3.93 | 1.70 | 5.63 |
| 2" x 14" (LB214) | BC@.052 | Ea | 4.28 | 1.92 | 6.20 |
| **Top flange hangers, heavy (HW)** | | | | | |
| 4" x 6" (HW46) | BC@.052 | Ea | 21.90 | 1.92 | 23.82 |
| 4" x 8" (HW48) | BC@.052 | Ea | 23.50 | 1.92 | 25.42 |
| 4" x 10" (HW410) | BC@.070 | Ea | 25.10 | 2.59 | 27.69 |
| 4" x 12" (HW412) | BC@.070 | Ea | 26.70 | 2.59 | 29.29 |
| 4" x 14" (HW414) | BC@.087 | Ea | 28.30 | 3.21 | 31.51 |
| 4" x 16" (HW416) | BC@.087 | Ea | 29.90 | 3.21 | 33.11 |
| 6" x 6" (HW66) | BC@.052 | Ea | 22.10 | 1.92 | 24.02 |
| 6" x 8" (HW68) | BC@.070 | Ea | 23.70 | 2.59 | 26.29 |
| 6" x 10" (HW610) | BC@.087 | Ea | 25.40 | 3.21 | 28.61 |
| 6" x 12" (HW612) | BC@.087 | Ea | 27.00 | 3.21 | 30.21 |
| 6" x 14" (HW614) | BC@.117 | Ea | 28.60 | 4.32 | 32.92 |
| 6" x 16" (HW616) | BC@.117 | Ea | 30.20 | 4.32 | 34.52 |
| 8" x 6" (HW86) | BC@.070 | Ea | 24.00 | 2.59 | 26.59 |
| 8" x 8" (HW88) | BC@.070 | Ea | 25.60 | 2.59 | 28.19 |
| 8" x 10" (HW810) | BC@.087 | Ea | 27.20 | 3.21 | 30.41 |
| 8" x 12" (HW812) | BC@.117 | Ea | 28.90 | 4.32 | 33.22 |
| 8" x 14" (HW814) | BC@.117 | Ea | 30.50 | 4.32 | 34.82 |
| 8" x 16" (HW816) | BC@.117 | Ea | 32.10 | 4.32 | 36.42 |

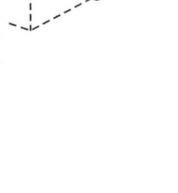

**"U"-type or formed seat type hangers ("U"). Standard duty, 16 gauge, galvanized**

| | Craft@Hrs | Unit | Material | Labor | Total |
|---|---|---|---|---|---|
| 2" x 4" (U24) | BC@.040 | Ea | .90 | 1.48 | 2.38 |
| 2" x 6" (U26) | BC@.046 | Ea | 1.03 | 1.70 | 2.73 |
| 2" x 10" (U210) | BC@.046 | Ea | 1.50 | 1.70 | 3.20 |
| 2" x 14" (U214) | BC@.052 | Ea | 2.51 | 1.92 | 4.43 |
| 3" x 4" (U34) | BC@.046 | Ea | 1.58 | 1.70 | 3.28 |
| 3" x 6" (U36) | BC@.046 | Ea | 1.74 | 1.70 | 3.44 |
| 3" x 10" (U310) | BC@.052 | Ea | 2.90 | 1.92 | 4.82 |
| 3" x 14" (U314) or 3" x 16" (U316) | BC@.070 | Ea | 3.94 | 2.59 | 6.53 |
| 4" x 4" (U44) | BC@.046 | Ea | 1.62 | 1.70 | 3.32 |
| 4" x 6" (U46) | BC@.052 | Ea | 1.81 | 1.92 | 3.73 |
| 4" x 10" (U410) | BC@.070 | Ea | 2.96 | 2.59 | 5.55 |
| 4" x 14" (U414) | BC@.087 | Ea | 4.35 | 3.21 | 7.56 |
| 6" x 6" (U66) | BC@.052 | Ea | 5.21 | 1.92 | 7.13 |
| 6" x 10" (U610) | BC@.087 | Ea | 5.64 | 3.21 | 8.85 |

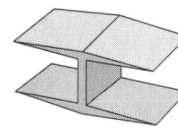

**Plywood clips (PSCL, extruded aluminum, purchased in cartons of 500)**

| | Craft@Hrs | Unit | Material | Labor | Total |
|---|---|---|---|---|---|
| 3/8", 7/16", 1/2", or 5/8" | BC@.006 | Ea | .10 | .22 | .32 |

# Framing Connectors

| | Craft@Hrs | Unit | Material | Labor | Total |
|---|---|---|---|---|---|
| **Post base for decks (base with standoff, PBS), 12 gauge, galvanized** | | | | | |
| 4" x 4" (PBS44) | BC@.046 | Ea | 6.62 | 1.70 | 8.32 |
| 4" x 6" (PBS46) | BC@.046 | Ea | 9.07 | 1.70 | 10.77 |
| 6" x 6" (PBS66) | BC@.070 | Ea | 11.90 | 2.59 | 14.49 |
| Add for rough lumber sizes | — | % | 35.0 | — | — |
| **Post base (AB), adjustable, 16 gauge, galvanized** | | | | | |
| 4" x 4" (AB44B) | BC@.046 | Ea | 5.98 | 1.70 | 7.68 |
| 4" x 6" (AB46) | BC@.046 | Ea | 13.00 | 1.70 | 14.70 |
| 6" x 6" (AB66) | BC@.070 | Ea | 15.80 | 2.59 | 18.39 |
| Add for rough lumber sizes | — | % | 35.0 | — | — |
| **Post bases (PB), 12 gauge, galvanized** | | | | | |
| 4" x 4" (PB44) | BC@.046 | Ea | 5.21 | 1.70 | 6.91 |
| 4" x 6" (PB46) | BC@.046 | Ea | 7.69 | 1.70 | 9.39 |
| 6" x 6" (PB66) | BC@.070 | Ea | 11.00 | 2.59 | 13.59 |
| **Post cap and base combination (BC), 18 gauge, galvanized** | | | | | |
| 4" x 4" (BC4) | BC@.037 | Ea | 2.98 | 1.37 | 4.35 |
| 4" x 6" (BC46) | BC@.037 | Ea | 6.74 | 1.37 | 8.11 |
| 6" x 6" (BC6) | BC@.054 | Ea | 7.44 | 1.99 | 9.43 |
| 8" x 8" (BC8) | BC@.054 | Ea | 26.90 | 1.99 | 28.89 |
| **Post caps and end post caps (PC), 12 gauge, galvanized. Beam size x post size** | | | | | |
| 4" x 4" (PC44) | BC@.052 | Ea | 17.00 | 1.92 | 18.92 |
| 4" x 6" (PC46) | BC@.052 | Ea | 22.70 | 1.92 | 24.62 |
| 4" x 8" (PC48) | BC@.052 | Ea | 33.50 | 1.92 | 35.42 |
| 6" x 4" (PC64) | BC@.052 | Ea | 22.30 | 1.92 | 24.22 |
| 6" x 6" (PC66) | BC@.077 | Ea | 35.10 | 2.84 | 37.94 |
| 6" x 8" (PC68) | BC@.077 | Ea | 24.10 | 2.84 | 26.94 |
| 8" x 8" (PC88) | BC@.077 | Ea | 48.60 | 2.84 | 51.44 |
| **Post top tie or anchor (AC), 18 gauge, galvanized** | | | | | |
| 4" x 4" (AC4) | BC@.040 | Ea | 2.43 | 1.48 | 3.91 |
| 6" x 6" (AC6) | BC@.040 | Ea | 2.98 | 1.48 | 4.46 |
| **Purlin anchors (PA), 12 gauge, 2" wide** | | | | | |
| 18" anchor (PA18) | BC@.165 | Ea | 4.86 | 6.10 | 10.96 |
| 23" anchor (PA23) | BC@.165 | Ea | 5.67 | 6.10 | 11.77 |
| 28" anchor (PA28) | BC@.185 | Ea | 5.79 | 6.83 | 12.62 |
| 35" anchor (PA35) | BC@.185 | Ea | 7.95 | 6.83 | 14.78 |
| **Split ring connectors** | | | | | |
| 2-5/8" ring | BC@.047 | Ea | 1.78 | 1.74 | 3.52 |
| 4" ring | BC@.054 | Ea | 2.48 | 1.99 | 4.47 |
| **"L"-straps, 12 gauge (L)** | | | | | |
| 5" x 5", 1" wide (55L) | BC@.132 | Ea | 1.02 | 4.88 | 5.90 |
| 6" x 6", 1-1/2" wide (66L) | BC@.132 | Ea | 2.10 | 4.88 | 6.98 |
| 8" x 8", 2" wide (88L) | BC@.132 | Ea | 3.21 | 4.88 | 8.09 |
| 12" x 12", 2" wide (1212L) | BC@.165 | Ea | 4.26 | 6.10 | 10.36 |
| Add for heavy duty straps, 7 gauge | — | Ea | 5.80 | — | 5.80 |
| **Long straps (MST), 12 gauge** | | | | | |
| 27" (MST27) | BC@.165 | Ea | 4.16 | 6.10 | 10.26 |
| 37" (MST37) | BC@.165 | Ea | 5.94 | 6.10 | 12.04 |
| 48" (MST48) | BC@.265 | Ea | 8.23 | 9.79 | 18.02 |
| 60" (MST60) | BC@.265 | Ea | 12.70 | 9.79 | 22.49 |
| 72" (MST72) | BC@.285 | Ea | 16.80 | 10.50 | 27.30 |

| | Craft@Hrs | Unit | Material | Labor | Total |
|---|---|---|---|---|---|
| **Plate and stud straps (FHA), galvanized, 1-1/2" wide, 12 gauge** | | | | | |
| 6" (FHA6) | BC@.100 | Ea | 1.37 | 3.69 | 5.06 |
| 9" (FHA9) | BC@.132 | Ea | 1.38 | 4.88 | 6.26 |
| 12" (FHA12) | BC@.132 | Ea | 1.53 | 4.88 | 6.41 |
| 18" (FHA18) | BC@.132 | Ea | 2.04 | 4.88 | 6.92 |
| 24" (FHA24) | BC@.165 | Ea | 2.67 | 6.10 | 8.77 |
| 30" (FHA30) | BC@.165 | Ea | 4.42 | 6.10 | 10.52 |
| **Tie straps (ST)** | | | | | |
| 2-1/16" x 9-5/16" 20 ga. (ST292) | BC@.132 | Ea | .86 | 4.88 | 5.74 |
| 3/4" x 16-5/16" 20 ga. (ST2115) | BC@.046 | Ea | .81 | 1.70 | 2.51 |
| 2-1/16" x 16-5/16" 20 ga. (ST2215) | BC@.132 | Ea | 1.06 | 4.88 | 5.94 |
| 2-1/16" x 23-5/16" 16 ga. (ST6224) | BC@.165 | Ea | 2.35 | 6.10 | 8.45 |
| 2-1/16" x 33-13/16" 14 ga. (ST6236) | BC@.165 | Ea | 2.84 | 6.10 | 8.94 |
| **"T"-straps, 14 gauge (T)** | | | | | |
| 6" x 6", 1-1/2" wide (66T) | BC@.132 | Ea | 2.10 | 4.88 | 6.98 |
| 12" x 8", 2" wide (128T) | BC@.132 | Ea | 4.77 | 4.88 | 9.65 |
| 12" x 12", 2" wide (1212T) | BC@.132 | Ea | 7.13 | 4.88 | 12.01 |
| Add for heavy-duty straps, 7 gauge | — | Ea | 5.80 | — | 5.80 |
| **Twist straps (TS), 16 gauge, galvanized, 1-1/4" wide** | | | | | |
| 9" (TS9) | BC@.132 | Ea | .76 | 4.88 | 5.64 |
| 11-5/8" (TS12) | BC@.132 | Ea | 1.03 | 4.88 | 5.91 |
| 17-3/4" (TS18) | BC@.132 | Ea | 1.48 | 4.88 | 6.36 |
| 21-5/8" (TS22) | BC@.165 | Ea | 1.74 | 6.10 | 7.84 |
| **Wall braces (WB), diagonal, 1-1/4" wide, 16 gauge** | | | | | |
| 10' long, corrugated or flat (WB106C or WB106) | BC@.101 | Ea | 6.52 | 3.73 | 10.25 |
| 12' long, flat (WB126) | BC@.101 | Ea | 7.68 | 3.73 | 11.41 |
| 14' long, flat (WB143C) | BC@.101 | Ea | 13.80 | 3.73 | 17.53 |
| **Wall braces (CWB), diagonal, 18 gauge, 1" wide** | | | | | |
| 10' long, L-shaped (CWB106) | BC@.125 | Ea | 6.55 | 4.62 | 11.17 |
| 12' long, L-shaped (CWB126) | BC@.125 | Ea | 7.66 | 4.62 | 12.28 |

**Garage Door Openers** Radio controlled, electric, single or double door, includes typical electric hookup to existing adjacent 110 volt outlet.

| | Craft@Hrs | Unit | Material | Labor | Total |
|---|---|---|---|---|---|
| Screw drive, 1/2 HP | BC@3.90 | Ea | 183.00 | 144.00 | 327.00 |
| Security, belt drive, 1/2 HP | BC@3.90 | Ea | 324.00 | 144.00 | 468.00 |
| Security, chain drive, 1/2 HP | BC@3.90 | Ea | 258.00 | 144.00 | 402.00 |
| Add for additional transmitter | — | Ea | 28.80 | — | 28.80 |
| Add for operator reinforcement kit | — | Ea | 23.50 | — | 23.50 |
| Add for low headroom kit | — | Ea | 73.10 | — | 73.10 |
| Add for emergency release | — | Ea | 21.50 | — | 21.50 |
| Belt rail extension kit | — | Ea | 48.50 | — | 48.50 |
| Chain glide extension kit | — | Ea | 16.50 | — | 16.50 |
| Chain rail extension kit | — | Ea | 49.60 | — | 49.60 |
| Screw drive extension kit | — | Ea | 21.60 | — | 21.60 |

# Garage Doors

|  | Craft@Hrs | Unit | Material | Labor | Total |
|---|---|---|---|---|---|

**Garage Doors**

Steel rollup garage door. Torsion springs. 2-layer insulated construction. 7/8" insulation. Steel frame with overlapped section joints. With hardware and lock.

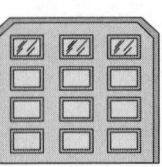

| | | | | | |
|---|---|---|---|---|---|
| 8' x 7' | B1@6.46 | Ea | 286.00 | 215.00 | 501.00 |
| 9' x 7' | B1@6.46 | Ea | 307.00 | 215.00 | 522.00 |
| 16' x 7' | B1@8.33 | Ea | 527.00 | 277.00 | 804.00 |

Raised panel steel rollup garage door. 2-layer construction. 7/8" thick polystyrene insulation. With hardware and lock.

| | | | | | |
|---|---|---|---|---|---|
| 8' x 7' | B1@6.46 | Ea | 307.00 | 215.00 | 522.00 |
| 9' x 7' | B1@6.46 | Ea | 328.00 | 215.00 | 543.00 |
| 16' x 7' | B1@8.33 | Ea | 568.00 | 277.00 | 845.00 |

Bonded steel garage door. 3-layer steel exterior. 2"-thick insulation. With torsion springs, hardware and lock.

| | | | | | |
|---|---|---|---|---|---|
| 8' x 7' | B1@6.46 | Ea | 445.00 | 215.00 | 660.00 |
| 8' x 7', molding face | B1@6.46 | Ea | 463.00 | 215.00 | 678.00 |
| 9' x 7' | B1@6.46 | Ea | 465.00 | 215.00 | 680.00 |
| 9' x 7', molding face | B1@6.46 | Ea | 484.00 | 215.00 | 699.00 |
| 16' x 7' | B1@6.46 | Ea | 725.00 | 215.00 | 940.00 |
| 16' x 7', molding face | B1@6.46 | Ea | 771.00 | 215.00 | 986.00 |
| 8' x 6'6", with glass window | B1@6.46 | Ea | 631.00 | 215.00 | 846.00 |
| 9' x 6'6", with glass window | B1@6.46 | Ea | 698.00 | 215.00 | 913.00 |
| 8' x 6'6", low headroom | B1@6.46 | Ea | 504.00 | 215.00 | 719.00 |
| 9' x 6'6", low headroom | B1@6.46 | Ea | 543.00 | 215.00 | 758.00 |

Non-insulated steel garage door. With rollup hardware and lock.

| | | | | | |
|---|---|---|---|---|---|
| 8' x 7' | B1@6.46 | Ea | 234.00 | 215.00 | 449.00 |
| 9' x 7' | B1@6.46 | Ea | 252.00 | 215.00 | 467.00 |
| 16' x 7' | B1@8.33 | Ea | 451.00 | 277.00 | 728.00 |

Wood garage doors. Unfinished, jamb hinge, with hardware and slide lock.

Plain panel uninsulated hardboard door

| | | | | | |
|---|---|---|---|---|---|
| 9' wide x 7' high | B1@4.42 | Ea | 309.00 | 147.00 | 456.00 |
| 16' wide x 7' high | B1@5.90 | Ea | 535.00 | 197.00 | 732.00 |

Styrofoam core hardboard door

| | | | | | |
|---|---|---|---|---|---|
| 9' wide x 7' high | B1@4.42 | Ea | 333.00 | 147.00 | 480.00 |
| 16' wide x 7' high | B1@5.90 | Ea | 586.00 | 197.00 | 783.00 |

Redwood door, raised panels

| | | | | | |
|---|---|---|---|---|---|
| 9' wide x 7' high | B1@4.42 | Ea | 614.00 | 147.00 | 761.00 |
| 16' wide x 7' high | B1@5.90 | Ea | 1,350.00 | 197.00 | 1,547.00 |

Hardboard door with two small lites

| | | | | | |
|---|---|---|---|---|---|
| 9' wide x 7' high | B1@4.42 | Ea | 339.00 | 147.00 | 486.00 |
| 16' wide x 7' high | B1@5.90 | Ea | 614.00 | 197.00 | 811.00 |

Hardboard door with sunburst lites

| | | | | | |
|---|---|---|---|---|---|
| 9' wide x 7' high | B1@4.42 | Ea | 493.00 | 147.00 | 640.00 |
| 16' wide x 7' high | B1@5.90 | Ea | 954.00 | 197.00 | 1,151.00 |

Hardboard door with cathedral-style lites

| | | | | | |
|---|---|---|---|---|---|
| 9' wide x 7' high | B1@4.42 | Ea | 436.00 | 147.00 | 583.00 |
| 16' wide x 7' high | B1@5.90 | Ea | 917.00 | 197.00 | 1,114.00 |

Garage door weather seal

| | | | | | |
|---|---|---|---|---|---|
| For 9' garage door | BC@.500 | Ea | 4.43 | 18.50 | 22.93 |
| For garage door to 18' wide | BC@1.00 | Ea | 7.18 | 36.90 | 44.08 |

| | Craft@Hrs | Unit | Material | Labor | Total |
|---|---|---|---|---|---|
| Garage door lock bar cylinder and lock assembly | | | | | |
|    Garage door lock | BC@.500 | Ea | 6.40 | 18.50 | 24.90 |
| "T"-handle for rollup garage door bar lock | | | | | |
|    Handle | BC@.250 | Ea | 4.48 | 9.24 | 13.72 |

## Gates, Ornamental Aluminum, Swing, Automatic

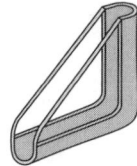

Welded aluminum driveway swing gates, double swing, arched up or down. Labor costs assume connection, but no wire runs. 2" square frame, 2" x 1-1/2" horizontal rails and 1" square pickets. Hoover Fence (J series, 4' to 5' high)

| | Craft@Hrs | Unit | Material | Labor | Total |
|---|---|---|---|---|---|
|    12' wide opening | BL@1.00 | Ea | 2,270.00 | 29.70 | 2,299.70 |
|    16' wide opening | BL@1.20 | Ea | 2,870.00 | 35.60 | 2,905.60 |
|    20' wide opening | BL@1.40 | Ea | 3,540.00 | 41.50 | 3,581.50 |
|    Add for custom width gates | — | % | 20.0 | — | — |
| Gate posts | | | | | |
|    4" x  9' post | BL@1.00 | Ea | 90.00 | 29.70 | 119.70 |
|    6" x 12' post | BL@1.00 | Ea | 218.00 | 29.70 | 247.70 |
|    Add for setting post in concrete | BL@.330 | Ea | 2.55 | 9.79 | 12.34 |
| 12 volt DC swing gate operator | BL@5.00 | Ea | 470.00 | 148.00 | 618.00 |
|    Dual gate unit | BL@5.00 | Ea | 748.00 | 148.00 | 896.00 |
|    Add for additional gate transmitters | — | Ea | 40.00 | — | 40.00 |
|    Add for low voltage wire | — | LF | .30 | — | .30 |

## Glass Material only. See labor costs below.

Cut window glass, clear DSB glass (double strength, B Grade)

| | Craft@Hrs | Unit | Material | Labor | Total |
|---|---|---|---|---|---|
|    8" x 10" | — | Ea | 1.67 | — | 1.67 |
|    10" x 12" | — | Ea | 2.25 | — | 2.25 |
|    11" x 14" | — | Ea | 2.69 | — | 2.69 |
|    12" x 16" | — | Ea | 3.42 | — | 3.42 |
|    12" x 36" | — | Ea | 6.11 | — | 6.11 |
|    16" x 20" | — | Ea | 4.77 | — | 4.77 |
|    18" x 36" | — | Ea | 9.10 | — | 9.10 |
|    24" x 30" | — | Ea | 10.10 | — | 10.10 |
|    24" x 36" | — | Ea | 12.00 | — | 12.00 |
|    30" x 36" | — | Ea | 15.20 | — | 15.20 |
|    36" x 48" | — | Ea | 27.90 | — | 27.90 |
| SSB glass (single strength, B Grade) | | | | | |
|    To 60" width plus length | — | SF | 2.02 | — | 2.02 |
|    Over 60" to 70" width plus length | — | SF | 2.51 | — | 2.51 |
|    Over 70" to 80" width plus length | — | SF | 2.22 | — | 2.22 |
| DSB glass (double strength, B Grade) | | | | | |
|    To 60" width plus length | — | SF | 2.83 | — | 2.83 |
|    Over 60" to 70" width plus length | — | SF | 2.87 | — | 2.87 |
|    Over 70" to 80" width plus length | — | SF | 2.91 | — | 2.91 |
|    Over 80" to 100" width plus length | — | SF | 3.12 | — | 3.12 |
| Non-glare glass | — | SF | 4.82 | — | 4.82 |
| Float glass, 1/4" thick | | | | | |
|    To 25 SF | — | SF | 3.16 | — | 3.16 |
|    25 to 35 SF | — | SF | 3.40 | — | 3.40 |
|    35 to 50 SF | — | SF | 3.69 | — | 3.69 |
| Insulating glass, clear | | | | | |
|    1/2" thick, 1/4" airspace, up to 25 SF | — | SF | 6.82 | — | 6.82 |
|    5/8" thick, 1/4" airspace, up to 35 SF | — | SF | 7.98 | — | 7.98 |
|    1" thick, 1/2" airspace, up to 50 SF | — | SF | 9.45 | — | 9.45 |

# Glass

| | Craft@Hrs | Unit | Material | Labor | Total |
|---|---|---|---|---|---|
| **Tempered glass, clear, 1/4" thick** | | | | | |
| To 25 SF | — | SF | 6.82 | — | 6.82 |
| 25 to 50 SF | — | SF | 8.17 | — | 8.17 |
| 50 to 75 SF | — | SF | 9.01 | — | 9.01 |
| **Tempered glass, clear, 3/16" thick** | | | | | |
| To 25 SF | — | SF | 6.72 | — | 6.72 |
| 25 to 50 SF | — | SF | 7.69 | — | 7.69 |
| 50 to 75 SF | — | SF | 8.51 | — | 8.51 |
| **Laminated safety glass, 1/4" thick, ASI safety rating** | | | | | |
| Up to 25 SF | — | SF | 8.76 | — | 8.76 |
| 25 to 35 SF | — | SF | 9.19 | — | 9.19 |
| **Polycarbonate sheet, .093" thick, clear** | | | | | |
| 8" x 10" | — | Ea | 5.00 | — | 5.00 |
| 12" x 24" | — | Ea | 11.80 | — | 11.80 |
| 18" x 24" | — | Ea | 18.80 | — | 18.80 |
| 36" x 48" | — | Ea | 83.60 | — | 83.60 |
| 36" x 72" | — | Ea | 103.00 | — | 103.00 |
| **Acrylic plastic sheet (Plexiglas), clear** | | | | | |
| 11" x 14" .093" thick | — | Ea | 3.52 | — | 3.52 |
| 12" x 24" .093" thick | — | Ea | 5.20 | — | 5.20 |
| 18" x 24" .093" thick | — | Ea | 10.10 | — | 10.10 |
| 18" x 24" .236" thick | — | Ea | 19.80 | — | 19.80 |
| 30" x 36" .093" thick | — | Ea | 21.40 | — | 21.40 |
| 30" x 36" .236" thick | — | Ea | 46.50 | — | 46.50 |
| 36" x 48" .093" thick | — | Ea | 33.20 | — | 33.20 |
| 36" x 72" .093" thick | — | Ea | 51.10 | — | 51.10 |
| 36" x 72" .236" thick | — | Ea | 106.00 | — | 106.00 |
| **Beveled edge mirrors, no frame included** | — | SF | 8.03 | — | 8.03 |
| **Labor setting glass in wood or metal sash with vinyl bed** | | | | | |
| To 15 SF, 1 hour charge | BG@1.77 | LS | — | 60.60 | 60.60 |
| 15 to 25 SF, 2 hour charge | BG@3.53 | LS | — | 121.00 | 121.00 |
| 25 to 35 SF, 3 hour charge | BG@5.30 | LS | — | 182.00 | 182.00 |
| **Labor to bevel edges, per linear inch** | | | | | |
| 1/2" beveled edge | BG@.013 | Inch | — | .45 | .45 |
| 1" beveled edge | BG@.063 | Inch | — | 2.16 | 2.16 |

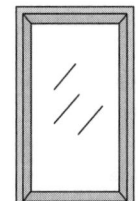

**Greenhouses, Sun Porches, Spa Enclosures** Freestanding or lean-to models. Anodized aluminum frame, acrylic glazed. Add the cost of excavation and concrete. Expect to see a wide variation in greenhouse costs for any given size.

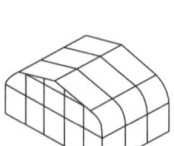

| Freestanding greenhouse enclosures, one door and four walls. 8' deep | Craft@Hrs | Unit | Material | Labor | Total |
|---|---|---|---|---|---|
| 7' high, 6' wide | B1@21.0 | Ea | 5,000.00 | 700.00 | 5,700.00 |
| 7-1/2' high, 9' wide | B1@33.0 | Ea | 6,240.00 | 1,100.00 | 7,340.00 |
| 8' high, 12' wide | B1@47.0 | Ea | 7,250.00 | 1,570.00 | 8,820.00 |
| 8' high, 15' wide | B1@59.0 | Ea | 8,470.00 | 1,970.00 | 10,440.00 |
| 9' high, 18' wide | B1@80.0 | Ea | 9,570.00 | 2,660.00 | 12,230.00 |
| Add for additional outswing door | — | Ea | 298.00 | — | 298.00 |
| Add for additional sliding window | — | Ea | 161.00 | — | 161.00 |

Lean-to greenhouse enclosures attached on long dimension to an existing structure, one door and three walls. 8' deep

| | Craft@Hrs | Unit | Material | Labor | Total |
|---|---|---|---|---|---|
| 7' high, 3' wide | B1@8.40 | Ea | 2,640.00 | 280.00 | 2,920.00 |
| 7' high, 6' wide | B1@16.8 | Ea | 3,980.00 | 560.00 | 4,540.00 |

| | Craft@Hrs | Unit | Material | Labor | Total |
|---|---|---|---|---|---|
| 7-1/2' high, 9' wide | B1@27.0 | Ea | 4,990.00 | 899.00 | 5,889.00 |
| 8' high, 12' wide | B1@38.4 | Ea | 6,050.00 | 1,280.00 | 7,330.00 |
| 8' high, 15' wide | B1@48.0 | Ea | 7,100.00 | 1,600.00 | 8,700.00 |
| Add for additional outswing door | — | Ea | 298.00 | — | 298.00 |
| Add for additional sliding window | — | Ea | 161.00 | — | 161.00 |

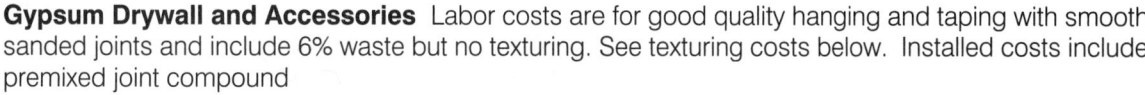

**Gutters and Downspouts**

Galvanized steel, 5" box type

| | Craft@Hrs | Unit | Material | Labor | Total |
|---|---|---|---|---|---|
| Gutter, 5", 10' lengths | SW@.070 | LF | 1.12 | 2.87 | 3.99 |
| Inside or outside miter corner | SW@.091 | Ea | 13.60 | 3.73 | 17.33 |
| Fascia support brackets | — | Ea | 4.37 | — | 4.37 |
| Spike and ferrule support, 7" | — | Ea | 1.44 | — | 1.44 |
| Gutter slip-joint connector | — | Ea | 4.61 | — | 4.61 |
| Add per 2" x 3" downspout, with drop outlet, 3 ells, strainer, 2 bands and 15" extension | | | | | |
| 10' downspout | SW@.540 | Ea | 63.40 | 22.20 | 85.60 |
| 20' downspout with coupling | SW@.810 | Ea | 83.40 | 33.20 | 116.60 |

Aluminum rain gutter, 5" box type, primed

| | Craft@Hrs | Unit | Material | Labor | Total |
|---|---|---|---|---|---|
| Gutter, 5", 10' lengths | SW@.070 | LF | .74 | 2.87 | 3.61 |
| Inside or outside corner | SW@.091 | Ea | 8.57 | 3.73 | 12.30 |
| Fascia support brackets | — | Ea | 2.38 | — | 2.38 |
| Gutter hanger and support | — | Ea | 1.78 | — | 1.78 |
| Section joint kit | — | Ea | 4.43 | — | 4.43 |

**Gypsum Drywall and Accessories** Labor costs are for good quality hanging and taping with smooth sanded joints and include 6% waste but no texturing. See texturing costs below. Installed costs include premixed joint compound
(5 gallons per 1,000 SF), 2" perforated joint tape (380 LF per 1,000 SF) and drywall screws
(4-1/2 pounds per 1,000 SF) at the following material prices.

| | Craft@Hrs | Unit | Material | Labor | Total |
|---|---|---|---|---|---|
| 1/4" plain board, per 4' x 8' sheet | — | Ea | 8.38 | — | 8.38 |
| 1/4" hi flex board, per 4' x 8' sheet | — | Ea | 11.10 | — | 11.10 |
| 3/8" plain board, per 4' x 8' sheet | — | Ea | 6.84 | — | 6.84 |
| 1/2" plain board, per 4' x 8' sheet | — | Ea | 7.33 | — | 7.33 |
| 1/2" plain board, per 4' x 9' sheet | — | Ea | 8.13 | — | 8.13 |
| 1/2" plain board, per 4' x 10' sheet | — | Ea | 8.49 | — | 8.49 |
| 1/2" plain board, per 4' x 12' sheet | — | Ea | 11.00 | — | 11.00 |
| 1/2" plain board, per 1/2" x 54" x 12' sheet | — | Ea | 15.80 | — | 15.80 |
| 1/2" water-resistant green board, per 4' x 8' sheet | — | Ea | 9.97 | — | 9.97 |
| 1/2" XP mold/moisture resistant, per 4' x 8' sheet | — | Ea | 11.00 | — | 11.00 |
| 1/2" mold-tough, per 4' x 12' sheet | — | Ea | 16.30 | — | 16.30 |
| 1/2" mold-tough, per 4' x 8' sheet | — | Ea | 11.30 | — | 11.30 |
| 5/8" fire-rated board, per 4' x 6' sheet | — | Ea | 6.56 | — | 6.56 |
| 5/8" fire-rated board, per 4' x 8' sheet | — | Ea | 8.28 | — | 8.28 |
| 5/8" fire-rated board, per 4' x 9' sheet | — | Ea | 13.80 | — | 13.80 |
| 5/8" fire-rated board, per 4' x 10' sheet | — | Ea | 10.90 | — | 10.90 |
| 5/8" fire-rated board, per 4' x 12' sheet | — | Ea | 12.60 | — | 12.60 |
| 5/8" XP mold/moisture resistant, per 4' x 8' sheet | — | Ea | 13.10 | — | 13.10 |
| 5/8" fire-rated, mold-tough, per 4' x 12' sheet | — | Ea | 14.80 | — | 14.80 |
| 5/8" foil-backed fire-rated X board, per 2' x 8' sheet | — | Ea | 7.08 | — | 7.08 |
| All-purpose joint compound, 5 gal | — | Ea | 14.20 | — | 14.20 |
| 2nd and 3rd coat joint finishing compound, 5 gal | — | Ea | 14.30 | — | 14.30 |
| Perforated joint tape, 500' roll | — | Ea | 3.41 | — | 3.41 |
| Self-adhesive fiberglass joint tape, 500' roll | — | Ea | 9.60 | — | 9.60 |
| Drywall screws, 5 pound box | — | Ea | 21.00 | — | 21.00 |

# Gypsum Drywall and Accessories

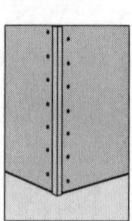

| | Craft@Hrs | Unit | Material | Labor | Total |
|---|---|---|---|---|---|
| **1/4" plain board** | | | | | |
| Walls | D1@.016 | SF | .29 | .57 | .86 |
| Ceilings | D1@.021 | SF | .29 | .75 | 1.04 |
| **3/8" plain board** | | | | | |
| Walls | D1@.017 | SF | .24 | .61 | .85 |
| Ceilings | D1@.023 | SF | .24 | .82 | 1.06 |
| **1/2" plain board** | | | | | |
| Walls | D1@.018 | SF | .26 | .64 | .90 |
| Ceilings | D1@.024 | SF | .26 | .86 | 1.12 |
| **1/2" moisture resistant green board** | | | | | |
| Walls | D1@.018 | SF | .35 | .64 | .99 |
| Ceilings | D1@.024 | SF | .39 | .86 | 1.25 |
| **1/2" fire-rated mold-tough** | | | | | |
| Walls | D1@.018 | SF | .43 | .64 | 1.07 |
| Ceilings | D1@.024 | SF | .43 | .86 | 1.29 |
| **5/8" fire-rated board** | | | | | |
| Walls | D1@.020 | SF | .29 | .71 | 1.00 |
| Ceilings | D1@.025 | SF | .29 | .89 | 1.18 |
| **Drywall casing and channel trim** | | | | | |
| Stop or casing or jamb casing | D1@.020 | LF | .26 | .71 | .97 |
| 7/8" hat channel | D1@.011 | LF | .62 | .39 | 1.01 |
| Narrow soundproofing channel, metal | D1@.011 | LF | .80 | .39 | 1.19 |
| Narrow soundproofing channel, PVC | D1@.011 | LF | 1.06 | .39 | 1.45 |
| Regular soundproofing channel, metal | D1@.011 | LF | .80 | .39 | 1.19 |
| Regular soundproofing channel, w/ padding tape | D1@.011 | LF | 1.08 | .39 | 1.47 |
| RC-1 resilient hat channel | D1@.011 | LF | .58 | .39 | .97 |
| RC-1 resilient hat channel, w/ padding tape | D1@.011 | LF | .88 | .39 | 1.27 |
| RC-2 resilient hat channel (for ceilings) | D1@.011 | LF | 1.10 | .39 | 1.49 |
| **Drywall arch and corner trim** | | | | | |
| Vinyl arch trim, 1-1/4", flexible | D1@.030 | LF | .22 | 1.07 | 1.29 |
| Vinyl outside corner, 1-1/4" | D1@.030 | LF | .22 | 1.07 | 1.29 |
| Corner guard, 3/4" x 3/4", nail on | D1@.030 | LF | .55 | 1.07 | 1.62 |
| Corner guard, 3/4" x 3/4", self-stick | D1@.030 | LF | .50 | 1.07 | 1.57 |
| Corner guard, bullnose, 2-1/2" x 2-1/2" | D1@.030 | LF | 2.38 | 1.07 | 3.45 |
| Bullnose, metal, 3/4" radius | D1@.030 | LF | .46 | 1.07 | 1.53 |
| Bullnose, plastic, 3/4" radius | D1@.030 | LF | .20 | 1.07 | 1.27 |
| Metal bullnose corner, 2-way | D1@.160 | Ea | 1.72 | 5.71 | 7.43 |
| Metal bullnose corner, 3-way | D1@.160 | Ea | 2.47 | 5.71 | 8.18 |
| Plastic bullnose corner, 2-way | D1@.160 | Ea | 1.30 | 5.71 | 7.01 |
| Plastic bullnose corner, 3-way | D1@.160 | Ea | 1.29 | 5.71 | 7.00 |
| **Metal corner bead, galvanized** | | | | | |
| Type "J" bead, 1/2" | D1@.016 | LF | .21 | .57 | .78 |
| Bullnose type corner, 1/2" | D1@.016 | LF | .47 | .57 | 1.04 |
| Bullnose type corner, 1-1/4" | D1@.016 | LF | .19 | .57 | .76 |
| Beadex casing B-9 | D1@.016 | LF | .22 | .57 | .79 |
| **Deduct for no taping or joint finishing** | | | | | |
| Walls | — | SF | -.04 | -.22 | -.26 |
| Ceilings | — | SF | -.04 | -.24 | -.28 |
| **Acoustic ceiling (popcorn), spray application** | | | | | |
| Light coverage | DT@.008 | SF | .05 | .29 | .34 |
| Medium coverage | DT@.011 | SF | .07 | .39 | .46 |
| Heavy coverage | DT@.016 | SF | .09 | .57 | .66 |
| Glitter application to acoustic ceiling texture | DT@.002 | SF | .01 | .07 | .08 |
| Remove acoustic ceiling texture | BL@.011 | SF | — | .33 | .33 |

| | Craft@Hrs | Unit | Material | Labor | Total |
|---|---|---|---|---|---|
| **Drywall texture** | | | | | |
| Light hand texture | DT@.005 | SF | .11 | .18 | .29 |
| Medium hand texture | DT@.006 | SF | .16 | .21 | .37 |
| Heavy hand texture | DT@.007 | SF | .20 | .25 | .45 |
| Skim coat, smooth | DT@.012 | SF | .13 | .43 | .56 |
| Knockdown, spray applied and troweled | DT@.020 | SF | .13 | .71 | .84 |

**Hardware** See also Bath Accessories, Door Chimes, Door Closers, Foundation Bolts, Garage Door Openers, Mailboxes, Nails, Screen Wire, Sheet Metal, Framing Connectors, and Weatherstripping. Rule of thumb: Cost of finish hardware (including door hardware, catches, stops, racks, brackets and pulls) per house.

| | Craft@Hrs | Unit | Material | Labor | Total |
|---|---|---|---|---|---|
| Low to medium cost house (1,600 SF) | BC@16.0 | LS | 1,120.00 | 591.00 | 1,711.00 |
| Better quality house (2,000 SF) | BC@20.0 | LS | 1,340.00 | 739.00 | 2,079.00 |

Keyed entry locksets. Meets ANSI Grade 2 standards for residential security. One-piece knob. Triple option faceplate fits square, radius-round, and drive-in door preps.

| | Craft@Hrs | Unit | Material | Labor | Total |
|---|---|---|---|---|---|
| Antique brass | BC@.250 | Ea | 29.90 | 9.24 | 39.14 |
| Antique pewter | BC@.250 | Ea | 33.30 | 9.24 | 42.54 |
| Bright brass | BC@.250 | Ea | 27.70 | 9.24 | 36.94 |
| Satin chrome | BC@.250 | Ea | 24.50 | 9.24 | 33.74 |

Lever locksets. Dual torque springs eliminate wobble.

| | Craft@Hrs | Unit | Material | Labor | Total |
|---|---|---|---|---|---|
| Antique brass | BC@.250 | Ea | 68.60 | 9.24 | 77.84 |
| Polished brass | BC@.250 | Ea | 66.10 | 9.24 | 75.34 |

Egg knob entry locksets

| | Craft@Hrs | Unit | Material | Labor | Total |
|---|---|---|---|---|---|
| Satin nickel finish | BC@.250 | Ea | 54.50 | 9.24 | 63.74 |
| Solid forged brass, polished brass finish | BC@.250 | Ea | 84.20 | 9.24 | 93.44 |

Colonial knob entry locksets

| | Craft@Hrs | Unit | Material | Labor | Total |
|---|---|---|---|---|---|
| Polished brass | BC@.250 | Ea | 79.40 | 9.24 | 88.64 |

Double cylinder high security lockset. Forged brass handle with thumb latch. Separate grade 1 security double cylinder deadbolt.

| | Craft@Hrs | Unit | Material | Labor | Total |
|---|---|---|---|---|---|
| Chrome finish | BC@.820 | Ea | 175.00 | 30.30 | 205.30 |

Construction keying system (minimum order of 20 keyed entry locks)

| | Craft@Hrs | Unit | Material | Labor | Total |
|---|---|---|---|---|---|
| Additional cost per lockset | — | Ea | 6.15 | — | 6.15 |

Deadbolt and lockset combo. Includes keyed alike entry knob and grade 2 security deadbolt with 4 keys. Single cylinder.

| | Craft@Hrs | Unit | Material | Labor | Total |
|---|---|---|---|---|---|
| Polished brass finish | BC@.750 | Ea | 50.60 | 27.70 | 78.30 |

Mortise lockset and deadbolt. Steel reinforced cast 1" throw deadbolt. Solid brass latch. Rocker switch locks and unlocks exterior knob. Labor includes mortising of door.

| | Craft@Hrs | Unit | Material | Labor | Total |
|---|---|---|---|---|---|
| Left or right hand throw | BC@1.05 | Ea | 177.00 | 38.80 | 215.80 |

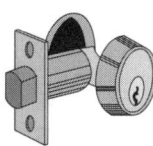

Deadbolts. All metal and brass. ANSI Grade 2 rating. 1" throw deadbolt. Includes 2 keys. Polished brass.

| | Craft@Hrs | Unit | Material | Labor | Total |
|---|---|---|---|---|---|
| Single cylinder | BC@.500 | Ea | 23.20 | 18.50 | 41.70 |
| Double cylinder | BC@.500 | Ea | 33.20 | 18.50 | 51.70 |

Knob passage latch

| | Craft@Hrs | Unit | Material | Labor | Total |
|---|---|---|---|---|---|
| Dummy, brass | BC@.250 | Ea | 14.10 | 9.24 | 23.34 |
| Dummy, satin nickel | BC@.250 | Ea | 15.60 | 9.24 | 24.84 |
| Hall or closet, brass | BC@.250 | Ea | 27.40 | 9.24 | 36.64 |
| Hall or closet, satin nickel | BC@.250 | Ea | 32.80 | 9.24 | 42.04 |

Egg knob privacy latchset

| | Craft@Hrs | Unit | Material | Labor | Total |
|---|---|---|---|---|---|
| Bright brass | BC@.250 | Ea | 29.30 | 9.24 | 38.54 |
| Satin nickel | BC@.250 | Ea | 36.30 | 9.24 | 45.54 |

Privacy lever latchset. Both levers locked or unlocked by turn button. Outside lever can be unlocked by emergency key (included).

| | Craft@Hrs | Unit | Material | Labor | Total |
|---|---|---|---|---|---|
| Rubbed bronze | BC@.250 | Ea | 22.30 | 9.24 | 31.54 |
| Polished brass | BC@.250 | Ea | 31.80 | 9.24 | 41.04 |
| Satin nickel | BC@.250 | Ea | 41.20 | 9.24 | 50.44 |

# Hardware

| | Craft@Hrs | Unit | Material | Labor | Total |
|---|---|---|---|---|---|
| **Door knockers. Polished brass.** | | | | | |
| Apartment type, brass, with viewer | BC@.250 | Ea | 17.00 | 9.24 | 26.24 |
| 4" x 8-1/2", ornate style | BC@.250 | Ea | 43.90 | 9.24 | 53.14 |
| **Door pull plates** | | | | | |
| Brass finish, 4" x 16" | BC@.333 | Ea | 22.20 | 12.30 | 34.50 |
| **Door push plates** | | | | | |
| Aluminum, 3" x 12" | BC@.333 | Ea | 5.44 | 12.30 | 17.74 |
| Solid brass, 4" x 16" | BC@.333 | Ea | 11.20 | 12.30 | 23.50 |
| **Door kick plates** | | | | | |
| 6" x 30", brass | BC@.333 | Ea | 47.20 | 12.30 | 59.50 |
| 8" x 34", brass | BC@.333 | Ea | 53.80 | 12.30 | 66.10 |
| 8" x 34", magnetic, for metal door, brass | BC@.333 | Ea | 29.10 | 12.30 | 41.40 |
| 8" x 34", satin nickel | BC@.333 | Ea | 37.10 | 12.30 | 49.40 |
| **Door stops** | | | | | |
| Spring type wall bumper, 3" long | BC@.116 | Ea | 2.01 | 4.29 | 6.30 |
| Wall bumper, 3" long, brass | BC@.116 | Ea | 4.48 | 4.29 | 8.77 |
| Floor mounted stop, satin nickel | BC@.333 | Ea | 5.65 | 12.30 | 17.95 |
| Hinge pin door stop, brass | BC@.250 | Ea | 5.98 | 9.24 | 15.22 |
| Kick down door holder, brass | BC@.250 | Ea | 9.29 | 9.24 | 18.53 |
| Magnetic wall stop | BC@.333 | Ea | 30.70 | 12.30 | 43.00 |
| Adhesive rubber cup knob strike | BC@.116 | Ea | 2.16 | 4.29 | 6.45 |
| **Door peep sights** | | | | | |
| 180-degree viewing angle | BC@.258 | Ea | 10.40 | 9.53 | 19.93 |
| **Pocket door frame and track.** Fully assembled. Designed to hold 150-pound standard 80" door. Door sold separately. | | | | | |
| 24" x 80" | BC@1.00 | Ea | 64.40 | 36.90 | 101.30 |
| 28" x 80" | BC@1.00 | Ea | 65.30 | 36.90 | 102.20 |
| 30" x 80" | BC@1.00 | Ea | 66.50 | 36.90 | 103.40 |
| 32" x 80" | BC@1.00 | Ea | 67.90 | 36.90 | 104.80 |
| 36" x 80" | BC@1.00 | Ea | 69.20 | 36.90 | 106.10 |
| **Pocket door edge pulls** | | | | | |
| 2-1/8" round brass | BC@.300 | Ea | 3.66 | 11.10 | 14.76 |
| 2-1/8" round brass dummy | BC@.300 | Ea | 5.43 | 11.10 | 16.53 |
| 1-3/8" x 3" oval brass | BC@.300 | Ea | 3.34 | 11.10 | 14.44 |
| 1-3/8" x 3" rectangular brass | BC@.300 | Ea | 3.36 | 11.10 | 14.46 |
| 3/4" round brass | BC@.300 | Ea | 2.34 | 11.10 | 13.44 |
| Brass plate | BC@.300 | Ea | 4.58 | 11.10 | 15.68 |
| **Ball bearing pocket door hangers** | | | | | |
| Hanger | BC@.175 | Ea | 10.10 | 6.46 | 16.56 |
| **Converging pocket door hardware kit.** For double pocket doors. | | | | | |
| Door kit | — | Ea | 4.28 | — | 4.28 |
| **Pocket door hangers** | | | | | |
| 3/8" offset, one wheel | BC@.175 | Ea | 2.92 | 6.46 | 9.38 |
| 1/16" offset, one wheel | BC@.175 | Ea | 2.92 | 6.46 | 9.38 |
| 7/8" offset, two wheel | BC@.175 | Ea | 4.45 | 6.46 | 10.91 |

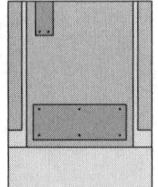

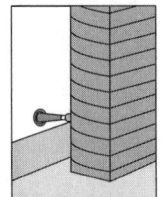

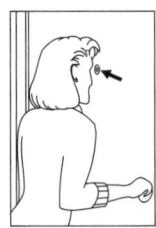

| | Craft@Hrs | Unit | Material | Labor | Total |
|---|---|---|---|---|---|

Touch bar exit hardware. For use with 1-3/4"-thick hollow metal and wood doors. Extruded anodized aluminum with stainless steel spring. 3/4" throw stainless steel latch bolts. Strike for 5/8" stop; shim for 1/2" stop; 2-3/4" backset, 4-1/4" minimum stile. Includes standard mounting with sheet metal and machine screws. Recommended mounting height: 40-5/16" from ceiling to finished floor. 36" maximum width. Rim-type cylinders with Schlage "C" keyway. Includes measure and mark, drill holes, assemble, set strike plate.

| | Craft@Hrs | Unit | Material | Labor | Total |
|---|---|---|---|---|---|
| Install, single door | BC@1.45 | Ea | 112.00 | 53.60 | 165.60 |
| Install, per set of double doors | BC@2.97 | Ea | 223.00 | 110.00 | 333.00 |
| Add for monitor switch for electrically operated locks | | | | | |
| Per lock | — | Ea | 67.10 | — | 67.10 |

**Electromagnetic security and life safety lock system** Surface mounted on door frame. Stainless steel finish. Includes magnet, strike plate, mounting fasteners and template. Costs include connection. Add for wire runs. Magnalock™.

| | Craft@Hrs | Unit | Material | Labor | Total |
|---|---|---|---|---|---|
| 600 lb holding force 12/24 VDC | E4@2.00 | Ea | 349.00 | 98.30 | 447.30 |
| 1,200 lb holding force 12/24 VDC | E4@2.00 | Ea | 441.00 | 98.30 | 539.30 |
| 1,800 lb holding force 12/24 VDC | E4@2.00 | Ea | 626.00 | 98.30 | 724.30 |
| LCD programmable timer, 12/24 AC/DC | E4@.500 | Ea | 123.00 | 24.60 | 147.60 |
| Exit delay timer, regulated DC voltage | E4@2.50 | Ea | 226.00 | 123.00 | 349.00 |
| Exit delay system documentation and pressure sensitive vinyl door label | | | | | |
| for NFPA 101 requirements | — | Ea | 20.50 | — | 20.50 |
| Tamper-resistant plunger sex bolt | — | Ea | 158.00 | — | 158.00 |
| Energy absorbing sex bolt | — | Ea | 267.00 | — | 267.00 |
| Replacement strike plate and fasteners | — | Ea | 120.00 | — | 120.00 |
| Field mounted stainless steel housing. Includes cover, end caps, | | | | | |
| mounting plate and fasteners | BC@1.50 | Ea | 564.00 | 55.40 | 619.40 |

## Hinges

Full mortise hinge, heavy duty, satin brass finish

| | Craft@Hrs | Unit | Material | Labor | Total |
|---|---|---|---|---|---|
| 3-1/2" x 3-1/2", 6 screws, .123 gauge | BC@.333 | Ea | 6.99 | 12.30 | 19.29 |
| 4" x 4", 8 screws, .130 gauge | BC@.333 | Ea | 11.90 | 12.30 | 24.20 |
| 4" x 4", spring hinge | BC@.333 | Ea | 15.00 | 12.30 | 27.30 |

Full mortise hinge, light duty, satin brass finish

| | Craft@Hrs | Unit | Material | Labor | Total |
|---|---|---|---|---|---|
| 3-1/2" x 3-1/2", loose pin | BC@.333 | Ea | 2.13 | 12.30 | 14.43 |
| 4" x 4", loose pin | BC@.333 | Ea | 2.55 | 12.30 | 14.85 |

Screen door hinge kit, 2 hinges, door pull and hook

| | Craft@Hrs | Unit | Material | Labor | Total |
|---|---|---|---|---|---|
| Bright zinc | BC@.750 | Ea | 4.33 | 27.70 | 32.03 |

House street numbers, per number

| | Craft@Hrs | Unit | Material | Labor | Total |
|---|---|---|---|---|---|
| 4" high, nail-on, aluminum | BC@.103 | Ea | 1.38 | 3.80 | 5.18 |
| 4" high, traditional brass | BC@.103 | Ea | 3.20 | 3.80 | 7.00 |
| 4" high, contemporary bronze | BC@.103 | Ea | 4.13 | 3.80 | 7.93 |
| 4" high, stainless steel | BC@.103 | Ea | 6.05 | 3.80 | 9.85 |
| 5" high, solid brass | BC@.103 | Ea | 12.20 | 3.80 | 16.00 |

Garage door hardware, jamb type, 7' height, with springs and slide bolt

| | Craft@Hrs | Unit | Material | Labor | Total |
|---|---|---|---|---|---|
| 8' to 10' wide door | BC@4.36 | Ea | 132.00 | 161.00 | 293.00 |
| 10' to 16' wide door, with truss | BC@5.50 | Ea | 154.00 | 203.00 | 357.00 |
| Truss assembly for 16' door | BC@.493 | Ea | 47.40 | 18.20 | 65.60 |

Garage door hardware, track type, 7' height, with springs and lock

| | Craft@Hrs | Unit | Material | Labor | Total |
|---|---|---|---|---|---|
| 7' to 9' wide door | BC@4.36 | Ea | 265.00 | 161.00 | 426.00 |
| 10' to 16' double track | BC@5.50 | Ea | 457.00 | 203.00 | 660.00 |

# Heating and Controls, Hydronic

| | Craft@Hrs | Unit | Material | Labor | Total |
|---|---|---|---|---|---|

**Hydronic Heating and Controls** A complete system will consist of a hydronic boiler, cast iron radiators or hydronic radiation baseboard heaters, boiler controls and accessories. This type system is engineered and designed to meet specific heating requirements. Once these design requirements are known, the following components may be used to estimate the cost of a complete installed system. For pipe, fittings, hangers and insulation costs, see the Commercial and Industrial section. For more detailed coverage, see the *National Plumbing & HVAC Estimator*, http://CraftsmanSiteLicense.com.

Hydronic boilers. U.S. Green Council and ASME "H" stamp certified to a rating of 95% AFUE (annual fuel usage efficiency) with EPA Tier II Nox emissions compliant burner, Kunkel safety valve, low-water cutoff, digital panel controls and automatic level controls. Weil-Mclain UE-3 series or equal.

| | Craft@Hrs | Unit | Material | Labor | Total |
|---|---|---|---|---|---|
| 37,500 Btu with standard gas pilot | P1@3.00 | Ea | 1,680.00 | 107.00 | 1,787.00 |
| 62,000 Btu with standard gas pilot | P1@3.50 | Ea | 2,000.00 | 125.00 | 2,125.00 |
| 96,000 Btu with standard gas pilot | P1@3.50 | Ea | 2,200.00 | 125.00 | 2,325.00 |
| 130,000 Btu with standard gas pilot | P1@4.00 | Ea | 2,530.00 | 143.00 | 2,673.00 |
| 164,000 Btu with standard gas pilot | P1@4.00 | Ea | 2,890.00 | 143.00 | 3,033.00 |
| 198,000 Btu with standard gas pilot | P1@4.00 | Ea | 3,310.00 | 143.00 | 3,453.00 |
| Add for electronic ignition | — | Ea | 320.00 | — | 320.00 |

LEED certification for hydronic boilers. Requires 88% boiler efficiency, recording controls, compliance with SCAQMD 1146.2 emission standards and zone thermostats.

| | Craft@Hrs | Unit | Material | Labor | Total |
|---|---|---|---|---|---|
| Add for a LEED certified boiler with USGBC rating | — | Ea | 15.0% | — | — |
| Add for LEED certification, per project | — | Ea | — | — | 2,000.00 |
| Add for central performance monitor/recorder | — | Ea | 1,100.00 | — | 1,100.00 |
| Add for zone recording thermostat | — | Ea | 150.00 | — | 150.00 |

Air eliminator-purger, threaded connections

| | Craft@Hrs | Unit | Material | Labor | Total |
|---|---|---|---|---|---|
| 3/4", cast iron, steam & water, 150 PSI | P1@.250 | Ea | 407.00 | 8.94 | 415.94 |
| 3/4", cast iron, 150 PSI water | P1@.250 | Ea | 131.00 | 8.94 | 139.94 |
| 1", cast iron, 150 PSI water | P1@.250 | Ea | 151.00 | 8.94 | 159.94 |
| 3/8", brass, 125 PSI steam | P1@.250 | Ea | 107.00 | 8.94 | 115.94 |
| 1/2", cast iron, 250 PSI steam | P1@.250 | Ea | 222.00 | 8.94 | 230.94 |
| 3/4", cast iron, 250 PSI steam | P1@.250 | Ea | 279.00 | 8.94 | 287.94 |
| Airtrol fitting, 3/4" | P1@.300 | Ea | 47.50 | 10.70 | 58.20 |
| Air eliminator vents | P1@.150 | Ea | 16.10 | 5.36 | 21.46 |

Atmospheric (anti-siphon) vacuum breakers, rough brass, threaded

| | Craft@Hrs | Unit | Material | Labor | Total |
|---|---|---|---|---|---|
| 1/2" vacuum breaker | P1@.210 | Ea | 23.30 | 7.51 | 30.81 |
| 3/4" vacuum breaker | P1@.250 | Ea | 25.00 | 8.94 | 33.94 |
| 1" vacuum breaker | P1@.300 | Ea | 39.70 | 10.70 | 50.40 |
| 1-1/4" vacuum breaker | P1@.400 | Ea | 65.80 | 14.30 | 80.10 |
| 1-1/2" vacuum breaker | P1@.450 | Ea | 77.50 | 16.10 | 93.60 |
| 2" vacuum breaker | P1@.500 | Ea | 116.00 | 17.90 | 133.90 |

Circuit balancing valves, bronze, threaded

| | Craft@Hrs | Unit | Material | Labor | Total |
|---|---|---|---|---|---|
| 1/2" circuit balancing valve | P1@.210 | Ea | 66.70 | 7.51 | 74.21 |
| 3/4" circuit balancing valve | P1@.250 | Ea | 70.80 | 8.94 | 79.74 |
| 1" circuit balancing valve | P1@.300 | Ea | 82.20 | 10.70 | 92.90 |
| 1-1/2" circuit balancing valve | P1@.450 | Ea | 129.00 | 16.10 | 145.10 |

Baseboard fin tube radiation, per linear foot, copper tube with aluminum fins, wall mounted

| | Craft@Hrs | Unit | Material | Labor | Total |
|---|---|---|---|---|---|
| 1/2" element and 8" cover | P1@.280 | LF | 10.20 | 10.00 | 20.20 |
| 3/4" element and 9"cover | P1@.280 | LF | 16.70 | 10.00 | 26.70 |
| 3/4" element, 10" cover, high capacity | P1@.320 | LF | 26.20 | 11.40 | 37.60 |
| 3/4" fin tube element only | P1@.140 | LF | 6.62 | 5.01 | 11.63 |
| 1" fin tube element only | P1@.150 | LF | 7.94 | 5.36 | 13.30 |
| 8" cover only | P1@.140 | LF | 6.17 | 5.01 | 11.18 |
| 9" cover only | P1@.140 | LF | 7.50 | 5.01 | 12.51 |
| 10" cover only | P1@.160 | LF | 8.21 | 5.72 | 13.93 |

| | Craft@Hrs | Unit | Material | Labor | Total |
|---|---|---|---|---|---|
| Add for pipe connection and control valve | P1@2.15 | Ea | 158.00 | 76.90 | 234.90 |
| Add for corners, fillers and caps, average per foot | — | % | 10.0 | — | — |
| **Flow check valves, brass, threaded, horizontal type** | | | | | |
| 3/4" check valve | P1@.250 | Ea | 12.70 | 8.94 | 21.64 |
| 1" check valve | P1@.300 | Ea | 16.80 | 10.70 | 27.50 |
| **Thermostatic mixing valves** | | | | | |
| 1/2", soldered | P1@.240 | Ea | 41.50 | 8.58 | 50.08 |
| 3/4", threaded | P1@.250 | Ea | 53.60 | 8.94 | 62.54 |
| 3/4", soldered | P1@.300 | Ea | 45.70 | 10.70 | 56.40 |
| 1", threaded | P1@.300 | Ea | 209.00 | 10.70 | 219.70 |
| **Liquid level gauges** | | | | | |
| 1/2", 175 PSI bronze | P1@.210 | Ea | 74.70 | 7.51 | 82.21 |
| **Wye pattern strainers, threaded, Class 125, bronze body valves** | | | | | |
| 3/4" strainer | P1@.260 | Ea | 36.40 | 9.30 | 45.70 |
| 1" strainer | P1@.330 | Ea | 44.60 | 11.80 | 56.40 |
| 1-1/4" strainer | P1@.440 | Ea | 62.30 | 15.70 | 78.00 |
| 1-1/2" strainer | P1@.495 | Ea | 93.80 | 17.70 | 111.50 |
| 2" strainer | P1@.550 | Ea | 162.00 | 19.70 | 181.70 |
| **Boiler expansion tank** | | | | | |
| 2.1 gallon | P1@.300 | Ea | 32.59 | 10.70 | 43.29 |
| 4.5 gallon | P1@.300 | Ea | 54.32 | 10.70 | 65.02 |

Installation of packaged hydronic boilers. Add the cost of control wiring, supply lines (electric, feedwater and gas), drain line, circulating pump, vent stack, permits, final inspection and rental of an appliance dolly ($14 per day), come-a-long ($16 per day), a ½ ton chain hoist ($21 per day) or a forklift, if required.

| | Craft@Hrs | Unit | Material | Labor | Total |
|---|---|---|---|---|---|
| Form and pour a 4' x 4' interior slab | CF@3.00 | Ea | 97.00 | 105.00 | 202.00 |
| Place 4' x 4' x 1/2" vibration pads | CF@.750 | Ea | 38.80 | 26.30 | 65.10 |
| Bolt down boiler | P1@1.00 | Ea | 35.00 | 35.80 | 70.80 |
| Connect gas and feedwater lines | P1@2.50 | Ea | — | 89.40 | 89.40 |
| Mount interior boiler drain | P1@.500 | Ea | 7.76 | 17.90 | 25.66 |
| Bore pipe hole in basement wall | P1@.250 | Ea | — | 8.94 | 8.94 |
| Bore burner stack vent in exterior wall | P1@.250 | Ea | — | 8.94 | 8.94 |
| Mount and edge-seal stack | P1@.500 | Ea | 40.00 | 17.90 | 57.90 |
| Mount circulating pump | P1@.450 | Ea | — | 16.10 | 16.10 |

**Hydronic radiant floor heating systems**  Includes PEX (cross-linked polyethylene) tube, the circulating pump, boiler, expansion tank, manifold, return piping, hangers, insulation, thermostatic controls, service fill and drain fittings, anti-crimp pipe support spacers, system commissioning and startup. Add the cost of control wiring, gas and electric connection, permit and post-installation inspection. By rated boiler capacity. Typically, 35,000 Btu per hour in boiler capacity will heat a 2,500 SF home.

Embedded system. 5/8" tube set 12" O.C. in a concrete slab. Includes 1" extruded polystyrene insulation under interior areas and 2" extruded polystyrene insulation at the building perimeter, forms, supports, mesh reinforcing and the concrete.

| | Craft@Hrs | Unit | Material | Labor | Total |
|---|---|---|---|---|---|
| 35,000 Btu per hour (2,500 SF of floor) | P1@120. | Ea | 5,460.00 | 4,290.00 | 9,750.00 |
| 70,000 Btu per hour (5,000 SF of floor) | P1@200. | Ea | 8,650.00 | 7,150.00 | 15,800.00 |
| 105,000 Btu per hour (7,500 SF of floor) | P1@280. | Ea | 13,400.00 | 10,000.00 | 23,400.00 |

Thin-slab system. 5/8" tube fastened 12" O.C. to the top of a wood-frame subfloor installed by others and then covered with 1-1/2" plasticized concrete.

| | Craft@Hrs | Unit | Material | Labor | Total |
|---|---|---|---|---|---|
| 35,000 Btu per hour (2,500 SF of floor) | P1@160. | Ea | 4,530.00 | 5,720.00 | 10,250.00 |
| 70,000 Btu per hour (5,000 SF of floor) | P1@300. | Ea | 7,730.00 | 10,700.00 | 18,430.00 |
| 105,000 Btu per hour (7,500 SF of floor) | P1@400. | Ea | 10,300.00 | 14,300.00 | 24,600.00 |
| Add per SF of floor for a poured gypsum cover | — | SF | 2.06 | — | 2.06 |
| Add for installation in an existing structure | — | % | — | — | 80.0 |

| | Craft@Hrs | Unit | Material | Labor | Total |
|---|---|---|---|---|---|

Plate-type system. 3/4" tube fastened 8" O.C. with aluminum heat transfer plates either above or below a wood-frame subfloor installed by others.

| | Craft@Hrs | Unit | Material | Labor | Total |
|---|---|---|---|---|---|
| 35,000 Btu per hour (2,500 SF of floor) | P1@100. | Ea | 3,910.00 | 3,580.00 | 7,490.00 |
| 70,000 Btu per hour (5,000 SF of floor) | P1@175. | Ea | 6,700.00 | 6,260.00 | 12,960.00 |
| 105,000 Btu per hour (7,500 SF of floor) | P1@240. | Ea | 9,790.00 | 8,580.00 | 18,370.00 |
| Add for installation in an existing structure | — | % | — | — | 50.0 |

**Hybrid central heating and cooling** High air velocity and narrow duct type. Attic or basement mounted modular system. Cooling is with a R22 refrigerant-charged A/C unit. Heating is with an electric hot water heating system. Rated at 13 SEER (Seasonal Energy Efficiency Rating). Provides precision zone control and low humidity for retrofit projects required to meet high energy-efficiency performance standards. Includes automatic mode switchover, individual zone digital thermostats, digital main controller module, transition-mounted duct fans and motor drive, hot water coils and a circulating pump mounted in the air handling unit. Designed for HVAC upgrades of historic buildings, condo conversions and for meeting stringent indoor air quality and occupant comfort standards. See central air filter systems in the section that follows. Add the cost of a hot water expansion tank. Add the cost of equipment rental, if required: an appliance dolly ($14 per day), a 3,000 lb. come-a-long ($16 per day) and a 1-ton chain hoist ($21 per day). Mount and bolt down hybrid A/C unit, hot water coils and air handler enclosure

| | Craft@Hrs | Unit | Material | Labor | Total |
|---|---|---|---|---|---|
| 2.0-ton | P1@.750 | Ea | 6,000.00 | 26.80 | 6,026.80 |
| 3.0-ton | P1@.750 | Ea | 7,100.00 | 26.80 | 7,126.80 |
| 3.5-ton | P1@.750 | Ea | 7,300.00 | 26.80 | 7,326.80 |
| 4.0-ton | P1@.750 | Ea | 8,500.00 | 26.80 | 8,526.80 |
| 5.0-ton | P1@.750 | Ea | 9,500.00 | 26.80 | 9,526.80 |

Installation for hybrid central heating

| | Craft@Hrs | Unit | Material | Labor | Total |
|---|---|---|---|---|---|
| Form and pour a 4' x 4' interior slab | CF@3.00 | Ea | 97.00 | 105.00 | 202.00 |
| Reinforce joists in an attic | B1@3.00 | LS | 450.00 | 99.90 | 549.90 |
| Place two vibration pads | CF@.750 | LS | 40.00 | 26.30 | 66.30 |
| Core-drill and mount A/C drain | P1@.150 | LF | 3.00 | 5.36 | 8.36 |
| Mount air mover in air handler duct | P1@.120 | Ea | — | 3.58 | 3.58 |
| Connect ductwork mains to manifold | P1@4.50 | Ea | 150.00 | 161.00 | 311.00 |
| Cut and braze copper A/C coil piping and run condensate drain | P1@4.00 | Ea | 8.00 | 143.00 | 151.00 |
| Mount high-velocity ductwork and silencer tubing | P1@0.10 | LF | 17.50 | 3.58 | 21.08 |
| Digital zone control thermostat, manual | BE@.750 | Ea | 125.00 | 29.70 | 154.70 |
| Digital master heat/cool controller unit with auto change-over | BE@.750 | Ea | 300.00 | 29.70 | 329.70 |
| Wiring of controls, typical | BE@3.00 | Ea | 1.50 | 119.00 | 120.50 |
| Calibration and test | BE@2.00 | Ea | — | 79.20 | 79.20 |

**Electric Heating** Labor includes installation and connecting only. Add the cost of wiring from the Commercial and Industrial section of this manual.

Electric baseboard heaters, convection type, surface mounted, 20 gauge steel. Labor includes installation and connecting only. Add the cost of wiring. 7" high, 3-1/4" deep, 3.41 Btu/hr/watt, 240 volt. Low or medium density.

| | Craft@Hrs | Unit | Material | Labor | Total |
|---|---|---|---|---|---|
| 2'6" long, 500 watt (1,700 Btu) | BE@1.70 | Ea | 30.80 | 67.30 | 98.10 |
| 3'0" long, 750 watt (2,225 Btu) | BE@1.81 | Ea | 36.10 | 71.70 | 107.80 |
| 4'0" long, 1,000 watt (3,400 Btu) | BE@1.81 | Ea | 42.40 | 71.70 | 114.10 |
| 6'0" long, 1,500 watt (5,100 Btu) | BE@1.89 | Ea | 57.10 | 74.80 | 131.90 |

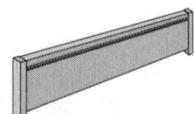

| | Craft@Hrs | Unit | Material | Labor | Total |
|---|---|---|---|---|---|
| 8'0" long, 2,000 watt (6,800 Btu) | BE@2.04 | Ea | 63.60 | 80.80 | 144.40 |
| Add for integral thermostat | BE@.094 | Ea | 20.80 | 3.72 | 24.52 |
| Add for wall mounted thermostat | BE@1.62 | Ea | 27.90 | 64.10 | 92.00 |

Fan-forced electric wall heaters, 12-3/4" x 6" rough opening. Add for 240 volt wiring and thermostat.

| | Craft@Hrs | Unit | Material | Labor | Total |
|---|---|---|---|---|---|
| 700/900/1600 watt, 50 CFM | BE@.650 | Ea | 148.00 | 25.70 | 173.70 |
| 500 watt, 6.25 amp, 5118 Btu, 83 CFM | BE@.650 | Ea | 116.00 | 25.70 | 141.70 |
| 1000 watt, 4.17 amp, 3412 Btu, 83 CFM | BE@.650 | Ea | 116.00 | 25.70 | 141.70 |
| 2000 watt, 8.33 amp, 6824 Btu, 60 CFM | BE@.650 | Ea | 148.00 | 25.70 | 173.70 |
| 2000 watt, 8.33 amp, 6824 Btu, 83 CFM | BE@.650 | Ea | 116.00 | 25.70 | 141.70 |
| 3000 watt,12.5 amp, 12,236 Btu, 65 CFM | BE@.650 | Ea | 163.00 | 25.70 | 188.70 |
| Double pole thermostat, 120 or 240 volt | BE@.440 | Ea | 24.90 | 17.40 | 42.30 |

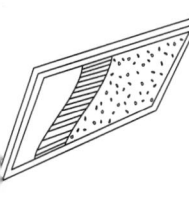

Surface mounted 1" thick radiant ceiling panels.120/208/240/277 volts AC.

| | Craft@Hrs | Unit | Material | Labor | Total |
|---|---|---|---|---|---|
| 250 watts, 24" x 24" | BE@.850 | Ea | 190.00 | 33.70 | 223.70 |
| 310 watts, 24" x 24" | BE@.850 | Ea | 205.00 | 33.70 | 238.70 |
| 375 watts, 24" x 24" | BE@.850 | Ea | 212.00 | 33.70 | 245.70 |
| 625 watts, 24" x 48" | BE@1.11 | Ea | 322.00 | 43.90 | 365.90 |
| 750 watts, 24" x 48" | BE@1.31 | Ea | 342.00 | 51.90 | 393.90 |
| Add for radiant line voltage wall thermostat | BE@.450 | Ea | 65.00 | 17.80 | 82.80 |
| Add for motion sensing setback thermostat | BE@.450 | Ea | 82.40 | 17.80 | 100.20 |

Floor units, drop-in type, 14" long x 7-1/4" wide, includes fan motor assembly, housing and grille

| | Craft@Hrs | Unit | Material | Labor | Total |
|---|---|---|---|---|---|
| 120 volts, 375 or 750 watts | BE@1.76 | Ea | 281.00 | 69.70 | 350.70 |
| 277 volts, 750 watts | BE@1.76 | Ea | 345.00 | 69.70 | 414.70 |
| Add for wall thermostat kit | BE@1.62 | Ea | 66.10 | 64.10 | 130.20 |
| Add for concrete accessory kit and housing | — | LS | 21.00 | — | 21.00 |

Infrared quartz tube heaters, indoor or outdoor, chromed guard, polished aluminum reflector. Includes modulating control.

| | Craft@Hrs | Unit | Material | Labor | Total |
|---|---|---|---|---|---|
| 3.2 KW, 24" long, 240 volts | BE@4.14 | Ea | 306.00 | 164.00 | 470.00 |
| 5 KW, 33" long, 240 volts | BE@4.14 | Ea | 404.00 | 164.00 | 568.00 |
| 7.3 KW, 46" long, 480 volts | BE@4.14 | Ea | 508.00 | 164.00 | 672.00 |
| 10.95 KW, 46" long, 480 volts | BE@4.14 | Ea | 843.00 | 164.00 | 1,007.00 |
| Add for wall or ceiling brackets (pair) | — | LS | 12.60 | — | 12.60 |
| Deduct for indoor use only | — | % | -30.0 | — | — |

Suspension blower heaters, 208 or 240 volts, single phase, propeller type, direct drive

| | Craft@Hrs | Unit | Material | Labor | Total |
|---|---|---|---|---|---|
| 1/4 HP | BE@4.80 | Ea | 335.00 | 190.00 | 525.00 |
| 1/3 HP | BE@4.80 | Ea | 390.00 | 190.00 | 580.00 |
| 1/2 HP | BE@4.80 | Ea | 545.00 | 190.00 | 735.00 |
| Add for wall mounting bracket | — | LS | 18.40 | — | 18.40 |
| Add for line voltage wall thermostat | | | | | |
| Single pole | — | LS | 30.00 | — | 30.00 |
| Double pole | — | LS | 45.00 | — | 45.00 |

Wall mounted fan-forced downflow insert heaters, heavy duty, built-in thermostat, 240 volts, 14" wide, 20" high, 4" deep rough-in, 3.41 Btu per hr per watt

| | Craft@Hrs | Unit | Material | Labor | Total |
|---|---|---|---|---|---|
| 1,500 watts | BE@3.95 | Ea | 158.00 | 156.00 | 314.00 |
| 2,000 watts | BE@3.95 | Ea | 172.00 | 156.00 | 328.00 |
| 3,000 watts | BE@3.95 | Ea | 180.00 | 156.00 | 336.00 |
| 4,500 watts | BE@3.95 | Ea | 218.00 | 156.00 | 374.00 |
| Add for surface mounting kit | — | LS | 18.50 | — | 18.50 |

# Heating System Accessories

| | Craft@Hrs | Unit | Material | Labor | Total |
|---|---|---|---|---|---|

**Central air filter for hybrid central heating system**  Drop-in module to fit in a high-velocity narrow duct air handling unit.  Complies with Mil-STD 282 (certified removal of 99% of dust particles over .3 microns in size) as well as ASHRAE specifications.  For residences requiring hypoallergenic air quality, medical clinics and pharmaceutical manufacturing/test facilities.  DOP filter (di-octyl phthalate) is a non-woven denier nylon/cellulose acetate composite with an activated carbon final layer. Optional Dwyer-type pressure differential switch and relay provides automatic shutdown of the air handling unit and a warning light when filter elements are clogged. Pressure drop across the filter is 1" to 2" W.C. (water column). Sized by the nominal tonnage of associated cooling/heating equipment.
Central air filter

| | Craft@Hrs | Unit | Material | Labor | Total |
|---|---|---|---|---|---|
| 2.5-ton unit, drop in filter | P1@.100 | Ea | 535.00 | 3.58 | 538.58 |
| 3.0-ton unit, drop in filter | P1@.100 | Ea | 565.00 | 3.58 | 568.58 |
| 3.5-ton unit, drop in filter | P1@.100 | Ea | 565.00 | 3.58 | 568.58 |
| 4.0-ton unit, drop in filter | P1@.100 | Ea | 597.00 | 3.58 | 600.58 |
| 5.0-ton unit, drop in filter | P1@.100 | Ea | 597.00 | 3.58 | 600.58 |
| Mount Dwyer switch on the air handler | BE@.500 | Ea | 150.00 | 19.80 | 169.80 |
| Wiring of Dwyer switch | BE@1.00 | Ea | 50.00 | 39.60 | 89.60 |
| Calibration and test | BE@.500 | Ea | — | 19.80 | 19.80 |

**Gas Furnaces**  Add the cost of piping, electrical connection and vent. See pipe costs in the Commercial and Industrial section of this manual. For more detailed coverage of gas heating, see the *National Plumbing & HVAC Estimator*, http://CraftsmanSiteLicense.com.
Gravity gas wall furnace. Natural gas or propane fired.

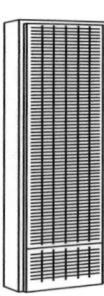

| | Craft@Hrs | Unit | Material | Labor | Total |
|---|---|---|---|---|---|
| 14,000 Btu | PM@4.67 | Ea | 484.00 | 195.00 | 679.00 |
| 30,000 Btu | PM@4.67 | Ea | 632.00 | 195.00 | 827.00 |
| 50,000 Btu | PM@4.67 | Ea | 680.00 | 195.00 | 875.00 |
| Wall register kit for 2nd room | PM@1.00 | Ea | 51.80 | 41.90 | 93.70 |
| Add for wall thermostat | PM@.972 | Ea | 32.60 | 40.70 | 73.30 |
| Add for blower | PM@.500 | Ea | 148.00 | 20.90 | 168.90 |
| Add for free standing vent trim kit | PM@.254 | LS | 193.00 | 10.60 | 203.60 |
| Add for gas valve for thermostat | PM@.250 | Ea | 109.00 | 10.50 | 119.50 |

Direct vent gas wall furnace. Includes wall thermostat. No electricity required. Includes vent kit to outside wall. By input Btu rating. Natural gas or propane fired.

| | Craft@Hrs | Unit | Material | Labor | Total |
|---|---|---|---|---|---|
| 10,000 Btu | PM@2.98 | Ea | 439.00 | 125.00 | 564.00 |
| 14,000 Btu | PM@2.98 | Ea | 484.00 | 125.00 | 609.00 |
| 22,000 Btu | PM@2.98 | Ea | 572.00 | 125.00 | 697.00 |
| 22,000 Btu, high altitude | PM@2.98 | Ea | 542.00 | 125.00 | 667.00 |
| 25,000 Btu | PM@2.98 | Ea | 569.00 | 125.00 | 694.00 |
| 30,000 Btu | PM@2.98 | Ea | 623.00 | 125.00 | 748.00 |
| 35,000 Btu | PM@2.98 | Ea | 651.00 | 125.00 | 776.00 |
| Add for blower | PM@.500 | Ea | 106.00 | 20.90 | 126.90 |
| Add for gas valve for thermostat | PM@.250 | Ea | 108.00 | 10.50 | 118.50 |

Gas floor furnace, including vent at $80, valve and wall thermostat

| | Craft@Hrs | Unit | Material | Labor | Total |
|---|---|---|---|---|---|
| 30 MBtu input, 24 MBtu output | PM@7.13 | Ea | 891.00 | 298.00 | 1,189.00 |
| 50 MBtu input, 35 MBtu output | PM@8.01 | Ea | 1,020.00 | 335.00 | 1,355.00 |
| Add for furnace serving two rooms | — | % | 30.0 | — | — |
| Add for spark ignition | — | LS | 59.10 | — | 59.10 |

Forced air upflow gas furnace with electronic ignition. Includes vent at $100, typical piping, at $50, thermostat at $40 and distribution plenum at $100. No ducting included.

| | Craft@Hrs | Unit | Material | Labor | Total |
|---|---|---|---|---|---|
| 50 MBtu input,  48 MBtu output | PM@7.13 | Ea | 2,900.00 | 298.00 | 3,198.00 |
| 75 MBtu input,  70 MBtu output | PM@7.13 | Ea | 3,050.00 | 298.00 | 3,348.00 |
| 100 MBtu input,  92 MBtu output | PM@7.80 | Ea | 3,240.00 | 326.00 | 3,566.00 |
| 125 MBtu input, 113 MBtu output | PM@7.80 | Ea | 3,450.00 | 326.00 | 3,776.00 |
| Add for liquid propane models | — | LS | 296.00 | — | 296.00 |
| Add for time control thermostat | — | Ea | 98.50 | — | 98.50 |

| | Craft@Hrs | Unit | Material | Labor | Total |
|---|---|---|---|---|---|

Ductwork for forced air furnace. Sheet metal, 6 duct and 2 air return residential system, with typical elbows and attaching boots.

| | Craft@Hrs | Unit | Material | Labor | Total |
|---|---|---|---|---|---|
| 4" x 12" registers, and 2" R-5 insulation | SW@22.1 | LS | 937.00 | 907.00 | 1,844.00 |
| Add for each additional duct run | SW@4.20 | Ea | 201.00 | 172.00 | 373.00 |
| Fiberglass flex duct, insulated, 8" dia. | SW@.012 | LF | 3.90 | .49 | 4.39 |

Suspended space heater, fan-forced, with tubular aluminum heat exchanger, mounting brackets and gas valve. Includes vent to 30' at $100, typical gas piping, at $50, and thermostat, at $40.

| | Craft@Hrs | Unit | Material | Labor | Total |
|---|---|---|---|---|---|
| 30 MBtu input, 24 MBtu output | PM@2.75 | Ea | 840.00 | 115.00 | 955.00 |
| 45 MBtu input, 36 MBtu output | PM@2.75 | Ea | 840.00 | 115.00 | 955.00 |
| 60 MBtu input, 48 MBtu output | PM@2.75 | Ea | 900.00 | 115.00 | 1,015.00 |
| 75 MBtu input, 60 MBtu output | PM@2.75 | Ea | 996.00 | 115.00 | 1,111.00 |

**Evaporative coolers,** for roof installation. Add $110 for an electrical switch and connection and $90 for a recirculation pump and mounting hardware.

| | Craft@Hrs | Unit | Material | Labor | Total |
|---|---|---|---|---|---|
| 3,000 CFM, down draft | PM@11.1 | Ea | 354.00 | 465.00 | 819.00 |
| 3,000 CFM, side draft | PM@11.1 | Ea | 353.00 | 465.00 | 818.00 |
| 4,000 CFM, side draft | PM@11.1 | Ea | 416.00 | 465.00 | 881.00 |
| 4,300 CFM, down draft | PM@11.1 | Ea | 446.00 | 465.00 | 911.00 |
| 5,500 CFM, down draft | PM@11.1 | Ea | 528.00 | 465.00 | 993.00 |
| 5,500 CFM, side draft | PM@11.1 | Ea | 354.00 | 465.00 | 819.00 |

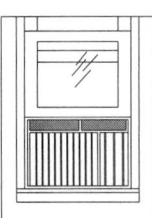

**Window type air conditioners** High-efficiency units, adjustable thermostat. Labor is for installation in an existing opening, connected to an existing electrical outlet.

| | Craft@Hrs | Unit | Material | Labor | Total |
|---|---|---|---|---|---|
| 5,000 Btu | PM@.751 | Ea | 219.00 | 31.40 | 250.40 |
| 8,000 Btu | PM@.751 | Ea | 319.00 | 31.40 | 350.40 |
| 10,000 Btu | PM@.751 | Ea | 356.00 | 31.40 | 387.40 |
| 15,000 Btu | PM@.751 | Ea | 521.00 | 31.40 | 552.40 |
| 25,000 Btu | PM@.751 | Ea | 665.00 | 31.40 | 696.40 |

**Packaged split system electric heat pumps** Window-wall units with galvanized steel cabinet and 35' of tubing, two thermostats and control package. Installed on concrete pad. Add for the electrical connection and concrete work.

| | Craft@Hrs | Unit | Material | Labor | Total |
|---|---|---|---|---|---|
| 8 MBtu cooling, 5 Mbtu heating | PM@10.9 | Ea | 1,110.00 | 456.00 | 1,566.00 |
| 15 MBtu cooling, 13 Mbtu heating | PM@10.9 | Ea | 1,590.00 | 456.00 | 2,046.00 |
| 18 Mbtu cooling, 16 Mbtu heating | PM@12.9 | Ea | 1,790.00 | 540.00 | 2,330.00 |
| Add for electrical connections, typical | PM@14.0 | LS | 231.00 | 586.00 | 817.00 |

**Residential air-to-air heat pump** Forced-air-aspirated heat pump, split system, slab mounted. Rated in nominal tonnage. Agency rating ARI and CSA. Installed in an existing air handling unit and with an optional filtration system. See the sections that follow. Trane or equal, rated at 13 SEER (Seasonal Energy Efficiency Rating), with automatic reversing valve for switching between heating and cooling modes, semiconductor thermocouple sensor for monitoring indoor/outdoor temperature differential, interior thermostat, controller relay, digital controller module, centrifugal compressor and motor drive. 208/230 V, 60 Hz, single phase. Requires two fan-aspirated pre-charged R22 refrigerant evaporator coils, one interior in the air handling unit and one outside in a weatherproof enclosure. See the air handler and evaporator coils that follow. Add, if required, for the cost of an appliance dolly ($14 per day), 3,000 lb. come-a-long (16 per day) and 1-ton chain hoist ($34 per day).
Mount and bolt interior heat pump unit in air handler enclosure

| | Craft@Hrs | Unit | Material | Labor | Total |
|---|---|---|---|---|---|
| 2.0-ton unit | P1@.750 | Ea | 2,100.00 | 26.80 | 2,126.80 |
| 3.0-ton unit | P1@.750 | Ea | 2,300.00 | 26.80 | 2,326.80 |
| 3.5-ton unit | P1@.750 | Ea | 2,800.00 | 26.80 | 2,826.80 |
| 4.0-ton unit | P1@.750 | Ea | 3,000.00 | 26.80 | 3,026.80 |
| 5.0-ton unit | P1@.750 | Ea | 3,500.00 | 26.80 | 3,526.80 |

# Heat Pumps

| | Craft@Hrs | Unit | Material | Labor | Total |
|---|---|---|---|---|---|
| Place 4' x 4' x 1/2" vibration pads | CF@.750 | Ea | 40.00 | 26.30 | 66.30 |
| Mount interior evaporator drain | P1@.500 | Ea | 8.00 | 17.90 | 25.90 |
| Bore piping hole through external coil | P1@.250 | Ea | — | 8.94 | 8.94 |
| Mount and bolt down the external coil | P1@.450 | Ea | 30.00 | 16.10 | 46.10 |
| Mount interior evaporator coil in air handler duct | P1@1.20 | Ea | — | 42.90 | 42.90 |
| Tie-in of connecting ductwork to house air ductwork manifold and exhaust | P1@4.50 | Ea | 150.00 | 161.00 | 311.00 |
| Cut and braze copper evaporator coil piping & drain | P1@4.00 | Ea | 8.00 | 143.00 | 151.00 |
| Wiring of controls | BE@3.00 | Ea | 1.50 | 119.00 | 120.50 |
| Calibration and test | BE@2.00 | Ea | — | 79.20 | 79.20 |
| Heat pump accessories | | | | | |
| Fossil fuel kit | P1@2.50 | Ea | 240.00 | 89.40 | 329.40 |
| High pressure cut-out kit | P1@.500 | Ea | 55.00 | 17.90 | 72.90 |
| Outdoor stat/Low ambient sensors | BE@.500 | Ea | 60.00 | 19.80 | 79.80 |
| Short cycle protection | BE@.500 | Ea | 75.00 | 19.80 | 94.80 |
| Digital heat/cool thermostat, manual | BE@.750 | Ea | 125.00 | 29.70 | 154.70 |
| Digital, programmable heat/cool thermostat with auto change-over | BE@.750 | Ea | 300.00 | 29.70 | 329.70 |
| Heat pump supplemental electric heating coil | | | | | |
| 5 KW, 208/230 Volt, 1-phase, 60 Hz | BE@1.50 | Ea | 140.00 | 59.40 | 199.40 |
| 7.5 KW 208/230 Volt, 1-phase, 60 Hz | BE@1.75 | Ea | 160.00 | 69.30 | 229.30 |
| 10 KW 208/230 Volt, 1-phase, 60 Hz | BE@2.00 | Ea | 175.00 | 79.20 | 254.20 |
| 12.5 KW, 208/230 Volt, 1-phase, 60 Hz | BE@2.25 | Ea | 250.00 | 89.10 | 339.10 |
| 15 KW, 208/230 Volt, 1-phase, 60 Hz | BE@2.75 | Ea | 350.00 | 109.00 | 459.00 |
| 20 KW, 208/230 Volt, 1-phase, 60 Hz | BE@3.50 | Ea | 470.00 | 139.00 | 609.00 |
| 25 KW 208/230 Volt, 1-phase, 60 Hz | BE@4.25 | Ea | 525.00 | 168.00 | 693.00 |

**LEED Certified Heating and Cooling** (Leadership in Energy & Environmental Design) offers an incentive to building owners to install energy-efficient and environmentally sensitive heating and cooling equipment. The HVAC system can earn a maximum of 17 points toward LEED certification by meeting ASHRAE 90 standards. A LEED certified air conditioning system must also meet ASHRAE 62.1-2004 standards for indoor air quality and include recording controls to measure system performance.

| | Craft@Hrs | Unit | Material | Labor | Total |
|---|---|---|---|---|---|
| Add for LEED certified residential systems with USGBC rating | — | Ea | 20.0% | — | — |
| Add for LEED certification, registration and inspection, per project | — | Ea | 2,000.00 | — | 2,000.00 |
| Add for LEED central digital performance monitor/recorder, ComfortView 3 or equal | — | Ea | 2,350.00 | — | 2,350.00 |
| Add for LEED certified zone recording thermostat, USB remote-ready | — | Ea | 350.00 | — | 350.00 |
| Add for IAQ (Indoor Air Quality) $CO_2$ sensor and duct-mounted aspirator box | — | Ea | 425.00 | — | 425.00 |

| | Craft@Hrs | Unit | Material | Labor | Total |
|---|---|---|---|---|---|

**Residential central air handling unit** Two-piece 16-gauge steel sheet metal enclosure with main inlet and exhaust ductwork tie-ins and inlet dust filter mount. With 208/230 volt single-phase 60 Hz variable-speed fan, relay and control and breaker box. Designed to accommodate an optional evaporator coil, "drop-in" oil or gas heater to fit an existing integral air-to-air heat exchanger, electric heater coil or hot water/steam coil and filter (not included). See the sections that follow. Add for LEED certification if required.

Mount and bolt down air handler

| | Craft@Hrs | Unit | Material | Labor | Total |
|---|---|---|---|---|---|
| 2.0-ton | P1@.750 | Ea | 1,100.00 | 26.80 | 1,126.80 |
| 3.0-ton | P1@.750 | Ea | 1,300.00 | 26.80 | 1,326.80 |
| 3.5-ton | P1@.750 | Ea | 1,500.00 | 26.80 | 1,526.80 |
| 4.0-ton | P1@.750 | Ea | 1,600.00 | 26.80 | 1,626.80 |
| 5.0-ton | P1@.750 | Ea | 1,800.00 | 26.80 | 1,826.80 |
| Mount internal evaporator coil in An air handler | P1@1.20 | Ea | 50.00 | 42.90 | 92.90 |
| Tie-in of connecting interior ductwork to house air ductwork manifold and exhaust | P1@4.50 | Ea | 100.00 | 161.00 | 261.00 |
| Cut and braze copper evaporator coil piping and drain | P1@4.00 | Ea | 8.00 | 143.00 | 151.00 |
| Wiring of controls | BE@3.00 | Ea | 1.50 | 119.00 | 120.50 |
| Calibration and test | BE@2.00 | Ea | — | 79.20 | 79.20 |

**Evaporator coils for a residential air handler** Uncased, pre-charged with R22 refrigerant, with flanged mounting base plate, connector tubing, refrigerant fill valve and check valve, twin-finned radiator baffles. For use in conjunction with heat pump units, to be mounted indoors in air handler or air handler ductwork; cased units are to be used for outdoor mounting. Sized by the associated nominal tonnage of the heat pump. Add for LEED certification if required.

| | Craft@Hrs | Unit | Material | Labor | Total |
|---|---|---|---|---|---|
| Form and pour a 4' x 4' x 2" exterior slab | CF@2.50 | Ea | 100.00 | 87.60 | 187.60 |
| Form and pour a 4' x 4' x 2" interior slab | CF@3.00 | Ea | 100.00 | 105.00 | 205.00 |
| Place two 4' x 4' x 1/2" (vibration pads) | CF@.300 | Ea | 50.00 | 10.50 | 60.50 |
| Mount interior evaporator drain | P1@.500 | Ea | 8.00 | 17.90 | 25.90 |
| Bore piping hole through exterior wall | P1@.250 | Ea | — | 8.94 | 8.94 |
| Mount external evaporator coil | | | | | |
| 2.0-ton | P1@.750 | Ea | 450.00 | 26.80 | 476.80 |
| 3.0-ton | P1@.750 | Ea | 480.00 | 26.80 | 506.80 |
| 3.5-ton | P1@.750 | Ea | 520.00 | 26.80 | 546.80 |
| 4.0-ton | P1@.750 | Ea | 700.00 | 26.80 | 726.80 |
| 5.0-ton | P1@.750 | Ea | 800.00 | 26.80 | 826.80 |
| Mount interior evaporator coil in air handler duct | | | | | |
| 2.0-ton | P1@1.75 | Ea | 400.00 | 62.60 | 462.60 |
| 3.0-ton | P1@1.75 | Ea | 450.00 | 62.60 | 512.60 |
| 3.5-ton | P1@1.75 | Ea | 500.00 | 62.60 | 562.60 |
| 4.0-ton | P1@1.75 | Ea | 550.00 | 62.60 | 612.60 |
| 5.0-ton | P1@1.75 | Ea | 600.00 | 62.60 | 662.60 |
| Tie-in of connecting interior ductwork manifold and exhaust | P1@4.50 | Ea | 100.00 | 161.00 | 261.00 |
| Cutting and brazing of copper evaporator coil piping and drain | P1@4.00 | Ea | 8.00 | 143.00 | 151.00 |
| Wiring of controls | BE@3.00 | Ea | 1.50 | 119.00 | 120.50 |
| Calibration and test | BE@2.00 | Ea | — | 79.20 | 79.20 |

# Insulation

| | Craft@Hrs | Unit | Material | Labor | Equipment | Total |
|---|---|---|---|---|---|---|

**Central air particulate removal filter system** Residential or light commercial DOP (di-octyl phthalate) type. Provided as drop-in module to manually fit into an air handling units. Complies with Mil-STD 282 (certified removal of 99% of dust particles over .3 microns in size) as well as ASHRAE specifications. For homes, medical clinics and pharmaceutical manufacturing/test facilities. DOP filter type is non-woven denier nylon/cellulose acetate composite with activated carbon final layer. Optional Dwyer-type pressure differential switch and relay provides for automatic shutdown of air handling unit and triggering of signal light when filter elements are clogged with particulates and require cleaning or replacement. Pressure drop across filter is 1" W.C. (water column) to 2" W.C. Sized according to nominal tonnage of associated cooling/heating capacity.

| | Craft@Hrs | Unit | Material | Labor | Equipment | Total |
|---|---|---|---|---|---|---|
| Dropping in of filter | | | | | | |
| 2.5-ton filter, | P1@.100 | Ea | 700.00 | 3.58 | — | 703.58 |
| 3.0-ton filter | P1@.100 | Ea | 800.00 | 3.58 | — | 803.58 |
| 3.5-ton filter | P1@.100 | Ea | 850.00 | 3.58 | — | 853.58 |
| 4.0-ton filter | P1@.100 | Ea | 900.00 | 3.58 | — | 903.58 |
| 5.0-ton filter | P1@.100 | Ea | 950.00 | 3.58 | — | 953.58 |
| Mount Dwyer switch on required air handler | BE@.500 | Ea | 150.00 | 19.80 | — | 169.80 |
| Wiring of Dwyer switch | BE@1.00 | Ea | 50.00 | 39.60 | — | 89.60 |
| Calibrate and test | BE@.500 | Ea | — | 19.80 | — | 19.80 |

| | Craft@Hrs | Unit | Material | Labor | Total |
|---|---|---|---|---|---|

**Insulation** See also Building Paper and Polyethylene Film. Coverage allows for studs and joists.
Fiberglass insulation, wall and ceiling application, coverage allows for framing

| | Craft@Hrs | Unit | Material | Labor | Total |
|---|---|---|---|---|---|
| Kraft-faced, 16" OC framing members | | | | | |
| 3-1/2" (R-11), 171 SF roll at $50.60 | BC@.007 | SF | .30 | .26 | .56 |
| 3-1/2" (R-13), 116 SF roll at $30.70 | BC@.007 | SF | .26 | .26 | .52 |
| 3-1/2" (R-15), 30 SF roll at $10.40 | BC@.007 | SF | .35 | .26 | .61 |
| 10" (R-30), 53 SF batts at $41.80 | BC@.008 | SF | .78 | .30 | 1.08 |
| 12" (R-38), 43 SF batts at $40.20 | BC@.008 | SF | .94 | .30 | 1.24 |
| Kraft-faced, 24" OC framing members | | | | | |
| 3-1/2" (R-15), 30 SF roll at $10.70 | BC@.005 | SF | .35 | .18 | .53 |
| 6-1/4" (R-19), 119 SF batts at $56.20 | BC@.005 | SF | .47 | .18 | .65 |
| 10" (R-30), 80 SF batts at $63.50 | BC@.006 | SF | .79 | .22 | 1.01 |
| 12" (R-38), 48 SF batts at $43.40 | BC@.006 | SF | .90 | .22 | 1.12 |
| Unfaced, 16" OC framing members | | | | | |
| 3-1/2" (R-13), 116 SF roll at $43.30 | BC@.004 | SF | .37 | .15 | .52 |
| 6-1/4" (R-19), 77.5 SF batts at $33.30 | BC@.005 | SF | .43 | .18 | .61 |
| 10" (R-30), 58 SF batts at $46.00 | BC@.006 | SF | .78 | .22 | 1.00 |
| Unfaced, 24" OC framing members | | | | | |
| 23" (R-30), 88 SF batts at $69.10 | BC@.003 | SF | .79 | .11 | .90 |
| 23" (R-30), 48 SF roll at $38.00 | BC@.003 | SF | .79 | .11 | .90 |
| Poly-faced insulation, vapor retarding twisted Miraflex fibers, ASTM C665, Type II, Class C, coverage allows for framing | | | | | |
| 3-1/2" (R-13), 16" or 24" OC framing | BC@.005 | SF | .42 | .18 | .60 |
| 8-3/4" (R-25), 16" or 24" OC framing | BC@.006 | SF | .63 | .22 | .85 |
| Encapsulated fiberglass roll insulation, Johns Manville Comfort Therm™, coverage allows for framing | | | | | |
| 3-1/2" (R-13) rolls, 16" OC framing | BC@.005 | SF | .32 | .18 | .50 |
| 6-1/4" (R-19) batts, 16" or 24" OC framing | BC@.006 | SF | .56 | .22 | .78 |
| 8-1/4" (R-25) rolls, 16" or 24" OC framing | BC@.006 | SF | .77 | .22 | .99 |
| 10-1/4" (R-30) batts, 16" or 24" OC framing | BC@.006 | SF | .84 | .22 | 1.06 |

# Insulation

| | Craft@Hrs | Unit | Material | Labor | Total |
|---|---|---|---|---|---|

Insulation board, Owens Corning Foamular XAE rigid extruded polystyrene foam board, film-faced, stapled in place, including taped joints, 4' x 8' or 9' panels, tongue and groove or square edge, including 5% waste

| | | | | | |
|---|---|---|---|---|---|
| 1/2" thick (R-2.5), $10.10 per 4' x 8' panel | BC@.010 | SF | .31 | .37 | .68 |
| 3/4" thick (R-3.8), $12.40 per 4' x 8' panel | BC@.011 | SF | .39 | .41 | .80 |
| 1" thick (R-5.0), $16.80 per 4 x 8 panel | BC@.011 | SF | .52 | .41 | .93 |
| 1-1/2" thick (R-7.5), $17.70 per 4' x 8' panel | BC@.015 | SF | .56 | .55 | 1.11 |
| 2" thick (R-10), $24.90 per 4' x 8' panel | BC@.015 | SF | .78 | .55 | 1.33 |

Extruded polystyrene insulation panel, Dow Blue Board, water resistant. By thermal resistance value, (R-). Density of 1.6 lbs. per CF. Gray Board density is 1.3 lbs. per CF.

| | | | | | |
|---|---|---|---|---|---|
| 3/4" x 2' x 8', tongue & groove, R-3.8 | BC@.011 | SF | .40 | .41 | .81 |
| 1" x 2' x 8', R-5.0 | BC@.011 | SF | .55 | .41 | .96 |
| 1" x 2' x 8', tongue & groove, R-5.0 | BC@.011 | SF | .34 | .41 | .75 |
| 1" x 4' x 8', R-5.0 | BC@.011 | SF | .52 | .41 | .93 |
| 1" x 4' x 8', tongue & groove, R-5.0 | BC@.011 | SF | .48 | .41 | .89 |
| 1-1/2" x 2' x 8', R-7.5 | BC@.015 | SF | .77 | .55 | 1.32 |
| 1-1/2" x 2' x 8', tongue & groove, R-7.5 | BC@.015 | SF | .85 | .55 | 1.40 |
| 1-1/2" 4' x 8', square edge, R-7.5 | BC@.015 | SF | .89 | .55 | 1.44 |
| 2" x 2' x 8', R-10 | BC@.015 | SF | 1.26 | .55 | 1.81 |
| 2" x 4' x 8', R-10 | BC@.015 | SF | 1.06 | .55 | 1.61 |
| 1" x 2' x 8' Gray Board (1.3 lbs. per CF) | BC@.015 | SF | .50 | .55 | 1.05 |
| 1-1/2" x 2' x 8' Gray Board | BC@.015 | SF | .71 | .55 | 1.26 |
| 2" x 2' x 8' Gray Board | BC@.015 | SF | .91 | .55 | 1.46 |

Rigid polyisocyanurate insulated sheathing, foil face on two sides, Dow/Celotex Tuff-R, 4' x 8' panels, including 5% waste

| | | | | | |
|---|---|---|---|---|---|
| 1/2" (R-3.3) | BC@.010 | SF | .29 | .37 | .66 |
| 3/4" (R-5.6) | BC@.011 | SF | .39 | .41 | .80 |
| 1" (R-7.2) | BC@.011 | SF | .55 | .41 | .96 |
| 2" (R-13) | BC@.015 | SF | .80 | .55 | 1.35 |

Fanfold extruded polystyrene insulation 4' x 50' panels, including 5% waste

| | | | | | |
|---|---|---|---|---|---|
| 1/4" (R-1) | BC@.011 | SF | 2.19 | .41 | 2.60 |
| 3/8" (R-1) | BC@.011 | SF | 2.44 | .41 | 2.85 |
| Add for nails, 50 lb cartons (large square heads) | | | | | |
| 2-1/2", for 1-1/2" boards | — | LS | 101.00 | — | 101.00 |
| 3", for 2" boards | — | LS | 101.00 | — | 101.00 |

Foil-faced urethane sheathing, 4' x 8' panels, including 5% waste,

| | | | | | |
|---|---|---|---|---|---|
| 1" (R-7.2) | BC@.011 | SF | .59 | .41 | 1.00 |

Asphalt impregnated sheathing board on walls, nailed, 4' x 8' panels

| | | | | | |
|---|---|---|---|---|---|
| 1/2", standard | BC@.013 | SF | .26 | .48 | .74 |
| 1/2", intermediate | BC@.013 | SF | .33 | .48 | .81 |
| 1/2", nail base | BC@.013 | SF | .46 | .48 | .94 |

Dens Glass Gold insulating exterior sheathing. Glass mat embedded in a water-resistant treated gypsum core with bond-enhancing gold primer coating. Square edge.

| | | | | | |
|---|---|---|---|---|---|
| 1/4" x 4' x 4' Dens-Shield | BC@.015 | SF | .63 | .55 | 1.18 |
| 1/2" x 32" x 4' Dens-Shield | BC@.015 | SF | .68 | .55 | 1.23 |
| 1/2" x 4' x 8' | BC@.016 | SF | .56 | .59 | 1.15 |
| 5/8" x 4' x 8' | BC@.016 | SF | .63 | .59 | 1.22 |

Polystyrene foam underlay, polyethylene covered, R-Gard.

| | | | | | |
|---|---|---|---|---|---|
| 3/8" 4' x 24' fanfold | BC@.011 | SF | .26 | .41 | .67 |
| 1/2" x 4' x 8' | BC@.011 | SF | .29 | .41 | .70 |
| 1" x 4' x 8' | BC@.011 | SF | .43 | .41 | .84 |
| 1-1/2" x 2' x 8' | BC@.015 | SF | .78 | .55 | 1.33 |
| 2" x 2' x 8' | BC@.015 | SF | .84 | .55 | 1.39 |

# Insulation

| | Craft@Hrs | Unit | Material | Labor | Total |
|---|---|---|---|---|---|
| **Cant strips for insulated roof board, one piece wood, tapered** | | | | | |
| 2" strip | RR@.018 | LF | .17 | .74 | .91 |
| 3" strip | RR@.020 | LF | .16 | .82 | .98 |
| 4" strip | RR@.018 | LF | .15 | .74 | .89 |
| **Sill plate gasket. Fills gaps between the foundation and the plate. Also used for sealing around windows and doors.** | | | | | |
| 1/4" x 50', 3-1/2" wide | RR@.003 | LF | .09 | .12 | .21 |
| 1/4" x 50', 5-1/2" wide | RR@.003 | LF | .13 | .12 | .25 |
| **Perimeter foundation insulation, expanded polystyrene with polyethylene skinned surface. Low moisture retention. Minimum compressive strength of 1,440 lbs. per SF.** | | | | | |
| 3/4" x 4' x 8' | BC@.011 | SF | .27 | .41 | .68 |
| 1-1/2" x 2' x 4' | BC@.015 | SF | .60 | .55 | 1.15 |
| 2" x 2' x 4' | BC@.015 | SF | .80 | .55 | 1.35 |
| 2" x 4' x 4' | BC@.015 | SF | .64 | .55 | 1.19 |
| **FBX 1240 industrial board insulation, rockwool, semi-rigid, no waste included, 24" OC framing members** | | | | | |
| 2" (R-8) | BC@.004 | SF | .49 | .15 | .64 |
| 3" (R-12) | BC@.004 | SF | .71 | .15 | .86 |
| 3-1/2" (R-14) | BC@.004 | SF | .83 | .15 | .98 |
| **Sound control insulation, fiberglass acoustically designed to absorb sound vibration, for use between interior walls, floors and ceilings, Quietzone™ batts, coverage allows for framing.** | | | | | |
| 3-1/2", 16" OC framing | BC@.006 | SF | .32 | .22 | .54 |
| 3-1/2", 24" OC framing | BC@.006 | SF | .32 | .22 | .54 |
| Acoustic floor mat, 140 SF roll at $71.50 | BC@.003 | SF | .51 | .11 | .62 |
| Acoustic caulk, 29 oz. tube | — | Ea | 5.78 | — | 5.78 |
| **Sound insulation board, 4' x 8' panels installed on walls** | | | | | |
| 1/2" Homasote | BC@.013 | SF | .77 | .48 | 1.25 |
| 1/2" Soundstop | BC@.013 | SF | .32 | .48 | .80 |
| 1/2" Sound deadening board | BC@.013 | SF | .32 | .48 | .80 |
| Add for installation on ceilings | BC@.004 | SF | — | .15 | .15 |
| **Sound Attenuation Fire Batt insulation (SAFB), rockwool, semi-rigid, no waste included, pressed between framing members** | | | | | |
| 16" OC framing members | | | | | |
| 2" (R-8) | BC@.004 | SF | .38 | .15 | .53 |
| 3" (R-12) | BC@.004 | SF | .58 | .15 | .73 |
| 4" (R-16) | BC@.005 | SF | .75 | .18 | .93 |
| 24" OC framing members | | | | | |
| 2" (R-8) | BC@.003 | SF | .38 | .11 | .49 |
| 3" (R-12) | BC@.003 | SF | .58 | .11 | .69 |
| 4" (R-16) | BC@.004 | SF | .75 | .15 | .90 |
| **Housewrap building envelope, wrapped on exterior walls and stapled, including 10% waste** | | | | | |
| 9' x 100' roll | BC@.006 | SF | .14 | .22 | .36 |
| Tyvek tape, 2" x 165' | — | Roll | 11.80 | — | 11.80 |
| **Stabilized cellulose insulation, for blown-in applications. Bag covers 40 SF of attic joists at 6" depth. Add the cost of blower rental.** | | | | | |
| 40 SF bag at $10.40 | BC@.010 | SF | .26 | .37 | .63 |
| 1" wood plugs | BC@.020 | Ea | .32 | .74 | 1.06 |
| 1" plastic plugs | BC@.020 | Ea | .28 | .74 | 1.02 |
| **Vermiculite insulation, poured over ceilings** | | | | | |
| Vermiculite, (4 CF) | — | Ea | 30.50 | — | 30.50 |
| At 3" depth (60 sacks per 1,000 SF) | BL@.007 | SF | 1.91 | .21 | 2.12 |
| At 4" depth (72 sacks per 1,000 SF) | BL@.007 | SF | 2.55 | .21 | 2.76 |

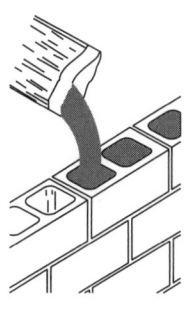

| | Craft@Hrs | Unit | Material | Labor | Total |
|---|---|---|---|---|---|
| Masonry batts, 1-1/2" x 15" x 48" | B9@.008 | SF | .26 | .26 | .52 |
| Masonry fill, poured in concrete block cores | | | | | |
|    Using 4 CF bags of perline, per bag | — | Ea | 27.70 | — | 27.70 |
|    4" wall, 8.1 SF per CF | B9@.006 | SF | .86 | .20 | 1.06 |
|    6" wall, 5.4 SF per CF | B9@.006 | SF | 1.28 | .20 | 1.48 |
|    8" wall, 3.6 SF per CF | B9@.006 | SF | 1.92 | .20 | 2.12 |

Natural fiber building insulation. EPA registered anti-microbial agent offers protection from mold, mildew, fungi and pests. Made from all natural cotton fibers. Friction fit. 94" long batting. Based on Ultra Touch.

| | Craft@Hrs | Unit | Material | Labor | Total |
|---|---|---|---|---|---|
|    16" long (R-13) | BC@.005 | SF | .56 | .18 | .74 |
|    24" long (R-13) | BC@.006 | SF | .56 | .22 | .78 |
|    16" long (R-19) | BC@.005 | SF | .82 | .18 | 1.00 |
|    24" long (R-19) | BC@.006 | SF | .82 | .22 | 1.04 |

**Insulation, Subcontract** New construction, coverage includes framing.

| | Craft@Hrs | Unit | Material | Labor | Total |
|---|---|---|---|---|---|
| Blown-in cellulose | | | | | |
|    3-1/2" (R-12) | — | MSF | — | — | 688.00 |
|    6" (R-21) | — | MSF | — | — | 977.00 |
| Sprayed-on urethane foam (rigid) | | | | | |
|    1" thick | — | MSF | — | — | 1,680.00 |
|    Add for acrylic waterproofing | — | MSF | — | — | 1,160.00 |
|    Add for urethane rubber waterproofing | — | MSF | — | — | 1,960.00 |
|    Add for cleaning rock off existing roof | — | MSF | — | — | 126.00 |

Insulation blown into walls in an existing structure, includes coring hole in exterior or interior of wall, filling cavity, sealing hole and paint to match existing surface. R-12 to R-14 rating depending on cavity thickness

| | Craft@Hrs | Unit | Material | Labor | Total |
|---|---|---|---|---|---|
|    Stucco (exterior application) | — | MSF | — | — | 2,000.00 |
|    Wallboard (interior application) | — | MSF | — | — | 2,000.00 |
|    Wood siding (exterior application) | — | MSF | — | — | 2,650.00 |

**Thermal Analysis** (Infrared thermography) Includes infrared video inspection of entire building, written report by qualified technician or engineer, and thermograms (infrared photographs) of all problem areas of building. Inspection is done using high-quality thermal sensitive instrumentation. Costs listed are for exterior facade shots, or interior (room by room) shots. All travel expenses for inspector are extra.

| | Craft@Hrs | Unit | Material | Labor | Total |
|---|---|---|---|---|---|
|    Residence, typically 4 hours required | — | Hr | — | — | 270.00 |
|    Commercial structure ($800.00 minimum) | — | Hr | — | — | 156.00 |

Add per day lost due to weather, lack of preparation by building occupants, etc., when inspector and equipment are at building site prepared to commence inspection

| | Craft@Hrs | Unit | Material | Labor | Total |
|---|---|---|---|---|---|
|    Typical cost per day | — | LS | — | — | 430.00 |

Add per night for accommodations when overnight stay is required. This is common when inspection must be performed at night to get accurate results

| | Craft@Hrs | Unit | Material | Labor | Total |
|---|---|---|---|---|---|
|    Typical cost, per night | — | LS | — | — | 150.00 |

**Insurance and Bonding** Typical rates. Costs vary by state and class of construction.

Rule of thumb: Employer's cost for payroll taxes, and insurance.

| | Craft@Hrs | Unit | Material | Labor | Total |
|---|---|---|---|---|---|
|    Per $100 (C$) of payroll | — | C$ | — | — | 30.00 |

Rule of thumb: Complete insurance program (comprehensive general liability policy, truck, automobile, and equipment floaters and fidelity bond), contractor's typical cost

| | Craft@Hrs | Unit | Material | Labor | Total |
|---|---|---|---|---|---|
|    Per $100 of payroll | — | C$ | — | — | 7.00 |

# Insurance, Liability

| | Craft@Hrs | Unit | Material | Labor | Total |
|---|---|---|---|---|---|

**Liability insurance** Comprehensive contractor's liability insurance, including operations, completed operations, bodily injury and property damage, protective and contractual coverages, $1,000,000 policy limit. Minimum annual premium will be between $2,500 and $10,000. Rates vary by state and with the contractor's loss experience. Typical costs per $100 (C$) of payroll for each trade employed. (Calculate and add for each category separately.)

| | Craft@Hrs | Unit | Material | Labor | Total |
|---|---|---|---|---|---|
| General contractors | — | C$ | — | — | 2.47 |
| Carpentry | — | C$ | — | — | 3.90 |
| Concrete, formed or flat | — | C$ | — | — | 5.00 |
| Drywall hanging and finishing | — | C$ | — | — | 3.37 |
| Electrical wiring | — | C$ | — | — | 3.40 |
| Floor covering installation | — | C$ | — | — | 4.50 |
| Glaziers | — | C$ | — | — | 3.37 |
| Heating, ventilating, air conditioning | — | C$ | — | — | 4.70 |
| Insulation | — | C$ | — | — | 3.90 |
| Masonry, tile | — | C$ | — | — | 3.70 |
| Painting | — | C$ | — | — | 5.10 |
| Plastering and stucco | — | C$ | — | — | 3.60 |
| Plumbing | — | C$ | — | — | 6.05 |

**Workers compensation coverage** Rates vary from state to state and by contractor's loss history. Coverage cost per $100 (C$) of base payroll excluding fringe benefits.

| | Craft@Hrs | Unit | Material | Labor | Total |
|---|---|---|---|---|---|
| Bricklayer | — | C$ | — | — | 9.69 |
| Carpenter | | | | | |
|   1 and 2 family dwellings | — | C$ | — | — | 15.80 |
|   Multiple units and commercial | — | C$ | — | — | 5.50 |
| Clerical (office worker) | — | C$ | — | — | .99 |
| Concrete | | | | | |
|   1 and 2 family dwellings | — | C$ | — | — | 6.20 |
|   Other concrete | — | C$ | — | — | 6.68 |
| Construction laborer | — | C$ | — | — | 15.80 |
| Drywall taper | — | C$ | — | — | 8.25 |
| Electrical wiring | — | C$ | — | — | 4.49 |
| Elevator erectors | — | C$ | — | — | 2.37 |
| Executive supervisors | — | C$ | — | — | 3.50 |
| Excavation, grading | — | C$ | — | — | 9.27 |
| Excavation, rock (no tunneling) | — | C$ | — | — | 12.10 |
| Glazing | — | C$ | — | — | 10.50 |
| Insulation work | — | C$ | — | — | 20.00 |
| Iron or steel erection work | — | C$ | — | — | 11.70 |
| Lathing | — | C$ | — | — | 5.43 |
| Operating engineers | — | C$ | — | — | 9.27 |
| Painting and paperhanging | — | C$ | — | — | 7.83 |
| Pile driving | — | C$ | — | — | 7.71 |
| Plastering and stucco | — | C$ | — | — | 13.00 |
| Plumbing | — | C$ | — | — | 6.27 |
| Reinforcing steel installation (concrete) | — | C$ | — | — | 11.70 |
| Roofing | — | C$ | — | — | 27.20 |
| Sewer construction | — | C$ | — | — | 9.10 |
| Sheet metal work (on site) | — | C$ | — | — | 9.36 |
| Steam and boiler work | — | C$ | — | — | 10.00 |
| Tile, stone, and terrazzo work | — | C$ | — | — | 5.71 |
| Truck driver | — | C$ | — | — | 9.27 |
| Tunneling | — | C$ | — | — | 20.00 |

# Insurance, Payment and Performance Bonds

| | Craft@Hrs | Unit | Material | Labor | Total |
|---|---|---|---|---|---|

**Payment and performance bonds** Cost per $1,000 (M$) of final contract price. Rates depend on the experience, credit, and net worth of the applicant.

Preferred rates, contract coverage for:

| | Craft@Hrs | Unit | Material | Labor | Total |
|---|---|---|---|---|---|
| First $ 100,000 | — | M$ | — | — | 13.70 |
| Next $2,400,000 | — | M$ | — | — | 8.27 |
| Next $2,500,000 | — | M$ | — | — | 6.55 |

Standard rates, contract coverage for:

| | | | | | |
|---|---|---|---|---|---|
| First $ 100,000 | — | M$ | — | — | 28.50 |
| Next $ 400,000 | — | M$ | — | — | 17.10 |
| Next $2,000,000 | — | M$ | — | — | 11.40 |
| Next $2,500,000 | — | M$ | — | — | 8.55 |
| Next $2,500,000 | — | M$ | — | — | 7.98 |
| Over $7,500,000 | — | M$ | — | — | 7.41 |

Substandard rates, contract coverage for:

| | | | | | |
|---|---|---|---|---|---|
| First $100,000 | — | M$ | — | — | 34.20 |
| Balance of contract | — | M$ | — | — | 22.80 |

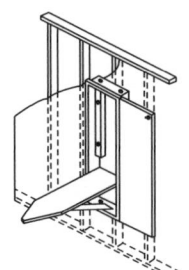

**Ironing Centers, Built-In** Wood frame with hinged door, can be recessed or surface mounted. Includes one 110 volt control panel where shown, receptacle outlet, timed shutoff switch, spotlight, storage shelf, and steel ironing board. No electrical work included. Iron-A-Way Products.

| | Craft@Hrs | Unit | Material | Labor | Total |
|---|---|---|---|---|---|
| Unit Type 1, no electric control panel, non-adjustable 42" board, 48" high, 15" wide, 6" deep | BC@.994 | Ea | 80.00 | 36.70 | 116.70 |
| Unit Type 2, with electric control panel, non-adjustable 42" board, 48" high, 15" wide, 6" deep | BC@.994 | Ea | 400.00 | 36.70 | 436.70 |
| Unit Type 3, with electric control panel, adjustable 46" board, 61" high, 15" wide, 6" deep | BC@.994 | Ea | 540.00 | 36.70 | 576.70 |

Added cost for any unit

| | | | | | |
|---|---|---|---|---|---|
| Add for sleeve board with holder | — | Ea | 40.00 | — | 40.00 |
| Add for door mirror | — | Ea | 115.00 | — | 115.00 |
| Add for oak paneled door front | — | Ea | 115.00 | — | 115.00 |

**Jackposts, Steel, Adjustable Building Columns** Permanent or temporary adjustable steel posts, 3" diameter.

| | Craft@Hrs | Unit | Material | Labor | Total |
|---|---|---|---|---|---|
| 19" To 36" extension | B1@.250 | Ea | 27.80 | 8.33 | 36.13 |
| 34" To 55" extension | B1@.250 | Ea | 33.20 | 8.33 | 41.53 |
| 54" To 93" extension | B1@.250 | Ea | 41.20 | 8.33 | 49.53 |
| 58" To 100" extension | B1@.250 | Ea | 41.20 | 8.33 | 49.53 |
| Heavy duty jack post | B1@.250 | Ea | 57.10 | 8.33 | 65.43 |

**Landscaping,** Seeding, level area

| | | | | | |
|---|---|---|---|---|---|
| Using tall fescue seed, per 20 lb bag | — | Ea | 37.10 | — | 37.10 |
| Seeding preparation, grade, rake, clean | BL@.003 | SY | .11 | .09 | .20 |
| Mechanical seeding (5 lbs per 1,000 SF) | BL@.020 | MSF | 9.30 | .59 | 9.89 |
| Hand seeding (10 lbs per 1,000 SF) | BL@.040 | MSF | 18.55 | 1.19 | 19.74 |

Fertilizer, pre-planting, 18-nitrogen, 24-phosphorus, 12-potassium

| | | | | | |
|---|---|---|---|---|---|
| (5 lb per 1,000 SF) | BL@.222 | MSF | 4.98 | 6.59 | 11.57 |
| Liming (70 lbs per 1,000 SF) | BL@.222 | MSF | 8.38 | 6.59 | 14.97 |

# Landscaping

| | Craft@Hrs | Unit | Material | Labor | Total |
|---|---|---|---|---|---|

Placing topsoil, delivered to site, 1 CY covers 81 SF at 4" depth. Add cost for equipment, where needed, from Excavation Equipment Costs section. Labor shown below is for operation of equipment.

| | Craft@Hrs | Unit | Material | Labor | Total |
|---|---|---|---|---|---|
| With equipment, level site | BL@.092 | CY | 21.20 | 2.73 | 23.93 |
| With equipment, sloped site | BL@.107 | CY | 21.20 | 3.17 | 24.37 |
| By hand, level site | BL@.735 | CY | 21.20 | 21.80 | 43.00 |
| By hand, sloped site | BL@.946 | CY | 21.20 | 28.10 | 49.30 |

Placing soil amendments, by hand spreading

| | | | | | |
|---|---|---|---|---|---|
| Wood chip mulch, pine bark | BL@1.27 | CY | 24.40 | 37.70 | 62.10 |
| Bale of straw, 25 bales per ton | BL@.200 | Ea | 3.24 | 5.93 | 9.17 |

Landscape stepping stones, concrete

| | | | | | |
|---|---|---|---|---|---|
| 12" round | BL@.133 | Ea | 2.64 | 3.95 | 6.59 |
| 14" round | BL@.133 | Ea | 3.55 | 3.95 | 7.50 |
| 18" round | BL@.153 | Ea | 6.15 | 4.54 | 10.69 |
| 24" round | BL@.153 | Ea | 10.40 | 4.54 | 14.94 |
| Add for square shapes | — | % | 20.00 | — | — |
| 18" or 24" diameter natural tree ring | BL@.133 | Ea | 2.23 | 3.95 | 6.18 |

Redwood benderboard, staked redwood

| | | | | | |
|---|---|---|---|---|---|
| 5/16" x 4" | BL@.011 | LF | .28 | .33 | .61 |
| 5/8" x 3-3/8" | BL@.016 | LF | 1.25 | .47 | 1.72 |
| 5/8" x 5-3/8" | BL@.016 | LF | 1.63 | .47 | 2.10 |

Roto-tilling light soil

| | | | | | |
|---|---|---|---|---|---|
| To 4" depth | BL@.584 | CSY | — | 17.30 | 17.30 |

Sodding, no soil preparation, nursery sod, Bermuda or Blue Rye mix

| | | | | | |
|---|---|---|---|---|---|
| Per SF (500 SF minimum charge) | BL@.013 | SF | .41 | .39 | .80 |

Hydroseeding subcontract (spray application of seed, binder and fertilizer slurry). Costs will vary based upon site conditions, quality/type of seed. For areas greater than 10,000 SF, see the hydroseeding cost in the commercial section.

| | | | | | |
|---|---|---|---|---|---|
| 1,000 – 2,000 SF job | — | MSF | — | — | 180.00 |
| 2,001 – 4,000 SF job | — | MSF | — | — | 150.00 |
| 4,001 – 6,000 SF job | — | MSF | — | — | 120.00 |
| 6,001 – 10,000 SF job | — | MSF | — | — | 110.00 |

## Lath

Imperial gypsum veneer base lath, nailed to walls and ceilings, including 10% waste.

| | | | | | |
|---|---|---|---|---|---|
| 3/8" x 4' x 8', tapered edge | BR@.076 | SY | 1.82 | 2.73 | 4.55 |
| 1/2" x 4' x 8', square edge | BR@.083 | SY | 1.96 | 2.98 | 4.94 |
| 1/2" x 4' x 10', square edge | BR@.083 | SY | 2.26 | 2.98 | 5.24 |
| 1/2" x 4' x 12', square edge | BR@.083 | SY | 1.64 | 2.98 | 4.62 |
| 5/8" x 4' x 10', square edge | BR@.083 | SY | 1.64 | 2.98 | 4.62 |

Steel lath, diamond pattern (junior mesh), 27" x 96" sheets, nailed to walls.

| | | | | | |
|---|---|---|---|---|---|
| 1.75 pound | BR@.076 | SY | 3.25 | 2.73 | 5.98 |
| 2-1/2 pound, galvanized | BR@.076 | SY | 3.83 | 2.73 | 6.56 |
| 2-1/2 pounds, paper back, dimpled | BR@.076 | SY | 4.97 | 2.73 | 7.70 |
| 3.4 pounds, galvanized | BR@.076 | SY | 5.27 | 2.73 | 8.00 |
| 3.4 pounds, paper back | BR@.076 | SY | 4.69 | 2.73 | 7.42 |
| 3/8" hi rib, 29 gauge | BR@.076 | SY | 6.44 | 2.73 | 9.17 |
| Add for ceiling applications | — | % | — | 25.0 | — |

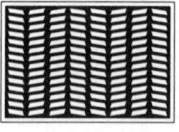

Galvanized stucco netting, 1-1/2" mesh, per roll, nailed to walls

| | | | | | |
|---|---|---|---|---|---|
| 36" x 150', 20 gauge | BR@2.80 | Ea | 55.30 | 100.00 | 155.30 |
| 36" x 150', 17-gauge | BR@2.80 | Ea | 64.60 | 100.00 | 164.60 |
| 36" x 100', 17-gauge paper back | BR@2.80 | Ea | 65.90 | 100.00 | 165.90 |
| 6" x 50' stucco repair net | — | Ea | 13.10 | — | 13.10 |

| | Craft@Hrs | Unit | Material | Labor | Total |
|---|---|---|---|---|---|
| **Corner bead, 26 gauge, 8' to 12' lengths.** | | | | | |
| Expanded corner bead, 8' long | BR@.027 | LF | .64 | .97 | 1.61 |
| Expanded corner bead, 10' long | BR@.027 | LF | .61 | .97 | 1.58 |
| Cornerite corner bead, 3" x 3" x 8' | BR@.027 | LF | .31 | .97 | 1.28 |
| Cornerite plaster corner, 2" x 2" x 4' | BR@.027 | LF | .61 | .97 | 1.58 |
| Cornerite stucco corner, 3" x 3" 8' long | BR@.027 | LF | .43 | .97 | 1.40 |
| Short flange corner bead, 8' long | BR@.027 | LF | 1.03 | .97 | 2.00 |
| Plaster J-trim, 1-3/8" wide x 10' long | BR@.027 | LF | .65 | .97 | 1.62 |
| **Casing beads** | | | | | |
| Short flange, 1/2" x 10' long | BR@.025 | LF | .59 | .90 | 1.49 |
| Short flange, 7/8" x 10' long | BR@.025 | LF | .58 | .90 | 1.48 |
| Expanded flange, 3/4" x 10' | BR@.025 | LF | .56 | .90 | 1.46 |
| **Expansion joint, 26 gauge, 1/2" ground** | BR@.050 | LF | 1.57 | 1.79 | 3.36 |

## Lawn Sprinkler Systems

PVC Schedule 40 pipe laid in an open trench and connected. Add the cost of trenching, valves and sprinkler heads below.

| | Craft@Hrs | Unit | Material | Labor | Total |
|---|---|---|---|---|---|
| **1/2" Schedule 40 PVC pressure pipe and fittings** | | | | | |
| 1/2" pipe | BL@.010 | LF | .16 | .30 | .46 |
| 1/2" 90 degree ell | BL@.100 | Ea | .30 | 2.97 | 3.27 |
| 1/2" tee | BL@.130 | Ea | .30 | 3.86 | 4.16 |
| **3/4" Schedule 40 PVC pressure pipe and fittings** | | | | | |
| 3/4" pipe | BL@.012 | LF | .20 | .36 | .56 |
| 3/4" 90 degree ell | BL@.115 | Ea | .36 | 3.41 | 3.77 |
| 3/4" tee | BL@.140 | Ea | .35 | 4.15 | 4.50 |
| **1" Schedule 40 PVC pressure pipe and fittings** | | | | | |
| 1" pipe | BL@.015 | LF | .30 | .45 | .75 |
| 1" 90 degree ell | BL@.120 | Ea | .51 | 3.56 | 4.07 |
| 1" tee | BL@.170 | Ea | .68 | 5.04 | 5.72 |
| **Deduct for Class 200 PVC pipe** | — | % | -40.0 | — | — |
| **Trenching for sprinkler pipe installation, by hand, including backfill and tamp** | | | | | |
| Light soil, 8" wide | | | | | |
| 12" deep | BL@.027 | LF | — | .80 | .80 |
| 18" deep | BL@.040 | LF | — | 1.19 | 1.19 |
| 24" deep | BL@.053 | LF | — | 1.57 | 1.57 |
| Average soil, 8" wide | | | | | |
| 12" deep | BL@.043 | LF | — | 1.28 | 1.28 |
| 18" deep | BL@.065 | LF | — | 1.93 | 1.93 |
| 24" deep | BL@.087 | LF | — | 2.58 | 2.58 |
| Heavy soil or loose rock, 8" wide | | | | | |
| 12" deep | BL@.090 | LF | — | 2.67 | 2.67 |
| 18" deep | BL@.100 | LF | — | 2.97 | 2.97 |
| 24" deep | BL@.134 | LF | — | 3.98 | 3.98 |

| | Craft@Hrs | Unit | Material | Labor | Total |
|---|---|---|---|---|---|
| **Impact sprinkler heads, riser mounted, adjustable circle** | | | | | |
| Brass | BL@.422 | Ea | 18.20 | 12.50 | 30.70 |
| Plastic | BL@.422 | Ea | 6.81 | 12.50 | 19.31 |
| Plastic, 3" pop-up | BL@.422 | Ea | 19.20 | 12.50 | 31.70 |
| **Gear drive rotor sprinkler, 3/4" inlet** | | | | | |
| Adjustable, 30' radius | BL@.422 | Ea | 14.90 | 12.50 | 27.40 |
| **Spray head pop-up type, plastic** | | | | | |
| 2-1/2" height | BL@.056 | Ea | 3.20 | 1.66 | 4.86 |
| 4" height | BL@.056 | Ea | 3.84 | 1.66 | 5.50 |
| 6" height | BL@.056 | Ea | 7.50 | 1.66 | 9.16 |

# Lawn Sprinkler Systems

| | Craft@Hrs | Unit | Material | Labor | Total |
|---|---|---|---|---|---|
| **Spray head, plastic with brass insert, for riser mounting** | | | | | |
| Adjustable 0 to 330 degrees | BL@.056 | Ea | 2.11 | 1.66 | 3.77 |
| Standard quarter, half or full pattern | BL@.056 | Ea | 1.56 | 1.66 | 3.22 |
| Shrub spray, all patterns | BL@.056 | Ea | 1.75 | 1.66 | 3.41 |

**Sprinkler head and riser, 1/2" PVC riser connected to branch supply with three threaded fittings and plastic 4" pop-up head**

| | Craft@Hrs | Unit | Material | Labor | Total |
|---|---|---|---|---|---|
| 4" riser | BL@.056 | Ea | 6.98 | 1.66 | 8.64 |
| 6" riser | BL@.056 | Ea | 7.12 | 1.66 | 8.78 |
| 12" riser | BL@.056 | Ea | 7.66 | 1.66 | 9.32 |
| 18" riser | BL@.056 | Ea | 7.86 | 1.66 | 9.52 |
| 24" riser | BL@.056 | Ea | 8.23 | 1.66 | 9.89 |

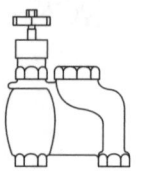

| | Craft@Hrs | Unit | Material | Labor | Total |
|---|---|---|---|---|---|
| **Valves, non-siphon, manual, brass** | | | | | |
| 3/4" valve | BL@1.03 | Ea | 33.70 | 30.60 | 64.30 |
| 1" valve | BL@1.38 | Ea | 59.70 | 40.90 | 100.60 |

**Valves, electric solenoid, plastic, with pressure-assisted flow control, hookup but no wire included**

| | Craft@Hrs | Unit | Material | Labor | Total |
|---|---|---|---|---|---|
| 3/4" valve | BL@1.26 | Ea | 16.70 | 37.40 | 54.10 |
| 1" valve | BL@1.26 | Ea | 19.50 | 37.40 | 56.90 |

**Valve control wire, 16 gauge low voltage plastic jacket direct burial cable, laid with pipe. No trenching or end connections included.**

| | Craft@Hrs | Unit | Material | Labor | Total |
|---|---|---|---|---|---|
| 2 conductor | BL@.003 | LF | .12 | .09 | .21 |
| 3 conductor | BL@.003 | LF | .16 | .09 | .25 |
| 5 conductor | BL@.003 | LF | .26 | .09 | .35 |
| 7 conductor | BL@.003 | LF | .40 | .09 | .49 |
| 8 conductor | BL@.003 | LF | .40 | .09 | .49 |

**Programmable irrigation control stations, no valve or wiring included, add electrical connection if required.**

| | Craft@Hrs | Unit | Material | Labor | Total |
|---|---|---|---|---|---|
| For up to 6 electric valves | BL@1.83 | LS | 42.80 | 54.30 | 97.10 |
| For up to 8 electric valves | BL@1.83 | LS | 58.80 | 54.30 | 113.10 |
| Add for 120 volt receptacle | — | LS | — | — | 85.00 |

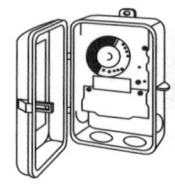

**Residential sprinkler system, typical labor and material costs for irrigating 1,700 SF using 400 LF of 3/4" Schedule 40 PVC pipe, 35 LF of 1" Schedule 40 PVC pipe, four electric 3/4" valves, pop-up sprinkler heads on 6" risers spaced 10', 100 LF of 5-conductor control wire and 235 LF of trenching 12" deep in average soil.**

| | Craft@Hrs | Unit | Material | Labor | Total |
|---|---|---|---|---|---|
| Large or regular area, 30 heads | BL@.023 | SF | .27 | .68 | .95 |
| Narrow or irregular area, 40 heads | BL@.025 | SF | .31 | .74 | 1.05 |
| Add for freezing zones | — | SF | — | — | .12 |

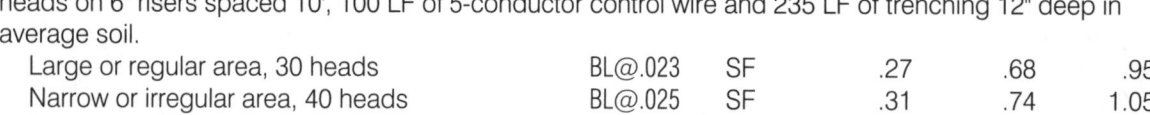

**Layout** Foundation layout, medium to large size residence, 6 to 8 outside corners, includes shooting elevation from nearby reference, stakes and batterboards as required. No surveying or clearing included.

| | Craft@Hrs | Unit | Material | Labor | Total |
|---|---|---|---|---|---|
| Typical cost per residence | B1@7.86 | LS | 65.10 | 262.00 | 327.10 |

**Lighting Fixtures** See also Electrical Work. Costs are to hang and connect fixtures only. Add the cost of wiring and fittings from the Commercial and Industrial section of this manual.

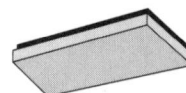

### Fluorescent lighting fixtures
Surface mounted fluorescent fixtures, including ballasts but no lamps.
Wraparound acrylic diffuser

| | Craft@Hrs | Unit | Material | Labor | Total |
|---|---|---|---|---|---|
| 24", two 20 watt tubes | BE@1.60 | Ea | 25.50 | 63.30 | 88.80 |
| 48", two 40 watt tubes | BE@1.60 | Ea | 39.10 | 63.30 | 102.40 |
| 48", four 40 watt tubes | BE@1.60 | Ea | 104.00 | 63.30 | 167.30 |
| 96" tandem, four 40 watt tubes | BE@1.60 | Ea | 106.00 | 63.30 | 169.30 |

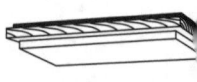

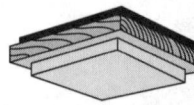

|  | Craft@Hrs | Unit | Material | Labor | Total |
|---|---|---|---|---|---|
| **Low profile wraparound model** | | | | | |
| 10" x 48" x 4", two 20 watt tubes | BE@1.60 | Ea | 66.90 | 63.30 | 130.20 |
| 15" x 48" x 4", four 40 watt tubes | BE@1.60 | Ea | 113.00 | 63.30 | 176.30 |
| 15" x 96" x 4", eight 40 watt tubes | BE@1.60 | Ea | 176.00 | 63.30 | 239.30 |
| 19" x 19" x 4", one 32/40 watt circline tube | BE@1.60 | Ea | 72.50 | 63.30 | 135.80 |
| 24" x 24" x 4", two 40 watt U/6 tubes | BE@1.60 | Ea | 93.20 | 63.30 | 156.50 |
| **Basic wraparound acrylic diffusers, solid oak wood ends** | | | | | |
| 10" x 24" x 3", two 20 watt tubes | BE@1.60 | Ea | 43.10 | 63.30 | 106.40 |
| 10" x 48" x 3", two 40 watt tubes | BE@1.60 | Ea | 52.60 | 63.30 | 115.90 |
| **Economy, wraparound acrylic diffuser, basic metal ends** | | | | | |
| 14" x 48" x 2", four 40 watt tubes | BE@1.60 | Ea | 78.80 | 63.30 | 142.10 |
| 8" x 48" x 2", two 40 watt tubes | BE@1.60 | Ea | 43.10 | 63.30 | 106.40 |
| **Electronic ballast troffer fixture, for mounting in a suspended ceiling** | | | | | |
| Four 32 watt T8 tubes | BE@.900 | Ea | 45.40 | 35.60 | 81.00 |
| Four 32 watt T8 tubes, with flex connection | BE@.800 | Ea | 49.50 | 31.70 | 81.20 |
| **Basic circular fixtures for circline tubes** | | | | | |
| Simulated glass crystal with antiqued metal band, | | | | | |
| 11" round x 3" high, one 22 watt tube | BE@.696 | Ea | 16.70 | 27.60 | 44.30 |
| Circular fixture | | | | | |
| 12" round x 3 high, 22 and 32 watt tubes | BE@.696 | Ea | 23.00 | 27.60 | 50.60 |
| Fixture with frosted glass diffuser | | | | | |
| 19" round x 4" high, 32 and 40 watt tubes | BE@.696 | Ea | 70.80 | 27.60 | 98.40 |
| **Low profile rectangular wall or ceiling mount fixture, metal housing, white acrylic diffuser** | | | | | |
| 9-5/8" x1-7/8", two 9 watt energy saver bulbs | BE@.696 | Ea | 40.70 | 27.60 | 68.30 |
| **Rectangular indoor/outdoor model with discoloration resistant/impact resistant acrylic diffuser and cold weather ballast** | | | | | |
| 10" x 10" x 5" deep, one 22 watt tube | BE@.696 | Ea | 65.80 | 27.60 | 93.40 |
| 12" x 12" x 5" deep, one 34 watt tube | BE@.696 | Ea | 85.20 | 27.60 | 112.80 |
| **Hall lantern type, including tubes** | | | | | |
| 4-1/4" wide x 12-1/8" long x 4" deep | BE@.696 | Ea | 41.50 | 27.60 | 69.10 |
| **Decorative valance type, with solid oak frame and acrylic diffuser** | | | | | |
| 12" x 25" x 4", two 20 watt tubes | BE@2.06 | Ea | 83.70 | 81.60 | 165.30 |
| 25" x 25" x 4", two 40 watt U/6 tubes | BE@1.59 | Ea | 104.00 | 62.90 | 166.90 |
| 12" x 49" x 4", two 40 watt tubes | BE@1.59 | Ea | 95.70 | 62.90 | 158.60 |
| **Economy valance type, with metal ends. Round or square models available** | | | | | |
| All models are 2-5/8" wide x 4-1/4" deep | | | | | |
| 24-3/4" L (requires one 15 watt tube) | BE@1.59 | Ea | 23.00 | 62.90 | 85.90 |
| 24-3/4" L (requires one 20 watt tube) | BE@1.59 | Ea | 24.70 | 62.90 | 87.60 |
| 36-3/4" L (requires one 30 watt tube) | BE@1.59 | Ea | 32.10 | 62.90 | 95.00 |
| 36-3/4" L (requires two 30 watt tubes) | BE@1.59 | Ea | 38.90 | 62.90 | 101.80 |
| 48-3/4" L (requires one 40 watt tube) | BE@1.59 | Ea | 43.90 | 62.90 | 106.80 |
| 48-3/4" L (requires two 40 watt tubes) | BE@1.59 | Ea | 49.40 | 62.90 | 112.30 |
| **Under shelf/cabinet type for wiring to wall switch** | | | | | |
| 18" x 5" x 2", one 15 watt tube | BE@.696 | Ea | 16.50 | 27.60 | 44.10 |
| 24" x 5" x 2", one 20 watt tube | BE@.696 | Ea | 22.00 | 27.60 | 49.60 |
| 36" x 5" x 2", one 30 watt tube | BE@.696 | Ea | 34.10 | 27.60 | 61.70 |
| 48" x 5" x 2", one 40 watt tube | BE@.696 | Ea | 48.10 | 27.60 | 75.70 |

# Lighting Fixtures

| | Cool white | Deluxe cool white | Energy saving cool white | Warm white |
|---|---|---|---|---|
| **Fluorescent tubes** | | | | |
| 8 watt (12" long) | 7.02 | 10.50 | — | 9.92 |
| 15 watt (18" long) | 7.63 | 9.06 | — | 7.94 |
| 20 watt (24" long) | 5.18 | 9.65 | 9.33 | 6.88 |
| 30 watt (36" long) | 6.34 | 12.10 | 6.44 | 10.40 |
| 40 watt (48" long) | 2.18 | — | — | — |
| **Slimline straight tubes** | | | | |
| 30 watt (24" or 36" long) | 16.50 | — | 13.10 | 17.60 |
| 40 watt (48" long) | 10.90 | 15.10 | 21.00 | 13.10 |
| **Circular (circline) tubes** | | | | |
| 20 watt (6-1/2" diameter) | 10.50 | — | — | 11.60 |
| 22 watt (8-1/4" diameter) | 7.90 | 11.00 | — | 11.40 |
| 32 watt (12" diameter) | 9.65 | 13.40 | — | 12.80 |
| 40 watt (16" diameter) | 14.00 | — | — | 16.90 |
| **"U"-shaped tubes (3" spacing x 12" long)** | | | | |
| U-3 40 watt | 14.30 | 19.20 | — | 14.30 |
| U-6 40 watt | 14.30 | 19.20 | — | 14.30 |

| | Craft@Hrs | Unit | Material | Labor | Total |
|---|---|---|---|---|---|
| Labor to install each tube | BE@.035 | Ea | — | 1.39 | 1.39 |

## Incandescent lighting fixtures

Outdoor post lantern fixtures, antique satin brass finish, no digging included

| | Craft@Hrs | Unit | Material | Labor | Total |
|---|---|---|---|---|---|
| Post lantern type fixture, cast aluminum | BE@2.00 | Ea | 64.00 | 79.20 | 143.20 |
| **Modern porch ceiling fixture** | | | | | |
| 7-1/2" x 7-1/2" x 3" drop, single light | BE@2.15 | Ea | 25.30 | 85.10 | 110.40 |
| **Flood light lampholders** | | | | | |
| Cadmium steel plate | BE@2.15 | Ea | 30.10 | 85.10 | 115.20 |
| 1/2" male swivel 7-5/16" long, aluminum | BE@2.15 | Ea | 28.70 | 85.10 | 113.80 |
| **Path lighting, kits with cable and timer** | | | | | |
| 20 light low voltage | BE@2.63 | LS | 57.10 | 104.00 | 161.10 |
| 6 flood lights, 11 watt | BE@2.98 | LS | 60.60 | 118.00 | 178.60 |
| 12 tier lights, 7 watt, programmable | BE@2.63 | LS | 81.40 | 104.00 | 185.40 |
| **Recessed fixtures, including housing, incandescent** | | | | | |
| **Integral thermally protected recessed (can) light housing** | | | | | |
| 4" recessed light housing, non-insulated | BE@1.00 | Ea | 17.90 | 39.60 | 57.50 |
| 5" recessed light housing, non-insulated | BE@1.00 | Ea | 22.60 | 39.60 | 62.20 |
| 6" recessed light housing, non-insulated | BE@1.00 | Ea | 11.30 | 39.60 | 50.90 |
| Add for insulated housing | — | % | 25.0 | — | — |
| Add for low voltage | — | % | 100.0 | — | — |
| Recessed baffle light trim | BE@0.25 | Ea | 10.20 | 9.90 | 20.10 |
| Recessed flat light trim | BE@0.25 | Ea | 10.20 | 9.90 | 20.10 |
| Recessed eyeball light trim | BE@0.25 | Ea | 33.90 | 9.90 | 43.80 |
| Recessed reflector light trim | BE@0.25 | Ea | 14.70 | 9.90 | 24.60 |
| Recessed shower light trim | BE@0.25 | Ea | 13.60 | 9.90 | 23.50 |
| Recessed open light trim | BE@0.25 | Ea | 4.52 | 9.90 | 14.42 |
| Recessed adjustable light trim | BE@0.25 | Ea | 25.50 | 9.90 | 35.40 |
| Recessed splay light trim | BE@0.25 | Ea | 11.20 | 9.90 | 21.10 |
| Utility 12" x 10", diffusing glass lens | BE@1.00 | Ea | 50.30 | 39.60 | 89.90 |

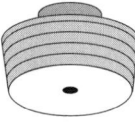

| | Craft@Hrs | Unit | Material | Labor | Total |
|---|---|---|---|---|---|
| **Ceiling fixtures, using 60 watt standard bulbs, pre-wired housing, white, canopy finish** | | | | | |
| Cloud 9-3/4" x 5-1/4" (2 bulbs) | BE@1.59 | Ea | 26.80 | 62.90 | 89.70 |
| Cloud 8-1/4" x 5-1/2" (1 bulb) | BE@1.59 | Ea | 33.90 | 62.90 | 96.80 |
| Standard globe 8-3/4" x 5" (2 bulbs) | BE@1.59 | Ea | 23.40 | 62.90 | 86.30 |
| Cloud globe, brass canopy 10" x 5" | BE@1.59 | Ea | 28.90 | 62.90 | 91.80 |

## Log Home Construction

Logs are cut from standing dead lodgepole pine or spruce. Add the foundation cost. Based on Authentic Log Homes.

Rule of thumb cost, including labor, material and hardware but no foundation, plumbing, HVAC, electrical work or interior finishing.

| | Craft@Hrs | Unit | Material | Labor | Total |
|---|---|---|---|---|---|
| Total cost per square foot of wall (shell only) | — | SF | — | — | 31.90 |
| **Lodgepole pine or spruce logs, including local delivery, per LF of log at 8" to 12" diameter** | | | | | |
| 8" machine peeled | — | LF | 12.30 | — | 12.30 |
| 8" hand peeled | — | LF | 13.20 | — | 13.20 |
| 10" machine peeled | — | LF | 16.60 | — | 16.60 |
| 10" hand peeled | — | LF | 18.40 | — | 18.40 |
| 12" machine peeled | — | LF | 21.60 | — | 21.60 |
| 12" hand peeled | — | LF | 23.40 | — | 23.40 |

Labor costs for log home construction. Using a crew experienced in log home construction. Add 10% to 50% for crews on their first log home project. Subfloor and porch joist hardware uses Type B Simpson Strong-Tie hangers. These labor estimates assume precut 8" x 6" log profiles. For 9" x 7" profiles, add 10%. For 10" x 8" profiles, add 20%. For random length logs, add 10%. The preferred crew is three carpenters and three helpers.

| | Craft@Hrs | Unit | Material | Labor | Total |
|---|---|---|---|---|---|
| **Subfloor** | | | | | |
| CCA treated sill plate, anchor bolts. | B1@.057 | LF | — | 1.90 | 1.90 |
| Built-up girders (using 2 x 8 lumber) | B1@.222 | LF | — | 7.39 | 7.39 |
| Posts (8" log), with anchor | B1@.500 | Ea | — | 16.70 | 16.70 |
| Rim joists and blocking (2 x 8) | B1@.065 | LF | — | 2.17 | 2.17 |
| Floor joists (2 x 8), hangers, bridging/blocking | B1@.025 | LF | — | .83 | .83 |
| Decking | B1@.014 | SF | — | .47 | .47 |
| **Porch** | | | | | |
| Rim joists | B1@.017 | LF | — | .57 | .57 |
| Posts (8" log), with anchor | B1@.500 | Ea | — | 16.70 | 16.70 |
| Joists with hangers | B1@.025 | SF | — | .83 | .83 |
| Decking (2 x 6 redwood) | B1@.025 | SF | — | .83 | .83 |
| Balcony railing | B1@.125 | LF | — | 4.16 | 4.16 |
| **Sill log** | | | | | |
| Flashing | B1@.010 | LF | — | .33 | .33 |
| Sill log spiked onto subfloor | B1@.050 | LF | — | 1.67 | 1.67 |
| **Laying wall courses** | | | | | |
| Courses 2 through 8 | B1@.025 | LF | — | .83 | .83 |
| Courses 9 through 16 | B1@.033 | LF | — | 1.10 | 1.10 |
| Course 17 (sill log) | B1@.040 | LF | — | 1.33 | 1.33 |
| Courses 18 through 24 | B1@.040 | LF | — | 1.33 | 1.33 |
| Courses 25 through 32 | B1@.050 | LF | — | 1.67 | 1.67 |
| **Doors and windows, per opening** | B1@4.00 | Ea | — | 133.00 | 133.00 |
| Interior trim | B1@2.00 | Ea | — | 66.60 | 66.60 |
| Exterior trim | B1@2.00 | Ea | — | 66.60 | 66.60 |
| **Mantle (10 feet long)** | | | | | |
| With bracing | B1@1.50 | Ea | — | 50.00 | 50.00 |

# Log Home Construction

| | Craft@Hrs | Unit | Material | Labor | Total |
|---|---|---|---|---|---|
| **Log floor system** | | | | | |
| Girder | B1@.050 | LF | — | 1.67 | 1.67 |
| Support posts | B1@.500 | Ea | — | 16.70 | 16.70 |
| Support wall (studs, top and bottom plates) | B1@.167 | LF | — | 5.56 | 5.56 |
| Cutting wall notch to accept girder | B1@1.50 | Ea | — | 50.00 | 50.00 |
| Floor joists | B1@.063 | LF | — | 2.10 | 2.10 |
| Cutting wall notch to accept floor joist | B1@0.50 | Ea | — | 16.70 | 16.70 |
| T&G decking | B1@.025 | SF | — | .83 | .83 |
| **Log staircase** | | | | | |
| Straight | B1@30.0 | Ea | — | 999.00 | 999.00 |
| "L"- or "U"-shaped | B1@36.0 | Ea | — | 1,200.00 | 1,200.00 |
| **Wall stiffener** | | | | | |
| 6" x 6" or 8" x 8" beam channeled into wall | B1@2.00 | Ea | — | 66.60 | 66.60 |
| **Log rafter system, including dormers** | | | | | |
| Ridge beam, main and secondary | B1@0.25 | LF | — | 8.33 | 8.33 |
| Notching gable to accept ridge beam | B1@1.50 | Ea | — | 50.00 | 50.00 |
| Built up truss pocket | B1@2.00 | Ea | — | 66.60 | 66.60 |
| Valley beam layout and installation | B1@0.33 | LF | — | 11.00 | 11.00 |
| Ridge beam support post | B1@0.50 | Ea | — | 16.70 | 16.70 |
| Rafter 20' or less | B1@1.00 | Ea | — | 33.30 | 33.30 |
| Rafter over 20' | B1@1.50 | Ea | — | 50.00 | 50.00 |
| Valley rafter 16' or less | B1@1.50 | Ea | — | 50.00 | 50.00 |
| Valley rafter between 16' and 20' | B1@2.00 | Ea | — | 66.60 | 66.60 |
| Valley rafter over 20' | B1@2.50 | Ea | — | 83.30 | 83.30 |
| Timber rafter 18' or less | B1@3.00 | Ea | — | 99.90 | 99.90 |
| Timber rafter over 18' | B1@4.00 | Ea | — | 133.00 | 133.00 |
| Rafter layout | B1@.250 | Ea | — | 8.33 | 8.33 |
| Snow block | B1@1.00 | Ea | — | 33.30 | 33.30 |
| **Log trusses, 32' and less** | | | | | |
| Hardware plate | B1@2.00 | Ea | — | 66.60 | 66.60 |
| Collar ties | B1@1.00 | Ea | — | 33.30 | 33.30 |
| Vertical or diagonal web | B1@3.00 | Ea | — | 99.90 | 99.90 |
| Add for each 5' of truss over 32' | — | % | — | 50.0 | — |
| **Cold roof construction** | | | | | |
| Secondary rafters | B1@0.50 | Ea | — | 16.70 | 16.70 |
| Jack rafters | B1@0.25 | Ea | — | 8.33 | 8.33 |
| Fly rafters | B1@0.50 | Ea | — | 16.70 | 16.70 |
| Blocking | B1@0.25 | Ea | — | 8.33 | 8.33 |
| **OSB roof sheathing** | | | | | |
| Less than or equal to 6/12 pitch | B1@1.72 | Sq | — | 57.30 | 57.30 |
| 7/12 pitch to 9/12 pitch | B1@2.00 | Sq | — | 66.60 | 66.60 |
| Less than or equal to 6/12 pitch | B1@2.50 | Sq | — | 83.30 | 83.30 |
| Felt | B1@1.00 | Sq | — | 33.30 | 33.30 |
| Asphalt shingles | B1@2.00 | Sq | — | 66.60 | 66.60 |
| Metal roof, ridge cap and closure strips | B1@1.33 | Sq | — | 44.30 | 44.30 |
| Ridge vents and soffit vents | B1@3.00 | Ea | — | 99.90 | 99.90 |
| Fascia | B1@0.02 | LF | — | .67 | .67 |
| Soffit | B1@0.03 | SF | — | 1.00 | 1.00 |
| **Caulking** | | | | | |
| Exterior only | B1@0.02 | LF | — | .67 | .67 |
| **Preservative** | | | | | |
| Exterior only | B1@.005 | SF | — | .17 | .17 |
| **Electrical chases** | | | | | |
| Cut in log ends and down tops of rafters | B1@.083 | LF | — | 2.76 | 2.76 |
| Holes drilled through ridge beams | B1@0.50 | Ea | — | 16.70 | 16.70 |

### Framing Lumber

MBF means per 1,000 board feet. One board foot is the volume of lumber in a nominal 1" x 12" board one foot long. A 2" x 6" joist 10 feet long has 10 board feet. To determine the board feet of lumber in any piece, multiply the width in inches by the depth in inches by the length in feet. Then divide by 12.

| | Unit | MBF | 8' L | 12' L | 16' L |
|---|---|---|---|---|---|
| **Douglas fir – #2 & Better** | | | | | |
| 2" x 4" – Green | Ea | 400.00 | 2.13 | 3.19 | 4.26 |
| 2" x 6" – KD | Ea | 450.00 | 3.60 | 5.40 | 7.20 |
| 2" x 8" – KD | Ea | 467.00 | 4.98 | 7.47 | 9.96 |
| 2" x 10" – KD | Ea | 491.00 | 6.54 | 9.81 | 13.10 |
| 2" x 12" – KD | Ea | 494.00 | 7.90 | 11.90 | 15.80 |
| **Douglas fir – kiln dried Standard & Better** | | | | | |
| 2" x 4" | Ea | 423.00 | 2.25 | 3.38 | 4.51 |
| 2" x 6" | Ea | 459.00 | 3.67 | 5.51 | 7.34 |
| 2" x 8" | Ea | 486.00 | 5.18 | 7.78 | 10.40 |
| 2" x 10" | Ea | 529.00 | 7.05 | 10.60 | 14.10 |
| 2" x 12" | Ea | 576.00 | 9.21 | 13.80 | 18.40 |
| **Douglas fir finish and clear – C & Better, vertical grain, S4S, dry. West Coast prices. Add 10% to 30% for transportation to other areas** | | | | | |
| 2" x 2" | Ea | 3,100.00 | 8.24 | 12.40 | 16.50 |
| 2" x 3" | Ea | 3,150.00 | 12.40 | 18.50 | 24.80 |
| 2" x 4" | Ea | 3,500.00 | 18.60 | 27.90 | 37.30 |
| 2" x 6" | Ea | 3,900.00 | 31.20 | 46.70 | 62.40 |
| 2" x 8" | Ea | 4,410.00 | 47.00 | 70.60 | 94.40 |
| 2" x 10" | Ea | 5,480.00 | 73.20 | 110.00 | 146.00 |
| 2" x 12" | Ea | 7,090.00 | 113.00 | 170.00 | 227.00 |
| **Douglas fir finish and clear – C & Better, flat grain, S4S, dry. West Coast prices. Add 10% to 30% for transportation to other areas** | | | | | |
| 2" x 2" | Ea | 2,700.00 | 7.20 | 10.80 | 14.50 |
| 2" x 3" | Ea | 2,700.00 | 10.80 | 16.20 | 21.50 |
| 2" x 4" | Ea | 2,730.00 | 14.60 | 21.90 | 29.10 |
| 2" x 6" | Ea | 3,160.00 | 25.30 | 37.90 | 50.60 |
| 2" x 8" | Ea | 3,160.00 | 33.80 | 50.60 | 67.40 |
| 2" x 10" | Ea | 3,160.00 | 42.20 | 63.20 | 84.30 |
| 2" x 12" | Ea | 4,200.00 | 67.20 | 102.00 | 134.00 |
| **Hemlock/fir, kiln dried** | | | | | |
| 2" x 4" – Standard & Better | Ea | 477.00 | 2.54 | 3.81 | 5.09 |
| 2" x 6" – #2 & Better | Ea | 472.00 | 3.78 | 5.67 | 7.56 |
| 2" x 8" – #2 & Better | Ea | 457.00 | 4.88 | 7.31 | 9.75 |
| 2" x 10" – #2 & Better | Ea | 449.00 | 5.98 | 8.97 | 12.00 |
| 2" x 12" – #2 & Better | Ea | 526.00 | 8.42 | 12.60 | 16.80 |
| **Spruce pine fir (SPF), kiln dried** | | | | | |
| 2" x 4", #2 and Better | Ea | 471.00 | 2.51 | 3.76 | 5.02 |
| 2" x 6", #2 and Better | Ea | 493.00 | 3.94 | 5.92 | 7.89 |
| 2" x 8", #2 and Better | Ea | 498.00 | 5.31 | 7.96 | 10.60 |
| 2" x 10", #2 and Better | Ea | 473.00 | 6.31 | 9.46 | 12.60 |
| 2" x 12", #2 and Better | Ea | 568.00 | 9.09 | 13.60 | 18.20 |
| 2" x 4", Standard and Better | Ea | 438.00 | 2.33 | 3.50 | 4.67 |
| 4" x 4", Standard and Better | Ea | 836.00 | 8.91 | 13.40 | 17.80 |

# Lumber, Framing

| | Unit | MBF | 8' L | 12' L | 16' L |
|---|---|---|---|---|---|
| **Red cedar – green  S4S** | | | | | |
| 2" x 4" | Ea | 1,420.00 | 7.58 | 11.40 | 15.20 |
| 2" x 6" | Ea | 1,760.00 | 14.10 | 21.10 | 28.20 |
| **Red cedar – green – D and Better  S4S** | | | | | |
| 2" x 2" | Ea | 2,470.00 | 6.58 | 9.88 | 13.20 |
| 2" x 4" | Ea | 3,060.00 | 16.30 | 24.50 | 32.60 |
| 2" x 6" | Ea | 3,470.00 | 27.70 | 41.60 | 55.50 |
| **Red cedar – rough – Standard and Better** | | | | | |
| 2" x 2" | Ea | 1,610.00 | 4.29 | 6.44 | 8.58 |
| 2" x 4" | Ea | 1,610.00 | 8.60 | 12.90 | 17.20 |
| 2" x 6" | Ea | 1,920.00 | 15.40 | 23.10 | 30.80 |
| 2" x 8" | Ea | 2,060.00 | 22.00 | 32.90 | 43.90 |
| **Southern yellow pine – #1 S4S** | | | | | |
| 2" x  4" | Ea | 690.00 | 3.68 | 5.52 | 7.36 |
| 2" x  6" | Ea | 571.00 | 4.57 | 6.85 | 9.13 |
| 2" x 10" | Ea | 525.00 | 7.00 | 10.50 | 14.00 |
| 2" x 12" | Ea | 632.00 | 10.10 | 15.20 | 20.20 |
| **Southern yellow pine – #2 S4S** | | | | | |
| 2" x  4" | Ea | 381.00 | 2.03 | 3.05 | 4.06 |
| 2" x  6" | Ea | 418.00 | 3.34 | 5.02 | 6.69 |
| 2" x  8" | Ea | 433.00 | 4.62 | 6.93 | 9.24 |
| 2" x 10" | Ea | 462.00 | 6.16 | 9.24 | 12.30 |
| 2" x 12" | Ea | 546.00 | 8.73 | 13.10 | 17.50 |
| **Southern yellow pine – Prime S4S** | | | | | |
| 2" x  4" | Ea | 507.00 | 2.70 | 4.05 | 5.41 |
| 2" x  6" | Ea | 390.00 | 3.12 | 4.69 | 6.25 |
| 2" x  8" | Ea | 450.00 | 4.79 | 7.19 | 9.59 |
| 2" x 10" | Ea | 461.00 | 6.14 | 9.21 | 12.30 |
| 2" x 12" | Ea | 598.00 | 9.56 | 14.40 | 19.10 |
| **Redwood. West Coast prices. Add 10% to 30% for transportation to other areas.** | | | | | |
| 2" x  2" – B Grade S4S | Ea | 2,340.00 | 6.24 | 9.36 | 12.50 |
| 2" x  2" – Green | Ea | 1,810.00 | 4.82 | 7.22 | 9.63 |
| 2" x  3" – Green | Ea | 1,350.00 | 5.42 | 8.12 | 10.80 |
| 2" x  4" – Construction Common | Ea | 1,510.00 | 8.07 | 12.10 | 16.10 |
| 2" x  6" – Construction Common | Ea | 1,220.00 | 9.80 | 14.70 | 19.60 |
| 2" x  4" – Construction Rough | Ea | 1,400.00 | 7.44 | 11.20 | 14.90 |
| 2" x  4" – Common S4S | Ea | 1,490.00 | 7.96 | 11.90 | 15.90 |
| 2" x  6" – Common S4S | Ea | 1,420.00 | 11.40 | 17.10 | 22.80 |
| 2" x  8" – Common S4S | Ea | 1,530.00 | 16.30 | 24.40 | 32.60 |
| 2" x 12" – Common S4S | Ea | 1,700.00 | 27.20 | 40.80 | 54.30 |
| 2" x  4" – Construction Heart, Rough | Ea | 2,030.00 | 10.80 | 16.20 | 21.60 |
| 2" x  6" – Construction Heart, Rough | Ea | 1,900.00 | 15.20 | 22.80 | 30.40 |
| 2" x  4" – Construction Heart, S4S | Ea | 2,180.00 | 11.60 | 17.40 | 23.20 |
| 2" x  6" – Construction Heart, S4S | Ea | 1,750.00 | 14.00 | 21.00 | 27.90 |
| 2" x  8" – Construction Heart, S4S | Ea | 1,910.00 | 20.40 | 30.60 | 40.70 |
| 2" x 12" – Construction Heart, S4S | Ea | 2,300.00 | 36.70 | 55.10 | 73.50 |

|  | Craft@Hrs | Unit | Material | Labor | Total |
|---|---|---|---|---|---|

**Studs** Typical costs for pre-cut white wood studding material, stud grade. Nominal sizes. Actual width and depth are about 1/2" less when dry. White wood species includes Engelmann spruce, all true firs, hemlocks, and pines (excluding Ponderosa pine).

|  | Craft@Hrs | Unit | Material | Labor | Total |
|---|---|---|---|---|---|
| 2" x 3" x 84" | B1@.192 | Ea | 1.95 | 6.40 | 8.35 |
| 2" x 4" x 84" | B1@.196 | Ea | 1.89 | 6.53 | 8.42 |
| 2" x 4" x 88" | B1@.200 | Ea | 2.43 | 6.66 | 9.09 |
| 2" x 4" x 92-5/8" | B1@.216 | Ea | 2.30 | 7.19 | 9.49 |
| 2" x 4" x 96" | B1@.224 | Ea | 2.31 | 7.46 | 9.77 |
| 2" x 4" x 104-5/8" | B1@.230 | Ea | 2.60 | 7.66 | 10.26 |
| 2" x 6" x 92-5/8" | B1@.308 | Ea | 3.28 | 10.30 | 13.58 |

**Boards**

Douglas fir finish and clear – C & Better (V/G) S4S, dry, vertical grain. West Coast prices. Add 10% to 30% for transportation to other areas.

|  | Unit | MBF | 8' L | 12' L | 16' L |
|---|---|---|---|---|---|
| 1" x 2" | Ea | 4,170.00 | 5.53 | 8.31 | 11.10 |
| 1" x 3" | Ea | 4,170.00 | 8.34 | 12.50 | 16.70 |
| 1" x 4" | Ea | 4,170.00 | 11.10 | 16.70 | 22.30 |
| 1" x 6" | Ea | 4,730.00 | 19.00 | 28.20 | 37.80 |
| 1" x 8" | Ea | 5,780.00 | 30.90 | 46.10 | 61.50 |
| 1" x 10" | Ea | 6,430.00 | 42.80 | 64.30 | 85.60 |
| 1" x 12" | Ea | 6,880.00 | 55.00 | 82.50 | 110.00 |

Douglas fir finish and clear – C & Better (F/G) S4S, dry, flat grain. West Coast prices. Add 10% to 30% for transportation to other areas.

|  | Unit | MBF | 8' L | 12' L | 16' L |
|---|---|---|---|---|---|
| 1" x 2" | Ea | 3,580.00 | 4.76 | 7.13 | 9.50 |
| 1" x 3" | Ea | 3,580.00 | 7.16 | 10.70 | 14.30 |
| 1" x 4" | Ea | 3,580.00 | 9.53 | 14.30 | 19.10 |
| 1" x 6" | Ea | 3,750.00 | 15.00 | 22.50 | 29.90 |
| 1" x 8" | Ea | 4,820.00 | 25.70 | 38.50 | 51.50 |
| 1" x 10" | Ea | 6,710.00 | 44.70 | 67.10 | 89.40 |
| 1" x 12" | Ea | 6,870.00 | 54.90 | 82.40 | 110.00 |

Southern pine boards – C & Better

|  | Unit | MBF | 8' L | 12' L | 16' L |
|---|---|---|---|---|---|
| 1" x 4" | Ea | 2,220.00 | 5.91 | 8.87 | 11.80 |
| 1" x 6" | Ea | 2,450.00 | 9.79 | 14.70 | 19.60 |
| 1" x 8" | Ea | 2,460.00 | 13.10 | 19.70 | 26.20 |
| 1" x 12" | Ea | 3,290.00 | 26.30 | 39.50 | 52.60 |

White pine boards, Select

|  | Unit | MBF | 8' L | 12' L | 16' L |
|---|---|---|---|---|---|
| 1" x 4" | Ea | 3,370.00 | 8.97 | 13.50 | 18.00 |
| 1" x 6" | Ea | 4,360.00 | 17.40 | 26.10 | 34.90 |
| 1" x 8" | Ea | 3,700.00 | 19.70 | 29.60 | 39.50 |
| 1" x 10" | Ea | 3,880.00 | 25.80 | 38.70 | 51.70 |
| 1" x 12" | Ea | 4,240.00 | 34.00 | 50.90 | 67.90 |

Eastern pine boards – Premium

|  | Unit | MBF | 8' L | 12' L | 16' L |
|---|---|---|---|---|---|
| 1" x 4" | Ea | 1,930.00 | 5.14 | 7.71 | 10.30 |
| 1" x 6" | Ea | 1,870.00 | 7.49 | 11.20 | 15.00 |
| 1" x 8" | Ea | 1,460.00 | 7.78 | 11.70 | 15.60 |
| 1" x 10" | Ea | 1,570.00 | 10.50 | 15.70 | 21.00 |
| 1" x 12" | Ea | 2,080.00 | 16.70 | 25.00 | 33.30 |

# Lumber, Boards

| | Unit | MBF | 8' L | 12' L | 16' L |
|---|---|---|---|---|---|
| Southern yellow pine flooring – D Grade | | | | | |
| 1" x 4" | Ea | 1,270.00 | 3.40 | 5.11 | 6.87 |
| 1" x 6" | Ea | 1,450.00 | 5.73 | 8.59 | 11.30 |
| 1" x 8" | Ea | 1,400.00 | 7.53 | 10.30 | 15.10 |
| 1" x 10" | Ea | 1,290.00 | 8.61 | 11.30 | 17.10 |
| 1" x 12" | Ea | 1,630.00 | 13.00 | 19.50 | 26.00 |
| Fir flooring – C & Better tongue & groove | | | | | |
| 1" x 3" vertical grain, C and Better | Ea | 2,750.00 | 5.50 | 8.25 | 11.00 |
| 1" x 4" vertical grain, C and Better | Ea | 2,750.00 | 7.33 | 11.00 | 14.60 |
| 1" x 3" flat grain, C and Better | Ea | 1,650.00 | 3.31 | 4.96 | 6.62 |
| 1" x 4" flat grain, C and Better | Ea | 1,650.00 | 4.40 | 6.62 | 8.81 |
| 1" x 3" vertical grain, D and Better | Ea | 1,960.00 | 3.96 | 5.92 | 7.90 |
| 1" x 4" vertical grain, D and Better | Ea | 1,960.00 | 5.27 | 7.90 | 10.10 |
| 1" x 3" flat grain, D and Better | Ea | 1,310.00 | 2.59 | 3.91 | 5.19 |
| 1" x 4" flat grain, D and Better | Ea | 1,310.00 | 3.47 | 5.19 | 6.92 |
| Ponderosa pine boards – #4 & Better | | | | | |
| 1" x 3" | Ea | 1,710.00 | 3.41 | 5.12 | 6.82 |
| 1" x 4" | Ea | 1,500.00 | 4.00 | 5.99 | 7.99 |
| 1" x 6" | Ea | 1,540.00 | 6.15 | 9.22 | 12.30 |
| 1" x 8" | Ea | 1,480.00 | 7.87 | 11.80 | 15.70 |
| 1" x 10" | Ea | 1,540.00 | 10.20 | 15.40 | 20.50 |
| White wood boards – Appearance Grade | | | | | |
| 1" x 4" | Ea | 1,780.00 | 4.74 | 7.11 | 9.48 |
| 1" x 6" | Ea | 1,910.00 | 7.66 | 11.50 | 15.30 |
| 1" x 8" | Ea | 1,950.00 | 10.40 | 15.60 | 20.70 |
| 1" x 12" | Ea | 2,430.00 | 19.50 | 29.20 | 38.90 |
| Cypress boards | | | | | |
| 1" x 4" | Ea | 1,760.00 | 4.70 | 7.05 | 9.40 |
| 1" x 6" | Ea | 2,130.00 | 8.53 | 12.80 | 17.10 |
| 1" x 8" | Ea | 2,390.00 | 12.70 | 19.10 | 25.40 |
| 1" x 10" | Ea | 2,420.00 | 16.10 | 24.20 | 32.30 |
| 1" x 12" | Ea | 2,790.00 | 22.40 | 33.50 | 44.70 |
| Hemlock boards – C & Better | | | | | |
| 1" x 2" | Ea | 3,920.00 | 5.22 | 7.84 | 10.50 |
| 1" x 3" | Ea | 3,810.00 | 7.61 | 11.40 | 15.20 |
| 1" x 4" | Ea | 3,870.00 | 10.30 | 15.50 | 20.60 |
| 1" x 6" | Ea | 4,730.00 | 18.90 | 28.40 | 37.80 |
| Pine boards – Select | | | | | |
| 1" x 2" | Ea | 5,210.00 | 6.95 | 10.40 | 13.90 |
| 1" x 3" | Ea | 6,060.00 | 12.10 | 18.20 | 24.20 |
| Clear Douglas fir – S4S | | | | | |
| 1" x 2" | Ea | 3,320.00 | 4.43 | 6.65 | 8.86 |
| 1" x 3" | Ea | 3,380.00 | 6.75 | 10.10 | 13.50 |
| 1" x 4" | Ea | 3,890.00 | 10.40 | 15.50 | 20.70 |
| 1" x 6" | Ea | 4,740.00 | 19.00 | 28.40 | 37.90 |
| 1" x 8" | Ea | 4,100.00 | 21.90 | 32.80 | 43.70 |
| Douglas fir / hemlock | | | | | |
| 1" x 4" | Ea | 1,150.00 | 3.06 | 4.59 | 6.12 |
| 1" x 6" | Ea | 1,230.00 | 4.92 | 7.38 | 9.84 |
| Southern yellow pine premium decking | | | | | |
| 5/4" x 6" | Ea | 2,500.00 | 12.50 | 18.80 | 25.00 |

| | Unit | 6' L | 8' L | 12' L | 16' L |
|---|---|---|---|---|---|
| **Red oak boards, S4S, East Coast and Midwest prices** | | | | | |
| 1" x 2" | Ea | 5.90 | 7.83 | 10.10 | 12.10 |
| 1" x 3" | Ea | 9.02 | 12.10 | 15.50 | 18.60 |
| 1" x 4" | Ea | 12.10 | 16.10 | 22.30 | 26.50 |
| 1" x 6" | Ea | 18.60 | 24.80 | 32.70 | 39.20 |
| 1" x 8" | Ea | 24.40 | 32.40 | 44.00 | 58.70 |
| 1" x 10" | Ea | 36.10 | 49.00 | 60.40 | 72.40 |
| 1" x 12" | Ea | 43.50 | 54.60 | 77.00 | 91.80 |

| | Unit | MBF | 8' L | 12' L | 16' L |
|---|---|---|---|---|---|
| **Cedar boards – S1S2E (surfaced 1 side, 2 ends)** | | | | | |
| 3/4" x 2" | Ea | 3,970.00 | 3.97 | 5.96 | 7.95 |
| 3/4" x 4" | Ea | 3,410.00 | 6.83 | 10.20 | 13.70 |
| 3/4" x 6" | Ea | 4,510.00 | 13.50 | 20.30 | 27.10 |
| 7/8" x 4" | Ea | 2,830.00 | 6.61 | 9.91 | 13.20 |
| 7/8" x 6" | Ea | 3,980.00 | 13.90 | 20.90 | 27.90 |
| 7/8" x 8" | Ea | 3,250.00 | 15.20 | 22.70 | 30.30 |
| 7/8" x 10" | Ea | 3,310.00 | 19.30 | 28.90 | 38.60 |
| 7/8" x 12" | Ea | 3,550.00 | 24.90 | 37.30 | 49.50 |
| 1" x 4" | Ea | 2,790.00 | 7.43 | 11.10 | 14.90 |
| 1" x 6" | Ea | 3,460.00 | 13.80 | 20.80 | 27.70 |
| **Clear cedar trim boards** | | | | | |
| 1" x 2" | Ea | 1,980.00 | 2.64 | 3.96 | 5.28 |
| 1" x 3" | Ea | 4,240.00 | 8.48 | 12.70 | 17.00 |
| 1" x 4" | Ea | 4,580.00 | 12.20 | 18.30 | 24.40 |
| 1" x 6" | Ea | 4,590.00 | 18.40 | 27.50 | 36.70 |
| 1" x 8" | Ea | 4,550.00 | 24.30 | 36.40 | 48.60 |
| **Red cedar boards, kiln dried** | | | | | |
| 1" x 4" | Ea | 2,490.00 | 6.64 | 9.95 | 13.30 |
| 1" x 6" | Ea | 3,500.00 | 14.00 | 21.00 | 28.00 |
| 1" x 8" | Ea | 3,240.00 | 17.30 | 25.90 | 34.60 |
| **Redwood boards, West Coast prices.** | | | | | |
| **Add 10% to 30% for transportation to other areas.** | | | | | |
| 1" x 2" – B Grade S4S | Ea | 3,520.00 | 4.70 | 7.04 | 9.39 |
| 1" x 3" – B Grade S4S | Ea | 3,480.00 | 6.96 | 10.40 | 13.90 |
| 1" x 4" – B Grade S4S | Ea | 3,890.00 | 10.40 | 15.60 | 20.70 |
| 1" x 6" – B Grade S4S | Ea | 3,770.00 | 15.10 | 22.70 | 30.20 |
| 1" x 4" – Green | Ea | 2,570.00 | 6.85 | 10.30 | 13.70 |
| 1" x 6" – Green | Ea | 3,300.00 | 13.20 | 19.80 | 26.40 |
| 1" x 8" – Select S1S2E | Ea | 2,270.00 | 12.10 | 18.10 | 24.20 |

| | Craft@Hrs | Unit | Material | Labor | Total |
|---|---|---|---|---|---|
| **Lath, Redwood "A" & Better, 50 per bundle, nailed to walls for architectural treatment.** | | | | | |
| 5/16" x 1-1/2" x 4' | BC@.500 | Bdle | 16.80 | 18.50 | 35.30 |
| 5/16" x 1-1/2" x 6' | BC@.625 | Bdle | 22.50 | 23.10 | 45.60 |
| 5/16" x 1-1/2" x 8' | BC@.750 | Bdle | 31.50 | 27.70 | 59.20 |

| | Craft@Hrs | Unit | Material | Labor | Total |
|---|---|---|---|---|---|

## Board Siding

Pine bevel siding, 1/2" thick by 6" wide. Each linear foot of 6" wide bevel siding covers .38 square feet of wall. Do not deduct for wall openings less than 10 square feet. Add 5% for typical waste and end cutting.

| | Craft@Hrs | Unit | Material | Labor | Total |
|---|---|---|---|---|---|
| 8' lengths | B1@.038 | SF | 1.08 | 1.27 | 2.35 |
| 12' lengths | B1@.036 | SF | 1.07 | 1.20 | 2.27 |
| 16' lengths | B1@.033 | SF | 1.09 | 1.10 | 2.19 |

Western red cedar bevel siding. A Grade and Better, 1/2" thick by 6" wide. Each linear foot of 6" wide bevel siding covers .38 square feet of wall. Do not deduct for wall openings less than 10 square feet. Add 5% for typical waste and end cutting.

| | Craft@Hrs | Unit | Material | Labor | Total |
|---|---|---|---|---|---|
| 8' lengths | B1@.038 | SF | 4.66 | 1.27 | 5.93 |
| 12' lengths | B1@.036 | SF | 4.50 | 1.20 | 5.70 |
| 16' lengths | B1@.033 | SF | 4.57 | 1.10 | 5.67 |

Clear cedar 1/2" bevel siding. Vertical grain, 1/2" thick by 6" wide, kiln dried. Each linear foot of 6" wide bevel siding covers .38 square feet of wall. Do not deduct for wall openings less than 10 square feet. Add 5% for typical waste and end cutting.

| | Craft@Hrs | Unit | Material | Labor | Total |
|---|---|---|---|---|---|
| 8' lengths | B1@.038 | SF | 4.26 | 1.27 | 5.53 |
| 12' lengths | B1@.036 | SF | 4.29 | 1.20 | 5.49 |
| 16' lengths | B1@.033 | SF | 4.34 | 1.10 | 5.44 |

Cedar 3/4" bevel siding, 3/4" thick by 8" wide. Each linear foot of 8" wide bevel siding covers .54 square feet of wall. Do not deduct for wall openings less than 10 square feet. Add 5% for typical waste and end cutting.

| | Craft@Hrs | Unit | Material | Labor | Total |
|---|---|---|---|---|---|
| 8' lengths | B1@.036 | SF | 2.48 | 1.20 | 3.68 |
| 12' lengths | B1@.034 | SF | 2.46 | 1.13 | 3.59 |
| 16' lengths | B1@.031 | SF | 2.50 | 1.03 | 3.53 |

Tongue and groove cedar siding, 1" x 6". Kiln dried, select. Each linear foot of 6" wide tongue and groove siding covers .43 square feet of wall. Do not deduct for wall openings less than 10 square feet. Add 5% for typical waste and end cutting.

| | Craft@Hrs | Unit | Material | Labor | Total |
|---|---|---|---|---|---|
| 8' lengths | B1@.036 | SF | 3.95 | 1.20 | 5.15 |
| 10' lengths | B1@.035 | SF | 3.24 | 1.17 | 4.41 |
| 12' lengths | B1@.034 | SF | 3.53 | 1.13 | 4.66 |

Redwood 5/8" tongue and groove siding, 5/8" thick by 5-3/8" wide. Each linear foot of 5-3/8" wide tongue and groove siding covers .42 square feet of wall. Do not deduct for wall openings less than 10 square feet. Add 5% for typical waste and end cutting.

| | Craft@Hrs | Unit | Material | Labor | Total |
|---|---|---|---|---|---|
| 8' lengths | B1@.042 | SF | 4.73 | 1.40 | 6.13 |
| 10' lengths | B1@.040 | SF | 4.62 | 1.33 | 5.95 |
| 12' lengths | B1@.038 | SF | 4.71 | 1.27 | 5.98 |

Redwood 5/8" shiplap siding, B Grade. 5/8" thick by 5-3/8" wide. Each linear foot of 5-3/8" wide shiplap siding covers .38 square feet of wall. Do not deduct for wall openings less than 10 square feet. Add 5% for typical waste and end cutting.

| | Craft@Hrs | Unit | Material | Labor | Total |
|---|---|---|---|---|---|
| 8' lengths | B1@.038 | SF | 5.23 | 1.27 | 6.50 |
| 10' lengths | B1@.037 | SF | 5.22 | 1.23 | 6.45 |
| 12' lengths | B1@.036 | SF | 5.21 | 1.20 | 6.41 |

Southern yellow pine, 6" pattern 105 siding, 1" x 6". Select Grade except as noted. Each linear foot of 6" wide siding covers .38 square feet of wall. Do not deduct for wall openings less than 10 square feet. Add 5% for typical waste and end cutting.

| | Craft@Hrs | Unit | Material | Labor | Total |
|---|---|---|---|---|---|
| 8' lengths | B1@.038 | SF | 3.65 | 1.27 | 4.92 |
| 10' lengths | B1@.037 | SF | 3.71 | 1.23 | 4.94 |
| 12' lengths, No. 2 | B1@.036 | SF | 2.09 | 1.20 | 3.29 |
| 12' lengths | B1@.036 | SF | 3.68 | 1.20 | 4.88 |

| | Craft@Hrs | Unit | Material | Labor | Total |
|---|---|---|---|---|---|

Southern yellow pine, 8" pattern 105 siding, 1" x 8". Select Grade except as noted. Each linear foot of 8" wide siding covers .54 square feet of wall. Do not deduct for wall openings less than 10 square feet. Add 5% for typical waste and end cutting.

| | Craft@Hrs | Unit | Material | Labor | Total |
|---|---|---|---|---|---|
| 8' lengths | B1@.036 | SF | 3.39 | 1.20 | 4.59 |
| 10' lengths | B1@.034 | SF | 3.28 | 1.13 | 4.41 |
| 12' lengths, No. 2 | B1@.031 | SF | 1.68 | 1.03 | 2.71 |
| 12' lengths | B1@.031 | SF | 3.27 | 1.03 | 4.30 |

Southern yellow pine, 6" pattern 116 siding, 1" x 6". Select Grade except as noted. Each linear foot of 6" wide siding covers .38 square feet of wall. Do not deduct for wall openings less than 10 square feet. Add 5% for typical waste and end cutting.

| | Craft@Hrs | Unit | Material | Labor | Total |
|---|---|---|---|---|---|
| 8' lengths | B1@.038 | SF | 3.70 | 1.27 | 4.97 |
| 10' lengths | B1@.037 | SF | 3.68 | 1.23 | 4.91 |
| 12' lengths | B1@.036 | SF | 3.73 | 1.20 | 4.93 |

Southern yellow pine, 6" pattern 117 siding, 1" x 6". Select Grade except as noted. Each linear foot of 6" wide siding covers .38 square feet of wall. Do not deduct for wall openings less than 10 square feet. Add 5% for typical waste and end cutting.

| | Craft@Hrs | Unit | Material | Labor | Total |
|---|---|---|---|---|---|
| 8' lengths | B1@.038 | SF | 3.62 | 1.27 | 4.89 |
| 10' lengths | B1@.037 | SF | 3.72 | 1.23 | 4.95 |
| 12' lengths | B1@.036 | SF | 3.62 | 1.20 | 4.82 |

Southern yellow pine, 8" pattern 131 siding, 1" x 8". Select Grade except as noted. Each linear foot of 8" wide siding covers .54 square feet of wall. Do not deduct for wall openings less than 10 square feet. Add 5% for typical waste and end cutting.

| | Craft@Hrs | Unit | Material | Labor | Total |
|---|---|---|---|---|---|
| 8' lengths | B1@.036 | SF | 3.31 | 1.20 | 4.51 |
| 10' lengths | B1@.034 | SF | 3.30 | 1.13 | 4.43 |
| 12' lengths | B1@.031 | SF | 3.29 | 1.03 | 4.32 |

Southern yellow pine, 8" pattern 122 siding, 1" x 8". "V"-joint. Select Grade except as noted. Each linear foot of 8" wide siding covers .54 square feet of wall. Do not deduct for wall openings less than 10 square feet. Add 5% for typical waste and end cutting.

| | Craft@Hrs | Unit | Material | Labor | Total |
|---|---|---|---|---|---|
| 8' lengths | B1@.036 | SF | 3.27 | 1.20 | 4.47 |
| 10' lengths | B1@.034 | SF | 3.24 | 1.13 | 4.37 |
| 12' lengths | B1@.031 | SF | 3.23 | 1.03 | 4.26 |

Southern yellow pine, 6" pattern 122 siding, 1" x 6". Select Grade except as noted. Each linear foot of 6" wide siding covers .38 square feet of wall. Do not deduct for wall openings less than 10 square feet. Add 5% for typical waste and end cutting.

| | Craft@Hrs | Unit | Material | Labor | Total |
|---|---|---|---|---|---|
| 8' lengths | B1@.038 | SF | 2.56 | 1.27 | 3.83 |
| 10' lengths | B1@.037 | SF | 2.55 | 1.23 | 3.78 |
| 12' lengths | B1@.036 | SF | 2.55 | 1.20 | 3.75 |

Southern yellow pine, 8" shiplap siding, 1" x 8". No. 2 Grade. Each linear foot of 8" wide siding covers .54 square feet of wall. Do not deduct for wall openings less than 10 square feet. Add 5% for typical waste and end cutting.

| | Craft@Hrs | Unit | Material | Labor | Total |
|---|---|---|---|---|---|
| 8' lengths | B1@.036 | SF | 1.55 | 1.20 | 2.75 |
| 10' lengths | B1@.034 | SF | 1.38 | 1.13 | 2.51 |
| 12' lengths | B1@.031 | SF | 1.54 | 1.03 | 2.57 |

Rustic 8" pine siding, 1" x 8". "V"-grooved. Each linear foot of 8" wide siding covers .54 square feet of wall. Do not deduct for wall openings less than 10 square feet. Add 5% for typical waste and end cutting.

| | Craft@Hrs | Unit | Material | Labor | Total |
|---|---|---|---|---|---|
| 16' lengths | B1@.031 | SF | 3.78 | 1.03 | 4.81 |
| 20' lengths | B1@.031 | SF | 3.85 | 1.03 | 4.88 |

# Lumber, Timbers

| | Craft@Hrs | Unit | Material | Labor | Total |
|---|---|---|---|---|---|

Log cabin siding, 1-1/2" thick. Each linear foot of 6" wide siding covers .41 square feet of wall. Each linear foot of 8" wide siding covers .55 square feet of wall. Do not deduct for wall openings less than 10 square feet. Add 5% for typical waste and end cutting.

| | Craft@Hrs | Unit | Material | Labor | Total |
|---|---|---|---|---|---|
| 6" x 8', T & G latia | B1@.038 | SF | 2.63 | 1.27 | 3.90 |
| 6" x 12', T & G latia | B1@.036 | SF | 2.68 | 1.20 | 3.88 |
| 8" x 12 | B1@.031 | SF | 2.91 | 1.03 | 3.94 |
| 8" x 16' | B1@.030 | SF | 2.90 | 1.00 | 3.90 |

Log lap spruce siding, 1-1/2" thick. Each linear foot of 6" wide siding covers .41 square feet of wall. Each linear foot of 8" wide siding covers .55 square feet of wall. Each linear foot of 10" wide siding covers .72 square feet of wall. Do not deduct for wall openings less than 10 square feet. Add 5% for typical waste and end cutting.

| | Craft@Hrs | Unit | Material | Labor | Total |
|---|---|---|---|---|---|
| 6" width | B1@.038 | SF | 4.29 | 1.27 | 5.56 |
| 8" width | B1@.031 | SF | 4.23 | 1.03 | 5.26 |
| 10" width | B1@.030 | SF | 3.42 | 1.00 | 4.42 |

## Timbers

| | Unit | MBF | 12' L | 16' L | 20' L |
|---|---|---|---|---|---|
| Green Douglas fir timbers – #1 Grade | | | | | |
| 4" x 8" | Ea | 843.00 | 18.00 | 27.00 | 36.00 |
| 4" x 10" | Ea | 827.00 | 22.00 | 33.10 | 44.10 |
| 4" x 12" | Ea | 830.00 | 26.60 | 39.80 | 53.10 |
| 6" x 6" | Ea | 942.00 | 22.60 | 33.90 | 45.20 |

| | Unit | 8' L | 10' L | 12' L | 16' L |
|---|---|---|---|---|---|
| Green Douglas fir timbers – #2 and Better | | | | | |
| 4" x 4" | Ea | 6.37 | 7.97 | 9.56 | 12.80 |
| 4" x 6" | Ea | 10.20 | 12.70 | 15.20 | 20.30 |
| 4" x 8" | Ea | 12.70 | 15.90 | 19.10 | 25.50 |
| 4" x 10" | Ea | 16.90 | 21.20 | 25.40 | 33.90 |
| 4" x 12" | Ea | 22.20 | 27.80 | 33.30 | 44.40 |

| | Unit | MBF | 8' L | 12' L | 16' L |
|---|---|---|---|---|---|
| Red cedar timbers | | | | | |
| 4" x 4" – Green S4S | Ea | 2,450.00 | 26.10 | 39.20 | 52.20 |
| 4" x 4" – Rough | Ea | 1,680.00 | 17.90 | 26.90 | 35.90 |
| 6" x 6" – Rough | Ea | 2,740.00 | 65.70 | 98.60 | 131.00 |

Redwood timbers  West Coast prices.

Add 10% to 30% for transportation to other areas.

| | Unit | MBF | 8' L | 12' L | 16' L |
|---|---|---|---|---|---|
| 4" x 4" – Construction Common | Ea | 1,570.00 | 16.80 | 25.10 | 33.50 |
| 4" x 4" – Construction Heart - Rough | Ea | 1,900.00 | 20.20 | 30.40 | 40.50 |
| 4" x 4" – Construction Heart – S4S | Ea | 1,960.00 | 20.90 | 31.40 | 41.80 |

## Pressure treated lumber, ACQ treatment

Pressure treated trim board S4S, .25 treatment, Appearance Grade

| | Unit | MBF | | | |
|---|---|---|---|---|---|
| 1" x 4" – .25 appearance grade | Ea | 1,270.00 | 3.39 | 5.08 | 6.78 |
| 1" x 6" – .25 appearance grade | Ea | 1,170.00 | 4.67 | 7.00 | 9.34 |
| 1" x 8" – .25 appearance grade | Ea | 1,190.00 | 6.34 | 9.51 | 12.70 |

# Lumber, Pressure Treated

| | Unit | MBF | 8' L | 12' L | 16' L |
|---|---|---|---|---|---|
| **Pressure treated trim board S4S, .25 treatment, D Grade** | | | | | |
| 1" x 4", – .25 D grade | Ea | 1,650.00 | 4.38 | 6.57 | 8.76 |
| 1" x 6", – .25 D grade | Ea | 1,870.00 | 7.49 | 11.20 | 15.00 |
| 1" x 8", – .25 D grade | Ea | 2,000.00 | 10.70 | 16.00 | 21.40 |
| 1" x 12", – .25 D grade | Ea | 1,990.00 | 15.90 | 23.90 | 31.80 |
| **Pressure treated lumber, .25 ACQ treatment** | | | | | |
| 2" x 3" | Ea | 601.00 | 3.20 | 4.80 | 6.40 |
| 2" x 4" | Ea | 843.00 | 4.49 | 6.74 | 8.98 |
| 2" x 6" | Ea | 820.00 | 6.56 | 9.84 | 13.10 |
| 2" x 8" | Ea | 911.00 | 9.71 | 14.60 | 19.40 |
| 2" x 10" | Ea | 868.00 | 11.60 | 17.40 | 23.10 |
| 2" x 12" | Ea | 900.00 | 14.40 | 21.60 | 28.80 |
| 4" x 4" | Ea | 1,040.00 | 11.10 | 16.70 | 22.30 |
| 4" x 6" | Ea | 1,090.00 | 17.40 | 26.10 | 34.70 |
| 4" x 8" | Ea | 1,190.00 | 25.40 | 38.20 | 50.90 |
| 4" x 10" | Ea | 1,220.00 | 32.70 | 49.00 | 65.30 |
| 4" x 12" | Ea | 1,280.00 | 41.10 | 61.60 | 82.20 |
| 6" x 6" | Ea | 876.00 | 21.00 | 31.50 | 42.10 |
| **Pressure treated southern yellow pine – .40 ACQ treatment, Prime** | | | | | |
| 1" x 4" | Ea | 1,000.00 | 2.68 | 4.01 | 5.35 |
| 1" x 6" | Ea | 892.00 | 3.57 | 5.35 | 7.13 |
| 2" x 4" | Ea | 861.00 | 4.59 | 6.88 | 9.17 |
| 2" x 6" | Ea | 791.00 | 6.33 | 9.49 | 12.70 |
| 2" x 8" | Ea | 823.00 | 8.77 | 13.20 | 17.50 |
| 2" x 10" | Ea | 781.00 | 10.40 | 15.60 | 20.80 |
| 2" x 12" | Ea | 1,180.00 | 19.00 | 28.40 | 37.90 |
| 4" x 4" | Ea | 1,810.00 | 19.30 | 29.00 | 38.70 |
| 4" x 6" | Ea | 1,690.00 | 27.00 | 40.50 | 54.00 |
| 6" x 6" | Ea | 1,910.00 | 45.90 | 68.90 | 91.90 |

**Pressure treated timbers .40 CCA pressure treatment. For uses with direct contact with water or the ground, or where moisture content exceeds 20%. Add 20% for .60 CCA treated southern yellow pine timber.**

| | Unit | MBF | 8' L | 12' L | 16' L |
|---|---|---|---|---|---|
| 4" x 4" | Ea | 861.00 | 9.18 | 13.80 | 18.40 |
| 4" x 6" | Ea | 909.00 | 14.60 | 21.80 | 29.10 |
| 6" x 6" | Ea | 922.00 | 22.10 | 33.20 | 44.30 |

| | Unit | 8' L | 10'L | 12' L | 16' L |
|---|---|---|---|---|---|
| **Hibor treated Douglas fir (Hawaii)** | | | | | |
| 2" x 4", Standard and Better | Ea | 3.88 | 5.00 | 6.12 | 7.79 |
| 2" x 6", No. 2 and Better | Ea | 6.12 | 7.79 | 9.47 | 12.30 |
| 2" x 8", No. 2 and Better | Ea | 10.00 | 12.30 | 14.50 | 19.00 |
| 2" x 10", No. 2 and Better | Ea | 12.30 | 15.60 | 19.00 | 24.60 |
| 2" x 12", No. 2 and Better | Ea | 14.50 | 17.90 | 22.30 | 29.00 |
| 4" x 4", No. 2 and Better | Ea | 8.91 | 12.30 | 14.50 | — |
| 4" x 6", No. 2 and Better | Ea | 15.60 | 20.10 | 23.50 | 44.70 |
| 4" x 8", No. 1 and Better, S4S | Ea | — | — | 49.20 | 65.90 |
| 4" x 10", No. 1 and Better, S4S | Ea | — | — | 60.40 | 82.70 |
| 4" x 12", No. 1 and Better, S4S | Ea | — | — | 78.20 | 106.00 |

**Identification Index:**
A set of two numbers separated by a slash that appears in the grade trademarks on standard sheathing, Structural I, Structural II and C-C Grades. Number on left indicates recommended maximum spacing in inches for supports when panel is used for roof decking. Number on right shows maximum recommended spacing in inches for supports when the panel is used for subflooring.

# Lumber, Plywood

## Plywood

| | Unit | MSF | 4' x 8' |
|---|---|---|---|
| **Sanded Exterior Grade Western plywood – AC** | | | |
| 1/4" thick | Ea | 788.00 | 25.20 |
| 3/8" thick | Ea | 827.00 | 26.50 |
| 1/2" thick | Ea | 1,300.00 | 41.70 |
| 5/8" thick | Ea | 1,300.00 | 41.60 |
| 3/4" thick | Ea | 1,670.00 | 53.30 |
| **Exterior Grade plywood – CDX** | | | |
| 3/8" thick | Ea | 435.00 | 13.90 |
| 1/2" thick | Ea | 489.00 | 15.70 |
| 1/2" thick, 4-ply | Ea | 507.00 | 16.20 |
| 1/2" thick, 5-ply | Ea | 542.00 | 17.30 |
| 5/8" thick, 4-ply | Ea | 616.00 | 19.70 |
| 5/8" thick, 5-ply | Ea | 627.00 | 20.10 |
| 3/4" thick | Ea | 746.00 | 23.90 |
| **Sanded Exterior Grade Western plywood – BC** | | | |
| 1/4" thick | Ea | 788.00 | 25.20 |
| 3/8" thick | Ea | 827.00 | 26.50 |
| 1/2" thick | Ea | 976.00 | 31.20 |
| 5/8" thick | Ea | 1,220.00 | 39.00 |
| 3/4" thick | Ea | 1,410.00 | 45.30 |

**A-A EXT-DFPA**
Designed for use in exposed applications where the appearance of both sides is important. Fences, wind screens, exterior cabinets and built-ins, signs, boats, commercial refrigerators, shipping containers, tanks, and ducts

A-A  G-2  EXT-DFPA PS 1-66
Typical edge-mark

**A-B EXT-DFPA**
For uses similar to A-A EXT but where the appearance of one side is a little less important than the face.

A-B  G-3  EXT-DFPA PS 1-66
Typical edge-mark

| | Craft@Hrs | Unit | Material | Labor | Total |
|---|---|---|---|---|---|
| **Texture 1-11 (T-1-11) siding, rough sawn. Per 100 square feet installed including 10% cutting waste.** | | | | | |
| 3/8" x 4' x 8' Southern pine, grooved 4" OC | B1@2.18 | Sq | 103.00 | 72.60 | 175.60 |
| 3/8" x 4' x 8' Decorative southern pine | B1@2.18 | Sq | 93.40 | 72.60 | 166.00 |
| 3/8" x 4' x 8' Hibor treated, plain (Hawaii) | B1@2.18 | Sq | 148.00 | 72.60 | 220.60 |
| 5/8" x 4' x 8' Premium pine, grooved | B1@2.45 | Sq | 127.00 | 81.60 | 208.60 |
| 5/8" x 4' x 8' Treated pine, Decorative Grade | B1@2.45 | Sq | 188.00 | 81.60 | 269.60 |
| 5/8" x 4' x 9' Fir, grooved | B1@2.45 | Sq | 231.00 | 81.60 | 312.60 |
| 5/8" x 4' x 8' Hibor treated, grooved (Hawaii) | B1@2.45 | Sq | 224.00 | 81.60 | 305.60 |
| 5/8" x 4' x 9' Hibor treated, grooved (Hawaii) | B1@2.45 | Sq | 358.00 | 81.60 | 439.60 |

| | Unit | MSF | 4' x 8' |
|---|---|---|---|
| **2-4-1 Sturdi-Floor subflooring. Tongue and groove attic decking** | | | |
| 5/8" | Ea | 897.00 | 28.70 |
| **Concrete form (Plyform), square edged** | | | |
| 5/8", 4 ply | Ea | 2,840.00 | 90.90 |
| 3/4", 4 ply | Ea | 2,890.00 | 92.50 |
| **Structural I Grade plywood. Price per panel** | | | |
| 3/8" | Ea | 556.00 | 17.80 |
| 1/2" | Ea | 647.00 | 20.70 |
| **Medium Density Overlay (MDO) plywood. Paintable paper surface bonded to plywood core.** | | | |
| 3/8" | Ea | 1,710.00 | 54.70 |
| 1/2" | Ea | 2,090.00 | 66.60 |
| 3/4" | Ea | 2,560.00 | 81.90 |
| **Marine Grade plywood** | | | |
| 1/2", AB Grade | Ea | 3,080.00 | 98.40 |
| 3/4", AB Grade | Ea | 3,920.00 | 125.00 |

| | Unit | MSF | 4' x 8' |
|---|---|---|---|
| SB strand board – square edged, oriented grain | | | |
| 1/4" | Ea | 256.00 | 8.20 |
| 3/8" | Ea | 274.00 | 8.78 |
| 7/16" | Ea | 265.00 | 8.48 |
| 1/2" | Ea | 306.00 | 9.80 |
| 3/4" | Ea | 544.00 | 17.40 |
| Underlayment "X"-Band, C Grade, tongue and groove | | | |
| 5/8", 4 ply | Ea | 783.00 | 25.10 |
| 3/4", 4 ply | Ea | 848.00 | 27.20 |
| Birch hardwood plywood | | | |
| 1/8" thick | Ea | 819.00 | 26.20 |
| 1/4" thick | Ea | 961.00 | 30.80 |
| 1/2" thick | Ea | 1,710.00 | 54.70 |
| 3/4" thick | Ea | 2,070.00 | 66.10 |
| Cedar hardwood plywood | | | |
| 1/4" thick | Ea | 1,240.00 | 39.60 |
| 3/8" thick | Ea | 1,500.00 | 47.90 |
| 1/2" thick | Ea | 2,080.00 | 66.70 |
| 5/8" thick | Ea | 2,410.00 | 77.00 |
| 3/4" thick | Ea | 2,800.00 | 89.50 |
| Oak hardwood plywood | | | |
| 1/8" thick | Ea | 735.00 | 23.50 |
| 1/4" thick | Ea | 1,020.00 | 32.70 |
| 1/2" thick | Ea | 2,030.00 | 64.90 |
| 3/4" thick | Ea | 2,400.00 | 76.70 |
| Lauan plywood | | | |
| 1/8" thick | Ea | 479.00 | 15.30 |
| 1/4" thick | Ea | 550.00 | 17.60 |
| 1/4" thick, moisture resistant | Ea | 493.00 | 15.80 |

| | Craft@Hrs | Unit | Material | Labor | Total |
|---|---|---|---|---|---|
| Hardboard, 4' x 8' panels | | | | | |
| 1/8" tempered hardboard | BC@1.02 | Ea | 11.00 | 37.70 | 48.70 |
| 1/8" tempered pegboard | BC@1.02 | Ea | 17.40 | 37.70 | 55.10 |
| 1/8" backer board | BC@1.02 | Ea | 19.50 | 37.70 | 57.20 |
| 1/4" tempered hardboard | BC@1.02 | Ea | 17.10 | 37.70 | 54.80 |
| 3/16" white garage liner pegboard | BC@1.02 | Ea | 23.00 | 37.70 | 60.70 |
| 1/4" white vinyl coated | BC@1.02 | Ea | 25.90 | 37.70 | 63.60 |
| 1/4" tempered perforated pegboard | BC@1.02 | Ea | 18.20 | 37.70 | 55.90 |
| Particleboard | | | | | |
| 3/8" x 4' x 8' | — | Ea | 15.50 | — | 15.50 |
| 1/2" x 4' x 8' | — | Ea | 16.30 | — | 16.30 |
| 5/8" x 4' x 8' | — | Ea | 19.90 | — | 19.90 |
| 3/4" x 4' x 8' | — | Ea | 22.20 | — | 22.20 |
| 3/4" x 25" x 97" countertop | — | Ea | 15.50 | — | 15.50 |
| 3/4" x 30" x 97", countertop | — | Ea | 19.50 | — | 19.50 |
| 3/4" x 25" x 121", countertop | — | Ea | 23.50 | — | 23.50 |
| 3/4" x 25" x 145", countertop | — | Ea | 24.40 | — | 24.40 |
| 3/4" x 49" x 97", industrial grade | — | Ea | 25.00 | — | 25.00 |

# Lumber, Recycled Plastic

| | Unit | MBF | 8' L | 12' L | 16' L |
|---|---|---|---|---|---|
| **Recycled Plastic Lumber** Produced from post-consumer plastics. Resistant to termites, rot, decay. | | | | | |
| 2" x 4" | Ea | 5,040.00 | 26.80 | 40.30 | 53.70 |
| 2" x 6" | Ea | 5,060.00 | 40.50 | 60.70 | 80.90 |
| 2" x 8" | Ea | 4,880.00 | 52.00 | 78.00 | 104.00 |
| 2" x 10" | Ea | 5,050.00 | 67.30 | 101.00 | 135.00 |
| 2" x 12" | Ea | 5,180.00 | 82.90 | 124.00 | 166.00 |
| 4" x 4" | Ea | 6,160.00 | 65.70 | 98.50 | 131.00 |
| 4" x 6" | Ea | 5,270.00 | 84.30 | 126.00 | 169.00 |
| 6" x 6" | Ea | 5,240.00 | 126.00 | 189.00 | 251.00 |
| 1" x 6" siding, smooth edge | Ea | 4,560.00 | 18.20 | 29.90 | 36.50 |
| 1" x 6" siding, tongue & groove | Ea | 4,980.00 | 19.90 | 26.00 | 39.80 |

**Poles** Select Appearance Grade building poles. Poles have an average taper of 1" in diameter per 10' of length. Ponderosa pine treated with Penta-Dow or ACZA (.60 lbs/CF), meets AWPA-C 23-77. Costs based on quantity of 8 to 10 poles. For delivery of smaller quantities, add $50.00 to the total cost. See installation labor below. Costs are per linear foot.

| | Unit | 8" butt size | 10" butt size | 12" butt size |
|---|---|---|---|---|
| To 15' long pole | LF | 10.20 | 13.40 | 17.60 |
| To 20' long pole | LF | 9.93 | 12.60 | 16.40 |
| To 25' long pole | LF | 9.63 | 10.50 | 15.00 |
| Add for Douglas fir | LF | 2.31 | 2.41 | 2.51 |
| Add for delivery, typical per trip | Ea | 85.00 | 85.00 | 85.00 |

| | Unit | 6' long | 7' long | 8' long | 10' long |
|---|---|---|---|---|---|
| **Posts** Sawn posts, construction common, redwood | | | | | |
| 4" x 4" | Ea | 12.20 | 14.10 | 16.10 | 20.10 |
| Yellow pine penta treated posts | | | | | |
| 3" x 3" | Ea | 6.21 | 7.52 | 8.75 | 10.10 |
| 3-1/2" x 3-1/2" | Ea | 6.99 | 8.22 | 9.28 | 10.60 |
| 4" x 4" | Ea | 8.63 | 10.10 | 11.00 | 12.50 |
| 4-1/2" x 4-1/2" | Ea | 10.40 | 11.10 | 11.40 | 12.90 |
| 5" x 5" | Ea | 10.20 | 12.60 | 13.20 | 15.50 |
| 6" x 6" | Ea | 12.50 | 14.00 | 14.70 | 16.50 |

| | Unit | 8' long | 10' long | 12' long | 16' long |
|---|---|---|---|---|---|
| Pressure treated southern yellow pine posts No. 2, 0.40 CCA treatment | | | | | |
| 4" x 4" | Ea | 19.90 | 24.80 | 29.80 | 39.70 |
| 4" x 6" | Ea | 27.60 | 34.50 | 41.50 | 55.30 |
| 6" x 6" | Ea | 47.10 | 58.80 | 70.70 | 94.10 |

| | Craft@Hrs | Unit | Material | Labor | Total |
|---|---|---|---|---|---|
| Labor to install poles and posts Costs shown are per post and include measuring, cutting one end, drilling two holes at the top and bottom and installation. Add for hardware and bolts as needed. | | | | | |
| 3" x 3", 4" x 4" or 5" x 5" posts | B1@.333 | Ea | — | 11.10 | 11.10 |
| 6" x 6" or 7" x 7" posts | B1@.500 | Ea | — | 16.70 | 16.70 |
| 8" or 10" butt diameter poles | B1@.666 | Ea | — | 22.20 | 22.20 |
| 12" butt diameter poles | B1@.750 | Ea | — | 25.00 | 25.00 |

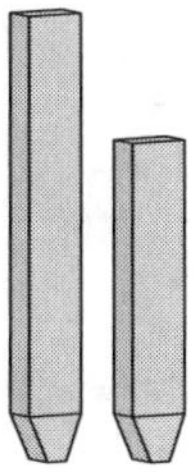

|  | Craft@Hrs | Unit | Material | Labor | Total |
|---|---|---|---|---|---|
| **Stakes** | | | | | |
| Survey stakes, pine or fir, 1" x 2" | | | | | |
| 18" | — | Ea | .23 | — | .23 |
| 24" | — | Ea | .28 | — | .28 |
| 36" | — | Ea | .57 | — | .57 |
| Ginnies, pine or fir, 1" x 1", per 1,000 | | | | | |
| 6" | — | M | 81.20 | — | 81.20 |
| 8" | — | M | 86.10 | — | 86.10 |
| Pointed lath, pine or fir, per 1,000, 1/4" x 1-1/2", S4S | | | | | |
| 18" | — | M | 239.00 | — | 239.00 |
| 24" | — | M | 258.00 | — | 258.00 |
| 36" | — | M | 357.00 | — | 357.00 |
| 48" | — | M | 487.00 | — | 487.00 |
| Foundation stakes, 1" x 3", per 1,000, 20 bundles per 1000 except for 48" (40 bundles per 1000) | | | | | |
| 12" or 14" | — | M | 295.00 | — | 295.00 |
| 18" | — | M | 378.00 | — | 378.00 |
| 24" | — | M | 436.00 | — | 436.00 |
| 36" | — | M | 615.00 | — | 615.00 |
| 48" | — | M | 994.00 | — | 994.00 |
| Hubs, pine or fir, 2" x 2", per 1,000 | | | | | |
| 8" | — | M | 515.00 | — | 515.00 |
| 10" | — | M | 575.00 | — | 575.00 |
| 12" | — | M | 628.00 | — | 628.00 |
| Form stakes (wedges), pine or fir, 5" point, per 1,000 | | | | | |
| 2" x 2" x 6" | — | M | 201.00 | — | 201.00 |
| 2" x 2" x 8" | — | M | 217.00 | — | 217.00 |
| 2" x 2" x 10" | — | M | 237.00 | — | 237.00 |
| 2" x 4" x 6" | — | M | 261.00 | — | 261.00 |
| 2" x 4" x 8" | — | M | 307.00 | — | 307.00 |
| 2" x 4" x 10" | — | M | 332.00 | — | 332.00 |
| 2" x 4" x 12" | — | M | 378.00 | — | 378.00 |
| Feather wedges | — | M | 282.00 | — | 282.00 |
| **Lumber Grading** Lumber grading services by West Coast Lumber Inspection Bureau. | | | | | |
| Inspection per MBF | — | MBF | — | — | .33 |
| Per month minimum – operating and shipping | — | Month | — | — | 330.00 |
| Per month min. – non-operating (No Shipments) | — | Month | — | — | 100.00 |

Reinspection, certificate inspection and grade stamping service for WCLIB Members. Costs do not include travel and subsistence charges.

|  | Craft@Hrs | Unit | Material | Labor | Total |
|---|---|---|---|---|---|
| On shipments west of the Rocky Mountains | — | Day | — | — | 450.00 |
| Overtime hours | — | Hour | — | — | 85.00 |
| Saturday, Sunday and holidays | — | Day | — | — | 700.00 |
| On shipments east of the Rocky Mountains | — | Day | — | — | 450.00 |
| Overtime hours | — | Hour | — | — | 85.00 |
| Saturday, Sunday and holidays | — | Day | — | — | 700.00 |

# Lumber Preservation Treatment

| | Craft@Hrs | Unit | Material | Labor | Total |
|---|---|---|---|---|---|

Reinspection, certificate inspection and grade stamping service for WCLIB Non-Members. Costs do not include travel and subsistence charges.

| | Craft@Hrs | Unit | Material | Labor | Total |
|---|---|---|---|---|---|
| On shipments west of the Rocky Mountains | — | Day | — | — | 600.00 |
| Overtime hours | — | Hour | — | — | 110.00 |
| Saturday, Sunday and holidays | — | Day | — | — | 900.00 |
| On shipments east of the Rocky Mountains | — | Day | — | — | 600.00 |
| Overtime hours | — | Hour | — | — | 210.00 |
| Saturday, Sunday and holidays | — | Day | — | — | 400.00 |

Certificate inspection service. Issuance of certificates to inspection subscribers employing WCLIB certified inspectors.

| | Craft@Hrs | Unit | Material | Labor | Total |
|---|---|---|---|---|---|
| Per thousand board feet, $30.00 minimum | — | MBF | — | — | 0.30 |

**Lumber Preservation Treatment** See also Soil Treatments. Costs assume treatment at a treatment plant, not on the construction site. Treatments listed below are designed to protect wood and wood products from decay and destruction caused by moisture (fungi), wood-boring insects and organisms. All retention rates listed (in parentheses) are in pounds per cubic foot and are the minimum recommended by the Society of American Wood Preservers. Costs are for treatment only and assume treating of 10,000 board feet or more. No lumber or transportation included. Add $70.00 per MBF for kiln-drying after treatment.

| | Craft@Hrs | Unit | Material | Labor | Total |
|---|---|---|---|---|---|
| Lumber submerged in or frequently exposed to salt water | | | | | |
| ACZA treatment (2.5) | — | MBF | — | — | 869.00 |
| ACZA treatment for structural lumber | | | | | |
| Piles and building poles (2.5) | — | CF | — | — | 9.80 |
| Board lumber and timbers (2.5) | — | MBF | — | — | 869.00 |
| Creosote treatment for structural lumber | | | | | |
| Piles and building poles (20.0) | — | CF | — | — | 8.69 |
| Board lumber and timbers (25.0) | — | MBF | — | — | 520.00 |
| Lumber in contact with fresh water or soil, ACZA treatment | | | | | |
| Building poles, structural (.60) | — | CF | — | — | 5.44 |
| Boards, timbers, structural (.60) | — | MBF | — | — | 250.00 |
| Boards, timbers, non-structural (.40) | — | MBF | — | — | 223.00 |
| Plywood, structural (.40), 1/2" | — | MSF | — | — | 250.00 |
| Posts, guardrail, signs (.60) | — | MBF | — | — | 314.00 |
| All weather foundation KD lumber (.60) | — | MBF | — | — | 350.00 |
| All weather foundation, 1/2" plywood (.60) | — | MSF | — | — | 278.30 |
| Lumber in contact with fresh water or soil, creosote treatment | | | | | |
| Piles, building poles, structural (12.0) | — | CF | — | — | 6.16 |
| Boards, timbers, structural (12.0) | — | MBF | — | — | 377.00 |
| Boards, timbers, non-structural (10.0) | — | MBF | — | — | 350.00 |
| Plywood, structural, 1/2" (10.0) | — | MSF | — | — | 302.00 |
| Posts (fence, guardrail, sign) (12.0) | — | MBF | — | — | 377.00 |
| Lumber used above ground, ACZA treatment | | | | | |
| Boards, timbers, structural (.25) | — | MBF | — | — | 206.00 |
| Boards, timbers, non-structural (.25) | — | MBF | — | — | 206.00 |
| Plywood, structural 1/2" (.25) | — | MSF | — | — | 230.00 |
| Lumber used above ground, creosote treatment | | | | | |
| Boards, timbers, structural (8.0) | — | MBF | — | — | 314.00 |
| Boards, timbers, non-structural (8.0) | — | MBF | — | — | 314.00 |
| Plywood, structural (8.0), 1/2" | — | MSF | — | — | 278.00 |
| Lumber used for decking, ACQ treatment | | | | | |
| Boards, structural (.41) | — | MBF | — | — | 271.00 |
| Boards, timbers, non-structural (.41) | — | MSF | — | — | 271.00 |

| | Craft@Hrs | Unit | Material | Labor | Total |
|---|---|---|---|---|---|
| Fire retardant treatments, structural use (interior only), "D-Blaze" | | | | | |
|    Surfaced lumber | — | MBF | — | — | 289.00 |
|    Rough lumber | — | MBF | — | — | 289.00 |
| Architectural use (interior and exterior), "NCX" | | | | | |
|    Surfaced lumber | — | MBF | — | — | 668.00 |
|    Rough lumber | — | MBF | — | — | 668.00 |

Wolmanized treatment. Copper chromated arsenic preservative is not used for residential applications. The term "Wolmanized" is now being applied to ACQ and CA preservatives which do not contain arsenic.

| | Craft@Hrs | Unit | Material | Labor | Total |
|---|---|---|---|---|---|
|    Lumber used above ground (.25) | — | MBF | — | — | 102.00 |
|    Lumber in ground contact or fresh water (.40) | — | MBF | — | — | 128.00 |
|    Lumber used as foundation (.60) | — | MBF | — | — | 139.00 |

## Mailboxes

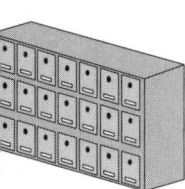

Apartment type, vertical, meets current postal regulations, tumbler locks, aluminum or gold finish, price per unit, recessed

| | Craft@Hrs | Unit | Material | Labor | Total |
|---|---|---|---|---|---|
|    3 box unit, 5-1/2" wide x 16" high | B1@1.88 | Ea | 115.00 | 62.60 | 177.60 |
|    4 box unit, 5-1/2" wide x 16" high | B1@1.88 | Ea | 145.00 | 62.60 | 207.60 |
|    5 box unit, 5-1/2" wide x 16" high | B1@1.88 | Ea | 175.00 | 62.60 | 237.60 |
|    6 box unit, 5-1/2" wide x 16" high | B1@1.88 | Ea | 205.00 | 62.60 | 267.60 |
|    7 box unit, 5-1/2" wide x 16" high | B1@1.88 | Ea | 235.00 | 62.60 | 297.60 |
|    Add for surface mounting collar | B1@.550 | Ea | 38.90 | 18.30 | 57.20 |

Office type, horizontal, meets postal regulations, tumbler locks, aluminum or brass finish

| | Craft@Hrs | Unit | Material | Labor | Total |
|---|---|---|---|---|---|
|    For front or rear load, per box | | | | | |
|       23" wide x 30" high x 12" deep | B1@1.02 | Ea | 525.00 | 34.00 | 559.00 |

Rural, aluminum finish, 19" long x 6-1/2" wide x 9" high, with flag

| | Craft@Hrs | Unit | Material | Labor | Total |
|---|---|---|---|---|---|
|    Black, does not include mounting post | B1@.776 | Ea | 18.60 | 25.80 | 44.40 |

Rural heavy gauge steel mailboxes, aluminum finish, 19" long x 7" wide x 10" high, with flag

| | Craft@Hrs | Unit | Material | Labor | Total |
|---|---|---|---|---|---|
|    Vandal-proof, does not include mounting post | B1@.776 | Ea | 73.20 | 25.80 | 99.00 |

Suburban style, finished, 16" long x 3" wide x 6" deep, with flag

| | Craft@Hrs | Unit | Material | Labor | Total |
|---|---|---|---|---|---|
|    Post mount bracket, does not include post | B1@.776 | Ea | 23.10 | 25.80 | 48.90 |
|    Add for 5' high aluminum post | — | Ea | 25.50 | — | 25.50 |

Cast aluminum mailbox and pedestal, 18" x 12" box, 4' pedestal,

| | Craft@Hrs | Unit | Material | Labor | Total |
|---|---|---|---|---|---|
|    With base, lockable | B1@.776 | Ea | 158.00 | 25.80 | 183.80 |

Bronze mailbox and post, 18" x 12" box, 5' post,

| | Craft@Hrs | Unit | Material | Labor | Total |
|---|---|---|---|---|---|
|    Steel anchor base | B1@.776 | Ea | 232.00 | 25.80 | 257.80 |

Plastic mailbox, mounting sleeve slides over 4' high post, lockable

| | Craft@Hrs | Unit | Material | Labor | Total |
|---|---|---|---|---|---|
|    Size 1 box, add for post | B1@.776 | Ea | 50.70 | 25.80 | 76.50 |
|    Add for treated wood or aluminum post | — | Ea | 20.00 | — | 20.00 |

Thru-the-wall type, 13" long x 3" high

| | Craft@Hrs | Unit | Material | Labor | Total |
|---|---|---|---|---|---|
|    Gold or aluminum finish | B1@1.25 | Ea | 26.50 | 41.60 | 68.10 |

Wall- or post-mount security mailboxes, lockable, 15" wide x 13" high x 8" deep

| | Craft@Hrs | Unit | Material | Labor | Total |
|---|---|---|---|---|---|
|    Black plastic and aluminum | B1@.450 | Ea | 62.80 | 15.00 | 77.80 |

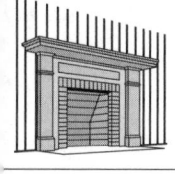

## Mantels, Ponderosa Pine  Unfinished.

| | Craft@Hrs | Unit | Material | Labor | Total |
|---|---|---|---|---|---|
| Modern design, 50" x 37" opening, | | | | | |
|    paint grade, 11" x 77" shelf | B1@3.81 | Ea | 651.00 | 127.00 | 778.00 |
| Colonial design, 50" x 39" opening, | | | | | |
|    stain grade, 7" x 72" shelf | B1@3.81 | Ea | 1,040.00 | 127.00 | 1,167.00 |
| Ornate design, 51-1/2" x 39" opening, | | | | | |
|    with detailing and scroll work, stain grade | | | | | |
|    11" x 68" shelf | B1@3.81 | Ea | 1,940.00 | 127.00 | 2,067.00 |

# Marble Setting, Subcontract

| | Craft@Hrs | Unit | Material | Labor | Total |
|---|---|---|---|---|---|
| Mantel moulding (simple trim for masonry fireplace openings) | | | | | |
| 15 LF, for opening 6'0" x 3'6" or less | B1@1.07 | Set | 230.00 | 35.60 | 265.60 |
| Mantel shelf only, with iron or wood support brackets, prefinished | | | | | |
| 4" x  8" x  8" | B1@.541 | Ea | 59.60 | 18.00 | 77.60 |
| 4" x  8" x  6" | B1@.541 | Ea | 45.50 | 18.00 | 63.50 |
| 4" x 10" x 10" | B1@.541 | Ea | 101.00 | 18.00 | 119.00 |

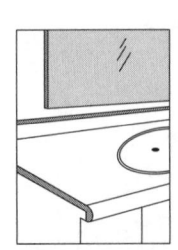

**Marble Setting, Subcontract**  See also Tile section. Marble prices vary with quality, quantity, and availability of materials. Check prices with local suppliers. These prices are based on 200 SF quantities.

| | Craft@Hrs | Unit | Material | Labor | Total |
|---|---|---|---|---|---|
| Marble tile flooring, tiles or slabs | | | | | |
| 3/8" x 12" x 12", thin set over concrete slab. | | | | | |
| Typical cost per SF of area covered | — | SF | — | — | 19.00 |
| 3/4" slabs set in mortar bed, cut to size. | | | | | |
| Typical cost per SF of area covered | — | SF | — | — | 70.00 |
| Marble bathroom vanity top, to 96" x 30" wide, with one sink cutout, bullnose or eased edge on 3 sides. | | | | | |
| Most colors and textures | — | Ea | — | — | 750.00 |
| Select colors and textures | — | Ea | — | — | 950.00 |
| Marble fireplace mantel and facing. 46" high to 67" wide overall. Including typical delivery. Custom-made mantels require several months for production and delivery. | | | | | |
| Minimum, plain design, white marble | — | Ea | — | — | 1,100.00 |
| Better quality, limestone | — | Ea | — | — | 2,000.00 |
| Ornate design, onyx | — | Ea | — | — | 3,300.00 |

**Markup**  Typical markup for light construction, percentage of gross contract price. No insurance or taxes included. See also Insurance and Taxes sections. Markup is the amount added to an estimate after all job costs have been accounted for: labor and materials, job fees, permits, bonds and insurance, job site supervision and similar job-related expenses. Profit is the return on money invested in the construction business.

| | Craft@Hrs | Unit | Material | Labor | Total |
|---|---|---|---|---|---|
| Contingency (allowance for unknown and unforeseen conditions) | | | | | |
| Varies widely from job to job, typically | — | % | — | — | 2.0 |
| Overhead (office, administrative and general business expense) | | | | | |
| Typical overhead expense | — | % | — | — | 10.0 |
| Profit (return on the money invested in the business, varies with competitive conditions) | | | | | |
| Typical profit for residential construction | — | % | — | — | 8.0 |
| Total markup (add 25% to the total job cost to attain 20% of gross contract price) | | | | | |
| Total for Contingency, Overhead & Profit | — | % | — | — | 20.0 |

Note that costs listed in this book do not include the general contractor's markup. However, sections identified as "subcontract" include the subcontractor's markup. Here's how to calculate markup:

**Project Costs**

| | |
|---|---|
| Material | $ 50,000 |
| Labor (with taxes & insurance) | 40,000 |
| Subcontract | 10,000 |
| Total job costs | 100,000 |
| **Markup (25% of $100,000)** | **$ 25,000** |
| Contract price | $125,000 |

Markup is $25,000 which is 25% of the total job costs ($100,000) but only 20% of the contract price ($25,000 divided by $125,000 is .20).

**Markup For General Contractors** On larger commercial and industrial projects, a markup of 24.8% on the total job cost is suggested. See the paragraph General Contractor Markup at the beginning of the section General Requirements in the Industrial and Commercial Division of this manual. For example, note how markup is calculated on a project with these costs:

| | |
|---|---|
| Material | $ 3,000,000 |
| Labor (with taxes and insurance) | 4,500,000 |
| Equipment | 1,000,000 |
| Subcontracts | 1,500,000 |
| Total job cost | $10,000,000 |
| Indirect overhead (8%) | 800,000 |
| Direct overhead (7.3%) | 730,000 |
| Contingency (2%) | 200,000 |
| Profit (7.5%) | 750,000 |
| Contract price | $12,480,000 |

Note that the total job cost includes both work done with the general contractor's crews and work done by subcontractor-specialists. Markup for the general contractor (overhead, including supervision expense, contingency and profit) is $2,480,000. This amounts to 24.8% of the total job costs but only 19.87% of the contract price ($2,480,000 divided by $12,480,000 is 0.1987).

| | Craft@Hrs | Unit | Material | Labor | Total |
|---|---|---|---|---|---|

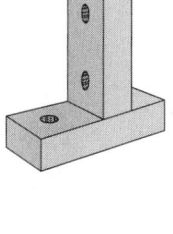

**Masonry, Brick Wall Assemblies** Typical costs for smooth red clay brick walls, laid in running bond with 3/8" concave joints. These costs include the bricks, mortar for bricks and cavities, typical ladder type reinforcing, wall ties and normal waste. Foundations are not included. Wall thickness shown is the nominal size based on using the type bricks described. "Wythe" means the quantity of bricks in the thickness of the wall. Costs shown are per square foot (SF) of wall measured on one face. Deduct for openings over 10 SF in size. The names and dimensions of bricks can be expected to vary, depending on the manufacturer. For more detailed coverage of brick and masonry see the *National Concrete & Masonry Estimator* at http://CraftsmanSiteLicense.com.

| | Craft@Hrs | Unit | Material | Labor | Total |
|---|---|---|---|---|---|
| **Standard brick, 3-3/4" wide x 2-1/4" high x 8" long** | | | | | |
| Common brick, per thousand | — | M | 470.00 | — | 470.00 |
| 4" thick wall, single wythe, veneer facing | B9@.211 | SF | 3.95 | 6.88 | 10.83 |
| 8" thick wall, double wythe, cavity filled | B9@.464 | SF | 7.75 | 15.10 | 22.85 |
| 12" thick wall, triple wythe, cavity filled | B9@.696 | SF | 11.60 | 22.70 | 34.30 |
| **Norman brick, 3-1/2" wide x 2-1/2" high x 11-1/2" long** | | | | | |
| Norman brick, per thousand | — | M | 1,330.00 | — | 1,330.00 |
| 4" thick wall, single wythe, veneer facing | B9@.211 | SF | 6.70 | 6.88 | 13.58 |
| 8" thick wall, double wythe, cavity filled | B9@.464 | SF | 13.40 | 15.10 | 28.50 |
| 12" thick wall, triple wythe, cavity filled | B9@.696 | SF | 20.00 | 22.70 | 42.70 |
| **Modular brick, 3" wide x 3-1/2" x 11-1/2" long** | | | | | |
| Modular brick, per thousand | — | M | 1,330.00 | — | 1,330.00 |
| 3-1/2" thick wall, single wythe, veneer face | B9@.156 | SF | 4.99 | 5.09 | 10.08 |
| 7-1/2" thick wall, double wythe, cavity filled | B9@.343 | SF | 9.86 | 11.20 | 21.06 |
| 11-1/2" thick wall, triple wythe, cavity filled | B9@.515 | SF | 14.70 | 16.80 | 31.50 |
| **Colonial brick, 3" wide x 3-1/2" x 10" long** | | | | | |
| Colonial brick, per thousand | — | M | 970.00 | — | 970.00 |
| 3-1/2" thick wall, single wythe, veneer face | B9@.177 | SF | 4.27 | 5.78 | 10.05 |
| 7-1/2" thick wall, double wythe, cavity filled | B9@.389 | SF | 8.40 | 12.70 | 21.10 |
| 11-1/2" thick wall, triple wythe, cavity filled | B9@.584 | SF | 12.50 | 19.10 | 31.60 |

| | Craft@Hrs | Unit | Material | Labor | Total |
|---|---|---|---|---|---|

**Masonry, Face Brick** Laid in running bond with 3/8" concave joints. Mortar, wall ties and foundations (footings) not included. Includes normal waste and delivery up to 30 miles with a 5 ton minimum. Coverage (units per SF) includes 3/8" mortar joints but the SF cost does not include mortar cost. Add for mortar below.

Standard brick, 2-1/4" high x 3-3/4" deep x 8" long (6.45 units per SF)

| | Craft@Hrs | Unit | Material | Labor | Total |
|---|---|---|---|---|---|
| Brown, per square foot | B9@.211 | SF | 5.28 | 6.88 | 12.16 |
| Brown, per thousand | B9@32.7 | M | 780.00 | 1,070.00 | 1,850.00 |
| Burnt oak, per square foot | B9@.211 | SF | 5.28 | 6.88 | 12.16 |
| Burnt oak, per thousand | B9@32.7 | M | 780.00 | 1,070.00 | 1,850.00 |
| Flashed, per square foot | B9@.211 | SF | 4.74 | 6.88 | 11.62 |
| Flashed, per thousand | B9@32.7 | M | 700.00 | 1,070.00 | 1,770.00 |
| Red, per square foot | B9@.211 | SF | 3.18 | 6.88 | 10.06 |
| Red, per thousand | B9@32.7 | M | 470.00 | 1,070.00 | 1,540.00 |

Norman brick, 2-1/2" high x 3-1/2" deep x 11-1/2" long (4.22 units per SF)

| | Craft@Hrs | Unit | Material | Labor | Total |
|---|---|---|---|---|---|
| Brown | B9@.211 | SF | 11.50 | 6.88 | 18.38 |
| Burnt oak | B9@.211 | SF | 11.50 | 6.88 | 18.38 |
| Flashed | B9@.211 | SF | 5.89 | 6.88 | 12.77 |
| Red | B9@.211 | SF | 5.89 | 6.88 | 12.77 |

Modular, 3-1/2" high x 3" deep x 11-1/2" long (3.13 units per SF)

| | Craft@Hrs | Unit | Material | Labor | Total |
|---|---|---|---|---|---|
| Brown | B9@.156 | SF | 4.37 | 5.09 | 9.46 |
| Burnt oak | B9@.156 | SF | 4.37 | 5.09 | 9.46 |
| Flashed | B9@.156 | SF | 4.37 | 5.09 | 9.46 |
| Red | B9@.156 | SF | 4.14 | 5.09 | 9.23 |

Jumbo (filler brick) 3-1/2" high x 3" deep x 11-1/2" long (3.13 units per SF)

| | Craft@Hrs | Unit | Material | Labor | Total |
|---|---|---|---|---|---|
| Fireplace, cored | B9@.156 | SF | 1.74 | 5.09 | 6.83 |
| Fireplace, solid | B9@.156 | SF | 1.54 | 5.09 | 6.63 |
| Mission, cored | B9@.156 | SF | 3.40 | 5.09 | 8.49 |
| Mission, solid | B9@.156 | SF | 3.49 | 5.09 | 8.58 |

Padre brick, 4" high x 3" deep

| | Craft@Hrs | Unit | Material | Labor | Total |
|---|---|---|---|---|---|
| 7-1/2" long cored, 4.2 per SF | B9@.156 | SF | 8.09 | 5.09 | 13.18 |
| 7-1/2" long solid, 4.2 per SF | B9@.156 | SF | 10.20 | 5.09 | 15.29 |
| 11-1/2" long cored, 2.8 per SF | B9@.177 | SF | 6.09 | 5.78 | 11.87 |
| 11-1/2" long solid, 2.8 per SF | B9@.177 | SF | 9.00 | 5.78 | 14.78 |
| 15-1/2" long cored, 2.1 per SF | B9@.211 | SF | 7.47 | 6.88 | 14.35 |
| 15-1/2" long solid, 2.1 per SF | B9@.211 | SF | 9.38 | 6.88 | 16.26 |

Commercial, 3-1/4" high x 3-1/4" deep x 10" long (3.6 per SF)

| | Craft@Hrs | Unit | Material | Labor | Total |
|---|---|---|---|---|---|
| Solid | B9@.156 | SF | 3.67 | 5.09 | 8.76 |
| Cored | B9@.177 | SF | 3.67 | 5.78 | 9.45 |
| Harlequin (manufactured "used" brick) | B9@.177 | SF | 3.89 | 5.78 | 9.67 |

**Mortar for brick** Based on an all-purpose Type S mortar mix used in single wythe walls laid with running bond. Includes normal waste. Type N mortar is used for severe conditions, such as chimneys. Costs are per SF of wall face using the brick sizes (width x height x length) listed.

| | Craft@Hrs | Unit | Material | Labor | Total |
|---|---|---|---|---|---|
| Type S mortar, per cubic foot | — | CF | 7.96 | — | 7.96 |
| Type S mortar, 60 lb. bag | — | Ea | 3.98 | — | 3.98 |
| 4" x 2-1/2" to 2-5/8" x 8" bricks | | | | | |
| 3/8" joints (5.5 CF per CSF) | — | SF | .44 | — | .44 |
| 1/2" joints (7.0 CF per CSF) | — | SF | .56 | — | .56 |
| 4" x 3-1/2" to 3-5/8" x 8" bricks | | | | | |
| 3/8" joints (4.8 CF per CSF) | — | SF | .38 | — | .38 |
| 1/2" joints (6.1 CF per CSF) | — | SF | .49 | — | .49 |

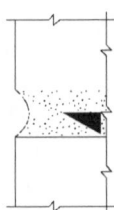

| | Craft@Hrs | Unit | Material | Labor | Total |
|---|---|---|---|---|---|
| 4" x 4" x 8" bricks | | | | | |
| 3/8" joints (4.2 CF per CSF) | — | SF | .33 | — | .33 |
| 1/2" joints (5.3 CF per CSF) | — | SF | .42 | — | .42 |
| 4" x 5-1/2" to 5-5/8" x 8" bricks | | | | | |
| 3/8" joints (3.5 CF per CSF) | — | SF | .28 | — | .28 |
| 1/2" joints (4.4 CF per CSF) | — | SF | .35 | — | .35 |
| 4" x 2" x 12" bricks | | | | | |
| 3/8" joints (6.5 CF per CSF) | — | SF | .52 | — | .52 |
| 1/2" joints (8.2 CF per CSF) | — | SF | .65 | — | .65 |
| 4" x 2-1/2" to 2-5/8" x 12" bricks | | | | | |
| 3/8" joints (5.1 CF per CSF) | — | SF | .41 | — | .41 |
| 1/2" joints (6.5 CF per CSF) | — | SF | .52 | — | .52 |
| 4" x 3-1/2" to 3-5/8" x 12" bricks | | | | | |
| 3/8" joints (4.4 CF per CSF) | — | SF | .35 | — | .35 |
| 1/2" joints (5.6 CF per CSF) | — | SF | .45 | — | .45 |
| 4" x 4" x 12" bricks | | | | | |
| 3/8" joints (3.7 CF per CSF) | — | SF | .29 | — | .29 |
| 1/2" joints (4.8 CF per CSF) | — | SF | .38 | — | .38 |
| 4" x 5-1/2" to 5-5/8" x 12" bricks | | | | | |
| 3/8" joints (3.0 CF per CSF) | — | SF | .24 | — | .24 |
| 1/2" joints (3.9 CF per CSF) | — | SF | .31 | — | .31 |

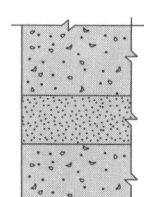

| | Craft@Hrs | Unit | Material | Labor | Total |
|---|---|---|---|---|---|
| 6" x 2-1/2" to 2-5/8" x 12" bricks | | | | | |
| 3/8" joints (7.9 CF per CSF) | — | SF | .63 | — | .63 |
| 1/2" joints (10.2 CF per CSF) | — | SF | .81 | — | .81 |
| 6" x 3-1/2" to 3-5/8" x 12" bricks | | | | | |
| 3/8" joints (6.8 CF per CSF) | — | SF | .54 | — | .54 |
| 1/2" joints (8.8 CF per CSF) | — | SF | .70 | — | .70 |
| 6" x 4" x 12" bricks | | | | | |
| 3/8" joints (5.6 CF per CSF) | — | SF | .45 | — | .45 |
| 1/2" joints (7.4 CF per CSF) | — | SF | .59 | — | .59 |

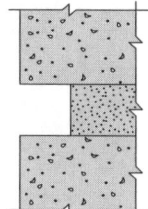

| | Craft@Hrs | Unit | Material | Labor | Total |
|---|---|---|---|---|---|
| Add for 6-11/16" 22 gauge galvanized wall ties | | | | | |
| 50 ties per 100 SF | B9@.011 | SF | .07 | .36 | .43 |
| Add for raked joints | B9@.027 | SF | — | .88 | .88 |
| Add for flush joints | B9@.072 | SF | — | 2.35 | 2.35 |
| Add for header course every 6th course | B9@.009 | SF | .61 | .29 | .90 |
| Add for Flemish bond | | | | | |
| With header course every 6th course | B9@.036 | SF | .48 | 1.17 | 1.65 |

**Mini-brick veneer strips** Wirecut texture. Figures in parentheses indicate coverage using 1/2" grouted joints. No waste included.

| | Craft@Hrs | Unit | Material | Labor | Total |
|---|---|---|---|---|---|
| 2-1/4" x 3/4" x 7-5/8" (6.7 units per SF) | | | | | |
| Unglazed, standard colors, $.49 each | B9@.300 | SF | 3.37 | 9.79 | 13.16 |
| Unglazed, flashed, $.63 each | B9@.300 | SF | 4.31 | 9.79 | 14.10 |
| Manufactured used, $.52 each | B9@.300 | SF | 3.57 | 9.79 | 13.36 |
| 3-5/8" x 1/2" x 7-5/8" (4.50 units per SF) | | | | | |
| Unglazed, standard colors, $.65 each | B9@.250 | SF | 3.33 | 8.16 | 11.49 |
| Unglazed, flashed, $.93 each | B9@.250 | SF | 4.23 | 8.16 | 12.39 |
| Manufactured used, $.80 each | B9@.250 | SF | 3.64 | 8.16 | 11.80 |

# Masonry

| | Craft@Hrs | Unit | Material | Labor | Total |
|---|---|---|---|---|---|
| Mini-brick veneer strip corners, | | | | | |
| Unglazed, standard colors | B9@.077 | Ea | 1.73 | 2.51 | 4.24 |
| Unglazed, flashed | B9@.077 | Ea | 1.95 | 2.51 | 4.46 |
| Manufactured used | B9@.077 | Ea | 1.87 | 2.51 | 4.38 |

**Firebrick** Used in chimneys and boilers, delivered to 30 miles, typical costs.

| | Craft@Hrs | Unit | Material | Labor | Total |
|---|---|---|---|---|---|
| Standard backs, 9" x 4-1/2" x 2-9/16" | B9@30.9 | M | 1,770.00 | 1,010.00 | 2,780.00 |
| Split backs, 9" x 4-1/2" x 1-1/4" | B9@23.3 | M | 1,580.00 | 760.00 | 2,340.00 |
| Fireclay, Dosch clay | | | | | |
| 50 lb. sack | — | Ea | 8.64 | — | 8.64 |

**Brick paver** Costs assume simple pattern with 1/2" mortar joints. Delivered to 30 miles, 10 ton minimum. Costs in material column include 8% for waste and breakage.

| | Craft@Hrs | Unit | Material | Labor | Total |
|---|---|---|---|---|---|
| Adobe pavers (Fresno) | | | | | |
| 6" x 12" (2 units per SF) | B9@.193 | SF | 2.00 | 6.30 | 8.30 |
| 12" x 12" (1 unit per SF) ) | B9@.135 | SF | 2.00 | 4.41 | 6.41 |
| California paver, 1-1/4" thick, brown | | | | | |
| 3-5/8" x 7-5/8" (4.3 units per SF) | B9@.340 | SF | 4.28 | 11.10 | 15.38 |
| Giant mission paver, 3" thick x 4" x 11-1/2" | | | | | |
| Red (2.67 units per SF) | B9@.193 | SF | 5.12 | 6.30 | 11.42 |
| Redskin (2.67 units per SF) | B9@.193 | SF | 5.78 | 6.30 | 12.08 |
| Mini-brick paver, 1/2" thick x 3-5/8" x 7-5/8" | | | | | |
| Sunset Red | | | | | |
| (4.5 units per SF) | B9@.340 | SF | 3.21 | 11.10 | 14.31 |
| Red, flashed | | | | | |
| (4.5 units per SF) | B9@.340 | SF | 4.57 | 11.10 | 15.67 |
| Manufactured | | | | | |
| (4.5 units per SF) | B9@.340 | SF | 3.94 | 11.10 | 15.04 |
| Sunset Red, Dieskin texture | | | | | |
| (4.5 units per SF) | B9@.340 | SF | 2.82 | 11.10 | 13.92 |
| Red, flashed, Dieskin texture | | | | | |
| (4.5 units per SF) | B9@.340 | SF | 3.50 | 11.10 | 14.60 |
| Manufactured, Dieskin texture | | | | | |
| (4.5 units per SF) | B9@.340 | SF | 3.79 | 11.10 | 14.89 |
| Re-roll texture, all colors | | | | | |
| (4.5 units per SF) | B9@.340 | SF | 3.40 | 11.10 | 14.50 |
| Norman paver, 2-3/16" thick x 3-1/2" x 11-1/2" (3 units per SF) | | | | | |
| Sunset Red | B9@.222 | SF | 4.82 | 7.24 | 12.06 |
| Red Flashed | B9@.222 | SF | 5.24 | 7.24 | 12.48 |
| Padre paver, 1-1/4" thick x 7-1/2" x 7-1/2" | | | | | |
| (2.5 units per SF) | B9@.211 | SF | 4.30 | 6.88 | 11.18 |
| Split paver, 1-1/8" thick x 3-7/8" x 8-1/8" (4.5 units per SF) | | | | | |
| Red | B9@.209 | SF | 2.43 | 6.82 | 9.25 |
| Manufactured, used | B9@.209 | SF | 2.92 | 6.82 | 9.74 |
| Bullnose pavers, 2-3/16" thick x 3-1/2" x 11-1/2" (3 units per LF). Common applications include perimeter of brick patio, walkway and wall caps | | | | | |
| Single bullnose, Sunset Red | B9@.250 | LF | 7.49 | 8.16 | 15.65 |
| Single bullnose, Brown Flashed | B9@.250 | LF | 8.89 | 8.16 | 17.05 |
| Single bullnose, Iron Spot | B9@.250 | LF | 9.27 | 8.16 | 17.43 |
| Double bullnose, Sunset Red | B9@.250 | LF | 10.60 | 8.16 | 18.76 |
| Double bullnose, Brown Flashed | B9@.250 | LF | 11.30 | 8.16 | 19.46 |
| Double bullnose, Iron Spot | B9@.250 | LF | 11.70 | 8.16 | 19.86 |
| Bullnose corners, all colors | B9@.242 | Ea | 38.90 | 7.90 | 46.80 |
| Mortar joints for pavers, 1/2" typical | B9@.010 | SF | .80 | .33 | 1.13 |

| | Craft@Hrs | Unit | Material | Labor | Total |
|---|---|---|---|---|---|

**Concrete block wall assemblies** Typical costs for standard natural gray medium weight masonry block walls including blocks, mortar, typical reinforcing and normal waste. Foundations (footings) are not included. See also individual component material and labor costs for blocks, mortar and reinforcing on the pages that follow. For more detailed coverage of concrete block see the *National Concrete and Masonry Estimator* at http://CraftsmanSiteLicense.com.

| | Craft@Hrs | Unit | Material | Labor | Total |
|---|---|---|---|---|---|
| **Walls constructed with 8" x 16" blocks laid in running bond** | | | | | |
| 4" thick wall | B9@.090 | SF | 2.12 | 2.94 | 5.06 |
| 6" thick wall | B9@.100 | SF | 2.62 | 3.26 | 5.88 |
| 8" thick wall | B9@.120 | SF | 3.12 | 3.92 | 7.04 |
| 12" thick wall | B9@.150 | SF | 4.61 | 4.89 | 9.50 |
| **Add for grouting cores at 36" intervals, poured by hand** | | | | | |
| 4" thick wall | B9@.007 | SF | .58 | .23 | .81 |
| 6" thick wall | B9@.007 | SF | .88 | .23 | 1.11 |
| 8" thick wall | B9@.007 | SF | 1.17 | .23 | 1.40 |
| 12" thick wall | B9@.010 | SF | 1.74 | .33 | 2.07 |
| **Add for 2" thick caps, natural gray concrete** | | | | | |
| 4" thick wall | B9@.027 | LF | .89 | .88 | 1.77 |
| 6" thick wall | B9@.038 | LF | 1.06 | 1.24 | 2.30 |
| 8" thick wall | B9@.046 | LF | 1.09 | 1.50 | 2.59 |
| 12" thick wall | B9@.065 | LF | 1.66 | 2.12 | 3.78 |
| **Add for detailed block, 3/8" score** | | | | | |
| Single score, one side | — | SF | .55 | — | .55 |
| Single score, two sides | — | SF | 1.21 | — | 1.21 |
| Multi-scores, one side | — | SF | .55 | — | .55 |
| Multi-scores, two sides | — | SF | 1.21 | — | 1.21 |
| **Add for color block, any wall assembly** | | | | | |
| Light colors | — | SF | .13 | — | .13 |
| Medium colors | — | SF | .20 | — | .20 |
| Dark colors | — | SF | .28 | — | .28 |
| **Add for other than running bond** | B9@.021 | SF | — | .69 | .69 |

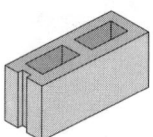

**Concrete block** Foundations (footings) not included. Includes local delivery, mortar and 8% waste. Natural gray, medium weight, no reinforcing included. See also concrete block wall assemblies. Costs shown as Ea are for the block only, the mortar from the adjacent block will be sufficient for these blocks.

| | Craft@Hrs | Unit | Material | Labor | Total |
|---|---|---|---|---|---|
| **4" wide units** | | | | | |
| 8" x 16", 1.125 per SF | B9@.085 | SF | 1.63 | 2.77 | 4.40 |
| 8" x 8", 2.25 per SF | B9@.093 | SF | 2.77 | 3.03 | 5.80 |
| 4" x 16", 2.25 per SF | B9@.093 | SF | 2.57 | 3.03 | 5.60 |
| 4" x 12", corner | B9@.052 | Ea | 1.05 | 1.70 | 2.75 |
| 4" x 8", 4.5 per SF, (half block) | B9@.106 | SF | 4.62 | 3.46 | 8.08 |
| **6" wide units** | | | | | |
| 8" x 16", 1.125 per SF | B9@.095 | SF | 2.02 | 3.10 | 5.12 |
| 8" x 16" corner | B9@.058 | Ea | 2.00 | 1.89 | 3.89 |
| 8" x 8", 2.25 per SF, (half block) | B9@.103 | SF | 3.58 | 3.36 | 6.94 |
| 4" x 16", 2.25 per SF | B9@.103 | SF | 3.25 | 3.36 | 6.61 |
| 4" x 14" corner | B9@.070 | Ea | .73 | 2.28 | 3.01 |
| 4" x 8", 4.5 per SF, (half block) | B9@.114 | SF | 5.94 | 3.72 | 9.66 |
| 4" x 12" (3/4 block) | B9@.052 | Ea | .69 | 1.70 | 2.39 |

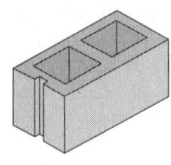

| | Craft@Hrs | Unit | Material | Labor | Total |
|---|---|---|---|---|---|
| **8" wide units** | | | | | |
| 4" x 16", 2.25 per SF | B9@.141 | SF | 3.88 | 4.60 | 8.48 |
| 8" x 16", 1.125 per SF | B9@.111 | SF | 2.41 | 3.62 | 6.03 |
| 6" x 8", 3 per SF | B9@.137 | SF | 6.22 | 4.47 | 10.69 |
| 8" x 12" (3/4 block) | B9@.074 | Ea | 1.69 | 2.41 | 4.10 |
| 8" x 8" (half) | B9@.052 | Ea | 1.61 | 1.70 | 3.31 |
| 8" x 8" (lintel) | B9@.052 | Ea | 1.90 | 1.70 | 3.60 |
| 6" x 16", 1.5 per SF | B9@.117 | SF | 3.46 | 3.82 | 7.28 |
| 4" x 8" (half block) | B9@.053 | Ea | 1.37 | 1.73 | 3.10 |
| **12" wide units** | | | | | |
| 8" x 16", 1.125 per SF | B9@.148 | SF | 3.75 | 4.83 | 8.58 |
| 8" x 8" (half) | B9@.082 | Ea | 2.67 | 2.68 | 5.35 |
| 8" x 8" (lintel) | B9@.082 | Ea | 3.00 | 2.68 | 5.68 |
| Add for high strength units | — | % | 27.0 | — | — |
| Add for lightweight block, typical | — | Ea | .26 | — | .26 |
| Deduct for lightweight block, labor | — | % | — | -7.0 | — |
| Add for ladder type reinforcement | B9@.004 | SF | .22 | .13 | .35 |
| Add for 1/2" rebars 24" OC | B9@.004 | SF | .22 | .13 | .35 |
| Add for other than running bond | — | % | — | 20.0 | — |
| **Add for 2" wall caps, natural color** | | | | | |
| 4" x 16" | B9@.027 | LF | .67 | .88 | 1.55 |
| 6" x 16" | B9@.038 | LF | .79 | 1.24 | 2.03 |
| 8" x 16" | B9@.046 | LF | .74 | 1.50 | 2.24 |
| **Add for detailed block, 3/8" score, typical prices** | | | | | |
| Single score, one side | — | SF | .55 | — | .55 |
| Single score, two sides | — | SF | 1.21 | — | 1.21 |
| Multi-scores, one side | — | SF | .55 | — | .55 |
| Multi-scores, two sides | — | SF | 1.21 | — | 1.21 |
| **Add for color block** | | | | | |
| Light colors | — | SF | .11 | — | .11 |
| Medium colors | — | SF | .17 | — | .17 |
| Dark colors | — | SF | .21 | — | .21 |
| **Add for grouting cores poured by hand, normal waste included** | | | | | |
| Using grout, per cubic foot | — | CF | 8.69 | — | 8.69 |
| **36" interval** | | | | | |
| 4" wide wall (16.1 SF/CF grout) | B9@.007 | SF | .58 | .23 | .81 |
| 6" wide wall (10.7 SF/CF grout) | B9@.007 | SF | .88 | .23 | 1.11 |
| 8" wide wall (8.0 SF/CF grout) | B9@.007 | SF | 1.17 | .23 | 1.40 |
| 12" wide wall (5.4 SF/CF grout) | B9@.010 | SF | 1.74 | .33 | 2.07 |
| **24" interval** | | | | | |
| 4" wide wall (16.1 SF/CF grout) | B9@.009 | SF | .87 | .29 | 1.16 |
| 6" wide wall (10.7 SF/CF grout) | B9@.009 | SF | 1.32 | .29 | 1.61 |
| 8" wide wall (8.0 SF/CF grout) | B9@.009 | SF | 1.76 | .29 | 2.05 |
| 12" wide wall (5.4 SF/CF grout) | B9@.013 | SF | 2.61 | .42 | 3.03 |

| | Craft@Hrs | Unit | Material | Labor | Total |
|---|---|---|---|---|---|

**Split face concrete block** Textured block, medium weight, split one side, structural. Includes 8% waste and local delivery.

| | Craft@Hrs | Unit | Material | Labor | Total |
|---|---|---|---|---|---|
| 8" wide, 4" x 16", 2.25 per SF | B9@.111 | SF | 6.13 | 3.62 | 9.75 |
| 8" wide, 8" x 16", 1.125 per SF | B9@.111 | SF | 3.17 | 3.62 | 6.79 |
| 12" wide, 8" x 8", 2.25 per SF | B9@.117 | SF | 8.70 | 3.82 | 12.52 |
| 12" wide, 8" x 16", 1.125 per SF | B9@.117 | SF | 4.50 | 3.82 | 8.32 |
| Add for lightweight block | — | % | 12.0 | — | — |
| Add for color | — | SF | .33 | — | .33 |
| Add for mortar (at 7 cubic feet, per 100 square feet of wall) | — | SF | .58 | — | .58 |

**Concrete pavers** Natural concrete 3/4" thick, including local delivery. Pavers installed in concrete bed but no concrete included.

| | Craft@Hrs | Unit | Material | Labor | Total |
|---|---|---|---|---|---|
| 6" x 6", 4 per SF | B9@.150 | SF | 16.80 | 4.89 | 21.69 |
| 6" x 12", 2 per SF | B9@.150 | SF | 8.55 | 4.89 | 13.44 |
| 12" x 12", 1 per SF | B9@.150 | SF | 3.99 | 4.89 | 8.88 |
| 12" x 18", .667 per SF | B9@.150 | SF | 2.81 | 4.89 | 7.70 |
| 18" x 18", .444 per SF | B9@.150 | SF | 1.92 | 4.89 | 6.81 |
| Add for light colors | — | SF | .54 | — | .54 |
| Add for dark colors | — | SF | 1.60 | — | 1.60 |
| Add for pavers with beach pebble surface | — | SF | 1.07 | — | 1.07 |

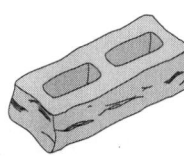

**Concrete screen block** Natural color. Foundations and supports not included. Material costs include 8% waste. (Quantity per SF is shown in parentheses)
Textured screen (no corebar defacements)

| | Craft@Hrs | Unit | Material | Labor | Total |
|---|---|---|---|---|---|
| 4" x 12" x 12" (1 per SF) | B9@.211 | SF | 2.59 | 6.88 | 9.47 |
| 4" x 6" x 6" (3.68 per SF) | B9@.407 | SF | 3.06 | 13.30 | 16.36 |
| 4" x 8" x 8" (2.15 per SF) | B9@.285 | SF | 4.46 | 9.30 | 13.76 |
| Four sided sculpture screens | | | | | |
| 4" x 12" x 12" (1 per SF) | B9@.312 | SF | 2.57 | 10.20 | 12.77 |
| Sculpture screens (patterned face, rug texture reverse) | | | | | |
| 4" x 12" x 12" (1 per SF) | B9@.211 | SF | 3.45 | 6.88 | 10.33 |
| 4" x 16" x 16" (.57 per SF) | B9@.285 | SF | 3.43 | 9.30 | 12.73 |
| 4" x 16" x 12" (.75 per SF) | B9@.285 | SF | 3.65 | 9.30 | 12.95 |
| Add for channel steel support (high rise) | | | | | |
| 2" x 2" x 3/16" | — | LF | 2.40 | — | 2.40 |
| 3" x 3" x 3/16" | — | LF | 2.85 | — | 2.85 |
| 4" x 4" x 1/4" | — | LF | 5.15 | — | 5.15 |
| 6" x 4" x 3/8" | — | LF | 10.90 | — | 10.90 |

**Concrete slump block** Natural concrete color, 4" widths are veneer, no foundation or reinforcing included. Costs in material column include 8% waste. (Quantity per SF is shown in parentheses.)

| | Craft@Hrs | Unit | Material | Labor | Total |
|---|---|---|---|---|---|
| 4" x 4" x 8" (4.5 per SF) | B9@.267 | SF | 4.96 | 8.71 | 13.67 |
| 4" x 4" x 12" (3 per SF) | B9@.267 | SF | 3.52 | 8.71 | 12.23 |
| 4" x 4" x 16" (2.25 per SF) | B9@.267 | SF | 2.64 | 8.71 | 11.35 |
| 6" x 4" x 16" (2.25 per SF) | B9@.243 | SF | 3.27 | 7.93 | 11.20 |
| 6" x 6" x 16" (1.5 per SF) | B9@.217 | SF | 2.23 | 7.08 | 9.31 |
| 8" x 4" x 16" (2.25 per SF) | B9@.264 | SF | 4.36 | 8.61 | 12.97 |
| 8" x 6" x 16" (1.5 per SF) | B9@.243 | SF | 3.20 | 7.93 | 11.13 |
| 12" x 4" x 16" (2.25 per SF) | B9@.267 | SF | 7.50 | 8.71 | 16.21 |
| 12" x 6" x 16" (1.5 per SF) | B9@.260 | SF | 5.34 | 8.48 | 13.82 |

# Masonry

| | Craft@Hrs | Unit | Material | Labor | Total |
|---|---|---|---|---|---|
| **Cap or slab block, 2" thick, 16" long, .75 units per LF. No waste included** | | | | | |
| 4" wide | B9@.160 | LF | .91 | 5.22 | 6.13 |
| 6" wide | B9@.160 | LF | 1.14 | 5.22 | 6.36 |
| 8" wide | B9@.160 | LF | 1.44 | 5.22 | 6.66 |
| **Add for curved slump block** | | | | | |
| 4" wide, 2' or 4' diameter circle | — | % | 125.0 | 25.0 | — |
| 8" wide, 2' or 4' diameter circle | — | % | 150.0 | 20.0 | — |
| Add for slump one side block | — | SF | .05 | — | .05 |
| Add for light colors for slump block | — | % | 13.0 | — | — |
| Add for darker colors for slump block | — | % | 25.0 | — | — |
| **Add for mortar for slump block. Cost includes normal waste** | | | | | |
| Using mortar, per cubic foot | — | CF | 7.96 | — | 7.96 |
| 4" x 4" x  8" (4 CF per CSF) | — | SF | .34 | — | .34 |
| 6" x 6" x 16" (4 CF per CSF) | — | SF | .34 | — | .34 |
| 8" x 6" x 16" (5 CF per CSF) | — | SF | .43 | — | .43 |
| 12" x 6" x 16" (7.5 CF per CSF) | — | SF | .64 | — | .64 |

**Adobe block**  Walls to 8', not including reinforcement or foundations, plus delivery, 4" high, 16" long, 2.25 blocks per SF, including 10% waste.

| | Craft@Hrs | Unit | Material | Labor | Total |
|---|---|---|---|---|---|
| 4" wide | B9@.206 | SF | 3.30 | 6.72 | 10.02 |
| 6" wide | B9@.233 | SF | 4.03 | 7.60 | 11.63 |
| 8" wide | B9@.238 | SF | 5.14 | 7.77 | 12.91 |
| 12" wide | B9@.267 | SF | 8.55 | 8.71 | 17.26 |
| Add for reinforcement (24" OC both ways) | — | SF | .36 | — | .36 |

**Flagstone**  Including concrete bed.

| | Craft@Hrs | Unit | Material | Labor | Total |
|---|---|---|---|---|---|
| Veneer on walls, 3" thick | B9@.480 | SF | 4.91 | 15.70 | 20.61 |
| Walks and porches, 2" thick | B9@.248 | SF | 2.43 | 8.09 | 10.52 |
| Coping, 4" x 12" | B9@.272 | SF | 7.75 | 8.88 | 16.63 |
| Steps, 6" risers | B9@.434 | LF | 2.78 | 14.20 | 16.98 |
| Steps, 12" treads | B9@.455 | LF | 5.57 | 14.80 | 20.37 |

**Flue lining**  Pumice, 1' lengths, delivered to 30 miles, including 8% waste.

| | Craft@Hrs | Unit | Material | Labor | Total |
|---|---|---|---|---|---|
| 8-1/2" round | B9@.148 | Ea | 5.86 | 4.83 | 10.69 |
| 8" x 13" oval | B9@.206 | Ea | 8.31 | 6.72 | 15.03 |
| 8" x 17" oval | B9@.228 | Ea | 9.01 | 7.44 | 16.45 |
| 10" x 17" oval | B9@.243 | Ea | 9.69 | 7.93 | 17.62 |
| 13" x 13" square | B9@.193 | Ea | 10.20 | 6.30 | 16.50 |
| 13" x 17" oval | B9@.193 | Ea | 10.20 | 6.30 | 16.50 |
| 13" x 21" oval | B9@.228 | Ea | 14.60 | 7.44 | 22.04 |
| 17" x 17" square | B9@.228 | Ea | 12.90 | 7.44 | 20.34 |
| 17" x 21" oval | B9@.254 | Ea | 16.30 | 8.29 | 24.59 |
| 21" x 21" square | B9@.265 | Ea | 24.90 | 8.65 | 33.55 |

**Glass block**  Costs include 6% for waste and local delivery. Add grout, spacer, sealant and reinforcing costs (accessories) and mortar from below. Labor cost includes laying the block, grouting and sealing. Costs assume a 100 SF job.3-7/8" thick block, smooth or irregular faces

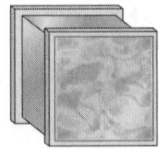

| | Craft@Hrs | Unit | Material | Labor | Total |
|---|---|---|---|---|---|
| 4" x 8", Decora, 4.5 per SF | B9@.392 | SF | 34.10 | 12.80 | 46.90 |
| 4" x 8", Icescapes, 4.5 per SF | B9@.392 | SF | 35.20 | 12.80 | 48.00 |
| 6" x 6", Decora, 4 per SF | B9@.376 | SF | 30.30 | 12.30 | 42.60 |
| 6" x 8", Decora, 3 per SF | B9@.336 | SF | 24.00 | 11.00 | 35.00 |
| 6" x 8", Icescapes, 3 per SF | B9@.336 | SF | 26.60 | 11.00 | 37.60 |

| | Craft@Hrs | Unit | Material | Labor | Total |
|---|---|---|---|---|---|
| 8" x 8", Decora, 2.25 per SF | B9@.296 | SF | 18.60 | 9.66 | 28.26 |
| 12" x 12", Decora, 1 per SF | B9@.211 | SF | 24.40 | 6.88 | 31.28 |
| 12" x 12", Icescapes, 1 per SF | B9@.211 | SF | 25.00 | 6.88 | 31.88 |
| 3-1/8" thick Thinline® block, smooth, irregular, or diamond relief faces | | | | | |
| 4" x 8" Decora, 4.5 per SF | B9@.391 | SF | 27.90 | 12.80 | 40.70 |
| 4" x 8" Delphi, 4 per SF | B9@.344 | SF | 24.80 | 11.20 | 36.00 |
| 6" x 8" Decora, 3 per SF | B9@.376 | SF | 18.90 | 12.30 | 31.20 |
| 8" x 8" Decora, 2.25 per SF | B9@.296 | SF | 14.40 | 9.66 | 24.06 |
| 8" x 8" Delphi, 2.25 per SF | B9@.296 | SF | 15.40 | 9.66 | 25.06 |
| 8" x 8", Vistabrick, 2.25 per SF each | B9@.296 | SF | 96.10 | 9.66 | 105.76 |
| Typical total cost of anchors, spacers, grout and sealer per SF | | | | | |
| 4" x 8" block | — | SF | 22.50 | — | 22.50 |
| 6" x 6" block | — | SF | 20.00 | — | 20.00 |
| 6" x 8" block | — | SF | 15.00 | — | 15.00 |
| 8" x 8" block | — | SF | 11.30 | — | 11.30 |
| 12" x 12" block | — | Ea | 5.01 | — | 5.01 |
| Corner units, most sizes and styles | B9@.125 | Ea | 26.30 | 4.08 | 30.38 |
| End units, most sizes and styles | B9@.125 | Ea | 26.30 | 4.08 | 30.38 |
| End and curve units, most sizes and styles | B9@.125 | Ea | 37.80 | 4.08 | 41.88 |
| 45-degree block, most sizes and styles | B9@.125 | Ea | 21.00 | 4.08 | 25.08 |
| Add for block with heat and glare reducing insert | — | % | 75.0 | — | — |
| Deduct for 100 SF to 500 SF job | — | % | -5.0 | — | — |
| Deduct for over 500 SF job to 1,000 SF job | — | % | -10.0 | — | — |
| Deduct for over 1,000 SF job | — | % | -12.0 | — | — |
| Accessories for 3-7/8" thick glass block | | | | | |
| Anchor pack (anchors, shims, fasteners), 2 per 100 block | | | | | |
| Per pack | — | Ea | 20.70 | — | 20.70 |
| Horizontal spacer pack (10 flat vertical spacers and 5 horizontal spacers) | | | | | |
| 2 packs per 100 block, per pack | — | Ea | 14.40 | — | 14.40 |
| Vertical spacer pack (ten 8" spacers) | | | | | |
| 10 packs per 100 block, per pack | — | Ea | 20.20 | — | 20.20 |
| Horizontal spacers (40" long), cut to length as needed | | | | | |
| 20 pieces per 100 block, per piece | — | Ea | 8.84 | — | 8.84 |
| Glass block surface grout, 15 pound bucket | | | | | |
| 30 pounds per 100 block, per bucket | — | Ea | 38.60 | — | 38.60 |
| Glass block silicone sealer, 10 ounce tube | | | | | |
| 16 tubes per 100 block, per tube | — | Ea | 9.45 | — | 9.45 |
| Glass block support channel, per LF | — | LF | 2.24 | — | 2.24 |
| Accessories for 3-1/8" thick glass block | | | | | |
| Rigid track silicone kit (rigid track, vertical spacers, sealant), 6 per 100 block | | | | | |
| Per kit | — | Ea | 30.20 | — | 30.20 |
| Horizontal spacer track, 36" long | — | Ea | 4.05 | — | 4.05 |
| Rigid vertical spacer track, per pack of five | | | | | |
| 8" spacers | — | Ea | 4.51 | — | 4.51 |
| Glass block spacers, per bag of 25 | — | Ea | 13.60 | — | 13.60 |
| White polyethylene foam expansion strips | | | | | |
| 3/8" x 4" x 24", per piece | — | Ea | 1.51 | — | 1.51 |
| Panel anchors, 1-3/4" x 24" steel strip | | | | | |
| Anchors block to the framing, per strip | — | Ea | 4.50 | — | 4.50 |
| Glass block support channel, per LF | — | LF | 3.31 | — | 3.31 |
| Panel reinforcing, 1-5/8" x 18", embedded in | | | | | |
| every other mortar joint, per strip | — | Ea | 4.99 | — | 4.99 |

# Masonry Accessories

| | Craft@Hrs | Unit | Material | Labor | Total |
|---|---|---|---|---|---|

Mortar for glass block. White cement mix (1 part white cement, 1/2 part hydrated lime, 2-1/4 to 3 parts #20 or #30 silica sand)

| | | | | | |
|---|---|---|---|---|---|
| White cement mix | — | CF | 26.80 | — | 26.80 |
| 1/4" mortar joints, 3-7/8" thick block, white cement mix | | | | | |
| 4" x 8" block | — | SF | 1.51 | — | 1.51 |
| 6" x 6" block | — | SF | 1.34 | — | 1.34 |
| 6" x 8" block | — | SF | 1.17 | — | 1.17 |
| 8" x 8" block | — | SF | 1.01 | — | 1.01 |
| 12" x 12" block | — | SF | .67 | — | .67 |
| 1/4" mortar joints, 3" and 3-1/8" thick block, white cement mix | | | | | |
| 4" x 8" block | — | SF | 1.22 | — | 1.22 |
| 6" x 8" block | — | SF | .95 | — | .95 |
| 8" x 8" block | — | SF | .90 | — | .90 |

## Masonry Accessories

Non-fibered masonry waterproofing for below-grade application

| | | | | | |
|---|---|---|---|---|---|
| Gallon covers 80 SF | B9@.688 | Gal | 10.00 | 22.40 | 32.40 |
| Glass reinforcing mesh | — | SF | .17 | — | .17 |

Lime

Hydrated (builders)

| | | | | | |
|---|---|---|---|---|---|
| 25 lb sack | — | Sack | 5.63 | — | 5.63 |
| 40 lb sack, Type S | — | Sack | 7.93 | — | 7.93 |
| 50 lb sack, white lime | — | Sack | 7.57 | — | 7.57 |
| 70 lb sack, portland lime | — | Sack | 9.07 | — | 9.07 |
| 50 lb sack, athletic field marker | — | Sack | 4.48 | — | 4.48 |
| Silica sand, #30, 100 lb sack | — | Sack | 9.90 | — | 9.90 |

Mortar colors

| | | | | | |
|---|---|---|---|---|---|
| Red, yellow, brown or black, 1 lb bag | — | Lb | 3.70 | — | 3.70 |
| Green, 1 lb bag | — | Lb | 7.56 | — | 7.56 |

Trowel-ready mortar (factory prepared, ready to use), 30 hour life. Deposit required on container, local delivery included. Type S mortar has high compressive strength (1,800 PSI) and high tensile bond strength. Type S is made from 2 parts portland cement, 1 part hydrated lime and 9 parts sand, and offers a longer board life. Type M mortar is made with 3 parts of portland cement, 1 part lime and 12 parts sand. Type M has a high compressive strength (at least 2,500 PSI) and is recommended for masonry in structural walls and masonry below grade such as foundations, retaining walls, sidewalks and driveways. Type N is a medium compressive-strength (750 PSI) mortar made of 1 part portland cement, 1 part lime and 6 parts sand. Type N is recommended for exterior above-grade walls exposed to severe weather, such as chimneys.

| | | | | | |
|---|---|---|---|---|---|
| Type S natural mortar, 1/3 CY | — | Ea | 99.00 | — | 99.00 |
| Type M natural mortar, 1/3 CY | — | Ea | 125.00 | — | 125.00 |
| Add to natural mortar for colored, 1/3 CY | — | Ea | 25.00 | — | 25.00 |
| Type S white mortar, 1/3 CY | — | Ea | 170.00 | — | 170.00 |

Trowel-ready mortar, cost per square foot of wall, 3/8" joint

| | | | | | |
|---|---|---|---|---|---|
| 4" x 4" x 8" brick wall | — | SF | .52 | — | .52 |
| 8" x 8" x 16" block wall | — | SF | .43 | — | .43 |

Dry mortar mix

| | | | | | |
|---|---|---|---|---|---|
| Type S, 80 lb sack | — | Sack | 5.41 | — | 5.41 |
| Type N, 70 lb sack | — | Sack | 9.11 | — | 9.11 |

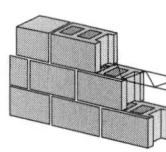

| | Craft@Hrs | Unit | Material | Labor | Total |
|---|---|---|---|---|---|
| **Hi-rise channel reinforcing steel (hat channel)** | | | | | |
| 2" x 2" x 3/16" | — | LF | 1.88 | — | 1.88 |
| 3" x 3" x 3/16" | — | LF | 2.23 | — | 2.23 |
| 4" x 4" x 1/4" | — | LF | 4.42 | — | 4.42 |
| 6" x 4" x 3/8" | — | LF | 8.64 | — | 8.64 |
| Welding | SW@.176 | Ea | — | 7.22 | 7.22 |
| Cutting | SW@.088 | Ea | — | 3.61 | 3.61 |
| Flanges, drill one hole, weld | SW@.272 | Ea | — | 11.20 | 11.20 |
| Additional holes | SW@.028 | Ea | — | 1.15 | 1.15 |
| **Expansion joints** | | | | | |
| 1/2" x 3-1/2" | — | LF | .39 | — | .39 |
| 1/4" x 3-1/2" | — | LF | .35 | — | .35 |
| 1/2" x 4" | — | LF | .49 | — | .49 |
| **Wall anchors, 1-3/4" x 24"** | — | Ea | 3.50 | — | 3.50 |
| **Dur-O-Wal reinforcing, Ladur or standard, 10' lengths** | | | | | |
| 3" wide or 4" wide wall | — | LF | .20 | — | .20 |
| 6" wide wall | — | LF | .21 | — | .21 |
| 8" wide wall | — | LF | .23 | — | .23 |
| 10" wide or 12" wide wall | — | LF | .25 | — | .25 |
| **Wired adobe blocks** | | | | | |
| 2" x 2" | — | Ea | .13 | — | .13 |
| 3" x 3" | — | Ea | .25 | — | .25 |
| **Tie wire, black, 16 gauge** | | | | | |
| Roll, 328' | — | Ea | 5.06 | — | 5.06 |
| **Wall ties, 22 gauge** | — | M | 100.00 | — | 100.00 |
| **Wall ties, 28 gauge** | — | M | 100.00 | — | 100.00 |
| **Angle iron** | | | | | |
| 3" x 3" x 3/16" | — | LF | 5.94 | — | 5.94 |
| 4" x 4" x 1/4" | — | LF | 10.40 | — | 10.40 |
| 6" x 4" x 3/8" | — | LF | 16.90 | — | 16.90 |
| **Post anchors (straps)** | | | | | |
| 4" x 4" x 10" | — | Ea | 5.50 | — | 5.50 |
| 4" x 4" x 16" | — | Ea | 15.40 | — | 15.40 |

**Masonry, Subcontract** Typical prices.

Fireplace chimneys, masonry only. Height above firebox shown.

| | Craft@Hrs | Unit | Material | Labor | Total |
|---|---|---|---|---|---|
| 5' high chimney (3' opening) | — | Ea | — | — | 2,670.00 |
| 13' high chimney (3' opening) | — | Ea | — | — | 3,150.00 |
| **Floors** | | | | | |
| Flagstone | — | SF | — | — | 12.50 |
| Patio tile | — | SF | — | — | 9.82 |
| Pavers | — | SF | — | — | 10.70 |
| Quarry tile, unglazed | — | SF | — | — | 13.80 |

**Mats, Runners and Treads** Koffler Products.

Mats

Vinyl link mats, closed link, 1/2" thick, including nosing

| | Craft@Hrs | Unit | Material | Labor | Total |
|---|---|---|---|---|---|
| Solid colors incl. black 28" x 36" | B1@.167 | Ea | 160.00 | 5.56 | 165.56 |
| For 9" to 12" letters add | — | Ea | 22.00 | — | 22.00 |

# Medicine Cabinets

| | Craft@Hrs | Unit | Material | Labor | Total |
|---|---|---|---|---|---|
| Steel link mats, 3/8" deep, 1" mesh | | | | | |
| 16" x 24" | B1@.501 | Ea | 24.00 | 16.70 | 40.70 |
| 22" x 36" | B1@.501 | Ea | 47.50 | 16.70 | 64.20 |
| 30" x 48" | B1@.501 | Ea | 81.20 | 16.70 | 97.90 |
| 36" x 54" | B1@.501 | Ea | 100.00 | 16.70 | 116.70 |
| 36" x 72" | B1@.501 | Ea | 127.00 | 16.70 | 143.70 |
| **Runners** | | | | | |
| Corrugated rubber runner, cut lengths, 2', 3', 4' widths | | | | | |
| 1/8" thick, black | B1@.060 | SF | 3.60 | 2.00 | 5.60 |
| Corrugated or round ribbed vinyl runner, 1/8" thick, 2', 3', or 4' wide, cut lengths, flame resistant | | | | | |
| Black corrugated | B1@.060 | SF | 2.68 | 2.00 | 4.68 |
| Black or brown, round ribbed | B1@.060 | SF | 2.75 | 2.00 | 4.75 |
| Deduct for full rolls (100 LF) | — | % | -10.0 | — | — |
| Safety strips, non-slip strips, press-on abrasive, black, 1/16" thick | | | | | |
| 3/4" x 24", box of 50 | B1@.080 | Ea | .90 | 2.66 | 3.56 |
| 6" x 24" cleats | B1@.080 | Ea | 6.11 | 2.66 | 8.77 |
| **Treads** | | | | | |
| Stair treads, homogeneous rubber, 9/64" thick, 1-1/2" nosing, 12-1/4" deep | | | | | |
| Diamond design, marbleized, medium duty | B1@.080 | LF | 13.10 | 2.66 | 15.76 |
| Plain, with abrasive strips | B1@.080 | LF | 14.40 | 2.66 | 17.06 |
| Matching cove riser | B1@.080 | LF | 7.10 | 2.66 | 9.76 |

## Medicine Cabinets

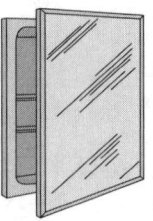

| | Craft@Hrs | Unit | Material | Labor | Total |
|---|---|---|---|---|---|
| Recessed or surface mount medicine cabinets, hinged door, no lighting | | | | | |
| Basic steel-frame mirror cabinet with polystyrene body, magnetic catch, three fixed shelves | | | | | |
| 14" x 18" | BC@.850 | Ea | 14.10 | 31.40 | 45.50 |
| Basic flat glass panel cabinet with enameled steel body, concealed brass hinges, one fixed shelf | | | | | |
| 15" x 19" | BC@.850 | Ea | 32.30 | 31.40 | 63.70 |
| Better quality etched and beveled mirror door cabinet, 2 adjustable shelves, polished brass knobs | | | | | |
| 25" x 26" | BC@.850 | Ea | 119.00 | 31.40 | 150.40 |
| Corner-mount frameless cabinet, one beveled mirror door, 2 adjustable shelves | | | | | |
| 14" x 26" | BC@.850 | Ea | 96.40 | 31.40 | 127.80 |
| Surface mount medicine cabinets, hinged door, no lighting | | | | | |
| Tri-view frame cabinet, flat glass panel, 2 adjustable shelves, brass exposed hinges | | | | | |
| 24" x 24" | BC@.850 | Ea | 53.20 | 31.40 | 84.60 |
| 30" x 29" | BC@.941 | Ea | 74.60 | 34.80 | 109.40 |
| 36" x 29" | BC@.941 | Ea | 86.20 | 34.80 | 121.00 |
| Beveled tri-view frame cabinet, flat glass panel, 2 adjustable shelves, concealed adjustable hinges | | | | | |
| 24" x 26" | BC@.850 | Ea | 96.30 | 31.40 | 127.70 |
| 29" x 25" | BC@.941 | Ea | 107.00 | 34.80 | 141.80 |
| 36" x 30" | BC@.941 | Ea | 139.00 | 34.80 | 173.80 |
| 48" x 30" | BC@.941 | Ea | 161.00 | 34.80 | 195.80 |
| Oval smoked bevel mirror, frameless swing door | | | | | |
| 14" x 18" | BC@.850 | Ea | 100.00 | 31.40 | 131.40 |
| All wood cabinets, 2 shelves, magnetic door latch, surface mount | | | | | |
| 15" x 25", oak finish | BC@.850 | Ea | 37.80 | 31.40 | 69.20 |
| 15" x 25", white finish | BC@.850 | Ea | 37.80 | 31.40 | 69.20 |
| Matching light bars for medicine cabinets, including electrical connection | | | | | |
| 4 lights | BE@1.50 | Ea | 44.70 | 59.40 | 104.10 |
| 5 lights | BE@1.50 | Ea | 55.60 | 59.40 | 115.00 |
| 6 lights | BE@1.50 | Ea | 66.90 | 59.40 | 126.30 |
| 7 lights | BE@1.50 | Ea | 93.10 | 59.40 | 152.50 |

| | Craft@Hrs | Unit | Material | Labor | Total |
|---|---|---|---|---|---|

**Meter Boxes** Precast concrete, FOB manufacturer, inside dimension, approximately 100 to 190 lbs each.

| | Craft@Hrs | Unit | Material | Labor | Total |
|---|---|---|---|---|---|
| 13-1/4" x 19-1/8" x 11" deep | | | | | |
| With 1 piece concrete cover | BL@.334 | Ea | 21.80 | 9.91 | 31.71 |
| With metal cover (cast iron) | BL@.334 | Ea | 30.60 | 9.91 | 40.51 |
| 14-5/8" x 19-3/4" x 11" deep | | | | | |
| With 1 piece concrete cover | BL@.334 | Ea | 17.60 | 9.91 | 27.51 |
| With metal cover (cast iron) | BL@.334 | Ea | 31.10 | 9.91 | 41.01 |
| 15-3/4" x 22-3/4" x 11" deep | | | | | |
| With 1 piece concrete cover | BL@.334 | Ea | 31.90 | 9.91 | 41.81 |
| With metal cover (cast iron) | BL@.334 | Ea | 56.50 | 9.91 | 66.41 |
| Add hinged or reading lid, concrete cover | — | Ea | 6.84 | — | 6.84 |
| Plastic meter and valve boxes, bolting plastic covers | | | | | |
| Standard 16" x 10-3/4" x 12" deep | BL@.167 | Ea | 21.70 | 4.95 | 26.65 |
| Jumbo 20-1/4" x 14-1/2" x 12" deep | BL@.167 | Ea | 28.20 | 4.95 | 33.15 |
| Round pit box, 10" diameter x 12" deep | BL@.167 | Ea | 11.80 | 4.95 | 16.75 |

**Mirrors** See also Bathroom Accessories.

| | Craft@Hrs | Unit | Material | Labor | Total |
|---|---|---|---|---|---|
| Door mirrors, polished edge | | | | | |
| 12" wide x 48" high | BG@.251 | Ea | 17.20 | 8.60 | 25.80 |
| 16" wide x 60" high | BG@.301 | Ea | 26.80 | 10.30 | 37.10 |
| 18" wide x 68" high | BG@.331 | Ea | 32.10 | 11.30 | 43.40 |
| Vanity mirrors, polished edge | | | | | |
| 24" wide x 36" high | BG@.251 | Ea | 27.00 | 8.60 | 35.60 |
| 36" wide x 36" high | BG@.301 | Ea | 37.70 | 10.30 | 48.00 |
| 36" wide x 42" high | BG@.331 | Ea | 43.00 | 11.30 | 54.30 |
| 48" wide x 36" high | BG@.331 | Ea | 49.80 | 11.30 | 61.10 |
| 60" wide x 36" high | BG@.331 | Ea | 63.10 | 11.30 | 74.40 |
| Frosted oval mirrors, floral design, polished edge | | | | | |
| 23" x 29" | BG@.251 | Ea | 42.70 | 8.60 | 51.30 |
| 23" x 36" | BG@.251 | Ea | 63.10 | 8.60 | 71.70 |
| 27" x 30", decorative chrome frame | BG@.251 | Ea | 124.00 | 8.60 | 132.60 |
| Mirror arches (radius, top edge) | | | | | |
| 23" x 35" high, frosted pattern | BG@.251 | Ea | 52.10 | 8.60 | 60.70 |
| Framed gallery mirrors | | | | | |
| 30" x 36", sandalwood frame | BG@.450 | Ea | 117.00 | 15.40 | 132.40 |
| 30" x 32", metal frame | BG@.450 | Ea | 138.00 | 15.40 | 153.40 |
| 36" x 32", metal frame | BG@.450 | Ea | 149.00 | 15.40 | 164.40 |
| 42" x 34", antique black frame | BG@.450 | Ea | 170.00 | 15.40 | 185.40 |

**Mouldings** Unfinished. Material costs include 5% waste. Labor costs are for typical installation only. (For installation costs of Hardwood Moulding, see Softwood Mouldings starting on next page under Mouldings.)

| | Unit | Oak | Poplar | Birch |
|---|---|---|---|---|
| **Hardwood moulding** | | | | |
| Base | | | | |
| 1/2" x 1-1/2" | LF | 1.26 | .90 | .91 |
| 1/2" x 2-1/2" | LF | 1.16 | 1.08 | 1.07 |
| Base shoe, 3/8" x 3/4" | LF | .58 | .45 | .45 |
| Casing | | | | |
| 1/2" x 1-1/2" | LF | 1.46 | 1.01 | .99 |
| 5/8" x 1-5/8" | LF | 1.53 | 1.11 | 1.10 |
| Cap, 1/2" x 1-1/2" | LF | 1.15 | .88 | .85 |

# Mouldings

| | Unit | Oak | Poplar | Birch |
|---|---|---|---|---|
| Crown, 1/2" x 2-1/4 | LF | 1.63 | 1.06 | .97 |
| Cove | | | | |
|   1/2" x 1/2" | LF | .86 | .51 | .52 |
|   3/4" x 1/4" | LF | 1.07 | .58 | .56 |
| Corner | | | | |
|   3/4" x 3/4" | LF | 1.10 | .57 | .54 |
|   1" x 1" x 6" | Ea | 2.83 | .80 | .83 |
| Round edge stop, 3/8" x 1-1/4" | LF | .79 | .48 | .56 |
| Chair rail, 1/2" x 2" | LF | 1.57 | 1.09 | 1.10 |
| Quarter round | | | | |
|   1/4" x 1/4" | LF | .69 | .49 | .49 |
|   1/2" x 1/2" | LF | .70 | .59 | .54 |
| Battens, 1/4" x 3/4" | LF | .61 | .37 | .36 |
| Door trim sets, 16'8" total length | | | | |
|   1/2" x 1-5/8" casing | Ea | 30.30 | 24.80 | 23.60 |
|   5/8" x 1-5/8" casing | Ea | 31.80 | 26.10 | 25.50 |
|   3/8" x 1-1/4" round edge stop oak birch | Ea | 21.20 | 16.80 | 15.30 |

| | Craft@Hrs | Unit | Material | Labor | Total |
|---|---|---|---|---|---|
| Oak threshold, LF of threshold | | | | | |
|   5/8" x 3-1/2" or 3/4" x 3-1/2" | BC@.174 | LF | 14.10 | 6.43 | 20.53 |
|   3/4" x 5-1/2" | BC@.174 | LF | 29.40 | 6.43 | 35.83 |
|   7/8" x 3-1/2" | BC@.174 | LF | 14.60 | 6.43 | 21.03 |
| **Redwood moulding** | | | | | |
|   Band moulding, 3/4" x 5/16" | BC@.030 | LF | .97 | 1.11 | 2.08 |
|   Battens and lattice | | | | | |
|     5/16" x 1-1/4" | BC@.026 | LF | .56 | .96 | 1.52 |
|     5/16" x 1-5/8" | BC@.026 | LF | .56 | .96 | 1.52 |
|     5/16" x 2-1/2" | BC@.026 | LF | .88 | .96 | 1.84 |
|     5/16" x 3-1/2" | BC@.026 | LF | .88 | .96 | 1.84 |
|     3/8" x 2-1/2" | BC@.026 | LF | 1.47 | .96 | 2.43 |
|   Brick mould, 1-1/2" x 1-1/2" | BC@.026 | LF | 1.87 | .96 | 2.83 |
|   Drip cap – water table | | | | | |
|     1-5/8" x 2-1/2" (no lip) | BC@.060 | LF | 2.45 | 2.22 | 4.67 |
|     1-5/8" x 2-1/2" | BC@.060 | LF | 2.45 | 2.22 | 4.67 |
|   Quarter round, 3/4" x 3/4" | BC@.026 | LF | 1.17 | .96 | 2.13 |
|   Rabbeted siding mould, 1-1/8" x 1-5/8" | BC@.026 | LF | 1.50 | .96 | 2.46 |
|   S4S (rectangular) | | | | | |
|     11/16" x 11/16" | BC@.026 | LF | .73 | .96 | 1.69 |
|     1" x 2" | BC@.026 | LF | 1.09 | .96 | 2.05 |
|     1" x 3" | BC@.026 | LF | 1.61 | .96 | 2.57 |
|     1" x 4" | BC@.026 | LF | 2.06 | .96 | 3.02 |
|   Stucco mould | | | | | |
|     13/16" x 1-3/8" | BC@.026 | LF | 1.07 | .96 | 2.03 |
|     13/16" x 1 1/2" | BC@.026 | LF | 1.07 | .96 | 2.03 |
|   Window sill, 2" x 8" | BC@.043 | LF | 12.30 | 1.59 | 13.89 |

| | Craft@Hrs | Unit | Material | Labor | Total |
|---|---|---|---|---|---|

**Softwood moulding, pine** (Labor costs here can also be used for hardwood mouldings in the previous section.)

| | Craft@Hrs | Unit | Material | Labor | Total |
|---|---|---|---|---|---|
| Astragal moulding | | | | | |
| For 1-3/8" x 7' doors | BC@.032 | LF | 2.76 | 1.18 | 3.94 |
| For 1-3/4" x 7' doors | BC@.032 | LF | 4.42 | 1.18 | 5.60 |
| Base (all patterns) | | | | | |
| 7/16" x 7/16" | BC@.016 | LF | .63 | .59 | 1.22 |
| 7/16" x 1-1/4" | BC@.016 | LF | .85 | .59 | 1.44 |
| 3/8" x 3" | BC@.016 | LF | 1.05 | .59 | 1.64 |
| 1/2" x 2-1/4" | BC@.016 | LF | 1.13 | .59 | 1.72 |
| 1/2" x 2-1/2" | BC@.016 | LF | 1.56 | .59 | 2.15 |
| 1/2" x 3-1/2" | BC@.016 | LF | 1.46 | .59 | 2.05 |
| Base (combination or cove) | | | | | |
| 7/16" x 1-1/4" | BC@.016 | LF | .80 | .59 | 1.39 |
| 1/2" x 2" | BC@.016 | LF | 1.27 | .59 | 1.86 |
| 1/2" x 3-1/4" | BC@.016 | LF | 1.69 | .59 | 2.28 |
| Base shoe | | | | | |
| 7/16" x 3/4" | BC@.016 | LF | .41 | .59 | 1.00 |
| Casing (all patterns) | | | | | |
| 1/2" x 1 5/8" | BC@.023 | LF | 1.09 | .85 | 1.94 |
| 5/8" x 1-5/8" | BC@.023 | LF | 1.52 | .85 | 2.37 |
| 7/16" x 2-1/2" | BC@.023 | LF | 1.34 | .85 | 2.19 |
| 9/16" x 3-1/2" | BC@.023 | LF | 1.82 | .85 | 2.67 |
| Chair rail | | | | | |
| 11/16" x 1-5/8" | BC@.021 | LF | 1.30 | .78 | 2.08 |
| 9/16" x 2-1/2" | BC@.021 | LF | 1.72 | .78 | 2.50 |
| Chamfer strip, 3/4" x 3/4" | BC@.016 | LF | .23 | .59 | .82 |
| Corner bead (outside corner moulding) | | | | | |
| 11/16" x 11/16" | BC@.016 | LF | .87 | .59 | 1.46 |
| 3/4" x 3/4" | BC@.016 | LF | .77 | .59 | 1.36 |
| 9/16" x 1-5/8" | BC@.016 | LF | 1.08 | .59 | 1.67 |
| Cove moulding, solid | | | | | |
| 3/8" x 3/8" | BC@.020 | LF | .43 | .74 | 1.17 |
| 1/2" x 1/2" | BC@.020 | LF | .67 | .74 | 1.41 |
| 11/16" x 11/16" | BC@.020 | LF | .94 | .74 | 1.68 |
| 3/4" x 3/4" | BC@.020 | LF | .83 | .74 | 1.57 |
| 1-1/2" x 1-1/2" | BC@.020 | LF | 1.18 | .74 | 1.92 |
| Cove moulding, sprung | | | | | |
| 3/4" x 1-5/8" | BC@.020 | LF | 1.11 | .74 | 1.85 |
| 3/4" x 3-1/2" | BC@.020 | LF | 2.15 | .74 | 2.89 |
| Crown or bed moulding | | | | | |
| 3/4" x 3/4" | BC@.044 | LF | .83 | 1.63 | 2.46 |
| 9/16" x 1-5/8" | BC@.044 | LF | .66 | 1.63 | 2.29 |
| 9/16" x 1-3/4" | BC@.044 | LF | .79 | 1.63 | 2.42 |
| 9/16" x 2-5/8" | BC@.044 | LF | 1.88 | 1.63 | 3.51 |
| 9/16" x 3-5/8" | BC@.044 | LF | 2.13 | 1.63 | 3.76 |

# Moulding, Softwood

| | Craft@Hrs | Unit | Material | Labor | Total |
|---|---|---|---|---|---|
| Add for special cutting, any crown moulding shown: | | | | | |
|   Inside corner, using coping saw | BC@.167 | Ea | — | 6.17 | 6.17 |
|   Outside corners, using miter saw | BC@.083 | Ea | — | 3.07 | 3.07 |
| Drip cap (water table moulding or bar nosing) | | | | | |
|   1-1/16" x 1-5/8" | BC@.060 | LF | 2.42 | 2.22 | 4.64 |
|   1-1/16" x 2" | BC@.060 | LF | 2.68 | 2.22 | 4.90 |
| Drip moulding | | | | | |
|   3/4" x 1-1/4" | BC@.030 | LF | .74 | 1.11 | 1.85 |
| Glass bead | | | | | |
|   1/2" x 9/16" | BC@.016 | LF | .72 | .59 | 1.31 |
| Handrail, fir | | | | | |
|   1-1/2" x 2-1/4" | BC@.060 | LF | 3.34 | 2.22 | 5.56 |
|   1-7/16" x 1-3/4" | BC@.060 | LF | 2.56 | 2.22 | 4.78 |
| Lattice | | | | | |
|   1/4" x 1-1/8" | BC@.016 | LF | .58 | .59 | 1.17 |
|   1/4" x 1-3/4" | BC@.016 | LF | .73 | .59 | 1.32 |
|   5/16" x 1-3/8" | BC@.016 | LF | .97 | .59 | 1.56 |
|   5/16" x 1-3/4" | BC@.016 | LF | 1.13 | .59 | 1.72 |
| Mullion casing | | | | | |
|   3/8" x 2" | BC@.016 | LF | 1.16 | .59 | 1.75 |
| Panel moulding | | | | | |
|   3/8" x 5/8" | BC@.030 | LF | .82 | 1.11 | 1.93 |
|   5/8" x 1-1/8" | BC@.030 | LF | .99 | 1.11 | 2.10 |
|   3/4" x 1" | BC@.030 | LF | 1.05 | 1.11 | 2.16 |
|   3/4" x 1-5/16" | BC@.030 | LF | 1.54 | 1.11 | 2.65 |
|   3/4" x 2" | BC@.030 | LF | 1.32 | 1.11 | 2.43 |
| Parting bead | | | | | |
|   3/8" x 3/4" | BC@.016 | LF | .67 | .59 | 1.26 |
| Picture frame mould | | | | | |
|   11/16" x 1 3/4" | BC@.040 | LF | 1.69 | 1.48 | 3.17 |
|   11/16" x 1-5/8" | BC@.040 | LF | 1.14 | 1.48 | 2.62 |
| Plaster ground | | | | | |
|   3/4" x 7/8" | BC@.010 | LF | .64 | .37 | 1.01 |
| Quarter round | | | | | |
|   1/4" x 1/4" | BC@.016 | LF | .39 | .59 | .98 |
|   3/8" x 3/8" | BC@.016 | LF | .44 | .59 | 1.03 |
|   1/2" x 1/2" | BC@.016 | LF | .62 | .59 | 1.21 |
|   11/16" x 11/16" | BC@.016 | LF | .73 | .59 | 1.32 |
|   3/4" x 3/4" | BC@.016 | LF | 1.10 | .59 | 1.69 |
|   1" x 1-11/16" | BC@.016 | LF | 1.08 | .59 | 1.67 |
| Half round | | | | | |
|   1/4" x 1/2" or 5/16" x 5/8" | BC@.016 | LF | .87 | .59 | 1.46 |
|   3/8" x 3/4" or 1/2" x 1" | BC@.016 | LF | .65 | .59 | 1.24 |
|   3/4" x 1-1/2" | BC@.016 | LF | 1.13 | .59 | 1.72 |
| Full round | | | | | |
|   1/2" or 3/4" diameter | BC@.016 | LF | .79 | .59 | 1.38 |
|   1-1/4" or 1-3/8" diameter | BC@.016 | LF | 1.73 | .59 | 2.32 |
|   1" diameter | BC@.016 | LF | 1.01 | .59 | 1.60 |
|   1-1/2" diameter | BC@.016 | LF | 1.97 | .59 | 2.56 |
|   2" diameter | BC@.016 | LF | 2.72 | .59 | 3.31 |

| | Craft@Hrs | Unit | Material | Labor | Total |
|---|---|---|---|---|---|
| **S4S (rectangular)** | | | | | |
| 7/16" x 2", pine | BC@.020 | Ea | 1.14 | .74 | 1.88 |
| 7/16" x 2-1/2", pine | BC@.020 | Ea | 1.46 | .74 | 2.20 |
| 7/16" x 2-5/8", pine | BC@.020 | Ea | 1.77 | .74 | 2.51 |
| 7/16" x 3-1/2", pine | BC@.020 | Ea | 2.48 | .74 | 3.22 |
| 1/2" x 3/4", pine | BC@.016 | Ea | .70 | .59 | 1.29 |
| 1/2" x 2-1/2", pine | BC@.020 | Ea | 1.59 | .74 | 2.33 |
| 1/2" x 3-1/2", pine | BC@.020 | Ea | 2.57 | .74 | 3.31 |
| 11/16" x 11/16", pine | BC@.016 | Ea | .88 | .59 | 1.47 |
| 11/16" x 1-5/8", pine | BC@.016 | Ea | 1.28 | .59 | 1.87 |
| 11/16" x 1-3/4", pine | BC@.016 | Ea | 1.31 | .59 | 1.90 |
| 11/16" x 2-1/2", pine | BC@.020 | Ea | 2.43 | .74 | 3.17 |
| 3/4" x 3/4", pine | BC@.016 | Ea | .95 | .59 | 1.54 |
| 1-1/16" x 1-3/4", pine | BC@.016 | Ea | 2.06 | .59 | 2.65 |
| **Sash bar** | | | | | |
| 7/8" x 1-3/8" | BC@.016 | LF | 1.13 | .59 | 1.72 |
| **Screen moulding** | | | | | |
| 1/4" x 3/4" beaded | BC@.016 | LF | .56 | .59 | 1.15 |
| 1/4" x 3/4" flat or insert | BC@.016 | LF | .51 | .59 | 1.10 |
| 3/8" x 3/4" clover leaf | BC@.016 | LF | .52 | .59 | 1.11 |
| **Square moulding** | | | | | |
| 1/2" x 1/2" | BC@.016 | LF | .49 | .59 | 1.08 |
| 11/16" x 11/16" | BC@.016 | LF | .93 | .59 | 1.52 |
| 3/4" x 3/4" | BC@.016 | LF | .87 | .59 | 1.46 |
| 2" x 2" | BC@.016 | LF | 2.17 | .59 | 2.76 |
| **Stops, round edge** | | | | | |
| 3/8" x 7/8" | BC@.025 | LF | .37 | .92 | 1.29 |
| 3/8" x 1-1/4" | BC@.025 | LF | .65 | .92 | 1.57 |
| 3/8" x 1-3/8" | BC@.025 | LF | .63 | .92 | 1.55 |
| 3/8" x 1-5/8" or 1/2" x 3/4" | BC@.025 | LF | .63 | .92 | 1.55 |
| 7/16" x 3/4" | BC@.025 | LF | .57 | .92 | 1.49 |
| 7/16" x 1-3/8" | BC@.025 | LF | .80 | .92 | 1.72 |
| 7/16" x 1-5/8" | BC@.025 | LF | 1.04 | .92 | 1.96 |
| **Wainscot cap** | | | | | |
| 5/8" x 1-1/4" or 1/2" x 1-1/2" | BC@.016 | LF | 1.18 | .59 | 1.77 |
| **Window stool, beveled** | | | | | |
| 11/16" x 2-1/4" | BC@.037 | LF | 2.45 | 1.37 | 3.82 |
| 11/16" x 2-1/2" | BC@.037 | LF | 1.03 | 1.37 | 2.40 |
| 1" x 3-1/2" | BC@.037 | LF | 3.81 | 1.37 | 5.18 |
| 1" x 5-1/2" | BC@.037 | LF | 5.87 | 1.37 | 7.24 |
| **Window stool, flat** | | | | | |
| 11/16" x 4-5/8" | BC@.037 | LF | 2.37 | 1.37 | 3.74 |
| 1" x 5-1/2" | BC@.037 | LF | 5.37 | 1.37 | 6.74 |
| **Medium density fiberboard moulding** Primed, ready for painting. | | | | | |
| Base, 9/16" x 3-1/16" | BC@.012 | LF | .89 | .44 | 1.33 |
| Base, 2-1/2" x 3-1/4" | BC@.012 | LF | .56 | .44 | 1.00 |
| Casing, 11/16" x 3" | BC@.016 | LF | 1.33 | .59 | 1.92 |
| Door casing set, 17' x 5/8" x 2-1/4" | BC@.272 | Ea | 7.40 | 10.00 | 17.40 |
| Chair rail, 9/16" x 3" | BC@.015 | LF | 1.56 | .55 | 2.11 |
| Corner, bullnose, 2-1/4" x 2-1/4" | BC@.025 | LF | .72 | .92 | 1.64 |
| Crown, 11/16" x 3-1/2" | BC@.025 | LF | 1.70 | .92 | 2.62 |

# Moulding, White Vinyl

| | Craft@Hrs | Unit | Material | Labor | Total |
|---|---|---|---|---|---|
| Crown inside or outside corner block, | | | | | |
| 3/4" x 6" | BC@.125 | Ea | 15.00 | 4.62 | 19.62 |
| Plinth block, 1" x 3-3/4" | BC@.125 | Ea | 3.43 | 4.62 | 8.05 |
| **White vinyl moulding** Unfinished. | | | | | |
| Backband casing | BC@.016 | LF | 2.44 | .59 | 3.03 |
| Base cap, 11/16" x 1-1/8" | BC@.009 | LF | .73 | .33 | 1.06 |
| Bead, 9/16" x 1-3/4" | BC@.016 | LF | .68 | .59 | 1.27 |
| Blind stop | BC@.016 | LF | .69 | .59 | 1.28 |
| Casing, 5/8" x 2-1/2" | BC@.023 | LF | 2.60 | .85 | 3.45 |
| Crown, 1/2" x 3-1/4" | BC@.044 | LF | 2.59 | 1.63 | 4.22 |
| Dentil, 27/32" x 5-7/8" | BC@.016 | LF | 4.06 | .59 | 4.65 |
| Dentil, 27/32" x 4-1/8" | BC@.016 | LF | 1.54 | .59 | 2.13 |
| Drip cap, 11/16" x 1-5/8" | BC@.016 | LF | 1.13 | .59 | 1.72 |
| Exterior bead 3/16" x 1/2" | BC@.016 | LF | .36 | .59 | .95 |
| Flat utility trim | BC@.023 | LF | 1.92 | .85 | 2.77 |
| Quarter round, 3/4" x 3/4" x 12' | BC@.016 | LF | .74 | .59 | 1.33 |
| Shingle moulding, 11/16" x 1-3/4" | BC@.016 | LF | 1.16 | .59 | 1.75 |
| **Extruded polymer moulding** Factory finished. | | | | | |
| Bead, Brite White | BC@.016 | LF | .30 | .59 | .89 |
| Bead, Cherry/Mahogany | BC@.016 | LF | .30 | .59 | .89 |
| Bead, Cinnamon Chestnut | BC@.016 | LF | .31 | .59 | .90 |
| Bed mould, Brite White | BC@.016 | LF | .60 | .59 | 1.19 |
| Bed mould, Cherry/Mahogany | BC@.016 | LF | .60 | .59 | 1.19 |
| Bed mould, Cinnamon Chestnut | BC@.016 | LF | .60 | .59 | 1.19 |
| Casing and base, Brite White | BC@.023 | LF | .60 | .85 | 1.45 |
| Casing and base, Cherry/Mahogany | BC@.023 | LF | .53 | .85 | 1.38 |
| Casing and base, Cinnamon Chestnut | BC@.023 | LF | .60 | .85 | 1.45 |
| Outside corner, Brite White | BC@.016 | LF | .30 | .59 | .89 |
| Outside corner, Cherry/Mahogany | BC@.016 | LF | .30 | .59 | .89 |
| Outside corner, Cinnamon Chestnut | BC@.016 | LF | .30 | .59 | .89 |
| Plywood cap, Brite White | BC@.016 | LF | .38 | .59 | .97 |
| Shoe, Brite White | BC@.016 | LF | .30 | .59 | .89 |
| Shoe, Cherry/Mahogany | BC@.016 | LF | .30 | .59 | .89 |
| Shoe, Cinnamon Chestnut | BC@.016 | LF | .30 | .59 | .89 |
| Stop, Brite White | BC@.025 | LF | .30 | .92 | 1.22 |
| Stop, Cinnamon Chestnut | BC@.025 | LF | .30 | .92 | 1.22 |

## Nails and Fasteners

Nail sizes are usually identified by the "penny," abbreviated "d." Use the following figures to estimate nail quantities for framing:

Sills and plates – 8 pounds of 10d, 16d and 20d nails per MBF

Walls and partition studs – 10 pounds of 10d and 16d nails per MBF

2" x 6" or 2" x 8" joists and rafters – 9 pounds of 16d nails per MBF

2" x 10" or 2" x 12" joist and rafters – 7 pounds of 16d nails per MBF

Sheathing to 3/8" thick – 10 pounds of 6d nails per MSF

Sheathing over 3/8" thick – 20 pounds of 8d nails per MSF

Siding – 25 pounds of 8d nails per MSF

| | Unit | Galvanized | Bright |
|---|---|---|---|
| **Common nails** Bright nails are tumbled to remove debris and have a polished appearance. Galvanized nails are zinc coated to protect against corrosion and have a dull sheen. Cost per box except as noted. | | | |
| 3 penny, 1-1/4", 1-pound box, 568 nails | Box | 4.41 | 3.59 |
| 4 penny, 1-1/2", 1-pound box, 316 nails | Box | 4.41 | 3.45 |
| 4 penny, 1-1/2", 5-pound box, 2,840 nails | Box | 15.40 | 12.60 |

| | Unit | Galvanized | Bright |
|---|---|---|---|
| 5 penny, 1-3/4", 1-pound box, 271 nails | Box | — | 3.85 |
| 6 penny, 2", 1-pound box, 181 nails | Box | 4.40 | — |
| 6 penny, 2", 5-pound box, 905 nails | Box | 15.40 | 12.10 |
| 6 penny, 2", 30-pound box, 5,430 nails | Box | 50.60 | 37.80 |
| 6 penny, 2", 50-pound box, 9,050 nails | Box | 58.70 | — |
| 7 penny, 2-1/2", 5-pound box, 805 nails | Box | 16.40 | — |
| 7 penny, 2-1/2", 50-pound box, 8,050 nails | Box | 66.90 | — |
| 8 penny, 2-1/2", 1-pound box, 106 nails | Box | 4.33 | 3.49 |
| 8 penny, 2-1/2", 5-pound box, 530 nails | Box | 15.40 | 12.10 |
| 8 penny, 2-1/2", 30-pound box, 3,180 nails | Box | 55.40 | 41.80 |
| 8 penny, 2-1/2", 50-pound box, 5,300 nails | Box | — | 50.60 |
| 10 penny, 3", 1-pound box, 69 nails | Box | 4.40 | 3.49 |
| 10 penny, 3", 5-pound box, 345 nails | Box | 15.40 | 12.10 |
| 10 penny, 3", 30-pound box, 2,070 nails | Box | 55.40 | 41.00 |
| 10 penny, 3", 50-pound box, 3,450 nails | Box | 65.90 | 48.70 |
| 12 penny, 3-1/4", 1-pound box, 64 nails | Box | 4.30 | — |
| 12 penny, 3-1/4", 5-pound box, 320 nails | Box | 15.40 | 12.00 |
| 12 penny, 3-1/4", 30-pound box, 1,920 nails | Box | 55.00 | 42.00 |
| 12 penny, 3-1/4", 50-pound box, 3,200 nails | Box | — | 51.70 |
| 16 penny, 3-1/2", 1-pound box, 49 nails | Box | 4.32 | 3.49 |
| 16 penny, 3-1/2", 5-pound box, 245 nails | Box | 15.40 | 12.20 |
| 16 penny, 3-1/2", 30-pound box, 1,470 nails | Box | 64.10 | 42.40 |
| 16 penny, 3-1/2", 50-pound box, 2,450 nails | Box | 55.60 | 51.30 |
| 20 penny, 4", 1-pound box, 31 nails | Box | 4.40 | 3.49 |
| 20 penny, 4", 5-pound box, 155 nails | Box | 15.40 | 12.20 |
| 40 penny, 5", per nail, 18 per pound | Ea | .06 | .05 |
| 60 penny, 6", per nail, 11 per pound | Ea | .17 | .08 |
| 60 penny, 6", 30-pound box, 330 nails | Box | — | 43.00 |

**Box nails** Box nails are thinner and about 1/8" shorter than common nails of the same nominal size. Points are blunter to reduce wood splitting. Use box nails for molding and trim and when the lumber is dry or brittle. Cost per box.

| | Unit | Galvanized | Bright |
|---|---|---|---|
| 3 penny, 1-1/4", 1-pound box, 1,010 nails | Box | 4.70 | 3.85 |
| 3 penny, 1-1/4", 5-pound box, 5,050 nails | Box | 16.40 | — |
| 4 penny, 1-1/2", 1-pound box, 473 nails | Box | 4.63 | 3.85 |
| 4 penny, 1-1/2", 5-pound box, 2,365 nails | Box | 17.00 | — |
| 5 penny, 1-3/4", 1-pound box, 406 nails | Box | 4.70 | — |
| 5 penny, 1-3/4", 5-pound box, 2,030 nails | Box | 16.50 | — |
| 6 penny, 2", 1-pound box, 236 nails | Box | 4.66 | 3.85 |
| 6 penny, 2", 5-pound box, 1,180 nails | Box | 16.20 | 13.20 |
| 7 penny, 2-1/4", 1-pound box, 210 nails | Box | 4.70 | — |
| 8 penny, 2-1/2", 1-pound box, 145 nails | Box | 4.66 | 3.85 |
| 8 penny, 2-1/2", 5-pound box, 725 nails | Box | 16.20 | 13.20 |
| 8 penny, 2-1/2", 30-pound box, 4,350 nails | Box | 56.60 | — |
| 10 penny, 3", 1-pound box, 132 nails | Box | 4.69 | 3.85 |
| 10 penny, 3", 5-pound box, 660 nails | Box | 16.40 | 13.30 |
| 12 penny, 3-1/4", 1-pound box, 94 nails | Box | 4.56 | 3.85 |
| 12 penny, 3-1/4", 5-pound box, 470 nails | Box | 16.21 | — |
| 16 penny, 3-1/2", 1-pound box, 71 nails | Box | 4.66 | 3.85 |
| 16 penny, 3-1/2", 5-pound box, 355 nails | Box | 16.21 | 13.24 |
| 16 penny, 3-1/2", 30-pound box, 2,130 nails | Box | 59.60 | 54.10 |
| 20 penny, 4", 1-pound box, 62 nails | Box | 4.70 | 3.85 |
| 20 penny, 4", 5-pound box, 310 nails | Box | 16.40 | — |

# Nails

| | Unit | Screw shank | Ring shank |
|---|---|---|---|
| **Deck nails** For use with ACQ or CA pressure-treated lumber. Cost per box. | | | |
| 6 penny, 2", 1-pound box | Box | 4.80 | 4.96 |
| 6 penny, 2", 5-pound box | Box | 16.70 | 17.30 |
| 8 penny, 2-1/2", 1-pound box | Box | 4.86 | 4.96 |
| 8 penny, 2-1/2", 5-pound box | Box | 16.90 | 16.70 |
| 10 penny, 3", 1-pound box | Box | 4.82 | 4.96 |
| 12 penny, 3-1/4", 1-pound box | Box | 4.81 | 4.95 |
| 12 penny, 3-1/4", 5-pound box | Box | 17.10 | 16.40 |
| 16 penny, 3-1/2", 1-pound box | Box | 4.86 | 4.96 |
| 16 penny, 3-1/2", 5-pound box | Box | 16.90 | 17.30 |

| | Craft@Hrs | Unit | Material | Labor | Total |
|---|---|---|---|---|---|
| **Vinyl/cement coated sinkers** Vinyl coating lubricates the wood and holds fast when driven. Head sinks flush with surface. | | | | | |
| 4 penny, 1-3/8", 1-pound box, 550 nails | — | Box | 2.93 | — | 2.93 |
| 6 penny, 1-7/8", 1-pound box, 275 nails | — | Box | 3.11 | — | 3.11 |
| 6 penny, 1-7/8", 5-pound box, 1,375 nails | — | Box | 10.90 | — | 10.90 |
| 8 penny, 2-3/8", 1-pound box, 142 nails, | — | Box | 3.11 | — | 3.11 |
| 8 penny, 2-3/8", 5-pound box, 710 nails | — | Box | 10.90 | — | 10.90 |
| 8 penny, 2-3/8", 50-pound box, 7,100 nails | — | Box | 48.20 | — | 48.20 |
| 10 penny, 2-7/8", 5-pound box, 104 nails | — | Box | 11.00 | — | 11.00 |
| 12 penny, 3-1/8", 1-pound box, 77 nails | — | Box | 3.11 | — | 3.11 |
| 12 penny, 3-1/8", 5-pound box, 385 nails | — | Box | 10.90 | — | 10.90 |
| 12 penny, 3-1/8", 50-pound box, 3,850 nails | — | Box | 51.40 | — | 51.40 |
| 16 penny, 3-3/8", 1-pound box, 61 nails | — | Box | 3.11 | — | 3.11 |
| 16 penny, 3-3/8", 5-pound box, 305 nails | — | Box | 10.90 | — | 10.90 |
| 16 penny, 3-3/8", 25-pound box, 1,525 nails | — | Box | 32.40 | — | 32.40 |
| 16 penny, 3-3/8", 50-pound box, 3,050 nails | — | Box | 48.10 | — | 48.10 |
| Plain shank stick collated framing nails | | | | | |
| 2" x 0.113", coated, box of 2,000 | — | Box | 33.30 | — | 33.30 |
| 2-1/2" x 0.120", coated, box of 2,000 | — | Box | 36.80 | — | 36.80 |
| 2-1/2" x 0.120", galvanized, box of 2,000 | — | Box | 59.60 | — | 59.60 |
| 3" x 0.120", coated, box of 2,000 | — | Box | 31.00 | — | 31.00 |
| 3" x 0.120", galvanized, box of 2,000 | — | Box | 70.60 | — | 70.60 |
| 3-1/4" x 0.120", coated, box of 2,000 | — | Box | 41.10 | — | 41.10 |
| 3-1/4" x 0.120", galvanized, box of 2,000 | — | Box | 71.90 | — | 71.90 |
| 3-1/2" x 0.131", coated, box of 2,000 | — | Box | 48.90 | — | 48.90 |
| 3 1/2" x 0.131", galvanized, box of 2,000 | — | Box | 68.60 | — | 68.60 |
| Plain shank coil framing nails, galvanized and coated | | | | | |
| 2" x 0.099", box of 3,600 | — | Box | 83.10 | — | 83.10 |
| 2-1/2" x 0.099", box of 3,600 | — | Box | 84.60 | — | 84.60 |
| 3" x 0.120", box of 2,700 | — | Box | 79.90 | — | 79.90 |
| 3-1/4" x 0.120", box of 2,700 | — | Box | 96.90 | — | 96.90 |
| Screw shank coil framing nails, plastic coated | | | | | |
| 1-5/8" x 0.099", box of 10,800, steel | — | Box | 195.00 | — | 195.00 |
| 1-7/8" x 0.099", box of 9,000, aluminum | — | Box | 206.00 | — | 206.00 |
| 1-3/4" x 0.086", box of 9,000, steel | — | Box | 149.00 | — | 149.00 |
| 2" x 0.099", box of 9,000, steel | — | Box | 118.00 | — | 118.00 |
| 2-1/4" x 0.099", box of 7,200, steel | — | Box | 97.50 | — | 97.50 |
| 2-1/2" x 0.105", box of 4,000, steel | — | Box | 71.80 | — | 71.80 |
| 2-7/8" x 0.105", box of 5,400, steel | — | Box | 108.00 | — | 108.00 |

| | Craft@Hrs | Unit | Material | Labor | Total |
|---|---|---|---|---|---|

**Brad nails for an impulse nailer, bright, 18-gauge**

| | Craft@Hrs | Unit | Material | Labor | Total |
|---|---|---|---|---|---|
| 1", box of 1,000 | — | Box | 4.72 | — | 4.72 |
| 1-1/4", box of 1,000 | — | Box | 4.41 | — | 4.41 |
| 2", box of 1,000 | — | Box | 5.42 | — | 5.42 |

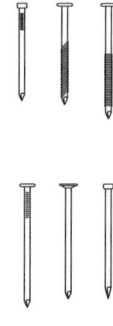

**Casing nails, galvanized.** Used where the nail head must be hidden. Casing nails have small, conical heads. The nail diameter is smaller than a common nail but larger than a finishing nail. Casing nails are used to attach trim.

| | | | | | |
|---|---|---|---|---|---|
| 6 penny, 2", 1-pound box, 236 nails | — | Box | 4.70 | — | 4.70 |
| 8 penny, 2-1/2", 1-pound box, 145 nails | — | Box | 4.47 | — | 4.47 |
| 10 penny, 3", 1-pound box, 94 nails | — | Box | 4.70 | — | 4.70 |
| 16 penny, 3-1/2", 1-pound box, 71 nails | — | Box | 4.46 | — | 4.46 |

**Drywall nails.** Allow 6 pounds of nails (1,300 nails) per MSF of gypsum wallboard

| | | | | | |
|---|---|---|---|---|---|
| 1-1/4", bright, ring shank, 5-pound box | — | Box | 13.50 | — | 13.50 |
| 1-3/8", electro galvanized 5-pound box | — | Box | 12.70 | — | 12.70 |
| 1-3/8", phosphate coated, 5-pound box | — | Box | 13.70 | — | 13.70 |
| 1-3/8", blued, threaded, 5-pound box | — | Box | 14.30 | — | 14.30 |
| 1-1/2", blued, threaded, 5-pound box | — | Box | 14.30 | — | 14.30 |
| 1-5/8", blued, threaded, 5-pound box | — | Box | 12.70 | — | 12.70 |
| 1-7/8", phosphate coated, 5-pound box | — | Box | 14.30 | — | 14.30 |

**Drywall screws.** Black phosphate coating. For hanging drywall and plasterboard on wood or metal studs. Allow 1,300 screws per MSF of gypsum wallboard.

| | | | | | |
|---|---|---|---|---|---|
| 1-1/4", 5-pound box of 1,330 | — | Box | 21.40 | — | 21.40 |
| 1-5/8", 5-pound box of 1,300 | — | Box | 21.40 | — | 21.40 |
| 2", 5-pound box of 1,070 | — | Box | 21.40 | — | 21.40 |
| 2-1/2", 5-pound box of 480 | — | Box | 21.40 | — | 21.40 |
| 3", 5-pound box of 415 | — | Box | 21.40 | — | 21.40 |

**Finishing nails, bright.** Finishing nails have a barrel-shaped head with a small diameter and a dimple on the top. Cost per box.

| | | | | | |
|---|---|---|---|---|---|
| 3 penny, 1-1/4", 1-pound box, 807 nails | — | Box | 3.42 | — | 3.42 |
| 4 penny, 1-1/2", 1-pound box, 584 nails | — | Box | 3.42 | — | 3.42 |
| 6 penny, 2", 1-pound box, 309 nails | — | Box | 3.42 | — | 3.42 |
| 8 penny, 2-1/2", 1-pound box, 189 nails | — | Box | 3.42 | — | 3.42 |
| 16 penny, 3-1/2", 1-pound box, 90 nails | — | Box | 3.42 | — | 3.42 |

**Finishing nail sticks for angled impulse nailer, 16 gauge, bright.** Cost per box.

| | | | | | |
|---|---|---|---|---|---|
| 1", box of 2,000 | — | Box | 9.80 | — | 9.80 |
| 1-1/4", box of 2,000 | — | Box | 11.50 | — | 11.50 |
| 1-1/2", box of 2,000 | — | Box | 14.00 | — | 14.00 |
| 1-3/4", box of 2,000 | — | Box | 14.40 | — | 14.40 |
| 2", box of 2,000 | — | Box | 15.60 | — | 15.60 |
| 2-1/2", box of 2,000 | — | Box | 19.90 | — | 19.90 |

**Bright joist hanger nails, 9 gauge except as noted.** Cost per box.

| | | | | | |
|---|---|---|---|---|---|
| 1-1/4", 1-pound box | — | Box | 4.07 | — | 4.07 |
| 1-1/2", 1-pound box | — | Box | 4.07 | — | 4.07 |
| 1-1/2", 1-pound box, 10 gauge | — | Box | 4.07 | — | 4.07 |
| 1-1/2", 5-pound box | — | Box | 14.30 | — | 14.30 |
| 1-1/2", 5-pound box, 10 gauge | — | Box | 14.30 | — | 14.30 |
| 1-1/2", 30-pound box | — | Box | 56.40 | — | 56.40 |

**Galvanized joist hanger nails. 9 gauge.** Cost per box.

| | | | | | |
|---|---|---|---|---|---|
| 1-1/4", 5-pound box | — | Box | 17.20 | — | 17.20 |
| 1-1/4", 25-pound box | — | Box | 54.40 | — | 54.40 |
| 1-1/2", 1-pound box | — | Box | 4.70 | — | 4.70 |
| 1-1/2", 50-pound box | — | Box | 75.30 | — | 75.30 |

# Nails

| | Craft@Hrs | Unit | Material | Labor | Total |
|---|---|---|---|---|---|
| Stainless steel joist hanger nails | | | | | |
| 1-1/2", 9 gauge, 1-pound box | — | Box | 10.40 | — | 10.40 |
| 1-1/2", 10 gauge, 5-pound box | — | Box | 47.00 | — | 47.00 |
| Masonry nails. Temper hardened, bright. | | | | | |
| 1", fluted nail, 1-pound box | — | Box | 5.73 | — | 5.73 |
| 1-1/2" fluted nail, 1-pound box | — | Box | 5.63 | — | 5.63 |
| 2" fluted nail, 1-pound box | — | Box | 5.61 | — | 5.61 |
| 2-1/2" cut nail, flat head, 1-pound box | — | Box | 5.61 | — | 5.61 |
| 3", cut nail, flat head, 5-pound box | — | Box | 20.70 | — | 20.70 |
| 4-1/2", cut nail, flat head, 1-pound box | — | Box | 5.81 | — | 5.81 |
| Plasterboard nails, 13 gauge, blued | | | | | |
| 3/8" head, 1-1/8" (449) or 1-1/4" (405) | — | Lb | 4.07 | — | 4.07 |
| 3/8" head, 1-5/8", 342 nails | — | Lb | 3.62 | — | 3.62 |
| Roofing nail coils, for pneumatic nailer | | | | | |
| 2 penny, 1", 7,200 nails, galvanized | — | Box | 65.20 | — | 65.20 |
| 3 penny, 1-1/4", 7,200 nails, galvanized | — | Box | 36.10 | — | 36.10 |
| 5 penny, 2", 3,600 nails, bright | — | Box | 51.50 | — | 51.50 |
| Roofing nails, electro galvanized | | | | | |
| 1/2", 1-pound box, 380 nails | — | Box | 2.89 | — | 2.89 |
| 3/4", 1-pound box, 315 nails | — | Box | 2.69 | — | 2.69 |
| 1-3/4" or 2", 5-pound box, 150 or 138 nails | — | Box | 10.80 | — | 10.80 |
| Roofing nails, galvanized. Ring shank. | | | | | |
| 1-1/4", 30-pound box | — | Box | 60.30 | — | 60.30 |
| 1-1/2", 30-pound box | — | Box | 64.40 | — | 64.40 |
| 1-3/4", 5-pound box | — | Box | 17.60 | — | 17.60 |
| 2-1/2", 50-pound box | — | Box | 60.80 | — | 60.80 |
| Roofing nails, plastic round cap. For securing felt on sandable roof decking or attaching foam board sheathing | | | | | |
| 7/8", 11-1/2 gauge, box of 3,000 | — | Box | 36.60 | — | 36.60 |
| 2" x 0.113", galvanized, box of 5,000 | — | Box | 87.20 | — | 87.20 |
| Galvanized roofing staples, wide crown. For shingles and polystyrene sheathing. | | | | | |
| 7/8", box of 9,000 | — | Box | 48.90 | — | 48.90 |
| 1-1/4", box of 9,000 | — | Box | 63.60 | — | 63.60 |
| Double head (duplex) nails for use on scaffolds, concrete forms and temporary construction | | | | | |
| 6 penny, 2", 1-pound box, 143 nails | — | Box | 4.41 | — | 4.41 |
| 6 penny, 2", 5-pound box, 715 nails | — | Box | 15.40 | — | 15.40 |
| 8 penny, 2-1/4", 1-pound box, 88 nails | — | Box | 4.39 | — | 4.39 |
| 8 penny, 2-1/4", 5-pound box, 440 nails | — | Box | 15.00 | — | 15.00 |
| 8 penny, 2-1/4", 30-pound box, 2,640 nails | — | Box | 49.80 | — | 49.80 |
| 16 penny, 3-1/4", 1-pound box, 44 nails | — | Box | 4.33 | — | 4.33 |
| 16 penny, 3-1/4", 5-pound box, 220 nails, | — | Box | 15.00 | — | 15.00 |
| 16 penny, 3-1/4", 30-pound box, 1,320 nails | — | Box | 49.80 | — | 49.80 |
| Siding nails | | | | | |
| 1-1/2", aluminum, 1-pound box | — | Box | 12.60 | — | 12.60 |
| 2", aluminum, 1-pound box | — | Box | 12.20 | — | 12.20 |
| 1-3/4", galvanized, 5-pound box | — | Box | 15.70 | — | 15.70 |
| 2-1/4", galvanized, 5-pound box | — | Box | 17.10 | — | 17.10 |
| 2-1/2", galvanized, 5-pound box | — | Box | 15.30 | — | 15.30 |
| 3", galvanized, 5-pound box | — | Box | 15.00 | — | 15.00 |
| 3" double galvanized, 5-pound box | — | Box | 18.70 | — | 18.70 |

| | Craft@Hrs | Unit | Material | Labor | Total |
|---|---|---|---|---|---|

Roofing nails, plastic round cap . For securing felt on sandable roof decking or attaching foam board sheathing

| | | | | | |
|---|---|---|---|---|---|
| 7/8", 11-1/2 gauge, box of 3,000 | — | Box | 36.60 | — | 36.60 |
| 2" x 0.113", galvanized, box of 5,000 | — | Box | 87.20 | — | 87.20 |

Spikes, galvanized. For landscape timbers, railroad ties, log home construction.

| | | | | | |
|---|---|---|---|---|---|
| 3/8" x 8", per spike | — | Ea | .50 | — | .50 |
| 3/8" x 10", per spike | — | Ea | .61 | — | .61 |
| 3/8" x 10", 5-pound box | — | Box | 14.30 | — | 14.30 |
| 3/8" x 12", per spike | — | Ea | .72 | — | .72 |
| 3/8" x 12", 5-pound box | — | Box | 14.30 | — | 14.30 |

Powder-driven fastening nails, zinc plated

| | | | | | |
|---|---|---|---|---|---|
| 3/4", per pack of 20 | — | Ea | 2.26 | — | 2.26 |
| 7/8", per pack of 20 | — | Ea | 4.09 | — | 4.09 |
| 1", per pack of 20 | — | Ea | 6.98 | — | 6.98 |
| 1-1/4", per pack of 20 | — | Ea | 7.08 | — | 7.08 |
| 1-1/2", per pack of 20 | — | Ea | 6.57 | — | 6.57 |
| 1-7/8", per pack of 20 | — | Ea | 7.25 | — | 7.25 |
| 2-1/2", per pack of 20 | — | Ea | 6.90 | — | 6.90 |
| 3", per pack of 20 | — | Ea | 7.26 | — | 7.26 |

Powder-driven 0.300 pins, with washers

| | | | | | |
|---|---|---|---|---|---|
| 1-1/2", per pack of 100 | — | Ea | 13.90 | — | 13.90 |
| 2-1/2", per pack of 100 | — | Ea | 14.20 | — | 14.20 |
| 3", per pack of 100 | — | Ea | 15.30 | — | 15.30 |

**Nuts, Bolts and Washers** Bolts, standard hex head, material cost per bolt.

| | | | Length | | |
|---|---|---|---|---|---|
| | 4" | 6" | 8" | 10" | 12" |
| Heavy duty lag bolts, galvanized, deduct 25% for carton quantities | | | | | |
| 1/2" | 1.09 | 1.60 | 2.41 | 2.85 | 3.30 |
| 5/8" | 2.63 | 3.89 | 4.62 | 5.75 | 6.04 |
| 3/4" | 5.25 | 6.97 | 7.54 | 9.43 | 11.29 |
| Machine bolts (hex bolts), galvanized, deduct 25% for carton quantities | | | | | |
| 1/2", 13 threads per inch | .85 | 1.32 | 1.88 | 2.15 | 2.57 |
| 5/8", 11 threads per inch | 2.12 | 3.19 | 3.90 | 4.52 | 5.25 |
| 3/4", 10 threads per inch | 2.70 | 3.71 | 5.09 | 5.97 | 7.08 |
| 7/8", 9 threads per inch | 5.65 | 6.84 | 7.97 | 9.74 | 11.20 |
| 1", 8 threads per inch | 6.48 | 7.37 | 7.63 | 8.07 | 9.15 |
| Carriage bolts (domed top and a square under the head), galvanized, deduct 25% for carton quantities | | | | | |
| 1/2", 13 threads per inch | .84 | 1.07 | 1.61 | 1.81 | 2.35 |
| 5/8", 11 threads per inch | 1.60 | 2.41 | 3.05 | 3.63 | 4.12 |
| 3/4", 10 threads per inch | 3.28 | 4.85 | 6.04 | 7.23 | 8.03 |

| | Craft@Hrs | Unit | Material | Labor | Total |
|---|---|---|---|---|---|

Labor to install machine bolts or lag bolts. Includes installation of nuts and washers and drilling as needed.

| | | | | | |
|---|---|---|---|---|---|
| 4" to 6" long | BC@.100 | Ea | — | 3.69 | 3.69 |
| 8" to 10" long | BC@.150 | Ea | — | 5.54 | 5.54 |
| 12" long | BC@.200 | Ea | — | 7.39 | 7.39 |

# Paint Removal

| | Craft@Hrs | Unit | Material | Labor | Total |
|---|---|---|---|---|---|
| Nuts, standard hex, galvanized | | | | | |
| 1/2" | — | Ea | .27 | — | .27 |
| 5/8" | — | Ea | .46 | — | .46 |
| 3/4" | — | Ea | .70 | — | .70 |
| 7/8" | — | Ea | 1.52 | — | 1.52 |
| 1" | — | Ea | 2.65 | — | 2.65 |
| Washers, standard cut steel, galvanized AKA Lock Washers | | | | | |
| 1/2" | — | Ea | .21 | — | .21 |
| 5/8" | — | Ea | .37 | — | .37 |
| 3/4" | — | Ea | .64 | — | .64 |
| 7/8" | — | Ea | 1.26 | — | 1.26 |
| 1" | — | Ea | 1.31 | — | 1.31 |
| Washers, USS Flat steel, stainless | | | | | |
| 1/2" | — | Ea | .20 | — | .20 |
| 5/8" | — | Ea | .36 | — | .36 |
| 3/4" | — | Ea | .55 | — | .55 |
| 7/8" | — | Ea | .79 | — | .79 |
| 1" | — | Ea | .86 | — | .86 |
| Lock washers, larger OD than flat washers, galvanized, OD typically 3" | | | | | |
| 1/2" | — | Ea | 1.98 | — | 1.98 |
| 5/8" | — | Ea | 1.98 | — | 1.98 |
| 3/4" | — | Ea | 1.98 | — | 1.98 |
| 1" | — | Ea | 1.98 | — | 1.98 |

**Paint Removal** Dustless removal of lead-based paints. The removal system is a mechanical process and no water, chemicals or abrasive grit are used. Containment or special ventilation is not required. Based on Pentek Inc. Costs shown are for a 5,000 SF or larger job. For estimating purposes figure one 55-gallon drum of contaminated waste per 2,500 square feet of coatings removed from surfaces with an average coating thickness of 12 mils. Waste disposal costs are not included. Use $5,000 as a minimum charge for work of this type. On pre-1978 homes, child care facilities and schools, rules adopted by the Environmental Protection Agency require that: (1) a trained supervisor be on site during paint removal, (2) those affected be given a copy of the EPA pamphlet on lead paint removal, (3) records be retained for three years and (4) steps be taken to prevent the spread of dust and debris. For more information, see http://www.epa.gov/lead/pubs/sbcomplianceguide.pdf.

| | Craft@Hrs | Unit | Material | Labor | Equipment | Total |
|---|---|---|---|---|---|---|
| **Removing paint from ceilings, doors, door & window frames, roofs, siding** Equipment includes three pneumatic needle scaler units attached to one air-powered vacuum-waste packing unit and one 150 CFM gas or diesel powered air compressor including all hoses and connectors. Add the one-time charge listed below for move-on, move-off and refurbishing of equipment. | | | | | | |
| Ceilings | | | | | | |
| Plaster, 70 SF per hour | PT@.042 | SF | — | 1.56 | .95 | 2.51 |
| Doors, wood or metal. Remove and re-hang doors, no carpentry included | | | | | | |
| To 3'0" x 6'8" | PT@1.00 | Ea | — | 37.30 | — | 37.30 |
| Over 3'0" x 6'8" to 6'0" x 8'0" | PT@1.50 | Ea | — | 55.90 | — | 55.90 |
| Over 6'0" x 8'0" to 12'0" x 20'0" | PT@2.00 | Ea | — | 74.50 | — | 74.50 |
| Paint removal (assumes doors are laying flat) | | | | | | |
| 80 SF per hour | PT@.038 | SF | — | 1.42 | .86 | 2.28 |
| Door or window frames | | | | | | |
| To 3'0" x 6'8" | PT@1.50 | Ea | — | 55.90 | 33.90 | 89.80 |
| Over 3'0" x 6'8" to 6'0" x 8'0" | PT@2.00 | Ea | — | 74.50 | 45.20 | 119.70 |
| Over 6'0" x 8'0" to 12'0" x 20'0" | PT@3.00 | Ea | — | 112.00 | 67.80 | 179.80 |

| | Craft@Hrs | Unit | Material | Labor | Equipment | Total |
|---|---|---|---|---|---|---|
| Lintels | | | | | | |
| Per SF of surface area | PT@.033 | SF | — | 1.23 | .75 | 1.98 |
| Roofs | | | | | | |
| Metal roofs to 3 in 12 pitch | | | | | | |
| 75 SF of surface per hour | PT@.040 | SF | — | 1.49 | .90 | 2.39 |
| Metal roofs over 3 in 12 pitch | | | | | | |
| 65 SF of surface per hour | PT@.046 | SF | — | 1.71 | 1.04 | 2.75 |
| Wood roofs to 3 in 12 pitch | | | | | | |
| 80 SF of surface per hour | PT@.038 | SF | — | 1.42 | .86 | 2.28 |
| Wood roofs over 3 in 12 pitch | | | | | | |
| 70 SF of surface per hour | PT@.042 | SF | — | 1.56 | .95 | 2.51 |
| Siding | | | | | | |
| Metal siding, 80 SF per hour | PT@.038 | SF | — | 1.42 | .86 | 2.28 |
| Wood siding, 85 SF per hour | PT@.035 | SF | — | 1.30 | .79 | 2.09 |
| Walls measured on one face of wall | | | | | | |
| Concrete, 80 SF per hour | PT@.038 | SF | — | 1.42 | .86 | 2.28 |
| Concrete block, 70 SF per hour | PT@.042 | SF | — | 1.56 | .95 | 2.51 |
| Plaster, 85 SF per hour | PT@.035 | SF | — | 1.30 | .79 | 2.09 |
| Add for mobilization and demobilization, per job | | | | | | |
| Move-on & off and refurbish | — | LS | — | — | — | 2,500.00 |

Removing paint from concrete floor and slab surfaces. Equipment includes three pneumatic manually-operated wheel-mounted scabblers attached to one air-powered vacuum-waste packing unit to scarify concrete floors and slabs and one 150 CFM gas- or diesel-powered air compressor, including all hoses and connectors. Add the one-time charge listed below for move-on, move-off and refurbishing of equipment.

| | Craft@Hrs | Unit | Material | Labor | Equipment | Total |
|---|---|---|---|---|---|---|
| Large unobstructed areas; floor slabs, driveways, sidewalks | | | | | | |
| 90 SF of surface per hour | PT@.033 | SF | — | 1.23 | .75 | 1.98 |
| Obstructed areas; narrow aisles, areas around equipment | | | | | | |
| 60 SF of surface per hour | PT@.050 | SF | — | 1.86 | 1.13 | 2.99 |
| Small areas; around decks | | | | | | |
| 40 SF of surface per hour | PT@.075 | SF | — | 2.79 | 1.69 | 4.48 |
| Add for mobilization and demobilization, per job | | | | | | |
| Move-on & off and refurbish | — | LS | — | — | — | 2,500.00 |

| | Craft@Hrs | Unit | Material | Labor | Total |
|---|---|---|---|---|---|
| **Paints, Coatings, and Supplies** Costs are for good to better quality coatings purchased in 1 to 4 gallon quantities. Add 30% for premium quality paints. General contractor prices may be discounted 10% to 15% on residential job quantities. Painting contractors receive discounts of 20% to 25% on steady volume accounts. Labor costs for painting are listed at the end of this section. For more complete coverage of painting costs, see the *National Painting Cost Estimator* at http://CraftsmanSiteLicense.com/ | | | | | |

**Primers**

Oil-based exterior primer-sealer, universal undercoat for use on wood, hardboard, aluminum and wrought iron. Prevents extractive bleeding.

| | Craft@Hrs | Unit | Material | Labor | Total |
|---|---|---|---|---|---|
| Quart (100 SF per Qt) | — | Qt | 12.40 | — | 12.40 |
| Gallon (400 SF per Gal) | — | Gal | 25.90 | — | 25.90 |
| Latex block filler | | | | | |
| (1,000 SF per 5 Gal) | — | 5 Gal | 52.90 | — | 52.90 |

# Paint, House

| | Craft@Hrs | Unit | Material | Labor | Total |
|---|---|---|---|---|---|
| Acrylic (vinyl acrylic) thermoplastic rust inhibiting metal primer | | | | | |
| (500 SF per Gal) | — | Gal | 66.00 | — | 66.00 |
| Acrylic latex interior primer and sealer, for drywall, plaster, masonry or wood | | | | | |
| (400 SF per Gal) | — | Gal | 25.90 | — | 25.90 |
| Linseed (oil base) exterior/ interior primer | | | | | |
| (300 SF per Gal) | — | Gal | 27.00 | — | 27.00 |
| Red oxide metal primer | | | | | |
| (500 SF per Gal) | — | Gal | 26.30 | — | 26.30 |
| Concrete bonding primer, clear | | | | | |
| (350 to 400 SF per Gal) | — | Gal | 19.40 | — | 19.40 |
| Zinc chromate primer | | | | | |
| (400 SF per Gal) | — | Qt | 16.40 | — | 16.40 |
| KILZ® exterior sealer, primer, stainblocker. Mildew-resistant. Blocks tannin bleed on redwood and cedar. | | | | | |
| Gallon (370 SF per gallon) | — | Gal | 18.70 | — | 18.70 |
| 5 gallons (1,850 SF per 5 gallons) | — | 5 Gal | 82.00 | — | 82.00 |

## House paint

Exterior acrylic latex house paint. Mildew and fade resistant. For wood, masonry, brick, stucco, vinyl and aluminum siding and metal. Typical coverage is 400 square feet per gallon.

| | Craft@Hrs | Unit | Material | Labor | Total |
|---|---|---|---|---|---|
| Flat, gallon | — | Gal | 23.90 | — | 23.90 |
| Flat, 5 gallons | — | 5 Gal | 116.00 | — | 116.00 |
| Satin, gallon | — | Gal | 26.50 | — | 26.50 |
| Satin, 5 gallons | — | 5 Gal | 119.00 | — | 119.00 |
| Semi-gloss, gallon | — | Gal | 25.97 | — | 25.97 |
| Semi-gloss, 5 gallons | — | 5 Gal | 120.20 | — | 120.20 |

Interior acrylic latex house paint. Typical coverage is 400 square feet per gallon.

| | Craft@Hrs | Unit | Material | Labor | Total |
|---|---|---|---|---|---|
| Flat pastel base, gallon | — | Gal | 21.50 | — | 21.50 |
| Flat pastel base, 5 gallons | — | 5 Gal | 94.60 | — | 94.60 |
| Flat ultra white, quart | — | Qt | 10.20 | — | 10.20 |
| Flat ultra white, gallon | — | Gal | 21.50 | — | 21.50 |
| Flat ultra white, 5 gallons | — | 5 Gal | 95.60 | — | 95.60 |
| Satin enamel, quart | — | Qt | 12.40 | — | 12.40 |
| Satin enamel, gallon | — | Gal | 25.80 | — | 25.80 |
| Satin enamel, 5 gallons | — | 5 Gal | 116.00 | — | 116.00 |
| High gloss enamel, quart | — | Qt | 14.00 | — | 14.00 |
| High gloss enamel, gallon | — | Gal | 29.70 | — | 29.70 |
| High gloss enamel, 5 gallons | — | 5 Gal | 132.00 | — | 132.00 |
| Roll-on fibered texture paint, 2 gallons | — | 2 Gal | 25.30 | — | 25.30 |

## Stucco, masonry and concrete paint

Masonry and stucco paint. Acrylic latex paint for interior or exterior. Smooth, rough or textured masonry surfaces

| | Craft@Hrs | Unit | Material | Labor | Total |
|---|---|---|---|---|---|
| Gallon | — | Gal | 19.40 | — | 19.40 |
| 5 gallons | — | 5 Gal | 85.30 | — | 85.30 |
| Elastomeric waterproofing, gallon | — | Gal | 28.00 | — | 28.00 |
| Elastomeric waterproofing, 5 gallons | — | 5 Gal | 125.00 | — | 125.00 |
| Penetrating sealer, gallon | — | Gal | 13.50 | — | 13.50 |
| Penetrating sealer, 5 gallons | — | 5 Gal | 58.00 | — | 58.00 |

Concrete stain. Forms a tough film on properly prepared concrete patios, basements, pool decks, pillars, brick and sidewalks. Natural slate color. 100 percent acrylic.

| | Craft@Hrs | Unit | Material | Labor | Total |
|---|---|---|---|---|---|
| Gallon | — | Gal | 27.00 | — | 27.00 |
| 5 gallons | — | 5 Gal | 120.00 | — | 120.00 |

| | Craft@Hrs | Unit | Material | Labor | Total |
|---|---|---|---|---|---|

Gloss porch and floor enamel. 100 percent acrylic. Mildew-resistant. Recoat in 24 to 48 hours. Full cure in 30 days

| | | | | | |
|---|---|---|---|---|---|
| Gloss, quart | — | Qt | 11.50 | — | 11.50 |
| Gloss, gallon | — | Gal | 25.90 | — | 25.90 |
| Low luster, slate gray, gallon | — | Gal | 25.90 | — | 25.90 |
| Low luster, slate gray, 5 gallons | — | 5 Gal | 115.00 | — | 115.00 |

Masonry waterproofer, for concrete, stucco and masonry, white when applied, dries clear

| | | | | | |
|---|---|---|---|---|---|
| 150 SF per gallon | — | Gal | 21.50 | — | 21.50 |
| 750 SF per 5 gallons | — | 5 Gal | 74.10 | — | 74.10 |

### Metal coatings

Clean metal primer. White. For use on bare, lightly rusted or previously painted surfaces. Creates a strong surface for better topcoat adhesion and weather resistance.

| | | | | | |
|---|---|---|---|---|---|
| Quart | — | Qt | 9.74 | — | 9.74 |

Rust converting metal primer. Non-toxic, USDA and FDA approved. Works on clean or rusty steel, zinc galvanized metal, aluminum, tin and previously painted surfaces

| | | | | | |
|---|---|---|---|---|---|
| Quart | — | Qt | 16.00 | — | 16.00 |
| Gallon | — | Gal | 55.90 | — | 55.90 |

Oil-based enamel. Industrial grade. Prevents rust, resists cracking, peeling, chipping, and fading. Scuff-, abrasion- and moisture-resistant. Lead-free. Not low volatile organic compound compliant.

| | | | | | |
|---|---|---|---|---|---|
| Gallon | — | Gal | 29.10 | — | 29.10 |

Stops Rust, Rust-Oleum®. Resists moisture and corrosion. For metal, wood, concrete and masonry.

| | | | | | |
|---|---|---|---|---|---|
| Half pint | — | Ea | 4.61 | — | 4.61 |
| Quart | — | Ea | 9.50 | — | 9.50 |

### Wood finishes

Exterior opaque stain. For wood siding, decks and fences. Water repellent and mildew-resistant. Gallon covers 400 square feet.

| | | | | | |
|---|---|---|---|---|---|
| Acrylic latex, tint or white, gallon | — | Gal | 27.90 | — | 27.90 |
| Acrylic latex, tint or white, 5 gallons | — | 5 Gal | 125.00 | — | 125.00 |
| Oil and latex, accent tint, gallon | — | Gal | 23.10 | — | 23.10 |
| Oil and latex, accent tint, 5 gallons | — | 5 Gal | 101.00 | — | 101.00 |

Exterior semi-opaque stain. For wood siding, decks and fences. Water repellent and mildew-resistant. Gallon covers 400 square feet.

| | | | | | |
|---|---|---|---|---|---|
| Oil-based stain, gallon | — | Gal | 26.70 | — | 26.70 |
| Oil-based stain, 5 gallons | — | 5 Gal | 122.00 | — | 122.00 |
| Water-based stain, gallon | — | Gal | 25.70 | — | 25.70 |
| Water-based stain, 5 gallons | — | 5 Gal | 82.60 | — | 82.60 |

Exterior clear wood finish. For exterior wood doors, windows, furniture and trim. Gallon covers 300 to 400 square feet.

| | | | | | |
|---|---|---|---|---|---|
| Tung oil spar varnish, quart | — | Qt | 8.62 | — | 8.62 |
| Tung oil spar varnish, gallon | — | Gal | 22.60 | — | 22.60 |
| Marine varnish, gloss or satin, quart | — | Qt | 17.60 | — | 17.60 |
| Marine varnish, gloss or satin, gallon | — | Gal | 45.20 | — | 45.20 |
| Urethane clear finish, gloss or satin, quart | — | Qt | 16.20 | — | 16.20 |
| Urethane clear finish, gloss or satin, gallon | — | Gal | 40.60 | — | 40.60 |
| Varathane® gloss, semi or satin, 13 ounces | — | Can | 7.45 | — | 7.45 |
| Varathane® gloss, semi or satin, quart | — | Qt | 17.30 | — | 17.30 |
| Varthane® gloss, semi or satin, gallon | — | Gal | 47.90 | — | 47.90 |
| Exterior polyurethane, gloss or satin, quart | — | Qt | 20.40 | — | 20.40 |

# Paint, Wood Finishes

| | Craft@Hrs | Unit | Material | Labor | Total |
|---|---|---|---|---|---|

Waterproofing wood finish. For wood siding, decks and fences. Water repellent and mildew resistant. Clear or tinted. Gallon covers 300 to 400 square feet.

| | Craft@Hrs | Unit | Material | Labor | Total |
|---|---|---|---|---|---|
| 6-way exterior protection, gallon | — | Gal | 22.50 | — | 22.50 |
| 6-way exterior protection, 5 gallons | — | 5 Gal | 55.50 | — | 55.50 |
| Premium wood weatherproofing, gallon | — | Gal | 32.30 | — | 32.30 |
| Premium wood weatherproofing, 5 gallons | — | 5 Gal | 144.00 | — | 144.00 |
| UV-resistant wood finish, gallon | — | Gal | 20.00 | — | 20.00 |
| UV-resistant wood finish, 5 gallons | — | 5 Gal | 89.30 | — | 89.30 |
| Penetrating wood finish, quart | — | Qt | 11.20 | — | 11.20 |
| Penetrating wood finish, gallon | — | Gal | 28.00 | — | 28.00 |
| Penetrating wood finish, 5 gallons | — | 5 Gal | 125.00 | — | 125.00 |
| Log home gloss finish, clear or tint, gallon | — | Gal | 35.60 | — | 35.60 |
| Log home gloss finish, clear or tint, 5 gallons | — | 5 Gal | 159.00 | — | 159.00 |
| Cuprinol Revive wood resurfacer, 2 gallons | — | 2 Gal | 56.70 | — | 56.70 |

Wood water sealer. Clear. Helps maintain wood's natural color. Protects against damage from water and sun. Provides a mildew-resistant coating. Gallon covers 225 to 325 square feet of smooth wood or as little as 125 square feet for a first application on rough-sawn wood.

| | Craft@Hrs | Unit | Material | Labor | Total |
|---|---|---|---|---|---|
| Thompson's WaterSeal, gallon | — | Gal | 15.00 | — | 15.00 |
| Thompson's WaterSeal, 5 gallons | — | 5 Gal | 64.60 | — | 64.60 |
| Waterproofing wood protector, gallon | — | Gal | 15.00 | — | 15.00 |
| Waterproofing wood protector, 5 gallons | — | 5 Gal | 67.00 | — | 67.00 |

Wood preservative. Helps maintain wood's natural color. Protects against wood rot, warping, swelling, termite damage, mildew and moisture. Gallon covers 100 to 300 square feet when applied by brush or roller.

| | Craft@Hrs | Unit | Material | Labor | Total |
|---|---|---|---|---|---|
| Dock and fence post preservative, gallon | — | Gal | 21.60 | — | 21.60 |
| Seasonite® new wood treatment, gallon | — | Gal | 14.60 | — | 14.60 |
| Termin-8 green preservative, quart | — | Qt | 10.00 | — | 10.00 |
| Termin-8 green preservative, gallon | — | Gal | 20.40 | — | 20.40 |
| Termin-8 green preservative, 5 gallons | — | 5 Gal | 75.90 | — | 75.90 |
| Copper-Green® preservative, 13.5 ounces | — | Can | 10.20 | — | 10.20 |
| Copper-Green® preservative, gallon | — | Gal | 19.00 | — | 19.00 |

Interior oil stain. Penetrates into unfinished or stripped wood to highlight and seal the grain.

| | Craft@Hrs | Unit | Material | Labor | Total |
|---|---|---|---|---|---|
| Tinted wiping stain, half pint | — | Ea | 4.80 | — | 4.80 |
| Tinted wiping stain, quart | — | Qt | 7.76 | — | 7.76 |
| Tinted wiping stain, gallon | — | Gal | 27.40 | — | 27.40 |
| Danish oil, quart | — | Qt | 11.30 | — | 11.30 |
| Danish oil, gallon | — | Gal | 29.50 | — | 29.50 |
| Gel stain, quart | — | Qt | 16.40 | — | 16.40 |

Interior water-based stain. Fast-drying and easy clean-up.

| | Craft@Hrs | Unit | Material | Labor | Total |
|---|---|---|---|---|---|
| White wash pickling stain, quart | — | Qt | 10.80 | — | 10.80 |
| Tinting stain, half-pint | — | Ea | 7.50 | — | 7.50 |
| Tinting stain, quart | — | Qt | 10.70 | — | 10.70 |

## Paint specialties

Asphalt-fiber roof and foundation coating, 75 SF per gallon on roof or concrete.

| | Craft@Hrs | Unit | Material | Labor | Total |
|---|---|---|---|---|---|
| Solvent type, 100 SF per gallon | — | Gal | 13.50 | — | 13.50 |

Driveway coating

| | Craft@Hrs | Unit | Material | Labor | Total |
|---|---|---|---|---|---|
| Coal tar emulsion (500 SF per 5 Gal) | — | 5 Gal | 33.70 | — | 33.70 |

Traffic paint, acrylic, 100 SF per gallon

| | Craft@Hrs | Unit | Material | Labor | Total |
|---|---|---|---|---|---|
| Blue, white, yellow or red, 1 gallon | — | Gal | 25.90 | — | 25.90 |
| Blue, white, yellow or red, 5 gallon | — | 5 Gal | 89.00 | — | 89.00 |

| | Craft@Hrs | Unit | Material | Labor | Total |
|---|---|---|---|---|---|
| Swimming pool enamel | | | | | |
| Sau-Sea Type R (rubber base) | | | | | |
| (350 SF per Gal) | — | 5 Gal | 290.00 | — | 290.00 |
| Sau-Sea plaster pool coating | | | | | |
| (350 SF per Gal) | — | 5 Gal | 297.00 | — | 297.00 |
| Sau-Sea thinner | | | | | |
| For plaster or rubber | — | Gal | 20.40 | — | 20.40 |
| Sau-Sea patching compound for plastic | | | | | |
| (150 LF per Gal) | — | Gal | 85.00 | — | 85.00 |

Multi-surface waterproofer. For wood, brick, concrete, stucco and masonry. Can be applied to new pressure-treated wood. Paintable. Allows wood to weather naturally. Goes on milky white, dries clear.

| | Craft@Hrs | Unit | Material | Labor | Total |
|---|---|---|---|---|---|
| Aerosol can, 12 ounces | — | Can | 4.66 | — | 4.66 |
| Gallon | — | Gal | 10.50 | — | 10.50 |
| 5 gallon | — | 5 Gal | 47.20 | — | 47.20 |

Paint and varnish remover, 150 SF per gallon

| | Craft@Hrs | Unit | Material | Labor | Total |
|---|---|---|---|---|---|
| Quart | — | Qt | 16.20 | — | 16.20 |
| Gallon | — | Gal | 60.40 | — | 60.40 |
| 5 gallon | — | 5 Gal | 204.00 | — | 204.00 |

Spackling compound (vinyl-paste)

| | Craft@Hrs | Unit | Material | Labor | Total |
|---|---|---|---|---|---|
| Interior, 8-ounce tube | — | Ea | 3.64 | — | 3.64 |
| Wood filler paste | — | Pint | 6.34 | — | 6.34 |

Thinners

| | Craft@Hrs | Unit | Material | Labor | Total |
|---|---|---|---|---|---|
| Acetone | — | Gal | 17.80 | — | 17.80 |
| Xylene epoxy thinner | — | Qt | 4.30 | — | 4.30 |
| Denatured alcohol | — | Gal | 16.20 | — | 16.20 |
| Mineral spirits | — | Gal | 14.00 | — | 14.00 |
| Paint thinner | — | Gal | 2.93 | — | 2.93 |
| Shellac or lacquer thinner | — | Gal | 16.60 | — | 16.60 |
| Turpentine | — | Qt | 6.25 | — | 6.25 |

**Painting, Labor** Single coat applications except as noted. Not including equipment rental costs. These figures are based on hand work (roller, and brush where required) on residential jobs. Where spray equipment can be used to good advantage, reduce these costs 30% to 40%. Spray painting will increase paint requirements by 30% to 60%. Protection of adjacent materials is included in these costs, but little or no surface preparation is assumed. Per SF of surface area to be painted. Figures in parentheses show how much work should be done in an hour. For more complete coverage of painting costs, see *National Painting Cost Estimator* at http://CraftsmanSiteLicense.com/

Exterior surfaces, per coat, paint grade

| | Craft@Hrs | Unit | Material | Labor | Total |
|---|---|---|---|---|---|
| Door and frame, 6 sides (2.5 doors/hour) | PT@.400 | Ea | — | 14.90 | 14.90 |
| Siding, smooth, wood (250 SF/hour) | PT@.004 | SF | — | .15 | .15 |
| Siding, rough or shingle (200 SF/hour) | PT@.005 | SF | — | .19 | .19 |
| Shutters (50 SF/hour) | PT@.020 | SF | — | .75 | .75 |
| Stucco (300 SF/hour) | PT@.003 | SF | — | .11 | .11 |
| Trim, posts, rails (20 LF/hour) | PT@.050 | LF | — | 1.86 | 1.86 |

Windows, including mullions, per SF of opening,

| | Craft@Hrs | Unit | Material | Labor | Total |
|---|---|---|---|---|---|
| one side only (67 SF/hour) | PT@.015 | SF | — | .56 | .56 |
| Add for light sanding (330 SF/hour) | PT@.003 | SF | — | .11 | .11 |

| | Craft@Hrs | Unit | Material | Labor | Total |
|---|---|---|---|---|---|
| Interior surfaces, per coat | | | | | |
| Ceiling, flat latex (250 SF/hour) | PT@.004 | SF | — | .15 | .15 |
| Ceiling, enamel (200 SF/hour) | PT@.005 | SF | — | .19 | .19 |
| Walls, flat latex (500 SF/hour) | PT@.002 | SF | — | .07 | .07 |
| Walls, enamel (400 SF/hour) | PT@.003 | SF | — | .11 | .11 |
| Baseboard (120 LF/hour) | PT@.008 | LF | — | .30 | .30 |
| Bookcases, cabinets (90 SF/hour) | PT@.011 | SF | — | .41 | .41 |
| Windows, including mullions and frame, per SF of opening, | | | | | |
| one side only (67 SF/hour) | PT@.015 | SF | — | .56 | .56 |
| Floors, wood | | | | | |
| Filling | PT@.007 | SF | — | .26 | .26 |
| Staining | PT@.006 | SF | — | .22 | .22 |
| Shellacking | PT@.006 | SF | — | .22 | .22 |
| Varnishing | PT@.006 | SF | — | .22 | .22 |
| Waxing, machine | PT@.006 | SF | — | .22 | .22 |
| Polishing, machine | PT@.008 | SF | — | .30 | .30 |
| Sanding, hand | PT@.012 | SF | — | .45 | .45 |
| Sanding, machine | PT@.007 | SF | — | .26 | .26 |
| Walls, brick, concrete and masonry | | | | | |
| Oiling and sizing (140 SF/hour) | PT@.007 | SF | — | .26 | .26 |
| Sealer coat (150 SF/hour) | PT@.007 | SF | — | .26 | .26 |
| Brick or masonry (165 SF/hour) | PT@.006 | SF | — | .22 | .22 |
| Concrete (200 SF/hour) | PT@.005 | SF | — | .19 | .19 |
| Concrete steps, 3 coats (38 SF/hour) | PT@.026 | SF | — | .97 | .97 |
| Window frames and sash, using primer to mask glass and prime sash, topcoats as shown | | | | | |
| 1 topcoat (100 SF/hour) | PT@.010 | SF | — | .37 | .37 |
| 2 topcoats (67 SF/hour) | PT@.015 | SF | — | .56 | .56 |
| Score and peel primer from glass (peels from glass only) | | | | | |
| Per lite | PT@.020 | Ea | — | .75 | .75 |

| | Day | Week | Month |
|---|---|---|---|
| **Paint Spraying Equipment Rental** Complete spraying outfit including paint cup, hose, spray gun. | | | |
| 8 CFM electric compressor | 65.00 | 210.00 | 590.00 |
| 44 CFM gas-driven compressor | 95.00 | 300.00 | 840.00 |
| Professional airless, complete outfit | 137.00 | 400.00 | 1,200.00 |

**Painting, Subcontract, Rule of Thumb** Including material, labor, equipment and the subcontractor's overhead and profit.

| | Craft@Hrs | Unit | Material | Labor | Total |
|---|---|---|---|---|---|
| Typical subcontract costs per square foot of floor area for painting the interior and exterior of residential buildings, including only minimum surface preparation | | | | | |
| Economy, 1 or 2 coats, little brushwork | — | SF | — | — | 2.28 |
| Good quality, 2 or 3 coats | — | SF | — | — | 5.29 |

**Painting, Subcontract, Cost Breakdown** The costs that follow are based on the area of the surface being prepared or painted. Typical subcontract costs, including labor, material, equipment and the subcontractor's overhead and profit.

**Surface preparation, subcontract** Work performed by hand.

| | | | | | |
|---|---|---|---|---|---|
| Water blasting (pressure washing), using wheel-mounted portable 2,200 PSI pressure washer | | | | | |
| Light blast (250 SF per hour) | — | SF | — | — | .23 |
| Most work (150 SF per hour) | — | SF | — | — | .37 |
| Rough surface, grime (75 SF per hour) | — | SF | — | — | .70 |

| | Craft@Hrs | Unit | Material | Labor | Total |
|---|---|---|---|---|---|
| **Light cleaning to remove surface dust and stains** | | | | | |
| Concrete or masonry surfaces | — | SF | — | — | .37 |
| Gypsum or plaster surfaces | — | SF | — | — | .37 |
| Wood surfaces | — | SF | — | — | .36 |
| **Normal preparation including scraping, patching and puttying** | | | | | |
| Concrete or masonry surface | | | | | |
| Painted | — | SF | — | — | .37 |
| Unpainted | — | SF | — | — | .30 |
| Gypsum or plaster surfaces | | | | | |
| Painted | — | SF | — | — | .37 |
| Unpainted | — | SF | — | — | .19 |
| Metal surfaces, light sanding | — | SF | — | — | .18 |
| Wood surfaces, surface preparation, subcontract | | | | | |
| Painted | — | SF | — | — | .36 |
| Unpainted | — | SF | — | — | .18 |
| Puttying | — | LF | — | — | 1.60 |

**Exterior painting, subcontract** The estimates below are for hand work (brush and roller). Costs will be 15% to 30% lower when spray equipment can be used to good advantage. These costs include only minimum surface preparation. Add additional surface preparation, if required, from the figures above. Typical subcontract costs.

| | Craft@Hrs | Unit | Material | Labor | Total |
|---|---|---|---|---|---|
| **Brick or concrete, including acrylic latex masonry primer and latex paint** | | | | | |
| 1 coat latex | — | SF | — | — | .71 |
| 2 coats latex | — | SF | — | — | .99 |
| Concrete floors, etch and epoxy enamel | — | SF | — | — | .41 |
| Columns and pilasters, per coat | — | SF | — | — | .58 |
| Cornices, per coat | — | SF | — | — | 1.47 |
| **Doors, including trim, exterior only** | | | | | |
| 3 coats | — | Ea | — | — | 46.90 |
| 2 coats | — | Ea | — | — | 33.80 |
| **Downspouts and gutters** | | | | | |
| Per coat | — | LF | — | — | 1.19 |
| **Eaves** | | | | | |
| No rafters, per coat | — | SF | — | — | .57 |
| With rafters, per coat, brush | — | SF | — | — | .99 |
| **Fences (gross area)** | | | | | |
| Plain, 2 coats (per side) | — | SF | — | — | 1.38 |
| Plain, 1 coat (per side) | — | SF | — | — | .82 |
| Picket, 2 coats (per side) | — | SF | — | — | 1.28 |
| Picket, 1 coat (per side) | — | SF | — | — | .73 |
| **Lattice work** | | | | | |
| 1 coat 1 side, gross area | — | SF | — | — | .63 |
| **Metal, typical** | | | | | |
| 1 coat | — | SF | — | — | .51 |
| 2 coats | — | SF | — | — | .89 |
| **Porch rail and balusters, per coat** | | | | | |
| Gross area | — | SF | — | — | 1.26 |
| Handrail only | — | LF | — | — | .86 |

# Painting, Subcontract

| | Craft@Hrs | Unit | Material | Labor | Total |
|---|---|---|---|---|---|
| **Roofs, wood shingle** | | | | | |
| 1 coat stain plus 1 coat sealer | | | | | |
| Flat roof | — | SF | — | — | .59 |
| 4 in 12 pitch | — | SF | — | — | .76 |
| 8 in 12 pitch | — | SF | — | — | .80 |
| 12 in 12 pitch | — | SF | — | — | .95 |
| 1 coat stain plus 2 coats sealer | | | | | |
| Flat roof | — | SF | — | — | .95 |
| 4 in 12 pitch | — | SF | — | — | 1.14 |
| 8 in 12 pitch | — | SF | — | — | 1.23 |
| 12 in 12 pitch | — | SF | — | — | 1.32 |
| **Siding, plain** | | | | | |
| Sanding and puttying, typical | — | SF | — | — | .39 |
| Siding and trim, 3 coats | — | SF | — | — | 1.20 |
| Siding and trim, 2 coats | — | SF | — | — | .89 |
| Trim only, 2 coats | — | SF | — | — | .94 |
| Trim only, 3 coats | — | SF | — | — | 1.28 |
| **Siding, shingle, including trim** | | | | | |
| 2 coats oil paint | — | SF | — | — | .91 |
| 1 coat oil paint | — | SF | — | — | .56 |
| 2 coats stain | — | SF | — | — | .86 |
| 1 coat stain | — | SF | — | — | .49 |
| **Steel sash** | | | | | |
| 3 coats (per side) | — | SF | — | — | 1.15 |
| **Stucco** | | | | | |
| 1 coat, smooth surface | — | SF | — | — | .57 |
| 2 coats, smooth surface | — | SF | — | — | .87 |
| 1 coat, rough surface | — | SF | — | — | .63 |
| 2 coats, rough surface | — | SF | — | — | 1.04 |
| **Window frames and sash, 3 coats** | | | | | |
| Per side, one lite | — | Ea | — | — | 35.80 |
| Per side, add for each additional lite | — | Ea | — | — | 2.00 |

**Interior painting, subcontract** The estimates below are for hand work (roller with some brush cut-in). Costs will be 15% to 30% lower when spray equipment can be used to good advantage. These costs include only minimum surface preparation. Add for additional surface preparation, if required, from the preceding pages. Typical subcontract costs.

| | Craft@Hrs | Unit | Material | Labor | Total |
|---|---|---|---|---|---|
| Cabinets, bookcases, cupboards, per SF of total surface area | | | | | |
| 3 coats, paint | — | SF | — | — | 2.85 |
| Stain, shellac, varnish or plastic | — | SF | — | — | 2.92 |
| Concrete block or brick | | | | | |
| Primer and 1 coat latex | — | SF | — | — | .66 |
| Doors, including trim | | | | | |
| 3 coats (per side) | — | SF | — | — | 1.75 |
| 2 coats (per side) | — | SF | — | — | 1.22 |
| Plaster or drywall walls, latex. Add for ceilings from below | | | | | |
| Prime coat or sealer | — | SF | — | — | .36 |
| 1 coat smooth surface | — | SF | — | — | .43 |
| 1 coat rough surface | — | SF | — | — | .48 |

|  | Craft@Hrs | Unit | Material | Labor | Total |
|---|---|---|---|---|---|
| **1 coat sealer, 1 coat flat** | | | | | |
| Smooth surface | — | SF | — | — | .58 |
| Rough surface | — | SF | — | — | .78 |
| **1 coat sealer, 2 coats flat** | | | | | |
| Smooth surface | — | SF | — | — | .79 |
| Rough surface | — | SF | — | — | 1.04 |
| **1 coat sealer, 2 coats gloss or semi-gloss** | | | | | |
| Smooth surface | — | SF | — | — | .75 |
| Rough surface | — | SF | — | — | 1.03 |
| Add for ceilings | — | % | — | — | 45.0 |
| **Stairs, including risers** | | | | | |
| 3 coats paint | — | SF | — | — | 2.00 |
| Stain, shellac, varnish | — | SF | — | — | 2.06 |
| **Woodwork, painting** | | | | | |
| Priming | — | SF | — | — | .47 |
| 2 coats | — | SF | — | — | .75 |
| 2 coats, top workmanship | — | SF | — | — | .93 |
| 3 coats | — | SF | — | — | 1.22 |
| 3 coats, top workmanship | — | SF | — | — | 1.61 |
| **Woodwork, staining** | | | | | |
| Apply stain | — | SF | — | — | .39 |
| Filling (paste wood filler) | — | SF | — | — | .83 |
| Sanding | — | SF | — | — | .49 |
| Apply wax and polish | — | SF | — | — | .49 |

**Paneling** See also Hardwood under Lumber and Hardboard.
Solid plank paneling, prefinished, "V"-joint or T&G, costs include nails and waste.

| Tennessee cedar plank paneling | Craft@Hrs | Unit | Material | Labor | Total |
|---|---|---|---|---|---|
| Clear 3/8" x 4" or 6" wide | BC@.050 | SF | 2.50 | 1.85 | 4.35 |
| Select tight knot, 3/4" x 4" or 6" wide | BC@.050 | SF | 2.61 | 1.85 | 4.46 |
| Knotty pine, 5/16" x 4" or 6" wide | BC@.050 | SF | 2.34 | 1.85 | 4.19 |
| Redwood, clear, 5/16" x 4" wide | BC@.050 | SF | 4.93 | 1.85 | 6.78 |

**Plywood and hardboard paneling** Prefinished 4' x 8' x 1/8" panels with hardboard back. Costs include nails, adhesive and 5% waste.

| | Craft@Hrs | Unit | Material | Labor | Total |
|---|---|---|---|---|---|
| Harbor white | BC@.030 | SF | .61 | 1.11 | 1.72 |
| Vintage maple | BC@.030 | SF | .53 | 1.11 | 1.64 |
| Traditional oak | BC@.030 | SF | .56 | 1.11 | 1.67 |
| Honey oak | BC@.030 | SF | .56 | 1.11 | 1.67 |
| Cherry oak | BC@.030 | SF | .56 | 1.11 | 1.67 |

**Hardwood plywood paneling** Unfinished 4' x 8' panels. Costs include nails, adhesive and 5% waste

| Birch plywood paneling | Craft@Hrs | Unit | Material | Labor | Total |
|---|---|---|---|---|---|
| 1/8" x 4' x 8' | BC@.032 | SF | .61 | 1.18 | 1.79 |
| 1/4" x 4' x 8' | BC@.032 | SF | .75 | 1.18 | 1.93 |
| 1/2" x 4' x 8' | BC@.040 | SF | 1.33 | 1.48 | 2.81 |
| 3/4" x 4' x 8' | BC@.042 | SF | 1.61 | 1.55 | 3.16 |
| **Oak plywood paneling** | | | | | |
| 1/4" x 4' x 8' | BC@.032 | SF | .80 | 1.18 | 1.98 |
| 1/2" x 4' x 8' | BC@.040 | SF | 1.57 | 1.48 | 3.05 |
| 3/4" x 4' x 8', red oak | BC@.042 | SF | 1.85 | 1.55 | 3.40 |

# Paneling

| | Craft@Hrs | Unit | Material | Labor | Total |
|---|---|---|---|---|---|
| Cedar hardwood paneling, natural | | | | | |
| 1/4" x 4' x 8' | BC@.032 | SF | .89 | 1.18 | 2.07 |
| 3/8" x 4' x 8' | BC@.040 | SF | 1.01 | 1.48 | 2.49 |
| 1/2" x 4' x 8' | BC@.040 | SF | 1.61 | 1.48 | 3.09 |
| 5/8" x 4' x 8' | BC@.042 | SF | 1.85 | 1.55 | 3.40 |
| 3/4" x 4' x 8' | BC@.042 | SF | 2.15 | 1.55 | 3.70 |
| Golden Virola paneling | | | | | |
| 1/4" x 2' x 2' | BC@.032 | SF | .99 | 1.18 | 2.17 |
| 1/4" x 2' x 4' | BC@.032 | SF | 1.00 | 1.18 | 2.18 |
| 1/4" x 4' x 4' | BC@.032 | SF | .66 | 1.18 | 1.84 |
| 1/4" x 4' x 8' | BC@.032 | SF | .71 | 1.18 | 1.89 |
| 3/8" x 4' x 8' | BC@.040 | SF | .96 | 1.48 | 2.44 |
| 1/2" x 2' x 2' | BC@.040 | SF | 1.51 | 1.48 | 2.99 |
| 5/8" x 4' x 8' | BC@.042 | SF | 1.35 | 1.55 | 2.90 |
| 3/4" x 2' x 2' | BC@.042 | SF | 2.12 | 1.55 | 3.67 |
| Maple paneling | | | | | |
| 1/2" x 4' x 8', C-2 grade | BC@.040 | SF | 1.24 | 1.48 | 2.72 |
| 3/4" x 4' x 8', B-3 grade | BC@.042 | SF | 2.06 | 1.55 | 3.61 |
| 3/4" x 4' x 8,' maple/birch | BC@.042 | SF | 1.68 | 1.55 | 3.23 |

**Vinyl-clad hardboard paneling** Vinyl-clad film over hardboard, 4' x 8' x 1/4" panels. Costs include nails, adhesive and 5% waste.

| | Craft@Hrs | Unit | Material | Labor | Total |
|---|---|---|---|---|---|
| White or Aegean Gold | BC@.032 | SF | 1.33 | 1.18 | 2.51 |
| Blue marble | BC@.032 | SF | .51 | 1.18 | 1.69 |
| Harvest pattern | BC@.032 | SF | 1.03 | 1.18 | 2.21 |

**FRP wall and ceiling paneling** Fiberglass-reinforced plastic, textured, 0.09" thick, Class C flame spread, 24" x 48" panels. Includes fasteners, adhesive and 5% waste.

| | Craft@Hrs | Unit | Material | Labor | Total |
|---|---|---|---|---|---|
| White or almond | BC@.032 | SF | 4.01 | 1.18 | 5.19 |

**Bamboo tambour paneling** 4' x 8' panels. Costs include nails, adhesive and 5% waste.

| | Craft@Hrs | Unit | Material | Labor | Total |
|---|---|---|---|---|---|
| Unfinished natural | BC@.032 | SF | 3.82 | 1.18 | 5.00 |
| Prefinished natural | BC@.032 | SF | 4.35 | 1.18 | 5.53 |
| Unfinished amber | BC@.032 | SF | 3.82 | 1.18 | 5.00 |
| Prefinished amber | BC@.032 | SF | 4.35 | 1.18 | 5.53 |
| Raw amber | BC@.032 | SF | 3.82 | 1.18 | 5.00 |
| Raw green | BC@.032 | SF | 3.82 | 1.18 | 5.00 |
| Raw black | BC@.032 | SF | 3.82 | 1.18 | 5.00 |

**Z-brick paneling** Non-ceramic mineral brick-like veneer. Interior or exterior. Coverage per carton is 3-1/2 square feet, 5% added for waste. Include the cost for adhesive from below.

| | Craft@Hrs | Unit | Material | Labor | Total |
|---|---|---|---|---|---|
| Red classic brick, 2-1/4" x 8" x 7/16" | BC@.050 | SF | 5.50 | 1.85 | 7.35 |
| Inca decorative brick, 2-1/4" x 8" x 7/16" | BC@.050 | SF | 4.89 | 1.85 | 6.74 |
| Used brick (Inca), 2-1/4" x 8" x 5/16" | BC@.050 | SF | 4.89 | 1.85 | 6.74 |
| Old Chicago brick, 2-3/8" x 8-1/8" x 5/16" | BC@.050 | SF | 4.89 | 1.85 | 6.74 |
| Liberty grey brick, 2-1/4" x 8" x 7/16" | BC@.050 | SF | 5.50 | 1.85 | 7.35 |
| Wheat brick, 2-1/4" x 8" x 7/16" | BC@.050 | SF | 4.89 | 1.85 | 6.74 |
| Burnt sienna brick, 2-1/4" x 8" x 7/16" | BC@.050 | SF | 6.11 | 1.85 | 7.96 |
| Mesa beige brick, 2-1/4" x 8" x 7/16" | BC@.050 | SF | 6.11 | 1.85 | 7.96 |
| Z-Ment mortar adhesive | — | SF | 1.75 | 1.50 | 3.25 |
| Z-Ment mortar adhesive, 16 to 20 SF per Gal | — | Gal | 30.60 | — | 30.60 |

| | Craft@Hrs | Unit | Material | Labor | Total |
|---|---|---|---|---|---|

**Paving, Subcontract** Small areas such as walks and driveways around residential buildings. Typical costs including material, labor, equipment and subcontractor's overhead and profit. No substrate preparation included. Use $1,800 as a minimum job charge.

Asphalt paving, including oil seal coat

| | Craft@Hrs | Unit | Material | Labor | Total |
|---|---|---|---|---|---|
| 2" asphalt | — | SF | — | — | 2.39 |
| 3" asphalt | — | SF | — | — | 2.68 |
| 4" asphalt | — | SF | — | — | 3.45 |
| 4" base and fine grading | — | SF | — | — | 1.46 |
| 6" base and fine grading | — | SF | — | — | 1.50 |

Asphalt seal coats

| | Craft@Hrs | Unit | Material | Labor | Total |
|---|---|---|---|---|---|
| Applied with broom or squeegee | — | SF | — | — | .40 |
| Fog seal, thin layer of diluted SS1H oil | — | SF | — | — | .18 |

Seal and sand

| | Craft@Hrs | Unit | Material | Labor | Total |
|---|---|---|---|---|---|
| Thin layer of oil with sand laid on top | — | SF | — | — | .40 |

Simco "walk-top" or other quick-dry applications

| | Craft@Hrs | Unit | Material | Labor | Total |
|---|---|---|---|---|---|
| Creamy oil | — | SF | — | — | .39 |

Brick paving

| | Craft@Hrs | Unit | Material | Labor | Total |
|---|---|---|---|---|---|
| On concrete, grouted, laid flat, including concrete base | — | SF | — | — | 15.90 |
| On concrete, laid solid on edge, 8" x 2-1/4" exposure, no grouting, including concrete base | — | SF | — | — | 17.30 |
| On sand bed, laid flat | — | SF | — | — | 12.30 |
| On sand bed, 8" x 2-1/4" exposure | — | SF | — | — | 13.90 |

Concrete (including fine grading) paving

| | Craft@Hrs | Unit | Material | Labor | Total |
|---|---|---|---|---|---|
| 3" concrete, unreinforced | — | SF | — | — | 3.34 |
| 4" concrete, unreinforced | — | SF | — | — | 4.01 |
| 4" concrete, mesh reinforcing | — | SF | — | — | 4.06 |
| 4" concrete, seeded aggregate finish | — | SF | — | — | 7.04 |
| Add for colored concrete, most colors | — | SF | — | — | .48 |

**Porous paving systems** Thin-walled HDPE plastic rings connected by an interlocking geogrid structure, installed on a porous base course for residential applications. Rings transfer loads from the surface to the grid structure to an engineered course base. Installed on a prepared base. www.invisiblestructures.com

Porous paving system with grass filler. Grid rolled out over base course and seeded with grass mixture. Includes fertilizer and soil polymer mix for spreading over base.

| | Craft@Hrs | Unit | Material | Labor | Total |
|---|---|---|---|---|---|
| 3.3' wide by 33' long rolls | BL@.003 | SF | 2.37 | .09 | 2.46 |
| Add for 1" sand fill | BL@.001 | SF | .04 | .03 | .07 |

Porous paving system with gravel fill. Grid rolled out over base course and covered with decorative fill gravel.

| | Craft@Hrs | Unit | Material | Labor | Total |
|---|---|---|---|---|---|
| 3.3' wide by 33' long rolls | BL@.004 | SF | 2.31 | .12 | 2.43 |
| Add for 1" gravel fill | BL@.001 | SF | .04 | .03 | .07 |

**Plastering, Subcontract** See also Lathing. Typical costs including material, labor, equipment and subcontractor's overhead and profit.

Acoustical plaster

| | Craft@Hrs | Unit | Material | Labor | Total |
|---|---|---|---|---|---|
| Including brown coat, 1-1/4" total thickness | — | SY | — | — | 35.50 |

Plaster on concrete or masonry, with bonding agent

| | Craft@Hrs | Unit | Material | Labor | Total |
|---|---|---|---|---|---|
| Gypsum plaster, 1/2" sanded | — | SY | — | — | 18.60 |

| | Craft@Hrs | Unit | Material | Labor | Total |
|---|---|---|---|---|---|
| **Plaster on interior wood frame walls, no lath included** | | | | | |
| Gypsum plaster, 7/8", 3 coats | — | SY | — | — | 28.00 |
| Keene's cement finish, 3 coats | — | SY | — | — | 29.70 |
| Cement plaster finish, 3 coats | — | SY | — | — | 31.40 |
| Gypsum plaster, 2 coats, tract work | — | SY | — | — | 24.50 |
| Add for ceiling work | — | % | — | — | 15.0 |
| **Plaster for tile** | | | | | |
| Scratch coat | — | SY | — | — | 13.70 |
| Brown coat and scratch coat | — | SY | — | — | 20.60 |
| **Exterior plaster (stucco)** | | | | | |
| **Walls, 3 coats, 7/8", no lath included** | | | | | |
| Textured finish | — | SY | — | — | 20.50 |
| Float finish | — | SY | — | — | 24.10 |
| Dash finish | — | SY | — | — | 27.00 |
| Add for soffits | — | % | — | — | 15.0 |
| **Portland cement plaster, including metal lath, exterior, 3 coats, 7/8" on wood frame** | | | | | |
| Walls | — | SY | — | — | 33.60 |
| Soffits | — | SY | — | — | 38.60 |
| Deduct for paper back lath | — | SY | — | — | -.18 |

**Plumbing Fixtures and Equipment** Costs are for good to better quality fixtures and trim installed with the accessories listed. All vitreous china, enameled steel, cast iron, and fiberglass fixtures are white except as noted. Colored acrylic fiberglass fixtures will cost about 6% more. Colored enameled steel, cast iron, or vitreous fixtures will cost about 25% more. No rough plumbing included. Rough plumbing is the drain, waste and vent (DWV) piping and water supply piping that isn't exposed when construction is complete. See costs for rough plumbing at the end of this section. Minimum quality fixtures and trim will cost 10% to 20% less, and deluxe fixtures will cost much more. Installation times assume standard quality work and uncomplicated layouts. For expanded coverage of plumbing cost estimates, see the *National Plumbing & HVAC Estimator* at http://CraftsmanSiteLicense.com/

## Bathtubs

Acrylic fiberglass one-piece bathtub and shower, 5' L, 32" W, 72" H. Includes polished chrome pop-up drain/overflow. Mixing valve, tub filler, and 2.5 GPM shower head

| | Craft@Hrs | Unit | Material | Labor | Total |
|---|---|---|---|---|---|
| Acrylic tub, shower and trim | P1@4.50 | Ea | 513.00 | 161.00 | 674.00 |

Enameled cast iron bathtub, 5' L, 32" W, 15" H for recessed installation. Includes polished chrome pop-up drain/overflow, mixing valve, tub filler and 2.5 GPM shower head

| | Craft@Hrs | Unit | Material | Labor | Total |
|---|---|---|---|---|---|
| Cast iron tub and trim | P1@3.50 | Ea | 590.00 | 125.00 | 715.00 |

Enameled steel bathtub, 5' L , 30" W, 15" H for recessed installations. With full wall flange, pop-up drain/overflow, chrome mixing valve, tub filler, shower flange and 2.5 GPM shower head

| | Craft@Hrs | Unit | Material | Labor | Total |
|---|---|---|---|---|---|
| Enameled steel tub and trim | P1@1.40 | Ea | 270.00 | 50.10 | 320.10 |

Enameled cast iron bathtub with chrome grab rails, 5' L, 32" W, 16" H. Includes polished chrome pop-up drain/overflow, mixing valve, tub filler, and 2.5 GPM shower head

| | Craft@Hrs | Unit | Material | Labor | Total |
|---|---|---|---|---|---|
| Cast iron tub with handrail and trim | P1@5.25 | Ea | 1,140.00 | 188.00 | 1,328.00 |

Recessed whirlpool bathtub with built-in 1.25 HP 115-volt motor and pump, integral drain. With eight multidirectional and adjustable jets. Acrylic with fiberglass reinforcement. White.

| | Craft@Hrs | Unit | Material | Labor | Total |
|---|---|---|---|---|---|
| Whirlpool tub and trim | P1@5.50 | Ea | 978.00 | 197.00 | 1,175.00 |
| Add for electrical connection | — | Ea | — | — | 240.00 |

| | Craft@Hrs | Unit | Material | Labor | Total |
|---|---|---|---|---|---|

**Indoor whirlpool tub** Acrylic fiberglass drop-in whirlpool bathtub, includes chrome pop-up drain/overflow, mixing valve and tub filler, with pump and air actuator switch. Add for the electrical connection and tub-support system

| | Craft@Hrs | Unit | Material | Labor | Total |
|---|---|---|---|---|---|
| 6 jets, 1.25 HP, 60" x 36" x 20" | P1@4.00 | Ea | 1,100.00 | 143.00 | 1,243.00 |
| 6 jets, 1.25 HP, 60" x 36" x 20", corner | P1@4.00 | Ea | 1,210.00 | 143.00 | 1,353.00 |
| 8 jets, 2 HP, 72" x 42" x 22" | P1@5.00 | Ea | 1,900.00 | 179.00 | 2,079.00 |
| 8 jets, 2 HP, 72" x 42" x 22", corner | P1@5.00 | Ea | 2,400.00 | 179.00 | 2,579.00 |
| Add for polished brass Roman tub filler | P1@1.50 | Ea | 289.00 | 53.60 | 342.60 |
| Add for chrome Roman tub hand shower | P1@1.50 | Ea | 186.00 | 53.60 | 239.60 |
| Add for auxiliary in-line water heater | P1@2.00 | Ea | 224.00 | 71.50 | 295.50 |
| Add for non-load-bearing tub apron | P1@1.00 | Ea | 227.00 | 35.80 | 262.80 |
| Add for electrical connection | — | Ea | — | — | 237.00 |

Whirlpool tub support system, including 150 pounds per square foot pre-tested bracing joists, galvanized cross members, vertical structural 2" pipe bracing, vertical support plate and tub grout pad.

| | Craft@Hrs | Unit | Material | Labor | Total |
|---|---|---|---|---|---|
| 1,500 pound capacity (60" x 36" x 20") | B1@8.00 | Ea | 510.00 | 266.00 | 776.00 |
| 2,500 pound capacity (72" x 42" x 22") | B1@8.00 | Ea | 615.00 | 266.00 | 881.00 |

**Hot tub** Whirlpool type, illuminated, corner or central, 24" platform mounted, with water temperature regulating heater, whirlpool re-circulating pump, 2" diameter drain, filter and cleanout, with marble-tile grouted hot tub ledge and sidewall fascia, custom plywood frame reinforcement, underfloor joist cross-bracing and joist support bracing, subfloor PEX piping to hot water and cold water lines, with valves, control wiring and ground fault protection. NSF, Underwriters Laboratories, and CSA rating.

| | Craft@Hrs | Unit | Material | Labor | Total |
|---|---|---|---|---|---|
| 56-gallon tub and trim | — | Ea | — | — | 5,100.00 |
| 85-gallon tub and trim | — | Ea | — | — | 6,120.00 |
| 110-gallon tub and trim | — | Ea | — | — | 6,600.00 |
| Prepping and sealing of floor pad | B9@8.00 | Ea | — | 261.00 | 261.00 |
| Installation of drain | P1@2.00 | Ea | 310.00 | 71.50 | 381.50 |
| Installation of PEX piping and valves | P1@2.00 | Ea | 200.00 | 71.50 | 271.50 |
| Cutting and assembly of wooden frame | B1@16.0 | Ea | 380.00 | 533.00 | 913.00 |

Tiling and grouting of sidewall and ledge frame

| | Craft@Hrs | Unit | Material | Labor | Total |
|---|---|---|---|---|---|
| Level and position frame | B1@16.0 | Ea | 760.00 | 533.00 | 1,293.00 |

Mounting and leveling of hot tub in frame

Final electrical connections to circulation pump

| | Craft@Hrs | Unit | Material | Labor | Total |
|---|---|---|---|---|---|
| and submerged tub lighting | CE@4.00 | Ea | 160.00 | 233.00 | 393.00 |

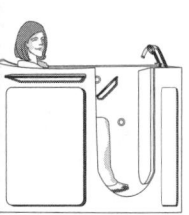

**Walk-in Doored Bathtub, ADA-compliant,** aerated hydrotherapy type, step-in with watertight sealing door, reinforced fiberglass, illuminated, corner or stand-alone, floor flush-mounted, with submersible seat, handrails and non-slip integral floor and seat surfaces, handheld flexible shower fixture with switchover valve, water temperature regulating 1,500 watt heater, whirlpool re-circulating pump, 2" diameter drain, filter and cleanout, under-floor joist cross-bracing and joist support vertical bracing, sub-floor PEX piping to hot water and cold water lines, with valves, control wiring and ground fault protection. NSF, Underwriters Laboratories, Americans with Disabilities Act, and CSA ratings.

| | Craft@Hrs | Unit | Material | Labor | Total |
|---|---|---|---|---|---|
| 55" long x 30" wide x 40" high (tub only) | —@.000 | Ea | 5,000.00 | — | 5,000.00 |
| Prepping and sealing of floor pad | B9@8.00 | Ea | — | 261.00 | 261.00 |
| Installation of drain | P1@2.00 | Ea | 300.00 | 71.50 | 371.50 |
| Installation of PEX piping, pump and valves | P1@2.00 | Ea | 190.00 | 71.50 | 261.50 |
| Mounting of tub, cross-joists, vertical column support bracing, and leveling | B1@16.0 | Ea | 750.00 | 533.00 | 1,283.00 |
| Final electrical connections to circulation pump, submerged tub lighting and grounding | CE@4.00 | Ea | 150.00 | 233.00 | 383.00 |

# Plumbing Fixtures and Equipment

| | Craft@Hrs | Unit | Material | Labor | Total |
|---|---|---|---|---|---|

## Shower stalls

Acrylic fiberglass shower stall, 36" W, 36" D, 72" H, includes mixing valve, shower head and door

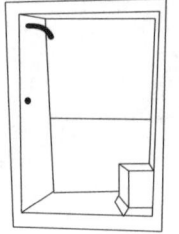

| | Craft@Hrs | Unit | Material | Labor | Total |
|---|---|---|---|---|---|
| Fiberglass 32" shower stall and trim | P1@4.15 | Ea | 649.00 | 148.00 | 797.00 |

Acrylic fiberglass shower stall, 48" W, 35" D, 72" H with integral soap dishes, grab bar and drain, mixing valve, shower head and door

| | | | | | |
|---|---|---|---|---|---|
| Fiberglass 48" shower stall and trim | P1@4.65 | Ea | 690.00 | 166.00 | 856.00 |

Acrylic fiberglass shower stall, 60" W, 35" D, 72" H with integral soap dishes, grab bar, drain and seat, mixing valve, shower head and door

| | | | | | |
|---|---|---|---|---|---|
| Fiberglass 60" shower stall and trim | P1@4.80 | Ea | 739.00 | 172.00 | 911.00 |
| Roll-in wheelchair stall and trim | P1@5.40 | Ea | 1,780.00 | 193.00 | 1,973.00 |
| Add for shower cap (roof) | P1@.500 | Ea | 175.00 | 17.90 | 192.90 |

## Handicapped tub and shower

Handicapped and elderly access fiberglass tub and shower, 60" x 30" H with integral soap dishes, seat and four grab bars, body washer shower, polished chrome pop-up drain/overflow, mixing valve and tub filler

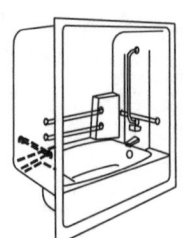

| | | | | | |
|---|---|---|---|---|---|
| Handicapped shower-tub and trim | P1@4.75 | Ea | 4,540.00 | 170.00 | 4,710.00 |
| Add for shower curtain rod | P1@.572 | Ea | 35.00 | 20.50 | 55.50 |

## Water closets

Floor-mounted, residential tank type 1.6 gal. vitreous china toilet, valved closet supply, and toilet seat with cover

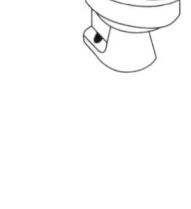

| | | | | | |
|---|---|---|---|---|---|
| Floor-mount tank type water closet and trim | P1@2.10 | Ea | 206.00 | 75.10 | 281.10 |

Floor-mounted, back outlet, residential tank type 1.6 gal. vitreous china toilet, valved closet supply, elongated bowl, and toilet seat with cover

| | | | | | |
|---|---|---|---|---|---|
| Floor-mount, back outlet tank type WC and trim | P1@2.35 | Ea | 250.00 | 84.00 | 334.00 |

Wall-mounted, tank type 1.6 gal. vitreous china toilet, toilet carrier, valved closet supply, elongated bowl, and toilet seat with cover

| | | | | | |
|---|---|---|---|---|---|
| Wall-mount, tank type WC, carrier and trim | P1@3.65 | Ea | 516.00 | 131.00 | 647.00 |

One piece, 1.6 gal. floor-mounted vitreous china toilet including toilet seat and valved closet supply

| | | | | | |
|---|---|---|---|---|---|
| One-piece water closet and trim | P1@2.35 | Ea | 300.00 | 84.00 | 384.00 |

Floor-mounted vitreous china 1.6 gal. pressure assist toilet, including toilet seat and valved closet supply

| | | | | | |
|---|---|---|---|---|---|
| Pressure assist water closet and trim | P1@2.25 | Ea | 399.00 | 80.50 | 479.50 |
| Add for elongated bowl models | — | Ea | 65.60 | — | 65.60 |

## Bidets

Standard vitreous china bidet with floor-mounted hardware, hot and cold valved supplies and polished chrome faucet

| | | | | | |
|---|---|---|---|---|---|
| Vitreous china bidet, vertical spray, and trim | P1@2.50 | Ea | 380.00 | 89.40 | 469.40 |
| Vitreous china bidet, over the rim, and trim | P1@2.50 | Ea | 354.00 | 89.40 | 443.40 |

## Lavatories

Oval self-rimming enameled steel lavatory, hot and cold valved supplies, single lever faucet and pop-up drain fitting and trap assembly

| | | | | | |
|---|---|---|---|---|---|
| Oval self-rim steel lavatory and trim | P1@2.50 | Ea | 304.00 | 89.40 | 393.40 |

Oval, self-rimming vitreous china lavatory, hot and cold valved supplies, single lever faucet and pop-up drain fitting and trap assembly

| | | | | | |
|---|---|---|---|---|---|
| Oval self-rim china lavatory and trim | P1@2.00 | Ea | 363.00 | 71.50 | 434.50 |

Oval, pedestal-mounted vitreous china lavatory, hot and cold valved supplies, single lever faucet and pop-up drain fitting and trap

| | | | | | |
|---|---|---|---|---|---|
| Oval pedestal-mount china lavatory and trim | P1@2.50 | Ea | 414.00 | 89.40 | 503.40 |

Rectangular, pedestal-mount vitreous china lavatory, hot and cold valved supplies, single lever faucet and pop-up drain fitting and trap

| | | | | | |
|---|---|---|---|---|---|
| Rectangular pedestal china lavatory and trim | P1@2.50 | Ea | 483.00 | 89.40 | 572.40 |

| | Craft@Hrs | Unit | Material | Labor | Total |
|---|---|---|---|---|---|

Rectangular, self-rimming vitreous china lavatory, hot and cold valved supplies, single lever faucet and pop-up drain fitting and trap assembly

| | Craft@Hrs | Unit | Material | Labor | Total |
|---|---|---|---|---|---|
| Counter-type rectangular china lavatory & trim | P1@2.00 | Ea | 423.00 | 71.50 | 494.50 |

Rectangular, wall-hung vitreous china lavatory, hot and cold valved supplies, single lever faucet and pop-up drain fitting and trap assembly

| | | | | | |
|---|---|---|---|---|---|
| Wall-hung china lavatory and trim | P1@2.50 | Ea | 395.00 | 89.40 | 484.40 |

Wheelchair access lavatory, wall-hung vitreous china, with carrier insulated hot and cold valved supplies, gooseneck faucet with wrist handles and pop-up drain and insulated offset trap

| | | | | | |
|---|---|---|---|---|---|
| Wall-hung wheelchair lavatory and trim | P1@2.75 | Ea | 680.00 | 98.30 | 778.30 |

## Bar sinks

Acrylic bar sink, 15" L, 15" W, 6" D, self-rimming, with chrome faucet, hot and cold water valved supplies and P-trap assembly

| | | | | | |
|---|---|---|---|---|---|
| Acrylic bar sink and trim | P1@2.00 | Ea | 328.00 | 71.50 | 399.50 |

Enameled cast iron bar sink, 16" L, 19" W, 6" D with chrome faucet, hot and cold water valved supplies and P-trap assembly

| | | | | | |
|---|---|---|---|---|---|
| Enameled cast iron bar sink and trim | P1@2.25 | Ea | 496.00 | 80.50 | 576.50 |

Stainless steel bar sink, 15" L, 15" W, 6" D, chrome faucet, hot and cold valved supplies, and trap assembly

| | | | | | |
|---|---|---|---|---|---|
| Stainless steel bar sink and trim | P1@2.00 | Ea | 409.00 | 71.50 | 480.50 |

## Kitchen sinks

Single bowl, enameled cast iron, self-rimming, 25" L, 22" W, 8" D with chrome faucet and sprayer, P-trap assembly and valved supplies

| | | | | | |
|---|---|---|---|---|---|
| Single bowl cast iron sink and trim | P1@1.75 | Ea | 345.00 | 62.60 | 407.60 |

Single bowl, stainless steel, self-rimming sink, 20" L, 20" W, 8" D with chrome faucet and sprayer, strainer, P-trap assembly and valved supplies

| | | | | | |
|---|---|---|---|---|---|
| Single bowl stainless sink and trim | P1@2.00 | Ea | 298.00 | 71.50 | 369.50 |

Double bowl, enameled cast iron, self-rimming sink, 33" L, 22" W, 8" D, chrome faucet and sprayer, two strainers, P-trap assembly and valved supplies

| | | | | | |
|---|---|---|---|---|---|
| Double bowl cast iron sink and trim | P1@2.00 | Ea | 462.00 | 71.50 | 533.50 |

Double bowl, stainless steel, self-rimming sink, 32" L, 22" W, 8" D with chrome faucet and sprayer, two strainers, P-trap assembly and valved supplies

| | | | | | |
|---|---|---|---|---|---|
| Double bowl stainless sink and trim | P1@2.25 | Ea | 355.00 | 80.50 | 435.50 |

Triple bowl, stainless steel, self-rimming sink, 43" L, 22" W with chrome faucet and sprayer, three strainers, P-trap assembly and valved supplies

| | | | | | |
|---|---|---|---|---|---|
| Triple bowl stainless sink and trim | P1@2.75 | Ea | 668.00 | 98.30 | 766.30 |

## Laundry sinks

Enameled cast iron laundry sink, 24" L, 20" D, 14" D, wall-mounted faucet with integral stops and vacuum breaker, strainer and P-trap assembly

| | | | | | |
|---|---|---|---|---|---|
| Cast iron laundry sink and trim | P1@2.76 | Ea | 510.00 | 98.70 | 608.70 |

Acrylic laundry sink, 24" L, 20" D, 14" D, deck-mounted faucet valved supplies, and P-trap assembly

| | | | | | |
|---|---|---|---|---|---|
| Acrylic laundry sink and trim | P1@2.00 | Ea | 156.00 | 71.50 | 227.50 |

## Service sinks

Enameled cast iron service sink, 22" L, 18" D, wall mounted faucet with integral stops and vacuum breaker, strainer and cast iron P-trap

| | | | | | |
|---|---|---|---|---|---|
| Cast iron service sink and trim | P1@3.00 | Ea | 1,020.00 | 107.00 | 1,127.00 |

## Drinking fountains and electric water coolers

Wall-hung vitreous china fountain, 14" W, 13" H, complete with fittings and 1-1/4" trap

| | | | | | |
|---|---|---|---|---|---|
| Wall hung vitreous fountain and trim | P1@2.00 | Ea | 487.00 | 71.50 | 558.50 |

# Plumbing Fixtures and Equipment

| | Craft@Hrs | Unit | Material | Labor | Total |
|---|---|---|---|---|---|
| Semi-recessed vitreous china, 15" W, 27" H, with fittings, wall cleanout and 1-1/4" trap | | | | | |
| Semi-recessed vitreous fountain and trim | P1@2.50 | Ea | 857.00 | 89.40 | 946.40 |
| Wall-hung electric water cooler, 13 gallon capacity, complete with trap and fittings (requires electrical outlet for plug-in connection, not included) | | | | | |
| Electric water cooler and trim | P1@2.75 | Ea | 1,140.00 | 98.30 | 1,238.30 |
| Wall-hung wheelchair access electric water cooler, front push bar, 7.5 gallon capacity, complete with trap and fittings (requires electrical outlet for plug-in connection, not included) | | | | | |
| Wheelchair access water cooler and trim | P1@2.50 | Ea | 893.00 | 89.40 | 982.40 |
| Fully recessed electric water cooler, 11.5 gallon capacity, complete with trap and fittings (requires electrical outlet for plug-in connection, not included) | | | | | |
| Recessed electric water cooler and trim | P1@2.95 | Ea | 2,510.00 | 105.00 | 2,615.00 |

## Garbage disposers

| | Craft@Hrs | Unit | Material | Labor | Total |
|---|---|---|---|---|---|
| Small, 1/3 HP kitchen disposer, including fittings, no waste pipe | | | | | |
| 1/3 HP kitchen disposer | P1@2.00 | Ea | 151.00 | 71.50 | 222.50 |
| Standard, 5/8 HP kitchen disposer, including fittings, no waste pipe | | | | | |
| 5/8 HP kitchen disposer | P1@2.00 | Ea | 180.00 | 71.50 | 251.50 |
| Standard, 3/4 HP kitchen disposer, including fittings, no waste pipe | | | | | |
| 3/4 HP kitchen disposer | P1@2.00 | Ea | 225.00 | 71.50 | 296.50 |
| Septic system, 3/4 HP kitchen disposer, including fittings, no waste pipe | | | | | |
| 3/4 HP kitchen disposer | P1@2.00 | Ea | 350.00 | 71.50 | 421.50 |
| Large capacity, 1 HP kitchen disposer, including fittings, no waste pipe | | | | | |
| 1 HP kitchen disposer | P1@2.15 | Ea | 380.00 | 76.90 | 456.90 |
| Add for electrical connection and switch | BE@1.00 | Ea | 35.30 | 39.60 | 74.90 |

## Dishwashers Under-counter installation, connected to existing drain. Add the cost of electrical work, if required.

| | Craft@Hrs | Unit | Material | Labor | Total |
|---|---|---|---|---|---|
| Deluxe, stainless steel, steam cleaning | P1@2.40 | Ea | 1,430.00 | 85.80 | 1,515.80 |
| Standard, 5-cycle | P1@2.40 | Ea | 640.00 | 85.80 | 725.80 |
| Economy, 3 wash levels | P1@2.40 | Ea | 295.00 | 85.80 | 380.80 |
| Add for electrical connection | BE@1.00 | Ea | 26.90 | 39.60 | 66.50 |

## Water heaters

Gas water heaters. Includes labor to connect to existing water supply piping and gas line. 6 year warranty. Labor column includes cost of installing heater and making the gas and water connections, but not the cost of the supply lines.

| | Craft@Hrs | Unit | Material | Labor | Total |
|---|---|---|---|---|---|
| 30 gallon tall, 53 GPH | P1@1.90 | Ea | 401.00 | 67.90 | 468.90 |
| 40 gallon short, 55 GPH | P1@2.00 | Ea | 410.00 | 71.50 | 481.50 |
| 40 gallon tall, 62 GPH | P1@2.00 | Ea | 340.00 | 71.50 | 411.50 |
| 50 gallon tall, 72 GPH | P1@2.15 | Ea | 502.00 | 76.90 | 578.90 |
| 75 gallon tall, 75 MBtu | P1@3.50 | Ea | 892.00 | 125.00 | 1,017.00 |
| 40 gallon, 40 MBtu, power vent | P1@2.00 | Ea | 750.00 | 71.50 | 821.50 |
| 50 gallon, 42 MBtu, power vent | P1@2.15 | Ea | 865.00 | 76.90 | 941.90 |
| Add for power vent electrical connection | BE@1.00 | Ea | 29.90 | 39.60 | 69.50 |

Propane water heaters. Includes labor to connect to existing water supply piping. 6 year warranty. Flammable vapor ignition resistant. Labor column includes cost of installing heater and making the water connections, but not the cost of the supply lines.

| | Craft@Hrs | Unit | Material | Labor | Total |
|---|---|---|---|---|---|
| 30 gallon, 27 MBtu | P1@2.00 | Ea | 489.00 | 71.50 | 560.50 |
| 40 gallon, 36 MBtu | P1@2.00 | Ea | 508.00 | 71.50 | 579.50 |
| 50 gallon, 36 MBtu | P1@2.15 | Ea | 590.00 | 76.90 | 666.90 |
| 40 gallon, 40 MBtu, power vent | P1@2.00 | Ea | 818.00 | 71.50 | 889.50 |
| 50 gallon, 36 MBtu, power vent | P1@2.15 | Ea | 920.00 | 76.90 | 996.90 |
| Add for power vent electrical connection | BE@1.00 | Ea | 29.90 | 39.60 | 69.50 |

| | Craft@Hrs | Unit | Material | Labor | Total |
|---|---|---|---|---|---|
| Electric water heaters. Includes labor to connect to existing water supply piping. Add for the cost of supply piping. | | | | | |
| 40 gallon 53 GPH, 12 year | P1@2.15 | Ea | 442.00 | 76.90 | 518.90 |
| 50 gallon 63 GPH, 12 year | P1@2.25 | Ea | 450.00 | 80.50 | 530.50 |
| 80 gallon 87 GPH, 12 year | P1@3.50 | Ea | 640.00 | 125.00 | 765.00 |
| 40 gallon 49 GPH, 9 year | P1@2.00 | Ea | 380.00 | 71.50 | 451.50 |
| 50 gallon 55 GPH, 9 year | P1@2.15 | Ea | 420.00 | 76.90 | 496.90 |
| 80 gallon tall, 80 GPH, 9 year | P1@3.50 | Ea | 435.00 | 125.00 | 560.00 |
| 30 gallon 38 GPH, 6 year | P1@1.90 | Ea | 305.00 | 67.90 | 372.90 |
| 40 gallon short, 42 GPH, 6 year | P1@2.00 | Ea | 310.00 | 71.50 | 381.50 |
| Thermal expansion tank. Controls thermal expansion of water in domestic hot water systems | | | | | |
| 2.1 gallon | P1@.300 | Ea | 33.60 | 10.70 | 44.30 |
| 4.5 gallon | P1@.300 | Ea | 56.00 | 10.70 | 66.70 |

**Solar water heating system**  Water storage tanks compatible with most closed-loop solar water heating systems, heat transfer by insulated all-copper heat exchanger, collector feed and return fittings. With pressure-temperature relief valve and back-up 4,500 watt electric heating element. Rheem.

| | Craft@Hrs | Unit | Material | Labor | Total |
|---|---|---|---|---|---|
| 80 gallon | P1@8.44 | Ea | 820.00 | 302.00 | 1,122.00 |
| Passive thermosiphon domestic hot water system with integral 80 gallon water tank, retrofitted to an existing single story structure, with 5 year limited warranty | | | | | |
| Passive solar water heater system (retrofit) | P1@6.50 | Ea | 3,380.00 | 232.00 | 3,612.00 |
| Active solar hot water system for 100 or 120 gallon water supply tank, fitted with automatic drain retrofitted to existing single story, single family unit, 5 year warranty | | | | | |
| Active solar water heater system (retrofit) | — | Ea | — | — | 5,590.00 |
| Add for strengthening existing roof structure under solar collectors | | | | | |
| Per SF of roof strengthened | — | SF | — | — | 1.36 |
| Active solar hot water system for new one-story home | | | | | |
| Including design and engineering (new work) | — | Ea | — | — | 4,850.00 |
| For designs requiring special cosmetic treatment | | | | | |
| Add for concealment of pipe, collectors, etc. | — | % | — | — | 15.5 |
| Solar pool heater, active heater for pools with freeze-tolerant collector, roof-mounted | | | | | |
| Per SF of collector area | — | SF | — | — | 6.25 |

**Water softening system**  24 volt electronic controls, demand-driven regeneration, 5,200 grams per gallon efficiency, 200 pound salt capacity, with bypass valve and drain tubing, 2 hour regeneration time.

| | Craft@Hrs | Unit | Material | Labor | Total |
|---|---|---|---|---|---|
| 27,000 grain capacity, 95 grams/gallon max hardness removal, 5 grams/gallon max iron reduction, 23 to 30 gallon regeneration | P1@4.00 | Ea | 525.00 | 143.00 | 668.00 |
| 38,500 grain capacity, 100 grams/gallon max hardness removal, 7 grams/gallon max iron reduction, 40 to 50 gallon regeneration | P1@4.00 | Ea | 583.00 | 143.00 | 726.00 |

**Water distillation unit**  Potable central household water system type, Polar Bear 200-L or equal. Underwriters Laboratories and CSA approved, Water Quality Association and certified performance tested to U.S. Environmental Protection Agency Water Quality Act testing standards. Permanent basement mounting, all-electric, 2 parts per million total dissolved solids warranted purity. 220/208/240 volts, 22 amps, 5,300 watts electrical rating. For residential single- and multi-family applications. Includes process condition and flow monitor and 75 gallon tank. Add $500.00 for optional auto-drain kit.

| | Craft@Hrs | Unit | Material | Labor | Total |
|---|---|---|---|---|---|
| 50 gal/day distilled water unit | — | Ea | 7,500.00 | — | 7,500.00 |
| Prepping and pouring of pad | P1@2.00 | Ea | 150.00 | 71.50 | 221.50 |
| Mounting of unit | P1@4.00 | Ea | — | 143.00 | 143.00 |
| Piping of unit | P1@3.00 | Ea | 200.00 | 107.00 | 307.00 |
| Wiring of controls | CE@4.00 | Ea | 150.00 | 233.00 | 383.00 |
| Piping of vents, pipe and drains | P1@1.00 | Ea | 150.00 | 35.80 | 185.80 |
| Startup and inspection | P1@1.00 | Ea | — | 35.80 | 35.80 |

# Plumbing Fixtures and Equipment

**Plumbing System Components**

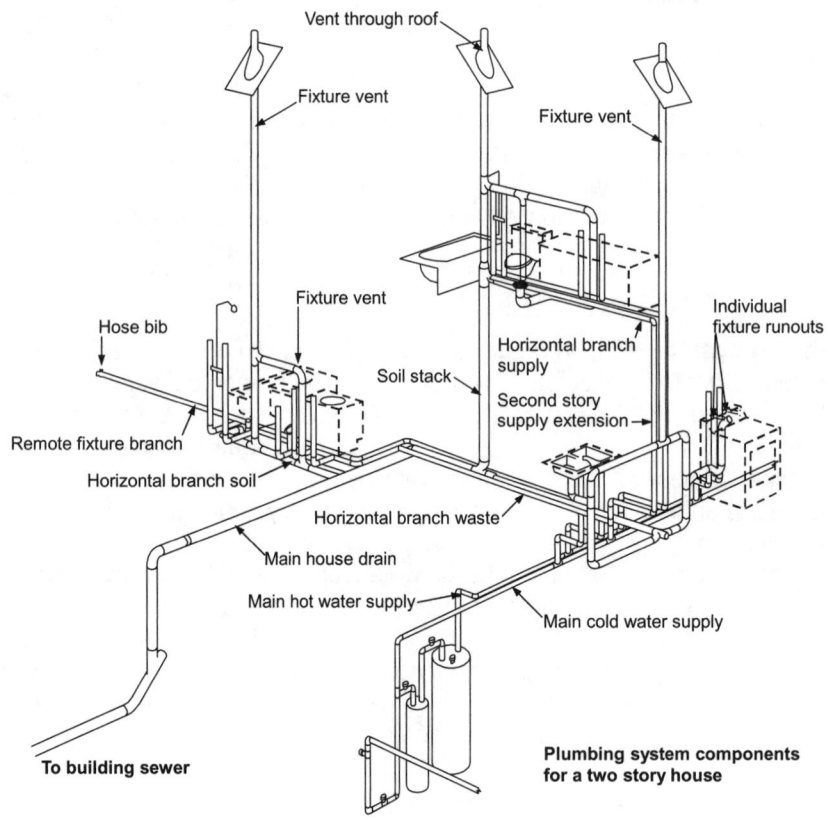

Vent through roof

Fixture vent

Fixture vent

Fixture vent

Hose bib

Horizontal branch supply

Individual fixture runouts

Soil stack

Second story supply extension

Remote fixture branch

Horizontal branch soil

Horizontal branch waste

Main house drain

Main hot water supply

Main cold water supply

To building sewer

**Plumbing system components for a two story house**

## Whole House Plumbing Rough-In

Rough plumbing is the drain, waste and vent (DWV) piping and water supply piping that isn't visible when construction is complete. Rough plumbing costs vary with the number of fixtures and length of the pipe runs. The figures in this section show rough plumbing costs for a typical single family home using plastic drainage and copper water supply materials. No plumbing fixtures, equipment or their final connections are included in these costs. See Plumbing Fixtures and Equipment for fixture and fixture connection costs. These costs include the subcontractor's markup. Add the general contractor's markup when applicable. For more complete information on plumbing and piping costs, see *National Plumbing & HVAC Estimator* at http://CraftsmanSiteLicense.com/

|  | Craft@Hrs | Unit | Material | Labor | Total |
|---|---|---|---|---|---|
| **Single story homes** | | | | | |
| Hot and cold potable water rough-in | P1@15.0 | Ea | 1,340.00 | 536.00 | 1,876.00 |
| Drainage and venting rough-in | P1@20.0 | Ea | 743.00 | 715.00 | 1,458.00 |
| Single story, total plumbing rough-in | P1@35.0 | Ea | 2,083.00 | 1,250.00 | 3,334.00 |
| **Two story homes** | | | | | |
| Hot and cold potable water rough-in | P1@20.0 | Ea | 1,970.00 | 715.00 | 2,685.00 |
| Drainage and venting rough-in | P1@25.0 | Ea | 917.00 | 894.00 | 1,811.00 |
| Two story, total plumbing rough-in | P1@45.0 | Ea | 2,887.00 | 1,609.00 | 4,496.00 |

| | Craft@Hrs | Unit | Material | Labor | Total |
|---|---|---|---|---|---|

**Plumbing Fixture Rough-In Assemblies** Includes labor and material costs for the installation of a typical water supply, drain and vent piping system on new residential construction projects. For renovation projects, add 20% to labor and 10% to material costs to allow for connection to existing service lines and working within existing spaces. Add the cost of any demolition or removal required before the new rough-in can be installed. No plumbing fixtures or fixture connections are included in these costs. See Plumbing Fixtures and Equipment on the following pages for fixture and fixture connection costs. These costs include the subcontractor's markup. Add the general contractor's mark-up.

**3-piece washroom group rough-in assembly** Includes the water supply, drain and vent piping to connect all three fixtures (tub, toilet and sink) in the group to one common waste, one vent, one hot and one cold water supply line that will be connected to each respective main service drain, vent or supply pipe. These figures are based on the main service piping being within 5' of the fixture group. Add the cost of main water supply, drain and vent lines to which the fixture group will connect.

**Hot and cold water supply:** 26' of 1/2" type M hard copper pipe (straight lengths), including associated soldered wrought copper fittings, solder, flux, and pipe clips.

**Drains and vents:** 5' of 3" drainage pipe on the water closet, 38' of 1-1/2" pipe serving all other drains and vents. 43' total plastic drain and vent pipe, with associated fittings, glue and strapping. Make additional allowances for final fixture connections. (See Plumbing Fixtures and Equipment for fixture and connection costs.)

| | Craft@Hrs | Unit | Material | Labor | Total |
|---|---|---|---|---|---|
| Hot and cold water supply | P1@2.50 | Ea | 28.30 | 89.40 | 117.70 |
| Drains and vents | P1@3.35 | Ea | 35.80 | 120.00 | 155.80 |
| **Total washroom group rough-in assembly** | **P1@5.85** | **Ea** | **64.10** | **209.40** | **273.50** |
| Add for shower head rough-in | P1@.350 | Ea | 6.52 | 12.50 | 19.02 |

### Kitchen sink rough-in assembly

Hot and cold water supply: 10' of 1/2" type M hard copper pipe, (straight lengths), including associated soldered wrought copper fittings, solder, flux, and pipe clips. Drains and vents: 15' of 1-1/2" plastic drain and vent pipe with associated fittings, glue and strapping. Make additional allowance for final fixture connection. (See Plumbing Fixtures and Equipment for fixture and connection costs.)

| | Craft@Hrs | Unit | Material | Labor | Total |
|---|---|---|---|---|---|
| Hot and cold water supply | P1@.750 | Ea | 10.90 | 26.80 | 37.70 |
| Drains and vents | P1@1.00 | Ea | 10.30 | 35.80 | 46.10 |
| **Total kitchen sink rough-in assembly** | **P1@1.75** | **Ea** | **21.20** | **62.60** | **83.80** |

### Laundry tub and washing machine rough-in assembly

Hot and cold water supply: 16' of 1/2" type M hard copper pipe (straight lengths) , including associated fittings, solder, flux, and pipe clips. Drains and vents: 21' of 1-1/2" plastic drain and vent pipe with associated fittings, glue and strapping.

| | Craft@Hrs | Unit | Material | Labor | Total |
|---|---|---|---|---|---|
| Hot and cold water supply | P1@1.00 | Ea | 17.40 | 35.80 | 53.20 |
| Drains and vents | P1@1.35 | Ea | 14.40 | 48.30 | 62.70 |
| **Total tub & washer rough-in assembly** | **P1@2.35** | **Ea** | **31.80** | **84.10** | **115.90** |

### Hot water tank rough-in assembly

Hot and cold water supply: 10' of 3/4" type M hard copper pipe (straight lengths), including associated soldered wrought copper fittings, solder, flux, and pipe clips

| | Craft@Hrs | Unit | Material | Labor | Total |
|---|---|---|---|---|---|
| Hot and cold water supply | P1@.500 | Ea | 17.90 | 17.90 | 35.80 |
| Add for 3/4" isolation valve | P1@.250 | Ea | 9.77 | 8.94 | 18.71 |

### Water softener rough-in assembly

Hot and cold water supply: 10' of 3/4" type M hard copper pipe (straight lengths), including associated soldered wrought copper fittings, solder, flux, and pipe clips. Regeneration drain: 10' of Schedule 40 PVC, including associated fittings, glue and strapping

| | Craft@Hrs | Unit | Material | Labor | Total |
|---|---|---|---|---|---|
| Cold water supply and return | P1@.500 | Ea | 17.90 | 17.90 | 35.80 |
| Regeneration drain | P1@.500 | Ea | 1.94 | 17.90 | 19.84 |

# Plumbing Fixture Rough-In Assemblies

| | Craft@Hrs | Unit | Material | Labor | Total |
|---|---|---|---|---|---|

### Natural gas appliance rough-in assembly

Includes labor and material to install and test 12' of 3/4" Schedule 40 black steel piping, associated 150 lb malleable iron fittings and a plug valve to serve a typical residential gas fired appliance. Add for the natural gas main supply line running through the building, meter connection and final connection to the appliance. Assembly assumes the appliance is within 5' of the main supply line.

| | Craft@Hrs | Unit | Material | Labor | Total |
|---|---|---|---|---|---|
| Gas fired domestic hot water heater | P1@.650 | Ea | 20.90 | 23.20 | 44.10 |
| Gas fired hot water heating boiler | P1@.650 | Ea | 20.90 | 23.20 | 44.10 |
| Gas fired forced air furnace | P1@.650 | Ea | 20.90 | 23.20 | 44.10 |
| Gas fired stove and oven | P1@.650 | Ea | 20.90 | 23.20 | 44.10 |
| Gas fired clothes dryer | P1@.650 | Ea | 20.90 | 23.20 | 44.10 |
| Gas main, 1" dia. with fittings and hangers | P1@.120 | LF | 6.73 | 4.29 | 11.02 |
| Gas meter connection (house side only) | P1@.750 | Ea | 36.20 | 26.80 | 63.00 |

### Natural gas appliance connections

Includes labor and material to connect and test a typical residential natural gas fired appliance. Make additional allowances for the natural gas main supply line running through the building, the appliance gas supply rough-in and labor to set the appliance in place.

| | Craft@Hrs | Unit | Material | Labor | Total |
|---|---|---|---|---|---|
| Gas fired domestic hot water heater | P1@.650 | Ea | 24.00 | 23.20 | 47.20 |
| Gas fired hot water heating boiler | P1@.650 | Ea | 24.00 | 23.20 | 47.20 |
| Gas fired forced air furnace | P1@.650 | Ea | 24.00 | 23.20 | 47.20 |
| Gas fired grill, stove or oven | P1@.650 | Ea | 60.40 | 23.20 | 83.60 |
| Gas fired clothes dryer | P1@.650 | Ea | 60.40 | 23.20 | 83.60 |

### Central hot water heating system

Includes 150' of 1-1/2" insulated supply and return mains, 150' of 3/4" insulated branch piping, 10 radiant baseboard convectors (50'), a 110 Mbh high efficiency gas fired hot water boiler, a circulation pump, electric controls, installed and tested. (Based on 1,200 SF of heated space)

| | Craft@Hrs | Unit | Material | Labor | Total |
|---|---|---|---|---|---|
| Distribution system | P1@46.0 | Ea | 1,120.00 | 1,640.00 | 2,760.00 |
| Distribution system (second floor) | P1@16.0 | Ea | 484.00 | 572.00 | 1,056.00 |
| Boiler, pump and controls | P1@16.0 | Ea | 4,300.00 | 572.00 | 4,872.00 |
| Radiant baseboard and valves | P1@16.0 | Ea | 2,600.00 | 572.00 | 3,172.00 |
| Radiant baseboard and valves (2nd floor) | P1@12.0 | Ea | 1,740.00 | 429.00 | 2,169.00 |
| Total hot water heating system, 1 story | P1@78.0 | Ea | 9,760.00 | 3,213.00 | 12,973.00 |
| Total hot water heating system, 2 story | P1@106 | Ea | 10,244.00 | 3,785.00 | 14,029.00 |

**Plumbing Access Doors** for plumbing, primer coated steel 14 gauge door, 18 gauge frame, with cam lock, for masonry, drywall, tile, wall or ceiling. L&L Louvers

Standard, wall or ceiling

| | Craft@Hrs | Unit | Material | Labor | Total |
|---|---|---|---|---|---|
| 8" x 8" | SW@.500 | Ea | 26.00 | 20.50 | 46.50 |
| 12" x 12" | SW@.500 | Ea | 28.00 | 20.50 | 48.50 |
| 14" x 14" | SW@.500 | Ea | 33.00 | 20.50 | 53.50 |
| 16" x 16" | SW@.500 | Ea | 36.00 | 20.50 | 56.50 |
| 18" x 18" | SW@.500 | Ea | 41.00 | 20.50 | 61.50 |
| 22" x 22" | SW@.500 | Ea | 55.00 | 20.50 | 75.50 |
| 22" x 30" | SW@.500 | Ea | 70.00 | 20.50 | 90.50 |
| 24" x 24" | SW@.500 | Ea | 58.00 | 20.50 | 78.50 |
| 24" x 36" | SW@.500 | Ea | 84.00 | 20.50 | 104.50 |
| 30" x 30" | SW@.500 | Ea | 114.00 | 20.50 | 134.50 |
| Add for cylinder lock | — | Ea | 15.00 | — | 15.00 |
| Add for louvers | — | % | 150.0 | — | — |
| Add for fire rating, non-insulated | — | % | 400.0 | — | — |
| Add for fire rating and insulated | — | % | 500.0 | — | — |
| Add for stainless steel, 304-2B | — | % | 300.0 | — | — |

| | Craft@Hrs | Unit | Material | Labor | Total |
|---|---|---|---|---|---|

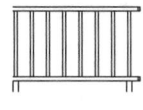

**Polyethylene Film, Clear or Black**  Material costs include 5% for waste and 10% for laps. Labor shown is for installation on grade.

| | Craft@Hrs | Unit | Material | Labor | Total |
|---|---|---|---|---|---|
| 4 mil (.004" thick) | | | | | |
|   50 LF rolls, 3' to 20' wide | BL@.002 | SF | .07 | .06 | .13 |
| 6 mil (.006" thick) | | | | | |
|   50 LF rolls, 3' to 40' wide | BL@.002 | SF | .08 | .06 | .14 |
| Add for installation on walls, ceilings and roofs, | | | | | |
|   tack stapled, typical | BL@.001 | SF | — | .03 | .03 |

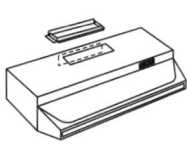

**Railings, Steel, Prefabricated**  See also Ornamental Iron. Porch or step rail, site fitted.

| | Craft@Hrs | Unit | Material | Labor | Total |
|---|---|---|---|---|---|
| 2'-10" high, 6" OC 1/2" solid twisted pickets, 4' or 6' sections | | | | | |
|   1" x 1/2" rails | RI@.171 | LF | 3.66 | 6.08 | 9.74 |
|   1" x 1" rails | RI@.180 | LF | 5.05 | 6.40 | 11.45 |
| Add for newel post 3' H with hardware | — | Ea | 13.90 | — | 13.90 |
| Add lamb's tongue for end post | — | Ea | 5.86 | — | 5.86 |

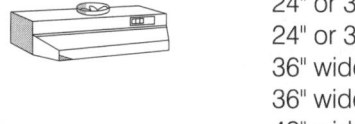

**Range Hoods**  Material costs are for ducted or ductless range hood only, no electrical wiring included. Add for ducting and electrical connection. Labor costs are for installation of range hood only. No carpentry work included. All models are steel construction with baked enamel finish in various colors. Economy ducted range hood.  Under-cabinet-mounted 190 CFM (cubic feet per minute) exhaust. 6.0 sones. 7" duct connector. 18" deep x 6" high. Washable charcoal filter with replaceable charcoal filter pad. Uses 75-watt incandescent lamp. Polymeric lens.

| | Craft@Hrs | Unit | Material | Labor | Total |
|---|---|---|---|---|---|
| 24" or 30" wide, enamel finish | SW@1.42 | Ea | 89.00 | 58.30 | 147.30 |
| 24" or 30" wide, stainless steel | SW@1.42 | Ea | 128.00 | 58.30 | 186.30 |
| 36" wide, enamel finish | SW@1.42 | Ea | 89.00 | 58.30 | 147.30 |
| 36" wide, stainless steel | SW@1.42 | Ea | 128.00 | 58.30 | 186.30 |
| 42" wide, enamel finish | SW@1.42 | Ea | 95.00 | 58.30 | 153.30 |
| 42" wide, stainless steel | SW@1.42 | Ea | 134.00 | 58.30 | 192.30 |
| Add for ducted hood with a duct damper | — | Ea | 15.00 | — | 15.00 |

Standard quality ducted range hood.  Under-cabinet-mounted 190 CFM (cubic feet per minute) 3-speed blower. 3.5 sones. Either vertical or horizontal 3-1/4" x 10" duct, 7" round duct. Powder-coated 22 gauge steel. 18" deep x 6" high. Uses one 40-watt incandescent lamp.

| | Craft@Hrs | Unit | Material | Labor | Total |
|---|---|---|---|---|---|
| 24" or 30" wide, ducted, black | SW@1.42 | Ea | 126.00 | 58.30 | 184.30 |
| 24" or 30" wide, ducted, stainless | SW@1.42 | Ea | 174.00 | 58.30 | 232.30 |
| 36" wide, ducted, black | SW@1.42 | Ea | 116.00 | 58.30 | 174.30 |
| 36" wide, ducted, stainless | SW@1.42 | Ea | 180.00 | 58.30 | 238.30 |
| 42" wide, ducted, black | SW@1.42 | Ea | 122.00 | 58.30 | 180.30 |
| 42" wide, ducted, stainless | SW@1.42 | Ea | 185.00 | 58.30 | 243.30 |
| Add for ductless range hood | — | Ea | 50.00 | — | 50.00 |

Better quality range hood, Allure III NuTone. Under-cabinet-mounted 400 CFM (cubic feet per minute). 0.4 sones at 100 CFM. Teflon-coated  bottom pan. Dual halogen lamps. Sensor detects excessive heat and adjusts vent speed. 20" deep x 7-1/4" high. Four ducting options: 3 1/4" x 10" horizontal or vertical vent, 7" round vertical vent or non-ducted.

| | Craft@Hrs | Unit | Material | Labor | Total |
|---|---|---|---|---|---|
| 30" wide, polyester finish | SW@1.42 | Ea | 460.00 | 58.30 | 518.30 |
| 30" wide, stainless steel finish | SW@1.42 | Ea | 505.00 | 58.30 | 563.30 |
| 36" wide, polyester finish | SW@1.42 | Ea | 470.00 | 58.30 | 528.30 |
| 36" wide, stainless steel finish | SW@1.42 | Ea | 520.00 | 58.30 | 578.30 |
| 42" wide, polyester finish | SW@1.42 | Ea | 480.00 | 58.30 | 538.30 |
| 42" wide, stainless steel finish | SW@1.42 | Ea | 535.00 | 58.30 | 593.30 |
| Add for filter for ducted application | — | Ea | 50.00 | — | 50.00 |
| Add for filter for non-ducted application | — | Ea | 29.50 | — | 29.50 |

# Range Hoods

| | Craft@Hrs | Unit | Material | Labor | Total |
|---|---|---|---|---|---|

Wall-mount range hood, Ballista NuTone. Stainless steel finish with hood-mounted blower. Three-speed control. Dual 20-watt halogen bulbs. Adjusts to fit under 8' to 10' ceiling. Stainless steel grease filter. Built-in backdraft damper. 20" deep. 33" to 54" high.

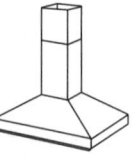

| | Craft@Hrs | Unit | Material | Labor | Total |
|---|---|---|---|---|---|
| 30" wide, 450 CFM | SW@2.50 | Ea | 1,530.00 | 103.00 | 1,633.00 |
| 36" wide, 450 CFM | SW@2.50 | Ea | 1,550.00 | 103.00 | 1,653.00 |
| 48" wide, 900 CFM | SW@2.50 | Ea | 1,730.00 | 103.00 | 1,833.00 |
| Deduct for exterior blower | — | Ea | -410.00 | — | -410.00 |

Ceiling-hung range hood, Provisa NuTone. Stainless steel finish with hood-mounted blower. Three-speed control. Four 20-watt halogen bulbs. Telescopic flue adjusts to fit under 8' to 9' ceiling. Stainless steel grease filter. Built-in backdraft damper. 27" deep. 33" high.

| | Craft@Hrs | Unit | Material | Labor | Total |
|---|---|---|---|---|---|
| 40" wide, 900 CFM | SW@3.50 | Ea | 2,660.00 | 144.00 | 2,804.00 |
| Deduct for exterior blower | — | Ea | -295.00 | — | -295.00 |

Custom range hood power package. Exhaust system components installed in a custom-built range hood 24" to 30" above the cooling surface. Stainless steel. With variable speed control, filters and halogen or incandescent lighting. Bulbs not included. No ductwork included. Add the cost of a decorative enclosure, ducting and electrical connection.

700 CFM, dual blowers, 7" duct, 1.5 to 4.6 sones

| | Craft@Hrs | Unit | Material | Labor | Total |
|---|---|---|---|---|---|
| 30" wide x 22" deep | SW@1.42 | Ea | 467.00 | 58.30 | 525.30 |
| 32" wide x 22" deep | SW@1.42 | Ea | 509.00 | 58.30 | 567.30 |
| 36" wide x 28" deep | SW@1.42 | Ea | 578.00 | 58.30 | 636.30 |
| 42" wide x 28" deep | SW@1.42 | Ea | 637.00 | 58.30 | 695.30 |
| 48" wide x 28" deep | SW@1.42 | Ea | 721.00 | 58.30 | 779.30 |

1,300 CFM, dual blowers, twin 7" ducts, 1.5 to 4.6 sones

| | Craft@Hrs | Unit | Material | Labor | Total |
|---|---|---|---|---|---|
| 30" wide x 22" deep | SW@1.42 | Ea | 758.00 | 58.30 | 816.30 |
| 32" wide x 22" deep | SW@1.42 | Ea | 835.00 | 58.30 | 893.30 |
| 36" wide x 28" deep | SW@1.42 | Ea | 939.00 | 58.30 | 997.30 |
| 42" wide x 28" deep | SW@1.42 | Ea | 1,030.00 | 58.30 | 1,088.30 |
| 48" wide x 28" deep | SW@1.42 | Ea | 1,170.00 | 58.30 | 1,228.30 |

Range splash plate, Broan. Mounts behind range with screws.

| | Craft@Hrs | Unit | Material | Labor | Total |
|---|---|---|---|---|---|
| 30" x 24", reversible white or almond | SW@.165 | Ea | 23.00 | 6.77 | 29.77 |
| 30" x 24", stainless steel | SW@.165 | Ea | 10.20 | 6.77 | 16.97 |
| 36" x 24", reversible white or almond | SW@.165 | Ea | 25.00 | 6.77 | 31.77 |
| 36" x 24", stainless steel | SW@.165 | Ea | 58.70 | 6.77 | 65.47 |

**Ranges, Built-In** Good to better quality counter-mounted cooktops. Labor costs are for carpentry installation only. Add for gas and electric runs and venting. Vented cooktops require exhaust venting. See Range Hoods above.

Gas, pilot-free ignition

| | Craft@Hrs | Unit | Material | Labor | Total |
|---|---|---|---|---|---|
| Four sealed burners, 30 " wide x 21" deep | | | | | |
|     Baked enamel top | P1@1.92 | Ea | 387.00 | 68.70 | 455.70 |
|     Stainless steel top | P1@1.92 | Ea | 640.00 | 68.70 | 708.70 |
| Five sealed burners, 36" wide x 21" deep | | | | | |
|     Stainless steel top | P1@1.92 | Ea | 860.00 | 68.70 | 928.70 |
|     Ceramic top | P1@1.92 | Ea | 1,510.00 | 68.70 | 1,578.70 |

Electric

Conventional coil type cooktop

| | Craft@Hrs | Unit | Material | Labor | Total |
|---|---|---|---|---|---|
| Two coils, 21" wide | BE@.903 | Ea | 215.00 | 35.70 | 250.70 |
| Four coils, 30" wide, black | BE@.903 | Ea | 450.00 | 35.70 | 485.70 |
| Four coils, 30" wide, stainless | BE@.903 | Ea | 517.00 | 35.70 | 552.70 |

Four element smooth glass-ceramic cooktop, 30" wide

| | Craft@Hrs | Unit | Material | Labor | Total |
|---|---|---|---|---|---|
| 6" and 9" elements | BE@.903 | Ea | 640.00 | 35.70 | 675.70 |
| 6"/9"/12" elements | BE@.903 | Ea | 1,020.00 | 35.70 | 1,055.70 |
| 6" and 12" elements, vented | BE@.903 | Ea | 1,290.00 | 35.70 | 1,325.70 |

| | Craft@Hrs | Unit | Material | Labor | Total |
|---|---|---|---|---|---|
| Five element smooth glass-ceramic cooktop, 36" wide | | | | | |
|     6"/9"/12" elements | BE@.903 | Ea | 1,450.00 | 35.70 | 1,485.70 |
|     6"/9"/12" elements, touch controls | BE@.903 | Ea | 1,450.00 | 35.70 | 1,485.70 |
| Five element induction smooth glass-ceramic cooktop, 36" wide | | | | | |
|     6"/7"/8"/11" elements, touch controls | BE@.903 | Ea | 2,690.00 | 35.70 | 2,725.70 |

**Ovens, Built-In** Material costs are for good to better quality 30" wide built-in self-cleaning ovens with stainless steel exterior and a clear window (except as noted). Labor costs are for setting the oven in a prepared opening and making the electrical connection. Add the cost of electric runs.

| | Craft@Hrs | Unit | Material | Labor | Total |
|---|---|---|---|---|---|
| Single oven, black on black | BE@1.67 | Ea | 1,080.00 | 66.10 | 1,146.10 |
| Single oven, stainless steel front | BE@1.67 | Ea | 1,380.00 | 66.10 | 1,446.10 |
| Single thermal oven, stainless steel | BE@1.67 | Ea | 1,670.00 | 66.10 | 1,736.10 |
| Double convection/thermal oven | BE@1.67 | Ea | 1,990.00 | 66.10 | 2,056.10 |
| Double convection/microwave oven | BE@1.67 | Ea | 2,800.00 | 66.10 | 2,866.10 |
| Double thermal oven, LCD controls | BE@1.67 | Ea | 3,450.00 | 66.10 | 3,516.10 |
| Double thermal oven, GE Monogram series | BE@1.67 | Ea | 4,440.00 | 66.10 | 4,506.10 |
| 30" warming drawer, stainless steel | BE@1.00 | Ea | 1,130.00 | 39.60 | 1,169.60 |

**Obtaining Roof Area from Plan Area**

| Rise | Factor | Rise | Factor |
|---|---|---|---|
| 3" | 1.031 | 8" | 1.202 |
| 3½" | 1.042 | 8½" | 1.225 |
| 4" | 1.054 | 9" | 1.250 |
| 4½" | 1.068 | 9½" | 1.275 |
| 5" | 1.083 | 10" | 1.302 |
| 5½" | 1.100 | 10½" | 1.329 |
| 6" | 1.118 | 11" | 1.357 |
| 6½" | 1.137 | 11½" | 1.385 |
| 7" | 1.158 | 12" | 1.414 |
| 7½" | 1.179 | | |

When a roof has to be figured from a plan only, and the roof pitch is known, the roof area may be fairly accurately computed from the table above. The horizontal or plan area (including overhangs) should be multiplied by the factor shown in the table opposite the rise, which is given in inches per horizontal foot. The result will be the roof area.

**Roofing Rule of Thumb** Typical subcontract prices per square (100 SF). Based on a 1,500 to 2,000 SF job not over two stories above ground level. Add the cost of flashing and deck preparation, if required. See detailed labor and material costs in the following sections. Many communities restrict use of wood roofing products.

| | Craft@Hrs | Unit | Material | Labor | Total |
|---|---|---|---|---|---|
| Composition shingles (class A, minimum) | — | Sq | — | — | 160.00 |
| Composition shingles (class C, premium) | — | Sq | — | — | 218.00 |
| Built-up roof, 3-ply and gravel | — | Sq | — | — | 295.00 |
| Cedar shingles, #3, 16" | — | Sq | — | — | 551.00 |
| Cedar shingles, #1, 18", fire treated | — | Sq | — | — | 709.00 |
| Clay mission tile | — | Sq | — | — | 385.00 |
| Concrete tile | — | Sq | — | — | 325.00 |

**Built-Up Roofing** Installed over existing suitable substrate. Add the cost of equipment rental (the kettle) and consumable supplies (mops, brooms, gloves). Typical cost for equipment and consumable supplies is $15 per 100 square feet applied.

Type 1: 2-ply felt and 1-ply 90-pound cap sheet, including 2 plies of hot mop asphalt

| | Craft@Hrs | Unit | Material | Labor | Total |
|---|---|---|---|---|---|
| Base sheet, nailed down, per 100 SF | R1@.250 | Sq | 7.16 | 8.86 | 16.02 |
| Hot mop asphalt, 30 lbs. per 100 SF | R1@.350 | Sq | 19.30 | 12.40 | 31.70 |
| Asphalt felt, 15 lbs. per 100 SF | R1@.100 | Sq | 7.16 | 3.54 | 10.70 |
| Hot mop asphalt, 30 lbs. per 100 SF | R1@.350 | Sq | 19.30 | 12.40 | 31.70 |
| Mineral surface | | | | | |
|     90 lb. cap sheet, per 100 SF | R1@.200 | Sq | 34.60 | 7.08 | 41.68 |
| **Type 1, total per 100 SF** | R1@1.25 | Sq | 87.52 | 44.28 | 131.80 |

Type 2: 3-ply asphalt, 1-ply 30 lb felt, 2 plies 15 lb felt, 3 coats hot mop asphalt and gravel

| | Craft@Hrs | Unit | Material | Labor | Total |
|---|---|---|---|---|---|
| Asphalt felt, 30 lbs. nailed down | R1@.300 | Sq | 14.30 | 10.60 | 24.90 |
| Hot mop asphalt, | | | | | |
|     30 lbs. per 100 SF | R1@.350 | Sq | 19.30 | 12.40 | 31.70 |
| Asphalt felt, 15 lbs. per 100 SF | R1@.100 | Sq | 7.16 | 3.54 | 10.70 |
| Hot mop asphalt, | | | | | |
|     30 lbs. per 100 SF | R1@.350 | Sq | 19.30 | 12.40 | 31.70 |
| Asphalt felt, 15 lbs. per 100 SF | R1@.100 | Sq | 7.16 | 3.54 | 10.70 |
| Hot mop asphalt, | | | | | |
|     30 lbs. per 100 SF | R1@.350 | Sq | 19.30 | 12.40 | 31.70 |
| Gravel, 400 lbs. per 100 SF | R1@.600 | Sq | 25.60 | 21.30 | 46.90 |
| **Type 2, total per 100 SF** | R1@2.15 | Sq | 112.12 | 76.18 | 188.30 |

# Roof Coatings and Adhesives

|  | Craft@Hrs | Unit | Material | Labor | Total |
|---|---|---|---|---|---|
| **Type 3: 4-ply asphalt, 4 plies 15 lb felt including 4 coats hot mop and gravel** | | | | | |
| Base sheet, nailed down, per 100 SF | R1@.250 | Sq | 7.16 | 8.86 | 16.02 |
| Hot mop asphalt, | | | | | |
|    30 lbs. per 100 SF | R1@.350 | Sq | 19.30 | 12.40 | 31.70 |
| Asphalt felt, 15 pound, per 100 SF | R1@.100 | Sq | 7.16 | 3.54 | 10.70 |
| Hot mop asphalt, | | | | | |
|    30 lbs. per 100 SF | R1@.350 | Sq | 19.30 | 12.40 | 31.70 |
| Asphalt felt, 15 pound, per 100 SF | R1@.100 | Sq | 7.16 | 3.54 | 10.70 |
| Hot mop asphalt, | | | | | |
|    30 lbs. per 100 SF | R1@.350 | Sq | 19.30 | 12.40 | 31.70 |
| Asphalt felt, 15 lbs. per 100 SF | R1@.100 | Sq | 7.16 | 3.54 | 10.70 |
| Hot mop asphalt, | | | | | |
|    30 lbs. per 100 SF | R1@.350 | Sq | 19.30 | 12.40 | 31.70 |
| Gravel, 400 lbs. per 100 SF | R1@.600 | Sq | 25.60 | 21.30 | 46.90 |
| **Type 3, total per 100 square feet** | **R1@2.55** | **Sq** | **131.44** | **90.28** | **221.72** |

## Roof Coatings and Adhesives

|  | Craft@Hrs | Unit | Material | Labor | Total |
|---|---|---|---|---|---|
| **Roofing asphalt, 100 lb carton** | | | | | |
| 145 to 165 degree (low melt) | — | Ea | 58.60 | — | 58.60 |
| 185 degree, certified (high melt) – Type IV | — | Ea | 47.20 | — | 47.20 |
| **Asphalt emulsion coating, (35 SF per gallon)** | | | | | |
| 5 gallons | — | Ea | 37.40 | — | 37.40 |
| **Asphalt primer, Henry #105, (150 to 300 SF per gallon)** | | | | | |
| 1 gallon | — | Ea | 15.40 | — | 15.40 |
| 5 gallons | — | Ea | 51.70 | — | 51.70 |
| 55 gallons | — | Ea | 436.00 | — | 436.00 |
| **Wet patch roof cement, Western Colloid #101, (12.5 SF per gallon at 1/8" thick)** | | | | | |
| 1 gallon | — | Ea | 18.10 | — | 18.10 |
| 5 gallons | — | Ea | 62.10 | — | 62.10 |
| **Roof patch elastomeric, seals vertical and horizontal seams (67 SF per gallon)** | | | | | |
| 1 gallon | — | Ea | 22.70 | — | 22.70 |
| **Elastomeric white roof coating, non-fibered, white acrylic (100 SF per gallon per coat)** | | | | | |
| 1 gallon | — | Ea | 28.10 | — | 28.10 |
| 5 gallons | — | Ea | 89.10 | — | 89.10 |
| **Fibered aluminum roof coating, non-asbestos fiber, (50 SF per gallon on built-up or smooth surface, 65 SF per gallon on metal surface)** | | | | | |
| 1 gallon | — | Ea | 9.10 | — | 9.10 |
| 5 gallons | — | Ea | 34.00 | — | 34.00 |
| **Aqua-Bright aluminum roof coating, aluminum flakes suspended in emulsified asphalt, 65 SF per gallon on built-up or smooth surface, Henry #558** | | | | | |
| 1 gallon | — | Ea | 28.20 | — | 28.20 |
| 5 gallons | — | Ea | 114.00 | — | 114.00 |
| **Flashing compound, Western Colloid #106, (12.5 SF per gallon at 1/8" thick)** | | | | | |
| 1 gallon | — | Ea | 15.00 | — | 15.00 |
| 3 gallons | — | Ea | 38.70 | — | 38.70 |
| 5 gallons | — | Ea | 55.90 | — | 55.90 |
| **Lap cement, Henry #108, (280 LF at 2" lap per gallon)** | | | | | |
| 1 gallon | — | Ea | 19.70 | — | 19.70 |
| 5 gallons | — | Ea | 55.00 | — | 55.00 |
| **Plastic roof cement (12-1/2 SF at 1/8" thick per gallon)** | | | | | |
| 1 gallon | — | Ea | 13.30 | — | 13.30 |
| 5 gallons | — | Ea | 39.40 | — | 39.40 |

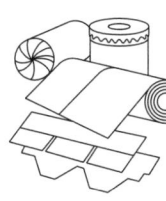

| | Craft@Hrs | Unit | Material | Labor | Total |
|---|---|---|---|---|---|
| Non-fibered roof coating, (80 SF per gallon on masonry surfaces or roofing, 100 SF per gallon on metal surfaces) | | | | | |
| 1 gallon | — | Ea | 14.00 | — | 14.00 |
| 5 gallons | — | Ea | 33.00 | — | 33.00 |
| Reflective fibered aluminum-flake roof coating (50 SF per gallon on built-up roof or smooth surface, 70 SF per gallon on metal surface) | | | | | |
| 1 gallon | — | Ea | 15.40 | — | 15.40 |
| 5 gallons | — | Ea | 59.10 | — | 59.10 |
| Wet-dry premium roof cement (12-1/2 SF per gallon at 1/8" thick) | | | | | |
| 11 oz cartridge for caulking gun | — | Ea | 2.23 | — | 2.23 |
| 1 gallon | — | Ea | 9.00 | — | 9.00 |
| 5 gallons | — | Ea | 31.40 | — | 31.40 |
| Cotton mat reinforcing for roof coatings | | | | | |
| 6" x 150' roll | — | Ea | 20.50 | — | 20.50 |
| Polyester fabric reinforcing for roof coatings, Henry | | | | | |
| 3 oz per SY, 36" x 375' | — | Sq | 21.20 | — | 21.20 |

## Roofing materials

| | Craft@Hrs | Unit | Material | Labor | Total |
|---|---|---|---|---|---|
| Mineral surface roll roofing. Roll covers 1 square (100 square feet). | | | | | |
| 90-pound fiberglass mat reinforced | R1@.200 | Sq | 47.00 | 7.08 | 54.08 |
| GAF Mineral Guard | R1@.200 | Sq | 46.90 | 7.08 | 53.98 |
| SBS-modified bitumen, Flintlastic | R1@.200 | Sq | 96.00 | 7.08 | 103.08 |
| Add for mopping in 30-pound hot asphalt | R1@.350 | Sq | 19.30 | 12.40 | 31.70 |
| Roofing felt. Asphalt saturated. Per square. 15-pound roll covers 400 square feet. 30-pound roll covers 200 square feet. Nailed down. | | | | | |
| 15-pound | R1@.140 | Sq | 6.07 | 4.96 | 11.03 |
| 15-pound, ASTM compliant | R1@.140 | Sq | 7.03 | 4.96 | 11.99 |
| 30-pound | R1@.150 | Sq | 12.10 | 5.31 | 17.41 |
| 30-pound, ASTM-D226 | R1@.150 | Sq | 14.10 | 5.31 | 19.41 |
| Add for mopping in 30-pound hot asphalt | R1@.350 | Sq | 19.30 | 12.40 | 31.70 |
| Fiberglass ply IV roll roofing. ASTM D-2178 Type IV. 500 square foot roll. | | | | | |
| Per 100 square feet | R1@.140 | Sq | 3.58 | 4.96 | 8.54 |
| Add for mopping in 30-pound hot asphalt | R1@.350 | Sq | 19.30 | 12.40 | 31.70 |
| Fiberglass base sheet ASTM D-4601. 300 square foot roll. Nailed down | | | | | |
| Per 100 square feet | R1@.300 | Sq | 4.72 | 10.60 | 15.32 |
| Modified bitumen adhesive. Rubberized. Use to adhere SBS modified bitumen membranes and rolled roofing. Blind nailing cement. Brush grade. | | | | | |
| 5 gallons | — | Ea | 62.40 | — | 62.40 |
| Cold process and lap cement. Bonds plies of built-up roofing and gravel to bare spots on existing gravel roof. Restores severely worn roof. Gallon covers 50 square feet. | | | | | |
| 1 gallon | — | Ea | 9.67 | — | 9.67 |
| 5 gallons | — | Ea | 30.20 | — | 30.20 |
| Cold-Ap® built-up roofing cement, Henry. Bonds layers of conventional built-up roofing or SBS base and cap sheets. Replaces hot asphalt in built-up roofing. For embedding polyester fabric, gravel and granules. Not recommended for saturated felt. ASTM D3019 Type III. Two gallons cover 100 square feet. | | | | | |
| 1 gallon | — | Ea | 15.70 | — | 15.70 |
| 5 gallons | — | Ea | 59.30 | — | 59.30 |
| Cold-process and lap cement. ASTM D3019 Type III. Two gallons cover 100 square feet. | | | | | |
| 5 gallons | — | Ea | 44.70 | — | 44.70 |

# Roofing Shingles

| | Craft@Hrs | Unit | Material | Labor | Total |
|---|---|---|---|---|---|

**Composition Roofing Shingles**  Material costs include 5% waste. Labor costs include roof loading and typical cutting and fitting for roofs of average complexity.

Architectural grade laminated shingles, Owens Corning. Three bundles per square (100 square feet). Class A fire rating. 70 MPH wind resistance.

| | Craft@Hrs | Unit | Material | Labor | Total |
|---|---|---|---|---|---|
| Oakridge, 40 year | R1@1.83 | Sq | 89.50 | 64.80 | 154.30 |
| Oakridge, 30 year, algae-resistant | R1@1.83 | Sq | 84.30 | 64.80 | 149.10 |

Hip and ridge shingles with sealant, Owens Corning. Per linear foot of hip or ridge.

| | Craft@Hrs | Unit | Material | Labor | Total |
|---|---|---|---|---|---|
| High Ridge | R1@.028 | LF | 2.59 | .99 | 3.58 |
| High Ridge, algae-resistant | R1@.028 | LF | 2.63 | .99 | 3.62 |
| High Style | R1@.028 | LF | 3.01 | .99 | 4.00 |

3-tab asphalt fiberglass® shingles, Owens Corning. Three bundles per square (100 square feet). Class A fire rating. 60 MPH wind resistance.

| | Craft@Hrs | Unit | Material | Labor | Total |
|---|---|---|---|---|---|
| Supreme, 25 year | R1@1.83 | Sq | 68.10 | 64.80 | 132.90 |
| Supreme, 25 year, algae-resistant | R1@1.83 | Sq | 66.80 | 64.80 | 131.60 |
| Classic, 20 year | R1@1.83 | Sq | 58.30 | 64.80 | 123.10 |

Hip and ridge shingles with sealant, Owens Corning. Per linear foot of hip or ridge.

| | Craft@Hrs | Unit | Material | Labor | Total |
|---|---|---|---|---|---|
| Asphalt-fiberglass | R1@.028 | LF | 1.21 | .99 | 2.20 |

Glaslock® interlocking shingles, Owens Corning. Three bundles per square (100 square feet). Class A fire rating. 60 MPH wind resistance.

| | Craft@Hrs | Unit | Material | Labor | Total |
|---|---|---|---|---|---|
| 25 year rating | R1@1.83 | Sq | 73.70 | 64.80 | 138.50 |

High definition laminated shingles, Timberline® Prestique®, GAF-Elk, Class A fire rating. Four bundles per square (100 square feet).

| | Craft@Hrs | Unit | Material | Labor | Total |
|---|---|---|---|---|---|
| Lifetime warranty, 110 MPH, brown | R1@1.83 | Sq | 102.00 | 64.80 | 166.80 |
| Lifetime warranty, 110 MPH, white | R1@1.83 | Sq | 127.00 | 64.80 | 191.80 |
| Lifetime warranty, 110 MPH, stain guard | R1@1.83 | Sq | 143.00 | 64.80 | 207.80 |
| 40 year warranty, 90 MPH, stain guard | R1@1.83 | Sq | 140.00 | 64.80 | 204.80 |
| 40 year warranty, 90 MPH | R1@1.83 | Sq | 96.70 | 64.80 | 161.50 |
| 30 year warranty, 80 MPH | R1@1.83 | Sq | 81.10 | 64.80 | 145.90 |

Architectural grade laminated shingles, Timberline® Natural Shadow, GAF-Elk Class A fire rating. Three bundles per square (100 square feet).

| | Craft@Hrs | Unit | Material | Labor | Total |
|---|---|---|---|---|---|
| 30 year warranty, 70 MPH, white | R1@1.83 | Sq | 74.00 | 64.80 | 138.80 |
| 30 year warranty, 70 MPH, most colors | R1@1.83 | Sq | 79.50 | 64.80 | 144.30 |
| 30 year warranty, 70 MPH, stain guard | R1@1.83 | Sq | 105.00 | 64.80 | 169.80 |

Hip and ridge shingles, GAF-Elk. Per linear foot of hip or ridge.

| | Craft@Hrs | Unit | Material | Labor | Total |
|---|---|---|---|---|---|
| Timbertex® | R1@.028 | LF | 2.64 | .99 | 3.63 |
| Pacific Ridge | R1@.028 | LF | 3.42 | .99 | 4.41 |
| Timber Ridge | R1@.028 | LF | 3.03 | .99 | 4.02 |

3-tab fiberglass asphalt shingles, GAF-Elk. Class A fire rating. 60-mph wind warranty. 3 bundles cover 100 square feet at 5" exposure. Per square (100 square feet).

| | Craft@Hrs | Unit | Material | Labor | Total |
|---|---|---|---|---|---|
| Royal Sovereign®, 25 year | R1@1.83 | Sq | 61.20 | 64.80 | 126.00 |
| Royal Sovereign®, 25 year, algae-resistant | R1@1.83 | Sq | 63.60 | 64.80 | 128.40 |
| Sentinel®, 20 year | R1@1.83 | Sq | 54.30 | 64.80 | 119.10 |

Grand Sequoia®, GAF-Elk 3-tab laminate shingles. Class A fire rating. 40-year limited warranty. 17" x 40". 4 bundles cover 100 square feet at 5" exposure. Hip and ridge bundle covers 20 linear feet.

| | Craft@Hrs | Unit | Material | Labor | Total |
|---|---|---|---|---|---|
| 40 year shingles, per square | R1@1.83 | Sq | 133.00 | 64.80 | 197.80 |
| Hip and ridge units, per linear foot | R1@.028 | LF | 2.42 | .99 | 3.41 |

Asphalt shingles, CertainTeed. Class A fire rated. Three bundles cover 100 square feet.

| | Craft@Hrs | Unit | Material | Labor | Total |
|---|---|---|---|---|---|
| Sealdon, 25 | R1@1.83 | Sq | 66.40 | 64.80 | 131.20 |

Shingle starter strip, Peel 'n' stick SBS membrane. Rubberized to seal around nails and resist wind lift.

| | Craft@Hrs | Unit | Material | Labor | Total |
|---|---|---|---|---|---|
| 7" x 33' roll | R1@.460 | Ea | 16.30 | 16.30 | 32.60 |
| Per linear foot | R1@.014 | LF | .50 | .50 | 1.00 |

Allowance for felt, flashing, fasteners, and vents.

| | Craft@Hrs | Unit | Material | Labor | Total |
|---|---|---|---|---|---|
| Typical cost | — | Sq | 16.70 | — | 16.70 |

|  | Craft@Hrs | Unit | Material | Labor | Total |
|---|---|---|---|---|---|

**Roofing Papers**  See also Building Paper. Labor laying paper on 3-in-12 pitch roof.

Asphalt roofing felt (115.5 SF covers 100 SF), labor to roll and mop

| | Craft@Hrs | Unit | Material | Labor | Total |
|---|---|---|---|---|---|
| 15 lb (432 SF roll) | R1@.490 | Sq | 6.07 | 17.40 | 23.47 |
| 30 lb (216 SF roll) | R1@.500 | Sq | 12.10 | 17.70 | 29.80 |

Glass base sheet (40" x 98' roll covers 300 SF)

| | Craft@Hrs | Unit | Material | Labor | Total |
|---|---|---|---|---|---|
| Base for torch applied membranes | R1@.360 | Sq | 4.72 | 12.80 | 17.52 |
| Roofing asphalt, 100 lbs/Sq | — | Sq | 55.60 | — | 55.60 |

Mineral surfaced roll roofing, 36" x 36' roll covers 100 square feet (1 square)

| | Craft@Hrs | Unit | Material | Labor | Total |
|---|---|---|---|---|---|
| 90 lb, black or white | R1@1.03 | Sq | 29.70 | 36.50 | 66.20 |
| 90 lb, tan | R1@1.03 | Sq | 33.20 | 36.50 | 69.70 |

Mineral surfaced selvage edge roll roofing, 39.4" x 32.6' roll covers 100 square feet (1 square) with 2" selvage edge overlap

| | Craft@Hrs | Unit | Material | Labor | Total |
|---|---|---|---|---|---|
| 90 lb, white or colors | R1@1.03 | Sq | 46.90 | 36.50 | 83.40 |

## Roofing Shakes and Shingles

Red cedar. No pressure treating. Labor includes typical flashing.

Taper split No. 1 medium cedar shakes, four bundles cover 100 square feet (1 square) at 10" exposure. Includes 10% for waste.

| | Craft@Hrs | Unit | Material | Labor | Total |
|---|---|---|---|---|---|
| No. 1 medium shakes, $60.70 per bundle | R1@3.52 | Sq | 243.00 | 125.00 | 368.00 |
| Add for pressure treated shakes | — | % | 30.0 | — | — |
| Deduct in the Pacific Northwest | — | % | -20.0 | — | — |
| Shake felt, 30 lb | R1@.500 | Sq | 14.40 | 17.70 | 32.10 |

Hip and ridge shakes, medium, 20 per bundle, covers 16.7 LF at 10" and 20 LF at 12" exposure

| | Craft@Hrs | Unit | Material | Labor | Total |
|---|---|---|---|---|---|
| Per bundle | R1@1.00 | Ea | 48.70 | 35.40 | 84.10 |
| Per linear foot, 10" exposure | R1@.060 | LF | 2.92 | 2.13 | 5.05 |
| Per linear foot, 12" exposure | R1@.058 | LF | 2.44 | 2.05 | 4.49 |

Sawn shakes, sawn 1 side, Class "C" fire retardant

| | Craft@Hrs | Unit | Material | Labor | Total |
|---|---|---|---|---|---|
| 1/2" to 3/4" x 24" (4 bdle/sq at 10" exp.) | R1@3.52 | Sq | 355.00 | 125.00 | 480.00 |
| 3/4" to 5/4" x 24" (5 bdle/sq at 10" exp.) | R1@4.16 | Sq | 492.00 | 147.00 | 639.00 |

Cedar roofing shingles, No. 1 Grade. Four bundles of Perfection Grade 18" shingles cover 100 square feet (1 square) at 5-1/2" exposure. Five Perfections are 2-1/4" thick. Four bundles of 5X Perfect Grade shingles cover 100 square feet (1 square) at 5" exposure. Five Perfects are 2" thick.

| | Craft@Hrs | Unit | Material | Labor | Total |
|---|---|---|---|---|---|
| Perfections, $60.50 per bundle (Florida) | R1@3.52 | Sq | 242.00 | 125.00 | 367.00 |
| Perfects, $54.25 per bundle | R1@3.52 | Sq | 217.00 | 125.00 | 342.00 |
| No. 2 Perfections, $67.25 per bundle | R1@3.52 | Sq | 269.00 | 125.00 | 394.00 |
| Add for pitch over 6 in 12 | — | % | — | 40.0 | — |

No. 2 cedar ridge shingles, medium, bundle covers 16 linear feet

| | Craft@Hrs | Unit | Material | Labor | Total |
|---|---|---|---|---|---|
| Per bundle | R1@1.00 | Ea | 46.20 | 35.40 | 81.60 |
| Per linear foot | R1@.062 | Ea | 2.89 | 2.20 | 5.09 |

Fire treated cedar shingles, 18", 5-1/2" exposure, 5 shingles are 2-1/4" thick, (5/2-1/4)

| | Craft@Hrs | Unit | Material | Labor | Total |
|---|---|---|---|---|---|
| No. 1 (houses) | R1@3.52 | Sq | 419.00 | 125.00 | 544.00 |
| No. 2, red label (houses or garages) | R1@1.00 | Sq | 349.00 | 35.40 | 384.40 |
| No. 3 (garages) | R1@1.00 | Sq | 300.00 | 35.40 | 335.40 |

Eastern white cedar shingles, smooth butt edge, four bundles cover 100 square feet (1 square) at 5" exposure

| | Craft@Hrs | Unit | Material | Labor | Total |
|---|---|---|---|---|---|
| White extras, $57.75 per bundle | R1@1.00 | Sq | 231.00 | 35.40 | 266.40 |
| White clears, $46.00 per bundle | R1@1.00 | Sq | 184.00 | 35.40 | 219.40 |
| White No. 2 clears, $31.25 per bundle | R1@1.00 | Sq | 125.00 | 35.40 | 160.40 |

Cedar shim shingles, tapered Western red cedar

| | Craft@Hrs | Unit | Material | Labor | Total |
|---|---|---|---|---|---|
| Builder's 12-pack, per pack | — | Ea | 4.09 | — | 4.09 |

## Roofing Sheets

| | Craft@Hrs | Unit | Material | Labor | Total |
|---|---|---|---|---|---|
| **Roofing Sheets** See also Fiberglass Panels. | | | | | |
| Metal sheet roofing, utility gauge | | | | | |
| 26" wide, 5-V crimp (includes 15% coverage loss) | | | | | |
| 6' to 12' lengths | R1@.027 | SF | .89 | .96 | 1.85 |
| 26" wide, corrugated (includes 15% coverage loss) | | | | | |
| 6' to 12' lengths | R1@.027 | SF | .88 | .96 | 1.84 |
| 27-1/2" wide, corrugated (includes 20% coverage loss) | | | | | |
| 6' to 12' lengths | R1@.027 | SF | .90 | .96 | 1.86 |
| Ridge roll, plain | | | | | |
| 10" wide | R1@.030 | LF | 2.56 | 1.06 | 3.62 |
| Ridge cap, formed, plain, 12" | | | | | |
| 12" x 10' roll | R1@.061 | LF | 1.22 | 2.16 | 3.38 |
| Sidewall flashing, plain | | | | | |
| 3" x 4" x 10' | R1@.035 | LF | .87 | 1.24 | 2.11 |
| V-5 crimp closure strips | | | | | |
| 24" long | R1@.035 | Ea | 1.33 | 1.24 | 2.57 |
| Endwall flashing, 2-1/2" corrugated | | | | | |
| 10" x 28" | R1@.035 | LF | 1.77 | 1.24 | 3.01 |
| Wood filler strip, corrugated | | | | | |
| 7/8" x 7/8" x 6' (at $1.68 each) | R1@.035 | LF | .28 | 1.24 | 1.52 |
| **Steel Panel Roofing** Simulated tile, Met-Tile™. 15% waste included. | | | | | |
| Roofing panels, 2' to 20' long | | | | | |
| 26 gauge .019" thick | R1@.027 | SF | 1.69 | .96 | 2.65 |
| End cap | | | | | |
| 4-3/4" | R1@.030 | Ea | 3.30 | 1.06 | 4.36 |
| 5-3/4" round | R1@.030 | Ea | 3.30 | 1.06 | 4.36 |
| 9-1/2" oversized | R1@.030 | Ea | 4.39 | 1.06 | 5.45 |
| Flashing, galvanized steel | | | | | |
| End wall | R1@.047 | LF | 1.69 | 1.66 | 3.35 |
| Plain side wall | R1@.047 | LF | 1.69 | 1.66 | 3.35 |
| Valley cover | R1@.047 | LF | 1.69 | 1.66 | 3.35 |
| Valley pan | R1@.047 | LF | 2.24 | 1.66 | 3.90 |
| Contoured end wall | R1@.047 | LF | 4.39 | 1.66 | 6.05 |
| Ridge-hip-gables | | | | | |
| Spanish 18" x 14-1/2" | R1@.035 | Ea | 4.77 | 1.24 | 6.01 |
| Standard 7-1/2" x 42" | R1@.030 | Ea | 9.57 | 1.06 | 10.63 |
| Trim | | | | | |
| Standard eave trim | R1@.035 | LF | 1.36 | 1.24 | 2.60 |
| Flat edge trim | R1@.035 | LF | 1.36 | 1.24 | 2.60 |
| Shake eave trim | R1@.035 | LF | 1.76 | 1.24 | 3.00 |
| Hood vents | | | | | |
| 18" x 22" | R1@.448 | Ea | 1.76 | 15.90 | 17.66 |

**Roofing Slate** Local delivery included. Costs will be higher where slate is not mined. Add freight cost at 800 to 1,000 pounds per 100 SF. Includes 20" long random width slate, 3/16" thick, 7-1/2" exposure. Meets Standard SS-S-451.

| | Craft@Hrs | Unit | Material | Labor | Total |
|---|---|---|---|---|---|
| Semi-weathering green and gray | R1@11.3 | Sq | 449.00 | 400.00 | 849.00 |
| Vermont black and gray black | R1@11.3 | Sq | 534.00 | 400.00 | 934.00 |
| China black or gray | R1@11.3 | Sq | 606.00 | 400.00 | 1,006.00 |
| Unfading and variegated purple | R1@11.3 | Sq | 606.00 | 400.00 | 1,006.00 |
| Unfading mottled green and purple | R1@11.3 | Sq | 606.00 | 400.00 | 1,006.00 |
| Unfading green | R1@11.3 | Sq | 496.00 | 400.00 | 896.00 |
| Red slate | R1@13.6 | Sq | 1,550.00 | 482.00 | 2,032.00 |
| Add for other specified widths and lengths | — | % | 20.0 | — | — |

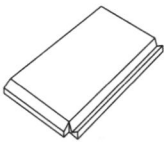

| | Craft@Hrs | Unit | Material | Labor | Total |
|---|---|---|---|---|---|

**Fiber Cement Slate Roofing** Eternit™, Stonit™, or Thrutone™, various colors. Non-asbestos formulation. Laid over 1/2" sheathing. Includes 30 lb. felt and one copper storm anchor per slate. Add for sheathing, flashing, hip, ridge and valley units.

23-5/8" x 11-7/8", English (420 lb, 30 year)

| | Craft@Hrs | Unit | Material | Labor | Total |
|---|---|---|---|---|---|
| 2" head lap (113 pieces per Sq) | R1@5.50 | Sq | 404.00 | 195.00 | 599.00 |
| 3" head lap (119 pieces per Sq) | R1@5.75 | Sq | 425.00 | 204.00 | 629.00 |
| 4" head lap (124 pieces per Sq) | R1@6.00 | Sq | 440.00 | 213.00 | 653.00 |
| Add for hip and ridge units, in mastic | R1@0.10 | LF | 8.26 | 3.54 | 11.80 |

15-3/4" x 10-5/8", Continental (420 lb, 30 year)

| | Craft@Hrs | Unit | Material | Labor | Total |
|---|---|---|---|---|---|
| 2" head lap (197 pieces per Sq) | R1@6.50 | Sq | 417.00 | 230.00 | 647.00 |
| 3" head lap (214 pieces per Sq) | R1@6.75 | Sq | 455.00 | 239.00 | 694.00 |
| 4" head lap (231 pieces per Sq) | R1@7.00 | Sq | 493.00 | 248.00 | 741.00 |
| Add for hip and ridge units, in mastic | R1@0.10 | LF | 7.58 | 3.54 | 11.12 |
| Copper storm anchors, box of 1,000 ($40.00) | — | Ea | .04 | — | .04 |
| Stainless steel slate hooks | — | Ea | .37 | — | .37 |
| Add for extra felt under 5-in-12 pitch | R1@0.20 | Sq | 11.70 | 7.08 | 18.78 |

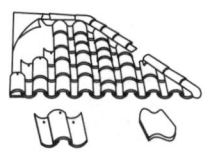

**Roofing Tile, Clay** Material costs include felt and flashing. No freight or waste included. Costs will be higher where clay tile is not manufactured locally. U.S. Tile Co.

Spanish tile, "S"-shaped, 88 pieces per square,
(800 lbs per square at 11" centers and 15" exposure)

| | Craft@Hrs | Unit | Material | Labor | Total |
|---|---|---|---|---|---|
| Red clay tile | R1@3.46 | Sq | 120.00 | 123.00 | 243.00 |
| Add for coloring | — | Sq | 13.10 | — | 13.10 |
| Red hip and ridge units | R1@.047 | LF | 1.00 | 1.66 | 2.66 |
| Color hip and ridge units | R1@.047 | LF | 1.43 | 1.66 | 3.09 |
| Red rake units | R1@.047 | LF | 1.90 | 1.66 | 3.56 |
| Color rake units | R1@.047 | LF | 2.08 | 1.66 | 3.74 |

Red clay mission tile 2-piece, 86 pans and 86 tops per square,
7-1/2" x 18" x 8-1/2" tiles at 11" centers and 15" exposure

| | Craft@Hrs | Unit | Material | Labor | Total |
|---|---|---|---|---|---|
| Red clay tile | R1@5.84 | Sq | 222.00 | 207.00 | 429.00 |
| Add for coloring (costs vary widely) | — | Sq | 49.90 | — | 49.90 |
| Red hip and ridge units | R1@.047 | LF | 1.05 | 1.66 | 2.71 |
| Color hip and ridge units | R1@.047 | LF | 1.43 | 1.66 | 3.09 |
| Red rake units | R1@.047 | LF | 1.90 | 1.66 | 3.56 |
| Color rake units | R1@.047 | LF | 2.08 | 1.66 | 3.74 |

**Concrete Roof Tile** Material includes felt, nails and flashing.
Approximately 90 pieces per square. Monier Lifetile

| | Craft@Hrs | Unit | Material | Labor | Total |
|---|---|---|---|---|---|
| Shake, slurry color | R1@3.25 | Sq | 134.00 | 115.00 | 249.00 |
| Slate, thru color | R1@3.25 | Sq | 162.00 | 115.00 | 277.00 |
| Espana, slurry coated | R1@3.25 | Sq | 162.00 | 115.00 | 277.00 |
| Monier 2000 | R1@3.25 | Sq | 174.00 | 115.00 | 289.00 |
| Vignette | R1@3.25 | Sq | 152.00 | 115.00 | 267.00 |
| Collage | R1@3.25 | Sq | 152.00 | 115.00 | 267.00 |
| Tapestry | R1@3.25 | Sq | 166.00 | 115.00 | 281.00 |
| Split shake | R1@3.25 | Sq | 182.00 | 115.00 | 297.00 |
| Villa, slurry coat | R1@3.25 | Sq | 151.00 | 115.00 | 266.00 |
| Villa, Roma, thru color | R1@3.25 | Sq | 156.00 | 115.00 | 271.00 |

| | Craft@Hrs | Unit | Material | Labor | Total |
|---|---|---|---|---|---|
| Trim tile | | | | | |
| Mansard "V"-ridge or rake, slurry coated | — | Ea | 2.14 | — | 2.14 |
| Mansard, ridge or rake, thru color | — | Ea | 2.23 | — | 2.23 |
| Hipstarters, slurry coated | — | Ea | 18.20 | — | 18.20 |
| Hipstarters, thru color | — | Ea | 23.30 | — | 23.30 |
| Accessories | | | | | |
| Eave closure or birdstop | R1@.030 | LF | 1.29 | 1.06 | 2.35 |
| Hurricane or wind clips | R1@.030 | Ea | .46 | 1.06 | 1.52 |
| Underlayment or felt | R1@.050 | Sq | 16.80 | 1.77 | 18.57 |
| Metal flashing and nails | R1@.306 | Sq | 9.72 | 10.80 | 20.52 |
| Pre-formed plastic flashings | | | | | |
| Hip (13" long) | R1@.020 | Ea | 33.60 | .71 | 34.31 |
| Ridge (39" long) | R1@.030 | Ea | 34.40 | 1.06 | 35.46 |
| Anti-ponding foam | R1@.030 | LF | .30 | 1.06 | 1.36 |
| Batten extenders | R1@.030 | Ea | 1.16 | 1.06 | 2.22 |
| Roof loading | | | | | |
| Add to load tile and accessories on roof | R1@.822 | Sq | — | 29.10 | 29.10 |

**Wall Safes, Residential** UL approved combination lock, flush combination dial, 1/2" thick door. Fits between 16" on center studs. Installed during new construction.

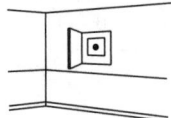

| | Craft@Hrs | Unit | Material | Labor | Total |
|---|---|---|---|---|---|
| For 2 x 4 wall thickness, outside dimensions 13.5" x 14" x 3.75" | | | | | |
| Capacity 400 cu. inches | BC@.788 | Ea | 425.00 | 29.10 | 454.10 |
| For 2 x 6 wall thickness, outside dimensions 13.5" x 14" x 5.75" | | | | | |
| Capacity 750 cu. inches | BC@.788 | Ea | 462.00 | 29.10 | 491.10 |
| For 2 x 12 wall thickness, outside dimensions 13.5" x 14" x 10.25" | | | | | |
| Capacity 1,350 cu. inches | BC@.950 | Ea | 498.00 | 35.10 | 533.10 |
| For 2 x 12 wall thickness, outside dimensions 14" x 14" x 10.25" | | | | | |
| Capacity 1,550 cu. inches | BC@.950 | Ea | 698.00 | 35.10 | 733.10 |
| Optional equipment | | | | | |
| Electronic push button lock | — | Ea | 165.00 | — | 165.00 |
| Dual key lock – in place of combination lock | — | Ea | 35.00 | — | 35.00 |
| Installation kit | — | Ea | 20.00 | — | 20.00 |
| Combination change kit | — | Ea | 10.00 | — | 10.00 |

**Floor Safes, Residential** 1/2" thick spring counter-balanced square door (13" x 13"), UL approved Group Two combination lock, metal cover plate – creates flush floor surface. Installed during new construction.

| | Craft@Hrs | Unit | Material | Labor | Total |
|---|---|---|---|---|---|
| Outside dimensions 14.5" x 16" x 10.25" | | | | | |
| Capacity 1,100 cu. inches | BC@1.10 | Ea | 451.00 | 40.60 | 491.60 |
| Outside dimensions 14.5" x 16" x 16.25" | | | | | |
| Capacity 2,325 cu. inches | BC@1.25 | Ea | 533.00 | 46.20 | 579.20 |
| Outside dimensions 14.5" x 16" x 21.75" | | | | | |
| Capacity 3,545 cu. inches | BC@1.35 | Ea | 620.00 | 49.90 | 669.90 |
| Outside dimensions 14.5" x 25" x 16.25" | | | | | |
| Capacity 3,970 cu. inches | BC@1.42 | Ea | 739.00 | 52.50 | 791.50 |
| Outside dimensions 25" x 25" x 16.75" | | | | | |
| Capacity 7,050 cu. inches | BC@1.55 | Ea | 875.00 | 57.30 | 932.30 |
| Outside dimensions 14.5" x 48" x 16.25" | | | | | |
| Capacity 7,775 cu. inches | BC@1.60 | Ea | 999.00 | 59.10 | 1,058.10 |
| Upgrade to CB-Rating, 1" thick steel door, 1/4" body | — | % | 15.0 | — | — |

| | Craft@Hrs | Unit | Material | Labor | Total |
|---|---|---|---|---|---|

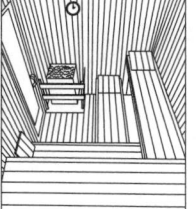

**Sandblasting, Waterblasting, Subcontract** The minimum charge for on-site (truck-mounted) sandblasting and waterblasting will usually be about $500.00. Add the cost of extra liability insurance and physical barriers to protect adjacent swimming pools, residences, process plants, etc. No scaffolding included. Based on sandblasting with 375 CFM, 125 PSI compressor. Portable hydroblast and sandblast equipment for worksites beyond 30 miles from base will incur a $50 transit time fee. Typical fuel surcharge is $100 per 8-hour shift.

| | Craft@Hrs | Unit | Material | Labor | Total |
|---|---|---|---|---|---|
| Sandblasting most surfaces | | | | | |
| Water soluble paints | — | SF | — | — | 2.60 |
| Oil paints | — | SF | — | — | 2.96 |
| Heavy mastic | — | SF | — | — | 3.28 |
| Sandblasting masonry | | | | | |
| Brick | — | SF | — | — | 2.85 |
| Block walls, most work | — | SF | — | — | 2.85 |
| Remove heavy surface grime | — | SF | — | — | 3.44 |
| Heavy, exposing aggregate | — | SF | — | — | 3.75 |
| Sandblasting concrete tilt-up panels | | | | | |
| Light blast | — | SF | — | — | 3.02 |
| Medium, exposing aggregate | — | SF | — | — | 3.85 |
| Heavy, exposing aggregate | — | SF | — | — | 4.33 |
| Sandblasting steel | | | | | |
| New, uncoated (commercial grade) | — | SF | — | — | 3.38 |
| New, uncoated (near white grade) | — | SF | — | — | 4.21 |
| Epoxy coated (near white grade) | — | SF | — | — | 4.85 |
| Sandblasting wood | | | | | |
| Medium blast, clean and texture | — | SF | — | — | 3.20 |
| Waterblast (hydroblast) with mild detergent | | | | | |
| To 5,000 PSI blast (4 hour minimum) | — | Hr | — | — | 379.00 |
| 5,000 to 10,000 PSI blast (8 hour min.) | — | Hr | — | — | 393.00 |
| Over 10,000 PSI blast (8 hour minimum) | — | Hr | — | — | 627.00 |
| Wet sandblasting | | | | | |
| (4 hour minimum) | — | Hr | — | — | 527.00 |

**Sauna Rooms** Costs listed are for labor and materials to install sauna rooms as described. Units are shipped as a disassembled kit complete with kiln dried softwood paneling boards, sauna heater, temperature and humidity controls and monitors, pre-assembled benches, prehung door, water bucket and dipper, and interior light. No electrical work included.

Pre-cut western red cedar sauna room package with heater, controls, duckboard floor and accessories, ready for nailing to existing framed and insulated walls with exterior finish.

| | Craft@Hrs | Unit | Material | Labor | Total |
|---|---|---|---|---|---|
| 4' x 4' room | B1@8.01 | Ea | 2,510.00 | 267.00 | 2,777.00 |
| 4' x 6' room | B1@8.01 | Ea | 3,050.00 | 267.00 | 3,317.00 |
| 6' x 6' room | B1@10.0 | Ea | 3,670.00 | 333.00 | 4,003.00 |
| 6' x 8' room | B1@12.0 | Ea | 4,360.00 | 400.00 | 4,760.00 |
| 8' x 8' room | B1@14.0 | Ea | 4,970.00 | 466.00 | 5,436.00 |
| 8' x 10' room | B1@15.0 | Ea | 5,560.00 | 500.00 | 6,060.00 |
| 8' x 12' room | B1@16.0 | Ea | 6,220.00 | 533.00 | 6,753.00 |
| Add for electrical connection wiring | — | LS | — | — | 335.00 |

# Sauna Rooms

| | Craft@Hrs | Unit | Material | Labor | Total |
|---|---|---|---|---|---|

Modular sauna room assembly. Includes the wired and insulated panels to construct a free-standing sauna on a waterproof floor. With heater, rocks, door, benches and accessories.

| | Craft@Hrs | Unit | Material | Labor | Total |
|---|---|---|---|---|---|
| 4' x 4' room | B1@4.00 | Ea | 4,040.00 | 133.00 | 4,173.00 |
| 4' x 6' room | B1@4.00 | Ea | 4,940.00 | 133.00 | 5,073.00 |
| 6' x 6' room | B1@6.00 | Ea | 5,960.00 | 200.00 | 6,160.00 |
| 6' x 8' room | B1@8.01 | Ea | 7,120.00 | 267.00 | 7,387.00 |
| 6' x 10' room | B1@8.01 | Ea | 8,120.00 | 267.00 | 8,387.00 |
| 8' x 8' room | B1@10.0 | Ea | 8,210.00 | 333.00 | 8,543.00 |
| 8' x 10' room | B1@11.0 | Ea | 9,240.00 | 366.00 | 9,606.00 |
| 8' x 12' room | B1@12.0 | Ea | 10,300.00 | 400.00 | 10,700.00 |
| Add for glass panel, 22-1/2" or 24" | — | Ea | 84.20 | — | 84.20 |
| Add for electrical connection wiring | — | LS | — | — | 260.00 |

Pre-cut 100% vertical grain redwood or clear western red cedar sauna room package with heater, controls and accessories (including non-skid removable rubberized flooring). Ready for nailing to existing framed and insulated walls with exterior

| | Craft@Hrs | Unit | Material | Labor | Total |
|---|---|---|---|---|---|
| 3' x 5' or 4' x 4' room | B1@6.68 | Ea | 2,080.00 | 223.00 | 2,303.00 |
| 4' x 6' room | B1@6.68 | Ea | 2,410.00 | 223.00 | 2,633.00 |
| 5' x 6' room | B1@6.68 | Ea | 2,650.00 | 223.00 | 2,873.00 |
| 6' x 6' room | B1@8.01 | Ea | 2,840.00 | 267.00 | 3,107.00 |
| 6' x 8' room | B1@10.7 | Ea | 3,470.00 | 356.00 | 3,826.00 |
| 8' x 8' room | B1@10.7 | Ea | 4,040.00 | 356.00 | 4,396.00 |
| 8' x 10' room | B1@10.7 | Ea | 4,670.00 | 356.00 | 5,026.00 |
| 8' x 12' room | B1@10.7 | Ea | 5,550.00 | 356.00 | 5,906.00 |
| 10' x 10' room | B1@10.7 | Ea | 5,440.00 | 356.00 | 5,796.00 |
| 10' x 12' room | B1@10.7 | Ea | 6,000.00 | 356.00 | 6,356.00 |
| Add for electrical connection wiring | — | LS | — | — | 335.00 |

Modular sauna room assembly. Completely pre-wired pre-insulated panels to construct a free-standing sauna on waterproof floor. Includes thermometer, sand timer, heater, rocks, door, benches and accessories.

| | Craft@Hrs | Unit | Material | Labor | Total |
|---|---|---|---|---|---|
| 4' x 4' room | B1@1.34 | Ea | 2,820.00 | 44.60 | 2,864.60 |
| 4' x 6' room | B1@1.34 | Ea | 3,290.00 | 44.60 | 3,334.60 |
| 5' x 6' room | B1@2.68 | Ea | 3,430.00 | 89.30 | 3,519.30 |
| 6' x 6' room | B1@2.68 | Ea | 3,880.00 | 89.30 | 3,969.30 |
| 6' x 8' room | B1@2.68 | Ea | 4,420.00 | 89.30 | 4,509.30 |
| 8' x 8' room | B1@4.02 | Ea | 5,460.00 | 134.00 | 5,594.00 |
| 8' x 10' room | B1@4.02 | Ea | 6,250.00 | 134.00 | 6,384.00 |
| 8' x 12' room | B1@4.02 | Ea | 7,010.00 | 134.00 | 7,144.00 |
| 10' x 10' room | B1@4.02 | Ea | 7,010.00 | 134.00 | 7,144.00 |
| 10' x 12' room | B1@4.02 | Ea | 7,680.00 | 134.00 | 7,814.00 |
| Add for electrical connection wiring | — | LS | — | — | 260.00 |
| Add for delivery (typical) | — | % | — | — | 5.0 |

| | 7-foot roll | 10-foot roll | 25-foot roll | 100-foot roll |
|---|---|---|---|---|
| **Screen Wire** Vinyl-coated fiberglass, noncombustible Charcoal or gray, cost per roll | | | | |
| 36" wide roll | 6.77 | — | 13.60 | 55.70 |
| 48" wide roll | 8.87 | — | 17.80 | 73.50 |
| 60" wide roll | — | 17.30 | 32.40 | 111.00 |
| 72" wide roll | — | 20.10 | 67.20 | 111.00 |
| 84" wide roll | — | 21.30 | 53.50 | 172.00 |
| 96" wide roll | — | 24.00 | 61.90 | 196.00 |

| | Craft@Hrs | Unit | Material | Labor | Total |
|---|---|---|---|---|---|
| **Labor to install screen wire** | | | | | |
| Includes measure, lay out, cut and attach screen wire to an existing metal or wood frame. | | | | | |
| Per linear foot of perimeter of frame | BC@.040 | LF | — | 1.48 | 1.48 |

**Security Alarms, Subcontract** Costs are for securing building perimeter and all accessible openings with electronic alarm system, including all labor and materials. Costs assume that no interior wall, floor, or ceiling finishes are in place. All equipment is good to better quality.

| | Craft@Hrs | Unit | Material | Labor | Total |
|---|---|---|---|---|---|
| Alarm control panel | — | Ea | — | — | 516.00 |
| Wiring, detectors per opening, with switch | — | Ea | — | — | 89.30 |
| Monthly monitoring charge | — | Mo | — | — | 36.70 |
| Alarm options (installed) | | | | | |
| Audio detectors | — | Ea | — | — | 129.00 |
| Communicator (central station service) | — | Ea | — | — | 286.00 |
| Digital touch pad control | — | Ea | — | — | 158.00 |
| Entry/exit delay | — | Ea | — | — | 111.00 |
| Fire/smoke detectors | — | Ea | — | — | 191.00 |
| Interior/exterior sirens | — | Ea | — | — | 85.50 |
| Motion detector | — | Ea | — | — | 261.00 |
| Panic button | — | Ea | — | — | 63.70 |
| Passive infrared detector | — | Ea | — | — | 357.00 |
| Pressure mat | — | LF | — | — | 102.00 |
| Security light control | — | Ea | — | — | 628.00 |
| Add for wiring detectors and controls in existing buildings | | | | | |
| With wall and ceiling finishes already in place | — | % | — | — | 25.0 |

**Security Guards, Subcontract** Construction site security guards, unarmed, in uniform with two-way radio, backup patrol car, bond, and liability insurance. Per manhour.

| | Craft@Hrs | Unit | Material | Labor | Total |
|---|---|---|---|---|---|
| Short term (1 night to 1 week) | — | Hr | — | — | 40.80 |
| Medium duration (1 week to 1 month) | — | Hr | — | — | 39.70 |
| Long term (1 to 6 months) | — | Hr | — | — | 27.00 |
| Add for licensed armed guard | — | Hr | — | — | 6.13 |
| Add for holidays | — | % | — | — | 55.0 |

Construction site guard dog service for sites with totally enclosed perimeter, 24 hour per day, 7 day per week service. Local business codes may restrict or prohibit unattended guard dogs. Includes dog handling training session, liability insurance, on-site kennel. Cost per month.

| | Craft@Hrs | Unit | Material | Labor | Total |
|---|---|---|---|---|---|
| Contractor feeding and tending dog | — | Mo | — | — | 702.00 |

Daily delivery of guard dog at quitting time and pickup at starting time by guard dog service (special hours require advance notice)

| | Craft@Hrs | Unit | Material | Labor | Total |
|---|---|---|---|---|---|
| Service feeding and tending dog | — | Mo | — | — | 735.00 |

Construction site man and dog guard team, includes two-way radio communication with guard service office, backup patrol car, and liability insurance. Guard service may require that a telephone and guard shack be provided by the contractor. Cost for man and dog per hour.

| | Craft@Hrs | Unit | Material | Labor | Total |
|---|---|---|---|---|---|
| Short term (1 week to 1 month) | — | Hr | — | — | 44.60 |
| Long term (1 to 6 months) | — | Hr | — | — | 31.80 |

**Septic Sewer Systems, Subcontract** Soils testing (percolation test) by qualified engineer (required by some communities when applying for septic system permit). Does not include application fee.

| | Craft@Hrs | Unit | Material | Labor | Total |
|---|---|---|---|---|---|
| Minimum cost | — | Ea | — | — | 1,800.00 |

Residential septic sewer tanks. Costs include excavation for tank with good site conditions, placing of tank, inlet and outlet fittings, and backfill after hookup.

| | Craft@Hrs | Unit | Material | Labor | Total |
|---|---|---|---|---|---|
| Steel reinforced concrete tanks | | | | | |
| 1,250 gallons (3 or 4 bedroom house) | — | Ea | — | — | 2,320.00 |
| 1,500 gallons (5 or 6 bedroom house) | — | Ea | — | — | 2,470.00 |

# Septic Sewer Systems, Subcontract

| | Craft@Hrs | Unit | Material | Labor | Total |
|---|---|---|---|---|---|
| Fiberglass tanks | | | | | |
| 1,000 gallons (3 bedroom house) | — | Ea | — | — | 2,290.00 |
| 1,250 gallons (4 bedroom house) | — | Ea | — | — | 2,340.00 |
| 1,500 gallons (5 or 6 bedroom house) | — | Ea | — | — | 2,490.00 |
| Polyethylene tanks | | | | | |
| 500 gallons | — | Ea | — | — | 610.00 |
| 750 gallons | — | Ea | — | — | 1,000.00 |
| 1,000 gallons | — | Ea | — | — | 1,100.00 |
| 1,250 gallons | — | Ea | — | — | 1,750.00 |
| 1,500 gallons | — | Ea | — | — | 1,950.00 |

Residential septic sewer drain fields (leach lines). Costs include labor, unsaturated paper, piping, gravel, excavation with good site conditions, backfill, and disposal of excess soil. 4" PVC pipe (ASTM D2729) laid in 3' deep by 1' wide trench.

| | Craft@Hrs | Unit | Material | Labor | Total |
|---|---|---|---|---|---|
| With 12" gravel base | — | LF | — | — | 9.00 |
| With 24" gravel base | — | LF | — | — | 10.60 |
| With 36" gravel base | — | LF | — | — | 11.90 |
| Add for pipe laid 6' deep | — | LF | — | — | 3.00 |

Add for piping from house to septic tank, and from septic tank to remote drain fields

| | Craft@Hrs | Unit | Material | Labor | Total |
|---|---|---|---|---|---|
| 4" PVC Schedule 40 | — | LF | — | — | 6.84 |

**Low cost pumping systems for residential wastes,** including fiberglass basin, pump, installation from septic tank or sewer line, 40' pipe run, and automatic float switch (no electric work or pipe included).

| | Craft@Hrs | Unit | Material | Labor | Total |
|---|---|---|---|---|---|
| To 15' head | — | LS | — | — | 3,100.00 |
| To 25' head | — | LS | — | — | 3,660.00 |
| To 30' head | — | LS | — | — | 4,750.00 |
| Add for high water or pump failure alarm | — | LS | — | — | 676.00 |

Better quality pump system with two alternating pumps, 700 to 800 gallon concrete or fiberglass basin, automatic float switch, indoor control panel, high water or pump failure alarm, explosion proof electrical system but no electric work or pipe.

| | Craft@Hrs | Unit | Material | Labor | Total |
|---|---|---|---|---|---|
| Per pump system | — | LS | — | — | 9,000.00 |

**Pipe locating service** to detect sewage and water leaks, breaks, and stoppages in pipes with diameter of 2" or more. Equipment can't be used where pipe has 90 degree turns. Maximum depth is 18" for cast iron and 30" for vitreous clay.

| | Craft@Hrs | Unit | Material | Labor | Total |
|---|---|---|---|---|---|
| Residential 1/2 day rate | — | LS | — | — | 459.00 |
| Commercial 1/2 day rate | — | LS | — | — | 615.00 |
| Add for each additional pipe entered | — | LS | — | — | 72.50 |
| Plus per LF of pipe length | — | LF | — | — | 2.65 |
| Add for travel over 15 miles, per mile | — | Ea | — | — | 1.00 |

**Sewer Connections, Subcontract** Including typical excavation and backfill.

| | Craft@Hrs | Unit | Material | Labor | Total |
|---|---|---|---|---|---|
| 4" vitrified clay pipeline, house to property line | | | | | |
| Long runs | — | LF | — | — | 21.00 |
| Short runs | — | LF | — | — | 22.50 |
| 6" vitrified clay pipeline | | | | | |
| Long runs | — | LF | — | — | 23.80 |
| Short runs | — | LF | — | — | 28.20 |
| Street work | — | LF | — | — | 59.00 |
| 4" PVC sewer pipe, city installed, property line to main line | | | | | |
| Up to 40 feet (connect in street) | — | LS | — | — | 2,150.00 |
| Up to 40 feet (connect in alley) | — | LS | — | — | 1,550.00 |
| Each foot over 40 feet | — | LF | — | — | 46.20 |

| | Craft@Hrs | Unit | Material | Labor | Total |
|---|---|---|---|---|---|
| 6" PVC sewer pipe, city installed, property line to main line | | | | | |
| Up to 40 feet (connect in street) | — | LS | — | — | 3,440.00 |
| Up to 40 feet (connect in alley) | — | LS | — | — | 2,460.00 |
| Each foot over 40 feet | — | LF | — | — | 75.00 |
| 8" PVC sewer pipe, city installed, property line to main line | | | | | |
| Up to 40 feet (connect in street) | — | LS | — | — | 4,790.00 |
| Up to 40 feet (connect in alley) | — | LS | — | — | 3,730.00 |
| Each foot over 40 feet | — | LF | — | — | 97.50 |

## Sheet Metal Access Doors

| | Craft@Hrs | Unit | Material | Labor | Total |
|---|---|---|---|---|---|
| Metal access doors, concealed metal hinge | | | | | |
| 8" x 8" | SW@.363 | Ea | 19.30 | 14.90 | 34.20 |
| 12" x 12" | SW@.363 | Ea | 23.90 | 14.90 | 38.80 |
| 18" x 18" | SW@.363 | Ea | 37.70 | 14.90 | 52.60 |
| 24" x 24" | SW@.363 | Ea | 53.60 | 14.90 | 68.50 |
| Wall access doors, nail-on type, galvanized or painted | | | | | |
| 6" x 9" | SW@.363 | Ea | 9.95 | 14.90 | 24.85 |
| 14" x 14" | SW@.363 | Ea | 15.00 | 14.90 | 29.90 |
| 24" x 18" | SW@.363 | Ea | 60.30 | 14.90 | 75.20 |
| 24" x 24" | SW@.363 | Ea | 71.10 | 14.90 | 86.00 |
| 24" x 18", screen only | — | Ea | 17.70 | — | 17.70 |
| 24" x 24", screen only | — | Ea | 20.20 | — | 20.20 |
| Tub access doors | | | | | |
| 6" x 9", white ABS | SW@.363 | Ea | 10.40 | 14.90 | 25.30 |
| 14" x 14", white ABS | SW@.363 | Ea | 15.30 | 14.90 | 30.20 |
| 12" x 12", steel | SW@.363 | Ea | 24.00 | 14.90 | 38.90 |

## Sheet Metal Area Walls  Provides light well for basement windows, galvanized, corrugated.

| | Craft@Hrs | Unit | Material | Labor | Total |
|---|---|---|---|---|---|
| 12" deep, 6" projection, 37" wide | SW@.410 | Ea | 30.60 | 16.80 | 47.40 |
| 18" deep, 6" projection, 37" wide | SW@.410 | Ea | 40.40 | 16.80 | 57.20 |
| 24" deep, 6" projection, 37" wide | SW@.410 | Ea | 52.70 | 16.80 | 69.50 |
| 30" deep, 6" projection, 37" wide | SW@.410 | Ea | 62.70 | 16.80 | 79.50 |
| Add for steel grille cover | — | Ea | 71.30 | — | 71.30 |

## Sheet Metal Flashing
**Aluminum flashing**  .0175" thickness is commercial grade. Mill finish except as noted.

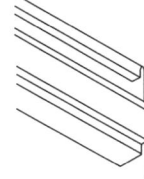

| | Craft@Hrs | Unit | Material | Labor | Total |
|---|---|---|---|---|---|
| Chimney flash kit (base, cricket and step flash) | SW@.900 | Ea | 34.30 | 36.90 | 71.20 |
| Drip cap, 1-5/8" x 10' x .0175, white | SW@.350 | Ea | 5.61 | 14.40 | 20.01 |
| Drip edge, 2-1/2" x 1" x .0145", 10' long | SW@.350 | Ea | 3.93 | 14.40 | 18.33 |
| Drip edge, 6-1/2" x 1", 10' long | SW@.350 | Ea | 12.40 | 14.40 | 26.80 |
| Flex corner, 2" x 2" x 3-1/2" radius | SW@.050 | Ea | 3.47 | 2.05 | 5.52 |
| Gravel stop, 10' x 5-1/2", "C"-shape | SW@.350 | Ea | 5.65 | 14.40 | 20.05 |
| Roll  6" x 10' x .011" | SW@.400 | Ea | 5.84 | 16.40 | 22.24 |
| Roll  6" x 10' x .0175" | SW@.400 | Ea | 15.30 | 16.40 | 31.70 |
| Roll  8" x 10' x .011" | SW@.400 | Ea | 7.32 | 16.40 | 23.72 |
| Roll  8" x 10' x .0175" | SW@.400 | Ea | 20.10 | 16.40 | 36.50 |
| Roll 10" x 10' x .011" | SW@.400 | Ea | 8.87 | 16.40 | 25.27 |
| Roll 10" x 10' x .0175" | SW@.400 | Ea | 12.80 | 16.40 | 29.20 |
| Roll 14" x 10' x .011" | SW@.400 | Ea | 11.10 | 16.40 | 27.50 |
| Roll 14" x 10' x .0175" | SW@.400 | Ea | 12.90 | 16.40 | 29.30 |
| Roll 20" x 10' x .011" | SW@.400 | Ea | 14.10 | 16.40 | 30.50 |
| Roll 20" x 10' x .0175" | SW@.400 | Ea | 38.20 | 16.40 | 54.60 |
| Roof edge, 2-1/2" x 1-1/2" x 10' | SW@.350 | Ea | 5.32 | 14.40 | 19.72 |

# Sheet Metal Flashing

|  | Craft@Hrs | Unit | Material | Labor | Total |
|---|---|---|---|---|---|
| Roof edge, 4-1/2" x 1-1/2" x 10' | SW@.350 | Ea | 14.20 | 14.40 | 28.60 |
| Sheet, 1' x 1', black | SW@.250 | Ea | 10.60 | 10.30 | 20.90 |
| Sheet, 1' x 1', mill finish | SW@.250 | Ea | 8.14 | 10.30 | 18.44 |
| Sheet, 1' x 2', black | SW@.250 | Ea | 18.80 | 10.30 | 29.10 |
| Sheet, 1' x 2' x .020", lincaine perforated | SW@.250 | Ea | 19.80 | 10.30 | 30.10 |
| Sheet, 2' x 3', mill finish | SW@.500 | Ea | 12.50 | 20.50 | 33.00 |
| Sheet, 3' x 3', mill finish | SW@.500 | Ea | 22.00 | 20.50 | 42.50 |
| Sheet, 3' x 3', lincaine perforated | SW@.500 | Ea | 36.20 | 20.50 | 56.70 |
| Step flashing bent, 2-1/2" x 2-1/2" x 7" | SW@.040 | Ea | .55 | 1.64 | 2.19 |
| Step flashing, bent, 4" x 4" x 8" | SW@.040 | Ea | .60 | 1.64 | 2.24 |
| Step flashing, bent, 5" x 7", pack of 10 | SW@.400 | Ea | 2.80 | 16.40 | 19.20 |
| Trim coil 24" x 50', white and black | SW@2.50 | Ea | 81.30 | 103.00 | 184.30 |
| Trim coil nails, box of 250 | — | Ea | 6.34 | — | 6.34 |

**Copper flashing** 16 ounce per square foot, .021 gauge, bright.

| | Craft@Hrs | Unit | Material | Labor | Total |
|---|---|---|---|---|---|
| Roll  8" x 15' | SW@.750 | Ea | 164.00 | 30.80 | 194.80 |
| Roll 12" x 15' | SW@.500 | Ea | 241.00 | 20.50 | 261.50 |
| Roll 12" x 20', 3 ounce | SW@1.00 | Ea | 62.40 | 41.00 | 103.40 |
| Roof edging, 1-1/2" x 2" x 10' | SW@.350 | Ea | 81.40 | 14.40 | 95.80 |
| Roof edging, 2" x 2" x 10' | SW@.350 | Ea | 90.40 | 14.40 | 104.80 |
| Roof-to-wall flashing, 4" x 6" x 10' | SW@.350 | Ea | 216.00 | 14.40 | 230.40 |
| Shingle flashing, 8" x 12" | SW@.040 | Ea | 25.30 | 1.64 | 26.94 |
| Step flashing, 5" x 7" | SW@.040 | Ea | 5.07 | 1.64 | 6.71 |
| W-valley flashing, 18" x 10' | SW@.500 | Ea | 318.00 | 20.50 | 338.50 |
| Copper nails, 1-1/4", box of 100 | — | Box | 5.38 | — | 5.38 |

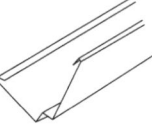

**Paper-backed copper flashing** 3 ounce per square foot. Compatible with ACQ treated lumber. Laminated to Kraft paper and reinforced with fiberglass scrim.

| | Craft@Hrs | Unit | Material | Labor | Total |
|---|---|---|---|---|---|
| Roll 8" x 20' | SW@.500 | Ea | 41.60 | 20.50 | 62.10 |
| Roll 10" x 20' | SW@.500 | Ea | 57.30 | 20.50 | 77.80 |
| Roll 12" x 20' | SW@.500 | Ea | 62.40 | 20.50 | 82.90 |

**Galvanized steel flashing** Galvanized mill (plain) finish except as noted. 26 gauge. Galvanized steel flashing has a zinc coating which resists corrosion. Bonderized steel flashing adds a colored phosphate coating which improves paint adhesion.

| | Craft@Hrs | Unit | Material | Labor | Total |
|---|---|---|---|---|---|
| ACQ compatible flashing, 8" x 20' | SW@1.00 | Ea | 28.50 | 41.00 | 69.50 |
| ACQ compatible flashing, 10" x 20' | SW@1.00 | Ea | 39.60 | 41.00 | 80.60 |
| Angle flashing, 3" x 5" x 10', 90-degree | SW@.500 | Ea | 7.91 | 20.50 | 28.41 |
| Angle flashing, 4" x 4" x 10', 90-degree | SW@.500 | Ea | 6.72 | 20.50 | 27.22 |
| Angle flashing, 6" x 6" x 10', 90-degree | SW@.500 | Ea | 9.70 | 20.50 | 30.20 |
| Angle hem flashing, 4" x 3-1/2" x 10' | SW@.400 | Ea | 7.56 | 16.40 | 23.96 |
| Chimney flash kit (base/cricket/step) | SW@.900 | Ea | 33.10 | 36.90 | 70.00 |
| Counter flash, 1/2" x 6" x 10', bonderized | SW@.400 | Ea | 10.00 | 16.40 | 26.40 |
| Deck ledger, 2" x 2" x 10', ACQ lumber | SW@.400 | Ea | 7.80 | 16.40 | 24.20 |
| Deck ledger, 2" x 2" x 8', galvanized | SW@.350 | Ea | 5.77 | 14.40 | 20.17 |
| Dormer flashing, 7", black | SW@.250 | Ea | 10.80 | 10.30 | 21.10 |
| Dormer flashing, 7", mill finish | SW@.250 | Ea | 8.16 | 10.30 | 18.46 |
| Drip cap, 1-1/2" x 10' | SW@.400 | Ea | 3.41 | 16.40 | 19.81 |
| Drip edge, 2-1/2" x 1" x 10', white | SW@.400 | Ea | 4.51 | 16.40 | 20.91 |
| Drip edge, 5" x 2' x 10', mill finish | SW@.400 | Ea | 6.44 | 16.40 | 22.84 |
| Drip edge, 6" x 10', Beige, | SW@.400 | Ea | 8.80 | 16.40 | 25.20 |
| Endwall flashing, 3" x 4" x 10' | SW@.400 | Ea | 9.90 | 16.40 | 26.30 |
| Fascia cap, 2" x 10', bonderized | SW@.400 | Ea | 7.92 | 16.40 | 24.32 |

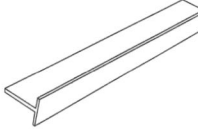

| | Craft@Hrs | Unit | Material | Labor | Total |
|---|---|---|---|---|---|
| Fungus preventer, 2-1/2" x 50' zinc strip | SW@1.25 | Ea | 22.10 | 51.30 | 73.40 |
| Girder shield flashing, 2" x 8" x 2" x 10' | SW@.400 | Ea | 10.60 | 16.40 | 27.00 |
| Gravel guard, 6" x 10', type "E" | SW@.400 | Ea | 5.80 | 16.40 | 22.20 |
| Gravel stop, 2" x 4" x 10', bonderized | SW@.400 | Ea | 9.43 | 16.40 | 25.83 |
| Gravel stop, 2" x 4" x 10', galvanized | SW@.400 | Ea | 8.05 | 16.40 | 24.45 |
| Gutter apron, 2' x 3' x 10', galvanized | SW@.400 | Ea | 6.10 | 16.40 | 22.50 |
| Gutter apron, 2" x 3" x 10', bonderized | SW@.400 | Ea | 6.94 | 16.40 | 23.34 |
| Half round square dormer, 12" x 12" | BC@.871 | Ea | 46.90 | 32.20 | 79.10 |
| L-flashing, 2" x 2" x 10', bonderized | SW@.400 | Ea | 6.35 | 16.40 | 22.75 |
| L-flashing, 2" x 2" x 10', galvanized | SW@.400 | Ea | 5.50 | 16.40 | 21.90 |
| L-flashing, 4" x 6" x 10', bonderized | SW@.400 | Ea | 20.70 | 16.40 | 37.10 |
| L-flashing, 4" x 6" x 10', galvanized | SW@.400 | Ea | 11.40 | 16.40 | 27.80 |
| Moss preventer, 2-2/3" x 50', zinc | SW@1.25 | Ea | 34.60 | 51.30 | 85.90 |
| Parapet coping, 2" x 8" x 2" x 10' | SW@.500 | Ea | 15.14 | 20.50 | 35.64 |
| Vent pipe flash cone, 1/2" pipe | SW@.250 | Ea | 5.35 | 10.30 | 15.65 |
| Vent pipe flash cone, 1" pipe | SW@.250 | Ea | 7.65 | 10.30 | 17.95 |
| Vent pipe flash cone, 3" pipe | SW@.311 | Ea | 8.56 | 12.80 | 21.36 |
| Vent pipe flash cone, 4" pipe | SW@.311 | Ea | 9.86 | 12.80 | 22.66 |
| Vent pipe flash cone, 5" pipe | SW@.311 | Ea | 12.20 | 12.80 | 25.00 |
| Rain diverter, 7-1/2" x 10' | SW@.400 | Ea | 8.15 | 16.40 | 24.55 |
| Reglet, 3/4" x 10', snap top, bonderized | SW@.500 | Ea | 12.50 | 20.50 | 33.00 |
| Reglet, 3/4" x 10', snap top, mill finish | SW@.500 | Ea | 6.99 | 20.50 | 27.49 |
| Roll  6" x 10', galvanized | SW@.500 | Ea | 6.48 | 20.50 | 26.98 |
| Roll  7" x 10', galvanized | SW@.500 | Ea | 7.66 | 20.50 | 28.16 |
| Roll  8" x 10', bonderized | SW@.500 | Ea | 13.00 | 20.50 | 33.50 |
| Roll  8" x 50', galvanized | SW@2.50 | Ea | 22.80 | 103.00 | 125.80 |
| Roll 10" x 10', galvanized | SW@.500 | Ea | 12.80 | 20.50 | 33.30 |
| Roll 14" x 10', galvanized | SW@.500 | Ea | 13.70 | 20.50 | 34.20 |
| Roll 20" x 10', galvanized | SW@.500 | Ea | 17.50 | 20.50 | 38.00 |
| Roll 20" x 50', galvanized | SW@.500 | Ea | 62.40 | 20.50 | 82.90 |
| Roll valley, 20" x 50', gray | SW@2.50 | Ea | 69.20 | 103.00 | 172.20 |
| Roll valley, 20" x 50', brown | SW@2.50 | Ea | 68.10 | 103.00 | 171.10 |
| Roll valley, 30" x 25', galvanized | SW@1.75 | Ea | 33.70 | 71.80 | 105.50 |
| Roll valley, 30" x 50', galvanized | SW@2.50 | Ea | 53.80 | 103.00 | 156.80 |
| Roof apron, 5" x 10', bonderized | SW@.400 | Ea | 4.68 | 16.40 | 21.08 |
| Roof edge, 1" x 2" x 10', brown | SW@.400 | Ea | 3.78 | 16.40 | 20.18 |
| Roof edge, 1" x 2" x 10', gray | SW@.400 | Ea | 3.78 | 16.40 | 20.18 |
| Roof edge, 3" x 3" x 10', brown | SW@.400 | Ea | 9.18 | 16.40 | 25.58 |
| Roof edge, 2" x 2" x 10', weatherwood | SW@.400 | Ea | 3.89 | 16.40 | 20.29 |
| Roof edge, 1-1/2" x 2" x 10', galvanized | SW@.400 | Ea | 5.81 | 16.40 | 22.21 |
| Roof edge, 1-1/2" x 2" x 10', bonderized | SW@.500 | Ea | 5.81 | 20.50 | 26.31 |
| Roof edge, 6" x 6" x 10', galvanized | SW@.400 | Ea | 15.80 | 16.40 | 32.20 |
| Roof grip edge, 2" x 3" x 10', bonderized | SW@.400 | Ea | 8.05 | 16.40 | 24.45 |
| Roof-to-wall flashing, 2" x 3" x 10' | SW@.400 | Ea | 7.52 | 16.40 | 23.92 |
| Roof-to-wall flashing, 4" x 5" x 10' | SW@.400 | Ea | 15.10 | 16.40 | 31.50 |
| Saddle stock, 4" x 14" x 10' | SW@.400 | Ea | 19.40 | 16.40 | 35.80 |
| Sheet 3' x 4', 26-gauge | SW@.500 | Ea | 16.70 | 20.50 | 37.20 |
| Shingle flashing, 5" x 7", box of 10 | SW@.400 | Box | 12.68 | 16.40 | 29.08 |
| Starter strip, 4-1/4" x 10', bonderized | SW@.500 | Ea | 5.09 | 20.50 | 25.59 |
| Step flashing, bent 2" x 3" x 7", .031" | SW@.040 | Ea | .41 | 1.64 | 2.05 |
| Step flashing, bent, 4" x 12", bonderized | SW@.040 | Ea | 1.56 | 1.64 | 3.20 |
| Step flashing, bent, 8" x 8", brown | SW@.040 | Ea | .89 | 1.64 | 2.53 |

# Sheet Metal Flashing

|  | Craft@Hrs | Unit | Material | Labor | Total |
|---|---|---|---|---|---|
| Trim coil, 24" x 50' | SW@2.50 | Ea | 89.60 | 103.00 | 192.60 |
| Tile roof pan, bent 3-1/2" x 6" x 1/2" x 10' | SW@.400 | Ea | 17.50 | 16.40 | 33.90 |
| Tin repair shingles, 5" x 7", galvanized | SW@.040 | Ea | .64 | 1.64 | 2.28 |
| Tin repair shingles, 8" x 12", bonderized | SW@.040 | Ea | 1.41 | 1.64 | 3.05 |
| Water table, 1-1/2" x 10', bonderized | SW@.400 | Ea | 7.51 | 16.40 | 23.91 |
| Water table, 1-1/2" x 10', galvanized | SW@.400 | Ea | 6.84 | 16.40 | 23.24 |
| W-valley, 18" x 10', 28 gauge | SW@.500 | Ea | 22.50 | 20.50 | 43.00 |
| W-valley, 20" x 10', 28 gauge, brown | SW@.500 | Ea | 20.90 | 20.50 | 41.40 |
| W-valley, 24" x 10', bonderized | SW@.500 | Ea | 32.70 | 20.50 | 53.20 |
| Z-bar flashing, 1/2" x 1-1/2" x 2" x 10' | SW@.400 | Ea | 2.92 | 16.40 | 19.32 |
| Z-bar flashing, 2" x 5/8" x 1" x 10' | SW@.400 | Ea | 2.99 | 16.40 | 19.39 |
| Z-bar flashing, 2" x 1" x 3" x 10' | SW@.400 | Ea | 9.48 | 16.40 | 25.88 |

## Lead flashing

|  | Craft@Hrs | Unit | Material | Labor | Total |
|---|---|---|---|---|---|
| Roll 8" wide, per linear foot | SW@.050 | Ea | 5.34 | 2.05 | 7.39 |
| Roll 10" wide, per linear foot | SW@.066 | Ea | 7.12 | 2.71 | 9.83 |
| Roll 12" wide, per linear foot | SW@.075 | Ea | 8.31 | 3.08 | 11.39 |
| Vent pipe flash cone, 1-1/2" pipe | SW@.311 | Ea | 16.90 | 12.80 | 29.70 |
| Vent pipe flash cone 2" pipe | SW@.311 | Ea | 18.30 | 12.80 | 31.10 |
| Vent pipe flash cone, 3" pipe | SW@.311 | Ea | 21.80 | 12.80 | 34.60 |
| Vent pipe flash cone, 4" pipe | SW@.311 | Ea | 12.80 | 12.80 | 25.60 |

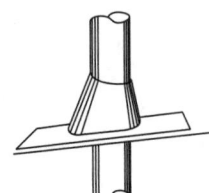

## Vinyl flashing  For windows, fascia, soffits, siding and posts in contact with ACQ treated lumber.

|  | Craft@Hrs | Unit | Material | Labor | Total |
|---|---|---|---|---|---|
| Deck ledger, J-channel, 4" x 10' x .035" | SW@.350 | Ea | 11.00 | 14.40 | 25.40 |
| Roll 20" x 50', tan | SW@3.50 | Ea | 44.00 | 144.00 | 188.00 |
| Roll 10" x 50', white | SW@2.50 | Ea | 25.50 | 103.00 | 128.50 |
| Roof edge, 1-7/8" x 10' | SW@.400 | Ea | 7.43 | 16.40 | 23.83 |
| Vent pipe flash cone, 1" to 3" pipe | SW@.250 | Ea | 5.26 | 10.30 | 15.56 |
| Vent pipe flash cone, 3" to 4" pipe | SW@.311 | Ea | 6.81 | 12.80 | 19.61 |

## Sheet Metal Hoods

Entrance hoods, galvanized

|  | Craft@Hrs | Unit | Material | Labor | Total |
|---|---|---|---|---|---|
| 48" x 22" x 12", double mold | SW@1.00 | Ea | 113.00 | 41.00 | 154.00 |
| 60" x 24" x 12", double mold | SW@1.00 | Ea | 107.00 | 41.00 | 148.00 |
| 48" x 22" x 12", scalloped edge | SW@1.00 | Ea | 139.00 | 41.00 | 180.00 |
| 60" x 24" x 12", scalloped edge | SW@1.00 | Ea | 146.00 | 41.00 | 187.00 |

Side hoods, galvanized, square

|  | Craft@Hrs | Unit | Material | Labor | Total |
|---|---|---|---|---|---|
| 4" x 4" | SW@.190 | Ea | 10.70 | 7.80 | 18.50 |
| 6" x 6" | SW@.216 | Ea | 18.20 | 8.86 | 27.06 |
| 7" x 7" | SW@.272 | Ea | 20.30 | 11.20 | 31.50 |

## Shelving

Wood shelving, using 1" thick #3 & Btr paint grade board lumber. Costs shown include an allowance for ledger boards, nails and normal waste. Painting not included.  Based on a six foot shelf. For other lengths up to ten feet, use the same labor cost but adjust for material.

Closet shelves, with 1" diameter clothes pole and pole support brackets each 3' OC

|  | Craft@Hrs | Unit | Material | Labor | Total |
|---|---|---|---|---|---|
| 10" wide shelf, 6'L | BC@.250 | Ea | 7.61 | 9.24 | 16.85 |
| 12" wide shelf, 6'L | BC@.250 | Ea | 11.40 | 9.24 | 20.64 |
| 18" wide shelf, 6'L | BC@.250 | Ea | 17.30 | 9.24 | 26.54 |

Linen cabinet shelves, cost of linen closet not included

|  | Craft@Hrs | Unit | Material | Labor | Total |
|---|---|---|---|---|---|
| 18" wide shelf, 6'L | BC@.250 | Ea | 6.08 | 9.24 | 15.32 |
| 24" wide shelf, 6'L | BC@.300 | Ea | 8.10 | 11.10 | 19.20 |

|  | Craft@Hrs | Unit | Material | Labor | Total |
|---|---|---|---|---|---|
| Utility shelves, laundry room walls, garage walls, etc. | | | | | |
| 10" wide shelf, 6'L | BC@.250 | Ea | 2.64 | 9.24 | 11.88 |
| 12" wide shelf, 6'L | BC@.250 | Ea | 4.04 | 9.24 | 13.28 |
| 18" wide shelf, 6'L | BC@.250 | Ea | 6.08 | 9.24 | 15.32 |
| 24" wide shelf, 6'L | BC@.300 | Ea | 8.10 | 11.10 | 19.20 |
| Wire shelving kits, vinyl coated steel wire 1" OC. Includes support braces and clamps. | | | | | |
| Heavy-duty garage shelf | | | | | |
| 3' x 16" | BC@.800 | Ea | 37.80 | 29.60 | 67.40 |
| 6' x 16" | BC@.800 | Ea | 53.30 | 29.60 | 82.90 |
| Linen shelf | | | | | |
| 4' x 12" | BC@.600 | Ea | 15.30 | 22.20 | 37.50 |
| Closet organizer. Wire shelf above and clothes hanging rack below | | | | | |
| 5' wide shelf to 8' high | BC@1.00 | Ea | 55.60 | 36.90 | 92.50 |
| 5' wide, with free-slide clothes | BC@1.00 | Ea | 42.30 | 36.90 | 79.20 |
| Glass shelving kits, includes wall support brackets | | | | | |
| 6" x 18", wall, brass support bracket | BC@.400 | Ea | 12.50 | 14.80 | 27.30 |
| 8" x 24", wall, brass support bracket | BC@.400 | Ea | 22.40 | 14.80 | 37.20 |
| 6" x 18", wall, white support bracket | BC@.400 | Ea | 11.50 | 14.80 | 26.30 |
| 8" x 24", wall, white support bracket | BC@.400 | Ea | 22.50 | 14.80 | 37.30 |
| 12" x 12", corner, brass support brackets | BC@.500 | Ea | 23.30 | 18.50 | 41.80 |
| 12" x 12", corner, white support brackets | BC@.500 | Ea | 20.00 | 18.50 | 38.50 |
| Glass shelf, 3/8" glass. Add shelf supports below. | | | | | |
| 24" x 8", clear | BC@.100 | Ea | 19.70 | 3.69 | 23.39 |
| 24" x 12", clear | BC@.100 | Ea | 22.00 | 3.69 | 25.69 |
| 36" x 8", clear | BC@.100 | Ea | 26.50 | 3.69 | 30.19 |
| 36" x 12", clear | BC@.100 | Ea | 33.10 | 3.69 | 36.79 |
| 48" x 8", clear | BC@.100 | Ea | 30.60 | 3.69 | 34.29 |
| 48" x 12", clear | BC@.100 | Ea | 45.80 | 3.69 | 49.49 |
| Brackets for glass shelf, per bracket | | | | | |
| Pelican bracket, any color | BC@.250 | Ea | 6.61 | 9.24 | 15.85 |
| Glacier bracket, 48" | BC@.250 | Ea | 18.10 | 9.24 | 27.34 |

## Shower and Tub Doors

| Swinging shower doors, with hardware, tempered safety glass | | | | | |
|---|---|---|---|---|---|
| Minimum quality pivot shower door, 64" high, silver finish frame, obscure glass | | | | | |
| 28" to 31" wide | BG@1.57 | Ea | 112.00 | 53.80 | 165.80 |
| 31" to 34" wide | BG@1.57 | Ea | 118.00 | 53.80 | 171.80 |
| Better quality pivot shower door 64" high, silver or gold tone frame, obscure glass | | | | | |
| 24" to 26" wide | BG@1.57 | Ea | 137.00 | 53.80 | 190.80 |
| 26" to 28" wide | BG@1.57 | Ea | 148.00 | 53.80 | 201.80 |
| 30" to 32" wide | BG@1.57 | Ea | 155.00 | 53.80 | 208.80 |
| 32" to 34" wide | BG@1.57 | Ea | 169.00 | 53.80 | 222.80 |
| 48" wide | BG@1.57 | Ea | 221.00 | 53.80 | 274.80 |
| Frameless swinging shower doors, to 60" wide by 70" high | | | | | |
| 1/4" clear glass | BG@1.57 | Ea | 233.00 | 53.80 | 286.80 |
| Silver solstice | BG@1.57 | Ea | 315.00 | 53.80 | 368.80 |
| Nickel solstice | BG@1.57 | Ea | 342.00 | 53.80 | 395.80 |
| Silver clear | BG@1.57 | Ea | 315.00 | 53.80 | 368.80 |

# Shower Stalls

| | Craft@Hrs | Unit | Material | Labor | Total |
|---|---|---|---|---|---|
| **Sliding shower doors, 2 panels, 5/32" tempered glass, anodized aluminum frame, 70" high, outside towel bar** | | | | | |
| Non-textured glass, 48" wide | BG@1.57 | Ea | 263.00 | 53.80 | 316.80 |
| Non-textured glass, 56" wide | BG@1.57 | Ea | 275.00 | 53.80 | 328.80 |
| Non-textured glass, 64" wide | BG@1.57 | Ea | 280.00 | 53.80 | 333.80 |
| Texture, opaque, 48" wide | BG@1.57 | Ea | 302.00 | 53.80 | 355.80 |
| Texture, opaque, 56" wide | BG@1.57 | Ea | 320.00 | 53.80 | 373.80 |
| Texture, opaque, 64" wide | BG@1.57 | Ea | 330.00 | 53.80 | 383.80 |
| One panel double mirrored, one bronze tinted | BG@1.57 | Ea | 341.00 | 53.80 | 394.80 |
| Snap-on plastic trim kit, per set of doors | BG@.151 | Ea | 31.20 | 5.17 | 36.37 |
| **Sliding shower doors, 2 panels, 1/4" tempered glass, anodized aluminum frame, 70" high, outside towel bar** | | | | | |
| Non-textured glass, 48" wide | BG@1.57 | Ea | 290.00 | 53.80 | 343.80 |
| Non-textured glass, 56" wide | BG@1.57 | Ea | 285.00 | 53.80 | 338.80 |
| Non-textured glass, 64" wide | BG@1.57 | Ea | 317.00 | 53.80 | 370.80 |
| Frosted opaque design, 48" wide | BG@1.57 | Ea | 379.00 | 53.80 | 432.80 |
| Frosted opaque design, 56" wide | BG@1.57 | Ea | 400.00 | 53.80 | 453.80 |
| Frosted, opaque design, 64" wide | BG@1.57 | Ea | 423.00 | 53.80 | 476.80 |
| **Tub doors, two sliding panels, 54" high, to 60" wide opening, with towel bar** | | | | | |
| Minimum quality, hammered glass | BG@1.77 | Ea | 112.00 | 60.60 | 172.60 |
| Better quality, textured glass | BG@1.77 | Ea | 163.00 | 60.60 | 223.60 |
| 5/32-inch glass, silk screen design | BG@1.77 | Ea | 320.00 | 60.60 | 380.60 |
| Frameless with 1/4" clear glass | BG@1.77 | Ea | 285.00 | 60.60 | 345.60 |
| Frameless with 1/4" box lite glass | BG@1.77 | Ea | 267.00 | 60.60 | 327.60 |
| Frameless with 1/4" delta frost glass | BG@1.77 | Ea | 313.00 | 60.60 | 373.60 |

**Shower Stalls**  No plumbing, piping or valves included.

Three-wall molded fiberglass shower stall with integrated soap shelf and textured bottom. White. 72" high. Add the cost of a shower door from the preceding section.

| | Craft@Hrs | Unit | Material | Labor | Total |
|---|---|---|---|---|---|
| 32" x 32" | P1@3.900 | Ea | 416.00 | 139.00 | 555.00 |
| 36" x 36" | P1@3.900 | Ea | 355.00 | 139.00 | 494.00 |
| 48" x 34" | P1@3.900 | Ea | 387.00 | 139.00 | 526.00 |
| 48" x 34" with bench seats | P1@3.900 | Ea | 427.00 | 139.00 | 566.00 |
| 60" x 34" with bench seats | P1@3.900 | Ea | 430.00 | 139.00 | 569.00 |

Neo-angle 2-wall molded fiberglass shower stalls with integrated soap shelf and textured bottom but no door or glass walls. White. 72" high.

| | Craft@Hrs | Unit | Material | Labor | Total |
|---|---|---|---|---|---|
| 38" x 38" | P1@2.50 | Ea | 420.00 | 89.40 | 509.40 |
| Add for neo-angle glass walls and pivot door | P1@1.40 | Ea | 232.00 | 50.10 | 282.10 |

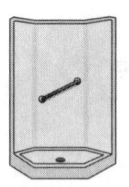

Neo-angle 4-wall molded fiberglass shower stalls with integrated soap shelf, textured bottom, glass walls and door. 72" high. Obscure glass in an aluminum frame.

| | Craft@Hrs | Unit | Material | Labor | Total |
|---|---|---|---|---|---|
| 38" x 38", chrome frame | P1@3.900 | Ea | 490.00 | 139.00 | 629.00 |
| 38" x 38", chrome frame, French glass | P1@3.900 | Ea | 492.00 | 139.00 | 631.00 |
| 42" x 42", chrome frame | P1@3.900 | Ea | 610.00 | 139.00 | 749.00 |
| 42" x 42", nickel frame | P1@3.900 | Ea | 695.00 | 139.00 | 834.00 |

Corner entry shower stall, 2 molded polystyrene walls and a two-panel aluminum-framed sliding glass door. With integrated corner shelves, textured base and drain. White with chrome frame.

| | Craft@Hrs | Unit | Material | Labor | Total |
|---|---|---|---|---|---|
| 32" x 32" x 74" | P1@3.250 | Ea | 292.00 | 116.00 | 408.00 |

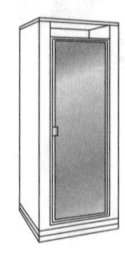

Round shower stall, white aluminum frame and walls, polystyrene sliding door and panel. With soap dishes, 6" textured base, drain and drain cover.

| | Craft@Hrs | Unit | Material | Labor | Total |
|---|---|---|---|---|---|
| 32" x 32" x 74" | P1@3.250 | Ea | 520.00 | 116.00 | 636.00 |

| | Craft@Hrs | Unit | Material | Labor | Total |
|---|---|---|---|---|---|

**Fiberglass shower receptors** One-piece molded fiberglass shower floor. Slip resistant with a starburst pattern. Drain molded-in with seal for 2" ABS, PVC, iron pipe and strainer. White. Single threshold 7" high.

| | Craft@Hrs | Unit | Material | Labor | Total |
|---|---|---|---|---|---|
| 36", neo-angle | PM@.600 | Ea | 150.00 | 25.10 | 175.10 |
| 38", neo-angle | PM@.800 | Ea | 170.00 | 33.50 | 203.50 |
| 32" x 32", square | PM@.600 | Ea | 96.00 | 25.10 | 121.10 |
| 36" x 36", square | PM@.600 | Ea | 105.00 | 25.10 | 130.10 |
| 32" x 48", rectangular | PM@.800 | Ea | 139.00 | 33.50 | 172.50 |
| 34" x 48", rectangular | PM@.800 | Ea | 150.00 | 33.50 | 183.50 |
| 34" x 60", rectangular | PM@.800 | Ea | 134.00 | 33.50 | 167.50 |
| Cultured marble receptor, 36" x 48" | PM@2.00 | Ea | 458.00 | 83.70 | 541.70 |

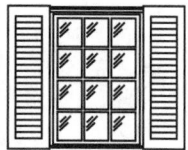

**Exterior shutters** Louvered co-polymer. Price per pair. If needed, add hinges and anchors from the section below.

| | Craft@Hrs | Unit | Material | Labor | Total |
|---|---|---|---|---|---|
| 15" x 39" | BC@.200 | Pr | 27.50 | 7.39 | 34.89 |
| 15" x 43" | BC@.200 | Pr | 29.50 | 7.39 | 36.89 |
| 15" x 48" | BC@.200 | Pr | 32.50 | 7.39 | 39.89 |
| 15" x 52" | BC@.320 | Pr | 34.60 | 11.80 | 46.40 |
| 15" x 55" | BC@.320 | Pr | 37.50 | 11.80 | 49.30 |
| 15" x 60" | BC@.320 | Pr | 40.50 | 11.80 | 52.30 |

**Installation of exterior shutters** Figures in the material column show the cost of anchors and hinges.

Installing shutters on hinges

| | Craft@Hrs | Unit | Material | Labor | Total |
|---|---|---|---|---|---|
| Wood frame construction, 1-3/8" throw | | | | | |
| 25" to 51" high | BC@.400 | Pr | 28.10 | 14.80 | 42.90 |
| 52" to 80" high | BC@.634 | Pr | 42.30 | 23.40 | 65.70 |
| Masonry construction, 4-1/4" throw | | | | | |
| 25" to 51" high | BC@.598 | Pr | 45.10 | 22.10 | 67.20 |
| 52" to 80" high | BC@.827 | Pr | 60.10 | 30.50 | 90.60 |
| Installing fixed shutters, any size | BC@.390 | Pr | 5.35 | 14.40 | 19.75 |

**Shutters, hurricane protection** Exterior shutters for residential and commercial applications. International Building Code Standard D-4355, Dade County Building Code standard E-1886. ASTM D-5621. Tested to ASTM specifications in excess of 150 MPH wind speed. Folding Shutter Corporation Armor Screen. Add $70 per day for a 30-foot rolling scissor lift, if required, for installation.

Roll-up hurricane shutters, corrugated metal. Includes retractor reel, enclosure box, hand crank, guide rails, above-window aluminum support bar with lag bolt plate and mounting bolts. Electric drive module includes drive motor, switch and 12-volt rechargeable battery pack for power out operation.

| | Craft@Hrs | Unit | Material | Labor | Total |
|---|---|---|---|---|---|
| Standard window and door sizes | B1@.015 | SF | 25.00 | .50 | 25.50 |
| Custom window and door sizes | B1@.015 | SF | 28.00 | .50 | 28.50 |
| Add for optional electric drive, battery & switch | — | Ea | 1,100.00 | — | 1,100.00 |
| Add for clear Lexan or equal ballistic plastic corrugation | — | % | 15.0 | — | — |
| Add for wind sensor module | — | % | 10.0 | — | — |

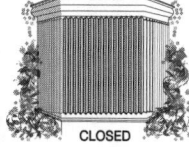

CLOSED

OPEN

Accordion-type hurricane shutters, flat metal or reinforced vinyl panels. Includes above-window guide-rail mount with bottom rail bearing track, vertical retractor reel, enclosure box, bearing-mounted guide rails, supporting mount plates with lag bolts and hand crank. Electric drive module includes drive motor, switch and 12-volt rechargeable battery pack for power out operation.

| | Craft@Hrs | Unit | Material | Labor | Total |
|---|---|---|---|---|---|
| Standard window and door sizes | B1@.01 | SF | 22.00 | .33 | 22.33 |
| Custom window and door sizes | B1@.01 | SF | 24.00 | .33 | 24.33 |
| Add for optional electric drive, battery & switch | — | Ea | 850.00 | — | 850.00 |
| Add for Lexan or equal ballistic plastic panels | — | % | 15.0 | — | — |
| Add for wind sensor module | — | % | 10.0 | — | — |

# Shutters, Hurricane Protection

| | Craft@Hrs | Unit | Material | Labor | Total |
|---|---|---|---|---|---|

Fixed corrugated panel hurricane shutters. Galvanized steel, ground anti-cut plastic-coated edges with F-shaped steel and paint-coated vertical mounting tracks. Includes studcons with wing nuts or lag bolt tie-downs and edge-seal water barrier stripping.

| | Craft@Hrs | Unit | Material | Labor | Total |
|---|---|---|---|---|---|
| Standard window and door sizes | B1@.015 | SF | 12.00 | .50 | 12.50 |
| Custom window and door sizes | B1@.015 | SF | 13.00 | .50 | 13.50 |
| Add for Lexan or equal ballistic plastic panels | — | % | 15.0 | — | — |

Bahama type coated metal, vinyl-coated wood or reinforced vinyl panels. Includes above-window single-panel swivel mount with bottom rail twin support outriggers.  Not suitable for covering doors. Secured with either lag bolts or studcons with wing nuts.

| | Craft@Hrs | Unit | Material | Labor | Total |
|---|---|---|---|---|---|
| Standard window sizes | B1@.012 | SF | 17.00 | .40 | 17.40 |
| Custom window sizes | B1@.012 | SF | 22.00 | .40 | 22.40 |
| Add for Lexan or equal ballistic plastic panels | — | % | 15.0 | — | — |

Colonial type hurricane shutters. Coated metal, vinyl-coated wood, or reinforced vinyl panels. Includes side swivel-mounted twin-panel with hold-open lag bolts or studcons with wing nuts. Deployed mode uses supplied hold-closed barrier bar with tie-down lag bolts or studcons with wing nuts.  Not suitable for covering doors.

| | Craft@Hrs | Unit | Material | Labor | Total |
|---|---|---|---|---|---|
| Standard window sizes | B1@.015 | SF | 11.00 | .50 | 11.50 |
| Custom window sizes | B1@.015 | SF | 13.00 | .50 | 13.50 |
| Add for Lexan or equal ballistic plastic panels | — | % | 15.2 | — | — |

Kevlar window and door hurricane protection. Padded flexible Kevlar or equal plastic fabric secured with lag bolts or studcons with wing nut tie-downs. Secured to wall studs or the window/door frame and removed.

| | Craft@Hrs | Unit | Material | Labor | Total |
|---|---|---|---|---|---|
| Standard window sizes | B1@.01 | SF | 24.00 | .33 | 24.33 |
| Custom window sizes | B1@.01 | SF | 26.00 | .33 | 26.33 |
| Add for transparent flexible panels | — | % | 10.0 | — | — |

Cover window or door with plywood panel. Single use, minimum 5/8 inch thickness. Installed with lag bolts or studcons with wing nut tie-downs. Secured to wall studs or the window/door frame and removed.

| | Craft@Hrs | Unit | Material | Labor | Total |
|---|---|---|---|---|---|
| Per SF or window or door covered | B1@.02 | SF | 0.50 | .67 | 1.17 |

**Siding**  See also Siding in the Lumber Section, Plywood and Hardboard.

**Fiber cement lap siding, HardiPlank®**  6-1/4" width requires 20 planks per 100 square feet of wall. 7-1/4" width requires 17 planks per 100 square feet of wall. 8-1/4" width requires 15 planks per 100 square feet of wall. 9-1/4" width requires 13 planks per 100 square feet of wall. 12" width requires 10 planks per 100 square feet of wall. Material price includes 10% for waste and fasteners.

| | Craft@Hrs | Unit | Material | Labor | Total |
|---|---|---|---|---|---|
| 6-1/4" x 12', cedarmill | B1@.048 | SF | 1.20 | 1.60 | 2.80 |
| 6-1/4" x 12', smooth | B1@.048 | SF | 1.24 | 1.60 | 2.84 |
| 7-1/4" x 12', cedarmill | B1@.048 | SF | 1.01 | 1.60 | 2.61 |
| 8-1/4" x 12', colonial roughsawn | B1@.048 | SF | 1.16 | 1.60 | 2.76 |
| 8-1/4" x 12', cedarmill | B1@.048 | SF | .99 | 1.60 | 2.59 |
| 8-1/4" x 12', beaded cedarmill | B1@.048 | SF | 1.25 | 1.60 | 2.85 |
| 8-1/4" x 12', beaded | B1@.048 | SF | 1.06 | 1.60 | 2.66 |
| 8-1/4" x 12', smooth | B1@.048 | SF | 1.21 | 1.60 | 2.81 |
| 9-1/4" x 12', cedarmill | B1@.046 | SF | 1.33 | 1.53 | 2.86 |
| 12" x 12', smooth | B1@.042 | SF | 1.20 | 1.40 | 2.60 |
| 12" x 12', cedarmill | B1@.042 | SF | 1.17 | 1.40 | 2.57 |

**Fiber cement panel siding**  Primed, HardiPanel®, 4' x 8' panels. Includes 10% for waste and fasteners.

| | Craft@Hrs | Unit | Material | Labor | Total |
|---|---|---|---|---|---|
| Sierra, Grooves 8" OC | BC@.042 | SF | .88 | 1.55 | 2.43 |
| Stucco | BC@.042 | SF | .93 | 1.55 | 2.48 |

| | Craft@Hrs | Unit | Material | Labor | Total |
|---|---|---|---|---|---|
| **Primed cement fiber trim, HardiTrim®, 12' lengths, per linear foot** | | | | | |
| Rustic, 3/4" x 4", 12' | BC@.032 | LF | .85 | 1.18 | 2.03 |
| Rustic, 3/4" x 6", 12' | BC@.032 | LF | 1.28 | 1.18 | 2.46 |
| **Primed cement fiber soffit, James Hardie** | | | | | |
| Vented, 1/4" x 12" wide | BC@.031 | LF | 1.50 | 1.15 | 2.65 |
| Solid, 1/4" x 4' x 8' panel | BC@.031 | SF | 1.04 | 1.15 | 2.19 |
| **Hardboard panel siding** Cost per square foot including 5% waste. 4' x 8' panels. | | | | | |
| 7/16" plain panel | B1@.022 | SF | .82 | .73 | 1.55 |
| 7/16" grooved 8" OC | B1@.022 | SF | 1.02 | .73 | 1.75 |
| 7/16" Sturdi-Panel grooved 8" OC | B1@.022 | SF | .65 | .73 | 1.38 |
| 7/16" cedar, grooved 8" OC | B1@.022 | SF | .76 | .73 | 1.49 |
| 7/16" cedar, plain, textured | B1@.022 | SF | .87 | .73 | 1.60 |
| 15/32" Duratemp, grooved 8" OC | B1@.022 | SF | 1.13 | .73 | 1.86 |
| Add for 4' x 9' panels | — | SF | .15 | — | .15 |
| **Hardboard lap siding** Primed plank siding. With 8% waste, 6" x 16' covers 5.6 square feet, 8" x 16' covers 7.4 square feet, 9-1/2" x 16' covers 9.7 square feet. | | | | | |
| 7/16" x 6" x 16', smooth lap | B1@.033 | SF | 1.02 | 1.10 | 2.12 |
| 7/16" x 6" x 16', textured cedar | B1@.033 | SF | 1.06 | 1.10 | 2.16 |
| 7/16" x 8" x 16', textured cedar | B1@.031 | SF | 1.37 | 1.03 | 2.40 |
| 7/16" x 8" x 16', smooth | B1@.031 | SF | 1.33 | 1.03 | 2.36 |
| 1/2" x 8" x 16', self-aligning | B1@.031 | SF | 1.88 | 1.03 | 2.91 |
| 1/2" x 8" x 16', old mill sure lock | B1@.031 | SF | 1.95 | 1.03 | 2.98 |
| 7/16" x 6" joint cover | — | Ea | .57 | — | .57 |
| 7/16" x 6" corner | — | Ea | .88 | — | .88 |
| 7/16" x 8" joint cover | — | Ea | 1.08 | — | 1.08 |
| 7/16" x 8" corner | — | Ea | 1.09 | — | 1.09 |
| 7/16" x 9-1/2" joint cover | — | Ea | 1.05 | — | 1.05 |
| 7/16" x 9-1/2" corner | — | Ea | 1.13 | — | 1.13 |
| **OSB siding** Oriented strand board. Includes 6% waste. | | | | | |
| Lap siding, smooth finish | | | | | |
| 3/8" x 8" x 16' (190 LF per Sq) | B1@.033 | SF | 1.22 | 1.10 | 2.32 |
| Lap siding, rough sawn finish | | | | | |
| 7/16" x 6" x 16' (250 LF per Sq) | B1@.033 | SF | 1.26 | 1.10 | 2.36 |
| 7/16" x 8" x 16' (190 LF per Sq) | B1@.031 | SF | 1.67 | 1.03 | 2.70 |
| Smart Panel II siding | | | | | |
| 3/8" x 4' x 8', smooth | B1@.022 | SF | .94 | .73 | 1.67 |
| 3/8" x 4' x 9', 8" grooved | B1@.022 | SF | 1.43 | .73 | 2.16 |
| 7/16" x 4' x 8', 4" grooved | B1@.022 | SF | 1.68 | .73 | 2.41 |
| 7/16" x 4' x 8', 8" grooved | B1@.022 | SF | 1.64 | .73 | 2.37 |
| 5/8" x 4' x 8', borate treated | B1@.022 | SF | 2.37 | .73 | 3.10 |
| Primed Smart Trim. Per linear foot of trim, including 10% waste. | | | | | |
| 4/4" x 4" x 16' | B1@.012 | LF | .90 | .40 | 1.30 |
| 4/4" x 6" x 16' | B1@.017 | LF | 1.39 | .57 | 1.96 |
| 4/4" x 8" x 16' | B1@.021 | LF | 1.86 | .70 | 2.56 |
| Pre-stained finish, Olympic machine coat, any of 85 colors, factory applied | | | | | |
| One-coat application, 5-year warranty | — | SF | .10 | — | .10 |
| Two-coat application, 10-year warranty | — | SF | .17 | — | .17 |

# Siding, Plywood

| | Craft@Hrs | Unit | Material | Labor | Total |
|---|---|---|---|---|---|
| Oriented strand board soffit panels. Textured Smart Soffit. Including 10% waste. | | | | | |
| 3/8" x 4" x 8' | B1@.033 | SF | 1.18 | 1.10 | 2.28 |
| OSB prefinished "splitless" and corrosion-resistant nails | | | | | |
| Siding nails (1.7 Lb per Sq) | — | SF | .05 | — | .05 |
| Trim nails (.85 Lb per 100 LF) | — | LF | .02 | — | .02 |
| Acrylic latex caulking, 1/4" bead | | | | | |
| Trim (1 tube per CLF) | B1@.001 | LF | .03 | .03 | .06 |
| Sheet metal flashing | | | | | |
| 7/8", 5/4", 1-5/8" x 10' | B1@.040 | LF | .62 | 1.33 | 1.95 |
| Brick mold, 10' lengths | B1@.040 | LF | 1.07 | 1.33 | 2.40 |
| Under eave soffit vents, screen type | | | | | |
| 4" or 8" x 16" | B1@.500 | Ea | 8.90 | 16.70 | 25.60 |
| Continuous aluminum soffit vents | | | | | |
| 2" x 8' | B1@.050 | LF | .83 | 1.67 | 2.50 |
| Touch-up stain, Behr, per coat, applied with brush or roller | | | | | |
| Prime coat (300 SF per gal) | PT@.006 | SF | .08 | .22 | .30 |
| Oil stain (400 SF per gal) | PT@.006 | SF | .06 | .22 | .28 |
| Penofin oil finish (400 SF per gal) | PT@.006 | SF | .10 | .22 | .32 |

**Plywood siding** 4' x 8' panels. Includes caulking, nails and 6% waste. For T1-11 siding, see Plywood in the Lumber section.

| | Craft@Hrs | Unit | Material | Labor | Total |
|---|---|---|---|---|---|
| 11/32" satin bead | B1@.025 | SF | .92 | .83 | 1.75 |
| 3/8" plybead, 16" grooved | B1@.025 | SF | .94 | .83 | 1.77 |
| 3/8" southern pine | B1@.025 | SF | .94 | .83 | 1.77 |
| 3/8" premium plain fir | B1@.025 | SF | 1.20 | .83 | 2.03 |
| 15/32" primed, 8" grooved | B1@.025 | SF | 1.15 | .83 | 1.98 |
| 15/32" fir, 8" grooved | B1@.025 | SF | 1.34 | .83 | 2.17 |
| 19/32" primed, 8" grooved | B1@.025 | SF | 1.60 | .83 | 2.43 |
| 5/8" rough sawn pine | B1@.025 | SF | 1.45 | .83 | 2.28 |
| 5/8" premium fir, #303-6 | B1@.025 | SF | 1.94 | .83 | 2.77 |

Yellow pine 4' x 8' plywood batten siding. Battens are 10" wide each 12" on center. Includes caulking, nails and 6% waste.

| | Craft@Hrs | Unit | Material | Labor | Total |
|---|---|---|---|---|---|
| 19/32" thick | B1@.025 | SF | 1.21 | .83 | 2.04 |
| 5/8" ACQ treated | B1@.025 | SF | 1.36 | .83 | 2.19 |

**Aluminum sheet siding**

Aluminum corrugated 4-V x 2-1/2", plain or embossed finish. Includes 15% waste and coverage loss.

| | Craft@Hrs | Unit | Material | Labor | Total |
|---|---|---|---|---|---|
| 17 gauge, 26" x 6' to 24' | B1@.034 | SF | 2.74 | 1.13 | 3.87 |
| 19 gauge, 26" x 6' to 24' | B1@.034 | SF | 2.84 | 1.13 | 3.97 |
| Rubber filler strip, 3/4" x 7/8" x 6' | — | LF | .39 | — | .39 |
| Flashing for corrugated aluminum siding, embossed | | | | | |
| End wall, 10" x 52" | B1@.048 | Ea | 4.90 | 1.60 | 6.50 |
| Side wall, 7-1/2" x 10' | B1@.015 | LF | 2.11 | .50 | 2.61 |
| Aluminum smooth 24 gauge, horizontal patterns, non-insulated | | | | | |
| 8" or double 4" widths, acrylic finish | B1@2.77 | Sq | 195.00 | 92.30 | 287.30 |
| 12" widths, bonded vinyl finish | B1@2.77 | Sq | 198.00 | 92.30 | 290.30 |
| Add for foam backing | — | Sq | 40.00 | — | 40.00 |
| Starter strip | B1@.030 | LF | .47 | 1.00 | 1.47 |
| Inside and outside corners | B1@.033 | LF | 1.41 | 1.10 | 2.51 |
| Casing and trim | B1@.033 | LF | .47 | 1.10 | 1.57 |
| Drip cap | B1@.044 | LF | .48 | 1.47 | 1.95 |

| | Craft@Hrs | Unit | Material | Labor | Total |
|---|---|---|---|---|---|
| **Galvanized steel siding** | | | | | |
| 28 gauge, 27-1/2" wide (includes 20% coverage loss) | | | | | |
|   6' to 12' standard lengths | B1@.034 | SF | 1.31 | 1.13 | 2.44 |
| 26 gauge, 26" wide (includes 15% coverage loss) | | | | | |
|   6' to 12' standard lengths | B1@.034 | SF | 1.60 | 1.13 | 2.73 |
| | | | | | |
| **Shingle siding** | | | | | |
| Cedar sidewall shingles, bundle covers 25 square feet, 16" long, 7-1/2" exposure, 5 shingles are 2" thick (5/2) | | | | | |
|   #2 Western red cedar | B1@3.86 | Sq | 200.00 | 129.00 | 329.00 |
|   Bleached cedar, rebutted and re-jointed | B1@3.86 | Sq | 307.00 | 129.00 | 436.00 |
|   Sanded cedar | B1@3.86 | Sq | 367.00 | 129.00 | 496.00 |
|   Grooved cedar | B1@3.86 | Sq | 366.00 | 129.00 | 495.00 |
|   Primed smooth cedar | B1@3.86 | Sq | 388.00 | 129.00 | 517.00 |
| | | | | | |
| **Panelized shingle siding**   Shakertown, KD, regraded shingle on plywood backing, no waste included. | | | | | |
| 8' long x 7" wide single course | | | | | |
|   Colonial 1 | B1@.033 | SF | 3.88 | 1.10 | 4.98 |
|   Cascade | B1@.033 | SF | 3.74 | 1.10 | 4.84 |
| 8' long x 14" wide, single course | | | | | |
|   Colonial 1 | B1@.033 | SF | 3.88 | 1.10 | 4.98 |
| 8' long x 14" wide, double course | | | | | |
|   Colonial 2 | B1@.033 | SF | 4.25 | 1.10 | 5.35 |
|   Cascade Classic | B1@.033 | SF | 3.74 | 1.10 | 4.84 |
| Decorative shingles, 5" wide x 18" long, Shakertown, fancy cuts, 9 hand-shaped patterns | | | | | |
|   7-1/2" exposure | B1@.035 | SF | 4.54 | 1.17 | 5.71 |
|   Deduct for 10" exposure | — | SF | -.49 | — | -.49 |
| | | | | | |
| **Vinyl siding**   Solid vinyl. Embossed wood grain texture. | | | | | |
| Double 4" traditional lap profile, 12' 6" long panels, 8" exposure. Each panel covers 8.33 SF before cutting waste. Case of 24 panels covers 200 square feet. Cost per square (100 square feet). | | | | | |
|   .040" thick, lighter colors | B1@3.34 | Sq | 59.10 | 111.00 | 170.10 |
|   .040" thick, darker colors and textures | B1@3.34 | Sq | 74.30 | 111.00 | 185.30 |
|   .042" thick, lighter colors | B1@3.34 | Sq | 78.40 | 111.00 | 189.40 |
|   .042" thick, darker colors and textures | B1@3.34 | Sq | 74.30 | 111.00 | 185.30 |
|   .044" thick, most colors and textures | B1@3.34 | Sq | 76.60 | 111.00 | 187.60 |
| Double 4.5" traditional lap profile, 12' 1" long panels, 9" exposure. Each panel covers 9.09 SF. Case of 22 panels covers 200 square feet. Cost per square (100 square feet). | | | | | |
|   .040" thick, most colors | B1@3.34 | Sq | 60.00 | 111.00 | 171.00 |
|   .040" thick, darker colors and textures | B1@3.34 | Sq | 63.20 | 111.00 | 174.20 |
| Double 5" traditional lap profile, 12' long panels, 10" exposure. Each panel covers 10 SF. Case of 20 panels covers 200 square feet. Cost per square (100 square feet). | | | | | |
|   .042" thick, lighter colors | B1@3.34 | Sq | 73.60 | 111.00 | 184.60 |
|   .042" thick, darker colors and textures | B1@3.34 | Sq | 81.90 | 111.00 | 192.90 |
| Trim for vinyl siding | | | | | |
|   Starter strip, 2-1/4" x 12.6' | B1@.030 | LF | .42 | 1.00 | 1.42 |
|   "J"-channel trim at wall openings, 12.5' long | B1@.033 | LF | .44 | 1.10 | 1.54 |
|   Outside or inside corner, 10' long | B1@.033 | LF | 1.86 | 1.10 | 2.96 |
|   Casing and trim, 12.5' long | B1@.033 | LF | .49 | 1.10 | 1.59 |
|   Mini mounting block, 4" x 6" | B1@.350 | Ea | 10.90 | 11.70 | 22.60 |
|   Vinyl "J"-block, 7" x 9" | B1@.350 | Ea | 11.70 | 11.70 | 23.40 |
| Add for R-3 insulated vinyl siding | B1@.350 | Sq | 33.20 | 11.70 | 44.90 |

# Skylights

| | Craft@Hrs | Unit | Material | Labor | Total |
|---|---|---|---|---|---|

**Thermoplastic resin siding** Nailite™ 9/10" thick panels installed over sheathing. Add for trim board, corners, starter strips, "J"-channels and mortar fill. Based on five nails per piece. Includes 10% waste.

| | Craft@Hrs | Unit | Material | Labor | Total |
|---|---|---|---|---|---|
| Hand-split shake, 41-3/8" x 18-3/4" | B1@.043 | SF | 2.32 | 1.43 | 3.75 |
| Hand-laid brick, 44-1/4" x 18-5/8" | B1@.043 | SF | 2.48 | 1.43 | 3.91 |
| Hand-cut stone, 44-1/4" x 18-5/8" | B1@.043 | SF | 2.48 | 1.43 | 3.91 |
| Rough sawn cedar, 59-1/4" x 15" | B1@.043 | SF | 2.58 | 1.43 | 4.01 |
| Perfection plus cedar, 36-1/8" x 15" | B1@.043 | SF | 2.58 | 1.43 | 4.01 |
| Ledge trim | | | | | |
| Brick ledge trim | B1@.033 | LF | 5.36 | 1.10 | 6.46 |
| Stone ledge trim | B1@.033 | LF | 5.36 | 1.10 | 6.46 |
| Corners | | | | | |
| Hand-split shake, 4" x 18" | B1@.030 | LF | 10.20 | 1.00 | 11.20 |
| Rough sawn cedar 3" x 26" | B1@.030 | LF | 10.80 | 1.00 | 11.80 |
| Hand-laid brick 4" x 18" | B1@.030 | LF | 10.80 | 1.00 | 11.80 |
| Hand-cut stone 4" x 18" | B1@.030 | LF | 10.80 | 1.00 | 11.80 |
| Perfection plus cedar 3" x 13" | B1@.030 | LF | 8.94 | 1.00 | 9.94 |
| Starter strips | | | | | |
| 12' universal | B1@.025 | LF | .55 | .83 | 1.38 |
| "J"-channels | | | | | |
| Hand-split shake 3/4" opening | B1@.025 | LF | .81 | .83 | 1.64 |
| Hand-split shake 1-1/4" flexible opening | B1@.027 | LF | 1.63 | .90 | 2.53 |
| Rough sawn cedar 3/4" opening | B1@.025 | LF | .61 | .83 | 1.44 |
| Rough sawn cedar 1-1/4" flexible opening | B1@.027 | LF | 1.63 | .90 | 2.53 |
| Hand-laid brick 3/4" opening | B1@.025 | LF | .61 | .83 | 1.44 |
| Hand-cut stone 3/4" opening | B1@.025 | LF | .61 | .83 | 1.44 |
| Perfection plus cedar 3/4" opening | B1@.025 | LF | .61 | .83 | 1.44 |
| Thermoplastic siding accessories | | | | | |
| Mortar fill, 12 tubes and 200 LF per carton | — | Ea | 47.10 | — | 47.10 |
| Touch up paint, 6 oz. aerosol can | — | Ea | 4.24 | — | 4.24 |

**Skylights** Polycarbonate dome, clear transparent or tinted, roof opening sizes, price each, Crestline Skylights, curb or flush mount, self-flashing. Comfort™

| | Craft@Hrs | Unit | Material | Labor | Total |
|---|---|---|---|---|---|
| Single dome, 2' x 2' | B1@2.58 | Ea | 69.00 | 85.90 | 154.90 |
| Single dome, 2' x 4' | B1@2.58 | Ea | 107.00 | 85.90 | 192.90 |
| Double dome, 4" x 4" | B1@2.84 | Ea | 199.00 | 94.60 | 293.60 |

Tempered over laminated low-E argon gas filled doubled insulated fixed glass skylights. Includes costs for flashing. Conforms to all building codes. For installation on roofs with 15- to 85-degree slope. Labor costs are for installation of skylight and flashing. Roll shade costs are for factory attached shades. No carpentry or roofing included.

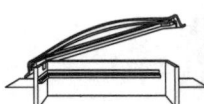

Operable skylight, Velux®, VS® Comfort™ models. Prices include EDL® flashing.

| | Craft@Hrs | Unit | Material | Labor | Total |
|---|---|---|---|---|---|
| 21-1/2" x 38-1/2", #104 | B1@3.42 | Ea | 815.00 | 114.00 | 929.00 |
| 21-1/2" x 46-3/8", #106 | B1@3.42 | Ea | 869.00 | 114.00 | 983.00 |
| 21-1/2" x 55", #108 | B1@3.42 | Ea | 931.00 | 114.00 | 1,045.00 |
| 30-5/8" x 38-1/2", #304 | B1@3.42 | Ea | 894.00 | 114.00 | 1,008.00 |
| 30-5/8" x 46-3/8", #306 | B1@3.42 | Ea | 931.00 | 114.00 | 1,045.00 |
| 30-5/8" x 55", #308 | B1@3.52 | Ea | 977.00 | 117.00 | 1,094.00 |
| 44-3/4" x 46-1/2", #606 | B1@3.52 | Ea | 1,050.00 | 117.00 | 1,167.00 |
| Add for electric operation | — | % | — | — | 200.0 |
| Add for ComfortPlus™ Laminated Low E | — | % | 10.0 | — | — |
| Add for roll shade, natural | | | | | |
| 21-1/2" x 27-1/2" to 55" width | — | Ea | 58.00 | — | 58.00 |
| 30-5/8" x 38-1/2" to 55" width | — | Ea | 66.70 | — | 66.70 |

| | Craft@Hrs | Unit | Material | Labor | Total |
|---|---|---|---|---|---|
| Fixed skylight, low-E laminated and tempered insulating glass. Wood frame. Aluminum exterior. | | | | | |
| 5-5/16" x 46-3/8" | B1@3.42 | Ea | 200.00 | 114.00 | 314.00 |
| 21-1/2" x 27-1/2" | B1@3.42 | Ea | 163.00 | 114.00 | 277.00 |
| 21-1/2" x 38-1/2" | B1@3.42 | Ea | 207.00 | 114.00 | 321.00 |
| 21-1/2" x 46-3/8" | B1@3.42 | Ea | 231.00 | 114.00 | 345.00 |
| 30-5/8" x 38-1/2" | B1@3.42 | Ea | 245.00 | 114.00 | 359.00 |
| 30-5/8" x 46-3/8" | B1@3.42 | Ea | 279.00 | 114.00 | 393.00 |
| 30-5/8" x 55" | B1@3.52 | Ea | 313.00 | 117.00 | 430.00 |
| Add for roll shade, natural | | | | | |
| 21-1/2" x 27-1/2" to 70-7/8" width | — | Ea | 62.00 | — | 62.00 |
| 30-5/8" x 38-1/2" to 55" width | — | Ea | 70.60 | — | 70.60 |

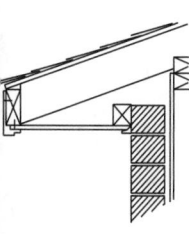

| | Craft@Hrs | Unit | Material | Labor | Total |
|---|---|---|---|---|---|
| Fixed ventilation skylight with removable filter on ventilation flap, Velux® model FSF® ComfortPlus™. Prices include EDL® flashing. | | | | | |
| 21-1/2" x 27-1/2", #101 | B1@3.14 | Ea | 226.00 | 105.00 | 331.00 |
| 21-1/2" x 38-1/2", #104 | B1@3.14 | Ea | 263.00 | 105.00 | 368.00 |
| 21-1/2" x 46-1/2", #106 | B1@3.14 | Ea | 298.00 | 105.00 | 403.00 |
| 21-1/2" x 70-7/8", #108 | B1@3.42 | Ea | 326.00 | 114.00 | 440.00 |
| 30-5/8" x 38-1/2", #304 | B1@3.14 | Ea | 321.00 | 105.00 | 426.00 |
| 30-5/8" x 46-3/8", #306 | B1@3.14 | Ea | 367.00 | 105.00 | 472.00 |
| 30-5/8" x 55", #308 | B1@3.39 | Ea | 399.00 | 113.00 | 512.00 |
| 44-3/4" x 27-1/2", #601 | B1@3.36 | Ea | 350.00 | 112.00 | 462.00 |
| 44-3/4" x 46-1/2", #606 | B1@3.39 | Ea | 461.00 | 113.00 | 574.00 |
| Add for roll shade, natural | | | | | |
| 21-1/2" x 27-1/2" to 70-7/8" width | — | Ea | 69.70 | — | 69.70 |
| 30-5/8" x 38-1/2" to 55" width | — | Ea | 78.80 | — | 78.80 |
| Curb mounted fixed skylight, Velux model FCM™ | | | | | |
| 22-1/2" x 22-1/2", #2222 | B1@3.14 | Ea | 124.00 | 105.00 | 229.00 |
| 22-1/2" x 34-1/2", #2234 | B1@3.14 | Ea | 164.00 | 105.00 | 269.00 |
| 22-1/2" x 46-1/2", #2246 | B1@3.14 | Ea | 178.00 | 105.00 | 283.00 |
| 30-1/2" x 30-1/2", #3030 | B1@3.14 | Ea | 197.00 | 105.00 | 302.00 |
| 46-1/2" x 46-1/2", #4646 | B1@3.14 | Ea | 302.00 | 105.00 | 407.00 |

**Top hinged roof windows** Aluminum-clad wood frame, double insulated tempered over laminated low-E gas filled double insulated glass, including prefabricated flashing, exterior awning, interior roller blind and insect screen. Sash rotates 180 degrees. For installation on roofs with 20- to 55-degree slope. Labor costs are for installation of roof window, flashing and sun screening accessories. No carpentry or roofing work other than curb included. Add for interior trim. Listed by actual unit dimensions, top hung, Velux® model GPL™ Comfort™ Roof Window.

| | Craft@Hrs | Unit | Material | Labor | Total |
|---|---|---|---|---|---|
| 30-5/8" wide x 55" high, #308 | B1@3.42 | Ea | 718.00 | 114.00 | 832.00 |
| 44-3/4" wide x 46-3/8" high, #606 | B1@3.52 | Ea | 800.00 | 117.00 | 917.00 |

**Soffit Systems**
Baked enamel finish, 6" fascia, "J"-channel, 8" perforated or solid soffit, .019 gauge aluminum

| | Craft@Hrs | Unit | Material | Labor | Total |
|---|---|---|---|---|---|
| 12" soffit | B1@.047 | LF | 3.98 | 1.57 | 5.55 |
| 18" soffit | B1@.055 | LF | 4.62 | 1.83 | 6.45 |
| 24" soffit | B1@.060 | LF | 5.25 | 2.00 | 7.25 |
| Vinyl soffit systems | | | | | |
| 12" soffit | B1@.030 | LF | 3.23 | 1.00 | 4.23 |
| 18" soffit | B1@.030 | LF | 3.71 | 1.00 | 4.71 |
| 24" soffit | B1@.030 | LF | 4.17 | 1.00 | 5.17 |
| Aluminum fascia alone | | | | | |
| 4" fascia | B1@.030 | LF | .97 | 1.00 | 1.97 |
| 6" fascia | B1@.030 | LF | 1.17 | 1.00 | 2.17 |
| 8" fascia | B1@.030 | LF | 1.38 | 1.00 | 2.38 |

# Soil Testing

| | Craft@Hrs | Unit | Material | Labor | Total |
|---|---|---|---|---|---|
| **Soil Testing** | | | | | |
| Field observation and testing | | | | | |
|    Soil technician (average) | — | Hr | — | — | 60.00 |
| Soil testing, per test | | | | | |
|    Moisture retention | — | Ea | — | — | 27.00 |
|    Bulk density (loose or core) | — | Ea | — | — | 25.00 |
|    Porosity | — | Ea | — | — | 40.00 |
|    Organic matter | — | Ea | — | — | 35.00 |
|    Plasticity (Liquid and Plastic Limit) | — | Ea | — | — | 50.00 |
|    Water holding capacity | — | Ea | — | — | 25.00 |
|    Herbicide detection | — | Ea | — | — | 50.00 |
|    Complete mineral analysis | — | Ea | — | — | 63.00 |

Foundation investigation (bearing capacity, lateral loads, piles, seismic, stability, settlement for design purposes) and preliminary soils investigation (depth of fill, classification and profile of soil, expansion, shrinkage, grading, drainage recommendations). Including drilling and sampling costs.

| | Craft@Hrs | Unit | Material | Labor | Total |
|---|---|---|---|---|---|
|    Drill rig (driller and helper) | — | Hr | — | — | 148.00 |
|    Staff engineer | — | Hr | — | — | 74.20 |
|    Project engineer, RCE | — | Hr | — | — | 83.70 |
|    Principal engineer | — | Hr | — | — | 140.00 |
|    Transportation | — | Mile | — | — | 1.10 |
|    Transportation | — | Hr | — | — | 10.00 |
|    Laboratory testing (general) | — | Hr | — | — | 70.00 |
|    Laboratory testing (varies on individual test basis) | — | Ea | — | — | 50.00 |
| Report preparation | | | | | |
|    Analysis, conclusions, recommendations | — | Ea | — | — | 2,120.00 |
|    Court preparation (Principal Engineer) | — | Hr | — | — | 160.00 |
|    Expert witness court appearance | — | Hr | — | — | 275.00 |
|    Add for hazardous studies | — | % | — | — | 20.0 |

**Soil Treatments, Subcontract** Costs based on total square footage inside of, and including, foundation. Includes inside and outside perimeter of foundation, interior foundation walls, and underneath all slabs. Three-plus year re-application guarantee against existence of termites. Use $300 as a minimum job charge.
Dursban T.C, Primus, or Equity. Mix applied per manufacturers' and state and federal specifications.

| | Craft@Hrs | Unit | Material | Labor | Total |
|---|---|---|---|---|---|
|    Complete application | — | SF | — | — | .42 |

**Solar Photovoltaic Electrical Systems, Subcontract** Costs are for solar electric generating systems for use where commercial power is not available. These estimates assume 8 clear sunlight hours per day. Costs will be higher where less sun time is available. Photocomm with Kyocera modules.
Small remote vacation home system (12 volt DC). 64 watt solar electric module (98 amp hours), digital voltage regulator, mounting rack and wiring.

| | Craft@Hrs | Unit | Material | Labor | Total |
|---|---|---|---|---|---|
|    384 watt-hours per day | — | LS | — | — | 1,420.00 |

Larger remote primary home system (120 volt AC). Includes eight L16 storage batteries (1,050 amp hours), voltage regulator, mounting rack, 2,400 watt DC to AC inverter.

| | Craft@Hrs | Unit | Material | Labor | Total |
|---|---|---|---|---|---|
|    2,880 watt-hours per day | — | LS | — | — | 8,430.00 |

Basic two room solar power package  Simpler Solar Systems
Includes three 48 watt PV panels, one 12 amp voltage regulator, one 220 Ah battery with hydrocaps, one 600 watt inverter, mounting frame and hardware, and two receptacles

| | Craft@Hrs | Unit | Material | Labor | Total |
|---|---|---|---|---|---|
|    12 amp regulator, with 600 watt inverter | — | LS | — | — | 3,400.00 |

|  | Craft@Hrs | Unit | Material | Labor | Total |
|---|---|---|---|---|---|

Small sized home solar power package. Simpler Solar Systems
90% solar powered system for 1,500 SF home includes six 60 watt PV panels, one 750 Ah battery, 20 amp voltage regulator and DC distribution box, and one 2.2 kW inverter. Costs also include one solar hot water system, one water source heat pump, one 19 cubic foot refrigerator freezer, one ringer type washing machine, six 12 volt lights and three 12 volt ceiling fans

| 20 amp regulator, 12 volt system | — | LS | — | — | 17,700.00 |
|---|---|---|---|---|---|

Medium sized home solar power package. Simpler Solar Systems. 80% solar powered system for 2,500 SF home includes ten 45 watt PV panels, one 750 Ah battery, one 30 amp voltage regulator, and one 2 kW inverter with charger. Costs also include one solar domestic hot water system and 22 assorted 12 volt light fixtures

| 30 amp regulator, 12 volt system | — | LS | — | — | 10,700.00 |
|---|---|---|---|---|---|

**Photovoltaic roofing systems** Photovoltaic panels and shingles. Converts sunlight into electric energy and distributes it directly to the building. The solar electric roofing products are configured in series or parallel on the roof deck to form an array. The array is then integrated into the roof following standard roofing specifications. Installation time does not include electrical connections.

| Architectural or structural standing photovoltaic seam panels | | | | | |
|---|---|---|---|---|---|
| 65 watt panel, 30" x 26" | RR@.035 | SF | 66.00 | 1.44 | 67.44 |
| 85 watt panel, 40" x 26" | RR@.035 | SF | 62.20 | 1.44 | 63.64 |
| 130 watt panel, 56" x 26" | RR@.033 | SF | 42.00 | 1.36 | 43.36 |
| 135 watt panel, 59" x 26" | RR@.033 | SF | 39.90 | 1.36 | 41.26 |
| Photovoltaic shingles, 12" x 86" panel | | | | | |
| 17 watt shingle panel, 3 SF exposure | RR@.030 | SF | 43.80 | 1.24 | 45.04 |
| Photovoltaic roofing system accessories | | | | | |
| Add for combiner box | — | Ea | 712.00 | — | 712.00 |
| Add for 2.8 kW inverter | — | Ea | 3,240.00 | — | 3,240.00 |
| Add for 3.6 kW inverter | — | Ea | 3,350.00 | — | 3,350.00 |
| Add for 6 volt DC 350 Ah storage batteries | — | Ea | 167.00 | — | 167.00 |
| Add for utility ground fault protection device | — | Ea | 346.00 | — | 346.00 |

**Solar pool heating system** Simpler Solar Systems

| Pool system, eight panels, valves, and piping | — | LS | — | — | 4,740.00 |
|---|---|---|---|---|---|
| Spa system, one panel, valves and piping | — | LS | — | — | 1,410.00 |

**Spas, fiberglass, outdoor** Includes pump and standard heater. Add the cost of a concrete slab base.

| 60" x 82", 31 jets | P1@3.50 | Ea | 3,200.00 | 125.00 | 3,325.00 |
|---|---|---|---|---|---|
| 73" x 73", 19 jets | P1@4.00 | Ea | 2,700.00 | 143.00 | 2,843.00 |
| 78" x 78", 25 jets | P1@4.00 | Ea | 3,000.00 | 143.00 | 3,143.00 |
| 92" x 92", 78 jets | P1@4.00 | Ea | 4,800.00 | 143.00 | 4,943.00 |
| Add for electrical connection | — | Ea | — | — | 230.00 |

**Stairs**

Factory cut and assembled straight closed box stairs, unfinished, 36" width

| Oak treads with 7-1/2" risers, price per riser | | | | | |
|---|---|---|---|---|---|
| 3'0" wide | B1@.324 | Ea | 112.00 | 10.80 | 122.80 |
| 3'6" wide | B1@.324 | Ea | 145.00 | 10.80 | 155.80 |
| 4'0" wide | B1@.357 | Ea | 157.00 | 11.90 | 168.90 |
| Deduct for pine treads and plywood risers | — | % | -22.0 | — | — |
| Add for prefinished assembled stair rail with balusters and newel, per riser | B1@.257 | Ea | 57.50 | 8.56 | 66.06 |
| Add for prefinished handrail, with brackets and balusters | B1@.135 | LF | 20.00 | 4.50 | 24.50 |

## Stairs

| | Craft@Hrs | Unit | Material | Labor | Total |
|---|---|---|---|---|---|
| **Basement stairs, open riser, delivered to job site assembled, 12 treads, 2' 11-3/4" wide** | | | | | |
| 13 risers at 7-13/16" riser height | B1@4.00 | Ea | 250.00 | 133.00 | 383.00 |
| 13 risers at 8" riser height | B1@4.00 | Ea | 256.00 | 133.00 | 389.00 |
| 13 risers at 8-3/16" riser height | B1@4.00 | Ea | 262.00 | 133.00 | 395.00 |
| Add for handrail | B1@.161 | LF | 10.40 | 5.36 | 15.76 |
| **Curved stair, clear oak, self-supporting, 8'6" radii, factory cut, unfinished, unassembled** | | | | | |
| **Open one side (1 side against wall)** | | | | | |
| 8'9" to 9'4" rise | B1@24.6 | Ea | 5,810.00 | 819.00 | 6,629.00 |
| 9'5" to 10'0" rise | B1@24.6 | Ea | 6,260.00 | 819.00 | 7,079.00 |
| 10'1" to 10'8" rise | B1@26.5 | Ea | 6,730.00 | 883.00 | 7,613.00 |
| 10'9" to 11'4" rise | B1@26.5 | Ea | 7,190.00 | 883.00 | 8,073.00 |
| **Open two sides** | | | | | |
| 8'9" to 9'4" rise | B1@30.0 | Ea | 10,400.00 | 999.00 | 11,399.00 |
| 9'5" to 10'0" rise | B1@30.0 | Ea | 11,200.00 | 999.00 | 12,199.00 |
| 10'1" to 10'8" rise | B1@31.9 | Ea | 11,800.00 | 1,060.00 | 12,860.00 |
| 10'9" to 11'4" rise | B1@31.9 | Ea | 12,600.00 | 1,060.00 | 13,660.00 |
| Add for newel posts | B1@.334 | Ea | 72.40 | 11.10 | 83.50 |
| **Spiral stairs, aluminum, delivered to job site unassembled, 7-3/4" or 8-3/4" riser heights, 5' diameter** | | | | | |
| 85-1/4" to 96-1/4" with 10 treads | B1@9.98 | Ea | 1,890.00 | 332.00 | 2,222.00 |
| 93" to 105" with 11 treads | B1@10.7 | Ea | 2,060.00 | 356.00 | 2,416.00 |
| 100-3/4" to 113-3/4" with 12 treads | B1@11.3 | Ea | 2,240.00 | 376.00 | 2,616.00 |
| 108-1/2" to 122-1/2" with 13 treads | B1@12.0 | Ea | 2,400.00 | 400.00 | 2,800.00 |
| 116-1/4" to 131-1/4" with 14 risers | B1@12.6 | Ea | 2,560.00 | 420.00 | 2,980.00 |
| 124" to 140" with 15 risers | B1@12.8 | Ea | 2,760.00 | 426.00 | 3,186.00 |
| 131-3/4" to 148-3/4" with 16 risers | B1@13.5 | Ea | 2,910.00 | 450.00 | 3,360.00 |
| 138-1/2" to 157-1/2" with 17 risers | B1@14.2 | Ea | 3,060.00 | 473.00 | 3,533.00 |
| 147-1/4" to 166-1/4" with 18 risers | B1@14.8 | Ea | 3,260.00 | 493.00 | 3,753.00 |
| Deduct for 4' diameter | — | % | -20.0 | — | — |
| Deduct for 4'6" diameter | — | % | -10.0 | — | — |
| Add for additional aluminum treads | — | Ea | 174.00 | — | 174.00 |
| Add for oak tread inserts | — | Ea | 72.20 | — | 72.20 |
| **Spiral stairs, red oak, FOB factory, double handrails, 5' diameter** | | | | | |
| 74" to 82" with 9 risers | B1@16.0 | Ea | 3,400.00 | 533.00 | 3,933.00 |
| 82" to 90" with 10 risers | B1@16.0 | Ea | 3,650.00 | 533.00 | 4,183.00 |
| 90" to 98" with 11 risers | B1@16.0 | Ea | 3,710.00 | 533.00 | 4,243.00 |
| 98" to 106" with 12 risers | B1@16.0 | Ea | 3,780.00 | 533.00 | 4,313.00 |
| 106" to 114" with 13 risers | B1@20.0 | Ea | 3,860.00 | 666.00 | 4,526.00 |
| 114" to 122" with 14 risers | B1@20.0 | Ea | 4,000.00 | 666.00 | 4,666.00 |
| 122" to 130" with 15 risers | B1@20.0 | Ea | 4,200.00 | 666.00 | 4,866.00 |
| Add for 6' diameter | — | % | 20.0 | — | — |
| Add for 8'6" diameter | — | % | 25.0 | — | — |
| Add for mahogany, any of above | — | % | 10.0 | — | — |
| Add for straight rail, unbored | B1@.060 | LF | 11.70 | 2.00 | 13.70 |
| Add for straight nosing, unbored | B1@.060 | LF | 7.40 | 2.00 | 9.40 |
| Add for balusters, 7/8" x 42" | B1@.334 | Ea | 4.76 | 11.10 | 15.86 |
| Add for newel posts 1-7/8" x 44" | B1@.334 | Ea | 28.60 | 11.10 | 39.70 |
| **Spiral stairs, steel, tubular steel handrail, composition board treads, delivered unassembled, 5' diameter** | | | | | |
| 86" to 95" with 10 risers | B1@9.98 | Ea | 1,140.00 | 332.00 | 1,472.00 |
| 95" to 105" with 11 risers | B1@10.7 | Ea | 1,270.00 | 356.00 | 1,626.00 |
| 105" to 114" with 12 risers | B1@11.3 | Ea | 1,340.00 | 376.00 | 1,716.00 |
| 114" to 124" with 13 risers | B1@12.0 | Ea | 1,480.00 | 400.00 | 1,880.00 |
| 124" to 133" with 14 risers | B1@12.7 | Ea | 1,700.00 | 423.00 | 2,123.00 |

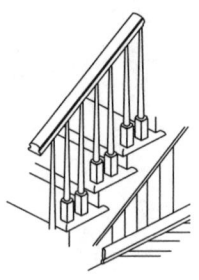

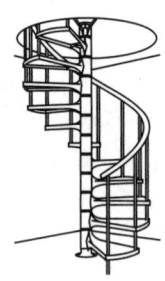

| | Craft@Hrs | Unit | Material | Labor | Total |
|---|---|---|---|---|---|
| Deduct for 4' diameter | — | % | -20.0 | — | — |
| Add for 6' diameter | — | % | 24.0 | — | — |
| Add for oak treads | — | % | 25.0 | — | — |
| Add for oak handrail | — | % | 45.0 | — | — |

Straight stairs, red oak, factory cut, unassembled , 36" tread, includes two 1-1/2" x 3-1/2" handrails and 7/8" round balusters.

| | Craft@Hrs | Unit | Material | Labor | Total |
|---|---|---|---|---|---|
| 70" to 77" with 10 risers | B1@12.0 | Ea | 1,560.00 | 400.00 | 1,960.00 |
| 77" to 85" with 11 risers | B1@12.0 | Ea | 1,760.00 | 400.00 | 2,160.00 |
| 85" to 93" with 12 risers | B1@12.0 | Ea | 2,070.00 | 400.00 | 2,470.00 |
| 93" to 100" with 13 risers | B1@14.0 | Ea | 2,080.00 | 466.00 | 2,546.00 |
| 100" to 108" with 14 risers | B1@14.0 | Ea | 2,270.00 | 466.00 | 2,736.00 |
| 108" to 116" with 15 risers | B1@14.0 | Ea | 2,370.00 | 466.00 | 2,836.00 |
| 116" to 124" with 16 risers | B1@16.0 | Ea | 2,590.00 | 533.00 | 3,123.00 |
| 124" to 131" with 17 risers | B1@16.0 | Ea | 2,670.00 | 533.00 | 3,203.00 |
| 131" to 139" with 18 risers | B1@16.0 | Ea | 2,990.00 | 533.00 | 3,523.00 |
| Add for 44" tread, any of above | — | Ea | 98.50 | — | 98.50 |
| Add for straight rail, unbored | B1@.060 | LF | 11.50 | 2.00 | 13.50 |
| Add for straight nosing | B1@.060 | LF | 7.24 | 2.00 | 9.24 |
| Add for round balusters, 7/8" x 42" | B1@.334 | Ea | 4.56 | 11.10 | 15.66 |
| Add for square balusters, 1-1/8" x 1-1/8" x 44" | B1@.334 | Ea | 9.38 | 11.10 | 20.48 |
| Add for newel posts, 3" x 3" x 44" | B1@.334 | Ea | 41.20 | 11.10 | 52.30 |

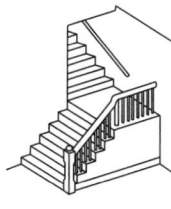

**Job-built stairways** 3 stringers cut from 2" x 12" Douglas fir material, treads and risers from 3/4" CDX plywood or OSB sheathing. Cost per 7-1/2" rise, 36" wide

| | Craft@Hrs | Unit | Material | Labor | Total |
|---|---|---|---|---|---|
| Using 2" Douglas fir, per MBF | — | MBF | 576.00 | — | 576.00 |
| Using 3/4" CDX plywood, per MSF | — | MSF | 746.00 | — | 746.00 |
| Using 3/4" OSB sheathing, per MSF | — | MSF | 544.00 | — | 544.00 |
| Using 1" x 6" white pine, per MBF | — | MBF | 1,870.00 | — | 1,870.00 |
| Straight run, 8'0" to 10'0" rise, plywood, per riser | B1@.530 | Ea | 12.00 | 17.70 | 29.70 |
| Straight run, 8'0" to 10'0" rise, OSB, per riser | B1@.530 | Ea | 10.30 | 17.70 | 28.00 |
| "L"- or "U"-shape, plywood, add for landings | B1@.625 | Ea | 13.30 | 20.80 | 34.10 |
| "L"- or "U"-shape, per riser, OSB, add for landings | B1@.625 | Ea | 11.60 | 20.80 | 32.40 |
| Semi-circular, repetitive work, plywood, per riser | B1@.795 | Ea | 13.30 | 26.50 | 39.80 |
| Semi-circular, repetitive work, OSB, per riser | B1@.795 | Ea | 11.60 | 26.50 | 38.10 |
| Landings, from 2" x 6" Douglas fir and 3/4" CDX plywood or 3/4" OSB, surfaced, per SF of landing surface, plywood | B1@.270 | SF | 2.40 | 8.99 | 11.39 |
| per SF of landing surface, OSB | B1@.270 | SF | 2.04 | 8.99 | 11.03 |

**Stair treads and risers** Unfinished. Including layout, cutting, fitting, and installation of treads, risers and typical tread nosing. Includes a miter return on one side of each tread. 9' rise is usually 14 treads and 15 risers

| | Craft@Hrs | Unit | Material | Labor | Total |
|---|---|---|---|---|---|
| Yellow pine treads, 11-1/2" x 48" | — | Ea | 13.20 | — | 13.20 |
| Oak treads, 11-1/2" x 48" | — | Ea | 21.90 | — | 21.90 |
| Oak curved first tread, 10-3/8" x 48" | — | Ea | 30.70 | — | 30.70 |
| Oak risers, 8" x 48" | — | Ea | 19.30 | — | 19.30 |
| Oak nosing | — | LF | 2.73 | — | 2.73 |
| Oak treads, risers and nosing for 9' rise stairway | BC@2.50 | Ea | 702.00 | 92.40 | 794.40 |
| Skirt board with dadoes for treads and risers | BC@3.75 | Ea | 29.30 | 139.00 | 168.30 |
| Skirt board on open side, mitered risers | BC@5.50 | Ea | 29.30 | 203.00 | 232.30 |
| Oak shoe rail with fillet, 1-1/4" plowed, 2-1/2" x 3/4" | BC@.030 | LF | 3.30 | 1.11 | 4.41 |
| Oak landing tread, 1" x 5-1/2" | BC@.180 | LF | 6.54 | 6.65 | 13.19 |

# Stair Rails, Posts and Balusters

| | Craft@Hrs | Unit | Material | Labor | Total |
|---|---|---|---|---|---|
| **False treads, 4-3/4" x 16-1/4" x 6"** | | | | | |
| Right end | BC@.180 | Ea | 14.00 | 6.65 | 20.65 |
| Left end | BC@.180 | Ea | 14.00 | 6.65 | 20.65 |
| Wall tread, 4-3/4" x 13-1/4" | BC@.180 | Ea | 16.80 | 6.65 | 23.45 |
| **Anti-skid adhesive strips. Peel and stick on wood, concrete or metal.** | | | | | |
| 2-3/4" x 14" | BC@.050 | Ea | 3.77 | 1.85 | 5.62 |
| 4" x 16" | BC@.050 | Ea | 5.92 | 1.85 | 7.77 |

**Stair rails, posts and balusters** Unfinished. Including layout, cutting, fitting and installation

| | Craft@Hrs | Unit | Material | Labor | Total |
|---|---|---|---|---|---|
| **Newel posts, 3" x 3" x 48" high** | | | | | |
| Half newel, ball top, hemlock | BC@2.50 | Ea | 26.00 | 92.40 | 118.40 |
| Half newel, ball top, oak | BC@2.50 | Ea | 32.40 | 92.40 | 124.80 |
| Starting newel, ball top, hemlock | BC@2.50 | Ea | 35.50 | 92.40 | 127.90 |
| Starting newel, ball top, oak | BC@2.50 | Ea | 49.80 | 92.40 | 142.20 |
| Starting newel, peg top, oak | BC@2.50 | Ea | 50.00 | 92.40 | 142.40 |
| Landing newel, ball top, hemlock | BC@2.50 | Ea | 43.30 | 92.40 | 135.70 |
| Landing newel, ball top, oak | BC@2.50 | Ea | 66.70 | 92.40 | 159.10 |
| Starting newel, peg top, oak | BC@2.50 | Ea | 70.50 | 92.40 | 162.90 |
| Starting newel, ball top, poplar | BC@2.50 | Ea | 42.30 | 92.40 | 134.70 |
| **Handrail, 2-3/8" D x 2-1/4" W, set between posts** | | | | | |
| Oak, set over newel posts | BC@.150 | LF | 6.33 | 5.54 | 11.87 |
| Oak, set between newel posts | BC@.250 | LF | 6.33 | 9.24 | 15.57 |
| Oak with fillet, set over newel posts | BC@.250 | LF | 7.47 | 9.24 | 16.71 |
| Hemlock, set between newel posts | BC@.150 | LF | 4.98 | 5.54 | 10.52 |
| Poplar, set between newel posts | BC@.150 | LF | 4.95 | 5.54 | 10.49 |
| Hemlock, set over newel posts | BC@.250 | LF | 4.95 | 9.24 | 14.19 |
| Poplar, set over newel posts | BC@.250 | LF | 4.95 | 9.24 | 14.19 |
| Flat bottom, set over newel posts | BC@.250 | LF | 3.09 | 9.24 | 12.33 |
| **Balusters, 1-1/4" D x 2-1/4" W, round top** | | | | | |
| Oak, 34" high | BC@.330 | Ea | 7.30 | 12.20 | 19.50 |
| Oak, 36" high | BC@.330 | Ea | 7.84 | 12.20 | 20.04 |
| Poplar, 34" high | BC@.330 | Ea | 4.95 | 12.20 | 17.15 |
| Poplar, 36" high | BC@.330 | Ea | 5.21 | 12.20 | 17.41 |
| Hemlock, 31" high, square top | BC@.330 | Ea | 4.73 | 12.20 | 16.93 |
| Hemlock, 34" high, tapered top | BC@.330 | Ea | 5.24 | 12.20 | 17.44 |
| Hemlock, 36" high, tapered top | BC@.330 | Ea | 5.89 | 12.20 | 18.09 |
| Hemlock, 36" high, square top | BC@.330 | Ea | 5.59 | 12.20 | 17.79 |
| Hemlock, 41" high, tapered top | BC@.330 | Ea | 6.54 | 12.20 | 18.74 |
| Hemlock, 41" high, square top | BC@.330 | Ea | 6.80 | 12.20 | 19.00 |
| **Handrail easings and transition fittings, oak** | | | | | |
| Gooseneck riser | BC@1.75 | LF | 79.90 | 64.60 | 144.50 |
| Upeasing | BC@1.75 | Ea | 23.60 | 64.60 | 88.20 |
| Opening cap | BC@1.75 | Ea | 19.90 | 64.60 | 84.50 |
| Quarter return | BC@1.75 | Ea | 19.70 | 64.60 | 84.30 |
| Volute | BC@1.75 | Ea | 86.80 | 64.60 | 151.40 |
| Turnout | BC@1.75 | Ea | 51.00 | 64.60 | 115.60 |
| Starting easing | BC@1.75 | Ea | 39.60 | 64.60 | 104.20 |
| Over easing | BC@1.75 | Ea | 24.90 | 64.60 | 89.50 |

| | Craft@Hrs | Unit | Material | Labor | Total |
|---|---|---|---|---|---|

## Stairs, attic

Economy folding stairway, tri-fold design, sanded plywood door, full width main hinge, fir door panel, 3/16" handrail, preassembled, 1" x 4" x 13" treads, 1" x 5" stringer, 22" x 54" opening size

| | Craft@Hrs | Unit | Material | Labor | Total |
|---|---|---|---|---|---|
| 7' to 8'9" ceiling height, 64" landing | B1@5.00 | Ea | 106.00 | 167.00 | 273.00 |
| 8' 9" to 10' ceiling height, 67" landing | B1@5.00 | Ea | 118.00 | 167.00 | 285.00 |

Economy all-aluminum tri-fold stairway, sanded plywood door, full width main hinge, double-riveted treads, 3/16" handrail, preassembled, 25" x 54" opening size

| | Craft@Hrs | Unit | Material | Labor | Total |
|---|---|---|---|---|---|
| 8'9" ceiling height, 64" landing | B1@5.00 | Ea | 172.00 | 167.00 | 339.00 |
| 10' ceiling height, 67" landing | B1@5.00 | Ea | 183.00 | 167.00 | 350.00 |

Good quality folding stairway, Bessler Stairways, full width main hinge, fir door panel, molded handrail, preassembled, 1" x 6" treads, 1" x 5" stringer and frame of select yellow pine, "space saver" for narrow openings, 22", 25-1/2", or 30" wide by 54" long

| | Craft@Hrs | Unit | Material | Labor | Total |
|---|---|---|---|---|---|
| 8'9" or 10' height | B1@5.00 | Ea | 384.00 | 167.00 | 551.00 |

Superior quality, Bessler Stairways, solid one-piece stringer sliding stairway, stringer and treads of clear yellow pine, unassembled, full-width piano hinge at head of stairs, with springs.

Compact model 26/26T, for a 22-1/2" x 54" to 22-1/2" x 74" opening, 7'7" to 9'3" height

| | Craft@Hrs | Unit | Material | Labor | Total |
|---|---|---|---|---|---|
| 1" x 4" stringers and treads | B1@6.00 | Ea | 686.00 | 200.00 | 886.00 |

Standard unit, for openings from 25-1/2" x 5'6" to 2'6" x 6'0", 7'7" to 10'10" heights

| | Craft@Hrs | Unit | Material | Labor | Total |
|---|---|---|---|---|---|
| 1" x 6" stringers and treads | B1@6.00 | Ea | 686.00 | 200.00 | 886.00 |

Heavy-duty unit, for openings from 2'6" x 5'10" to 2'6" x 8'0", 7'7" to 12'10" heights

| | Craft@Hrs | Unit | Material | Labor | Total |
|---|---|---|---|---|---|
| 1" x 8" stringers and treads | B1@9.00 | Ea | 1,310.00 | 300.00 | 1,610.00 |

**Steam Bath Generators** Steam bath equipment, 240 volts AC, single phase, UL approved. Costs include steam generator and controls. Labor costs include installation of steam unit and connections to enclosure, no rough plumbing or electrical included. Installed in existing, adequately sealed, tub enclosure. Maximum enclosure size as shown. www.amerec.com

| | Craft@Hrs | Unit | Material | Labor | Total |
|---|---|---|---|---|---|
| 100 cubic feet max | P1@10.0 | Ea | 947.00 | 358.00 | 1,305.00 |
| 160 cubic feet max | P1@10.0 | Ea | 1,030.00 | 358.00 | 1,388.00 |
| 220 cubic feet max | P1@10.0 | Ea | 1,040.00 | 358.00 | 1,398.00 |
| 300 cubic feet max | P1@10.0 | Ea | 1,150.00 | 358.00 | 1,508.00 |
| 400 cubic feet max | P1@10.0 | Ea | 1,190.00 | 358.00 | 1,548.00 |

## Tarpaulins

Polyethylene, heavy duty. Water repellant, with rustproof grommets, cut sizes

| | Craft@Hrs | Unit | Material | Labor | Total |
|---|---|---|---|---|---|
| 6' x  8' | — | Ea | 5.18 | — | 5.18 |
| 9' x  9' | — | Ea | 10.50 | — | 10.50 |
| 9' x 12' | — | Ea | 13.60 | — | 13.60 |
| 12" x 16' | — | Ea | 24.10 | — | 24.10 |
| 16' x 20' | — | Ea | 47.20 | — | 47.20 |
| 20' x 30' | — | Ea | 63.00 | — | 63.00 |
| 40' x 60' | — | Ea | 168.00 | — | 168.00 |

Canvas, 8 oz.

| | Craft@Hrs | Unit | Material | Labor | Total |
|---|---|---|---|---|---|
| 8' x 10' | — | Ea | 30.40 | — | 30.40 |
| 10' x 12' | — | Ea | 45.60 | — | 45.60 |
| 10' x 16' | — | Ea | 60.80 | — | 60.80 |
| 14' x 20' | — | Ea | 106.00 | — | 106.00 |
| 20' x 20' | — | Ea | 152.00 | — | 152.00 |

|  | Craft@Hrs | Unit | Material | Labor | Total |
|---|---|---|---|---|---|

**Taxes, Payroll**

Employer's cost for payroll taxes and insurance, expressed as a percent of total payroll. See also, Insurance for more detailed information.

Rule of thumb: When actual payroll taxes and insurance are not known, estimate that for each $1 paid in wages the employer must pay an additional $.30

| | | | | | |
|---|---|---|---|---|---|
| Payroll taxes and insurance | — | % | — | — | 30.0 |

Specific taxes, percent of total payroll

| | | | | | |
|---|---|---|---|---|---|
| Typical SUI, State unemployment insurance | — | % | — | — | 5.40 |
| FICA Social Security 6.20%, Medicare 1.45% | — | % | — | — | 7.65 |
| FUTA, Federal Unemployment Insurance | — | % | — | — | .80 |

**Thresholds**

Aluminum threshold with vinyl feet

| | | | | | |
|---|---|---|---|---|---|
| 3-3/8" x 5/8" x 36", economy | BC@.515 | Ea | 8.08 | 19.00 | 27.08 |
| 3-3/4" x 1" x 36", bronze | BC@.515 | Ea | 10.90 | 19.00 | 29.90 |
| 3-3/4" x 1" x 36", silver | BC@.515 | Ea | 16.30 | 19.00 | 35.30 |
| 3-1/2" x 5/8" x 72", brown | BC@.730 | Ea | 28.20 | 27.00 | 55.20 |
| 3-1/2" x 5/8" x 72", silver | BC@.730 | Ea | 24.30 | 27.00 | 51.30 |
| 3-1/2" x 5/8" x 72", gold | BC@.730 | Ea | 28.40 | 27.00 | 55.40 |

Aluminum bumper threshold with replaceable vinyl insert and vinyl feet

| | | | | | |
|---|---|---|---|---|---|
| 3-3/16" x 5/8" x 36", silver | BC@.515 | Ea | 15.20 | 19.00 | 34.20 |
| 3-3/8" x 1" x 36" silver | BC@.515 | Ea | 12.50 | 19.00 | 31.50 |
| 3-3/8" x 1" x 36", gold | BC@.515 | Ea | 18.70 | 19.00 | 37.70 |
| 3-1/2" x 3/4" x 36", silver | BC@.515 | Ea | 10.50 | 19.00 | 29.50 |
| 3-1/2" x 3/4" x 36", gold | BC@.515 | Ea | 19.40 | 19.00 | 38.40 |
| 3-3/8" x 1" x 72", silver | BC@.730 | Ea | 28.20 | 27.00 | 55.20 |

Aluminum saddle interior threshold, smooth flat top, covers floor seams

| | | | | | |
|---|---|---|---|---|---|
| 1-3/4" x 1/8" x 36", silver | BC@.515 | Ea | 9.67 | 19.00 | 28.67 |
| 1-3/4" x 1/8" x 36", gold | BC@.515 | Ea | 11.90 | 19.00 | 30.90 |
| 2-1/2" x 1/4" x 36", gold | BC@.515 | Ea | 9.87 | 19.00 | 28.87 |

Aluminum commercial fluted saddle threshold, UL fire labeled

| | | | | | |
|---|---|---|---|---|---|
| 5" x 1/2" x 36", mill finish | BC@.515 | Ea | 21.20 | 19.00 | 40.20 |
| 5" x 1/2" x 72", mill finish | BC@.730 | Ea | 36.90 | 27.00 | 63.90 |

Aluminum adjustable threshold, height adjusts from 7/8" to 1-5/16", vinyl top insert and vinyl feet

| | | | | | |
|---|---|---|---|---|---|
| 3-1/2" x 36", silver | BC@.515 | Ea | 23.50 | 19.00 | 42.50 |

Aluminum saddle exterior threshold, 1" minimum door clearance, for under-door and drip cap door bottoms.

| | | | | | |
|---|---|---|---|---|---|
| 3-1/2" x 1" x 36", silver | BC@.515 | Ea | 12.40 | 19.00 | 31.40 |
| 4" x 1" x 36", silver | BC@.515 | Ea | 17.50 | 19.00 | 36.50 |

Aluminum saddle exterior threshold, with vinyl bottom insert

| | | | | | |
|---|---|---|---|---|---|
| 5" x 36", chrome | BC@.515 | Ea | 21.00 | 19.00 | 40.00 |
| 5" x 36", gold | BC@.515 | Ea | 21.50 | 19.00 | 40.50 |

Oak and aluminum adjustable threshold, height adjusts from 1-1/8" to 1-3/8"

| | | | | | |
|---|---|---|---|---|---|
| 5-5/8" x 36", silver | BC@.515 | Ea | 29.90 | 19.00 | 48.90 |

Oak bumper threshold, clear vinyl insert

| | | | | | |
|---|---|---|---|---|---|
| 3-1/2" x 1" x 36" | BC@.515 | Ea | 22.70 | 19.00 | 41.70 |

Oak low boy threshold

| | | | | | |
|---|---|---|---|---|---|
| 3-1/2" x 3/4" x 36" | BC@.515 | Ea | 13.20 | 19.00 | 32.20 |
| 3-1/2" x 3/4" x 72" | BC@.730 | Ea | 25.20 | 27.00 | 52.20 |

Aluminum under-door threshold kit, with vinyl insert.

| | | | | | |
|---|---|---|---|---|---|
| 1-1/4" x 36", silver | BC@.515 | Ea | 7.75 | 19.00 | 26.75 |
| 3-3/4" x 36", gold | BC@.515 | Ea | 21.90 | 19.00 | 40.90 |

| | Craft@Hrs | Unit | Material | Labor | Total |
|---|---|---|---|---|---|
| **Vinyl replacement threshold insert** | | | | | |
| 1-7/8" x 36", gray | BC@.100 | Ea | 3.11 | 3.69 | 6.80 |
| 1-1/2" x 36", brown | BC@.100 | Ea | 3.36 | 3.69 | 7.05 |
| **Oak sill, adjustable, commercial grade, adjusts to accommodate gaps** | | | | | |
| 1-1/8" x 4-9/16" x 36", in-swing | BC@.750 | Ea | 25.40 | 27.70 | 53.10 |
| 1-1/8" x 4-9/16" x 36", out-swing | BC@.750 | Ea | 26.70 | 27.70 | 54.40 |
| **Aluminum sill nosing** | | | | | |
| 1-1/2" x 2-3/4" x 36", silver | BC@.381 | Ea | 6.72 | 14.10 | 20.82 |
| 2-3/4" x 4-1/2" x 36", bronze | BC@.381 | Ea | 10.50 | 14.10 | 24.60 |
| 2-3/4" x 1-1/2" x 72", mill | BC@.540 | Ea | 9.86 | 19.90 | 29.76 |
| **Aluminum sill cover, for use over rough concrete sills** | | | | | |
| 2-3/4" x 1-1/2" x 36" | BC@.381 | Ea | 13.20 | 14.10 | 27.30 |
| 2-3/4" x 1-1/2" x 72" | BC@.540 | Ea | 13.60 | 19.90 | 33.50 |
| **Aluminum sill edging** | | | | | |
| 2-3/4" x 1-1/2" x 36" | BC@.381 | Ea | 12.10 | 14.10 | 26.20 |
| **Aluminum door shoe, with vinyl insert** | | | | | |
| 1-3/8" x 36", brown | BC@.381 | Ea | 12.30 | 14.10 | 26.40 |
| 1-3/4" x 36" brown, with drip cap | BC@.381 | Ea | 15.20 | 14.10 | 29.30 |
| **Aluminum "L"-shape door bottom, with drip cap and vinyl weather seal** | | | | | |
| 1-3/4" x 36", gold | BC@.381 | Ea | 12.00 | 14.10 | 26.10 |
| **Aluminum "U"-shape door bottom, for exterior doors, with vinyl insert** | | | | | |
| 1-3/4" x 36" | BC@.381 | Ea | 12.40 | 14.10 | 26.50 |
| 1-3/4" x 36", with drip cap | BC@.381 | Ea | 15.20 | 14.10 | 29.30 |
| **Aluminum commercial door bottom, UL fire labeled** | | | | | |
| 36", mill finish | BC@.381 | Ea | 13.40 | 14.10 | 27.50 |
| 36", bronze anodized | BC@.381 | Ea | 15.00 | 14.10 | 29.10 |
| **Brush door bottom, with brush sweep** | | | | | |
| 1-3/4" x 36", vinyl | BC@.381 | Ea | 8.94 | 14.10 | 23.04 |
| 2-1/8" x 36", rubber | BC@.381 | Ea | 8.48 | 14.10 | 22.58 |
| 2" x 36", wood | BC@.381 | Ea | 13.50 | 14.10 | 27.60 |
| **Aluminum drip cap door bottom** | | | | | |
| 1-3/4" x 36", silver | BC@.381 | Ea | 12.80 | 14.10 | 26.90 |
| 1-1/2" x 36", heavy duty | BC@.381 | Ea | 13.40 | 14.10 | 27.50 |
| **Vinyl slide-on door bottom** | | | | | |
| 1-3/4" x 36" | BC@.381 | Ea | 11.60 | 14.10 | 25.70 |
| **Aluminum and vinyl door bottom** | | | | | |
| 36", aluminum | BC@.381 | Ea | 12.50 | 14.10 | 26.60 |
| **Aluminum door bottom and sweep, interior or exterior doors** | | | | | |
| 2" x 36", heavy duty | BC@.381 | Ea | 13.10 | 14.10 | 27.20 |
| **Oak door sweep, unfinished** | | | | | |
| 36" wide | BC@.381 | Ea | 12.50 | 14.10 | 26.60 |

**Tile, Ceramic** Costs shown are typical costs for the tile only, including 10% allowance for breakage and cutting waste. Add costs for mortar or adhesive, grout and labor to install tile from the section that follows tile material costs.

Glazed ceramic wall tile (5/16" thickness)

| | | | | | |
|---|---|---|---|---|---|
| 4-1/4" x 4-1/4", 8 tiles cover 1 SF | | | | | |
| Minimum quality | — | SF | 2.00 | — | 2.00 |
| Good quality | — | SF | 3.20 | — | 3.20 |
| Glass tile (typical) | — | SF | 16.00 | — | 16.00 |

# Tile, Ceramic

| | Craft@Hrs | Unit | Material | Labor | Total |
|---|---|---|---|---|---|
| **6" x 6", 4 tiles cover 1 SF** | | | | | |
| Minimum quality | — | SF | 2.30 | — | 2.30 |
| Good quality | — | SF | 2.85 | — | 2.85 |
| Better quality | — | SF | 4.00 | — | 4.00 |
| **Ceramic mosaic tile, face-mounted sheets, 1/4" x 12" x 12"** | | | | | |
| 1" hexagonal | | | | | |
| Minimum quality | — | SF | 2.12 | — | 2.12 |
| Good quality | — | SF | 2.87 | — | 2.87 |
| Iridescent glass | — | SF | 5.00 | — | 5.00 |
| 2" hexagonal or square | | | | | |
| Minimum quality, white | — | SF | 2.70 | — | 2.70 |
| Good quality, pergamo | — | SF | 3.63 | — | 3.63 |
| Better quality, travertine | — | SF | 5.00 | — | 5.00 |
| Mixed mosaic, montagna | — | SF | 10.00 | — | 10.00 |
| Metal sheet | — | SF | 15.60 | — | 15.60 |
| 3/4" square | | | | | |
| Light brown iridescent | — | SF | 5.00 | — | 5.00 |
| Square cobalt blue | — | SF | 6.85 | — | 6.85 |
| Beige multi mosaic | — | SF | 12.50 | — | 12.50 |
| **Ceramic fixtures** | | | | | |
| 5-piece porcelain fixture set (soap dish, toothbrush holder, tub soap dish, toilet paper holder, 24" towel bar) | TL@1.55 | Set | 23.20 | 54.90 | 78.10 |
| Toothbrush/tumbler holder | TL@.250 | SF | 19.80 | 8.86 | 28.66 |
| Soap dish | TL@.250 | SF | 7.56 | 8.86 | 16.42 |
| Towel bar assembly, 24" long | TL@.550 | SF | 15.50 | 19.50 | 35.00 |
| Bath corner shelf, 8" x 8" | TL@.500 | SF | 20.90 | 17.70 | 38.60 |
| Toilet tissue holder, 4-1/4" x 6" | TL@.250 | SF | 10.20 | 8.86 | 19.06 |
| **Glazed ceramic floor tile** | | | | | |
| 8" x 8", 2.3 tiles cover 1 square foot | | | | | |
| Minimum quality | — | SF | 1.58 | — | 1.58 |
| Better quality | — | SF | 3.22 | — | 3.22 |
| 12" x 12", 1 tile per SF | | | | | |
| Minimum quality | — | SF | 1.02 | — | 1.02 |
| Good quality – Porcelain | — | SF | 1.87 | — | 1.87 |
| Better quality – Alabaster | — | SF | 2.25 | — | 2.25 |

**Tile, Stone and Marble** Prices and availability can vary widely. Costs include a 5% allowance for breakage and cutting waste. Add costs for mortar or adhesive, grout and labor to install tile from the section that follows.

| | Craft@Hrs | Unit | Material | Labor | Total |
|---|---|---|---|---|---|
| **Slate tile, .40" x 12" x 12"** | | | | | |
| Minimum quality gauged slate | — | SF | 2.01 | — | 2.01 |
| Good quality gauged slate | — | SF | 2.41 | — | 2.41 |
| Better quality gauged slate, 16" x 16" | — | SF | 2.81 | — | 2.81 |
| Minimum quality natural slate | — | SF | 1.89 | — | 1.89 |
| Better quality natural slate | — | SF | 6.06 | — | 6.06 |
| **Natural stone tile, 3/8" x 12" x 12"** | | | | | |
| Minimum quality | — | SF | 1.66 | — | 1.66 |
| Better quality, polished | — | SF | 6.43 | — | 6.43 |
| **Paver tile, 3/4" x 12" x 12"** | | | | | |
| Mexican red pavers | — | SF | 1.23 | — | 1.23 |

| | Craft@Hrs | Unit | Material | Labor | Total |
|---|---|---|---|---|---|
| **Marble tile, 3/8" x 12" x 12", polished one face** | | | | | |
| Minimum quality | — | SF | 2.76 | — | 2.76 |
| Good quality | — | SF | 4.27 | — | 4.27 |
| Top quality | — | SF | 17.00 | — | 17.00 |
| **Travertine beige marble tile, 3/8" x 18" x 18", 2.25 SF per tile** | | | | | |
| Minimum quality | — | SF | 4.27 | — | 4.27 |
| Better quality | — | SF | 7.13 | — | 7.13 |
| **Carrera white marble tile, 3/8" x 12" x 12"** | | | | | |
| White, imported | — | SF | 5.50 | — | 5.50 |
| **Granite 3/8" tile, 12" x 12", polished on one face** | | | | | |
| Minimum quality | — | SF | 3.21 | — | 3.21 |
| Good quality | — | SF | 7.43 | — | 7.43 |
| Better quality black granite | — | SF | 10.00 | — | 10.00 |
| Best quality black granite | — | SF | 13.60 | — | 13.60 |
| **Granite, 3/4", cut to size, imported** | — | SF | 26.80 | — | 26.80 |

**Tile installation** The costs below are for setting and grouting tile only. Add the cost of preparing the surface for installation (tile backer) and the tile. See Floor Leveling for related costs to patch and level floor; See Membrane Ditra for detachable membrane costs.

Ceramic tile adhesive. Premixed Type 1, Acryl-4000. 1 gallon covers 70 SF applied with #1 trowel (3/16" x 5/32" "V"-notch), 50 SF applied with #2 trowel (3-/16" x 1/4" "V"-notch) or 40 SF applied with #3 trowel (1/4" x 1/4" square-notch).

| | Craft@Hrs | Unit | Material | Labor | Total |
|---|---|---|---|---|---|
| 1 quart | — | Ea | 6.06 | — | 6.06 |
| 1 gallon | — | Ea | 12.60 | — | 12.60 |
| 3-1/2 gallons | — | Ea | 32.50 | — | 32.50 |
| Tile adhesive applied with #2 trowel | — | SF | .23 | — | .23 |

Thin-set tile mortar. 50 pound bag covers 100 SF applied with #1 trowel (1/4" x 1/4" square-notch), 80 SF applied with #2 trowel (1/4" x 3/8" square-notch), 45 SF applied with #3 trowel (1/2" x 1/2" square notch). Cost per 50 pound bag.

| | Craft@Hrs | Unit | Material | Labor | Total |
|---|---|---|---|---|---|
| Standard, gray | — | Ea | 5.93 | — | 5.93 |
| Standard, white | — | Ea | 8.37 | — | 8.37 |
| VersaBond, gray | — | Ea | 14.70 | — | 14.70 |
| VersaBond, white | — | Ea | 16.80 | — | 16.80 |
| FlexiBond, gray | — | Ea | 28.30 | — | 28.30 |
| Marble and granite mix | — | Ea | 23.90 | — | 23.90 |
| Master-Blend applied with #3 trowel | — | SF | .25 | — | .25 |

Tile grout. Polymer-modified dry grout for joints 1/16" to 1/2". Coverage varies with tile and joint size. Cost per 25 pound bag.

| | Craft@Hrs | Unit | Material | Labor | Total |
|---|---|---|---|---|---|
| Unsanded | — | Ea | 12.40 | — | 12.40 |
| Sanded, light colors | — | Ea | 13.90 | — | 13.90 |
| Sanded, dark colors | — | Ea | 13.90 | — | 13.90 |

**Installation of tile in adhesive** Includes adhesive and grout. Add the cost of preparing the surface for installation (tile backer) and the cost of tile.
Countertops

| | Craft@Hrs | Unit | Material | Labor | Total |
|---|---|---|---|---|---|
| Ceramic mosaic field tile | TL@.203 | SF | .28 | 7.19 | 7.47 |
| 4-1/4" x 4-1/4" to 6" x 6" glazed field tile | TL@.180 | SF | .28 | 6.38 | 6.66 |
| Countertop trim pieces and edge tile | TL@.180 | LF | .05 | 6.38 | 6.43 |

# Tile Installation

| | Craft@Hrs | Unit | Material | Labor | Total |
|---|---|---|---|---|---|
| **Floors** | | | | | |
| Ceramic mosaic field tile | TL@.121 | SF | .28 | 4.29 | 4.57 |
| 4-1/4" x 4-1/4" to 6" x 6" glazed field tile | TL@.110 | SF | .28 | 3.90 | 4.18 |
| Floor trim pieces and edge tile | TL@.110 | LF | .05 | 3.90 | 3.95 |
| **Walls** | | | | | |
| Ceramic mosaic field tile | TL@.143 | SF | .28 | 5.07 | 5.35 |
| 4-1/4" x 4-1/4" to 6" x 6" glazed field tile | TL@.131 | SF | .28 | 4.64 | 4.92 |
| Wall trim pieces and edge tile | TL@.131 | LF | .05 | 4.64 | 4.69 |

**Installation of tile in thin-set mortar**  Includes grout. Costs are for mortar and grout. Add the cost of surface preparation (backerboard) and tile.

| | Craft@Hrs | Unit | Material | Labor | Total |
|---|---|---|---|---|---|
| **Countertops** | | | | | |
| Ceramic mosaic field tile | TL@.407 | SF | .35 | 14.40 | 14.75 |
| 4-1/4" x 4-1/4" to 6" x 6" glazed field tile | TL@.352 | SF | .35 | 12.50 | 12.85 |
| Countertop trim pieces and edge tile | TL@.352 | LF | .06 | 12.50 | 12.56 |
| **Floors** | | | | | |
| Ceramic mosaic field tile | TL@.238 | SF | .35 | 8.43 | 8.78 |
| 4-1/4" x 4-1/4" to 6" x 6" glazed field tile | TL@.210 | SF | .35 | 7.44 | 7.79 |
| 8" x 8" to 12" x 12" glazed field tile | TL@.200 | SF | .35 | 7.09 | 7.44 |
| Quarry or paver tile | TL@.167 | SF | .35 | 5.92 | 6.27 |
| Marble or granite, 3/8" thick | TL@.354 | SF | .35 | 12.50 | 12.85 |
| Marble or granite, 3/4" thick | TL@.591 | SF | .35 | 20.90 | 21.25 |
| Floor trim pieces and edge tile | TL@.210 | LF | .06 | 7.44 | 7.50 |
| **Walls** | | | | | |
| Ceramic mosaic field tile | TL@.315 | SF | .35 | 11.20 | 11.55 |
| 4-1/4" x 4-1/4" to 6" x 6" glazed field tile | TL@.270 | SF | .35 | 9.57 | 9.92 |
| Wall trim pieces and edge tile | TL@.270 | LF | .06 | 9.57 | 9.63 |

**Tile backerboard** (Durock™ or Wonderboard™).  Water-resistant underlayment for ceramic tile on floors, walls, countertops and other interior wet areas. Material cost for 100 square feet (CSF) of board includes waterproofing membrane, the backerboard (and 10% waste), 50 pounds of job mixed latex-fortified mortar for the joints and surface skim coat, 75 linear feet of fiberglass joint tape, 200 1-1/4" backerboard screws. For scheduling purposes, estimate that a crew of 2 can install, tape and apply the skim coat on the following quantity of backerboard in an 8-hour day: countertops 180 SF, floors 525 SF, and walls 350 SF. Use $250.00 as a minimum charge for this type work.

| | Craft@Hrs | Unit | Material | Labor | Total |
|---|---|---|---|---|---|
| Using 15 Lb. felt waterproofing, per 400 SF roll | — | Roll | 14.00 | — | 14.00 |
| Using 1/4" or 1/2" backerboard, per 100 SF | — | CSF | 76.40 | — | 76.40 |
| Using 50 Lb. of latex-fortified mortar each 100 SF, per 50 Lb. sack | — | Sack | 14.70 | — | 14.70 |
| Using 75 LF of backerboard tape each 100 SF per 50 LF roll | — | Roll | 4.17 | — | 4.17 |
| Using 200 backerboard screws for each 100 SF, per pack of 200 screws | — | Pack | 4.19 | — | 4.19 |
| Backerboard with waterproofing, mortar, tape and screws | | | | | |
| Floors | T1@.030 | SF | 1.16 | .98 | 2.14 |
| Walls | T1@.045 | SF | 1.16 | 1.47 | 2.63 |
| Countertops | T1@.090 | SF | 1.16 | 2.93 | 4.09 |

| | Craft@Hrs | Unit | Material | Labor | Total |
|---|---|---|---|---|---|

**Vacuum Cleaning Systems, Central** Subcontract. Costs listed are for better quality central vacuum power unit, piping to 4 remote outlets, hose and attachments (adequate for typical 2,000 SF home). Assumes that all vacuum tubes are installed before any interior finishes are applied.

| | Craft@Hrs | Unit | Material | Labor | Total |
|---|---|---|---|---|---|
| Basic package with 23' hose | — | LS | — | — | 1,600.00 |
| Additional remote outlets | — | Ea | — | — | 150.00 |
| Add for electric or turbo power head | — | Ea | — | — | 155.00 |

**Vanity cabinets** Sink base with no countertop, no lavatory bowl and no plumbing included. Prefinished. See also Countertops and Plumbing. 3/4" furniture-grade sides with veneer interior, overlay doors and drawers, and concealed hinges, 31-1/2" high.

Door only sink base vanity cabinets, white exterior finish. Maple interior finish, concealed hinges.

| | Craft@Hrs | Unit | Material | Labor | Total |
|---|---|---|---|---|---|
| 18" x 16", 1 door | BC@.400 | Ea | 66.70 | 14.80 | 81.50 |
| 24" x 18", 2 doors | BC@.400 | Ea | 88.30 | 14.80 | 103.10 |
| 30" x 18", 2 doors | BC@.400 | Ea | 98.60 | 14.80 | 113.40 |

Standard quality vanity cabinets, veneer doors, drawer front(s) and face frame

| | | | | | |
|---|---|---|---|---|---|
| 24" x 18", 1 door, 2 drawers | BC@.400 | Ea | 150.00 | 14.80 | 164.80 |
| 30" x 18", 1 door, 2 drawers | BC@.400 | Ea | 167.00 | 14.80 | 181.80 |
| 36" x 18", 2 doors, 2 drawers | BC@.400 | Ea | 189.00 | 14.80 | 203.80 |

Good quality vanity cabinets, white or hardwood exterior finish, ball bearing drawer glides

| | | | | | |
|---|---|---|---|---|---|
| 24" x 21", 1 door, 1-drawer | BC@.400 | Ea | 259.00 | 14.80 | 273.80 |
| 30" x 21", 1 door, 4-drawers | BC@.400 | Ea | 267.00 | 14.80 | 281.80 |
| 36" x 21", 2 doors, 4 drawers | BC@.400 | Ea | 298.00 | 14.80 | 312.80 |
| 48" x 21", 2 doors, 7 drawers | BC@.500 | Ea | 352.00 | 18.50 | 370.50 |

Premium quality vanity cabinets, cinnamon oak doors and drawer fonts, adjustable hinges

| | | | | | |
|---|---|---|---|---|---|
| 24" x 21", 1 door, 2 drawer | BC@.400 | Ea | 232.00 | 14.80 | 246.80 |
| 30" x 21", 1 door, 2 drawer | BC@.400 | Ea | 309.00 | 14.80 | 323.80 |
| 36" x 21", 1 door, 2 drawers | BC@.400 | Ea | 366.00 | 14.80 | 380.80 |
| 48" x 21", 1 door, 4 drawers | BC@.500 | Ea | 484.00 | 18.50 | 502.50 |
| 60" x 21", 2 doors, 4 drawers | BC@.500 | Ea | 530.00 | 18.50 | 548.50 |

Over the toilet vanity cabinet, 8" deep x 24" high

| | | | | | |
|---|---|---|---|---|---|
| 24" wide, 2 doors, 1 shelf | B1@1.37 | Ea | 106.00 | 45.60 | 151.60 |
| 30" wide, 2 doors, 1 shelf | B1@1.37 | Ea | 143.00 | 45.60 | 188.60 |

## Vents and Louvers

Attic and gable louver vents, UV stabilized copolymer, screened, rough opening sizes

| | | | | | |
|---|---|---|---|---|---|
| 12" x 12", 44 sq. inch vent area | SW@.448 | Ea | 35.30 | 18.40 | 53.70 |
| 12" x 18", 86 sq. inch vent area | SW@.448 | Ea | 37.30 | 18.40 | 55.70 |
| 12" x 24", 140 sq. inch vent area | SW@.871 | Ea | 53.00 | 35.70 | 88.70 |
| 21" x 27" oval | SW@.871 | Ea | 89.00 | 35.70 | 124.70 |
| 16" x 24", half round, 100 sq. inch vent area | SW@.871 | Ea | 74.20 | 35.70 | 109.90 |
| 16" round, 50 sq. inch vent area | SW@.448 | Ea | 47.70 | 18.40 | 66.10 |
| 12" x 18", round top 50 sq. inch vent area | SW@.448 | Ea | 44.40 | 18.40 | 62.80 |
| 16" octagon, 54 sq. inch vent area | SW@.871 | Ea | 47.70 | 35.70 | 83.40 |

Chimney caps

| | | | | | |
|---|---|---|---|---|---|
| 9" x 9" rectangular | SW@.324 | Ea | 29.70 | 13.30 | 43.00 |
| 9" x 13" rectangular | SW@.324 | Ea | 33.30 | 13.30 | 46.60 |
| 13" x 13" rectangular | SW@.324 | Ea | 34.50 | 13.30 | 47.80 |
| 13" x 13" rectangular, stainless | SW@.324 | Ea | 43.00 | 13.30 | 56.30 |
| 8" x 8" to 13" x 13" adjustable | SW@.324 | Ea | 32.30 | 13.30 | 45.60 |
| 13" x 18" fixed, black | SW@.324 | Ea | 34.90 | 13.30 | 48.20 |

## Vents and Louvers

| | Craft@Hrs | Unit | Material | Labor | Total |
|---|---|---|---|---|---|
| **Clothes dryer vent set, aluminum, with hood, inside plate, 8' of flexible duct pipe, clamps and collar** | | | | | |
| 4" diameter vent kit | SW@.440 | Ea | 13.40 | 18.10 | 31.50 |
| Flex hose, 4" diameter, 8' length | — | Ea | 16.60 | — | 16.60 |
| Flex hose clamps, 4" diameter | — | Ea | 1.71 | — | 1.71 |
| **Code caps** | | | | | |
| 4" diameter, top only | SW@.324 | Ea | 23.30 | 13.30 | 36.60 |
| 6" diameter, top only | SW@.324 | Ea | 29.60 | 13.30 | 42.90 |
| 8" diameter, top only | SW@.324 | Ea | 36.00 | 13.30 | 49.30 |
| 6" diameter shake base | SW@.324 | Ea | 27.70 | 13.30 | 41.00 |
| 8" diameter shake base | SW@.324 | Ea | 28.90 | 13.30 | 42.20 |
| **Dormer louver vents, galvanized, 1/4" mesh screen** | | | | | |
| 19" x 3", low profile, rectangular | BC@.871 | Ea | 39.10 | 32.20 | 71.30 |
| 24" x 12", half round | BC@.871 | Ea | 40.30 | 32.20 | 72.50 |
| Add for 1/8" mesh | — | % | 3.0 | — | — |
| **Under-eave louver vents, rectangular, with 1/4" mesh screen, 16" wide** | | | | | |
| 16" x 4", mill finish | SW@.255 | Ea | 1.52 | 10.50 | 12.02 |
| 16" x 8", mill finish | SW@.255 | Ea | 1.92 | 10.50 | 12.42 |
| 16" x 8", brown finish | SW@.255 | Ea | 1.96 | 10.50 | 12.46 |
| 16" x 4", white finish | SW@.255 | Ea | 1.67 | 10.50 | 12.17 |
| 16" x 8", white finish | SW@.255 | Ea | 1.96 | 10.50 | 12.46 |
| **Foundation screen vents, galvanized, with screen, no louvers** | | | | | |
| 14" x 6" | SW@.255 | Ea | 3.38 | 10.50 | 13.88 |
| 16" x 4" | SW@.255 | Ea | 2.87 | 10.50 | 13.37 |
| 16" x 6" | SW@.255 | Ea | 3.12 | 10.50 | 13.62 |
| 16" x 8" | SW@.255 | Ea | 3.51 | 10.50 | 14.01 |

| | Craft@Hrs | Unit | Material | Labor | Total |
|---|---|---|---|---|---|
| **Automatic foundation vent, fits 8" x 16" concrete block opening, aluminum mesh on front and molded plastic screen on back. Opens at 72 degrees, closes at 38 degrees. 1" frame each side** | | | | | |
| Automatic open and close | SW@.255 | Ea | 16.70 | 10.50 | 27.20 |
| **Foundation vent with manual damper, sheet metal** | | | | | |
| 4" x 16" | SW@.255 | Ea | 6.89 | 10.50 | 17.39 |
| 6" x 16" | SW@.255 | Ea | 7.53 | 10.50 | 18.03 |
| 8" x 16" | SW@.255 | Ea | 7.97 | 10.50 | 18.47 |
| **Foundation access door** | | | | | |
| 24" x 24" | BC@.363 | Ea | 30.00 | 13.40 | 43.40 |
| 32" x 24" | BC@.363 | Ea | 35.50 | 13.40 | 48.90 |
| **Heater closet door vents, louver** | | | | | |
| 70 square inches, louvers only | SW@.220 | Ea | 6.11 | 9.03 | 15.14 |
| 70 square inches, louvers and screen | SW@.255 | Ea | 11.30 | 10.50 | 21.80 |
| 38 square inches, louvers only | SW@.190 | Ea | 5.93 | 7.80 | 13.73 |
| 38 square inches, louvers and screen | SW@.210 | Ea | 8.98 | 8.62 | 17.60 |
| **Circular louver vents, for moisture control in sidewalls, aluminum mill finish** | | | | | |
| 2" circular vent | SW@.255 | Ea | 1.54 | 10.50 | 12.04 |
| 3" circular vent | SW@.255 | Ea | 3.18 | 10.50 | 13.68 |
| 4" circular vent | SW@.255 | Ea | 2.88 | 10.50 | 13.38 |
| **Rafter eve vents, galvanized steel with nailing flange, 1/8" mesh, galvanized, louver** | | | | | |
| 14" x 5" | SW@.220 | Ea | 4.17 | 9.03 | 13.20 |
| 14" x 6" | SW@.220 | Ea | 5.16 | 9.03 | 14.19 |
| 22" x 3" | SW@.220 | Ea | 4.06 | 9.03 | 13.09 |
| 22" x 6" | SW@.220 | Ea | 5.36 | 9.03 | 14.39 |
| 22" x 7" | SW@.220 | Ea | 5.36 | 9.03 | 14.39 |

|  | Craft@Hrs | Unit | Material | Labor | Total |
|---|---|---|---|---|---|
| Rafter bay vent channel, keeps attic insulation away from vents, promotes circulation. | | | | | |
| Rafters 16" on center, | | | | | |
| 14" wide, 15" net-free vent area | BC@.220 | Ea | 1.93 | 8.13 | 10.06 |
| Rafters 24" on center, | | | | | |
| 22" wide, 26" net-free vent area | BC@.220 | Ea | 2.46 | 8.13 | 10.59 |

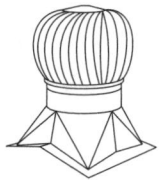

| Ridge ventilators, 1/8" louver opening, double baffle, aluminum | | | | | |
|---|---|---|---|---|---|
| 10' long, black | BC@.650 | Ea | 17.80 | 24.00 | 41.80 |
| 10' long, brown | BC@.650 | Ea | 17.10 | 24.00 | 41.10 |
| 10' long, white | BC@.650 | Ea | 17.20 | 24.00 | 41.20 |
| 4' long, hinged for steep roof | BC@.255 | Ea | 11.60 | 9.42 | 21.02 |
| Joint strap | BC@.010 | LF | 1.30 | .37 | 1.67 |
| End connector plug | BC@.010 | Ea | 2.34 | .37 | 2.71 |

| Roof vents, with mesh screen, for roof venting of kitchen and bath exhaust fans | | | | | |
|---|---|---|---|---|---|
| Square cap type, 26" L x 23" D x 5" H | SW@.448 | Ea | 53.80 | 18.40 | 72.20 |
| Dormer type, 20" L x 10" D x 6" H | SW@.448 | Ea | 35.40 | 18.40 | 53.80 |

| Rotary roof ventilators, with base, by turbine diameter | | | | | |
|---|---|---|---|---|---|
| 12", galvanized | SW@.694 | Ea | 28.30 | 28.50 | 56.80 |
| 12", brown | SW@.694 | Ea | 35.60 | 28.50 | 64.10 |
| 14", galvanized | SW@.694 | Ea | 36.90 | 28.50 | 65.40 |
| 14", brown | SW@.694 | Ea | 38.70 | 28.50 | 67.20 |
| 16", galvanized | SW@.930 | Ea | 111.00 | 38.20 | 149.20 |
| 18", galvanized | SW@1.21 | Ea | 128.00 | 49.70 | 177.70 |

| Round louver vents, 1/8" mesh, galvanized | | | | | |
|---|---|---|---|---|---|
| 12" or 14" diameter | SW@.440 | Ea | 46.50 | 18.10 | 64.60 |
| 16" diameter | SW@.440 | Ea | 53.50 | 18.10 | 71.60 |
| 18" diameter | SW@.440 | Ea | 59.00 | 18.10 | 77.10 |
| 24" diameter | SW@.661 | Ea | 107.00 | 27.10 | 134.10 |

| Soffit vents, aluminum louvers, 8' long, reversible | | | | | |
|---|---|---|---|---|---|
| 2-5/8" wide, white finish | SW@.440 | Ea | 5.20 | 18.10 | 23.30 |
| 2-5/8" wide, mill finish | SW@.440 | Ea | 3.49 | 18.10 | 21.59 |

| Wall louvers, reversible for flush or recessed mounting. White finish. Screened | | | | | |
|---|---|---|---|---|---|
| 12" x 12", aluminum | SW@.440 | Ea | 10.30 | 18.10 | 28.40 |
| 12" x 18", aluminum | SW@.440 | Ea | 12.50 | 18.10 | 30.60 |
| 14" x 24", aluminum | SW@.440 | Ea | 19.20 | 18.10 | 37.30 |
| 18" x 24", aluminum | SW@.440 | Ea | 22.10 | 18.10 | 40.20 |
| 12" x 18", plastic | SW@.440 | Ea | 8.60 | 18.10 | 26.70 |
| 18" x 24", plastic | SW@.440 | Ea | 14.60 | 18.10 | 32.70 |

**Wallboard Partitions, Subcontract** Costs for interior partitions in one- and two-story buildings. 1/4" gypsum wallboard on 2 sides with "V"-grooved edges and plastic moulding at corners fastened to 25 gauge metal framing at 24" OC, including framing, per LF of wall.

|  | Craft@Hrs | Unit | Material | Labor | Total |
|---|---|---|---|---|---|
| Vinyl covered 8' high | — | LF | — | — | 95.10 |
| Vinyl covered 10' high | — | LF | — | — | 128.00 |
| Paintable paper-covered 8' high | — | LF | — | — | 69.30 |
| Paintable paper-covered 10' high | — | LF | — | — | 103.00 |
| Add for glued vinyl base on two sides | — | LF | — | — | 3.75 |

# Wallcoverings

| | Craft@Hrs | Unit | Material | Labor | Total |
|---|---|---|---|---|---|

**Wallcoverings** Costs listed are for good to better quality materials installed on a clean, smooth surface and include adhesive, edge trimming, pattern matching, and normal cutting and fitting waste. Surface preparation such as patching and priming are extra. Typical roll is 8 yards long and 18" wide (approximately 36 square feet).

| | Craft@Hrs | Unit | Material | Labor | Total |
|---|---|---|---|---|---|
| Scrape off old wallpaper, up to 2 layers | PP@.015 | SF | — | .50 | .50 |
| Paper stripper (1,000 SF per gallon) | — | Gal | 24.20 | — | 24.20 |
| Patch holes, scrape marks | PP@.010 | SF | .03 | .33 | .36 |
| Blank stock (underliner) | | | | | |
| Bath, kitchen, and laundry room | PP@.668 | Roll | 8.08 | 22.40 | 30.48 |
| Other rooms | PP@.558 | Roll | 8.08 | 18.70 | 26.78 |
| Textured basket weave | | | | | |
| Bath, kitchen, and laundry rooms | PP@.995 | Roll | 19.60 | 33.30 | 52.90 |
| Other rooms | PP@.830 | Roll | 19.60 | 27.80 | 47.40 |
| Papers, vinyl-coated papers, trimmed, and pre-pasted | | | | | |
| Bath, kitchen, and laundry rooms | PP@.884 | Roll | 20.20 | 29.60 | 49.80 |
| Other rooms | PP@.780 | Roll | 20.20 | 26.10 | 46.30 |
| Misc. patterns | | | | | |
| Bath, kitchen, and laundry rooms | PP@1.12 | Roll | 11.10 | 37.50 | 48.60 |
| Other rooms | PP@.844 | Roll | 11.10 | 28.20 | 39.30 |
| Designer paper and hand-painted prints | | | | | |
| Bath, kitchen, and laundry rooms | PP@1.68 | Roll | 55.60 | 56.20 | 111.80 |
| Other rooms | PP@1.41 | Roll | 55.60 | 47.20 | 102.80 |
| Murals, no pattern | | | | | |
| Solid vinyl | PP@.844 | Yd | 19.20 | 28.20 | 47.40 |
| Coordinating borders, vinyl coated | | | | | |
| roll is 4" to 6" x 12' to 15' | PP@.017 | LF | .76 | .57 | 1.33 |
| Wallpaper adhesive | | | | | |
| Clear, strippable (300 SF per gallon) | — | Gal | 7.22 | — | 7.22 |
| Clear, strippable, per SF | — | SF | .02 | — | .02 |
| Mildew-resistant (250 SF per gallon) | — | Gal | 13.60 | — | 13.60 |
| Adhesive, per SF | — | SF | .05 | — | .05 |

## Waterproofing

Below grade waterproofing, applied to concrete or masonry walls, no surface preparation, excavation, or backfill included, typical job

| | Craft@Hrs | Unit | Material | Labor | Total |
|---|---|---|---|---|---|
| Bentonite waterproofing | | | | | |
| One layer, nailed in place | BL@.079 | SF | 1.25 | 2.34 | 3.59 |
| Cementitious waterproofing with protection board | | | | | |
| Spray or brush applied, 2 coats | BL@.077 | SF | 1.21 | 2.28 | 3.49 |
| Crystalline waterproofing | | | | | |
| Spray or brush applied, 3 coats | BL@.110 | SF | 1.63 | 3.26 | 4.89 |
| Elastomeric waterproofing 30 mm per coat | | | | | |
| Sprayed, troweled, or rolled, 2 coats | BL@.065 | SF | 1.30 | 1.93 | 3.23 |

Above grade waterproofing, applied to smooth concrete or plywood deck, no surface preparation included, typical job

| | Craft@Hrs | Unit | Material | Labor | Total |
|---|---|---|---|---|---|
| Between slabs membrane | | | | | |
| Sprayed, troweled or rolled, 2 coats | PT@.047 | SF | .92 | 1.75 | 2.67 |
| Pedestrian walking surface, elastomeric membrane, | | | | | |
| Applied 4 coats thick with aggregate | PT@.086 | SF | 1.18 | 3.20 | 4.38 |
| Chloroprene/Neoprene latex deck | | | | | |
| Surfacing | PT@.144 | SF | 2.49 | 5.37 | 7.86 |

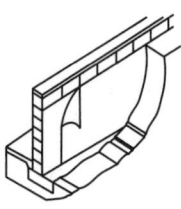

| | Craft@Hrs | Unit | Material | Labor | Total |
|---|---|---|---|---|---|

**Waterproofing system for foundations** Wrapped over concrete foundation footing. System should be tied into a perimeter drain system at or around the footing. Cover top of system with at least 6 inches of soil and backfill immediately after installation. No surface preparation, excavation, or backfill included.

| | Craft@Hrs | Unit | Material | Labor | Total |
|---|---|---|---|---|---|
| MiraDRI™ 860 | | | | | |
| rubberized asphalt membrane | B1@.075 | SF | .62 | 2.50 | 3.12 |
| Latex-based primer (400 SF per gal) | B1@.075 | SF | .16 | 2.50 | 2.66 |
| LM-800 liquid membrane | B1@.034 | SF | .13 | 1.13 | 1.26 |
| Carlisle 704 mastic, | | | | | |
| 1/8" x 3/4", 130 LF per gallon | B1@.034 | LF | .21 | 1.13 | 1.34 |
| Mira Drain™ 6200 (200 SF Roll) | B1@.075 | SF | .86 | 2.50 | 3.36 |
| Quick Drain™ (16" x 50' Roll) | B1@.075 | SF | 3.10 | 2.50 | 5.60 |

**Weatherstripping Materials**

| | Craft@Hrs | Unit | Material | Labor | Total |
|---|---|---|---|---|---|
| Door bottom seals, brass | | | | | |
| Brass "U"-shaped door bottom with adjustment for gaps | | | | | |
| 36" x 1-3/8" | BC@.250 | Ea | 17.30 | 9.24 | 26.54 |
| 36" x 1-3/4" | BC@.250 | Ea | 19.10 | 9.24 | 28.34 |
| Door frame seal sets | | | | | |
| Spring bronze, nailed every 1-3/8 inch | | | | | |
| 36" x 6'8" door | BC@.500 | Ea | 19.40 | 18.50 | 37.90 |
| 36" x 7'0" door | BC@.500 | Ea | 20.60 | 18.50 | 39.10 |
| Rigid PVC frame, with silicon bulb / screws | BC@.008 | Ea | 13.70 | .30 | 14.00 |
| Header and jamb side stripping, adhesive (Peel N Stick®) | | | | | |
| 36" x 6'8" door, brass stripping | BC@.500 | Ea | 27.00 | 18.50 | 45.50 |
| Window seal sets, spring brass, adhesive (Peel N Stick®), linear feet per side | | | | | |
| Double hung window, head and sill | BC@.008 | LF | 1.59 | .30 | 1.89 |
| Add to remove & replace window sash | BC@.250 | Ea | — | 9.24 | 9.24 |
| Casement window, head and sill | BC@.008 | LF | .70 | .30 | 1.00 |
| Gray or black pile gasket | BC@.008 | LF | .35 | .30 | .65 |
| Gray or black pile gasket with adhesive | BC@.006 | LF | .75 | .22 | .97 |
| Garage door top and side seal | | | | | |
| Kry-O-Gem™ fiber gasket in aluminum or vinyl frame (KEL-EEZ®) | | | | | |
| 9' x 7' door / screws | BC@.500 | Ea | 42.60 | 18.50 | 61.10 |
| 16' x 7' door / screws | BC@.600 | Ea | 55.50 | 22.20 | 77.70 |
| Flexible EPDM in aluminum frame | BC@.050 | LF | 1.65 | 1.85 | 3.50 |
| Garage door bottom seal. Set with adhesive | | | | | |
| 9', neoprene or EPDM | BC@.383 | Ea | 9.45 | 14.10 | 23.55 |
| 16', neoprene or EPDM | BC@.500 | Ea | 29.60 | 18.50 | 48.10 |
| Soft cushion foam | BC@.025 | LF | 1.05 | .92 | 1.97 |
| EPDM Rubber | BC@.025 | LF | 1.95 | .92 | 2.87 |

**Well Drilling, Subcontract** Typical costs shown, based on well depth including welded steel casing (or continuous PVC casing). Actual costs will depend on specific conditions, i.e. hard-rock, alluvial riverbed or a combination of conditions. The amount of casing needed depends on soil conditions. Some types of rock require casings at the well surface only. Your local health department can probably provide information on the average well depth and subsurface conditions in the community. Note that unforeseen subsurface conditions can increase drilling costs substantially. Most subcontractors will estimate costs on an hourly basis.

| | Craft@Hrs | Unit | Material | Labor | Total |
|---|---|---|---|---|---|
| Well hole with 4" ID F480 PVC casing | — | LF | — | — | 33.60 |
| Well hole with 6" ID steel casing | — | LF | — | — | 57.90 |
| Well hole with 8" ID steel casing | — | LF | — | — | 79.60 |
| 6" well hole in sturdy rock (no casing) | — | LF | — | — | 21.00 |

# Window and Door Mantels, Aluminum

| | Craft@Hrs | Unit | Material | Labor | Total |
|---|---|---|---|---|---|
| Add per well for 6" diameter bottom filter screen, if needed due to site conditions | | | | | |
| Stainless steel | — | LF | — | — | 163.00 |
| Low carbon steel | — | LF | — | — | 97.40 |
| Add per well for drive shoe | — | LS | — | — | 105.00 |
| Add for 6" surface seal, if needed due to local code or site conditions | | | | | |
| 20' sanitary seal – for individual well | — | LS | — | — | 882.00 |
| 50' sanitary seal – for community well | — | LS | — | — | 2,880.00 |
| Add per well for well drilling permit | | | | | |
| Individual water well | — | LS | — | — | 521.00 |
| Community water well with Source Assessment | — | LS | — | — | 1,380.00 |
| Add for automatic electric pumping system with pressure tank(s) domestic use, no electrical work included. | | | | | |
| Individual (18 GPM @ 400 TDH), typical | — | LS | 19,800.00 | — | 19,800.00 |
| Community (60 GPM @ 400 TDH), typical | — | LS | 19,800.00 | — | 19,800.00 |
| Add for electrical service drop pole | | | | | |
| Typical cost 100 amp 230 volt 1 phase service | — | LS | — | — | 1,430.00 |

**Window and Door Mantels, Aluminum** Includes hardware, aluminum crown mould and roof. Modern design, bright copper or primed aluminum finish.

| | Craft@Hrs | Unit | Material | Labor | Total |
|---|---|---|---|---|---|
| Up to 24" width | SW@.500 | Ea | 115.00 | 20.50 | 135.50 |
| 25" to 36" width | SW@.500 | Ea | 154.00 | 20.50 | 174.50 |
| 37" to 48" width | SW@.500 | Ea | 193.00 | 20.50 | 213.50 |
| 49" to 60" width | SW@.600 | Ea | 231.00 | 24.60 | 255.60 |
| 61" to 72" width | SW@.600 | Ea | 271.00 | 24.60 | 295.60 |
| 73" to 84" width | SW@.600 | Ea | 310.00 | 24.60 | 334.60 |
| 85" to 96" width | SW@.700 | Ea | 350.00 | 28.70 | 378.70 |
| 97" to 108" width | SW@.700 | Ea | 389.00 | 28.70 | 417.70 |
| 109" to 120" width | SW@.700 | Ea | 429.00 | 28.70 | 457.70 |
| Add for aged copper or brass aluminum finish | — | % | 10.0 | — | — |
| Add for units over 120" | — | LF | 52.10 | — | 52.10 |

**Window Sills, Cultured Stone**

| | Craft@Hrs | Unit | Material | Labor | Total |
|---|---|---|---|---|---|
| Marble sill 40" or less | B1@.687 | Ea | 60.00 | 22.90 | 82.90 |
| Marble sill 41" to 60" | B1@.730 | Ea | 80.00 | 24.30 | 104.30 |
| Onyx sill 40" or less | B1@.687 | Ea | 80.00 | 22.90 | 102.90 |
| Onyx sill 41" to 60" | B1@.730 | Ea | 115.00 | 24.30 | 139.30 |
| Granite sill 40" or less | B1@.687 | Ea | 98.00 | 22.90 | 120.90 |
| Granite sill 41" to 60" | B1@.730 | Ea | 145.00 | 24.30 | 169.30 |

**Window Sills, Vinyl** Solid vinyl. Set with latex or solvent base adhesive. Hard, scratch resistant surface. Ultraviolet protection to prevent yellowing. Gloss finish

| | Craft@Hrs | Unit | Material | Labor | Total |
|---|---|---|---|---|---|
| 3-1/2" x 3' | BC@.129 | Ea | 16.20 | 4.77 | 20.97 |
| 3-1/2" x 4' | BC@.172 | Ea | 20.50 | 6.35 | 26.85 |
| 3-1/2" x 5' | BC@.215 | Ea | 27.00 | 7.94 | 34.94 |
| 3-1/2" x 6' | BC@.258 | Ea | 30.20 | 9.53 | 39.73 |
| 4-1/4" x 3' | BC@.129 | Ea | 20.50 | 4.77 | 25.27 |
| 4-1/4" x 4' | BC@.172 | Ea | 25.90 | 6.35 | 32.25 |
| 4-1/4" x 5' | BC@.215 | Ea | 32.40 | 7.94 | 40.34 |
| 4-1/4" x 6' | BC@.258 | Ea | 37.80 | 9.53 | 47.33 |
| 5-1/2" x 3' | BC@.129 | Ea | 22.70 | 4.77 | 27.47 |
| 5-1/2" x 4' | BC@.172 | Ea | 30.20 | 6.35 | 36.55 |

| | Craft@Hrs | Unit | Material | Labor | Total |
|---|---|---|---|---|---|
| 5-1/2" x 5' | BC@.215 | Ea | 35.60 | 7.94 | 43.54 |
| 5-1/2" x 6' | BC@.258 | Ea | 43.20 | 9.53 | 52.73 |
| 6-1/4" x 3' | BC@.129 | Ea | 25.20 | 4.77 | 29.97 |
| 6-1/4" x 4' | BC@.172 | Ea | 34.20 | 6.35 | 40.55 |
| 6-1/4" x 5' | BC@.215 | Ea | 40.90 | 7.94 | 48.84 |
| 6-1/4" x 6' | BC@.258 | Ea | 50.90 | 9.53 | 60.43 |

**Window Treatments** Blinds and shades. See also Draperies.
Horizontal blinds. Per square foot of blind, width x height. Includes head rail and wand tilter. Standard colors and custom sizes.

| | Craft@Hrs | Unit | Material | Labor | Total |
|---|---|---|---|---|---|
| 1" wide PVC | BC@.083 | SF | 1.21 | 3.07 | 4.28 |
| 1" wide aluminum | BC@.083 | SF | 2.05 | 3.07 | 5.12 |
| 1" wide basswood | BC@.083 | SF | 6.64 | 3.07 | 9.71 |
| 2" wide faux wood | BC@.083 | SF | 2.86 | 3.07 | 5.93 |
| 2" wide basswood | BC@.083 | SF | 5.09 | 3.07 | 8.16 |
| 2" wide American hardwood | BC@.083 | SF | 5.68 | 3.07 | 8.75 |
| 2-1/2" wide faux wood | BC@.083 | SF | 3.70 | 3.07 | 6.77 |
| 2-1/2" wide beveled basswood | BC@.083 | SF | 6.80 | 3.07 | 9.87 |

Roll-up blinds. Per square foot of blind, width x height. Includes aluminum head rail, valance and roller chain lift. Standard colors and custom sizes.

| | Craft@Hrs | Unit | Material | Labor | Total |
|---|---|---|---|---|---|
| Woven wood (rattan and jute) | BC@.083 | SF | 6.64 | 3.07 | 9.71 |
| Bamboo | BC@.083 | SF | 8.84 | 3.07 | 11.91 |
| Vinyl-fiberglass, translucent | BC@.083 | SF | 2.98 | 3.07 | 6.05 |
| Room darkening fabric | BC@.083 | SF | 9.50 | 3.07 | 12.57 |
| Denim fabric | BC@.083 | SF | 4.00 | 3.07 | 7.07 |
| Decorative woven cotton fabric | BC@.083 | SF | 5.00 | 3.07 | 8.07 |
| Sun control vinyl fiberglass sheer weave | BC@.083 | SF | 4.98 | 3.07 | 8.05 |

Vertical blinds. 3-1/2" vanes. Per square foot of blind, width x height. Includes aluminum head rail, tension pulley and dust cover valance. Standard colors and custom sizes. Left or right pull.

| | Craft@Hrs | Unit | Material | Labor | Total |
|---|---|---|---|---|---|
| Smooth surface PVC | BC@.083 | SF | 4.93 | 3.07 | 8.00 |
| Textured surface PVC | BC@.083 | SF | 6.13 | 3.07 | 9.20 |
| Premium textured surface PVC | BC@.083 | SF | 6.38 | 3.07 | 9.45 |
| Faux wood | BC@.083 | SF | 15.50 | 3.07 | 18.57 |
| Add for center pull, per blind | — | Ea | 21.60 | — | 21.60 |
| Add for tile or molding cutout, per blind | — | Ea | 21.60 | — | 21.60 |
| Add for motorized remote control, per blind | — | Ea | 108.00 | — | 108.00 |

Cellular shades, polyester. With head rail and mounting hardware.

| | Craft@Hrs | Unit | Material | Labor | Total |
|---|---|---|---|---|---|
| 3/8" double cell light filter | BC@.083 | SF | 5.56 | 3.07 | 8.63 |
| 3/8" single cell translucent | BC@.083 | SF | 7.67 | 3.07 | 10.74 |
| 3/8" double cell room darkening | BC@.083 | SF | 7.77 | 3.07 | 10.84 |
| 3/8" single cell blackout | BC@.083 | SF | 8.67 | 3.07 | 11.74 |
| 1/2" double cell light filter | BC@.083 | SF | 8.22 | 3.07 | 11.29 |
| 1/2" single cell light filter | BC@.083 | SF | 5.56 | 3.07 | 8.63 |
| 3/4" single cell extreme light | BC@.083 | SF | 5.67 | 3.07 | 8.74 |
| 3/4" single cell blackout | BC@.083 | SF | 9.22 | 3.07 | 12.29 |
| Add for width over 8' | — | % | 20.0 | — | — |
| Add for continuous loop lift cord | — | % | 25.0 | — | — |
| Add for no lift cord (child safety), per blind | — | Ea | 40.00 | — | 40.00 |
| Add for tile or molding cutout, per blind | — | Ea | 20.00 | — | 20.00 |

# Windows

| | Craft@Hrs | Unit | Material | Labor | Total |
|---|---|---|---|---|---|

**Windows**  See also Skylights and Sky Windows.

Single-hung insulating glass vinyl windows. 5/8" insulating glass with grid between the lites. 4/4 grid means the upper lite is divided into four panes and the lower lite is divided into four panes. Tilt sash for easy cleaning. Includes half screen. Dimensions are rough opening sizes, width x height. White finish.

| | Craft@Hrs | Unit | Material | Labor | Total |
|---|---|---|---|---|---|
| 1'8" x 4'2", 4/4 grid | BC@.500 | Ea | 96.60 | 18.50 | 115.10 |
| 2'0" x 3'0", 4/4 grid | BC@.500 | Ea | 85.50 | 18.50 | 104.00 |
| 2'0" x 6'0", 6/4 grid | BC@.500 | Ea | 104.00 | 18.50 | 122.50 |
| 2'8" x 3'0", 6/6 grid | BC@.500 | Ea | 94.60 | 18.50 | 113.10 |
| 2'8" x 4'4", 6/6 grid | B1@1.00 | Ea | 110.00 | 33.30 | 143.30 |
| 2'8" x 5'0", 6/6 grid | B1@1.00 | Ea | 113.00 | 33.30 | 146.30 |
| 2'8" x 6'0", 6/6 grid | B1@1.00 | Ea | 116.00 | 33.30 | 149.30 |
| 3'0" x 3'0", 6/6 grid | BC@.500 | Ea | 98.80 | 18.50 | 117.30 |
| 3'0" x 4'0", 6/6 grid | B1@1.00 | Ea | 112.00 | 33.30 | 145.30 |
| 3'0" x 4'4", 6/6 grid | B1@1.00 | Ea | 112.00 | 33.30 | 145.30 |
| 3'0" x 5'0", 6/6 grid | B1@1.50 | Ea | 124.00 | 50.00 | 174.00 |
| 3'0" x 6'0", 9/6 grid | B1@1.50 | Ea | 137.00 | 50.00 | 187.00 |

Single-hung insulated low-E glass vinyl windows. Low-E (energy efficient) 5/8" insulating glass with grids between the glass. 4/4 grid means the upper light is divided into four panes and the lower light is divided into four panes. White finish. Tilt sash for easy cleaning. Includes half screen. Dimensions are rough opening sizes, width x height.

| | Craft@Hrs | Unit | Material | Labor | Total |
|---|---|---|---|---|---|
| 2'0" x 3'0", 4/4 grid | BC@.500 | Ea | 85.50 | 18.50 | 104.00 |
| 2'0" x 5'0", 4/4 grid | BC@.500 | Ea | 121.00 | 18.50 | 139.50 |
| 2'0" x 6'0", 6/4 grid | B1@1.00 | Ea | 138.00 | 33.30 | 171.30 |
| 2'8" x 3'0", 6/6 grid | BC@.500 | Ea | 102.00 | 18.50 | 120.50 |
| 2'8" x 4'4", 6/6 grid | B1@1.00 | Ea | 137.00 | 33.30 | 170.30 |
| 2'8" x 5'0", 6/6 grid | B1@1.00 | Ea | 124.00 | 33.30 | 157.30 |
| 2'8" x 6'0", 9/6 grid | B1@1.50 | Ea | 150.00 | 50.00 | 200.00 |
| 3'0" x 3'0", 6/6 grid | BC@.500 | Ea | 95.70 | 18.50 | 114.20 |
| 3'0" x 4'0", 6/6 grid | B1@1.00 | Ea | 125.00 | 33.30 | 158.30 |
| 3'0" x 4'4", 6/6 grid | B1@1.00 | Ea | 142.00 | 33.30 | 175.30 |
| 3'0" x 5'0", 6/6 grid | B1@1.50 | Ea | 144.00 | 50.00 | 194.00 |
| 3'0" x 6'0", 9/6 grid | B1@1.50 | Ea | 158.00 | 50.00 | 208.00 |

Double-hung insulating glass vinyl windows with low-E glass insulating glass. Upper and lower lites are divided into panes by a vinyl grid between the sheets of glass. Sash tilts for easy cleaning. With full screen. 4-9/16" white jamb. Built-in "J"-channel. Designed for new construction. Dimensions are rough opening sizes, width x height.

| | Craft@Hrs | Unit | Material | Labor | Total |
|---|---|---|---|---|---|
| 2'4" x 3'2" | BC@.500 | Ea | 137.00 | 18.50 | 155.50 |
| 2'4" x 3'10" | BC@.500 | Ea | 145.00 | 18.50 | 163.50 |
| 2'4" x 4'2" | BC@.500 | Ea | 156.00 | 18.50 | 174.50 |
| 2'4" x 4'6" | BC@.500 | Ea | 161.00 | 18.50 | 179.50 |
| 2'8" x 3'2" | BC@.500 | Ea | 142.00 | 18.50 | 160.50 |
| 2'8" x 3'10" | BC@.500 | Ea | 155.00 | 18.50 | 173.50 |
| 2'8" x 4'2" | BC@.500 | Ea | 161.00 | 18.50 | 179.50 |
| 2'8" x 4'6" | B1@1.00 | Ea | 165.00 | 33.30 | 198.30 |
| 3'0" x 4'2" | B1@1.50 | Ea | 166.00 | 50.00 | 216.00 |
| 3'0" x 4'6" | B1@1.00 | Ea | 180.00 | 33.30 | 213.30 |

Double-hung insulating glass vinyl windows. White, with full screen. Low-E (energy-efficient) insulating glass. By rough opening size, width x height. Actual size is 1/2" less. Upper and lower lites are divided into six panes by a vinyl grid between the sheets of glass. 4-9/16" wide jamb.

| | Craft@Hrs | Unit | Material | Labor | Total |
|---|---|---|---|---|---|
| 2'0" x 3'2" | BC@.500 | Ea | 133.00 | 18.50 | 151.50 |
| 2'4" x 2'10" | BC@.500 | Ea | 123.00 | 18.50 | 141.50 |

| | Craft@Hrs | Unit | Material | Labor | Total |
|---|---|---|---|---|---|
| 2'4" x 3'2" | BC@.500 | Ea | 131.00 | 18.50 | 149.50 |
| 2'4" x 3'10" | BC@.500 | Ea | 136.00 | 18.50 | 154.50 |
| 2'4" x 4'6" | B1@1.00 | Ea | 151.00 | 33.30 | 184.30 |
| 2'8" x 2'10" | BC@.500 | Ea | 136.00 | 18.50 | 154.50 |
| 2'8" x 3'2" | BC@.500 | Ea | 131.00 | 18.50 | 149.50 |
| 2'8" x 3'10" | BC@.500 | Ea | 140.00 | 18.50 | 158.50 |
| 2'8" x 4'2" | B1@1.00 | Ea | 138.00 | 33.30 | 171.30 |
| 2'8" x 4'6" | B1@1.00 | Ea | 149.00 | 33.30 | 182.30 |
| 3'0" x 3'10" | B1@1.00 | Ea | 169.00 | 33.30 | 202.30 |
| 3'0" x 4'6" | B1@1.00 | Ea | 179.00 | 33.30 | 212.30 |

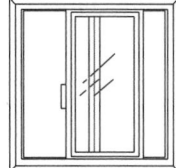

Horizontal sliding single-glazed vinyl windows. By nominal size, width x height. Actual size is 1/2" less. With screen.

| | Craft@Hrs | Unit | Material | Labor | Total |
|---|---|---|---|---|---|
| 2'0" x 2'0" | BC@.500 | Ea | 66.00 | 18.50 | 84.50 |
| 3'0" x 2'0" | BC@.500 | Ea | 83.50 | 18.50 | 102.00 |
| 3'0" x 3'0" | BC@.500 | Ea | 98.70 | 18.50 | 117.20 |
| 4'0" x 3'0" | B1@1.00 | Ea | 114.00 | 33.30 | 147.30 |
| 4'0" x 4'0" | B1@1.50 | Ea | 132.00 | 50.00 | 182.00 |
| 5'0" x 4'0" | B1@1.50 | Ea | 147.00 | 50.00 | 197.00 |

Horizontal sliding insulated low-E glass vinyl windows. Argon filled. For new construction. By rough opening size, width x height. With screen.

| | Craft@Hrs | Unit | Material | Labor | Total |
|---|---|---|---|---|---|
| 2'0" x 2'0" | BC@.500 | Ea | 76.40 | 18.50 | 94.90 |
| 2'0" x 3'0" | BC@.500 | Ea | 83.50 | 18.50 | 102.00 |
| 2'4" x 4'6" | BC@.500 | Ea | 135.00 | 18.50 | 153.50 |
| 2'8" x 4'6" | B1@1.00 | Ea | 144.00 | 33.30 | 177.30 |
| 3'0" x 2'0" | BC@.500 | Ea | 96.70 | 18.50 | 115.20 |
| 3'0" x 3'0" | BC@.500 | Ea | 98.70 | 18.50 | 117.20 |
| 3'0" x 4'0" | B1@1.00 | Ea | 136.00 | 33.30 | 169.30 |
| 3'0" x 5'2" | B1@1.50 | Ea | 153.00 | 50.00 | 203.00 |
| 3'0" x 6'0" | B1@1.50 | Ea | 153.00 | 50.00 | 203.00 |
| 4'0" x 2'0" | BC@.500 | Ea | 138.00 | 18.50 | 156.50 |
| 4'0" x 3'0" | B1@1.00 | Ea | 134.00 | 33.30 | 167.30 |
| 4'0" x 4'0" | B1@1.50 | Ea | 135.00 | 50.00 | 185.00 |
| 5'0" x 4'0" | B1@1.50 | Ea | 187.00 | 50.00 | 237.00 |

Casement insulated low-E glass vinyl windows. Argon filled. By window size, width x height. With screen. Sash opening direction when viewed from the interior.

| | Craft@Hrs | Unit | Material | Labor | Total |
|---|---|---|---|---|---|
| 28" x 48", left opening | BC@.500 | Ea | 201.00 | 18.50 | 219.50 |
| 28" x 48", right opening | BC@.500 | Ea | 201.00 | 18.50 | 219.50 |
| 30" x 36", left opening | BC@.500 | Ea | 241.00 | 18.50 | 259.50 |
| 30" x 36", right opening | B1@1.00 | Ea | 241.00 | 33.30 | 274.30 |
| 30" x 48", right opening | B1@1.00 | Ea | 274.00 | 33.30 | 307.30 |
| 30" x 48", right opening | B1@1.00 | Ea | 274.00 | 33.30 | 307.30 |
| 36" x 42", right opening | B1@1.50 | Ea | 263.00 | 50.00 | 313.00 |
| 36" x 42", left opening | B1@1.50 | Ea | 263.00 | 50.00 | 313.00 |

Casement bay insulating glass vinyl windows. Fixed center lite and two casement flankers. White vinyl. Factory assembled. By nominal (opening) size, width x height. Actual size is 1/2" less in both width and height. Includes two screens.

| | Craft@Hrs | Unit | Material | Labor | Total |
|---|---|---|---|---|---|
| 69" x 50" | B1@4.00 | Ea | 1,050.00 | 133.00 | 1,183.00 |
| 74" x 50" | B1@4.00 | Ea | 1,060.00 | 133.00 | 1,193.00 |
| 93" x 50" | B1@4.50 | Ea | 1,200.00 | 150.00 | 1,350.00 |
| 98" x 50" | B1@4.50 | Ea | 1,260.00 | 150.00 | 1,410.00 |

# Windows

| | Craft@Hrs | Unit | Material | Labor | Total |
|---|---|---|---|---|---|

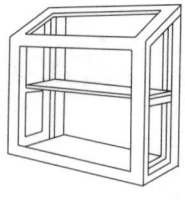

Garden insulating glass vinyl windows. Single front awning vent. White vinyl construction with wood frame. Tempered glass shelf. 5/4" vinyl laminated head and seat board. With hardware and screen. By nominal (opening) size, width x height. Replacement window. Labor includes removing the old stops and sash and setting new "J"-trim as needed.

| | Craft@Hrs | Unit | Material | Labor | Total |
|---|---|---|---|---|---|
| 36" x 36" | B1@2.75 | Ea | 569.00 | 91.60 | 660.60 |
| 36" x 36", low-E glass | B1@2.75 | Ea | 602.00 | 91.60 | 693.60 |
| 36" x 48", low-E glass | B1@2.75 | Ea | 652.00 | 91.60 | 743.60 |

Fixed half circle (picture) insulating glass vinyl windows. White frame. Low-E (energy-efficient) glass. Stock units for new construction. By width x height.

| | Craft@Hrs | Unit | Material | Labor | Total |
|---|---|---|---|---|---|
| 20" x 10" | B1@2.75 | Ea | 474.00 | 91.60 | 565.60 |
| 30" x 16" | B1@2.75 | Ea | 474.00 | 91.60 | 565.60 |
| 40" x 20" | B1@2.75 | Ea | 632.00 | 91.60 | 723.60 |
| 50" x 26" | B1@2.75 | Ea | 748.00 | 91.60 | 839.60 |
| 60" x 30" | B1@2.75 | Ea | 850.00 | 91.60 | 941.60 |
| 80" x 40" | B1@2.75 | Ea | 1,230.00 | 91.60 | 1,321.60 |

Fixed (picture) low-E insulating glass vinyl windows. Argon filled. White frame. Dimensions are rough opening sizes, width x height.

| | Craft@Hrs | Unit | Material | Labor | Total |
|---|---|---|---|---|---|
| 24" x 36" | BC@.500 | Ea | 79.40 | 18.50 | 97.90 |
| 24" x 48" | BC@.500 | Ea | 103.00 | 18.50 | 121.50 |
| 24" x 60" | BC@.500 | Ea | 114.00 | 18.50 | 132.50 |
| 30" x 36" | BC@.500 | Ea | 109.00 | 18.50 | 127.50 |
| 30" x 48" | BC@.500 | Ea | 111.00 | 18.50 | 129.50 |
| 36" x 48" | B1@1.00 | Ea | 114.00 | 33.30 | 147.30 |
| 36" x 60" | B1@1.00 | Ea | 137.00 | 33.30 | 170.30 |
| 48" x 48" | B1@1.50 | Ea | 135.00 | 50.00 | 185.00 |
| 48" x 60" | B1@1.50 | Ea | 149.00 | 50.00 | 199.00 |
| 60" x 48" | B1@1.50 | Ea | 154.00 | 50.00 | 204.00 |

Jalousie (louver) vinyl windows. Positive locking handle for security. By opening size, width x height.

| | Craft@Hrs | Unit | Material | Labor | Total |
|---|---|---|---|---|---|
| 24" x 35-1/2", clear | BC@.500 | Ea | 268.00 | 18.50 | 286.50 |
| 24" x 35-1/2", obscure glass | BC@.500 | Ea | 305.00 | 18.50 | 323.50 |
| 30" x 35-1/2", clear | BC@.500 | Ea | 285.00 | 18.50 | 303.50 |
| 30" x 35-1/2", obscure glass | BC@.500 | Ea | 328.00 | 18.50 | 346.50 |
| 36" x 35-1/2", clear | BC@.500 | Ea | 295.00 | 18.50 | 313.50 |
| 36" x 35-1/2", obscure glass | BC@.500 | Ea | 350.00 | 18.50 | 368.50 |

Hopper insulating glass vinyl windows. Low-E (energy-efficient) insulating glass. Argon filled. Frame depth 3-5/16". By opening size, width x height.

| | Craft@Hrs | Unit | Material | Labor | Total |
|---|---|---|---|---|---|
| 32" x 15" | BC@.500 | Ea | 83.70 | 18.50 | 102.20 |
| 32" x 17" | BC@.500 | Ea | 88.90 | 18.50 | 107.40 |
| 32" x 19" | BC@.500 | Ea | 91.80 | 18.50 | 110.30 |
| 32" x 23" | BC@.500 | Ea | 96.00 | 18.50 | 114.50 |

Vinyl-clad double-hung insulating glass wood windows. Pine interior. White vinyl exterior. Tilt sash. By nominal (opening) size, width x height. Actual width is 1-5/8" more. Actual height is 3-1/4" more. Add the cost of screens and colonial grilles.

| | Craft@Hrs | Unit | Material | Labor | Total |
|---|---|---|---|---|---|
| 2'0" x 3'0" | BC@.500 | Ea | 160.00 | 18.50 | 178.50 |
| 2'4" x 3'6" | BC@.500 | Ea | 188.00 | 18.50 | 206.50 |
| 2'4" x 4'0" | BC@.500 | Ea | 204.00 | 18.50 | 222.50 |
| 2'4" x 4'6" | B1@1.00 | Ea | 219.00 | 33.30 | 252.30 |
| 2'4" x 4'9" | B1@1.00 | Ea | 229.00 | 33.30 | 262.30 |
| 2'8" x 3'0" | B1@1.00 | Ea | 183.00 | 33.30 | 216.30 |
| 2'8" x 3'6" | B1@1.00 | Ea | 200.00 | 33.30 | 233.30 |
| 2'8" x 4'0" | B1@1.00 | Ea | 211.00 | 33.30 | 244.30 |
| 2'8" x 4'6" | B1@1.00 | Ea | 232.00 | 33.30 | 265.30 |

| | Craft@Hrs | Unit | Material | Labor | Total |
|---|---|---|---|---|---|
| 2'8" x 4'9" | B1@1.00 | Ea | 240.00 | 33.30 | 273.30 |
| 3'0" x 3'6" | B1@1.00 | Ea | 210.00 | 33.30 | 243.30 |
| 3'0" x 4'0" | B1@1.00 | Ea | 228.00 | 33.30 | 261.30 |
| 3'0" x 4'9" | B1@1.00 | Ea | 249.00 | 33.30 | 282.30 |

Screens for vinyl-clad wood windows. By window nominal size.

| | Craft@Hrs | Unit | Material | Labor | Total |
|---|---|---|---|---|---|
| 2'0" x 3'0" | BC@.050 | Ea | 20.80 | 1.85 | 22.65 |
| 2'4" x 3'0" | BC@.050 | Ea | 23.00 | 1.85 | 24.85 |
| 2'4" x 3'6" | BC@.050 | Ea | 25.20 | 1.85 | 27.05 |
| 2'4" x 4'0" | BC@.050 | Ea | 26.30 | 1.85 | 28.15 |
| 2'4" x 4'6" | BC@.050 | Ea | 28.50 | 1.85 | 30.35 |
| 2'4" x 4'9" | BC@.050 | Ea | 29.50 | 1.85 | 31.35 |
| 2'8" x 3'0" | BC@.050 | Ea | 23.40 | 1.85 | 25.25 |
| 2'8" x 3'6" | BC@.050 | Ea | 26.10 | 1.85 | 27.95 |
| 2'8" x 4'0" | BC@.050 | Ea | 28.60 | 1.85 | 30.45 |
| 2'8" x 4'6" | BC@.050 | Ea | 29.60 | 1.85 | 31.45 |
| 2'8" x 4'9" | BC@.050 | Ea | 30.50 | 1.85 | 32.35 |
| 3'0" x 3'6" | BC@.050 | Ea | 26.20 | 1.85 | 28.05 |
| 3'0" x 4'0" | BC@.050 | Ea | 29.70 | 1.85 | 31.55 |
| 3'0" x 4'6" | BC@.050 | Ea | 31.90 | 1.85 | 33.75 |
| 3'0" x 4'9" | BC@.050 | Ea | 32.00 | 1.85 | 33.85 |

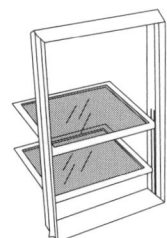

Vinyl-clad insulating glass wood awning windows. Prefinished pine interior. White vinyl exterior. Add the cost of screens. By nominal (opening) size, width x height.

| | Craft@Hrs | Unit | Material | Labor | Total |
|---|---|---|---|---|---|
| 2'1" x 2'0" | BC@.500 | Ea | 195.00 | 18.50 | 213.50 |
| 3'1" x 2'0" | BC@.500 | Ea | 235.00 | 18.50 | 253.50 |
| 4'1" x 2'0" | BC@.500 | Ea | 246.00 | 18.50 | 264.50 |

White vinyl-clad insulating glass wood casement windows, Andersen. Prefinished pine interior. White vinyl exterior. Tilt sash. With nailing flange for new construction. By nominal (opening) size, width x height. Add the cost of screens. Hinge operation as viewed from the exterior.

| | Craft@Hrs | Unit | Material | Labor | Total |
|---|---|---|---|---|---|
| 2'1" x 3'0", right hinge | BC@.500 | Ea | 228.00 | 18.50 | 246.50 |
| 2'1" x 3'0", left hinge | BC@.500 | Ea | 228.00 | 18.50 | 246.50 |
| 2'1" x 3'6", right hinge | BC@.500 | Ea | 256.00 | 18.50 | 274.50 |
| 2'1" x 3'6", left hinge | BC@.500 | Ea | 260.00 | 18.50 | 278.50 |
| 2'1" x 4'0", right hinge | BC@.500 | Ea | 271.00 | 18.50 | 289.50 |
| 2'1" x 4'0", left hinge | BC@.500 | Ea | 266.00 | 18.50 | 284.50 |
| 2'1" x 5'0", right hinge | B1@1.00 | Ea | 301.00 | 33.30 | 334.30 |
| 2'1" x 5'0", left hinge | B1@1.00 | Ea | 301.00 | 33.30 | 334.30 |
| 4'1" x 3'0", left and right hinge | B1@1.00 | Ea | 336.00 | 33.30 | 369.30 |
| 4'1" x 3'5", left and right hinge | B1@1.00 | Ea | 458.00 | 33.30 | 491.30 |
| 4'1" x 4'0", left and right hinge | B1@1.00 | Ea | 560.00 | 33.30 | 593.30 |
| 2'1" x 3'6", right hinge, egress | BC@.500 | Ea | 500.00 | 18.50 | 518.50 |
| 2'1" x 3'6", left hinge, egress | BC@.500 | Ea | 260.00 | 18.50 | 278.50 |
| 2'1" x 4'0", right hinge, egress | BC@.500 | Ea | 291.00 | 18.50 | 309.50 |
| 3'5" x 3'5", left hinge, egress | B1@1.00 | Ea | 507.00 | 33.30 | 540.30 |

Vinyl-clad insulating glass wood casement windows. Vinyl exterior. Unfinished wood interior. With hardware. Add the cost of grilles and screens. By nominal size, width x height.

| | Craft@Hrs | Unit | Material | Labor | Total |
|---|---|---|---|---|---|
| 2'1" x 3'6", single opening, egress | BC@.500 | Ea | 232.00 | 18.50 | 250.50 |
| 2'1" x 4'0", single opening, egress | BC@.500 | Ea | 293.00 | 18.50 | 311.50 |
| 3'5" x 3'5", single opening | BC@.500 | Ea | 411.00 | 18.50 | 429.50 |
| 4'1" x 3'0", double opening | B1@1.50 | Ea | 427.00 | 50.00 | 477.00 |
| 4'1" x 3'5"", double opening | B1@1.50 | Ea | 458.00 | 50.00 | 508.00 |
| 4'1" x 4'0", double opening | B1@1.50 | Ea | 529.00 | 50.00 | 579.00 |

# Windows

| | Craft@Hrs | Unit | Material | Labor | Total |
|---|---|---|---|---|---|

Vinyl-clad insulating glass wood basement windows. Prefinished pine interior. White vinyl exterior. Tilt sash. With nailing flange for new construction. By nominal (opening) size, width x height. With screen.

| | Craft@Hrs | Unit | Material | Labor | Total |
|---|---|---|---|---|---|
| 2'8" x 1'3" | BC@.500 | Ea | 86.90 | 18.50 | 105.40 |
| 2'8" x 1'7" | BC@.500 | Ea | 97.90 | 18.50 | 116.40 |
| 2'8" x 2'0" | BC@.500 | Ea | 109.00 | 18.50 | 127.50 |

Aluminum single-hung vertical sliding windows. Removable sash. Dual sash weatherstripping. Includes screen. By nominal (opening) size, width x height. With nailing flange for new construction.
Mill finish, single glazed

| | Craft@Hrs | Unit | Material | Labor | Total |
|---|---|---|---|---|---|
| 2'0" x 3'0" | BC@.500 | Ea | 66.30 | 18.50 | 84.80 |
| 2'8" x 3'0" | BC@.500 | Ea | 74.90 | 18.50 | 93.40 |
| 2'8" x 4'4" | BC@.500 | Ea | 90.50 | 18.50 | 109.00 |
| 2'8" x 5'0" | BC@.500 | Ea | 101.00 | 18.50 | 119.50 |
| 3'0" x 3'0" | BC@.500 | Ea | 77.10 | 18.50 | 95.60 |
| 3'0" x 4'0" | BC@.500 | Ea | 90.10 | 18.50 | 108.60 |
| 3'0" x 4'4" | BC@.500 | Ea | 90.10 | 18.50 | 108.60 |
| 3'0" x 5'0" | BC@.500 | Ea | 105.00 | 18.50 | 123.50 |

White finish, insulated glass, with grid

| | Craft@Hrs | Unit | Material | Labor | Total |
|---|---|---|---|---|---|
| 2'0" x 3'0" | BC@.500 | Ea | 80.40 | 18.50 | 98.90 |
| 2'8" x 3'0" | BC@.500 | Ea | 90.10 | 18.50 | 108.60 |
| 2'8" x 4'4" | BC@.500 | Ea | 113.00 | 18.50 | 131.50 |
| 2'8" x 5'0" | BC@.500 | Ea | 124.00 | 18.50 | 142.50 |
| 3'0" x 3'0" | BC@.500 | Ea | 97.80 | 18.50 | 116.30 |
| 3'0" x 4'0" | B1@1.00 | Ea | 113.00 | 33.30 | 146.30 |
| 3'0" x 4'4" | B1@1.00 | Ea | 119.00 | 33.30 | 152.30 |
| 3'0" x 5'0" | B1@1.50 | Ea | 128.00 | 50.00 | 178.00 |
| 3'0" x 6'0" | B1@1.50 | Ea | 146.00 | 50.00 | 196.00 |

Solar bronze insulating glass

| | Craft@Hrs | Unit | Material | Labor | Total |
|---|---|---|---|---|---|
| 2'0" x 2'0" | BC@.500 | Ea | 94.50 | 18.50 | 113.00 |
| 2'0" x 3'0" | BC@.500 | Ea | 94.50 | 18.50 | 113.00 |
| 2'0" x 4'0" | BC@.500 | Ea | 114.00 | 18.50 | 132.50 |
| 3'0" x 3'0" | BC@.500 | Ea | 122.00 | 18.50 | 140.50 |
| 3'0" x 4'0" | B1@1.00 | Ea | 165.00 | 33.30 | 198.30 |

Impact-resistant aluminum single hung windows. Glass bonded with clear polyvinyl. With screen.

| | Craft@Hrs | Unit | Material | Labor | Total |
|---|---|---|---|---|---|
| 26" x 38", obscure | BC@.500 | Ea | 213.00 | 18.50 | 231.50 |
| 37" x 26" | BC@.500 | Ea | 239.00 | 18.50 | 257.50 |
| 37" x 50" | B1@1.00 | Ea | 309.00 | 33.30 | 342.30 |
| 37" x 63" | B1@1.50 | Ea | 370.00 | 50.00 | 420.00 |
| 53" x 38" | B1@1.50 | Ea | 351.00 | 50.00 | 401.00 |
| 53" x 50" | B1@1.50 | Ea | 437.00 | 50.00 | 487.00 |

Aluminum horizontal sliding windows. Right sash is operable when viewed from the interior. With screen, 7/8" nailing fin setback for 3-coat stucco or wood siding. By nominal (opening) size, width x height. Obscure glass is usually available in smaller (1' and 2') windows at no additional cost.
Mill finish windows, single glazed

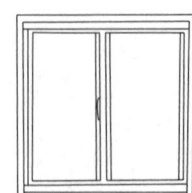

| | Craft@Hrs | Unit | Material | Labor | Total |
|---|---|---|---|---|---|
| 2'0" x 2'0" | BC@.500 | Ea | 36.50 | 18.50 | 55.00 |
| 3'0" x 2'0" | BC@.500 | Ea | 41.60 | 18.50 | 60.10 |
| 3'0" x 3'0" | BC@.500 | Ea | 52.10 | 18.50 | 70.60 |
| 3'0" x 4'0" | B1@1.00 | Ea | 62.80 | 33.30 | 96.10 |
| 4'0" x 3'0" | B1@1.00 | Ea | 62.80 | 33.30 | 96.10 |
| 4'0" x 4'0" | BC@1.50 | Ea | 73.50 | 55.40 | 128.90 |
| 6'0" x 3'0" | BC@1.50 | Ea | 84.10 | 55.40 | 139.50 |
| 6'0" x 4'0" | BC@1.50 | Ea | 94.60 | 55.40 | 150.00 |

| | Craft@Hrs | Unit | Material | Labor | Total |
|---|---|---|---|---|---|
| **White finish windows, insulating glass** | | | | | |
| 2'0" x 2'0" | BC@.500 | Ea | 52.10 | 18.50 | 70.60 |
| 3'0" x 1'0" | BC@.500 | Ea | 60.80 | 18.50 | 79.30 |
| 3'0" x 2'0" | BC@.500 | Ea | 67.30 | 18.50 | 85.80 |
| 3'0" x 3'0" | BC@.500 | Ea | 84.20 | 18.50 | 102.70 |
| 3'0" x 4'0" | B1@1.00 | Ea | 101.00 | 33.30 | 134.30 |
| 4'0" x 3'0" | B1@1.00 | Ea | 101.00 | 33.30 | 134.30 |
| 4'0" x 4'0" | BC@1.50 | Ea | 106.00 | 55.40 | 161.40 |
| **White finish windows, insulating glass, with grid** | | | | | |
| 3'0" x 3'0" | BC@.500 | Ea | 108.00 | 18.50 | 126.50 |
| 3'0" x 4'0" | B1@1.00 | Ea | 131.00 | 33.30 | 164.30 |
| 4'0" x 3'0" | B1@1.00 | Ea | 130.00 | 33.30 | 163.30 |
| 4'0" x 4'0" | B1@1.50 | Ea | 151.00 | 50.00 | 201.00 |
| 5'0" x 3'0" | B1@1.50 | Ea | 150.00 | 50.00 | 200.00 |
| 5'0" x 4'0" | B1@1.50 | Ea | 162.00 | 50.00 | 212.00 |
| 6'0" x 3'0" | B1@1.50 | Ea | 172.00 | 50.00 | 222.00 |
| 6'0" x 4'0" | B1@1.50 | Ea | 184.00 | 50.00 | 234.00 |
| **Solar bronze insulating glass windows** | | | | | |
| 2'0" x 2'0" | BC@.500 | Ea | 69.50 | 18.50 | 88.00 |
| 3'0" x 3'0" | BC@.500 | Ea | 112.00 | 18.50 | 130.50 |
| 3'0" x 4'0" | B1@1.00 | Ea | 116.00 | 33.30 | 149.30 |
| 4'0" x 3'0" | B1@1.00 | Ea | 116.00 | 33.30 | 149.30 |
| 4'0" x 4'0" | B1@1.50 | Ea | 138.00 | 50.00 | 188.00 |
| 5'0" x 4'0" | B1@1.50 | Ea | 154.00 | 50.00 | 204.00 |
| 6'0" x 4'0" | B1@1.50 | Ea | 180.00 | 50.00 | 230.00 |
| **White finish windows, insulating low-E glass** | | | | | |
| 3'0" x 3'0" | BC@.500 | Ea | 108.00 | 18.50 | 126.50 |
| 3'0" x 4'0" | B1@1.00 | Ea | 135.00 | 33.30 | 168.30 |
| 4'0" x 3'0" | B1@1.00 | Ea | 130.00 | 33.30 | 163.30 |
| 4'0" x 4'0" | B1@1.50 | Ea | 145.00 | 50.00 | 195.00 |
| 5'0" x 4'0" | B1@1.50 | Ea | 162.00 | 50.00 | 212.00 |
| 6'0" x 4'0" | BC@1.50 | Ea | 194.00 | 55.40 | 249.40 |

**Aluminum single glazed awning windows. With hardware and screen. Painted powder polyurethane finish. Solar bronze glass. Opening (nominal) sizes, width x height.**

| | Craft@Hrs | Unit | Material | Labor | Total |
|---|---|---|---|---|---|
| 24" x 20-1/8", one lite high | BC@.500 | Ea | 47.00 | 18.50 | 65.50 |
| 24" x 28-3/4", two lites high | BC@.500 | Ea | 67.30 | 18.50 | 85.80 |
| 24" x 37-3/8", three lites high | B1@1.00 | Ea | 70.50 | 33.30 | 103.80 |
| 24" x 46", four lites high | B1@1.00 | Ea | 70.50 | 33.30 | 103.80 |
| 24" x 54-5/8", five lites high | B1@1.50 | Ea | 89.80 | 50.00 | 139.80 |
| 24" x 63-1/4", six lites high | B1@1.50 | Ea | 95.00 | 50.00 | 145.00 |
| 30" x 20-1/8", one lite high | BC@.500 | Ea | 57.60 | 18.50 | 76.10 |
| 30" x 28-3/4", two lites high | BC@.500 | Ea | 66.00 | 18.50 | 84.50 |
| 30" x 37-3/8", three lites high | B1@1.00 | Ea | 77.00 | 33.30 | 110.30 |
| 30" x 46", four lites high | B1@1.00 | Ea | 86.10 | 33.30 | 119.40 |
| 30" x 54-5/8", five lites high | B1@1.50 | Ea | 100.00 | 50.00 | 150.00 |
| 30" x 63-1/4", six lites high | B1@1.50 | Ea | 116.00 | 50.00 | 166.00 |
| 36" x 20-1/8", one lite high | BC@.500 | Ea | 57.40 | 18.50 | 75.90 |
| 36" x 28-3/4", two lites high | BC@.500 | Ea | 73.80 | 18.50 | 92.30 |
| 36" x 37-3/8", three lites high | B1@1.00 | Ea | 78.60 | 33.30 | 111.90 |
| 36" x 46", four lites high | B1@1.00 | Ea | 89.60 | 33.30 | 122.90 |
| 36" x 54-5/8", five lites high | B1@1.50 | Ea | 109.00 | 50.00 | 159.00 |
| 36" x 63-1/4", six lites high | B1@1.50 | Ea | 154.00 | 50.00 | 204.00 |

# Windows

|  | Craft@Hrs | Unit | Material | Labor | Total |
|---|---|---|---|---|---|

Double-hung insulating glass wood windows. Treated western pine. Natural wood interior. 4-9/16" jamb width. 5/8" insulating glass. Weatherstripped. Tilt-in, removable sash. With hardware. Nominal sizes, width x height. With screen.

| | Craft@Hrs | Unit | Material | Labor | Total |
|---|---|---|---|---|---|
| 2'0" x 3'2" | BC@.500 | Ea | 132.00 | 18.50 | 150.50 |
| 2'4" x 3'2" | BC@.500 | Ea | 144.00 | 18.50 | 162.50 |
| 2'4" x 4'6" | BC@.500 | Ea | 165.00 | 18.50 | 183.50 |
| 2'8" x 3'2" | BC@.500 | Ea | 155.00 | 18.50 | 173.50 |
| 2'8" x 3'10" | BC@.500 | Ea | 160.00 | 18.50 | 178.50 |
| 2'8" x 4'6" | B1@1.00 | Ea | 172.00 | 33.30 | 205.30 |
| 2'8" x 5'2" | B1@1.00 | Ea | 183.00 | 33.30 | 216.30 |
| 2'8" x 6'2" | B1@1.00 | Ea | 202.00 | 33.30 | 235.30 |
| 3'0" x 3'2" | BC@.500 | Ea | 160.00 | 18.50 | 178.50 |
| 3'0" x 3'10" | B1@1.00 | Ea | 158.00 | 33.30 | 191.30 |
| 3'0" x 4'6" | B1@1.00 | Ea | 180.00 | 33.30 | 213.30 |
| 3'0" x 5'2" | B1@1.00 | Ea | 187.00 | 33.30 | 220.30 |
| 3'0" x 6'2" | B1@1.00 | Ea | 227.00 | 33.30 | 260.30 |
| 2'8" x 4'6", twin | B1@1.00 | Ea | 320.00 | 33.30 | 353.30 |
| 3'0" x 4'6", twin | B1@1.00 | Ea | 374.00 | 33.30 | 407.30 |

Fixed octagon wood windows. Stock units for new construction.

| | Craft@Hrs | Unit | Material | Labor | Total |
|---|---|---|---|---|---|
| 24", 9-lite beveled glass | BC@1.50 | Ea | 129.00 | 55.40 | 184.40 |
| 24", insulating glass | BC@1.50 | Ea | 111.00 | 55.40 | 166.40 |
| 24", beveled glass | BC@1.50 | Ea | 89.70 | 55.40 | 145.10 |
| 24", single glazed | BC@1.50 | Ea | 86.30 | 55.40 | 141.70 |
| 24", venting insulating glass | BC@1.50 | Ea | 149.00 | 55.40 | 204.40 |

Angle bay windows. Fixed center lite and two double-hung flankers. 30-degree angle. Pine with aluminum cladding insulating glass. With screen. White or bronze finish. 4-9/16" wall thickness. Opening (nominal) sizes, width x height. Add the cost of interior trim and installing or framing the bay window roof. See the figures that follow.

| | Craft@Hrs | Unit | Material | Labor | Total |
|---|---|---|---|---|---|
| 6'4" x 4'8" | B1@4.00 | Ea | 1,500.00 | 133.00 | 1,633.00 |
| 7'2" x 4'8" | B1@4.00 | Ea | 1,510.00 | 133.00 | 1,643.00 |
| 7'10" x 4'8" | B1@4.50 | Ea | 1,620.00 | 150.00 | 1,770.00 |
| 8'6" x 4'8" | B1@4.50 | Ea | 1,870.00 | 150.00 | 2,020.00 |
| 8'10" x 4'8" | B1@4.50 | Ea | 1,970.00 | 150.00 | 2,120.00 |
| 9'2" x 4'8" | B1@5.00 | Ea | 1,930.00 | 167.00 | 2,097.00 |
| 9'10" x 4'8" | B1@5.00 | Ea | 2,040.00 | 167.00 | 2,207.00 |
| Add for 45 degree angle windows | — | Ea | 76.50 | — | 76.50 |
| Add for roof framing kit | — | Ea | 136.00 | — | 136.00 |

Bay window roof cover. Architectural copper roof system for standard angle bay windows. Includes hardware. Stock units for new construction.

| | Craft@Hrs | Unit | Material | Labor | Total |
|---|---|---|---|---|---|
| Up to 48" width | SW@.500 | Ea | 801.00 | 20.50 | 821.50 |
| 49" to 60" width | SW@.500 | Ea | 910.00 | 20.50 | 930.50 |
| 61" to 72" width | SW@.600 | Ea | 1,020.00 | 24.60 | 1,044.60 |
| 73" to 84" width | SW@.600 | Ea | 1,120.00 | 24.60 | 1,144.60 |
| 85" to 96" width | SW@.600 | Ea | 1,240.00 | 24.60 | 1,264.60 |
| 97" to 108" width | SW@.700 | Ea | 1,340.00 | 28.70 | 1,368.70 |
| 109" to 120" width | SW@.700 | Ea | 1,470.00 | 28.70 | 1,498.70 |
| 121" to 132" width | SW@.800 | Ea | 1,670.00 | 32.80 | 1,702.80 |
| 133" to 144" width | SW@.800 | Ea | 1,780.00 | 32.80 | 1,812.80 |
| 145" to 156" width | SW@1.00 | Ea | 1,900.00 | 41.00 | 1,941.00 |
| 157" to 168" width | SW@1.00 | Ea | 2,020.00 | 41.00 | 2,061.00 |
| Add for aged copper, brass or aluminum finish | — | % | 20.0 | — | — |
| Add for 24" to 26" projection | — | % | 10.0 | — | — |

|  | Craft@Hrs | Unit | Material | Labor | Total |
|---|---|---|---|---|---|
| Add for 26" to 28" projection | — | % | 15.0 | — | — |
| Add for 28" to 30" projection | — | % | 20.0 | — | — |
| Add for brick soffit kit | — | % | 45.0 | — | — |
| Add for 90-degree bays | — | Ea | 74.40 | — | 74.40 |

Casement bow windows. Insulating glass, assembled, with screens, weatherstripped, unfinished pine, 16" projection, 22" x 56" lites, 4-7/16" jambs. Add the cost of interior trim and installing or framing the bow window roof. Stock units for new construction. Generally a roof kit is required for projections of 12" or more. See the figures that follow.

|  | Craft@Hrs | Unit | Material | Labor | Total |
|---|---|---|---|---|---|
| 8'1" wide x 4'8" (4 lites, 2 venting) | B1@4.75 | Ea | 1,620.00 | 158.00 | 1,778.00 |
| 8'1" wide x 5'4" (4 lites, 2 venting) | B1@4.75 | Ea | 1,750.00 | 158.00 | 1,908.00 |
| 8 '1" wide x 6'0" (4 lites, 2 venting) | B1@4.75 | Ea | 1,880.00 | 158.00 | 2,038.00 |
| 10' wide x 4'8" (5 lites, 2 venting) | B1@4.75 | Ea | 1,980.00 | 158.00 | 2,138.00 |
| 10' wide x 5'4" (5 lites, 2 venting) | B1@4.75 | Ea | 2,130.00 | 158.00 | 2,288.00 |
| 10' wide x 6'0" (5 lites, 2 venting) | B1@4.75 | Ea | 2,340.00 | 158.00 | 2,498.00 |
| Add for roof framing kit, 5 lites | — | Ea | 121.00 | — | 121.00 |
| Add for interior and exterior trim |  |  |  |  |  |
| 3, 4 or 5 lites | B1@2.00 | Set | 37.90 | 66.60 | 104.50 |

Bow or bay window roof framed on site. 1/2" plywood sheathing, fiberglass batt insulation, fascia board, cove moulding, roof flashing, 15-pound felt, 240-pound seal-tab asphalt shingles, and cleanup.

|  | Craft@Hrs | Unit | Material | Labor | Total |
|---|---|---|---|---|---|
| Up to 120" width | B1@3.50 | Ea | 75.50 | 117.00 | 192.50 |
| Add per linear foot of width over 10' | B1@.350 | LF | 7.55 | 11.70 | 19.25 |

Basement storm windows. White aluminum frame with removable plexiglass window. Clips on over aluminum or wood basement window. Screen included. Nominal window sizes, width x height.

|  | Craft@Hrs | Unit | Material | Labor | Total |
|---|---|---|---|---|---|
| 32" x 14" | BC@.420 | Ea | 23.30 | 15.50 | 38.80 |
| 32" x 18" | BC@.420 | Ea | 23.30 | 15.50 | 38.80 |
| 32" x 22" | BC@.420 | Ea | 22.30 | 15.50 | 37.80 |

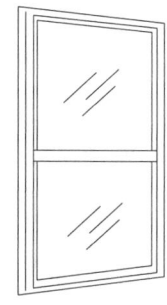

Aluminum storm windows. Removable glass inserts. Removable screen. Double track. Includes weatherstripping. Adaptable to aluminum single-hung windows.

|  | Craft@Hrs | Unit | Material | Labor | Total |
|---|---|---|---|---|---|
| 24" x 39", mill finish | BC@.420 | Ea | 33.80 | 15.50 | 49.30 |
| 24" x 39", white finish | BC@.420 | Ea | 35.90 | 15.50 | 51.40 |
| 24" x 55", mill finish | BC@.420 | Ea | 47.00 | 15.50 | 62.50 |
| 24" x 55", white finish | BC@.420 | Ea | 47.00 | 15.50 | 62.50 |
| 28" x 39", mill finish | BC@.420 | Ea | 36.50 | 15.50 | 52.00 |
| 28" x 39", white finish | BC@.420 | Ea | 38.40 | 15.50 | 53.90 |
| 28" x 47", white finish | BC@.420 | Ea | 43.80 | 15.50 | 59.30 |
| 28" x 55", mill finish | BC@.420 | Ea | 42.60 | 15.50 | 58.10 |
| 28" x 55", white finish | BC@.420 | Ea | 48.00 | 15.50 | 63.50 |
| 32" x 39", mill finish | BC@.420 | Ea | 36.90 | 15.50 | 52.40 |
| 32" x 39", white finish | BC@.420 | Ea | 41.60 | 15.50 | 57.10 |
| 32" x 47", white finish | BC@.420 | Ea | 50.40 | 15.50 | 65.90 |
| 32" x 55", mill finish | BC@.420 | Ea | 43.90 | 15.50 | 59.40 |
| 32" x 55", white finish | BC@.420 | Ea | 47.00 | 15.50 | 62.50 |
| 32" x 63", mill finish | BC@.420 | Ea | 52.40 | 15.50 | 67.90 |
| 32" x 63", white finish | BC@.420 | Ea | 52.40 | 15.50 | 67.90 |
| 32" x 75", white finish | BC@.420 | Ea | 63.10 | 15.50 | 78.60 |
| 36" x 39", mill finish | BC@.420 | Ea | 37.60 | 15.50 | 53.10 |
| 36" x 39", white finish | BC@.420 | Ea | 43.90 | 15.50 | 59.40 |
| 36" x 55", mill finish | BC@.420 | Ea | 44.80 | 15.50 | 60.30 |
| 36" x 55", white finish | BC@.420 | Ea | 50.00 | 15.50 | 65.50 |
| 36" x 63", mill finish | BC@.420 | Ea | 51.30 | 15.50 | 66.80 |
| 36" x 63", white finish | BC@.420 | Ea | 56.10 | 15.50 | 71.60 |

# Window Wells

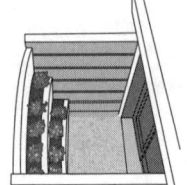

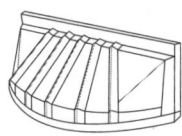

| | Craft@Hrs | Unit | Material | Labor | Total |
|---|---|---|---|---|---|

**Window wells** ScapeWel egress window, meets code requirements for basement escape window. Snaps into place on site. For new construction and remodeling, attaches directly to foundation. Bilco.

| | Craft@Hrs | Unit | Material | Labor | Total |
|---|---|---|---|---|---|
| Model 4048-42, 42" x 48" two tier | B1@1.50 | Ea | 733.00 | 50.00 | 783.00 |
| 4042C polycarbonate window well cover | | | | | |
| Fits Model 4048-42 | — | Ea | 366.00 | — | 366.00 |
| Model 4048-54, 54" x 48" two tier | B1@1.50 | Ea | 784.00 | 50.00 | 834.00 |
| 4054C polycarbonate window well cover | | | | | |
| Fits Model 4048-54 | — | Ea | 397.00 | — | 397.00 |
| Model 4048-66, 66" x 48" two tier | B1@1.50 | Ea | 814.00 | 50.00 | 864.00 |
| 4066C polycarbonate window well cover | | | | | |
| Fits Model 4048-66 | — | Ea | 422.00 | — | 422.00 |
| Model 4862-42, 42" x 62" three tier | B1@1.50 | Ea | 957.00 | 50.00 | 1,007.00 |
| 4842C polycarbonate window well cover | | | | | |
| Fits Model 4862-42 | — | Ea | 361.00 | — | 361.00 |
| Model 4862-54, 54" x 62" three tier | B1@1.50 | Ea | 1,000.00 | 50.00 | 1,050.00 |
| 4854C polycarbonate window well cover | | | | | |
| Fits Model 4862-54 | — | Ea | 397.00 | — | 397.00 |
| Model 4862-66, 66" x 62" three tier | B1@1.50 | Ea | 1,060.00 | 50.00 | 1,110.00 |
| 4866C polycarbonate window well cover | | | | | |
| Fits Model 4862-66 | — | Ea | 443.00 | — | 443.00 |

# Single Family Home Costs

Use the costs below to check your estimates and to verify bids submitted by subcontractors. These costs are based on a 1,600 square foot home (as described on page 306) built during the third quarter of 2010 at a cost of $99.88 per square foot. Figures in the column "% of Total" are the same as those listed in the Residential Rule of Thumb Construction Costs on page 306. These are averages and will vary widely with the type and quality of construction.

| Item | Total manpower required | % of Total | Unit | $ Material | $ Labor | $ Equipment | Total $ per SF | Total $ 1600 SF |
|---|---|---|---|---|---|---|---|---|
| Excavation | 3 men 2 days | 1.2% | SF | — | $0.90 | $0.30 | $1.20 | $1,920 |
| Foundation, slab, piers | 3 men 4 days | 3.7% | SF | $1.48 | 1.85 | 0.36 | 3.69 | 5,904 |
| Flatwork (drive and walk) | 3 men 3 days | 2.4% | SF | 0.96 | 1.20 | 0.24 | 2.40 | 3,840 |
| Brick hearth & veneer | 2 men 1 day | 0.7% | SF | 0.28 | 0.35 | 0.07 | 0.70 | 1,120 |
| Rough hardware | | 0.6% | SF | 0.24 | 0.30 | 0.06 | 0.60 | 960 |
| Finish hardware | | 0.2% | SF | 0.12 | 0.08 | — | 0.20 | 320 |
| Rough lumber | | 8.4% | SF | 8.39 | — | — | 8.39 | 13,424 |
| Finish lumber | | 0.5% | SF | 0.50 | — | — | 0.50 | 800 |
| Rough carpentry labor | 2 men 20 days | 8.9% | SF | — | 8.88 | — | 8.88 | 14,208 |
| Finish carpentry labor | 1 man 7 days | 1.7% | SF | — | 1.70 | — | 1.70 | 2,720 |
| Countertops | 1 man 2 days | 1.5% | SF | 0.90 | 0.60 | — | 1.50 | 2,400 |
| Cabinets | 2 men 2 days | 3.7% | SF | 2.95 | 0.74 | — | 3.69 | 5,904 |
| Insulation (R19 ceiling) | 1 man 3 days | 2.3% | SF | 1.50 | 0.81 | — | 2.31 | 3,696 |
| Roofing | 4 men 3 days | 5.5% | SF | 3.29 | 2.20 | — | 5.49 | 8,784 |
| Painting | 2 men 3 days | 3.6% | SF | 1.26 | 2.33 | — | 3.59 | 5,744 |
| Shower & tub enclosure | 1 man 1 day | 0.5% | SF | 0.30 | 0.20 | — | 0.50 | 800 |
| Prefabricated fireplace | 1 man 1 day | 0.9% | SF | 0.68 | 0.18 | 0.04 | 0.90 | 1,440 |
| Bath accessories | 1 man 1 day | 0.7% | SF | 0.47 | 0.23 | — | 0.70 | 1,120 |
| Built-in appliances | 1 man 1 day | 1.6% | SF | 1.44 | 0.16 | — | 1.60 | 2,560 |
| Heating and ducting | 2 men 5 days | 2.9% | SF | 1.16 | 1.74 | — | 2.90 | 4,640 |
| Plmbg & sewer connc'tn | 3 men 5 days | 7.3% | SF | 2.92 | 3.65 | 0.72 | 7.29 | 11,664 |
| Doors | 2 men 2 days | 1.9% | SF | 1.24 | 0.67 | — | 1.90 | 3,040 |
| Garage door | 1 man 1 day | 0.4% | SF | 0.30 | 0.10 | — | 0.40 | 640 |
| Alum windows/slidg drs | 1 man 2 days | 1.2% | SF | 0.72 | 0.48 | — | 1.20 | 1,920 |
| Exterior stucco | 3 men 3 days | 6.4% | SF | 4.15 | 1.92 | 0.32 | 6.39 | 10,224 |
| Gysum wallbord | 2 men 6 days | 4.7% | SF | 2.10 | 2.58 | — | 4.69 | 7,504 |
| Resilient flooring | 1 man 1 day | 2.0% | SF | 0.94 | 1.06 | — | 2.00 | 3,200 |
| Carpeting | 2 men 1 day | 2.4% | SF | 1.92 | 0.48 | — | 2.40 | 3,840 |
| Wiring (Romex) | 1 man 10 days | 3.2% | SF | 1.28 | 1.91 | — | 3.19 | 5,104 |
| Lighting fixtures | | 1.2% | SF | 0.96 | 0.24 | — | 1.20 | 1,920 |
| **Subtotal, DIRECT JOB COSTS** | | **82.2%** | | **$42.45** | **$37.54** | **$2.11** | **$82.10** | **$131,360** |
| Amount above, as a percent of total direct costs | | | | 52% | 46% | 3% | 100% | — |

| Indirect Costs | % of Total | Unit | $ Material | $ Labor | $ Equipment | Total $ per SF | Total $ 1600 SF |
|---|---|---|---|---|---|---|---|
| Insurance, payroll tax | 2.8% | SF | $2.80 | — | — | $2.80 | $4,416 |
| Plans & specs | 0.4% | SF | 0.40 | — | — | 0.40 | 624 |
| Permits & utilities | 1.7% | SF | 1.70 | — | — | 1.70 | 2,688 |
| Final cleanup | 0.4% | SF | — | 0.40 | — | 0.40 | 624 |
| **Subtotal, INDIRECT JOB COSTS** | **5.3%** | | **$4.90** | **$0.40** | **—** | **$5.30** | **$8,352** |
| **Overhead & profit** | **12.5%** | SF | **$12.48** | **—** | **—** | **$12.48** | **$19,718** |
| **Total all costs, rounded (based on 1600 SF)** | **100.0%** | SF | **$59.83** | **$37.94** | **$2.11** | **$99.88** | **$159,430** |
| Amount above, as a percent of total costs = | | | 59.8% | 38.0% | 2.2% | 100% | — |

The "Manpower" column assumes work is done in phases. Elapsed time, from breaking ground to final cleanup, will be about 10 weeks, plus delays due to weather, shortages and schedule conflicts.

# Construction Economics Division
## Construction Cost Index for New Single Family Homes

| Period | % of 1967 Cost | $ per SF of Floor | Period | % of 1967 Cost | $ per SF of Floor | Period | % of 1967 Cost | $ per SF of Floor |
|---|---|---|---|---|---|---|---|---|
| **1993** | | | **1999** | | | **2005** | | |
| 1st quarter | 443.9 | 62.15 | 1st quarter | 504.3 | 70.56 | 1st quarter | 608.70 | 85.17 |
| 2nd quarter | 450.7 | 63.10 | 2nd quarter | 505.2 | 70.69 | 2nd quarter | 616.87 | 86.31 |
| 3rd quarter | 456.4 | 63.90 | 3rd quarter | 506.5 | 70.85 | 3rd quarter | 621.21 | 87.18 |
| 4th quarter | 458.0 | 64.12 | 4th quarter | 511.1 | 71.50 | 4th quarter | 631.35 | 88.34 |
| **1994** | | | **2000** | | | **2006** | | |
| 1st quarter | 459.5 | 64.33 | 1st quarter | 516.8 | 72.30 | 1st quarter | 641.63 | 89.77 |
| 2nd quarter | 460.7 | 64.50 | 2nd quarter | 520.4 | 72.80 | 2nd quarter | 641.73 | 89.79 |
| 3rd quarter | 461.4 | 64.60 | 3rd quarter | 524.2 | 73.33 | 3rd quarter | 644.78 | 89.87 |
| 4th quarter | 461.6 | 64.62 | 4th quarter | 525.7 | 73.55 | 4th quarter | 655.91 | 91.77 |
| **1995** | | | **2001** | | | **2007** | | |
| 1st quarter | 462.1 | 64.68 | 1st quarter | 527.4 | 73.78 | 1st quarter | 656.08 | 91.80 |
| 2nd quarter | 463.6 | 64.90 | 2nd quarter | 528.8 | 73.98 | 2nd quarter | 653.57 | 91.45 |
| 3rd quarter | 464.3 | 65.00 | 3rd quarter | 530.3 | 74.19 | 3rd quarter | 655.00 | 91.67 |
| 4th quarter | 464.8 | 65.07 | 4th quarter | 533.1 | 74.58 | 4th quarter | 658.00 | 92.07 |
| **1996** | | | **2002** | | | **2008** | | |
| 1st quarter | 469.4 | 65.72 | 1st quarter | 530.11 | 74.16 | 1st quarter | 665.00 | 93.04 |
| 2nd quarter | 474.1 | 66.37 | 2nd quarter | 530.29 | 74.18 | 2nd quarter | 677.06 | 94.73 |
| 3rd quarter | 478.6 | 67.00 | 3rd quarter | 532.80 | 74.52 | 3rd quarter | 699.06 | 95.17 |
| 4th quarter | 480.6 | 672.5 | 4th quarter | 535.31 | 74.89 | 4th quarter | 720.45 | 100.80 |
| **1997** | | | **2003** | | | **2009** | | |
| 1st quarter | 482.0 | 67.53 | 1st quarter | 539.22 | 75.45 | 1st quarter | 707.78 | 99.03 |
| 2nd quarter | 487.4 | 68.20 | 2nd quarter | 540.58 | 75.63 | 2nd quarter | 704.72 | 98.60 |
| 3rd quarter | 491.9 | 68.84 | 3rd quarter | 541.11 | 75.71 | 3rd quarter | 704.69 | 98.59 |
| 4th quarter | 493.5 | 69.09 | 4th quarter | 554.32 | 77.56 | 4th quarter | 704.85 | 98.62 |
| **1998** | | | **2004** | | | **2010** | | |
| 1st quarter | 495.2 | 69.30 | 1st quarter | 557.54 | 78.01 | 1st quarter | — | 99.42 |
| 2nd quarter | 498.4 | 69.76 | 2nd quarter | 578.42 | 80.93 | 2nd quarter | — | 99.75 |
| 3rd quarter | 501.7 | 70.22 | 3rd quarter | 587.05 | 82.10 | 3rd quarter | — | 99.83 |
| 4th quarter | 503.5 | 70.45 | 4th quarter | 611.14 | 85.51 | 4th quarter | — | — |

The figures under the column "$ per SF of Floor" show construction costs for building a good quality home in a suburban area under competitive conditions in each calendar quarter since 1993. This home is described below under the section "Residential Rule of Thumb." These costs include the builder's overhead and profit and a 450 square foot garage but no basement. The cost of a finished basement per square foot will be approximately 40% of the square foot cost of living area. If the garage area is more than 450 square feet, use 50% of the living area cost to adjust for the larger or smaller garage. To find the total construction cost of the home, multiply the living area (excluding the garage) by the cost in the column "$ per SF of Floor."

| | |
|---|---|
| Deduct for rural areas | 5.0% |
| Add for 1,800 SF house (better quality) | 4.0 |
| Add for 2,000 SF house (better quality) | 3.0 |
| Deduct for over 2,400 SF house | 3.0 |
| Add for split level house | 3.0 |
| Add for 3-story house | 10.0 |
| Add for masonry construction | 9.0 |

Construction costs are higher in some cities and lower in others. Square foot costs listed in the table above are national averages. To modify these costs to your job site, apply the appropriate area modification factor from pages 12 through 15 of this manual. But note that area modifications on pages 12 through 15 are based on recent construction and may not apply to work that was completed many years ago.

## Residential Rule of Thumb

### Construction Costs

The following figures are percentages of total construction cost for a good quality single family residence: 1,600 square foot single story non-tract 3-bedroom, 1¾ bath home plus attached two car (450 SF) garage. Costs assume a conventional floor, stucco exterior, wallboard interior, shake roof, single fireplace, forced air heat, copper water supply, ABS drain lines, range with double oven, disposer, dishwasher and 900 SF of concrete flatwork.

| Item | Percent | Item | Percent | Item | Percent | Item | Percent | Item | Percent |
|---|---|---|---|---|---|---|---|---|---|
| Excavation | 1.2 | Finish lumber | .5 | Painting | 3.6 | Doors | 1.9 | Wiring (Romex) | 3.2 |
| Flatwork (drive & walk) | 2.4 | Rough carpentry labor | 8.9 | Shower & tub enclosure | .5 | Garage door | .4 | Lighting fixtures | 1.2 |
| Foundation, slab, piers | 3.7 | Finish carpentry labor | 1.7 | Prefabricated fireplace | .9 | Alum. windows, door | 1.2 | Insurance, tax | 2.8 |
| Brick hearth & veneer | .7 | Countertops | 1.5 | Bath accessories | .7 | Exterior stucco | 6.4 | Plans and specs | .4 |
| Rough hardware | .6 | Cabinets | 3.7 | Built-in appliances | 1.6 | Gypsum wallboard | 4.7 | Permits & utilities | 1.7 |
| Finish hardware | .2 | Insulation (R19 ceiling) | 2.3 | Heating and ducting | 2.9 | Resilient flooring | 2.0 | Final cleanup | .4 |
| Rough lumber | 8.4 | Roofing | 5.5 | Plumb. & sewer conn. | 7.3 | Carpeting | 2.4 | Overhead & profit | 12.5 |

# Industrial and Commercial Division Contents
## A complete index begins on page 646

# Industrial and Commercial Division
## Hourly Labor Costs

The hourly labor costs shown in the column headed "Hourly Cost" have been used to compute the manhour costs for the crews on pages 8 and 9 and the costs in the "Labor" column on pages 309 to 645 of this book. All figures are in U.S. dollars per hour.

"Hourly Wage and Benefits" includes the wage, welfare, pension, vacation, apprentice and other mandatory contributions. The "Employer's Burden" is the cost to the contractor for Unemployment Insurance (FUTA), Social Security and Medicare (FICA), state unemployment insurance, workers' compensation insurance and liability insurance. Tax and insurance expense included in these labor-hour costs are itemized in the sections beginning on pages 181 and 284.

These hourly labor costs will apply within a few percent on many jobs. But wages may be much higher or lower on the job you are estimating. If the hourly cost on this page is not accurate, use the labor cost adjustment procedure on page 11.

If your hourly labor cost is not known and can't be estimated, use both labor and material figures in this book without adjustment. When all material and labor costs have been compiled, multiply the total by the appropriate figure in the area modification table on pages 12 through 15.

| Craft | Base Wage ($) | Typical Employer Burden (%) | Employer's Burden Per Hour ($) | Hourly Cost ($) |
|---|---|---|---|---|
| Air Tool Operator | 31.10 | 36.14% | 11.24 | 42.34 |
| Asbestos Worker | 42.94 | 37.35% | 16.04 | 58.98 |
| Boilermaker | 44.28 | 36.59% | 16.20 | 60.48 |
| Bricklayer | 39.97 | 32.25% | 12.89 | 52.86 |
| Bricklayer Tender | 30.54 | 32.22% | 9.84 | 40.38 |
| Building Laborer | 30.13 | 32.36% | 9.75 | 39.88 |
| Carpenter | 39.45 | 31.25% | 12.33 | 51.78 |
| Cement Mason | 37.57 | 33.51% | 12.59 | 50.16 |
| Crane Operator | 40.54 | 36.19% | 14.67 | 55.21 |
| Drywall Installer | 39.09 | 29.44% | 11.51 | 50.60 |
| Electrician | 45.42 | 28.51% | 12.95 | 58.37 |
| Elevator Constructor | 42.09 | 27.23% | 11.46 | 53.55 |
| Floor Layer | 37.76 | 33.71% | 12.73 | 50.49 |
| Glazier | 39.74 | 31.10% | 12.36 | 52.10 |
| Iron Worker (Structural) | 43.15 | 47.30% | 20.41 | 63.56 |
| Lather | 36.16 | 31.17% | 11.27 | 47.43 |
| Marble Setter | 39.49 | 31.25% | 12.34 | 51.83 |
| Millwright | 40.12 | 31.16% | 12.50 | 52.62 |
| Mosaic & Terrazzo Worker | 37.12 | 31.25% | 11.60 | 48.72 |
| Painter | 39.23 | 31.10% | 12.20 | 51.43 |
| Pile Driver | 39.77 | 35.60% | 14.16 | 53.93 |
| Pipefitter | 45.30 | 31.43% | 14.24 | 59.54 |
| Plasterer | 40.67 | 33.59% | 13.66 | 54.33 |
| Plasterer Helper | 32.33 | 33.56% | 10.85 | 43.18 |
| Plumber | 46.06 | 31.46% | 14.49 | 60.55 |
| Reinforcing Ironworker | 42.09 | 38.54% | 16.22 | 58.31 |
| Roofer | 35.12 | 37.22% | 13.07 | 48.19 |
| Sheet Metal Worker | 44.79 | 30.54% | 13.68 | 58.47 |
| Sprinkler Fitter | 47.48 | 30.12% | 14.30 | 61.78 |
| Tractor Operator | 40.40 | 36.11% | 14.59 | 54.99 |
| Truck Driver | 32.75 | 36.12% | 11.83 | 44.58 |

**General Contractor's Markup**  Costs in this manual do not include the general contractor's markup. Costs in sections identified as *subcontract* are representative of prices quoted by subcontractors and include the subcontractor's markup. See pages 3 through 5 for more on subcontracted work and markup. Typical markup for general contractors handling commercial and industrial projects is shown at the bottom of this page. The two sections that follow, Indirect Overhead and Direct Overhead, give a more detailed breakdown.

**Indirect Overhead** (home office overhead)
This cost, usually estimated as a percentage, is calculated by dividing average annual indirect overhead costs by average annual receipts. Indirect overhead usually doesn't increase or decrease as quickly as gross volume increases or decreases. Home office expense may be about the same even if annual sales double or drop by one-half or more.

The figures below show typical indirect overhead costs per $1,000 of estimated project costs for a general contractor handling $600,000 to $1,000,000 commercial and industrial projects. Indirect overhead costs vary widely but will be about 8% of gross for most profitable firms. Total indirect overhead cost ($20.00 per $1,000 for materials and $60.00 per $1,000 for labor) appears as 2.0% and 6.0% under Total General Contractor's Markup at the bottom of this page.

|  | Craft@Hrs | Unit | Material | Labor | Total |
|---|---|---|---|---|---|
| Rent, office supplies, utilities, equipment, advertising, etc. | — | M$ | 20.00 | — | 20.00 |
| Office salaries and professional fees | — | M$ | — | 60.00 | 60.00 |
| General Contractor's indirect overhead | — | M$ | 20.00 | 60.00 | 80.00 |

**Direct Overhead** (job site overhead)
The figures below show typical direct overhead costs per $1,000 of a project's total estimated costs for a general contractor handling a $600,000 to $1,000,000 job that requires 6 to 9 months for completion. Use these figures for preliminary estimates and to check final bids. Add the cost of mobilization, watchmen, fencing, hoisting, permits, bonds, scaffolding and testing. Total direct overhead cost ($14.30 per $1,000 for materials and $58.70 per $1,000 for labor) appears as 1.4% and 5.9% under Total General Contractor's Markup at the bottom of this page. These costs can be expected to vary and will usually be lower on larger projects.

|  | Craft@Hrs | Unit | Material | Labor | Total |
|---|---|---|---|---|---|
| Job supervision | — | M$ | — | 44.00 | 44.00 |
| Truck and fuel for superintendent ($450 per month) | — | M$ | 4.50 | — | 4.50 |
| Temporary power, light and heat ($250 per month) | — | M$ | 2.50 | — | 2.50 |
| Temporary water ($100 per month) | — | M$ | 1.00 | — | 1.00 |
| Cellular phone ($50 per month) | — | M$ | .50 | — | .50 |
| Job site toilets ($100 per month) | — | M$ | 1.00 | — | 1.00 |
| Job site office trailer 8' x 30' ($150 per month) | — | M$ | 1.50 | — | 1.50 |
| Storage bin and tool shed 8' x 24' ($80 per month) | — | M$ | .80 | — | .80 |
| Job site cleanup & debris removal ($300 per month) | — | M$ | — | 3.00 | 3.00 |
| Job signs & first aid equipment ($50 per month) | — | M$ | .50 | — | .50 |
| Small tools, supplies ($200 per month) | — | M$ | 2.00 | — | 2.00 |
| Taxes and insurance on wages (at 24.9%) | — | M$ | — | 11.70 | 11.70 |
| Total General Contractor's direct overhead | — | M$ | 14.30 | 58.70 | 73.00 |

**Total General Contractor's Markup**  Typical costs for commercial and industrial projects.

|  | Craft@Hrs | Unit | Material | Labor | Total |
|---|---|---|---|---|---|
| Indirect overhead (home office overhead) | — | % | 2.0 | 6.0 | 8.0 |
| Direct overhead (job site overhead) | — | % | 1.4 | 5.9 | 7.3 |
| Contingency (allowance for unknown conditions) | — | % | — | — | 2.0 |
| Profit (varies widely, mid-range shown) | — | % | — | — | 7.5 |
| Total General Contractor's markup | — | % | — | — | 24.8 |

**Project Financing** Construction loans are usually made for a term of up to 18 months. The maximum loan will be based on the value of the land and building when completed. Typically this maximum is 75% for commercial, industrial and apartment buildings and 77.5% for tract housing.

The initial loan disbursement will be about 50% of the land cost. Periodic disbursements are based on the percentage of completion as verified by voucher or inspection. The last 20% of loan proceeds is disbursed when an occupancy permit is issued. Fund control fees cover the cost of monitoring disbursements. Typical loan fees, loan rates and fund control fees are listed below.

| | Craft@Hrs | Unit | Material | Labor | Total |
|---|---|---|---|---|---|
| Loan origination fee, based on amount of loan | | | | | |
| Tracts | — | % | — | — | 2.5 |
| Apartments | — | % | — | — | 2.0 |
| Commercial, industrial buildings | — | % | — | — | 3.0 |
| Loan interest rate, prime rate plus | | | | | |
| Tracts, commercial, industrial | — | % | — | — | 2.0 |
| Apartments | — | % | — | — | 1.5 |
| Fund control fees (costs per $1,000 disbursed) | | | | | |
| To $200,000 | — | LS | — | — | 1,000.00 |
| Over $200,000 to $500,000, $1,000 plus | — | M$ | — | — | 1.50 |
| Over $500,000 to $1,000,000, $1,450 plus | — | M$ | — | — | 1.00 |
| Over $1,000,000 to $3,000,000, $1,250 plus | — | M$ | — | — | .85 |
| Over $3,000,000 to $5,000,000, $1,150 plus | — | M$ | — | — | .75 |
| Over $5,000,000, $1,000 plus | — | M$ | — | — | .55 |

**Project Scheduling by CPM** (Critical Path Method). Includes consultation, review of construction documents, development of construction logic, and graphic schedule. Project scheduling software can be purchased for home use. Costs typically will be between $150.00 and $600.00

| | Craft@Hrs | Unit | Material | Labor | Total |
|---|---|---|---|---|---|
| Wood frame buildings, one or two stories | | | | | |
| Simple schedule | — | LS | — | — | 855.00 |
| Complex schedule | — | LS | — | — | 2,060.00 |
| Tilt-up concrete buildings, 10,000 to 50,000 SF | | | | | |
| Simple schedule | — | LS | — | — | 1,160.00 |
| Complex schedule | — | LS | — | — | 1,750.00 |
| Two to five story buildings (mid rise) | | | | | |
| Simple schedule | — | LS | — | — | 2,120.00 |
| Complex schedule | — | LS | — | — | 3,000.00 |
| Five to ten story buildings (high rise) | | | | | |
| Simple schedule | — | LS | — | — | 3,430.00 |
| Complex schedule | — | LS | — | — | 8,610.00 |
| Manufacturing plants and specialized use low rise buildings | | | | | |
| Simple schedule | — | LS | — | — | 1,520.00 |
| Complex schedule | — | LS | — | — | 2,110.00 |
| Project scheduling software for in-house production | | | | | |
| Complete package | — | LS | — | — | 7,470.00 |

Comprehensive schedules to meet government or owner specifications may cost 20% to 40% more. Daily schedule updates can increase costs by 40% to 50%.

**Sewer Connection Fees** Check with the local sanitation district for actual charges. These costs are typical for work done by city or county crews and include excavation, re-compaction and repairs to the street. Note that these costs are set by local government and will vary widely. These costs do not include capacity fees or other charges which are often levied on connection to new construction. The capacity fee per living unit may be $3,000 or more. For commercial buildings, figure 20 fixture units equal one living unit. A 6" sewer connection will be required for buildings that include more than 216 fixture units. (Bathtubs, showers and sinks are 2 fixture units, water closets are 4, clothes washers are 3 and lavatories are 1.) The costs shown assume that work will be done by sanitation district crews. Similar work done by a private contractor under a public improvement permit may cost 50% less. Typical charges based on main up to 11' deep and lateral up to 5' deep.

| | Craft@Hrs | Unit | Material | Labor | Total |
|---|---|---|---|---|---|
| 4" connection and 40' run to main in street | — | LS | — | — | 1,700.00 |
| 4" pipe runs over 40' to street | — | LF | — | — | 38.40 |
| 6" connection and 40' run to main in street | — | LS | — | — | 2,770.00 |
| 6" pipe runs over 40' to street | — | LF | — | — | 65.60 |
| 4" connection and 15' run to main in alley | — | LS | — | — | 1,230.00 |
| 6" connection and 15' run to main in alley | — | LS | — | — | 1,990.00 |
| Manhole cut-in | — | Ea | — | — | 338.00 |
| Add for main depths over 5' | | | | | |
| Over 5' to 8' | — | % | — | — | 30.0 |
| Over 8' to 11' | — | % | — | — | 60.0 |
| Over 11' | — | % | — | — | 100.0 |

**Water Meters** Check with the local water district for actual charges. These costs are typical for work done by city or county crews and include excavation and pipe to 40' from the main, meter, vault, re-compaction and repairs to the street. These costs do not include capacity fees or other charges which are often levied on connection to new construction. The capacity fee per living unit is typically $5,000. For commercial buildings, figure 20 fixture units equal one living unit. (Bathtubs, showers and sinks are 2 fixture units, water closets are 4, clothes washers are 3 and lavatories are 1.) The cost for discontinuing service will usually be about the same as the cost for starting new service, but no capacity fee will be charged. Add the backflow device and capacity fee, if required.

| | Craft@Hrs | Unit | Material | Labor | Total |
|---|---|---|---|---|---|
| 1" service, 3/4" meter | — | LS | — | — | 2,620.00 |
| 1" service, 1" meter | — | LS | — | — | 2,710.00 |
| Add for 1" service per LF over 40' | — | LF | — | — | 62.60 |
| 2" service, 1-1/2" meter | — | LS | — | — | 3,030.00 |
| 2" service, 2" meter | — | LS | — | — | 3,150.00 |
| Add for 2" service per LF over 40' | — | LF | — | — | 67.40 |
| Two 2" service lines and meter manifold | — | LS | — | — | 4,980.00 |
| Add for two 2" lines per LF over 40' | — | LF | — | — | 91.90 |

Hydrant meter, typical deposit is $1,000 to $2,000, typical water cost is $5 per 1,000 gallons

| | Craft@Hrs | Unit | Material | Labor | Total |
|---|---|---|---|---|---|
| Rental per day | — | Day | — | — | 11.10 |

Installation of meter only. Connected to existing water service if correct size line exists from the water main to the meter box location

| | Craft@Hrs | Unit | Material | Labor | Total |
|---|---|---|---|---|---|
| 3/4" meter | — | LS | — | — | 115.00 |
| 1" x 3/4" meter | — | LS | — | — | 110.00 |
| 1" meter | — | LS | — | — | 131.00 |
| 1-1/2" meter | — | LS | — | — | 424.00 |
| 2" meter | — | LS | — | — | 616.00 |
| 2" meter (double line) | — | LS | — | — | 1,090.00 |

Backflow preventer with single check valve. Includes pressure vacuum breaker (PVB), two ball valves, one air inlet and one check valve

| | Craft@Hrs | Unit | Material | Labor | Total |
|---|---|---|---|---|---|
| 1" pipe | — | LS | — | — | 260.00 |
| 2" pipe | — | LS | — | — | 513.00 |

Backflow preventer with double check valves. Includes pressure vacuum breaker (PVB), two ball or gate valves, one air inlet and two check valves

| | Craft@Hrs | Unit | Material | Labor | Total |
|---|---|---|---|---|---|
| 3/4" pipe | — | LS | — | — | 342.00 |
| 1" pipe | — | LS | — | — | 342.00 |
| 1-1/2" pipe | — | LS | — | — | 418.00 |
| 2" pipe | — | LS | — | — | 560.00 |

Reduced pressure backflow preventer. Includes pressure reducing vacuum breaker (PRVB), two ball or gate valves, one air inlet and two check valves and a relief valve

| | Craft@Hrs | Unit | Material | Labor | Total |
|---|---|---|---|---|---|
| 1" pipe | — | LS | — | — | 904.00 |
| 2" pipe | — | LS | — | — | 1,100.00 |
| Two 2" pipes (double installation) | — | LS | — | — | 2,160.00 |

# 01 General Requirements

| | Craft@Hrs | Unit | Material | Labor | Total |
|---|---|---|---|---|---|
| **Surveying** | | | | | |
| Surveying party, 2 or 3 technicians | | | | | |
|    Typical cost | — | Day | — | — | 1,240.00 |
|    Higher cost | — | Day | — | — | 1,600.00 |
| Data reduction and drafting | — | Hr | — | — | 75.80 |

**Surveys** Including wood hubs or pipe markers as needed. These figures assume that recorded monuments are available adjacent to the site.

| | Craft@Hrs | Unit | Material | Labor | Total |
|---|---|---|---|---|---|
| Residential lot, tract work, 4 corners | — | LS | — | — | 754.00 |
| Residential lot, individual, 4 corners | — | LS | — | — | 1,280.00 |
| Commercial lot, based on 30,000 SF lot | — | LS | — | — | 1,380.00 |
|    Over 30,000 SF, add per acre | — | Acre | — | — | 220.00 |

Lots without recorded markers cost more, depending on distance to the nearest recorded monument.

| | Craft@Hrs | Unit | Material | Labor | Total |
|---|---|---|---|---|---|
|    Add up to | — | % | — | — | 104.0 |

Some states require that a corner record be prepared and filed with the county surveyor's office when new monuments are set. The cost of preparing and filing a corner record will be about $350.00 to $400.00.

**Aerial Mapping** Typical costs based on a scale of 1" to 40'.

| | Craft@Hrs | Unit | Material | Labor | Total |
|---|---|---|---|---|---|
| Back up ground survey including flagging model | — | Ea | — | — | 3,540.00 |

(Typical "model" is approximately 750' by 1,260' or 25 acres)
Aerial photo flight, plane and crew, per local flight. One flight can cover several adjacent models

| | Craft@Hrs | Unit | Material | Labor | Total |
|---|---|---|---|---|---|
|    Per local flight | — | Ea | — | — | 440.00 |
| Vertical photos (2 required per model) | — | Ea | — | — | 48.00 |
| Data reduction and drafting, per hour | — | Hr | — | — | 66.00 |
| Complete survey with finished drawing, per model, including both field work and flying | | | | | |
|    Typical cost per model | — | Ea | — | — | 3,670.00 |
|    Deduct if ground control is marked on the ground by others | | | | | |
|    Per model | — | LS | — | — | -1,100.00 |
| Typical mapping cost per acre including aerial control, flight and compilation | | | | | |
|    Based on 25 acre coverage | — | Acre | — | — | 172.00 |
| Aerial photos, oblique (minimum 4 per job), one 8" x 10" print | | | | | |
|    Black and white | — | Ea | — | — | 46.80 |
|    Color | — | Ea | — | — | 100.00 |
| Quantitative studies for earthmoving operations to evaluate soil quantities moved | | | | | |
|    Ground control and one flight | — | Ea | — | — | 2,990.00 |
|    Each additional flight | — | Ea | — | — | 2,050.00 |

**Engineering Fees** Typical billing rates.

| | Craft@Hrs | Unit | Material | Labor | Total |
|---|---|---|---|---|---|
| Registered engineers (calculations, design and drafting) | | | | | |
|    Assistant engineer | — | Hr | — | — | 83.30 |
|    Senior engineer | — | Hr | — | — | 141.00 |
| Principal engineers (cost control, project management, client contact) | | | | | |
|    Assistant principal engineer | — | Hr | — | — | 149.00 |
|    Senior principal engineer | — | Hr | — | — | 200.00 |
| Job site engineer (specification compliance) | — | Hr | — | — | 83.30 |

**Estimating Cost** Compiling detailed labor and material cost estimate for commercial and industrial construction, in percent of job cost. Actual fees are based on time spent plus cost of support services.

| | Craft@Hrs | Unit | Material | Labor | Total |
|---|---|---|---|---|---|
| Most types of buildings, typical fees | | | | | |
|    Under $100,000 (range of .75 to 1.5%) | — | % | — | — | 1.1 |
|    Over $100,000 to $500,000 (range of .5 to 1.25%) | — | % | — | — | 0.9 |
|    Over $500,000 to $1,000,000 (range of .5 to 1.0%) | — | % | — | — | 0.8 |
|    Over $1,000,000 (range of .25 to .75%) | — | % | — | — | 0.5 |

# 01 General Requirements

|  | Craft@Hrs | Unit | Material | Labor | Total |
|---|---|---|---|---|---|
| Complex, high-tech, research or manufacturing facilities or other buildings with many unique design features, typical fees |  |  |  |  |  |
| Under $100,000 (range of 2.3 to 3%) | — | % | — | — | 2.7 |
| Over $100,000 to $500,000 (range of 1.5 to 2.5%) | — | % | — | — | 2.0 |
| Over $500,000 to $1,000,000 |  |  |  |  |  |
| (range of 1.0 to 2.0%) | — | % | — | — | 1.5 |
| Over $1,000,000 (range of .5 to 1.0%) | — | % | — | — | 0.8 |
| Typical estimator's fee per hour | — | Hr | — | — | 50.00 |

**Specifications Fee** Preparing detailed construction specifications to meet design requirements for commercial and industrial buildings, in percent of job cost. Actual fees are based on time spent plus cost of support services, if any.

|  | Craft@Hrs | Unit | Material | Labor | Total |
|---|---|---|---|---|---|
| Typical fees for most jobs | — | % | — | — | 0.5 |
| Small or complex jobs |  |  |  |  |  |
| Typical cost | — | % | — | — | 1.0 |
| High cost | — | % | — | — | 1.5 |
| Large jobs or repetitive work |  |  |  |  |  |
| Typical cost | — | % | — | — | 0.2 |
| High cost | — | % | — | — | 0.5 |

**Job Layout** Setting layout stakes for the foundation from monuments identified on the plans and existing on site. Per square foot of building first floor area. Typical costs. Actual costs will be based on time spent.

|  | Craft@Hrs | Unit | Material | Labor | Total |
|---|---|---|---|---|---|
| Smaller or more complex jobs | — | SF | — | — | .89 |
| Large job, easy work | — | SF | — | — | .56 |

**General Non-Distributable Supervision and Expense**

|  | Craft@Hrs | Unit | Material | Labor | Total |
|---|---|---|---|---|---|
| Project manager, typical | — | Mo | — | — | 6,590.00 |
| Job superintendent | — | Mo | — | — | 5,100.00 |
| Factory representative | — | Day | — | — | 787.00 |
| Timekeeper | — | Mo | — | — | 2,120.00 |
| Job site engineer for layout | — | Mo | — | — | 3,940.00 |
| Office manager | — | Mo | — | — | 2,380.00 |
| Office clerk | — | Mo | — | — | 1,750.00 |

**Mobilization** Typical costs for mobilization and demobilization within a 50 mile radius as a percent of total contract price. Includes cost of moving general contractor-owned equipment and job office to the site, setting up a fenced material yard, and removing equipment, office and fencing at job completion.

|  | Craft@Hrs | Unit | Material | Labor | Total |
|---|---|---|---|---|---|
| Allow, as a percentage of total contract price | — | % | — | — | 0.5 |

**Aggregate Testing**

|  | Craft@Hrs | Unit | Material | Labor | Total |
|---|---|---|---|---|---|
| Add for sample prep | — | Hr | — | — | 49.50 |
| Abrasion, L.A. Rattler, 100/500 cycles, ASTM C 131 | — | Ea | — | — | 121.00 |
| Absorption, ASTM C 127, 128 | — | Ea | — | — | 60.00 |
| Acid solubility | — | Ea | — | — | 35.00 |
| Aggregate test for mix design, including sieve analysis, specific gravity, number 200 wash, |  |  |  |  |  |
| organic impurities | — | Ea | — | — | 49.50 |
| Clay lumps and friable particles, ASTM C 142 | — | Ea | — | — | 66.00 |
| Coal and lignite, ASTM C 123 | — | Ea | — | — | 105.00 |
| Material finer than #200 sieve, ASTM C 117 | — | Ea | — | — | 41.00 |
| Organic impurities, ASTM C 40 | — | Ea | — | — | 52.00 |
| Percent crushed particles, CAL 205 | — | Ea | — | — | 69.00 |
| Percent flat or elongated particles, CRD C 119 | — | Ea | — | — | 110.00 |
| Potential reactivity, chemical method, 3 determinations per series, |  |  |  |  |  |
| ASTM C 289 | — | Ea | — | — | 347.00 |
| Potential reactivity, mortar bar method, ASTM C 227 | — | Ea | — | — | 330.00 |

| | Craft@Hrs | Unit | Material | Labor | Total |
|---|---|---|---|---|---|
| Sieve analysis, pit run aggregate | — | Ea | — | — | 69.00 |
| Lightweight aggregate ASTM C 123 | — | Ea | — | — | 77.00 |
| Sand equivalent | — | Ea | — | — | 88.00 |
| Sieve analysis, processed (each size), ASTM C 136 | — | Ea | — | — | 80.00 |
| Soft particles, ASTM C 235 | — | Ea | — | — | 69.00 |
| Soundness, sodium or magnesium, 5 cycle, per series, ASTM C 88 | — | Ea | — | — | 149.00 |
| Specific gravity, coarse, ASTM C 127 | — | Ea | — | — | 60.00 |
| Specific gravity, fine, ASTM C 127 | — | Ea | — | — | 83.00 |
| Unit weight per cubic foot, ASTM C 29 | — | Ea | — | — | 35.00 |

**Asphaltic Concrete Testing**  Minimum fee is usually $40.00.

| | Craft@Hrs | Unit | Material | Labor | Total |
|---|---|---|---|---|---|
| Asphaltic concrete core drilling, field or laboratory | | | | | |
| Per linear inch | — | Ea | — | — | 5.00 |
| Add for equipment and technician for core drilling | — | Hr | — | — | 70.00 |
| Asphaltic core density | — | Ea | — | — | 25.00 |
| Extraction, percent asphalt (Method B), excluding ash correction, ASTM D 2172 | — | Ea | — | — | 60.00 |
| Gradation on extracted sample (including wash), ASTM C 136 | — | Ea | — | — | 75.00 |
| Maximum density, lab mixed, Marshall, 1559 | — | Ea | — | — | 175.00 |
| Maximum density, premixed, Marshall, 1559 | — | Ea | — | — | 125.00 |
| Maximum theoretical unit weight (Rice Gravity), ASTM 2041 | — | Ea | — | — | 75.00 |
| Penetration, ASTM D 5 | — | Ea | — | — | 60.00 |
| Stability, lab mixed, Marshall, ASTM D 1559 | — | Ea | — | — | 250.00 |
| Stability, premixed, Marshall, ASTM D 1559 | — | Ea | — | — | 150.00 |

**Concrete Testing**  Minimum fee is usually $40.00.

| | Craft@Hrs | Unit | Material | Labor | Total |
|---|---|---|---|---|---|
| Aggregate tests for concrete mix designs only, including sieve analysis, specific gravity, number 200 wash, organic impurities, weight per cubic foot | | | | | |
| Per aggregate size | — | Ea | — | — | 72.00 |
| Amend or re-type existing mix designs | | | | | |
| Not involving calculations | — | Ea | — | — | 42.00 |
| Compression test, 2", 4", or 6" cores (excludes sample preparation) | | | | | |
| ASTM C 42 | — | Ea | — | — | 37.00 |
| Compression test, 6" x 12" cylinders, including mold | | | | | |
| ASTM C 39 | — | Ea | — | — | 13.00 |
| Core cutting in laboratory | — | Ea | — | — | 42.00 |
| Cylinder pickup within 40 miles of laboratory | | | | | |
| Cost per cylinder (minimum of 3) | — | Ea | — | — | 9.00 |
| Flexure test, 6" x 6" beams, ASTM C 78 | — | Ea | — | — | 53.00 |
| Laboratory trial, concrete batch, ASTM C 192 | — | Ea | — | — | 345.00 |
| Length change (3 bars, 4 readings, up to 90 days) | | | | | |
| ASTM C 157 modified | — | LS | — | — | 240.00 |
| Additional readings, per set of 3 bars | — | LS | — | — | 37.00 |
| Storage over 90 days, per set of 3 bars, per month | — | Mo | — | — | 37.00 |
| Modulus of elasticity test, static, ASTM C 469 | — | Ea | — | — | 73.00 |
| Mix design, determination of proportions | — | Ea | — | — | 79.00 |
| Pick up aggregate sample | | | | | |
| Within 40 mile radius of laboratory, per trip | — | Ea | — | — | 47.00 |
| Pickup or delivery of shrinkage molds, per trip | — | Ea | — | — | 32.00 |
| Prepare special strength documentation for mix design | — | Ea | — | — | 89.00 |

| | Craft@Hrs | Unit | Material | Labor | Total |
|---|---|---|---|---|---|
| Proportional analysis, cement factor and percent of aggregate | | | | | |
| Per analysis | — | Ea | — | — | 230.00 |
| Review mix design prepared by others | — | Ea | — | — | 68.00 |
| Splitting tensile, 6" x 12" cylinder, ASTM C 496 | — | Ea | — | — | 32.00 |
| Weight per cubic foot determination of lightweight concrete | | | | | |
| cylinders | — | Ea | — | — | 6.00 |

**Masonry and Tile Testing**

Brick, ASTM C 67

| | Craft@Hrs | Unit | Material | Labor | Total |
|---|---|---|---|---|---|
| Modulus of rupture (flexure) or compressive strength | — | Ea | — | — | 27.00 |
| Absorption initial rate, 5 hour or 24 hour | — | Ea | — | — | 22.00 |
| Boil, 1, 2, or 5 hours | — | Ea | — | — | 22.00 |
| Efflorescence | — | Ea | — | — | 27.00 |
| Dimensions, overall, coring, shell and web | — | Ea | — | — | 22.00 |
| Compression of core | — | Ea | — | — | 37.00 |
| Cores, shear, 6" and 8" diameter, 2 faces, per core | — | Ea | — | — | 47.00 |
| Coefficient of friction (slip test) | — | Ea | — | — | 84.00 |

Concrete block, ASTM C 140

| | Craft@Hrs | Unit | Material | Labor | Total |
|---|---|---|---|---|---|
| Moisture content as received | — | Ea | — | — | 22.00 |
| Absorption | — | Ea | — | — | 27.00 |
| Compression | — | Ea | — | — | 32.00 |
| Shrinkage, modified British, ASTM C 426 | — | Ea | — | — | 68.00 |
| Compression, 4", 6", 8" cores | — | Ea | — | — | 37.00 |
| Fireproofing, oven dry density | — | Ea | — | — | 32.00 |

Gunite

| | Craft@Hrs | Unit | Material | Labor | Total |
|---|---|---|---|---|---|
| Compression, 2", 4", 6" cores, ASTM C 42 | — | Ea | — | — | 37.00 |
| Pickup gunite field sample (40 mile trip maximum) | — | Ea | — | — | 53.00 |

Building stone

| | Craft@Hrs | Unit | Material | Labor | Total |
|---|---|---|---|---|---|
| Compression | — | Ea | — | — | 32.00 |
| Flex strength | — | Ea | — | — | 32.00 |
| Modules/rupture | — | Ea | — | — | 32.00 |
| Specific gravity | — | Ea | — | — | 32.00 |
| Water absorption | — | Ea | — | — | 32.00 |

Masonry prisms, ASTM E 447

| | Craft@Hrs | Unit | Material | Labor | Total |
|---|---|---|---|---|---|
| Compression test, grouted prisms | — | Ea | — | — | 136.00 |
| Pickup prisms, within 40 miles of lab | — | Ea | — | — | 37.00 |

Mortar and grout, UBC Standard 24-22 & 24-28

| | Craft@Hrs | Unit | Material | Labor | Total |
|---|---|---|---|---|---|
| Compression test, 2" x 4" mortar cylinder | — | Ea | — | — | 18.00 |
| Compression test, 3" x 6" grout prisms | — | Ea | — | — | 24.00 |
| Compression test, 2" cubes, ASTM C 109 | — | Ea | — | — | 47.00 |

Roof fill, lightweight ASTM C 495

| | Craft@Hrs | Unit | Material | Labor | Total |
|---|---|---|---|---|---|
| Compression test | — | Ea | — | — | 22.00 |
| Density | — | Ea | — | — | 16.00 |

Roofing tile

| | Craft@Hrs | Unit | Material | Labor | Total |
|---|---|---|---|---|---|
| Roofing tile breaking strength, UBC | — | Ea | — | — | 22.00 |
| Roofing tile absorption | — | Ea | — | — | 22.00 |

**Reinforcement Testing**

Sampling at fabricator's plant (within 40 miles of lab), $50.00 minimum

| | Craft@Hrs | Unit | Material | Labor | Total |
|---|---|---|---|---|---|
| During normal business hours, per sample | — | Ea | — | — | 18.00 |
| Other than normal business hours, per sample | — | Ea | — | — | 27.00 |
| Tensile test, mechanically spliced bar | — | Ea | — | — | 100.00 |
| Tensile test, number 11 bar or smaller, | | | | | |
| ASTM 615 | — | Ea | — | — | 26.00 |

# 01 General Requirements

| | Craft@Hrs | Unit | Material | Labor | Total |
|---|---|---|---|---|---|
| Tensile test, number 14 bar or smaller, ASTM 615 | — | Ea | — | — | 58.00 |
| Tensile test, number 18 bar, ASTM 615 | — | Ea | — | — | 68.00 |
| Tensile test, welded number 11 bar or smaller | — | Ea | — | — | 37.00 |
| Tensile test, welded number 14 bar | — | Ea | — | — | 68.00 |
| Tensile test, welded number 18 bar | — | Ea | — | — | 84.00 |
| Tensile and elongation test for pre-stress strands | | | | | |
|     24" test, ASTM A 416 | — | Ea | — | — | 84.00 |
|     10" test, ASTM A 421 | — | Ea | — | — | 47.00 |

**Soils and Aggregate Base Testing**  Minimum fee is usually $400.

| | Craft@Hrs | Unit | Material | Labor | Total |
|---|---|---|---|---|---|
| Bearing ratio (excluding moisture-density curve) | — | Ea | — | — | 110.00 |
| Consolidation test (single point) | — | Ea | — | — | 75.00 |
| Consolidation test (without rate data) | — | Ea | — | — | 95.00 |
| Direct shear test | | | | | |
|     Unconsolidated, undrained | — | Ea | — | — | 65.00 |
|     Consolidated, undrained | — | Ea | — | — | 100.00 |
|     Consolidated, drained | — | Ea | — | — | 125.00 |
| Durability index – coarse and fine | — | Ea | — | — | 140.00 |
| Expansion index test, UBC 29-2 | — | Ea | — | — | 110.00 |
| Liquid limit or plastic limit, Atterberg limits | — | Ea | — | — | 100.00 |
| Mechanical analysis – ASTM D1140 (wash 200 sieve) | — | Ea | — | — | 41.00 |
| Mechanical analysis – sand and gravel (wash sieve) | — | Ea | — | — | 116.00 |
| Moisture content | — | Ea | — | — | 15.00 |
| Moisture-density curve for compacted fill, ASTM 1557 | | | | | |
|     4-inch mold | — | Ea | — | — | 130.00 |
|     6-inch mold | — | Ea | — | — | 140.00 |
| Permeability (falling head) | — | Ea | — | — | 225.00 |
| Permeability (constant head) | — | Ea | — | — | 175.00 |
| Resistance value | — | Ea | — | — | 220.00 |
| Sand equivalent | — | Ea | — | — | 65.00 |
| Specific gravity – fine-grained soils | — | Ea | — | — | 60.00 |
| Unconfined compression test (undisturbed sample) | — | Ea | — | — | 70.00 |
| Unit dry weight and moisture content | | | | | |
|     (undisturbed sample) | — | Ea | — | — | 21.00 |

**Steel Testing**  $100.00 minimum for machine time.

| | Craft@Hrs | Unit | Material | Labor | Total |
|---|---|---|---|---|---|
| Charpy impact test, reduced temperature | — | Ea | — | — | 26.00 |
| Charpy impact test, room temperature | — | Ea | — | — | 13.00 |
| Electric extensometer | — | Ea | — | — | 63.00 |
| Hardness tests, Brinell or Rockwell | — | Ea | — | — | 22.00 |
| Photo micrographs | | | | | |
|     20× to 2,000× | — | Ea | — | — | 22.00 |
|     With negative | — | Ea | — | — | 26.00 |
| Tensile test, under 100,000 lbs | — | Ea | — | — | 37.00 |
| Welder Qualification Testing. Add $36 minimum per report | | | | | |
|     Tensile test | — | Ea | — | — | 37.00 |
|     Bend test | — | Ea | — | — | 26.00 |
|     Macro etch | — | Ea | — | — | 53.00 |
|     Fracture test | — | Ea | — | — | 32.00 |
|     Machining for weld tests, 1/2" and less | — | Ea | — | — | 37.00 |
|     Machining for weld tests, over 1/2" | — | Ea | — | — | 53.00 |

| | Craft@Hrs | Unit | Material | Labor | Total |
|---|---|---|---|---|---|

**Construction Photography** Digital, professional quality work, hand printed, labeled, numbered and dated. Also high resolution photographs on CD-ROM or lower resolution for emailing purposes.
Job progress photos, includes shots of all building corners, two 8" x 10" color prints of each shot, per visit

| | Craft@Hrs | Unit | Material | Labor | Total |
|---|---|---|---|---|---|
| Typical cost | — | LS | — | — | 550.00 |
| High cost | — | LS | — | — | 630.00 |

Periodic survey visits to job site to figure angles (for shots of all building corners)

| | | | | | |
|---|---|---|---|---|---|
| Minimum two photos, per visit | — | Ea | — | — | 355.00 |

Exterior completion photographs, two 8" x 10" prints (two angles)

| | | | | | |
|---|---|---|---|---|---|
| Black and white | — | Ea | — | — | 550.00 |
| Color | — | Ea | — | — | 550.00 |

Interior completion photographs, with lighting setup, two 8" x 10" prints of each shot (minimum three shots)

| | | | | | |
|---|---|---|---|---|---|
| Black and white | — | Ea | — | — | 630.00 |
| Color | — | Ea | — | — | 630.00 |

Reprints or enlargements from existing digital file. Hand (custom) prints

| | | | | | |
|---|---|---|---|---|---|
| 8" x 10", black and white | — | Ea | — | — | 25.00 |
| 11" x 14", black and white | — | Ea | — | — | 35.00 |
| 8" x 10", color | — | Ea | — | — | 25.00 |
| 11" x 14", color | — | Ea | — | — | 35.00 |

Retouching services (taking out phone lines, power poles, autos, signs, etc.) will run about $100.00 per hour of computer time.

**Signs and Markers** Pipe markers, self-stick vinyl, by overall outside diameter of pipe or covering, meets OSHA and ANSI requirements. Costs are based on standard wording and colors.

| | Craft@Hrs | Unit | Material | Labor | Total |
|---|---|---|---|---|---|
| 3/4" to 1-3/8" diameter, 1/2" letters, 8" long | PA@.077 | Ea | 2.90 | 3.96 | 6.86 |
| 1-1/2" to 2 3/8" diameter, 3/4" letters, 8" long | PA@.077 | Ea | 2.90 | 3.96 | 6.86 |
| 2 1/2" to 7 7/8" diameter, 1-1/4" letters, 12" long | PA@.077 | Ea | 4.60 | 3.96 | 8.56 |
| 8" to 10" diameter, 2-1/2" letters, 24" long | PA@.095 | Ea | 6.50 | 4.89 | 11.39 |
| Flow arrows, 2" wide tape x 108' roll, perforated every 8" | | | | | |
| (use $50.00 as minimum material cost) | PA@.095 | LF | .52 | 4.89 | 5.41 |

Barrier tapes, 3" wide x 1,000' long. "Caution," "Danger," "Open Trench," and other stock wording
(Use $35.00 as minimum material cost.) Labor shown is based on one man installing 500 LF in 1 hour.

| | Craft@Hrs | Unit | Material | Labor | Total |
|---|---|---|---|---|---|
| 3 mil thick, short-term or indoor use | PA@.002 | LF | .06 | .10 | .16 |
| 4 mil thick, long-term or outdoor use | PA@.002 | LF | .07 | .10 | .17 |

Traffic control signage, "No Parking Between Signs," "Customer Parking" and other stock traffic control signs and graphics, reflective aluminum signs except as noted.

| | Craft@Hrs | Unit | Material | Labor | Total |
|---|---|---|---|---|---|
| 12" x 18" | PA@.455 | Ea | 54.00 | 23.40 | 77.40 |
| 18" x 24" | PA@.455 | Ea | 94.50 | 23.40 | 117.90 |
| 24" x 8" Diamond grade (highly reflective) | PA@.455 | Ea | 100.00 | 23.40 | 123.40 |
| 24" x 12", Diamond grade (highly reflective) | PA@.455 | Ea | 114.10 | 23.40 | 137.50 |
| 24" x 18" | PA@.455 | Ea | 94.00 | 23.40 | 117.40 |
| 24" x 24" | PA@.455 | Ea | 106.00 | 23.40 | 129.40 |
| 30" x 30" | PA@.455 | Ea | 168.00 | 23.40 | 191.40 |
| Add for 8' breakaway pole and mounting | CL@1.00 | Ea | 92.00 | 39.90 | 131.90 |
| Add for aluminum signs | — | Ea | 5.00 | — | 5.00 |

**Temporary Utilities** No permit fees included.
Drop pole for electric service
Power pole installation, wiring, 6 months rent and pickup

| | | | | | |
|---|---|---|---|---|---|
| 100 amp meter pole | — | Ea | — | — | 471.00 |
| 200 amp meter pole | — | Ea | — | — | 619.00 |

| | Craft@Hrs | Unit | Material | Labor | Total |
|---|---|---|---|---|---|
| Electric drop line and run to contractor's panel (assuming proper phase, voltage and service are available within 100' of site) | | | | | |
| Overhead run to contractor's "T" pole (including removal) | | | | | |
| Single phase, 100 amps | — | LS | — | — | 378.00 |
| Single phase, over 100 to 200 amps | — | LS | — | — | 511.00 |
| Three phase, to 200 amps | — | LS | — | — | 588.00 |
| Underground run in contractor's trench and conduit (including removal) | | | | | |
| Single phase, 100 amps, "USA" wire | — | LS | — | — | 545.00 |
| Single phase, 101-200 amps, "USA" wire | — | LS | — | — | 705.00 |
| Three phase, to 200 amps, "USA" wire | — | LS | — | — | 781.00 |
| Typical power cost | | | | | |
| Per kilowatt hour ($5.00 minimum) | — | Ea | — | — | .13 |
| Per 1,000 SF of building per month of construction | — | MSF | — | — | 2.15 |
| Temporary water service, meter installed on fire hydrant. Add the cost of distribution hose or piping. | | | | | |
| Meter fee, 2" meter (excluding $400 deposit) | — | LS | — | — | 78.60 |
| Water cost, per 1,000 gallons | — | M | — | — | 1.75 |
| Job phone, typical costs including message unit charges | | | | | |
| Cellular phone and one year service | — | LS | — | — | 482.00 |
| Portable job site toilets, including delivery, weekly service, and pickup, per 28 day month, six month contract | | | | | |
| Good quality (polyethylene) | — | Mo | — | — | 70.00 |
| Deluxe quality, with sink, soap dispenser, towel dispenser, mirror | — | Mo | — | — | 110.00 |
| Trailer-mounted male and female compartment double, 2 stalls in each, sink, soap dispenser, mirror and towel dispenser in each compartment | | | | | |
| Rental per week | — | Wk | — | — | 299.00 |

**Construction Elevators and Hoists**

| | Craft@Hrs | Unit | Material | Labor | Total |
|---|---|---|---|---|---|
| Hoisting towers to 200', monthly rent | | | | | |
| Material hoist | CO@176. | Mo | 3,300.00 | 9,720.00 | 13,020.00 |
| Personnel hoist | CO@176. | Mo | 8,000.00 | 9,720.00 | 17,720.00 |
| Tower cranes, self-climbing (electric), monthly rent | | | | | |
| 5 ton lift at 180' radius, one operator | CO@176. | Mo | 19,600.00 | 9,720.00 | 29,320.00 |
| With two operators | CO@352. | Mo | 19,600.00 | 19,400.00 | 39,000.00 |
| 7 ton lift at 170' radius, one operator | CO@176. | Mo | 27,600.00 | 9,720.00 | 37,320.00 |
| With two operators | CO@352. | Mo | 27,600.00 | 19,400.00 | 47,000.00 |
| Mobilize, erect and dismantle self-climbing tower cranes | | | | | |
| Good access | — | LS | — | — | 55,000.00 |
| Average access | — | LS | — | — | 64,200.00 |
| Poor access | — | LS | — | — | 100,000.00 |

**Hydraulic Truck Cranes** Includes equipment rental and operator. Four hour minimum. Time charge begins at the rental yard. By rated crane capacity.

| | Craft@Hrs | Unit | Material | Labor | Total |
|---|---|---|---|---|---|
| 12-1/2 ton | — | Hr | — | — | 182.00 |
| 15 ton | — | Hr | — | — | 189.00 |
| 25 ton | — | Hr | — | — | 194.00 |
| 50 ton | — | Hr | — | — | 253.00 |
| 70 ton | — | Hr | — | — | 281.00 |
| 80 ton | — | Hr | — | — | 313.00 |
| 100 ton | — | Hr | — | — | 342.00 |

| | Craft@Hrs | Unit | Material | Labor | Total |
|---|---|---|---|---|---|
| 125 ton | — | Hr | — | — | 365.00 |
| 150 ton | — | Hr | — | — | 387.00 |
| Add for clamshell or breaker ball work | — | Hr | — | — | 46.00 |

**Conventional Cable Cranes** Including equipment rental and operator. Four hour minimum. Time charge begins at the rental yard. By rated crane capacity with 80' boom.

| | Craft@Hrs | Unit | Material | Labor | Total |
|---|---|---|---|---|---|
| 50 tons | — | Hr | — | — | 289.00 |
| 70 tons | — | Hr | — | — | 318.00 |
| 90 tons | — | Hr | — | — | 365.00 |
| 125 tons | — | Hr | — | — | 508.00 |
| 140 tons | — | Hr | — | — | 396.00 |

**Equipment Rental Rates** Costs are for equipment in good condition but exclude accessories. Includes the costs of damage waiver (about 10% of the rental cost), insurance, fuel, and oil. Add the cost of an equipment operator, pickup, return to yard and repairs, if necessary. Delivery and pickup within 30 miles of the rental yard are usually offered at an additional cost of one day's rent. A rental "day" is assumed to be one 8-hour shift and begins when the equipment leaves the rental yard. A "week" is 40 hours in five consecutive days. A "month" is 176 hours in 30 consecutive days.

| | Day | Week | Month |
|---|---|---|---|
| **Air equipment rental** | | | |
| Air compressors, wheel-mounted | | | |
| 16 CFM, shop type, electric | 54.00 | 161.00 | 481.00 |
| 30 CFM, shop type, electric | 74.00 | 221.00 | 663.00 |
| 80 CFM, shop type, electric | 86.00 | 261.00 | 782.00 |
| 100 CFM, gasoline unit | 107.00 | 321.00 | 903.00 |
| 125 CFM, gasoline unit | 133.00 | 401.00 | 1,200.00 |
| Air compressors wheel-mounted diesel units | | | |
| to 159 CFM | 101.00 | 300.00 | 903.00 |
| 160 - 249 CFM | 133.00 | 401.00 | 1,200.00 |
| 250 - 449 CFM | 202.00 | 603.00 | 1,810.00 |
| 450 - 749 CFM | 268.00 | 802.00 | 2,410.00 |
| 750 - 1,200 CFM | 534.00 | 1,610.00 | 4,810.00 |
| 1,400 CFM and over | 669.00 | 2,010.00 | 6,820.00 |
| Paving breakers (no bits included) hand-held, pneumatic | | | |
| to 40 lb | 30.00 | 87.00 | 235.00 |
| 41 - 55 lb | 35.00 | 90.00 | 250.00 |
| 56 - 70 lb | 35.00 | 110.00 | 290.00 |
| 71 - 90 lb | 35.00 | 110.00 | 300.00 |
| Paving breakers jackhammer bits | | | |
| Moil points, 15" to 18" | 6.00 | 18.00 | 38.00 |
| Chisels, 3" | 7.00 | 20.00 | 40.00 |
| Clay spades, 5-1/2" | 15.00 | 45.00 | 90.00 |
| Asphalt cutters, 5" | 9.00 | 25.00 | 50.00 |
| Rock drills, jackhammers, hand-held, pneumatic | | | |
| to 29 lb | 25.00 | 80.00 | 215.00 |
| 30 - 39 lb | 40.00 | 130.00 | 360.00 |
| 40 - 49 lb | 35.00 | 130.00 | 340.00 |
| 50 lb and over | 30.00 | 105.00 | 305.00 |
| Rock drill bits, hexagonal | | | |
| 7/8", 2' long | 10.00 | 30.00 | 55.00 |
| 1", 6' long | 8.00 | 30.00 | 60.00 |

| | Day | Week | Month |
|---|---|---|---|
| Pneumatic chippers, medium weight, 10 lb | 35.00 | 105.00 | 290.00 |
| Pneumatic tampers, medium weight, 30 lb | 33.00 | 110.00 | 315.00 |
| Pneumatic grinders, hand-held, 7 lb (no stone) | 25.00 | 95.00 | 280.00 |
| Air hose, 50 LF section | | | |
|     5/8" air hose | 6.00 | 15.00 | 40.00 |
|     3/4" air hose | 7.00 | 20.00 | 65.00 |
|     1" air hose | 10.00 | 30.00 | 70.00 |
|     1-1/2" air hose | 21.00 | 70.00 | 165.00 |
| **Compaction equipment rental** | | | |
| Vibro plate, 300 lb, 24" plate width, gas | 80.00 | 265.00 | 775.00 |
| Vibro plate, 600 lb, 32" plate width, gas | 140.00 | 460.00 | 1,350.00 |
| Rammer, 60 CPM, 200 lb, gas powered | 70.00 | 218.00 | 610.00 |
| Rollers, two axle steel drum, road type | | | |
|     1 ton, gasoline | 270.00 | 855.00 | 2,500.00 |
|     6 - 8 ton, gasoline | 410.00 | 1,250.00 | 3,720.00 |
|     8 - 10 ton | 525.00 | 1,580.00 | 4,680.00 |
| Rollers, rubber tired, self propelled | | | |
|     12 ton,  9 wheel, diesel | 331.00 | 998.00 | 2,940.00 |
|     15 ton,  9 wheel, diesel | 457.00 | 1,950.00 | 4,140.00 |
|     30 ton, 11 wheel, diesel | 462.00 | 1,570.00 | 4,430.00 |
| Rollers, towed, sheepsfoot type | | | |
|     40" diameter, double drum, 28" wide | 126.00 | 382.00 | 1,150.00 |
|     60" diameter, double drum, 54" wide | 234.00 | 695.00 | 2,080.00 |
| Rollers, self-propelled vibratory dual sheepsfoot drum, gas | | | |
|     25 HP, 26" wide roll | 370.00 | 1,250.00 | 3,780.00 |
|     100 HP, 36" wide roll | 590.00 | 2,450.00 | 7,850.00 |
| Rollers, vibrating steel drum, gas powered, walk behind | | | |
|     7 HP, 1,000 lb, one drum, 2' wide | 90.00 | 310.00 | 835.00 |
|     10 HP, 2,000 lb, two drum, 2'-6" wide | 130.00 | 480.00 | 1,300.00 |
| Rollers, self-propelled riding type, vibrating steel drum | | | |
|     9,000 lb, gasoline, two drum | 225.00 | 690.00 | 1,920.00 |
|     15,000 lb, diesel, two drum | 445.00 | 1,290.00 | 3,830.00 |
|     25,000 lb, diesel, two drum | 550.00 | 1,610.00 | 4,700.00 |
| **Concrete equipment rental** | | | |
| Buggies, push type, 7 CF | 60.00 | 220.00 | 510.00 |
| Buggies, walking type, 12 CF | 75.00 | 225.00 | 750.00 |
| Buggies, riding type, 14 CF | 85.00 | 300.00 | 810.00 |
| Buggies, riding type, 18 CF | 115.00 | 360.00 | 725.00 |
| Vibrator, electric, 3 HP, flexible shaft | 40.00 | 130.00 | 335.00 |
| Vibrator, gasoline, 6 HP, flexible shaft | 63.00 | 184.00 | 551.00 |
| Troweling machine, 36", 4 paddle | 50.00 | 185.00 | 475.00 |
| Cement mixer, 6 CF | 55.00 | 200.00 | 590.00 |
| Towable mixer (10+ CF) | 70.00 | 155.00 | 640.00 |
| Power trowel 36", gas, with standard handle | 50.00 | 185.00 | 475.00 |
| Bull float | 15.00 | 45.00 | 110.00 |
| Concrete saws, gas powered, excluding blade cost | | | |
|     10 HP, push type | 65.00 | 215.00 | 590.00 |
|     20 HP, self propelled | 95.00 | 325.00 | 910.00 |
|     37 HP, self propelled | 140.00 | 510.00 | 1,450.00 |

|  | Day | Week | Month |
|---|---|---|---|
| Concrete bucket, bottom dump 1 CY | 45.00 | 135.00 | 380.00 |
| Plaster mixer, gas, portable, 5 CF | 55.00 | 200.00 | 590.00 |
| Concrete conveyor, belt type, portable, gas powered | | | |
|    All sizes | 130.00 | 500.00 | 1,325.00 |
| Vibrating screed, 16', single beam | 65.00 | 220.00 | 640.00 |
| Column clamps, to 48", per set | — | — | 6.50 |

**Crane rental** Add for clamshell, dragline, or pile driver attachments from below.

| Cable controlled crawler-mounted lifting cranes, diesel | Day | Week | Month |
|---|---|---|---|
|    45 ton | — | 2,420 | 6,830.00 |
|    60 ton | — | 3,670.00 | 10,200.00 |
|    100 ton (125 ton) | 1,300.00 | 5,100.00 | 15,000.00 |
|    150 ton | 2,000.00 | 5,900.00 | 18,000.00 |
| Cable controlled, truck-mounted lifting cranes, diesel | | | |
|    50 ton | 675.00 | 1,990.00 | 6,800.00 |
|    75 ton | 860.00 | 2,570.00 | 7,750.00 |
|    100 ton | 1,250.00 | 3,740.00 | 9,750.00 |
| Hydraulic – rough terrain | | | |
|    10 ton, gasoline | 570.00 | 1,400.00 | 3,800.00 |
|    17 ton, diesel | 485.00 | 1,550.00 | 4,100.00 |
|    35 ton, diesel | 780.00 | 2,250.00 | 6,500.00 |
|    50 ton, diesel | 1,350.00 | 4,000.00 | 11,200.00 |

**Clamshell, dragline, and pile driver attachments** Add to crane rental costs shown above.

| Clamshell buckets | Day | Week | Month |
|---|---|---|---|
|    Rehandling type, 1 CY | — | 320.00 | 970.00 |
|    General purpose, 2 CY | 80.00 | 510.00 | 1,370.00 |
| Dragline buckets | | | |
|    1 CY bucket | 46.00 | 116.00 | 325.00 |
|    2 CY bucket | 87.00 | 221.00 | 662.00 |
| Pile drivers, hammers | | | |
|    25,000 - 49,999 ft-lb | 515.00 | 2,390.00 | 4,270.00 |
|    75 - 99 tons vibratory | 1,300.00 | 3,100.00 | 8,000.00 |
|    1,000 - 1,999 ft-lb extractor | — | 710.00 | 1,840.00 |

**Excavation equipment rental**

| Backhoe/loaders, crawler mounted, diesel | Day | Week | Month |
|---|---|---|---|
|    3/8 CY backhoe | 350.00 | 1,000.00 | 3,000.00 |
|    3/4 CY backhoe | 410.00 | 1,480.00 | 3,350.00 |
|    1-1/2 CY backhoe | 485.00 | 1,480.00 | 4,350.00 |
|    2 CY backhoe | 725.00 | 2,280.00 | 7,050.00 |
|    2-1/2 CY backhoe | 860.00 | 2,720.00 | 8,100.00 |
|    3 CY backhoe | 1,510.00 | 4,500.00 | 13,500.00 |
| Backhoe/loaders, wheel mounted, diesel or gasoline | | | |
|    1/2 CY bucket capacity, 55 HP | 185.00 | 650.00 | 1,850.00 |
|    1 CY bucket capacity, 65 HP | 270.00 | 820.00 | 2,450.00 |
|    1-1/4 CY bucket capacity, 75 HP | 280.00 | 720.00 | 2,100.00 |
|    1-1/2 CY bucket capacity, 100 HP | 265.00 | 830.00 | 2,410.00 |
|    Mounted with 750-lb. breaker | 480.00 | 1,670.00 | 5,000.00 |
|    Mounted with 2,000-lb. breaker | 714.00 | 2,500.00 | 7,510.00 |

|  | Day | Week | Month |
|---|---|---|---|
| **Crawler dozers, with angle dozer or bulldozer blade, diesel** | | | |
| 65 HP, D-3 | 340.00 | 1,060.00 | 3,100.00 |
| 90-105 HP, D-4 or D-5 | 460.00 | 1,320.00 | 3,870.00 |
| 140 HP, D-6 | 710.00 | 2,150.00 | 6,400.00 |
| 200 HP, D-7 | 1,160.00 | 3,300.00 | 10,200.00 |
| 335 HP, D-8 | 1,650.00 | 4,920.00 | 15,100.00 |
| 460 HP, D-9 | 2,590.00 | 7,100.00 | 21,300.00 |
| **Crawler loaders, diesel** | | | |
| 3/4 CY loader | 410.00 | 1,150.00 | 3,400.00 |
| 1-1/2 CY loader | 485.00 | 1,480.00 | 4,350.00 |
| 2 CY loader | 725.00 | 2,280.00 | 7,050.00 |
| **Hydraulic excavators, crawler mounted with backhoe type arm, diesel** | | | |
| .2 - .4 CY bucket | 350.00 | 1,060.00 | 3,200.00 |
| .4 - .6 CY bucket | 385.00 | 1,130.00 | 3,350.00 |
| .6 - .8 CY bucket | 500.00 | 1,350.00 | 4,030.00 |
| **Gradall, truck mounted, diesel** | | | |
| 3/4 CY bucket | 1,000.00 | 3,000.00 | 8,600.00 |
| 1 CY bucket | 1,150.00 | 3,900.00 | 10,100.00 |
| **Front-end loader, diesel or gas (Skiploader)** | | | |
| 40-hp | 120.00 | 450.00 | 1,300.00 |
| 60-hp | 205.00 | 600.00 | 1,800.00 |
| 80-hp | 215.00 | 715.00 | 2,100.00 |
| **Skid steer loaders** | | | |
| Bobcat 643, 1,000-lb capacity | 155.00 | 535.00 | 1,410.00 |
| Bobcat 753, 1,350-lb capacity | 180.00 | 600.00 | 1,680.00 |
| Bobcat 763, 1,750-lb capacity | 200.00 | 640.00 | 1,830.00 |
| Bobcat 863, 1,900-lb capacity | 220.00 | 680.00 | 2,000.00 |
| **Skid steer attachments** | | | |
| Auger | 90.00 | 280.00 | 780.00 |
| Hydraulic breaker | 170.00 | 535.00 | 1,600.00 |
| Backhoe | 105.00 | 360.00 | 2,000.00 |
| Sweeper | 115.00 | 380.00 | 1,080.00 |
| Grapple bucket | 60.00 | 166.00 | 515.00 |
| **Wheel loaders, front-end load and dump, diesel** | | | |
| 3/4 CY bucket, 4WD, articulated | 235.00 | 690.00 | 2,000.00 |
| 1 CY bucket, 4WD, articulated | 290.00 | 850.00 | 2,480.00 |
| 2 CY bucket, 4WD, articulated | 340.00 | 1,050.00 | 3,150.00 |
| 3-1/4 CY bucket, 4WD, articulated | 700.00 | 2,200.00 | 6,500.00 |
| 5 CY bucket, 4WD, articulated | 1,000.00 | 3,050.00 | 9,200.00 |
| **Motor scraper-hauler, 2 wheel tractor** | | | |
| Under 18 CY capacity, single engine | 1,260.00 | 4,450.00 | 14,200.00 |
| 18 CY and over, single engine | 1,780.00 | 6,650.00 | 20,400.00 |
| Under 18 CY capacity, Two engine | 2,300.00 | 6,300.00 | 20,000.00 |
| 18 CY and over, Two engine | 2,400.00 | 7,000.00 | 24,200.00 |
| **Graders, diesel, pneumatic tired, articulated** | | | |
| 100 HP | 460.00 | 1,280.00 | 4,100.00 |
| 150 HP | 800.00 | 2,300.00 | 7,100.00 |
| 200 HP | 930.00 | 2,950.00 | 8,800.00 |
| **Trenchers, inclined chain boom type, pneumatic tired** | | | |
| 15 HP, 12" wide, 48" max. depth, walking | 190.00 | 660.00 | 1,850.00 |
| 20 HP, 12" wide, 60" max. depth, riding | 230.00 | 775.00 | 2,200.00 |
| 55 HP, 18" wide, 96" max. depth, riding | 450.00 | 1,550.00 | 4,600.00 |
| 100 HP, 24" x 96" deep, crawler mount | 700.00 | 2,600.00 | 7,700.00 |

|  | Day | Week | Month |
|---|---|---|---|
| Trenching shields, steel, 8' high x 12' long | | | |
|    Single wall | 110.00 | 350.00 | 1,000.00 |
|    Hydraulic trench braces, 8' long, to 42" wide | 21.00 | 62.00 | 190.00 |
| Truck-mounted wagon drills, self propelled, crawler mounted | | | |
|    4" drifter, straight boom | 505.00 | 1,330.00 | 3,730.00 |
|    4-1/2" drifter, hydraulic swing boom | 734.00 | 2,240.00 | 6,620.00 |
| Truck rental | | | |
|    1-1/2 ton truck | 87.00 | 260.00 | 781.00 |
|    Flatbed with boom | 470.00 | 1,430.00 | 4,140.00 |
| Dump trucks, rental rate plus mileage | | | |
|    3 CY | 210.00 | 550.00 | 1,570.00 |
|    5 CY | 230.00 | 820.00 | 2,550.00 |
|    10 CY | 280.00 | 1,070.00 | 3,010.00 |
|    25 CY off highway | — | 3,770.00 | 11,300.00 |

**Forklift rental**

| Rough terrain, towable, gas powered | | | |
|---|---|---|---|
|    4,000 lb capacity | 250.00 | 700.00 | 1,800.00 |
|    8,000 lb capacity | 235.00 | 775.00 | 2,250.00 |
| Extension boom, 2 wheel steering | | | |
|    4,000 lb, 30' lift | 240.00 | 725.00 | 2,100.00 |
|    6,000 lb, 35' lift | 220.00 | 650.00 | 1,900.00 |
|    8,000 lb, 40' lift, 4 wheel steering | 310.00 | 700.00 | 2,610.00 |

**Aerial platform rental**

| Rolling scissor lifts, 2' x 3' platform, 650 lb capacity, push around, electric powered | | | |
|---|---|---|---|
|    30' high | 65.00 | 185.00 | 550.00 |
|    40' high | 140.00 | 380.00 | 980.00 |
|    50' high | 204.00 | 555.00 | 1,510.00 |
| Rolling scissor lifts, self-propelled, hydraulic, electric powered | | | |
|    To 20' | 85.00 | 225.00 | 580.00 |
|    21' - 30' | 110.00 | 290.00 | 750.00 |
|    31' - 40' | 155.00 | 440.00 | 1,260.00 |
| Rolling scissor lifts, self-propelled, hydraulic, diesel powered | | | |
|    To 20' | 110.00 | 280.00 | 675.00 |
|    21' - 30' | 165.00 | 475.00 | 1,300.00 |
|    31' - 40' | 180.00 | 575.00 | 1,550.00 |
| Boomlifts, telescoping and articulating booms, self-propelled, gas or diesel powered, 2-wheel drive | | | |
|    21' - 30' | 225.00 | 675.00 | 1,890.00 |
|    31' - 40' | 230.00 | 630.00 | 1,780.00 |
|    41' - 50' | 240.00 | 650.00 | 1,800.00 |
|    51' - 60' | 275.00 | 900.00 | 2,500.00 |

**Pump rental** Hoses not included.

| Submersible pump rental | | | |
|---|---|---|---|
|    1-1/2" to 2", electric | 34.00 | 105.00 | 290.00 |
|    3" to 4", electric | 75.00 | 250.00 | 675.00 |
|    All diameters, hydraulic | 240.00 | 720.00 | 2,230.00 |
| Trash pumps, self-priming, gas | | | |
|    1-1/2" connection | 39.00 | 130.00 | 330.00 |
|    4" connection | 65.00 | 210.00 | 520.00 |
|    6" connection | 170.00 | 500.00 | 1,500.00 |

| | Day | Week | Month |
|---|---|---|---|
| Diaphragm pumps | | | |
| 2", gas, single action | 40.00 | 130.00 | 365.00 |
| 4", gas, double action | 90.00 | 290.00 | 750.00 |
| Hose, coupled, 25 LF sections | | | |
| 2" suction line | 12.00 | 32.00 | 87.00 |
| 1-1/2" discharge line | 7.00 | 20.00 | 55.00 |
| 3" discharge line | 12.00 | 35.00 | 90.00 |
| **Sandblasting equipment rental** | | | |
| Compressor and hopper | | | |
| To 250 PSI | 80.00 | 270.00 | 710.00 |
| Over 250 to 300 PSI | 85.00 | 300.00 | 865.00 |
| Over 600 PSI to 1,000 PSI | 130.00 | 420.00 | 1,180.00 |
| Sandblasting accessories | | | |
| Hoses, 50', coupled | | | |
| 3/4" sandblast hose or air hose | 8.00 | 30.00 | 80.00 |
| 1" air hose | 12.00 | 40.00 | 105.00 |
| 1-1/4" air hose or sandblast hose | 18.00 | 65.00 | 180.00 |
| Nozzles, all types | 27.00 | 83.00 | 250.00 |
| Valve, remote control (deadman), all sizes | 39.00 | 48.00 | 144.00 |
| Air-fed hood | 22.00 | 66.00 | 198.00 |
| **Roadway equipment rental** | | | |
| Boom mounted pavement breaker, hydraulic 1,500 lb | 390.00 | 1,230.00 | 3,550.00 |
| Paving breaker, backhoe mount, pneumatic, 1,000 lb | 245.00 | 800.00 | 2,250.00 |
| Asphalt curbing machine | 105.00 | 320.00 | 966.00 |
| Paving machine diesel, 10' width, self propelled | 1,300.00 | 5,250.00 | 15,700.00 |
| Distribution truck for asphalt prime coat, 3,000 gallon | 1,100.00 | 3,310.00 | 9,800.00 |
| Pavement striper, 1 line, walk behind | 50.00 | 200.00 | 550.00 |
| Water blast truck for 1,000 PSI pavement cleaning | 1,390.00 | 3,800.00 | 12,500.00 |
| Router-groover for joint or crack sealing in pavement | 224.00 | 761.00 | 2,290.00 |
| Sealant pot for joint or crack sealing | 270.00 | 842.00 | 2,530.00 |
| Self-propelled street sweeper, vacuum (7 CY max.) | 1,100.00 | 3,300.00 | 9,800.00 |
| Water trucks/tanks | | | |
| 1,800-gallon water truck | 220.00 | 760.00 | 2,260.00 |
| 3,500-gallon water truck | 378.00 | 1,530.00 | 4,360.00 |
| Water tower | 200.00 | 575.00 | 1,650.00 |
| Tow behind trailer, 500 gal. | 80.00 | 210.00 | 600.00 |
| Off-road water tanker, 10,000 gal. | 330.00 | 1,140.00 | 3,390.00 |
| **Traffic control equipment rental** | | | |
| Arrow board – 25 lights | 65.00 | 240.00 | 690.00 |
| Barricades | 12.00 | 25.00 | 50.00 |
| Barricades with flashers | 8.00 | 22.00 | 65.00 |
| Delineators | 6.00 | 10.00 | 17.00 |

|  | Day | Week | Month |
|---|---|---|---|
| **Drill rental** | | | |
| 1/2" cordless drill, 18 volt | 11.00 | 42.00 | 125.00 |
| 3/8" or 1/2", electric drill | 11.00 | 42.00 | 125.00 |
| 3/4" electric drill | 26.00 | 105.00 | 315.00 |
| 1/2" right angle electric drill | 28.00 | 106.00 | 315.00 |
| 1/2" electric hammer drill | 50.00 | 185.00 | 550.00 |
| **Electric generator rental** | | | |
| 2.5 kw, gasoline | 40.00 | 170.00 | 500.00 |
| 5 kw, gasoline | 60.00 | 235.00 | 700.00 |
| 15 kw, diesel | 115.00 | 365.00 | 1,040.00 |
| 60 kw, diesel | 265.00 | 750.00 | 2,170.00 |
| **Floor equipment rental** | | | |
| Carpet blower | 30.00 | 120.00 | 360.00 |
| Carpet cleaner | 25.00 | 105.00 | 300.00 |
| Carpet kicker | 15.00 | 60.00 | 175.00 |
| Carpet power stretcher | 28.00 | 105.00 | 315.00 |
| Carpet stapler, electric | 16.00 | 63.00 | 189.00 |
| Carpet seamer, electric | 15.00 | 60.00 | 170.00 |
| Floor edger, 7", electric | 30.00 | 120.00 | 350.00 |
| Floor maintainer | 42.00 | 165.00 | 500.00 |
| Grinder 9" | 25.00 | 100.00 | 290.00 |
| Mini grinder, 4-1/2" | 19.00 | 63.00 | 189.00 |
| Pergo installation kit | 32.00 | 105.00 | 315.00 |
| Tile roller | 20.00 | 75.00 | 220.00 |
| Tile stripper, electric | 55.00 | 210.00 | 624.00 |
| Tile stripper, manual | 9.00 | 34.00 | 101.00 |
| Vacuum, 10 gallons, wet/dry | 45.00 | 180.00 | 550.00 |
| **Hammer rental** | | | |
| Electric brute breaker | 65.00 | 250.00 | 750.00 |
| Gas breaker | 65.00 | 255.00 | 760.00 |
| Demolition hammer, electric | 50.00 | 190.00 | 565.00 |
| Roto hammer, 7/8", electric | 50.00 | 205.00 | 610.00 |
| Roto hammer, 1-1/2", electric | 45.00 | 170.00 | 490.00 |
| **Heater rental**, fan forced | | | |
| Kerosene heater, to 200 MBtu | 38.00 | 140.00 | 390.00 |
| Kerosene heater, over 200 to 300 MBtu | 30.00 | 155.00 | 370.00 |
| Kerosene heater, over 500 to 1,000 MBtu | 65.00 | 250.00 | 680.00 |
| Salamanders, LP gas (no powered fan) | 40.00 | 140.00 | 365.00 |
| **Ladder rental** | | | |
| Step ladder, 6' or 8' | 15.00 | 55.00 | 160.00 |
| Step ladder, 10' or 14' | 20.00 | 75.00 | 220.00 |
| Extension ladder, 24' | 32.00 | 125.00 | 365.00 |
| Extension ladder, 28' or 32' | 33.00 | 124.00 | 365.00 |
| Extension ladder, 40' | 39.00 | 148.00 | 440.00 |
| Stairway step ladder | 16.00 | 63.00 | 189.00 |

# 01 General Requirements

| | Day | Week | Month |
|---|---|---|---|
| **Landscape equipment rental** | | | |
| Aerator, gas | 66.00 | 270.00 | 790.00 |
| Air broom backpack, gas | 34.00 | 132.00 | 400.00 |
| Auger, 2 man (gas) | 65.00 | 250.00 | 750.00 |
| 6" chipper, gas | 180.00 | 550.00 | 1,620.00 |
| Brush cutter, gas | 40.00 | 150.00 | 440.00 |
| Brush hog, 24" | 58.00 | 230.00 | 675.00 |
| Clamshovel, manual | 11.00 | 36.00 | 106.00 |
| Digging bar, 5' | 11.00 | 42.00 | 126.00 |
| Edger, lawn, gas | 27.00 | 110.00 | 330.00 |
| Fertilizer spreader | 10.00 | 36.00 | 106.00 |
| Hand auger, 6" | 13.00 | 45.00 | 131.00 |
| Hedge trimmer, 24", electric | 16.00 | 65.00 | 195.00 |
| Hedge trimmer, 30", electric | 25.00 | 100.00 | 300.00 |
| Hedge trimmer, 30", gas | 25.00 | 100.00 | 300.00 |
| Lawn roller | 15.00 | 65.00 | 185.00 |
| Log splitter, gas | 90.00 | 350.00 | 1,100.00 |
| Overseeder, gas | 58.00 | 200.00 | 600.00 |
| Power rake, gas | 52.00 | 210.00 | 630.00 |
| Sod cutter | 70.00 | 285.00 | 850.00 |
| Stump grinder, 9 HP | 105.00 | 410.00 | 1,210.00 |
| Tiller, light duty, gas | 68.00 | 265.00 | 770.00 |
| Tiller, medium duty, gas | 65.00 | 250.00 | 730.00 |
| Tiller, heavy duty, gas | 80.00 | 310.00 | 950.00 |
| Tree pruning shears | 15.00 | 50.00 | 150.00 |
| Weed eater, gas | 32.00 | 125.00 | 375.00 |
| Wheelbarrow | 12.00 | 45.00 | 135.00 |
| | | | |
| **Level rental** | | | |
| Laser level, small | 45.00 | 160.00 | 505.00 |
| Level kit | 60.00 | 210.00 | 630.00 |
| Transit level kit | 41.00 | 170.00 | 500.00 |
| | | | |
| **Light tower rental** | | | |
| Floodlight, 1,000 watt, on stand | 35.00 | 115.00 | 340.00 |
| To 7,000 watt trailer mounted light set, gas | 100.00 | 300.00 | 840.00 |
| Over 7,000 watt trailer mounted light set, gas | 120.00 | 390.00 | 1,000.00 |
| Extension cord, 50' | 10.00 | 32.00 | 100.00 |
| | | | |
| **Nailer and stapler rental** | | | |
| Drywall screwdriver | 15.00 | 60.00 | 170.00 |
| Finish nailer, pneumatic or cordless | 32.00 | 125.00 | 365.00 |
| Floor stapler, pneumatic | 35.00 | 130.00 | 390.00 |
| Framing nailer, cordless | 32.00 | 125.00 | 370.00 |
| Framing nailer, pneumatic, 1-1/2" to 3-1/4" | 32.00 | 125.00 | 370.00 |
| Hardwood floor nailer, 3/4" | 20.00 | 85.00 | 250.00 |
| Low velocity nail gun, single | 16.00 | 63.00 | 189.00 |
| Low velocity gun, multi | 33.00 | 125.00 | 370.00 |
| Roofing nailer, pneumatic, 1/2" to 2" | 40.00 | 150.00 | 440.00 |
| Screw gun | 37.00 | 147.00 | 440.00 |
| Underlay spot nailer | 16.00 | 63.00 | 189.00 |

|  | Day | Week | Month |
|---|---|---|---|
| **Painting equipment rental** | | | |
| Airless paint sprayer – large | 75.00 | 265.00 | 790.00 |
| Heat gun blower | 10.00 | 34.00 | 101.00 |
| Mud mixing paddle | 9.00 | 34.00 | 100.00 |
| Pressure washer – cold, 1,500 PSI | 58.00 | 210.00 | 700.00 |
| Pressure washer – cold, 2,000 PSI | 70.00 | 270.00 | 800.00 |
| Pressure washer – cold, 3,500 PSI | 85.00 | 325.00 | 1,000.00 |
| Texture sprayer with compressor | 58.00 | 350.00 | 1,050.00 |
| Wallpaper steamer, electric | 30.00 | 110.00 | 315.00 |
| **Plumbing and piping equipment rental** | | | |
| Basin wrench, 3/8" to 1-1/4" | 13.00 | 38.00 | 105.00 |
| Chain wrench | 10.00 | 36.00 | 106.00 |
| Conduit & pipe cutter | 16.00 | 53.00 | 158.00 |
| Drain cleaner, corded drill, 3/8" | 35.00 | 140.00 | 410.00 |
| Laser pipe level | 85.00 | 300.00 | 850.00 |
| Pipe reamer, 1/2" to 2" | 12.00 | 45.00 | 125.00 |
| Pipe stand | 14.00 | 50.00 | 151.00 |
| Pipe threader, 1/2" to 2" | 15.00 | 50.00 | 150.00 |
| Pipe wrench, 18" or 24" | 15.00 | 50.00 | 151.00 |
| Pipe wrench, 36" | 15.00 | 50.00 | 150.00 |
| Sewer snake, 100' x 5/8", electric | 60.00 | 240.00 | 730.00 |
| Sewer snake, 25' x 5/16", manual | 21.00 | 85.00 | 255.00 |
| Sewer snake, 50' x 1/2", electric | 50.00 | 195.00 | 580.00 |
| Sewer snake, 50' x 1/2", manual | 32.00 | 125.00 | 365.00 |
| Soil pipe cutter, ratchet, 6" | 15.00 | 50.00 | 150.00 |
| Toilet auger | 11.00 | 43.00 | 125.00 |
| **Sanding equipment rental** | | | |
| Belt sander, 3" x 21" | 17.00 | 63.00 | 189.00 |
| Belt sander, 4" x 24" | 25.00 | 100.00 | 290.00 |
| Drywall sander and vacuum | 50.00 | 190.00 | 560.00 |
| Floor sander, drum type | 45.00 | 180.00 | 530.00 |
| Floor sander, square | 45.00 | 170.00 | 510.00 |
| Orbit sander | 20.00 | 75.00 | 220.00 |
| Palm sander | 15.00 | 50.00 | 150.00 |
| Vibrator sander | 20.00 | 75.00 | 220.00 |
| **Saw rental** | | | |
| Band saw, electric | 30.00 | 120.00 | 360.00 |
| Bolt cutters, 24" | 11.00 | 35.00 | 110.00 |
| Bolt cutters, 36" | 10.00 | 45.00 | 125.00 |
| Chain saws, 18", gasoline | 50.00 | 195.00 | 580.00 |
| Chop saw, 14", electric | 37.00 | 150.00 | 440.00 |
| Circular saw, 8-1/4" (electric) | 18.00 | 60.00 | 180.00 |
| Compound miter saw | 38.00 | 150.00 | 440.00 |
| Jamb saw | 28.00 | 110.00 | 315.00 |
| Jig saw, electric | 22.00 | 75.00 | 220.00 |
| Masonry table saw, 2 HP | 68.00 | 250.00 | 750.00 |
| Reciprocating saw, electric | 20.00 | 75.00 | 220.00 |
| Router, 2 HP, fixed base | 32.00 | 126.00 | 378.00 |
| Sliding compound miter saw | 43.00 | 160.00 | 490.00 |
| Table saw, 10", electric | 43.00 | 165.00 | 490.00 |
| Tile cutter, manual | 12.00 | 44.00 | 132.00 |
| Tile saw and stand, wet cut | 55.00 | 210.00 | 625.00 |

# 01 General Requirements

|  | Day | Week | Month |
|---|---|---|---|
| **Scaffolding rental** | | | |
| 5' high, with casters | 32.00 | 32.00 | 95.00 |
| 10' high, with casters | 45.00 | 45.00 | 135.00 |
| 15' high, with casters | 65.00 | 65.00 | 190.00 |
| 5' high, with leg jacks | 30.00 | 30.00 | 80.00 |
| 10' high with leg jacks | 45.00 | 45.00 | 130.00 |
| 15' high, with leg jacks | 60.00 | 60.00 | 176.00 |
| 14' wide with base plates | 40.00 | 40.00 | 118.00 |
| 21' wide with base plates | 55.00 | 55.00 | 170.00 |
| 14' wide with leg jacks | 45.00 | 45.00 | 130.00 |
| 21' wide with leg jacks | 60.00 | 60.00 | 178.00 |
| 14' wide with casters | 45.00 | 45.00 | 135.00 |
| 21' wide with casters | 65.00 | 65.00 | 190.00 |
| Drywall panel lift, 5 piece | 26.00 | 105.00 | 115.00 |
| Drywall scaffold | 30.00 | 115.00 | 340.00 |
| Scissor lift | 60.00 | 190.00 | 550.00 |
| | | | |
| **Welding equipment rental,** with helmet | | | |
| To 200 amp | 55.00 | 200.00 | 520.00 |
| 201 to 300 amp | 75.00 | 240.00 | 615.00 |
| | | | |
| **Miscellaneous equipment rental** | | | |
| Aluminum break, 10' | 53.00 | 210.00 | 630.00 |
| Appliance dolly | 14.00 | 50.00 | 150.00 |
| Chain hoist, 1 ton, 20' lift | 21.00 | 84.00 | 252.00 |
| Chain hoist, 5 ton, 20' lift | 34.00 | 126.00 | 378.00 |
| Come-a-long, 3,000 pounds | 16.00 | 63.00 | 189.00 |
| Dehumidifier, electric | 25.00 | 100.00 | 290.00 |
| Dolly, 4 wheel | 12.00 | 43.00 | 125.00 |
| Fan, low profile | 15.00 | 45.00 | 125.00 |
| Fan, 30", with pedestal | 30.00 | 100.00 | 260.00 |
| Fishtape, 200', 1/8" | 12.00 | 46.00 | 139.00 |
| Hydraulic jack, 12 ton | 13.00 | 50.00 | 151.00 |
| Loading ramps | 9.00 | 21.00 | 63.00 |
| Metal detector | 16.00 | 63.00 | 189.00 |
| Pavement line marker | 16.00 | 63.00 | 189.00 |
| Screwjack, 10 ton | 11.00 | 42.00 | 126.00 |

**Scaffolding Rental** Exterior scaffolding, tubular steel, 60" wide, with 6'4" open end frames set at 7', with cross braces, base plates, mud sills, adjustable legs, post-mounted guardrail, climbing ladders and landings, brackets, clamps and building ties. Including two 2" x 10" scaffold planks on all side brackets and six 2" x 10" scaffold planks on scaffold where indicated. Add local delivery and pickup. Minimum rental is one month.

| | Craft@Hrs | Unit | Material | Labor | Total |
|---|---|---|---|---|---|
| Costs are per square foot of building wall covered, per month | | | | | |
| With plank on scaffold only | — | SF | .96 | — | .96 |
| With plank on side brackets and scaffold | — | SF | 1.20 | — | 1.20 |
| Add for erection and dismantling, level ground, truck accessible | | | | | |
| With plank on scaffold only | CL@.008 | SF | — | .32 | .32 |
| With planks on side brackets and scaffold | CL@.012 | SF | — | .48 | .48 |

| | Craft@Hrs | Unit | Material | Labor | Total |
|---|---|---|---|---|---|
| Caster-mounted scaffolding, 30" wide by 7' or 10' long, rental. Minimum charge is one month | | | | | |
| 6' high | — | Mo | 120.00 | — | 120.00 |
| 7' high | — | Mo | 130.00 | — | 130.00 |
| 10' high | — | Mo | 140.00 | — | 140.00 |
| 13' high | — | Mo | 150.00 | — | 150.00 |
| 15' high | — | Mo | 160.00 | — | 160.00 |
| 20' high | — | Mo | 170.00 | — | 170.00 |
| Hook-end scaffold plank, rental, each | — | Mo | 15.00 | — | 15.00 |
| Caster-mounted interior scaffold, purchase. Light duty | | | | | |
| 51" high x 4' long x 22" wide | — | Ea | 192.00 | — | 192.00 |
| 72" high x 56" long x 25" wide | — | Ea | 425.00 | — | 425.00 |
| Plain-end micro-lam scaffold plank, rental each | | | | | |
| 9' length | — | Mo | 3.50 | — | 3.50 |
| 12' length | — | Mo | 4.50 | — | 4.50 |
| Aluminum extension scaffold plank, purchase. 250-pound load capacity. 2"-deep box section extrusions. Plank length extends in 12" increments. OSHA compliant | | | | | |
| 8' to 13' long, 14" wide | — | Ea | 207.00 | — | 207.00 |
| Swinging stage, 10', complete, motor operated, purchase | CL@4.00 | Ea | 1,100.00 | 160.00 | 1,260.00 |

**Heavy Duty Shoring** Adjustable vertical tower type shoring. Rated capacity 5.5 tons per leg. Base frames are 4' wide x 6' high or 2' wide x 5' high. Extension frames are 4' wide x 5'4" high or 2' wide x 4'4" high. Frames are erected in pairs using cross-bracing to form a tower. Combinations of base frames and extension frames are used to reach the height required. Screw jacks with base plates are installed in the bottom of each leg. Similar screw jacks with "U"-bracket heads are installed in the top of each leg. Material costs shown are rental rates based on a 1-month minimum. Add the cost of delivery and pickup. For scheduling purposes, estimate that a crew of 2 can unload, handle and erect 8 to 10 frames (including typical jacks and heads) per hour. Dismantling, handling and moving or loading will require nearly the same time.

| | Craft@Hrs | Unit | Material | Labor | Total |
|---|---|---|---|---|---|
| Frames (any size), including braces and assembly hardware | | | | | |
| Rental per frame per month | — | Mo | 11.60 | — | 11.60 |
| Screw jacks (four required per frame with base plate or "U"-head) | | | | | |
| Rental per jack per month | — | Mo | 4.85 | — | 4.85 |
| Add for erecting each frame | CL@.245 | Ea | — | 9.77 | 9.77 |
| Add for dismantling each frame | CL@.197 | Ea | — | 7.86 | 7.86 |

**Temporary Structures** Portable job site office trailers with electric air conditioners and baseboard heat, built-in desks, plan tables, insulated and paneled walls, tiled floors, locking doors and windows with screens, closet, fluorescent lights, and electrical receptacles. Add $30 per month for units with toilet and lavatory. Monthly rental based on 6 month minimum. Dimensions are length, width and height overall and floor area (excluding hitch area).

| | Craft@Hrs | Unit | Material | Labor | Total |
|---|---|---|---|---|---|
| 16' x  8' x  7' high (108 SF) | — | Mo | — | — | 130.00 |
| 20' x  8' x  7' high (160 SF) | — | Mo | — | — | 140.00 |
| 24' x  8' x  7' high (192 SF) | — | Mo | — | — | 150.00 |
| 28' x  8' x  7' high (240 SF) | — | Mo | — | — | 160.00 |
| 28' x 10' x 10' high (360 SF) | — | Mo | — | — | 200.00 |
| 40' x 10' x 10' high (440 SF) | — | Mo | — | — | 232.00 |
| 48' x 10' x 10' high (500 SF) | — | Mo | — | — | 263.00 |
| 40' x 12' x 10' high (552 SF) | — | Mo | — | — | 263.00 |
| 60' x 12' x 10' high (720 SF) | — | Mo | — | — | 345.00 |
| 60' x 14' x 10' high (840 SF) | — | Mo | — | — | 355.00 |
| Add for skirting, per LF of perimeter | — | LF | — | — | 9.30 |
| Deduct for unit without heating & cooling | — | Mo | — | — | -25.00 |

| | Craft@Hrs | Unit | Material | Labor | Total |
|---|---|---|---|---|---|
| Add for delivery, setup, and dismantling of office trailers within 15 miles | | | | | |
|   Delivery and setup | — | LS | — | — | 200.00 |
|   Dismantle and pickup | — | LS | — | — | 200.00 |
| Add for delivery or pickup over 15 miles, per mile | | | | | |
|   8' wide units | — | Mile | — | — | 2.00 |
|   10' wide units | — | Mile | — | — | 2.90 |
|   12' wide units | — | Mile | — | — | 3.50 |
|   14' wide units | — | Mile | — | — | 3.90 |
| Portable steel storage containers (lockable), suitable for job site storage of tools and materials. Rental per month, based on 6 month rental | | | | | |
|   8' x 8' x 20' | — | Mo | — | — | 69.00 |
|   8' x 8' x 24' | — | Mo | — | — | 89.60 |
|   8' x 8' x 40' | — | Mo | — | — | 139.60 |
| Add for delivery and pickup, within 15 miles | | | | | |
|   One time charge | — | LS | — | — | 190.00 |
| Portable job site shacks with lights, power receptacles, locking door and window. Rental per month, based on 6 month rental | | | | | |
|   12' x 8' x 8' | — | Mo | — | — | 70.00 |
|   8' x 8' x 8' | — | Mo | — | — | 60.00 |
| Add for typical delivery and pickup, within 15 miles | | | | | |
|   One time charge | — | LS | — | — | 70.00 |

**Temporary Enclosures** Rented chain link fence and accessories. Costs are a one-time charge for up to six months usage on a rental basis. Costs include installation and one trip for removal, and assume level site with truck access along pre-marked fence line. Add for gates and barbed wire as shown. Minimum charge is $350.

| | Craft@Hrs | Unit | Material | Labor | Total |
|---|---|---|---|---|---|
| Chain link fence, 6' high | | | | | |
|   Less than 250 feet | — | LF | — | — | 1.86 |
|   250 to 500 feet | — | LF | — | — | 1.80 |
|   501 to 750 feet | — | LF | — | — | 1.70 |
|   751 to 1,000 feet | — | LF | — | — | 1.43 |
|   Over 1,000 feet | — | LF | — | — | 1.25 |
| Add for gates | | | | | |
|   6' x 10' single | — | Ea | — | — | 116.00 |
|   6' x 12' single | — | Ea | — | — | 146.00 |
|   6' x 15' single | — | Ea | — | — | 190.00 |
|   6' x 20' double | — | Ea | — | — | 220.00 |
|   6' x 24' double | — | Ea | — | — | 294.00 |
|   6' x 30' double | — | Ea | — | — | 366.00 |
| Add for barbed wire | | | | | |
|   per strand, per linear foot | — | LF | — | — | .21 |
| Contractor furnished items, installed and removed, based on single use and no salvage value | | | | | |
|   Railing on stairway, two sides, 2" x 4" | CL@.121 | LF | .53 | 4.83 | 5.36 |
| Guardrail at second and higher floors | | | | | |
|   toe rail, mid rail and top rail | CL@.124 | LF | 3.18 | 4.95 | 8.13 |
| Plywood barricade fence | | | | | |
|   Bolted to pavement, 8' high | CL@.430 | LF | 12.70 | 17.10 | 29.80 |
|   Post in ground, 8' high | CL@.202 | LF | 10.60 | 8.06 | 18.66 |
|   8' high with 4' wide sidewalk cover | CL@.643 | LF | 33.90 | 25.60 | 59.50 |

# 01 General Requirements

| | Craft@Hrs | Unit | Material | Labor | Total |
|---|---|---|---|---|---|
| **Cleanup** | | | | | |
| Progressive "broom clean" cleanup, per 1,000 SF of floor per cleaning | | | | | |
|    Typical cost | CL@.183 | MSF | — | 7.30 | 7.30 |
| Final, total floor area (no glass cleaning) | CL@1.65 | MSF | — | 65.80 | 65.80 |

Glass cleaning, per 1,000 square feet of glass cleaned on one side. (Double these figures when both sides are cleaned.) Add the cost of staging or scaffolding when needed

| | Craft@Hrs | Unit | Material | Labor | Total |
|---|---|---|---|---|---|
|    Cleaning glass with sponge and squeegee | CL@1.34 | MSF | 16.10 | 53.40 | 69.50 |
|    Cleaning glass and window trim | | | | | |
|       with solvent, towel, sponge and squeegee | CL@5.55 | MSF | 20.20 | 221.00 | 241.20 |
|    Removing paint and concrete splatter from windows and trim | | | | | |
|       by scraping and solvent | CL@15.9 | MSF | 50.50 | 634.00 | 684.50 |
|    Mop resilient floor by hand | CL@1.20 | MSF | 1.85 | 47.90 | 49.75 |

## Waste Disposal

Dump fees. Tippage charges for solid waste disposal at the dump vary from $30 to $120 per ton. For planning purposes, estimate waste disposal at $75 per ton plus the hauling cost. Call the dump or trash disposal company for actual charges. Typical costs are shown below.

| | Craft@Hrs | Unit | Material | Labor | Total |
|---|---|---|---|---|---|
|    Dumpster, 3 CY trash bin, emptied weekly | — | Mo | — | — | 350.00 |
|    Dumpster, 40 CY solid waste bin (lumber, drywall, roofing) | | | | | |
|       Hauling cost, per load | — | Ea | — | — | 250.00 |
|       Add to per load charge, per ton | — | Ton | — | — | 60.00 |
|    Low-boy, 14 CY solid waste container (asphalt, dirt, masonry, concrete) | | | | | |
|       Hauling cost, per load (use 7 CY as maximum load) | — | Ea | — | — | 250.00 |
|       Add to per load charge, per ton | — | Ton | — | — | 60.00 |

Recycler fees. Recycling construction waste materials can substantially reduce disposal costs. Recycling charges vary from $95 to $120 per load, depending on the type material and the size of the load. Call the recycling company for actual charges. Typical recycler fees are shown below. Add the cost for hauling.

| | Craft@Hrs | Unit | Material | Labor | Total |
|---|---|---|---|---|---|
|    Green waste | — | Ton | — | — | 30.00 |
|    Asphalt, per load (7 CY) | — | Ea | — | — | 100.00 |
|    Concrete, masonry or rock, per load (7 CY) | — | Ea | — | — | 100.00 |
|    Dirt, per load (7 CY) | — | Ea | — | — | 100.00 |
|    Mixed loads, per load (7 CY) | — | Ea | — | — | 100.00 |

# 02 Existing Conditions

**Paving and Curb Demolition** No salvage of materials. These costs include the cost of loading and hauling to a legal dump within 6 miles. Dump fees are not included. See Waste Disposal. Equipment cost includes one wheel-mounted air compressor, one paving breaker and jackhammer bits, one 55 HP wheel loader with integral backhoe and one 5 CY dump truck. The figures in parentheses give the approximate loose volume of the materials (volume after being demolished). Add the cost of saw cutting, if required. Use $500.00 as a minimum charge.

| | Craft@Hrs | Unit | Material | Labor | Equipment | Total |
|---|---|---|---|---|---|---|
| Demolish bituminous paving, depths to 3" (10 SY per CY) | | | | | | |
|    Large area, with a wheel loader | C3@.038 | SY | — | 1.77 | 1.02 | 2.79 |
|    Strips 24" wide for utility lines | C3@.044 | SY | — | 2.05 | 1.18 | 3.23 |
|    Add for jobs under 50 SY | C3@.010 | SY | — | .46 | .27 | .73 |
| Demolish bituminous curbs, | | | | | | |
|    to 12" width (65 LF per CY) | C3@.028 | LF | — | 1.30 | .75 | 2.05 |

# 02 Existing Conditions

| | Craft@Hrs | Unit | Material | Labor | Equipment | Total |
|---|---|---|---|---|---|---|
| **Demolish concrete sidewalk or paving** | | | | | | |
| 4" concrete with no reinforcing | | | | | | |
| (1 CY of waste per 75 SF demolished) | C3@.045 | SF | — | 2.09 | 1.21 | 3.30 |
| 4" concrete with mesh but no rebars | | | | | | |
| (1 CY of waste per 65 SF demolished) | C3@.050 | SF | — | 2.32 | 1.34 | 3.66 |
| 4" concrete with rebars | | | | | | |
| (1 CY of waste per 55 SF demolished) | C3@.058 | SF | — | 2.70 | 1.56 | 4.26 |
| 5" concrete with mesh but no rebars | | | | | | |
| (1 CY of waste per 55 SF demolished) | C3@.063 | SF | — | 2.93 | 1.69 | 4.62 |
| 5" concrete with rebars | | | | | | |
| (1 CY of waste per 46 SF demolished) | C3@.073 | SF | — | 3.39 | 1.96 | 5.35 |
| 6" concrete with mesh but no rebars | | | | | | |
| (1 CY of waste per 45 SF demolished) | C3@.078 | SF | — | 3.63 | 2.10 | 5.73 |
| 6" concrete with rebars | | | | | | |
| (1 CY of waste per 38 SF demolished) | C3@.089 | SF | — | 4.14 | 2.39 | 6.53 |
| 8" concrete with mesh but no rebars | | | | | | |
| (1 CY of waste per 32 SF demolished) | C3@.109 | SF | — | 5.07 | 2.93 | 8.00 |
| 8" concrete with rebars | | | | | | |
| (1 CY of waste per 27 SF demolished) | C3@.125 | SF | — | 5.81 | 3.36 | 9.17 |
| **Concrete over 8" to 12" thick** | | | | | | |
| Per CY without rebars | | | | | | |
| (2 CY of waste per 1 CY demolished) | C3@4.60 | CY | — | 214.00 | 124.00 | 338.00 |
| Per CY with rebars | | | | | | |
| (2.4 CY of waste per 1 CY demolished) | C3@5.29 | CY | — | 246.00 | 142.00 | 388.00 |
| **Demolish concrete curb** | | | | | | |
| Curb and 24" monolithic gutter | | | | | | |
| (7 LF per CY) | C3@.100 | LF | — | 4.65 | 2.69 | 7.34 |
| Planter and batter type curbs, 6" wide | | | | | | |
| (30 LF per CY) | C3@.043 | LF | — | 2.00 | 1.16 | 3.16 |
| **Remove pavement markings by water blasting. Use $100.00 as a minimum charge** | | | | | | |
| Water blaster, per hour | — | Hour | — | — | 10.70 | 10.70 |
| 4" wide strips | CL@.032 | LF | — | 1.28 | .34 | 1.62 |
| Per square foot | CL@.098 | SF | — | 3.91 | 1.05 | 4.96 |

**Fencing demolition** Equipment includes one 5 CY dump truck. No salvage except as noted. These costs include loading and hauling to a legal dump within 6 miles. Dump fees are not included. See Waste Disposal. Use $250.00 as a minimum charge.

| | Craft@Hrs | Unit | Material | Labor | Equipment | Total |
|---|---|---|---|---|---|---|
| Remove chain link fence, | | | | | | |
| To 4' high | C4@.030 | LF | — | 1.24 | .35 | 1.59 |
| 5' to 8' high | C4@.055 | LF | — | 2.28 | .64 | 2.92 |
| Remove and salvage chain link fence, | | | | | | |
| To 8' high | C4@.073 | LF | — | 3.03 | .86 | 3.89 |
| Remove and dispose wood fence | | | | | | |
| Picket or board fence to 4' high | C4@.056 | LF | — | 2.32 | .66 | 2.98 |
| Picket or board fence 5' to 8' high | C4@.080 | LF | — | 3.32 | .94 | 4.26 |
| Split rail fence, 2 or 3 rail | C4@.035 | LF | — | 1.45 | .41 | 1.86 |
| Remove and dispose wood or chain link fence gate | | | | | | |
| To 8' high, including one gate post | C4@.333 | LF | — | 13.80 | 3.90 | 17.70 |

| | Craft@Hrs | Unit | Material | Labor | Equipment | Total |
|---|---|---|---|---|---|---|

**Highway type guardrail demolition** These costs include loading and hauling to a legal dump within 6 miles. Dump fees are not included. See Waste Disposal. No salvage except as noted. Equipment includes one wheel-mounted air compressor, one paving breaker and jackhammer bits, one 55 HP wheel loader with integral backhoe and one 5 CY dump truck. Use $500.00 as a minimum charge.

| | Craft@Hrs | Unit | Material | Labor | Equipment | Total |
|---|---|---|---|---|---|---|
| Remove and dispose guardrail | C4@.043 | LF | — | 1.78 | 1.16 | 2.94 |
| Remove guardrail in salvage condition | C4@.074 | LF | — | 3.07 | 1.99 | 5.06 |
| Remove and dispose guardrail posts | C3@.245 | Ea | — | 11.40 | 6.59 | 17.99 |

**Manhole, Piping and Underground Tank Demolition** No salvage of materials except as noted. These costs include loading and hauling to a legal dump within 6 miles. Dump fees are not included. See Waste Disposal. Equipment includes one wheel-mounted air compressor, one paving breaker and jackhammer bits, one 55 HP wheel loader with integral backhoe and one 5 CY dump truck. Use $400.00 as a minimum charge.
Manholes and catch basins, demolition, to 10' deep. Break below collar and plug

| | Craft@Hrs | Unit | Material | Labor | Equipment | Total |
|---|---|---|---|---|---|---|
| Brick | C3@5.16 | Ea | — | 240.00 | 139.00 | 379.00 |
| Masonry | C3@5.33 | Ea | — | 248.00 | 143.00 | 391.00 |
| Precast concrete | C3@6.64 | Ea | — | 309.00 | 179.00 | 488.00 |
| Add for sand fill, any of above | C3@.125 | CY | 23.20 | 5.81 | 3.36 | 32.37 |

Frame and cover from manhole or catch basin

| | Craft@Hrs | Unit | Material | Labor | Equipment | Total |
|---|---|---|---|---|---|---|
| Remove in salvage condition | C3@1.45 | Ea | — | 67.40 | 39.00 | 106.40 |
| Remove and reset | C3@3.99 | Ea | — | 185.00 | 107.00 | 292.00 |

Fire hydrant demolition

| | Craft@Hrs | Unit | Material | Labor | Equipment | Total |
|---|---|---|---|---|---|---|
| Remove and dispose | C3@5.57 | Ea | — | 259.00 | 150.00 | 409.00 |

Break out storm or sewer pipe, non-salvageable. Excavation or backfill not included

| | Craft@Hrs | Unit | Material | Labor | Equipment | Total |
|---|---|---|---|---|---|---|
| Up to 12" diameter | C3@.114 | LF | — | 5.30 | 3.07 | 8.37 |
| 15" to 18" | C3@.134 | LF | — | 6.23 | 3.60 | 9.83 |
| 21" to 24" | C3@.154 | LF | — | 7.16 | 4.14 | 11.30 |
| 27" to 36" | C3@.202 | LF | — | 9.39 | 5.43 | 14.82 |

Remove welded steel pipe for salvage. Excavation or backfill not included

| | Craft@Hrs | Unit | Material | Labor | Equipment | Total |
|---|---|---|---|---|---|---|
| 4" diameter or smaller | C3@.111 | LF | — | 5.16 | 2.98 | 8.14 |
| 6" to 10" diameter | C3@.202 | LF | — | 9.39 | 5.43 | 14.82 |

Remove and haul away empty underground liquid storage tanks. (Draining and disposing of hazardous liquid in a tank may require special waste handling equipment.) Cost of draining not included. Includes excavation and backfill to 6' deep.

| | Craft@Hrs | Unit | Material | Labor | Equipment | Total |
|---|---|---|---|---|---|---|
| 50 to 250 gallon tank | C3@4.24 | Ea | — | 197.00 | 114.00 | 311.00 |
| Over 250 to 600 gallon tank | C3@11.7 | Ea | — | 544.00 | 315.00 | 859.00 |
| Over 600 to 1,000 gallon tank | C3@22.4 | Ea | — | 1,040.00 | 602.00 | 1,642.00 |
| Add for sand fill, any of above | C3@.125 | CY | 13.50 | 5.81 | 3.36 | 22.67 |

**Miscellaneous Sitework Demolition** No salvage of materials except as noted. These costs include loading and hauling to a legal dump within 6 miles. Dump fees are not included. See Waste Disposal.
Railroad demolition, siding quantities. Equipment includes one wheel-mounted air compressor, one paving breaker and jackhammer bits, one 55 HP wheel loader with integral backhoe and one 5 CY dump truck

| | Craft@Hrs | Unit | Material | Labor | Equipment | Total |
|---|---|---|---|---|---|---|
| Remove track and ties for scrap | C5@.508 | LF | — | 22.80 | 10.30 | 33.10 |
| Remove and dispose of ballast stone | C5@.111 | CY | — | 4.98 | 2.24 | 7.22 |
| Remove wood ties alone | C5@.143 | Ea | — | 6.41 | 2.88 | 9.29 |

Remove light standards, flagpoles, playground poles, up to 30' high, including foundations. Equipment includes one wheel-mounted air compressor, one pneumatic breaker and jackhammer bits and one 5 CY dump truck. Use $400.00 as a minimum charge

| | Craft@Hrs | Unit | Material | Labor | Equipment | Total |
|---|---|---|---|---|---|---|
| Remove item, in salvage condition | C4@6.24 | Ea | — | 259.00 | 104.00 | 363.00 |
| Remove item, no salvage | C4@.746 | Ea | — | 30.90 | 12.50 | 43.40 |

Torch cutting steel plate. Equipment is an oxygen-acetylene manual welding and cutting torch with gases, regulator and goggles. Use $200.00 as a minimum charge

| | Craft@Hrs | Unit | Material | Labor | Equipment | Total |
|---|---|---|---|---|---|---|
| To 3/8" thick | CL@.073 | LF | — | 2.91 | .55 | 3.46 |

|  | Craft@Hrs | Unit | Material | Labor | Equipment | Total |
|---|---|---|---|---|---|---|

**Building Demolition** Costs for demolishing an entire building. Includes loading and hauling up to 6 miles but no dump fees. See Waste Disposal. Costs are by square foot of floor area based on 8' ceiling height. No salvage value assumed. Figures in parentheses give the approximate "loose" volume of the materials (volume after being demolished).

**Light wood-frame structures** Up to three stories in height. Based on 2,500 SF job. No basements included. Estimate each story separately. Equipment includes one 55 HP wheel loader with integral backhoe and one 5 CY dump truck. Use $3,200.00 as a minimum charge.

| | Craft@Hrs | Unit | Material | Labor | Equipment | Total |
|---|---|---|---|---|---|---|
| First story (8 SF per CY) | C5@.038 | SF | — | 1.70 | .62 | 2.32 |
| Second story (8 SF per CY) | C5@.053 | SF | — | 2.38 | .87 | 3.25 |
| Third story (8 SF per CY) | C5@.070 | SF | — | 3.14 | 1.15 | 4.29 |

**Building demolition with pneumatic tools** Equipment includes one wheel-mounted air compressor, three breakers and jackhammer bits, two 55 HP wheel loaders with integral backhoes and two 5 CY dump trucks. Use $10,000.00 as a minimum charge.

| | Craft@Hrs | Unit | Material | Labor | Equipment | Total |
|---|---|---|---|---|---|---|
| Concrete building (30 SF per CY) | C6@.089 | SF | — | 3.90 | 1.36 | 5.26 |
| Reinforced concrete building | | | | | | |
| (20 SF per CY) | C6@.101 | SF | — | 4.43 | 1.55 | 5.98 |
| Masonry building (50 SF per CY) | C6@.074 | SF | — | 3.24 | 1.13 | 4.37 |

**Building demolition with crane and headache ball** Equipment includes one wheel-mounted air compressor, one pneumatic paving breaker and jackhammer bits, one 55 HP wheel loader with integral backhoe and backhoe mounted paving breaker, one 15-ton hydraulic crane with headache ball and three 5 CY dump trucks. Use $10,000.00 as a minimum charge.

| | Craft@Hrs | Unit | Material | Labor | Equipment | Total |
|---|---|---|---|---|---|---|
| Concrete building (30 SF per CY) | C7@.031 | SF | — | 1.43 | 1.01 | 2.44 |
| Reinforced concrete building | | | | | | |
| (20 SF per CY) | C7@.037 | SF | — | 1.71 | 1.21 | 2.92 |
| Masonry building (50 SF per CY) | C7@.026 | SF | — | 1.20 | .85 | 2.05 |

**Concrete foundation and footing demolition** with D-6 crawler dozer and pneumatic tools. Equipment includes one wheel-mounted air compressor, one pneumatic paving breaker and jackhammer bits, one D-6 crawler dozer with attachments and one 5 CY dump truck. Use $5,000.00 as a minimum charge.

| | Craft@Hrs | Unit | Material | Labor | Equipment | Total |
|---|---|---|---|---|---|---|
| Non-reinforced concrete | | | | | | |
| (1 CY yields 1.33 loose CY) | C3@1.10 | CY | — | 51.10 | 44.70 | 95.80 |
| Reinforced concrete | | | | | | |
| (1 CY yields 1.66 loose CY) | C3@1.58 | CY | — | 73.40 | 64.20 | 137.60 |
| Chip out concrete using paving breaker | | | | | | |
| No dozer used, (.8 CF yields 1 loose CF) | CL@.203 | CF | — | 8.10 | 3.40 | 11.50 |

**Gutting a building** Interior finishes stripped back to the structural walls. Building structure to remain. No allowance for salvage value. These costs include loading and hauling up to 6 miles. Dump fees are not included. See Waste Disposal. Costs shown are per square foot of floor area based on 8' ceiling height. Costs will be about 50% less if a small tractor can be used and up to 25% higher if debris must be carried to ground level in an elevator. Equipment includes one air compressor, two pneumatic breakers with jackhammer bits and one 5 CY dump truck. Figures in parentheses give the approximate "loose" volume of the materials (volume after being demolished). Use $3,000.00 as a minimum charge.

| | Craft@Hrs | Unit | Material | Labor | Equipment | Total |
|---|---|---|---|---|---|---|
| Residential buildings (125 SF per CY) | C4@.110 | SF | — | 4.56 | 1.84 | 6.40 |
| Commercial buildings (140 SF per CY) | C4@.100 | SF | — | 4.15 | 1.67 | 5.82 |

**Partition wall demolition** Building structure to remain. No allowance for salvage value. Dump fees are not included. See Waste Disposal. Costs shown are per square foot of wall removed (as measured on one side). Knock down with pneumatic jackhammers and pile adjacent to site but no removal. Equipment includes one air compressor and two pneumatic breakers with jackhammer bits. Figures in parentheses give the approximate "loose" volume of the materials (volume after being demolished). Use $350.00 as a minimum charge.

| | Craft@Hrs | Unit | Material | Labor | Equipment | Total |
|---|---|---|---|---|---|---|
| Brick or block partition demolition | | | | | | |
| 4" thick partition (60 SF per CY) | CL@.040 | SF | — | 1.60 | .74 | 2.34 |
| 8" thick partition (30 SF per CY) | CL@.057 | SF | — | 2.27 | 1.06 | 3.33 |
| 12" thick partition (20 SF per CY) | CL@.073 | SF | — | 2.91 | 1.36 | 4.27 |
| Concrete partition demolition | | | | | | |
| Non-reinforced (18 CF per CY) | CL@.273 | CF | — | 10.90 | 5.07 | 15.97 |
| Reinforced (20 CF per CY) | CL@.363 | CF | — | 14.50 | 6.74 | 21.24 |
| Knock down with hand tools and pile adjacent to site but no removal. | | | | | | |
| Stud partition demolition (typically 75 to 100 SF per CY) | | | | | | |
| Gypsum or terra cotta on metal lath | CL@.027 | SF | — | 1.08 | — | 1.08 |
| Drywall on metal or wood studs | CL@.028 | SF | — | 1.12 | — | 1.12 |
| Plaster on metal studs | CL@.026 | SF | — | 1.04 | — | 1.04 |

**Ceiling demolition** Building structure to remain. No allowance for salvage value. Dump fees are not included. See Waste Disposal. Costs shown are per square foot of ceiling (as measured on one side). Figures in parentheses give the approximate "loose" volume of the materials (volume after being demolished). Use $300.00 as a minimum charge. Knock down with hand tools to remove ceiling finish on framing at heights to 9'.

| | Craft@Hrs | Unit | Material | Labor | Equipment | Total |
|---|---|---|---|---|---|---|
| Plaster ceiling (typically 175 to 200 SF per CY) | | | | | | |
| Including lath and furring | CL@.025 | SF | — | 1.00 | — | 1.00 |
| Including suspended grid | CL@.020 | SF | — | .80 | — | .80 |
| Acoustic tile ceiling (typically 200 to 250 SF per CY) | | | | | | |
| Including suspended grid | CL@.010 | SF | — | .40 | — | .40 |
| Including grid in salvage condition | CL@.019 | SF | — | .76 | — | .76 |
| Including strip furring | CL@.014 | SF | — | .56 | — | .56 |
| Tile glued to ceiling | CL@.015 | SF | — | .60 | — | .60 |
| Drywall ceiling (typically 250 to 300 SF per CY) | | | | | | |
| Nailed or attached with screws to joists | CL@.012 | SF | — | .48 | — | .48 |
| Including strip furring | CL@.023 | SF | — | .92 | — | .92 |

**Roof demolition** Building structure to remain. No allowance for salvage value. Dump fees are not included, refer to construction materials disposal section and add for same. Costs shown are per square foot of area removed (as measured on one side).
Roof surface removal, using hand tools. Figures in parentheses give the approximate "loose" volume of the materials (volume after being demolished). Use $900.00 as a minimum charge.

| | Craft@Hrs | Unit | Material | Labor | Equipment | Total |
|---|---|---|---|---|---|---|
| Asphalt shingles (2.50 Sq per CY) | CL@.850 | Sq | — | 33.90 | — | 33.90 |
| Built-up roofing, including sheathing and gravel (1.25 Sq per CY) | CL@2.91 | Sq | — | 116.00 | — | 116.00 |
| Clay or concrete tile (.70 Sq per CY) | CL@1.03 | Sq | — | 41.10 | — | 41.10 |
| Concrete plank, no covering (.80 Sq per CY) | CL@1.10 | Sq | — | 43.90 | — | 43.90 |
| Gypsum plank, no covering (.70 Sq per CY) | CL@.820 | Sq | — | 32.70 | — | 32.70 |
| Metal deck, no covering (.50 Sq per CY) | CL@2.27 | Sq | — | 90.50 | — | 90.50 |
| Wood shingles or flashing (1.66 Sq per CY) | CL@.756 | Sq | — | 30.10 | — | 30.10 |
| Remove gravel stop | CL@.070 | LF | — | 2.79 | — | 2.79 |

| | Craft@Hrs | Unit | Material | Labor | Equipment | Total |
|---|---|---|---|---|---|---|

**Floor slab demolition** Using pneumatic jackhammers. Piled adjacent to site but no removal. Equipment includes one air compressor and two pneumatic breakers with jackhammer bits. Use $350.00 as a minimum charge.

| | Craft@Hrs | Unit | Material | Labor | Equipment | Total |
|---|---|---|---|---|---|---|
| Slab on grade, 4" to 6" thick, reinforced | | | | | | |
| With wire mesh (55 SF per CY) | CL@.057 | SF | — | 2.27 | 1.06 | 3.33 |
| With number 4 bars (45 SF per CY) | CL@.063 | SF | — | 2.51 | 1.17 | 3.68 |
| Slab, suspended, 6" to 8" thick, free fall | | | | | | |
| (35 SF per CY) | CL@.092 | SF | — | 3.67 | 1.71 | 5.38 |
| Slab fill, lightweight concrete fill | | | | | | |
| On metal deck (45 SF per CY) | CL@.029 | SF | — | 1.16 | .54 | 1.70 |
| Topping, insulating (150 SF per CY) | CL@.023 | SF | — | .92 | .43 | 1.35 |

**Floor covering demolition** Using hand tools. Debris piled adjacent to site but no removal. Figures in parentheses give the approximate "loose" volume of the materials (volume after being demolished). Use $350.00 as a minimum charge.

| | Craft@Hrs | Unit | Material | Labor | Equipment | Total |
|---|---|---|---|---|---|---|
| Ceramic or quarry tile, brick | | | | | | |
| (200 SF per CY) | CL@.029 | SF | — | 1.16 | — | 1.16 |
| Resilient materials only | | | | | | |
| (270 SF per CY) | CL@.018 | SF | — | .72 | — | .72 |
| Terrazzo tile | | | | | | |
| (225 SF per CY) | CL@.032 | SF | — | 1.28 | — | 1.28 |
| Wood block floor, laid in mastic | | | | | | |
| (200 SF per CY) | CL@.051 | SF | — | 2.03 | — | 2.03 |
| Wood, residential strip floor | | | | | | |
| (225 SF per CY) | CL@.032 | SF | — | 1.28 | — | 1.28 |

**Cutting openings in frame walls** Using hand tools. Cost per square foot of wall measured on one face.

| | Craft@Hrs | Unit | Material | Labor | Equipment | Total |
|---|---|---|---|---|---|---|
| Metal stud wall with stucco or plaster | CL@.094 | SF | — | 3.75 | — | 3.75 |
| Wood stud wall with drywall | CL@.051 | SF | — | 2.03 | — | 2.03 |
| Dust control partitions, 6 mil plastic | C8@.011 | SF | .30 | .50 | — | .80 |

**Debris removal** Break demolition debris into manageable size with hand tools, load into a 5 CF wheelbarrow, move to chute and dump. Costs shown are per cubic foot of material dumped.

| | Craft@Hrs | Unit | Material | Labor | Equipment | Total |
|---|---|---|---|---|---|---|
| Wheelbarrow, 50' | | | | | | |
| to trash chute and dump | CL@.018 | CF | — | .72 | — | .72 |
| Wheelbarrow, 100' | | | | | | |
| to trash chute and dump | CL@.022 | CF | — | .88 | — | .88 |
| Wheelbarrow, 50' to elevator, descend 10 floors | | | | | | |
| to trash chute and dump | CL@.024 | CF | — | .96 | — | .96 |
| Trash chutes. Prefabricated steel chute installed and removed in a multi-story building. | | | | | | |
| One-time charge for up to six-month's usage. Use 50 LF as minimum charge | | | | | | |
| 18" diameter | — | LS | — | — | — | 1,160.00 |
| 36" diameter | — | LS | — | — | — | 1,735.00 |

**Load demolition debris on a truck** and haul 6 miles to a dump site. Equipment includes one 5 CY dump truck and one 1 CY articulating wheel loader. Includes truck cost but dump fees are not included. See Waste Disposal. Per cubic yard of debris.

| | Craft@Hrs | Unit | Material | Labor | Equipment | Total |
|---|---|---|---|---|---|---|
| Truck loaded by hand | C4@.828 | CY | — | 34.30 | 9.70 | 44.00 |
| Truck loaded using a wheel loader | C5@.367 | CY | — | 16.50 | 6.51 | 23.01 |

**Removal of interior items** Using hand tools. Items removed in salvageable condition do not include an allowance for salvage value. Figures in parentheses give the approximate "loose" volume of the materials (volume after being demolished.) Removal in non-salvageable condition (demolished) except as noted.

| | Craft@Hrs | Unit | Material | Labor | Equipment | Total |
|---|---|---|---|---|---|---|
| Hollow metal door and frame in a masonry wall. (2 doors per CY) | | | | | | |
| Single door to 4' x 7' | CL@1.00 | Ea | — | 39.90 | — | 39.90 |
| Two doors, per opening to 8' x 7' | CL@1.50 | Ea | — | 59.80 | — | 59.80 |

# 02 Existing Conditions

| | Craft@Hrs | Unit | Material | Labor | Equipment | Total |
|---|---|---|---|---|---|---|
| Wood door and frame in a wood-frame wall (2 doors per CY) | | | | | | |
| Single door to 4' x 7' | CL@.500 | Ea | — | 19.90 | — | 19.90 |
| Two doors, per opening to 8' x 7' | CL@.750 | Ea | — | 29.90 | — | 29.90 |
| Remove door and frame in masonry wall, salvage condition | | | | | | |
| Hollow metal door to 4' x 7' | C8@2.00 | Ea | — | 91.70 | — | 91.70 |
| Wood door to 4' x 7' | C8@1.00 | Ea | — | 45.80 | — | 45.80 |
| Frame for resilient mat, metal | | | | | | |
| Per SF of mat area | C8@.054 | SF | — | 2.47 | — | 2.47 |
| Lockers, metal 12" W, 60" H, 15" D | C8@.500 | Ea | — | 22.90 | — | 22.90 |
| Sink and soap dispenser, wall-hung | C8@.500 | Ea | — | 22.90 | — | 22.90 |
| Toilet partitions, wood or metal | | | | | | |
| Per partition | C8@.750 | Ea | — | 34.40 | — | 34.40 |
| Urinal screens, wood or metal | | | | | | |
| Per partition | C8@.500 | Ea | — | 22.90 | — | 22.90 |
| Window and frame, wood or metal | C8@.076 | SF | — | 3.48 | — | 3.48 |
| Remove and reset airlock doors | C8@5.56 | Ea | — | 255.00 | — | 255.00 |

**Structure Moving** Up to 5 mile haul, not including new foundation costs, utility hookup or finishing. These figures assume two stories maximum, no obstruction from trees or utility lines and that adequate right of way is available. Fees and charges imposed by government are not included. Costs shown are per SF of total floor area. Equipment cost includes one tow-truck equipped with dollies, jacks, support beams and hand tools. For estimating purpose, use 1,000 SF as a minimum job charge.

| | Craft@Hrs | Unit | Material | Labor | Equipment | Total |
|---|---|---|---|---|---|---|
| Concrete or masonry structures, 12" maximum thick walls or floors | | | | | | |
| 1,000 to 2,000 SF | C1@.168 | SF | — | 6.86 | 10.20 | 17.06 |
| 2,000 to 4,000 SF | C1@.149 | SF | — | 6.08 | 9.06 | 15.14 |
| Wood frame | | | | | | |
| 1,000 to 2,000 SF | C1@.142 | SF | — | 5.80 | 8.64 | 14.44 |
| 2,000 to 4,000 SF | C1@.131 | SF | — | 5.35 | 7.97 | 13.32 |
| Steel frame | | | | | | |
| 1,000 to 2,000 SF | C1@.215 | SF | — | 8.78 | 13.08 | 21.86 |
| 2,000 to 4,000 SF | C1@.199 | SF | — | 8.12 | 12.10 | 20.22 |

**Asbestos Hazard Surveys** Building inspection, hazard identification, sampling and ranking of asbestos risk.

| | Craft@Hrs | Unit | Material | Labor | Equipment | Total |
|---|---|---|---|---|---|---|
| Asbestos hazard survey and sample collection (10,000 SF per hour) | | | | | | |
| 4 hour minimum | — | Hour | — | — | — | 90.00 |
| Sample analysis (usually one sample per 1,000 SF) | | | | | | |
| 10 sample minimum | — | Ea | — | — | — | 55.00 |
| Report writing (per 1,000 SF of floor) | | | | | | |
| $200 minimum | — | MSF | — | — | — | 28.50 |

**Asbestos Removal, Subcontract** Typical costs including site preparation, monitoring, equipment, and removal of waste. Disposal of hazardous waste materials vary widely, consult your local waste disposal facility concerning prices and procedures.

| | Craft@Hrs | Unit | Material | Labor | Equipment | Total |
|---|---|---|---|---|---|---|
| Ceiling insulation in containment structure | | | | | | |
| 500 to 5,000 SF job | — | SF | — | — | — | 40.20 |
| 5,000 to 20,000 SF job | — | SF | — | — | — | 28.50 |
| Pipe insulation in containment structure | | | | | | |
| 100 to 1,000 LF of 6" pipe | — | LF | — | — | — | 70.90 |
| 1,000 to 3,000 LF of 6" pipe | — | LF | — | — | — | 45.50 |
| Pipe insulation using glove bags | | | | | | |
| 100 to 1,000 LF of 6" pipe | — | LF | — | — | — | 60.30 |
| 1,000 to 3,000 LF of 6" pipe | — | LF | — | — | — | 45.50 |

|  | Craft@Hrs | Unit | Material | Labor | Equipment | Total |
|---|---|---|---|---|---|---|

**Excavation for Concrete Work** These costs do not include shoring or disposal. Typical soil conditions.

Batterboards, lay out for footings,

| per corner | C8@1.24 | Ea | 7.00 | 56.80 | — | 63.80 |

Trenching with a 1/2 CY utility backhoe/loader, small jobs, good soil conditions, no backfill included

18" x 24" depth,

| .111 CY per LF (135 LF/Hr) | S1@.015 | LF | — | .71 | .27 | .98 |

24" x 36" depth,

| .222 CY per LF (68 LF/Hr) | S1@.029 | LF | — | 1.38 | .52 | 1.90 |

36" x 48" depth,

| .444 CY per LF (34 LF/Hr) | S1@.059 | LF | — | 2.80 | 1.05 | 3.85 |

48" x 60" depth,

| .741 CY per LF (20 LF/Hr) | S1@.099 | LF | — | 4.70 | 1.76 | 6.46 |

Per cubic yard,

| 15 CY per hour | S1@.133 | CY | — | 6.31 | 2.36 | 8.67 |

Hand labor

Clearing trench shoulders

| of obstructions | CL@.017 | LF | — | .68 | — | .68 |

Hard pan or rock outcropping

| Breaking and stacking | CL@3.23 | CY | — | 129.00 | — | 129.00 |

Column pads and small piers,

| average soil | CL@1.32 | CY | — | 52.60 | — | 52.60 |

Backfilling against foundations, no import or export of materials

| Hand, including hand tamp | CL@.860 | CY | — | 34.30 | — | 34.30 |

Backfill with a 1/2 CY utility backhoe/loader

| Minimum compaction, wheel-rolled | S1@.055 | CY | — | 2.61 | .98 | 3.59 |

Backfill with 1/2 CY utility backhoe/loader, 150 CFM compressor & pneumatic tamper

| Spec grade compaction | S1@.141 | CY | — | 6.69 | 4.36 | 11.05 |

Disposal of excess backfill material with a 1/2 CY backhoe, no compaction

| Spot spread, 100' haul | S1@.034 | CY | — | 1.61 | .60 | 2.21 |

Area spread, 4" deep

| (covers 81 SF per CY) | S1@.054 | CY | — | 2.56 | .96 | 3.52 |

Grading for slabs

Using a D-4 crawler tractor with dozer blade

| (1000 SF per hour) | S1@.001 | SF | — | .05 | .02 | .07 |

Fine grading, by hand, light to medium soil

| (165 to 170 SF per hour) | CL@.006 | SF | — | .24 | — | .24 |

Capillary fill, one CY covers 81 SF at 4" deep

4", hand graded and rolled,

| small job | CL@.014 | SF | .31 | .56 | — | .87 |

4", machine graded and rolled,

| larger job | S1@.006 | SF | .31 | .28 | .13 | .72 |
| Add for each extra 1" hand graded | CL@.004 | SF | .09 | .16 | — | .25 |
| Add for each extra 1" machine graded | S1@.001 | SF | .09 | .05 | .02 | .16 |

Sand fill, 1 CY covers 162 SF at 2" deep

| 2" sand cushion, hand spread, 1 throw | CL@.003 | SF | .14 | .12 | — | .26 |
| Add for each additional 1" | CL@.002 | SF | .07 | .08 | — | .15 |

Waterproof membrane, polyethylene, over sand bed, including 20% lap and waste

| .004" (4 mil) clear or black | CL@.001 | SF | .05 | .04 | — | .09 |
| .006" (6 mil) clear or black | CL@.001 | SF | .07 | .04 | — | .11 |

**Formwork for Concrete** Labor costs include the time needed to prepare formwork sketches at the job site, measure for the forms, fabricate, erect, align, brace, strip, clean and stack the forms. Costs for reinforcing and concrete are not included. Multiple use of forms shows the cost per use when a form is used several times on the same job without being totally disassembled or discarded. Normal cleaning and repairs are included in the labor cost on lines that show multiple use of forms. Costs will be higher if plans are not detailed enough to identify the work to be performed. No salvage value is assumed except as noted. Costs listed are per square foot of form in contact with the concrete (SFCA). These costs assume Standard and Better lumber price of $544 per MBF and $1,900 per MSF for 3/4" plyform before waste allowance. For more detailed coverage of concrete formwork, see *National Concrete & Masonry Estimator*, http://CraftsmanSiteLicense.com

| | Craft@Hrs | Unit | Material | Labor | Total |
|---|---|---|---|---|---|

**Wall footing, grade beam or tie beam forms** These figures assume nails, stakes and form oil costing $.25 per square foot and 2.5 board feet of lumber are used for each square foot of form in contact with the concrete (SFCA). To calculate the quantity of formwork required, multiply the depth of footing in feet by the length of footing in feet. Then double the results if two sides will be formed. For scheduling purposes, estimate that a crew of 5 can lay out, fabricate and erect 600 to 700 SF of footing, grade beam or tie beam forms in an 8-hour day.

| | Craft@Hrs | Unit | Material | Labor | Total |
|---|---|---|---|---|---|
| 1 use | F5@.070 | SFCA | 1.42 | 3.29 | 4.71 |
| 3 uses | F5@.050 | SFCA | .84 | 2.35 | 3.19 |
| 5 uses | F5@.040 | SFCA | .72 | 1.88 | 2.60 |
| Add for stepped footings | F5@.028 | SFCA | .51 | 1.32 | 1.83 |
| Add for keyed joint, 1 use | F5@.020 | LF | .56 | .94 | 1.50 |

Reinforcing bar supports for footing, grade beam and tie beam forms. Bars suspended from 2" x 4" lumber, including stakes. Based on .86 BF of lumber per LF

| | Craft@Hrs | Unit | Material | Labor | Total |
|---|---|---|---|---|---|
| 1 use | F5@.060 | LF | .65 | 2.82 | 3.47 |
| 3 uses | F5@.050 | LF | .45 | 2.35 | 2.80 |
| 5 uses | F5@.040 | LF | .41 | 1.88 | 2.29 |

Integral starter wall forms (stem walls) formed monolithic with footings, heights up to 4'0", with 3 BF of lumber per SFCA plus an allowance of $0.25 for nails, stakes and form oil

| | Craft@Hrs | Unit | Material | Labor | Total |
|---|---|---|---|---|---|
| 1 use | F5@.100 | SFCA | 1.66 | 4.70 | 6.36 |
| 3 uses | F5@.075 | SFCA | .95 | 3.53 | 4.48 |
| 5 uses | F5@.067 | SFCA | .81 | 3.15 | 3.96 |

Bulkheads or pour-stops. When integral starter walls are formed, forms will usually be required at the ends of each wall or footing. These are usually called bulkheads or pour-stops. These figures assume the use of 1.1 SF of 3/4" plyform per SFCA plus an allowance of $0.25 per SFCA for nails, stakes and bracing

| | Craft@Hrs | Unit | Material | Labor | Total |
|---|---|---|---|---|---|
| 1 use | F5@.150 | SFCA | 3.43 | 7.05 | 10.48 |
| 3 uses | F5@.120 | SFCA | 1.84 | 5.64 | 7.48 |
| 5 uses | F5@.100 | SFCA | 1.52 | 4.70 | 6.22 |

**Column footing or pile cap forms** These figures include nails, stakes and form oil and 3.5 board feet of lumber for each square foot of form in contact with the concrete. For scheduling purposes, estimate that a crew of 5 can lay out, fabricate and erect 500 to 600 SF of square or rectangular footing forms in an 8-hour day or 350 to 450 SF of octagonal, hexagonal or triangular footing forms in an 8-hour day.

Square or rectangular column forms

| | Craft@Hrs | Unit | Material | Labor | Total |
|---|---|---|---|---|---|
| 1 use | F5@.090 | SFCA | 1.89 | 4.23 | 6.12 |
| 3 uses | F5@.070 | SFCA | 1.07 | 3.29 | 4.36 |
| 5 uses | F5@.060 | SFCA | .91 | 2.82 | 3.73 |

Octagonal, hexagonal or triangular forms

| | Craft@Hrs | Unit | Material | Labor | Total |
|---|---|---|---|---|---|
| 1 use | F5@.117 | SFCA | 1.89 | 5.50 | 7.39 |
| 3 uses | F5@.091 | SFCA | 1.07 | 4.28 | 5.35 |
| 5 uses | F5@.078 | SFCA | .91 | 3.67 | 4.58 |

Reinforcing bar supports for column forms. Bars suspended from 2" x 4" lumber, including stakes. Based on .86 BF of lumber per LF

| | Craft@Hrs | Unit | Material | Labor | Total |
|---|---|---|---|---|---|
| 1 use | F5@.060 | LF | .40 | 2.82 | 3.22 |

| | Craft@Hrs | Unit | Material | Labor | Total |
|---|---|---|---|---|---|

Anchor bolt templates or dowel supports for column forms. Includes cost of nails, stakes, bracing, and form oil and 1.1 SF of 3/4" plyform for each SF of column base or dowel support. No anchor bolts included

| | Craft@Hrs | Unit | Material | Labor | Total |
|---|---|---|---|---|---|
| 1 use | F5@.150 | SF | 3.43 | 7.05 | 10.48 |
| 3 uses | F5@.100 | SF | 1.84 | 4.70 | 6.54 |
| 5 uses | F5@.090 | SF | 1.52 | 4.23 | 5.75 |

**Slab-on-grade forms** These costs assume nails, stakes, form oil, and accessories are used for each board foot of lumber. Note that costs listed below are per linear foot (LF) of form. Figures in parentheses show the average square feet of contact area (SFCA) per linear foot of form. Use this figure to convert costs and man hours per linear foot to costs and manhours per square foot. For scheduling purposes, estimate that a crew of 5 will lay out, fabricate and erect the following quantities of slab-on-grade edge forms in an 8-hour day:

Slabs to 6" high at 615 to 720 LF. Slabs over 6" to 12" high at 450 to 500 LF

Slabs over 12" to 24" high at 320 to 350 LF. Slabs over 24" to 36" high, 240 to 260 LF

Edge forms to 6" high, 1.5 BF per LF of form (.5 SFCA per LF)

| | Craft@Hrs | Unit | Material | Labor | Total |
|---|---|---|---|---|---|
| 1 use | F5@.065 | LF | .95 | 3.06 | 4.01 |
| 3 uses | F5@.061 | LF | .60 | 2.87 | 3.47 |
| 5 uses | F5@.055 | LF | .53 | 2.59 | 3.12 |

Edge forms over 6" high to 12" high, 2.25 BF per LF of form (average .75 SFCA per LF)

| | Craft@Hrs | Unit | Material | Labor | Total |
|---|---|---|---|---|---|
| 1 use | F5@.090 | LF | 1.31 | 4.23 | 5.54 |
| 3 uses | F5@.086 | LF | .78 | 4.04 | 4.82 |
| 5 uses | F5@.080 | LF | .67 | 3.76 | 4.43 |

Edge forms over 12" high to 24" high, 4.5 BF per LF of form (average 1.5 SFCA per LF)

| | Craft@Hrs | Unit | Material | Labor | Total |
|---|---|---|---|---|---|
| 1 use | F5@.124 | LF | 2.36 | 5.83 | 8.19 |
| 3 uses | F5@.119 | LF | 1.31 | 5.60 | 6.91 |
| 5 uses | F5@.115 | LF | 1.09 | 5.41 | 6.50 |

Edge forms over 24" high to 36" high, 7.5 BF per LF of forms (average 2.5 SFCA per LF)

| | Craft@Hrs | Unit | Material | Labor | Total |
|---|---|---|---|---|---|
| 1 use | F5@.166 | LF | 3.77 | 7.81 | 11.58 |
| 3 uses | F5@.160 | LF | 2.01 | 7.52 | 9.53 |
| 5 uses | F5@.155 | LF | 1.66 | 7.29 | 8.95 |

Add for 2" x 2" tapered keyed joint, one-piece, .38 BF per LF

| | Craft@Hrs | Unit | Material | Labor | Total |
|---|---|---|---|---|---|
| 1 use | F5@.044 | LF | .18 | 2.07 | 2.25 |
| 3 uses | F5@.040 | LF | .09 | 1.88 | 1.97 |
| 5 uses | F5@.035 | LF | .07 | 1.65 | 1.72 |

Blockout and slab depression forms. Figure the linear feet required Use the linear foot costs for slab edge forms for the appropriate height and then add per linear foot

| | Craft@Hrs | Unit | Material | Labor | Total |
|---|---|---|---|---|---|
| Blockouts | F5@.030 | LF | .54 | 1.41 | 1.95 |

**Wall forms** Formwork over 6' high includes an allowance for a work platform and handrail built on one side of the form for use by the concrete placing crew.

Heights to 4', includes 1.1 SF of plyform, 1.5 BF of lumber and allowance for nails, ties and oil per SFCA

| | Craft@Hrs | Unit | Material | Labor | Total |
|---|---|---|---|---|---|
| 1 use | F5@.119 | SFCA | 4.13 | 5.60 | 9.73 |
| 3 uses | F5@.080 | SFCA | 2.19 | 3.76 | 5.95 |
| 5 uses | F5@.069 | SFCA | 1.80 | 3.24 | 5.04 |
| Add for 1 side battered | F5@.016 | SFCA | .41 | .75 | 1.16 |
| Add for 2 sides battered | F5@.024 | SFCA | .83 | 1.13 | 1.96 |

Heights over 4' to 6', includes 1.1 SF of plyform, 2.0 BF of lumber and allowance for nails, ties and oil

| | Craft@Hrs | Unit | Material | Labor | Total |
|---|---|---|---|---|---|
| 1 use | F5@.140 | SFCA | 4.37 | 6.58 | 10.95 |
| 3 uses | F5@.100 | SFCA | 2.31 | 4.70 | 7.01 |
| 5 uses | F5@.080 | SFCA | 1.90 | 3.76 | 5.66 |
| Add for 1 side battered | F5@.016 | SFCA | .44 | .75 | 1.19 |
| Add for 2 sides battered | F5@.024 | SFCA | .87 | 1.13 | 2.00 |

| | Craft@Hrs | Unit | Material | Labor | Total |
|---|---|---|---|---|---|
| **Heights over 6' to 12', includes 1.2 SF of plyform, 2.5 BF of lumber and allowance for nails, ties and oil** | | | | | |
| 1 use | F5@.160 | SFCA | 4.89 | 7.52 | 12.41 |
| 3 uses | F5@.110 | SFCA | 2.57 | 5.17 | 7.74 |
| 5 uses | F5@.100 | SFCA | 2.11 | 4.70 | 6.81 |
| Add for 1 side battered | F5@.016 | SFCA | .49 | .75 | 1.24 |
| Add for 2 sides battered | F5@.024 | SFCA | .98 | 1.13 | 2.11 |
| **Heights over 12' to 16', includes 1.2 SF of plyform, 3.0 BF of lumber and allowance for nails, ties and oil** | | | | | |
| 1 use | F5@.180 | SFCA | 5.13 | 8.46 | 13.59 |
| 3 uses | F5@.130 | SFCA | 2.69 | 6.11 | 8.80 |
| 5 uses | F5@.110 | SFCA | 2.20 | 5.17 | 7.37 |
| Add for 1 side battered | F5@.016 | SFCA | .51 | .75 | 1.26 |
| Add for 2 sides battered | F5@.024 | SFCA | 1.03 | 1.13 | 2.16 |
| **Heights over 16', includes 1.3 SF of plyform, 3.5 BF of lumber and allowance for nails, ties and oil** | | | | | |
| 1 use | F5@.199 | SFCA | 5.65 | 9.36 | 15.01 |
| 3 uses | F5@.140 | SFCA | 2.95 | 6.58 | 9.53 |
| 5 uses | F5@.119 | SFCA | 2.41 | 5.60 | 8.01 |
| Add for 1 side battered | F5@.016 | SFCA | .57 | .75 | 1.32 |
| Add for 2 sides battered | F5@.024 | SFCA | 1.13 | 1.13 | 2.26 |
| **Architectural form liner for wall forms** | | | | | |
| Low cost liners | F5@.020 | SFCA | 1.56 | .94 | 2.50 |
| Average cost liners | F5@.100 | SFCA | 2.49 | 4.70 | 7.19 |
| **Reveal strips 1" deep by 2" wide, one-piece, wood** | | | | | |
| 1 use | F5@.069 | LF | .10 | 3.24 | 3.34 |
| 3 uses | F5@.050 | LF | .06 | 2.35 | 2.41 |
| 5 uses | F5@.020 | LF | .05 | .94 | .99 |

**Blockouts for openings. Form area is the opening perimeter times the depth. These figures assume that nails and form oil costing $.25 per square foot, 1.2 SF of plyform and .5 board feet of lumber are used per SF of form. Blockouts usually can be used only once.**

| | Craft@Hrs | Unit | Material | Labor | Total |
|---|---|---|---|---|---|
| 1 use | F5@.250 | SF | 3.95 | 11.80 | 15.75 |

**Bulkheads or pour-stops. These costs assume that nails, form oil and accessories, 1.1 SF of plyform and 1 board foot of lumber are used per SF of form**

| | Craft@Hrs | Unit | Material | Labor | Total |
|---|---|---|---|---|---|
| 1 use | F5@.220 | SF | 3.90 | 10.30 | 14.20 |
| 3 uses | F5@.149 | SF | 2.08 | 7.01 | 9.09 |
| 5 uses | F5@.130 | SF | 1.71 | 6.11 | 7.82 |
| Add for keyed wall joint, two-piece, tapered, 1.2 BF per SF | | | | | |
| 1 use | F5@.100 | SF | .81 | 4.70 | 5.51 |
| 3 uses | F5@.069 | SF | .53 | 3.24 | 3.77 |
| 5 uses | F5@.061 | SF | .48 | 2.87 | 3.35 |
| **Curved wall forms** | | | | | |
| Smooth radius, add to straight wall cost | F5@.032 | SFCA | 1.22 | 1.50 | 2.72 |
| 8' chord sections, add to straight wall cost | F5@.032 | SFCA | .73 | 1.50 | 2.23 |

**Haunches or ledges. Area is the width of the ledge times the length. These costs assume nails, form oil and accessories, 2 SF of plyform, and .5 board foot of lumber are used per SF of ledge**

| | Craft@Hrs | Unit | Material | Labor | Total |
|---|---|---|---|---|---|
| 1 use | F5@.300 | SF | 6.27 | 14.10 | 20.37 |
| 3 uses | F5@.210 | SF | 3.26 | 9.87 | 13.13 |
| 5 uses | F5@.180 | SF | 2.66 | 8.46 | 11.12 |

**Steel framed plywood forms** Rented steel framed plywood forms can reduce forming costs on many jobs. Where rented forms are used 3 times a month, use the wall forming costs shown for 3 uses at the appropriate height but deduct 25% from the material cost and 50% from the labor manhours and labor costs. Savings will be smaller where layouts change from one use to the next, when form penetrations must be made and repaired before returning a form, where form delivery costs are high, and where non-standard form sizes are needed. For more detailed coverage of steel framed plywood forms, see *National Concrete & Masonry Estimator*, http://CraftsmanSiteLicense.com

**Column forms for square or rectangular columns** Quantities in parentheses show the cross section area of a square column. When estimating a rectangular column, use the costs for the square column with closest cross section area. For scheduling purposes estimate that a crew of 5 can lay out, fabricate and erect about 550 to 600 SFCA of square or rectangular column forms in an 8-hour day.

| | Craft@Hrs | Unit | Material | Labor | Total |
|---|---|---|---|---|---|
| Up to 12" x 12" (144 square inches), using nails, snap ties, oil and column clamps with 1.15 SF of plyform and 2 BF of lumber per SFCA | | | | | |
| 1 use | F5@.130 | SFCA | 4.51 | 6.11 | 10.62 |
| 3 uses | F5@.090 | SFCA | 2.38 | 4.23 | 6.61 |
| 5 uses | F5@.080 | SFCA | 1.96 | 3.76 | 5.72 |
| Over 12" x 12" to 16" x 16" (256 square inches), using nails, snap ties, oil and column clamps, with 1.125 SF of plyform and 2.1 BF of lumber per SFCA | | | | | |
| 1 use | F5@.100 | SFCA | 4.49 | 4.70 | 9.19 |
| 3 uses | F5@.069 | SFCA | 2.37 | 3.24 | 5.61 |
| 5 uses | F5@.061 | SFCA | 1.95 | 2.87 | 4.82 |
| Over 16" x 16" to 20" x 20" (400 square inches), using nails, snap ties, oil and column clamps, with 1.1 SF of plyform and 2.2 BF of lumber per SFCA | | | | | |
| 1 use | F5@.080 | SFCA | 4.46 | 3.76 | 8.22 |
| 3 uses | F5@.061 | SFCA | 2.36 | 2.87 | 5.23 |
| 5 uses | F5@.050 | SFCA | 1.94 | 2.35 | 4.29 |
| Over 20" x 20" to 24" x 24" (576 square inches), using nails, snap ties, oil and column clamps, with 1.1 SF of plyform and 2.4 BF of lumber per SFCA | | | | | |
| 1 use | F5@.069 | SFCA | 4.56 | 3.24 | 7.80 |
| 3 uses | F5@.050 | SFCA | 2.40 | 2.35 | 4.75 |
| 5 uses | F5@.040 | SFCA | 1.97 | 1.88 | 3.85 |
| Over 24" x 24" to 30" x 30" (900 square inches), using nails, snap ties, oil and column clamps, with 1.1 SF of plyform and 2.4 BF of lumber per SFCA | | | | | |
| 1 use | F5@.061 | SFCA | 4.56 | 2.87 | 7.43 |
| 3 uses | F5@.040 | SFCA | 2.40 | 1.88 | 4.28 |
| 5 uses | F5@.031 | SFCA | 1.97 | 1.46 | 3.43 |
| Over 30" x 30" to 36" x 36" (1,296 square inches), using nails, snap ties, oil and column clamps, with 1.1 SF of plyform and 2.4 BF of lumber per SFCA | | | | | |
| 1 use | F5@.050 | SFCA | 4.56 | 2.35 | 6.91 |
| 3 uses | F5@.040 | SFCA | 2.40 | 1.88 | 4.28 |
| 5 uses | F5@.031 | SFCA | 1.97 | 1.46 | 3.43 |
| Over 36" x 36" to 48" x 48" (2,304 square inches), using nails, snap ties, oil and column clamps, with 1.1 SF of plyform and 2.5 BF of lumber per SFCA | | | | | |
| 1 use | F5@.040 | SFCA | 4.60 | 1.88 | 6.48 |
| 3 uses | F5@.031 | SFCA | 2.43 | 1.46 | 3.89 |
| 5 uses | F5@.020 | SFCA | 2.05 | .94 | 2.99 |

**Column capitals for square columns** Capital forms for square columns usually have four symmetrical sides. Length and width at the top of the capital are usually twice the length and width at the bottom of the capital. Height is usually the same as the width at the capital base. These costs assume capitals installed not over 12' above floor level, use of nails, form oil, shores and accessories with 1.5 SF of plyform and 2 BF of lumber per SF of contact area (SFCA). Complexity of these forms usually makes more than 3 uses impractical.

For scheduling purposes estimate that a crew of 5 can lay out, fabricate and erect the following quantities of capital formwork in an 8-hour day:

70 to 80 SF for 12" to 16" columns      90 to 100 SF for 20" to 24" columns

110 to 120 SF for 30" to 36" columns    160 to 200 SF for 48" columns

| | Craft@Hrs | Unit | Material | Labor | Total |
|---|---|---|---|---|---|
| Up to 12" x 12" column, 6.0 SFCA | | | | | |
|   1 use of forms | F5@4.01 | Ea | 33.20 | 189.00 | 222.20 |
|   3 uses of forms | F5@3.00 | Ea | 16.60 | 141.00 | 157.60 |
| Over 12" x 12" to 16" x 16" column, 10.7 SFCA | | | | | |
|   1 use of forms | F5@6.00 | Ea | 59.10 | 282.00 | 341.10 |
|   3 uses of forms | F5@4.01 | Ea | 29.60 | 189.00 | 218.60 |
| Over 16" x 16" to 20" x 20" column, 16.6 SFCA | | | | | |
|   1 use of forms | F5@8.31 | Ea | 91.70 | 391.00 | 482.70 |
|   3 uses of forms | F5@6.00 | Ea | 45.90 | 282.00 | 327.90 |
| Over 20" x 20" to 24" x 24" column, 24.0 SFCA | | | | | |
|   1 use of forms | F5@12.0 | Ea | 133.00 | 564.00 | 697.00 |
|   3 uses of forms | F5@9.00 | Ea | 66.30 | 423.00 | 489.30 |
| Over 24" x 24" to 30" x 30" column, 37.5 SFCA | | | | | |
|   1 use of forms | F5@16.0 | Ea | 207.00 | 752.00 | 959.00 |
|   3 uses of forms | F5@12.0 | Ea | 104.00 | 564.00 | 668.00 |
| Over 30" x 30" to 36" x 36" column, 54.0 SFCA | | | | | |
|   1 use of forms | F5@20.0 | Ea | 298.00 | 940.00 | 1,238.00 |
|   3 uses of forms | F5@16.0 | Ea | 149.00 | 752.00 | 901.00 |
| Over 36" x 36" to 48" x 48", 96.0 SFCA | | | | | |
|   1 use of forms | F5@24.0 | Ea | 530.00 | 1,130.00 | 1,660.00 |
|   3 uses of forms | F5@18.0 | Ea | 265.00 | 846.00 | 1,111.00 |

**Column forms for round columns** Use the costs below to estimate round fiber tube forms (Sonotube is one manufacturer). Add the cost of column footings or foundations. These forms are peeled off when the concrete has cured. Costs shown include setting, aligning, bracing and stripping and assume that bracing and collars at top and bottom can be used 3 times. Column forms over 12'0" long will cost more per linear foot. Costs are per linear foot for standard weight column forms. Plastic lined forms have one vertical seam. For scheduling purposes, estimate that a crew of 5 can lay out, erect and brace the following quantities of 10' to 12' high round fiber tube forms in an 8-hour day: twenty 8" to 12" diameter columns; sixteen 14" to 20" columns and twelve 24" to 48" columns. For more detailed coverage of round column forms, see *National Concrete & Masonry Estimator*, http://CraftsmanSiteLicense.com.

| | Craft@Hrs | Unit | Material | Labor | Total |
|---|---|---|---|---|---|
| Bracing per LF (add to each cost below) | — | LF | .47 | — | .47 |
|   8" spiral type | F5@.166 | LF | 2.63 | 7.81 | 10.44 |
| 10" spiral type | F5@.182 | LF | 3.75 | 8.56 | 12.31 |
| 12" spiral type | F5@.207 | LF | 2.15 | 9.73 | 11.88 |
| 14" spiral type | F5@.224 | LF | 6.51 | 10.50 | 17.01 |
| 16" spiral type | F5@.232 | LF | 6.91 | 10.90 | 17.81 |
| 20" spiral type | F5@.250 | LF | 10.90 | 11.80 | 22.70 |
| 24" spiral type | F5@.275 | LF | 9.82 | 12.90 | 22.72 |
| 30" spiral type | F5@.289 | LF | 17.60 | 13.60 | 31.20 |
| 36" spiral type | F5@.308 | LF | 21.00 | 14.50 | 35.50 |
| 48" spiral type | F5@.334 | LF | 65.30 | 15.70 | 81.00 |
|   8" plastic lined | F5@.166 | LF | 11.70 | 7.81 | 19.51 |
| 10" plastic lined | F5@.177 | LF | 12.30 | 8.32 | 20.62 |
| 12" plastic lined | F5@.207 | LF | 16.50 | 9.73 | 26.23 |
| 14" plastic lined | F5@.224 | LF | 18.10 | 10.50 | 28.60 |
| 16" plastic lined | F5@.232 | LF | 21.20 | 10.90 | 32.10 |
| 20" plastic lined | F5@.250 | LF | 28.00 | 11.80 | 39.80 |
| 24" plastic lined | F5@.275 | LF | 33.60 | 12.90 | 46.50 |
| 30" plastic lined | F5@.289 | LF | 39.90 | 13.60 | 53.50 |
| 36" plastic lined | F5@.308 | LF | 43.30 | 14.50 | 57.80 |
| 48" plastic lined | F5@.334 | LF | 65.30 | 15.70 | 81.00 |

| | Craft@Hrs | Unit | Material | Labor | Total |
|---|---|---|---|---|---|

## Column capitals for round columns

Use these figures to estimate costs for rented prefabricated steel conical shaped capital forms. Costs assume a minimum rental period of 1 month and include an allowance for nails and form oil. Labor shown assumes the capital will be supported from the deck formwork above the capital and includes laying out and cutting the hole in the deck to receive the capital. Capital bottom diameter is sized to fit into the column tube form. The column form size (diameter) does not significantly affect the capital form rental cost. Dimensions shown below are the top diameter. Form weights are shown in parentheses. Only 3 uses of these forms are usually possible in a 30-day period. Setting forms and installing reinforcing steel takes one day, concrete placing takes another day, curing requires 7 to 10 days, removing and cleaning forms takes another day. For scheduling purposes estimate that a crew of 5 can lay out, cut openings in floor deck formwork and install an average of 8 or 9 round column capital forms in an 8-hour day.

| | Craft@Hrs | Unit | Material | Labor | Total |
|---|---|---|---|---|---|
| 3'6" (100 pounds) 1 use | F5@4.01 | Ea | 208.00 | 189.00 | 397.00 |
| 3'6" (100 pounds) 3 uses | F5@4.01 | Ea | 69.30 | 189.00 | 258.30 |
| 4'0" (125 pounds) 1 use | F5@4.01 | Ea | 224.00 | 189.00 | 413.00 |
| 4'0" (125 pounds) 3 uses | F5@4.01 | Ea | 74.70 | 189.00 | 263.70 |
| 4'6" (150 pounds) 1 use | F5@4.30 | Ea | 232.00 | 202.00 | 434.00 |
| 4'6" (150 pounds) 3 uses | F5@4.30 | Ea | 77.40 | 202.00 | 279.40 |
| 5'0" (175 pounds) 1 use | F5@4.51 | Ea | 244.00 | 212.00 | 456.00 |
| 5'0" (175 pounds) 3 uses | F5@4.51 | Ea | 87.00 | 212.00 | 299.00 |
| 5'6" (200 pounds) 1 use | F5@4.51 | Ea | 280.00 | 212.00 | 492.00 |
| 5'6" (200 pounds) 3 uses | F5@4.51 | Ea | 93.80 | 212.00 | 305.80 |
| 6'0" (225 pounds) 1 use | F5@4.70 | Ea | 293.00 | 221.00 | 514.00 |
| 6'0" (225 pounds) 3 uses | F5@4.70 | Ea | 92.60 | 221.00 | 313.60 |

## Beam and girder forms

**Beam and girder forms** Using nails, snap ties and oil, 1.3 SF of 3/4" plyform and 2.1 BF of lumber per SFCA. For scheduling purposes, estimate that a crew of 5 can lay out, fabricate and erect 250 to 300 SF of beam and girder forms in an 8-hour day. For more detailed coverage of beam and girder forms, see *National Concrete & Masonry Estimator*, http://CraftsmanSiteLicense.com.

| | Craft@Hrs | Unit | Material | Labor | Total |
|---|---|---|---|---|---|
| 1 use | F5@.149 | SFCA | 4.99 | 7.01 | 12.00 |
| 3 uses | F5@.141 | SFCA | 2.62 | 6.63 | 9.25 |
| 5 uses | F5@.120 | SFCA | 2.15 | 5.64 | 7.79 |

## Shores for beams and girders

**Shores for beams and girders** Using 4" x 4" wooden posts. These figures assume an average shore height of 12'0" with a 2'0" long horizontal 4" x 4" head and 2" x 6" diagonal brace (30 BF of lumber per shore) and include nails, adjustable post clamps and accessories. For scheduling purposes estimate that a crew of 5 can lay out, fabricate and install 30 to 35 adjustable wood shores in an 8-hour day when shores have an average height of 12'0". Quantity and spacing of shores is dictated by the size of the beams and girders. Generally, allow one shore for each 6 LF of beam or girder. Costs for heavy duty shoring are listed under General Requirements, Heavy Duty Shoring.

| | Craft@Hrs | Unit | Material | Labor | Total |
|---|---|---|---|---|---|
| 1 use | F5@1.50 | Ea | 19.60 | 70.50 | 90.10 |
| 3 uses | F5@1.00 | Ea | 11.30 | 47.00 | 58.30 |
| 5 uses | F5@.748 | Ea | 9.64 | 35.20 | 44.84 |

## Elevated slab forms

**Elevated slab forms** These are costs for forms per square foot of finished slab area and include shoring and stripping at heights to 12'. Bays are assumed to be 20' x 20' bays.

## Waffle slab-joist pans, monolithic

**Waffle slab-joist pans, monolithic** Using rented forms (30-day rental period minimum). For scheduling purposes, estimate that a crew of 5 can lay out, fabricate and install 500 to 600 SF of waffle slab-joist pan forms and shoring in an 8-hour day. Two uses of the forms in a 30-day period is usually the maximum.

| | Craft@Hrs | Unit | Material | Labor | Total |
|---|---|---|---|---|---|
| Metal pans, 10,000 to 20,000 SF | | | | | |
| 1 use | F5@.100 | SF | 5.25 | 4.70 | 9.95 |
| 2 uses | F5@.066 | SF | 2.94 | 3.10 | 6.04 |
| Fiberglass pans, 10,000 to 20,000 SF | | | | | |
| 1 use | F5@.077 | SF | 5.35 | 3.62 | 8.97 |
| 2 uses | F5@.061 | SF | 3.06 | 2.87 | 5.93 |

| | Craft@Hrs | Unit | Material | Labor | Total |
|---|---|---|---|---|---|

**Flat slab with beams, monolithic** Including nails, braces, stiffeners and oil 1.2 SF of plyform and 3.2 BF of lumber per SFCA for one-way beam systems. Two-way beam systems require approximately 40% more forming materials than one-way beam systems. For scheduling purposes, estimate that a crew of 5 can lay out, fabricate and erect 350 to 400 SF of one-way beam systems and 250 to 300 SF of two-way beam systems in an 8-hour day, including shoring.

| | Craft@Hrs | Unit | Material | Labor | Total |
|---|---|---|---|---|---|
| One-way beam 1 use | F5@.110 | SFCA | 5.22 | 5.17 | 10.39 |
| One-way beam 2 uses | F5@.080 | SFCA | 4.55 | 3.76 | 8.31 |
| Two-way beam 1 use | F5@.152 | SFCA | 7.21 | 7.15 | 14.36 |
| Two-way beam 2 uses | F5@.100 | SFCA | 4.67 | 4.70 | 9.37 |

**Structural flat slab** Using nails, braces, stiffeners and oil costing $.25, 1.2 SF of plyform and 2.25 BF of lumber per SFCA. For scheduling purposes, estimate that a crew of 5 can lay out, fabricate and erect 500 to 600 SFCA of structural flat slab forms in an 8-hour day, including shoring.

| | Craft@Hrs | Unit | Material | Labor | Total |
|---|---|---|---|---|---|
| Edge forms | F5@.072 | LF | 1.31 | 3.39 | 4.70 |
| Slab forms 1 use | F5@.069 | SFCA | 4.78 | 3.24 | 8.02 |
| Slab forms 3 uses | F5@.048 | SFCA | 2.51 | 2.26 | 4.77 |
| Slab forms 5 uses | F5@.043 | SFCA | 2.06 | 2.02 | 4.08 |

**Additional forming costs for elevated slabs, if required** Based on single use of plyform and lumber. For scheduling purposes estimate that a 5 man crew can install 500 SFCA per 8-hour day.

| | Craft@Hrs | Unit | Material | Labor | Total |
|---|---|---|---|---|---|
| Control joints | F5@.031 | LF | 1.40 | 1.46 | 2.86 |
| Curbs & pads | F5@.083 | SFCA | 4.78 | 3.90 | 8.68 |
| Depressions | F5@.083 | SFCA | 1.43 | 3.90 | 5.33 |
| Keyed joints | F5@.031 | LF | .59 | 1.46 | 2.05 |

**Miscellaneous formwork** Based on single use of plyform and lumber. For scheduling purposes, estimate that a crew of 5 can lay out, fabricate and erect 225 to 250 SFCA of these types of forms in an 8-hour day.

| | Craft@Hrs | Unit | Material | Labor | Total |
|---|---|---|---|---|---|
| Flat soffits & landings | F5@.166 | SFCA | 7.40 | 7.81 | 15.21 |
| Sloping soffits | F5@.166 | SFCA | 7.64 | 7.81 | 15.45 |
| Stair risers (steps) | F5@.120 | SFCA | 2.87 | 5.64 | 8.51 |

**Reinforcing Steel Bars** 20' lengths, material costs only. Add the cost of delivery, shop fabrication, cutting, bending, setting, lap and tying bars.

ASTM-615 grade 60

| | Craft@Hrs | Unit | Material | Labor | Total |
|---|---|---|---|---|---|
| 3/8" x 20' | — | Ea | 4.62 | — | 4.62 |
| 1/2" x 20' | — | Ea | 7.24 | — | 7.24 |
| 5/8" x 20' | — | Ea | 11.30 | — | 11.30 |
| 3/4" x 20' | — | Ea | 13.20 | — | 13.20 |
| 1" x 20' | — | Ea | 31.30 | — | 31.30 |

ASTM-775 epoxy coated

| | Craft@Hrs | Unit | Material | Labor | Total |
|---|---|---|---|---|---|
| 1/2" x 20' | — | Ea | 15.70 | — | 15.70 |
| 5/8" x 20' | — | Ea | 22.60 | — | 22.60 |
| 3/4" x 20' | — | Ea | 32.30 | — | 32.30 |

**Reinforcing for Cast-in-Place Concrete** Steel reinforcing bars (rebar), ASTM A615 Grade 60. Material costs are for deformed steel reinforcing rebar, including 10% lap allowance, cutting and bending. These costs also include detailed shop drawings and delivery to jobsite with identity tags per shop drawings. Add for epoxy or galvanized coating of rebar, chairs, splicing, spiral caissons and round column reinforcing, if required, from the sections following the data below. Costs per pound (Lb) and per linear foot (LF) including tie wire and tying.

Reinforcing steel placed and tied in footings, foundations and grade beams

| | Craft@Hrs | Unit | Material | Labor | Total |
|---|---|---|---|---|---|
| 1/4" diameter, #2 rebar | RB@.015 | Lb | 1.20 | .87 | 2.07 |
| 1/4" diameter, #2 rebar (.17 lb per LF) | RB@.003 | LF | .22 | .17 | .39 |
| 3/8" diameter, #3 rebar | RB@.011 | Lb | .62 | .64 | 1.26 |

# 03 Concrete

| | Craft@Hrs | Unit | Material | Labor | Total |
|---|---|---|---|---|---|
| 3/8" diameter, #3 rebar (.38 lb per LF) | RB@.004 | LF | .26 | .23 | .49 |
| 1/2" diameter, #4 rebar | RB@.010 | Lb | .55 | .58 | 1.13 |
| 1/2" diameter, #4 rebar (.67 lb per LF) | RB@.007 | LF | .40 | .41 | .81 |
| 5/8" diameter, #5 rebar | RB@.009 | Lb | .47 | .52 | .99 |
| 5/8" diameter, #5 rebar (1.04 lb per LF) | RB@.009 | LF | .54 | .52 | 1.06 |
| 3/4" diameter, #6 rebar | RB@.008 | Lb | .44 | .47 | .91 |
| 3/4" diameter, #6 rebar (1.50 lb per LF) | RB@.012 | LF | .73 | .70 | 1.43 |
| 7/8" diameter, #7 rebar | RB@.008 | Lb | .45 | .47 | .92 |
| 7/8" diameter, #7 rebar (2.04 lb per LF) | RB@.016 | LF | 1.01 | .93 | 1.94 |
| 1" diameter, #8 rebar | RB@.008 | Lb | .47 | .47 | .94 |
| 1" diameter, #8 rebar (2.67 lb per LF) | RB@.021 | LF | 1.39 | 1.22 | 2.61 |
| 1-1/8" diameter, #9 rebar | RB@.008 | Lb | .65 | .47 | 1.12 |
| 1-1/8" diameter, #9 rebar (3.40 lb per LF) | RB@.027 | LF | 2.43 | 1.57 | 4.00 |
| 1-1/4" diameter, #10 rebar | RB@.007 | Lb | .65 | .41 | 1.06 |
| 1-1/4" diameter, #10 rebar (4.30 lb per LF) | RB@.030 | LF | 3.08 | 1.75 | 4.83 |
| 1-3/8" diameter, #11 rebar | RB@.007 | Lb | .65 | .41 | 1.06 |
| 1-3/8" diameter, #11 rebar (5.31 lb per LF) | RB@.037 | LF | 3.80 | 2.16 | 5.96 |

Reinforcing steel placed and tied in structural slabs

| | Craft@Hrs | Unit | Material | Labor | Total |
|---|---|---|---|---|---|
| 1/4" diameter, #2 rebar | RB@.014 | Lb | 1.20 | .82 | 2.02 |
| 1/4" diameter, #2 rebar (.17 lb per LF) | RB@.002 | LF | .22 | .12 | .34 |
| 3/8" diameter, #3 rebar | RB@.010 | Lb | .62 | .58 | 1.20 |
| 3/8" diameter, #3 rebar (.38 lb per LF) | RB@.004 | LF | .26 | .23 | .49 |
| 1/2" diameter, #4 rebar | RB@.009 | Lb | .55 | .52 | 1.07 |
| 1/2" diameter, #4 rebar (.67 lb per LF) | RB@.006 | LF | .40 | .35 | .75 |
| 5/8" diameter, #5 rebar | RB@.008 | Lb | .47 | .47 | .94 |
| 5/8" diameter, #5 rebar (1.04 lb per LF) | RB@.008 | LF | .54 | .47 | 1.01 |
| 3/4" diameter, #6 rebar | RB@.007 | Lb | .44 | .41 | .85 |
| 3/4" diameter, #6 rebar (1.50 lb per LF) | RB@.011 | LF | .73 | .64 | 1.37 |
| 7/8" diameter, #7 rebar | RB@.007 | Lb | .45 | .41 | .86 |
| 7/8" diameter, #7 rebar (2.04 lb per LF) | RB@.014 | LF | 1.01 | .82 | 1.83 |
| 1" diameter, #8 rebar | RB@.007 | Lb | .47 | .41 | .88 |
| 1" diameter, #8 rebar (2.67 lb per LF) | RB@.019 | LF | 1.39 | 1.11 | 2.50 |
| 1-1/8" diameter, #9 rebar | RB@.007 | Lb | .65 | .41 | 1.06 |
| 1-1/8" diameter, #9 rebar (3.40 lb per LF) | RB@.024 | LF | 2.43 | 1.40 | 3.83 |
| 1-1/4" diameter, #10 rebar | RB@.006 | Lb | .65 | .35 | 1.00 |
| 1-1/4" diameter, #10 rebar (4.30 lb per LF) | RB@.026 | LF | 3.08 | 1.52 | 4.60 |
| 1-3/8" diameter, #11 rebar | RB@.006 | Lb | .65 | .35 | 1.00 |
| 1-3/8" diameter, #11 rebar (5.31 lb per LF) | RB@.032 | LF | 3.80 | 1.87 | 5.67 |

Reinforcing steel placed and tied in columns, stairs and walls

| | Craft@Hrs | Unit | Material | Labor | Total |
|---|---|---|---|---|---|
| 1/4" diameter, #2 rebar | RB@.017 | Lb | 1.20 | .99 | 2.19 |
| 1/4" diameter, #2 rebar (.17 lb per LF) | RB@.003 | LF | .22 | .17 | .39 |
| 3/8" diameter, #3 rebar | RB@.012 | Lb | .62 | .70 | 1.32 |
| 3/8" diameter, #3 rebar (.38 lb per LF) | RB@.005 | LF | .26 | .29 | .55 |
| 1/2" diameter, #4 rebar | RB@.011 | Lb | .55 | .64 | 1.19 |
| 1/2" diameter, #4 rebar (.67 lb per LF) | RB@.007 | LF | .40 | .41 | .81 |
| 5/8" diameter, #5 rebar | RB@.010 | Lb | .47 | .58 | 1.05 |
| 5/8" diameter, #5 rebar (1.04 lb per LF) | RB@.010 | LF | .54 | .58 | 1.12 |
| 3/4" diameter, #6 rebar | RB@.009 | Lb | .44 | .52 | .96 |
| 3/4" diameter, #6 rebar (1.50 lb per LF) | RB@.014 | LF | .73 | .82 | 1.55 |
| 7/8" diameter, #7 rebar | RB@.009 | Lb | .45 | .52 | .97 |
| 7/8" diameter, #7 rebar (2.04 lb per LF) | RB@.018 | LF | 1.01 | 1.05 | 2.06 |
| 1" diameter, #8 rebar | RB@.009 | Lb | .47 | .52 | .99 |
| 1" diameter, #8 rebar (2.67 lb per LF) | RB@.024 | LF | 1.39 | 1.40 | 2.79 |
| 1-1/8" diameter, #9 rebar | RB@.009 | Lb | .65 | .52 | 1.17 |

# 03 Concrete

| | Craft@Hrs | Unit | Material | Labor | Total |
|---|---|---|---|---|---|
| 1-1/8" diameter, #9 rebar (3.40 lb per LF) | RB@.031 | LF | 2.43 | 1.81 | 4.24 |
| 1-1/4" diameter, #10 rebar | RB@.008 | Lb | .65 | .47 | 1.12 |
| 1-1/4" diameter, #10 rebar (4.30 lb per LF) | RB@.034 | LF | 3.08 | 1.98 | 5.06 |
| 1-3/8" diameter, #11 rebar | RB@.008 | Lb | .65 | .47 | 1.12 |
| 1-3/8" diameter, #11 rebar (5.31 lb per LF) | RB@.037 | LF | 3.80 | 2.16 | 5.96 |
| Galvanized coating, add to uncoated bars, per pound | | | | | |
| Any #2 rebar, #3 rebar or #4 rebar, add | — | Lb | .28 | — | .28 |
| Any #5 rebar or #6 rebar, add | — | Lb | .35 | — | .35 |
| Any #7 rebar or #8 rebar, add | — | Lb | .33 | — | .33 |
| Any #9 rebar, #10 rebar or #11 rebar, add | — | Lb | .31 | — | .31 |
| Pre-bent rebar, 90 degree except as noted | | | | | |
| 3/8" diameter, #3 rebar, 24" x 24" | RB@.014 | Ea | 1.56 | .82 | 2.38 |
| 1/2" diameter, #4 rebar, 10" x 3' | RB@.021 | Ea | 3.40 | 1.22 | 4.62 |
| 1/2" diameter, #4 rebar, 2' x 3' | RB@.021 | Ea | 4.37 | 1.22 | 5.59 |
| 5/8" diameter, #5 rebar, 24" x 24" | RB@.014 | Ea | 4.91 | .82 | 5.73 |
| 5/8" diameter, #5 rebar, 1' x 4' | RB@.035 | Ea | 6.25 | 2.04 | 8.29 |
| 5/8" diameter, #5 rebar, 3' x 3' | RB@.043 | Ea | 6.73 | 2.51 | 9.24 |
| 5/8" diameter, #5 rebar, 30" x 9" J-bar | RB@.028 | Ea | 2.65 | 1.63 | 4.28 |
| 3/4" diameter, #6 rebar, 24" x 24" | RB@.043 | Ea | 6.85 | 2.51 | 9.36 |
| Rebar stirrups | | | | | |
| 3/8" #3 bar, 6" x 12" | RB@.014 | Ea | 1.65 | .82 | 2.47 |
| 3/8" #3 bar, 6" x 14" | RB@.014 | Ea | 1.89 | .82 | 2.71 |
| 3/8" #3 bar, 6" x 18" | RB@.014 | Ea | 2.15 | .82 | 2.97 |
| 3/8" #3 bar, 6" x 24" | RB@.014 | Ea | 2.15 | .82 | 2.97 |
| 5/8" #8 bar, 6" x 48" | RB@.014 | Ea | 5.05 | .82 | 5.87 |
| Rebar rings | | | | | |
| Rectangular, 6" x 3-1/2" | RB@.014 | Ea | 1.28 | .82 | 2.10 |
| Rectangular, 9" x 5" | RB@.014 | Ea | 1.65 | .82 | 2.47 |
| Rectangular, 11" x 5" | RB@.014 | Ea | 1.54 | .82 | 2.36 |
| Rectangular, 12" x 3-1/2" | RB@.014 | Ea | 1.66 | .82 | 2.48 |
| Rectangular, 14" x 4" | RB@.014 | Ea | 1.66 | .82 | 2.48 |
| Rectangular, 14" x 5" | RB@.014 | Ea | 1.56 | .82 | 2.38 |
| Round, 6" diameter | RB@.014 | Ea | 1.44 | .82 | 2.26 |
| Round, 8" diameter | RB@.014 | Ea | 1.33 | .82 | 2.15 |
| Round, 10" diameter | RB@.014 | Ea | 1.67 | .82 | 2.49 |
| Square, 8" | RB@.014 | Ea | 1.62 | .82 | 2.44 |
| Reinforcing steel chairs. Labor for placing rebar shown above includes time for setting chairs. | | | | | |
| Chairs, individual type, cost each | | | | | |
| 3" high | | | | | |
|   Steel | — | Ea | .56 | — | .56 |
|   Plastic | — | Ea | .21 | — | .21 |
|   Galvanized steel | — | Ea | .86 | — | .86 |
| 5" high | | | | | |
|   Steel | — | Ea | .86 | — | .86 |
|   Plastic | — | Ea | .53 | — | .53 |
|   Galvanized steel | — | Ea | 1.13 | — | 1.13 |
| 8" high | | | | | |
|   Steel | — | Ea | 1.81 | — | 1.81 |
|   Plastic | — | Ea | 1.09 | — | 1.09 |
|   Galvanized steel | — | Ea | 2.43 | — | 2.43 |
| 12" high | | | | | |
|   Steel | — | Ea | 3.53 | — | 3.53 |
|   Plastic | — | Ea | 1.87 | — | 1.87 |
|   Galvanized steel | — | Ea | 4.32 | — | 4.32 |

| | Craft@Hrs | Unit | Material | Labor | Total |
|---|---|---|---|---|---|
| **Chairs, continuous type, with legs 8" OC, cost per linear foot** | | | | | |
| 3" high | | | | | |
|     Steel | — | LF | .78 | — | .78 |
|     Plastic | — | LF | .44 | — | .44 |
|     Galvanized steel | — | LF | 1.00 | — | 1.00 |
| 6" high | | | | | |
|     Steel | — | LF | 1.06 | — | 1.06 |
|     Plastic | — | LF | .62 | — | .62 |
|     Galvanized steel | — | LF | 1.41 | — | 1.41 |
| 8" high | | | | | |
|     Steel | — | LF | 1.64 | — | 1.64 |
|     Plastic | — | LF | 1.14 | — | 1.14 |
|     Galvanized steel | — | LF | 1.98 | — | 1.98 |
| 12" high | | | | | |
|     Steel | — | LF | 3.57 | — | 3.57 |
|     Plastic | — | LF | 1.84 | — | 1.84 |
|     Galvanized steel | — | LF | 4.08 | — | 4.08 |
| **Reinforcing bar weld splicing. Cost per splice** | | | | | |
|   #8 bars, #9 bars or #10 bars | RB@.527 | Ea | 3.20 | 30.70 | 33.90 |
| **Reinforcing bar clip splicing, sleeve and wedge, with hand-held hydraulic ram. Cost per splice** | | | | | |
|   Number 4 bars, 1/2" | RB@.212 | Ea | 5.98 | 12.40 | 18.38 |
|   Number 5 bars, 5/8" | RB@.236 | Ea | 14.00 | 13.80 | 27.80 |
|   Number 6 bars, 3/4" | RB@.258 | Ea | 26.20 | 15.00 | 41.20 |
| **Waterstop, 3/8"** | | | | | |
|   Rubber, 6" | B2@.059 | LF | 3.37 | 2.04 | 5.41 |
|   Rubber, 9" | B2@.059 | LF | 4.79 | 2.04 | 6.83 |
| **Spiral caisson and round column reinforcing, 3/8" diameter hot rolled steel spirals with main vertical bars as shown.** | | | | | |
| **Costs include typical engineering and shop drawings. Shop fabricated spirals delivered, tagged, ready to install.** | | | | | |
| **Cost per vertical linear foot (VLF)** | | | | | |
|   16" diameter with 6 #6 bars, 16.8 lbs per VLF | RB@.041 | VLF | 30.90 | 2.39 | 33.29 |
|   24" diameter with 6 #6 bars, 28.5 lbs per VLF | RB@.085 | VLF | 52.80 | 4.96 | 57.76 |
|   36" diameter with 8 #10 bars, 51.2 lbs per VLF | RB@.216 | VLF | 96.10 | 12.60 | 108.70 |
| **Welded wire mesh steel, electric weld, including 15% waste and overlap** | | | | | |
|   2" x 2" W.9 x W.9 (#12 x #12), slabs | RB@.004 | SF | .80 | .23 | 1.03 |
|   2" x 2" W.9 x W.9 (#12 x #12), beams and columns | RB@.020 | SF | .77 | 1.17 | 1.94 |
|   4" x 4" W1.4 x W1.4 (#10 x #10), slabs | RB@.003 | SF | .37 | .17 | .54 |
|   4" x 4" W2.0 x W2.0 (#8 x #8), slabs | RB@.004 | SF | .43 | .23 | .66 |
|   4" x 4" W2.9 x W2.9 (#6 x #6), slabs | RB@.005 | SF | .48 | .29 | .77 |
|   4" x 4" W4.0 x W4.0 (#4 x #4), slabs | RB@.006 | SF | .63 | .35 | .98 |
|   6" x 6" W1.4 x W1.4 (#10 x #10), slabs | RB@.003 | SF | .17 | .17 | .34 |
|   6" x 6" W2.0 x W2.0 (#8 x #8), slabs | RB@.004 | SF | .28 | .23 | .51 |
|   6" x 6" W2.9 x W2.9 (#6 x #6), slabs | RB@.004 | SF | .32 | .23 | .55 |
|   6" x 6" W4.0 x W4.0 (#4 x #4), slabs | RB@.005 | SF | .45 | .29 | .74 |
|   Add for lengthwise cut, LF of cut | RB@.002 | LF | — | .12 | .12 |
| **Welded wire mesh, galvanized steel, electric weld, including 15% waste and overlap** | | | | | |
|   2" x 2" W.9 x W.9 (#12 x #12), slabs | RB@.004 | SF | 1.55 | .23 | 1.78 |
|   2" x 2" W.9 x W.9 (#12 x #12), beams and columns | RB@.020 | SF | 1.55 | 1.17 | 2.72 |
|   4" x 4" W1.4 x W1.4 (#10 x #10), slabs | RB@.003 | SF | .74 | .17 | .91 |
|   4" x 4" W2.0 x W2.0 (#8 x #8), slabs | RB@.004 | SF | .86 | .23 | 1.09 |
|   4" x 4" W2.9 x W2.9 (#6 x #6), slabs | RB@.005 | SF | .97 | .29 | 1.26 |
|   4" x 4" W4.0 x W4.0 (#4 x #4), slabs | RB@.006 | SF | 1.25 | .35 | 1.60 |
|   6" x 6" W1.4 x W1.4 (#10 x #10), slabs | RB@.003 | SF | .40 | .17 | .57 |

| | Craft@Hrs | Unit | Material | Labor | Total |
|---|---|---|---|---|---|
| 6" x 6" W2.0 x W2.0 (#8 x #8), slabs | RB@.004 | SF | .52 | .23 | .75 |
| 6" x 6" W2.9 x W2.9 (#6 x #6), slabs | RB@.004 | SF | .63 | .23 | .86 |
| 6" x 6" W4.0 x W4.0 (#4 x #4), slabs | RB@.005 | SF | .91 | .29 | 1.20 |
| Add for lengthwise cut, LF of cut | RB@.002 | LF | — | .12 | .12 |
| Welded wire mesh, epoxy-coated steel, electric weld, including 15% waste and overlap | | | | | |
| 2" x 2" W.9 x W.9 (#12 x #12), slabs | RB@.004 | SF | 3.19 | .23 | 3.42 |
| 2" x 2" W.9 x W.9 (#12 x #12), beams and columns | RB@.020 | SF | 3.19 | 1.17 | 4.36 |
| 4" x 4" W1.4 x W1.4 (#10 x #10), slabs | RB@.003 | SF | 2.76 | .17 | 2.93 |
| 4" x 4" W2.0 x W2.0 (#8 x #8), slabs | RB@.004 | SF | 2.82 | .23 | 3.05 |
| 4" x 4" W2.9 x W2.9 (#6 x #6), slabs | RB@.005 | SF | 2.86 | .29 | 3.15 |
| 4" x 4" W4.0 x W4.0 (#4 x #4), slabs | RB@.006 | SF | 3.02 | .35 | 3.37 |
| 6" x 6" W1.4 x W1.4 (#10 x #10), slabs | RB@.003 | SF | 2.58 | .17 | 2.75 |
| 6" x 6" W2.0 x W2.0 (#8 x #8), slabs | RB@.004 | SF | 2.65 | .23 | 2.88 |
| 6" x 6" W2.9 x W2.9 (#6 x #6), slabs | RB@.004 | SF | 2.70 | .23 | 2.93 |
| 6" x 6" W4.0 x W4.0 (#4 x #4), slabs | RB@.005 | SF | 2.85 | .29 | 3.14 |
| Add for lengthwise cut, LF of cut | RB@.002 | LF | — | .12 | .12 |
| Reinforcing mesh mini-mats, electric weld, including 15% waste and overlap | | | | | |
| 6" x 6" W1.4 x W1.4 (#10 x #10), 4' x 8' | RB@.003 | SF | .30 | .17 | .47 |

**Ready-Mix Concrete** Ready-mix delivered by truck. Typical prices for most cities. Includes delivery up to 20 miles for 10 CY or more, 3" to 4" slump. Material cost only, no placing or pumping included. All concrete material costs in this manual are based on these figures.

| | | Unit | Material | Labor | Total |
|---|---|---|---|---|---|
| 5.0 sack mix, 2,000 PSI | — | CY | 96.00 | — | 96.00 |
| 6.0 sack mix, 3,000 PSI | — | CY | 100.00 | — | 100.00 |
| 6.6 sack mix, 3,500 PSI | — | CY | 102.00 | — | 102.00 |
| 7.1 sack mix, 4,000 PSI | — | CY | 104.00 | — | 104.00 |
| 8.5 sack mix, 5,000 PSI | — | CY | 119.00 | — | 119.00 |
| Extra costs for ready-mix concrete | | | | | |
| Add for less than 10 CY per load | — | CY | 25.00 | — | 25.00 |
| Add for delivery over 20 miles | — | Mile | 8.00 | — | 8.00 |
| Add for standby charge in excess of 5 minutes per CY delivered, per minute of extra time | — | Ea | 1.50 | — | 1.50 |
| Add for super plasticized mix, 7"-8" slump | — | % | 8.0 | — | — |
| Add for high early strength concrete | | | | | |
| 5 sack mix | — | CY | 12.50 | — | 12.50 |
| 6 sack mix | — | CY | 16.50 | — | 16.50 |
| Add for lightweight aggregate, typical | — | CY | 44.00 | — | 44.00 |
| Add for pump mix (pea-gravel aggregate) | — | CY | 10.00 | — | 10.00 |
| Add for granite aggregate, typical | — | CY | 6.90 | — | 6.90 |
| Add for white cement (architectural) | — | CY | 57.00 | — | 57.00 |
| Add for 1% calcium chloride | — | CY | 1.60 | — | 1.60 |
| Add for chemical compensated shrinkage | — | CY | 21.00 | — | 21.00 |

Add for colored concrete, typical prices. The ready-mix supplier charges for cleanup of the ready-mix truck used for delivery of colored concrete. The usual practice is to provide one truck per day for delivery of all colored concrete for a particular job. Add the cost below to the cost per cubic yard for the design mix required. Also add the cost for truck cleanup.

| | | Unit | Material | Labor | Total |
|---|---|---|---|---|---|
| Colored concrete, truck cleanup, per day | — | LS | 78.80 | — | 78.80 |
| Adobe | — | CY | 20.00 | — | 20.00 |
| Black | — | CY | 31.50 | — | 31.50 |
| Blended red | — | CY | 17.90 | — | 17.90 |
| Brown | — | CY | 23.10 | — | 23.10 |
| Green | — | CY | 35.70 | — | 35.70 |
| Yellow | — | CY | 24.20 | — | 24.20 |

# 03 Concrete

| | Craft@Hrs | Unit | Material | Labor | Total |
|---|---|---|---|---|---|

**Concrete Pumping, Subcontract**  Small jobs, including trailer mounted pump and operator at 15 CY per hour. Based on 6 sack concrete mix designed for pumping. No concrete included. $125.00 minimum charge.

3/8" aggregate mix (pea gravel), using hose to 200'

| | Craft@Hrs | Unit | Material | Labor | Total |
|---|---|---|---|---|---|
| Cost per cubic yard (pumping only) | — | CY | — | — | 14.00 |
| Add for hose over 200', per LF | — | LF | — | — | 2.00 |

3/4" aggregate mix, using hose to 200'

| | Craft@Hrs | Unit | Material | Labor | Total |
|---|---|---|---|---|---|
| Cost per cubic yard (pumping only) | — | CY | — | — | 14.50 |
| Add for hose over 200', per LF | — | LF | — | — | 1.75 |

**Concrete Pumping with a Boom Truck, Subcontract**  Includes truck rent, operator, local travel but no concrete. Add costs equal to 1 hour for equipment setup and 1 hour for cleanup. Use 4 hours as the minimum cost for 23, 28 and 32 meter boom trucks and 5 hours as the minimum cost for 36 through 52 meter boom trucks. Estimate the actual pour rate at 70% of the rated capacity on thicker slabs and 50% of the capacity on most other work. Costs include operator and oiler where necessary. Costs shown include subcontractor's markup.

| | Craft@Hrs | Unit | Material | Labor | Total |
|---|---|---|---|---|---|
| 23 meter boom (75'), 70 CY per hour rating | — | Hr | — | — | 127.00 |
| Add per CY pumped with 23 meter boom | — | CY | — | — | 2.89 |
| 28 meter boom (92'), 70 CY per hour rating | — | Hr | — | — | 145.00 |
| Add per CY pumped with 28 meter boom | — | CY | — | — | 2.90 |
| 32 meter boom (105'), 90 CY per hour rating | — | Hr | — | — | 170.00 |
| Add per CY pumped with 32 meter boom | — | CY | — | — | 2.95 |
| 36 meter boom (120'), 90 CY per hour rating | — | Hr | — | — | 210.00 |
| Add per CY pumped with 36 meter boom | — | CY | — | — | 2.95 |
| 42 meter boom (138'), 100 CY per hour rating | — | Hr | — | — | 278.00 |
| Add per CY pumped with 42 meter boom | — | CY | — | — | 3.50 |
| 52 meter boom (170'), 100 CY per hour rating | — | Hr | — | — | 325.00 |
| Add per CY pumped with 52 meter boom | — | CY | — | — | 4.00 |

| | Craft@Hrs | Unit | Material | Labor | Equipment | Total |
|---|---|---|---|---|---|---|

**Ready-Mix Concrete and Placing**  No forms, finishing or reinforcing included. Material cost is based on 3,000 PSI concrete. Use $900 minimum charge for boom truck and $125 minimum for trailer mounted pump.

Columns

| | Craft@Hrs | Unit | Material | Labor | Equipment | Total |
|---|---|---|---|---|---|---|
| By crane (60 ton crawler crane) | H3@.873 | CY | 100.00 | 39.50 | 16.50 | 156.00 |
| By pump (28 meter boom truck and operator) | M2@.062 | CY | 115.00 | 2.72 | 5.60 | 123.32 |

Slabs-on-grade

| | Craft@Hrs | Unit | Material | Labor | Equipment | Total |
|---|---|---|---|---|---|---|
| Direct from chute | CL@.430 | CY | 100.00 | 17.10 | — | 117.10 |
| By crane (60 ton crawler crane) | H3@.594 | CY | 100.00 | 26.80 | 11.20 | 138.00 |
| By pump (trailer mounted, 20 CY/ hour) | M2@.188 | CY | 115.00 | 8.24 | 4.55 | 127.79 |
| With buggy | CL@.543 | CY | 115.00 | 21.70 | 7.15 | 143.85 |

Elevated slabs

| | Craft@Hrs | Unit | Material | Labor | Equipment | Total |
|---|---|---|---|---|---|---|
| By crane (60 ton crawler crane) | H3@.868 | CY | 100.00 | 39.20 | 16.40 | 155.60 |
| By pump (28 meter boom truck and operator) | M2@.054 | CY | 115.00 | 2.37 | 5.13 | 122.50 |

Footings, pile caps, foundations

| | Craft@Hrs | Unit | Material | Labor | Equipment | Total |
|---|---|---|---|---|---|---|
| Direct from chute | CL@.564 | CY | 100.00 | 22.50 | — | 122.50 |
| By pump (trailer mounted, 30 CY/ hour) | M2@.125 | CY | 115.00 | 5.48 | 1.90 | 122.38 |
| With buggy | CL@.701 | CY | 115.00 | 28.00 | 9.24 | 152.24 |

Beams and girders

| | Craft@Hrs | Unit | Material | Labor | Equipment | Total |
|---|---|---|---|---|---|---|
| By pump (28 meter boom truck and operator) | M2@.062 | CY | 115.00 | 2.72 | 5.60 | 123.32 |
| By crane (60 ton crawler crane) | H3@.807 | CY | 100.00 | 36.50 | 15.30 | 151.80 |

Stairs

| | Craft@Hrs | Unit | Material | Labor | Equipment | Total |
|---|---|---|---|---|---|---|
| Direct from chute | CL@.814 | CY | 100.00 | 32.50 | — | 132.50 |
| By crane (60 ton crawler crane) | H3@1.05 | CY | 100.00 | 47.50 | 19.90 | 167.40 |

| | Craft@Hrs | Unit | Material | Labor | Equipment | Total |
|---|---|---|---|---|---|---|
| By pump (28 meter boom truck and operator) | M2@.108 | CY | 115.00 | 4.74 | 8.26 | 128.00 |
| With buggy | CL@.948 | CY | 115.00 | 37.80 | 12.50 | 165.30 |
| **Walls and grade beams to 4' high** | | | | | | |
| Direct from chute | CL@.479 | CY | 100.00 | 19.10 | — | 119.10 |
| By pump (trailer mounted. 20 CY/ hour) | M2@.188 | CY | 115.00 | 8.24 | 1.52 | 124.76 |
| With buggy | CL@.582 | CY | 115.00 | 23.20 | 7.67 | 145.87 |
| **Walls over 4' to 8' high** | | | | | | |
| By crane (60 ton crawler crane) | H3@.821 | CY | 100.00 | 37.10 | 15.50 | 152.60 |
| By pump (28 meter boom truck and operator) | M2@.062 | CY | 115.00 | 2.72 | 3.60 | 121.32 |
| **Walls over 8' to 16' high** | | | | | | |
| By crane (60 ton crawler crane) | H3@.895 | CY | 100.00 | 40.50 | 16.90 | 157.40 |
| By pump (28 meter boom truck and operator) | M2@.078 | CY | 115.00 | 3.42 | 6.52 | 124.94 |
| **Walls over 16' high** | | | | | | |
| By crane (60 ton crawler crane) | H3@.986 | CY | 100.00 | 44.60 | 18.70 | 163.30 |
| By pump (28 meter boom truck and operator) | M2@.097 | CY | 115.00 | 4.25 | 7.63 | 126.88 |

**Concrete Slab-on-Grade Assemblies** Typical costs including fine grading, edge forms, 3,000 PSI concrete, wire mesh, finishing and curing.

| | Craft@Hrs | Unit | Material | Labor | Equipment | Total |
|---|---|---|---|---|---|---|
| **4" thick 4' x 6' equipment pad** | | | | | | |
| Hand trimming and shaping, 24 SF | CL@.512 | LS | — | 20.40 | — | 20.40 |
| Lay out, set and strip edge forms, 20 LF, 5 uses | C8@1.30 | LS | 10.60 | 59.60 | — | 70.20 |
| Place W2.9 x W2.9 x 6" x 6" mesh, 24 SF | RB@.097 | LS | 7.68 | 5.66 | — | 13.34 |
| Place ready-mix concrete, from chute .3 CY | CL@.127 | LS | 30.00 | 5.06 | — | 35.06 |
| Finish concrete, broom finish | CM@.591 | LS | — | 29.60 | — | 29.60 |
| Cure concrete, curing paper | CL@.124 | LS | 1.00 | 4.95 | — | 5.95 |
| **Total job cost for 4" thick 24 SF pad** | —@2.75 | LS | 49.28 | 125.27 | — | 174.55 |
| Cost per SF for 4" thick 24 SF pad | —@.115 | SF | 2.05 | 5.22 | — | 7.27 |
| Cost per CY of 4" thick concrete, .3 CY job | —@9.17 | CY | 164.30 | 417.60 | — | 581.90 |
| **8" thick 10' x 10' equipment pad** | | | | | | |
| Hand trimming and shaping, 100 SF | CL@2.79 | LS | — | 111.00 | — | 111.00 |
| Lay out, set and strip edge forms, 40 LF, 5 uses | C8@3.28 | LS | 26.80 | 150.00 | — | 176.80 |
| Place W2.9 x W2.9 x 6" x 6" mesh, 100 SF | RB@.400 | LS | 32.00 | 23.30 | — | 55.30 |
| Place ready-mix concrete, from chute 2.5 CY | CL@1.08 | LS | 250.00 | 43.10 | — | 293.10 |
| Finish concrete, broom finish | CM@2.11 | LS | — | 106.00 | — | 106.00 |
| Cure concrete, curing paper | CL@.515 | LS | 4.00 | 20.50 | — | 24.50 |
| **Total job cost for 8" thick 100 SF pad** | —@10.2 | LS | 312.80 | 453.90 | — | 766.70 |
| Cost per SF for 8" thick 100 SF job | —@.102 | SF | 3.13 | 4.54 | — | 7.67 |
| Cost per CY of 8" thick concrete, 2.5 CY job | —@4.08 | CY | 125.00 | 182.00 | — | 307.00 |

**Note:** Ready-mix concrete used in both examples assumes a minimum of 10 cubic yards will be delivered (per load) with the excess used elsewhere on the same job.

| | | Unit | Material | Labor | Equipment | Total |
|---|---|---|---|---|---|---|
| Add for each CY less than 10 delivered | — | CY | 25.00 | — | — | 25.00 |

| | Craft@Hrs | Unit | Material | Labor | Equipment | Total |
|---|---|---|---|---|---|---|

**Floor Slab Assemblies** Typical reinforced concrete slab-on-grade including excavation, gravel fill, forms, vapor barrier, wire mesh, 3,000 PSI concrete, finishing and curing.

| | Craft@Hrs | Unit | Material | Labor | Equipment | Total |
|---|---|---|---|---|---|---|
| 3,000 PSI concrete | — | CY | 100.00 | — | — | 100.00 |
| 4" thick slab | —@.024 | SF | 2.30 | 1.19 | .11 | 3.60 |
| 5" thick slab | —@.025 | SF | 2.61 | 1.24 | .11 | 3.96 |
| 6" thick slab | —@.026 | SF | 2.92 | 1.29 | .12 | 4.33 |

Detailed cost breakdown slab as described above:

**4" thick 100' x 75' slab**

| | Craft@Hrs | Unit | Material | Labor | Equipment | Total |
|---|---|---|---|---|---|---|
| Grade sandy loam site using a D-4 tractor, 140 CY, balanced job, (no import or export) | S1@2.66 | LS | — | 126.00 | 65.00 | 191.00 |
| Buy and spread 6" crushed rock base using a D-4 tractor, 140 CY | S1@15.8 | LS | 4,210.00 | 750.00 | 386.00 | 5,346.00 |
| Lay out, set and strip edge forms, 350 LF, 5 uses | C8@19.6 | LS | 186.00 | 898.00 | — | 1,084.00 |
| Place .006" polyethylene vapor barrier 7,500 SF | CL@8.84 | LS | 637.00 | 353.00 | — | 990.00 |
| Place W2.9 x W2.9 x 6" x 6" mesh, 7,500 SF | RB@30.0 | LS | 2,400.00 | 1,750.00 | — | 4,150.00 |
| Place and remove 2" x 4" keyway, 200 LF, 1 use | C8@8.80 | LS | 36.00 | 403.00 | — | 439.00 |
| Place 4" concrete, 93 CY, from chute | CL@39.9 | LS | 9,300.00 | 1,590.00 | — | 10,890.00 |
| Float finish 7,500 SF | CM@52.4 | LS | — | 2,630.00 | 359.00 | 2,989.00 |
| Cure with slab curing paper | CM@5.58 | LS | 450.00 | 280.00 | — | 730.00 |
| **Total job cost 4" thick 100' x 75' floor slab** | **—@184** | **LS** | **17,219.00** | **8,780.00** | **810.00** | **26,809.00** |
| Cost per SF for 4" thick 7,500 SF job | —@.025 | SF | 2.30 | 1.17 | .11 | 3.58 |
| Cost per CY for 4" thick 93 CY job | —@1.98 | CY | 185.00 | 94.40 | 8.71 | 288.11 |
| Cost per each additional 1" of concrete | —@.001 | SF | .58 | .29 | .03 | .90 |

**Cast-in-Place Concrete, Subcontract** Typical costs including forms, concrete, reinforcing, finishing and curing. These costs include the subcontractor's overhead and profit but no excavation, shoring or backfill. Use these figures only for preliminary estimates and on jobs that require a minimum of 1,000 CY of concrete.

Foundations

| | Craft@Hrs | Unit | Material | Labor | Equipment | Total |
|---|---|---|---|---|---|---|
| Institutional or office buildings | — | CY | — | — | — | 507.00 |
| Heavy engineered structures | — | CY | — | — | — | 552.00 |

Structural concrete walls

| | Craft@Hrs | Unit | Material | Labor | Equipment | Total |
|---|---|---|---|---|---|---|
| Single story, 8" wall | — | CY | — | — | — | 677.00 |
| Single story, 10" wall | — | CY | — | — | — | 762.00 |
| Multi-story, 8" wall | — | CY | — | — | — | 902.00 |
| Multi-story, 10" | — | CY | — | — | — | 952.00 |
| Slip formed 8" wall | — | CY | — | — | — | 788.00 |

Structural slabs, including shoring,

| | Craft@Hrs | Unit | Material | Labor | Equipment | Total |
|---|---|---|---|---|---|---|
| 6", 1 way beams, 100 pounds reinforcing | — | CY | — | — | — | 818.00 |
| 6", 2 way beams, 225 pounds reinforcing | — | CY | — | — | — | 755.00 |
| 8", flat, with double mat reinforcing over steel frame by others, 100 pounds reinforcing | — | CY | — | — | — | 682.00 |
| 12" flat, 250 pounds reinforcing bars per CY | — | CY | — | — | — | 618.00 |

| | Craft@Hrs | Unit | Material | Labor | Equipment | Total |
|---|---|---|---|---|---|---|
| 7", post-tensioned, | | | | | | |
|   to 1 pound tempered bars | — | CY | — | — | — | 769.00 |
| 6" to 8" including beam jacketing | — | CY | — | — | — | 736.00 |
| 6" to 8" on permanent metal form | — | CY | — | — | — | 664.00 |
| 8" concurrent | | | | | | |
|   with slipform construction | — | CY | — | — | — | 709.00 |
| Lift slab construction 6" to 8" | — | SF | — | — | — | 15.00 |
| Pan and joist slabs 3" thick, 30" pan, including shoring | | | | | | |
|   6" x 16" joist | — | CY | — | — | — | 782.00 |
|   6" x 12" joist | — | CY | — | — | — | 692.00 |
|   6" x 10" joist | — | CY | — | — | — | 651.00 |
| Pan and joist slabs 4-1/2" thick, 20" pan, including shoring | | | | | | |
|   6" x 20" joist | — | CY | — | — | — | 654.00 |
|   6" x 16" joist | — | CY | — | — | — | 640.00 |
|   6" x 12" joist | — | CY | — | — | — | 635.00 |
| Parapet and fascia, 6" thick | — | CY | — | — | — | 992.00 |
| Loading docks, | | | | | | |
|   based on 8" walls and 8" slab | — | CY | — | — | — | 448.00 |
| Beams and girders, not slab integrated | | | | | | |
|   12" x 24", typical | — | CY | — | — | — | 1,340.00 |
|   18" x 24", typical | — | CY | — | — | — | 1,040.00 |
| Columns, average reinforcing | | | | | | |
|   Square, wood formed, with chamfer | | | | | | |
|     12" column | — | CY | — | — | — | 1,690.00 |
|     16" column | — | CY | — | — | — | 1,560.00 |
|     18" column | — | CY | — | — | — | 1,400.00 |
|     20" column | — | CY | — | — | — | 1,280.00 |
|     24" column | — | CY | — | — | — | 1,110.00 |
|   Round, fiber tube formed, 12" to 18" | — | CY | — | — | — | 1,130.00 |

**Cast-in-Place Concrete Steps on Grade Assemblies** Typical in-place costs, including layout, fabrication and placing forms, setting and tying steel reinforcing, installation of steel nosing, placing 2,000 PSI concrete directly from the chute of a ready-mix truck, finishing, stripping forms and curing. Costs assume excavation, back filling and compaction have been completed by others. Crew is a carpenter, laborer and finisher. Formwork is based on using standard & better lumber and concrete and includes a 5% waste allowance. For scheduling purposes, estimate that a crew of 3 can set forms and place steel for and pour 4 to 5 CY of concrete cast-in-place steps in an 8-hour day. Costs shown are for concrete steps with a 6" riser height and 12" tread depth supported by a 6" thick monolithic slab. The "tricks of the trade" below explain how to use these unit prices to estimate stairways of various widths and heights. Note: Total cost per CY has been rounded. For more detailed coverage of formed concrete stairs, see the *National Concrete & Masonry Estimator*, http://CraftsmanSiteLicense.com

| | Craft@Hrs | Unit | Material | Labor | Total |
|---|---|---|---|---|---|
| Fabricate and erect forms, 2 uses of forms, includes stakes, braces, form oil and nails at $.13 per LF of forms | | | | | |
|   Side forms, 1.7 BF per LF of tread | P9@.080 | LF | .58 | 3.78 | 4.36 |
|   Riser forms, 1.5 BF per LF of tread | P9@.040 | LF | .51 | 1.89 | 2.40 |
|   Rebar, #3, Grade 60, 1 pound per LF of tread | P9@.009 | LF | .60 | .43 | 1.03 |
| Embedded steel step nosing, 2-1/2" x 2-1/2" x 1/4" black mild steel | | | | | |
|   Angle iron, 4.5 pounds per LF of tread. | P9@.004 | LF | 4.18 | .19 | 4.37 |
| Concrete, .03 CY per LF of tread | P9@.030 | LF | 3.00 | 1.42 | 4.42 |
|   Total cost per LF of risers | P9@.163 | LF | 8.87 | 7.71 | 16.58 |
|   Total cost per CY of concrete | P9@5.43 | CY | 296.00 | 257.00 | 553.00 |

# 03 Concrete

Tricks of the trade:

To quickly estimate the cost of cast-in-place concrete steps on grade, do this:

    Multiply the width of the steps (in feet) times their height in feet and multiply the result times 2.

    This will give the total linear feet (LF) of risers in the steps. Multiply the total LF by the costs per LF.

Sample estimates for cast-in-place concrete steps on grade using the procedure and costs from above:

    2'6" wide by 3' high (6 steps) would be: 2.5' x 3' x 2 = 15 LF x $16.58 = $248.70

    3' wide by 3' high (6 steps) would be: 3' x 3' x 2 = 18 LF x $16.58 = $298.44

    4' wide by 4' high (8 steps) would be: 4' x 4' x 2 = 32 LF x $16.58 = $530.56

    7' wide by 4'6" high (9 steps) would be: 7' x 4.5' x 2 = 63 LF x $16.58 = $1,044.54

Tip for using these estimates with The National Estimator computer estimating program: Calculate the linear feet of stair, copy and paste the line "Total cost per LF of risers" and type the quantity.

| Concrete Accessories | Craft@Hrs | Unit | Material | Labor | Total |
|---|---|---|---|---|---|
| Bentonite | | | | | |
|   50 pound bag, granules | — | Ea | 43.60 | — | 43.60 |
|   3 gallon Volclay Bentoseal | — | Ea | 148.00 | — | 148.00 |
|   4' x 15' Bentonite geotextile roll | — | Ea | 85.10 | — | 85.10 |
| Bond breaker and form release, Thompson's CBA, spray on | | | | | |
|   Lumber & plywood form release, 200 SF/gal | CL@.006 | SF | .06 | .24 | .30 |
|   Metal form release, 600 SF per gallon | CL@.003 | SF | .02 | .12 | .14 |
|   Casting bed form breaker, 300 SF per gallon | CL@.005 | SF | .04 | .20 | .24 |
|   Cure coat, 300 SF per gallon | CL@.005 | SF | .04 | .20 | .24 |
| Chamfer strip | | | | | |
|   3/4" x 3/4" x 1", 10' lengths | CC@.015 | LF | .22 | .78 | 1.00 |
| Chamfer corner forms | | | | | |
|   3/4" corner forms | CC@.015 | LF | 10.40 | .78 | 11.18 |
| Column clamps for plywood forms, per set, by column size, Econ-o-clamp | | | | | |
|   8" x 8" to 24" x 24", purchase | — | Set | 133.00 | — | 133.00 |
|   24" x 24" to 48" x 48", purchase | — | Set | 174.00 | — | 174.00 |
|   Monthly rental cost as % of purchase price | — | % | 5.0 | — | — |
| Column clamps for lumber forms, per set, by column size | | | | | |
|   10" x 10" to 28" x 28", purchase | — | Set | 164.00 | — | 164.00 |
|   16" x 16" to 40" x 40", purchase | — | Set | 206.00 | — | 206.00 |
|   21" x 21" to 50" x 50", purchase | — | Set | 344.00 | — | 344.00 |
|   10" x 10" to 28" x 28", rent per month, per set | — | Mo | 5.67 | — | 5.67 |
|   16" x 16" to 40" x 40", rent per month, per set | — | Mo | 6.27 | — | 6.27 |
|   21" x 21" to 50" x 50", rent per month, per set | — | Mo | 10.40 | — | 10.40 |
| Concrete sleeves, circular polyethylene liner, any slab thickness | | | | | |
|   1-1/2" to 3" diameter | — | Ea | 2.40 | — | 2.40 |
|   4" to 6" diameter | — | Ea | 4.60 | — | 4.60 |
| Embedded iron, installation only | C8@.045 | Lb | — | 2.06 | 2.06 |
| Expansion joints, cane fiber, asphalt impregnated, pre-molded | | | | | |
|   1/2" x 4", slabs | C8@.028 | LF | .69 | 1.28 | 1.97 |
|   1/2" x 6", in walls | C8@.035 | LF | .96 | 1.60 | 2.56 |
|   1/2" x 8", in walls | C8@.052 | LF | 1.23 | 2.38 | 3.61 |
| Concrete footing form wedge ties | | | | | |
|   6" wall thickness, pack of 50 | — | Pack | 26.20 | — | 26.20 |
|   8" wall thickness, pack of 50 | — | Pack | 28.90 | — | 28.90 |
|   Tie wedge, pack of 100 | — | Pack | 19.50 | — | 19.50 |
| Form spreaders, 4" to 10" | — | Ea | .40 | — | .40 |
| Form oil, 800 to 1,200 SF per gallon | — | Gal | 6.75 | — | 6.75 |

| | Craft@Hrs | Unit | Material | Labor | Total |
|---|---|---|---|---|---|
| **Grout for machine bases mixed and placed (add for edge forms if required, from below)** | | | | | |
| Five Star epoxy | | | | | |
|   per SF for each 1" of thickness | C8@.170 | SF | 14.10 | 7.79 | 21.89 |
| Five Star NBEC non-shrink, non-metallic | | | | | |
|   per SF for each 1" of thickness | C8@.170 | SF | 3.23 | 7.79 | 11.02 |
| Embeco 636, metallic aggregate | | | | | |
|   per SF for each 1" of thickness | C8@.163 | SF | 7.36 | 7.47 | 14.83 |
| Forms for grout, based on using 2" x 4" lumber, | | | | | |
|   1 use | C8@.195 | LF | 2.00 | 8.94 | 10.94 |
| Inserts, Unistrut, average | C8@.068 | Ea | 8.78 | 3.12 | 11.90 |
| **She bolt form clamps, with cathead and bolt, purchase** | | | | | |
| 17" rod | — | Ea | 5.02 | — | 5.02 |
| 21" rod | — | Ea | 8.36 | — | 8.36 |
| 24" rod | — | Ea | 9.00 | — | 9.00 |
| Monthly rental as % of purchase price | — | % | 15.0 | — | — |
| Snap ties, average, 4,000 pound, 6" to 10" | — | Ea | .45 | — | .45 |
| **Water stop, 3/8"** | | | | | |
| Rubber, 6" | C8@.059 | LF | 3.32 | 2.70 | 6.02 |
| Rubber, 9" | C8@.059 | LF | 4.72 | 2.70 | 7.42 |
| **Concrete waler brackets** | | | | | |
| 2" x 4" | — | Ea | 3.20 | — | 3.20 |
| 2" x 6" | — | Ea | 3.68 | — | 3.68 |
| Waler jacks (scaffold support on form), purchase | — | Ea | 50.00 | — | 50.00 |
| Rental per month, each | — | Mo | 6.25 | — | 6.25 |

## Concrete Slab Finishes

| | Craft@Hrs | Unit | Material | Labor | Total |
|---|---|---|---|---|---|
| Float finish | CM@.009 | SF | — | .45 | .45 |
| **Trowel finishing** | | | | | |
| Steel, machine work | CM@.014 | SF | — | .70 | .70 |
| Steel, hand work | CM@.017 | SF | — | .85 | .85 |
| Broom finish | CM@.012 | SF | — | .60 | .60 |
| Scoring concrete surface, hand work | CM@.005 | LF | — | .25 | .25 |
| Exposed aggregate (washed, including finishing), | | | | | |
|   no disposal of slurry | CM@.015 | SF | .33 | .75 | 1.08 |
| **Color hardener, nonmetallic, trowel application** | | | | | |
| Standard colors | — | Lb | .87 | — | .87 |
| Non standard colors | — | Lb | 1.13 | — | 1.13 |
| Light duty floors (40 pounds per 100 SF) | CM@.008 | SF | — | .40 | .40 |
| Pastel shades (60 pounds per 100 SF) | CM@.012 | SF | — | .60 | .60 |
| Heavy duty (90 pounds per 100 SF) | CM@.015 | SF | — | .75 | .75 |
| Add for colored wax, 800 SF per gal | CM@.003 | SF | .03 | .15 | .18 |
| **Sidewalk grits, abrasive, 25 pounds per 100 SF** | | | | | |
| Aluminum oxide 100 pound bag | CM@.008 | SF | .26 | .40 | .66 |
| Silicon carbide, 100 pound bag | CM@.008 | SF | .45 | .40 | .85 |
| **Metallic surface hardener, dry shake, Masterplate 200** | | | | | |
| Moderate duty, 1 pound per SF | CM@.008 | SF | .85 | .40 | 1.25 |
| Heavy duty, 1.5 pounds per SF | CM@.009 | SF | 1.31 | .45 | 1.76 |
| Concentrated traffic, 2 pounds per SF | CM@.010 | SF | 1.74 | .50 | 2.24 |
| Liquid curing and sealing compound, sprayed-on, | | | | | |
|   Mastercure, 400 SF | CL@.004 | SF | .09 | .16 | .25 |

# 03 Concrete

| | Craft@Hrs | Unit | Material | Labor | Total |
|---|---|---|---|---|---|
| Sweep, scrub and wash down | CL@.006 | SF | .02 | .24 | .26 |
| Finish treads and risers | | | | | |
|   No abrasives, no plastering, per SF of tread | CM@.038 | SF | — | 1.91 | 1.91 |
|   With abrasives, plastered, per SF of tread | CM@.058 | SF | .80 | 2.91 | 3.71 |
| Wax coating | CL@.002 | SF | .05 | .08 | .13 |
| **Concrete Wall Finishes** | | | | | |
| Cut back ties and patch | CM@.010 | SF | .13 | .50 | .63 |
| Remove fins | CM@.007 | LF | .06 | .35 | .41 |
| Grind smooth | CM@.020 | SF | .10 | 1.00 | 1.10 |
| Sack, simple | CM@.012 | SF | .07 | .60 | .67 |
| Bush hammer light, green concrete | CM@.020 | SF | .14 | 1.00 | 1.14 |
| Bush hammer standard, cured concrete | CM@.029 | SF | .14 | 1.45 | 1.59 |
| Bush hammer, heavy, cured concrete | CM@.077 | SF | .14 | 3.86 | 4.00 |
| Needle gun treatment, large areas | CM@.058 | SF | .14 | 2.91 | 3.05 |
| Wire brush, green | CM@.014 | SF | .06 | .70 | .76 |
| Wash with acid and rinse | CM@.004 | SF | .31 | .20 | .51 |
| Break fins, patch voids, Carborundum rub | CM@.032 | SF | .08 | 1.61 | 1.69 |
| Break fins, patch voids, burlap grout rub | CM@.024 | SF | .08 | 1.20 | 1.28 |
| Sandblasting, Subcontract. No scaffolding included, flat vertical surfaces | | | | | |
|   Light sandblast | — | SF | — | — | 2.06 |
|   Medium, to expose aggregate | — | SF | — | — | 2.35 |
|   Heavy, with dust control requirements | — | SF | — | — | 3.31 |
|   Note: Minimum charge for sandblasting | — | LS | — | — | 950.00 |
| **Miscellaneous Concrete Finishes** | | | | | |
| Monolithic natural aggregate topping | | | | | |
|   1/16" topping | P9@.006 | SF | .20 | .28 | .48 |
|   3/16" topping | P9@.018 | SF | .45 | .85 | 1.30 |
|   1/2" topping | P9@.021 | SF | 1.16 | .99 | 2.15 |
| Integral colors, | | | | | |
|   Typically 8 pounds per sack of cement | — | Lb | 2.35 | — | 2.35 |
| Mono rock, 3/8" wear course only, typical cost | — | SF | 1.37 | — | 1.37 |
| Kalman 3/4", wear course only, typical cost | — | SF | 2.08 | — | 2.08 |
| Acid etching, 5% muriatic acid, typical cost | P9@.005 | SF | .40 | .24 | .64 |
| Felton sand, for use with white cement | — | CY | 43.20 | — | 43.20 |
| **Specially Placed Concrete**  Typical subcontract prices. | | | | | |
| Gunite, no forms and no reinforcing included | | | | | |
|   Flat plane, vertical areas, 1" thick | — | SF | — | — | 3.86 |
|   Curved arch, barrel, per 1" thickness | — | SF | — | — | 4.80 |
|   Pool or retaining walls, per 1" thickness | — | SF | — | — | 3.70 |
|   Add for architectural finish | — | SF | — | — | 1.47 |
|   Wire mesh for gunite 2" x 2", W.9 x W.9 | — | SF | — | — | 1.62 |
| Pressure grouting, large quantity, 50/50 mix | — | CY | — | — | 548.00 |

**Precast Concrete**  Panel costs are based on 3,000 PSI natural gray concrete with a smooth or board finish on one side. Placing (lifting) costs include handling, hoisting, alignment, bracing and permanent connections. Erection costs vary with the type of crane. Heavier lifts over a longer radius (reach) require larger cranes. Equipment costs include the crane and a 300 amp gas powered welding machine. Add the cost of grouting or caulking joints. Costs will be higher on small jobs (less than 35,000 SF).

| | Craft@Hrs | Unit | Material | Labor | Equipment | Total |
|---|---|---|---|---|---|---|

**Precast wall panels** Costs include delivery to 40 miles, typical reinforcing steel and embedded items but no lifting. Use 2,500 SF as a minimum job charge. See placing costs below.

| | Craft@Hrs | Unit | Material | Labor | Equipment | Total |
|---|---|---|---|---|---|---|
| 4" thick wall panels (52 lbs per SF) | — | SF | 12.00 | — | — | 12.00 |
| 5" thick wall panels (65 lbs per SF) | — | SF | 13.20 | — | — | 13.20 |
| 6" thick wall panels (78 lbs per SF) | — | SF | 14.30 | — | — | 14.30 |
| 8" thick wall panels (105 lbs per SF) | — | SF | 15.20 | — | — | 15.20 |
| Add for insulated sandwich wall panels | | | | | | |
| 2-1/2" of cladding and 2" of fiberboard | — | SF | 4.73 | — | — | 4.73 |
| Add for high strength concrete | | | | | | |
| 3,500 lb concrete mix | — | % | 2.0 | — | — | — |
| 4,000 lb concrete mix | — | % | 3.0 | — | — | — |
| 4,500 lb concrete mix | — | % | 4.0 | — | — | — |
| 5,000 lb concrete mix | — | % | 6.0 | — | — | — |
| Add for white facing or integral white mix | — | % | 20.0 | — | — | — |
| Add for broken rib | | | | | | |
| or fluted surface design | — | % | 20.0 | — | — | — |
| Add for exposed aggregate finish | — | % | 17.0 | — | — | — |
| Add for sandblasted surface finish | — | % | 12.0 | — | — | — |

**Placing precast structural wall panels** Figures in parentheses show the panels placed per day with a 100 ton crawler-mounted crane and welding machine. Use $9,000 as a minimum job charge.

| | Craft@Hrs | Unit | Material | Labor | Equipment | Total |
|---|---|---|---|---|---|---|
| 1 ton panels to 180' reach (23 per day) | H4@2.79 | Ea | — | 168.00 | 71.50 | 239.50 |
| 1 to 3 tons, to 121' reach (21 per day) | H4@3.05 | Ea | — | 183.00 | 78.10 | 261.10 |
| 3 to 5 tons, to 89' reach (18 per day) | H4@3.54 | Ea | — | 213.00 | 90.70 | 303.70 |
| 5 to 7 tons, to 75' reach (17 per day) | H4@3.76 | Ea | — | 226.00 | 96.30 | 322.30 |
| 7 to 9 tons, to 65' reach (16 per day) | H4@4.00 | Ea | — | 241.00 | 103.00 | 344.00 |
| 9 to 11 tons, to 55' reach (15 per day) | H4@4.28 | Ea | — | 257.00 | 110.00 | 367.00 |
| 11 to 13 tons, to 50' reach (14 per day) | H4@4.56 | Ea | — | 274.00 | 117.00 | 391.00 |
| 13 to 15 tons, to 43' reach (13 per day) | H4@4.93 | Ea | — | 296.00 | 126.00 | 422.00 |
| 15 to 20 tons, to 36' reach (11 per day) | H4@5.81 | Ea | — | 349.00 | 149.00 | 498.00 |
| 20 to 25 tons to 31' reach (10 per day) | H4@6.40 | Ea | — | 385.00 | 164.00 | 549.00 |
| For placing panels with cladding, add | — | % | — | 5.0 | 5.0 | — |
| For placing sandwich wall panels, add | — | % | — | 10.0 | 10.0 | — |
| If a 45 ton crane can be used, deduct | — | % | — | — | -22.0 | — |

**Placing precast partition wall panels** Figures in parentheses show the panels placed per day with a 45 ton truck-mounted crane and welding machine. Use $6,000 as a minimum job charge.

| | Craft@Hrs | Unit | Material | Labor | Equipment | Total |
|---|---|---|---|---|---|---|
| 1 ton panels to 100' reach (23 per day) | H4@2.79 | Ea | — | 168.00 | 41.40 | 209.40 |
| 1 to 2 tons, to 82' reach (21 per day) | H4@3.05 | Ea | — | 183.00 | 45.30 | 228.30 |
| 2 to 3 tons, to 65' reach (18 per day) | H4@3.54 | Ea | — | 213.00 | 52.50 | 265.50 |
| 3 to 4 tons, to 57' reach (16 per day) | H4@4.00 | Ea | — | 241.00 | 59.40 | 300.40 |
| 4 to 5 tons, to 50' reach (15 per day) | H4@4.28 | Ea | — | 257.00 | 63.50 | 320.50 |
| 5 to 6 tons, to 44' reach (14 per day) | H4@4.56 | Ea | — | 274.00 | 67.70 | 341.70 |
| 6 to 7 tons, to 40' reach (13 per day) | H4@4.93 | Ea | — | 296.00 | 73.10 | 369.10 |
| 7 to 8 tons, to 36' reach (11 per day) | H4@5.81 | Ea | — | 349.00 | 86.20 | 435.20 |
| 8 to 10 tons, to 31' reach (10 per day) | H4@6.40 | Ea | — | 385.00 | 95.00 | 480.00 |
| 10 to 12 tons to 26' reach (9 per day) | H4@7.11 | Ea | — | 428.00 | 106.00 | 534.00 |
| 12 to 14 tons, to 25' reach (8 per day) | H4@8.02 | Ea | — | 482.00 | 119.00 | 601.00 |
| 14 to 16 tons, to 21' reach (7 per day) | H4@9.14 | Ea | — | 550.00 | 136.00 | 686.00 |
| If a 35 ton crane can be used, deduct | — | % | — | — | -5.0 | — |

# 03 Concrete

|  | Craft@Hrs | Unit | Material | Labor | Equipment | Total |
|---|---|---|---|---|---|---|

**Precast flat floor slab or floor plank** Costs include delivery to 40 miles, 3,000 PSI concrete, typical reinforcing steel and embedded items. Use $2,000 as a minimum job charge. See placing costs below.

|  | Craft@Hrs | Unit | Material | Labor | Equipment | Total |
|---|---|---|---|---|---|---|
| 4" thick floor panels (52 lbs per SF) | — | SF | 10.00 | — | — | 10.00 |
| 5" thick floor panels (65 lbs per SF) | — | SF | 11.90 | — | — | 11.90 |
| 6" thick floor panels (78 lbs per SF) | — | SF | 15.40 | — | — | 15.40 |
| 8" thick floor panels (105 lbs per SF) | — | SF | 15.90 | — | — | 15.90 |

Placing panels with a 100 ton crawler-mounted crane and welding machine at $775 per day. Figures in parentheses show the number of panels placed per day. Use $8,500 as a minimum job charge

| | Craft@Hrs | Unit | Material | Labor | Equipment | Total |
|---|---|---|---|---|---|---|
| 2 ton panels to 150' reach (23 per day) | H4@2.79 | Ea | — | 168.00 | 56.80 | 224.80 |
| 2 to 3 tons, to 121' reach (21 per day) | H4@3.05 | Ea | — | 183.00 | 62.00 | 245.00 |
| 3 to 4 tons, to 105' reach (19 per day) | H4@3.37 | Ea | — | 203.00 | 68.60 | 271.60 |
| 4 to 5 tons, to 88' reach (18 per day) | H4@3.56 | Ea | — | 214.00 | 72.40 | 286.40 |
| 5 to 6 tons, to 82' reach (17 per day) | H4@3.76 | Ea | — | 226.00 | 76.50 | 302.50 |
| 6 to 7 tons, to 75' reach (16 per day) | H4@4.00 | Ea | — | 241.00 | 81.40 | 322.40 |
| 7 to 8 tons, to 70' reach (15 per day) | H4@4.28 | Ea | — | 257.00 | 87.10 | 344.10 |
| 8 to 9 tons, to 65' reach (14 per day) | H4@4.56 | Ea | — | 274.00 | 93.00 | 367.00 |
| 9 to 10 tons, to 60' reach (11 per day) | H4@5.82 | Ea | — | 350.00 | 118.00 | 468.00 |
| 20 to 25 tons to 31' reach (10 per day) | H4@6.40 | Ea | — | 385.00 | 130.00 | 515.00 |

**Precast combination beam and slab units** Beam and slab units with beam on edge of slab. Weight of beams and slabs will be about 4,120 lbs per CY. Use 40 CY as a minimum job charge. See placing costs below.

| | Craft@Hrs | Unit | Material | Labor | Equipment | Total |
|---|---|---|---|---|---|---|
| Slabs with no intermediate beam | — | CY | 810.00 | — | — | 810.00 |
| Slabs with one intermediate beam | — | CY | 857.00 | — | — | 857.00 |
| Slabs with two intermediate beams | — | CY | 935.00 | — | — | 935.00 |

Placing deck units with a 100 ton crawler-mounted crane and a welding machine. Figures in parentheses show the number of units placed per day. Use $8,500 as a minimum job charge.

| | Craft@Hrs | Unit | Material | Labor | Equipment | Total |
|---|---|---|---|---|---|---|
| 5 ton units to 88' reach (21 per day) | H4@3.05 | Ea | — | 183.00 | 62.00 | 245.00 |
| 5 to 10 tons, to 60' reach (19 per day) | H4@3.37 | Ea | — | 203.00 | 68.60 | 271.60 |
| 10 to 15 tons, to 43' reach (17 per day) | H4@3.76 | Ea | — | 226.00 | 76.50 | 302.50 |
| 15 to 20 tons, to 36' reach (15 per day) | H4@4.28 | Ea | — | 257.00 | 87.00 | 344.00 |
| 20 to 25 tons, to 31' reach (14 per day) | H4@4.57 | Ea | — | 275.00 | 93.00 | 368.00 |
| If a 150 ton crane is required, add | — | % | — | — | 16.0 | — |

**Precast beams, girders and joists** Beams, girders and joists, costs per linear foot of span, including 3,000 PSI concrete, reinforcing steel, embedded items and delivery to 40 miles. Use 400 LF as a minimum job charge. See placing costs below.

| | Craft@Hrs | Unit | Material | Labor | Equipment | Total |
|---|---|---|---|---|---|---|
| 1,000 lb load per foot, 15' span | — | LF | 97.80 | — | — | 97.80 |
| 1,000 lb load per foot, 20' span | — | LF | 109.00 | — | — | 109.00 |
| 1,000 lb load per foot, 30' span | — | LF | 124.00 | — | — | 124.00 |
| 3,000 lb load per foot, 10' span | — | LF | 97.80 | — | — | 97.80 |
| 3,000 lb load per foot, 20' span | — | LF | 124.00 | — | — | 124.00 |
| 3,000 lb load per foot, 30' span | — | LF | 151.00 | — | — | 151.00 |
| 5,000 lb load per foot, 10' span | — | LF | 97.80 | — | — | 97.80 |
| 5,000 lb load per foot, 20' span | — | LF | 137.00 | — | — | 137.00 |
| 5,000 lb load per foot, 30' span | — | LF | 161.00 | — | — | 161.00 |

Placing beams, girders and joists with a 100 ton crawler-mounted crane and a welding machine. Figures in parentheses show the number of units placed per day. Use $8,500 as a minimum job charge.

| | Craft@Hrs | Unit | Material | Labor | Equipment | Total |
|---|---|---|---|---|---|---|
| 5 ton units to 88' reach (18 per day) | H4@3.56 | Ea | — | 214.00 | 72.40 | 286.40 |
| 5 to 10 tons, to 60' reach (17 per day) | H4@3.76 | Ea | — | 226.00 | 76.50 | 302.50 |

# 03 Concrete

|  | Craft@Hrs | Unit | Material | Labor | Equipment | Total |
|---|---|---|---|---|---|---|
| 10 to 15 tons, to 43' reach (16 per day) | H4@4.00 | Ea | — | 241.00 | 81.40 | 322.40 |
| 15 to 20 tons, to 36' reach (15 per day) | H4@4.27 | Ea | — | 257.00 | 86.90 | 343.90 |
| If a 150 ton crane is required, add | — | % | — | — | 16.0 | — |

**Precast concrete columns** Precast concrete columns, 12" x 12" to 36" x 36" including concrete, reinforcing steel, embedded items and delivery to 40 miles. Use 40 CY as a minimum job charge. See placing costs below.

| | | | | | | |
|---|---|---|---|---|---|---|
| Columns with | | | | | | |
| 500 lbs of reinforcement per CY | — | CY | 843.00 | — | — | 843.00 |
| Columns with | | | | | | |
| 350 lbs of reinforcement per CY | — | CY | 826.00 | — | — | 826.00 |
| Columns with | | | | | | |
| 200 lbs of reinforcement per CY | — | CY | 810.00 | — | — | 810.00 |

Placing concrete columns with a 100 ton crawler-mounted crane and a welding machine. Figures in parentheses show the number of units placed per day. Use $8,500 as a minimum job charge.

| | | | | | | |
|---|---|---|---|---|---|---|
| 3 ton columns | | | | | | |
| to 121' reach (29 per day) | H4@2.21 | Ea | — | 133.00 | 45.00 | 178.00 |
| 3 to 5 tons, to 89' reach (27 per day) | H4@2.37 | Ea | — | 143.00 | 48.20 | 191.20 |
| 5 to 7 tons, to 75' reach (25 per day) | H4@2.55 | Ea | — | 153.00 | 51.90 | 204.90 |
| 7 to 9 tons, to 65' reach (23 per day) | H4@2.79 | Ea | — | 168.00 | 56.80 | 224.80 |
| 9 to 11 tons, to 55' reach (22 per day) | H4@2.92 | Ea | — | 176.00 | 59.40 | 235.40 |
| 11 to 13 tons, to 50' reach (20 per day) | H4@3.20 | Ea | — | 192.00 | 65.10 | 257.10 |
| 13 to 15 tons, to 43' reach (18 per day) | H4@3.56 | Ea | — | 214.00 | 72.40 | 286.40 |

**Precast concrete stairs** Costs include concrete, reinforcing, embedded steel, nosing, and delivery to 40 miles. Use $500 as a minimum job charge. See placing costs below.
Stairs, 44" to 48" wide, 10' to 12' rise, "U"- or "L"-shape, including typical landings,

| | | | | | | |
|---|---|---|---|---|---|---|
| Cost per each step (7" rise) | — | Ea | 48.20 | — | — | 48.20 |

Placing precast concrete stairs with a 45 ton truck-mounted crane and a welding machine. Figures in parentheses show the number of stair flights placed per day. Use $6,000 as a minimum job charge.

| | | | | | | |
|---|---|---|---|---|---|---|
| 2 ton stairs to 82' reach (21 per day) | H4@3.05 | Ea | — | 183.00 | 45.30 | 228.30 |
| 2 to 3 tons, to 65' reach (19 per day) | H4@3.37 | Ea | — | 203.00 | 50.00 | 253.00 |
| 3 to 4 tons, to 57' reach (18 per day) | H4@3.54 | Ea | — | 213.00 | 52.50 | 265.50 |
| 4 to 5 tons, to 50' reach (17 per day) | H4@3.76 | Ea | — | 226.00 | 55.80 | 281.80 |
| 5 to 6 tons, to 44' reach (16 per day) | H4@4.00 | Ea | — | 241.00 | 59.40 | 300.40 |
| 6 to 7 tons, to 40' reach (15 per day) | H4@4.28 | Ea | — | 257.00 | 63.50 | 320.50 |
| 7 to 8 tons, to 36' reach (14 per day) | H4@4.56 | Ea | — | 274.00 | 67.70 | 341.70 |
| 8 to 9 tons, to 33' reach (10 per day) | H4@6.40 | Ea | — | 385.00 | 95.00 | 480.00 |
| If a 35 ton crane can be used, deduct | — | % | — | — | -10.0 | — |

**Precast double tees** 8' wide, 24" deep. Design load is shown in pounds per square foot (PSF). Costs include lifting and placing but no slab topping. Equipment cost is for a 45 ton truck crane and a welding machine. Use $6,000 as a minimum job charge.

| | | | | | | |
|---|---|---|---|---|---|---|
| 30' to 35' span, 115 PSF | H4@.019 | SF | 11.10 | 1.14 | .28 | 12.52 |
| 30' to 35' span, 141 PSF | H4@.019 | SF | 12.10 | 1.14 | .28 | 13.52 |
| 35' to 40' span, 78 PSF | H4@.018 | SF | 11.10 | 1.08 | .27 | 12.45 |
| 35' to 40' span, 98 to 143 PSF | H4@.018 | SF | 12.10 | 1.08 | .27 | 13.45 |
| 40' to 45' span, 53 PSF | H4@.017 | SF | 11.10 | 1.02 | .25 | 12.37 |
| 40' to 45' span, 69 to 104 PSF | H4@.017 | SF | 12.10 | 1.02 | .25 | 13.37 |
| 40' to 45' span, 134 PSF | H4@.017 | SF | 12.70 | 1.02 | .25 | 13.97 |
| 45' to 50' span, 77 PSF | H4@.016 | SF | 12.10 | .96 | .24 | 13.30 |
| 45' to 50' span, 101 PSF | H4@.016 | SF | 12.70 | .96 | .24 | 13.90 |

| | Craft@Hrs | Unit | Material | Labor | Equipment | Total |
|---|---|---|---|---|---|---|

Concrete topping, based on 350 CY job 3,000 PSI design mix with lightweight aggregate, no forms or finishing included.

| | Craft@Hrs | Unit | Material | Labor | Equipment | Total |
|---|---|---|---|---|---|---|
| By crane | H3@.868 | CY | 100.00 | 39.20 | 16.40 | 155.60 |
| By pump (28 meter boom truck and operator) | M2@.080 | CY | 115.00 | 3.51 | 6.64 | 125.15 |

Hollow core roof planks, not including topping, 35 PSF live load, 10' to 15' span

| | Craft@Hrs | Unit | Material | Labor | Equipment | Total |
|---|---|---|---|---|---|---|
| 4" thick | H4@.018 | SF | 5.86 | 1.08 | .26 | 7.20 |
| 6" thick | H4@.018 | SF | 7.16 | 1.08 | .26 | 8.50 |
| 8" thick | H4@.020 | SF | 8.39 | 1.20 | .28 | 9.87 |
| 10" thick | H4@.026 | SF | 8.74 | 1.56 | .37 | 10.67 |

**Prestressed concrete support poles** Costs shown are per linear foot for sizes and lengths shown. Equipment cost is for a 45 ton truck crane. These costs do not include base or excavation. Lengths other than as shown will cost more per foot. Use $6,000 as a minimum job charge.

10" x 10" poles with four 7/16" 270 KSI ultimate tendon strand strength

| | Craft@Hrs | Unit | Material | Labor | Equipment | Total |
|---|---|---|---|---|---|---|
| 10', 14', 20', 25' or 30' long | H4@.028 | LF | 15.70 | 1.68 | .38 | 17.76 |

12" x 12" poles with six 7/16" 270 KSI ultimate tendon strand strength

| | Craft@Hrs | Unit | Material | Labor | Equipment | Total |
|---|---|---|---|---|---|---|
| 10', 14' or 20' long | H4@.028 | LF | 18.90 | 1.68 | .38 | 20.96 |
| 25' or 30' long | H4@.028 | LF | 21.00 | 1.68 | .38 | 23.06 |

**Tilt-up concrete construction** Costs will be higher if access is not available from all sides or obstructions delay the work. Except as noted, no grading, compacting, engineering, design, permit or inspection fees are included.

**Foundations and footings for tilt-up** Costs assume normal soil conditions with no forming or shoring required and are based on concrete poured directly from the chute of a ready-mix truck. Soil excavated and not needed for backfill can be stored on site and used by others. For scheduling purposes, estimate that a crew of 5 can set forms for and pour 40 to 50 CY of concrete footings and foundations in an 8-hour day.

Lay out foundation from known point on site, set stakes and mark for excavation.

| | Craft@Hrs | Unit | Material | Labor | Equipment | Total |
|---|---|---|---|---|---|---|
| Per SF of floor slab | C8@.001 | SF | — | .05 | — | .05 |

Excavation with a 1/2 CY utility backhoe

| | Craft@Hrs | Unit | Material | Labor | Equipment | Total |
|---|---|---|---|---|---|---|
| 27 CY per hour | S1@.075 | CY | — | 3.56 | 1.33 | 4.89 |

Set and tie grade A60 reinforcing bars, typical foundation has 50 to 60 pounds of reinforcing bar per CY of concrete

| | Craft@Hrs | Unit | Material | Labor | Equipment | Total |
|---|---|---|---|---|---|---|
| | RB@.008 | Lb | .54 | .47 | — | 1.01 |

Place stringer supports for reinforcing bars (at 1.1 LF of support per LF of rebar), using 2" x 4" or 2" x 6" lumber. Includes typical stakes and hardware

| | Craft@Hrs | Unit | Material | Labor | Equipment | Total |
|---|---|---|---|---|---|---|
| Based on 3 uses of supports | C8@.017 | LF | .51 | .78 | — | 1.29 |

(Inspection of reinforcing may be required before concrete is placed.)

Make and place steel column anchor bolt templates (using 1.1 SF of plywood per SF of column base). Based on 2 uses of 3/4" plyform. If templates are furnished by the steel column fabricator, include only the labor cost to place the templates as shown

| | Craft@Hrs | Unit | Material | Labor | Equipment | Total |
|---|---|---|---|---|---|---|
| Make and place templates | C8@.160 | SF | 1.95 | 7.33 | — | 9.28 |
| Place templates furnished by others | C8@.004 | SF | — | .18 | — | .18 |

Blockout form at foundation, step footings, columns, etc., per SF of contact area, using 2" x 6" lumber. Includes allowance for waste, typical stakes and hardware.

| | Craft@Hrs | Unit | Material | Labor | Equipment | Total |
|---|---|---|---|---|---|---|
| Based on 3 uses of forms | C8@.018 | SFCA | 3.10 | .82 | — | 3.92 |

Column anchor bolts, usually 4 per column. "J"-hook bolt includes flat washer and two nuts

| | Craft@Hrs | Unit | Material | Labor | Equipment | Total |
|---|---|---|---|---|---|---|
| 1/2" x  6" | C8@.037 | Ea | .95 | 1.70 | — | 2.65 |
| 3/4" x 12" | C8@.037 | Ea | 3.29 | 1.70 | — | 4.99 |

Concrete, 2,000 PSI design mix, including 3% allowance for waste,

| | Craft@Hrs | Unit | Material | Labor | Equipment | Total |
|---|---|---|---|---|---|---|
| Placed directly from chute | CL@.552 | CY | 98.90 | 22.00 | — | 120.90 |

| | Craft@Hrs | Unit | Material | Labor | Equipment | Total |
|---|---|---|---|---|---|---|
| Remove, clean and stack reinforcing supports | CL@.020 | LF | — | .80 | — | .80 |
| Dry pack non-shrink grout under column bases. Typical grout area is 2' x 2' x 4" (1.33 CF of grout) | CM@.625 | Ea | 11.60 | 31.40 | — | 43.00 |
| Panel support pads. (Supports panel when grout is placed.) One needed per panel. Based on 3 CF of 2,000 PSI concrete per pad | C8@.141 | Ea | 10.70 | 6.46 | — | 17.16 |
| Backfill at foundation with 1/2 CY utility backhoe. 90 CY per hour | S1@.022 | CY | — | 1.04 | .39 | 1.43 |
| Dispose of excess soil on site, using backhoe | S1@.011 | CY | — | .52 | .20 | .72 |
| Concrete test cylinders, typically 5 required per 100 CY of concrete | — | CY | 1.45 | — | — | 1.45 |

**Floor slabs for tilt-up** These costs assume that the site has been graded and compacted by others before slab work begins. The costs below include fine grading the site, placing a 2" layer of sand, a vapor barrier and a second 2" layer of sand over the barrier. Each layer of sand should be screeded level. Reinforcing may be either wire mesh or bars. When sand and vapor barrier are used, concrete may have to be pumped in place. If inspection of the reinforcing is required, the cost will usually be included in the building permit fee. Floor slabs are usually poured 6" deep in 14' to 18' wide strips with edge forms on both sides and ends. Edge forms are used as supports for the mechanical screed which consolidates the concrete with pneumatic vibrators as the surface is leveled. The screed is pulled forward by a winch and cable. Edge forms should have an offset edge to provide a groove which is filled by the adjacent slab. The slab perimeter should end 2' to 3' inside the finished building walls. This "pour strip" is filled in with concrete when the walls are up and braces have been removed. About 2 days after pouring, the slab should be saw-cut 1-1/2" to 2" deep and 1/4" wide each 20' both ways. This controls cracking as the slab cures. For scheduling purposes, estimate that a finishing crew of 10 workers can place and finish 10,000 to 12,000 square feet of slab in 8 hours.

| | Craft@Hrs | Unit | Material | Labor | Equipment | Total |
|---|---|---|---|---|---|---|
| Fine grade the building pad with a utility tractor with blade and bucket 2,500 SF per hour | S6@.001 | SF | — | .04 | .01 | .05 |
| Spread sand 4" deep in 2" layers, (1.35 tons is 1 cubic yard and covers 81 SF at 4" depth), using a utility tractor | | | | | | |
|   Cost of sand | — | Ton | 10.00 | — | — | 10.00 |
|   60 SF per hour | S6@.006 | SF | .17 | .27 | .06 | .50 |
| Vapor barrier, 6 mil polyethylene | CL@.001 | SF | .08 | .04 | — | .12 |
| Welded wire mesh, 6" x 6", W1.4 x W1.4 including 10% for overlap and waste | RB@.004 | SF | .17 | .23 | — | .40 |
| Reinforcing bars, grade 60, typically #4 bars on 2' centers each way with 1.05 LF bars per SF | RB@.004 | SF | .42 | .23 | — | .65 |
| Dowel out for #4 bars at construction joints | RB@.004 | LF | .28 | .23 | — | .51 |
| Edge forms with keyed joint, rented | F5@.013 | LF | — | .61 | .09 | .70 |
| Ready-mix concrete for the slab (with 3% waste allowance) | | | | | | |
|   Placed from the chute of the ready-mix truck | CL@.421 | CY | 103.00 | 16.80 | — | 119.80 |
|   Pump mix, including pumping cost | M2@.188 | CY | 115.00 | 8.24 | 10.90 | 134.14 |
| Finish the floor slab, using power trowels | CM@.003 | SF | — | .15 | .03 | .18 |
| Vibrating screed and accessories, (includes screeds, fuel, air compressor and hoses) rented | | | | | | |
|   Per SF of slab | — | SF | — | — | .13 | .13 |
| Curing compound, spray applied, 250 SF per gallon | CL@.001 | SF | .03 | .04 | — | .07 |

# 03 Concrete

| | Craft@Hrs | Unit | Material | Labor | Equipment | Total |
|---|---|---|---|---|---|---|
| Saw cut green concrete, | | | | | | |
| 1-1/2" to 2" deep | C8@.010 | LF | — | .46 | .09 | .55 |
| Strip, clean and stack edge forms | CL@.011 | LF | — | .44 | — | .44 |
| Ready-mix concrete for the pour strip including 3% waste | | | | | | |
| Placed from the chute of the | | | | | | |
| ready-mix truck | CL@.421 | CY | 122.00 | 16.80 | — | 138.80 |
| Pump mix, including pumping cost | M2@.188 | CY | 129.00 | 8.24 | 10.90 | 148.14 |
| Finish the pour strip | CM@.004 | SF | — | .20 | .03 | .23 |
| Special inspection of the concrete pour (if required) | | | | | | |
| Based on 10,000 SF in 8 hours | — | SF | — | — | — | .04 |
| Clean and caulk the saw joints, based on 50 LF per hour and | | | | | | |
| caulk at 1,500 LF per gallon | CL@.020 | LF | .04 | .80 | — | .84 |
| Clean slab (broom clean) | CL@.001 | SF | — | .04 | — | .04 |

**Tilt-up wall panels** Wall thickness is usually a nominal 6" (5-1/2" actual) or a nominal 8" (7-1/2"). Panel heights over 40' are uncommon. Typical panel width is 20'. When the floor area isn't large enough to form all wall panels on the floor slab, stacking will be required. Construction period will be longer and the costs higher than the figures shown below when panels must be poured and cured in stacked multiple layers. When calculating the volume of concrete required, the usual practice is to multiply overall panel dimensions (width, length and thickness) and deduct only for panel openings that exceed 25 square feet. When pilasters (thickened wall sections) are formed as part of the panel, add the cost of extra formwork and the additional concrete. Use the production rates that follow for scheduling a job.

## Production rates for tilt-up

Lay out and form panels, place reinforcing bars and install embedded lifting devices:
2,000 to 3,000 SF of panel face (measured one side) per 8-hour day for a crew of 5. Note that reinforcing may have to be inspected before concrete is poured.

Place and finish concrete: 4,000 to 5,000 SF of panel per 8-hour day for a crew of 5.

Continuous inspection of the concrete pour by a licensed inspector may be required.
The minimum charge will usually be 4 hours at $50 per hour plus travel cost.

Install ledgers with embedded anchor bolts on panels before the concrete hardens:
150 to 200 LF of ledger per 8-hour day for a crew of 2.

Sack and patch exposed panel face before panel is erected:
2,000 to 3,000 SF of panel face per 8-hour day for a crew of 2.

Install panel braces before lifting panels:
30 to 40 braces per 8-hour day for a crew of 2. Panels usually need 3 or 4 braces. Braces are usually rented for 30 days.

**Tilt up and set panels** 20 to 24 panels per 8-hour day for a crew of 11 (6 workers, the crane operator, oiler and 3 riggers). If the crew sets 22 panels averaging 400 SF each in an 8-hour day, the crew is setting 1,100 SF per hour. These figures include time for setting and aligning panels, drilling the floor slab for brace bolts, and securing the panel braces. Panel tilt-up will go slower if the crane can't work from the floor slab when lifting panels. Good planning of the panel lifting sequence and easy exit of the crane from the building interior will increase productivity. Size and weight of the panel has little influence on the setting time if the crane is capable of making the lift and is equipped with the right rigging equipment, spreader bars and strongbacks. Panels will usually have to cure for 14 days before lifting unless high early strength concrete is used.

Weld or bolt steel embedded in panels: 16 to 18 connections per 8-hour day for a crew of 2 working from ladders. Work more than 30 feet above the floor may take more time.

Sack and patch panel faces after erection: 400 SF of face per 8-hour day for a crew of 2.

# 03 Concrete

| | Craft@Hrs | Unit | Material | Labor | Equipment | Total |
|---|---|---|---|---|---|---|
| **Tilt-up panel costs** | | | | | | |
| Engineering fee, per panel, typical | — | Ea | — | — | — | 50.00 |
| Fabricate and lay out forms for panel edges and blockouts, assumes 2 uses and includes | | | | | | |
| typical hardware and chamfer | | | | | | |
| 2" x 6" per MBF | — | MBF | 465.00 | — | — | 465.00 |
| 2" x 8" per MBF | — | MBF | 465.00 | — | — | 465.00 |
| 2" x 6" per LF | F5@.040 | LF | .46 | 1.88 | — | 2.34 |
| 2" x 8" per LF | F5@.040 | LF | .62 | 1.88 | — | 2.50 |
| Reveals (1" x 4" accent strips on exterior panel face) | | | | | | |
| based on 1 use of lumber, | | | | | | |
| nailed to slab | C8@.018 | LF | .59 | .82 | — | 1.41 |
| Clean and prepare the slab prior to placing the concrete for the panels, | | | | | | |
| 1,250 SF per hour | CL@.001 | SF | — | .04 | — | .04 |
| Releasing agent (bond breaker), sprayed on, | | | | | | |
| 250 SF and $7.50 per gallon | CL@.001 | SF | .05 | .04 | — | .09 |
| Place and tie grade 60 reinforcing bars and supports. Based on using #3 and #4 bars, including waste and laps. A | | | | | | |
| typical panel has 80 to 100 pounds of reinforcing steel per CY of concrete in the panel | | | | | | |
| Reinforcing per CY of concrete | RB@.500 | CY | 44.10 | 29.20 | — | 73.30 |
| Reinforcing per SF of 6" thick panel | RB@.009 | SF | .81 | .52 | — | 1.33 |
| Reinforcing per SF of 8" thick panel | RB@.012 | SF | 1.09 | .70 | — | 1.79 |
| Ledger boards with anchor bolts. Based on 4" x 12" treated lumber and 3/4" x 12" bolts placed each 2 feet, installed | | | | | | |
| prior to lifting panel | | | | | | |
| Using 4" x 12" treated lumber | — | MBF | 1,290.00 | — | — | 1,290.00 |
| Typical ledgers and anchor bolts | C8@.353 | LF | 5.27 | 16.20 | — | 21.47 |
| Embedded steel for pickup and brace point hardware. Usually 10 to 12 points per panel, quantity and type depend | | | | | | |
| on panel size | | | | | | |
| Embedded pickup & brace hardware, | | | | | | |
| typical | C8@.001 | SF | .68 | .05 | — | .73 |
| Note: Inspection of reinforcing may be required before concrete is placed. | | | | | | |
| Concrete, 3,000 PSI, placed direct from chute | | | | | | |
| Concrete only (before waste allowance) | — | CY | 100.00 | — | — | 100.00 |
| 6" thick panels (54 SF per CY) | CL@.008 | SF | 1.91 | .32 | .07 | 2.30 |
| 8" thick panels (40.5 SF per CY) | CL@.010 | SF | 2.54 | .40 | .09 | 3.03 |
| Concrete, 3,000 PSI, pumped in place | | | | | | |
| Concrete only (before waste allowance) | — | CY | 115.00 | — | — | 115.00 |
| 6" thick panels (54 SF per CY) | M2@.008 | SF | 2.19 | .35 | .02 | 2.56 |
| 8" thick panels (40.5 SF per CY) | M2@.011 | SF | 2.93 | .48 | .03 | 3.44 |
| Finish concrete panels, | | | | | | |
| one face before lifting | P8@.004 | SF | — | .18 | .02 | .20 |
| Continuous inspection of pour, if required, based on 5,000 to 6,000 SF being placed per 8-hour day | | | | | | |
| Typical cost | | | | | | |
| for continuous inspection of pour | — | SF | — | — | — | .08 |
| Curing wall panels with spray-on curing compound, per SF of face, | | | | | | |
| based on 250 SF per gallon | CL@.001 | SF | .04 | .04 | — | .08 |
| Strip, clean and stack 2" edge forms | CL@.030 | LF | — | 1.20 | — | 1.20 |

**Braces for panels**

Install rented panel braces, based on an average of three braces per panel for panels averaging 400 to 500 SF each.
See also, Remove panel braces

| | Craft@Hrs | Unit | Material | Labor | Equipment | Total |
|---|---|---|---|---|---|---|
| Typical cost for braces, | | | | | | |
| per SF of panel | F5@.005 | SF | — | .24 | .13 | .37 |

| | Craft@Hrs | Unit | Material | Labor | Equipment | Total |
|---|---|---|---|---|---|---|
| **Miscellaneous equipment rental, typical costs per SF of wall panel** | | | | | | |
| Rent power trowels and vibrators | — | SF | — | — | .08 | .08 |
| Rent compressors, hoses, rotohammers, fuel | — | SF | — | — | .08 | .08 |
| **Lifting panels into place, with crane and riggers, using a 140 ton truck crane** | | | | | | |
| Move crane on and off site, typical | — | LS | — | — | 2,940.00 | 2,940.00 |
| Hourly cost (4 hr minimum applies), per hour | T2@11.0 | Hr | — | 565.00 | 396.00 | 961.00 |
| Set, brace and align panels, per SF of panel measured on one face, Based on 1,100 SF per hour | T2@.010 | SF | — | .51 | .41 | .92 |
| **Allowance for shims, bolts for panel braces and miscellaneous hardware, labor cost is included with associated items** | | | | | | |
| Typical cost per SF of panel | — | SF | .35 | — | — | .35 |
| **Weld or bolt panels together, typically 2 connections are required per panel, equipment is a welding machine.** | | | | | | |
| Typical cost per panel connection | T3@.667 | Ea | 89.00 | 32.70 | 3.29 | 124.99 |
| **Continuous inspection of welding, if required.** | | | | | | |
| Inspection of welds, per connection (3 per hour) | — | Ea | — | — | — | 27.50 |
| **Set grout forms at exterior wall perimeter. Based on using 2" x 12" boards. Costs shown include stakes and assume 3 uses of the forms** | | | | | | |
| Using 2" x 12" boards (per MBF) | — | MBF | 563.00 | — | — | 563.00 |
| Cost of forms per linear foot (LF) | F5@.019 | LF | .74 | .89 | — | 1.63 |
| **Pour grout under panel bottoms. Place 2,000 PSI concrete (grout) directly from the chute of ready-mix truck.** | | | | | | |
| Typically 1.85 CY per 100 LF of wall | M2@.005 | LF | 1.83 | .22 | — | 2.05 |
| **Sack and patch wall, per SF of area (measure area to be sacked and patched, both inside and outside),** | | | | | | |
| per SF of wall face(s), before tilt-up | P8@.007 | SF | .09 | .32 | — | .41 |
| per SF of wall face(s), after tilt-up | P8@.040 | SF | .09 | 1.80 | — | 1.89 |
| **Caulking and backing strip for panel joints. Cost per LF for caulking both inside & outside face of joint,** | | | | | | |
| per linear foot (LF) of joint | PA@.055 | LF | .24 | 2.83 | — | 3.07 |
| **Remove panel braces,** | | | | | | |
| stack & load onto trucks | CL@.168 | Ea | — | 6.70 | — | 6.70 |
| **Final cleanup & patch floor,** | | | | | | |
| per SF of floor | P8@.003 | SF | .01 | .14 | — | .15 |
| **Additional costs for tilt-up panels, if required** | | | | | | |
| Architectural form liner, minimum cost | F5@.001 | SF | 1.95 | .05 | — | 2.00 |
| Architectural form liner, typical cost | F5@.003 | SF | 2.43 | .14 | — | 2.57 |
| Sandblast, light and medium, subcontract | — | SF | — | — | — | 1.50 |
| Sandblast, heavy, subcontract | — | SF | — | — | — | 2.25 |
| Truck dock bumpers, installation only | C8@2.00 | Ea | — | 91.70 | — | 91.70 |
| Dock levelers, installation only | C8@7.46 | Ea | — | 342.00 | — | 342.00 |
| Angle iron guard around door opening, 2" x 2" x 1/4", mild steel (4' lengths) | C8@.120 | LF | 2.45 | 5.50 | — | 7.95 |
| Pipe guard at door opening, 6" black steel, concrete filled | C8@.281 | LF | 29.90 | 12.90 | — | 42.80 |
| Embedded reglet at parapet wall, aluminum | F5@.019 | LF | 1.15 | .89 | — | 2.04 |
| Integral color in concrete, light colors, add to concrete. Price per CY of concrete to be colored | — | CY | 27.20 | — | — | 27.20 |
| Backfill outside perimeter wall with a utility tractor. 90 CY per hour | S1@.022 | CY | — | 1.04 | .27 | 1.31 |

# 03 Concrete

| | Craft@Hrs | Unit | Material | Labor | Equipment | Total |
|---|---|---|---|---|---|---|

**Cast-in-place columns for tilt-up** Concrete columns are often designed as part of the entryway of a tilt-up building or to support lintel panels over a glass storefront. Both round and square freestanding columns are used. They must be formed and poured before the lintel panels are erected. Round columns are formed with plastic-treated fiber tubes (Sonotube is one manufacturer) that are peeled off when the concrete has cured. Square columns are usually formed with plywood and 2" x 4" wood braces or adjustable steel column clamps.

Round plastic-lined fiber tube forms, set and aligned, based on 12' long tubes. Longer tubes will cost more per LF

| | Craft@Hrs | Unit | Material | Labor | Equipment | Total |
|---|---|---|---|---|---|---|
| 12" diameter tubes | M2@.207 | LF | 16.50 | 9.08 | — | 25.58 |
| 20" diameter tubes | M2@.250 | LF | 28.00 | 11.00 | — | 39.00 |
| 24" diameter tubes | M2@.275 | LF | 33.60 | 12.10 | — | 45.70 |

Rectangular plywood forms, using 1.1 SF of 3/4" plyform at $92.50 per 4' x 8' sheet ($2.89 per SF), and 2" x 4" (at $469.00 per MBF) bracing 3' OC

| | Craft@Hrs | Unit | Material | Labor | Equipment | Total |
|---|---|---|---|---|---|---|
| Per SF of contact area | | | | | | |
| based on 3 uses | F5@.083 | SF | 1.65 | 3.90 | — | 5.55 |

Column braces, 2" x 4" at $469.00 per MBF, with 5 BF per LF of column height

| | Craft@Hrs | Unit | Material | Labor | Equipment | Total |
|---|---|---|---|---|---|---|
| Per LF of column height | | | | | | |
| based on 3 uses | F5@.005 | LF | 1.11 | .24 | — | 1.35 |

Column clamps rented at $4.50 for 30 days, 1 per LF of column height, cost per LF of column height

| | Craft@Hrs | Unit | Material | Labor | Equipment | Total |
|---|---|---|---|---|---|---|
| Per LF of column | | | | | | |
| based on 2 uses a month | F5@.040 | LF | — | 1.88 | 3.25 | 5.13 |

Reinforcing bars, grade 60, placed and tied, typically 150 to 200 pounds per CY of concrete

| | Craft@Hrs | Unit | Material | Labor | Equipment | Total |
|---|---|---|---|---|---|---|
| Cages, round or rectangular | RB@1.50 | CY | 84.20 | 87.50 | — | 171.70 |
| Vertical bars | | | | | | |
| attached to cages | RB@1.50 | CY | 84.20 | 87.50 | — | 171.70 |

Ready-mix concrete, 3,000 PSI pump mix

| | Craft@Hrs | Unit | Material | Labor | Equipment | Total |
|---|---|---|---|---|---|---|
| Placed by pump (includes 3% waste) | M2@.094 | CY | 118.00 | 4.12 | 7.45 | 129.57 |

Strip off fiber forms and dismantle bracing, per LF of height

| | Craft@Hrs | Unit | Material | Labor | Equipment | Total |
|---|---|---|---|---|---|---|
| 12" or 16" diameter | CL@.075 | LF | — | 2.99 | — | 2.99 |
| 24" diameter | CL@.095 | LF | — | 3.79 | — | 3.79 |

Strip wood forms and dismantle bracing

| | Craft@Hrs | Unit | Material | Labor | Equipment | Total |
|---|---|---|---|---|---|---|
| Per SF of contact area, no salvage | CL@.011 | SF | — | .44 | — | .44 |
| Clean and stack column clamps | CL@.098 | Ea | — | 3.91 | — | 3.91 |
| Sack and patch concrete column face | P8@.011 | SF | .15 | .50 | — | .65 |

Allowance for miscellaneous equipment rental (vibrators, compressors, hoses, etc.)

| | Craft@Hrs | Unit | Material | Labor | Equipment | Total |
|---|---|---|---|---|---|---|
| Per CY of concrete in column | — | CY | — | — | 2.50 | 2.50 |

**Trash enclosures for tilt-up** These usually consist of a 6" concrete slab measuring 5' x 10' and precast concrete tilt-up panels 6' high. A 6" ledger usually runs along the wall interior. Overall wall length is usually 20 feet. Cost for the trash enclosure foundation, floor slab and wall will be approximately the same as for the tilt-up foundation, floor slab and wall. The installed cost of a two-leaf steel gate measuring 6' high and 8' long overall will be about $400 including hardware.

**Equipment Foundations** Except as noted, no grading, compacting, engineering, design, permit or inspection fees are included. Costs assume normal soil conditions with no shoring or dewatering required and are based on concrete poured directly from the chute of a ready-mix truck. Costs will be higher if access is not available from all sides or if obstructions delay the work.

Labor costs include the time needed to prepare formwork sketches at the job site, measure for the forms, fabricate, erect, align and brace the forms, place and tie the reinforcing steel and embedded steel items, pour and finish the concrete and strip, clean and stack the forms.

Reinforcing steel costs assume an average of #3 and #4 bars at 12" on center each way. Weights shown include an allowance for waste and laps.

Form costs assume 3 uses of the forms and show the cost per use when a form is used on the same job without being totally disassembled. Normal cleaning and repairs are included. No salvage value is assumed. Costs listed are per square foot of form in contact with the concrete, assume using #2 & Btr lumber

For scheduling purposes, estimate that a crew of 5 can set the forms and place and tie the reinforcing steel for and pour 8 to 10 CY of concrete equipment foundations in an 8-hour day.

**Rule-of-thumb estimates** The costs below are based on the volume of concrete in the foundation. If plans for the foundation have not been prepared, estimate the quantity of concrete required from the weight of the equipment using the rule-of-thumb estimating data below. Design considerations may alter the actual quantities required. Based on the type foundation required, use the cost per CY given in the detailed cost breakdown section which follows the table on the next page.

Estimating concrete quantity from the weight of equipment:

1. Rectangular pad type foundations —
   Boilers and similar equipment mounted at grade:
       Allow 1 CY per ton
   Centrifugal pumps and compressors, including weight of motor:
       Up to 2 HP, allow .5 CY per ton
       Over 2 to 10 HP, allow 1 CY per ton
       Over 10 to 30 HP, allow 2 CY per ton
       Over 30 HP, allow 3 CY per ton
   Reciprocating pumps and compressors, including weight of motor:
       Up to 2 HP, allow .75 CY per ton
       Over 2 to 10 HP, allow 1.5 CY per ton
       Over 10 to 30 HP, allow 2.5 CY per ton
       Over 30 HP, allow 4.5 CY per ton

2. Rectangular pad and stem wall type foundations — Horizontal tanks or shell and tube heat exchangers, allow 6 CY per ton

3. Square pad and pedestal type foundations — Vertical tanks not over 3' in diameter, allow 6 CY per ton

4. Octagonal pad and pedestal type foundations — Vertical tanks over 3' in diameter, allow 4 CY per ton of empty weight

| | Craft@Hrs | Unit | Material | Labor | Equipment | Total |
|---|---|---|---|---|---|---|

**Rectangular pad type foundations** For pumps, compressors, boilers and similar equipment mounted at grade. Form costs assume nails, stakes and form oil costing $.50 per square foot and 1.1 SF of 3/4" plyform plus 2.2 BF of lumber per square foot of contact surface. Costs shown are per cubic yard (CY) of concrete within the foundation.

| | Craft@Hrs | Unit | Material | Labor | Equipment | Total |
|---|---|---|---|---|---|---|
| Lay out foundation from known point on site, set stakes and mark for excavation, with 1 CY per CY of concrete | C8@.147 | CY | .06 | 6.74 | — | 6.80 |
| Excavation with utility backhoe, at 28 CY per hour, with 1.5 CY per CY of concrete | S1@.108 | CY | — | 5.12 | 3.84 | 8.96 |
| Formwork, material (3 uses) and labor at .150 manhours per SF with 15 SF per CY of concrete | F5@2.25 | CY | 27.70 | 106.00 | — | 133.70 |
| Reinforcing bars, Grade 60, set and tied, material and labor .006 manhours per lb with 25 lb per CY of concrete | RB@.150 | CY | 12.60 | 8.75 | — | 21.35 |
| Anchor bolts, sleeves, embedded steel, material and labor at .040 manhours per lb with 5 lb per CY of concrete | C8@.200 | CY | 6.80 | 9.17 | — | 15.97 |
| Ready-mix concrete, 3,000 PSI mix after allowance for 10% waste with 1 CY per CY of concrete | C8@.385 | CY | 110.00 | 17.60 | — | 127.60 |
| Concrete test cylinders including test reports, with 5 per 100 CY of concrete | — | CY | .65 | — | — | .65 |
| Finish, sack and patch concrete, material and labor at .010 manhours per SF with 15 SF per CY of concrete | P8@.150 | CY | .92 | 6.75 | — | 7.67 |
| Backfill around foundation with utility backhoe at 40 CY per hour, with .5 CY per CY of concrete | S1@.025 | CY | — | 1.19 | .44 | 1.63 |
| Dispose of excess soil on site with utility backhoe at 80 CY per hour with 1 CY per CY of concrete | S1@.025 | CY | — | 1.19 | .44 | 1.63 |
| **Rectangular pad type foundation** Total cost per CY of concrete | —@3.44 | CY | 158.73 | 162.51 | 4.72 | 325.96 |

| | Craft@Hrs | Unit | Material | Labor | Equipment | Total |
|---|---|---|---|---|---|---|

**Rectangular pad and stem wall type foundations** For horizontal cylindrical or shell and tube heat exchangers mounted above grade. Form costs assume nails, stakes and form oil costing $.60 per square foot and 1.1 SF of 3/4" plyform plus 4.5 BF of lumber per square foot of contact surface. Costs shown are per cubic yard (CY) of concrete within the foundations.

| | Craft@Hrs | Unit | Material | Labor | Equipment | Total |
|---|---|---|---|---|---|---|
| Lay out foundation from known point on site, set stakes and mark for excavation, with 1 CY per CY of concrete | C8@.147 | CY | .05 | 6.74 | — | 6.79 |
| Excavation with utility backhoe, at $28 per hour, with 1 CY per CY of concrete | S1@.072 | CY | — | 3.42 | 1.28 | 4.70 |
| Formwork, material (3 uses) per SF and labor at .200 manhours per SF with 60 SF per CY of concrete | F5@12.0 | CY | 172.00 | 564.00 | — | 736.00 |
| Reinforcing bars, Grade 60, set and tied, material per lb and labor at .006 manhours per lb with 45 lb per CY of concrete | RB@.270 | CY | 22.70 | 15.70 | — | 38.40 |
| Anchor bolts, sleeves, embedded steel, material and labor at .040 manhours per lb with 5 lb per CY of concrete | C8@.200 | CY | 6.80 | 9.17 | — | 15.97 |
| Ready-mix concrete, 3,000 PSI mix, after 10% waste allowance with 1 CY per CY | C8@.385 | CY | 110.00 | 17.60 | — | 127.60 |
| Concrete test cylinders including test reports, with 5 per 100 CY | — | CY | .65 | — | — | .65 |
| Finish, sack and patch concrete, material per SF and labor at .010 manhours per SF with 60 SF per CY of concrete | P8@.600 | CY | 3.69 | 27.00 | — | 30.69 |
| Backfill around foundation with utility backhoe, at 40 CY per hour, with .5 CY per CY of concrete | S1@.025 | CY | — | 1.19 | .44 | 1.63 |
| Dispose of excess soil on site with utility backhoe, at 80 CY per hour, with .5 CY per CY of concrete | S1@.025 | CY | — | 1.19 | .44 | 1.63 |
| **Rectangular pad and stem wall type foundation** Total cost per CY of concrete | —@13.7 | CY | 315.89 | 646.01 | 2.16 | 964.06 |

**Square pad and pedestal type foundations** For vertical cylindrical tanks not over 3' in diameter. Form costs assume nails, stakes and form oil costing $.50 per square foot and 1.1 SF of 3/4" plyform plus 3 BF of lumber per square foot of contact surface. Costs shown are per cubic yard (CY) of concrete within the foundations.

| | Craft@Hrs | Unit | Material | Labor | Equipment | Total |
|---|---|---|---|---|---|---|
| Lay out foundation from known point on site, set stakes and mark for excavation, with 1 CY per CY of concrete | C8@.147 | CY | .06 | 6.74 | — | 6.80 |
| Excavation with utility backhoe, at 28 CY per hour, with 1 CY per CY of concrete | S1@.072 | CY | — | 3.42 | 1.28 | 4.70 |
| Formwork, material and labor at .192 manhours per SF with 13 SF per CY of concrete | F5@2.50 | CY | 31.50 | 118.00 | — | 149.50 |
| Reinforcing bars, Grade 60, set and tied, material and labor at .006 manhours per lb with 20 lb per CY of concrete | RB@.120 | CY | 10.10 | 7.00 | — | 17.10 |
| Anchor bolts, sleeves, embedded steel, material and labor at .040 manhours per lb with 2 lb per CY of concrete | C8@.080 | CY | 2.81 | 3.67 | — | 6.48 |
| Ready-mix concrete, 3,000 PSI mix, after 10% waste allowance with 1 CY per CY | C8@.385 | CY | 110.00 | 17.60 | — | 127.60 |
| Concrete test cylinders including test reports, with 5 per 100 CY | — | CY | .65 | — | — | .65 |
| Finish, sack and patch concrete, material per SF and labor at .010 manhours per SF with 60 SF per CY of concrete | P8@.600 | CY | 3.69 | 27.00 | — | 30.69 |
| Backfill around foundation with utility backhoe, at 40 CY per hour, with .5 CY per CY of concrete | S1@.025 | CY | — | 1.19 | .44 | 1.63 |
| Dispose of excess soil on site with utility backhoe, at 80 CY per hour, with .5 CY per CY of concrete | S1@.025 | CY | — | 1.19 | .44 | 1.63 |
| **Square pad and pedestal type foundation** Total cost per CY of concrete | —@3.95 | CY | 158.81 | 185.81 | 2.16 | 346.78 |

| | Craft@Hrs | Unit | Material | Labor | Equipment | Total |
|---|---|---|---|---|---|---|

**Octagonal pad and pedestal type foundations** For vertical cylindrical tanks over 3' in diameter. Form costs assume nails, stakes and form oil costing $.50 per square foot and 1.1 SF of 3/4" plyform plus 4.5 BF of lumber per square foot of contact surface. Costs shown are per cubic yard (CY) of concrete within the foundation.

| | Craft@Hrs | Unit | Material | Labor | Equipment | Total |
|---|---|---|---|---|---|---|
| Lay out foundation from known point on site, set stakes and mark for excavation, | | | | | | |
| with 1 CY per CY of concrete | C8@.147 | CY | .06 | 6.74 | — | 6.80 |
| Excavation with utility backhoe, at $28 CY per hour, | | | | | | |
| with 1 CY per CY of concrete | S1@.072 | CY | — | 3.42 | 1.28 | 4.70 |
| Formwork, material SF and labor at .250 manhours per SF | | | | | | |
| with 10 SF per CY of concrete | F5@2.50 | CY | 27.70 | 118.00 | — | 145.70 |
| Reinforcing bars, Grade 60, set and tied, material and labor at .006 manhours per lb | | | | | | |
| with 20 lb per CY of concrete | RB@.120 | CY | 10.10 | 7.00 | — | 17.10 |
| Anchor bolts, sleeves, embedded steel, material and labor at .040 manhours per lb | | | | | | |
| with 2 lb per CY of concrete | C8@.080 | CY | 2.65 | 3.67 | — | 6.32 |
| Ready-mix concrete, 3,000 PSI mix, after 10% waste allowance | | | | | | |
| with 1 CY per CY of concrete | C8@.385 | CY | 110.00 | 17.60 | — | 127.60 |
| Concrete test cylinders including test reports, | | | | | | |
| with 5 per 100 CY of concrete | — | CY | .65 | — | — | .65 |
| Finish, sack and patch concrete, material and labor at .010 manhours per SF | | | | | | |
| with 60 SF per CY of concrete | P8@.600 | CY | 3.69 | 27.00 | — | 30.69 |
| Backfill around foundation with utility backhoe, at 40 CY per hour, | | | | | | |
| with .5 CY per CY of concrete | S1@.025 | CY | — | 1.19 | .44 | 1.63 |
| Dispose of excess soil on site with utility backhoe, at 80 CY per hour, | | | | | | |
| with .5 CY per CY of concrete | S1@.025 | CY | — | 1.19 | .44 | 1.63 |
| **Octagonal pad and pedestal type foundation** | | | | | | |
| Total cost per CY of concrete | —@3.95 | CY | 154.85 | 185.81 | 2.16 | 342.82 |

| | Craft@Hrs | Unit | Material | Labor | Total |
|---|---|---|---|---|---|

**Cementitious Decks, Subcontract** Typical costs including the subcontractor's overhead and profit.

| | Craft@Hrs | Unit | Material | Labor | Total |
|---|---|---|---|---|---|
| Insulating concrete deck, poured on existing substrate, 1-1/2" thick | | | | | |
| Under 3,000 SF | — | SF | — | — | 2.16 |
| Over 3,000 SF to 10,000 SF | — | SF | — | — | 1.87 |
| Over 10,000 SF | — | SF | — | — | 1.57 |
| Poured on existing metal deck, 2-5/8" thick, lightweight | | | | | |
| Concrete fill, trowel finish | — | SF | — | — | 2.54 |
| Gypsum decking, poured over | | | | | |
| form board, screed and float | — | SF | — | — | 2.30 |
| Vermiculite, | | | | | |
| screed and float, field mixed | — | SF | — | — | 1.88 |
| Exterior concrete deck for walking service, troweled | | | | | |
| Lightweight, 2-5/8", 110 lbs per CF | — | SF | — | — | 3.27 |
| Add for 4,000 PSI mix | — | SF | — | — | .24 |
| Add for 5,000 PSI mix | — | SF | — | — | .40 |
| Pool deck at pool side, | | | | | |
| 1/8" to 3/16" thick, colors, finished | — | SF | — | — | 2.96 |
| Fiber deck, tongue and groove, cementitious planks, complete | | | | | |
| 2" thick | — | SF | — | — | 2.49 |
| 2-1/2" thick | — | SF | — | — | 3.03 |
| 3" thick | — | SF | — | — | 3.31 |

**Concrete Surfacing, Subcontract** Load-bearing, architectural, three-part, acrylic polymer cementitious resurfacing system. 6,000 PSI compressive strength. Applied over existing concrete, aggregate, masonry, steel, asphalt, foam or wood surfaces. Installed on surfaces that do not require crack repair, brick or tile patterns or the removal of heavy grease contaminants. Includes stain sealer and colored texture coat. Based on a typical 1,000 square foot project. www.permacrete.com

| | Craft@Hrs | Unit | Material | Labor | Total |
|---|---|---|---|---|---|
| Applied over new surface | — | SF | — | — | .89 |
| Applied over old concrete, includes cleaning | — | SF | — | — | 1.59 |
| Applied over pool interior, includes cleaning | — | SF | — | — | 1.89 |
| Applied over vertical surface | — | SF | — | — | 2.18 |

**Surface Imprinted Concrete Paving, Subcontract** Patterned and colored to simulate slate, granite, limestone, sandstone, wood or cobblestone. 3 to 4 inch depth. Costs include imprinting and texturing, concrete placement and finish, color hardener, release agent and sealer application. Add the cost of grading, forms and form stripping. www.bomanite.com

| | Craft@Hrs | Unit | Material | Labor | Total |
|---|---|---|---|---|---|
| Regular grade | — | SF | — | — | 17.60 |
| Heavy grade (for heavy road traffic and freeze areas) | — | SF | — | — | 21.80 |

**Troweled-on Cementitious Coating, Subcontract** For direct use on interior flooring, accent areas or walls. Bonds to concrete, wood, metal, plastic or asphalt.

| | Craft@Hrs | Unit | Material | Labor | Total |
|---|---|---|---|---|---|
| Cementitious coating | — | SF | — | — | 13.10 |

**Wall Sawing, Subcontract** Using an electric, hydraulic or air saw. Per LF of cut including overcuts at each corner equal to the depth of the cut. Costs shown assume electric power is available. Minimum cost will be $400.

| | Unit | Brick or block | Concrete w/#4 bar | Concrete w/#7 bar |
|---|---|---|---|---|
| To 4" depth | LF | 14.80 | 21.10 | 20.40 |
| To 5" depth | LF | 18.40 | 23.00 | 23.70 |
| To 6" depth | LF | 20.30 | 24.80 | 26.80 |
| To 7" depth | LF | 25.60 | 30.20 | 30.80 |
| To 8" depth | LF | 31.20 | 35.40 | 37.00 |
| To 10" depth | LF | 38.80 | 43.80 | 44.00 |
| To 12" depth | LF | 47.80 | 52.60 | 52.40 |
| To 14" depth | LF | 58.50 | 63.70 | 60.90 |
| To 16" depth | LF | 70.00 | 72.80 | 69.30 |
| To 18" depth | LF | 83.10 | 83.10 | 82.80 |

**Concrete Core Drilling, Subcontract** Using an electric, hydraulic or air drill. Costs shown are per LF of depth for diameter shown. Prices are based on one mat reinforcing with 3/8" to 5/8" bars and include cleanup. Difficult drill setups are higher. Prices assume availability of 110 volt electricity. Figure travel time at $110 per hour. Minimum cost will be $250.

| | Craft@Hrs | Unit | Material | Labor | Total |
|---|---|---|---|---|---|
| 1" or 1-1/2" diameter | — | LF | — | — | 64.80 |
| 2" or 2-1/2" diameter | — | LF | — | — | 67.10 |
| 3" or 3-1/2"diameter | — | LF | — | — | 80.80 |
| 4" or 5" diameter | — | LF | — | — | 97.80 |
| 6" or 8" diameter | — | LF | — | — | 135.00 |
| 10" diameter | — | LF | — | — | 184.00 |
| 12" diameter | — | LF | — | — | 302.00 |
| 14" diameter | — | LF | — | — | 346.00 |
| 16" diameter | — | LF | — | — | 393.00 |
| 18" diameter | — | LF | — | — | 439.00 |
| 20" diameter | — | LF | — | — | 487.00 |
| 24" diameter | — | LF | — | — | 583.00 |
| 30" diameter | — | LF | — | — | 715.00 |
| 36" diameter | — | LF | — | — | 859.00 |
| Hourly rate, 2 hour minimum | — | Hr | — | — | 175.00 |

**Concrete Slab Sawing, Subcontract** Using a gasoline powered saw. Costs per linear foot for cured concrete assuming a level surface with good access and saw cut lines laid out and pre-marked by others. Costs include local travel time. Minimum cost will be $250. Electric powered slab sawing will be approximately 40% higher for the same depth.

| Depth | Unit | Under 200' | 200'-1000' | Over 1000' |
|---|---|---|---|---|
| 1" deep | LF | .78 | .70 | .58 |
| 1-1/2" deep | LF | 1.16 | 1.06 | .89 |
| 2" deep | LF | 1.54 | 1.35 | 1.16 |
| 2-1/2" deep | LF | 1.94 | 1.74 | 1.45 |
| 3" deep | LF | 2.35 | 2.02 | 1.74 |
| 3-1/2" deep | LF | 2.69 | 2.41 | 2.02 |
| 4" deep | LF | 3.07 | 2.69 | 2.33 |
| 5" deep | LF | 3.85 | 3.39 | 2.88 |
| 6" deep | LF | 4.63 | 4.05 | 3.46 |
| 7" deep | LF | 5.59 | 4.80 | 4.16 |
| 8" deep | LF | 6.36 | 5.58 | 4.80 |
| 9" deep | LF | 7.23 | 6.36 | 5.50 |
| 10" deep | LF | 8.19 | 7.23 | 6.26 |
| 11" deep | LF | 9.18 | 8.08 | 7.03 |
| 12" deep | LF | 10.20 | 8.93 | 7.80 |

Green concrete (2 days old) sawing will usually cost 15 to 20% less. Work done on an hourly basis will cost $125 per hour for slabs up to 4" thick and $135 per hour for 5" or 6" thick slabs. A two hour minimum charge will apply on work done on an hourly basis.

**Asphalt Sawing, Subcontract** Using a gasoline powered saw. Minimum cost will be $200. Cost per linear foot of green or cured asphalt, assuming level surface with good access and saw cut lines laid out and pre-marked by others. Costs include local travel time. Work done on an hourly basis will cost $115 per hour for asphalt up to 4" thick and $125 per hour for 5" or 6" thick asphalt. A two hour minimum charge will apply on work done on an hourly basis.

| Depth | Unit | Under 450' | 450'-1000' | Over 1000' |
|---|---|---|---|---|
| 1" or 1-1/2" deep | LF | .49 | .39 | .26 |
| 2" deep | LF | .78 | .58 | .43 |
| 2-1/2" deep | LF | .95 | .78 | .50 |
| 3" deep | LF | 1.16 | .89 | .60 |
| 3-1/2" deep | LF | 1.35 | 1.06 | .69 |
| 4" deep | LF | 1.54 | 1.16 | .77 |
| 5" deep | LF | 1.94 | 1.45 | .93 |
| 6" deep | LF | 2.33 | 1.74 | 1.10 |

**Cleaning and Pointing Masonry** These costs assume the masonry surface is in fair to good condition with no unusual damage. Add the cost of protecting adjacent surfaces such as trim or the base of the wall and the cost of scaffolding, if required. Labor required to presoak or saturate the area cleaned is included in the labor cost. Work more than 12' above floor level will increase costs. Brushing (hand cleaning) masonry includes the cost of detergent or chemical solution.

| | Craft@Hrs | Unit | Material | Labor | Total |
|---|---|---|---|---|---|
| Light cleanup (100 SF per manhour) | M1@.010 | SF | .03 | .47 | .50 |
| Medium scrub (75 SF per manhour) | M1@.013 | SF | .04 | .61 | .65 |
| Heavy (50 SF per manhour) | M1@.020 | SF | .06 | .93 | .99 |

| | Craft@Hrs | Unit | Material | Labor | Equipment | Total |
|---|---|---|---|---|---|---|
| Water blasting masonry, using rented 400 to 700 PSI power washer with 3 to 8 gallon per minute flow rate, includes blaster rental at $65 per day | | | | | | |
| Smooth face (250 SF per manhour) | M1@.004 | SF | — | .19 | .03 | .22 |
| Rough face (200 SF per manhour) | M1@.005 | SF | — | .23 | .03 | .26 |
| Sandblasting masonry, using 150 PSI compressor (with accessories and sand) at $129.00 per day | | | | | | |
| Smooth face (50 SF per manhour) | M1@.020 | SF | .31 | .93 | .21 | 1.45 |
| Rough face (40 SF per manhour) | M1@.025 | SF | .34 | 1.17 | .51 | 2.02 |
| Steam cleaning masonry, using rented steam cleaning rig (with accessories) at $112.00 per day | | | | | | |
| Smooth face (75 SF per manhour) | M1@.013 | SF | — | .61 | .14 | .75 |
| Rough face (55 SF per manhour) | M1@.018 | SF | — | .84 | .19 | 1.03 |

| | Craft@Hrs | Unit | Material | Labor | Total |
|---|---|---|---|---|---|
| Add for masking adjacent surfaces | — | % | — | 5.0 | — |
| Add for difficult stain removal | — | % | 50.0 | 50.0 | — |
| Add for working from scaffold | — | % | — | 20.0 | — |
| Repointing brick, cut out joint, mask (blend-in), and regrout (tuck pointing) | | | | | |
| 30 SF per manhour | M1@.033 | SF | .08 | 1.54 | 1.62 |

**Masonry Cleaning** Costs to clean masonry surfaces using commercial cleaning agents, add for pressure washing and scaffolding equipment.

| | Craft@Hrs | Unit | Material | Labor | Total |
|---|---|---|---|---|---|
| Typical cleaning of surfaces | | | | | |
| Granite, sandstone, terra cotta, brick | CL@.015 | SF | .30 | .60 | .90 |
| Cleaning surfaces of heavily carbonated | | | | | |
| Limestone or cast stone | CL@.045 | SF | .40 | 1.79 | 2.19 |
| Typical wash with acid and rinse | CL@.004 | SF | .46 | .16 | .62 |

**Masonry Reinforcing and Flashing** Reinforcing bars for concrete block, ASTM A615 grade 60 bars, cost per linear foot including 10% overlap and waste.

| | Craft@Hrs | Unit | Material | Labor | Total |
|---|---|---|---|---|---|
| #3 bars (3/8", 5,319 LF per ton), horizontal | M1@.005 | LF | .23 | .23 | .46 |
| #3 bars placed vertically | M1@.006 | LF | .23 | .28 | .51 |
| #3 galvanized bars, horizontal | M1@.005 | LF | .33 | .23 | .56 |
| #3 galvanized bars placed vertically | M1@.006 | LF | .33 | .28 | .61 |
| #4 bars (1/2", 2,994 LF per ton), horizontal | M1@.008 | LF | .39 | .37 | .76 |
| #4 bars placed vertically | M1@.010 | LF | .39 | .47 | .86 |
| #4 galvanized bars, horizontal | M1@.008 | LF | .63 | .37 | 1.00 |
| #4 galvanized bars placed vertically | M1@.010 | LF | .63 | .47 | 1.10 |
| #5 bars (5/8", 1,918 LF per ton), horizontal | M1@.009 | LF | .54 | .42 | .96 |
| #5 bars placed vertically | M1@.012 | LF | .54 | .56 | 1.10 |
| #5 galvanized bars, horizontal | M1@.009 | LF | .99 | .42 | 1.41 |
| #5 galvanized bars placed vertically | M1@.012 | LF | .99 | .56 | 1.55 |
| #6 bars (3/4", 1,332 LF per ton), horizontal | M1@.010 | LF | .75 | .47 | 1.22 |
| #6 bars placed vertically | M1@.013 | LF | .75 | .61 | 1.36 |
| #6 galvanized bars, horizontal | M1@.014 | LF | 1.53 | .65 | 2.18 |
| #6 galvanized bars placed vertically | M1@.014 | LF | 1.53 | .65 | 2.18 |
| Wall ties | | | | | |
| Rectangular, galvanized | M1@.005 | Ea | .36 | .23 | .59 |
| Rectangular, copper coated | M1@.005 | Ea | .42 | .23 | .65 |
| Cavity "Z"-type, galvanized | M1@.005 | Ea | .34 | .23 | .57 |
| Cavity "Z"-type, copper coated | M1@.005 | Ea | .38 | .23 | .61 |
| Masonry anchors, steel, including bolts | | | | | |
| 3/16" x 18" long | M1@.173 | Ea | 4.24 | 8.07 | 12.31 |
| 3/16" x 24" long | M1@.192 | Ea | 5.32 | 8.95 | 14.27 |
| 1/4" x 18" long | M1@.216 | Ea | 5.47 | 10.10 | 15.57 |
| 1/4" x 24" long | M1@.249 | Ea | 6.66 | 11.60 | 18.26 |

# 04 Masonry

| | Craft@Hrs | Unit | Material | Labor | Total |
|---|---|---|---|---|---|
| Wall reinforcing | | | | | |
| Truss type, plain | | | | | |
| 4" wall | M1@.002 | LF | .23 | .09 | .32 |
| 6" wall | M1@.002 | LF | .27 | .09 | .36 |
| 8" wall | M1@.003 | LF | .29 | .14 | .43 |
| 10" wall | M1@.003 | LF | .32 | .14 | .46 |
| 12" wall | M1@.004 | LF | .34 | .19 | .53 |
| Truss type, galvanized | | | | | |
| 4" wall | M1@.002 | LF | .28 | .09 | .37 |
| 6" wall | M1@.002 | LF | .31 | .09 | .40 |
| 8" wall | M1@.003 | LF | .37 | .14 | .51 |
| 10" wall | M1@.003 | LF | .40 | .14 | .54 |
| 12" wall | M1@.004 | LF | .44 | .19 | .63 |
| Ladder type, galvanized, Class 3 | | | | | |
| 2" wide, for 4" wall | M1@.003 | LF | .17 | .14 | .31 |
| 4" wide | M1@.003 | LF | .32 | .14 | .46 |
| 6" wide | M1@.003 | LF | .30 | .14 | .44 |
| 8" wide | M1@.003 | LF | .36 | .14 | .50 |
| Control joints | | | | | |
| Cross-shaped PVC | M1@.024 | LF | 2.05 | 1.12 | 3.17 |
| 8" wide PVC | M1@.047 | LF | 2.68 | 2.19 | 4.87 |
| Closed cell 1/2" joint filler | M1@.012 | LF | .35 | .56 | .91 |
| Closed cell 3/4" joint filler | M1@.012 | LF | .63 | .56 | 1.19 |
| Through the wall flashing | | | | | |
| 5 ounce copper | M1@.023 | SF | 3.00 | 1.07 | 4.07 |
| .030" elastomeric sheeting | M1@.038 | SF | .55 | 1.77 | 2.32 |

**Expansion Shields** Masonry anchors set in block or concrete including drilling. With nut and washer. Length indicates drilled depth.

| | Craft@Hrs | Unit | Material | Labor | Total |
|---|---|---|---|---|---|
| One wedge anchors, 1/4" diameter | | | | | |
| 1-3/4" long | CL@.079 | Ea | .75 | 3.15 | 3.90 |
| 2-1/4" long | CL@.114 | Ea | .90 | 4.55 | 5.45 |
| 3-1/4" long | CL@.171 | Ea | 1.20 | 6.82 | 8.02 |
| One wedge anchors, 1/2" diameter | | | | | |
| 4-1/4" long | CL@.227 | Ea | 1.90 | 9.05 | 10.95 |
| 5-1/2" long | CL@.295 | Ea | 2.52 | 11.80 | 14.32 |
| 7" long | CL@.409 | Ea | 3.09 | 16.30 | 19.39 |
| One wedge anchors, 3/4" diameter | | | | | |
| 5-1/2" long | CL@.286 | Ea | 5.18 | 11.40 | 16.58 |
| 8-1/2" long | CL@.497 | Ea | 7.40 | 19.80 | 27.20 |
| 10" long | CL@.692 | Ea | 8.74 | 27.60 | 36.34 |
| One wedge anchors, 1" diameter | | | | | |
| 6" long | CL@.295 | Ea | 12.60 | 11.80 | 24.40 |
| 9" long | CL@.515 | Ea | 15.80 | 20.50 | 36.30 |
| 12" long | CL@.689 | Ea | 18.20 | 27.50 | 45.70 |
| Flush self-drilling anchors | | | | | |
| 1/4" x 1-1/4" | CL@.094 | Ea | .75 | 3.75 | 4.50 |
| 5/16" x 1-1/4" | CL@.108 | Ea | .90 | 4.31 | 5.21 |
| 3/8" x 1-1/2" | CL@.121 | Ea | 1.20 | 4.83 | 6.03 |
| 1/2" x 1-1/2" | CL@.159 | Ea | 1.77 | 6.34 | 8.11 |
| 5/8" x 2-3/8" | CL@.205 | Ea | 2.81 | 8.18 | 10.99 |
| 3/4" x 2-3/8" | CL@.257 | Ea | 4.59 | 10.20 | 14.79 |

# 04 Masonry

| | Craft@Hrs | Unit | Material | Labor | Total |
|---|---|---|---|---|---|

**Brick Wall Assemblies** Typical costs for smooth red clay brick walls, laid in running bond with 3/8" concave joints. These costs include the bricks, mortar for bricks and cavities, typical ladder type reinforcing, wall ties and normal waste. Foundations are not included. Wall thickness shown is the nominal size based on using the type bricks described. "Wythe" means the quantity of bricks in the thickness of the wall. Costs shown are per square foot (SF) of wall measured on one face. Deduct for openings over 10 SF in size. The names and dimensions of bricks can be expected to vary, depending on the manufacturer. For more detailed coverage of brick wall assemblies, see *National Concrete & Masonry Estimator*, http://CraftsmanSiteLicense.com

Standard bricks, 3-3/4" wide x 2-1/4" high x 8" long

| | | | | | |
|---|---|---|---|---|---|
| 4" thick wall, single wythe, veneer facing | M1@.211 | SF | 3.95 | 9.84 | 13.79 |
| 8" thick wall, double wythe, cavity filled | M1@.464 | SF | 7.75 | 21.60 | 29.35 |
| 12" thick wall, triple wythe, cavity filled | M1@.696 | SF | 11.60 | 32.40 | 44.00 |

Modular bricks, 3" wide x 3-1/2" x 11-1/2" long

| | | | | | |
|---|---|---|---|---|---|
| 3-1/2" thick wall, single wythe, veneer facing | M1@.156 | SF | 4.99 | 7.27 | 12.26 |
| 7-1/2" thick wall, double wythe, cavity filled | M1@.343 | SF | 9.86 | 16.00 | 25.86 |
| 11-1/2" thick wall, triple wythe, cavity filled | M1@.515 | SF | 14.70 | 24.00 | 38.70 |

Colonial bricks, 3" wide x 3-1/2" x 10" long (at $719 per M)

| | | | | | |
|---|---|---|---|---|---|
| 3-1/2" thick wall, single wythe, veneer facing | M1@.177 | SF | 4.27 | 8.25 | 12.52 |
| 7-1/2" thick wall, double wythe, cavity filled | M1@.389 | SF | 8.40 | 18.10 | 26.50 |
| 11-1/2" thick wall, triple wythe, cavity filled | M1@.584 | SF | 12.50 | 27.20 | 39.70 |

Add to any brick wall assembly above for:

| | | | | | |
|---|---|---|---|---|---|
| American bond or stacked bond | — | % | 5.0 | 10.0 | — |
| Basketweave bond or soldier bond | — | % | 10.0 | 10.0 | — |
| Flemish bond | — | % | 10.0 | 15.0 | — |
| Herringbone pattern | — | % | 15.0 | 15.0 | — |
| Institutional inspection | — | % | — | 20.0 | — |
| Short runs or cut-up work | — | % | 5.0 | 15.0 | — |

**Fireplaces** Common brick, including the chimney foundation, damper, flue lining and stack to 15'

| | | | | | |
|---|---|---|---|---|---|
| 30" box, brick to 5', large quantity | M1@30.3 | Ea | 1,020.00 | 1,410.00 | 2,430.00 |
| 36" box, brick to 5', custom | M1@33.7 | Ea | 1,240.00 | 1,570.00 | 2,810.00 |
| 42" box, brick to 8', custom | M1@48.1 | Ea | 1,440.00 | 2,240.00 | 3,680.00 |
| 48" box, brick to 8', custom | M1@53.1 | Ea | 1,660.00 | 2,480.00 | 4,140.00 |

Flue lining, 1' lengths

| | | | | | |
|---|---|---|---|---|---|
| 8-1/2" round | M1@.148 | LF | 5.86 | 6.90 | 12.76 |
| 10" x 17" oval | M1@.206 | LF | 9.69 | 9.60 | 19.29 |
| 13" x 13" square | M1@.193 | LF | 10.20 | 9.00 | 19.20 |

Prefabricated cast concrete

| | | | | | |
|---|---|---|---|---|---|
| 30" box and 15' chimney | M1@5.43 | Ea | 1,320.00 | 253.00 | 1,573.00 |
| Face brick, mantle to ceiling | M1@.451 | SF | 7.67 | 21.00 | 28.67 |

**Clay Backing Tile** Load bearing, 12" x 12"

| | | | | | |
|---|---|---|---|---|---|
| 4" thick | M1@.066 | SF | 4.24 | 3.08 | 7.32 |
| 6" thick | M1@.079 | SF | 4.78 | 3.68 | 8.46 |
| 8" thick | M1@.091 | SF | 5.62 | 4.24 | 9.86 |
| Deduct for non-load bearing tile | — | % | -8.0 | -10.0 | — |

**Structural Glazed Tile** Heavy-duty, fire-safe, structural glazed facing tile for interior surfaces. No foundations or reinforcing included.

Glazed one side, all colors, 5-1/3" x 12"

| | | | | | |
|---|---|---|---|---|---|
| 2" thick | M1@.125 | SF | 4.90 | 5.83 | 10.73 |
| 4" thick | M1@.136 | SF | 5.91 | 6.34 | 12.25 |
| 6" thick | M1@.141 | SF | 8.76 | 6.57 | 15.33 |
| 8" thick | M1@.172 | SF | 11.10 | 8.02 | 19.12 |

| | Craft@Hrs | Unit | Material | Labor | Total |
|---|---|---|---|---|---|
| Glazed one side, all colors, 8" x 16" block | | | | | |
| 2" thick | M1@.072 | SF | 5.51 | 3.36 | 8.87 |
| 4" thick | M1@.072 | SF | 5.91 | 3.36 | 9.27 |
| 6" thick | M1@.079 | SF | 8.15 | 3.68 | 11.83 |
| 8" thick | M1@.085 | SF | 9.55 | 3.96 | 13.51 |

Acoustical glazed facing tile, with small holes on a smooth ceramic glazed facing, and one mineral fiberglass sound batt located in the hollow core behind perforated face

| | Craft@Hrs | Unit | Material | Labor | Total |
|---|---|---|---|---|---|
| 8" x 8" face size unit, 4" thick | M1@.136 | SF | 9.68 | 6.34 | 16.02 |
| 8" x 16" face size unit, 4" thick | M1@.141 | SF | 8.61 | 6.57 | 15.18 |

Reinforced glazed security tile, with vertical cores and knockouts for horizontal reinforcing, 8" x 16" block

| | Craft@Hrs | Unit | Material | Labor | Total |
|---|---|---|---|---|---|
| 4" thick | M1@.092 | SF | 7.54 | 4.29 | 11.83 |
| 6" thick | M1@.095 | SF | 9.75 | 4.43 | 14.18 |
| 8" thick | M1@.101 | SF | 11.70 | 4.71 | 16.41 |
| Add to 8" security tile for horizontal and vertical reinforcing and grout cores | | | | | |
| to meet high security standards | M1@.047 | SF | 3.48 | 2.12 | 5.60 |
| Add for glazed 2 sides | — | % | 55.0 | 10.0 | — |
| Base, glazed 1 side | M1@.190 | LF | 7.31 | 8.86 | 16.17 |
| Cap, glazed 2 sides | M1@.193 | LF | 12.20 | 9.00 | 21.20 |
| Deduct for large areas | — | % | -5.0 | -5.0 | — |
| Add for institutional inspection | — | % | — | 10.0 | — |

**Chemical Resistant Brick or Tile, Subcontract** (also known as acid-proof brick).Typical prices for chemical resistant brick or tile meeting ASTM C-279, types I, II, and III requirements. Installed with 1/8" thick bed and mortar joints over a membrane of either elastomeric bonded to the floors or walls, sheet membrane with a special seam treatment, squeegee hot-applied asphalt, cold-applied adhesive troweled on, or special spray-applied material. The setting bed and joint materials can be either 100% carbon-filled furan or phenolic, sodium or potassium silicates, silica-filled epoxy, polyester, vinyl ester or other special mortars. Expansion joints are required in floor construction, particularly where construction joints exist in the concrete. Additional expansion joints may be required, depending on the service conditions to which the floor will be subjected. Brick or tile will be either 3-3/4" x 8" face, 1-1/8" thick (splits); 3-3/4" x 8" face, 2-1/4" thick (singles); or 4" x 8" face, 1-3/8" thick (pavers). Costs are based on a 2,000 SF job. For smaller projects, increase the cost 5% for each 500 SF or portion of 500 SF less than 2,000 SF. For scheduling purposes estimate that a crew of 4 can install 75 to 100 SF of membrane and brick in an 8-hour day. These costs assume that the membrane is laid on a suitable surface prepared by others and include the subcontractor's overhead and profit.

| Typical installed costs for chemical resistant brick or tile (acid-proof brick), installed | | | | | |
|---|---|---|---|---|---|
| Less complex jobs | — | SF | — | — | 30.90 |
| More complex jobs | — | SF | — | — | 37.50 |
| Minimum job cost | — | LS | — | — | 20,500.00 |

**Concrete Block Retaining Walls** Includes mortar, regular block, U-block, inspection block (for walls six courses high or higher) grout, cavity caps and No. 5 reinforcing steel. Equipment is a trailer-mounted 15 CY per hour grout pump with operator. Add the cost of wall foundations from the section that follows.

| | Craft@Hrs | Unit | Material | Labor | Equipment | Total |
|---|---|---|---|---|---|---|
| Retaining wall with one lintel block course and grouted and reinforced vertical cells at 16" O.C. | | | | | | |
| 3' - 4" high wall | M1@.504 | LF | 14.20 | 23.50 | 2.90 | 40.60 |
| 4' - 0" high wall | M1@.576 | LF | 17.40 | 26.90 | 3.32 | 47.62 |
| 4' - 8" high wall | M1@.648 | LF | 20.30 | 30.20 | 3.73 | 54.23 |
| 5' - 4" high wall | M1@.720 | LF | 23.20 | 33.60 | 4.15 | 60.95 |
| 6' - 0" high wall | M1@.793 | LF | 26.10 | 37.00 | 4.57 | 67.67 |
| 6' - 8" high wall | M1@.864 | LF | 29.10 | 40.30 | 4.98 | 74.38 |
| 7' - 4" high wall | M1@.936 | LF | 32.10 | 43.60 | 5.39 | 81.09 |
| 8' - 0" high wall | M1@1.01 | LF | 35.10 | 47.10 | 5.82 | 88.02 |

|  | Craft@Hrs | Unit | Material | Labor | Equipment | Total |
|---|---|---|---|---|---|---|
| **Retaining wall with one lintel block course and grouted and reinforced vertical cells at 32" O.C.** |  |  |  |  |  |  |
| 3' - 4" high wall | M1@.454 | LF | 11.20 | 21.20 | 2.62 | 35.02 |
| 4' - 0" high wall | M1@.526 | LF | 13.60 | 24.50 | 3.03 | 41.13 |
| 4' - 8" high wall | M1@.598 | LF | 15.70 | 27.90 | 3.45 | 47.05 |
| 5' - 4" high wall | M1@.670 | LF | 17.90 | 31.20 | 3.86 | 52.96 |
| 6' - 0" high wall | M1@.743 | LF | 20.10 | 34.60 | 4.28 | 58.98 |
| 6' - 8" high wall | M1@.814 | LF | 22.40 | 37.90 | 4.69 | 64.99 |
| 7' - 4" high wall | M1@.886 | LF | 24.50 | 41.30 | 5.11 | 70.91 |
| 8' - 0" high wall | M1@.958 | LF | 26.70 | 44.70 | 5.52 | 76.92 |
| **Retaining wall with one lintel block course and grouted and reinforced vertical cells at 48" O.C.** |  |  |  |  |  |  |
| 3' - 4" high wall | M1@.436 | LF | 10.20 | 20.30 | 2.51 | 33.01 |
| 4' - 0" high wall | M1@.508 | LF | 12.30 | 23.70 | 2.93 | 38.93 |
| 4' - 8" high wall | M1@.580 | LF | 14.20 | 27.00 | 3.34 | 44.54 |
| 5' - 4" high wall | M1@.652 | LF | 16.10 | 30.40 | 3.76 | 50.26 |
| 6' - 0" high wall | M1@.725 | LF | 18.10 | 33.80 | 4.18 | 56.08 |
| 6' - 8" high wall | M1@.797 | LF | 20.00 | 37.20 | 4.59 | 61.79 |
| 7' - 4" high wall | M1@.869 | LF | 21.90 | 40.50 | 5.01 | 67.41 |
| 8' - 0" high wall | M1@.941 | LF | 23.90 | 43.90 | 5.42 | 73.22 |
| **Retaining wall with two lintel block courses and grouted and reinforced vertical cells at 16" O.C.** |  |  |  |  |  |  |
| 3' - 4" high wall | M1@.546 | LF | 14.20 | 25.50 | 3.15 | 42.85 |
| 4' - 0" high wall | M1@.617 | LF | 17.40 | 28.80 | 3.56 | 49.76 |
| 4' - 8" high wall | M1@.689 | LF | 20.30 | 32.10 | 3.39 | 55.79 |
| 5' - 4" high wall | M1@.761 | LF | 23.20 | 35.50 | 4.39 | 63.09 |
| 6' - 0" high wall | M1@.833 | LF | 26.10 | 38.80 | 4.80 | 69.70 |
| 6' - 8" high wall | M1@.906 | LF | 29.10 | 42.20 | 5.22 | 76.52 |
| 7' - 4" high wall | M1@.978 | LF | 32.10 | 45.60 | 5.64 | 83.34 |
| 8' - 0" high wall | M1@1.05 | LF | 35.10 | 49.00 | 6.05 | 90.15 |
| **Retaining wall with two lintel block courses and grouted and reinforced vertical cells at 32" O.C.** |  |  |  |  |  |  |
| 3' - 4" high wall | M1@.495 | LF | 11.70 | 23.10 | 2.85 | 37.65 |
| 4' - 0" high wall | M1@.567 | LF | 14.00 | 26.40 | 3.27 | 43.67 |
| 4' - 8" high wall | M1@.639 | LF | 16.20 | 29.80 | 3.68 | 49.68 |
| 5' - 4" high wall | M1@.711 | LF | 18.40 | 33.10 | 4.10 | 55.60 |
| 6' - 0" high wall | M1@.783 | LF | 20.60 | 36.50 | 4.51 | 61.61 |
| 6' - 8" high wall | M1@.856 | LF | 22.80 | 39.90 | 4.93 | 67.63 |
| 7' - 4" high wall | M1@.928 | LF | 25.00 | 43.30 | 5.35 | 73.65 |
| 8' - 0" high wall | M1@1.00 | LF | 27.20 | 46.60 | 5.76 | 79.56 |
| **Retaining wall with two lintel block courses and grouted and reinforced vertical cells at 48" O.C.** |  |  |  |  |  |  |
| 3' - 4" high wall | M1@.478 | LF | 10.80 | 22.30 | 2.75 | 35.85 |
| 4' - 0" high wall | M1@.550 | LF | 12.90 | 25.60 | 3.17 | 41.67 |
| 4' - 8" high wall | M1@.621 | LF | 14.80 | 29.00 | 3.58 | 47.38 |
| 5' - 4" high wall | M1@.693 | LF | 16.70 | 32.30 | 3.99 | 52.99 |
| 6' - 0" high wall | M1@.766 | LF | 18.60 | 35.70 | 4.41 | 58.71 |
| 6' - 8" high wall | M1@.839 | LF | 20.60 | 39.10 | 4.84 | 64.54 |
| 7' - 4" high wall | M1@.910 | LF | 22.50 | 42.40 | 5.24 | 70.14 |
| 8' - 0" high wall | M1@.982 | LF | 24.50 | 45.80 | 5.66 | 75.96 |
| **Retaining wall with three lintel block courses and grouted and reinforced vertical cells at 16" O.C.** |  |  |  |  |  |  |
| 3' - 4" high wall | M1@.587 | LF | 14.50 | 27.40 | 3.38 | 45.28 |
| 4' - 0" high wall | M1@.660 | LF | 17.40 | 30.80 | 3.80 | 52.00 |
| 4' - 8" high wall | M1@.732 | LF | 20.40 | 34.10 | 4.22 | 58.72 |
| 5' - 4" high wall | M1@.804 | LF | 23.20 | 37.50 | 4.63 | 65.33 |
| 6' - 0" high wall | M1@.876 | LF | 26.10 | 40.80 | 5.05 | 71.95 |
| 6' - 8" high wall | M1@.948 | LF | 30.00 | 44.20 | 5.46 | 79.66 |
| 7' - 4" high wall | M1@1.02 | LF | 32.10 | 47.60 | 5.88 | 85.58 |
| 8' - 0" high wall | M1@1.09 | LF | 35.10 | 50.80 | 6.28 | 92.18 |

|  | Craft@Hrs | Unit | Material | Labor | Equipment | Total |
|---|---|---|---|---|---|---|
| Retaining wall with three lintel block courses and grouted and reinforced vertical cells at 32" O.C. | | | | | | |
| 3' - 4" high wall | M1@.537 | LF | 12.10 | 25.00 | 3.09 | 40.19 |
| 4' - 0" high wall | M1@.610 | LF | 14.50 | 28.40 | 3.52 | 46.42 |
| 4' - 8" high wall | M1@.682 | LF | 16.70 | 31.80 | 3.93 | 52.43 |
| 5' - 4" high wall | M1@.754 | LF | 18.80 | 35.20 | 4.35 | 58.35 |
| 6' - 0" high wall | M1@.826 | LF | 21.00 | 38.50 | 4.76 | 64.26 |
| 6' - 8" high wall | M1@.898 | LF | 23.20 | 41.90 | 5.18 | 70.28 |
| 7' - 4" high wall | M1@.971 | LF | 25.50 | 45.30 | 5.60 | 76.40 |
| 8' - 0" high wall | M1@1.04 | LF | 27.70 | 48.50 | 5.99 | 82.19 |
| Retaining wall with three lintel block course and grouted and reinforced vertical cells at 48" O.C. | | | | | | |
| 3' - 4" high wall | M1@.520 | LF | 11.40 | 24.20 | 3.00 | 38.60 |
| 4' - 0" high wall | M1@.593 | LF | 13.50 | 27.60 | 3.42 | 44.52 |
| 4' - 8" high wall | M1@.664 | LF | 15.40 | 31.00 | 3.83 | 50.23 |
| 5' - 4" high wall | M1@.736 | LF | 17.30 | 34.30 | 4.24 | 55.84 |
| 6' - 0" high wall | M1@.808 | LF | 19.30 | 37.70 | 4.66 | 61.66 |
| 6' - 8" high wall | M1@.880 | LF | 21.20 | 41.00 | 5.07 | 67.27 |
| 7' - 4" high wall | M1@.953 | LF | 23.10 | 44.40 | 5.49 | 72.99 |
| 8' - 0" high wall | M1@1.03 | LF | 25.10 | 48.00 | 5.94 | 79.04 |

**Concrete Foundations for Retaining Walls** Includes hand excavation in normal soil, backfilling, board or plywood forming as noted, grade 60 reinforcing steel and 3,000 P.S.I. concrete poured from the chute. Concrete waste when poured against earth is assumed to be 1" of concrete at the bottom of the footing and 2" along the walls. Concrete waste when poured against forms is assumed to be 1" at the bottom of the pad. When forms are used, estimate that excavation will be required 18" beyond the forms.

|  | Craft@Hrs | Unit | Material | Labor | Total |
|---|---|---|---|---|---|
| 36" wide x 12" deep foundation with 4 No. 5 bars, 2' deep | | | | | |
| Hand excavated, poured against earth | | | | | |
| Excavation | CL@.412 | LF | — | 16.40 | 16.40 |
| Reinforcing steel | RB@.072 | LF | 3.17 | 4.20 | 7.37 |
| Concrete | CL@.075 | LF | 13.20 | 2.99 | 16.19 |
| Hand excavated, poured against forms | | | | | |
| Excavation | CL@.894 | LF | — | 35.70 | 35.70 |
| Formwork | M2@.251 | LF | 1.77 | 11.00 | 12.77 |
| Reinforcing steel | RB@.072 | LF | 3.17 | 4.20 | 7.37 |
| Concrete | CL@.068 | LF | 11.90 | 2.71 | 14.61 |
| 36" wide x 12" deep with 4 No. 5 bars, 3' deep | | | | | |
| Hand excavated, poured against earth | | | | | |
| Excavation | CL@.635 | LF | — | 25.30 | 25.30 |
| Reinforcing steel | RB@.072 | LF | 3.17 | 4.20 | 7.37 |
| Concrete | CL@.075 | LF | 13.20 | 2.99 | 16.19 |
| Hand excavated, poured against forms | | | | | |
| Excavation | CL@1.36 | LF | — | 54.20 | 54.20 |
| Formwork | M2@.251 | LF | 2.08 | 11.00 | 13.08 |
| Reinforcing steel | RB@.072 | LF | 3.17 | 4.20 | 7.37 |
| Concrete | CL@.068 | LF | 11.90 | 2.71 | 14.61 |
| 36" wide x 12" deep with 6 No. 5 bars, 2' deep | | | | | |
| Hand excavated, poured against earth | | | | | |
| Excavation | CL@.412 | LF | — | 16.40 | 16.40 |
| Reinforcing steel | RB@.108 | LF | 4.59 | 6.30 | 10.89 |
| Concrete | CL@.075 | LF | 13.20 | 2.99 | 16.19 |

| | Craft@Hrs | Unit | Material | Labor | Total |
|---|---|---|---|---|---|
| Hand excavated, poured against forms | | | | | |
| Excavation | CL@.894 | LF | — | 35.70 | 35.70 |
| Formwork | M2@.251 | LF | 2.08 | 11.00 | 13.08 |
| Reinforcing steel | RB@.108 | LF | 4.59 | 6.30 | 10.89 |
| Concrete | CL@.068 | LF | 12.20 | 2.71 | 14.91 |
| **36" wide x 12" deep with 6 No. 5 bars, 3' deep** | | | | | |
| Hand excavated, poured against earth | | | | | |
| Excavation | CL@.635 | LF | — | 25.30 | 25.30 |
| Reinforcing steel | RB@.108 | LF | 4.59 | 6.30 | 10.89 |
| Concrete | CL@.075 | LF | 13.20 | 2.99 | 16.19 |
| Hand excavated, poured against forms | | | | | |
| Excavation | CL@1.36 | LF | — | 54.20 | 54.20 |
| Formwork | M2@.251 | LF | 2.08 | 11.00 | 13.08 |
| Reinforcing steel | RB@.108 | LF | 4.59 | 6.30 | 10.89 |
| Concrete | CL@.068 | LF | 11.90 | 2.71 | 14.61 |
| **36" wide x 16" deep with 6 No. 5 bars, 2' deep, No. 4 transverse bars at 2' O.C.** | | | | | |
| Hand excavated, poured against earth | | | | | |
| Excavation | CL@.400 | LF | — | 16.00 | 16.00 |
| Reinforcing steel | RB@.108 | LF | 5.15 | 6.30 | 11.45 |
| Concrete | CL@.096 | LF | 16.90 | 3.83 | 20.73 |
| Hand excavated, poured against forms | | | | | |
| Excavation | CL@.882 | LF | — | 35.20 | 35.20 |
| Formwork | M2@.337 | LF | 2.77 | 14.80 | 17.57 |
| Reinforcing steel | RB@.108 | LF | 5.15 | 6.30 | 11.45 |
| Concrete | CL@.089 | LF | 15.60 | 3.55 | 19.15 |
| **36" wide x 16" deep with 6 No. 5 bars, 3' deep, No. 4 transverse bars at 2' O.C.** | | | | | |
| Hand excavated, poured against earth | | | | | |
| Excavation | CL@.624 | LF | — | 24.90 | 24.90 |
| Reinforcing steel | RB@.108 | LF | 5.15 | 6.30 | 11.45 |
| Concrete | CL@.096 | LF | 16.90 | 3.83 | 20.73 |
| Hand excavated, poured against forms | | | | | |
| Excavation | CL@1.35 | LF | — | 53.80 | 53.80 |
| Formwork | M2@.337 | LF | 2.77 | 14.80 | 17.57 |
| Reinforcing steel | RB@.108 | LF | 5.15 | 6.30 | 11.45 |
| Concrete | CL@.089 | LF | 15.60 | 3.55 | 19.15 |

**Concrete Block Wall Assemblies** Typical costs for standard natural gray medium weight masonry block walls including blocks, mortar, typical reinforcing and normal waste. Foundations are not included. For more detailed coverage of concrete block masonry, see *National Concrete & Masonry Estimator*, http://CraftsmanSiteLicense.com
Walls constructed with 8" x 16" blocks laid in running bond

| | Craft@Hrs | Unit | Material | Labor | Total |
|---|---|---|---|---|---|
| 4" thick wall | M1@.090 | SF | 2.12 | 4.20 | 6.32 |
| 6" thick wall | M1@.100 | SF | 2.62 | 4.66 | 7.28 |
| 8" thick wall | M1@.120 | SF | 3.12 | 5.59 | 8.71 |
| 12" thick wall | M1@.150 | SF | 4.61 | 6.99 | 11.60 |

**Additional costs for concrete block wall assemblies**
Add for grouting concrete block cores at 36" intervals, poured by hand

| | Craft@Hrs | Unit | Material | Labor | Total |
|---|---|---|---|---|---|
| 4" thick wall | M1@.007 | SF | .53 | .33 | .86 |
| 6" thick wall | M1@.007 | SF | .80 | .33 | 1.13 |
| 8" thick wall | M1@.007 | SF | 1.07 | .33 | 1.40 |
| 12" thick wall | M1@.010 | SF | 1.59 | .47 | 2.06 |

| | Craft@Hrs | Unit | Material | Labor | Total |
|---|---|---|---|---|---|
| Add for 2" thick caps, natural gray concrete | | | | | |
| 4" thick wall | M1@.027 | LF | .89 | 1.26 | 2.15 |
| 6" thick wall | M1@.038 | LF | 1.06 | 1.77 | 2.83 |
| 8" thick wall | M1@.046 | LF | 1.16 | 2.14 | 3.30 |
| 12" thick wall | M1@.065 | LF | 1.66 | 3.03 | 4.69 |
| Add for detailed block, 3/8" score | | | | | |
| Single score, one side | — | SF | .55 | — | .55 |
| Single score, two sides | — | SF | 1.21 | — | 1.21 |
| Multi-scores, one side | — | SF | .55 | — | .55 |
| Multi-scores, two sides | — | SF | 1.21 | — | 1.21 |
| Add for color block, any of above | | | | | |
| Light colors | — | % | 12.0 | — | — |
| Medium colors | — | % | 18.0 | — | — |
| Dark colors | — | % | 25.0 | — | — |
| Add for other than running bond | — | % | — | 20.0 | — |

**Glazed Concrete Block** Prices are per square foot of wall and include allowance for mortar and waste but no foundation or excavation. 8" high x 16" long, 3/8" joints, no reinforcing or grout included.
Based on 1.05 blocks per square foot (SF).

| | Craft@Hrs | Unit | Material | Labor | Total |
|---|---|---|---|---|---|
| Glazed 1 side | | | | | |
| 2" wide | M1@.102 | SF | 11.60 | 4.76 | 16.36 |
| 4" wide | M1@.107 | SF | 13.50 | 4.99 | 18.49 |
| 6" wide | M1@.111 | SF | 13.40 | 5.17 | 18.57 |
| 8" wide | M1@.118 | SF | 14.20 | 5.50 | 19.70 |
| Glazed 2 sides | | | | | |
| 4" wide | M1@.122 | SF | 20.30 | 5.69 | 25.99 |
| 6" wide | M1@.127 | SF | 21.50 | 5.92 | 27.42 |
| 8" wide | M1@.136 | SF | 21.80 | 6.34 | 28.14 |
| Glazed cove base, glazed 1 side, 8" high | | | | | |
| 4" length | M1@.127 | LF | 39.80 | 5.92 | 45.72 |
| 6" length | M1@.137 | LF | 30.00 | 6.39 | 36.39 |
| 8" length | M1@.147 | LF | 23.10 | 6.85 | 29.95 |

**Decorative Concrete Block** Prices are per square foot of wall and include allowance for mortar and waste but no foundation, reinforcing, grout or excavation. Regular weight units, 1.05 blocks per SF.

| | Craft@Hrs | Unit | Material | Labor | Total |
|---|---|---|---|---|---|
| Split face hollow block, 8" high x 16" long, 3/8" joints | | | | | |
| 4" wide | M1@.110 | SF | 1.64 | 5.13 | 6.77 |
| 6" wide | M1@.121 | SF | 1.73 | 5.64 | 7.37 |
| 8" wide | M1@.133 | SF | 1.99 | 6.20 | 8.19 |
| 10" wide | M1@.140 | SF | 2.84 | 6.53 | 9.37 |
| 12" wide | M1@.153 | SF | 2.80 | 7.13 | 9.93 |
| Split rib (combed) hollow block, 8" high x 16" long, 3/8" joints | | | | | |
| 4" wide | M1@.107 | SF | 2.45 | 4.99 | 7.44 |
| 6" wide | M1@.115 | SF | 2.55 | 5.36 | 7.91 |
| 8" wide | M1@.121 | SF | 2.82 | 5.64 | 8.46 |
| 10" wide | M1@.130 | SF | 3.68 | 6.06 | 9.74 |
| 12" wide | M1@.146 | SF | 3.63 | 6.81 | 10.44 |
| Screen block, 3/8" joints | | | | | |
| 6" x 6" x 4" wide | M1@.210 | SF | 3.30 | 9.79 | 13.09 |
| 8" x 8" x 4" wide (2.25 per SF) | M1@.140 | SF | 4.98 | 6.53 | 11.51 |
| 12" x 12" x 4" wide (1 per SF) | M1@.127 | SF | 2.53 | 5.92 | 8.45 |
| 8" x 16" x 8" wide (1.11 per SF) | M1@.115 | SF | 2.62 | 5.36 | 7.98 |

| | Craft@Hrs | Unit | Material | Labor | Total |
|---|---|---|---|---|---|
| Acoustical slotted block, by rated Noise Reduction Coefficient (NRC) | | | | | |
| 4" wide, with metal baffle, NRC, .45 to .55 | M1@.121 | SF | 4.07 | 5.64 | 9.71 |
| 6" wide, with metal baffle, NRC, .45 to .55 | M1@.130 | SF | 4.82 | 6.06 | 10.88 |
| 8" wide, with metal baffle, NRC, .45 to .55 | M1@.121 | SF | 5.99 | 5.64 | 11.63 |
| 8" wide, with fiber filler, NRC, .65 to .75 | M1@.143 | SF | 7.74 | 6.67 | 14.41 |

**Concrete Block Bond Beams**  For door and window openings, no grout or reinforcing included.
Cost per linear foot of bond beam block

| | Craft@Hrs | Unit | Material | Labor | Total |
|---|---|---|---|---|---|
| 6" wide,   8" high, 16" long | M1@.068 | LF | 2.40 | 3.17 | 5.57 |
| 8" wide,   8" high, 16" long | M1@.072 | LF | 2.72 | 3.36 | 6.08 |
| 10" wide,   8" high, 16" long | M1@.079 | LF | 4.00 | 3.68 | 7.68 |
| 12" wide,   8" high, 16" long | M1@.089 | LF | 3.80 | 4.15 | 7.95 |
| 12" wide,   8" high double bond beam, 16" long | M1@.102 | LF | 3.80 | 4.76 | 8.56 |
| 8" wide, 16" high, 16" long | M1@.127 | LF | 6.29 | 5.92 | 12.21 |

**Grout for Concrete Block**  Grout for block cores, Type "S". Costs shown are based on 2 cores per 16" block and include a 3% allowance for waste.
Grout per SF of wall, all cells filled, pumped in place and rodded

| | Craft@Hrs | Unit | Material | Labor | Total |
|---|---|---|---|---|---|
| 4" wide block, .11 CF per SF | M1@.018 | SF | .88 | .84 | 1.72 |
| 6" wide block, .22 CF per SF | M1@.025 | SF | 1.75 | 1.17 | 2.92 |
| 8" wide block, .32 CF per SF | M1@.035 | SF | 2.55 | 1.63 | 4.18 |
| 10" wide block, .45 CF per SF | M1@.035 | SF | 3.58 | 1.63 | 5.21 |
| 12" wide block, .59 CF per SF | M1@.039 | SF | 4.70 | 1.82 | 6.52 |

Grout per LF of cell filled measured vertically

| | Craft@Hrs | Unit | Material | Labor | Total |
|---|---|---|---|---|---|
| 4" wide block, .06 CF per LF | M1@.010 | LF | .48 | .47 | .95 |
| 6" wide block, .11 CF per LF | M1@.015 | LF | .88 | .70 | 1.58 |
| 8" wide block, .16 CF per LF | M1@.021 | LF | 1.27 | .98 | 2.25 |
| 10" wide block, .23 CF per LF | M1@.022 | LF | 1.83 | 1.03 | 2.86 |
| 12" wide block, .30 CF per LF | M1@.024 | LF | 2.39 | 1.12 | 3.51 |
| Add for type M mortar (2,500 PSI) | — | % | 6.0 | — | — |

Grout for bond beams, 3,000 PSI concrete, by block size and linear foot of beam, no block or reinforcing included, based on block cavity at 50% of block volume

| | Craft@Hrs | Unit | Material | Labor | Total |
|---|---|---|---|---|---|
| 8" high x   4" wide, .10 CF per LF | M1@.005 | LF | .80 | .23 | 1.03 |
| 8" high x   6" wide, .15 CF per LF | M1@.010 | LF | 1.19 | .47 | 1.66 |
| 8" high x   8" wide, .22 CF per LF | M1@.017 | LF | 1.75 | .79 | 2.54 |
| 8" high x 10" wide, .30 CF per LF | M1@.023 | LF | 2.39 | 1.07 | 3.46 |
| 8" high x 12" wide, .40 CF per LF | M1@.032 | LF | 3.18 | 1.49 | 4.67 |
| 16" high x   8" wide, .44 CF per LF | M1@.061 | LF | 3.50 | 2.84 | 6.34 |
| 16" high x 10" wide, .60 CF per LF | M1@.058 | LF | 4.78 | 2.70 | 7.48 |
| 16" high x 12" wide, .80 CF per LF | M1@.079 | LF | 6.37 | 3.68 | 10.05 |

Grout hollow metal door frame in a masonry wall

| | Craft@Hrs | Unit | Material | Labor | Total |
|---|---|---|---|---|---|
| Single door | M1@.668 | Ea | 9.94 | 31.10 | 41.04 |
| Double door | M1@.851 | Ea | 13.20 | 39.70 | 52.90 |

**Additional Costs for Concrete Block**

| | Craft@Hrs | Unit | Material | Labor | Total |
|---|---|---|---|---|---|
| Sill blocks | M1@.087 | LF | 1.58 | 4.06 | 5.64 |
| Cutting blocks | M1@.137 | LF | — | 6.39 | 6.39 |
| Stack bond | — | % | 5.0 | 10.0 | — |
| Add for cut-up jobs | — | % | 5.0 | 5.0 | — |
| Add for institutional grade work | — | % | 5.0 | 15.0 | — |
| Add for colored block | — | % | 15.0 | — | — |

# 04 Masonry

| | Craft@Hrs | Unit | Material | Labor | Total |
|---|---|---|---|---|---|

**Glass Block** Clear 3-7/8" thick block installed in walls to 8' high. Costs shown include 1/4" mortar joints, caulking, wall ties and 6% waste. Heights over 8' to 10' require additional scaffolding and handling and may increase labor costs up to 40%. Mortar is 1 part portland cement, 1/2 part lime, and 4 parts sand, plus integral waterproofer. Costs for continuous panel reinforcing in horizontal joints are listed below.

| | Craft@Hrs | Unit | Material | Labor | Total |
|---|---|---|---|---|---|
| Flat glass block, under 1,000 SF job | | | | | |
|    6" x 6", 4 per SF | M1@.323 | SF | 31.80 | 15.10 | 46.90 |
|    8" x 8", 2.25 per SF | M1@.245 | SF | 18.80 | 11.40 | 30.20 |
|    12" x 12", 1 per SF | M1@.202 | SF | 25.50 | 9.42 | 34.92 |
| Deduct for jobs over 1,000 SF to 5,000 SF | — | % | -4.0 | -4.0 | — |
| Deduct for jobs over 5,000 SF | — | % | -10.0 | -10.0 | — |
| Deduct for Thinline interior glass block | — | % | -10.0 | — | — |
| Add for thermal control fiberglass inserts | — | % | 5.0 | 5.0 | — |
| Replacing 8" x 8" x 4" clear glass block, 10 unit minimum | | | | | |
|    Per block replaced | M1@.851 | Ea | 9.10 | 39.70 | 48.80 |
| Additional costs for glass block | | | | | |
|    Fiberglass or polymeric expansion joints | — | LF | .37 | — | .37 |
|    Reinforcing steel mesh | — | LF | .52 | — | .52 |
|    Wall anchors, 2' long | | | | | |
|      Used where blocks aren't set into a wall | — | Ea | 2.01 | — | 2.01 |
|    Asphalt emulsion, 600 LF per gallon | — | Gal | 12.20 | — | 12.20 |
|    Waterproof sealant, 180 LF per gallon | — | Gal | 14.90 | — | 14.90 |
|    Cleaning block after installation, using sponge and water only | | | | | |
|      Per SF cleaned | CL@.023 | SF | — | .92 | .92 |

**Stone Work** The delivered price of stone can be expected to vary widely. These costs include mortar, based on 1/2" joints, but no scaffolding. Figures in parentheses show typical quantities installed per 8-hour day by a crew of two.

| | Craft@Hrs | Unit | Material | Labor | Total |
|---|---|---|---|---|---|
| Rough stone veneer, 4", placed over stud wall | | | | | |
|    Lava stone (79 SF) | M4@.202 | SF | 6.66 | 9.26 | 15.92 |
|    Rubble stone (40 SF) | M4@.400 | SF | 9.43 | 18.30 | 27.73 |
|    Most common veneer (60 SF) | M4@.267 | SF | 11.80 | 12.20 | 24.00 |
| Cut stone | | | | | |
|    Thinset granite tile, 1/2" thick (120 SF) | M4@.133 | SF | 13.90 | 6.10 | 20.00 |
|    Granite, 7/8" thick, interior (70 SF) | M4@.229 | SF | 22.70 | 10.50 | 33.20 |
|    Granite, 1-1/4" thick, exterior (60 SF) | M4@.267 | SF | 16.90 | 12.20 | 29.10 |
|    Limestone, 2" thick (70 SF) | M4@.229 | SF | 21.00 | 10.50 | 31.50 |
|    Limestone, 3" thick (60 SF) | M4@.267 | SF | 25.60 | 12.20 | 37.80 |
|    Marble, 7/8" thick, interior (70 SF) | M4@.229 | SF | 26.80 | 10.50 | 37.30 |
|    Marble, 1-1/4" thick, interior (70 SF) | M4@.229 | SF | 33.00 | 10.50 | 43.50 |
|    Sandstone, 2" thick (70 SF) | M4@.229 | SF | 18.90 | 10.50 | 29.40 |
|    Sandstone, 3" thick (60 SF) | M4@.267 | SF | 24.50 | 12.20 | 36.70 |
|    Add for honed finish | — | SF | 2.20 | — | 2.20 |
|    Add for polished finish | — | SF | 3.85 | — | 3.85 |
| Marble specialties, typical prices | | | | | |
|    Floors, 7/8" thick, exterior (70 SF) | M1@.229 | SF | 20.20 | 10.70 | 30.90 |
|    Thresholds, 1-1/4" thick (70 LF) | M1@.229 | LF | 10.70 | 10.70 | 21.40 |
|    Bases, 7/8" x 6" (51 LF) | M1@.314 | LF | 9.61 | 14.60 | 24.21 |
|    Columns, plain (15 CF) | M1@1.06 | CF | 113.00 | 49.40 | 162.40 |

| | Craft@Hrs | Unit | Material | Labor | Total |
|---|---|---|---|---|---|
| Columns, fluted (13 CF) | M1@1.23 | CF | 157.00 | 57.30 | 214.30 |
| Window stools, 5" x 1" (60 LF) | M1@.267 | LF | 11.80 | 12.40 | 24.20 |
| Toilet partitions (1.5 Ea) | M1@10.6 | Ea | 625.00 | 494.00 | 1,119.00 |
| Stair treads, 1-1/4" x 11" (50 LF) | M1@.320 | LF | 26.40 | 14.90 | 41.30 |
| Limestone, rough cut large blocks (39 CF) | M1@.410 | CF | 32.50 | 19.10 | 51.60 |

Manufactured stone veneer, including mortar, Cultured Stone®

| | Craft@Hrs | Unit | Material | Labor | Total |
|---|---|---|---|---|---|
| Prefitted types (160 SF) | M1@.100 | SF | 6.75 | 4.66 | 11.41 |
| Brick veneer type (120 SF) | M1@.133 | SF | 5.60 | 6.20 | 11.80 |
| Random cast type (120 SF) | M1@.133 | SF | 6.17 | 6.20 | 12.37 |

**Quarry Tile** Unglazed natural red tile set in a portland cement bed with mortar joints.

Quarry floor tile

| | Craft@Hrs | Unit | Material | Labor | Total |
|---|---|---|---|---|---|
| 4" x 4" x 1/2", 1/8" straight joints | M1@.112 | SF | 3.54 | 5.22 | 8.76 |
| 6" x 6" x 1/2", 1/4" straight joints | M1@.101 | SF | 3.31 | 4.71 | 8.02 |
| 6" x 6" x 3/4", 1/4" straight joints | M1@.105 | SF | 4.33 | 4.90 | 9.23 |
| 8" x 8" x 3/4", 3/8" straight joints | M1@.095 | SF | 4.10 | 4.43 | 8.53 |
| 8" x 8" x 1/2", 1/4" hexagon joints | M1@.112 | SF | 4.66 | 5.22 | 9.88 |

Quarry wall tile

| | Craft@Hrs | Unit | Material | Labor | Total |
|---|---|---|---|---|---|
| 4" x 4" x 1/2", 1/8" straight joints | M1@.125 | SF | 3.52 | 5.83 | 9.35 |
| 6" x 6" x 3/4", 1/4" straight joints | M1@.122 | SF | 4.33 | 5.69 | 10.02 |

Quarry tile stair treads

| | Craft@Hrs | Unit | Material | Labor | Total |
|---|---|---|---|---|---|
| 6" x 6" x 3/4", 12" wide tread | M1@.122 | LF | 4.77 | 5.69 | 10.46 |

Quarry tile window sill

| | Craft@Hrs | Unit | Material | Labor | Total |
|---|---|---|---|---|---|
| 6" x 6" x 3/4" tile on 6" wide sill | M1@.078 | LF | 4.58 | 3.64 | 8.22 |

Quarry tile trim or cove base

| | Craft@Hrs | Unit | Material | Labor | Total |
|---|---|---|---|---|---|
| 5" x 6" x 1/2" straight top | M1@.083 | LF | 3.52 | 3.87 | 7.39 |
| 6" x 6" x 3/4" round top | M1@.087 | LF | 4.10 | 4.06 | 8.16 |
| Deduct for tile set in epoxy bed without grout joints | — | SF | — | -1.91 | -1.91 |
| Add for tile set in epoxy bed with grout joints | — | SF | 1.24 | — | 1.24 |

**Pavers and Floor Tile**

Brick, excluding platform cost

| | Craft@Hrs | Unit | Material | Labor | Total |
|---|---|---|---|---|---|
| Plate, glazed | M1@.114 | SF | 2.87 | 5.31 | 8.18 |
| Laid in sand | M1@.089 | SF | 2.19 | 4.15 | 6.34 |
| Pavers, on concrete, grouted | M1@.107 | SF | 3.87 | 4.99 | 8.86 |
| Add for special patterns | — | % | 25.0 | 35.0 | — |
| Slate | M1@.099 | SF | 3.38 | 4.62 | 8.00 |

Terrazzo tiles, 1/4" thick

| | Craft@Hrs | Unit | Material | Labor | Total |
|---|---|---|---|---|---|
| Standard | M1@.078 | SF | 5.45 | 3.64 | 9.09 |
| Granite | M1@.085 | SF | 8.88 | 3.96 | 12.84 |
| Brick steps | M1@.186 | SF | 3.44 | 8.67 | 12.11 |

**Interlocking Paving Stones** These costs include fine sand to fill joints and machine vibration, but no excavation.

| | Craft@Hrs | Unit | Material | Labor | Total |
|---|---|---|---|---|---|
| Interlocking pavers, rectangular, 60mm thick, Orion | M1@.083 | SF | 1.57 | 3.87 | 5.44 |
| Interlocking pavers, hexagonal, 80mm thick, Omni | M1@.085 | SF | 1.93 | 3.96 | 5.89 |
| Interlocking pavers, multi-angle, 80mm thick, Venus | M1@.081 | SF | 1.65 | 3.78 | 5.43 |
| Concrete masonry grid pavers (erosion control) | M1@.061 | SF | 2.77 | 2.84 | 5.61 |
| Add for a 1" sand cushion | M1@.001 | SF | .04 | .05 | .09 |
| Add for a 2" sand cushion | M1@.002 | SF | .08 | .09 | .17 |
| Deduct for over 5,000 SF | — | % | -3.0 | -7.0 | — |

# 04 Masonry

|  | Craft@Hrs | Unit | Material | Labor | Total |
|---|---|---|---|---|---|
| **Masonry Lintels and Coping** | | | | | |
| Reinforced precast concrete lintels | | | | | |
| 4" wide x 8" high, to 6'6" long | M1@.073 | LF | 7.96 | 3.40 | 11.36 |
| 6" wide x 8" high, to 6'6" long | M1@.078 | LF | 14.20 | 3.64 | 17.84 |
| 8" wide x 8" high, to 6'6" long | M1@.081 | LF | 14.80 | 3.78 | 18.58 |
| 10" wide x 8" high, to 6'6" long | M1@.091 | LF | 19.00 | 4.24 | 23.24 |
| 4" wide x 8" high, over 6'6" to 10' long | M1@.073 | LF | 14.90 | 3.40 | 18.30 |
| 6" wide x 8" high, over 6'6" to 10' long | M1@.077 | LF | 17.50 | 3.59 | 21.09 |
| 8" wide x 8" high, over 6'6" to 10' long | M1@.081 | LF | 18.20 | 3.78 | 21.98 |
| 10" wide x 8" high, over 6'6" to 10' long | M1@.091 | LF | 25.20 | 4.24 | 29.44 |
| Precast concrete coping, 5" average thickness, 4' to 8' long | | | | | |
| 10" wide, gray concrete | M1@.190 | LF | 13.30 | 8.86 | 22.16 |
| 12" wide, gray concrete | M1@.200 | LF | 14.90 | 9.32 | 24.22 |
| 12" wide, white concrete | M1@.226 | LF | 36.80 | 10.50 | 47.30 |
| **Pargeting, Waterproofing and Dampproofing** | | | | | |
| Pargeting, 2 coats, 1/2" thick | M3@.070 | SF | .23 | 3.41 | 3.64 |
| Pargeting, 2 coats, waterproof, 3/4" thick | M3@.085 | SF | .32 | 4.14 | 4.46 |
| Dampproofing, asphalt primer at 1 gallon per 100 SF, Dampproof coat at 30 pounds per 100 SF | CL@.029 | SF | .12 | 1.16 | 1.28 |

# 05 Metals

**Structural Steel** Costs shown below are for 60 to 100 ton jobs up to six stories high where the fabricated structural steel is purchased from a steel fabricator located within 50 miles of the job site. Material costs are for ASTM A36 steel, fabricated according to plans, specifications and AISC recommendations and delivered to the job site with connections attached, piece marked and ready for erection. Labor includes time to unload and handle the steel, erect and plumb, install permanent bolts, weld and all other work usually associated with erecting the steel, including field touch-up painting. Equipment includes a 60 ton crawler mounted crane, plus an electric powered welding machine, an oxygen/acetylene welding and cutting torch, a diesel powered 100 CFM air compressor with pneumatic grinders, ratchet wrenches, hoses, and other tools usually associated with work of this type. Structural steel is usually estimated by the tons of steel in the project. Labor for erection will vary with the weight of steel per linear foot. For example a member weighing 25 pounds per foot will require more manhours per ton than one weighing 60 pounds per foot.

**Structural shapes** Estimate the total weight of "bare" fabricated structural shapes and then add 15% to allow for the weight of connections.

|  | Craft@Hrs | Unit | Material | Labor | Equipment | Total |
|---|---|---|---|---|---|---|
| Beams, purlins, and girts | | | | | | |
| Under 20 pounds per LF | H8@14.0 | Ton | 2,580.00 | 826.00 | 339.00 | 3,745.00 |
| From 20 to 50 lbs per LF | H8@11.0 | Ton | 1,650.00 | 649.00 | 266.00 | 2,565.00 |
| Over 50 to 75 lbs per LF | H8@9.40 | Ton | 1,570.00 | 555.00 | 227.00 | 2,352.00 |
| Over 75 to 100 lbs per LF | H8@8.00 | Ton | 1,530.00 | 472.00 | 194.00 | 2,196.00 |
| Columns | | | | | | |
| Under 20 pounds per LF | H8@12.0 | Ton | 2,400.00 | 708.00 | 290.00 | 3,398.00 |
| From 20 to 50 lbs per LF | H8@10.0 | Ton | 1,570.00 | 590.00 | 241.00 | 2,401.00 |
| Over 50 to 75 lbs per LF | H8@9.00 | Ton | 1,490.00 | 531.00 | 218.00 | 2,239.00 |
| Over 75 to 100 lbs per LF | H8@7.00 | Ton | 1,490.00 | 413.00 | 169.00 | 2,072.00 |
| From 1 to 3 lbs per LF | H8@48.0 | Ton | 3,620.00 | 2,830.00 | 1,160.00 | 7,610.00 |
| Over 3 to 6 lbs per LF | H8@34.8 | Ton | 3,370.00 | 2,050.00 | 842.00 | 6,262.00 |

|  | Craft@Hrs | Unit | Material | Labor | Equipment | Total |
|---|---|---|---|---|---|---|
| Sag rods and X-type bracing |  |  |  |  |  |  |
| Under 1 pound per LF | H8@82.1 | Ton | 3,900.00 | 4,840.00 | 1,990.00 | 10,730.00 |
| Over 6 to 10 lbs per LF | H8@28.0 | Ton | 3,090.00 | 1,650.00 | 677.00 | 5,417.00 |
| Trusses and girders |  |  |  |  |  |  |
| Under 20 pounds per LF | H8@10.0 | Ton | 2,390.00 | 590.00 | 242.00 | 3,222.00 |
| From 20 to 50 lbs per LF | H8@7.50 | Ton | 1,570.00 | 443.00 | 182.00 | 2,195.00 |
| Over 50 to 75 lbs per LF | H8@6.00 | Ton | 1,490.00 | 354.00 | 145.00 | 1,989.00 |
| Over 75 to 100 lbs per LF | H8@5.00 | Ton | 1,440.00 | 295.00 | 121.00 | 1,856.00 |

**Structural Steel, Subcontract, by Square Foot of Floor** Use these figures only for preliminary estimates on buildings with 20,000 to 80,000 SF per story, 100 PSF live load and 14' to 16' story height. A "bay" is the center-to-center distance between columns. Costs assume steel averaging 2,550 per ton including installation labor and material and the subcontractor's overhead and profit. Include all area under the roof when calculating the total square footage.

|  | Craft@Hrs | Unit | Material | Labor | Equipment | Total |
|---|---|---|---|---|---|---|
| Single story |  |  |  |  |  |  |
| 20' x 20' bays (8 to 9 lbs per SF) | — | SF | — | — | — | 18.70 |
| 30' x 30' bays (10 to 11 lbs per SF) | — | SF | — | — | — | 23.00 |
| 40' x 40' bays (12 to 13 lbs per SF) | — | SF | — | — | — | 27.40 |
| Two story to six story |  |  |  |  |  |  |
| 20' x 20' bays (9 to 10 lbs per SF) | — | SF | — | — | — | 20.90 |
| 30' x 30' bays (11 to 12 lbs per SF) | — | SF | — | — | — | 25.20 |
| 40' x 40' bays (13 to 14 lbs per SF) | — | SF | — | — | — | 29.60 |

**Tubular Columns** Rectangular, round, and square structural columns. When estimating total weight of columns, figure the column weight and add 30% to allow for a shop attached "U"-bracket on the top, a square base plate on the bottom and four anchor bolts. This should be sufficient for columns 15' to 20' in length. Equipment cost includes a 10 ton hydraulic truck-mounted crane plus an electric powered welding machine, an oxygen/acetylene welding and cutting torch, a diesel powered 100 CFM air compressor with pneumatic grinders, ratchet wrenches, hoses, and other tools usually associated with work of this type.

|  | Craft@Hrs | Unit | Material | Labor | Equipment | Total |
|---|---|---|---|---|---|---|
| Rectangular tube columns, in pounds per LF |  |  |  |  |  |  |
| 3" x 2" to 12" x 4", to 20 lbs per LF | H7@9.00 | Ton | 2,370.00 | 547.00 | 248.00 | 3,165.00 |
| 6" x 4" to 16" x 8", 21 to 50 lbs per LF | H7@8.00 | Ton | 1,900.00 | 486.00 | 221.00 | 2,607.00 |
| 12 " x 6" to 20" x 12", |  |  |  |  |  |  |
| 51 to 75 lbs per LF | H7@6.00 | Ton | 1,860.00 | 365.00 | 166.00 | 2,391.00 |
| 14" x 10" to 20" x 12", 76 to 100 lbs/LF | H7@5.00 | Ton | 1,800.00 | 304.00 | 138.00 | 2,242.00 |
| Round columns, in pounds per LF |  |  |  |  |  |  |
| 3" to 6" diameter pipe, to 20 lbs per LF | H7@9.00 | Ton | 2,520.00 | 547.00 | 248.00 | 3,315.00 |
| 4" to 12" pipe, 20 to 50 lbs per LF | H7@8.00 | Ton | 2,030.00 | 486.00 | 221.00 | 2,737.00 |
| 6" to 12" pipe, 51 to 75 lbs per LF | H7@6.00 | Ton | 1,950.00 | 365.00 | 166.00 | 2,481.00 |
| Square tube columns, in pounds per LF |  |  |  |  |  |  |
| 2" x 2" to 8" x 8", to 20 lbs per LF | H7@9.00 | Ton | 2,280.00 | 547.00 | 248.00 | 3,075.00 |
| 4" x 4" to 12" x 12", 21 to 50 lbs per LF | H7@8.00 | Ton | 1,860.00 | 486.00 | 221.00 | 2,567.00 |
| 8" x 8" to 16" x 16", 51 to 75 lbs per LF | H7@6.00 | Ton | 1,810.00 | 365.00 | 166.00 | 2,341.00 |
| 10" x 10" to 16" x 16", 76 to 100 lbs/LF | H7@5.00 | Ton | 1,790.00 | 304.00 | 138.00 | 2,232.00 |
| Add for concrete fill in columns, 3,000 PSI design pump mix, placed by pump |  |  |  |  |  |  |
| (including 10% waste) | M2@.738 | CY | 127.00 | 32.40 | 5.95 | 165.35 |

**Space Frame System** 10,000 SF or more. Equipment cost shown is for a 15 ton truck crane and a 2 ton truck equipped for work of this type.

|  | Craft@Hrs | Unit | Material | Labor | Equipment | Total |
|---|---|---|---|---|---|---|
| 5' module 4.5 lb live load | H5@.146 | SF | 18.00 | 7.65 | 2.36 | 28.01 |
| 4' module, add | H5@.025 | SF | 2.61 | 1.31 | .40 | 4.32 |

# 05 Metals

| | Craft@Hrs | Unit | Material | Labor | Equipment | Total |
|---|---|---|---|---|---|---|

**Open web steel joists** Estimate the total weight of "bare" joists and then add 10% to allow for connections and accessories

"H" and "J" Series, 8H3 thru 26H10 or 8J3 thru 30J8

| | Craft@Hrs | Unit | Material | Labor | Equipment | Total |
|---|---|---|---|---|---|---|
| Under 10 lbs per LF | H8@10.0 | Ton | 1,620.00 | 590.00 | 245.00 | 2,455.00 |
| Over 10 to 20 lbs per LF | H8@8.00 | Ton | 1,600.00 | 472.00 | 194.00 | 2,266.00 |

"LH" and "LJ" Series, 18LH02 thru 48LH10 or 18LJ02 thru 48LJ11

| | | | | | | |
|---|---|---|---|---|---|---|
| Under 20 lbs per LF | H8@8.00 | Ton | 1,480.00 | 472.00 | 194.00 | 2,146.00 |
| Over 20 to 50 lbs per LF | H8@7.00 | Ton | 1,440.00 | 413.00 | 169.00 | 2,022.00 |
| Over 50 lbs per LF | H8@6.00 | Ton | 1,370.00 | 354.00 | 145.00 | 1,869.00 |

"DLH" or "DLJ" Series, 52DLH10 thru 72DLH18 or 52DLJ12 thru 52DLJ20

| | | | | | | |
|---|---|---|---|---|---|---|
| 20 to 50 lbs per LF | H8@8.00 | Ton | 1,600.00 | 472.00 | 194.00 | 2,266.00 |
| Over 50 lbs per LF | H8@6.00 | Ton | 1,520.00 | 354.00 | 145.00 | 2,019.00 |

**Steel Floor and Roof Decking** These costs assume a 40,000 SF job not over six stories high. Weight and cost include 10% for waste and connections. Equipment cost shown is for a 15 ton truck crane and a 2 ton truck equipped for work of this type.
Floor and roof decking, 1-1/2" standard rib, simple non-composite section, non-cellular, shop prime painted finish

| | | | | | | |
|---|---|---|---|---|---|---|
| 22 gauge, 2.1 lbs per SF | H5@.013 | SF | 1.55 | .68 | .21 | 2.44 |
| 20 gauge, 2.5 lbs per SF | H5@.013 | SF | 1.62 | .68 | .21 | 2.51 |
| 18 gauge, 3.3 lbs per SF | H5@.013 | SF | 2.04 | .68 | .21 | 2.93 |
| 16 gauge, 4.1 lbs per SF | H5@.013 | SF | 2.26 | .68 | .21 | 3.15 |
| Add for galvanized finish | — | SF | .28 | — | — | .28 |

Steel floor deck system. Consists of 18 gauge, 1-1/2" standard rib, shop prime painted steel decking, installed over a structural steel beam and girder support system (cost of support structure not included) covered with 3" thick 6.0 sack mix, 3,000 PSI lightweight ready-mix concrete, placed by pump, reinforced with 6" x 6", W1.4 x W1.4 welded wire mesh. Costs shown assume shoring is not required and include curing and finishing the concrete.

| | | | | | | |
|---|---|---|---|---|---|---|
| 18 gauge decking | H5@.013 | SF | 1.62 | .68 | .21 | 2.51 |
| Welded wire mesh | RB@.004 | SF | .14 | .23 | — | .37 |
| Lightweight concrete | M2@.006 | SF | 1.63 | .26 | .10 | 1.99 |
| Cure and finish concrete | P8@.011 | SF | .15 | .50 | — | .65 |
| **Total floor deck system** | —@.034 | SF | 3.54 | 1.67 | .31 | 5.52 |
| Add for 16 gauge decking | — | SF | .29 | — | — | .29 |

Steel roof deck system. Consists of 18 gauge, 1-1/2" standard rib, shop prime painted steel decking, installed over open web steel joists 4' OC (costs of joists not included) covered with 1-1/4" R-8.3 urethane insulation board overlaid with .045" loose laid membrane EPDM (ethylene propylene diene monomer) elastomeric roofing. Joist pricing assumes "H" or "J" Series at 9.45 lbs per LF spanning 20'0", spaced 4' OC.

| | | | | | | |
|---|---|---|---|---|---|---|
| 18 gauge decking | H5@.013 | SF | 2.04 | .68 | .21 | 2.93 |
| Insulation board | A1@.011 | SF | 1.58 | .54 | — | 2.12 |
| EPDM roofing | R3@.006 | SF | 1.64 | .27 | — | 1.91 |
| **Total roof deck system** | —@.030 | SF | 5.26 | 1.49 | .21 | 6.96 |
| Add for 16 gauge decking | — | SF | .29 | — | — | .29 |
| Add open web steel joists | H5@.005 | SF | .81 | .26 | .08 | 1.15 |

| | Craft@Hrs | Unit | Material | Labor | Total |
|---|---|---|---|---|---|

**Checkered-steel plate** Shop prime painted or galvanized after fabrication as shown
Checkered steel floor plate,

| | Craft@Hrs | Unit | Material | Labor | Total |
|---|---|---|---|---|---|
| 1/4", 11-1/4 lbs per SF, painted | IW@.025 | SF | 13.20 | 1.59 | 14.79 |
| 3/8", 16-1/2 lbs per SF, painted | IW@.025 | SF | 18.80 | 1.59 | 20.39 |
| 1/4", 11-1/4 lbs per SF, galvanized | IW@.025 | SF | 14.40 | 1.59 | 15.99 |
| 3/8", 16-1/2 lbs per SF, galvanized | IW@.025 | SF | 21.00 | 1.59 | 22.59 |

# 05 Metals

|  | Craft@Hrs | Unit | Material | Labor | Total |
|---|---|---|---|---|---|
| Checkered-steel trench cover plate |  |  |  |  |  |
| 1/4", 11-1/4 lbs per SF, primed | IW@.045 | SF | 17.20 | 2.86 | 20.06 |
| 1/4", 11-1/4 lbs per SF, galvanized | IW@.045 | SF | 20.00 | 2.86 | 22.86 |
| 3/8", 16-1/2 lbs per SF, primed | IW@.045 | SF | 24.90 | 2.86 | 27.76 |
| 3/8", 16-1/2 lbs per SF, galvanized | IW@.045 | SF | 30.20 | 2.86 | 33.06 |

|  | Craft@Hrs | Unit | Material | Labor | Equipment | Total |
|---|---|---|---|---|---|---|
| **Spiral stairways** Cost per LF of vertical rise, floor to floor. Costs assume a 6" diameter round tubular steel center column, 12 gauge steel pan type 10" treads with 7-1/2" open risers, and 2" x 2" tubular steel handrail with 1" x 1" tubular steel posts at each riser, shop prime painted. Add for railings at top of stairways as required. |  |  |  |  |  |  |
| 4'0" diameter | H7@.151 | LF | 277.00 | 9.18 | 4.17 | 290.35 |
| 6'0" diameter | H7@.151 | LF | 378.00 | 9.18 | 4.17 | 391.35 |
| 8'0" diameter | H7@.151 | LF | 477.00 | 9.18 | 4.17 | 490.35 |

**Straight stairways** Cost per LF based on sloping length. Rule of thumb: the sloping length of stairways is the vertical rise times 1.6. Costs assume standard "C"-shape structural steel channels for stringers, 12 gauge steel pan 10" treads with 7-1/2" closed risers. Posts are 2" diameter tubular steel, 5' OC. Handrails are 3'6" high, 1-1/2" diameter 2-rail type, on both sides of the stairs. Installation assumes stairs and landings are attached to an existing structure on two sides.

|  | Craft@Hrs | Unit | Material | Labor | Equipment | Total |
|---|---|---|---|---|---|---|
| 3'0" wide | H7@.469 | LF | 196.00 | 28.50 | 12.90 | 237.40 |
| 4'0" wide | H7@.630 | LF | 260.00 | 38.30 | 17.40 | 315.70 |
| 5'0" wide | H7@.781 | LF | 326.00 | 47.50 | 21.60 | 395.10 |

Ladders, vertical type. 1'6" wide, attached to existing structure

|  | Craft@Hrs | Unit | Material | Labor | Equipment | Total |
|---|---|---|---|---|---|---|
| Without safety cage | H7@.100 | LF | 36.50 | 6.08 | 2.76 | 45.34 |
| With steel safety cage | H7@.150 | LF | 43.30 | 9.12 | 4.14 | 56.56 |
| Add for galvanized finish | — | LF | 4.26 | — | — | 4.26 |

Landings. Costs assume standard "C"-shape structural steel channels for joists and frame, 10 gauge steel pan deck areas, shop prime painted. Add for support legs from Structural Steel Section.

|  | Craft@Hrs | Unit | Material | Labor | Equipment | Total |
|---|---|---|---|---|---|---|
| Cost per square foot | H7@.250 | SF | 132.00 | 15.20 | 6.90 | 154.10 |
| Add for galvanized finish | — | SF | 15.00 | — | — | 15.00 |

## Railings, Aluminum, Brass and Wrought Iron

Aluminum railings. 3'6" high with 1-1/2" square horizontal rails top and bottom, 5/8" square vertical bars at 4" OC, and 1-1/2" square posts spaced 4' OC. Welded construction with standard mill finish.

|  | Craft@Hrs | Unit | Material | Labor | | Total |
|---|---|---|---|---|---|---|
| Installed on balconies | H6@.092 | LF | 28.90 | 4.76 | — | 33.66 |
| Installed on stairways | H6@.115 | LF | 32.50 | 5.95 | — | 38.45 |
| Add for anodized finish, bronze or black | — | LF | 14.10 | — | — | 14.10 |

Brass railings. Round tubular type, bright finish, .050" wall commercial quality with supports at 3' OC attached to existing structure using brass screws. Labor includes layout, measure, cut and end preparation

|  | Craft@Hrs | Unit | Material | Labor | | Total |
|---|---|---|---|---|---|---|
| 1-1/2" tubing | H6@.058 | LF | 10.80 | 3.00 | — | 13.80 |
| 2" tubing | H6@.058 | LF | 14.30 | 3.00 | — | 17.30 |
| Fittings, slip-on type, no soldering or brazing required |  |  |  |  |  |  |
| 1-1/2" elbows | H6@.092 | Ea | 23.10 | 4.76 | — | 27.86 |
| 1-1/2" bracket | H6@.166 | Ea | 12.10 | 8.59 | — | 20.69 |
| 2" elbows | H6@.115 | Ea | 20.10 | 5.95 | — | 26.05 |
| 2" bracket | H6@.166 | Ea | 12.10 | 8.59 | — | 20.69 |

Wrought iron railings. Standard 3'6" high, welded steel construction, shop prime painted, attached to existing structure

|  | Craft@Hrs | Unit | Material | Labor | | Total |
|---|---|---|---|---|---|---|
| Stock patterns | H6@.171 | LF | 14.80 | 8.84 | — | 23.64 |
| Custom patterns | H6@.287 | LF | 22.20 | 14.80 | — | 37.00 |
| Stairway handrails | H6@.343 | LF | 24.40 | 17.70 | — | 42.10 |

# 05 Metals

| | Craft@Hrs | Unit | Material | Labor | Equipment | Total |
|---|---|---|---|---|---|---|

**Steel Handrails, Kickplates, and Railings** Material costs assume shop fabrication and assembly, prime painted and erection bolts. Equipment cost includes a 10-ton hydraulic truck-mounted crane, an electric powered welding machine, an oxygen/acetylene welding rig, a diesel powered 100 CFM air compressor and accessories.

Handrails. Wall mounted, with brackets 5' OC, based on 12' lengths, welded steel, shop prime painted

| | Craft@Hrs | Unit | Material | Labor | Equipment | Total |
|---|---|---|---|---|---|---|
| 1-1/4" diameter rail | H7@.092 | LF | 6.50 | 5.59 | 2.54 | 14.63 |
| 1-1/2" diameter rail | H7@.102 | LF | 7.60 | 6.20 | 2.81 | 16.61 |
| Add for galvanized finish | — | LF | 1.39 | — | — | 1.39 |

Kickplates, 1/4" thick flat steel plate, shop prime painted, attached to steel railings, stairways or landings

| | Craft@Hrs | Unit | Material | Labor | Equipment | Total |
|---|---|---|---|---|---|---|
| 4" high | H7@.058 | LF | 5.22 | 3.53 | 1.60 | 10.35 |
| 6" high | H7@.069 | LF | 5.67 | 4.19 | 1.90 | 11.76 |
| Add for galvanized finish | — | LF | 1.03 | — | — | 1.03 |

Pipe and chain railings. Welded steel pipe posts with U-brackets attached to accept 1/4" chain, shop prime painted. Installed in existing embedded sleeves. Posts are 5'0" long (1'6" in the sleeve and 3'6" above grade). Usual spacing is 4' OC. Add chain per lineal foot

| | Craft@Hrs | Unit | Material | Labor | Equipment | Total |
|---|---|---|---|---|---|---|
| 1-1/4" diameter post | H7@.287 | Ea | 10.50 | 17.40 | — | 27.90 |
| 1-1/2" diameter post | H7@.287 | Ea | 12.80 | 17.40 | — | 30.20 |
| 2" diameter post | H7@.343 | Ea | 17.20 | 20.80 | — | 38.00 |
| 3" diameter post | H7@.343 | Ea | 34.10 | 20.80 | — | 54.90 |
| Add for galvanized finish | — | Ea | 4.44 | — | — | 4.44 |
| Chain, 1/4" diameter galvanized | H7@.058 | LF | 3.47 | 3.53 | — | 7.00 |

Railings. Floor mounted, 3'6" high with posts 5' OC, based on 12' length, welded steel, shop prime painted
1-1/4" diameter rails and posts

| | Craft@Hrs | Unit | Material | Labor | Equipment | Total |
|---|---|---|---|---|---|---|
| 2-rail type | H7@.115 | LF | 43.80 | 6.99 | 3.17 | 53.96 |
| 3-rail type | H7@.125 | LF | 56.40 | 7.60 | 3.45 | 67.45 |

1-1/2" diameter rails and posts

| | Craft@Hrs | Unit | Material | Labor | Equipment | Total |
|---|---|---|---|---|---|---|
| 2-rail type | H7@.125 | LF | 52.20 | 7.60 | 3.45 | 63.25 |
| 3-rail type | H7@.137 | LF | 69.20 | 8.33 | 3.78 | 81.31 |
| Add for galvanized finish | — | LF | 11.10 | — | — | 11.10 |

| | Craft@Hrs | Unit | Material | Labor | Total |
|---|---|---|---|---|---|
| **Fabricated metals** | | | | | |
| Catch basin grating and frame | | | | | |
| 24" x 24", standard duty | CL@8.68 | Ea | 326.00 | 346.00 | 672.00 |
| Manhole rings with cover | | | | | |
| 24", 330 lbs | CL@6.80 | Ea | 314.00 | 271.00 | 585.00 |
| 30", 400 lbs | CL@6.80 | Ea | 326.00 | 271.00 | 597.00 |
| 38", 730 lbs | CL@6.80 | Ea | 899.00 | 271.00 | 1,170.00 |

**Welded steel grating** Weight shown is approximate. Costs shown are per square foot of grating, shop prime painted or galvanized after fabrication as shown

Grating bars as shown, 1/2" cross bars at 4" OC

| | Craft@Hrs | Unit | Material | Labor | Total |
|---|---|---|---|---|---|
| 1-1/4" x 3/16" at 1-1/4" OC, 9 lbs per SF, painted | IW@.019 | SF | 7.05 | 1.21 | 8.26 |
| 1-1/4" x 3/16" at 1-1/4" OC, 9 lbs per SF, galvanized | IW@.019 | SF | 8.70 | 1.21 | 9.91 |
| Add for banding on edges | — | LF | 3.69 | — | 3.69 |
| 2" x 1/4" at 1-1/4" OC, 18.7 lbs per SF, painted | IW@.026 | SF | 16.60 | 1.65 | 18.25 |
| 2" x 1/4" at 1-1/4" OC, 18.7 lbs per SF, galvanized | IW@.026 | SF | 22.30 | 1.65 | 23.95 |
| Add for banding on edges | — | LF | 5.50 | — | 5.50 |

Add for toe-plates, steel, shop prime painted

| | Craft@Hrs | Unit | Material | Labor | Total |
|---|---|---|---|---|---|
| 1/4" thick x 4" high attached after grating is in place | IW@.048 | LF | 7.08 | 3.05 | 10.13 |

# 05 Metals

| | Craft@Hrs | Unit | Material | Labor | Total |
|---|---|---|---|---|---|
| **Steel fabrications** | | | | | |
| Canopy framing, steel | IW@.009 | Lb | 1.06 | .57 | 1.63 |
| Miscellaneous supports, steel | IW@.019 | Lb | 1.78 | 1.21 | 2.99 |
| Miscellaneous supports, aluminum | IW@.061 | Lb | 11.50 | 3.88 | 15.38 |
| Embedded steel (embedded in concrete) | | | | | |
| Light, to 20 lbs per LF | T3@.042 | Lb | 2.17 | 2.06 | 4.23 |
| Medium, over 20 to 50 lbs per LF | T3@.035 | Lb | 1.95 | 1.72 | 3.67 |
| Heavy, over 50 lbs per LF | T3@.025 | Lb | 1.65 | 1.23 | 2.88 |
| Channel sill, 1/8" steel, 1-1/2" x 8" wide | T3@.131 | LF | 17.00 | 6.43 | 23.43 |
| Angle sill, 1/8" steel, 1-1/2" x 1-1/2" | T3@.095 | LF | 5.34 | 4.66 | 10.00 |
| Fire escapes, ladder and balcony, per floor | IW@16.5 | Ea | 2,000.00 | 1,050.00 | 3,050.00 |
| Grey iron foundry items | T3@.014 | Lb | .98 | .69 | 1.67 |
| Stair nosing, 2-1/2" x 2-1/2", set in concrete | T3@.139 | LF | 8.38 | 6.82 | 15.20 |
| Cast iron | | | | | |
| Lightweight sections | T3@.012 | Lb | 1.03 | .59 | 1.62 |
| Heavy section | T3@.010 | Lb | .84 | .49 | 1.33 |
| Cast column bases | | | | | |
| Iron, 16" x 16", custom | T3@.759 | Ea | 89.60 | 37.30 | 126.90 |
| Iron, 32" x 32", custom | T3@1.54 | Ea | 374.00 | 75.60 | 449.60 |
| Aluminum, 8" column, stock | T3@.245 | Ea | 20.70 | 12.00 | 32.70 |
| Aluminum, 10" column, stock | T3@.245 | Ea | 26.10 | 12.00 | 38.10 |
| Wheel type corner guards, per LF of height | T3@.502 | LF | 65.00 | 24.60 | 89.60 |
| Angle corner guards, with anchors | T3@.018 | Lb | 1.94 | .88 | 2.82 |
| Channel door frames, with anchors | T3@.012 | Lb | 1.80 | .59 | 2.39 |
| Steel lintels, painted | T3@.013 | Lb | 1.07 | .64 | 1.71 |
| Steel lintels, galvanized | T3@.013 | Lb | 1.42 | .64 | 2.06 |
| | | | | | |
| **Anodizing** Add to cost of fabrication | | | | | |
| Aluminum | — | % | 15.0 | — | — |
| Bronze | — | % | 30.0 | — | — |
| Black | — | % | 50.0 | — | — |
| Galvanizing, add to cost of fabrication | — | Lb | .57 | — | .57 |
| | | | | | |
| **Ornamental Metal** | | | | | |
| Ornamental screens, aluminum | H6@.135 | SF | 35.20 | 6.98 | 42.18 |
| Ornamental screens extruded metal | H6@.151 | SF | 29.50 | 7.81 | 37.31 |
| Gates, wrought iron, 6' x 7', typical installed cost | — | Ea | — | — | 1,490.00 |
| Sun screens, aluminum stock design, manual | H6@.075 | SF | 37.90 | 3.88 | 41.78 |
| Sun screens, aluminum stock design, motorized | H6@.102 | SF | 68.90 | 5.28 | 74.18 |
| Aluminum work, general fabrication | H6@.160 | Lb | 18.80 | 8.28 | 27.08 |
| Brass and bronze work, general fabrication | H6@.079 | Lb | 27.30 | 4.09 | 31.39 |
| Stainless steel, general fabrication | H6@.152 | Lb | 22.20 | 7.86 | 30.06 |
| | | | | | |
| **Security Gratings** | | | | | |
| Door grating, 1" tool steel vertical bars at 4" OC, 3/8" x 2-1/4" steel frame and horizontal bars at 18" OC, | | | | | |
| 3' x 7' overall | H6@10.5 | Ea | 1,670.00 | 543.00 | 2,213.00 |
| Add for butt hinges and locks, per door | H6@2.67 | Ea | 510.00 | 138.00 | 648.00 |
| Add for field painting, per door | H6@1.15 | Ea | 7.77 | 59.50 | 67.27 |
| Window grating, tool steel vertical bars at 4" OC, 3/8" x 2-1/4" steel frame, horizontal bars 6" OC, | | | | | |
| 20" x 30" overall set with 1/2" bolts at 6" OC | H6@3.21 | Ea | 224.00 | 166.00 | 390.00 |
| Louver vent grating, 1" x 3/16", "I" bearing bars 1-3/16" OC with 3/16" spaced bars 4" OC and banded ends, | | | | | |
| set with 1/2" bolts at 1" OC, prime painted | H6@.114 | SF | 27.30 | 5.90 | 33.20 |
| Steel plate window covers, hinged, 3/8" plate steel, 32" W x 48" H overall, 10 lbs per SF, | | | | | |
| including hardware | H6@.150 | SF | 12.50 | 7.76 | 20.26 |

|  | Craft@Hrs | Unit | Material | Labor | Equipment | Total |
|---|---|---|---|---|---|---|

**Steel fabrication on site** These costs are for work at the job site and assume engineering and design, including detail drawings sufficient to do the work, have been prepared by others at no cost to the installing contractor. Costs are based on ASTM A36 all-purpose carbon steel, which is widely used in building construction. Equipment cost includes a 1-1/4-ton flatbed truck equipped with a diesel powered welding machine, an oxygen/acetylene welding and cutting torch, a diesel powered 100 CFM air compressor with pneumatic grinders, ratchet wrenches, hoses and hand tools appropriate for this type of work. Material cost for the steel is not included.

## Flat plate, cutting and drilling holes

Flame cutting. Labor shown includes layout, marking, cutting and grinding of finished cut

| Square cut, thickness of plate as shown | Craft@Hrs | Unit | Material | Labor | Equipment | Total |
|---|---|---|---|---|---|---|
| 1/4", 5/16" or 3/8" | IW@.020 | LF | — | 1.27 | .96 | 2.23 |
| 1/2", 5/8" or 3/4" | IW@.025 | LF | — | 1.59 | 1.20 | 2.79 |
| 7/8", 1" or 1-1/4" | IW@.030 | LF | — | 1.91 | 1.44 | 3.35 |
| Bevel cut 30 to 45 degrees, thickness of plate as shown | | | | | | |
| 1/4", 5/16" or 3/8" | IW@.030 | LF | — | 1.91 | 1.44 | 3.35 |
| 1/2", 5/8" or 3/4" | IW@.035 | LF | — | 2.22 | 1.69 | 3.91 |
| 7/8", 1" or 1-1/4" | IW@.043 | LF | — | 2.73 | 2.07 | 4.80 |

Drill holes. Using a hand-held drill motor. Labor shown includes layout, drill and ream holes with plate thickness and hole size as shown (note: hole size is usually 1/16" greater than bolt diameter)

| | Craft@Hrs | Unit | Material | Labor | Equipment | Total |
|---|---|---|---|---|---|---|
| 1/4", 5/16" or 3/8" plate thickness, 5/16", 3/8" or 1/2" hole | IW@.084 | Ea | — | 5.34 | 4.05 | 9.39 |
| 1/2", 5/8" or 3/4" plate thickness, 9/16", 3/4" or 13/16" hole | IW@.134 | Ea | — | 8.52 | 6.45 | 14.97 |
| 7/8", 1" or 1-1/4" plate thickness, 1", 1-1/16" or 1-5/16" hole | IW@.200 | Ea | — | 12.70 | 9.63 | 22.33 |

**Welding** Using manual inert gas shielded arc electric welding and coated electrodes (rods) of the appropriate size and type. Equipment is an electric welding machine with amperage suitable for the welding rod being used. Labor includes layout, fit-up and weld.

Fillet welds for attaching base plates, clip angles, gussets. Welding in flat position

| Size of continuous fillet weld | Craft@Hrs | Unit | Material | Labor | Equipment | Total |
|---|---|---|---|---|---|---|
| 3/16" | IW@.300 | LF | — | 19.10 | 14.50 | 33.60 |
| 1/4" | IW@.400 | LF | — | 25.40 | 19.30 | 44.70 |
| 3/8" | IW@.585 | LF | — | 37.20 | 28.20 | 65.40 |
| 1/2" | IW@.810 | LF | — | 51.50 | 39.00 | 90.50 |
| 5/8" | IW@1.00 | LF | — | 63.60 | 48.20 | 111.80 |
| 3/4" | IW@1.20 | LF | — | 76.30 | 57.80 | 134.10 |

| Buttweld joints, 100% penetration, flat plates | | | | | | |
|---|---|---|---|---|---|---|
| Thickness at joint | | | | | | |
| 3/16" | IW@.200 | LF | — | 12.70 | 9.63 | 22.33 |
| 1/4" | IW@.250 | LF | — | 15.90 | 12.04 | 27.94 |
| 3/8" | IW@.440 | LF | — | 28.00 | 21.20 | 49.20 |
| 1/2" | IW@.660 | LF | — | 41.90 | 31.80 | 73.70 |
| 5/8" | IW@.825 | LF | — | 52.40 | 39.70 | 92.10 |
| 3/4" | IW@1.05 | LF | — | 66.70 | 50.60 | 117.30 |

# 06 Wood and Composites

| | Craft@Hrs | Unit | Material | Labor | Total |
|---|---|---|---|---|---|
| **Rough Carpentry**  Industrial and commercial work. Costs include 10% waste.Using standard and better dimension lumber at the following prices: | | | | | |
| 2" x  4" | — | MBF | 473.00 | — | 473.00 |
| 2" x  6" | — | MBF | 465.00 | — | 465.00 |
| 2" x  8" | — | MBF | 465.00 | — | 465.00 |
| 2" x 10" | — | MBF | 484.00 | — | 484.00 |
| 2" x 12" | — | MBF | 563.00 | — | 563.00 |
| 4" x  4" | — | MBF | 598.00 | — | 598.00 |
| 4" x  6" | — | MBF | 635.00 | — | 635.00 |
| 4" x  8" | — | MBF | 597.00 | — | 597.00 |
| 4" x 10" | — | MBF | 635.00 | — | 635.00 |
| 4" x 12" | — | MBF | 694.00 | — | 694.00 |
| 6" x  6" | — | MBF | 942.00 | — | 942.00 |
| 6" x  8" | — | MBF | 942.00 | — | 942.00 |
| 6" x 10" | — | MBF | 942.00 | — | 942.00 |
| 6" x 12" | — | MBF | 942.00 | — | 942.00 |
| Add for ACQ treatment | — | MBF | 250.00 | — | 250.00 |
| Stud walls, including plates, blocking, headers and diagonal bracing, per SF of wall, 16" OC | | | | | |
| 2" x 4",   8' high, 1.1 BF per SF | C8@.021 | SF | .57 | .96 | 1.53 |
| 2" x 4", 10' high, 1.0 BF per SF | C8@.021 | SF | .52 | .96 | 1.48 |
| 2" x 6",   8' high, 1.6 BF per SF | C8@.031 | SF | .82 | 1.42 | 2.24 |
| 2" x 6", 10' high, 1.5 BF per SF | C8@.030 | SF | .77 | 1.37 | 2.14 |
| 12" OC, 2" x 6", 8' high, 2.0 BF per SF | C8@.037 | SF | 1.02 | 1.70 | 2.72 |
| 8" OC, 2" x 4", staggered on 2" x 6" plates, 1.8 BF per SF | C8@.034 | SF | .94 | 1.56 | 2.50 |
| Pier caps, .60 ACQ treated | | | | | |
| 2" x 6" | C8@.024 | LF | .71 | 1.10 | 1.81 |
| Mud sills, treated material, add bolts or fastener | | | | | |
| Bolted, 2" x 4" | C8@.023 | LF | .72 | 1.05 | 1.77 |
| Bolted, 2" x 6" | C8@.024 | LF | .71 | 1.10 | 1.81 |
| Shot, 2" x 4" | C8@.018 | LF | .85 | .82 | 1.67 |
| Shot, 2" x 6" | C8@.020 | LF | .71 | .92 | 1.63 |
| Plates (add for .60 ACQ treatment if required) | | | | | |
| 2" x 4" | C8@.013 | LF | .35 | .60 | .95 |
| 2" x 6" | C8@.016 | LF | .51 | .73 | 1.24 |
| Posts, #1 structural | | | | | |
| 4" x 4" | C8@.110 | LF | .88 | 5.04 | 5.92 |
| 6" x 6" | C8@.145 | LF | 3.11 | 6.65 | 9.76 |
| Headers, 4" x 12 | C8@.067 | LF | 3.05 | 3.07 | 6.12 |
| Diagonal bracing, 1" x 4", let in | C8@.021 | LF | .42 | .96 | 1.38 |
| Blocking and nailers, 2" x 6" | C8@.034 | LF | .51 | 1.56 | 2.07 |
| Blocking, solid diagonal bracing, 2" x 4" | C8@.023 | LF | .35 | 1.05 | 1.40 |
| Beams and girders, #2 & better | | | | | |
| 4" x 8" to 4" x 12" | C8@.094 | LF | 2.79 | 4.31 | 7.10 |
| 6" x 8" to 6" x 12" | C8@.115 | LF | 6.22 | 5.27 | 11.49 |
| Ceiling joists and blocking, including rim joists, per LF of joist | | | | | |
| 2" x  6" | C8@.028 | LF | .51 | 1.28 | 1.79 |
| 2" x  8" | C8@.037 | LF | .68 | 1.70 | 2.38 |
| 2" x 10" | C8@.046 | LF | .89 | 2.11 | 3.00 |
| 2" x 12" | C8@.057 | LF | 1.24 | 2.61 | 3.85 |

# 06 Wood and Composites

| | Craft@Hrs | Unit | Material | Labor | Total |
|---|---|---|---|---|---|
| **Floor joists and blocking including rim joists and 6% waste, per LF of joist** | | | | | |
| 2" x  6", 1.06 BF per LF | C8@.017 | LF | .51 | .78 | 1.29 |
| 2" x  8", 1.41 BF per LF | C8@.018 | LF | .68 | .82 | 1.50 |
| 2" x 10", 1.77 BF per LF | C8@.019 | LF | .89 | .87 | 1.76 |
| 2" x 12", 2.12 BF per LF | C8@.021 | LF | 1.24 | .96 | 2.20 |
| **Purlins, struts, ridge boards** | | | | | |
| 4" x 10" | C8@.050 | LF | 2.33 | 2.29 | 4.62 |
| 4" x 12" | C8@.055 | LF | 3.05 | 2.52 | 5.57 |
| 6" x 10" | C8@.066 | LF | 5.18 | 3.02 | 8.20 |
| 6" x 12" | C8@.066 | LF | 6.22 | 3.02 | 9.24 |
| **Rafters and outriggers, to 4 in 12 pitch** | | | | | |
| 2" x  4" to 2" x  8" | C8@.033 | LF | .68 | 1.51 | 2.19 |
| 2" x 10" to 2" x 12" | C8@.036 | LF | 1.24 | 1.65 | 2.89 |
| Add for over 4 in 12 pitch roof | — | % | — | 30.0 | — |
| **Rafters, including frieze blocks, ribbon, strongbacks, 24" OC, per SF of roof plan area** | | | | | |
| 2" x  4",  .44 BF per SF | C8@.015 | SF | .21 | .69 | .90 |
| 2" x  6",  .65 BF per SF | C8@.021 | SF | .30 | .96 | 1.26 |
| 2" x  8",  .86 BF per SF | C8@.027 | SF | .40 | 1.24 | 1.64 |
| 2" x 10", 1.10 BF per SF | C8@.031 | SF | .53 | 1.42 | 1.95 |
| 2" x 12", 1.40 BF per SF | C8@.041 | SF | .79 | 1.88 | 2.67 |

**Sills and Ledgers Bolted to Concrete or Masonry Walls** Costs are per linear foot of Standard and Better pressure treated sill or ledger with anchor bolt holes drilled each 2'6" center. Labor costs include handling, measuring, cutting to length and drilling holes at floor level and installation working on a scaffold 10' to 20' above floor level. Figures in parentheses show the board feet per linear foot of sill or ledger. Costs shown include 5% waste.

For scheduling purposes, estimate that a crew of 2 can install 150 to 200 LF of sills with a single row of anchor bolts in an 8-hour day when working from scaffolds. Estimate that a crew of 2 can install 150 to 175 LF of ledgers with a single row of anchor bolt holes or 50 to 60 LF of ledgers with a double row of anchor bolt holes in an 8-hour day working from scaffolds. Add the cost of anchor bolts and scaffold, if required.

| | Craft@Hrs | Unit | Material | Labor | Total |
|---|---|---|---|---|---|
| **Sills with a single row of anchor bolt holes** | | | | | |
| 2" x  4" (0.70 BF per LF) | C8@.040 | LF | .59 | 1.83 | 2.42 |
| 2" x  6" (1.05 BF per LF) | C8@.040 | LF | .86 | 1.83 | 2.69 |
| 2" x  8" (1.40 BF per LF) | C8@.040 | LF | 1.28 | 1.83 | 3.11 |
| 2" x 10" (1.75 BF per LF) | C8@.040 | LF | 1.52 | 1.83 | 3.35 |
| 2" x 12" (2.10 BF per LF) | C8@.040 | LF | 1.89 | 1.83 | 3.72 |
| 4" x  6" (2.10 BF per LF) | C8@.051 | LF | 2.28 | 2.34 | 4.62 |
| 4" x  8" (2.80 BF per LF) | C8@.051 | LF | 3.34 | 2.34 | 5.68 |
| 4" x 10" (3.33 BF per LF) | C8@.059 | LF | 4.08 | 2.70 | 6.78 |
| 4" x 12" (4.20 BF per LF) | C8@.059 | LF | 5.39 | 2.70 | 8.09 |
| **Ledgers with a single row of anchor bolt holes** | | | | | |
| 2" x 4" (0.70 BF per LF) | C8@.059 | LF | .59 | 2.70 | 3.29 |
| 4" x 6" (2.10 BF per LF) | C8@.079 | LF | 2.28 | 3.62 | 5.90 |
| 4" x 8" (2.80 BF per LF) | C8@.079 | LF | 3.34 | 3.62 | 6.96 |
| **Ledgers with a double row of anchor bolt holes** | | | | | |
| 4" x  4" (1.40 BF per LF) | C8@.120 | LF | 1.46 | 5.50 | 6.96 |
| 4" x  6" (2.10 BF per LF) | C8@.160 | LF | 2.28 | 7.33 | 9.61 |
| 4" x  8" (2.80 BF per LF) | C8@.160 | LF | 3.34 | 7.33 | 10.67 |
| 4" x 12" (4.20 BF per LF) | C8@.250 | LF | 5.39 | 11.50 | 16.89 |

# 06 Wood and Composites

| | Craft@Hrs | Unit | Material | Labor | Total |
|---|---|---|---|---|---|
| **Sheathing** Includes nails and normal waste | | | | | |
| Floors, hand nailed, underlayment C-D Grade plywood | | | | | |
| 1/2" | C8@.012 | SF | .53 | .55 | 1.08 |
| 5/8" | C8@.012 | SF | .65 | .55 | 1.20 |
| 3/4" | C8@.013 | SF | .78 | .60 | 1.38 |
| Floors, machine nailed OSB | | | | | |
| 7/16" x 4' x 8' | C8@.009 | SF | .28 | .41 | .69 |
| 7/16" x 4' x 9' | C8@.009 | SF | .39 | .41 | .80 |
| 7/16" x 4' x 10' | C8@.009 | SF | .46 | .41 | .87 |
| 15/32" 4' x 8' | C8@.009 | SF | .34 | .41 | .75 |
| 1/2" x 4' x 8' | C8@.009 | SF | .32 | .41 | .73 |
| 5/8" x 4' x 8' | C8@.010 | SF | .43 | .46 | .89 |
| 5/8" x 4' x 8', tongue and groove | C8@.010 | SF | .55 | .46 | 1.01 |
| 23/32" x 4' x 8' | C8@.010 | SF | .64 | .46 | 1.10 |
| 23/32" 4' x 8' subfloor | C8@.010 | SF | .87 | .46 | 1.33 |
| 3/4" x 4' x 8', tongue and groove | C8@.011 | SF | .52 | .50 | 1.02 |
| 1-1/8" x 4' x 8', tongue and groove | C8@.015 | SF | 1.39 | .69 | 2.08 |
| 15/32" x 4' x 8', hibor treated | C8@.011 | SF | 1.05 | .50 | 1.55 |
| Walls, B-D sanded, standard, Interior Grade plywood | | | | | |
| 1/4" | C8@.015 | SF | .83 | .69 | 1.52 |
| 3/8" | C8@.015 | SF | .87 | .69 | 1.56 |
| 1/2" | C8@.016 | SF | 1.02 | .73 | 1.75 |
| 5/8" | C8@.018 | SF | 1.28 | .82 | 2.10 |
| 3/4" | C8@.018 | SF | 1.49 | .82 | 2.31 |
| 5/8" fire retardant, A-B, screwed to metal frame | C8@.026 | SF | .59 | 1.19 | 1.78 |
| Walls, Exterior Grade C-D plywood | | | | | |
| 1/2" | C8@.016 | SF | .53 | .73 | 1.26 |
| 5/8" | C8@.018 | SF | .66 | .82 | 1.48 |
| 3/4" | C8@.020 | SF | .78 | .92 | 1.70 |
| 3/8" rough sawn T-1-11 | C8@.018 | SF | .92 | .82 | 1.74 |
| 5/8" rough sawn T-1-11 | C8@.018 | SF | 1.36 | .82 | 2.18 |
| 5/8" rough sawn RBB | C8@.018 | SF | 1.36 | .82 | 2.18 |
| Walls, OSB | | | | | |
| 11/32" or 3/8" x 4' x 8' | C8@.013 | SF | .28 | .60 | .88 |
| 7/16" x 4' x 8' | C8@.015 | SF | .28 | .69 | .97 |
| 7/16" x 4' x 9' | C8@.015 | SF | .39 | .69 | 1.08 |
| 7/16" x 4' x 10' | C8@.015 | SF | .45 | .69 | 1.14 |
| 15/32" 4' x 8' | C8@.015 | SF | .34 | .69 | 1.03 |
| 1/2" x 4' x 8' | C8@.018 | SF | .32 | .82 | 1.14 |
| 5/8" x 4' x 8' | C8@.018 | SF | .43 | .82 | 1.25 |
| 5/8" x 4' x 8', tongue and groove | C8@.018 | SF | .53 | .82 | 1.35 |
| 23/32" x 4' x 8' | C8@.020 | SF | .64 | .92 | 1.56 |
| 15/32" x 4' x 10', Structural 1 | C8@.020 | SF | .55 | .92 | 1.47 |
| Add for nailing shear walls | — | % | — | 20.0 | — |
| Roofs, C-D Exterior plywood, machine nailed | | | | | |
| 3/8" | C8@.009 | SF | .46 | .41 | .87 |
| 1/2" | C8@.009 | SF | .53 | .41 | .94 |
| 5/8" | C8@.009 | SF | .66 | .41 | 1.07 |
| 3/4" | C8@.011 | SF | .78 | .50 | 1.28 |

# 06 Wood and Composites

| | Craft@Hrs | Unit | Material | Labor | Total |
|---|---|---|---|---|---|
| Roofs, OSB, machine nailed | | | | | |
| 11/32" or 3/8" x 4' x 8' | C8@.009 | SF | .28 | .41 | .69 |
| 7/16" x 4' x 8' | C8@.009 | SF | .28 | .41 | .69 |
| 7/16" x 4' x 9' | C8@.009 | SF | .39 | .41 | .80 |
| 7/16" x 4' x 10' | C8@.009 | SF | .38 | .41 | .79 |
| 15/32" 4' x 8' | C8@.009 | SF | .34 | .41 | .75 |
| 1/2" x 4' x 8' | C8@.009 | SF | .32 | .41 | .73 |
| 5/8" x 4' x 8' | C8@.010 | SF | .43 | .46 | .89 |
| 5/8" x 4' x 8', tongue and groove | C8@.010 | SF | .52 | .46 | .98 |
| 23/32" x 4' x 8' | C8@.010 | SF | .64 | .46 | 1.10 |
| 3/4" x 4' x 8', tongue and groove | C8@.011 | SF | .52 | .50 | 1.02 |
| Blocking for panel joints, | | | | | |
| 2" x 4" | C8@.010 | LF | .35 | .46 | .81 |
| Lumber floor sheathing | | | | | |
| Diagonal, 1" x 6" Standard and Better | | | | | |
| 1.14 BF per SF | C8@.018 | SF | 1.40 | .82 | 2.22 |
| Straight 2" x 6" Commercial Grade T&G fir | | | | | |
| 2.28 BF per SF | C8@.032 | SF | 1.06 | 1.47 | 2.53 |
| Straight 2" x 6" Commercial Grade T&G cedar | | | | | |
| 2.28 BF per SF | C8@.032 | SF | 7.90 | 1.47 | 9.37 |
| Timber deck, Southern Pine, 5/4" x 6", Premium Grade | | | | | |
| 1.25 BF per SF | C8@.012 | SF | 3.13 | .55 | 3.68 |
| Lumber wall sheathing, diagonal, 1" x 6" | | | | | |
| Standard and Better, 1.14 BF per SF | C8@.024 | SF | 2.12 | 1.10 | 3.22 |
| Lumber roof sheathing, diagonal, 1" x 6" | | | | | |
| Standard and Better, 1.14 BF per SF | C8@.018 | SF | 2.12 | .82 | 2.94 |

**Panelized Wood Roof Systems** The costs shown will apply on 20,000 to 30,000 SF jobs where roof panels are fabricated near the point of installation and moved into place with a forklift. Costs will be higher on smaller jobs and jobs with poor access. 4' x 8' and 4' x 10' roof panels are made of plywood or OSB nailed to 2" x 4" or 2" x 6" stiffeners spaced 24" on centers and running the long dimension of the panel. Based on fabricated panels lifted by forklift (with metal hangers attached) and nailed to 4" x 12" or 4" x 14" roof purlins spaced 8' on center. Equipment cost is for a 4-ton forklift (40' lift), a truck with hand tools, a gas powered air compressor with air hoses and two pneumatic-operated nailing guns. Costs are per 1,000 square feet (MSF) of roof surface and do not include roof supports except as noted.

| | Craft@Hrs | Unit | Material | Labor | Equipment | Total |
|---|---|---|---|---|---|---|

For scheduling purposes, estimate that a crew of 5 can fabricate 6,000 to 6,400 SF of 4' x 8' or 4' x 10' panels in an 8-hour day. Estimate that the same crew of 5 can install and nail down 7,600 to 8,000 SF of panels in an 8-hour day. Based on the following material costs:

| | Craft@Hrs | Unit | Material | Labor | Equipment | Total |
|---|---|---|---|---|---|---|
| 2" x 4" stiffeners | — | MBF | 473.00 | — | — | 473.00 |
| 2" x 6" stiffeners | — | MBF | 465.00 | — | — | 465.00 |
| Purlins | — | MBF | 694.00 | — | — | 694.00 |
| 1/2" CDX plywood | — | MSF | 507.00 | — | — | 507.00 |
| 1/2" OSB sheathing | — | MSF | 306.00 | — | — | 306.00 |

## 4' x 8' roof panels

With 1/2" CDX structural #1 plywood attached to 2" x 4" Construction Grade fir stiffeners

| | Craft@Hrs | Unit | Material | Labor | Equipment | Total |
|---|---|---|---|---|---|---|
| Panels fabricated, ready to install | F6@6.60 | MSF | 767.00 | 315.00 | 90.20 | 1,172.20 |
| Equipment & labor | | | | | | |
| to install and nail panels | F6@5.21 | MSF | — | 248.00 | 71.20 | 319.20 |
| Total installed cost, 4' x 8' panels | F6@11.8 | MSF | 767.00 | 562.00 | 161.40 | 1,490.40 |

# 06 Wood and Composites

| | Craft@Hrs | Unit | Material | Labor | Equipment | Total |
|---|---|---|---|---|---|---|

**With 1/2" OSB structural sheathing attached to 2" x 4" Construction Grade fir stiffeners**

| | Craft@Hrs | Unit | Material | Labor | Equipment | Total |
|---|---|---|---|---|---|---|
| Panels fabricated, ready to install | F6@6.60 | MSF | 557.00 | 315.00 | 90.20 | 962.20 |
| Equipment & labor | | | | | | |
|   to install and nail panels | F6@5.21 | MSF | — | 248.00 | 71.20 | 319.20 |
|     Total installed cost, 4' x 8' panels | F6@11.8 | MSF | 557.00 | 562.00 | 161.40 | 1,280.40 |

## 4' x 10' roof panels

**With 1/2" CDX structural #1 plywood attached to 2" x 6" Construction Grade fir stiffeners**

| | Craft@Hrs | Unit | Material | Labor | Equipment | Total |
|---|---|---|---|---|---|---|
| Panels fabricated, ready to install | F6@6.20 | MSF | 865.00 | 295.00 | 84.70 | 1,244.70 |
| Equipment & labor | | | | | | |
|   to install and nail panels | F6@4.90 | MSF | — | 234.00 | 67.10 | 301.10 |
|     Total installed cost, 4' x 10' panels | F6@11.1 | MSF | 865.00 | 529.00 | 151.8 | 1,545.80 |

**With 1/2" OSB structural sheathing attached to 2" x 6" Construction Grade fir stiffeners**

| | Craft@Hrs | Unit | Material | Labor | Equipment | Total |
|---|---|---|---|---|---|---|
| Panels fabricated, ready to install | F6@6.20 | MSF | 655.00 | 295.00 | 85.00 | 1,035.00 |
| Equipment & labor | | | | | | |
|   to install and nail panels | F6@4.90 | MSF | — | 234.00 | 67.10 | 301.10 |
|     Total installed cost, 4' x 10' panels | F6@11.1 | MSF | 655.00 | 529.00 | 152.10 | 1,336.10 |

## 8' x 20' and 8' x 32' roof panels

These panels can be fabricated on site using the same crew and equipment as shown in the preceding section. However, the stiffeners run across the short dimension of the panel and one purlin is attached to the outside edge of the long dimension of each panel before lifting into place. Figures below include the cost of purlins but not supporting beams. For scheduling purposes, estimate that a crew of 5 can fabricate 6,400 to 7,000 SF of 8' x 20' or 8' x 32' panels in an 8-hour day. Estimate that the same crew of 5 can install and nail down 10,000 to 11,000 SF of panels in an 8-hour day.

## 8' x 20' roof panels with purlins attached

**4' x 10' x 1/2" CDX structural #1 plywood attached to 2" x 4" Const. Grade fir stiffeners and 4" x 12" #1 Grade fir purlins**

| | Craft@Hrs | Unit | Material | Labor | Equipment | Total |
|---|---|---|---|---|---|---|
| Panels fabricated, ready to install | F6@6.23 | MSF | 1,470.00 | 297.00 | 85.00 | 1,852.00 |
| Equipment & labor | | | | | | |
|   to install and nail panels | F6@3.96 | MSF | — | 189.00 | 54.10 | 243.10 |
|     Total installed cost, 8' x 20' panels | F6@10.2 | MSF | 1,470.00 | 486.00 | 139.10 | 2,095.10 |

**8' x 20' x 1/2" OSB structural sheathing attached to 2" x 4" Const. Grade fir stiffeners and 4" x 12" #1 Grade fir purlins**

| | Craft@Hrs | Unit | Material | Labor | Equipment | Total |
|---|---|---|---|---|---|---|
| Panels fabricated, ready to install | F6@6.23 | MSF | 1,260.00 | 297.00 | 85.00 | 1,642.00 |
| Equipment & labor | | | | | | |
|   to install and nail panels | F6@3.96 | MSF | — | 189.00 | 54.10 | 243.10 |
|     Total installed cost, 8' x 20' panels | F6@10.2 | MSF | 1,260.00 | 486.00 | 139.10 | 1,885.10 |

## 8' x 32' panel with purlins attached

**4' x 8' x 1/2" CDX structural #1 plywood attached to 2" x 4" Const. Grade fir stiffeners and 4" x 12" #1 Grade fir purlins**

| | Craft@Hrs | Unit | Material | Labor | Equipment | Total |
|---|---|---|---|---|---|---|
| Panels fabricated, ready to install | F6@5.82 | MSF | 1,590.00 | 277.00 | 79.40 | 1,946.40 |
| Equipment & labor | | | | | | |
|   to install and nail panels | F6@3.54 | MSF | — | 169.00 | 48.50 | 217.50 |
|     Total installed cost, 8' x 32' panels | F6@9.36 | MSF | 1,590.00 | 446.00 | 127.90 | 2,163.90 |

**4' x 8' x 1/2" OSB structural sheathing attached to 2" x 4" Const. Grade fir stiffeners and 4" x 12" #1 Grade fir purlins**

| | Craft@Hrs | Unit | Material | Labor | Equipment | Total |
|---|---|---|---|---|---|---|
| Panels fabricated, ready to install | F6@5.82 | MSF | 1,380.00 | 277.00 | 79.40 | 1,736.40 |
| Equipment & labor | | | | | | |
|   to install and nail panels | F6@3.54 | MSF | — | 169.00 | 48.50 | 217.50 |
|     Total installed cost, 8' x 32' panels | F6@9.36 | MSF | 1,380.00 | 446.00 | 127.90 | 1,953.90 |
| Adjust to any of the panel costs for 5/8" CDX #1 structural plywood, instead of 1/2" attached before | | | | | | |
|   panels are erected | — | MSF | 171.00 | — | — | 171.00 |

| | Craft@Hrs | Unit | Material | Labor | Equipment | Total |
|---|---|---|---|---|---|---|
| Adjust to any of the panel costs for 15/32" OSB Structural I sheathing, instead of 1/2" attached before | | | | | | |
|    panels are erected | — | MSF | 221.00 | — | — | 221.00 |
| Add for roof hatch opening, includes reinforced opening and curb. Does not include roof hatch, see below | | | | | | |
|    2" x 4" roof system | F6@.055 | SF | 2.84 | 2.62 | — | 5.46 |
|    2" x 6" roof system | F6@.060 | SF | 3.68 | 2.86 | — | 6.54 |

**Panelized Wood Wall Panels** These costs will apply on jobs with 20,000 to 30,000 SF of wall panels. Panel material costs are for prefabricated panels purchased from a modular plant located within 50 miles of the job site. Costs will be higher on smaller jobs and on jobs with poor access. Costs assume wall panels are fabricated according to job plans and specifications, delivered piece-marked, ready to install and finish. Labor costs for installation of panels are listed at the end of this section and include unloading and handling, sorting, moving the panels into position using a forklift, installing and nailing. Maximum panel height is usually 12'. Equipment cost includes a forklift, hand tools, a portable air compressor with air hoses and two pneumatic-operated nailing guns. Costs are per square foot of wall, measured on one face. For scheduling purposes, estimate that a crew of 4 can place and fasten 3,500 to 4,500 SF of panels in an 8-hour day.

| | Craft@Hrs | Unit | Material | Labor | Equipment | Total |
|---|---|---|---|---|---|---|
| 2" x 4" studs 16" OC with gypsum wallboard on both sides | | | | | | |
|    1/2" wallboard | — | SF | 1.07 | — | — | 1.07 |
|    5/8" wallboard | — | SF | 1.17 | — | — | 1.17 |
| 2" x 4" studs 16" OC with gypsum wallboard inside and plywood or OSB outside | | | | | | |
|    1/2" wallboard and 1/2" CDX plywood | — | SF | 1.33 | — | — | 1.33 |
|    1/2" wallboard and 1/2" OSB sheathing | — | SF | 1.13 | — | — | 1.13 |
|    1/2" wallboard and 5/8" CDX plywood | — | SF | 1.44 | — | — | 1.44 |
|    1/2" wallboard and 5/8" OSB sheathing | — | SF | 1.24 | — | — | 1.24 |
|    5/8" wallboard and 1/2" CDX plywood | — | SF | 1.38 | — | — | 1.38 |
|    5/8" wallboard and 1/2" OSB sheathing | — | SF | 1.18 | — | — | 1.18 |
|    5/8" wallboard and 5/8" CDX plywood | — | SF | 1.49 | — | — | 1.49 |
|    5/8" wallboard and 5/8" OSB sheathing | — | SF | 1.29 | — | — | 1.29 |
| 2" x 6" studs 12" OC with gypsum wallboard on both sides | | | | | | |
|    1/2" wallboard | — | SF | 1.27 | — | — | 1.27 |
|    5/8" wallboard | — | SF | 1.32 | — | — | 1.32 |
| 2" x 6" studs 12" OC with gypsum wallboard inside and plywood or OSB outside | | | | | | |
|    1/2" wallboard and 1/2" CDX plywood | — | SF | 1.78 | — | — | 1.78 |
|    1/2" wallboard and 1/2" OSB sheathing | — | SF | 1.58 | — | — | 1.58 |
|    1/2" wallboard and 5/8" CDX plywood | — | SF | 1.89 | — | — | 1.89 |
|    1/2" wallboard and 5/8" OSB sheathing | — | SF | 1.69 | — | — | 1.69 |
|    5/8" wallboard and 1/2" CDX plywood | — | SF | 1.83 | — | — | 1.83 |
|    5/8" wallboard and 1/2" OSB sheathing | — | SF | 1.63 | — | — | 1.63 |
|    5/8" wallboard and 5/8" CDX plywood | — | SF | 1.94 | — | — | 1.94 |
|    5/8" wallboard and 5/8" OSB sheathing | — | SF | 1.74 | — | — | 1.74 |
| 2" x 6" studs 16" OC with gypsum wallboard on both sides | | | | | | |
|    1/2" wallboard | — | SF | 1.07 | — | — | 1.07 |
|    5/8" wallboard | — | SF | 1.12 | — | — | 1.12 |
| 2" x 6" studs 16" OC with gypsum wallboard inside and plywood or OSB outside | | | | | | |
|    1/2" wallboard and 1/2" CDX plywood | — | SF | 1.58 | — | — | 1.58 |
|    1/2" wallboard and 1/2" OSB sheathing | — | SF | 1.37 | — | — | 1.37 |
|    1/2" wallboard and 5/8" CDX plywood | — | SF | 1.69 | — | — | 1.69 |
|    1/2" wallboard and 5/8" OSB sheathing | — | SF | 1.49 | — | — | 1.49 |
|    5/8" wallboard and 1/2" CDX plywood | — | SF | 1.63 | — | — | 1.63 |
|    5/8" wallboard and 1/2" OSB sheathing | — | SF | 1.42 | — | — | 1.42 |
|    5/8" wallboard and 5/8" CDX plywood | — | SF | 1.74 | — | — | 1.74 |
|    5/8" wallboard and 5/8" OSB sheathing | — | SF | 1.54 | — | — | 1.54 |

# 06 Wood and Composites

| | Craft@Hrs | Unit | Material | Labor | Equipment | Total |
|---|---|---|---|---|---|---|
| Add to any of the above for the following, installed by panel fabricator: | | | | | | |
| R-19 insulation | — | SF | 1.01 | — | — | 1.01 |
| R-11 insulation | — | SF | .66 | — | — | .66 |
| Vapor barrier | — | SF | .45 | — | — | .45 |
| Door or window openings to 16 SF each | — | SF | 1.36 | — | — | 1.36 |
| Door or window openings over 16 SF each | — | SF | 1.24 | — | — | 1.24 |
| Electrical receptacles | — | Ea | 49.20 | — | — | 49.20 |
| Electrical switches | — | Ea | 52.40 | — | — | 52.40 |
| Labor and equipment cost to install panels | | | | | | |
| Up to 8' high | F7@.003 | SF | — | .15 | .04 | .19 |
| Over 8' to 12' high | F7@.002 | SF | — | .10 | .02 | .12 |

| | Craft@Hrs | Unit | Material | Labor | Total |
|---|---|---|---|---|---|
| **Miscellaneous Rough Carpentry** | | | | | |
| Furring, Utility Grade, unit costs incl. 5% allowance for waste and nails | | | | | |
| 2" x 4" on masonry walls | C8@.024 | LF | .33 | 1.10 | 1.43 |
| 1" x 4" machine nailed to ceiling | C8@.017 | LF | .40 | .78 | 1.18 |
| 1" x 4" nailed on concrete | C8@.016 | LF | .40 | .73 | 1.13 |
| Skip sheathing, on roof rafters, machine nailed, #2 and Better, unit costs incl. 7% allowance for waste & nails | | | | | |
| 1" x 6" at 9" centers (.67 BF per SF) | | | | | |
| Per SF of roof area | C8@.020 | SF | .88 | .92 | 1.80 |
| 1" x 4" at 7" centers (.50 BF per SF) | | | | | |
| Per SF of roof area | C8@.030 | SF | .61 | 1.37 | 1.98 |
| Particleboard, 4' x 8', used as underlayment, including 5% cutting waste. | | | | | |
| 3/8" x 4' x 8' sheets | C8@.011 | SF | .69 | .50 | 1.19 |
| 1/2" x 4' x 8' sheets | C8@.011 | SF | .73 | .50 | 1.23 |
| 5/8" x 4' x 8' sheets | C8@.011 | SF | .88 | .50 | 1.38 |
| 3/4" x 4' x 8' sheets | C8@.011 | SF | .99 | .50 | 1.49 |
| Add for attaching to metal frame | C8@.005 | SF | — | .23 | .23 |
| 15 pound felt on walls ($27.60 per 432 SF roll) | C8@.003 | SF | .07 | .14 | .21 |
| Super Jumbo Tex asphalt saturated Kraft (162 SF roll) | C8@.004 | SF | .11 | .18 | .29 |

| | Craft@Hrs | Unit | Material | Labor | Equipment | Total |
|---|---|---|---|---|---|---|
| **Wood Trusses** Material costs are for prefabricated wood trusses shipped to job site fully assembled ready for installation. Equipment is a 6,000 lb capacity fork lift. Labor includes unloading and handling and installing the trusses not over 20' above grade. | | | | | | |
| Truss joists, TJI Type at 50 PSF load design, at 16" OC | | | | | | |
| 9-1/2" TJI/15 | C8@.017 | SF | 1.75 | .78 | .26 | 2.79 |
| 11-7/8" TJI/15 | C8@.017 | SF | 1.92 | .78 | .26 | 2.96 |
| 14" TJI/35 | C8@.018 | SF | 2.71 | .82 | .27 | 3.80 |
| 16" TJI/35 | C8@.018 | SF | 2.95 | .82 | .27 | 4.04 |
| Add for job under 3,000 SF | C8@.004 | SF | .47 | .18 | .06 | .71 |
| Roof trusses, 24" OC, any slope from 3 in 12 to 12 in 12, total height not to exceed 12' high from bottom chord to highest point on truss. Prices for trusses over 12' high will be up to 100% higher. Square foot (SF) costs, where shown, are per square foot of roof area to be covered. | | | | | | |
| Scissor truss, 2" x 4" top and bottom chords | | | | | | |
| Up to 38' span | F5@.022 | SF | 2.15 | 1.03 | .13 | 3.31 |
| 40' to 50' span | F5@.028 | SF | 2.70 | 1.32 | .17 | 4.19 |

# 06 Wood and Composites

| | Craft@Hrs | Unit | Material | Labor | Equipment | Total |
|---|---|---|---|---|---|---|
| **Fink truss "W" (conventional roof truss), 2" x 4" top and bottom chords** | | | | | | |
| Up to 38' span | F5@.017 | SF | 1.87 | .80 | .10 | 2.77 |
| 40' to 50' span | F5@.022 | SF | 2.20 | 1.03 | .13 | 3.36 |
| **Fink truss with 2" x 6" top and bottom chords** | | | | | | |
| Up to 38' span | F5@.020 | SF | 2.24 | .94 | .12 | 3.30 |
| 40' to 50' span | F5@.026 | SF | 2.71 | 1.22 | .16 | 4.09 |
| **Fink truss with gable fill at 16" OC** | | | | | | |
| 28' span, 5 in 12 slope | F5@.958 | Ea | 153.00 | 45.00 | 5.84 | 203.84 |
| 32' span, 5 in 12 slope | F5@1.26 | Ea | 189.00 | 59.20 | 7.68 | 255.88 |
| 40' span, 5 in 12 slope | F5@1.73 | Ea | 270.00 | 81.30 | 10.54 | 361.84 |

**Laminated Glued Beams (Glu-lam beams)** Typical costs for horizontal glu-lam beams. Figures in parentheses show the approximate weight per linear foot of beam. Equipment cost is based on using a 4,000 lb capacity forklift with 30' lift.

| | Craft@Hrs | Unit | Material | Labor | Equipment | Total |
|---|---|---|---|---|---|---|
| Steel connectors and shoes for glu-lam beams, including bolts, | | | | | | |
| Typical per end or splice connection | — | Ea | 96.40 | — | — | 96.40 |

Drill holes in glu-lam beams and attach connectors at grade before installing beam, based on number of holes in the connector, with work performed at job site

| | Craft@Hrs | Unit | Material | Labor | Equipment | Total |
|---|---|---|---|---|---|---|
| Labor per each hole | F6@.236 | Ea | — | 11.20 | — | 11.20 |
| Install beams with connections attached | F6@.150 | LF | — | 7.15 | .85 | 8.00 |

West Coast Douglas fir, 24F-V4, H=240 PSI, exterior glue, camber is 3,500' R for beams. Industrial grade, without wrap. For other grades see the adjustments that follow.

| | Craft@Hrs | Unit | Material | Labor | Equipment | Total |
|---|---|---|---|---|---|---|
| 3-1/8" x 6" (4.4 pounds per LF) | — | LF | 4.88 | — | — | 4.88 |
| 3-1/8" x 9" (6.6 pounds per LF) | — | LF | 7.67 | — | — | 7.67 |
| 3-1/8" x 10-1/2" (7.5 pounds per LF) | — | LF | 8.34 | — | — | 8.34 |
| 3-1/8" x 12" (8.9 pounds per LF) | — | LF | 9.55 | — | — | 9.55 |
| 3-1/8" x 15" (11.1 pounds per LF) | — | LF | 11.90 | — | — | 11.90 |
| 3-1/8" x 18" (13.2 pounds per LF) | — | LF | 14.10 | — | — | 14.10 |
| 5-1/8" x 6" (7.3 pounds per LF) | — | LF | 7.00 | — | — | 7.00 |
| 5-1/8" x 9" (10.7 pounds per LF) | — | LF | 10.50 | — | — | 10.50 |
| 5-1/8" x 12" (14.5 pounds per LF) | — | LF | 14.00 | — | — | 14.00 |
| 5-1/8" x 15" (18.1 pounds per LF) | — | LF | 17.50 | — | — | 17.50 |
| 5-1/8" x 18" (21.7 pounds per LF) | — | LF | 21.00 | — | — | 21.00 |
| 5-1/8" x 21" (25.4 pounds per LF) | — | LF | 24.50 | — | — | 24.50 |
| 5-1/8" x 24" (29.0 pounds per LF) | — | LF | 27.90 | — | — | 27.90 |
| 6-3/4" x 12" (19.1 pounds per LF) | — | LF | 18.60 | — | — | 18.60 |
| 6-3/4" x 15" (23.8 pounds per LF) | — | LF | 23.30 | — | — | 23.30 |
| 6-3/4" x 18" (28.6 pounds per LF) | — | LF | 28.10 | — | — | 28.10 |
| 6-3/4" x 21" (33.3 pounds per LF) | — | LF | 32.60 | — | — | 32.60 |
| 6-3/4" x 24" (38.1 pounds per LF) | — | LF | 37.40 | — | — | 37.40 |

| Add to any of the beam costs shown in this section for: | Craft@Hrs | Unit | Material | Labor | Equipment | Total |
|---|---|---|---|---|---|---|
| Industrial grade, clean, wrapped, add | — | % | 12.5 | — | — | — |
| Architectural grade, wrapped, add | — | % | 20.0 | — | — | — |
| One-coat rez-latex stain, add | — | % | 10.0 | — | — | — |
| Transportation to job site, per delivery, maximum 50 miles one way, add | — | LS | 325.00 | — | — | 325.00 |

## 06 Wood and Composites

| | Craft@Hrs | Unit | Material | Labor | Total |
|---|---|---|---|---|---|

**Stairs** Job-built stairways, 3 stringers cut from 2" x 12" material, treads and risers from either 3/4" CDX plywood or 3/4" OSB sheathing.

Material costs based on:

| | Craft@Hrs | Unit | Material | Labor | Total |
|---|---|---|---|---|---|
| 2" x 12" | — | MBF | 576.00 | — | 576.00 |
| 1" x 6" | — | MSF | 1,870.00 | — | 1,870.00 |
| 3/4" CDX plywood | — | MSF | 746.00 | — | 746.00 |
| 3/4" OSB sheathing | — | MSF | 544.00 | — | 544.00 |
| Plywood sheathing cost per 7-1/2" rise, 36" wide | | | | | |
|   Straight run, 8'0" to 10'0" rise, per riser | C8@.530 | Ea | 12.00 | 24.30 | 36.30 |
|   "L"- or "U"-shape, per riser, add for landings | C8@.625 | Ea | 13.30 | 28.60 | 41.90 |
| OSB sheathing cost per 7-1/2" rise, 36" wide | | | | | |
|   Straight run, 8'0" to 10'0" rise, per riser | C8@.530 | Ea | 10.30 | 24.30 | 34.60 |
|   "L"- or "U"-shape, per riser, add for landings | C8@.625 | Ea | 11.60 | 28.60 | 40.20 |
| Landings, framed and 3/4" CDX plywood surface | | | | | |
|   Per SF of landing surface | C8@.270 | SF | 2.40 | 12.40 | 14.80 |
| Landings, framed and 3/4" OSB sheathing surface | | | | | |
|   Per SF of landing surface | C8@.270 | SF | 2.04 | 12.40 | 14.44 |
| Factory cut and assembled straight closed box stairs, unfinished, 36" width | | | | | |
|   Oak treads with 7-1/2" risers, price per riser | | | | | |
|     3'0" to 3'6" wide, pre riser | C8@.324 | Ea | 135.00 | 14.80 | 149.80 |
|     4'0" wide, per riser | C8@.357 | Ea | 163.00 | 16.40 | 179.40 |
|   Add for prefinished assembled stair rail with | | | | | |
|   balusters and newel, per riser | C8@.257 | Ea | 59.00 | 11.80 | 70.80 |
|   Add for prefinished handrail, brackets and balusters | C8@.135 | LF | 18.90 | 6.19 | 25.09 |
| Spiral stairs, 5' diameter. Steel, tubular handrail, composition board treads, prefabricated, shipped unassembled | | | | | |
|   86" to 95" with 10 risers | C8@9.98 | Ea | 1,100.00 | 457.00 | 1,557.00 |
|   95" to 105" with 11 risers | C8@10.7 | Ea | 1,220.00 | 490.00 | 1,710.00 |
|   105" to 114" with 12 risers | C8@11.3 | Ea | 1,270.00 | 518.00 | 1,788.00 |
|   114" to 124" with 13 risers | C8@12.0 | Ea | 1,400.00 | 550.00 | 1,950.00 |
|   124" to 133" with 14 risers | C8@12.7 | Ea | 1,610.00 | 582.00 | 2,192.00 |
|   Add for 6' diameter | — | % | 24.0 | — | — |
|   Add for oak treads | — | % | 25.0 | — | — |
|   Add for oak handrail | — | % | 45.0 | — | — |

## 07 Thermal and Moisture Protection

**Dampproofing**

| | Craft@Hrs | Unit | Material | Labor | Total |
|---|---|---|---|---|---|
| Asphalt wall primer, per coat, gallon | | | | | |
|   covers 250 SF | CL@.010 | SF | .03 | .40 | .43 |
| Asphalt emulsion, wall, per coat, gallon | | | | | |
|   covers 33 SF, brush on | CL@.013 | SF | .23 | .52 | .75 |
| Hot mop concrete wall, 2 coats and glass fabric | CL@.033 | SF | .38 | 1.32 | 1.70 |
| Hot mop deck, 3 ply, felt, typical subcontract price | — | Sq | — | — | 167.00 |
| Hot mop deck, 4 ply, felt, typical subcontract price | — | Sq | — | — | 214.00 |
|   Add for 1/2" asphalt fiberboard, subcontract price | — | Sq | — | — | 112.00 |
| Bituthene waterproofing membrane, | | | | | |
|   60 mil plain surface | CL@.025 | SF | .88 | 1.00 | 1.88 |
| Waterproof baths in high rise | CL@.025 | SF | 1.87 | 1.00 | 2.87 |
| Iron compound, internal coating, 2 coats | CL@.012 | SF | 1.74 | .48 | 2.22 |
| Roof vapor barrier, .004" polyethylene | CL@.002 | SF | .08 | .08 | .16 |

# 07 Thermal and Moisture Protection

| | Craft@Hrs | Unit | Material | Labor | Total |
|---|---|---|---|---|---|
| **Thermal Insulation** | | | | | |
| Fiberglass batts, placed between or over framing members | | | | | |
| 3-1/2" Kraft faced, R-11, between ceiling joists | A1@.006 | SF | .32 | .30 | .62 |
| 3-1/2" Kraft faced, R-11, on suspended ceiling, working off ladders from below | A1@.018 | SF | .32 | .89 | 1.21 |
| 3-1/2" Kraft faced, R-11, in crawl space | A1@.020 | SF | .32 | .99 | 1.31 |
| 3-1/2" Kraft faced, R-13, between ceiling joists | A1@.006 | SF | .28 | .30 | .58 |
| 3-1/2" Kraft faced, R-13, on suspended ceiling, working off ladders from below | A1@.018 | SF | .28 | .89 | 1.17 |
| 3-1/2" Kraft faced, R-13, in crawl space | A1@.020 | SF | .28 | .99 | 1.27 |
| 6-1/4" Kraft faced, R-19, between ceiling joists | A1@.006 | SF | .46 | .30 | .76 |
| 6-1/4" Kraft faced, R-19, on suspended ceiling, working off ladders from below | A1@.018 | SF | .46 | .89 | 1.35 |
| 6-1/4" Kraft faced, R-19, in crawl space | A1@.020 | SF | .50 | .99 | 1.49 |
| 6-1/4" Kraft faced, R-19, between studs | A1@.006 | SF | .50 | .30 | .80 |
| 8-1/4" Kraft faced, R-21, between ceiling joists | A1@.006 | SF | .70 | .30 | 1.00 |
| 8-1/4" Kraft faced, R-21, on suspended ceiling, working off ladders from below | A1@.018 | SF | .70 | .89 | 1.59 |
| 8-1/4" Kraft faced, R-21, in crawl space | A1@.020 | SF | .70 | .99 | 1.69 |
| 8-1/4" Kraft faced, R-21, between studs | A1@.006 | SF | .70 | .30 | 1.00 |
| 10-1/4" Kraft faced, R-30, between ceiling joists | A1@.006 | SF | .84 | .30 | 1.14 |
| 10-1/4" Kraft faced, R-30, on suspended ceiling, working off ladders from below | A1@.018 | SF | .84 | .89 | 1.73 |
| 10-1/4" Kraft faced, R-30, in crawl space | A1@.020 | SF | .84 | .99 | 1.83 |
| 10-1/4" Kraft faced, R-30, between studs | A1@.006 | SF | .84 | .30 | 1.14 |
| 12-1/4" Kraft faced, R-38, between ceiling joists | A1@.006 | SF | .97 | .30 | 1.27 |
| 12-1/4" Kraft faced, R-38, on suspended ceiling, working off ladders from below | A1@.018 | SF | .97 | .89 | 1.86 |
| 12-1/4" Kraft faced, R-38, in crawl space | A1@.020 | SF | .97 | .99 | 1.96 |
| 12-1/4" Kraft faced, R-38, between studs | A1@.006 | SF | .97 | .30 | 1.27 |
| Add for supporting batts on wire rods, 1 per SF | A1@.001 | SF | .04 | .05 | .09 |
| Add for foil one side | — | SF | .04 | — | .04 |
| Add for foil two sides | — | SF | .06 | — | .06 |
| Deduct for unfaced one side | — | SF | - .03 | — | - .03 |
| Add for Kraft paper two sides | — | SF | .03 | — | .03 |
| Fasteners for batts placed under a ceiling deck, including flat washers | | | | | |
| 2-1/2" long | A1@.025 | Ea | .22 | 1.24 | 1.46 |
| 4-1/2" long | A1@.025 | Ea | .25 | 1.24 | 1.49 |
| 6-1/2" long | A1@.025 | Ea | .36 | 1.24 | 1.60 |
| Domed self-locking washers | A1@.005 | Ea | .18 | .25 | .43 |
| Blown fiberglass or mineral wool, over ceiling joists. Add blowing equipment cost at $300 per day | | | | | |
| R-11, 5" | A1@.008 | SF | .21 | .40 | .61 |
| R-13, 6" | A1@.010 | SF | .25 | .49 | .74 |
| R-19, 9" | A1@.014 | SF | .35 | .69 | 1.04 |
| Insulation board on walls, 4' x 8' Foamular® 150 polystyrene | | | | | |
| 1/2" thick (R-2.5), $9.49 per 4' x 8' panel | C8@.010 | SF | .33 | .46 | .79 |
| 3/4" thick (R-3.8), $11.70 per 4' x 8' panel | C8@.011 | SF | .40 | .50 | .90 |
| 1" thick (R-5.0), $16.70 per 4' x 8' panel | C8@.011 | SF | .58 | .50 | 1.08 |
| 1-1/2" thick (R-7.5), $16.90 per 4' x 8' panel | C8@.015 | SF | .59 | .69 | 1.28 |
| 2" thick (R-10), $24.20 per 4' x 8' panel | C8@.015 | SF | .83 | .69 | 1.52 |

# 07 Thermal and Moisture Protection

| | Craft@Hrs | Unit | Material | Labor | Total |
|---|---|---|---|---|---|
| **Expanded polystyrene board installed on perimeter walls** | | | | | |
| 3/4", R-2.9 | C8@.010 | SF | .25 | .46 | .71 |
| 1", R-4.20 | C8@.011 | SF | .40 | .50 | .90 |
| 1", R-5.9 | C8@.011 | SF | .47 | .50 | .97 |
| 1-1/2", R-5.9 | C8@.015 | SF | .39 | .69 | 1.08 |
| 1-3/8", R-9.1 | C8@.015 | SF | .88 | .69 | 1.57 |
| 2", R-8 | C8@.015 | SF | .41 | .69 | 1.10 |
| 2", R-13 | C8@.015 | SF | 1.03 | .69 | 1.72 |
| **Polyisocyanurate foam core sheathing, Super Tuff-R®, foil face, painted back, 4' x 8' panels on perimeter walls** | | | | | |
| 1/2", R-3.3 | C8@.010 | SF | .38 | .46 | .84 |
| 3/4", R-5.5 | C8@.010 | SF | .41 | .46 | .87 |
| 1", R-6.5 | C8@.011 | SF | .58 | .50 | 1.08 |
| 1-3/8", R-9.1 | C8@.015 | SF | .88 | .69 | 1.57 |
| 1-1/2", R-9.8 | C8@.015 | SF | .78 | .69 | 1.47 |
| 2", R-13.0 commercial, | C8@.015 | SF | 1.02 | .69 | 1.71 |
| 2", R-13.0 | C8@.015 | SF | 1.03 | .69 | 1.72 |
| **Perlite or vermiculite, poured in concrete block cores** | | | | | |
| Perlite only | — | CF | 2.69 | — | 2.69 |
| 4" wall, 8.1 SF per CF | M1@.004 | SF | .35 | .19 | .54 |
| 6" wall, 5.4 SF per CF | M1@.007 | SF | .51 | .33 | .84 |
| 8" wall, 3.6 SF per CF | M1@.009 | SF | .77 | .42 | 1.19 |
| 10" wall, 3.0 SF per CF | M1@.011 | SF | .93 | .51 | 1.44 |
| 12" wall, 2.1 SF per CF | M1@.016 | SF | 1.33 | .75 | 2.08 |
| **Poured perlite or vermiculite in wall cavities** | M1@.032 | CF | 2.66 | 1.49 | 4.15 |
| **Sound board** | | | | | |
| 1/2", on walls | C8@.010 | SF | .34 | .46 | .80 |
| 1/2", on floors | C8@.009 | SF | .34 | .41 | .75 |
| **Cold box insulation** | | | | | |
| 2" polystyrene | A1@.012 | SF | .39 | .59 | .98 |
| 1" cork | A1@.013 | SF | 1.18 | .64 | 1.82 |
| Add for overhead work or poor access areas | — | SF | — | .08 | .08 |
| Add for enclosed areas | — | SF | — | .08 | .08 |
| Add for 1 hour fire rating | — | SF | .24 | — | .24 |

**Roof Insulation** Roof insulation. Local regulation can affect the cost and availability of urethanes, isocyanurates, phenolics and polystyrene roof insulation board.

| | Craft@Hrs | Unit | Material | Labor | Total |
|---|---|---|---|---|---|
| **Fiberglass board roof insulation** | | | | | |
| 3/4", R-2.80, C-0.36 | R3@.570 | Sq | 51.00 | 25.90 | 76.90 |
| 1", R-4.20, C-0.24 | R3@.636 | Sq | 60.20 | 28.90 | 89.10 |
| 1-3/8", R-5.30, C-0.19 | R3@.684 | Sq | 77.90 | 31.10 | 109.00 |
| 1-5/8", R-6.70, C-0.15 | R3@.706 | Sq | 88.20 | 32.10 | 120.30 |
| 2-1/4", R-8.30, C-0.12 | R3@.797 | Sq | 108.00 | 36.20 | 144.20 |
| **Polystyrene rigid foam roof insulation. R-19 rating when used with batt insulation. Including 5% waste.** | | | | | |
| 1/2" 4' x 8' panels | R3@1.10 | Sq | 25.60 | 50.00 | 75.60 |
| 3/4" 4' x 8' panels | R3@1.10 | Sq | 30.10 | 50.00 | 80.10 |
| 1" 4' x 8' panels | R3@1.10 | Sq | 46.90 | 50.00 | 96.90 |
| 1-1/2" 4' x 8' panels | R3@1.10 | Sq | 52.70 | 50.00 | 102.70 |
| 2" 4' x 8' panels | R3@1.10 | Sq | 70.60 | 50.00 | 120.60 |

| | Craft@Hrs | Unit | Material | Labor | Total |
|---|---|---|---|---|---|
| Phenolic board roof insulation | | | | | |
| 1-1/4", R-10.0, C-0.10 | R3@.500 | Sq | 64.40 | 22.70 | 87.10 |
| 1-1/2", R-12.5, C-0.08 | R3@.592 | Sq | 72.20 | 26.90 | 99.10 |
| 1-3/4", R-14.6, C-0.07 | R3@.592 | Sq | 78.60 | 26.90 | 105.50 |
| 2", R-16.7, C-0.06 | R3@.684 | Sq | 90.50 | 31.10 | 121.60 |
| 2-1/2", R-20.0, C-0.05 | R3@.728 | Sq | 99.30 | 33.10 | 132.40 |
| 3", R-25.0, C-0.04 | R3@.889 | Sq | 131.00 | 40.40 | 171.40 |
| 3-1/4", R-28.0, C-0.035 | R3@1.05 | Sq | 169.00 | 47.70 | 216.70 |
| Expanded polystyrene board roof insulation | | | | | |
| 3/4", R-5.30, C-0.19 | R3@.478 | Sq | 30.10 | 21.70 | 51.80 |
| 1", R-6.70, C-0.15 | R3@.478 | Sq | 36.00 | 21.70 | 57.70 |
| 1-1/2", R-11.1, C-0.09 | R3@.570 | Sq | 52.70 | 25.90 | 78.60 |
| 2", R-14.3, C-0.07 | R3@.684 | Sq | 88.40 | 31.10 | 119.50 |
| 2-1/4", R-16.7, C-0.06 | R3@.684 | Sq | 93.10 | 31.10 | 124.20 |
| Perlite board roof insulation | | | | | |
| 1", R-2.80, C-0.36 | R3@.639 | Sq | 54.30 | 29.00 | 83.30 |
| 1-1/2", R-4.20, C-0.24 | R3@.728 | Sq | 79.70 | 33.10 | 112.80 |
| 2", R-5.30, C-0.19 | R3@.820 | Sq | 97.80 | 37.20 | 135.00 |
| 2-1/2", R-6.70, C-0.15 | R3@.956 | Sq | 122.00 | 43.40 | 165.40 |
| 3", R-8.30, C-0.12 | R3@1.03 | Sq | 148.00 | 46.80 | 194.80 |
| 4", R-10.0, C-0.10 | R3@1.23 | Sq | 196.00 | 55.90 | 251.90 |
| 5-1/4", R-14.3, C-0.07 | R3@1.57 | Sq | 260.00 | 71.30 | 331.30 |
| Cellular Foamglass (isocyanurates) board insulation, on roofs | | | | | |
| 1-1/2" | R3@.775 | Sq | 198.00 | 35.20 | 233.20 |
| Tapered 3/16" per foot | R3@.775 | Sq | 181.00 | 35.20 | 216.20 |
| Polystyrene board roof insulation | | | | | |
| 1", R-4.20, C-0.24, foil one side | R3@.545 | Sq | 39.70 | 24.80 | 64.50 |
| 1-1/2", R-5.9, C-0.16 | R3@.639 | Sq | 33.30 | 29.00 | 62.30 |
| 2", R-8.30, C-0.12 | R3@.797 | Sq | 40.40 | 36.20 | 76.60 |
| Fasteners for board type insulation on roofs | | | | | |
| Typical 2-1/2" to 3-1/2" screws with 3" metal disks | | | | | |
| I-60 | R3@.053 | Sq | 3.38 | 2.41 | 5.79 |
| I-90 | R3@.080 | Sq | 5.11 | 3.63 | 8.74 |
| Add for plastic disks | — | Sq | .85 | — | .85 |

**Exterior Wall Insulation and Finish System (EIFS)** Polymer based exterior non-structural wall finish applied to concrete, masonry, stucco or exterior grade gypsum sheathing (such as Dryvit). System consists of a plaster base coat approximately 1/4" thick, and a glass fiber mesh embedded in the adhesive base coat. The base coat is applied over polystyrene insulation boards bonded to an existing flat vertical wall surface with adhesive. Costs shown are per SF of finished wall surface. These costs do not include surface preparation or wall costs. Add the cost of scaffolding if required. Add 50% to the labor costs when work is done from scaffolding.
For scheduling purposes, estimate that a crew of three can field apply 400 SF of exterior wall insulation and finish systems in an 8-hour day.

| | Craft@Hrs | Unit | Material | Labor | Total |
|---|---|---|---|---|---|
| Adhesive mixture, 2 coats | | | | | |
| 2.1 SF per SF of wall area including waste | F8@.013 | SF | .60 | .66 | 1.26 |
| Glass fiber mesh | | | | | |
| 1.1 SF per SF of wall area including waste | F8@.006 | SF | .30 | .30 | .60 |
| Insulation board, 1" thick | | | | | |
| 1.05 SF per SF of wall area including waste | F8@.013 | SF | .36 | .66 | 1.02 |
| Textured finish coat, with integral color | | | | | |
| 1.05 SF per SF of wall area including waste | F8@.024 | SF | .55 | 1.21 | 1.76 |
| Total with 1" thick insulation | F8@.056 | SF | 1.92 | 2.83 | 4.75 |
| Add for 2" thick insulation board | — | SF | .32 | — | .32 |
| Add for 3" thick insulation board | — | SF | .61 | — | .61 |
| Add for 4" thick insulation board | — | SF | .93 | — | .93 |

| | Craft@Hrs | Unit | Material | Labor | Total |
|---|---|---|---|---|---|

**Aluminum Wall Cladding System, Exterior** These costs assume installation by a manufacturer-trained contractor, on a framed and insulated exterior building surface, including track system, panels and trim. Alply, Inc.
Techwall solid aluminum panel, 1/8" thick, joints sealed with silicon sealer

| | Craft@Hrs | Unit | Material | Labor | Total |
|---|---|---|---|---|---|
| Kynar 500 paint finish | SM@.158 | SF | 19.50 | 9.24 | 28.74 |
| Anodized aluminum finish | SM@.158 | SF | 19.20 | 9.24 | 28.44 |
| Add for radius panels | SM@.032 | SF | 3.96 | 1.87 | 5.83 |

**Cladding, Preformed Roofing and Siding** Applied on metal framing.
Architectural insulated metal panels with smooth .040 aluminum exterior face with Kynar 500 finish, 2" thick isocyanurate insulation, .040 mill finish aluminum interior sheet, shop applied extrusion and dry seal gasketed joint system, shop fabricated. Alply, Inc.

| | Craft@Hrs | Unit | Material | Labor | Total |
|---|---|---|---|---|---|
| Kynar 500 paint finish | SM@.136 | SF | 14.00 | 7.95 | 21.95 |
| Add for radius panels | SM@.020 | SF | 1.99 | 1.17 | 3.16 |
| Corrugated or ribbed roofing, colored galvanized steel, 9/16" deep, uninsulated | | | | | |
| 18 gauge | SM@.026 | SF | 2.41 | 1.52 | 3.93 |
| 20 gauge | SM@.026 | SF | 2.11 | 1.52 | 3.63 |
| 22 gauge | SM@.026 | SF | 1.88 | 1.52 | 3.40 |
| 24 gauge | SM@.026 | SF | 1.60 | 1.52 | 3.12 |
| Corrugated siding, colored galvanized steel, 9/16" deep, uninsulated | | | | | |
| 18 gauge | SM@.034 | SF | 2.41 | 1.99 | 4.40 |
| 20 gauge | SM@.034 | SF | 2.11 | 1.99 | 4.10 |
| 22 gauge | SM@.034 | SF | 1.88 | 1.99 | 3.87 |
| Ribbed siding, colored galvanized steel, 1-3/4" deep, box rib | | | | | |
| 18 gauge | SM@.034 | SF | 2.54 | 1.99 | 4.53 |
| 20 gauge | SM@.034 | SF | 2.23 | 1.99 | 4.22 |
| 22 gauge | SM@.034 | SF | 1.96 | 1.99 | 3.95 |
| Corrugated aluminum roofing, natural finish | | | | | |
| .020" aluminum | SM@.026 | SF | 1.56 | 1.52 | 3.08 |
| .032" aluminum | SM@.026 | SF | 1.92 | 1.52 | 3.44 |
| Add for factory painted finish | — | SF | .20 | — | .20 |
| Corrugated aluminum siding, natural finish | | | | | |
| .020" aluminum | SM@.034 | SF | 1.56 | 1.99 | 3.55 |
| .032" aluminum | SM@.034 | SF | 1.92 | 1.99 | 3.91 |
| Add for factory painted finish | — | SF | .20 | — | .20 |
| Vinyl siding | | | | | |
| Solid vinyl, .042", horizontal siding | SM@.026 | SF | .72 | 1.52 | 2.24 |
| Add for insulated siding backer panel | SM@.001 | SF | .34 | .06 | .40 |
| Vinyl window and door trim | SM@.040 | LF | .40 | 2.34 | 2.74 |
| Vinyl siding and fascia system | SM@.032 | SF | .96 | 1.87 | 2.83 |
| Laminated sandwich panel siding, enameled 22 gauge aluminum face one side, polystyrene core | | | | | |
| 1" core | SM@.069 | SF | 4.98 | 4.03 | 9.01 |
| 1-1/2" core | SM@.069 | SF | 5.56 | 4.03 | 9.59 |

**Membrane Roofing** (One square [Sq] = 100 square feet)
Built-up roofing, asphalt felt

| | Craft@Hrs | Unit | Material | Labor | Total |
|---|---|---|---|---|---|
| 3-ply, smooth surface top sheet | R3@1.61 | Sq | 56.00 | 73.10 | 129.10 |
| 4-ply, base sheet, 3-ply felt, smooth surface top sheet | R3@1.85 | Sq | 83.40 | 84.00 | 167.40 |
| 5-ply, 20 year | R3@2.09 | Sq | 95.50 | 94.90 | 190.40 |
| Built-up fiberglass felt roof | | | | | |
| Base and 3-ply felt | R3@1.93 | Sq | 120.00 | 87.70 | 207.70 |
| Add for each additional ply | R3@.188 | Sq | 5.78 | 8.54 | 14.32 |
| Add for light rock dress off (250 lbs per Sq) | R3@.448 | Sq | 22.70 | 20.30 | 43.00 |
| Add for heavy rock dress off (400 lbs per Sq) | R3@.718 | Sq | 36.00 | 32.60 | 68.60 |

|  | Craft@Hrs | Unit | Material | Labor | Total |
|---|---|---|---|---|---|
| Remove and replace rock (400 lbs per Sq) | | | | | |
| Including new flood coat | R3@1.32 | Sq | 42.40 | 60.00 | 102.40 |
| Aluminized coating for built-up roofing | R3@.599 | Sq | 13.40 | 27.20 | 40.60 |
| Fire rated type | R3@.609 | Sq | 14.10 | 27.70 | 41.80 |
| Modified APP, SBS flashing membrane | R3@.033 | SF | 1.61 | 1.50 | 3.11 |
| Cap sheet | R3@.277 | Sq | 43.50 | 12.60 | 56.10 |
| Modified asphalt, APP, | | | | | |
| SBS base sheet and membrane | R3@1.97 | Sq | 162.00 | 89.50 | 251.50 |
| Roll roofing | | | | | |
| 90 lb mineral surface | R3@.399 | Sq | 34.40 | 18.10 | 52.50 |
| 19" wide selvage edge mineral surface (110 lb) | R3@.622 | Sq | 137.00 | 28.30 | 165.30 |
| Double coverage selvage edge roll (140 lb) | R3@.703 | Sq | 148.00 | 31.90 | 179.90 |
| Asphalt impregnated walkway for built-up roofing | | | | | |
| 1/2" | R3@.016 | SF | 4.00 | .73 | 4.73 |
| 3/4" | R3@.017 | SF | 5.05 | .77 | 5.82 |
| 1" | R3@.020 | SF | 6.12 | .91 | 7.03 |
| Strip off existing 4-ply roof, no disposal included | R3@1.32 | Sq | — | 60.00 | 60.00 |

**Elastomeric Roofing**

|  | Craft@Hrs | Unit | Material | Labor | Total |
|---|---|---|---|---|---|
| Butyl | | | | | |
| 1/16" | R3@1.93 | Sq | 146.00 | 87.70 | 233.70 |
| 1/32" | R3@1.93 | Sq | 125.00 | 87.70 | 212.70 |
| Neoprene, 1/16" | R3@2.23 | Sq | 217.00 | 101.00 | 318.00 |
| Acrylic-urethane foam roof system | | | | | |
| Remove gravel and prepare existing built-up roof | R3@1.26 | Sq | 5.56 | 57.20 | 62.76 |
| Apply base and top coats of acrylic (.030") | R3@.750 | Sq | 120.00 | 34.10 | 154.10 |
| Urethane foam 1" thick (R-7.1) | R3@.280 | Sq | 113.00 | 12.70 | 125.70 |
| Urethane foam 2" thick (R-14.3) | R3@1.10 | Sq | 190.00 | 50.00 | 240.00 |
| Spray-on mineral granules, 40 lbs per CSF | R3@.600 | Sq | 17.30 | 27.30 | 44.60 |
| EPDM roofing system (Ethylene Propylene Diene Monomer) | | | | | |
| .045" loose laid membrane | R3@.743 | Sq | 99.00 | 33.70 | 132.70 |
| .060" adhered membrane (excluding adhesive) | R3@1.05 | Sq | 113.00 | 47.70 | 160.70 |
| Bonding adhesive | R3@.305 | Sq | 46.00 | 13.90 | 59.90 |
| Lap splice cement, per 100 LF | R3@.223 | CLF | 27.50 | 10.10 | 37.60 |
| Ballast rock, 3/4" to 1-1/2" gravel | R3@.520 | Sq | 22.50 | 23.60 | 46.10 |
| Neoprene sheet flashing, .060" | R3@.021 | SF | 2.46 | .95 | 3.41 |
| CSPE (Chloro-sulfonated Polyethylene) | | | | | |
| .045" single ply membrane | R3@.743 | Sq | 157.00 | 33.70 | 190.70 |
| Add for mechanical fasteners and washers | | | | | |
| 6" long, 12" OC at laps | — | Ea | .77 | — | .77 |
| 12" long, 12" OC at laps | — | Ea | 1.55 | — | 1.55 |

**Cementitious Roofing System** Tectum™ roof deck systems for use on open beam ceilings with exposed joists. Provides a nailable top surface, a core of Styrofoam brand insulation and an exposed interior Tectum™ ceiling plank substrate. Tectum™ roof deck is highly moisture resistant and possesses high compressive strength, excellent sound absorption, insulative properties and a finished interior. Costs shown include nails, screws, and adhesive. Includes normal cutting waste.
Tectum I with T&G edges and square ends. Standard roof deck with foam insulation. Available in lengths up to 14' and widths to 3'

|  | Craft@Hrs | Unit | Material | Labor | Total |
|---|---|---|---|---|---|
| 1-1/2" thick | C8@.015 | SF | 2.06 | .69 | 2.75 |
| 2" thick | C8@.017 | SF | 2.40 | .78 | 3.18 |
| 2-1/2" thick | C8@.019 | SF | 2.95 | .87 | 3.82 |
| 3" thick | C8@.021 | SF | 3.56 | .96 | 4.52 |

|  | Craft@Hrs | Unit | Material | Labor | Total |
|---|---|---|---|---|---|

Tectum III with T&G edges and square ends. Composite roof deck used where greater insulative properties are required. Tectum III provides a finished interior surface, and foam insulation. Width is 47" and lengths are available from 4" to 16"

|  | Craft@Hrs | Unit | Material | Labor | Total |
|---|---|---|---|---|---|
| 3-1/2" thick | C8@.015 | SF | 4.53 | .69 | 5.22 |
| 4" thick | C8@.021 | SF | 5.13 | .96 | 6.09 |
| 5" thick | C8@.027 | SF | 5.58 | 1.24 | 6.82 |
| 6" thick | C8@.033 | SF | 6.42 | 1.51 | 7.93 |
| 7" thick | C8@.039 | SF | 7.37 | 1.79 | 9.16 |
| 8" thick | C8@.045 | SF | 8.21 | 2.06 | 10.27 |
| 9" thick | C8@.051 | SF | 9.06 | 2.34 | 11.40 |
| 10" thick | C8@.057 | SF | 11.00 | 2.61 | 13.61 |
| 11" thick | C8@.063 | SF | 12.00 | 2.89 | 14.89 |

**Cant Strips**

|  | Craft@Hrs | Unit | Material | Labor | Total |
|---|---|---|---|---|---|
| 3" fiber | R3@.017 | LF | .14 | .77 | .91 |
| 6" fiber | R3@.017 | LF | .28 | .77 | 1.05 |
| 4" wood | R3@.017 | LF | .70 | .77 | 1.47 |
| 6" wood | R3@.017 | LF | .70 | .77 | 1.47 |

**Flashing and Sheet Metal**

Prefinished sheet metal fascia and mansards, standing beam and batten

|  | Craft@Hrs | Unit | Material | Labor | Total |
|---|---|---|---|---|---|
| Straight or simple | SM@.054 | SF | 3.96 | 3.16 | 7.12 |
| Curved or complex | SM@.082 | SF | 5.97 | 4.79 | 10.76 |
| Coping, gravel stop and flashing | SM@.037 | SF | 2.89 | 2.16 | 5.05 |
| Sheet metal wainscot, galvanized | SM@.017 | SF | .80 | .99 | 1.79 |

Aluminum flashing .032",

|  | Craft@Hrs | Unit | Material | Labor | Total |
|---|---|---|---|---|---|
| Coping and wall cap, 16" girth | SM@.069 | LF | 3.28 | 4.03 | 7.31 |
| Counter flash, 8' | SM@.054 | LF | 2.89 | 3.16 | 6.05 |
| Reglet flashing, 8" | SM@.058 | LF | 1.19 | 3.39 | 4.58 |
| Neoprene gasket for flashing | SM@.015 | LF | 1.00 | .88 | 1.88 |
| Gravel stop and fascia, 10" | SM@.084 | LF | 3.44 | 4.91 | 8.35 |
| Side wall flashing, 9" | SM@.027 | LF | 1.80 | 1.58 | 3.38 |
| Gravel stop |  |  |  |  |  |
| 4" | SM@.058 | LF | 3.90 | 3.39 | 7.29 |
| 6" | SM@.063 | LF | 4.43 | 3.68 | 8.11 |
| 8" | SM@.068 | LF | 5.36 | 3.98 | 9.34 |
| 12" | SM@.075 | LF | 8.88 | 4.39 | 13.27 |
| Valley, 24" | SM@.070 | LF | 2.29 | 4.09 | 6.38 |

Copper flashing 16 oz,

|  | Craft@Hrs | Unit | Material | Labor | Total |
|---|---|---|---|---|---|
| Roll 8" x 15' | SW@.750 | Ea | 165.00 | 30.80 | 195.80 |
| Roof edging, 2" x 2" x 10' | SW@.350 | Ea | 90.80 | 14.40 | 105.20 |
| Gravel stop |  |  |  |  |  |
| 4" | SM@.058 | LF | 18.70 | 3.39 | 22.09 |
| 6" | SM@.063 | LF | 21.70 | 3.68 | 25.38 |
| 8" | SM@.068 | LF | 25.30 | 3.98 | 29.28 |
| 10" | SM@.075 | LF | 31.10 | 4.39 | 35.49 |
| Reglet flashing, 8" | SM@.054 | LF | 8.66 | 3.16 | 11.82 |
| Neoprene gasket for flashing | SM@.015 | LF | 1.00 | .88 | 1.88 |
| Side wall flashing, 6" | SM@.027 | LF | 8.22 | 1.58 | 9.80 |
| Base, 20 oz | SM@.072 | LF | 16.90 | 4.21 | 21.11 |
| Valley, 24" | SM@.069 | LF | 15.20 | 4.03 | 19.23 |

# 07 Thermal and Moisture Protection

| | Craft@Hrs | Unit | Material | Labor | Total |
|---|---|---|---|---|---|
| **Sheet lead flashing** | | | | | |
| Plumbing vents, soil stacks, 4 lbs per square foot | | | | | |
| 4" pipe size | P6@.709 | Ea | 31.80 | 35.60 | 67.40 |
| 6" pipe size | P6@.709 | Ea | 36.80 | 35.60 | 72.40 |
| 8" pipe size | P6@.709 | Ea | 45.10 | 35.60 | 80.70 |
| Roof drains, flat pan type, 36" x 36" | | | | | |
| 2.5 lbs/SF material | P6@1.49 | Ea | 38.70 | 74.80 | 113.50 |
| 4 lbs/SF material | P6@1.49 | Ea | 65.30 | 74.80 | 140.10 |
| 6 lbs/SF material | P6@1.49 | Ea | 92.90 | 74.80 | 167.70 |
| **Stainless steel flashing** | | | | | |
| Fascia roof edge, .018" | SM@.054 | LF | 4.50 | 3.16 | 7.66 |
| Base, .018" | SM@.054 | LF | 5.39 | 3.16 | 8.55 |
| Counter flash, .015" | SM@.054 | LF | 4.24 | 3.16 | 7.40 |
| Valley, .015", 24" | SM@.069 | LF | 4.22 | 4.03 | 8.25 |
| Reglets, .020", 8" | SM@.035 | LF | 3.67 | 2.05 | 5.72 |
| **Galvanized sheet metal flashing** | | | | | |
| Cap and counter flash, 8" | SM@.046 | LF | 1.31 | 2.69 | 4.00 |
| Coping and wall cap, 16" girth | SM@.069 | LF | 1.70 | 4.03 | 5.73 |
| Gravel stop, 6" | SM@.060 | LF | .51 | 3.51 | 4.02 |
| Neoprene gasket | SM@.005 | LF | 1.00 | .29 | 1.29 |
| Pitch pockets, 24 gauge, filled | | | | | |
| 4" x 4" | SM@.584 | Ea | 19.60 | 34.10 | 53.70 |
| 6" x 6" | SM@.584 | Ea | 24.70 | 34.10 | 58.80 |
| 8" x 8" | SM@.584 | Ea | 29.10 | 34.10 | 63.20 |
| 8" x 10" | SM@.777 | Ea | 46.00 | 45.40 | 91.40 |
| 8" x 12" | SM@.777 | Ea | 40.50 | 45.40 | 85.90 |
| Reglet flashing, 8" | SM@.054 | LF | 1.26 | 3.16 | 4.42 |
| Shingles (chimney flash) | RF@.064 | LF | .99 | 3.08 | 4.07 |
| Side wall flashing, 9" | SM@.027 | LF | .96 | 1.58 | 2.54 |
| "W" valley, 24" | SM@.069 | LF | 1.93 | 4.03 | 5.96 |
| Plumbers counter flash cone, 4" diameter | P8@.204 | Ea | 6.14 | 9.18 | 15.32 |
| **Roof safes and caps** | | | | | |
| 4" galvanized sheet metal | SM@.277 | Ea | 12.40 | 16.20 | 28.60 |
| 4" aluminum | SM@.277 | Ea | 14.10 | 16.20 | 30.30 |
| **Pitch pockets, 16 oz copper, filled** | | | | | |
| 4" x 4" | SM@.547 | Ea | 117.00 | 32.00 | 149.00 |
| 6" x 6" | SM@.547 | Ea | 148.00 | 32.00 | 180.00 |
| 8" x 8" | SM@.547 | Ea | 175.00 | 32.00 | 207.00 |
| 8" x 10" | SM@.679 | Ea | 205.00 | 39.70 | 244.70 |
| 8" x 12" | SM@.679 | Ea | 241.00 | 39.70 | 280.70 |
| Galvanized corner guards, 4" x 4" | SM@.039 | LF | 12.30 | 2.28 | 14.58 |
| **Copper sheet metal roofing, 16 oz** | | | | | |
| Batten seam | SM@.065 | SF | 12.70 | 3.80 | 16.50 |
| Standing seam | SM@.056 | SF | 14.70 | 3.27 | 17.97 |
| **Gutters and Downspouts** | | | | | |
| Aluminum, .032", including hangers | | | | | |
| Fascia gutter, 5" | SM@.050 | LF | 1.00 | 2.92 | 3.92 |
| Box gutter, 4" | SM@.050 | LF | .94 | 2.92 | 3.86 |
| Dropouts, elbows, for either of above | SM@.101 | Ea | 3.34 | 5.91 | 9.25 |

| | Craft@Hrs | Unit | Material | Labor | Total |
|---|---|---|---|---|---|
| **Downspouts, to 24' height** | | | | | |
| 2" x 3" | SM@.037 | LF | .83 | 2.16 | 2.99 |
| 3" x 4" | SM@.047 | LF | 1.00 | 2.75 | 3.75 |
| 4" diameter, round, 12' to 24' high | SM@.047 | LF | 1.29 | 2.75 | 4.04 |
| Add for height over 24' | SM@.017 | LF | — | .99 | .99 |
| **Copper, 16 oz, including hangers** | | | | | |
| Box gutter, 4" | SM@.054 | LF | 20.00 | 3.16 | 23.16 |
| Half round gutter | | | | | |
| 4" gutter | SM@.039 | LF | 17.20 | 2.28 | 19.48 |
| 5" gutter | SM@.051 | LF | 21.70 | 2.98 | 24.68 |
| 6" gutter | SM@.056 | LF | 26.90 | 3.27 | 30.17 |
| Dropouts, elbows, | | | | | |
| for either of above | SM@.154 | Ea | 43.40 | 9.00 | 52.40 |
| **Downspouts, to 24' height** | | | | | |
| 2" x 3" | SM@.036 | LF | 18.90 | 2.10 | 21.00 |
| 3" x 4" | SM@.047 | LF | 25.80 | 2.75 | 28.55 |
| 3" diameter | SM@.037 | LF | 12.80 | 2.16 | 14.96 |
| 4" diameter | SM@.047 | LF | 19.30 | 2.75 | 22.05 |
| 5" diameter | SM@.049 | LF | 23.50 | 2.87 | 26.37 |
| Add for heights over 24' | SM@.017 | LF | — | .99 | .99 |
| Scuppers, copper, 16 oz, | | | | | |
| 8" x 8" | SM@.972 | Ea | 493.00 | 56.80 | 549.80 |
| 10" x 10" | SM@.972 | Ea | 598.00 | 56.80 | 654.80 |
| **Galvanized steel, 26 gauge, including hangers** | | | | | |
| Box gutter 4" | SM@.054 | LF | 1.41 | 3.16 | 4.57 |
| Dropouts, elbows | SM@.082 | Ea | 2.17 | 4.79 | 6.96 |
| Fascia gutter | | | | | |
| 5" face | SM@.054 | LF | 1.68 | 3.16 | 4.84 |
| 7" face | SM@.062 | LF | 2.24 | 3.63 | 5.87 |
| Dropouts, elbows, for either of above | SM@.088 | Ea | 3.30 | 5.15 | 8.45 |
| Downspouts, with all fittings, height to 24' | | | | | |
| 2" x 3" | SM@.036 | LF | 1.20 | 2.10 | 3.30 |
| 3" x 4" | SM@.047 | LF | 1.81 | 2.75 | 4.56 |
| 3" round | SM@.037 | LF | 1.81 | 2.16 | 3.97 |
| 6" round | SM@.055 | LF | 1.70 | 3.22 | 4.92 |
| Add for heights over 24' | SM@.017 | LF | — | .99 | .99 |
| Roof sump, 18" x 18" x 5" | SM@.779 | Ea | 9.23 | 45.50 | 54.73 |
| Scupper, 6" x 6" x 8" | SM@.591 | Ea | 27.60 | 34.60 | 62.20 |
| **Stainless steel, .015" thick, including hangers** | | | | | |
| Box gutter, 5" wide, 4-1/2" high | SM@.054 | LF | 8.69 | 3.16 | 11.85 |
| Dropouts, elbows | SM@.076 | Ea | 12.70 | 4.44 | 17.14 |
| Downspout, 4" x 5", to 24' high | SM@.051 | LF | 7.78 | 2.98 | 10.76 |
| Add for heights over 24' | SM@.017 | LF | — | .99 | .99 |
| **Vents, Louvers and Screens** | | | | | |
| Fixed louvers, with screen, typical | SM@.402 | SF | 11.90 | 23.50 | 35.40 |
| Door louvers, typical | SM@.875 | Ea | 25.60 | 51.20 | 76.80 |
| Manual operating louvers, typical | SM@.187 | SF | 18.90 | 10.90 | 29.80 |
| Cooling tower screens | SM@.168 | SF | 10.60 | 9.82 | 20.42 |
| Bird screens with frame | SM@.028 | SF | 1.58 | 1.64 | 3.22 |

# 07 Thermal and Moisture Protection

| | Craft@Hrs | Unit | Material | Labor | Total |
|---|---|---|---|---|---|
| Concrete block vents, aluminum | M1@.370 | SF | 14.00 | 17.20 | 31.20 |
| Frieze vents, with screen, 14" x 4" | C8@.198 | Ea | 2.91 | 9.07 | 11.98 |
| Foundation vents, 6" x 14", ornamental | C8@.317 | Ea | 7.86 | 14.50 | 22.36 |
| Attic vents, with louvers, 14" x 24" | C8@.377 | Ea | 22.90 | 17.30 | 40.20 |
| Architectural facade screen, aluminum | SM@.187 | SF | 21.90 | 10.90 | 32.80 |
|    Add for enamel or light anodized | — | SF | 2.34 | — | 2.34 |
|    Add for porcelain or heavy anodized | — | SF | 5.67 | — | 5.67 |

**Rain Dispersing Gutters**  Rustproof aluminum, installed on mounting brackets. Disperses rain into a 4 foot area. No downspouts needed. Resists clogging by leaves and debris. Rainhandler™

| | Craft@Hrs | Unit | Material | Labor | Total |
|---|---|---|---|---|---|
| Rain disperser | RF@.100 | LF | 4.87 | 4.82 | 9.69 |
| Mounting brackets (use one each 18") | — | Ea | 1.09 | — | 1.09 |
| Horizontal adapter for roof edges that extend outward more than 2-1/2" from building (use one each 18") | RF@.020 | Ea | 2.17 | .96 | 3.13 |
| Vertical adapter for vertical fascias, under 3" from building (use one each 5') | RF@.020 | Ea | 2.17 | .96 | 3.13 |
| "Doorbrella" for installation above doors and windows | RF@.250 | Ea | 16.70 | 12.00 | 28.70 |
| Drip edge extension for roof edges that extend less than 1-1/2" from vertical fascia (use one each 18") | RF@.020 | Ea | 4.39 | .96 | 5.35 |
| Roof valley rain diverter | RF@.050 | Ea | 6.63 | 2.41 | 9.04 |
| Fascia adapter for angled fascia boards | RF@.050 | Ea | 2.17 | 2.41 | 4.58 |

## Roof Accessories

| | Craft@Hrs | Unit | Material | Labor | Total |
|---|---|---|---|---|---|
| Roof hatches, steel, not including ladder | | | | | |
|    30" x 30" | C8@3.09 | Ea | 466.00 | 142.00 | 608.00 |
|    30" x 72" | C8@4.69 | Ea | 1,120.00 | 215.00 | 1,335.00 |
|    36" x 36" | C8@3.09 | Ea | 533.00 | 142.00 | 675.00 |
|    36" x 72" | C8@4.69 | Ea | 1,190.00 | 215.00 | 1,405.00 |
| Ceiling access hatches | | | | | |
|    30" x 30" | C8@1.59 | Ea | 281.00 | 72.90 | 353.90 |
|    42" x 42" | C8@1.89 | Ea | 336.00 | 86.60 | 422.60 |
| Smoke vents | | | | | |
|    Aluminum, 48" x 48" | C8@3.09 | Ea | 1,330.00 | 142.00 | 1,472.00 |
|    Galvanized, 48" x 48" | C8@3.09 | Ea | 1,170.00 | 142.00 | 1,312.00 |
|    Fusible "shrink-out" heat and smoke vent (PVC dome in aluminum frame) | | | | | |
|       4' x 8' | C8@5.15 | Ea | 1,070.00 | 236.00 | 1,306.00 |
| Roof scuttle, aluminum, 2'6" x 3'0" | C8@3.50 | Ea | 525.00 | 160.00 | 685.00 |

## Ventilators

| | Craft@Hrs | Unit | Material | Labor | Total |
|---|---|---|---|---|---|
| Rotary, wind driven | | | | | |
|    6" diameter | SM@.913 | Ea | 70.20 | 53.40 | 123.60 |
|    12" diameter | SM@.913 | Ea | 85.60 | 53.40 | 139.00 |
|    24" diameter | SM@1.36 | Ea | 243.00 | 79.50 | 322.50 |
| Ventilators, mushroom type, motorized, single speed, including damper and bird screen but no electrical work | | | | | |
|    8", 180 CFM | SM@2.93 | Ea | 334.00 | 171.00 | 505.00 |
|    12", 1,360 CFM | SM@2.93 | Ea | 553.00 | 171.00 | 724.00 |
|    18", 2,000 CFM | SM@4.00 | Ea | 792.00 | 234.00 | 1,026.00 |
|    24", 4,000 CFM | SM@4.86 | Ea | 1,220.00 | 284.00 | 1,504.00 |
|    Add for two-speed motor | — | % | 35.0 | — | — |
|    Add for explosive proof units | — | Ea | 457.00 | — | 457.00 |

| | Craft@Hrs | Unit | Material | Labor | Total |
|---|---|---|---|---|---|

**Windstorm Certified Roof Turbine Fans** Complies with ASHRAE/American National Standards Institute specification 62.2-2007 and Air Conditioning Contractors of America standard ACCA 5QI-2007, "HVAC Indoor Air Quality Installation Specification 2007." U.S. Green Building Council and Underwriters Laboratory approved. Texas Department of Insurance report # RV-12. Air Vent, Empire, Lomanco or equal.

| | Craft@Hrs | Unit | Material | Labor | Total |
|---|---|---|---|---|---|
| 12" dia., galvanized steel | R1@2.50 | Ea | 244.00 | 88.60 | 332.60 |
| 14" dia., galvanized steel | R1@4.50 | Ea | 381.00 | 159.00 | 540.00 |
| 12" dia., high-impact polypropylene | R1@7.00 | Ea | 518.00 | 248.00 | 766.00 |
| 14" dia., high-impact polypropylene | R1@7.50 | Ea | 684.00 | 266.00 | 950.00 |
| Add for duct hangers | R1@.250 | Ea | 3.42 | 8.86 | 12.28 |
| Add for sheet metal duct | R1@.350 | LF | 5.42 | 12.40 | 17.82 |
| Add for edge flashing sealant | R1@.150 | LF | 2.35 | 5.31 | 7.66 |

**Cast-in-Place Concrete Fireproofing** Costs per linear foot (LF) of steel beam, column or girder. Use these figures to estimate the cubic feet (CF) of concrete per linear foot and square feet (SF) of form required per linear foot for each type of beam, column or girder. Quantities include concrete required to fill the void between the web and flange and provide 2" protection at the flanges. A small change in concrete thickness at flanges has very little effect on cost. Use $7,500.00 as a minimum job cost for work of this type. Concrete thickness for fireproofing will usually be:

Members at least 6" x 6" but less than 8" x 8" — 3" for a 4 hour rating, 2" for a 3 hour rating, 1-1/2" for a 2 hour rating, and 1" for a 1 hour rating.

Members at least 8" x 8" but less than 12" x 12" — 2-1/2" for a 4 hour rating, 2" for a 3 hour rating, 1" for a 1 or 2 hour rating.

Members 12" x 12" or greater — 2" for a 4 hour rating and 1" for a 1, 2, or 3 hour rating.

Labor costs include the time needed to prepare formwork sketches at the job site, measure for the forms, fabricate, erect, align and brace the forms, cut, bend, place and tie the reinforcing steel, install embedded steel items, place and finish the concrete and strip, clean and stack the forms. Labor costs assume a concrete forming, placing and finishing crew of 1 carpenter, 1 laborer and 1 cement finisher. Concrete placing equipment assumes the use of a concrete pump.

For scheduling purposes, estimate that a crew of 3 will fireproof the following quantities of steel in an 8-hour day:

15 LF of W21 through W36 members,

25 LF of W8 through W18 members, 22 LF of S12 through S24 members, and 42 LF of M4 through W6 or S6 through S8 members.

Material costs assume reinforcing steel is #3 and #4 bars at 12" on center each way and includes waste and laps. Form cost assumes 3 uses without completely disassembling the form. Normal cleaning and repairs are included. No salvage value is assumed. Cost for forms, before allowance for waste, is based on Standard & Better lumber and 3/4" plyform.

| | Craft@Hrs | Unit | Material | Labor | Equipment | Total |
|---|---|---|---|---|---|---|

**Concrete and accessories for fireproofing**

| | Craft@Hrs | Unit | Material | Labor | Equipment | Total |
|---|---|---|---|---|---|---|
| Concrete, 3,000 PSI pump mix with 5% allowance for waste including pump cost | P9@.600 | CY | 128.00 | 28.40 | 9.13 | 165.53 |
| Grade A60 reinforcing bars, set and tied, with 50 lbs per CY of concrete | P9@.790 | CY | 22.50 | 37.30 | — | 59.80 |
| Hangers, snap-ties, misc. embedded steel with 10 lb per CY of concrete | P9@.660 | CY | 52.30 | 31.20 | — | 83.50 |
| Concrete test cylinders including test reports, with 5 per 100 CY | — | CY | .90 | — | — | .90 |
| Total for concrete and accessories | P9@2.05 | CY | 203.70 | 96.90 | 9.13 | 309.73 |

**Forms for fireproofing** These costs assume the following materials are used per square foot of contact area (SFCA): Nails, clamps and form oil costing $.25 per board foot of lumber, 1.2 SF of 3/4" plyform and 3.7 BF of lumber per SFCA.

| | Craft@Hrs | Unit | Material | Labor | Equipment | Total |
|---|---|---|---|---|---|---|
| Make, erect, align and strip forms, 3 uses | P9@.150 | SF | 3.07 | 7.09 | — | 10.16 |
| Sack and patch concrete SF | P9@.010 | SF | .09 | .47 | — | .56 |
| Total for forms and finishing, per use | P9@.160 | SF | 3.16 | 7.56 | — | 10.72 |

|  | Craft@Hrs | Unit | Material | Labor | Equipment | Total |
|---|---|---|---|---|---|---|

**Sample fireproofing estimate** for ten W24 x 55 beams 20' long. Note from the table below that this beam requires 2.7 cubic feet of concrete per linear foot and 6.9 square feet of form per linear foot. Using labor and material prices from above to estimate 200 linear feet of beam. (Total costs have been rounded)

|  | Craft@Hrs | Unit | Material | Labor | Equipment | Total |
|---|---|---|---|---|---|---|
| Concrete (200 × 2.7 = 540 CF or 20 CY) | P9@41.0 | LS | 4,074.00 | 1,940.00 | 304.00 | 6,318.00 |
| Form and finish (200 × 6.9 = 1,380 SF) | P9@221 | LS | 4,360.00 | 10,400.00 | — | 14,760.00 |
| Total job cost shown above | P9@262 | LS | 8,434.00 | 12,340.00 | 304.00 | 21,078.00 |
| Cost per linear foot of beam | P9@1.31 | LF | 42.20 | 61.90 | 1.52 | 105.62 |

**Cast-in-place concrete fireproofing**  Cost per LF, including concrete and forming.

| Beam / CF and SF per LF | Craft@Hrs | Unit | Material | Labor | Equipment | Total |
|---|---|---|---|---|---|---|
| W36 x 135 to W36 x 300 |  |  |  |  |  |  |
| 5.13 CF and 10.07 SF per LF | P9@2.00 | LF | 70.50 | 94.50 | 1.73 | 166.73 |
| W33 x 118 to W30 x 241 |  |  |  |  |  |  |
| 4.59 CF and 9.35 SF per LF | P9@1.84 | LF | 64.20 | 87.00 | 1.55 | 152.75 |
| W30 x 99 to W30 x 211 |  |  |  |  |  |  |
| 3.78 CF and 8.4 SF per LF | P9@1.63 | LF | 55.10 | 77.10 | 1.28 | 133.48 |
| W27 x 84 to W27 x 178 |  |  |  |  |  |  |
| 3.51 CF and 8.19 SF per LF | P9@1.57 | LF | 52.40 | 74.20 | 1.19 | 127.79 |
| W24 x 55 to W24 x 162 |  |  |  |  |  |  |
| 2.7 CF and 6.9 SF per LF | P9@1.31 | LF | 42.20 | 61.90 | .91 | 105.01 |
| W21 x 44 to W21 x 147 |  |  |  |  |  |  |
| 2.43 CF and 6.75 SF per LF | P9@1.26 | LF | 39.70 | 59.60 | .82 | 100.12 |
| W18 x 37 to W18 x 71 |  |  |  |  |  |  |
| 1.62 CF and 5.34 SF per LF | P9@.974 | LF | 29.10 | 46.00 | .55 | 75.65 |
| W16 x 26 to W16 x 100 |  |  |  |  |  |  |
| 1.62 CF and 5.4 SF per LF | P9@.987 | LF | 29.30 | 46.70 | .55 | 76.55 |
| W14 x 90 to W14 x 730 |  |  |  |  |  |  |
| 2.97 CF and 7.15 SF per LF | P9@1.37 | LF | 45.00 | 64.80 | 1.00 | 110.80 |
| W14 x 22 to W14 x 82 |  |  |  |  |  |  |
| 1.35 CF and 4.7 SF per LF | P9@.854 | LF | 25.00 | 40.40 | .46 | 65.86 |
| W12 x 65 to W12 x 336 |  |  |  |  |  |  |
| 2.16 CF and 6.08 SF per LF | P9@1.13 | LF | 35.50 | 53.40 | .73 | 89.63 |
| W12 x 40 to W12 x 58 |  |  |  |  |  |  |
| 1.62 CF and 5.4 SF per LF | P9@.987 | LF | 29.30 | 46.70 | .55 | 76.55 |
| W12 x 14 to W12 x 35 |  |  |  |  |  |  |
| 1.08 CF and 4.44 SF per LF | P9@.790 | LF | 22.20 | 37.30 | .37 | 59.87 |
| W10 x 33 to W10 x 112 |  |  |  |  |  |  |
| 1.35 CF and 4.75 SF per LF | P9@.862 | LF | 25.20 | 40.70 | .46 | 66.36 |
| W10 x 12 to W10 x 30 |  |  |  |  |  |  |
| 1.08 CF and 4.56 SF per LF | P9@.808 | LF | 22.60 | 38.20 | .37 | 61.17 |
| W8 x 24 to W8 x 67 |  |  |  |  |  |  |
| 1.08 CF and 4.4 SF per LF | P9@.784 | LF | 22.10 | 37.10 | .37 | 59.57 |
| W8 x 10 to W8 x 21 |  |  |  |  |  |  |
| 0.81 CF and 3.87 SF per LF | P9@.678 | LF | 18.30 | 32.00 | .27 | 50.57 |
| W6 x 15 to W6 x 25 |  |  |  |  |  |  |
| 0.81 CF and 3.84 SF per LF | P9@.672 | LF | 18.30 | 31.80 | .27 | 50.37 |
| W6 x 9 to W6 x 16 |  |  |  |  |  |  |
| 0.54 CF and 2.92 SF per LF | P9@.507 | LF | 13.30 | 24.00 | .18 | 37.48 |
| W5 x 16 to W5 x 19 |  |  |  |  |  |  |
| 0.54 CF and 2.86 SF per LF | P9@.499 | LF | 13.10 | 23.60 | .18 | 36.88 |
| W4 x 13 |  |  |  |  |  |  |
| 0.54 CF and 3.2 SF per LF | P9@.552 | LF | 14.20 | 26.10 | .18 | 40.48 |

| | Craft@Hrs | Unit | Material | Labor | Equipment | Total |
|---|---|---|---|---|---|---|
| M4 x 13 to M14 x 18 | | | | | | |
|    0.54 CF and 2.9 SF per LF | P9@.504 | LF | 13.20 | 23.80 | .18 | 37.18 |
| S24 x 80 to S24 x 121 | | | | | | |
|    2.43 CF and 7.11 SF per LF | P9@1.32 | LF | 40.80 | 62.40 | .82 | 104.02 |
| S20 x 66 to S20 x 96 | | | | | | |
|    1.89 CF and 6.09 SF per LF | P9@1.12 | LF | 33.50 | 52.90 | .64 | 87.04 |
| S15 x 43 to S18 x 70 | | | | | | |
|    1.62 CF and 5.82 SF per LF | P9@1.05 | LF | 30.60 | 49.60 | .55 | 80.75 |
| S12 x 35 to S12 x 50 | | | | | | |
|    1.08 CF and 4.44 SF per LF | P9@.790 | LF | 22.20 | 37.30 | .37 | 59.87 |
| S8 x 23 to S10 x 35 | | | | | | |
|    0.81 CF and 3.81 SF per LF | P9@.670 | LF | 18.20 | 31.70 | .27 | 50.17 |
| S6 x 13 to S7 x 20 | | | | | | |
|    0.54 CF and 2.91 SF per LF | P9@.504 | LF | 13.30 | 23.80 | .18 | 37.28 |

**Spray-Applied Fireproofing, Subcontract** Fire endurance coating made from inorganic vermiculite and portland cement. Costs assume a 10,000 board foot job. (One BF is one square foot covered 1" thick.) For smaller jobs, increase the cost by 5% for each 500 BF less than 10,000 BF. Use $2,000 as a minimum subcontract price. For thickness other than 1", adjust these costs proportionately. For scheduling purposes, estimate that a crew of 2 plasterers and 1 helper can apply 200 to 250 board feet per hour.

| | Craft@Hrs | Unit | Material | Labor | Total |
|---|---|---|---|---|---|
| Structural steel columns | — | BF | — | — | 2.05 |
| Structural steel beams | — | BF | — | — | 1.83 |
| Purlins, girts, and miscellaneous members | — | BF | — | — | 1.69 |
| Decks, ceilings or walls | — | BF | — | — | 1.55 |
| Add for 18 gauge 2" hex mesh reinforcing | — | BF | — | — | .36 |
| Add for key coat bonded on primed surfaces | — | SF | — | — | .42 |

Rule-of-thumb method for estimating spray-applied fireproofing on bare structural steel by member size:

**Beams and Columns**

| | Beams | Cost LF | Columns | Cost LF |
|---|---|---|---|---|
| W36 x 135 to 300 | 10 BF/LF | 18.30 | 11 BF/LF | 22.60 |
| W33 x 118 to 241 | 9 BF/LF | 16.50 | 10 BF/LF | 20.50 |
| W30 x 99 to 211 | 8 BF/LF | 14.60 | 9 BF/LF | 18.50 |
| W27 x 84 to 178 | 7 BF/LF | 12.80 | 8 BF/LF | 16.40 |
| W24 x 55 to 182 | 6.5 BF/LF | 11.90 | 7.5 BF/LF | 15.40 |
| W21 x 44 to 147 | 6 BF/LF | 11.00 | 7 BF/LF | 14.40 |
| W18 x 35 to 118 | 5 BF/LF | 9.15 | 6 BF/LF | 12.30 |
| W16 x 28 to 57 | 3.5 BF/LF | 6.41 | 4.5 BF/LF | 9.23 |
| W14 x 61 to 132 | 5.5 BF/LF | 10.10 | 6.5 BF/LF | 13.30 |
| W14 x 22 to 63 | 4 BF/LF | 7.32 | 5 BF/LF | 10.30 |
| W12 x 65 to 190 | 5 BF/LF | 9.15 | 6 BF/LF | 12.30 |
| W12 x 40 to 58 | 4 BF/LF | 7.32 | 5 BF/LF | 10.30 |
| W12 x 14 to 35 | 3 BF/LF | 5.49 | 4 BF/LF | 8.20 |
| W10 x 49 to 112 | 4 BF/LF | 7.32 | 5 BF/LF | 10.30 |
| W10 x 22 to 45 | 3 BF/LF | 5.49 | 4 BF/LF | 8.20 |
| W10 x 12 to 19 | 2 BF/LF | 3.66 | 3 BF/LF | 6.15 |
| W8 x 24 to 67 | 3 BF/LF | 5.49 | 4 BF/LF | 8.20 |
| W6 x 9 to 25 | 2 BF/LF | 3.66 | 3 BF/LF | 6.15 |
| W5 x 16 to 19 | 2 BF/LF | 3.66 | 2.5 BF/LF | 5.13 |
| W4 x 13 | 1 BF/LF | 1.83 | 2 BF/LF | 4.10 |

**Purlins and Girts**

|  | Purlins or Girts | Cost LF |
|---|---|---|
| MC 18 members | 4 BF/LF | 6.76 |
| MC 10 to MC 13 members | 3 BF/LF | 5.07 |
| MC 8 members | 2 BF/LF | 3.38 |
| C 10 to C 15 members | 3 BF/LF | 5.07 |
| C 7 to C 9 members | 2 BF/LF | 3.38 |

**Sealants and Caulking** By joint size. Costs assume 10% waste. Twelve 10.5 ounce cartridges equal one gallon and yield 231 cubic inches of caulk or sealant. Productivity shown assumes one roofer using a bulk dispenser. Add the cost of joint cleaning, if required.

| | Craft@Hrs | Unit | Material | Labor | Total |
|---|---|---|---|---|---|
| Silicone latex caulk | | | | | |
| 1/4" x 1/4", 351 LF/gallon, 63 LF per hour | RF@.016 | LF | .10 | .77 | .87 |
| 1/4" x 3/8", 234 LF/gallon, 60 LF per hour | RF@.017 | LF | .15 | .82 | .97 |
| 1/4" x 1/2", 176 LF/gallon, 58 LF per hour | RF@.017 | LF | .20 | .82 | 1.02 |
| 3/8" x 3/8", 156 LF/gallon, 58 LF per hour | RF@.017 | LF | .23 | .82 | 1.05 |
| 3/8" x 1/2", 117 LF/gallon, 57 LF per hour | RF@.017 | LF | .30 | .82 | 1.12 |
| 3/8" x 5/8", 94 LF/gallon, 56 LF per hour | RF@.018 | LF | .38 | .87 | 1.25 |
| 3/8" x 3/4", 76 LF/gallon, 52 LF per hour | RF@.019 | LF | .47 | .92 | 1.39 |
| 1/2" x 1/2", 87 LF/gallon, 56 LF per hour | RF@.018 | LF | .41 | .87 | 1.28 |
| 1/2" x 5/8", 70 LF/gallon, 50 LF per hour | RF@.020 | LF | .51 | .96 | 1.47 |
| 1/2" x 3/4", 58 LF/gallon, 48 LF per hour | RF@.021 | LF | .62 | 1.01 | 1.63 |
| 1/2" x 7/8", 51 LF/gallon, 46 LF per hour | RF@.022 | LF | .70 | 1.06 | 1.76 |
| 1/2" x 1", 44 LF/gallon, 45 LF per hour | RF@.022 | LF | .81 | 1.06 | 1.87 |
| 3/4" x 3/4", 39 LF/gallon, 44 LF per hour | RF@.023 | LF | .91 | 1.11 | 2.02 |
| 1" x 1", 22 LF/gallon, 30 LF per hour | RF@.033 | LF | 1.62 | 1.59 | 3.21 |
| Acrylic latex plus silicone caulk | | | | | |
| 1/4" x 1/4", 351 LF/gallon, 63 LF per hour | RF@.016 | LF | .09 | .77 | .86 |
| 1/4" x 3/8", 234 LF/gallon, 60 LF per hour | RF@.017 | LF | .14 | .82 | .96 |
| 1/4" x 1/2", 176 LF/gallon, 58 LF per hour | RF@.017 | LF | .18 | .82 | 1.00 |
| 3/8" x 3/8", 156 LF/gallon, 58 LF per hour | RF@.017 | LF | .21 | .82 | 1.03 |
| 3/8" x 1/2", 117 LF/gallon, 57 LF per hour | RF@.017 | LF | .28 | .82 | 1.10 |
| 3/8" x 5/8", 94 LF/gallon, 56 LF per hour | RF@.018 | LF | .35 | .87 | 1.22 |
| 3/8" x 3/4", 76 LF/gallon, 52 LF per hour | RF@.019 | LF | .44 | .92 | 1.36 |
| 1/2" x 1/2", 87 LF/gallon, 56 LF per hour | RF@.018 | LF | .38 | .87 | 1.25 |
| 1/2" x 5/8", 70 LF/gallon, 50 LF per hour | RF@.020 | LF | .47 | .96 | 1.43 |
| 1/2" x 3/4", 58 LF/gallon, 48 LF per hour | RF@.021 | LF | .57 | 1.01 | 1.58 |
| 1/2" x 7/8", 51 LF/gallon, 46 LF per hour | RF@.022 | LF | .65 | 1.06 | 1.71 |
| 1/2" x 1", 44 LF/gallon, 45 LF per hour | RF@.022 | LF | .75 | 1.06 | 1.81 |
| 3/4" x 3/4", 39 LF/gallon, 44 LF per hour | RF@.023 | LF | .84 | 1.11 | 1.95 |
| 1" x 1", 22 LF/gallon, 30 LF per hour | RF@.033 | LF | 1.49 | 1.59 | 3.08 |
| 100 percent silicone caulk | | | | | |
| 1/4" x 1/4", 351 LF/gallon, 63 LF per hour | RF@.016 | LF | .21 | .77 | .98 |
| 1/4" x 3/8", 234 LF/gallon, 60 LF per hour | RF@.017 | LF | .32 | .82 | 1.14 |
| 1/4" x 1/2", 176 LF/gallon, 58 LF per hour | RF@.017 | LF | .49 | .82 | 1.31 |
| 3/8" x 3/8", 156 LF/gallon, 58 LF per hour | RF@.017 | LF | .49 | .82 | 1.31 |
| 3/8" x 1/2", 117 LF/gallon, 57 LF per hour | RF@.017 | LF | .65 | .82 | 1.47 |
| 3/8" x 5/8", 94 LF/gallon, 56 LF per hour | RF@.018 | LF | .80 | .87 | 1.67 |
| 3/8" x 3/4", 76 LF/gallon, 52 LF per hour | RF@.019 | LF | .99 | .92 | 1.91 |
| 1/2" x 1/2", 87 LF/gallon, 56 LF per hour | RF@.018 | LF | .87 | .87 | 1.74 |
| 1/2" x 5/8", 70 LF/gallon, 50 LF per hour | RF@.020 | LF | 1.09 | .96 | 2.05 |
| 1/2" x 3/4", 58 LF/gallon, 48 LF per hour | RF@.021 | LF | 1.31 | 1.01 | 2.32 |
| 1/2" x 7/8", 51 LF/gallon, 46 LF per hour | RF@.022 | LF | 1.48 | 1.06 | 2.54 |

| | Craft@Hrs | Unit | Material | Labor | Total |
|---|---|---|---|---|---|
| 1/2" x 1", 44 LF/gallon, 45 LF per hour | RF@.022 | LF | 1.72 | 1.06 | 2.78 |
| 3/4" x 3/4", 39 LF/gallon, 44 LF per hour | RF@.023 | LF | 1.94 | 1.11 | 3.05 |
| 1" x 1", 22 LF/gallon, 30 LF per hour | RF@.033 | LF | 3.44 | 1.59 | 5.03 |
| Polyurethane sealant | | | | | |
| 1/4" x 1/4", 351 LF/gallon, 63 LF per hour | RF@.016 | LF | .24 | .77 | 1.01 |
| 1/4" x 3/8", 234 LF/gallon, 60 LF per hour | RF@.017 | LF | .37 | .82 | 1.19 |
| 1/4" x 1/2", 176 LF/gallon, 58 LF per hour | RF@.017 | LF | .49 | .82 | 1.31 |
| 3/8" x 3/8", 156 LF/gallon, 58 LF per hour | RF@.017 | LF | .56 | .82 | 1.38 |
| 3/8" x 1/2", 117 LF/gallon, 57 LF per hour | RF@.017 | LF | .74 | .82 | 1.56 |
| 3/8" x 5/8", 94 LF/gallon, 56 LF per hour | RF@.018 | LF | .92 | .87 | 1.79 |
| 3/8" x 3/4", 76 LF/gallon, 52 LF per hour | RF@.019 | LF | 1.14 | .92 | 2.06 |
| 1/2" x 1/2", 87 LF/gallon, 56 LF per hour | RF@.018 | LF | .99 | .87 | 1.86 |
| 1/2" x 5/8", 70 LF/gallon, 50 LF per hour | RF@.020 | LF | 1.24 | .96 | 2.20 |
| 1/2" x 3/4", 58 LF/gallon, 48 LF per hour | RF@.021 | LF | 1.49 | 1.01 | 2.50 |
| 1/2" x 7/8", 51 LF/gallon, 46 LF per hour | RF@.022 | LF | 1.69 | 1.06 | 2.75 |
| 1/2" x 1", 44 LF/gallon, 45 LF per hour | RF@.022 | LF | 1.97 | 1.06 | 3.03 |
| 3/4" x 3/4", 39 LF/gallon, 44 LF per hour | RF@.023 | LF | 2.22 | 1.11 | 3.33 |
| 1" x 1", 22 LF/gallon, 30 LF per hour | RF@.033 | LF | 3.93 | 1.59 | 5.52 |
| Backing rods for sealant | | | | | |
| Closed cell, 1/4" rod | RF@.010 | LF | .08 | .48 | .56 |
| Closed cell, 3/8" rod | RF@.011 | LF | .14 | .53 | .67 |
| Closed cell, 1/2" rod | RF@.011 | LF | .18 | .53 | .71 |
| Closed cell, 5/8" rod | RF@.012 | LF | .22 | .58 | .80 |
| Closed cell, 7/8" rod | RF@.012 | LF | .35 | .58 | .93 |
| Closed cell, 1" rod | RF@.014 | LF | .42 | .67 | 1.09 |
| Closed cell, 1-1/4" rod | RF@.014 | LF | .73 | .67 | 1.40 |
| Closed cell, 1-1/2" rod | RF@.015 | LF | 1.09 | .72 | 1.81 |
| Closed cell, 2" rod | RF@.015 | LF | 1.63 | .72 | 2.35 |

**Expansion Joints, Preformed**

| | Craft@Hrs | Unit | Material | Labor | Total |
|---|---|---|---|---|---|
| Wall and ceiling, aluminum cover | | | | | |
| Drywall or panel type | CC@.067 | LF | 13.30 | 3.47 | 16.77 |
| Plaster type | CC@.067 | LF | 13.30 | 3.47 | 16.77 |
| Floor to wall, 3" x 3" | CC@.038 | LF | 7.28 | 1.97 | 9.25 |
| Floor, 3" | CC@.038 | LF | 9.08 | 1.97 | 11.05 |
| Neoprene joint with aluminum cover | CC@.079 | LF | 19.40 | 4.09 | 23.49 |
| Wall to roof joint | CC@.094 | LF | 7.80 | 4.87 | 12.67 |
| Bellows type expansion joints, butyl with neoprene backer | | | | | |
| 16 oz copper, 4" wide | CC@.046 | LF | 13.60 | 2.38 | 15.98 |
| 28 gauge stainless steel, 4" wide | CC@.046 | LF | 14.00 | 2.38 | 16.38 |
| 26 gauge galvanized sheet metal, 4" wide | CC@.046 | LF | 8.24 | 2.38 | 10.62 |
| 16 oz copper, 6" wide | CC@.046 | LF | 15.70 | 2.38 | 18.08 |
| 28 gauge stainless steel, 6" wide | CC@.046 | LF | 16.20 | 2.38 | 18.58 |
| 26 gauge galvanized sheet metal, 6" wide | CC@.046 | LF | 9.99 | 2.38 | 12.37 |
| Neoprene gaskets, closed cell | | | | | |
| 1/8" x 2" | CC@.019 | LF | .74 | .98 | 1.72 |
| 1/8" x 6" | CC@.023 | LF | 1.75 | 1.19 | 2.94 |
| 1/4" x 2" | CC@.022 | LF | .86 | 1.14 | 2.00 |
| 1/4" x 6" | CC@.024 | LF | 1.84 | 1.24 | 3.08 |
| 1/2" x 6" | CC@.025 | LF | 2.75 | 1.29 | 4.04 |
| 1/2" x 8" | CC@.028 | LF | 4.03 | 1.45 | 5.48 |

# 07 Thermal and Moisture Protection

|  | Craft@Hrs | Unit | Material | Labor | Total |
|---|---|---|---|---|---|
| Acoustical caulking, drywall or plaster type | CC@.038 | LF | .30 | 1.97 | 2.27 |
| Polyisobutylene tapes, non-drying |  |  |  |  |  |
| Polybutene | CC@.028 | LF | .28 | 1.45 | 1.73 |
| Polyisobutylene/butyl, preformed | CC@.020 | LF | .15 | 1.04 | 1.19 |
| **Expansion Joint Covers** |  |  |  |  |  |
| Surface type, aluminum |  |  |  |  |  |
| Floor, 1-1/2" | CC@.194 | LF | 25.20 | 10.00 | 35.20 |
| Wall and ceiling, 1-1/2" | CC@.185 | LF | 14.20 | 9.58 | 23.78 |
| Gymnasium base | CC@.099 | LF | 11.60 | 5.13 | 16.73 |
| Roof, typical | CC@.145 | LF | 24.20 | 7.51 | 31.71 |

# 08 Openings

**Complete Hollow Metal Door Assembly** These figures show the costs normally associated with installing an exterior hollow core steel door.

| Hollow metal exterior 3' x 7' flush door, with frame, hardware and trim, complete | Craft@Hrs | Unit | Material | Labor | Total |
|---|---|---|---|---|---|
| Stock hollow metal flush door, |  |  |  |  |  |
| 16 gauge, 3' x 7' x 1-3/4", non-rated | C8@.721 | Ea | 398.00 | 33.00 | 431.00 |
| Closer plate reinforcing on door | — | Ea | 15.70 | — | 15.70 |
| Stock 18 gauge frame, 6" jamb, non-rated | C8@.944 | Ea | 121.00 | 43.30 | 164.30 |
| Closer plate on frame | — | Ea | 5.98 | — | 5.98 |
| Three hinges, 4-1/2" x 4-1/2" | C8@.578 | LS | 100.00 | 26.50 | 126.50 |
| Lockset, mortise type | C8@.949 | Ea | 265.00 | 43.50 | 308.50 |
| Saddle type threshold, aluminum, 3' | C8@.276 | Ea | 19.30 | 12.60 | 31.90 |
| Standard duty closer | C8@.787 | Ea | 115.00 | 36.10 | 151.10 |
| Weatherstripping, bronze and neoprene | C8@2.58 | LS | 51.60 | 118.00 | 169.60 |
| Paint with primer and 2 coats enamel | C8@.781 | LS | 11.80 | 35.80 | 47.60 |
| **Total cost door, frame and trim as described above** | C8@7.62 | Ea | 1,103.38 | 349.00 | 1,452.38 |

**Hollow Metal Exterior Doors** Commercial quality doors.
These costs do not include the frame, hinges, lockset, trim or finishing.

| 6'8" high hollow flush doors, 1-3/8" thick, non-rated | Craft@Hrs | Unit | Material | Labor | Total |
|---|---|---|---|---|---|
| 2'6" wide, 20 gauge | C8@.681 | Ea | 225.00 | 31.20 | 256.20 |
| 2'8" wide, 20 gauge | C8@.683 | Ea | 232.00 | 31.30 | 263.30 |
| 3'0" wide, 20 gauge | C8@.721 | Ea | 248.00 | 33.00 | 281.00 |
| 6'8" high hollow flush doors, 1-3/4" thick, non-rated |  |  |  |  |  |
| 2'6" wide, 20 gauge | C8@.681 | Ea | 273.00 | 31.20 | 304.20 |
| 2'8" wide, 20 gauge | C8@.697 | Ea | 281.00 | 31.90 | 312.90 |
| 2'8" wide, 18 gauge | C8@.697 | Ea | 326.00 | 31.90 | 357.90 |
| 2'8" wide, 16 gauge | C8@.697 | Ea | 380.00 | 31.90 | 411.90 |
| 3'0" wide, 20 gauge | C8@.721 | Ea | 295.00 | 33.00 | 328.00 |
| 3'0" wide, 18 gauge | C8@.721 | Ea | 340.00 | 33.00 | 373.00 |
| 3'0" wide, 16 gauge | C8@.721 | Ea | 398.00 | 33.00 | 431.00 |
| 7'0" high hollow flush doors, 1-3/4" thick, non-rated |  |  |  |  |  |
| 2'6" or 2'8" wide, 20 gauge | C8@.757 | Ea | 289.00 | 34.70 | 323.70 |
| 2'8" or 3'0" wide, 18 gauge | C8@.757 | Ea | 345.00 | 34.70 | 379.70 |
| 2'8" or 3'0" wide, 16 gauge | C8@.757 | Ea | 404.00 | 34.70 | 438.70 |
| 3'0" wide, 20 gauge | C8@.816 | Ea | 308.00 | 37.40 | 345.40 |

| | Craft@Hrs | Unit | Material | Labor | Total |
|---|---|---|---|---|---|
| Additional costs for hollow metal doors, cost per door | | | | | |
| Add for steel astragal set on site | C8@.496 | Ea | 40.50 | 22.70 | 63.20 |
| Add for factory applied steel astragal | — | Ea | 40.50 | — | 40.50 |
| Add for 90-minute "B" fire label rating | — | Ea | 45.50 | — | 45.50 |
| Add for R-7 polyurethane foam core | — | Ea | 110.00 | — | 110.00 |
| Add for 10" x 10" wired glass panel | — | Ea | 79.20 | — | 79.20 |
| Add for closer reinforcing plate | — | Ea | 15.70 | — | 15.70 |
| Add for chain or bolt reinforcing plate | — | Ea | 117.00 | — | 117.00 |
| Add for rim exit latch reinforcing | — | Ea | 33.50 | — | 33.50 |
| Add for vertical exit latch reinforcing | — | Ea | 60.80 | — | 60.80 |
| Add for pull plate reinforcing | — | Ea | 24.50 | — | 24.50 |
| Add for galvanizing 1-3/8" doors | — | Ea | 30.80 | — | 30.80 |
| Add for galvanizing 1-3/4" door | — | Ea | 37.60 | — | 37.60 |
| Add for cutouts to 4 SF | — | Ea | 63.40 | — | 63.40 |
| Add for cutouts over 4 SF | — | SF | 13.40 | — | 13.40 |
| Add for stainless steel doors | — | Ea | 1,070.00 | — | 1,070.00 |
| Add for baked enamel finish | — | Ea | 80.50 | — | 80.50 |
| Add for porcelain enamel finish | — | Ea | 191.00 | — | 191.00 |
| Add for larger sizes | — | SF | 4.38 | — | 4.38 |
| Add for special dapping | — | Ea | 34.30 | — | 34.30 |

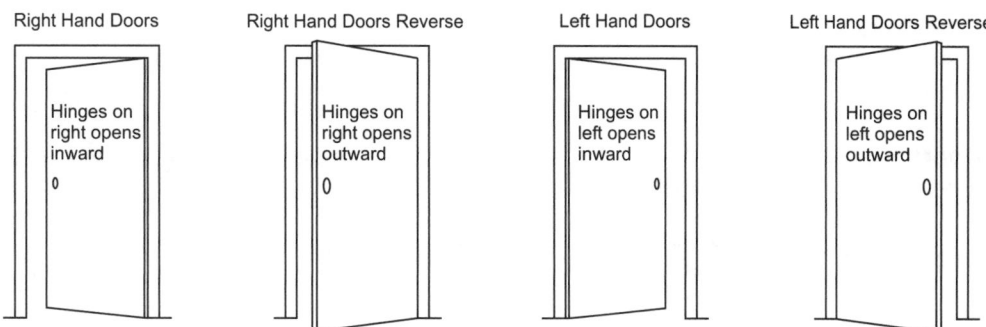

**Hollow Metal Door Frames** 18 gauge prefinished hollow metal door frames, non-rated, stock sizes, for 1-3/4" or 1-3/8" doors. These costs do not include the door, hinges, lockset, trim or finishing.

| | Craft@Hrs | Unit | Material | Labor | Total |
|---|---|---|---|---|---|
| 6'8" high | | | | | |
| To 3'6" wide, to 4-1/2" jamb | C8@.944 | Ea | 106.00 | 43.30 | 149.30 |
| To 3'6" wide, 4-3/4" to 6" jamb | C8@.944 | Ea | 111.00 | 43.30 | 154.30 |
| To 3'6" wide, over 6" jamb | C8@.944 | Ea | 148.00 | 43.30 | 191.30 |
| Over 4' wide, 4-1/2" jamb | C8@.944 | Ea | 126.00 | 43.30 | 169.30 |
| Over 4' wide, 4-3/4" to 6" jamb | C8@.944 | Ea | 134.00 | 43.30 | 177.30 |
| Over 4' wide, over 6" jamb | C8@.944 | Ea | 172.00 | 43.30 | 215.30 |
| 7' high | | | | | |
| To 3'6" wide, to 4-1/2" jamb | C8@.944 | Ea | 115.00 | 43.30 | 158.30 |
| To 3'6" wide, 4-3/4" to 6" jamb | C8@.944 | Ea | 121.00 | 43.30 | 164.30 |
| To 3'6" wide, over 6" jamb | C8@.944 | Ea | 158.00 | 43.30 | 201.30 |
| Over 4' wide, 4-1/2" jamb | C8@.944 | Ea | 136.00 | 43.30 | 179.30 |
| Over 4' wide, 4-3/4" to 6" jamb | C8@.944 | Ea | 141.00 | 43.30 | 184.30 |
| Over 4' wide, over 6" jamb | C8@.944 | Ea | 179.00 | 43.30 | 222.30 |

|  | Craft@Hrs | Unit | Material | Labor | Total |
|---|---|---|---|---|---|
| 8' high | | | | | |
| To 3'6" wide, to 4-1/2" jamb | C8@1.05 | Ea | 132.00 | 48.10 | 180.10 |
| To 3'6" wide, 4-3/4" to 6" jamb | C8@1.05 | Ea | 139.00 | 48.10 | 187.10 |
| To 3'6" wide, over 6" jamb | C8@1.05 | Ea | 178.00 | 48.10 | 226.10 |
| Over 4' wide, 4-1/2" jamb | C8@1.05 | Ea | 153.00 | 48.10 | 201.10 |
| Over 4' wide, 4-3/4" to 6" jamb | C8@1.05 | Ea | 157.00 | 48.10 | 205.10 |
| Over 4' wide, over 6" jamb | C8@1.05 | Ea | 201.00 | 48.10 | 249.10 |
| Cost additions or deductions for hollow metal door frames, per frame | | | | | |
| Add for concrete filled frames | C8@.382 | Ea | 8.22 | 17.50 | 25.72 |
| Add for borrowed lite | C8@.086 | SF | 7.90 | 3.94 | 11.84 |
| Add for fixed transom lite | C8@.099 | SF | 10.50 | 4.54 | 15.04 |
| Add for movable transom lite | C8@.107 | SF | 12.00 | 4.90 | 16.90 |
| Add for window frame sections | C8@.071 | LF | 6.40 | 3.25 | 9.65 |
| Add for wall frame sections | C8@.071 | LF | 7.28 | 3.25 | 10.53 |
| Add for 90-minute UL label | — | Ea | 31.70 | — | 31.70 |
| Add for aluminum casing | — | Ea | 26.40 | — | 26.40 |
| Add for communicator frame (back-to-back doors) | — | Ea | 19.30 | — | 19.30 |
| Add for lengthening, 7' to 8'10" | — | Ea | 17.00 | — | 17.00 |
| Add for extra hinge reinforcing | — | Ea | 5.77 | — | 5.77 |
| Add for stainless steel frames | — | Ea | 555.00 | — | 555.00 |
| Add for porcelain enamel finish | — | Ea | 179.00 | — | 179.00 |
| Add for reinforcing for chain and bolt | — | Ea | 7.39 | — | 7.39 |
| Add for closer plate reinforcing | — | Ea | 6.53 | — | 6.53 |
| Add for exit latch reinforcing | — | Ea | 12.10 | — | 12.10 |
| Deduct for 22 gauge frames | — | Ea | - 14.10 | — | - 14.10 |
| Add for galvanized finish | — | % | 12.0 | — | — |

**Prehung Steel Doors** 18 gauge primed insulated 1-3/4" thick entry doors with sweep and 18 gauge steel frame. Costs shown include three 4" x 4" x 1/4" hinges but no lockset.

| Flush doors | | | | | |
|---|---|---|---|---|---|
| Jamb to 5-1/2" wide, 2'8" x 6'8", or 3'0" x 6'8" | C8@1.05 | Ea | 458.00 | 48.10 | 506.10 |
| Jamb over 5-1/2" wide, 2'8" x 6'8", or 3'0" x 6'8" | C8@1.05 | Ea | 469.00 | 48.10 | 517.10 |
| 6 or 8 panel doors | | | | | |
| Jamb to 5-1/2" wide, 2'8" x 6'8", or 3'0" x 6'8" | C8@1.05 | Ea | 474.00 | 48.10 | 522.10 |
| Jamb over 5-1/2" wide, 2'8" x 6'8", or 3'0" x 6'8" | C8@1.05 | Ea | 480.00 | 48.10 | 528.10 |
| Additional costs for prehung steel doors | | | | | |
| 90-minute Factory Mutual fire rating | — | Ea | 23.40 | — | 23.40 |
| 60-minute Factory Mutual fire rating | — | Ea | 18.00 | — | 18.00 |
| Add for 16 gauge door and frame | — | Ea | 14.30 | — | 14.30 |
| Deduct for 22 gauge and door | — | Ea | - 3.76 | — | - 3.76 |
| Add for galvanized steel finish | — | Ea | 12.40 | — | 12.40 |
| Deduct for unfinished door | — | Ea | - 14.60 | — | - 14.60 |
| Add for installed weatherstripping | — | Ea | 36.20 | — | 36.20 |
| Add for aluminum threshold | — | Ea | 35.90 | — | 35.90 |

**Doors and Frames** Commercial and institutional quality. Unfinished, no hardware included.

| Wood hollow core 1-3/8" thick prehung flush interior doors | | | | | |
|---|---|---|---|---|---|
| 2'6" x 6'8", hardboard face | C8@.936 | Ea | 64.80 | 42.90 | 107.70 |
| 2'8" x 6'8", hardboard face | C8@.936 | Ea | 66.00 | 42.90 | 108.90 |
| 3'0" x 6'8", hardboard face | C8@.982 | Ea | 68.20 | 45.00 | 113.20 |
| 2'6" x 6'8", hardwood face | C8@.936 | Ea | 77.50 | 42.90 | 120.40 |
| 2'8" x 6'8", hardwood face | C8@.936 | Ea | 79.40 | 42.90 | 122.30 |
| 3'0" x 6'8", hardwood face | C8@.982 | Ea | 81.70 | 45.00 | 126.70 |

| | Craft@Hrs | Unit | Material | Labor | Total |
|---|---|---|---|---|---|
| **Wood solid core 1-3/4" thick flush prehung exterior doors** | | | | | |
| 2'8" x 6'8", hardboard face | C8@1.10 | Ea | 106.00 | 50.40 | 156.40 |
| 3'0" x 6'8", hardboard face | C8@1.10 | Ea | 108.00 | 50.40 | 158.40 |
| 2'8" x 6'8", hardwood face | C8@1.10 | Ea | 108.00 | 50.40 | 158.40 |
| 3'0" x 6'8", hardwood face | C8@1.10 | Ea | 108.00 | 50.40 | 158.40 |
| 2'6" x 6'8", hardboard face, B fire label | C8@1.10 | Ea | 151.00 | 50.40 | 201.40 |
| 2'8" x 6'8", hardboard face, B fire label | C8@1.10 | Ea | 156.00 | 50.40 | 206.40 |
| 3'0" x 6'8", hardboard face, B fire label | C8@1.10 | Ea | 159.00 | 50.40 | 209.40 |
| **Steel foam core 1-3/4" thick prehung exterior doors** | | | | | |
| 2'6" x 6'8", flush face | C8@.982 | Ea | 151.00 | 45.00 | 196.00 |
| 2'8" x 6'8", flush face | C8@.982 | Ea | 151.00 | 45.00 | 196.00 |
| 3'0" x 6'8", flush face | C8@1.01 | Ea | 151.00 | 46.30 | 197.30 |
| 2'6" x 6'8", colonial panel face | C8@.982 | Ea | 151.00 | 45.00 | 196.00 |
| 2'8" x 6'8", colonial panel face | C8@.982 | Ea | 151.00 | 45.00 | 196.00 |
| 3'0" x 6'8", colonial panel face | C8@1.01 | Ea | 151.00 | 46.30 | 197.30 |
| **Fiberglass foam core 1-3/4" thick prehung exterior doors** | | | | | |
| 2'8" x 6'8", colonial panel face | C8@.982 | Ea | 183.00 | 45.00 | 228.00 |
| 3'0" x 6'8", colonial panel face | C8@1.01 | Ea | 183.00 | 46.30 | 229.30 |
| 2'8" x 6'8", French 15-lite | C8@.982 | Ea | 270.00 | 45.00 | 315.00 |
| 3'0" x 6'8", French 15-lite | C8@1.01 | Ea | 270.00 | 46.30 | 316.30 |

| | Craft@Hrs | Unit | Material | Labor | Equipment | Total |
|---|---|---|---|---|---|---|
| **Special Doors** These costs do not include electrical work. Equipment is a 4,000 lb. forklift. | | | | | | |
| **Sliding metal clad fire doors including electric motor** | | | | | | |
| Light duty, minimum | H5@.249 | SF | 37.20 | 13.10 | 1.59 | 51.89 |
| Light duty, maximum | H5@.405 | SF | 64.90 | 21.20 | 2.58 | 88.68 |
| Heavy duty, minimum | H5@.600 | SF | 74.40 | 31.50 | 3.82 | 109.72 |
| Heavy duty, maximum | H5@.750 | SF | 128.00 | 39.30 | 4.78 | 172.08 |
| **Fire doors, overhead roll-up type, sectional steel, fusible link, UL label** | | | | | | |
| 6' x 7', manual operation | F7@13.4 | Ea | 2,160.00 | 665.00 | 107.00 | 2,932.00 |
| 5' x 8', manual operation | F7@13.4 | Ea | 1,460.00 | 665.00 | 107.00 | 2,232.00 |
| 14' x 14', chain hoist operated | F7@26.2 | Ea | 4,240.00 | 1,300.00 | 209.00 | 5,749.00 |
| 18' x 14', chain hoist operated | F7@32.9 | Ea | 5,850.00 | 1,630.00 | 262.00 | 7,742.00 |
| **Sliding metal fire doors. Motor operated, 3-hour rating, with fusible link. Add the cost of electrical hookup** | | | | | | |
| 3' x 6'8" | H5@16.0 | Ea | 3,450.00 | 839.00 | 102.00 | 4,391.00 |
| 3'8" x 6'8" | H5@16.0 | Ea | 3,680.00 | 839.00 | 102.00 | 4,621.00 |
| 4' x 8' | H5@16.0 | Ea | 3,810.00 | 839.00 | 102.00 | 4,751.00 |
| 5' x 8' | H5@16.0 | Ea | 3,960.00 | 839.00 | 102.00 | 4,901.00 |
| 6' x 8' | H5@16.0 | Ea | 4,730.00 | 839.00 | 102.00 | 5,671.00 |
| **Refrigerator doors, manually operated, 4", with hardware and standard frame** | | | | | | |
| Galvanized, cooler, 3' x 7', hinged | F7@6.60 | Ea | 1,820.00 | 327.00 | 52.60 | 2,199.60 |
| Stainless, cooler, 3' x 7', hinged | F7@6.60 | Ea | 2,100.00 | 327.00 | 52.60 | 2,479.60 |
| Stainless, freezer, 3' x 7', hinged | F7@6.60 | Ea | 2,730.00 | 327.00 | 52.60 | 3,109.60 |
| Stainless, cooler, 4' x 7', hinged | F7@6.60 | Ea | 2,360.00 | 327.00 | 52.60 | 2,739.60 |
| Galvanized, cooler, 5' x 7', sliding | F7@7.93 | Ea | 2,670.00 | 393.00 | 63.10 | 3,126.10 |
| Galvanized, freezer, 5' x 7' sliding | F7@7.93 | Ea | 3,110.00 | 393.00 | 63.10 | 3,566.10 |
| **Hangar doors, including hardware and motorized opener. Costs based on 150' wide doors, installed with a 25 ton hydraulic truck crane.** | | | | | | |
| To 20' high | H5@.213 | SF | 23.60 | 11.20 | 9.62 | 44.42 |
| 20' to 40' high | H5@.142 | SF | 30.50 | 7.44 | 6.41 | 44.35 |
| 40' to 60' high | H5@.085 | SF | 36.10 | 4.46 | 3.84 | 44.40 |
| 60' to 80' high | H5@.061 | SF | 42.90 | 3.20 | 2.75 | 48.85 |
| More than 80' high | H5@.047 | SF | 71.00 | 2.46 | 2.12 | 75.58 |

| | Craft@Hrs | Unit | Material | Labor | Equipment | Total |
|---|---|---|---|---|---|---|
| **Revolving doors, 6'6" to 7' diameter, manual controls** | | | | | | |
| Aluminum | F7@50.3 | Ea | 26,200.00 | 2,500.00 | 401.00 | 29,101.00 |
| Stainless steel | F7@50.3 | Ea | 33,100.00 | 2,500.00 | 401.00 | 36,001.00 |
| Bronze | F7@69.0 | Ea | 38,400.00 | 3,420.00 | 549.00 | 42,369.00 |
| Add for automatic door controls | — | % | 40.0 | — | — | — |
| **Security grilles, aluminum, overhead roll-up type, horizontal rods at 2" OC, chain hoist operated** | | | | | | |
| 8' x  8', clear anodized aluminum | F7@18.5 | Ea | 2,460.00 | 918.00 | 147.00 | 3,525.00 |
| 8' x  8', medium bronze aluminum | F7@18.5 | Ea | 3,480.00 | 918.00 | 147.00 | 4,545.00 |
| 10' x 10', anodized aluminum | F7@21.8 | Ea | 3,300.00 | 1,080.00 | 174.00 | 4,554.00 |
| 18' x  8', anodized aluminum | F7@26.3 | Ea | 5,080.00 | 1,300.00 | 209.00 | 6,589.00 |
| 18' x  8', anodized, motor operated | F7@28.7 | Ea | 5,930.00 | 1,420.00 | 229.00 | 7,579.00 |
| **Service doors, aluminum, overhead roll-up type** | | | | | | |
| 4' x  4', manual counter shutter | F7@13.6 | Ea | 1,010.00 | 675.00 | 108.00 | 1,793.00 |
| 8' x  4', manual operation | F7@12.7 | Ea | 1,460.00 | 630.00 | 101.00 | 2,191.00 |
| 10' x 10', chain hoist operated | F7@12.7 | Ea | 2,250.00 | 630.00 | 101.00 | 2,981.00 |
| **Service doors, steel, overhead roll-up, chain hoist operated** | | | | | | |
| 8' x  8' | F7@11.7 | Ea | 1,470.00 | 580.00 | 93.20 | 2,143.20 |
| 10' x 10' | F7@11.7 | Ea | 1,800.00 | 580.00 | 93.20 | 2,473.20 |
| 12' x 12' | F7@14.7 | Ea | 2,430.00 | 729.00 | 117.00 | 3,276.00 |
| 14' x 14' | F7@18.2 | Ea | 2,660.00 | 903.00 | 145.00 | 3,708.00 |
| 18' x 18' | F7@22.0 | Ea | 3,600.00 | 1,090.00 | 175.00 | 4,865.00 |
| 18' x 14', 240 volt motor operated | F7@25.0 | Ea | 4,680.00 | 1,240.00 | 199.00 | 6,119.00 |
| **Sliding glass doors with 1/4" tempered float glass including frame, trim, and hardware** | | | | | | |
| 6-0' wide economy grade | C8@2.75 | Ea | 916.00 | 126.00 | — | 1,042.00 |
| 6-0' wide premium grade | C8@3.00 | Ea | 1,060.00 | 137.00 | — | 1,197.00 |
| 12-0' wide economy grade | C8@4.80 | Ea | 1,450.00 | 220.00 | — | 1,670.00 |
| 12-0' wide premium grade | C8@5.00 | Ea | 2,130.00 | 229.00 | — | 2,359.00 |
| **Sliding glass doors with 5/8" insulating glass including frame, trim, and hardware** | | | | | | |
| 6-0' wide economy grade | C8@2.75 | Ea | 1,300.00 | 126.00 | — | 1,426.00 |
| 6-0' wide premium grade | C8@3.00 | Ea | 1,580.00 | 137.00 | — | 1,717.00 |
| 12-0' wide economy grade | C8@4.80 | Ea | 1,740.00 | 220.00 | — | 1,960.00 |
| 12-0' wide premium grade | C8@5.00 | Ea | 2,550.00 | 229.00 | — | 2,779.00 |
| **Telescoping doors, with frame, enamel finish, electric operated** | | | | | | |
| 10' high x 10' wide | H5@15.0 | Ea | 9,660.00 | 786.00 | 95.60 | 10,541.60 |
| 20' high x 10' wide | H5@19.3 | Ea | 15,500.00 | 1,010.00 | 123.00 | 16,633.00 |
| 20' high x 12' wide | H5@24.0 | Ea | 16,800.00 | 1,260.00 | 153.00 | 18,213.00 |
| 20' high x 16' wide | H5@26.4 | Ea | 20,200.00 | 1,380.00 | 168.00 | 21,748.00 |
| **Accordion folding doors, 4-1/2" wide, nylon wheels hung from straight track, with fittings, 6'8" to 12' high, manual operation, per SF of opening** | | | | | | |
| Vinyl surface | C8@.051 | SF | 20.30 | 2.34 | — | 22.64 |
| Birch or oak veneer | C8@.051 | SF | 32.80 | 2.34 | — | 35.14 |
| Walnut, cherry or maple veneer | C8@.051 | SF | 39.40 | 2.34 | — | 41.74 |
| Clear or bronze acrylic | C8@.051 | SF | 37.80 | 2.34 | — | 40.14 |
| Clear or bronze acrylic, with security | C8@.051 | SF | 46.20 | 2.34 | — | 48.54 |
| Anodized aluminum, solid | C8@.051 | SF | 45.20 | 2.34 | — | 47.54 |
| Anodized aluminum, perforated | C8@.051 | SF | 56.70 | 2.34 | — | 59.04 |
| Add for curved track | C8@.102 | LF | 16.80 | 4.67 | — | 21.47 |
| Add for hat channel (suspended ceiling) mount | C8@.102 | LF | 8.40 | 4.67 | — | 13.07 |
| Add for sound rated folding doors | — | SF | 6.30 | — | — | 6.30 |
| Add for locking bolt, per side | — | Ea | 36.80 | — | — | 36.80 |
| Add for fire retardant accordion folding doors | — | % | 40.0 | — | — | — |
| Deduct for finished one side (closets) doors | — | % | 10.0 | — | — | — |

# 08 Openings

|  | Craft@Hrs | Unit | Material | Labor | Equipment | Total |
|---|---|---|---|---|---|---|

Air curtains. Add the cost of motor starters, transformers, door switches and temperature controls. Unheated, with minimal wind stoppage. Per linear foot of opening width.

|  | Craft@Hrs | Unit | Material | Labor | Equipment | Total |
|---|---|---|---|---|---|---|
| 8-0' high, multiples of 3-0' wide | C8@3.70 | LF | 253.00 | 170.00 | — | 423.00 |
| 10-0' high, multiples of 4-0' wide | C8@3.80 | LF | 262.00 | 174.00 | — | 436.00 |
| 12-0' high, multiples of 3'-6" wide | C8@3.90 | LF | 272.00 | 179.00 | — | 451.00 |
| 16-0' high, multiples of 3'-6" wide | C8@4.00 | LF | 327.00 | 183.00 | — | 510.00 |
| Add for high wind stoppage | — | % | 20.0 | — | — | — |
| Add for heated units | — | % | 75.0 | — | — | — |

Door louvers. Non-vision design recommended for schools, class "A" and institutional buildings. Suitable for both wood and metal doors. Manufactured from 18 gauge Jetkote (steel) supplied with #8-3/4" phillips (pan head) L&L Louvers.

|  | Craft@Hrs | Unit | Material | Labor | Equipment | Total |
|---|---|---|---|---|---|---|
| 12" x 12" | C8@.500 | Ea | 39.00 | 22.90 | — | 61.90 |
| 18" x 12" | C8@.500 | Ea | 48.00 | 22.90 | — | 70.90 |
| 18" x 18" | C8@.500 | Ea | 55.00 | 22.90 | — | 77.90 |
| 24" x 12" | C8@.500 | Ea | 55.00 | 22.90 | — | 77.90 |
| 24" x 18" | C8@.500 | Ea | 64.00 | 22.90 | — | 86.90 |
| 24" x 24" | C8@.500 | Ea | 82.00 | 22.90 | — | 104.90 |

See-thru door frames. (Vision Lite). Recommended for applications where total visibility, light transmission and security are required. The beveled style frame provides a wider field of vision. Manufactured from 18 gauge Jetkote (steel). Installation screws are #8 x 2" phillips flat head screws. Glass not included. L&L Louvers.

|  | Craft@Hrs | Unit | Material | Labor | Equipment | Total |
|---|---|---|---|---|---|---|
| 6" x 27" | C8@.500 | Ea | 40.00 | 22.90 | — | 62.90 |
| 7" x 22" | C8@.500 | Ea | 36.00 | 22.90 | — | 58.90 |
| 12" x 12" | C8@.500 | Ea | 26.00 | 22.90 | — | 48.90 |
| 18" x 18" | C8@.500 | Ea | 43.00 | 22.90 | — | 65.90 |
| 24" x 24" | C8@.500 | Ea | 47.00 | 22.90 | — | 69.90 |
| 24" x 30" | C8@.500 | Ea | 51.00 | 22.90 | — | 73.90 |
| 24" x 36" | C8@.500 | Ea | 54.00 | 22.90 | — | 76.90 |

1/4" Wire glass for see-thru door frames

|  | Craft@Hrs | Unit | Material | Labor | Equipment | Total |
|---|---|---|---|---|---|---|
| 6" x 27" | — | Ea | 40.00 | — | — | 40.00 |
| 7" x 22" | — | Ea | 40.00 | — | — | 40.00 |
| 12" x 12" | — | Ea | 40.00 | — | — | 40.00 |
| 18" x 18" | — | Ea | 55.00 | — | — | 55.00 |
| 24" x 24" | — | Ea | 65.00 | — | — | 65.00 |
| 24" x 30" | — | Ea | 70.00 | — | — | 70.00 |
| 24" x 36" | — | Ea | 75.00 | — | — | 75.00 |

|  | Craft@Hrs | Unit | Material | Labor | Total |
|---|---|---|---|---|---|

Steel access doors, 1" frame, concealed hinge, screwdriver access

|  | Craft@Hrs | Unit | Material | Labor | Total |
|---|---|---|---|---|---|
| 12" x 12", galvanized | CC@.250 | Ea | 23.60 | 12.90 | 36.50 |
| 14" x 14", galvanized | CC@.250 | Ea | 25.00 | 12.90 | 37.90 |
| 18" x 18" | CC@.275 | Ea | 34.40 | 14.20 | 48.60 |
| 14" x 14", vinyl | CC@.250 | Ea | 14.60 | 12.90 | 27.50 |

|  | Craft@Hrs | Unit | Material | Labor | Equipment | Total |
|---|---|---|---|---|---|---|

Stock room doors, double-action .063" aluminum, with hardware and bumper strips both sides, 7' high doors, plastic laminate finish, 12" high baseplate in each door

|  | Craft@Hrs | Unit | Material | Labor | Equipment | Total |
|---|---|---|---|---|---|---|
| 2' wide opening, 1 door | F7@1.70 | Ea | 233.00 | 84.30 | 13.50 | 330.80 |
| 2'6" wide opening, 1 door | F7@1.70 | Ea | 261.00 | 84.30 | 13.50 | 358.80 |
| 3' wide opening, 1 door | F7@1.80 | Ea | 288.00 | 89.30 | 14.30 | 391.60 |
| 3'6" wide opening, 1 door | F7@1.90 | Ea | 320.00 | 94.30 | 15.10 | 429.40 |
| 4' wide opening, 2 doors, per pair | F7@3.36 | Pr | 466.00 | 167.00 | 26.80 | 659.80 |
| 5' wide opening, 2 doors, per pair | F7@3.36 | Pr | 522.00 | 167.00 | 26.80 | 715.80 |
| 6' wide opening, 2 doors, per pair | F7@3.98 | Pr | 574.00 | 197.00 | 31.70 | 802.70 |
| 7' wide opening, 2 doors, per pair | F7@3.98 | Pr | 622.00 | 197.00 | 31.70 | 850.70 |

417

# 08 Openings

| | Craft@Hrs | Unit | Material | Labor | Equipment | Total |
|---|---|---|---|---|---|---|
| Vault doors, minimum security | | | | | | |
| 3' x 7', 2 hour fire rating | F7@22.0 | Ea | 2,730.00 | 1,090.00 | 175.00 | 3,995.00 |
| 4' x 7', 2 hour fire rating | F7@23.4 | Ea | 3,900.00 | 1,160.00 | 186.00 | 5,246.00 |
| 3' x 7', 4 hour fire rating | F7@27.2 | Ea | 2,970.00 | 1,350.00 | 217.00 | 4,537.00 |
| 4' x 7', 4 hour fire rating | F7@28.7 | Ea | 4,050.00 | 1,420.00 | 229.00 | 5,699.00 |

| | Craft@Hrs | Unit | Material | Labor | Total |
|---|---|---|---|---|---|
| **Flexible strip doors** Suitable for interior or exterior use. High strength USDA-approved clear transparent PVC plastic strips, with aluminum mounting hardware. Typical uses include truck dock door openings, railway car entrances, freezer doors and as positive draft protection to check the flow of unwanted air at +140 degrees to –40 degrees F. Costs shown per SF are per square foot of flexible strip area. To calculate the flexible strip area, add 2 linear feet to the opening width and height to allow for coverage at the sides and top. Costs shown per LF are per linear foot of mounting hardware. To calculate the length of the mounting hardware, add 2 LF to the opening width. Labor costs are based on work performed using ladders. Add for scaffolding costs if required. | | | | | |
| Openings to 8' in height, 8" wide strips x .080" thick | | | | | |
| With 1/2 strip overlap | C8@.058 | SF | 4.29 | 2.66 | 6.95 |
| With 3/4 strip overlap | C8@.076 | SF | 4.75 | 3.48 | 8.23 |
| With full strip overlap | C8@.096 | SF | 6.63 | 4.40 | 11.03 |
| Openings over 8' to 14' in height, 12" wide strips x .120" thick | | | | | |
| With 2/3 strip overlap | C8@.100 | SF | 7.59 | 4.58 | 12.17 |
| With full strip overlap | C8@.115 | SF | 9.09 | 5.27 | 14.36 |
| Openings over 14' to 20' in height, 16" wide strips x .120" thick | | | | | |
| With 1/2 strip overlap | C8@.075 | SF | 8.11 | 3.44 | 11.55 |
| With 3/4 strip overlap | C8@.100 | SF | 10.70 | 4.58 | 15.28 |
| With full strip overlap | C8@.125 | SF | 12.20 | 5.73 | 17.93 |
| Add for mounting hardware, aluminum, pre-drilled, complete with nuts, bolts, washers and flex strip mounting pins | | | | | |
| Ceiling or header mounted (1.5 lbs per LF) | C8@.400 | LF | 10.90 | 18.30 | 29.20 |
| Wall mounted (2.5 lbs per LF) | C8@.500 | LF | 40.10 | 22.90 | 63.00 |
| **Curtain Walls** Typical costs | | | | | |
| Structural sealant curtain wall systems with aluminum or steel frame | | | | | |
| Regular weight, float glass | — | SF | — | — | 35.80 |
| Heavy duty, float glass | — | SF | — | — | 39.20 |
| Corning glasswall glass | — | SF | — | — | 41.40 |
| Add for spandrel glass | — | SF | — | — | 5.23 |
| Add for heat absorbing glass | — | SF | — | — | 9.39 |
| Add for low transmission glass | — | SF | — | — | 6.12 |
| Add for 1" insulating glass | — | SF | — | — | 8.85 |
| Add for bronze anodizing | — | % | — | — | 15.0 |
| Add for black anodizing | — | % | — | — | 18.0 |
| Flush panel doors for curtain walls | | | | | |
| 3' x 7' | H5@7.14 | Ea | 1,790.00 | 374.00 | 45.50 | 2,209.50 |
| 3'6" x 7' | H5@7.14 | Ea | 1,870.00 | 374.00 | 45.50 | 2,289.50 |
| Pair 2'6" x 7', per opening | H5@10.5 | Ea | 2,640.00 | 550.00 | 66.90 | 3,256.90 |
| Pair 3' x 7', per opening | H5@10.5 | Ea | 3,010.00 | 550.00 | 66.90 | 3,626.90 |
| Add for bronze anodized aluminum | — | % | 15.0 | — | — | — |

|  | Craft@Hrs | Unit | Material | Labor | Total |
|---|---|---|---|---|---|
| **Commercial and Industrial Grade Steel Windows,** glazed | | | | | |
| Industrial grade, fixed 100% | G1@.086 | SF | 16.00 | 3.96 | 19.96 |
| Industrial grade, vented 50% | G1@.086 | SF | 20.70 | 3.96 | 24.66 |
| Projected, vented 50% | G1@.086 | SF | 23.10 | 3.96 | 27.06 |
| Add for screen, SF of screen | — | SF | 2.79 | — | 2.79 |

**Hardware**  Commercial and industrial quality. Installation costs include drilling and routing of doors where required. Rule of thumb. Use these costs for preliminary estimates.

|  | Craft@Hrs | Unit | Material | Labor | Total |
|---|---|---|---|---|---|
| Hardware for commercial buildings, per square foot of floor area | | | | | |
| Economy grade, for doors, cabinets, | | | | | |
| and toilet rooms | CC@.002 | SF | .21 | .10 | .31 |
| Standard grade, for doors, cabinets, | | | | | |
| and toilet rooms | CC@.004 | SF | .44 | .21 | .65 |
| Hospital, not including panic hardware, per door | CC@2.46 | Ea | 568.00 | 127.00 | 695.00 |
| Office, not including panic hardware, per door | CC@1.57 | Ea | 272.00 | 81.30 | 353.30 |
| School, not including panic hardware, per door | CC@1.94 | Ea | 380.00 | 100.00 | 480.00 |
| Detail cost for individual hardware components | | | | | |
| Cabinet hardware, per LF of face, typical prices | | | | | |
| Commercial grade | CC@.145 | LF | 12.40 | 7.51 | 19.91 |
| Institutional grade | CC@.181 | LF | 15.60 | 9.37 | 24.97 |
| Closer, surface mounted. Based on Schlage Lock | | | | | |
| Interior doors | CC@.616 | Ea | 124.00 | 31.90 | 155.90 |
| Exterior doors | CC@.709 | Ea | 117.00 | 36.70 | 153.70 |
| Exterior, heavy duty | CC@.894 | Ea | 137.00 | 46.30 | 183.30 |
| Floor mounted | CC@1.16 | Ea | 236.00 | 60.10 | 296.10 |
| Deadbolts and deadlatches, chrome finish | | | | | |
| Standard deadbolt, single cylinder | CC@.745 | Ea | 36.00 | 38.60 | 74.60 |
| Standard deadbolt, double cylinder | CC@.745 | Ea | 49.40 | 38.60 | 88.00 |
| Standard turnbolt, no cylinder | CC@.745 | Ea | 27.70 | 38.60 | 66.30 |
| Nightlatch, single cylinder | CC@.745 | Ea | 87.10 | 38.60 | 125.70 |
| Nightlatch, double cylinder | CC@.745 | Ea | 96.30 | 38.60 | 134.90 |
| Nightlatch, no cylinder, turnbolt | CC@.745 | Ea | 75.10 | 38.60 | 113.70 |
| Heavy duty deadbolt, single cylinder | CC@.745 | Ea | 67.50 | 38.60 | 106.10 |
| Heavy duty deadbolt, double cylinder | CC@.745 | Ea | 84.00 | 38.60 | 122.60 |
| Heavy duty turnbolt, no cylinder | CC@.745 | Ea | 67.50 | 38.60 | 106.10 |
| Extra heavy deadbolt, single cylinder | CC@.745 | Ea | 107.00 | 38.60 | 145.60 |
| Extra heavy deadbolt, double cylinder | CC@.745 | Ea | 126.00 | 38.60 | 164.60 |
| Add for brass or bronze finish | — | Ea | 3.07 | — | 3.07 |
| Door stops | | | | | |
| Rubber tip, screw base | CC@.060 | Ea | 4.78 | 3.11 | 7.89 |
| Floor mounted, with holder | CC@.279 | Ea | 19.80 | 14.40 | 34.20 |
| Wall mounted, with holder | CC@.279 | Ea | 19.40 | 14.40 | 33.80 |
| Overhead mounted | CC@.528 | Ea | 113.00 | 27.30 | 140.30 |
| Hinges, butt type | | | | | |
| 3-1/2" x 3-1/2" | CC@.357 | Pr | 11.10 | 18.50 | 29.60 |
| 5" x 5", hospital swing clear | CC@.547 | Pr | 147.00 | 28.30 | 175.30 |
| 4", spring, single acting | CC@.243 | Pr | 36.40 | 12.60 | 49.00 |
| 4-1/2", heavy duty | CC@.347 | Pr | 68.00 | 18.00 | 86.00 |
| 7", spring, double acting | CC@.491 | Pr | 89.20 | 25.40 | 114.60 |
| Floor-mounted, interior, per door | CC@.665 | Ea | 298.00 | 34.40 | 332.40 |
| Floor-mounted, exterior, per 3' door | CC@.665 | Ea | 322.00 | 34.40 | 356.40 |
| Floor-mounted, exterior, over 3' wide door | CC@.799 | Ea | 679.00 | 41.40 | 720.40 |
| Invisible, blind doors, interior, per door | CC@.574 | Ea | 35.30 | 29.70 | 65.00 |
| Invisible, for blind doors, exterior | CC@.574 | Pr | 45.10 | 29.70 | 74.80 |

| | Craft@Hrs | Unit | Material | Labor | Total |
|---|---|---|---|---|---|
| **Kick plates** (see also Plates below) | | | | | |
| 10" x 34" 16 gauge, bronze | CC@.452 | Ea | 47.80 | 23.40 | 71.20 |
| 10" x 34" 18 gauge, stainless | CC@.452 | Ea | 39.60 | 23.40 | 63.00 |
| **Letter drop**, for wood doors, | | | | | |
| brass plated | CC@.369 | Ea | 33.20 | 19.10 | 52.30 |
| **Locksets** | | | | | |
| Heavy duty residential or light duty commercial locksets, chrome finish | | | | | |
| Classroom or storeroom, key lock | CC@.745 | Ea | 107.00 | 38.60 | 145.60 |
| Dummy knob | CC@.247 | Ea | 20.10 | 12.80 | 32.90 |
| Entrance lockset, key lock | CC@.745 | Ea | 97.90 | 38.60 | 136.50 |
| Passage latch, no lock | CC@.689 | Ea | 46.40 | 35.70 | 82.10 |
| Privacy latch, button lock | CC@.689 | Ea | 55.60 | 35.70 | 91.30 |
| Add for lever handle (handicapped) | — | Ea | 38.20 | — | 38.20 |
| Add for brass or bronze finish | — | Ea | 6.67 | — | 6.67 |
| Heavy duty commercial, chrome finish | | | | | |
| Bored entrance lockset | CC@.745 | Ea | 246.00 | 38.60 | 284.60 |
| Grip handle, with trim | CC@.855 | Ea | 270.00 | 44.30 | 314.30 |
| Single deadbolt-lockset | CC@.745 | Ea | 119.00 | 38.60 | 157.60 |
| Double deadbolt-lockset | CC@.745 | Ea | 157.00 | 38.60 | 195.60 |
| Mortise deadbolt-lockset | CC@.855 | Ea | 270.00 | 44.30 | 314.30 |
| Add for lever handle (handicapped) | — | Ea | 25.50 | — | 25.50 |
| Add for brass or bronze finish | — | Ea | 9.37 | — | 9.37 |
| **Panic type exit door hardware**, with trim | | | | | |
| Mortise type, for aluminum, steel or wood doors | | | | | |
| Satin aluminum finish | CC@2.41 | Ea | 494.00 | 125.00 | 619.00 |
| Dark bronze finish | CC@2.41 | Ea | 510.00 | 125.00 | 635.00 |
| Polished chrome finish | CC@2.41 | Ea | 605.00 | 125.00 | 730.00 |
| Add for external lockset | CC@.401 | Ea | 67.50 | 20.80 | 88.30 |
| Rim lock type, for aluminum, steel or wood doors | | | | | |
| Satin aluminum finish | CC@1.50 | Ea | 363.00 | 77.70 | 440.70 |
| Dark bronze finish | CC@1.50 | Ea | 381.00 | 77.70 | 458.70 |
| Polished chrome finish | CC@1.50 | Ea | 468.00 | 77.70 | 545.70 |
| Add for external lockset | CC@.401 | Ea | 51.00 | 20.80 | 71.80 |
| Vertical rod type, satin aluminum finish, with trim | | | | | |
| Aluminum doors, concealed rod | CC@2.14 | Ea | 481.00 | 111.00 | 592.00 |
| Metal or wood doors, external rod | CC@1.79 | Ea | 503.00 | 92.70 | 595.70 |
| Wood doors, concealed rod | CC@2.14 | Ea | 592.00 | 111.00 | 703.00 |
| Add for external lockset | CC@.401 | Ea | 65.50 | 20.80 | 86.30 |
| Add for bronze or black finish | — | % | 37.5 | — | — |
| Add for chrome finish | — | % | 70.0 | — | — |
| **Plates** (see also Kick Plates above) | | | | | |
| Pull plates, bronze, 4" x 16" | CC@.190 | Ea | 39.50 | 9.84 | 49.34 |
| Push plates bronze, 4" x 16" | CC@.190 | Ea | 30.60 | 9.84 | 40.44 |
| **Surface bolts** | | | | | |
| 4" bolt | CC@.308 | Ea | 6.09 | 15.90 | 21.99 |
| 6" bolt | CC@.310 | Ea | 8.30 | 16.10 | 24.40 |
| **Threshold** | | | | | |
| Aluminum, 36" | CC@.241 | Ea | 19.60 | 12.50 | 32.10 |
| Bronze, 36" | CC@.241 | Ea | 80.20 | 12.50 | 92.70 |

# 08 Openings

| | Craft@Hrs | Unit | Material | Labor | Total |
|---|---|---|---|---|---|
| **Weatherstripping** | | | | | |
| Bronze and neoprene, 3' x 7' door | | | | | |
|    Wood door | CC@1.53 | Ea | 22.10 | 79.20 | 101.30 |
|    Steel door, adjustable | CC@2.32 | Ea | 53.30 | 120.00 | 173.30 |
| Astragal, mortise mounted, bronze, adjustable | CC@.064 | LF | 13.70 | 3.31 | 17.01 |
| Glue-back foam for 3' x 7' door | CC@.006 | LF | .39 | .31 | .70 |
| **Glazing** | | | | | |
| Sheet (window) glass | | | | | |
|   Single strength "B" | | | | | |
|     To 60" width plus length | G1@.057 | SF | 1.96 | 2.62 | 4.58 |
|     Over 60" to 70" width plus length | G1@.057 | SF | 2.02 | 2.62 | 4.64 |
|     Over 70" to 80" width plus length | G1@.057 | SF | 2.19 | 2.62 | 4.81 |
|     Over 80" width plus length | G1@.057 | SF | 2.23 | 2.62 | 4.85 |
| Double strength window glass | | | | | |
|   Double strength "B" | | | | | |
|     To 60" width plus length | G1@.057 | SF | 2.55 | 2.62 | 5.17 |
|     Over 60" to 70" width plus length | G1@.057 | SF | 2.65 | 2.62 | 5.27 |
|     Over 70" to 80" width plus length | G1@.057 | SF | 2.74 | 2.62 | 5.36 |
|     Over 80" width plus length | G1@.057 | SF | 2.82 | 2.62 | 5.44 |
|   Tempered "B" | G1@.057 | SF | 4.16 | 2.62 | 6.78 |
|   Obscure | G1@.057 | SF | 3.93 | 2.62 | 6.55 |
| 3/16" glass | | | | | |
|   Crystal (clear) | G1@.068 | SF | 3.40 | 3.13 | 6.53 |
|   Tempered | G1@.068 | SF | 5.61 | 3.13 | 8.74 |
| 1/4" glass | | | | | |
|   Float, clear, quality 3 | G1@.102 | SF | 3.24 | 4.69 | 7.93 |
|   Float, bronze or gray, quality 3 | G1@.102 | SF | 3.41 | 4.69 | 8.10 |
|   Float, obscure | G1@.102 | SF | 4.92 | 4.69 | 9.61 |
|   Float, heat absorbing | G1@.102 | SF | 5.38 | 4.69 | 10.07 |
|   Float, safety, laminated, clear | G1@.102 | SF | 5.94 | 4.69 | 10.63 |
|   Float, tempered | G1@.102 | SF | 5.38 | 4.69 | 10.07 |
|   Spandrel, plain | G1@.102 | SF | 7.24 | 4.69 | 11.93 |
|   Spandrel, tinted gray or bronze | G1@.102 | SF | 7.97 | 4.69 | 12.66 |
|   Tempered, reflective | G1@.102 | SF | 8.28 | 4.69 | 12.97 |
| 3/8" glass | | | | | |
|   Float, clear | G1@.128 | SF | 5.52 | 5.89 | 11.41 |
|   Float, tinted | G1@.128 | SF | 7.75 | 5.89 | 13.64 |
|   Obscure, tempered | G1@.128 | SF | 8.28 | 5.89 | 14.17 |
|   Corrugated glass | G1@.128 | SF | 7.11 | 5.89 | 13.00 |
| 1/2" float, tempered | G1@.157 | SF | 20.40 | 7.22 | 27.62 |
| 1" bullet resistant, 12" x 12" panels | G1@1.32 | SF | 79.10 | 60.70 | 139.80 |
| 2" bullet resistant, 15 to 20 SF | G1@1.62 | SF | 47.30 | 74.50 | 121.80 |
| Insulating glass, 2 layers of 1/8", 1/2" overall, 10 to 15 square foot lites | | | | | |
|   "B" quality sheet glass | G1@.131 | SF | 6.28 | 6.02 | 12.30 |
|   Float, polished bronze or gray | G1@.131 | SF | 8.75 | 6.02 | 14.77 |
| Insulating glass, 2 layers of 1/4" float, 1" overall, 30 to 40 square foot lites | | | | | |
|   Clear float glass | G1@.136 | SF | 9.61 | 6.25 | 15.86 |

# 08 Openings

| | Craft@Hrs | Unit | Material | Labor | Total |
|---|---|---|---|---|---|
| Acrylic plastic sheet (Plexiglas), clear except as noted | | | | | |
| 11" x 14" .093" thick | — | Ea | 3.28 | — | 3.28 |
| 12" x 24" .093" thick | — | Ea | 5.03 | — | 5.03 |
| 18" x 24" .093" thick | — | Ea | 9.77 | — | 9.77 |
| 24" x 48" .093" thick | — | Ea | 25.60 | — | 25.60 |
| 30" x 36" .093" thick | — | Ea | 20.90 | — | 20.90 |
| 36" x 48" .093" thick | — | Ea | 32.70 | — | 32.70 |
| 36" x 72" .093" thick | — | Ea | 49.70 | — | 49.70 |
| 18" x 24" .236" thick | — | Ea | 19.10 | — | 19.10 |
| 24" x 48" .236" thick | — | Ea | 56.90 | — | 56.90 |
| 30" x 36" .236" thick | — | Ea | 45.50 | — | 45.50 |
| 36" x 72" .236" thick | — | Ea | 104.00 | — | 104.00 |
| 30" x 36" .093" thick, bronze | — | Ea | 17.00 | — | 17.00 |
| Wired glass, 1/4", type 3 | | | | | |
| Clear | G1@.103 | SF | 7.40 | 4.74 | 12.14 |
| Hammered | G1@.103 | SF | 6.92 | 4.74 | 11.66 |
| Obscure | G1@.103 | SF | 6.65 | 4.74 | 11.39 |
| Mirrors, unframed | | | | | |
| Sheet glass, 3/16" | G1@.103 | SF | 5.31 | 4.74 | 10.05 |
| Float glass, 1/4" | G1@.106 | SF | 8.22 | 4.87 | 13.09 |
| Reflective, 1 way, in wood stops | G1@.118 | SF | 15.10 | 5.43 | 20.53 |
| Small lites, adhesive mount | G1@.057 | SF | 3.46 | 2.62 | 6.08 |
| Polycarbonate sheet glazing. 4' x 8' panels, Cyro | | | | | |
| Impact resistant, weather resistant sheets | | | | | |
| Clear 1/8" | G1@.027 | SF | 4.60 | 1.24 | 5.84 |
| Clear 3/16" | G1@.027 | SF | 4.82 | 1.24 | 6.06 |
| Clear 1/4" | G1@.027 | SF | 6.24 | 1.24 | 7.48 |
| Clear 1/2" | G1@.035 | SF | 11.40 | 1.61 | 13.01 |
| Add for tints | — | % | 17.0 | — | — |
| Chemical and abrasion resistant sheets | | | | | |
| Clear 1/8" | G1@.027 | SF | 7.04 | 1.24 | 8.28 |
| Clear 3/16" | G1@.027 | SF | 8.00 | 1.24 | 9.24 |
| Clear 1/4" | G1@.027 | SF | 11.70 | 1.24 | 12.94 |
| Clear 1/2" | G1@.035 | SF | 17.90 | 1.61 | 19.51 |
| Add for tints | — | % | 17.0 | — | — |
| Shock absorbent, high light transmission sheets (greenhouse applications) | | | | | |
| Clear 5/16" | G1@.027 | SF | 7.04 | 1.24 | 8.28 |
| Clear 5/8" | G1@.035 | SF | 8.00 | 1.61 | 9.61 |

# 09 Finishes

**Gypsum Wallboard** Commercial grade work. Material costs include 6% for waste.
Costs per square foot of area covered.
Gypsum wallboard nailed or screwed to wood framing or wood furring, no taping or finishing included

| | Craft@Hrs | Unit | Material | Labor | Total |
|---|---|---|---|---|---|
| 3/8" on walls | CD@.007 | SF | .20 | .35 | .55 |
| 3/8" on ceilings | CD@.009 | SF | .20 | .46 | .66 |
| 3/8" on furred columns and beams | CD@.013 | SF | .20 | .66 | .86 |
| 1/2" on walls | CD@.007 | SF | .21 | .35 | .56 |
| 1/2" on ceilings | CD@.009 | SF | .21 | .46 | .67 |
| 1/2" on furred columns and beams | CD@.015 | SF | .21 | .76 | .97 |

| | Craft@Hrs | Unit | Material | Labor | Total |
|---|---|---|---|---|---|
| 5/8" on walls | CD@.007 | SF | .26 | .35 | .61 |
| 5/8" on ceilings | CD@.010 | SF | .26 | .51 | .77 |
| 5/8" on furred columns and beams | CD@.015 | SF | .26 | .76 | 1.02 |
| Add for taping and finishing wall joints | CD@.007 | SF | .04 | .35 | .39 |
| Add for taping and finishing ceiling joints | CD@.009 | SF | .04 | .46 | .50 |

Gypsum wallboard clipped to metal furring, no taping or finishing included. Add the cost of furring from the sections that follow.

| | Craft@Hrs | Unit | Material | Labor | Total |
|---|---|---|---|---|---|
| 3/8" on wall furring | CD@.008 | SF | .20 | .40 | .60 |
| 3/8" on ceiling furring | CD@.011 | SF | .20 | .56 | .76 |
| 1/2" on wall furring | CD@.009 | SF | .21 | .46 | .67 |
| 1/2" on ceiling furring | CD@.011 | SF | .21 | .56 | .77 |
| 1/2" on column or beam furring | CD@.018 | SF | .21 | .91 | 1.12 |
| 5/8" on wall furring | CD@.009 | SF | .26 | .46 | .72 |
| 5/8" on ceiling furring | CD@.011 | SF | .26 | .56 | .82 |
| 5/8" on column or beam furring | CD@.018 | SF | .26 | .91 | 1.17 |
| 1/2", two layers on ceiling furring | CD@.022 | SF | .42 | 1.11 | 1.53 |
| 1/2", two layers on furring | CD@.017 | SF | .42 | .86 | 1.28 |
| 1/4" sound board | CD@.008 | SF | .25 | .40 | .65 |
| Add for taping and finishing wall joints | CD@.007 | SF | .04 | .35 | .39 |
| Add for taping and finishing ceiling joints | CD@.009 | SF | .04 | .46 | .50 |
| Add for taping only, no joint finishing | CD@.003 | SF | .01 | .15 | .16 |

Additional costs for gypsum wallboard

| | Craft@Hrs | Unit | Material | Labor | Total |
|---|---|---|---|---|---|
| Add for foil-backed board | — | SF | .06 | — | .06 |
| Add for fire resistant board | — | SF | .05 | — | .05 |
| Add for water resistant board | — | SF | .15 | — | .15 |
| Add for 10' or 12' wall heights | CD@.001 | SF | .02 | .05 | .07 |
| Add for school jobs | CD@.002 | SF | — | .10 | .10 |
| Add for trowel textured finish | CD@.009 | SF | .40 | .46 | .86 |
| Add for adhesive application, 1/4" bead | | | | | |
| Studs 16" on center | CD@.200 | CSF | .89 | 10.10 | 10.99 |
| Joists 16" on center | CD@.200 | CSF | .89 | 10.10 | 10.99 |

Bead, casing and channels

| | Craft@Hrs | Unit | Material | Labor | Total |
|---|---|---|---|---|---|
| Corner bead, 1-1/4" x 1-1/4" | CD@.017 | LF | .20 | .86 | 1.06 |
| Stop or casing | CD@.021 | LF | .26 | 1.06 | 1.32 |
| Jamb casing | CD@.023 | LF | .26 | 1.16 | 1.42 |
| RC-1 channel or 7/8" hat channel | CD@.011 | LF | .62 | .56 | 1.18 |

Vinyl clad gypsum board, adhesive or clip application on walls

| | Craft@Hrs | Unit | Material | Labor | Total |
|---|---|---|---|---|---|
| 1/2", no mouldings included | CD@.010 | SF | .58 | .51 | 1.09 |
| 5/8", no mouldings included | CD@.011 | SF | .61 | .56 | 1.17 |

**Furring and Lathing** For commercial applications. See plaster costs in the following section.

Furring on walls, cold rolled, galvanized

| | Craft@Hrs | Unit | Material | Labor | Total |
|---|---|---|---|---|---|
| 3/4", 12" on center | C8@.306 | SY | 3.99 | 14.00 | 17.99 |
| 3/4", 16" on center | C8@.270 | SY | 3.48 | 12.40 | 15.88 |
| 3/4", 24" on center | C8@.207 | SY | 2.63 | 9.49 | 12.12 |
| 1-1/2", 12" on center | C8@.342 | SY | 5.66 | 15.70 | 21.36 |
| 1-1/2", 16" on center | C8@.297 | SY | 4.85 | 13.60 | 18.45 |
| 1-1/2", 24" on center | C8@.234 | SY | 4.04 | 10.70 | 14.74 |
| 3-5/8", 12" on center, 25 gauge | C8@.360 | SY | 8.38 | 16.50 | 24.88 |
| 3-5/8", 16" on center, 25 gauge | C8@.315 | SY | 6.59 | 14.40 | 20.99 |
| 3-5/8", 24" on center, 25 gauge | C8@.261 | SY | 6.25 | 12.00 | 18.25 |

| | Craft@Hrs | Unit | Material | Labor | Total |
|---|---|---|---|---|---|
| **Furring on ceilings, cold rolled, galvanized** | | | | | |
| 3/4", 12" on center | C8@.413 | SY | 3.99 | 18.90 | 22.89 |
| 3/4", 16" on center | C8@.365 | SY | 3.48 | 16.70 | 20.18 |
| 3/4", 24" on center | C8@.280 | SY | 2.63 | 12.80 | 15.43 |
| 1-1/2", 12" on center | C8@.462 | SY | 5.66 | 21.20 | 26.86 |
| 1-1/2", 16" on center | C8@.400 | SY | 4.85 | 18.30 | 23.15 |
| 1-1/2", 24" on center | C8@.316 | SY | 4.04 | 14.50 | 18.54 |
| 1-1/2" x 3/4", coffered | C8@.887 | SY | 6.55 | 40.70 | 47.25 |
| **Furring on beams and columns, per linear foot of galvanized channel** | | | | | |
| 7/8" channel | C8@.059 | LF | .45 | 2.70 | 3.15 |
| 1-5/8" channel | C8@.059 | LF | .55 | 2.70 | 3.25 |
| **Resilient channel on beams and columns, per linear foot of channel** | | | | | |
| Narrow soundproofing channel, metal | C8@.011 | LF | .80 | .50 | 1.30 |
| Narrow soundproofing channel, PVC | C8@.011 | LF | 1.06 | .50 | 1.56 |
| Regular soundproofing channel, metal | C8@.011 | LF | .80 | .50 | 1.30 |
| Regular soundproofing channel, with padding tape | C8@.011 | LF | 1.08 | .50 | 1.58 |
| RC-1 resilient hat channel | C8@.011 | LF | .58 | .50 | 1.08 |
| RC-1 resilient hat channel, with padding tape | C8@.011 | LF | .88 | .50 | 1.38 |
| RC-2 resilient hat channel (for ceilings) | C8@.011 | LF | .75 | .50 | 1.25 |
| **Ceiling suspension system for lath, installed 7' to 9' above floor level with rod hangers attached to structural steel or metal decking, 1-1/2" main at 24" OC. Add the cost of lath.** | | | | | |
| With 3/4" cross at 12" OC | C8@.378 | SY | 4.75 | 17.30 | 22.05 |
| With 3/4" cross at 24" OC | C8@.342 | SY | 4.23 | 15.70 | 19.93 |
| With 1-1/2" cross at 16" OC | C8@.432 | SY | 5.20 | 19.80 | 25.00 |
| With 2-1/2" cross at 24" OC | C8@.360 | SY | 4.55 | 16.50 | 21.05 |
| **Metal lath on walls and ceilings, nailed in place** | | | | | |
| 2.5 lb., diamond lath | F8@.107 | SY | 6.09 | 5.42 | 11.51 |
| 2.5 lb., paper-backed | F8@.107 | SY | 6.87 | 5.42 | 12.29 |
| 3.4 lb., diamond lath | F8@.119 | SY | 6.48 | 6.02 | 12.50 |
| 3.4 lb., paper-backed | F8@.114 | SY | 7.24 | 5.77 | 13.01 |
| 3.4 lb., 3/8" high ribbed lath | F8@.121 | SY | 8.39 | 6.12 | 14.51 |
| 3.4 lb., 1/8" low ribbed lath | F8@.124 | SY | 8.84 | 6.28 | 15.12 |
| Add for galvanized metal lath | — | % | 10.0 | — | — |
| Add for metal lath wired in place | — | % | — | 15.0 | — |
| Add for metal lath on columns and beams | — | % | — | 100.0 | — |
| **Stucco mesh, reverse twist, self-furring, galvanized, includes 5% waste** | | | | | |
| 1" mesh, 18 gauge | F8@.079 | SY | 1.23 | 4.00 | 5.23 |
| 1-1/2" mesh, 17 gauge | F8@.079 | SY | 1.44 | 4.00 | 5.44 |
| 1-1/2" mesh, 17 gauge, paper back | F8@.079 | SY | 2.20 | 4.00 | 6.20 |
| **Gypsum lath, perforated or plain, clipped to metal studs** | | | | | |
| 3/8" regular | F8@.056 | SY | 8.48 | 2.83 | 11.31 |
| 3/8" firestop | F8@.058 | SY | 9.35 | 2.94 | 12.29 |
| 3/8" foil back | F8@.058 | SY | 9.42 | 2.94 | 12.36 |
| 1/2" regular | F8@.075 | SY | 9.35 | 3.80 | 13.15 |
| 1/2" firestop | F8@.075 | SY | 10.20 | 3.80 | 14.00 |
| 1/2" foil back | F8@.075 | SY | 10.30 | 3.80 | 14.10 |
| Add for application to ceilings | — | % | — | 50.0 | — |
| Add for application to columns or beams | — | % | — | 80.0 | — |
| Deduct for gypsum lath screwed in place | — | % | -5.0 | -5.0 | — |

|  | Craft@Hrs | Unit | Material | Labor | Total |
|---|---|---|---|---|---|
| **Plaster bead, 1-3/4" perforated flange** | | | | | |
| 1" casing bead | F8@.025 | LF | .39 | 1.27 | 1.66 |
| 1-1/4" casing bead | F8@.025 | LF | .50 | 1.27 | 1.77 |
| 1-1/2" casing bead | F8@.025 | LF | .61 | 1.27 | 1.88 |
| 3/4" casing bead, expanded wing | F8@.037 | LF | 1.22 | 1.87 | 3.09 |
| 2" casing bead | F8@.025 | LF | .67 | 1.27 | 1.94 |
| Square nose stop and casing | F8@.025 | LF | .66 | 1.27 | 1.93 |
| 3/4" radius corner bead | F8@.025 | LF | .72 | 1.27 | 1.99 |
| Expanded nose corner bead | F8@.025 | LF | 1.39 | 1.27 | 2.66 |
| **Control joints, 4" perforated flange** | | | | | |
| 1/2" joint | F8@.035 | LF | .83 | 1.77 | 2.60 |
| 5/8" joint | F8@.035 | LF | .87 | 1.77 | 2.64 |
| 3/4" joint | F8@.035 | LF | .91 | 1.77 | 2.68 |
| 3/4" joint, expanded wing | F8@.037 | LF | 2.92 | 1.87 | 4.79 |
| 7/8" joint | F8@.035 | LF | 1.13 | 1.77 | 2.90 |
| **Expansion joints, slip type** | | | | | |
| 3/4", 26 gauge, 1 piece | F8@.025 | LF | 1.39 | 1.27 | 2.66 |
| 1/4" to 5/8" adjustment, 1/2" to 3/4" deep | F8@.035 | LF | 2.09 | 1.77 | 3.86 |
| 1/4" to 5/8" adjustment, 7/8" deep | F8@.035 | LF | 2.26 | 1.77 | 4.03 |
| 5/8" to 1-1/8" adjustment, 1/2" to 3/4" deep | F8@.035 | LF | 2.19 | 1.77 | 3.96 |
| 5/8" to 1-1/8" adjustment, 7/8" deep | F8@.035 | LF | 2.26 | 1.77 | 4.03 |
| 1-1/8" to 1-5/8" adjustment, 1/2" to 3/4" deep | F8@.035 | LF | 2.19 | 1.77 | 3.96 |
| 1-1/8" to 1-5/8" adjustment, 7/8" deep | F8@.035 | LF | 2.27 | 1.77 | 4.04 |
| **Lathing accessories** | | | | | |
| Base screed, 1/2", 26 gauge | F8@.032 | LF | .88 | 1.62 | 2.50 |
| Vents, 1-1/2", galvanized | F8@.042 | LF | 2.50 | 2.13 | 4.63 |
| Vents, 4", galvanized | F8@.061 | LF | 2.93 | 3.09 | 6.02 |
| Archbead, plastic nose | F8@.028 | LF | 1.50 | 1.42 | 2.92 |
| Stud clips for gypsum lath, field clip | F8@.004 | Ea | .12 | .20 | .32 |
| Joist clips for lath | F8@.005 | Ea | .12 | .25 | .37 |
| 2-1/2" metal base, galvanized | F8@.042 | LF | .67 | 2.13 | 2.80 |
| Tie wire, 18 gauge, galvanized, case of 5 reels | — | Ea | 37.60 | — | 37.60 |
| **Plastering** These costs do not include the furring or lath | | | | | |
| **Gypsum interior plaster, lime putty trowel finish** | | | | | |
| Two coat application, on ceilings | F8@.381 | SY | 6.65 | 19.30 | 25.95 |
| Two coat application, on walls | F8@.330 | SY | 6.65 | 16.70 | 23.35 |
| Three coat application, on ceilings | F8@.448 | SY | 8.74 | 22.70 | 31.44 |
| Three coat application, on walls | F8@.405 | SY | 8.74 | 20.50 | 29.24 |
| **Gypsum interior vermiculite plaster, trowel finish** | | | | | |
| Two coat application, on ceilings | F8@.381 | SY | 6.99 | 19.30 | 26.29 |
| Two coat application, on walls | F8@.330 | SY | 6.99 | 16.70 | 23.69 |
| Three coat application, on ceilings | F8@.448 | SY | 9.10 | 22.70 | 31.80 |
| Three coat application, on walls | F8@.405 | SY | 9.10 | 20.50 | 29.60 |
| **Keene's cement plaster, troweled lime putty, medium hard finish** | | | | | |
| Two coat application, on ceilings | F8@.435 | SY | 10.10 | 22.00 | 32.10 |
| Two coat application, on walls | F8@.381 | SY | 10.10 | 19.30 | 29.40 |
| Three coat application, on ceilings | F8@.521 | SY | 13.10 | 26.40 | 39.50 |
| Three coat application, on walls | F8@.459 | SY | 13.10 | 23.20 | 36.30 |

| | Craft@Hrs | Unit | Material | Labor | Total |
|---|---|---|---|---|---|
| Portland cement stucco on exterior walls, 3 coats totaling 1" thick | | | | | |
| Natural gray, sand float finish | F8@.567 | SY | 8.08 | 28.70 | 36.78 |
| Natural gray, trowel finish | F8@.640 | SY | 8.08 | 32.40 | 40.48 |
| White cement, sand float finish | F8@.575 | SY | 9.10 | 29.10 | 38.20 |
| White cement, trowel finish | F8@.650 | SY | 9.10 | 32.90 | 42.00 |
| Portland cement stucco on soffits, 3 coats totaling 1" thick | | | | | |
| Natural gray, sand float finish | F8@.678 | SY | 8.08 | 34.30 | 42.38 |
| Natural gray, trowel finish | F8@.829 | SY | 8.08 | 42.00 | 50.08 |
| White cement, sand float finish | F8@.878 | SY | 9.10 | 44.40 | 53.50 |
| White cement, trowel finish | F8@1.17 | SY | 9.10 | 59.20 | 68.30 |
| Brown and scratch coat base for tile | F8@.308 | SY | 5.05 | 15.60 | 20.65 |
| Scratch coat only for tile | F8@.152 | SY | 2.23 | 7.69 | 9.92 |
| Thin coat plaster, 2 coats, with rock lath but not including studs | | | | | |
| 3/8" rock lath (gypsum lath) | F8@.265 | SY | 10.30 | 13.40 | 23.70 |
| Simulated acoustic texture, sometimes called "popcorn" or "cottage cheese" | | | | | |
| On ceilings | F8@.052 | SY | 1.30 | 2.63 | 3.93 |
| Texture coat on exterior walls and soffits | F8@.045 | SY | 1.30 | 2.28 | 3.58 |
| Patching gypsum plaster, including lath repair but no studding | | | | | |
| To 5 SF repairs | P5@.282 | SF | 1.51 | 12.40 | 13.91 |
| Over 5 SF repairs | P5@.229 | SF | 1.51 | 10.00 | 11.51 |
| Repair cracks only ($150 per job minimum) | P5@.055 | LF | .26 | 2.41 | 2.67 |
| Plaster moldings, ornate designs | | | | | |
| 2" | CC@.090 | LF | 4.19 | 4.66 | 8.85 |
| 4" | CC@.115 | LF | 8.55 | 5.95 | 14.50 |
| 6" | CC@.132 | LF | 12.40 | 6.83 | 19.23 |

**Metal Framed Shaft Walls** Shaft walls are used to enclose vertical shafts that surround plumbing and electrical chases, elevators and stairwells. A 2-hour wall can be made from 1" gypsum wallboard screwed to a 2-1/2" metal stud partition on the shaft side with two layers of 5/8" gypsum wallboard screwed to the partition exterior face. The wall cavity is filled with fiberglass batt insulation and the drywall is taped and finished on both sides. This wall will be 4-3/4" thick. Labor includes installing metal studs, installing the insulation, hanging, taping and finishing the drywall, and cleanup. Metal studs and insulation include a 10% allowance for waste. Wallboard costs include corner and edge trim and a 15% allowance for waste. Costs shown are per square foot of wall measured on one side. For scheduling purposes, estimate that a crew of 4 can install metal studs and insulation, hang, tape and finish 320 SF of shaft wall per 8-hour day.

| | Craft@Hrs | Unit | Material | Labor | Total |
|---|---|---|---|---|---|
| Metal studs, "C" section, 2-1/2" wide, 20 gauge, 24" on center | | | | | |
| Complete with top runner and bottom plate | C8@.036 | SF | .36 | 1.65 | 2.01 |
| 1" Type "X" gypsum shaftboard | CD@.029 | SF | .53 | 1.47 | 2.00 |
| 2 layers 5/8" Type "X" gypsum wallboard | CD@.027 | SF | .58 | 1.37 | 1.95 |
| Fiberglass insulation, foil faced | A1@.006 | SF | .44 | .30 | .74 |
| **Total for metal framed shaft wall as described** | —@.098 | **SF** | **1.91** | **4.79** | **6.70** |

**Ceramic Tile** Set in adhesive and grouted. Add the cost of adhesive below. No scratch or brown coat included. Based on standard U.S. grades and stock colors. Custom colors, designs and imported tile will cost more.

| | Craft@Hrs | Unit | Material | Labor | Total |
|---|---|---|---|---|---|
| 4-1/4" x 4-1/4" glazed wall tile | | | | | |
| Smooth gloss glaze, minimum quality | T4@.131 | SF | 2.15 | 5.80 | 7.95 |
| Smooth gloss glaze, standard quality | T4@.131 | SF | 4.75 | 5.80 | 10.55 |
| Matte finish, better quality | T4@.131 | SF | 6.24 | 5.80 | 12.04 |
| High gloss finish | T4@.131 | SF | 8.31 | 5.80 | 14.11 |
| Commercial quality | | | | | |
| Group 1 (matte glaze) | T4@.131 | SF | 3.80 | 5.80 | 9.60 |
| Group 2 (bright glaze) | T4@.131 | SF | 4.65 | 5.80 | 10.45 |
| Group 3 (crystal glaze) | T4@.131 | SF | 5.35 | 5.80 | 11.15 |

# 09 Finishes

| | Craft@Hrs | Unit | Material | Labor | Total |
|---|---|---|---|---|---|
| 6" wall tile, smooth glaze | | | | | |
| 6" x 6", gray or brown | T4@.131 | SF | 2.54 | 5.80 | 8.34 |
| 6" x 8", quilted look | T4@.131 | SF | 5.17 | 5.80 | 10.97 |
| 6" x 8", fume look | T4@.131 | SF | 4.55 | 5.80 | 10.35 |
| Trim pieces for glazed wall tile | | | | | |
| Surface bullnose | T4@.077 | LF | 1.74 | 3.41 | 5.15 |
| Surface bullnose corner | T4@.039 | Ea | 1.13 | 1.73 | 2.86 |
| Sink rail or cap | T4@.077 | LF | 5.03 | 3.41 | 8.44 |
| Radius bullnose | T4@.077 | LF | 2.29 | 3.41 | 5.70 |
| Quarter round or bead | T4@.077 | LF | 2.71 | 3.41 | 6.12 |
| Outside corner or bead | T4@.039 | Ea | .89 | 1.73 | 2.62 |
| Base | T4@.077 | LF | 6.16 | 3.41 | 9.57 |
| Soap holder, soap dish | T4@.265 | Ea | 12.20 | 11.70 | 23.90 |
| Tissue holder, towel bar | T4@.268 | Ea | 17.90 | 11.90 | 29.80 |
| Glazed floor tile, 1/4" thick | | | | | |
| Minimum quality, standard colors | T4@.210 | SF | 2.45 | 9.30 | 11.75 |
| Better quality, patterns | T4@.210 | SF | 2.93 | 9.30 | 12.23 |
| Commercial quality, 6" x 6" to 12" x 12" | | | | | |
| Standard colors and reds and browns | T4@.210 | SF | 2.81 | 9.30 | 12.11 |
| Blues and greens | T4@.210 | SF | 3.44 | 9.30 | 12.74 |
| Add for abrasive surface | — | SF | .29 | — | .29 |
| 1" x 1" mosaic tile, back-mounted, natural clays | | | | | |
| Group I colors (browns, reds) | T4@.143 | SF | 2.44 | 6.33 | 8.77 |
| Group II colors (tans, grays) | T4@.143 | SF | 2.67 | 6.33 | 9.00 |
| Group III colors (charcoal, blues) | T4@.143 | SF | 2.92 | 6.33 | 9.25 |
| Ceramic tile adhesive, premixed Type 1, 1 gallon covers 70 SF applied with #1 trowel (3/16" x 5/32" "V"-notch), 50 SF applied with #2 trowel (3-/16" x 1/4" "V"-notch) or 40 SF applied with #3 trowel (1/4" x 1/4" square-notch). | | | | | |
| 1 quart | — | Ea | 6.06 | — | 6.06 |
| 1 gallon | — | Ea | 12.60 | — | 12.60 |
| 3-1/2 gallons | — | Ea | 32.50 | — | 32.50 |
| Tile adhesive applied with #2 trowel | — | SF | .23 | — | .23 |
| Thin-set tile mortar. 50 pound bag covers 100 SF applied with #1 trowel (1/4" x 1/4" square-notch), 80 SF applied with #2 trowel (1/4" x 3/8" square-notch), 45 SF applied with #3 trowel (1/2" x 1/2" square notch). Cost per 50 pound bag. | | | | | |
| Standard, gray | — | Ea | 5.93 | — | 5.93 |
| Master-Blend, white | — | Ea | 8.37 | — | 8.37 |
| FlexiBond, gray | — | Ea | 28.30 | — | 28.30 |
| Marble and granite mix | — | Ea | 23.90 | — | 23.90 |
| Master-Blend applied with #3 trowel | — | SF | .25 | — | .25 |
| Tile grout, polymer-modified dry grout for joints 1/16" to 1/2". Coverage varies with tile and joint size. Cost per 25 pound bag. | | | | | |
| Unsanded | — | Ea | 12.40 | — | 12.40 |
| Sanded, light colors | — | Ea | 13.90 | — | 13.90 |
| Sanded, dark colors | — | Ea | 13.90 | — | 13.90 |
| Decorator tile panels, institutional quality, 6" x 9" x 3/4", typical costs | | | | | |
| Unglazed domestic | T4@.102 | SF | 5.71 | 4.52 | 10.23 |
| Wash-glazed domestic | T4@.102 | SF | 9.49 | 4.52 | 14.01 |
| Full glazed domestic | T4@.102 | SF | 10.00 | 4.52 | 14.52 |
| Glazed imported, decorative | T4@.102 | SF | 23.80 | 4.52 | 28.32 |
| Deduct for imported tile | — | % | -20.0 | — | — |
| Plastic tile | | | | | |
| 12" x 12" | T4@.010 | Ea | 2.95 | .44 | 3.39 |
| Ramp edges | T4@.010 | Ea | 1.29 | .44 | 1.73 |

| | Craft@Hrs | Unit | Material | Labor | Total |
|---|---|---|---|---|---|
| Aluminum tile, 12" x 12" | T4@.102 | SF | 35.00 | 4.52 | 39.52 |
| Copper on aluminum, 4-1/4" x 4-1/4" | T4@.102 | SF | 75.00 | 4.52 | 79.52 |
| Stainless steel tile, 4-1/4" x 4-1/4" | T4@.102 | SF | 75.00 | 4.52 | 79.52 |
| Quarry tile set in portland cement | | | | | |
| 5" high base tile | T4@.109 | LF | 3.22 | 4.83 | 8.05 |
| 4" x 4" x 1/2" floor tile | T4@.096 | SF | 3.80 | 4.25 | 8.05 |
| 6" x 6" x 1/2" floor tile | T4@.080 | SF | 3.12 | 3.54 | 6.66 |
| Quarry tile set in furan resin | | | | | |
| 5" high base tile | T4@.096 | LF | 2.24 | 4.25 | 6.49 |
| 6" x 6" x 3/4" floor tile | T4@.093 | SF | 5.37 | 4.12 | 9.49 |
| Add for abrasive finish | — | SF | .49 | — | .49 |

**Acoustical Treatment**

Tile board and panels, no suspension grid included, applied to ceilings with staples, not including furring.
Decorative (non-acoustic) Armstrong ceiling tile,
12" x 12" x 1/2", T&G

| | Craft@Hrs | Unit | Material | Labor | Total |
|---|---|---|---|---|---|
| Washable white | C8@.017 | SF | .72 | .78 | 1.50 |
| Grenoble | C8@.017 | SF | 1.02 | .78 | 1.80 |
| Tin design | C8@.017 | SF | 1.77 | .78 | 2.55 |
| Glenwood | C8@.017 | SF | .79 | .78 | 1.57 |
| Tundra Fire Guard, 2' x 2' x 5/8" | C8@.017 | SF | 1.81 | .78 | 2.59 |

Wood fiber ceiling tile, Advantage Series, USG. Class C panels for non-acoustical and non fire-rated applications. Tongue-and-groove edging hides staples for a smooth, clean look. 12" x 12" x 1/2" with staple flange. Stapled or set with adhesive.

| | Craft@Hrs | Unit | Material | Labor | Total |
|---|---|---|---|---|---|
| Lace | C8@.017 | SF | 0.85 | .78 | 1.63 |
| Tivoli | C8@.017 | SF | 0.98 | .78 | 1.76 |
| White wood fiber | C8@.017 | SF | 0.96 | .78 | 1.74 |

Acoustical ceiling tile adhesive, 237, Henry. For use on 12" x 12" or 12" x 24" tile. Bonds to concrete, concrete block, drywall, plaster and brick. 30-minute working time. One gallon covers 60 square feet.

| | Craft@Hrs | Unit | Material | Labor | Total |
|---|---|---|---|---|---|
| Per square foot | — | SF | .25 | — | .25 |
| Sound-absorbing fabric covered wall panels for attaching to structural walls | | | | | |
| AlphaSorb™, style 2100, 1" thick x 4' x 8' | C8@.038 | SF | 6.81 | 1.74 | 8.55 |
| AlphaSorb™, style 2100, 2" thick x 4' x 8' | C8@.038 | SF | 8.75 | 1.74 | 10.49 |
| Anchorage™, panels, 1" thick x 4' x 8' | C8@.038 | SF | 7.69 | 1.74 | 9.43 |
| Anchorage™, panels, 2" thick x 4' x 8' | C8@.038 | SF | 9.56 | 1.74 | 11.30 |
| Impaling clip fastener, per panel | — | Ea | 5.00 | — | 5.00 |
| Add for radius, mitered, or beveled edges, per panel | — | Ea | 10.00 | — | 10.00 |
| "Z"-Clip fasteners, per panel | — | Ea | 20.00 | — | 20.00 |

**Suspended Ceiling Grid Systems** 5,000 SF job. Add for ceiling tile below.
"T"-bar suspension system, no tile included, suspended from ceiling joists with wires

| | Craft@Hrs | Unit | Material | Labor | Total |
|---|---|---|---|---|---|
| 2' x 2' grid | C8@.011 | SF | .73 | .50 | 1.23 |
| 2' x 4' grid | C8@.010 | SF | .56 | .46 | 1.02 |
| 4' x 4' grid | C8@.009 | SF | .37 | .41 | .78 |
| Add for jobs under 5,000 SF | | | | | |
| Less than 2,500 SF | C8@.007 | SF | .46 | .32 | .78 |
| 2,500 SF to 4,500 SF | C8@.005 | SF | .36 | .23 | .59 |

**Ceiling Tile** Tile laid in suspended ceiling grid. No suspended ceiling grid system included. Labor column shows the cost of laying tile in a suspended ceiling grid.
Acoustical rated suspended ceiling panels

| | Craft@Hrs | Unit | Material | Labor | Total |
|---|---|---|---|---|---|
| 2' x 2', fissured, "A" fire rated | C8@.012 | SF | .58 | .55 | 1.13 |
| 2' x 2', random texture, "A" fire rated | C8@.012 | SF | .93 | .55 | 1.48 |
| 2' x 2' "A" fire rated, with Bioguard™ | C8@.012 | SF | 1.30 | .55 | 1.85 |
| 2' x 2', "A" fire rated, Armstrong "Sahara" | C8@.012 | SF | 1.36 | .55 | 1.91 |

|  | Craft@Hrs | Unit | Material | Labor | Total |
|---|---|---|---|---|---|
| 2' x 2', Armstrong "Classic" | C8@.012 | SF | 1.83 | .55 | 2.38 |
| 2' x 2' x 3/4", foil back, USG "Cheyenne" | C8@.012 | SF | 1.15 | .55 | 1.70 |
| 2' x 2' x 3/4", USG "Luna", fine textured | C8@.012 | SF | 1.65 | .55 | 2.20 |
| 2' x 4' x 5/8", USG "5th Avenue" | C8@.007 | SF | .50 | .32 | .82 |
| 2' x 4' x 5/8", fissured fire code | C8@.007 | SF | .74 | .32 | 1.06 |
| 2' x 4', fiberglass class "A" fire rated | C8@.007 | SF | .73 | .32 | 1.05 |
| 2' x 4' x 5/8", "A" fire rated, USG "5th Ave" | C8@.007 | SF | .65 | .32 | .97 |
| Non-acoustical suspended ceiling panels | | | | | |
| 2' x 4' x 1/2", Sheetrock® Gypsum Lay-In Ceiling Panel | C8@.012 | SF | .83 | .55 | 1.38 |
| 2' x 2' x 5/8", USG "Ceramic ClimaPlus" | C8@.012 | SF | 1.20 | .55 | 1.75 |
| 2' x 2' x 3/4", USG "Saville Row" | C8@.012 | SF | 1.47 | .55 | 2.02 |
| 2' x 2', grid pattern, Armstrong "Prestige" | C8@.012 | SF | .36 | .55 | .91 |
| 2' x 4', mineral fiber, USG "Plateau" | C8@.007 | SF | .29 | .32 | .61 |
| 2' x 4' x 9/16", mineral, USG "Stonehurst" | C8@.007 | SF | .49 | .32 | .81 |
| 2' x 4' random texture, "A" fire rated | C8@.007 | SF | .74 | .32 | 1.06 |
| 2' x 4' clean room, non-perforated | C8@.007 | SF | 1.80 | .32 | 2.12 |
| Translucent panels, 2' x 4' | | | | | |
| Clear or frosted prismatic | C8@.004 | SF | .64 | .18 | .82 |
| Arctic opal or cracked ice | C8@.004 | SF | 1.13 | .18 | 1.31 |
| Clear acrylic | C8@.004 | SF | .62 | .18 | .80 |
| White acrylic | C8@.004 | SF | .63 | .18 | .81 |

## Wood Flooring

|  | Craft@Hrs | Unit | Material | Labor | Total |
|---|---|---|---|---|---|
| Oak, 3/4" x 2-1/4", unfinished | | | | | |
| Select grade, white | F9@.062 | SF | 6.52 | 2.80 | 9.32 |
| Select, grade, red | F9@.062 | SF | 7.91 | 2.80 | 10.71 |
| Select, grade, ash | F9@.062 | SF | 7.91 | 2.80 | 10.71 |
| Parquet, prefinished, 5/16" | | | | | |
| Ash or quartered oak, select | F9@.053 | SF | 16.90 | 2.40 | 19.30 |
| Ash or quartered oak, natural | F9@.053 | SF | 12.30 | 2.40 | 14.70 |
| Cherry or walnut select grade | F9@.053 | SF | 19.50 | 2.40 | 21.90 |
| Cherry or walnut natural grade | F9@.053 | SF | 14.70 | 2.40 | 17.10 |
| Oak, select plain | F9@.053 | SF | 14.00 | 2.40 | 16.40 |
| Oak, natural plain | F9@.053 | SF | 12.00 | 2.40 | 14.40 |
| Concrete moisture vapor barriers | | | | | |
| Koester VAP primer (3 coat system) | FL@.026 | SF | 3.02 | 1.31 | 4.33 |
| Koester VAP primer (4 coat system | FL@.033 | SF | 4.67 | 1.67 | 6.34 |
| Gym floors, 3/4", maple, including finishing | | | | | |
| On sleepers and membrane | F9@.086 | SF | 6.29 | 3.89 | 10.18 |
| On steel spring system | F9@.114 | SF | 7.47 | 5.15 | 12.62 |
| Connected with steel channels | F9@.097 | SF | 6.56 | 4.38 | 10.94 |
| Softwood floors, fir, unfinished | | | | | |
| 1" x 2" | F9@.031 | SF | 4.27 | 1.40 | 5.67 |
| 1" x 3" | F9@.031 | SF | 7.44 | 1.40 | 8.84 |
| Finishing wood flooring | | | | | |
| Fill and stain | F9@.010 | SF | .09 | .45 | .54 |
| Two coats urethane | F9@.047 | SF | .15 | 2.12 | 2.27 |
| Wax coating | F9@.001 | SF | .03 | .05 | .08 |
| Sand, scrape and edge hardwood floor | | | | | |
| Large room or hall, using 12" drum sander and 7" disc sander | | | | | |
| Three cuts | F9@.009 | SF | .05 | .41 | .46 |
| Four cuts | F9@.011 | SF | .06 | .50 | .56 |
| Closet or 4' x 10' area, using 7" disc sander | | | | | |
| Three cuts | F9@.010 | SF | .04 | .45 | .49 |
| Four cuts | F9@.012 | SF | .04 | .54 | .58 |

# 09 Finishes

| | Craft@Hrs | Unit | Material | Labor | Total |
|---|---|---|---|---|---|
| **Resilient Flooring** | | | | | |
| Asphalt tile | | | | | |
| 9" x 9" x 1/8" | F9@.018 | SF | 1.24 | .81 | 2.05 |
| Vinyl composition and vinyl tile, 12" x 12" | | | | | |
| No wax, self-stick, Metro, 3 year warranty | F9@.016 | SF | .73 | .72 | 1.45 |
| No wax, self-stick, Themes, 5 year warranty | F9@.016 | SF | 1.20 | .72 | 1.92 |
| No wax, self-stick, urethane, 25 year warranty | F9@.016 | SF | 1.59 | .72 | 2.31 |
| | | | | | |
| **Sheet vinyl flooring**, Armstrong. Includes 10% waste and adhesive. | | | | | |
| Medley series, .08" thick, urethane no-wax wear layer. Includes 10% waste and adhesive. | | | | | |
| Chalk White | FL@.300 | SY | 17.50 | 15.10 | 32.60 |
| Sage Stone | FL@.300 | SY | 18.10 | 15.10 | 33.20 |
| White, Edinburgh | FL@.300 | SY | 17.90 | 15.10 | 33.00 |
| Sentinel series, vinyl wear layer, .04" overall, ten year warranty. Includes 10% waste and adhesive. | | | | | |
| Most colors and textures | FL@.300 | SY | 9.81 | 15.10 | 24.91 |
| Select designs | FL@.300 | SY | 10.40 | 15.10 | 25.50 |
| Themes series, .08" thick, 2.9 Performance Appearance Rating | | | | | |
| Beige Cameo | FL@.300 | SY | 11.90 | 15.10 | 27.00 |
| Everest White | FL@.300 | SY | 12.60 | 15.10 | 27.70 |
| Green and Natural, Alexandria V | FL@.300 | SY | 11.80 | 15.10 | 26.90 |
| Green, Florentino | FL@.300 | SY | 12.60 | 15.10 | 27.70 |
| Light Shale, Leed's Landing | FL@.300 | SY | 12.60 | 15.10 | 27.70 |
| Natural White, Prescott | FL@.300 | SY | 12.80 | 15.10 | 27.90 |
| Sandstone | FL@.300 | SY | 11.90 | 15.10 | 27.00 |
| White Crystal, Marble Wisp | FL@.300 | SY | 12.80 | 15.10 | 27.90 |
| Sundial series, .098 thick. No adhesive needed | | | | | |
| Black/White, Florence 2 | FL@.300 | SY | 14.00 | 15.10 | 29.10 |
| Emerald and White, Florence 2 | FL@.300 | SY | 14.10 | 15.10 | 29.20 |
| Light Neutral, Antigua 2 | FL@.300 | SY | 13.70 | 15.10 | 28.80 |
| Light Oak, Myrtlewood | FL@.300 | SY | 13.30 | 15.10 | 28.40 |
| Light Wood, Parquet de Luxe | FL@.300 | SY | 14.00 | 15.10 | 29.10 |
| Limestone, Quinault Light | FL@.300 | SY | 13.20 | 15.10 | 28.30 |
| Mushroom, Bologna 2 | FL@.300 | SY | 14.10 | 15.10 | 29.20 |
| Rust, Chestnut Corner | FL@.300 | SY | 13.20 | 15.10 | 28.30 |
| Stone/White, Romanesque 2 | FL@.300 | SY | 13.90 | 15.10 | 29.00 |
| Metro series, .045" thick. vinyl no-wax wear layer. 2.3 Performance Appearance Rating | | | | | |
| Adobe Tan | FL@.300 | SY | 7.07 | 15.10 | 22.17 |
| Blue Green, colorweave | FL@.300 | SY | 10.20 | 15.10 | 25.30 |
| Bluelake | FL@.300 | SY | 7.54 | 15.10 | 22.64 |
| Cocoa | FL@.300 | SY | 7.03 | 15.10 | 22.13 |
| Jewel Flower, colorweave | FL@.300 | SY | 9.97 | 15.10 | 25.07 |
| Media Heights, Sandstone | FL@.300 | SY | 8.02 | 15.10 | 23.12 |
| Navy | FL@.300 | SY | 9.89 | 15.10 | 24.99 |
| Sand | FL@.300 | SY | 7.49 | 15.10 | 22.59 |
| Royelle series, .05" thick. 1.3 Performance Appearance Rating | | | | | |
| Black and White, Sheffley | FL@.300 | SY | 4.69 | 15.10 | 19.79 |
| Dark Oak, Augusta | FL@.300 | SY | 5.06 | 15.10 | 20.16 |
| Seashell, Kaley's Korner | FL@.300 | SY | 5.05 | 15.10 | 20.15 |
| | | | | | |
| **Plastic Interlocking Safety Floor Tiles** Based on Duragrid for a 1,000 to 5,000 SF job. | | | | | |
| Firm, smooth tiles, for athletic courts, gym floors, industrial areas | | | | | |
| 12" x 12" x 1/2" | F9@.003 | SF | 3.79 | .14 | 3.93 |
| 2" x 12" x 1/2" line strips | F9@.003 | LF | 2.19 | .14 | 2.33 |

| | Craft@Hrs | Unit | Material | Labor | Total |
|---|---|---|---|---|---|
| Firm, non-slip (nippled) tile for tennis courts, wet-weather surface, | | | | | |
|    12" x 12" x 1/2" | F9@.003 | SF | 3.35 | .14 | 3.49 |
|      2" x 12" x 1/2" line strips | F9@.003 | LF | 2.19 | .14 | 2.33 |
| Soft tiles, for anti-fatigue matting, household or deck areas | | | | | |
|    12" x 12" x 1/2" | F9@.003 | SF | 3.53 | .14 | 3.67 |
|      2" x 12" x 1/2" line strips | F9@.003 | LF | 2.19 | .14 | 2.33 |
| Mitered edging strip, 2" x 12" x 1/2" | F9@.003 | LF | 2.19 | .14 | 2.33 |
|    Monogramming/graphics pegs | F9@.085 | C | 3.72 | 3.84 | 7.56 |
|    Add for jobs under 1,000 SF | — | % | 10.0 | 20.0 | — |
|    Deduct for jobs over 5,000 SF | — | % | -5.0 | -5.0 | — |

**Rubber Flooring and Accessories**

| | Craft@Hrs | Unit | Material | Labor | Total |
|---|---|---|---|---|---|
| Rubber tile flooring | | | | | |
|    17-13/16" x 17-13/16" x 1/8" circular pattern | F9@.021 | SF | 9.48 | .95 | 10.43 |
| Base, top set rubber, 1/8" thick | | | | | |
|    2-1/2" high | F9@.018 | LF | .96 | .81 | 1.77 |
|    4" high | F9@.018 | LF | 1.21 | .81 | 2.02 |
|    6" high | F9@.018 | LF | 1.56 | .81 | 2.37 |
| Base corners, top set rubber, 1/8" thick | | | | | |
|    2-1/2" or 4" high | F9@.028 | Ea | 1.56 | 1.27 | 2.83 |
|    6" high | F9@.028 | Ea | 1.97 | 1.27 | 3.24 |
| Stair treads, molded rubber, 12-1/2" tread width, per LF of tread length | | | | | |
|    1/4" circle design, colors | F9@.067 | LF | 14.50 | 3.03 | 17.53 |
|    Rib design, light duty | F9@.067 | LF | 15.80 | 3.03 | 18.83 |
|    Tread with abrasive strip, light duty | F9@.071 | LF | 17.50 | 3.21 | 20.71 |
|    Tread with abrasive strip, heavy duty | F9@.071 | LF | 22.70 | 3.21 | 25.91 |
|    Stair risers, molded rubber, 7" x 1/8" | F9@.038 | LF | 4.59 | 1.72 | 6.31 |
|    Stringer cover, 0.100", 10" high | F9@.106 | LF | 4.59 | 4.79 | 9.38 |
|    Add for project under 2,000 SF | F9@.010 | LF | .26 | .45 | .71 |
|    Underlayment, plywood, 5/8" | C8@.009 | SF | .78 | .41 | 1.19 |

**Shock-absorbing resilient tile, rubber** Installed over asphalt, concrete or compacted gravel base. Indoor or outdoor use, for playgrounds, factory floors, aerobic centers, recreational decks and docks. Unit costs below include single component polyurethane adhesive (also listed separately below). CSSI. www.carlsurf.com

| | Craft@Hrs | Unit | Material | Labor | Total |
|---|---|---|---|---|---|
| 2-1/4" thickness, meets CPSC guidelines for 6' drop heights | | | | | |
|    Full tile, 24" x 24" | F9@.014 | SF | 9.95 | .63 | 10.58 |
|    Diagonal tile | F9@.014 | Ea | 23.60 | .63 | 24.23 |
|    Transition ramp, 48" x 8" | F9@.056 | Ea | 28.50 | 2.53 | 31.03 |
|    Accessible ramp, 24" x 24" | F9@.056 | Ea | 44.20 | 2.53 | 46.73 |
|    90-degree molded outside corner 16" x 8" | F9@.112 | Ea | 21.60 | 5.06 | 26.66 |
|    90-degree molded inside corner 12" x 8" | F9@.110 | Ea | 21.20 | 4.97 | 26.17 |
|    45-degree molded outside corner 8" x 8" | F9@.110 | Ea | 20.90 | 4.97 | 25.87 |
|    45-degree molded inside corner 8" x 8" | F9@.109 | Ea | 19.30 | 4.93 | 24.23 |
| 3-3/4" thickness, meets CPSC guidelines for 11' drop heights | | | | | |
|    Full tile, 24" x 24" | F9@.017 | SF | 17.10 | .77 | 17.87 |
|    Ramp, 48" x 12" | F9@.056 | Ea | 67.00 | 2.53 | 69.53 |
|    Add for one part polyurethane adhesive | | | | | |
|    (30 SF per gal) | — | Gal | 47.80 | — | 47.80 |

**Composition Flooring** 1/4" thick

| | Craft@Hrs | Unit | Material | Labor | Total |
|---|---|---|---|---|---|
|    Acrylic | F9@.075 | SF | 3.75 | 3.39 | 7.14 |
|    Epoxy terrazzo | F9@.105 | SF | 4.25 | 4.74 | 8.99 |
|    Conductive epoxy terrazzo | F9@.132 | SF | 5.89 | 5.97 | 11.86 |
|    Troweled neoprene | F9@.082 | SF | 4.80 | 3.71 | 8.51 |
|    Polyester | F9@.065 | SF | 3.20 | 2.94 | 6.14 |

# 09 Finishes

| | Craft@Hrs | Unit | Material | Labor | Total |
|---|---|---|---|---|---|
| **Terrazzo** | | | | | |
| Floors, 1/2" terrazzo topping on an underbed bonded to an existing concrete slab, | | | | | |
| #1 and #2 chips in gray portland cement. Add divider strips below | | | | | |
| 2" epoxy or polyester | T4@.128 | SF | 3.47 | 5.67 | 9.14 |
| 2" conductive | T4@.134 | SF | 4.50 | 5.94 | 10.44 |
| 2" with #3 and larger chips | T4@.134 | SF | 5.96 | 5.94 | 11.90 |
| 3" with 15 lb felt and sand cushion | T4@.137 | SF | 4.50 | 6.07 | 10.57 |
| 2-3/4" with mesh, felt and sand cushion | T4@.137 | SF | 4.38 | 6.07 | 10.45 |
| Additional costs for terrazzo floors. Add for: | | | | | |
| White portland cement | — | SF | .57 | — | .57 |
| Non-slip abrasive, light | — | SF | 1.24 | — | 1.24 |
| Non-slip abrasive, heavy | — | SF | 1.89 | — | 1.89 |
| Countertops, mud set | T4@.250 | SF | 11.00 | 11.10 | 22.10 |
| Wainscot, precast, mud set | T4@.198 | SF | 10.40 | 8.77 | 19.17 |
| Cove base, mud set | T4@.198 | LF | 2.53 | 8.77 | 11.30 |
| Add for 2,000 SF job | — | % | 11.0 | 15.0 | — |
| Add for 1,000 SF job | — | % | 44.0 | 60.0 | — |
| Add for waterproof membrane | — | SF | 1.36 | — | 1.36 |
| Divider strips | | | | | |
| Brass, 12 gauge | T4@.009 | LF | 2.72 | .40 | 3.12 |
| White metal, 12 gauge | T4@.009 | LF | 1.25 | .40 | 1.65 |
| Brass 4' OC each way | T4@.004 | SF | 1.43 | .18 | 1.61 |
| Brass 2' OC each way | T4@.007 | SF | 2.77 | .31 | 3.08 |
| **Carpeting** Glue-down carpet installation includes sweeping the floor and applying adhesive. Carpet over pad | | | | | |
| installation includes laying the pad and stapling or spot adhesive. | | | | | |
| Nylon, cut pile, with 3/8" nova pad | | | | | |
| 30 oz | F9@.114 | SY | 15.10 | 5.15 | 20.25 |
| 36 oz | F9@.114 | SY | 17.60 | 5.15 | 22.75 |
| 40 oz | F9@.114 | SY | 20.00 | 5.15 | 25.15 |
| Nylon, level loop, with 3/8" nova pad | | | | | |
| 20 oz | F9@.114 | SY | 15.10 | 5.15 | 20.25 |
| 24 oz | F9@.114 | SY | 17.60 | 5.15 | 22.75 |
| Base grade, level loop, with 50 oz pad | | | | | |
| 18 oz | F9@.114 | SY | 7.83 | 5.15 | 12.98 |
| 20 oz | F9@.114 | SY | 8.77 | 5.15 | 13.92 |
| 26 oz | F9@.114 | SY | 8.77 | 5.15 | 13.92 |
| Antron nylon, 20 oz, anti-static, no pad | F9@.068 | SY | 10.60 | 3.07 | 13.67 |
| Lee's "Faculty IV" with unibond back, direct glue | F9@.114 | SY | 20.60 | 5.15 | 25.75 |
| **Carpet Tile** | | | | | |
| Modular carpet tiles | | | | | |
| "Faculty II" tiles, 26 oz face weight | F9@.170 | SY | 16.30 | 7.68 | 23.98 |
| "Surfaces" tiles, 38 oz face weight | F9@.170 | SY | 17.30 | 7.68 | 24.98 |
| ESD 26 oz loop pile for computer floors | F9@.170 | SY | 18.90 | 7.68 | 26.58 |
| **Access Flooring** Standard 24" x 24" steel flooring panels with 1/8" high pressure laminate surface. Costs shown | | | | | |
| include the floor panels and the supporting understructures listed. Add for accessories below. Raised Floor | | | | | |
| Installation, Inc. | | | | | |
| Stringerless system, up to 10" finish floor height | | | | | |
| Under 1,000 SF job | D3@.058 | SF | 14.80 | 3.02 | 17.82 |
| 1,000 to 5,000 SF job | D3@.046 | SF | 12.70 | 2.39 | 15.09 |
| Over 5,000 SF job | D3@.039 | SF | 11.10 | 2.03 | 13.13 |

432

| | Craft@Hrs | Unit | Material | Labor | Total |
|---|---|---|---|---|---|
| Snap-on stringer system, up to 10" finish floor height | | | | | |
| Under 1,000 SF job | D3@.058 | SF | 16.20 | 3.02 | 19.22 |
| 1,000 to 5,000 SF job | D3@.046 | SF | 13.70 | 2.39 | 16.09 |
| Over 5,000 SF job | D3@.039 | SF | 12.30 | 2.03 | 14.33 |
| Rigid bolted stringer system, up to 18" finish floor height | | | | | |
| Under 1,000 SF job | D3@.058 | SF | 17.10 | 3.02 | 20.12 |
| 1,000 to 5,000 SF job | D3@.046 | SF | 14.40 | 2.39 | 16.79 |
| Over 5,000 SF job | D3@.039 | SF | 13.10 | 2.03 | 15.13 |
| Corner lock system, under 2,500 SF office areas, 6" to 10" finish floor height, blank panels | D3@.036 | SF | 7.83 | 1.87 | 9.70 |
| Used flooring, most floor system types | | | | | |
| Unrefurbished, typical | D3@.058 | SF | 10.70 | 3.02 | 13.72 |
| Refurbished, new panel surface and trim | D3@.058 | SF | 14.60 | 3.02 | 17.62 |
| Extras for all floor systems, add to the costs above | | | | | |
| Carpeted panels | — | SF | 1.46 | — | 1.46 |
| Concrete-filled steel panels | — | SF | .85 | — | .85 |
| Ramps, per SF of actual ramp area | D3@.104 | SF | 9.38 | 5.41 | 14.79 |
| Non-skid tape on ramp surface | D3@.033 | SF | 5.74 | 1.72 | 7.46 |
| Portable ramp for temporary use | — | SF | 63.50 | — | 63.50 |
| Steps, per SF of actual tread area | D3@.208 | SF | 15.50 | 10.80 | 26.30 |
| Fascia at steps and ramps, to 18" high | D3@.117 | LF | 10.80 | 6.09 | 16.89 |
| Cove base, 4" high, black or brown | D3@.026 | LF | 1.78 | 1.35 | 3.13 |
| Guardrail | D3@.573 | LF | 64.00 | 29.80 | 93.80 |
| Handrail | D3@.573 | LF | 29.50 | 29.80 | 59.30 |
| Cable cutouts | | | | | |
| Standard "L"-type trim | D3@.407 | Ea | 10.20 | 21.20 | 31.40 |
| Type "F" trim, under floor plenum systems | D3@.407 | Ea | 11.80 | 21.20 | 33.00 |
| Perforated air distribution panels | — | SF | 4.10 | — | 4.10 |
| Air distribution grilles, 6" x 18" | | | | | |
| With adjustable damper, including cutout | D3@.673 | Ea | 76.20 | 35.00 | 111.20 |
| Fire extinguishing systems, smoke detectors, halon type, subcontract | | | | | |
| With halon gas suppression | — | SF | — | — | 5.26 |
| Automatic fire alarm | — | SF | — | — | .82 |

**Wallcoverings** Based on 2,000 SF job. These costs include 10% waste allowance. For estimating purposes figure an average roll of wallpaper will cover 40 SF. The cost of wallpaper will vary greatly.

| | Craft@Hrs | Unit | Material | Labor | Total |
|---|---|---|---|---|---|
| Paperhanging, average quality, labor only | PA@.013 | SF | — | .67 | .67 |
| Paperhanging, good quality, labor only | PA@.016 | SF | — | .82 | .82 |
| Vinyl wallcovering, typical costs | | | | | |
| 7 oz, light grade | PA@.014 | SF | .36 | .72 | 1.08 |
| 14 oz, medium grade | PA@.015 | SF | .61 | .77 | 1.38 |
| 22 oz, heavy grade, premium quality | PA@.019 | SF | .95 | .98 | 1.93 |
| Vinyl, 14 oz, on aluminum backer | PA@.035 | SF | .96 | 1.80 | 2.76 |
| Flexwood, many veneers, typical job | PA@.053 | SF | 2.52 | 2.73 | 5.25 |
| Cloth wallcovering | | | | | |
| Linen, acrylic backer, scotch guarded | PA@.021 | SF | 1.20 | 1.08 | 2.28 |
| Grasscloth | PA@.021 | SF | 1.90 | 1.08 | 2.98 |
| Felt | PA@.027 | SF | 2.41 | 1.39 | 3.80 |
| Cork sheathing, 1/4", typical | PA@.027 | SF | 1.50 | 1.39 | 2.89 |
| Blank stock (underliner) | PA@.013 | SF | 2.77 | .67 | 3.44 |
| Laminated cork sheets, typical | PA@.015 | SF | 3.86 | .77 | 4.63 |

# 09 Finishes

| | Craft@Hrs | Unit | Material | Labor | Total |
|---|---|---|---|---|---|
| Laminated cork sheets, colors | PA@.035 | SF | 6.08 | 1.80 | 7.88 |

Gypsum impregnated jute fabric Flexi-wall™, with #500 adhesive. Use 360 SF as minimum job size

| | Craft@Hrs | Unit | Material | Labor | Total |
|---|---|---|---|---|---|
| Medium weight plaster wall liner #605 | PA@.014 | SF | 1.28 | .72 | 2.00 |
| Heavy duty plaster wall liner #609 | PA@.014 | SF | 1.47 | .72 | 2.19 |
| Classics line | PA@.016 | SF | 1.51 | .82 | 2.33 |
| Images line | PA@.016 | SF | 1.88 | .82 | 2.70 |
| Trimmed paper, hand prints, foils | PA@.012 | SF | 1.17 | .62 | 1.79 |
| Add for job less than 200 SF (5 rolls) | — | % | — | 20.0 | — |
| Add for small rooms (kitchen, bath) | — | % | 10.0 | 20.0 | — |
| Add for patterns | — | % | — | 15.0 | — |
| Deduct for job over 2,000 SF (50 rolls) | — | % | — | -5.0 | — |

**Vinyl Wallcovering and Borders** Solid vinyl, prepasted, standard patterns and colors.

| | Craft@Hrs | Unit | Material | Labor | Total |
|---|---|---|---|---|---|
| Gold floral on green | PA@.028 | SF | .56 | 1.44 | 2.00 |
| Red, green and gold | PA@.028 | SF | .54 | 1.44 | 1.98 |
| Ivy and berries | PA@.035 | SF | .54 | 1.80 | 2.34 |
| Pastel shells on sand | PA@.017 | SF | .44 | .87 | 1.31 |
| Coordinated border | PA@.028 | SF | .35 | 1.44 | 1.79 |

**Hardboard Wallcovering** Vinyl clad, printed.

| | Craft@Hrs | Unit | Material | Labor | Total |
|---|---|---|---|---|---|
| 3/16" pegboard | C8@.025 | SF | .52 | 1.15 | 1.67 |
| 1/4" perforated white garage liner | C8@.025 | SF | .43 | 1.15 | 1.58 |
| Add for adhesive, 1/4" bead, 16" OC | — | SF | .13 | — | .13 |
| Add for metal trim | — | LF | .46 | — | .46 |

**Decontamination and Surface Preparation** Dustless removal of undesirable surface contamination and paint from steel and concrete substrates. Applications include PCBs, radioactivity, toxic chemicals and lead-based paints. The removal system is a mechanical process and no water, chemicals or abrasive grit are used. Containment or special ventilation is not required. Based on Pentek Inc. For estimating purposes figure one 55-gallon drum of contaminated waste per 2,500 square feet of coatings removed from surfaces with an average coating thickness of 12 mils. Waste disposal costs are not included. Unit prices shown are based on a crew of three. Use $4,500 as a minimum charge for work of this type. Equipment is based on using three pneumatic needle scaler units attached to one air-powered vacuum-waste packing unit. Equipment cost includes one vacuum-waste packing unit, three needle scalers, and one 150 CFM gas or diesel powered air compressor including all hoses and connectors. Add the one-time charge for move-on, move-off and refurbishing of equipment as shown.

| | Craft@Hrs | Unit | Material | Labor | Equipment | Total |
|---|---|---|---|---|---|---|

**Preparation of steel surfaces to bare metal.** Steel Structures Painting Council Surface Preparation (SSPC-SP) specification codes are shown

Steel surfaces, SSPC-SP-11 (coating mechanically removed, bare metal exposed)
Overhead surfaces, box girders, obstructed areas

| | Craft@Hrs | Unit | Material | Labor | Equipment | Total |
|---|---|---|---|---|---|---|
| 30 SF of surface area per hour | PA@.100 | SF | — | 5.14 | 3.85 | 8.99 |

Large unobstructed structural steel members

| | | | | | | |
|---|---|---|---|---|---|---|
| 40 SF of surface area per hour | PA@.075 | SF | — | 3.86 | 2.89 | 6.75 |

Above ground storage tanks

| | | | | | | |
|---|---|---|---|---|---|---|
| 50 SF of surface area per hour | PA@.060 | SF | — | 3.09 | 2.31 | 5.40 |

Steel surfaces, SSPC-SP-3 (coating mechanically removed, mill scale left in place)
Overhead surfaces, box girders, obstructed areas

| | | | | | | |
|---|---|---|---|---|---|---|
| 50 SF of surface area per hour | PA@.060 | SF | — | 3.09 | 2.31 | 5.40 |

Large unobstructed structural steel members

| | | | | | | |
|---|---|---|---|---|---|---|
| 70 SF of surface area per hour | PA@.042 | SF | — | 2.16 | 1.62 | 3.78 |

| | Craft@Hrs | Unit | Material | Labor | Equipment | Total |
|---|---|---|---|---|---|---|
| Above ground storage tanks | | | | | | |
| 90 SF of surface area per hour | PA@.033 | SF | — | 1.70 | 1.27 | 2.97 |
| Door or window frames | | | | | | |
| To 3'0" x 6'8" | PA@1.50 | Ea | — | 77.10 | 57.70 | 134.80 |
| Over 3'0" x 6'8" to 6'0" x 8'0" | PA@2.00 | Ea | — | 103.00 | 76.80 | 179.80 |
| Over 6'0" x 8'0" to 12'0 x 20'0" | PA@3.00 | Ea | — | 154.00 | 115.00 | 269.00 |

## Preparation of concrete walls and lintels

| | Craft@Hrs | Unit | Material | Labor | Equipment | Total |
|---|---|---|---|---|---|---|
| Concrete lintels, per SF of area | PA@.033 | SF | — | 1.70 | 1.27 | 2.97 |
| Concrete walls, measured on one face of wall | | | | | | |
| 80 SF of surface area per hour | PA@.038 | SF | — | 1.95 | 1.46 | 3.41 |
| Concrete block walls, measured on one face of wall | | | | | | |
| 70 SF of surface area per hour | PA@.042 | SF | — | 2.16 | 1.62 | 3.78 |
| Add to total estimated costs arrived at using above, one-time cost | | | | | | |
| Move-on, move-off | | | | | | |
| and refurbish equipment | — | LS | — | — | — | 3,400.00 |

## Preparation of concrete floors and slabs

Equipment is based on using three pneumatic manually-operated wheel-mounted scabblers attached to one air-powered vacuum-waste packing unit to scarify concrete floors and slabs. Equipment cost includes one vacuum-waste packing unit, three scabblers, and one 150 CFM gas or diesel powered air compressor including all hoses and connectors. Add the one-time charge for move-on, move-off and refurbishing of equipment as shown.

| | Craft@Hrs | Unit | Material | Labor | Equipment | Total |
|---|---|---|---|---|---|---|
| Large unobstructed areas; warehouses and aircraft hanger floor slabs | | | | | | |
| 90 SF of surface area per hour | PA@.033 | SF | — | 1.70 | 1.52 | 3.22 |
| Obstructed areas; narrow aisles, areas in pits | | | | | | |
| 60 SF of surface area per hour | PA@.050 | SF | — | 2.57 | 2.30 | 4.87 |
| Small areas; around equipment, piping and conduit | | | | | | |
| 40 SF of surface area per hour | PA@.075 | SF | — | 3.86 | 3.45 | 7.31 |
| Add to total estimated costs arrived at using above, one-time cost | | | | | | |
| Move-on, move-off | | | | | | |
| and refurbish equipment | — | LS | — | — | — | 4,450.00 |

**Painting — commercial and industrial work** Painting costs in this section assume work can be done from the floor or from ladders. Add the cost of scaffolding or staging when if required. Painting costs in this section include no surface preparation and no masking or protection of adjacent surfaces. Estimates for surface preparation and protection of adjacent surfaces are shown separately below. For more complete information on commercial and industrial painting costs, see *National Painting Cost Estimator* at http://CraftsmanSiteLicense.com/

| | Craft@Hrs | Unit | Material | Labor | Total |
|---|---|---|---|---|---|
| **Preparing metals for painting** SSPC is Steel Structures Painting Council. | | | | | |
| Brush, scrape, sand by hand | | | | | |
| (SSPC-2), 50 SF per hour | PA@.020 | SF | — | 1.03 | 1.03 |
| Power tool cleaning (SSPC-3), 100 SF per hour | PA@.010 | SF | — | .51 | .51 |
| Pressure washing (SSPC-12), 500 SF per hour | PA@.002 | SF | — | .10 | .10 |
| Steam cleaning, 330 SF per hour | PA@.003 | SF | — | .15 | .15 |
| Structural metals, machinery or equipment, paint one coat | | | | | |
| Brush work, light to medium coating | PA@.006 | SF | — | .31 | .31 |
| Brush work, heavy coat | PA@.007 | SF | — | .36 | .36 |
| Mitt, or glove, light to medium coating | PA@.007 | SF | — | .36 | .36 |
| Roll, heavy coating | PA@.003 | SF | — | .15 | .15 |
| Spray light or medium coat | PA@.002 | SF | — | .10 | .10 |
| Spray heavy coat | PA@.003 | SF | — | .15 | .15 |
| Spray machinery or equipment | PA@.003 | SF | — | .15 | .15 |

# 09 Finishes

|  | Craft@Hrs | Unit | Material | Labor | Total |
|---|---|---|---|---|---|
| **Exterior metals, paint one coat** | | | | | |
| Brush metal siding | PA@.006 | SF | — | .31 | .31 |
| Roll metal siding | PA@.003 | SF | — | .15 | .15 |
| Spray metal siding | PA@.002 | SF | — | .10 | .10 |
| Spray metal decking | PA@.003 | SF | — | .15 | .15 |
| Brush metal trim | PA@.004 | SF | — | .21 | .21 |
| Spray chain link fencing | PA@.003 | SF | — | .15 | .15 |

Pipe or HVAC duct, paint one coat. Consider each linear foot of conduit under 4" in diameter as having one square foot of surface regardless of the pipe diameter. Consider each valve as having 2 square feet of surface regardless of the size.

|  | Craft@Hrs | Unit | Material | Labor | Total |
|---|---|---|---|---|---|
| Brush, no insulation | PA@.007 | SF | — | .36 | .36 |
| Brush, with insulation | PA@.006 | SF | — | .31 | .31 |
| Roll, mitt or glove, no insulation | PA@.005 | SF | — | .26 | .26 |
| Roll, mitt or glove, with insulation | PA@.007 | SF | — | .36 | .36 |
| Spray pipe or small ductwork | PA@.003 | SF | — | .15 | .15 |
| Spray larger ductwork | PA@.003 | SF | — | .15 | .15 |
| **Conduit, hangers or fasteners, paint one coat** | | | | | |
| Brush application | PA@.007 | SF | — | .36 | .36 |
| Mitt or glove application | PA@.006 | SF | — | .31 | .31 |
| Spray application | PA@.003 | SF | — | .15 | .15 |
| **Exterior metals coated with a single coat of epoxy or mastic, 8 to 10 mil film** | | | | | |
| Brush epoxy coating | PA@.016 | SF | — | .82 | .82 |
| Roll epoxy coating | PA@.006 | SF | — | .31 | .31 |
| Spray epoxy coating | PA@.004 | SF | — | .21 | .21 |
| Brush mastic coating | PA@.019 | SF | — | .98 | .98 |
| Roll mastic coating | PA@.008 | SF | — | .41 | .41 |
| Spray mastic coating | PA@.005 | SF | — | .26 | .26 |
| **Interior metals, paint one coat** | | | | | |
| Spray joists and decking | PA@.003 | SF | — | .15 | .15 |
| Brush hollow metal items | PA@.013 | SF | — | .67 | .67 |
| Roll hollow metal items | PA@.006 | SF | — | .31 | .31 |
| Spray hollow metal items | PA@.006 | SF | — | .31 | .31 |
| **Tanks or spheres, paint one coat** | | | | | |
| Roll the exterior shell | PA@.004 | SF | — | .21 | .21 |
| Roll the exterior roof | PA@.004 | SF | — | .21 | .21 |
| Spray the exterior shell | PA@.003 | SF | — | .15 | .15 |
| Spray the exterior roof | PA@.003 | SF | — | .15 | .15 |
| **Surface preparation for masonry or concrete** | | | | | |
| Acid etching, 250 SF per hour | PA@.004 | SF | — | .21 | .21 |
| Water blasting, 110 SF per hour | PA@.009 | SF | — | .46 | .46 |
| **Paint concrete, one coat** | | | | | |
| Roll tilt up walls | PA@.004 | SF | — | .21 | .21 |
| Spray tilt up walls | PA@.003 | SF | — | .15 | .15 |
| Roll stucco walls | PA@.005 | SF | — | .26 | .26 |
| Spray stucco walls | PA@.003 | SF | — | .15 | .15 |
| Roll form-poured concrete walls | PA@.004 | SF | — | .21 | .21 |
| Spray form-poured concrete walls | PA@.003 | SF | — | .15 | .15 |
| Brush elastomeric or mastic, walls | PA@.019 | SF | — | .98 | .98 |
| Roll elastomeric or mastic, walls | PA@.008 | SF | — | .41 | .41 |
| Spray elastomeric or mastic, walls | PA@.005 | SF | — | .26 | .26 |
| Brush epoxy coating on walls | PA@.014 | SF | — | .72 | .72 |
| Roll epoxy coating on walls | PA@.007 | SF | — | .36 | .36 |

| | Craft@Hrs | Unit | Material | Labor | Total |
|---|---|---|---|---|---|
| Spray epoxy coating on walls | PA@.005 | SF | — | .26 | .26 |
| Roll ceiling | PA@.005 | SF | — | .26 | .26 |
| Spray ceiling | PA@.003 | SF | — | .15 | .15 |
| Spray concrete deck | PA@.003 | SF | — | .15 | .15 |
| Brush concrete floor or steps | PA@.006 | SF | — | .31 | .31 |
| Roll concrete floor or steps | PA@.005 | SF | — | .26 | .26 |
| Masonry walls, paint one coat | | | | | |
| Brush concrete block | PA@.008 | SF | — | .41 | .41 |
| Roll concrete block | PA@.004 | SF | — | .21 | .21 |
| Spray concrete block | PA@.003 | SF | — | .15 | .15 |
| Roll scored concrete block or brick | PA@.008 | SF | — | .41 | .41 |
| Spray scored concrete block or brick | PA@.004 | SF | — | .21 | .21 |
| Spray fluted masonry | PA@.006 | SF | — | .31 | .31 |
| Brush block filler | PA@.019 | SF | — | .98 | .98 |
| Roll block filler | PA@.008 | SF | — | .41 | .41 |
| Spray block filler | PA@.005 | SF | — | .26 | .26 |
| Swimming pool paint, elastomerics and mastics | | | | | |
| Brush application | PA@.008 | SF | — | .41 | .41 |
| Roller application | PA@.005 | SF | — | .26 | .26 |
| Spray application | PA@.003 | SF | — | .15 | .15 |
| Polyurethane epoxy floor system | | | | | |
| Etch and neutralize concrete | PA@.006 | SF | .03 | .31 | .34 |
| Roll base coat | PA@.006 | SF | .07 | .31 | .38 |
| Roll first flood coat | PA@.006 | SF | .09 | .31 | .40 |
| Apply chip coat, by hand | PA@.003 | SF | .05 | .15 | .20 |
| Sanding and wiping, by machine | PA@.003 | SF | .02 | .15 | .17 |
| Roll second flood coat | PA@.002 | SF | .09 | .10 | .19 |
| Roll first finish glaze coat | PA@.002 | SF | .09 | .10 | .19 |
| Roll second finish glaze coat | PA@.002 | SF | .07 | .10 | .17 |
| Roll third finish glaze coat | PA@.002 | SF | .07 | .10 | .17 |
| Wood floors and decks, paint one coat | | | | | |
| Brush porch floor | PA@.006 | SF | — | .31 | .31 |
| Roll or spray porch floor | PA@.005 | SF | — | .26 | .26 |
| Brush wood steps, risers and stringers | PA@.008 | SF | — | .41 | .41 |
| Brush rail or baluster, per LF of rail or baluster | PA@.008 | LF | — | .41 | .41 |
| Brush large post | PA@.010 | SF | — | .51 | .51 |
| Shingles or siding, paint one coat | | | | | |
| Brush shingles or rough siding | PA@.010 | SF | — | .51 | .51 |
| Roll shingles or rough siding | PA@.006 | SF | — | .31 | .31 |
| Spray shingles or rough siding | PA@.003 | SF | — | .15 | .15 |
| Brush mineral fiber shingles | PA@.010 | SF | — | .51 | .51 |
| Roll mineral fiber shingles | PA@.008 | SF | — | .41 | .41 |
| Spray mineral fiber shingles | PA@.003 | SF | — | .15 | .15 |
| Brush shake siding | PA@.010 | SF | — | .51 | .51 |
| Spray shake siding | PA@.006 | SF | — | .31 | .31 |
| Brush rustic or lap siding | PA@.009 | SF | — | .46 | .46 |
| Roll rustic or lap siding | PA@.007 | SF | — | .36 | .36 |
| Spray rustic or lap siding | PA@.004 | SF | — | .21 | .21 |
| Roll board and batten siding | PA@.006 | SF | — | .31 | .31 |
| Spray board and batten siding | PA@.003 | SF | — | .15 | .15 |
| Roll smooth siding | PA@.004 | SF | — | .21 | .21 |
| Spray smooth siding | PA@.003 | SF | — | .15 | .15 |

# 09 Finishes

| | Craft@Hrs | Unit | Material | Labor | Total |
|---|---|---|---|---|---|
| **Preparing interior surfaces for painting** | | | | | |
| Lay and secure drop cloth, 1,300 SF per hour | PA@.078 | CSF | .14 | 4.01 | 4.15 |
| Masking with paper, 75 LF per hour | PA@.013 | LF | .06 | .67 | .73 |
| Cover up with plastic or paper, 3,200 SF per hour | PA@.031 | CSF | .04 | 1.59 | 1.63 |
| Light sanding of wood trim, 75 SF per hour | PA@.013 | LF | .05 | .67 | .72 |
| Wire brush surface, 100 SF per hour | PA@.010 | SF | .05 | .51 | .56 |
| Sand paneling, 110 SF per hour | PA@.009 | SF | .05 | .46 | .51 |
| Light hand wash, 250 SF per hour | PA@.004 | SF | .03 | .21 | .24 |
| Remove calcimine, 125 SF per hour | PA@.008 | SF | .06 | .41 | .47 |
| **Interior walls, paint one coat** | | | | | |
| Brush smooth plaster or drywall | PA@.006 | SF | — | .31 | .31 |
| Roll smooth plaster or drywall | PA@.004 | SF | — | .21 | .21 |
| Spray smooth plaster or drywall | PA@.002 | SF | — | .10 | .10 |
| Roll sand finish plaster or drywall | PA@.004 | SF | — | .21 | .21 |
| Spray sand finish plaster or drywall | PA@.003 | SF | — | .15 | .15 |
| Roll rough sand finish wall | PA@.005 | SF | — | .26 | .26 |
| Spray rough sand finish wall | PA@.003 | SF | — | .15 | .15 |
| Brush wood paneling | PA@.006 | SF | — | .31 | .31 |
| Spray wood paneling | PA@.003 | SF | — | .15 | .15 |
| **Interior ceilings, paint one coat** | | | | | |
| Roll acoustical ceiling | PA@.007 | SF | — | .36 | .36 |
| Spray acoustical ceiling | PA@.004 | SF | — | .21 | .21 |
| Roll acoustical metal pan ceiling | PA@.004 | SF | — | .21 | .21 |
| Spray acoustical metal pan ceiling | PA@.002 | SF | — | .10 | .10 |
| Roll smooth drywall ceiling | PA@.004 | SF | — | .21 | .21 |
| Spray smooth drywall ceiling | PA@.003 | SF | — | .15 | .15 |
| Spray open wood or ceiling | PA@.003 | SF | — | .15 | .15 |
| Roll tongue and groove ceiling | PA@.005 | SF | — | .26 | .26 |
| Spray tongue and groove ceiling | PA@.003 | SF | — | .15 | .15 |
| **Cabinets or shelving, paint one coat** | | | | | |
| Brush one coat | PA@.008 | SF | — | .41 | .41 |
| Spray one coat | PA@.006 | SF | — | .31 | .31 |

**Material cost for painting** Cost per coat based on purchase in gallon quantities. Paint coverage per gallon is based on second and later coats applied by roller or brush. Add 10% for first coats. Add 20% when paint is applied with a sprayer.

| | Craft@Hrs | Unit | Material | Labor | Total |
|---|---|---|---|---|---|
| Acrylic deck and siding stain | | | | | |
| $27.70 and 350 SF per gallon | — | SF | .08 | — | .08 |
| Penetrating tinted natural stain | | | | | |
| $27.80 and 250 SF per gallon | — | SF | .11 | — | .11 |
| Oil-based wiping stain | | | | | |
| $27.80 and 225 SF per gallon | — | SF | .12 | — | .12 |
| Water-based urethane exterior clear finish | | | | | |
| $47.50 and 400 SF per gallon | — | SF | .12 | — | .12 |
| Oil-based urethane interior clear finish | | | | | |
| $47.50 and 400 SF per gallon | — | SF | .12 | — | .12 |
| Interior latex primer | | | | | |
| $9.10 and 400 SF per gallon | — | SF | .02 | — | .02 |
| Interior flat latex | | | | | |
| $16.00 and 400 SF per gallon | — | SF | .04 | — | .04 |
| Interior latex gloss enamel | | | | | |
| $31.70 and 400 SF per gallon | — | SF | .08 | — | .08 |
| Exterior oil-based primer-sealer | | | | | |
| $25.70 and 350 SF per gallon | — | SF | .07 | — | .07 |

# 09 Finishes

| | Craft@Hrs | Unit | Material | Labor | Total |
|---|---|---|---|---|---|
| Concrete and masonry bonding epoxy primer | | | | | |
| $19.20 and 350 SF per gallon | — | SF | .05 | — | .05 |
| Exterior flat acrylic latex primer-sealer | | | | | |
| $22.50 and 350 SF per gallon | — | SF | .06 | — | .06 |
| Exterior gloss acrylic latex primer-sealer | | | | | |
| $28.30 and 350 SF per gallon | — | SF | .09 | — | .09 |
| Rust converting metal primer, 3 mil dry film thickness | | | | | |
| $42.60 and 300 SF per gallon | — | SF | .14 | — | .14 |
| Oil-based industrial enamel | | | | | |
| $28.90 and 300 SF per gallon | — | SF | .10 | — | .10 |
| Chlorinated rubber swimming pool paint | | | | | |
| $28.30 and 350 SF per gallon | — | SF | .09 | — | .09 |
| Waterborne epoxy architectural gloss coating | | | | | |
| $37.40 and 350 SF per gallon | — | SF | .11 | — | .11 |

# 10 Specialties

**Chalkboards, Whiteboards, and Tackboards**

| | Craft@Hrs | Unit | Material | Labor | Total |
|---|---|---|---|---|---|
| Chalkboard with 1/4" hardboard backing, full-length chalk rail | | | | | |
| 4' x 6' to 4' x 12' | CC@.034 | SF | 7.16 | 1.76 | 8.92 |
| Chalkboard with 1/2" particleboard backing, full-length chalk rail | | | | | |
| 4' x 6' to 4' x 16' | CC@.034 | SF | 9.41 | 1.76 | 11.17 |
| Melamine whiteboard with aluminum frame and tray | | | | | |
| 3' x 4' | CC@.408 | Ea | 32.70 | 21.10 | 53.80 |
| 3' x 5' | CC@.510 | Ea | 75.20 | 26.40 | 101.60 |
| 4' x 5' | CC@.680 | Ea | 88.40 | 35.20 | 123.60 |
| 4' x 6' | CC@.816 | Ea | 96.50 | 42.30 | 138.80 |
| 4' x 8' | CC@1.22 | Ea | 112.00 | 63.20 | 175.20 |
| 4' x 10' | CC@1.36 | Ea | 215.00 | 70.40 | 285.40 |
| 4' x 12' | CC@1.63 | Ea | 220.00 | 84.40 | 304.40 |
| Melamine whiteboard with wood frame and tray | | | | | |
| 3' x 4' | CC@.408 | Ea | 45.90 | 21.10 | 67.00 |
| 3' x 5' | CC@.510 | Ea | 79.60 | 26.40 | 106.00 |
| 4' x 5' | CC@.680 | Ea | 93.80 | 35.20 | 129.00 |
| 4' x 6' | CC@.816 | Ea | 98.20 | 42.30 | 140.50 |
| 4' x 8' | CC@1.22 | Ea | 120.00 | 63.20 | 183.20 |
| Porcelain on steel markerboard, aluminum frame and tray | | | | | |
| 3' x 4' | CC@.408 | Ea | 127.00 | 21.10 | 148.10 |
| 3' x 5' | CC@.510 | Ea | 202.00 | 26.40 | 228.40 |
| 4' x 5' | CC@.680 | Ea | 205.00 | 35.20 | 240.20 |
| 4' x 6' | CC@.816 | Ea | 214.00 | 42.30 | 256.30 |
| 4' x 8' | CC@1.09 | Ea | 272.00 | 56.40 | 328.40 |
| 4' x 10' | CC@1.36 | Ea | 315.00 | 70.40 | 385.40 |
| 4' x 12' | CC@1.63 | Ea | 367.00 | 84.40 | 451.40 |
| Porcelain on steel markerboard, wood frame and tray | | | | | |
| 3' x 4' | CC@.408 | Ea | 132.00 | 21.10 | 153.10 |
| 3' x 5' | CC@.510 | Ea | 202.00 | 26.40 | 228.40 |
| 4' x 5' | CC@.680 | Ea | 209.00 | 35.20 | 244.20 |
| 4' x 8' | CC@.816 | Ea | 348.00 | 42.30 | 390.30 |

|  | Craft@Hrs | Unit | Material | Labor | Total |
|---|---|---|---|---|---|
| **Natural corkboard with aluminum frame and tray** | | | | | |
| 3' x 5' | CC@.510 | Ea | 71.20 | 26.40 | 97.60 |
| 4' x 5' | CC@.680 | Ea | 80.50 | 35.20 | 115.70 |
| 4' x 6' | CC@.816 | Ea | 69.20 | 42.30 | 111.50 |
| 4' x 8' | CC@1.09 | Ea | 84.80 | 56.40 | 141.20 |
| 4' x 10' | CC@1.36 | Ea | 121.00 | 70.40 | 191.40 |
| 4' x 12' | CC@1.63 | Ea | 166.00 | 84.40 | 250.40 |

## Identifying Devices

Directory boards, felt-covered changeable letter boards, indoor mount

|  | Craft@Hrs | Unit | Material | Labor | Total |
|---|---|---|---|---|---|
| Open faced, aluminum frame | | | | | |
| 18" x 2', 1 door | C8@.635 | Ea | 151.00 | 29.10 | 180.10 |
| 2' x 3', 1 door | C8@1.31 | Ea | 216.00 | 60.00 | 276.00 |
| 4' x 3', 2 doors | C8@1.91 | Ea | 340.00 | 87.50 | 427.50 |
| Open faced, wood frame | | | | | |
| 18" x 2', 1 door | C8@.635 | Ea | 129.00 | 29.10 | 158.10 |
| 2' x 3', 1 door | C8@1.31 | Ea | 179.00 | 60.00 | 239.00 |
| 4' x 3', 2 doors | C8@1.91 | Ea | 266.00 | 87.50 | 353.50 |
| Acrylic door, aluminum frame | | | | | |
| 18" x 2', 1 door | C8@.635 | Ea | 340.00 | 29.10 | 369.10 |
| 2' x 3', 1 door | C8@1.31 | Ea | 477.00 | 60.00 | 537.00 |
| 4' x 3', 2 doors | C8@1.91 | Ea | 890.00 | 87.50 | 977.50 |
| **Hand-painted lettering, typical subcontract prices** | | | | | |
| Standard, per square foot | — | SF | — | — | 24.20 |
| Gold leaf, per square foot | — | SF | — | — | 48.70 |

Architectural signage, engraved, self-adhesive plastic signs with high contrast white lettering

|  | Craft@Hrs | Unit | Material | Labor | Total |
|---|---|---|---|---|---|
| 4" x 12", stock wording | PA@.191 | Ea | 32.10 | 9.82 | 41.92 |
| 4" x 12", custom wording | PA@.191 | Ea | 37.80 | 9.82 | 47.62 |
| 3" x 8", stock wording | PA@.191 | Ea | 24.90 | 9.82 | 34.72 |
| 3" x 8", custom wording | PA@.191 | Ea | 29.60 | 9.82 | 39.42 |
| Add for aluminum mounting frame | | | | | |
| 4" x 12" | PA@.096 | Ea | 26.30 | 4.94 | 31.24 |
| 3" x 8" | PA@.096 | Ea | 15.90 | 4.94 | 20.84 |

Illuminated letters, porcelain enamel finish, stock letters, aluminum with Plexiglas face, add electrical work

|  | Craft@Hrs | Unit | Material | Labor | Total |
|---|---|---|---|---|---|
| 12" high x 2-1/2" wide, 1 tube per letter | C8@.849 | Ea | 160.00 | 38.90 | 198.90 |
| 24" high x 5" wide, 2 tubes per letter | C8@1.61 | Ea | 309.00 | 73.80 | 382.80 |
| 36" high x 6-1/2" wide, 2 tubes per letter | C8@3.28 | Ea | 484.00 | 150.00 | 634.00 |

Fabricated back-lighted letters, aluminum with neon illumination, baked enamel finish, script, add electrical work

|  | Craft@Hrs | Unit | Material | Labor | Total |
|---|---|---|---|---|---|
| 12" high x 3" deep, 1 tube | G1@.834 | Ea | 87.90 | 38.40 | 126.30 |
| 24" high x 6" deep, 2 tubes | G1@1.41 | Ea | 185.00 | 64.80 | 249.80 |
| 36" high x 8" deep, 2 tubes | G1@2.53 | Ea | 334.00 | 116.00 | 450.00 |

Cast aluminum letters, various styles, for use on building interior or exterior

|  | Craft@Hrs | Unit | Material | Labor | Total |
|---|---|---|---|---|---|
| 4" high | C8@.215 | Ea | 22.60 | 9.85 | 32.45 |
| 8" high | C8@.320 | Ea | 36.80 | 14.70 | 51.50 |
| 12" high | C8@.320 | Ea | 56.70 | 14.70 | 71.40 |
| 15" high | C8@.456 | Ea | 89.30 | 20.90 | 110.20 |
| 18" high | C8@.456 | Ea | 116.00 | 20.90 | 136.90 |
| 24" high | C8@.456 | Ea | 184.00 | 20.90 | 204.90 |
| Deduct for injection molded plastic letters | — | % | 70.0 | — | — |
| Add for cast bronze letters | — | % | 100.0 | — | — |

# 10 Specialties

| | Craft@Hrs | Unit | Material | Labor | Total |
|---|---|---|---|---|---|
| Braille signage, meets ADA and CABO/ANSI A117.1 requirements, various colors, interior (See ADA section) | | | | | |
| 8" x 8" | PA@.250 | Ea | 62.60 | 12.90 | 75.50 |
| 2" x 8" | PA@.250 | Ea | 36.10 | 12.90 | 49.00 |
| Plaques, including standard lettering, expansion or toggle bolt applied | | | | | |
| Aluminum, 12" x 4" | G1@.900 | Ea | 206.00 | 41.40 | 247.40 |
| Aluminum, 24" x 18" | G1@2.70 | Ea | 1,040.00 | 124.00 | 1,164.00 |
| Aluminum, 24" x 36" | G1@2.91 | Ea | 1,900.00 | 134.00 | 2,034.00 |
| Bronze, 12" x 4" | G1@.900 | Ea | 281.00 | 41.40 | 322.40 |
| Bronze, 24" x 18" | G1@2.70 | Ea | 1,560.00 | 124.00 | 1,684.00 |
| Bronze, 24" x 36" | G1@2.91 | Ea | 2,900.00 | 134.00 | 3,034.00 |
| Safety and warning signs, stock signs ("High Voltage," "Fire Exit," etc.). Indoor, high performance plastic | | | | | |
| 10" x 7" | PA@.099 | Ea | 19.20 | 5.09 | 24.29 |
| 14" x 10" | PA@.099 | Ea | 27.70 | 5.09 | 32.79 |
| 20" x 14" | PA@.099 | Ea | 40.40 | 5.09 | 45.49 |
| Add for aluminum | — | % | 40.0 | — | — |
| Vandal resistant 14" x 10" | PA@.099 | Ea | 51.30 | 5.09 | 56.39 |
| Indoor, pressure sensitive vinyl signs | | | | | |
| 10" x 7" | PA@.099 | Ea | 15.20 | 5.09 | 20.29 |
| 14" x 10" | PA@.099 | Ea | 18.40 | 5.09 | 23.49 |

## Compartments and Cubicles

| | Craft@Hrs | Unit | Material | Labor | Total |
|---|---|---|---|---|---|
| Toilet partitions, floor mounted. Standard sizes, include 1 panel, 1 door and all pilasters and hardware. | | | | | |
| Powder coated metal | T5@2.66 | Ea | 449.00 | 131.00 | 580.00 |
| Solid plastic (polymer) | T5@4.00 | Ea | 855.00 | 197.00 | 1,052.00 |
| Laminated plastic | T5@2.66 | Ea | 450.00 | 131.00 | 581.00 |
| Stainless steel | T5@2.66 | Ea | 1,250.00 | 131.00 | 1,381.00 |
| Add for wheel chair units | — | Ea | 65.00 | — | 65.00 |
| Add for ceiling hung units | T5@.500 | Ea | 27.00 | 24.60 | 51.60 |
| Add for free-standing units (no adjacent wall) | T5@.900 | Ea | 160.00 | 44.30 | 204.30 |
| Urinal screens, wall mounted with brackets, 18" wide | | | | | |
| Standard quality, powder coated | T5@1.95 | Ea | 96.00 | 95.90 | 191.90 |
| Solid plastic (polymer) | T5@1.50 | Ea | 121.00 | 73.80 | 194.80 |
| Laminated plastic | T5@1.03 | Ea | 77.00 | 50.70 | 127.70 |
| Stainless steel | T5@1.03 | Ea | 282.00 | 50.70 | 332.70 |
| Add for floor mounted, 24" wide | T5@.476 | Ea | 34.50 | 23.40 | 57.90 |
| Accessories mounted on partitions | | | | | |
| Coat hooks and door stop, chrome | T5@.166 | Ea | 9.30 | 8.16 | 17.46 |
| Purse shelf, chrome, 5" x 14" | T5@.166 | Ea | 91.90 | 8.16 | 100.06 |
| Dressing cubicles, 80" high, with curtain, floor mounted, overhead braced | | | | | |
| Powder coated steel | T5@3.41 | Ea | 317.00 | 168.00 | 485.00 |
| Solid plastic (polymer) | T5@4.52 | Ea | 475.00 | 222.00 | 697.00 |
| Laminated plastic | T5@3.41 | Ea | 379.00 | 168.00 | 547.00 |
| Stainless steel | T5@3.41 | Ea | 775.00 | 168.00 | 943.00 |
| Shower compartments, industrial type, with receptor but no door or plumbing | | | | | |
| Powder coated steel | T5@2.68 | Ea | 822.00 | 132.00 | 954.00 |
| Solid plastic (polymer) | T5@4.04 | Ea | 922.00 | 199.00 | 1,121.00 |
| Laminated plastic | T5@2.68 | Ea | 913.00 | 132.00 | 1,045.00 |
| Stainless steel | T5@2.68 | Ea | 1,930.00 | 132.00 | 2,062.00 |
| Add for soap dish | T5@.166 | Ea | 9.93 | 8.16 | 18.09 |
| Add for curtain rod | T5@.166 | Ea | 6.95 | 8.16 | 15.11 |

| | Craft@Hrs | Unit | Material | Labor | Total |
|---|---|---|---|---|---|
| **Doors for industrial shower compartments** | | | | | |
| Wire or tempered glass, 24" x 72" | G1@1.03 | Ea | 105.00 | 47.40 | 152.40 |
| Plastic, 24" x 72" | G1@1.03 | Ea | 72.30 | 47.40 | 119.70 |
| Plastic, 1 panel door and 1 side panel | G1@1.03 | Ea | 147.00 | 47.40 | 194.40 |
| **Shower stalls, with receptor, 32" x 32", with door but no plumbing** | | | | | |
| Fiberglass | P6@2.65 | Ea | 273.00 | 133.00 | 406.00 |
| Painted metal | P6@2.65 | Ea | 284.00 | 133.00 | 417.00 |
| Hospital cubicle curtain track, ceiling mount | MW@.060 | LF | 6.28 | 3.16 | 9.44 |

**Movable Office Partitions** Prefabricated units. Cost per SF of partition wall measured one side. Use 100 SF as minimum job size.

| | Craft@Hrs | Unit | Material | Labor | Total |
|---|---|---|---|---|---|
| **Gypsum board on 2-1/2" metal studs (includes finish on both sides)** | | | | | |
| 1/2" board, unpainted, STC 38 | C8@.028 | SF | 4.27 | 1.28 | 5.55 |
| 5/8" board, unpainted, STC 40 | C8@.028 | SF | 4.36 | 1.28 | 5.64 |
| 1/2" or 5/8", unpainted, STC 45 | C8@.028 | SF | 4.79 | 1.28 | 6.07 |
| 1/2" vinyl-covered board, STC 38 | C8@.028 | SF | 5.16 | 1.28 | 6.44 |
| 5/8" vinyl-covered board, STC 40 | C8@.028 | SF | 5.42 | 1.28 | 6.70 |
| Metal-covered, baked enamel finish | C8@.028 | SF | 11.40 | 1.28 | 12.68 |
| Add for factory vinyl wrap, 15 to 22 oz | — | SF | .43 | — | .43 |
| Windows, including frame and glass, add to above, factory installed | | | | | |
| 3'6" x 2'0" | — | Ea | 91.70 | — | 91.70 |
| 3'6" x 4'0" | — | Ea | 135.00 | — | 135.00 |
| Wood doors, hollow core, prefinished, add to above, factory installed | | | | | |
| 1-3/4", 3'0" x 7'0" | — | Ea | 288.00 | — | 288.00 |
| **Additional costs for gypsum board partitions** | | | | | |
| Metal door jambs, 3'6" x 7'0" | — | Ea | 173.00 | — | 173.00 |
| Metal door jambs, 6'0" x 7'0" | — | Ea | 292.00 | — | 292.00 |
| Passage hardware set | — | Ea | 68.10 | — | 68.10 |
| Lockset hardware | — | LS | 92.00 | — | 92.00 |
| Corner for gypsum partitions | — | Ea | 104.00 | — | 104.00 |
| Corner for metal-covered partitions | — | Ea | 206.00 | — | 206.00 |
| Starter, gypsum partitions | — | Ea | 34.80 | — | 34.80 |
| Starter, metal-covered partitions | — | Ea | 74.50 | — | 74.50 |
| Metal base | — | LF | 2.11 | — | 2.11 |
| Cubicles and welded booths, 5' x 5' x 5' | C8@.028 | SF | 11.90 | 1.28 | 13.18 |
| **Banker type divider partition, subcontract** | | | | | |
| 5'6" high including 18" glass top, 2" thick, vinyl finish, not including door or gates | — | SF | — | — | 12.90 |

**Demountable Partitions** Prefinished units. Cost per SF of partition measured one side. Use 100 SF as minimum job size. Labor shown applies to installing or removing.

| | Craft@Hrs | Unit | Material | Labor | Total |
|---|---|---|---|---|---|
| Modular panels, spring pressure, mounted to floor and ceiling with removable tracks | | | | | |
| Standard panel, non-rated | C8@.124 | SF | 15.50 | 5.68 | 21.18 |

**Accordion Folding Partitions**
Vinyl-covered particleboard core, top hung, including structural supports and track. STC = Sound Transmission Coefficient.

| | Craft@Hrs | Unit | Material | Labor | Total |
|---|---|---|---|---|---|
| Economy, to 8' high, STC 36 | — | SF | 11.50 | — | 11.50 |
| Economy, to 30' wide x 17' high, STC 41 | — | SF | 15.80 | — | 15.80 |
| Good quality, to 30' W x 17' H, STC 43 | — | SF | 18.50 | — | 18.50 |

| | Craft@Hrs | Unit | Material | Labor | Total |
|---|---|---|---|---|---|
| Better quality, to 30' W x 17' H, STC 44 | — | SF | 19.00 | — | 19.00 |
| Better quality, large openings, STC 45 | — | SF | 17.70 | — | 17.70 |
| Better quality, extra large openings, STC 47 | — | SF | 19.00 | — | 19.00 |
| Installation of folding partitions, per LF of track | C8@.647 | LF | — | 29.70 | 29.70 |
| Add for prefinished birch or ash wood slats accordion folding partitions | | | | | |
| With vinyl hinges, 15' W x 8' H maximum | — | SF | 7.58 | — | 7.58 |

Folding fire partitions, accordion, with automatic closing system, baked enamel finish, with standard hardware and motor, cost per 5' x 7' module

| | Craft@Hrs | Unit | Material | Labor | Total |
|---|---|---|---|---|---|
| 20 minute rating | C8@22.6 | Ea | 4,890.00 | 1,040.00 | 5,930.00 |
| 60 minute rating | C8@22.6 | Ea | 6,210.00 | 1,040.00 | 7,250.00 |
| 90 minute rating | C8@22.6 | Ea | 6,670.00 | 1,040.00 | 7,710.00 |
| 2 hour rating | C8@22.6 | Ea | 7,850.00 | 1,040.00 | 8,890.00 |

Folding leaf partitions, typical subcontract prices. Not including header or floor track recess, vinyl covered.

| | Craft@Hrs | Unit | Material | Labor | Total |
|---|---|---|---|---|---|
| Metal panels, 7.5 lbs per SF, top hung, hinged, | | | | | |
| (STC 52), 16' high x 60' wide maximum | — | SF | — | — | 59.50 |
| Metal panels, 7.5 lbs per SF, top hung, single panel pivot, | | | | | |
| (STC 48), 18' high x 60' wide maximum | — | SF | — | — | 60.80 |
| Wood panels, hinged, floor or ceiling hung, 6 lbs per SF, | | | | | |
| (STC 40), 12' high x 36' wide maximum | — | SF | — | — | 49.20 |
| Add for laminated plastic | — | SF | — | — | 2.76 |
| Add for unfinished wood veneer | — | SF | — | — | 3.33 |
| Add for chalkboard, large area | — | SF | — | — | 2.67 |

**Woven Wire Partitions** 10 gauge painted, 1-1/2" steel mesh.

Wall panels, 8' high, installed on a concrete slab

| | Craft@Hrs | Unit | Material | Labor | Total |
|---|---|---|---|---|---|
| 1' wide | H6@.437 | Ea | 106.00 | 22.60 | 128.60 |
| 2' wide | H6@.491 | Ea | 125.00 | 25.40 | 150.40 |
| 3' wide | H6@.545 | Ea | 142.00 | 28.20 | 170.20 |
| 4' wide | H6@.545 | Ea | 162.00 | 28.20 | 190.20 |
| 5' wide | H6@.598 | Ea | 164.00 | 30.90 | 194.90 |
| Add for galvanized mesh | — | SF | .67 | — | .67 |
| Corner posts for 8' high walls | H6@.270 | Ea | 36.90 | 14.00 | 50.90 |
| Stiffener channel posts for 8' walls | H6@.277 | Ea | 72.60 | 14.30 | 86.90 |
| Service window panel, 8' H x 5' W panel | H6@.836 | Ea | 339.00 | 43.20 | 382.20 |

Sliding doors, 8' high wall, 7' H door with 1' transom

| | Craft@Hrs | Unit | Material | Labor | Total |
|---|---|---|---|---|---|
| 3' wide | H6@2.94 | Ea | 553.00 | 152.00 | 705.00 |
| 4' wide | H6@3.29 | Ea | 581.00 | 170.00 | 751.00 |
| 5' wide | H6@3.29 | Ea | 619.00 | 170.00 | 789.00 |
| 6' wide | H6@3.58 | Ea | 731.00 | 185.00 | 916.00 |

Hinged doors, 8' high wall, 7' H door with 1' transom

| | Craft@Hrs | Unit | Material | Labor | Total |
|---|---|---|---|---|---|
| 3' wide | H6@2.85 | Ea | 431.00 | 147.00 | 578.00 |
| 4' wide | H6@3.14 | Ea | 491.00 | 162.00 | 653.00 |

Dutch doors, 8' high

| | Craft@Hrs | Unit | Material | Labor | Total |
|---|---|---|---|---|---|
| 3' wide | H6@2.85 | Ea | 619.00 | 147.00 | 766.00 |
| 4' wide | H6@3.14 | Ea | 669.00 | 162.00 | 831.00 |

Complete wire partition installation including posts and typical doors, per SF of wall and ceiling panel

| | Craft@Hrs | Unit | Material | Labor | Total |
|---|---|---|---|---|---|
| Painted wall panels | H6@.029 | SF | 4.48 | 1.50 | 5.98 |
| Galvanized wall panels | H6@.029 | SF | 5.14 | 1.50 | 6.64 |
| Painted ceiling panels | H6@.059 | SF | 4.15 | 3.05 | 7.20 |
| Galvanized ceiling panels | H6@.059 | SF | 4.70 | 3.05 | 7.75 |

| | Craft@Hrs | Unit | Material | Labor | Total |
|---|---|---|---|---|---|

**Wall Protection Systems** .063" extruded aluminum retainer with a .100" extruded plastic (Acrovyn) cover. Installed with adhesive or mechanical fasteners over existing drywall, doors or door frames. No substrate costs are included.

| | Craft@Hrs | Unit | Material | Labor | Total |
|---|---|---|---|---|---|
| Flush-mounted corner guards, 3" x 3" wide legs | CC@.194 | LF | 11.70 | 10.00 | 21.70 |
| Surface-mounted corner guards, 3" x 3" wide legs | CC@.069 | LF | 6.60 | 3.57 | 10.17 |
| Handrail | CC@.111 | LF | 13.00 | 5.75 | 18.75 |
| Bumper guards, bed bumpers | CC@.138 | LF | 8.83 | 7.15 | 15.98 |
| Door frame protectors | CC@.056 | LF | 13.20 | 2.90 | 16.10 |
| Door protectors | CC@.083 | SF | 4.24 | 4.30 | 8.54 |
| Protective wallcovering (vinyl only) | PA@.087 | SF | 3.34 | 4.47 | 7.81 |

| | Unit | Commercial Industrial Grade | Institutional Grade |
|---|---|---|---|

**Toilet and Bath Accessories** Material cost only. See labor following this section.

| | Unit | Commercial Industrial Grade | Institutional Grade |
|---|---|---|---|
| Combination towel/waste units | | | |
| Recessed paper towel dispenser | Ea | 170.00 | 210.00 |
| Recessed roll towel dispenser | Ea | 298.00 | 460.00 |
| Surface paper towel dispenser | Ea | 48.00 | 66.40 |
| Surface mounted roll towel dispenser | Ea | 183.00 | 277.00 |
| Recessed roll towel dispenser with waste bin | Ea | 482.00 | 561.00 |
| Recessed paper towel dispenser with waste bin | Ea | 133.00 | 398.00 |
| Waste receptacles | | | |
| Recessed, 12 gallon | Ea | 174.00 | 210.00 |
| Surface mounted | Ea | 67.50 | 166.00 |
| Floor standing | Ea | 140.00 | 226.00 |
| Countertop | Ea | — | 310.00 |
| Wall urns | | | |
| Recessed | Ea | 136.00 | 158.00 |
| Surface mounted | Ea | 91.00 | 125.00 |
| Floor standing with waste | Ea | — | 148.00 |
| Sanitary napkin dispenser, recessed | Ea | 293.00 | 460.00 |
| Soap dispensers | | | |
| Liquid, 20 oz, lavatory mounted | Ea | 34.90 | 34.90 |
| Liquid, 34 oz, lavatory mounted | Ea | 40.50 | 40.50 |
| Powder | Ea | 37.10 | 51.80 |
| Under-counter reservoir system, with 3 stations | Ea | — | 1,010.00 |
| Add for additional stations | Ea | — | 86.70 |
| Liquid, surface mounted | Ea | 16.40 | 34.90 |
| Powdered, surface mounted | Ea | 12.90 | 81.00 |
| Toilet paper dispensers | | | |
| Multi roll, recessed | Ea | 69.70 | 106.00 |
| Multi roll, surface mounted | Ea | 54.00 | 82.00 |
| Toilet seat cover dispensers | | | |
| Recessed | Ea | 60.70 | 108.00 |
| Surfaced mounted | Ea | 31.50 | 45.10 |
| Soap dishes | | | |
| With grab bar, recessed | Ea | 20.80 | 25.90 |
| With grab bar, surface mounted | Ea | 33.70 | 41.60 |
| Grab bars, stainless steel, wall mounted | | | |
| 1-1/2" x 12" | Ea | — | 23.70 |
| 1-1/2" x 18" | Ea | — | 22.50 |
| 1-1/2" x 24" | Ea | — | 25.90 |

| | Unit | Commercial Industrial Grade | Institutional Grade |
|---|---|---|---|
| 1-1/2" x 30" | Ea | — | 27.00 |
| 1-1/2" x 36" | Ea | — | 28.10 |
| 1-1/2" x 48" | Ea | — | 30.50 |
| 1-1/4" x 18" | Ea | — | 54.00 |
| 1-1/4" x 24" | Ea | — | 58.50 |
| 1-1/4" x 36" | Ea | — | 72.00 |
| 1", 90 degree angle, 30" | Ea | — | 84.40 |
| 1-1/4", 90 degree angle, 30" | Ea | — | 104.00 |
| Towel bars | | | |
| 18" chrome | Ea | — | 37.10 |
| 24" chrome | Ea | — | 38.30 |
| 18" stainless steel | Ea | — | 39.30 |
| 24" stainless steel | Ea | — | 41.60 |
| Medicine cabinets, with mirror | | | |
| Swing door, 16" x 22", wall hung | Ea | 70.31 | 268.00 |
| Swing door, 16" x 22", recessed | Ea | 82.31 | 229.00 |
| Mirrors, unframed, polished stainless steel back | | | |
| 16" x 20" | SF | 42.70 | — |
| 16" x 24" | SF | 48.60 | — |
| 18" x 24" | SF | 51.60 | — |
| 18" x 30" | SF | 61.50 | — |
| 24" x 30" | SF | 79.30 | — |
| 24" x 36" | SF | 101.00 | — |
| Mirrors, stainless steel frame, 1/4" glass | | | |
| 16" x 24" | Ea | — | 37.80 |
| 18" x 24" | Ea | — | 40.10 |
| 18" x 30" | Ea | — | 48.10 |
| 18" x 36" | Ea | — | 63.10 |
| 24" x 30" | Ea | — | 66.50 |
| 24" x 36" | Ea | — | 77.90 |
| 24" x 48" | Ea | — | 96.80 |
| 24" x 60" | Ea | — | 137.00 |
| 72" x 36" | Ea | — | 231.00 |
| Shelves, stainless steel, 6" deep | | | |
| 18" long | Ea | 49.50 | 57.40 |
| 24" long | Ea | 51.80 | 59.70 |
| 36" long | Ea | 66.40 | 76.40 |
| 72" long | Ea | 109.00 | 143.00 |
| Shower rod, end flanges, vinyl curtain | | | |
| 6' long, straight | Ea | 52.10 | 65.00 |
| Robe hooks | | | |
| Single | Ea | 7.53 | 11.10 |
| Double | Ea | 9.22 | 12.20 |
| Single with door stop | Ea | 8.56 | 14.00 |
| Straddle bar, stainless steel | | | |
| 24" x 24" x 20" | Ea | 111.00 | 128.00 |
| Urinal bar, 24" long, stainless steel | Ea | 90.00 | 108.00 |
| Hot air hand dryer, 115 volt | | | |
| add electric connection | Ea | 573.00 | 691.00 |

# 10 Specialties

| | Craft@Hrs | Unit | Material | Labor | Total |
|---|---|---|---|---|---|
| **Labor to install toilet and bath accessories** | | | | | |
| Toilet paper dispensers | | | | | |
|     Surface mounted | CC@.182 | Ea | — | 9.42 | 9.42 |
|     Recessed | CC@.310 | Ea | — | 16.10 | 16.10 |
| Toilet seat cover dispensers | | | | | |
|     Surface mounted | CC@.413 | Ea | — | 21.40 | 21.40 |
|     Recessed | CC@.662 | Ea | — | 34.30 | 34.30 |
| Feminine napkin dispenser, surface mounted | CC@.760 | Ea | — | 39.40 | 39.40 |
| Feminine napkin disposer, recessed | CC@.882 | Ea | — | 45.70 | 45.70 |
| Soap dishes | CC@.230 | Ea | — | 11.90 | 11.90 |
| Soap dispensers, surface mounted | CC@.254 | Ea | — | 13.20 | 13.20 |
| Soap dispensers, recessed | CC@.628 | Ea | — | 32.50 | 32.50 |
| Soap dispenser system | CC@2.91 | Ea | — | 151.00 | 151.00 |
| Paper towel dispensers and waste receptacles | | | | | |
|     Semi-recessed | CC@1.96 | Ea | — | 101.00 | 101.00 |
|     Recessed | CC@2.57 | Ea | — | 133.00 | 133.00 |
| Medicine cabinets with mirror, swing door | | | | | |
|     Wall hung | CC@.943 | Ea | — | 48.80 | 48.80 |
| Mirrors | | | | | |
|     To 5 SF | CG@.566 | Ea | — | 29.50 | 29.50 |
|     5 to 10 SF | CG@.999 | Ea | — | 52.00 | 52.00 |
|     Over 10 SF | CG@.125 | SF | — | 6.51 | 6.51 |
| Towel bars and grab bars | CC@.286 | Ea | — | 14.81 | 14.81 |
| Shelves, stainless steel, to 72" | CC@.569 | Ea | — | 29.50 | 29.50 |
| Robe hooks | CC@.182 | Ea | — | 9.42 | 9.42 |
| Hot air hand dryer, add for electric circuit wiring | CE@.908 | Ea | — | 53.00 | 53.00 |

**Baby Changing Stations** Polyethylene with steel on steel hinges. Holds up to 350 lbs. Changing stations require 22" horizontal wall space. Costs include sanitary bed liner dispenser, safety straps, necessary hardware and metal door plaques.

| | Craft@Hrs | Unit | Material | Labor | Total |
|---|---|---|---|---|---|
| Horizontal baby changing station, 20" H x 35" W | CL@.750 | Ea | 220.00 | 29.90 | 249.90 |
| Add for sanitary bed liners, pack of 500 | — | Ea | 60.00 | — | 60.00 |
| Add for changing stations replacement strap | — | Ea | 22.00 | — | 22.00 |

**Child Protection Seat** Polyethylene with steel on steel hinges. Holds up to 150 lbs. Requires 19 square inches wall space. Costs include safety straps and necessary hardware.

| | Craft@Hrs | Unit | Material | Labor | Total |
|---|---|---|---|---|---|
| Child protection seat, 19" H x 12" W | CL@.550 | Ea | 90.40 | 21.90 | 112.30 |
| Add for replacement safety strap | — | Ea | 10.50 | — | 10.50 |

**Lockers, Metal** Baked enamel finish, key lock 12" wide x 60" high x 15" deep overall dimensions

| | Craft@Hrs | Unit | Material | Labor | Total |
|---|---|---|---|---|---|
| Single tier, cost per locker | D4@.571 | Ea | 148.00 | 26.40 | 174.40 |
| Double tier, cost per 2 lockers | D4@.571 | Ea | 168.00 | 26.40 | 194.40 |
| Five tier, cost per 5 lockers | D4@.571 | Ea | 204.00 | 26.40 | 230.40 |
| Six tier, cost per 6 lockers | D4@.571 | Ea | 235.00 | 26.40 | 261.40 |
| Add for 72" height | — | % | 10.0 | — | — |
| Add for 15" width | — | % | 25.0 | — | — |
| Subtract for unassembled | — | % | -20.0 | — | — |

| | Craft@Hrs | Unit | Material | Labor | Total |
|---|---|---|---|---|---|

## Postal Specialties

Letter boxes, cost per each, recessed mounted

Aluminum

| | Craft@Hrs | Unit | Material | Labor | Total |
|---|---|---|---|---|---|
| 15" x 19" private access | D4@1.10 | Ea | 150.00 | 50.90 | 200.90 |
| 15" x 19" USPS access | D4@1.10 | Ea | 125.00 | 50.90 | 175.90 |

Brass

| | Craft@Hrs | Unit | Material | Labor | Total |
|---|---|---|---|---|---|
| 15" x 19" private access | D4@1.10 | Ea | 150.00 | 50.90 | 200.90 |
| 15" x 19" USPS access | D4@1.10 | Ea | 125.00 | 50.90 | 175.90 |

Brass indoor mailboxes for private use. Solid brass die-cast doors with lock, 1/4" glass window and name plate

| | Craft@Hrs | Unit | Material | Labor | Total |
|---|---|---|---|---|---|
| Rear loading 30 door unit, 3-1/2" W x 5" H door | C8@1.10 | Ea | 580.00 | 50.40 | 630.40 |
| Rear loading 14 door unit, 3-1/2" W x 5" H door | C8@1.00 | Ea | 540.00 | 45.80 | 585.80 |
| Add for front loading unit | — | % | 20.0 | — | — |

Aluminum rack ladder system, private use, 1/4" extruded aluminum doors. 20 gauge cold rolled steel compartments, open back. Unit dimensions 23-1/4" W x 12-1/8" H x 15-1/2" D

| | Craft@Hrs | Unit | Material | Labor | Total |
|---|---|---|---|---|---|
| Rear loading 12 door unit, 3-1/2" W x 5-1/2" H door | C8@1.50 | Ea | 160.00 | 68.70 | 228.70 |
| Rear loading 8 door unit, 5-1/2" W x 5-1/2" H door | C8@1.50 | Ea | 150.00 | 68.70 | 218.70 |
| Rear loading 4 door unit, 10-3/4" W x 5-1/2" H door | C8@1.50 | Ea | 140.00 | 68.70 | 208.70 |
| Rack ladder, 61-7/8" H x 15-1/2" D | C8@.500 | Ea | 70.00 | 22.90 | 92.90 |
| Steel rotary mail center, 30" W x 74" H x 24" D | C8@1.00 | Ea | 975.00 | 45.80 | 1,020.80 |

Vertical mailboxes, aluminum, USPS approved, surface mounted or recessed. Door size 5-1/2" W x 15-3/4" H, front loading

| | Craft@Hrs | Unit | Material | Labor | Total |
|---|---|---|---|---|---|
| 7 door unit, 40-3/4" W x 19-1/2" H x 7-1/4" D | C8@1.00 | Ea | 210.00 | 45.80 | 255.80 |
| 5 door unit, 29-3/4" W x 19-1/2" H x 7-1/4" D | C8@1.00 | Ea | 150.00 | 45.80 | 195.80 |
| 3 door unit, 18-3/4" W x 19-1/2" H x 7-1/4" D | C8@1.00 | Ea | 90.00 | 45.80 | 135.80 |

Pedestal mailboxes, aluminum neighborhood delivery and collection box units with weather protection hood. USPS approved, rear access. Doors are 6-3/8" W x 5-1/4" H

| | Craft@Hrs | Unit | Material | Labor | Total |
|---|---|---|---|---|---|
| 16 door unit, 26-1/2" W x 22-1/2" H x 22" D | C8@1.10 | Ea | 1,250.00 | 50.40 | 1,300.40 |
| 12 door unit, 20" W x 22-1/2" H x 22" D | C8@1.10 | Ea | 1,200.00 | 50.40 | 1,250.40 |
| 32" replacement pedestal | C8@1.00 | Ea | 75.00 | 45.83 | 120.83 |

## Storage Shelving

Retail store gondolas (supermarket shelving units), material prices vary widely, check local source

| | Craft@Hrs | Unit | Material | Labor | Total |
|---|---|---|---|---|---|
| Install only | D4@.333 | LF | — | 15.40 | 15.40 |

Steel industrial shelving, enamel finish, 2 sides & back, 12" deep x 30" wide x 75" high

| | Craft@Hrs | Unit | Material | Labor | Total |
|---|---|---|---|---|---|
| 5 shelves high | D4@.758 | Ea | 108.00 | 35.10 | 143.10 |
| 8 shelves high | D4@1.21 | Ea | 149.00 | 56.00 | 205.00 |

Steel industrial shelving, enamel finish, open on sides, 12" deep x 36" wide x 75" high

| | Craft@Hrs | Unit | Material | Labor | Total |
|---|---|---|---|---|---|
| 5 shelves high | D4@.686 | Ea | 63.20 | 31.70 | 94.90 |
| 8 shelves high | D4@1.06 | Ea | 96.30 | 49.00 | 145.30 |

Stainless steel modular shelving, 20" deep x 36" wide x 63" high

| | Craft@Hrs | Unit | Material | Labor | Total |
|---|---|---|---|---|---|
| 3 shelves, 16 gauge | D4@1.19 | Ea | 1,050.00 | 55.00 | 1,105.00 |

| | Craft@Hrs | Unit | Material | Labor | Total |
|---|---|---|---|---|---|
| Stainless steel wall shelf, 12" deep, bracket mount | D4@.675 | LF | 48.70 | 31.20 | 79.90 |

Stainless steel work counters, with under-counter storage shelf, 16 gauge

| | Craft@Hrs | Unit | Material | Labor | Total |
|---|---|---|---|---|---|
| 4' wide x 2' deep x 3' high | D4@1.36 | Ea | 1,140.00 | 62.90 | 1,202.90 |
| 6' wide x 2' deep x 3' high | D4@2.01 | Ea | 1,730.00 | 93.00 | 1,823.00 |

Library type metal shelving, 90" high, 8" deep, 5 shelves

| | Craft@Hrs | Unit | Material | Labor | Total |
|---|---|---|---|---|---|
| 30" long units | D4@1.17 | Ea | 189.00 | 54.10 | 243.10 |
| 36" long units | D4@1.30 | Ea | 204.00 | 60.10 | 264.10 |

| | Craft@Hrs | Unit | Material | Labor | Total |
|---|---|---|---|---|---|

**Flagpoles** Including erection cost but no foundation. See foundation costs below. By pole height above ground, except as noted. Buried dimension is usually 10% of pole height. Costs include typical shipping cost.

Fiberglass, tapered, external halyard, commercial grade, with hardware and ground sleeve, white

| | Craft@Hrs | Unit | Material | Labor | Total |
|---|---|---|---|---|---|
| 25' high | D4@8.19 | Ea | 396.00 | 379.00 | 775.00 |
| 30' high | D4@9.21 | Ea | 758.00 | 426.00 | 1,184.00 |
| 35' high | D4@9.96 | Ea | 950.00 | 461.00 | 1,411.00 |
| 40' high | D4@10.2 | Ea | 1,120.00 | 472.00 | 1,592.00 |
| 50' high | D4@15.1 | Ea | 3,500.00 | 698.00 | 4,198.00 |
| 60' high | D4@18.5 | Ea | 3,650.00 | 856.00 | 4,506.00 |
| 70' high | D4@21.7 | Ea | 9,570.00 | 1,000.00 | 10,570.00 |

Aluminum, satin finish, tapered, external halyard, sectional, architectural grade

| | Craft@Hrs | Unit | Material | Labor | Total |
|---|---|---|---|---|---|
| 20' high | D4@8.19 | Ea | 689.00 | 379.00 | 1,068.00 |
| 25' high | D4@8.19 | Ea | 922.00 | 379.00 | 1,301.00 |
| 30' high | D4@9.21 | Ea | 1,050.00 | 426.00 | 1,476.00 |
| 35' high | D4@9.96 | Ea | 1,370.00 | 461.00 | 1,831.00 |
| 40' high | D4@10.2 | Ea | 1,780.00 | 472.00 | 2,252.00 |
| 45' high | D4@10.5 | Ea | 1,960.00 | 486.00 | 2,446.00 |
| 50' high | D4@15.1 | Ea | 2,670.00 | 698.00 | 3,368.00 |
| 60' high | D4@18.5 | Ea | 4,740.00 | 856.00 | 5,596.00 |
| 70' high | D4@21.7 | Ea | 5,290.00 | 1,000.00 | 6,290.00 |
| 80' high | D4@25.2 | Ea | 7,100.00 | 1,170.00 | 8,270.00 |
| Additional costs for tapered aluminum flagpoles | | | | | |
| Add for nautical yardarm mast | — | % | 20.0 | — | — |
| Add for internal halyard (vandal resistant) | — | % | 20.0 | — | — |
| Add for clear finish | — | % | 15.0 | — | — |
| Add for bronze finish | — | % | 20.0 | — | — |
| Add for black finish | — | % | 30.0 | — | — |

Wall-mounted aluminum poles, including wall mount on concrete or masonry structure

| | Craft@Hrs | Unit | Material | Labor | Total |
|---|---|---|---|---|---|
| 8' high | D4@7.95 | Ea | 598.00 | 368.00 | 966.00 |
| 10' high | D4@9.69 | Ea | 701.00 | 448.00 | 1,149.00 |
| 17' with wall support, vertical | D4@11.1 | Ea | 405.00 | 513.00 | 918.00 |
| 20' with wall support, vertical | D4@14.4 | Ea | 413.00 | 666.00 | 1,079.00 |
| Add for yardarm nautical mast | | | | | |
| 6'6" aluminum yardarm | — | Ea | 418.00 | — | 418.00 |
| 13'6" aluminum yardarm | — | Ea | 500.00 | — | 500.00 |

Add for eagle mounted on spindle rod

| | Craft@Hrs | Unit | Material | Labor | Total |
|---|---|---|---|---|---|
| 10" H gold finish, 24" wingspan | — | Ea | 155.00 | — | 155.00 |
| 12" H gold finish, 11" wingspan | — | Ea | 80.70 | — | 80.70 |
| 18" H gold finish, 15" wingspan | — | Ea | 91.20 | — | 91.20 |

Poured concrete flagpole foundations, including excavation, metal collar, steel casing, base plate, sand fill and ground spike, by pole height above ground, typical costs

| | Craft@Hrs | Unit | Material | Labor | Total |
|---|---|---|---|---|---|
| 30' high or less | P9@10.5 | Ea | 77.20 | 496.00 | 573.20 |
| 35' high | P9@10.8 | Ea | 84.00 | 511.00 | 595.00 |
| 40' high | P9@12.5 | Ea | 90.90 | 591.00 | 681.90 |
| 45' high | P9@13.2 | Ea | 98.10 | 624.00 | 722.10 |
| 50' high | P9@15.8 | Ea | 119.00 | 747.00 | 866.00 |
| 60' high | P9@19.2 | Ea | 140.00 | 908.00 | 1,048.00 |
| 70' high | P9@22.5 | Ea | 181.00 | 1,060.00 | 1,241.00 |
| 80' high | P9@26.0 | Ea | 225.00 | 1,230.00 | 1,455.00 |
| 90' to 100' high | P9@30.1 | Ea | 280.00 | 1,420.00 | 1,700.00 |

Direct embedded pole base, augered in place, no concrete required, by pole height

| | Craft@Hrs | Unit | Material | Labor | Total |
|---|---|---|---|---|---|
| 20' to 25' pole | D4@1.88 | Ea | 341.00 | 87.00 | 428.00 |
| 25' to 40' pole | D4@2.71 | Ea | 488.00 | 125.00 | 613.00 |
| 45' to 55' pole | D4@2.99 | Ea | 549.00 | 138.00 | 687.00 |
| 60' to 70' pole | D4@3.42 | Ea | 610.00 | 158.00 | 768.00 |

| | Craft@Hrs | Unit | Material | Labor | Total |
|---|---|---|---|---|---|
| **Service Station Equipment** These costs do not include excavation, concrete work, piping or electrical work | | | | | |
| Air compressor, 1-1/2 HP with receiver | PF@5.54 | Ea | 1,910.00 | 330.00 | 2,240.00 |
| Air compressor, 3 HP with receiver | PF@5.54 | Ea | 2,280.00 | 330.00 | 2,610.00 |
| Hose reels with hose | | | | | |
|   Air hose, 50', heavy duty | PF@8.99 | Ea | 962.00 | 535.00 | 1,497.00 |
|   Water hose, 50' and valve | PF@8.01 | Ea | 979.00 | 477.00 | 1,456.00 |
|   Lube rack, 3 hose, remote pump | PF@48.4 | Ea | 5,540.00 | 2,880.00 | 8,420.00 |
|   Lube rack, 5 hose, remote pump | PF@48.4 | Ea | 10,200.00 | 2,880.00 | 13,080.00 |
| Air-operated pumps for hose reels, fit 55 gallon drum | | | | | |
|   Motor oil or gear oil | PF@1.00 | Ea | 965.00 | 59.50 | 1,024.50 |
|   Lube oil | PF@1.00 | Ea | 2,060.00 | 59.50 | 2,119.50 |
| Gasoline dispensers, commercial type, two-sided display, add the cost of hose and fill nozzles. | | | | | |
|   Mechanical display, single product, one hose, low cabinet, standard suction pump | PF@9.94 | Ea | 3,720.00 | 592.00 | 4,312.00 |
|   Electronic display, two product, two hose, low cabinet, standard suction pump | PF@9.94 | Ea | 7,400.00 | 592.00 | 7,992.00 |
|   Electronic display, three product, four hose, high cabinet, PCI-compliant card payment terminal | PF@9.94 | Ea | 16,400.00 | 592.00 | 16,992.00 |
|   Remote control station, connects up to 16 dispensers to a point of sale terminal, add the cost of wire and conduit | PF@4.00 | Ea | 2,420.00 | 238.00 | 2,658.00 |
|   Vapor recovery fill nozzle, 8' hose with breakaway fitting, per hose and nozzle | PF@1.00 | LS | 550.00 | 59.50 | 609.50 |
| Air and water service, post type with auto-inflator and retracting hoses, hoses included | PF@9.58 | Ea | 1,640.00 | 570.00 | 2,210.00 |
|   Add for electric thermal unit, factory installed | — | Ea | 96.20 | — | 96.20 |
| Auto hoist. Prices vary with the rating and capacity | | | | | |
|   Hoist, single post frame contact | | | | | |
|     8,000 lb semi hydraulic | D4@71.3 | Ea | 3,940.00 | 3,300.00 | 7,240.00 |
|   Hoist, single post frame contact | | | | | |
|     8,000 lb fully hydraulic | D4@71.3 | Ea | 4,210.00 | 3,300.00 | 7,510.00 |
|   Hoist, two post, pneumatic, 11,000 lb | D4@114. | Ea | 7,660.00 | 5,270.00 | 12,930.00 |
|   Hoist, two post, pneumatic, 24,000 lb | D4@188. | Ea | 10,900.00 | 8,700.00 | 19,600.00 |
| Two-post fully hydraulic hoists. Prices vary with the rating and capacity | | | | | |
|   11,000 lb capacity | D4@188. | Ea | 6,220.00 | 8,700.00 | 14,920.00 |
|   13,000 lb capacity | D4@188. | Ea | 6,340.00 | 8,700.00 | 15,040.00 |
|   18,500 lb capacity | D4@188. | Ea | 8,490.00 | 8,700.00 | 17,190.00 |
|   24,000 lb capacity | D4@188. | Ea | 11,400.00 | 8,700.00 | 20,100.00 |
|   26,000 lb capacity | D4@188. | Ea | 12,400.00 | 8,700.00 | 21,100.00 |
| Cash box and pedestal stand | D4@2.55 | Ea | 294.00 | 118.00 | 412.00 |
| Tire changer, 120 PSI pneumatic, auto tire | D4@10.6 | Ea | 2,870.00 | 490.00 | 3,360.00 |
| Tire changer, hydraulic, truck tire | D4@21.4 | Ea | 8,890.00 | 990.00 | 9,880.00 |
| Exhaust fume system, underground, complete | | | | | |
|   Per station | D5@9.52 | Ea | 609.00 | 478.00 | 1,087.00 |
| **Parking Control Equipment** Costs are for parking equipment only, no concrete work, paving, or electrical work included. Labor costs include installation and testing. | | | | | |
| Parking gates, controllable by attendant, vehicle detector, ticket dispenser, card reader, or coin/token machine | | | | | |
|   Barrier gate operator with 10' arm | D4@9.13 | Ea | 2,010.00 | 422.00 | 2,432.00 |
|   Barrier gate operator with 14' arm | D4@9.13 | Ea | 2,110.00 | 422.00 | 2,532.00 |
|   Barrier gate operator with counter-balanced arm | D4@9.13 | Ea | 3,660.00 | 422.00 | 4,082.00 |

# 11 Equipment

| | Craft@Hrs | Unit | Material | Labor | Total |
|---|---|---|---|---|---|
| Heavy-duty swing gate opener | D4@9.13 | Ea | 1,600.00 | 422.00 | 2,022.00 |
| 1/2 HP commercial slide gate opener | D4@9.13 | Ea | 3,300.00 | 422.00 | 3,722.00 |
| Add for battery backup | — | Ea | 500.00 | — | 500.00 |
| Vehicle detector, for installation in approach lane | | | | | |
| including cutting and grouting | D4@4.96 | Ea | 392.00 | 229.00 | 621.00 |
| Ticket dispenser (spitter), with date/time imprinter | | | | | |
| actuated by vehicle detector or hand button | D4@9.13 | Ea | 5,910.00 | 422.00 | 6,332.00 |
| Card readers. Non-programmable, parking areas with long term parking privileges | | | | | |
| Per station | D4@3.39 | Ea | 614.00 | 157.00 | 771.00 |
| Programmable, parking areas for users paying a fee to renew parking privileges | | | | | |
| Per reader | D4@3.39 | Ea | 4,760.00 | 157.00 | 4,917.00 |
| Access controls, keypad type | | | | | |
| Wall, post or gooseneck mounting, weatherproof | D4@1.00 | Ea | 350.00 | 46.30 | 396.30 |
| Lane spikes, dragon teeth, installation includes cutting and grouting as needed | | | | | |
| 6' wide, surface mounted | D4@6.69 | Ea | 1,380.00 | 309.00 | 1,689.00 |
| 6' wide, flush mounted | D4@9.69 | Ea | 1,380.00 | 448.00 | 1,828.00 |
| Lighted warning sign for lane spikes | D4@5.71 | Ea | 573.00 | 264.00 | 837.00 |

| | Craft@Hrs | Unit | Material | Labor | Equipment | Total |
|---|---|---|---|---|---|---|

**Loading Dock Equipment** Costs shown assume that equipment is set in prepared locations. No excavation, concrete work or electrical work included. Equipment cost is for a 4,000 lb forklift.

| | Craft@Hrs | Unit | Material | Labor | Equipment | Total |
|---|---|---|---|---|---|---|
| Move equipment on or off job | | | | | | |
| Allow, per move | — | LS | — | — | 100.00 | 100.00 |
| Dock levelers, Systems, Inc. | | | | | | |
| Edge-of-dock leveler, manual | D6@2.25 | Ea | 890.00 | 120.00 | 23.90 | 1,033.90 |
| Mechanical platform dock leveler, recessed, 25,000 lb capacity | | | | | | |
| 6' wide x 6' long | D6@3.00 | Ea | 2,520.00 | 160.00 | 31.90 | 2,711.90 |
| 6' wide x 8' long | D6@3.00 | Ea | 2,720.00 | 160.00 | 31.90 | 2,911.90 |
| 6' wide x 10' long | D6@3.00 | Ea | 4,060.00 | 160.00 | 31.90 | 4,251.90 |
| Hydraulic, pit type | | | | | | |
| 6' wide x 6' long | D6@4.50 | Ea | 3,750.00 | 240.00 | 47.80 | 4,037.80 |
| 6' wide x 8' long | D6@4.50 | Ea | 3,990.00 | 240.00 | 47.80 | 4,277.80 |
| Dock seals, foam pad type, 12" projection, with anchor brackets (bolts not included), door opening size as shown, Systems, Inc. | | | | | | |
| 8' wide x 8' high | D6@3.75 | Ea | 629.00 | 200.00 | 39.80 | 868.80 |
| 8' wide x 10' high | D6@3.75 | Ea | 639.00 | 200.00 | 39.80 | 878.80 |
| Dock lifts, portable, electro-hydraulic scissor lifts, Systems, Inc. | | | | | | |
| 4,000 lb capacity, 6' x 6' dock | D6@2.25 | Ea | 6,930.00 | 120.00 | 23.90 | 7,073.90 |
| Dock lifts, pit recessed, electro-hydraulic scissor lifts, Advance Lifts | | | | | | |
| 5,000 lb capacity, 6' x 8' platform | D6@12.0 | Ea | 5,650.00 | 641.00 | 127.00 | 6,418.00 |
| 6,000 lb capacity, 6' x 8' platform | D6@12.0 | Ea | 5,780.00 | 641.00 | 127.00 | 6,548.00 |
| 8,000 lb capacity | | | | | | |
| 6' x 8' platform | D6@12.0 | Ea | 8,370.00 | 641.00 | 127.00 | 9,138.00 |
| 8' x 10' platform | D6@12.0 | Ea | 9,440.00 | 641.00 | 127.00 | 10,208.00 |
| 10,000 lb capacity | | | | | | |
| 6' x 8' platform | D6@12.0 | Ea | 9,530.00 | 641.00 | 127.00 | 10,298.00 |
| 8' x 10' platform | D6@12.0 | Ea | 10,500.00 | 641.00 | 127.00 | 11,268.00 |

| | Craft@Hrs | Unit | Material | Labor | Equipment | Total |
|---|---|---|---|---|---|---|
| 12,000 lb capacity | | | | | | |
| 6' x 10' platform | D6@15.0 | Ea | 12,300.00 | 801.00 | 159.00 | 13,260.00 |
| 8' x 12' platform | D6@15.0 | Ea | 13,100.00 | 801.00 | 159.00 | 14,060.00 |
| 15,000 lb capacity | | | | | | |
| 6' x 10' platform | D6@15.0 | Ea | 13,700.00 | 801.00 | 159.00 | 14,660.00 |
| 8' x 12' platform | D6@15.0 | Ea | 14,900.00 | 801.00 | 159.00 | 15,860.00 |
| 18,000 lb capacity | | | | | | |
| 6' x 10' platform | D6@15.0 | Ea | 14,400.00 | 801.00 | 159.00 | 15,360.00 |
| 8' x 12' platform | D6@15.0 | Ea | 15,300.00 | 801.00 | 159.00 | 16,260.00 |
| 20,000 lb capacity | | | | | | |
| 6' x 12' platform | D6@15.0 | Ea | 16,000.00 | 801.00 | 159.00 | 16,960.00 |
| 8' x 12' platform | D6@15.0 | Ea | 17,400.00 | 801.00 | 159.00 | 18,360.00 |
| Combination dock lift and dock leveler, electro-hydraulic scissor lifts, pit recessed, 5,000 lb capacity, Advance Lifts | | | | | | |
| 6' x 7'2" platform | D6@12.0 | Ea | 6,500.00 | 641.00 | 127.00 | 7,268.00 |
| 6' x 8' platform | D6@12.0 | Ea | 5,270.00 | 641.00 | 127.00 | 6,038.00 |
| Dock bumpers, laminated rubber, not including masonry anchors | | | | | | |
| 10" high x 18" wide x 4-1/2" thick | D4@.248 | Ea | 42.50 | 11.50 | — | 54.00 |
| 10" high x 24" wide x 4-1/2" thick | D4@.299 | Ea | 55.60 | 13.80 | — | 69.40 |
| 10" high x 36" wide x 4-1/2" thick | D4@.269 | Ea | 96.90 | 12.40 | — | 109.30 |
| 10" high x 24" wide x 6" thick | D4@.358 | Ea | 69.80 | 16.60 | — | 86.40 |
| 10" high x 36" wide x 6" thick | D4@.416 | Ea | 152.00 | 19.20 | — | 171.20 |
| 20" high x 11" wide x 4-1/2" thick | D4@.499 | Ea | 58.60 | 23.10 | — | 81.70 |
| 20" high x 11" wide x 6" thick | D4@.259 | Ea | 83.90 | 12.00 | — | 95.90 |
| Add for 4 masonry anchors | D4@1.19 | LS | 2.49 | 55.00 | — | 57.49 |
| Barrier posts, 8' long concrete-filled steel posts set 4' underground in concrete | | | | | | |
| 3" diameter | P8@.498 | Ea | 89.90 | 22.40 | — | 112.30 |
| 4" diameter | P8@.498 | Ea | 128.00 | 22.40 | — | 150.40 |
| 6" diameter | P8@.752 | Ea | 160.00 | 33.90 | — | 193.90 |
| 8" diameter | P8@1.00 | Ea | 193.00 | 45.00 | — | 238.00 |
| Loading dock shelter, fabric covered, steel frame, | | | | | | |
| 24" extension for 10' x 10' door | D4@9.91 | Ea | 937.00 | 458.00 | — | 1,395.00 |
| Perimeter door seal, | | | | | | |
| 12" x 12", vinyl cover | D4@.400 | LF | 38.10 | 18.50 | — | 56.60 |

| | Craft@Hrs | Unit | Material | Labor | Total |
|---|---|---|---|---|---|
| **Truck Scales** Dual platforms, single to multi-axle vehicles. These costs do not include concrete work or excavation. | | | | | |
| 40 ton, 11' x 32" x 3.6" platform | D4@112. | Ea | 23,200.00 | 5,180.00 | 28,380.00 |
| **Turnstiles** | | | | | |
| Access control turnstiles, 3' high | | | | | |
| Non-register type, manual | D4@2.78 | Ea | 766.00 | 129.00 | 895.00 |
| Register type, manual | D4@2.78 | Ea | 1,090.00 | 129.00 | 1,219.00 |
| Add for coin, token, card or electronic control | — | Ea | 276.00 | — | 276.00 |
| Register type, portable | — | Ea | 1,670.00 | — | 1,670.00 |
| Security-type rotary turnstiles, 7'0" high, indoor or outdoor | | | | | |
| Free access, Type B (vertical bar outer cage) | D4@15.6 | Ea | 5,670.00 | 722.00 | 6,392.00 |
| One way, Type B (vertical bar outer cage) | D4@15.6 | Ea | 6,140.00 | 722.00 | 6,862.00 |
| Free access, Type AA (horizontal bar outer cage) | D4@18.4 | Ea | 7,330.00 | 851.00 | 8,181.00 |
| One way, Type AA (horizontal bar outer cage) | D4@18.4 | Ea | 8,030.00 | 851.00 | 8,881.00 |
| Add for electric control | — | Ea | 633.00 | — | 633.00 |

# 11 Equipment

| | Craft@Hrs | Unit | Material | Labor | Total |
|---|---|---|---|---|---|

**Recessed Wall and Floor Safes** Safes with 1" thick steel door, 1/4" hardplate protecting locking mechanism, welded deadbolts, key or combination lock and internal relock device.

In-floor safes

| | Craft@Hrs | Unit | Material | Labor | Total |
|---|---|---|---|---|---|
| 15" L x 16" W x 11" D, floor safe | D4@2.33 | Ea | 544.00 | 108.00 | 652.00 |
| 15" L x 16" W x 22" D, floor safe | D4@2.33 | Ea | 723.00 | 108.00 | 831.00 |
| 25" L x 25" W x 17" D, floor safe | D4@2.33 | Ea | 989.00 | 108.00 | 1,097.00 |
| 14" L x 14" W x 4" D, wall safe | D4@2.00 | Ea | 425.00 | 92.50 | 517.50 |
| 14" L x 14" W x 11" D, wall safe | D4@2.00 | Ea | 698.00 | 92.50 | 790.50 |
| Add for installation in existing building | D4@2.00 | Ea | — | 92.50 | 92.50 |

| | Craft@Hrs | Unit | Material | Labor | Equipment | Total |
|---|---|---|---|---|---|---|
| **Bank Equipment** | | | | | | |
| Fire rated vault doors | | | | | | |
| 32" wide, 2 hour rating | D4@11.9 | Ea | 3,890.00 | 550.00 | 448.00 | 4,888.00 |
| 40" wide, 2 hour rating | D4@20.8 | Ea | 5,100.00 | 962.00 | 783.00 | 6,845.00 |
| 32" wide, 4 hour rating | D4@11.9 | Ea | 4,400.00 | 550.00 | 448.00 | 5,398.00 |
| 40" wide, 4 hour rating | D4@20.8 | Ea | 5,710.00 | 962.00 | 783.00 | 7,455.00 |
| 32" wide, 6 hour rating | D4@11.9 | Ea | 5,140.00 | 550.00 | 448.00 | 6,138.00 |
| 40" wide, 6 hour rating | D4@20.8 | Ea | 6,150.00 | 962.00 | 783.00 | 7,895.00 |
| Teller windows | | | | | | |
| Drive-up, motorized drawer, projected | D4@26.8 | Ea | 12,100.00 | 1,240.00 | — | 13,340.00 |
| Walk-up, one teller | D4@17.6 | Ea | 6,210.00 | 814.00 | — | 7,024.00 |
| Walk-up, two teller | D4@17.6 | Ea | 7,420.00 | 814.00 | — | 8,234.00 |
| After hours depository doors | | | | | | |
| Envelope and bag, whole chest, complete | D4@37.1 | Ea | 10,500.00 | 1,720.00 | — | 12,220.00 |
| Envelope only | D4@4.72 | Ea | 1,240.00 | 218.00 | — | 1,458.00 |
| Flush mounted, bag only | D4@5.44 | Ea | 1,880.00 | 252.00 | — | 2,132.00 |
| Teller counter, modular components | C8@.532 | LF | 115.00 | 24.40 | — | 139.40 |
| Square check desks, 48" x 48", 4 person | C8@2.00 | Ea | 1,380.00 | 91.70 | — | 1,471.70 |
| Round check desks, 48" diameter, 4 person | C8@2.00 | Ea | 1,520.00 | 91.70 | — | 1,611.70 |
| Rectangular check desks, 72" x 24", 4 person | C8@2.00 | Ea | 1,290.00 | 91.70 | — | 1,381.70 |
| Rectangular check desks, 36" x 72", 8 person | C8@2.00 | Ea | 2,530.00 | 91.70 | — | 2,621.70 |
| Bank partition, bullet resistant, 2.5" thick | C8@.093 | SF | 87.10 | 4.26 | — | 91.36 |
| Safe deposit boxes, modular units | | | | | | |
| 18 openings, 2" x 5" | D4@3.42 | Ea | 1,350.00 | 158.00 | — | 1,508.00 |
| 30 openings, 2" x 5" | D4@3.42 | Ea | 1,710.00 | 158.00 | — | 1,868.00 |
| 42 openings, 2" x 5" | D4@3.42 | Ea | 2,190.00 | 158.00 | — | 2,348.00 |
| Base, 32" x 24" x 3" | D4@1.11 | Ea | 130.00 | 51.30 | — | 181.30 |
| Canopy top | D4@.857 | Ea | 74.00 | 39.60 | — | 113.60 |
| Cash dispensing automatic teller, with phone | CE@84.3 | Ea | 48,500.00 | 4,920.00 | — | 53,420.00 |
| Video camera surveillance system, 3 cameras | CE@15.6 | Ea | 3,130.00 | 911.00 | — | 4,041.00 |
| Perimeter alarm system, typical cost | CE@57.7 | Ea | 7,840.00 | 3,370.00 | — | 11,210.00 |

| | Craft@Hrs | Unit | Material | Labor | Total |
|---|---|---|---|---|---|
| **Checkroom Equipment** | | | | | |
| Automatic electric checkroom conveyor, 350 coat capacity | | | | | |
| Conveyor and controls | D4@15.2 | Ea | 3,370.00 | 703.00 | 4,073.00 |
| Add for electrical connection for checkroom conveyor, typical subcontract price | | | | | |
| 40' wire run with conduit, circuit breaker, and junction box | — | LS | — | — | 420.00 |

# 11 Equipment

| | Craft@Hrs | Unit | Material | Labor | Total |
|---|---|---|---|---|---|

**Coat and Hat Racks** Assembly and installation of prefabricated coat and hat racks.
Wall-mounted coat and hat rack with pilfer-resistant hangers, aluminum

| | Craft@Hrs | Unit | Material | Labor | Total |
|---|---|---|---|---|---|
| 24" length,   8 hangers | CC@.665 | Ea | 156.00 | 34.40 | 190.40 |
| 36" length, 12 hangers | CC@.748 | Ea | 183.00 | 38.70 | 221.70 |
| 48" length, 16 hangers | CC@.833 | Ea | 233.00 | 43.10 | 276.10 |
| 60" length, 20 hangers | CC@.916 | Ea | 272.00 | 47.40 | 319.40 |
| 72" length, 24 hangers | CC@.997 | Ea | 294.00 | 51.60 | 345.60 |

**Food Service Equipment** These costs do not include electrical work or plumbing. Add the cost of hookup when needed. Based on Superior Products Company. (Figure in parentheses is equipment weight.)
Food stations, vinyl clad steel, with stainless steel work area and clear acrylic breath/sneeze guard,
24" D x 35" H, mounted on 4" casters

| | Craft@Hrs | Unit | Material | Labor | Total |
|---|---|---|---|---|---|
| Hot food station, 60" long (182 lbs) | D4@1.00 | Ea | 2,150.00 | 46.30 | 2,196.30 |
| Hot food station, 46" long (156 lbs) | D4@1.00 | Ea | 1,930.00 | 46.30 | 1,976.30 |
| Cold food station, 60" long (166 lbs) | D4@1.00 | Ea | 1,930.00 | 46.30 | 1,976.30 |
| Cold food station, 46" long (147 lbs) | D4@1.00 | Ea | 1,700.00 | 46.30 | 1,746.30 |
| Plate rest with mounting kit (10 lbs) | D4@.250 | Ea | 350.00 | 11.60 | 361.60 |
| Incandescent light assembly (20 lbs) | D4@.250 | Ea | 660.00 | 11.60 | 671.60 |

Food warmers, stainless steel top, hardwood cutting board, 12" x 20" openings, waterless

| | Craft@Hrs | Unit | Material | Labor | Total |
|---|---|---|---|---|---|
| 58-1/2", natural gas, 4 openings (135 lbs) | P6@1.00 | Ea | 1,150.00 | 50.20 | 1,200.20 |
| 58-1/2", LP gas, 4 openings (135 lbs) | P6@1.00 | Ea | 1,150.00 | 50.20 | 1,200.20 |
| 58-1/2", electric, 3000 watt, 4 openings, (135 lbs) | E4@1.00 | Ea | 1,280.00 | 49.10 | 1,329.10 |

Countertop one-compartment steamer, stainless steel. Holds multiple pan inserts. Thermostatically controlled 208 volt electric heater element. Holds 12" x 20" x 2-1/2" deep pans. Underwriters Laboratories and NSF listed. Add the cost of 3/4" water valve connection and 2" drain valve connection.

| | Craft@Hrs | Unit | Material | Labor | Total |
|---|---|---|---|---|---|
| 3-pan inserts, 8.1 KW | P1@16.0 | Ea | 8,600.00 | 572.00 | 9,172.00 |
| 5-pan inserts, 11 KW | P1@20.0 | Ea | 12,100.00 | 715.00 | 12,815.00 |

Meat and deli case, self-contained with gravity cooling, stainless steel top, 54" high x 34" deep

| | Craft@Hrs | Unit | Material | Labor | Total |
|---|---|---|---|---|---|
| 48-1/2" long (425 lbs) | D4@3.00 | Ea | 3,150.00 | 139.00 | 3,289.00 |
| 72-1/2" long (595 lbs) | D4@4.00 | Ea | 3,740.00 | 185.00 | 3,925.00 |

Freestanding bench type meat saw, 5/8" saw blade, belt-drive, toggle switch, 9.56" x 10.3" throat size. Includes a wheel and blade scraper, stainless steel carriage table with easy-slide moving height and a hose down cleanable epoxy enamel cabinet. 1.5 HP motor. Grade 304 stainless steel throughout. Underwriters Laboratories and NSF rated.

| | Craft@Hrs | Unit | Material | Labor | Total |
|---|---|---|---|---|---|
| 14" x 14" table | D4@2.0 | Ea | 2,020.00 | 92.50 | 2,112.50 |
| 24" x 24" table | D4@2.0 | Ea | 3,550.00 | 92.50 | 3,642.50 |

Sandwich preparation units, free standing, stainless steel top, sides and back, 115 V AC, refrigerated 30" D x 43" H, full length 12" deep cutting board

| | Craft@Hrs | Unit | Material | Labor | Total |
|---|---|---|---|---|---|
| 6.5 cubic feet, 27-1/2" long | D4@2.00 | Ea | 1,390.00 | 92.50 | 1,482.50 |
| 12 cubic feet, 48" long | D4@2.00 | Ea | 2,050.00 | 92.50 | 2,142.50 |

Pizza preparation units, free standing, stainless steel top, sides and back, 115 V AC, refrigerated 35-1/2" D x 41-1/2" H, full length 19-1/2" deep cutting board

| | Craft@Hrs | Unit | Material | Labor | Total |
|---|---|---|---|---|---|
| 11.4 cubic feet, 44-1/2" long, one door | D4@2.00 | Ea | 2,420.00 | 92.50 | 2,512.50 |
| 20.6 cubic feet, 67" long, two doors | D4@2.00 | Ea | 3,400.00 | 92.50 | 3,492.50 |
| 30.9 cubic feet, 93-1/4" long, three doors | D4@2.00 | Ea | 4,590.00 | 92.50 | 4,682.50 |
| Pan kit for prep tables | — | Ea | 75.00 | — | 75.00 |

Sushi serving table, custom made with multiple sealed wells for wet or dry operation. Grade 304 stainless steel with understorage, 14 gauge base supports, individual drains and valves and adjustable legs. White polyethylene top, glass customer product screen and cutting board. Thermostatically controlled electric heater elements. Add for customer seating. Underwriters Laboratories and NSF rated.

| | Craft@Hrs | Unit | Material | Labor | Total |
|---|---|---|---|---|---|
| 5-customer seating | T5@40.0 | Ea | 11,600.00 | 1,970.00 | 13,570.00 |
| 10-customer seating | T5@75.0 | Ea | 25,100.00 | 3,690.00 | 28,790.00 |

# 11 Equipment

| | Craft@Hrs | Unit | Material | Labor | Total |
|---|---|---|---|---|---|

Four-station ice cream and shake serving and dispenser, heat-treatment freezer system type, multiple quart per station capacity, with syrup dispensers, product pumps, water valve connections, cleaning drain connections, thermostatic controls, product digital temperature and serve rate metering. Grade 304 stainless steel throughout. Underwriters Laboratories and NSF rated. Equipment cost includes a forklift for one work day.

| | Craft@Hrs | Unit | Material | Labor | Total |
|---|---|---|---|---|---|
| 14 quart total capacity | D4@30.0 | Ea | 23,100.00 | 1,390.00 | 24,490.00 |
| 28 quart total capacity | D4@30.0 | Ea | 35,100.00 | 1,390.00 | 36,490.00 |

Shelving, aluminum, free-standing, 2,000 lb capacity lower shelf, 400 lb capacity uppers

| | Craft@Hrs | Unit | Material | Labor | Total |
|---|---|---|---|---|---|
| 18" x 36" x 68" high, 3 tier | C8@2.00 | Ea | 280.00 | 91.70 | 371.70 |
| 18" x 48" x 68" high, 3 tier | C8@2.00 | Ea | 280.00 | 91.70 | 371.70 |
| 18" x 36" x 76" high, 4 tier | C8@2.00 | Ea | 340.00 | 91.70 | 431.70 |
| 18" x 48" x 76" high, 4 tier | C8@2.00 | Ea | 370.00 | 91.70 | 461.70 |

Work tables, 16 gauge type 430 stainless steel, 30" high. Includes two 12" deep over-shelves, full depth under-shelf and pot and ladle rack.

| | Craft@Hrs | Unit | Material | Labor | Total |
|---|---|---|---|---|---|
| 48" long (118 lbs) | D4@1.00 | Ea | 1,020.00 | 46.30 | 1,066.30 |
| 60" long (135 lbs) | D4@1.00 | Ea | 1,110.00 | 46.30 | 1,156.30 |
| 72" long (154 lbs) | D4@1.00 | Ea | 1,200.00 | 46.30 | 1,246.30 |
| Add to any work table above | | | | | |
| 20" x 20" x 5" pull flange drawer (13 lbs) | D4@.250 | Ea | 115.00 | 11.60 | 126.60 |
| Stainless steel pot hooks | D4@.250 | Ea | 10.40 | 11.60 | 22.00 |

Exhaust hoods, 16 gauge grade 304 stainless steel with #3 polished finish. Conforms to NFPA code #96 and the National Sanitation Foundation requirements. Add for exhaust fans from the next section.

| | Craft@Hrs | Unit | Material | Labor | Total |
|---|---|---|---|---|---|
| 48" deep angled front hood, 24" high with length as listed below | | | | | |
| 6' long (205 lbs) | T5@2.00 | Ea | 1,170.00 | 98.40 | 1,268.40 |
| 8' long (250 lbs) | T5@3.00 | Ea | 1,530.00 | 148.00 | 1,678.00 |
| 10' long (285 lbs) | T5@4.00 | Ea | 1,880.00 | 197.00 | 2,077.00 |
| Additives to any hood above | | | | | |
| Installation kit | — | Ea | 195.00 | — | 195.00 |
| Grease filters, aluminum, cleanable, 2' deep | T5@.125 | LF | 42.40 | 6.15 | 48.55 |

Exhaust fans, 2,800 CFM, 1,265 RPM. Wall or roof mounted, for use with the exhaust hoods above. Heavy duty aluminum with 115 volt AC motor in a weathertight enclosure.
These costs are based on installing the fan in an existing opening, electrical hookup not included

| | Craft@Hrs | Unit | Material | Labor | Total |
|---|---|---|---|---|---|
| Fan for 6' long hood, 18-1/2" square curb size | | | | | |
| 1/2 HP (75 lbs) | T5@2.00 | Ea | 1,020.00 | 98.40 | 1,118.40 |
| Fan for 8' or 10' long hood, 23-1/2" square curb size | | | | | |
| 1/2 HP (100 lbs) | T5@2.00 | Ea | 1,080.00 | 98.40 | 1,178.40 |
| Additives to either fan above | | | | | |
| Roof curb | T5@1.00 | Ea | 306.00 | 49.20 | 355.20 |
| Variable speed motor controller, 115 volt AC | — | Ea | 137.00 | — | 137.00 |
| Lights, incandescent, pre-wired, 115 volt AC | — | Ea | 78.20 | — | 78.20 |
| Ductwork, aluminized steel | T5@.100 | LF | 37.40 | 4.92 | 42.32 |
| 90-degree elbow for ductwork, aluminized steel | T5@.500 | Ea | 249.00 | 24.60 | 273.60 |

Electrostatic precipitators for restaurant exhaust air purification. Removes kitchen exhaust grease, smoke and odors. Meets EPA, state and most local air quality standards. Multi-stage, self-contained filtration unit with electrostatic precipitator ionizer module, 95% DOP (dioctyl phthalate) Mil-Std 282 media filter, and carbon modules for odor control. For applications in commercial and institutional kitchens such as restaurants, food courts, hospitals, and schools.

| | Craft@Hrs | Unit | Material | Labor | Total |
|---|---|---|---|---|---|
| 1,200 CFM, 1.5 Amperes | P1@16.0 | Ea | 8,550.00 | 572.00 | 9,122.00 |
| 2,400 CFM, 5.0 Amperes | P1@20.0 | Ea | 12,100.00 | 715.00 | 12,815.00 |

| | Craft@Hrs | Unit | Material | Labor | Total |
|---|---|---|---|---|---|

Removable oven and stove exhaust hood filters with mounting flue enclosure. Seamless welded design, finished in marine-grade 316 stainless steel, 27" deep hood, with internal blower meeting Home Ventilating Institute certified performance standards. Fully enclosed with baffle filters, halogen lamps, marine grade controls to resist high moisture and heat in flue gases. With backsplash panel and soffit flue venting. Commercial and institutional grade. NSF and NFPA and Underwriters Laboratories approved.

| | Craft@Hrs | Unit | Material | Labor | Total |
|---|---|---|---|---|---|
| 1,100 CFM | P6@20.0 | Ea | 4,500.00 | 1,000.00 | 5,500.00 |
| 1,200 CFM | P6@20.0 | Ea | 6,500.00 | 1,000.00 | 7,500.00 |
| 1,500 CFM | P6@20.0 | Ea | 7,000.00 | 1,000.00 | 8,000.00 |

Restaurant cooking area fire protection wet chemical system. Exceeds the standards of Underwriters Laboratories (UL-300, Fire Extinguishing Systems for Protection of Restaurant Cooking Areas) and complies with NFPA 17A (Wet Chemical Extinguishing Systems), NFPA 96 (Commercial Cooking Operations, Ventilation Control and Fire) and major insurance company guidelines.

| | Craft@Hrs | Unit | Material | Labor | Total |
|---|---|---|---|---|---|
| 5' wide range hood | P6@16.0 | Ea | 1,650.00 | 804.00 | 2,454.00 |
| 14' wide range hood | P6@24.0 | Ea | 2,450.00 | 1,210.00 | 3,660.00 |

Charbroiler, lava rock, countertop model, burner and control every 12", natural or LP gas

| | Craft@Hrs | Unit | Material | Labor | Total |
|---|---|---|---|---|---|
| 40,000 Btu, 15" W (121 lbs) | P6@2.00 | Ea | 690.00 | 100.00 | 790.00 |
| 80,000 Btu, 24" W (186 lbs) | P6@2.00 | Ea | 820.00 | 100.00 | 920.00 |
| 120,000 Btu, 36" W (230 lbs) | P6@2.00 | Ea | 1,140.00 | 100.00 | 1,240.00 |

Charbroiler, radiant, countertop model, burner and control every 12", natural or LP gas

| | Craft@Hrs | Unit | Material | Labor | Total |
|---|---|---|---|---|---|
| 80,000 Btu, 24" W (155 lbs) | P6@2.00 | Ea | 2,030.00 | 100.00 | 2,130.00 |
| 120,000 Btu, 36" W (211 lbs) | P6@2.00 | Ea | 2,700.00 | 100.00 | 2,800.00 |
| 160,000 Btu, 48" W (211 lbs) | P6@2.00 | Ea | 3,250.00 | 100.00 | 3,350.00 |

Deep fat fryer, free standing floor model, electric, 45" height, stainless steel front and top

| | Craft@Hrs | Unit | Material | Labor | Total |
|---|---|---|---|---|---|
| 40 lbs capacity (200 lbs) | P6@2.00 | Ea | 1,680.00 | 100.00 | 1,780.00 |
| 60 lbs capacity (225 lbs) | P6@2.00 | Ea | 2,060.00 | 100.00 | 2,160.00 |
| Add for filter ready | — | Ea | 325.00 | — | 325.00 |

Gas ranges, free standing, floor mounted, 29-3/4" D x 59" H x 62-3/4" W, with twin ovens and 24" raised griddle/broiler

| | Craft@Hrs | Unit | Material | Labor | Total |
|---|---|---|---|---|---|
| 6 burners, 35,000 Btu each (835 lbs) | P6@4.00 | Ea | 5,450.00 | 201.00 | 5,651.00 |
| Add for stainless steel shelf above range | — | Ea | 120.00 | — | 120.00 |

Restaurant range, gas, 6 burners, 2 ovens, raised griddle/broiler, open burners with black porcelain top grates, stainless steel front, sides, back riser, high shelf and 6" legs. 3/4" rear gas connection with gas pressure regulator. NSF listed.

| | Craft@Hrs | Unit | Material | Labor | Total |
|---|---|---|---|---|---|
| 50 MBtu/Hr | P6@8.00 | Ea | 6,600.00 | 402.00 | 7,002.00 |
| 75 MBtu/Hr | P6@8.00 | Ea | 7,700.00 | 402.00 | 8,102.00 |
| 100 MBtu/Hr | P6@8.00 | Ea | 11,500.00 | 402.00 | 11,902.00 |

Griddles, gas, 27,500 Btu, 1" thick x 24" deep griddle plate, countertop mounted

| | Craft@Hrs | Unit | Material | Labor | Total |
|---|---|---|---|---|---|
| 24" wide (295 lbs) | P6@2.00 | Ea | 750.00 | 100.00 | 850.00 |
| 36" wide (385 lbs) | P6@2.00 | Ea | 910.00 | 100.00 | 1,010.00 |
| 48" wide (520 lbs) | P6@3.00 | Ea | 1,280.00 | 151.00 | 1,431.00 |

Ovens, free standing, floor mounted

Convection oven, natural gas, stainless steel, porcelain interior, 38" W x 36" D x 54-1/4" H

| | Craft@Hrs | Unit | Material | Labor | Total |
|---|---|---|---|---|---|
| 54,000 Btu (550 lbs) | D4@4.00 | Ea | 4,600.00 | 185.00 | 4,785.00 |

Conveyer oven, electric, stainless steel body, 18" H x 50" W x 31-3/8" D, 208 or 240 Volt AC

| | Craft@Hrs | Unit | Material | Labor | Total |
|---|---|---|---|---|---|
| 16" W conveyor, 6 KW, UL listed (190 lbs) | D4@4.00 | Ea | 4,790.00 | 185.00 | 4,975.00 |
| Entry & exit shelves | — | Ea | 220.00 | — | 220.00 |

Professional microwave oven, single shelf, permanent mount, stainless steel throughout, with 6 minute dial timer, bottom energy feed, see-thru side hinged door, digital controls, cavity or wall mount, 120 Volt, 60 Hertz, single phase, Underwriters Laboratories and NSF listed.

| | Craft@Hrs | Unit | Material | Labor | Total |
|---|---|---|---|---|---|
| 750 watt | E4@4.00 | Ea | 562.00 | 197.00 | 759.00 |
| 1,000 watt | E4@4.00 | Ea | 473.00 | 197.00 | 670.00 |
| 1,500 watt | E4@4.00 | Ea | 683.00 | 197.00 | 880.00 |
| 2,000 watt | E4@4.00 | Ea | 893.00 | 197.00 | 1,090.00 |

# 11 Equipment

| | Craft@Hrs | Unit | Material | Labor | Total |
|---|---|---|---|---|---|

Pastry oven, gas, 6 burners, 2 ovens, raised griddle/broiler, open burners with black porcelain top grates. Stainless steel front, sides, back riser, high shelf and 6" legs. 3/4" rear gas connection with gas pressure regulator. NSF listed.

| | Craft@Hrs | Unit | Material | Labor | Total |
|---|---|---|---|---|---|
| 50 MBtu/Hr | P6@8.00 | Ea | 6,300.00 | 402.00 | 6,702.00 |
| 75 MBtu/Hr | P6@8.00 | Ea | 7,880.00 | 402.00 | 8,282.00 |
| 100 MBtu/Hr | P6@8.00 | Ea | 11,600.00 | 402.00 | 12,002.00 |

Dough roller, production type, for bread, pastry, donut and bagel manufacture. Two-stage machine with stainless steel construction. Quick change handles, front infeed and front discharge. Rolls up to 18" diameter doughs. NSF 8 listed. 120 Volt , 60 Hertz, single phase, UL listed.

| | | | | | |
|---|---|---|---|---|---|
| 450 pastry blanks per hour | E4@4.00 | Ea | 4,520.00 | 197.00 | 4,717.00 |
| 1,000 pastry blanks per hour | E4@4.00 | Ea | 6,830.00 | 197.00 | 7,027.00 |
| 2,000 pastry blanks per hour | E4@4.00 | Ea | 15,800.00 | 197.00 | 15,997.00 |

Bread slicer, hard or soft crusted breads. With stainless steel bagging trough. 3/8" (10 mm), 7/16" (11 mm), 1/2" (13 mm), or 3/4" (19 mm) pre-set slice thickness. Independently mounted heat-treated alloy carbon steel blades. Single phase 115 volt, 60 Hz AC power. ETL listed for safety.

| | | | | | |
|---|---|---|---|---|---|
| 6.2 Amp/hr, 420 loaves per hour | E4@4.00 | Ea | 4,520.00 | 197.00 | 4,717.00 |
| 10 Amp/hr,  700 loaves per hour | E4@4.00 | Ea | 7,880.00 | 197.00 | 8,077.00 |
| 18 Amp/hr,  1,000 loaves per hour | E4@4.00 | Ea | 11,600.00 | 197.00 | 11,797.00 |

Toaster, conveyor type, stainless steel, fused quartz sheathing, countertop model

| | | | | | |
|---|---|---|---|---|---|
| 350 slices per hour, 115 volt AC, 1,400 watts | D4@3.00 | Ea | 629.00 | 139.00 | 768.00 |

Pot and pan racks, galvanized steel, pot hooks placed 4 per linear foot

| | | | | | |
|---|---|---|---|---|---|
| Ceiling mounted, 5' | C8@2.00 | Ea | 365.00 | 91.70 | 456.70 |
| Wall mounted, 5' | C8@2.00 | Ea | 245.00 | 91.70 | 336.70 |

Ice-maker, free standing, floor mounted model

| | | | | | |
|---|---|---|---|---|---|
| 147 lbs per day capacity (165 lbs) | P6@2.00 | Ea | 1,820.00 | 100.00 | 1,920.00 |
| 220 lbs per day capacity (165 lbs) | P6@2.00 | Ea | 2,130.00 | 100.00 | 2,230.00 |
| 290 lbs per day capacity (215 lbs) | P6@2.00 | Ea | 2,390.00 | 100.00 | 2,490.00 |
| 395 lbs per day capacity (185 lbs) | P6@2.00 | Ea | 3,200.00 | 100.00 | 3,300.00 |

Ice storage unit, self service for bagged ice, 40 cubic foot capacity, floor mounted. Includes refrigeration unit.

| | | | | | |
|---|---|---|---|---|---|
| Indoor type | E4@3.00 | Ea | 2,700.00 | 147.00 | 2,847.00 |
| Outdoor type, automatic defrost | E4@3.00 | Ea | 2,200.00 | 147.00 | 2,347.00 |

Refrigerators, CFC-free refrigeration system 115 volt AC

Under counter, stainless steel front and top

| | | | | | |
|---|---|---|---|---|---|
| 6.2 cubic foot (157 lbs) | P6@3.00 | Ea | 1,190.00 | 151.00 | 1,341.00 |
| 11.8 cubic foot (230 lbs) | P6@4.00 | Ea | 1,930.00 | 201.00 | 2,131.00 |

Beverage backbar refrigerator, 2 section glass doors, 34" high, condensing unit on right, galvanized top, stainless steel floors, galvanized interior walls, epoxy coated steel shelves.  208 volt, 1- or 3-phase, 60 Hz AC, Underwriters Laboratories and NSF rated.

| | | | | | |
|---|---|---|---|---|---|
| 5 MBtu/Hr | E4@4.00 | Ea | 3,680.00 | 197.00 | 3,877.00 |
| 10 MBtu/Hr | E4@4.00 | Ea | 5,780.00 | 197.00 | 5,977.00 |
| 20 MBtu/Hr | E4@4.00 | Ea | 7,880.00 | 197.00 | 8,077.00 |

Freezers, under counter type, stainless steel front and top, CFC-free refrigeration system

| | | | | | |
|---|---|---|---|---|---|
| 6.2 cubic foot (162 lbs) | P6@.500 | Ea | 1,620.00 | 25.10 | 1,645.10 |
| 11.8 cubic foot (234 lbs) | P6@1.00 | Ea | 2,060.00 | 50.20 | 2,110.20 |

Reach-in refrigerator, bottom mounted, stainless steel doors, anodized aluminum sides

| | | | | | |
|---|---|---|---|---|---|
| Single door, 23 cubic foot (276 lbs) | P6@4.00 | Ea | 2,260.00 | 201.00 | 2,461.00 |
| Double door, 49 cubic foot (410 lbs) | P6@4.00 | Ea | 2,880.00 | 201.00 | 3,081.00 |
| Triple door, 72 cubic foot (600 lbs) | P6@4.00 | Ea | 3,860.00 | 201.00 | 4,061.00 |

Custom walk-in coolers and freezers, in an existing building. Including thermostatically controlled external remote condenser and temperature recorder. UL and NSF rated compound thermoplastic internal insulation, food-grade grouted tile floor. Fluorescent illumination. High-traffic stainless steel vault door with safety panic bar-style exit latch override. Shelving not included. Per square foot of floor.

| | | | | | |
|---|---|---|---|---|---|
| Cooler (to +35 degrees Fahrenheit) | — | SF | — | — | 52.00 |
| Freezer (to −35 degrees Fahrenheit) | — | SF | — | — | 77.00 |

|  | Craft@Hrs | Unit | Material | Labor | Total |
|---|---|---|---|---|---|
| Sinks, stainless steel, free standing, floor mounted, with 7" high backsplash and faucet holes drilled on 8" centers, no trim included | | | | | |
| Scullery sink, 16" x 20" tubs, 12" deep | | | | | |
| 40" long, 1 tub, 1 drainboard (65 lbs) | P6@2.00 | Ea | 1,130.00 | 100.00 | 1,230.00 |
| 58" long, 2 tubs, 1 drainboard (80 lbs) | P6@2.00 | Ea | 1,360.00 | 100.00 | 1,460.00 |
| 103" long, 3 tubs, 2 drainboards (155 lbs) | P6@2.00 | Ea | 2,040.00 | 100.00 | 2,140.00 |
| 126" long, 4 tubs, 2 drainboards (175 lbs) | P6@2.00 | Ea | 2,510.00 | 100.00 | 2,610.00 |
| Mop sink, faucet not included | | | | | |
| 21" x 21" x 41", single compartment | P6@2.00 | Ea | 289.00 | 100.00 | 389.00 |
| Dining facility maintenance and cleaning station, stainless steel throughout. Includes floor-mounted mop sink, hose and cleaning fluid dispenser and connections for mop and bucket fill, shelving and mop racks, shielded lighting fixture, grounded high-mount electrical outlet and floor drain. UL and NSF rated. Built into an existing structure. Add the cost of plumbing and electric rough-in. Per square foot of floor. | | | | | |
| Installed cost | — | SF | — | — | 26.00 |
| Food mixing machines, floor mounted | | | | | |
| 20 quart, bench mount (260 lbs) | D4@2.00 | Ea | 2,760.00 | 92.50 | 2,852.50 |
| 30 quart (360 lbs) | D4@3.00 | Ea | 4,000.00 | 139.00 | 4,139.00 |
| Garbage disposers, includes control switch, solenoid valve, siphon breaker and flow control valve | | | | | |
| 1/2 HP (56 lbs) | P6@1.00 | Ea | 1,140.00 | 50.20 | 1,190.20 |
| 3/4 HP (57 lbs) | P6@1.00 | Ea | 1,390.00 | 50.20 | 1,440.20 |
| 1 HP (60 lbs) | P6@1.00 | Ea | 1,480.00 | 50.20 | 1,530.20 |
| 1-1/2 HP (76 lbs) | P6@2.00 | Ea | 1,880.00 | 100.00 | 1,980.00 |
| 2 HP (81 lbs) | P6@2.00 | Ea | 2,020.00 | 100.00 | 2,120.00 |
| 3 HP (129 lbs) | P6@3.00 | Ea | 2,450.00 | 151.00 | 2,601.00 |
| Dishwashers | | | | | |
| Under counter, 24 rack capacity per hour. Universal timer | | | | | |
| High temp 70° booster (245 lbs) | P6@4.00 | Ea | 4,120.00 | 201.00 | 4,321.00 |
| Low temp, chemical sanitizer (245 lbs) | P6@4.00 | Ea | 3,600.00 | 201.00 | 3,801.00 |
| Upright, heavy duty, semi-automatic, 58 racks per hour. Can be used for corner or straight-through applications | | | | | |
| With booster heater (420 lbs) | P6@6.00 | Ea | 9,620.00 | 301.00 | 9,921.00 |
| Without booster heater (420 lbs) | P6@6.00 | Ea | 7,920.00 | 301.00 | 8,221.00 |
| Cappuccino dispenser, for cappuccino, hot chocolate or other hot beverages, 1/4" water line required | | | | | |
| Two flavor, 42 servings | D4@1.00 | Ea | 1,430.00 | 46.30 | 1,476.30 |
| Three flavor, 62 servings | D4@1.00 | Ea | 1,630.00 | 46.30 | 1,676.30 |
| Multi-station decanter coffee brewer, stainless steel, with water feed tank valve connections, permanent countertop mounted, 115/230 volt single phase 60 Hertz, with warmer plates. UL and NSF rated. | | | | | |
| Single brewer, three warmers | P6@4.00 | Ea | 1,550.00 | 201.00 | 1,751.00 |
| Single brewer, five warmers | P6@4.00 | Ea | 2,060.00 | 201.00 | 2,261.00 |
| Twin brewers, eight warmers | P6@8.00 | Ea | 3,710.00 | 402.00 | 4,112.00 |
| Multi-station urn coffee percolator brewer, stainless steel, permanent countertop mounted, with water feed tank valve connections, 115/230 volt single phase, 60 Hz, with warmer plates. UL and NSF rated. | | | | | |
| Single percolator urn, three warmers | P6@4.00 | Ea | 1,550.00 | 201.00 | 1,751.00 |
| Twin percolator urn, five warmers | P6@8.00 | Ea | 2,060.00 | 402.00 | 2,462.00 |
| Beverage dispenser, non-carbonated, refrigerated, countertop model | | | | | |
| 5 gallon, 1 bowl | P6@2.00 | Ea | 750.00 | 100.00 | 850.00 |
| 5 gallon, 2 bowls | P6@2.00 | Ea | 1,060.00 | 100.00 | 1,160.00 |
| 2.4 gallon, 2 mini bowls | P6@2.00 | Ea | 865.00 | 100.00 | 965.00 |
| 2.4 gallon, 4 mini bowls | P6@2.00 | Ea | 1,430.00 | 100.00 | 1,530.00 |

| | Craft@Hrs | Unit | Material | Labor | Total |
|---|---|---|---|---|---|

**Educational Equipment** Labor includes unpacking, layout and complete installation of prefinished furniture and equipment.

Wardrobes, 40" x 78" x 26"

| | Craft@Hrs | Unit | Material | Labor | Total |
|---|---|---|---|---|---|
| Teacher type | C8@2.00 | Ea | 1,270.00 | 91.70 | 1,361.70 |
| Student type | C8@2.00 | Ea | 641.00 | 91.70 | 732.70 |

Seating, per seat

| | Craft@Hrs | Unit | Material | Labor | Total |
|---|---|---|---|---|---|
| Lecture room, pedestal, with folding arm | C8@2.00 | Ea | 167.00 | 91.70 | 258.70 |
| Horizontal sections | C8@2.00 | Ea | 80.10 | 91.70 | 171.80 |

Tables, fixed pedestal

| | Craft@Hrs | Unit | Material | Labor | Total |
|---|---|---|---|---|---|
| 48" x 16" | C8@2.00 | Ea | 362.00 | 91.70 | 453.70 |
| With chairs, 48" x 16" | C8@2.00 | Ea | 520.00 | 91.70 | 611.70 |

Projection screen

| | Craft@Hrs | Unit | Material | Labor | Total |
|---|---|---|---|---|---|
| Ceiling mounted, pull down | D4@.019 | SF | 3.85 | .88 | 4.73 |
| Videotape recorder | CE@.500 | Ea | 140.00 | 29.20 | 169.20 |
| T.V. camera | CE@.800 | Ea | 556.00 | 46.70 | 602.70 |
| T.V. monitor, wall mounted | CE@2.00 | Ea | 325.00 | 117.00 | 442.00 |

Language laboratory study carrels, plastic laminated wood

| | Craft@Hrs | Unit | Material | Labor | Total |
|---|---|---|---|---|---|
| Individual, 48" x 30" x 54" | C8@1.43 | Ea | 228.00 | 65.50 | 293.50 |
| Two position, 73" x 30" x 47" | C8@1.93 | Ea | 610.00 | 88.50 | 698.50 |
| Island, 4 position, 66" x 66" x 47" | C8@3.17 | Ea | 688.00 | 145.00 | 833.00 |

**Library equipment,** educational and institutional grade. Meets recommendations of the America Library Association. Complies with FEMA 356 standards and ANSI/NISO Z39.73-1994 seismic safety standards. Each 5-level shelf unit is 72" high by 36" wide by 21" deep. Floor-mounted units require four seismic assemblies. Wall-mounted units require two seismic assemblies.

| | Craft@Hrs | Unit | Material | Labor | Total |
|---|---|---|---|---|---|
| Torch-cut or concrete-saw hole in steel or concrete structural frame for seismic mounts (floor or wall) | D4@1.50 | Ea | — | 69.40 | 69.40 |
| Assemble, bolt down and level each seismic spring module | D4@0.25 | Ea | 125.00 | 11.60 | 136.60 |
| Assemble and position bookshelf | D4@1.25 | Ea | 650.00 | 57.80 | 707.80 |
| Bolt down and level completed assembly | D4@0.50 | Ea | — | 23.10 | 23.10 |
| Load and vibration test | D4@1.00 | Ea | — | 46.30 | 46.30 |
| Computer work station, 29" x 59" x 24", duplex unit, freestanding, Achieva or equal, including assembly and two seismic anchors | C8@1.00 | Ea | 370.00 | 45.80 | 415.80 |
| Computer work center, hexagonal, 26" x 80" x 35", seats six, Paragon or equal, including assembly and three seismic anchors | C8@1.00 | Ea | 750.00 | 45.80 | 795.80 |
| Double desk computer work station, 36" W x 30" D, MyOffice or equal, including assembly and two seismic anchors | C8@1.00 | Ea | 1,200.00 | 45.80 | 1,245.80 |

**Rolling Ladders** Rolling ladders for mercantile, library or industrial use.
Oak rolling ladder with top and bottom rolling fixtures, non-slip treads, pre-finished. By floor to track height. Painted aluminum hardware. Add cost of track from below

| | Craft@Hrs | Unit | Material | Labor | Total |
|---|---|---|---|---|---|
| Up to 9'6" high | D4@.819 | Ea | 605.00 | 37.90 | 642.90 |
| 9'6" to 10'6" high | D4@.819 | Ea | 615.00 | 37.90 | 652.90 |
| 10'6" to 11'6" high | D4@1.09 | Ea | 665.00 | 50.40 | 715.40 |
| 11'6" to 12'6" high | D4@1.09 | Ea | 701.00 | 50.40 | 751.40 |
| 12'6" to 13'6" high | D4@1.09 | Ea | 742.00 | 50.40 | 792.40 |
| 13' to 14'6" high | D4@1.09 | Ea | 773.00 | 50.40 | 823.40 |
| Add for bend at ladder top or bottom | — | Ea | 87.40 | — | 87.40 |
| Add for brass or chrome hardware | — | Ea | 98.70 | — | 98.70 |
| Add for ladder work shelf | — | Ea | 68.20 | — | 68.20 |
| Add for handrail | — | Ea | 46.50 | — | 46.50 |

| | Craft@Hrs | Unit | Material | Labor | Total |
|---|---|---|---|---|---|
| **Rolling ladder track, 7/8" slotted steel, per foot of length, 9' to 14' high** | | | | | |
| Painted aluminum finish | D4@.089 | LF | 6.10 | 4.12 | 10.22 |
| Painted black finish | D4@.089 | LF | 5.95 | 4.12 | 10.07 |
| Brass or chrome plated finish | D4@.089 | LF | 14.10 | 4.12 | 18.22 |
| Corner bend track, 90 degree, 30" radius | | | | | |
| Painted finish | D4@.363 | Ea | 50.50 | 16.80 | 67.30 |
| Chrome or brass finish | D4@.363 | Ea | 80.50 | 16.80 | 97.30 |

## Educational Computer Network Wiring

These estimates assume installation in steel frame interior walls and suspended ceilings before drywall and ceiling finish are installed. All cable runs terminate to either a telephone closet or a server room. All runs are to a punch block at both the distribution end and central interconnect. Underground cable runs assume communications duct has been installed by others. Central interconnect locations are usually a classroom located near or in a cluster of classroom buildings. One-hundred-pair cable is suitable for 12 network connections with 2 spares.

Campus network wiring pulled in underground communications duct from a central interconnection location to a distribution point. Based on a 1,000' cable run.

| | Craft@Hrs | Unit | Material | Labor | Total |
|---|---|---|---|---|---|
| **Category 5E cable** | | | | | |
| 25 pair cable | E4@.043 | LF | 1.62 | 2.11 | 3.73 |
| 100 pair cable | E4@.043 | LF | 5.90 | 2.11 | 8.01 |
| **Cable pulling come-along, 3,000 pound rating** | | | | | |
| Rental per day | — | Day | 15.00 | — | 15.00 |
| Rental per week | — | Wk | 60.00 | — | 60.00 |
| Rental per month | — | Mo | 180.00 | — | 180.00 |

Plywood backboard mount for a wall-mount rack supporting the patch panel and network switches. Set on a masonry wall.

| | Craft@Hrs | Unit | Material | Labor | Total |
|---|---|---|---|---|---|
| 4' x 4' x 3/4" | E4@.938 | Ea | 11.90 | 46.10 | 58.00 |
| 4' x 8' x 3/4" | E4@1.10 | Ea | 23.90 | 54.00 | 77.90 |

Classroom punch down assembly, wall mounted, includes frame and 110 style punch down. Does not include labor to punch down cable connections.

| | Craft@Hrs | Unit | Material | Labor | Total |
|---|---|---|---|---|---|
| Wall mounting frame with 110 style punch down | E4@.250 | Ea | 155.00 | 12.30 | 167.30 |

Classroom network station wiring, category 5E twisted pair 350 mbps cable pulled in conduit from a classroom to a central interconnect point. Based on 1,000' cable run with conduit installed by others. Add the cost of cable testing below.

| | Craft@Hrs | Unit | Material | Labor | Total |
|---|---|---|---|---|---|
| 4 pair UTP plenum-rated solid wire cable | E4@.008 | LF | 0.25 | .39 | .64 |

Network cable wall-mount rack with category 6 rack mount patch panels and 110 style punch down blocks. A wall-mount rack will hold eight 6" rack mount devices. Add the cost of cable punch down below.

| | Craft@Hrs | Unit | Material | Labor | Total |
|---|---|---|---|---|---|
| 24 port patch panel | E4@.250 | Ea | 370.00 | 12.30 | 382.30 |
| 48 port patch panel | E4@.350 | Ea | 565.00 | 17.20 | 582.20 |
| 96 port patch panel | E4@.450 | Ea | 896.00 | 22.10 | 918.10 |

Cable punch down. Labor for punch down (connection) on a 110 style block mounted in a classroom, central interconnect point or at server room.

| | Craft@Hrs | Unit | Material | Labor | Total |
|---|---|---|---|---|---|
| Per four pair of cable connections | E4@.020 | Ea | — | .98 | .98 |

Network cable testing, using a dedicated cable tester, shielded or unshielded cable

| | Craft@Hrs | Unit | Material | Labor | Total |
|---|---|---|---|---|---|
| Per twisted pair | E4@.125 | Ea | — | 6.14 | 6.14 |

Distribution of patch panel connections to a network switch. 96 port uses 2 DES-3550s. Includes 10/100 network switch and patch cables

| | Craft@Hrs | Unit | Material | Labor | Total |
|---|---|---|---|---|---|
| 24 station connection | CE@0.33 | Ea | 399.00 | 19.30 | 418.30 |
| 48 station connection | CE@0.50 | Ea | 652.00 | 29.20 | 681.20 |
| 96 station connection | CE@1.00 | Ea | 1,510.00 | 58.40 | 1,568.40 |

# 11 Equipment

|  | Craft@Hrs | Unit | Material | Labor | Total |
|---|---|---|---|---|---|

Wireless access point. Use the network wire pulling and termination costs above. Access points are usually installed 12" below the finished ceiling on a small shelf. Add the cost of an electrical outlet at the access point. Count each access point as one station termination and connection.

| | Craft@Hrs | Unit | Material | Labor | Total |
|---|---|---|---|---|---|
| Wireless access point | CE@.100 | Ea | 73.40 | 5.84 | 79.24 |

High gain antenna for extended range on a wireless access point. Add the cost of a RG-6U cable run to the wireless access point.

| | Craft@Hrs | Unit | Material | Labor | Total |
|---|---|---|---|---|---|
| 18Dbi high gain panel antenna | CE@.100 | Ea | 127.00 | 5.84 | 132.84 |

**Ecclesiastical Equipment**  Labor includes unpacking, layout and complete installation of prefinished furniture and equipment.

| | Craft@Hrs | Unit | Material | Labor | Total |
|---|---|---|---|---|---|
| Lecterns, 16" x 24" | | | | | |
|    Plain | C8@4.48 | Ea | 671.00 | 205.00 | 876.00 |
|    Deluxe | C8@4.48 | Ea | 1,230.00 | 205.00 | 1,435.00 |
|    Premium | C8@4.48 | Ea | 2,340.00 | 205.00 | 2,545.00 |
| Pulpits | | | | | |
|    Plain | C8@4.77 | Ea | 761.00 | 219.00 | 980.00 |
|    Deluxe | C8@4.77 | Ea | 1,190.00 | 219.00 | 1,409.00 |
|    Premium | C8@4.77 | Ea | 2,950.00 | 219.00 | 3,169.00 |
| Arks, with curtain | | | | | |
|    Plain | C8@5.97 | Ea | 788.00 | 274.00 | 1,062.00 |
|    Deluxe | C8@5.97 | Ea | 1,030.00 | 274.00 | 1,304.00 |
|    Premium | C8@5.97 | Ea | 1,710.00 | 274.00 | 1,984.00 |
| Arks, with door | | | | | |
|    Plain | C8@8.95 | Ea | 893.00 | 410.00 | 1,303.00 |
|    Deluxe | C8@8.95 | Ea | 1,520.00 | 410.00 | 1,930.00 |
|    Premium | C8@8.95 | Ea | 2,340.00 | 410.00 | 2,750.00 |
| Pews, bench type | | | | | |
|    Plain | C8@.382 | LF | 49.40 | 17.50 | 66.90 |
|    Deluxe | C8@.382 | LF | 53.80 | 17.50 | 71.30 |
|    Premium | C8@.382 | LF | 64.40 | 17.50 | 81.90 |
| Pews, seat type | | | | | |
|    Plain | C8@.464 | LF | 62.00 | 21.30 | 83.30 |
|    Deluxe | C8@.491 | LF | 80.80 | 22.50 | 103.30 |
|    Premium | C8@.523 | LF | 96.00 | 24.00 | 120.00 |
| Kneelers | | | | | |
|    Plain | C8@.185 | LF | 12.10 | 8.48 | 20.58 |
|    Deluxe | C8@.185 | LF | 15.60 | 8.48 | 24.08 |
|    Premium | C8@.185 | LF | 24.10 | 8.48 | 32.58 |
| Cathedral chairs, shaped wood | | | | | |
|    Without accessories | C8@.235 | Ea | 158.00 | 10.80 | 168.80 |
|    With book rack | C8@.235 | Ea | 171.00 | 10.80 | 181.80 |
|    With book rack and kneeler | C8@.235 | Ea | 190.00 | 10.80 | 200.80 |
| Cathedral chairs, upholstered | | | | | |
|    Without accessories | C8@.285 | Ea | 164.00 | 13.10 | 177.10 |
|    With book rack | C8@.285 | Ea | 177.00 | 13.10 | 190.10 |
|    With book rack and kneeler | C8@.285 | Ea | 214.00 | 13.10 | 227.10 |
| Adaptable fabric upholstered chair seating (Sauder) | | | | | |
|    Hardwood laminated frame, interlocking legs, book rack, stackable, "Modlok" | | | | | |
|       Standard | C8@.066 | Ea | 135.00 | 3.02 | 138.02 |
|       Deluxe | C8@.066 | Ea | 141.00 | 3.02 | 144.02 |

# 11 Equipment

|  | Craft@Hrs | Unit | Material | Labor | Total |
|---|---|---|---|---|---|
| Solid oak frame, stackable, "Oaklok" | C8@.066 | Ea | 110.00 | 3.02 | 113.02 |
| Laminated beech frame, stackable, "Plylok" | C8@.066 | Ea | 119.00 | 3.02 | 122.02 |
| Laminated beech frame, stacks & folds, "Plyfold" | C8@.066 | Ea | 104.00 | 3.02 | 107.02 |
| Confessionals, single, with curtain |  |  |  |  |  |
| Plain | C8@9.11 | Ea | 2,850.00 | 418.00 | 3,268.00 |
| Deluxe | C8@9.11 | Ea | 3,420.00 | 418.00 | 3,838.00 |
| Premium | C8@9.11 | Ea | 3,900.00 | 418.00 | 4,318.00 |
| Add for door in place of curtain | — | Ea | 455.00 | — | 455.00 |
| Confessionals, double, with curtain |  |  |  |  |  |
| Plain | C8@13.7 | Ea | 4,510.00 | 628.00 | 5,138.00 |
| Deluxe | C8@13.7 | Ea | 5,210.00 | 628.00 | 5,838.00 |
| Premium | C8@13.7 | Ea | 6,430.00 | 628.00 | 7,058.00 |
| Communion rail, hardwood, with standards |  |  |  |  |  |
| Plain | C8@.315 | LF | 52.00 | 14.40 | 66.40 |
| Deluxe | C8@.315 | LF | 65.40 | 14.40 | 79.80 |
| Premium | C8@.315 | LF | 78.10 | 14.40 | 92.50 |
| Communion rail, carved oak, with standards |  |  |  |  |  |
| Embellished | C8@.315 | LF | 84.80 | 14.40 | 99.20 |
| Ornate | C8@.315 | LF | 113.00 | 14.40 | 127.40 |
| Premium | C8@.315 | LF | 187.00 | 14.40 | 201.40 |
| Communion rail, bronze or stainless steel, |  |  |  |  |  |
| With standards | H6@.285 | LF | 121.00 | 14.70 | 135.70 |
| Altars, hardwood, sawn |  |  |  |  |  |
| Plain | C8@5.97 | Ea | 915.00 | 274.00 | 1,189.00 |
| Deluxe | C8@5.97 | Ea | 1,080.00 | 274.00 | 1,354.00 |
| Premium | C8@5.97 | Ea | 1,470.00 | 274.00 | 1,744.00 |
| Altars, carved hardwood |  |  |  |  |  |
| Plain | C8@5.97 | Ea | 2,730.00 | 274.00 | 3,004.00 |
| Deluxe | C8@6.73 | Ea | 6,310.00 | 308.00 | 6,618.00 |
| Premium | C8@7.30 | Ea | 7,180.00 | 335.00 | 7,515.00 |
| Altars, with Nimmerillium fonts, marble | D4@27.8 | Ea | 4,660.00 | 1,290.00 | 5,950.00 |
| Altars, marble or granite |  |  |  |  |  |
| Plain | D4@29.1 | Ea | 4,860.00 | 1,350.00 | 6,210.00 |
| Deluxe | D4@29.1 | Ea | 7,060.00 | 1,350.00 | 8,410.00 |
| Premium | D4@29.1 | Ea | 10,500.00 | 1,350.00 | 11,850.00 |
| Altars, marble legs and base |  |  |  |  |  |
| Deluxe | D4@29.1 | Ea | 2,920.00 | 1,350.00 | 4,270.00 |
| Premium | D4@29.1 | Ea | 4,170.00 | 1,350.00 | 5,520.00 |

## Theater & Stage Curtains

Straight track for manual curtains, see lifting and drawing equipment below

|  | Craft@Hrs | Unit | Material | Labor | Total |
|---|---|---|---|---|---|
| Standard duty |  |  |  |  |  |
| 24' wide stage | D4@.486 | LF | 33.60 | 22.50 | 56.10 |
| 40' wide stage | D4@.315 | LF | 33.60 | 14.60 | 48.20 |
| 60' wide stage | D4@.233 | LF | 33.60 | 10.80 | 44.40 |
| Medium duty |  |  |  |  |  |
| 24' wide stage | D4@.486 | LF | 45.80 | 22.50 | 68.30 |
| 40' wide stage | D4@.315 | LF | 45.80 | 14.60 | 60.40 |
| 60' wide stage | D4@.233 | LF | 45.80 | 10.80 | 56.60 |

# 11 Equipment

|  | Craft@Hrs | Unit | Material | Labor | Total |
|---|---|---|---|---|---|
| Heavy duty | | | | | |
| 24' wide stage | D4@.486 | LF | 46.50 | 22.50 | 69.00 |
| 40' wide stage | D4@.315 | LF | 46.50 | 14.60 | 61.10 |
| 60' wide stage | D4@.233 | LF | 46.50 | 10.80 | 57.30 |
| **Curved track, manual operation, see lifting and drawing equipment below** | | | | | |
| Standard duty | | | | | |
| 24' wide stage | D4@.486 | LF | 62.40 | 22.50 | 84.90 |
| 40' wide stage | D4@.371 | LF | 62.40 | 17.20 | 79.60 |
| 60' wide stage | D4@.315 | LF | 62.40 | 14.60 | 77.00 |
| Heavy duty | | | | | |
| 24' wide stage | D4@.584 | LF | 175.00 | 27.00 | 202.00 |
| 40' wide stage | D4@.427 | LF | 150.00 | 19.70 | 169.70 |
| 60' wide stage | D4@.342 | LF | 150.00 | 15.80 | 165.80 |
| **Cyclorama track, manual, see lifting and drawing equipment below** | | | | | |
| 24' wide stage | D4@.486 | LF | 39.50 | 22.50 | 62.00 |
| 40' wide stage | D4@.315 | LF | 39.50 | 14.60 | 54.10 |
| 60' wide stage | D4@.233 | LF | 39.50 | 10.80 | 50.30 |

**Stage curtains, Subcontract** Complete costs for custom fabricated, flameproof stage curtains including labor to hang. No track or curtain lifting or drawing equipment included. By percent of added fullness.

|  | Craft@Hrs | Unit | Material | Labor | Total |
|---|---|---|---|---|---|
| Heavyweight velour | | | | | |
| 0% fullness | — | SF | — | — | 5.61 |
| 50% fullness | — | SF | — | — | 6.92 |
| 75% fullness | — | SF | — | — | 7.13 |
| 100% fullness | — | SF | — | — | 7.66 |
| Lightweight velour | | | | | |
| 0% fullness | — | SF | — | — | 4.46 |
| 50% fullness | — | SF | — | — | 4.98 |
| 75% fullness | — | SF | — | — | 5.51 |
| 100% fullness | — | SF | — | — | 6.77 |
| Muslin, 0% fullness | — | SF | — | — | 3.30 |
| Repp cloth | | | | | |
| 0% fullness | — | SF | — | — | 3.57 |
| 50% fullness | — | SF | — | — | 5.40 |
| 75% fullness | — | SF | — | — | 5.35 |
| 100% fullness | — | SF | — | — | 5.93 |

**Stage curtain lifting and drawing equipment** Labor is equipment installation, see electrical connection below

|  | Craft@Hrs | Unit | Material | Labor | Total |
|---|---|---|---|---|---|
| Lift curtain equipment | | | | | |
| 1/2 HP | D4@7.50 | Ea | 5,690.00 | 347.00 | 6,037.00 |
| 3/4 HP | D4@7.50 | Ea | 6,120.00 | 347.00 | 6,467.00 |
| Draw curtain equipment | | | | | |
| 1/3 HP | D4@7.50 | Ea | 4,110.00 | 347.00 | 4,457.00 |
| 1/2 HP | D4@7.50 | Ea | 4,320.00 | 347.00 | 4,667.00 |
| 3/4 HP | D4@7.50 | Ea | 4,850.00 | 347.00 | 5,197.00 |

Add for electrical connection. Typical subcontract price for 40' wire run with conduit, circuit breaker, junction box and starter switch.

|  | Craft@Hrs | Unit | Material | Labor | Total |
|---|---|---|---|---|---|
| Per electrical connection | — | Ea | — | — | 1,750.00 |

# 11 Equipment

| | Craft@Hrs | Unit | Material | Labor | Total |
|---|---|---|---|---|---|

**Fresnel-type spotlights and general stage lighting** for small to medium (1,500 seat) venues, specialty dimmer rheostats, and breaker panels. Cost includes fixed directional mounting pedestals, prewired electrical conduit, fusing, ground fault protection to *National Electrical Code* standards and main control panel. Equipment cost is a self-propelled electric accordion-style work platform. All electrical components are to Underwriters Laboratories standards.

| | Craft@Hrs | Unit | Material | Labor | Total |
|---|---|---|---|---|---|
| 1,000 watt Fresnel lens light, 6" | E4@5.00 | Ea | 270.00 | 246.00 | 516.00 |
| 5,000 watt Fresnel lens light, 10" | E4@5.00 | Ea | 1,120.00 | 246.00 | 1,366.00 |
| Main breaker | E4@1.00 | Ea | 153.00 | 49.10 | 202.10 |
| Dimmer rheostat | E4@1.50 | Ea | 76.50 | 73.70 | 150.20 |
| Spotlight control panel | E4@3.00 | Ea | 357.00 | 147.00 | 504.00 |
| Pre-wired conduit | E4@5.00 | CLF | 51.00 | 246.00 | 297.00 |
| Calibration and test | E4@2.00 | Ea | — | 98.30 | 98.30 |

**Stage entertainment sound system** Appropriate for a small to medium (1,500 seat) venue. Includes prewired electrical conduit, 8-track digital 80 Gig sound recorder, high-definition 1080i-compliant digital motion picture camera and tripod, 64-channel sound mixing board, sound balancing system, 16-channel linear amplifier, filtered power supply, portable stage monitors, fixed acoustically tuned speaker system, fixed suspended stage microphones, ground fault protection to American Federation of Musicians safety standards and steel security cabinet with heavy-duty lockset for equipment storage. Equipment cost is for a self-propelled electric accordion-style work platform.

| | Craft@Hrs | Unit | Material | Labor | Total |
|---|---|---|---|---|---|
| Pre-wired conduit | E4@5.00 | CLF | 51.00 | 246.00 | 297.00 |
| Power supply and amplifier | E4@.500 | Ea | 816.00 | 24.60 | 840.60 |
| Fixed speakers | B1@.500 | Ea | 459.00 | 16.70 | 475.70 |
| Ground fault protection | E4@1.00 | Ea | 306.00 | 49.10 | 355.10 |
| Ceiling microphones | E4@1.00 | Ea | 255.00 | 49.10 | 304.10 |
| Digital 1080i-compliant camcorder | — | Ea | 6,000.00 | — | 6,000.00 |
| Mixing and recording table | E4@2.50 | Ea | 2,550.00 | 123.00 | 2,673.00 |
| Security cabinet for mixing and recording table | B1@1.50 | Ea | 1,220.00 | 50.00 | 1,270.00 |
| Calibration and test | E4@4.00 | Ea | — | 197.00 | 197.00 |

**Public address systems** For small to medium (1,500 seat) venues such as school assembly halls or civic centers. Includes a linear amplifier, filtered power supply, stage-mounted cable fixtures with covers, pedestal stage microphones, ground fault protection to American Federation of Musicians safety standards, fixed wall-hung speakers, pre-wired permanent conduit runs, removable stage monitors, and security cabinet with lockset for equipment storage.

| | Craft@Hrs | Unit | Material | Labor | Total |
|---|---|---|---|---|---|
| Pre-wired conduit | E4@5.00 | CLF | 51.00 | 246.00 | 297.00 |
| Stage cable fixtures | E4@2.00 | Ea | 10.20 | 98.30 | 108.50 |
| Ground fault protection | E4@1.00 | Ea | 306.00 | 49.10 | 355.10 |
| Fixed speakers | B1@.500 | Ea | 459.00 | 16.70 | 475.70 |
| Amplifier and power supply | E4@.500 | Ea | 816.00 | 24.60 | 840.60 |
| Security cabinet | B1@1.50 | Ea | 1,220.00 | 50.00 | 1,270.00 |
| Calibration and test | E4@4.00 | Ea | — | 197.00 | 197.00 |

**Stage entertainment video systems** Includes digital halogen element 1,500 watt back projector, prefab frame-supported Mylar back projection screens, medium-resolution stand-mounted digital CCTV cameras, operator's control panel and TV monitor array, digital 1080i-compliant panel-mounted video/audio recorder, ground fault protection to American Federation of Musicians safety standards and security cabinet with lockset for storage of equipment.

| | Craft@Hrs | Unit | Material | Labor | Total |
|---|---|---|---|---|---|
| Pre-wired conduit | E4@5.00 | CLF | 51.00 | 246.00 | 297.00 |
| Back-projection mounts for screen, amp, and monitors | B1@4.00 | Ea | 1,530.00 | 133.00 | 1,663.00 |
| Projector, screen, and cameras | B1@10.0 | Ea | 4,590.00 | 333.00 | 4,923.00 |
| Ground fault equipment | E4@1.00 | Ea | 306.00 | 49.10 | 355.10 |
| Calibration and test | E4@2.00 | Ea | — | 98.30 | 98.30 |

# 11 Equipment

|  | Craft@Hrs | Unit | Material | Labor | Total |
|---|---|---|---|---|---|

**Acoustically engineered sound reduction paneling**  Cost includes vertical support studs, spacers, fasteners and trim. Complies with American Institute of Sound Engineering specifications. Per square foot of wall or ceiling covered.

| | Craft@Hrs | Unit | Material | Labor | Total |
|---|---|---|---|---|---|
| Sidewall coverage | D2@.040 | SF | 1.50 | 1.81 | 3.31 |
| Ceiling coverage | D2@.060 | SF | 1.50 | 2.71 | 4.21 |

**Specialty mobile HVAC** for performing stage zone "spot" cooling to American Federation of Musicians and Screen Actors Guild safety standards.  For below-stage mounting in live theaters and for shooting stages for cinema and television work.  Includes integral ground fault protection. Material only.  Directional ducting costs not included.

| | | Unit | Material | Labor | Total |
|---|---|---|---|---|---|
| 250,000 Btu Unit | — | Ea | — | — | 27,500.00 |

**Specialty theater panic exit double doors**  Includes pre-hung frame, vertical center bar main brace, overhead illuminated light emitting diode "fire exit only" sign, fire alarm system interlocking switch and digital alarm horn, external-only safety lockset and panic bar. To National Fire Protection Association standards and specifications.

| | Craft@Hrs | Unit | Material | Labor | Total |
|---|---|---|---|---|---|
| Door frame and vertical bar | B1@8.00 | Ea | 765.00 | 266.00 | 1,031.00 |
| Double doors, per pair | B1@.500 | Ea | 1,430.00 | 16.70 | 1,446.70 |
| Wiring of interlock alarm and sign | E4@2.00 | Ea | 561.00 | 98.30 | 659.30 |
| Calibration and test | E4@3.00 | Ea | — | 147.00 | 147.00 |

## Gymnasium and Athletic Equipment
Basketball backstops

| | Craft@Hrs | Unit | Material | Labor | Total |
|---|---|---|---|---|---|
| Fixed, wall mounted | D4@8.00 | Ea | 716.00 | 370.00 | 1,086.00 |
| Swing-up, wall mounted | D4@10.0 | Ea | 932.00 | 463.00 | 1,395.00 |
| Ceiling suspended, to 20', swing-up manual | D4@16.0 | Ea | 2,050.00 | 740.00 | 2,790.00 |
| Add for glass backstop, fan shape | — | Ea | 627.00 | — | 627.00 |
| Add for glass backstop, rectangular | — | Ea | 704.00 | — | 704.00 |
| Add for power operation | — | Ea | — | — | 610.00 |

Gym floors, typical subcontract prices, not including subfloor or base

| | | Unit | Material | Labor | Total |
|---|---|---|---|---|---|
| Synthetic gym floor, 3/16" | — | SF | — | — | 5.84 |
| Synthetic gym floor, 3/8" | — | SF | — | — | 7.67 |
| Built-up wood "spring system" maple floor | — | SF | — | — | 10.90 |
| Rubber cushioned maple floor | — | SF | — | — | 9.70 |
| Maple floor on sleepers and membrane | | | | | |
|    Unfinished | — | SF | — | — | 7.56 |
| Apparatus inserts | — | Ea | — | — | 40.70 |
| Bleachers, telescoping, manual, per seat | D4@.187 | Ea | 45.80 | 8.65 | 54.45 |
| Bleachers, hydraulic operated, per seat | D4@.235 | Ea | 62.40 | 10.90 | 73.30 |

Scoreboards, basketball, single face

| | Craft@Hrs | Unit | Material | Labor | Total |
|---|---|---|---|---|---|
| Economy | CE@7.01 | Ea | 1,350.00 | 409.00 | 1,759.00 |
| Good | CE@13.0 | Ea | 1,860.00 | 759.00 | 2,619.00 |
| Premium | CE@21.6 | Ea | 2,930.00 | 1,260.00 | 4,190.00 |

**Weight and exercise equipment room,** commercial and educational facility grade, on a 4" concrete slab supporting special composite low-impact padded exercise room flooring.  Costs include labor for bolting of equipment to the floor but do not include the slab or flooring costs, electrical connections where required or cost of equipment assembly.

| | Craft@Hrs | Unit | Material | Labor | Total |
|---|---|---|---|---|---|
| Weight machine | B9@.500 | Ea | 2,900.00 | 16.30 | 2,916.30 |
| Rowing machine | B9@.250 | Ea | 700.00 | 8.16 | 708.16 |
| Treadmill | B9@.250 | Ea | 2,300.00 | 8.16 | 2,308.16 |
| Stationary bicycle | B9@.250 | Ea | 945.00 | 8.16 | 953.16 |
| Boxing ring | B9@4.00 | Ea | 8,500.00 | 131.00 | 8,631.00 |
| Trampoline | B9@2.00 | Ea | 600.00 | 65.30 | 665.30 |
| Parallel bars | B9@.500 | Ea | 1,750.00 | 16.30 | 1,766.30 |
| Balance bar | B9@.250 | Ea | 1,250.00 | 8.16 | 1,258.16 |

| | Craft@Hrs | Unit | Material | Labor | Total |
|---|---|---|---|---|---|
| **Playing Fields** Typical subcontract costs. Use 3,000 SF as a minimum job charge. | | | | | |
| Playing fields, synthetic surface | | | | | |
| Turf, including base and turf | — | SF | — | — | 12.30 |
| Turf, 3 layer, base foam, turf | — | SF | — | — | 13.30 |
| Uniturf, embossed running surface, 3/8" | — | SF | — | — | 9.06 |
| Running track | | | | | |
| Volcanic cinder, 7" | — | SF | — | — | 3.04 |
| Bituminous and cork, 2" | — | SF | — | — | 4.12 |

**Artificial grass turf,** glue-down carpet installation includes cleaning application surface, adhesive, seam seal and rolling. Based on 100 SY job. Challenger Industries, www.challengerind.com.

| | Craft@Hrs | Unit | Material | Labor | Total |
|---|---|---|---|---|---|
| Economy grade non-directional turf | F9@.055 | SY | 5.60 | 2.49 | 8.09 |
| Mid grade non-directional | | | | | |
| weather resistant turf | F9@.055 | SY | 6.71 | 2.49 | 9.20 |
| High performance non-directional | | | | | |
| UV stabilized turf | F9@.055 | SY | 7.64 | 2.49 | 10.13 |
| Miniature golf style weather resistant turf | F9@.055 | SY | 7.84 | 2.49 | 10.33 |
| Premium spike-proof non-directional | | | | | |
| UV stabilized turf | F9@.065 | SY | 19.30 | 2.94 | 22.24 |
| Padded spike-proof non-directional | | | | | |
| UV stabilized turf | F9@.065 | SY | 21.50 | 2.94 | 24.44 |
| Carpet adhesive (25 SY per gallon) | — | Gal | 13.50 | — | 13.50 |
| Seam sealer (800 LF per gallon) | — | Gal | 46.60 | — | 46.60 |
| Spike-proof indoor carpeting | | | | | |
| Economy grade patterned carpet | F9@.060 | SY | 15.30 | 2.71 | 18.01 |
| Mid grade patterned carpet | F9@.060 | SY | 16.90 | 2.71 | 19.61 |
| Mid grade insignia carpet | F9@.060 | SY | 22.90 | 2.71 | 25.61 |
| Premium patterned carpet | F9@.060 | SY | 26.80 | 2.71 | 29.51 |

| | Craft@Hrs | Unit | Material | Labor | Equipment | Total |
|---|---|---|---|---|---|---|
| **Athletic Equipment** Installed with a 4,000 lb. capacity forklift. | | | | | | |
| Football goals, costs shown are for two goals for one field. Galvanized steel pipe. | | | | | | |
| 4-1/2" center support post | S6@22.0 | Pr | 2,530.00 | 988.00 | 234.00 | 3,752.00 |
| 6-5/8" center support post | S6@24.0 | Pr | 4,990.00 | 1,080.00 | 255.00 | 6,325.00 |
| Combination football/soccer goal, regulation size, galvanized | | | | | | |
| steel pipe, dual uprights | S6@24.0 | Pr | 2,260.00 | 1,080.00 | 255.00 | 3,595.00 |
| Soccer goals, regulation size, costs shown are for two goals for one field | | | | | | |
| Aluminum frame with net | S6@24.0 | Pr | 2,560.00 | 1,080.00 | 255.00 | 3,895.00 |
| Outdoor, playground style basketball goals, single support, adjustable, 6' backboard extension aluminum fan backboard, hoop and net | | | | | | |
| Single goal | S6@12.0 | Ea | 875.00 | 539.00 | 127.00 | 1,541.00 |
| Double goal, back to back | S6@15.0 | Ea | 1,220.00 | 674.00 | 159.00 | 2,053.00 |
| Volleyball nets, with galvanized posts net, and ground sleeves | | | | | | |
| Per net | S6@18.0 | Ea | 934.00 | 809.00 | 191.00 | 1,934.00 |

**Tennis Courts** Typical costs for championship size single playing court installation including normal fine grading. These costs do not include site preparation, excavation, drainage, or retaining walls. Equipment, where shown, is a 4,000 lb. capacity forklift. Court is 60' x 120' with asphaltic concrete or portland cement slab (4" thickness built up to 6" thickness at perimeter). Includes compacted base material and wire mesh reinforcing in concrete.

| | Craft@Hrs | Unit | Material | Labor | Equipment | Total |
|---|---|---|---|---|---|---|
| Typical cost per court, slab-on-grade | — | LS | — | — | — | 21,000.00 |
| Court surface seal coat (7,200 SF) | | | | | | |
| One color plus 2" wide boundary lines | — | LS | — | — | — | 2,700.00 |

# 11 Equipment

| | Craft@Hrs | Unit | Material | Labor | Equipment | Total |
|---|---|---|---|---|---|---|
| **Fence, 10' high galvanized chain link (360 LF) at court perimeter** | | | | | | |
| Fencing including gates | — | LS | — | — | — | 10,000.00 |
| **Lighting** | | | | | | |
| Typical installation | — | LS | — | — | — | 11,500.00 |
| **Net posts, with tension reel** | | | | | | |
| Galvanized steel, 3-1/2" diameter | S6@6.00 | Pr | 336.00 | 270.00 | 63.70 | 669.70 |
| Galvanized steel, 4-1/2" diameter | S6@6.00 | Pr | 379.00 | 270.00 | 63.70 | 712.70 |
| **Tennis nets** | | | | | | |
| Nylon | S6@1.50 | Ea | 270.00 | 67.40 | 15.90 | 353.30 |
| Steel | S6@1.50 | Ea | 600.00 | 67.40 | 15.90 | 683.30 |
| **Ball wall, fiberglass, 8' x 5'** | | | | | | |
| Attached to existing wall | S6@1.50 | Ea | 730.00 | 67.40 | 15.90 | 813.30 |
| **Windbreak, 9' high, closed mesh polypropylene** | | | | | | |
| Lashed to existing chain link fence | S6@.003 | SF | .20 | .13 | .03 | .36 |

| | Craft@Hrs | Unit | Material | Labor | Total |
|---|---|---|---|---|---|
| **Athletic Benches** Backed athletic benches, galvanized pipe frame, stationary | | | | | |
| Aluminum seat and back | | | | | |
| 7' long | C8@.930 | Ea | 273.00 | 42.60 | 315.60 |
| 8' long | C8@1.16 | Ea | 316.00 | 53.20 | 369.20 |
| 15' long | C8@1.31 | Ea | 504.00 | 60.00 | 564.00 |
| Fiberglass seat and back | | | | | |
| 10' long | C8@.930 | Ea | 392.00 | 42.60 | 434.60 |
| 12' long | C8@1.16 | Ea | 451.00 | 53.20 | 504.20 |
| 16' long | C8@1.31 | Ea | 604.00 | 60.00 | 664.00 |
| Galvanized steel seat and back | | | | | |
| 15' long | C8@1.31 | Ea | 361.00 | 60.00 | 421.00 |
| Backless athletic bench, galvanized pipe frame, stationary, 12" wide x 10' long | | | | | |
| Wood seat | C8@.776 | Ea | 192.00 | 35.60 | 227.60 |
| Steel seat | C8@.776 | Ea | 211.00 | 35.60 | 246.60 |
| Aluminum seat | C8@.776 | Ea | 243.00 | 35.60 | 278.60 |
| **Surface Preparation and Striping for Athletic Courts** Use $150 as a minimum job charge. | | | | | |
| Surface preparation | | | | | |
| Acid etch concrete surface | PA@.013 | SY | .21 | .67 | .88 |
| Primer on asphalt | PA@.006 | SY | .29 | .31 | .60 |
| Primer on concrete | PA@.006 | SY | .44 | .31 | .75 |
| Asphalt emulsion filler, 1 coat | PA@.008 | SY | .61 | .41 | 1.02 |
| Acrylic filler, 1 coat | PA@.008 | SY | .75 | .41 | 1.16 |
| Rubber/acrylic compound, per coat | PA@.012 | SY | 1.96 | .62 | 2.58 |
| Acrylic emulsion texture, per coat | | | | | |
| Sand filled | PA@.006 | SY | 1.27 | .31 | 1.58 |
| Rubber filled | PA@.006 | SY | 1.33 | .31 | 1.64 |
| Acrylic emulsion color coat, per coat | | | | | |
| Tan, brown, or red | PA@.006 | SY | .42 | .31 | .73 |
| Greens | PA@.006 | SY | .47 | .31 | .78 |
| Blue | PA@.006 | SY | .52 | .31 | .83 |
| Rubber cushioned concrete pavers | | | | | |
| 18" x 18", dry joints | FL@.039 | SF | 16.00 | 1.97 | 17.97 |

# 11 Equipment

| | Craft@Hrs | Unit | Material | Labor | Total |
|---|---|---|---|---|---|
| **Striping for athletic courts** | | | | | |
| Marked with white or yellow striping paint | | | | | |
| Lines 3" wide | PA@.005 | LF | .06 | .26 | .32 |
| Lines 4" wide | PA@.005 | LF | .08 | .26 | .34 |
| Lines 4" wide, reflectorized | PA@.002 | LF | .17 | .10 | .27 |
| Lines 8" wide, reflectorized | PA@.004 | LF | .34 | .21 | .55 |
| Marked with reflectorized thermoplastic tape | | | | | |
| Lines 4" wide | PA@.004 | LF | 1.26 | .21 | 1.47 |
| Lines 6" wide | PA@.006 | LF | 1.92 | .31 | 2.23 |
| Symbols, lettering | | | | | |
| Within 2'-0" x 8'-0" template | PA@2.50 | Ea | 49.10 | 129.00 | 178.10 |

| | Craft@Hrs | Unit | Material | Labor | Equipment | Total |
|---|---|---|---|---|---|---|
| **Playground Equipment** Equipment cost is for a 4,000 lb. capacity forklift. Use $2,500 as a minimum job charge. | | | | | | |
| Balance beams. All are supported 12" above the ground by 2-3/8" OD galvanized pipe uprights. | | | | | | |
| 10' long "Z"-shaped beam | S6@6.00 | Ea | 390.00 | 270.00 | 63.70 | 723.70 |
| 10' long curved beam | S6@6.00 | Ea | 507.00 | 270.00 | 63.70 | 840.70 |
| 10' long straight beam | S6@6.00 | Ea | 383.00 | 270.00 | 63.70 | 716.70 |
| 15' long straight beam | S6@6.00 | Ea | 307.00 | 270.00 | 63.70 | 640.70 |
| 30' long "Z"-shaped beam | S6@9.00 | Ea | 597.00 | 404.00 | 95.60 | 1,096.60 |
| Geodesic dome beehive climber, 1" galvanized pipe rungs | | | | | | |
| 8' diameter, 4' high | S6@24.0 | Ea | 529.00 | 1,080.00 | 255.00 | 1,864.00 |
| 13' diameter, 5' high | S6@36.0 | Ea | 1,590.00 | 1,620.00 | 382.00 | 3,592.00 |
| 18' diameter, 7' high | S6@48.0 | Ea | 1,980.00 | 2,160.00 | 510.00 | 4,650.00 |
| Horizontal ladders, 20" wide with 1-1/16" OD galvanized pipe rungs 12" OC and 2-3/8" OD galvanized pipe headers and top rails | | | | | | |
| 12' long, 6' high | S6@6.00 | Ea | 675.00 | 270.00 | 63.70 | 1,008.70 |
| 18' long, 7' high | S6@9.00 | Ea | 831.00 | 404.00 | 95.60 | 1,330.60 |
| Rope and pole climbs. Horizontal vertical end posts are 3-1/2" OD galvanized steel tubing. | | | | | | |
| 10' H x 10' W with three 1-5/8" diameter poles | S6@24.0 | Ea | 996.00 | 1,080.00 | 255.00 | 2,331.00 |
| 10' H x 10' W with three 1" diameter ropes | S6@24.0 | Ea | 1,070.00 | 1,080.00 | 255.00 | 2,405.00 |
| 16' H x 12' W with three 1-5/8" diameter poles | S6@24.0 | Ea | 1,330.00 | 1,080.00 | 255.00 | 2,665.00 |
| 16' H x 12' W with three 1-1/2" diameter ropes | S6@24.0 | Ea | 1,520.00 | 1,080.00 | 255.00 | 2,855.00 |
| Sandboxes, prefabricated. Painted galvanized steel, 12" high, pine seats, sand not included | | | | | | |
| 6' x 6' | S6@9.00 | Ea | 419.00 | 404.00 | 95.60 | 918.60 |
| 12' x 12' | S6@12.0 | Ea | 595.00 | 539.00 | 127.00 | 1,261.00 |
| Masonry sand | — | CY | 31.70 | — | — | 31.70 |
| Washed plaster sand | — | CY | 40.60 | — | — | 40.60 |
| Concrete sand | — | CY | 52.70 | — | — | 52.70 |
| White silica sand | — | CY | 69.80 | — | — | 69.80 |
| Sand scoopers, crane simulator. Heavy-duty metal construction. Two handles control crane arm height and sand scooper, set in concrete. | | | | | | |
| 3' high miniature scooper with seat | CL@2.15 | Ea | 461.00 | 85.70 | 68.50 | 615.20 |
| 4' high rotating stand up scooper | CL@2.75 | Ea | 475.00 | 110.00 | 87.60 | 672.60 |
| Masonry sand | — | CY | 31.70 | — | — | 31.70 |

# 11 Equipment

|  | Craft@Hrs | Unit | Material | Labor | Equipment | Total |
|---|---|---|---|---|---|---|

Seesaws, galvanized pipe beam with polyethylene saddle or animal seats. Seesaw beam pivots between two coil springs. Double units have seesaw beams spaced 90 degrees apart.

| Single unit, 7' metal base | | | | | | |
|---|---|---|---|---|---|---|
| set in concrete | S6@5.00 | Ea | 404.00 | 225.00 | 53.10 | 682.10 |
| Double unit, 4 saddle seats | S6@9.00 | Ea | 1,350.00 | 404.00 | 95.60 | 1,849.60 |
| Double unit, 2 animal & 2 saddle seats | S6@9.00 | Ea | 1,930.00 | 404.00 | 95.60 | 2,429.60 |
| Double unit, 4 animal seats | S6@9.00 | Ea | 2,510.00 | 404.00 | 95.60 | 3,009.60 |

Spring saws, steel frame with aluminum seats, set in concrete

| 2' x 12' with two saddle seats | S6@6.00 | Ea | 938.00 | 270.00 | 63.70 | 1,271.70 |
|---|---|---|---|---|---|---|
| 12' diameter with two animal seats | S6@6.00 | Ea | 636.00 | 270.00 | 63.70 | 969.70 |
| 12' diameter with four saddle seats | S6@8.50 | Ea | 1,910.00 | 382.00 | 90.20 | 2,382.20 |
| 12' diameter with four animal seats | S6@8.50 | Ea | 815.00 | 382.00 | 90.20 | 1,287.20 |
| 12' diameter with six saddle seats | S6@9.00 | Ea | 2,830.00 | 404.00 | 95.60 | 3,329.60 |
| 12' diameter with six animal seats | S6@9.00 | Ea | 1,110.00 | 404.00 | 95.60 | 1,609.60 |

Slides

Traditional slide with 18" wide 16 gauge stainless steel slide and galvanized steel stairway

| Chute 5' H x 10' L, overall length is 13' | S6@12.0 | Ea | 1,220.00 | 539.00 | 127.00 | 1,886.00 |
|---|---|---|---|---|---|---|
| Chute 6' H x 12' L, overall length is 15' | S6@12.0 | Ea | 1,330.00 | 539.00 | 127.00 | 1,996.00 |
| Chute 8' H x 16' L, overall length is 20' | S6@12.0 | Ea | 1,800.00 | 539.00 | 27.00 | 2,366.00 |

Swings, belt seats, galvanized pipe frame

8' high two-leg ends

| 2 seats wood framed | S6@8.00 | Ea | 1,370.00 | 359.00 | 84.90 | 1,813.90 |
|---|---|---|---|---|---|---|
| 2 seats | S6@12.0 | Ea | 587.00 | 539.00 | 127.00 | 1,253.00 |
| 4 seats | S6@18.0 | Ea | 1,000.00 | 809.00 | 191.00 | 2,000.00 |
| 6 seats | S6@24.0 | Ea | 1,400.00 | 1,080.00 | 255.00 | 2,735.00 |

10' high two-leg ends

| 2 seats | S6@18.0 | Ea | 638.00 | 809.00 | 191.00 | 1,638.00 |
|---|---|---|---|---|---|---|
| 4 seats | S6@24.0 | Ea | 1,090.00 | 1,080.00 | 255.00 | 2,425.00 |
| 6 seats | S6@30.0 | Ea | 1,540.00 | 1,350.00 | 319.00 | 3,209.00 |

10' high three-leg ends

| 4 seats | S6@24.0 | Ea | 1,370.00 | 1,080.00 | 255.00 | 2,705.00 |
|---|---|---|---|---|---|---|
| 6 seats | S6@30.0 | Ea | 1,690.00 | 1,350.00 | 319.00 | 3,359.00 |
| 8 seats | S6@36.0 | Ea | 2,410.00 | 1,620.00 | 382.00 | 4,412.00 |

12' high three-leg ends

| 3 seats | S6@30.0 | Ea | 1,140.00 | 1,350.00 | 319.00 | 2,809.00 |
|---|---|---|---|---|---|---|
| 6 seats | S6@36.0 | Ea | 1,720.00 | 1,620.00 | 382.00 | 3,722.00 |
| 8 seats | S6@42.0 | Ea | 2,380.00 | 1,890.00 | 446.00 | 4,716.00 |

| Add for platform for wheelchair, any of above | S6@12.0 | Ea | 664.00 | 539.00 | 127.00 | 1,330.00 |
|---|---|---|---|---|---|---|

Add for seat replacements

| Seat replacement | — | Ea | 40.80 | — | — | 40.80 |
|---|---|---|---|---|---|---|
| Child's rubber bucket seat | — | Ea | 93.70 | — | — | 93.70 |

Tetherball sets, galvanized pipe post, ground sleeve and chain,

| complete with nylon rope and ball | S6@3.00 | Ea | 170.00 | 135.00 | 31.90 | 336.90 |
|---|---|---|---|---|---|---|
| Sand for pit (masonry sand) | — | CY | 31.70 | — | — | 31.70 |

# 11 Equipment

| | Craft@Hrs | Unit | Material | Labor | Equipment | Total |
|---|---|---|---|---|---|---|

Tower climber. Complete with 8' high fireman's slide pole, "loft" platform, cargo net, horizontal ladders and vertical ladders. Frame is 1-1/4" square steel tubing with painted finish, bracing, ladders and slide pole are 1" OD steel pipe with painted finish. The platform consists of two 30" x 96" fiberglass planks with integral color located 5'0" above the ground. Available in mixed or matched colors.

| | Craft@Hrs | Unit | Material | Labor | Equipment | Total |
|---|---|---|---|---|---|---|
| 8' L x 6' W x 8' H | S6@36.0 | Ea | 3,220.00 | 1,620.00 | 382.00 | 5,222.00 |
| Arch climbers, 1.9" OD galvanized steel pipe, set in concrete | | | | | | |
| 2'2" x 10'7" | S6@6.75 | Ea | 597.00 | 303.00 | 29.20 | 929.20 |
| 2'2" x 13'11" | S6@7.00 | Ea | 661.00 | 314.00 | 74.30 | 1,049.30 |
| 2'2" x 15'1" | S6@7.00 | Ea | 760.00 | 314.00 | 74.30 | 1,148.30 |
| 15' x 20' four-way climber | S6@10.0 | Ea | 1,010.00 | 449.00 | 106.00 | 1,565.00 |
| Funnel ball, 3-1/2" OD galvanized steel post with yellow polyethylene hopper | | | | | | |
| 10' high, set in concrete | S6@4.00 | Ea | 769.00 | 180.00 | 42.50 | 991.50 |
| Chin up bars, 1-1/2" OD galvanized steel post with 54" and 66" and 84" bar height levels | | | | | | |
| 12' width, set in concrete | S6@4.75 | Ea | 447.00 | 213.00 | 50.40 | 710.40 |
| Parallel bars, 1-7/8" OD galvanized steel post, 2'9" high. | | | | | | |
| 2' x 9'2" set in concrete | S6@6.00 | Ea | 273.00 | 270.00 | 63.70 | 606.70 |
| Add for motion beam attachment | S6@.250 | Ea | 186.00 | 11.20 | 2.65 | 199.85 |

| | Craft@Hrs | Unit | Material | Labor | Total |
|---|---|---|---|---|---|
| **Park Grills** Steel grill, set in concrete. Add for cost of concrete. | | | | | |
| Rotating adjustable grill, 40" high with | | | | | |
| 300 sq. inch cooking area | CM@2.50 | Ea | 181.00 | 125.00 | 306.00 |
| Large grill with two adjustable grates, | | | | | |
| 40" high with 1,008 sq. inch cooking area | CM@2.75 | Ea | 359.00 | 138.00 | 497.00 |
| Hooded rotating grill, with hood vents, | | | | | |
| 40" high with 500 sq. inch cooking area | CM@3.00 | Ea | 468.00 | 150.00 | 618.00 |
| Shelterhouse grill with shelves and two grates, | | | | | |
| 40" high with 1,368 sq. inch cooking area | CM@3.50 | Ea | 663.00 | 176.00 | 839.00 |
| Campfire rings with 300 sq. inch cooking area and security brackets set in concrete | | | | | |
| Hinged grate | CM@2.25 | Ea | 156.00 | 113.00 | 269.00 |
| Adjustable grate | CM@2.25 | Ea | 181.00 | 113.00 | 294.00 |

**Medical Refrigeration Equipment** Labor costs include unpacking, assembly and installation. Jewett Refrigerator. Blood bank refrigerators, upright enameled steel cabinets with locking glass doors, 7 day recording thermometer, LED temperature display, failure alarm, interior lights, and removable drawers. 2 to 4 degree C operating range.

| | Craft@Hrs | Unit | Material | Labor | Total |
|---|---|---|---|---|---|
| 5 drawer, 240 bag, 16.9 CF, 29" wide x 30" deep x 74" high, #BBR17 | D4@2.25 | Ea | 6,290.00 | 104.00 | 6,394.00 |
| 6 drawer, 360 bag, 24.8 CF, 29" wide x 36" deep x 83" high, #BBR25 | D4@2.25 | Ea | 6,570.00 | 104.00 | 6,674.00 |
| 10 drawer, 480 bag, 37.4 CF, 59" wide x 30" deep x 74" high, #BBR37 | D4@3.42 | Ea | 9,100.00 | 158.00 | 9,258.00 |
| 12 drawer, 720 bag, 55.0 CF, 59" wide x 36" deep x 83" high, #BBR55 | D4@3.42 | Ea | 9,360.00 | 158.00 | 9,518.00 |

Blood plasma storage freezers, upright enameled steel cabinets with locking steel doors, 7 day recording thermometer, LED temperature display, failure alarm, and removable drawers. −30 degree C operating temperature.

| | Craft@Hrs | Unit | Material | Labor | Total |
|---|---|---|---|---|---|
| 6 drawer, 265 pack, 13.2 CF, 34" wide x 30" deep x 74" high, #BPL13 | D4@2.25 | Ea | 10,700.00 | 104.00 | 10,804.00 |
| 8 drawer, 560 pack, 24.8 CF, 29" wide x 36" deep x 83" high, #BPL25 | D4@2.25 | Ea | 9,360.00 | 104.00 | 9,464.00 |
| 16 drawer, 1,120 pack, 55 CF, 59" wide x 36" deep x 83" high, #BPL55 | D4@3.42 | Ea | 14,900.00 | 158.00 | 15,058.00 |

# 11 Equipment

| | Craft@Hrs | Unit | Material | Labor | Total |
|---|---|---|---|---|---|
| Laboratory refrigerators, upright enameled steel cabinets with locking steel doors, enameled steel interior, and adjustable stainless steel shelves. 2 to 4 degree C operating temperature. | | | | | |
| 4 shelf, 16.9 CF, 29" wide x 30" deep x 74" high #LR17B | D4@2.25 | Ea | 4,290.00 | 104.00 | 4,394.00 |
| 8 shelf, 37.4 CF, 59" wide x 30" deep x 74" high #LR37B | D4@3.42 | Ea | 6,570.00 | 158.00 | 6,728.00 |
| Pharmacy refrigerators, upright enameled steel cabinets with locking steel doors, enameled steel interior, and adjustable stainless steel drawers and shelves. −2 degree C operating temperature. | | | | | |
| 6 drawer, 1 shelf, 24.8 CF, 29" wide x 36" deep x 83" high, #PR25B | D4@2.25 | Ea | 4,810.00 | 104.00 | 4,914.00 |
| 9 drawer, 5 shelf, 55.0 CF, 59" wide x 36" deep x 83" high, #PR55B | D4@3.42 | Ea | 7,650.00 | 158.00 | 7,808.00 |
| Hospital/Lab/Pharmacy undercounter refrigerators, stainless steel interior and exterior, locking door. 24" wide x 24" deep x 34-1/2" high. 2 degree C operating temperature. | | | | | |
| 5.4 CF, with blower coil cooling system, stainless steel racks, and auto-defrost, #UC5B | D4@1.67 | Ea | 2,200.00 | 77.20 | 2,277.20 |
| 5.4 CF with cold wall cooling system, stainless steel racks, and auto-defrost, #UC5C | D4@1.69 | Ea | 2,820.00 | 78.20 | 2,898.20 |
| Hospital/Lab/Pharmacy undercounter freezers, stainless steel interior and exterior, locking door. 24" wide x 24" deep x 34-1/2" high. −20 degree C operating temperature. | | | | | |
| 4.6 CF, stainless steel racks, auto-defrost system, #UCF5B | D4@1.69 | Ea | 3,120.00 | 78.20 | 3,198.20 |
| 4.6 CF, stainless steel racks, manual hot gas defrost system, #UCF5C | D4@1.69 | Ea | 3,680.00 | 78.20 | 3,758.20 |
| Morgue refrigerators | | | | | |
| 1 or 2 body roll-in type, 39" wide x 96" deep x 76" high, with 1/2 HP, 4,380 Btu/hour condenser, #1SPEC | D4@15.8 | Ea | 9,910.00 | 731.00 | 10,641.00 |
| 2 body sliding tray type, 39" wide x 96" deep x 76" high, with 1/2 HP, 4,380 Btu/hour condenser, #2EC | D4@18.0 | Ea | 9,790.00 | 833.00 | 10,623.00 |
| 2 body side opening type, 96" wide x 39" deep x 76" high, with 1/2 HP, 4,380 Btu/hour condenser, #2SC | D4@18.0 | Ea | 10,900.00 | 833.00 | 11,733.00 |
| 4 or 6 body with roll-in type lower compartments, sliding tray type upper compartments, 73" wide x 96" deep x 102" high, 6,060 Btu/hr, 3/4 HP condenser, #4SPEC2W | D4@22.5 | Ea | 16,900.00 | 1,040.00 | 17,940.00 |

# 12 Furnishings

| | Craft@Hrs | Unit | Material | Labor | Total |
|---|---|---|---|---|---|
| **Church Glass** Material cost includes fabrication of the lite. Labor cost includes setting in a prepared opening only. Minimum labor cost is usually $150 per job. Add the cost of scaffold or a hoist, if needed. | | | | | |
| Stained glass | | | | | |
| Simple artwork | G1@.184 | SF | 69.10 | 8.46 | 77.56 |
| Moderate artwork | G1@.318 | SF | 175.00 | 14.60 | 189.60 |
| Elaborate artwork | G1@.633 | SF | 555.00 | 29.10 | 584.10 |
| Faceted glass | | | | | |
| Simple artwork | G1@.184 | SF | 61.30 | 8.46 | 69.76 |
| Moderate artwork | G1@.318 | SF | 120.00 | 14.60 | 134.60 |
| Elaborate artwork | G1@.633 | SF | 166.00 | 29.10 | 195.10 |

# 12 Furnishings

|  | Craft@Hrs | Unit | Material | Labor | Total |
|---|---|---|---|---|---|
| Colored glass, no artwork |  |  |  |  |  |
| Single pane | G1@.056 | SF | 6.56 | 2.58 | 9.14 |
| Patterned | G1@.184 | SF | 13.80 | 8.46 | 22.26 |
| Small pieces | G1@.257 | SF | 35.40 | 11.80 | 47.20 |

## Window Treatment

Lead mesh fabrics, measure glass size only. Includes allowance for overlap, pleating and all hardware.

|  | Craft@Hrs | Unit | Material | Labor | Total |
|---|---|---|---|---|---|
| Lead mesh (soundproofing) | — | SY | — | — | 25.00 |
| Lead mesh (x-ray) 2.5 lbs per SF | — | SY | — | — | 28.60 |

Draperies, 8' height, fabric quality will cause major variation in prices, per LF of opening, typical installed prices

|  | Craft@Hrs | Unit | Material | Labor | Total |
|---|---|---|---|---|---|
| Standard quality, window or sliding door | — | LF | — | — | 21.60 |
| Better quality, window | — | LF | — | — | 71.40 |
| Better quality, sliding door | — | LF | — | — | 68.10 |
| Drapery lining | — | LF | — | — | 8.13 |
| Fiberglass fabric | — | LF | — | — | 69.50 |
| Filter light control type | — | LF | — | — | 62.30 |
| Flameproofed | — | LF | — | — | 67.40 |
| Velour, grand | — | LF | — | — | 89.00 |

FREE STANDING

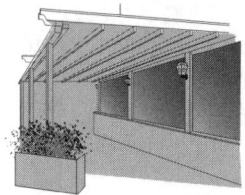

WALL BRACED

**Commercial grade awnings,** for restaurant patios, cafes, office outdoor break areas and for solar thermal influx mitigation above glazing. These units meet Western Canvas Products Association standards. Standard fabric is weatherproofed woven, solution-dyed polyester. Complies with Professional Awning Manufacturers Association standards. Arquati Roma, Milano or equal. Add the cost of electric connection and backlighting, if required.

Retractable wall-mounted fabric awnings. Lateral arm-type retractable units in lengths to 40 feet and projections to 17 feet. With pitch adjustment arm shoulders, aluminum frame and stainless-steel fasteners.

|  | Craft@Hrs | Unit | Material | Labor | Total |
|---|---|---|---|---|---|
| Reinforcing soffit/wall mount support and attaching transverse mount support to soffit or wall | H1@2.00 | Ea | 180.00 | 105.00 | 285.00 |
| Lifting and mounting mechanical assembly | H1@2.00 | Ea | 100.00 | 105.00 | 205.00 |
| Mounting and leveling frame, arm and fabric assembly, per SF of fabric | H1@.010 | SF | 24.30 | .53 | 24.83 |
| Adjusting fabric tautness, lateral arm tensioning and adjusting arm pulleys | H1@0.50 | Ea | — | 26.30 | 26.30 |

Retractable hood metal-frame lean-to-type awnings. With pre-assembled roller tube and hood assembly, vertical anodized aluminum support hood roller guide columns, curved lintel hood retractor roller guide and transverse wall or soffit mounting support plate assembly. Lengths to 42 feet and projections to 22 feet. Includes pitch adjustment arm shoulders, certified aluminum frame construction and stainless-steel fasteners.

|  | Craft@Hrs | Unit | Material | Labor | Total |
|---|---|---|---|---|---|
| Reinforcing soffit/wall mount support and attaching transverse mount support to soffit or wall | H1@2.00 | Ea | 200.00 | 105.00 | 305.00 |
| Lifting and mounting mechanical assembly | H1@3.00 | Ea | 125.00 | 158.00 | 283.00 |
| Mounting and leveling frame, vertical supports, reel, drive and fabric assembly, per SF of fabric | H1@.010 | SF | 25.73 | .53 | 26.26 |
| Adjusting fabric tautness, lateral arm tensioning and adjusting arm pulleys | H1@0.50 | Ea | — | 26.30 | 26.30 |

# 12 Furnishings

| | Craft@Hrs | Unit | Material | Labor | Total |
|---|---|---|---|---|---|

Free-standing or wall-braced canopy-type awning. With pre-assembled roller tube and hood assembly, vertical anodized aluminum support hood roller guide columns, transverse and parallel anodized aluminum girders with hood expander/retractor roller guide and mounting vertical support base-plate assemblies. Lengths to 42 feet and projections up to 22 feet. Aluminum frame construction, stainless-steel fasteners, motor and controls, rainproof hood, Kevlar pulley cord, roller-bearing pulley and hood guides.

| | Craft@Hrs | Unit | Material | Labor | Total |
|---|---|---|---|---|---|
| Reinforcing soffit/wall mount support and attaching transverse mount support to soffit or wall | H1@.010 | Ea | 300.00 | .53 | 300.53 |
| Lifting and mounting mechanical assembly | H1@1.45 | Ea | 200.00 | 76.20 | 276.20 |
| Mounting and leveling frame, arm and fabric assembly, per SF of fabric | H1@.015 | SF | 31.00 | .79 | 31.79 |
| Adjusting fabric tautness and lateral arm tensioning and adjusting arm pulleys | H1@0.80 | Ea | — | 42.00 | 42.00 |

Additional costs for retractable awnings. Wind sensor, motorized reel, automatic actuator and safety interlock are mandatory for compliance with Dade County (Florida) construction code.

| | Craft@Hrs | Unit | Material | Labor | Total |
|---|---|---|---|---|---|
| Add for one day forklift rental | — | Day | 250.00 | — | — |
| Add for woven flame-retardant fiber | — | % | 5.0 | — | — |
| Add for expanded Teflon coated glass fiber (PTFE) | — | % | 8.0 | — | — |
| Add for painted or laser-printed custom logo and lettering | — | % | 3.0 | — | — |
| Add for decal-mounted logo and text | — | % | 2.0 | — | — |
| Add for wind sensor, auto crank and safety interlock | — | % | 5.0 | — | — |
| Add for motorized crank | — | % | 10.0 | — | — |
| Add for calibrating and testing auto crank, wind sensor and automatic actuator | H1@0.80 | Ea | — | 42.00 | 42.00 |

**Cabinets** Cost per linear foot of face or back dimension. Base cabinets are 34" high x 24" deep, wall cabinets are 42" high x 12" deep and full height cabinets are 94" high x 24" deep. Labor shown is for installing shop fabricated units. Add for countertops at the end of this section.

Metal cabinets, shop and commercial type, with hardware but no countertop

| | Craft@Hrs | Unit | Material | Labor | Total |
|---|---|---|---|---|---|
| Base cabinet, drawer, door and shelf | D4@.379 | LF | 106.00 | 17.50 | 123.50 |
| Wall cabinet with door and 2 shelves | D4@.462 | LF | 97.00 | 21.40 | 118.40 |
| Full height cabinet, 5 shelves | D4@.595 | LF | 198.00 | 27.50 | 225.50 |
| Library shelving, 48" high x 8" deep modules | D4@.336 | LF | 69.10 | 15.50 | 84.60 |
| Wardrobe units, 72" high x 42" deep | D4@.398 | LF | 171.00 | 18.40 | 189.40 |

Custom made wood cabinets, no hardware or countertops included, unfinished, 36" wide modules. For more complete coverage of custom-built cabinets, see *National Framing and Finish Carpentry Estimator* at http://CraftsmanSiteLicense.com.

| | Craft@Hrs | Unit | Material | Labor | Total |
|---|---|---|---|---|---|
| Standard grade, mahogany or laminated plastic face | | | | | |
| Base | C8@.256 | LF | 79.70 | 11.70 | 91.40 |
| Wall | C8@.380 | LF | 60.80 | 17.40 | 78.20 |
| Full height | C8@.426 | LF | 138.00 | 19.50 | 157.50 |
| Custom grade, birch | | | | | |
| Base | C8@.361 | LF | 122.00 | 16.50 | 138.50 |
| Wall | C8@.529 | LF | 105.00 | 24.20 | 129.20 |
| Full height | C8@.629 | LF | 191.00 | 28.80 | 219.80 |
| Sink fronts | C8@.361 | LF | 101.00 | 16.50 | 117.50 |
| Drawer units, 4 drawers | C8@.361 | LF | 201.00 | 16.50 | 217.50 |
| Apron (knee space) | C8@.148 | LF | 25.90 | 6.78 | 32.68 |
| Additional costs for custom wood cabinets | | | | | |
| Ash face | — | LF | 4.33 | — | 4.33 |
| Walnut face | — | LF | 65.30 | — | 65.30 |
| Edge banding | — | LF | 7.06 | — | 7.06 |

# 12 Furnishings

| | Craft@Hrs | Unit | Material | Labor | Total |
|---|---|---|---|---|---|
| Prefinished exterior | — | LF | 6.42 | — | 6.42 |
| Prefinished interior | — | LF | 10.90 | — | 10.90 |
| Standard hardware, installed | — | LF | 5.12 | — | 5.12 |
| Institutional grade hardware, installed | — | LF | 7.45 | — | 7.45 |
| Drawer roller guides, per drawer | — | Ea | 8.70 | — | 8.70 |
| Drawer roller guides, suspension | — | Ea | 27.10 | — | 27.10 |
| Extra drawers | | | | | |
|    12" wide | — | Ea | 55.30 | — | 55.30 |
|    18" wide | — | Ea | 64.60 | — | 64.60 |
|    24" wide | — | Ea | 70.60 | — | 70.60 |
| **Classroom type wood cabinets, laminated plastic face, with hardware but no countertop** | | | | | |
| Base cabinet, drawer, door and shelf | C8@.391 | LF | 186.00 | 17.90 | 203.90 |
| Wall cabinet with door and 2 shelves | C8@.475 | LF | 162.00 | 21.80 | 183.80 |
| Full height cabinet with doors | C8@.608 | LF | 263.00 | 27.90 | 290.90 |
| **Hospital type wood cabinets, laminated plastic face, with hardware but no countertop** | | | | | |
| Base cabinet, drawer, door and shelf | C8@.982 | LF | 184.00 | 45.00 | 229.00 |
| Wall, with 2 shelves and door | C8@1.15 | LF | 169.00 | 52.70 | 221.70 |
| Full height with doors | C8@1.35 | LF | 251.00 | 61.90 | 312.90 |
| **Laboratory type metal cabinets, with hardware but no countertop** | | | | | |
| Base cabinet with door | D4@.785 | LF | 154.00 | 36.30 | 190.30 |
| Base cabinet with drawers | D4@.785 | LF | 191.00 | 36.30 | 227.30 |
| Base cabinets, island base | D4@.785 | LF | 191.00 | 36.30 | 227.30 |
| Wall cabinet with metal door | D4@.785 | LF | 144.00 | 36.30 | 180.30 |
| Wall cabinet with framed glass door | D4@.785 | LF | 188.00 | 36.30 | 224.30 |
| Wall cabinets with no doors | D4@.785 | LF | 121.00 | 36.30 | 157.30 |
| Wardrobe or storage cabinet, 80" high, with door | D4@.939 | LF | 388.00 | 43.40 | 431.40 |
| Storage cabinet, 80" high, no door | D4@.939 | LF | 233.00 | 43.40 | 276.40 |
| Fume hoods, steel, without duct, typical cost | T5@3.88 | Ea | 773.00 | 191.00 | 964.00 |
|   Add for duct, without electric work, typical cost | T5@16.6 | LF | 1,170.00 | 816.00 | 1,986.00 |

**Countertops,** shop fabricated, delivered ready for installation.
Laminated plastic tops 2'0" wide over plywood base

| | Craft@Hrs | Unit | Material | Labor | Total |
|---|---|---|---|---|---|
| Custom work, square edge front, 4" splash | C8@.187 | LF | 28.70 | 8.57 | 37.27 |
|   Add for square edge 9" backsplash | — | LF | 9.31 | — | 9.31 |
|   Add for solid colors | — | LF | 2.77 | — | 2.77 |
|   Add for acid proof tops | — | LF | 21.30 | — | 21.30 |
| Stainless steel, 2'0" wide with backsplash | C8@.366 | LF | 73.80 | 16.80 | 90.60 |

**Cigarette Disposal Units**
20 gauge stainless steel with grate. For cigarettes and cigars

| | Craft@Hrs | Unit | Material | Labor | Total |
|---|---|---|---|---|---|
| 5-1/4" x 5-3/16" x 8-3/8" wall mounted | CM@.500 | Ea | 150.00 | 25.10 | 175.10 |
| 5-1/4" x 5-3/16" x 8-3/8" pole mounted | CM@.750 | Ea | 160.00 | 37.60 | 197.60 |
| 13" x 16" x 43" wall mounted | CM@.600 | Ea | 183.00 | 30.10 | 213.10 |
| 13" x 16" x 43" pole mounted | CM@.900 | Ea | 206.00 | 45.10 | 251.10 |
| 15" x 16" x 43" wall mounted with trash can | CM@.750 | Ea | 431.00 | 37.60 | 468.60 |
| 15" x 16" x 43" pole mounted with trash can | CM@1.25 | Ea | 450.00 | 62.70 | 512.70 |

**Recessed Open Link Entry Mats** These costs assume that mats are mounted in an aluminum frame recessed into a concrete walkway and do not include concrete work, special designs or lettering

| | Craft@Hrs | Unit | Material | Labor | Total |
|---|---|---|---|---|---|
| Polypropylene, 3/8" thick | D4@.560 | SF | 20.00 | 25.90 | 45.90 |
| Coconut husk fiber, 3/8" thick | D4@.560 | SF | 10.00 | 25.90 | 35.90 |
| Recessed rubber | D4@.560 | SF | 6.50 | 25.90 | 32.40 |

# 12 Furnishings

| | Craft@Hrs | Unit | Material | Labor | Total |
|---|---|---|---|---|---|

**Entrance Mats and Recessed Foot Grilles**  These costs assume installation of an extruded aluminum tread foot grille on a level surface or over a catch basin with or without a drain.

Pedigrid/Pedimat extruded aluminum entrance mat

| | Craft@Hrs | Unit | Material | Labor | Total |
|---|---|---|---|---|---|
| Grid mat with carpet treads | D4@.150 | SF | 34.30 | 6.94 | 41.24 |
| Grid mat with vinyl treads | D4@.150 | SF | 37.30 | 6.94 | 44.24 |
| Grid mat with pool or shower vinyl treads | D4@.150 | SF | 41.60 | 6.94 | 48.54 |
| Grid mat with abrasive serrated aluminum treads | D4@.150 | SF | 41.80 | 6.94 | 48.74 |
| Recessed catch basin with 2" drain | D4@.302 | SF | 46.20 | 14.00 | 60.20 |
| Recessed catch basin without drain | D4@.242 | SF | 43.50 | 11.20 | 54.70 |
| Level base recessed mount | D4@.150 | SF | 41.50 | 6.94 | 48.44 |
| Surface-mount frame, vinyl | — | LF | 4.15 | — | 4.15 |
| Recessed-mount frame, bronze aluminum | D4@.092 | LF | 9.50 | 4.26 | 13.76 |

**Seating**  These costs do not include concrete work

| | Craft@Hrs | Unit | Material | Labor | Total |
|---|---|---|---|---|---|
| Theater style, economy | D4@.424 | Ea | 123.00 | 19.60 | 142.60 |
| Theater style, lodge, rocking type | D4@.470 | Ea | 245.00 | 21.70 | 266.70 |

Bleachers, closed riser, extruded aluminum, cost per 18" wide seat

| | Craft@Hrs | Unit | Material | Labor | Total |
|---|---|---|---|---|---|
| Portable bleachers | D4@.090 | Ea | 53.60 | 4.16 | 57.76 |
| Portable bleachers with guardrails | D4@.136 | Ea | 59.50 | 6.29 | 65.79 |
| Team bleachers, stationary | D4@.090 | Ea | 46.00 | 4.16 | 50.16 |

Stadium seating, on concrete foundation by others, extruded aluminum, cost per 18" seat width

| | Craft@Hrs | Unit | Material | Labor | Total |
|---|---|---|---|---|---|
| 10" flat bench seating, seatboard only | D4@.090 | Ea | 15.50 | 4.16 | 19.66 |
| 12" contour bench seating, seatboard only | D4@.090 | Ea | 16.60 | 4.16 | 20.76 |
| Chair seating, molded plastic | D4@.310 | Ea | 89.40 | 14.30 | 103.70 |

Replacement bench seating, stadium or bleacher, extruded aluminum

Cost per 18" seat width

| | Craft@Hrs | Unit | Material | Labor | Total |
|---|---|---|---|---|---|
| 10" wide flat seatboard only | D4@.069 | Ea | 9.81 | 3.19 | 13.00 |
| 10" wide flat seatboard and footboard | D4@.136 | Ea | 21.00 | 6.29 | 27.29 |
| 12" contour seatboard only | D4@.069 | Ea | 11.60 | 3.19 | 14.79 |

**Benches**

Floor-mounted wood locker room bench, metal pedestal bolted to a concrete floor.

| | Craft@Hrs | Unit | Material | Labor | Total |
|---|---|---|---|---|---|
| Hardwood top, 6' long, 12" wide | D4@.753 | Ea | 208.00 | 34.80 | 242.80 |

Park bench, for public and commercial use. Cast aluminum with treated 1-1/2" by 2-1/4" pine wood slats. Central Park or Victorian style.

| | Craft@Hrs | Unit | Material | Labor | Total |
|---|---|---|---|---|---|
| 4' long | D4@.750 | Ea | 239.00 | 34.70 | 273.70 |
| 5' long | D4@.750 | Ea | 259.00 | 34.70 | 293.70 |
| 6' long | D4@.750 | Ea | 295.00 | 34.70 | 329.70 |

**Bicycle, Moped and Motorcycle Security Racks**  Installation kit includes tamper proof bolts and quick setting cement. Locks supplied by users. Labor is to install units on concrete surface. Cost per bicycle or motorcycle of capacity. Add cost for drilling holes in concrete as needed. These costs are based on a minimum purchase of six racks.

Basic bicycle rack

| | Craft@Hrs | Unit | Material | Labor | Total |
|---|---|---|---|---|---|
| 5' long, looped steel, holds 8 bikes | C8@1.00 | Ea | 255.00 | 45.80 | 300.80 |
| 7' long, looped steel, holds 12 bikes | C8@1.25 | Ea | 313.00 | 57.30 | 370.30 |
| 10' long, single sided, holds 9 bikes | C8@.862 | Ea | 275.00 | 39.50 | 314.50 |
| 20' long, single sided, holds 18 bikes | C8@1.50 | Ea | 490.00 | 68.70 | 558.70 |
| 10' long, double sided, holds 18 bikes | C8@1.50 | Ea | 354.00 | 68.70 | 422.70 |
| 20' long, double sided, holds 36 bikes | C8@1.00 | Ea | 620.00 | 45.80 | 665.80 |

Moped or motorcycle rack

| | Craft@Hrs | Unit | Material | Labor | Total |
|---|---|---|---|---|---|
| With 24" security chain | C8@.575 | Ea | 126.00 | 26.40 | 152.40 |

# 12 Furnishings

| | Craft@Hrs | Unit | Material | Labor | Total |
|---|---|---|---|---|---|
| Add for mounting track for use on asphalt or non-concrete surfaces. No foundation work included. Each 10' track section holds 6 bicycle racks or 4 moped or motorcycle racks. | | | | | |
| Per 10' track section, including drilling holes | C8@2.50 | Ea | 131.00 | 115.00 | 246.00 |
| Add for purchasing less than six mounting racks | — | Ea | 15.40 | — | 15.40 |
| Add for drilling mounting holes in concrete, | | | | | |
| 2 required per rack, cost per rack | C8@.125 | Ea | — | 5.73 | 5.73 |
| Bicycle racks, galvanized pipe rack, embedded or surface mounted to concrete, cost per rack | | | | | |
| Double entry | | | | | |
| 5' long, 7 bikes | C8@3.31 | Ea | 286.00 | 152.00 | 438.00 |
| 10' long, 14 bikes | C8@3.31 | Ea | 414.00 | 152.00 | 566.00 |
| 20' long, 28 bikes | C8@4.88 | Ea | 752.00 | 224.00 | 976.00 |
| 30' long, 42 bikes | C8@6.46 | Ea | 1,090.00 | 296.00 | 1,386.00 |
| Single side entry | | | | | |
| 5' long, 4 bikes | C8@2.41 | Ea | 275.00 | 110.00 | 385.00 |
| 10' long, 8 bikes | C8@2.41 | Ea | 311.00 | 110.00 | 421.00 |
| 20' long, 16 bikes | C8@3.55 | Ea | 561.00 | 163.00 | 724.00 |
| Wood and metal rack, embedded or surface mounted | | | | | |
| Double entry | | | | | |
| 10' long, 14 bikes | C8@3.31 | Ea | 450.00 | 152.00 | 602.00 |
| 20' long, 28 bikes | C8@4.88 | Ea | 843.00 | 224.00 | 1,067.00 |
| Single entry | | | | | |
| 10' long, 8 bikes | C8@2.41 | Ea | 375.00 | 110.00 | 485.00 |
| 20' long, 10 bikes | C8@3.55 | Ea | 674.00 | 163.00 | 837.00 |
| Bike post bollard, | | | | | |
| dimensional wood, concrete footing | C8@.223 | Ea | 204.00 | 10.20 | 214.20 |
| Single tube galvanized steel, looped wave pattern, 30" high, surface mounted | | | | | |
| 3' long, 5 bikes | C8@3.45 | Ea | 669.00 | 158.00 | 827.00 |
| 5' long, 7 bikes | C8@3.45 | Ea | 767.00 | 158.00 | 925.00 |
| 7' long, 9 bikes | C8@4.61 | Ea | 999.00 | 211.00 | 1,210.00 |
| 9' long, 11 bikes | C8@4.61 | Ea | 1,180.00 | 211.00 | 1,391.00 |
| Bonded color, add | — | Ea | 102.00 | — | 102.00 |
| Add for embedded mounting | C8@1.77 | Ea | 25.50 | 81.10 | 106.60 |

# 13 Special Construction

**Swimming Pools** Complete in-ground pool including typical excavation, reinforcing, gunite, filter, pump, pop-up cleaning heads and circulation manifold, skimmer, PVC supply and backwash piping, float valve, ladder, electrical controls and one pool light. Add the cost of deck beyond the edge of coping, electrical supply, a pool equipment slab and enclosure, if required. Typical subcontract prices per SF of water surface.

| | | Unit | Material | Labor | Total |
|---|---|---|---|---|---|
| Apartment building | — | SF | — | — | 62.70 |
| Community | — | SF | — | — | 80.70 |
| Hotel or resort | — | SF | — | — | 75.90 |
| School, training (42' x 75') | — | SF | — | — | 85.00 |
| School, Olympic (42' x 165') | — | SF | — | — | 129.00 |

Swimming pool covers, retractable motor-operated, track mounted, including the motor, gearing, retracting reel, pulley system and remote switch. For square or rectangular pools. Add the cost of concrete work for the winding reel enclosure, electric supply and the cover.

| | | Unit | Material | Labor | Total |
|---|---|---|---|---|---|
| 3/4 HP drive motor and reel (to 600 SF water surface) | — | Ea | — | — | 4,410.00 |
| 1-1/4 HP drive motor and reel | | | | | |
| (to 1,200 SF water surface) | — | Ea | — | — | 5,520.00 |
| Add for the vinyl cover, per SF of water surface | — | SF | — | — | 3.31 |

# 13 Special Construction

|  | Craft@Hrs | Unit | Material | Labor | Equipment | Total |
|---|---|---|---|---|---|---|

**Access for the Disabled**

Use the following cost estimates when altering an existing building to meet requirements of the Americans with Disabilities Act (ADA). The ADA requires that existing public and private buildings be altered to accommodate the physically handicapped when that can be done without unreasonable difficulty or expense.

**Ramps, flared aprons and landings** Concrete access ramps, flared aprons and landings, 2,000 PSI concrete placed directly from the chute of a ready-mix truck, vibrated and finished. Costs include typical wire mesh reinforcing and an allowance for waste but no excavation, backfill or foundations.

Access ramps, based on ramps with 1" rise per 1' of run, 4" thick walls and 4" thick slab. Includes pipe sleeves for railings, sand fill in cavity and non-slip surface finish

|  | Craft@Hrs | Unit | Material | Labor | Equipment | Total |
|---|---|---|---|---|---|---|
| 24" vertical rise (4' wide x 25' long) | C8@40.0 | Ea | 1,100.00 | 1,830.00 | — | 2,930.00 |
| 36" vertical rise (4' wide x 38' long) | C8@60.0 | Ea | 2,120.00 | 2,750.00 | — | 4,870.00 |
| 48" vertical rise (4' wide x 50' long) | C8@96.0 | Ea | 3,120.00 | 4,400.00 | — | 7,520.00 |

Access aprons at curbs, flared and inclined, 2' long x 8' wide

|  | Craft@Hrs | Unit | Material | Labor | Equipment | Total |
|---|---|---|---|---|---|---|
| 4" thick (80 SF per CY) | P9@.024 | SF | 2.71 | 1.13 | — | 3.84 |
| 6" thick (54 SF per CY) | P9@.032 | SF | 4.07 | 1.51 | — | 5.58 |

Landings, 4" thick walls on three sides and 4" thick top slab cast against an existing structure on one side. Costs include pipe sleeves for railings, sand fill in cavity and an allowance for waste.

|  | Craft@Hrs | Unit | Material | Labor | Equipment | Total |
|---|---|---|---|---|---|---|
| 3'0" high with 6' x 6' square slab | C8@32.0 | Ea | 916.00 | 1,470.00 | — | 2,386.00 |
| Add per vertical foot of height over 3'0" | — | VLF | 173.00 | — | — | 173.00 |

Steel access ramps, 4140 carbon steel, 1" rise per foot of run, 8 gauge diamond pattern steel plate and 2" diameter handrail with electrostatic vinyl paint coating. Includes fabrication and assembly but no excavation. Add the cost of post foundations.

|  | Craft@Hrs | Unit | Material | Labor | Equipment | Total |
|---|---|---|---|---|---|---|
| Per square foot of ramp surface | T3@.150 | SF | 3.90 | 7.37 | — | 11.27 |

Redwood access ramps, S4S kiln-dried or better redwood, based on ramps with 1-inch rise foot of run, 6" x 6" posts supporting 2" x 12" beams with steel cross-bracing 36" OC. 2" x 6" decking with 2" diameter railing. Includes fabrication and assembly but no excavation. Add the cost of post foundations.

|  | Craft@Hrs | Unit | Material | Labor | Equipment | Total |
|---|---|---|---|---|---|---|
| Per square foot of ramp surface | B9@.150 | SF | 2.29 | 4.89 | — | 7.18 |
| Per square foot of ramp surface | B9@.150 | SF | 2.29 | 4.89 | — | 7.18 |

Snow melting cable for ramps and landings. Includes 208-277 volt AC self-regulating heater cable designed to be encased in concrete. Based on cable connected to an existing power source adjacent to the installation

|  | Craft@Hrs | Unit | Material | Labor | Equipment | Total |
|---|---|---|---|---|---|---|
| Heater cable |  |  |  |  |  |  |
| (1.1 LF per SF of pavement) | CE@.015 | SF | 9.28 | .88 | — | 10.16 |
| Power connection kit | CE@.500 | Ea | 30.00 | 29.20 | — | 59.20 |
| Thermostat | CE@.500 | Ea | 245.00 | 29.20 | — | 274.20 |

**Railing for ramps and landings** Welded steel. Equipment cost shown is for a flatbed truck and a welding machine. Wall-mounted handrails with brackets 5' OC, based on 12' lengths

|  | Craft@Hrs | Unit | Material | Labor | Equipment | Total |
|---|---|---|---|---|---|---|
| 1-1/4" diameter rail, shop prime painted | H6@.092 | LF | 6.60 | 4.76 | .45 | 11.81 |
| 1-1/2" diameter rail, shop prime painted | H6@.102 | LF | 7.73 | 5.28 | .50 | 13.51 |
| Add for galvanized finish | — | % | 20.0 | — | — | — |

Floor-mounted railings 3'6" high with posts 5' OC, based on 12' length

|  | Craft@Hrs | Unit | Material | Labor | Equipment | Total |
|---|---|---|---|---|---|---|
| 1-1/4" diameter rails and posts, shop prime painted |  |  |  |  |  |  |
| 2-rail type | H6@.115 | LF | 27.10 | 5.95 | .57 | 33.62 |
| 3-rail type | H6@.125 | LF | 35.10 | 6.47 | .62 | 42.19 |
| 1-1/2" diameter rails and posts, shop prime painted |  |  |  |  |  |  |
| 2-rail type | H6@.125 | LF | 32.70 | 6.47 | .62 | 39.79 |
| 3-rail type | H6@.137 | LF | 43.00 | 7.09 | .68 | 50.77 |
| Add for galvanized finish | — | % | 20.0 | — | — | — |

# 13 Special Construction

| | Craft@Hrs | Unit | Material | Labor | Equipment | Total |
|---|---|---|---|---|---|---|

## Drinking fountains and water coolers for the disabled

Remove an existing drinking fountain or water cooler and disconnect waste and water piping at the wall adjacent to the fixture for use with the new fixture. No salvage value assumed.

| | Craft@Hrs | Unit | Material | Labor | Equipment | Total |
|---|---|---|---|---|---|---|
| Floor or wall-mounted fixture | P6@1.00 | Ea | — | 50.20 | — | 50.20 |

Install a wall-hung white vitreous china fountain with chrome plated spout, 14" wide, 13" high, non-electric, self-closing dispensing valve and automatic volume control

| | | | | | | |
|---|---|---|---|---|---|---|
| Drinking fountain, trim and valves | P6@2.80 | Ea | 800.00 | 141.00 | — | 941.00 |

Wheelchair wall-hung water cooler, vitreous china, refrigerated, (no electrical work included)

| | | | | | | |
|---|---|---|---|---|---|---|
| Wall-hung water cooler and trim (add rough-in) | P6@2.50 | Ea | 886.00 | 126.00 | — | 1,012.00 |

Install floor or wall-mounted electric water cooler, 8 gallons per hour, cold water only, connected to existing electrical outlet adjacent to the cooler (no electrical work included)

| | | | | | | |
|---|---|---|---|---|---|---|
| Water cooler, trim and valves | P6@2.80 | Ea | 1,560.00 | 141.00 | — | 1,701.00 |

## Parking stall marking for the disabled

Remove existing pavement marking by water blasting

| | | | | | | |
|---|---|---|---|---|---|---|
| 4" wide stripes (allow 40 LF per stall) | CL@.032 | LF | — | 1.28 | .45 | 1.73 |
| Markings, per square foot (minimum 10 SF) | CL@.098 | SF | — | 3.91 | 1.35 | 5.26 |

Mark parking stall with handicapped symbol painted on, including layout
Reflectorized stripes and symbol, one color. Equipment to include a compressor, hose and spray gun, minimum daily rental $75.00.

| | | | | | | |
|---|---|---|---|---|---|---|
| Per stall, no striping | PA@.425 | Ea | 11.30 | 21.90 | — | 33.20 |

Parking lot handicapped sign, 12" x 18" 10 gauge metal, reflective lettering and handicapped symbol
Sign on 2" galvanized steel pipe post 10' long, set 2' into the ground. Includes digging of hole

| | | | | | | |
|---|---|---|---|---|---|---|
| Using a manual auger and backfill | CL@1.25 | Ea | 141.00 | 49.90 | — | 190.90 |
| On walls with mechanical fasteners | PA@.455 | Ea | 60.60 | 23.40 | — | 84.00 |

## Paving and curb removal

These costs include breaking out the paving or curb for ADA improvements, loading and hauling debris to a legal dump within 6 miles but no dump fees. Equipment cost per hour is for one 90 CFM air compressor, one paving breaker with jackhammer bits, one 55 HP wheel loader with integral backhoe and one 5 CY dump truck. The figures in parentheses show the approximate "loose" volume of the materials (volume after being demolished). Use $250 as a minimum charge for this type work and add dump charges as a separate item.

Asphalt paving, depths to 3"

| | | | | | | |
|---|---|---|---|---|---|---|
| (27 SF per CY) | C3@.040 | SF | — | 1.86 | 1.04 | 2.90 |

Asphalt curbs, to 12" width

| | | | | | | |
|---|---|---|---|---|---|---|
| (18 LF per CY) | C3@.028 | LF | — | 1.30 | .73 | 2.03 |

Concrete paving and slabs on grade
4" concrete without rebars

| | | | | | | |
|---|---|---|---|---|---|---|
| (54 SF per CY) | C3@.060 | SF | — | 2.79 | 1.56 | 4.35 |

6" concrete without rebars

| | | | | | | |
|---|---|---|---|---|---|---|
| (45 SF per CY) | C3@.110 | SF | — | 5.11 | 2.87 | 7.98 |
| Add for removal of bar reinforced paving | C3@.200 | SF | — | 9.30 | 5.21 | 14.51 |

## Curb, gutter and paving for the disabled

These costs assume 3 uses of forms and 2,000 PSI concrete placed directly from the chute of a ready-mix truck. Use $250 as a minimum charge for this type work. See also Snow Melting Cable in the Electrical section.
Concrete cast-in-place vertical curb, including 2% waste, forms and finishing. These costs include typical excavation and backfill with excess excavation spread on site.

| | | | | | | |
|---|---|---|---|---|---|---|
| 6" x 12" straight curb | P9@.136 | LF | 4.18 | 6.43 | — | 10.61 |
| 6" x 18" straight curb | P9@.149 | LF | 6.20 | 7.04 | — | 13.24 |
| 6" x 24" straight curb | P9@.154 | LF | 8.23 | 7.28 | — | 15.51 |
| Add for curved sections | — | % | — | 40.0 | — | — |
| Curb and 24" monolithic gutter (7 LF per CY) | C3@.100 | LF | 8.71 | 4.65 | — | 13.36 |
| Concrete paving and walkway, 4" thick | P8@.018 | SF | 1.99 | .81 | — | 2.80 |
| Add for integral colors, most pastels | — | SF | .94 | — | — | .94 |

# 13 Special Construction

|  | Craft@Hrs | Unit | Material | Labor | Equipment | Total |
|---|---|---|---|---|---|---|

## Plumbing fixtures for the disabled

Remove existing plumbing fixtures. Disconnect from existing waste and water piping at the wall adjacent to the fixture. Waste and water piping to remain in place for connection to a new fixture. No salvage value assumed.

| | Craft@Hrs | Unit | Material | Labor | Equipment | Total |
|---|---|---|---|---|---|---|
| Lavatories | P6@1.00 | Ea | — | 50.20 | — | 50.20 |
| Urinals | P6@1.00 | Ea | — | 50.20 | — | 50.20 |
| Water closets, floor or wall mounted | P6@1.00 | Ea | — | 50.20 | — | 50.20 |

Install new fixtures. White vitreous china fixtures with typical fittings, hangers, trim and valves. Connected to waste and supply piping from lines adjacent to the fixture.

| | Craft@Hrs | Unit | Material | Labor | Equipment | Total |
|---|---|---|---|---|---|---|
| Lavatories, wall hung | P6@1.86 | Ea | 600.00 | 93.40 | — | 693.40 |

Water closets, tank type, with elongated bowl, toilet seat, trim and valves

| | Craft@Hrs | Unit | Material | Labor | Equipment | Total |
|---|---|---|---|---|---|---|
| Floor mounted, with bowl ring | P6@1.86 | Ea | 334.00 | 93.40 | — | 427.40 |
| Wall hung, with carrier | P6@2.80 | Ea | 705.00 | 141.00 | — | 846.00 |

## Plumbing partitions and grab bars for the disabled

Remove a standard toilet partition, no salvage value included. Floor or ceiling mounted

| | Craft@Hrs | Unit | Material | Labor | Equipment | Total |
|---|---|---|---|---|---|---|
| Remove one wall and one door | T5@1.00 | Ea | — | 49.20 | — | 49.20 |

Install toilet partition designed for wheelchair access. Floor or ceiling mounted. These costs include one wall and one door, mechanical fasteners and drilling of holes

| | Craft@Hrs | Unit | Material | Labor | Equipment | Total |
|---|---|---|---|---|---|---|
| Powder coated metal | T5@2.66 | Ea | 514.00 | 131.00 | — | 645.00 |
| Solid plastic (polymer) | T5@4.00 | Ea | 881.00 | 197.00 | — | 1,078.00 |
| Laminated plastic | T5@2.66 | Ea | 498.00 | 131.00 | — | 629.00 |
| Stainless steel | T5@2.66 | Ea | 1,260.00 | 131.00 | — | 1,391.00 |

Grab bars, stainless steel, wall mounted, commercial grade. Set of two required for each stall. Including mechanical fasteners and drilling of holes

| | Craft@Hrs | Unit | Material | Labor | Equipment | Total |
|---|---|---|---|---|---|---|
| 1-1/4" x 24" urinal bars | CC@1.00 | Set | 100.00 | 51.80 | — | 151.80 |
| 1-1/4" x 36" toilet partition bars | CC@1.00 | Set | 123.00 | 51.80 | — | 174.80 |

| | Craft@Hrs | Unit | Material | Labor | Total |
|---|---|---|---|---|---|

**Signage in Braille** Standard signs (Rest Room, Elevator, Telephone, Stairway, etc.) mounted with self-stick foam, velcro or in frame. Add the cost of frames, if required, from below.

5/8" lettering raised 1/32". Grade 2 Braille raised 1/32"

| | Craft@Hrs | Unit | Material | Labor | Total |
|---|---|---|---|---|---|
| 2" x 8" sign, raised letters and tactile Braille | PA@0.25 | Ea | 44.90 | 12.90 | 57.80 |

8" x 8" sign with words, graphic symbols and tactile Braille

| | Craft@Hrs | Unit | Material | Labor | Total |
|---|---|---|---|---|---|
| standard graphic symbols | PA@0.25 | Ea | 44.90 | 12.90 | 57.80 |

Sign frames. Signs snap into frames mounted with mechanical fasteners

| | Craft@Hrs | Unit | Material | Labor | Total |
|---|---|---|---|---|---|
| 2" x 8" frames, black, almond or grey | PA@0.25 | Ea | 29.80 | 12.90 | 42.70 |
| 8" x 8" frames, black, almond or grey | PA@0.25 | Ea | 55.60 | 12.90 | 68.50 |

Elevator plates. Sign with graphic symbols and tactile Braille, standard colors

| | Craft@Hrs | Unit | Material | Labor | Total |
|---|---|---|---|---|---|
| Door jamb plates, 7" x 11", two per opening | PA@0.75 | Set | 94.70 | 38.60 | 133.30 |

Operation plates, Open Door, Close Door, Emergency Stop, Alarm,

| | Craft@Hrs | Unit | Material | Labor | Total |
|---|---|---|---|---|---|
| 1-1/4" x 1-1/4", set of four required per car | PA@1.00 | Set | 85.20 | 51.40 | 136.60 |

Operation plates, Floor Number, 1-1/4" x 1-1/4"

| | Craft@Hrs | Unit | Material | Labor | Total |
|---|---|---|---|---|---|
| One required for each floor per car | PA@0.25 | Ea | 21.30 | 12.90 | 34.20 |

**Handicapped automatic remote door opener,** 60 Hz, single phase, 110 volts, with manual override, electro-pneumatic actuator arm assembly , manual panic bar, door open-position locks, prehung twin metal doors, digital operation cycle setting module, actuator button panel, power/service panel. Americans With Disabilities Act compliant.

| | Craft@Hrs | Unit | Material | Labor | Total |
|---|---|---|---|---|---|
| Twin door frame | C8@4.00 | Ea | 318.00 | 183.00 | 501.00 |
| Twin doors, per pair | C8@.250 | Ea | 2,440.00 | 11.50 | 2,451.50 |
| Actuator arms for twin doors | C8@0.50 | Ea | 371.00 | 22.90 | 393.90 |
| Automatic remote door controls | C8@4.00 | Ea | 223.00 | 183.00 | 406.00 |
| Actuator mechanism | — | Ea | 2,760.00 | — | 2,760.00 |

# 13 Special Construction

| | Craft@Hrs | Unit | Material | Labor | Total |
|---|---|---|---|---|---|

**Bus stop shelters** Prefabricated structure systems, shipped knocked-down. Complete with all hardware and foundation anchors. Labor cost is for assembly on site. Add the cost of foundation and floor slab. Dimensions are approximate outside measurements. Based on Columbia Equipment Company.

Three-sided open front type. Clear satin silver anodized aluminum structure and fascia, 1/4" clear acrylic sheet glazing in panels and white baked enamel finish aluminum "V"-beam roof.

| | Craft@Hrs | Unit | Material | Labor | Total |
|---|---|---|---|---|---|
| 9'2" L x 5'3" W x 7'5" H | | | | | |
| Two rear panels, single panel each side | C8@8.00 | Ea | 4,590.00 | 367.00 | 4,957.00 |
| Four rear panels, two panels each side | C8@12.0 | Ea | 4,830.00 | 550.00 | 5,380.00 |
| 11'3" L x 5'3" W x 7'5" H | | | | | |
| Four rear panels, two panels each side | C8@12.0 | Ea | 5,640.00 | 550.00 | 6,190.00 |
| Add to three-sided open front shelters for the following: | | | | | |
| Wind break front entrance panel | | | | | |
| 5'3" W x 7'5" H, with two glazed panels | C8@2.00 | Ea | 556.00 | 91.70 | 647.70 |

| | Craft@Hrs | Unit | Material | Labor | Equipment | Total |
|---|---|---|---|---|---|---|

Bus stop enclosures, aluminum frame with Plexiglas sidewall elements, 12' wide by 30' long by 10' high, barrel roof with diamond pattern grille work, aluminum bench, solar lighting and bronze anodized finish.

| | Craft@Hrs | Unit | Material | Labor | Equipment | Total |
|---|---|---|---|---|---|---|
| Shelter assembly, bolted to the foundation | B5@8.00 | Ea | 19,600.00 | 284.00 | 250.00 | 20,134.00 |
| Structure leveling | B5@2.00 | Ea | 250.00 | 70.90 | — | 320.90 |

| | Craft@Hrs | Unit | Material | Labor | Total |
|---|---|---|---|---|---|
| Bus stop bench, all aluminum vandal-resistant seat and backrest | | | | | |
| 8'0" long | C8@1.00 | Ea | 459.00 | 45.80 | 504.80 |
| 10'0" long | C8@1.50 | Ea | 585.00 | 68.70 | 653.70 |

Light fixture with unbreakable Lexan diffuser, wiring not included

| | Craft@Hrs | Unit | Material | Labor | Total |
|---|---|---|---|---|---|
| 100 watt incandescent, photoelectric cell control | C8@2.00 | Ea | 230.00 | 91.70 | 321.70 |

Radiant heater, with vandal-resistant electric heating element in heavy duty enclosure and metal mesh guard over opening. Wiring not included.

| | Craft@Hrs | Unit | Material | Labor | Total |
|---|---|---|---|---|---|
| 2,500 watt unit (no controls) | C8@1.50 | Ea | 800.00 | 68.70 | 868.70 |
| 5,000 watt unit (no controls) | C8@1.50 | Ea | 950.00 | 68.70 | 1,018.70 |
| Electronic timer control, add to above | C8@1.00 | Ea | 420.00 | 45.80 | 465.80 |

Integrated map/schedule display panel, anodized aluminum frame, tamper-proof fasteners and "spanner-head" tool for opening.

| | Craft@Hrs | Unit | Material | Labor | Total |
|---|---|---|---|---|---|
| Full panel width x 30" high | | | | | |
| 3/16" clear Plexiglas | C8@1.00 | Ea | 154.00 | 45.80 | 199.80 |
| 3/16" polycarbonate with mar-resistant coating | C8@1.00 | Ea | 215.00 | 45.80 | 260.80 |
| Graphics, words "Bus Stop" in white letters, on fascia | | | | | |
| First side | — | LS | 115.00 | — | 115.00 |
| Each additional side | — | Ea | 40.10 | — | 40.10 |

**Metal shelter structures** Prefabricated structure systems, shipped knocked-down, complete with all necessary hardware. Labor cost is for assembly on site. Typical use is for picnic pavilions. These costs do not include foundation or floor slab. Dimensions are approximate outside measurements. Based on Americana Building Products. www.americana.com

Four-sided hip roof shelter. Steel structural rafters and columns, aluminum roof decking with stucco embossed, painted finish, 8' height at perimeter.

| | Craft@Hrs | Unit | Material | Labor | Total |
|---|---|---|---|---|---|
| 12' L x 12' W | S4@32.0 | Ea | 4,710.00 | 1,280.00 | 5,990.00 |
| 16' L x 16' W | S4@32.0 | Ea | 5,770.00 | 1,280.00 | 7,050.00 |
| 16' L x 32' W | S4@36.0 | Ea | 7,790.00 | 1,440.00 | 9,230.00 |

| | Craft@Hrs | Unit | Material | Labor | Total |
|---|---|---|---|---|---|
| 20' L x 20' W | S4@36.0 | Ea | 7,250.00 | 1,440.00 | 8,690.00 |
| 20' L x 30' W | S4@36.0 | Ea | 9,270.00 | 1,440.00 | 10,710.00 |
| 20' L x 40' W | S4@36.0 | Ea | 11,400.00 | 1,440.00 | 12,840.00 |
| 24' L x 24' W | S4@36.0 | Ea | 9,270.00 | 1,440.00 | 10,710.00 |
| 24' L x 48' W | S4@40.0 | Ea | 13,500.00 | 1,600.00 | 15,100.00 |
| Add for 48" x 48" cupola with tubular metal frame | S4@1.50 | Ea | 939.00 | 59.80 | 998.80 |
| Add for 6' x 12' cupola with tubular metal frame | S4@3.50 | Ea | 3,890.00 | 140.00 | 4,030.00 |

Gable roof shelter. Steel structural rafters and columns, aluminum roof and trim. 8' height, 3 in 12 roof pitch.

| | Craft@Hrs | Unit | Material | Labor | Total |
|---|---|---|---|---|---|
| 20' L x 20' W | S4@36.0 | Ea | 4,290.00 | 1,440.00 | 5,730.00 |
| 20' L x 24' W | S4@36.0 | Ea | 6,190.00 | 1,440.00 | 7,630.00 |
| 20' L x 30' W | S4@36.0 | Ea | 6,580.00 | 1,440.00 | 8,020.00 |
| 20' L x 50' W | S4@40.0 | Ea | 11,200.00 | 1,600.00 | 12,800.00 |
| 22' L x 20' W | S4@36.0 | Ea | 4,560.00 | 1,440.00 | 6,000.00 |
| 22' L x 30' W | S4@36.0 | Ea | 6,980.00 | 1,440.00 | 8,420.00 |
| 22' L x 40' W | S4@36.0 | Ea | 9,270.00 | 1,440.00 | 10,710.00 |
| 22' L x 50' W | S4@40.0 | Ea | 11,500.00 | 1,600.00 | 13,100.00 |

**Modular Buildings** (Manufactured housing)
Relocatable structures and institutional type housing for remote areas. Costs are per square foot under the roof. Based on 2" x 4" wood studs 16" OC, 1/2" gypsum wallboard inside and 1/2" CDX plywood outside, with minimum plumbing, heating and electrical systems for the intended use. Costs include factory assembly, delivery by truck to within 50 miles and setup on site. Add 1.5% for each additional 50 miles of delivery. No site preparation, earthwork, foundations or furnishings included. Cost for smaller, more complex structures will be higher.

Portable structures (temporary offices, school classrooms, etc.)

| | | Unit | Material | Labor | Total |
|---|---|---|---|---|---|
| Medium quality | — | SF | — | — | 53.90 |
| Better quality | — | SF | — | — | 66.70 |

Institutional housing, four-plex modules

| | | Unit | Material | Labor | Total |
|---|---|---|---|---|---|
| Medium quality | — | SF | — | — | 69.70 |
| Better quality | — | SF | — | — | 70.30 |

Construction or mining camp barracks, mess halls, kitchens, etc.

| | | Unit | Material | Labor | Total |
|---|---|---|---|---|---|
| Single story structures | — | SF | — | — | 61.20 |

**Air-Supported Structures** Air-supported pool or tennis court enclosures, polyester reinforced vinyl dome, with zippered entry air lock doors, cable tiedowns and blower, typical cost per square foot of area covered.

| | Craft@Hrs | Unit | Material | Labor | Total |
|---|---|---|---|---|---|
| Under 12,000 SF | D4@.030 | SF | 11.70 | 1.39 | 13.09 |
| 12,000 to 24,000 SF | D4@.030 | SF | 10.90 | 1.39 | 12.29 |
| 24,000 to 30,000 SF | D4@.030 | SF | 10.10 | 1.39 | 11.49 |
| 30,000 to 36,000 SF | D4@.030 | SF | 9.58 | 1.39 | 10.97 |
| 36,000 to 45,000 SF | D4@.030 | SF | 9.46 | 1.39 | 10.85 |
| 45,000 to 80,000 SF | D4@.030 | SF | 9.34 | 1.39 | 10.73 |
| 80,000 to 110,000 SF | D4@.030 | SF | 9.07 | 1.39 | 10.46 |

**Sunrooms, solariums and conservatories**. Custom-designed three-wall building additions supplied in kit form for assembly on site. Aluminum frame mounted on an existing 6" concrete slab with 12" wide x 18" deep footing. Moving glass panels have two adjustable tandem nylon wheel assemblies. Size is limited to 20'-wide roof span, 100' length and 18' height. Meets Architectural Aluminum Manufacturers Association (AAMA), Insulating Glass Certification Council (IGCC) and National Sunroom Association (NSA) standards. Costs include supervision of installation by a manufacturer's representative. Sunrooms, solariums and conservatories have distinct definitions under most building codes. Smaller units in high wind or snow load climates will cost more than larger units built in moderate climates.

# 13 Special Construction

| | Craft@Hrs | Unit | Material | Labor | Total |
|---|---|---|---|---|---|

Three-season sunrooms and solariums have one twin-door single-pivot or double-hung external door and a movable sliding glass egress panel, fixed glass roof and sidewall panels and one movable screen panel. Structural main frame sections are 4-1/4" extruded 6060, 6063, 6061, 6005 or 6105 aluminum alloy with "teardrop" drip channels. Glazed with either Lexan, GE Plastics polycarbonate or equal bronze- or blue-tinted or clear glass.

Four-season sunrooms, solariums and conservatories have, in addition, minimum 3" thick compound foam sealing inside structural tubular elements and twin-pane tempered glass either gas-filled or Dupont Butacite-interlayered.

Sunrooms (residential lean-to patio enclosure), three-season, no engineering required. Costs will vary with the wind load and snow load. Cost per square foot of floor.

| | Craft@Hrs | Unit | Material | Labor | Total |
|---|---|---|---|---|---|
| Minimum cost, per SF | — | SF | — | — | 125.00 |
| Typical maximum cost per SF | — | SF | — | — | 175.00 |
| Add for four-season sunrooms | — | % | — | — | 20.0 |

Residential three-season solarium. Custom-designed and engineered for the site. Cost per square foot of floor.

| | Craft@Hrs | Unit | Material | Labor | Total |
|---|---|---|---|---|---|
| Minimum cost per SF | — | SF | — | — | 240.00 |
| Typical maximum cost per SF | — | SF | — | — | 300.00 |

Residential four-season solarium, with electric service, HVAC, fireproofing and abrasion-resistant protective coating. Meets code requirements for a room addition. Cost per square foot of floor.

| | Craft@Hrs | Unit | Material | Labor | Total |
|---|---|---|---|---|---|
| Minimum cost per SF | — | SF | — | — | 400.00 |
| Typical maximum cost per SF | — | SF | — | — | 600.00 |

Residential conservatory (room addition). Four season. Stepped, bull-nose or gabled roof meets code requirements for glazed roof deflection and load. With electric service, HVAC, fireproofing, abrasion-resistant protective coating.

| | Craft@Hrs | Unit | Material | Labor | Total |
|---|---|---|---|---|---|
| Minimum cost per SF | — | SF | — | — | 500.00 |
| Typical maximum cost per SF | — | SF | — | — | 900.00 |

Commercial solarium, four season, with electric service, HVAC, fireproofing, abrasion-resistant protective coating, store-type doors, locksets and closers. A sidewalk café or retail facility will usually require high-impact (falling object) glazing, fire sprinklers and alarms. Cost per square foot of floor.

| | Craft@Hrs | Unit | Material | Labor | Total |
|---|---|---|---|---|---|
| Typical cost per SF | — | SF | — | — | 700.00 |
| Sidewalk café or retail application, cost per SF | — | SF | — | — | 1,500.00 |

Residential polycarbonate sheet glazing. 4' x 8' panels, Lexan, GE Plastics, or equal, for sunroom and 3-season service only

| | Craft@Hrs | Unit | Material | Labor | Total |
|---|---|---|---|---|---|
| Clear 1/4" | G1@.027 | SF | 11.20 | 1.24 | 12.44 |
| Clear 1/2" | G1@.035 | SF | 17.30 | 1.61 | 18.91 |
| Add for tints | — | % | 17.0 | — | — |
| Add for hurricane and/or heavy snow load zones | — | % | 25.0 | — | — |
| Add for continuous-length custom glazing | — | % | 20.0 | — | — |

Tempered glass, residential argon gas-filled twin-sheet Low-E glazing. PPG or equal, for 3-season and 4-season service.

| | Craft@Hrs | Unit | Material | Labor | Total |
|---|---|---|---|---|---|
| Clear, 5/8" | G1@.027 | SF | 18.00 | 1.24 | 19.24 |
| Clear, 1" | G1@.035 | SF | 19.70 | 1.61 | 21.31 |
| Add for tints | — | % | 17.0 | — | — |
| Add for hurricane and/or heavy snow load zones | — | % | 25.0 | — | — |
| Add for continuous-length custom glazing | — | % | 20.0 | — | — |

High-impact sheet glazing, four-season, for conservatory and urban commercial service (tempered Low-E, PPG or equal) twin pane with Dupont Butacite interlayer. Certified for hurricane and heavy snow load.

| | Craft@Hrs | Unit | Material | Labor | Total |
|---|---|---|---|---|---|
| Clear, 5/8" | G1@.035 | SF | 25.30 | 1.61 | 26.91 |
| Clear, 1" | G1@.045 | SF | 27.60 | 2.07 | 29.67 |
| Add for tints | — | % | 17.0 | — | — |
| Add for curved eaves | — | % | 5.0 | — | — |
| Add for continuous-length custom glazing | — | % | 20.0 | — | — |

# 13 Special Construction

|  | Craft@Hrs | Unit | Material | Labor | Total |
|---|---|---|---|---|---|
| Architectural aluminum structural elements | | | | | |
| Framing and joiner tube, 11 ga. x 1" x 3", | IW@.009 | Lb | 1.32 | .57 | 1.89 |
| Horizontal roof supports, 11 ga. x 4-1/4" x 4-1/4", | | | | | |
| with galvanized steel internal bracing | IW@.019 | Lb | 2.20 | 1.21 | 3.41 |
| Vertical sidewall supports, 10 ga. x 4-1/4 x 4-1/4", | | | | | |
| with galvanized steel internal bracing | IW@.061 | Lb | 14.30 | 3.88 | 18.18 |
| Add for 3" insulating foam (4-season units) | — | % | 15.0 | — | — |
| Concrete-embedded aluminum | | | | | |
| Light, to 20 lbs per LF | T3@.042 | Lb | 2.68 | 2.06 | 4.74 |
| Medium, over 20 to 50 lbs per LF | T3@.035 | Lb | 2.42 | 1.72 | 4.14 |
| Heavy, over 50 lbs per LF | T3@.025 | Lb | 2.05 | 1.23 | 3.28 |
| Channel sill, 1/8" aluminum, 1-1/2" x 8" wide | T3@.131 | LF | 21.00 | 6.43 | 27.43 |
| Angle sill, 1/8" aluminum, 1-1/2" x 1-1/2" | T3@.095 | LF | 6.61 | 4.66 | 11.27 |
| Solariums or conservatory foundation, including forms, 3,000 p.s.i. concrete and finishing | | | | | |
| Footing 12" wide x 18" deep (0.058 CY per LF) | B4@.104 | LF | 8.50 | 3.65 | 12.15 |
| Slab 6" thick, per SF or floor area | B5@.071 | SF | 3.41 | 2.52 | 5.93 |
| Mount pre-fabricated aluminum frame | IW@.061 | Lb | 14.30 | 3.88 | 18.18 |
| Anchor bolt and leveling each 10' wall section | D4@4.00 | Ea | 200.00 | 185.00 | 385.00 |
| Solarium or conservatory HVAC ducting | | | | | |
| Duct to the attached building, including plenums, crosses, tees, | | | | | |
| and air handler tie-in, per LF of duct | T6@0.03 | LF | .75 | 1.57 | 2.32 |
| Mount and connect electric service and lighting for a solarium or conservatory, including main breaker box tie-in | | | | | |
| Internal lighting, electric service, | | | | | |
| and wall switches, per LF of conduit | E2@.184 | LF | .35 | 9.04 | 9.39 |
| Fourth wall fabrication for a solarium or conservatory, including support frame fabrication and joist bolstering, with caulking and galvanized rain seal flashing | | | | | |
| Cut and fabricate wall opening | P9@.05 | SF | .75 | 2.36 | 3.11 |
| Add for permitting and inspection | — | % | — | 3.0 | — |

| | Craft@Hrs | Unit | Material | Labor | Equipment | Total |
|---|---|---|---|---|---|---|
| **Pre-Engineered Steel Buildings** 26 gauge colored galvanized steel roof and siding with 4 in 12 (20 pound live load) roof. Cost per SF of floor area. These costs do not include foundation or floor slab. Add delivery cost to the site. Equipment is a 15 ton hydraulic crane and a 2-ton truck equipped for work of this type. | | | | | | |
| 40' x 100' (4,000 SF) | | | | | | |
| 14' eave height | H5@.074 | SF | 4.62 | 3.88 | 1.20 | 9.70 |
| 16' eave height | H5@.081 | SF | 5.20 | 4.25 | 1.31 | 10.76 |
| 20' eave height | H5@.093 | SF | 5.87 | 4.88 | 1.50 | 12.25 |
| 60' x 100' (6,000 SF) | | | | | | |
| 14' eave height | H5@.069 | SF | 4.47 | 3.62 | 1.11 | 9.20 |
| 16' eave height | H5@.071 | SF | 4.56 | 3.72 | 1.15 | 9.43 |
| 20' eave height | H5@.081 | SF | 5.29 | 4.25 | 1.31 | 10.85 |
| 80' x 100' (8,000 SF) | | | | | | |
| 14' eave height | H5@.069 | SF | 4.38 | 3.62 | 1.11 | 9.11 |
| 16' eave height | H5@.071 | SF | 4.55 | 3.72 | 1.15 | 9.42 |
| 20' eave height | H5@.080 | SF | 5.29 | 4.19 | 1.29 | 10.77 |
| 100' x 100' (10,000 SF) | | | | | | |
| 14' eave height | H5@.067 | SF | 4.34 | 3.51 | 1.08 | 8.93 |
| 16' eave height | H5@.071 | SF | 4.55 | 3.72 | 1.15 | 9.42 |
| 20' eave height | H5@.080 | SF | 5.09 | 4.19 | 1.29 | 10.57 |
| 100' x 150' (15,000 SF) | | | | | | |
| 14' eave height | H5@.065 | SF | 3.72 | 3.41 | 1.05 | 8.18 |
| 16' eave height | H5@.069 | SF | 3.96 | 3.62 | 1.11 | 8.69 |
| 20' eave height | H5@.074 | SF | 4.39 | 3.88 | 1.20 | 9.47 |

# 13 Special Construction

| | Craft@Hrs | Unit | Material | Labor | Equipment | Total |
|---|---|---|---|---|---|---|
| **100' x 200' (20,000 SF)** | | | | | | |
| 14' eave height | H5@.063 | SF | 3.63 | 3.30 | 1.02 | 7.95 |
| 16' eave height | H5@.066 | SF | 3.80 | 3.46 | 1.07 | 8.33 |
| 20' eave height | H5@.071 | SF | 4.10 | 3.72 | 1.15 | 8.97 |
| **140' x 150' (21,000 SF)** | | | | | | |
| 14' eave height | H5@.058 | SF | 3.36 | 3.04 | .94 | 7.34 |
| 16' eave height | H5@.059 | SF | 3.48 | 3.09 | .95 | 7.52 |
| 20' eave height | H5@.066 | SF | 3.88 | 3.46 | 1.07 | 8.41 |
| **140' x 175' (24,500 SF)** | | | | | | |
| 14' eave height | H5@.057 | SF | 3.24 | 2.99 | .92 | 7.15 |
| 16' eave height | H5@.058 | SF | 3.36 | 3.04 | .94 | 7.34 |
| 20' eave height | H5@.065 | SF | 3.72 | 3.41 | 1.05 | 8.18 |
| **160' x 200' (32,000 SF)** | | | | | | |
| 14' eave height | H5@.055 | SF | 3.16 | 2.88 | 1.89 | 7.93 |
| 16' eave height | H5@.056 | SF | 3.24 | 2.94 | 1.90 | 8.08 |
| 20' eave height | H5@.061 | SF | 3.57 | 3.20 | .99 | 7.76 |
| **200' x 200' (40,000 SF)** | | | | | | |
| 14' eave height | H5@.052 | SF | 3.05 | 2.73 | .84 | 6.62 |
| 16' eave height | H5@.053 | SF | 3.13 | 2.78 | .86 | 6.77 |
| 20' eave height | H5@.058 | SF | 3.36 | 3.04 | .94 | 7.34 |
| **Additional costs:** | | | | | | |
| Personnel door, 3'0" x 6'8" | H5@2.83 | Ea | 580.00 | 148.00 | 45.70 | 773.70 |
| Overhead door, 8'0" x 10'0" | H5@8.02 | Ea | 1,420.00 | 420.00 | 130.00 | 1,970.00 |
| Interior liner panel, 26 gauge<br>  Per SF of wall | H5@.012 | SF | 1.56 | .63 | .19 | 2.38 |
| Foam insulated sandwich panel roofing and siding<br>  Per SF of roof and sidewall | H5@.005 | SF | 5.53 | .26 | .08 | 5.87 |
| Plastic skylight roof panel<br>  Per SF of panel | H5@.013 | SF | 4.79 | .68 | .21 | 5.68 |
| R-11 blanket insulation,<br>  3-1/2", wall or floor | H5@.005 | SF | .37 | .26 | .08 | .71 |
| Ridge ventilator, 9" throat | H5@.193 | LF | 46.60 | 10.10 | 3.12 | 59.82 |
| 24 gauge colored eave gutter | H5@.028 | LF | 7.47 | 1.47 | .45 | 9.39 |
| 4" x 4" colored downspouts | H5@.014 | LF | 6.66 | .73 | .23 | 7.62 |
| 30 pound live load roof design | — | % | 12.0 | — | — | — |

| | Craft@Hrs | Unit | Material | Labor | Total |
|---|---|---|---|---|---|

**X-Ray Viewing Panels, Clear Lead-Plastic** Based on CLEAR-Pb, by Cardinal Health. Based on SF of panel (rounded up to the next higher whole square foot). For panels 12 square feet or larger, add a crating charge. Panels larger than 72" x 96" are special order items. Weights shown are approximate.

| | Craft@Hrs | Unit | Material | Labor | Total |
|---|---|---|---|---|---|
| Crating charge | — | Ea | — | — | 95.10 |
| 7mm thick, 0.3mm lead equivalence, 2.3 lbs/SF | G1@.115 | SF | 144.00 | 5.29 | 149.29 |
| 12mm thick, 0.5mm lead equivalence, 3.9 lbs/SF | G1@.195 | SF | 194.00 | 8.97 | 202.97 |
| 18mm thick, 0.8mm lead equivalence, 5.9 lbs/SF | G1@.294 | SF | 210.00 | 13.50 | 223.50 |
| 22mm thick, 1.0mm lead equivalence, 7.2 lbs/SF | G1@.361 | SF | 223.00 | 16.60 | 239.60 |
| 35mm thick, 1.5mm lead equivalence, 11.5 lbs/SF | G1@.574 | SF | 242.00 | 26.40 | 268.40 |
| 46mm thick, 2.0mm lead equivalence, 15.0 lbs/SF | G1@.751 | SF | 328.00 | 34.50 | 362.50 |
| 70mm thick, 3.0mm lead equivalence, 18.0 lbs/SF | G1@1.00 | SF | 463.00 | 46.00 | 509.00 |

|  | Craft@Hrs | Unit | Material | Labor | Total |
|---|---|---|---|---|---|

**Mobile X-Ray Barriers** Clear lead-plastic window panels on the upper portion and opaque panels on the lower portion, mounted within a framework with casters on the bottom. Based on Cardinal Health. Labor shown is to uncrate factory assembled barrier and attach casters.

30" W x 75" H overall

| | | | | | |
|---|---|---|---|---|---|
| 0.5mm lead equiv. window panel 30" x 24" and 0.8mm lead equiv. opaque panel 30" x 48" | MW@.501 | Ea | 2,150.00 | 26.40 | 2,176.40 |

48" W x 75" H overall

| | | | | | |
|---|---|---|---|---|---|
| 0.5mm lead equiv. window panel 48" x 36" and 0.8mm lead equiv. opaque panel 48" x 36" | MW@.501 | Ea | 4,040.00 | 26.40 | 4,066.40 |
| 1.0mm lead equiv. window panel 48" x 36" and 1.5mm lead equiv. opaque panel 48" x 36" | MW@.501 | Ea | 4,780.00 | 26.40 | 4,806.40 |

72" W x 75" H, overall

| | | | | | |
|---|---|---|---|---|---|
| 0.5mm lead equiv. window panel 72" x 36" and 0.8mm lead equiv. opaque panel 72" x 36" | MW@.751 | Ea | 4,770.00 | 39.50 | 4,809.50 |
| 1.0mm lead equiv. window panel 72" x 36" and 1.5mm lead equiv. opaque panel 72" x 36" | MW@.751 | Ea | 5,960.00 | 39.50 | 5,999.50 |

**Modular X-Ray Barriers** Panels are mounted within a framework for attaching to floor, wall or ceiling. Shipped unassembled. Based on Cardinal Health. Structural supports not included. Costs shown are based on typical 36" wide x 84" high panel sections. Clear lead-plastic window panels 48" high are the upper portion and opaque leaded panels 36" high are the bottom portion.

1-section barrier, 36" W x 84" H overall

| | | | | | |
|---|---|---|---|---|---|
| 0.5mm lead equiv. panels | G1@2.78 | Ea | 3,320.00 | 128.00 | 3,448.00 |
| 0.8mm lead equiv. panels | G1@2.78 | Ea | 3,500.00 | 128.00 | 3,628.00 |
| 1.0mm lead equiv. panels | G1@3.34 | Ea | 3,590.00 | 154.00 | 3,744.00 |
| 1.5mm lead equiv. panels | G1@4.46 | Ea | 3,880.00 | 205.00 | 4,085.00 |

2-section barrier, 72" W x 84" H overall

| | | | | | |
|---|---|---|---|---|---|
| 0.5mm lead equiv. panels | G1@5.56 | Ea | 7,050.00 | 256.00 | 7,306.00 |
| 0.8mm lead equiv. panels | G1@5.56 | Ea | 7,430.00 | 256.00 | 7,686.00 |
| 1.0mm lead equiv. panels | G1@6.68 | Ea | 7,590.00 | 307.00 | 7,897.00 |
| 1.5mm lead equiv. panels | G1@8.90 | Ea | 8,180.00 | 409.00 | 8,589.00 |

3-section barrier, 108" W x 84" H overall

| | | | | | |
|---|---|---|---|---|---|
| 0.5mm lead equiv. panels | G1@8.34 | Ea | 10,800.00 | 384.00 | 11,184.00 |
| 0.8mm lead equiv. panels | G1@8.34 | Ea | 11,400.00 | 384.00 | 11,784.00 |
| 1.0mm lead equiv. panels | G1@10.0 | Ea | 11,700.00 | 460.00 | 12,160.00 |
| 1.5mm lead equiv. panels | G1@13.4 | Ea | 12,600.00 | 616.00 | 13,216.00 |

Larger than 3-section barriers, add to the cost of 1-section barriers for each section over 3

| | | | | | |
|---|---|---|---|---|---|
| Add per 18" W x 84" H section | — | % | 50.0 | — | — |
| Add per 36" W x 84" H section | — | % | 100.0 | — | — |

**Convenience Store Specialties**
Convenience store display shelving, open glass with stainless steel frame, shell and shelves. Fluorescent indirect lighting on unit rear wall, 60 Hz, single phase, 110 Volt. Prehung slide-in label and pricing panels. Underwriters Laboratories and CSA approved.

| | | | | | |
|---|---|---|---|---|---|
| 38" wide 2-tower unit | C8@.250 | Ea | 957.00 | 11.50 | 968.50 |
| 72" wide 4-tower unit | C8@.350 | Ea | 1,320.00 | 16.00 | 1,336.00 |

| | Craft@Hrs | Unit | Material | Labor | Total |
|---|---|---|---|---|---|

Convenience store external signage, all-weather internally illuminated, freestanding, with internal fluorescent lighting fixtures, custom painted Lucite 48" x 72" twin-sided sign panel, structural aluminum welded frame and retainer moldings, 30 foot high by 16" diameter structural brushed and anodized aluminum vertical mounting post and reinforced base plate. Includes photocell-actuated night switch and in-store manual actuation override fused control panel. Add the cost of concrete pad, trenching, underground wiring from the electrical panel and crane rental.

| | Craft@Hrs | Unit | Material | Labor | Total |
|---|---|---|---|---|---|
| Metal base plate | C8@1.50 | Ea | 600.00 | 68.70 | 668.70 |
| Post and sign, installed | C8@3.00 | Ea | 5,780.00 | 137.00 | 5,917.00 |
| Sign controls | C8@8.00 | Ea | 300.00 | 367.00 | 667.00 |

High-traffic convenience store aluminum twin-door portal, steel frame, pre-hung, with pneumatic adjustable closers, security deadbolt lockset, push bars both sides or "U"-section push/pull handgrips, swing-down swivel hold-open stanchions, clear glass single pane high impact tempered door glass.

| | Craft@Hrs | Unit | Material | Labor | Total |
|---|---|---|---|---|---|
| Store glass double door, per pair of doors | SW@8.00 | Ea | 2,420.00 | 328.00 | 2,748.00 |

Convenience store display window security grillwork. After-hours anti-forced entry lockable grillwork, painted carbon steel horizontally-expanding accordion type, shop fabricated, 8' high by 20' long, with anchored tracking bars top and bottom.

| | Craft@Hrs | Unit | Material | Labor | Total |
|---|---|---|---|---|---|
| Security grillwork, 8' x 20' | B9@8.00 | Ea | 845.00 | 261.00 | 1,106.00 |
| Security grillwork, per SF | B9@0.05 | SF | 5.28 | 1.63 | 6.91 |

Convenience store bulletproof cashier booth (enclosure only installed in an existing building), single panel, counter-mounted, with product/register transfer tray ("lazy Susan" type), glazed with ballistic Lucite, 5' high by 8' long.

| | Craft@Hrs | Unit | Material | Labor | Total |
|---|---|---|---|---|---|
| Bulletproof cashier booth | B9@4.00 | Ea | 1,160.00 | 131.00 | 1,291.00 |

**Modular prefabricated church steeple**  For houses of worship seating from 500 to 1,500 congregants. Includes 16' x 16' steeple base bolted and reinforced on a steel or wood frame, matched base plate steeple element, site-specific roof-mounted 1-1/2" diameter bolted joining plate with roof rafter and purlin support joists, lightning arrestor and #8 aluminum grounding cable with dielectric zinc grounding anchor. Labor cost includes laser leveling, vendor-supplied wind studies and roof-related load-bearing studies. Equipment cost is for a five-ton tracked spider crane and flatbed trailer truck with operator and truck driver. Add the cost of carillon speakers, special lighting, anti-bird/anti-bat nesting measures and internal access ladders, if required.

| | Craft@Hrs | Unit | Material | Labor | Equipment | Total |
|---|---|---|---|---|---|---|
| Building of roof base plate and plate support | F5@40.0 | Ea | 3,680.00 | 1,880.00 | — | 5,560.00 |
| Church steeple installation | | | | | | |
| Steeple, bolted in and shimmed | S5@24.0 | Ea | 15,800.00 | 1,030.00 | 1,800.00 | 18,630.00 |
| Roof flashing and roof edge seal | R3@16.0 | Ea | 840.00 | 727.00 | — | 1,567.00 |
| Painting of steeple and trim | D7@40.0 | Ea | 315.00 | 1,830.00 | — | 2,145.00 |

**Decorative ornamental fountains,** interior (for shopping center and office complex atriums) and exterior (for public parks and large educational and industrial campuses). Costs are for nominal 1,300 gallon capacity 16' diameter by 1' deep precast concrete circular water reservoir pond with 6-foot diameter fountain centerpiece and spigot matrix. Costs include 110 volt, 3 phase 60 Hz 5 GPM pumps, integrated drains and recirculation piping with digital adjustable flow regulator. Add the cost of foundation and gravel sub-foundation, custom decorative bas relief fascia for reservoir pond sidewalls, fencing, associated decorative ironwork and circular walkway esplanade. Equipment cost is for a forklift for two days. No electrical service or foundation costs included.

| | Craft@Hrs | Unit | Material | Labor | Equipment | Total |
|---|---|---|---|---|---|---|
| Positioning, anchoring and Gunite sealing of pond sidewalls | CM@40.0 | Ea | 14,300.00 | 2,010.00 | 250.00 | 16,560.00 |
| Bolting and wiring of pump and controller | CE@8.00 | Ea | 1,470.00 | 467.00 | — | 1,937.00 |
| Startup and adjustment of fountain pump | CE@2.00 | Ea | — | 117.00 | — | 117.00 |

# 13 Special Construction

| | Craft@Hrs | Unit | Material | Labor | Equipment | Total |
|---|---|---|---|---|---|---|

**Traffic signal** 20-foot horizontal trombone mount, with controller, emergency default override, uninterruptible power system battery pack, LED low-voltage signal lighting elements. 15-foot vertical aluminum pole mount and cast aluminum base. Meets NHTSA standards. Add the cost of breaking out existing concrete and running electric power to the site. Equipment cost is for a backhoe and stinger truck crane. Per traffic signal installed. No electrical service or foundation costs included.

| | Craft@Hrs | Unit | Material | Labor | Equipment | Total |
|---|---|---|---|---|---|---|
| Craning in, bolting and shimming | | | | | | |
|    Traffic light pole and base | B5@4.00 | Ea | 14,700.00 | 142.00 | 250.00 | 15,092.00 |
|    Electrical wiring and controls | CE@8.00 | Ea | 1,890.00 | 467.00 | — | 2,357.00 |
| Calibration and test | CE@4.00 | Ea | — | 233.00 | — | 233.00 |

**Pedestrian street crossing signal,** right angle bi-directional, with LED low-wattage signal lamps, timing sequence box, SCADA or equal interconnect with central traffic controller module, with aluminum post and cast aluminum base, pedestrian high-impact manual button switches, to NHTSA standards. Add the cost of removing the existing concrete and bringing electric power to the site. Equipment is a backhoe and a stinger truck crane. Per signal installed. No electrical service or foundation costs included.

| | Craft@Hrs | Unit | Material | Labor | Equipment | Total |
|---|---|---|---|---|---|---|
| Craning in, bolting and shimming | | | | | | |
|    of crosswalk light and base | B5@8.00 | Ea | 8,400.00 | 284.00 | 250.00 | 8,934.00 |
|    Fixtures plus assembly | CE@4.00 | Ea | 1,370.00 | 233.00 | — | 1,603.00 |
|    Wiring of main breaker and timer | CE@4.00 | Ea | 1,260.00 | 233.00 | — | 1,493.00 |
|    Calibration and test | CE@4.00 | Ea | — | 233.00 | — | 233.00 |

**Street lighting,** metal halide and high-pressure sodium twin luminaries mounted on twin 15-foot trombone horizontal extensions. Aluminum post with cast aluminum base and photocell switch actuator. Add the cost removing any existing concrete surface and excavation. Equipment is a backhoe and a stinger truck crane. Add the cost of the foundation, electrical wiring and conduit.

| | Craft@Hrs | Unit | Material | Labor | Equipment | Total |
|---|---|---|---|---|---|---|
| Craning in and positioning | | | | | | |
|    of pole and base | B5@2.00 | Ea | 14,200.00 | 70.90 | 250.00 | 14,520.90 |
|    Wiring of controls and power | CE@8.00 | Ea | 1,580.00 | 467.00 | — | 2,047.00 |
|    Startup and adjustment | CE@4.00 | Ea | — | 233.00 | — | 233.00 |

**Public park surveillance** CCTV configuration, street lamp mounted, with 4-station DVD/CD recorder, 4-station low-light digital camera units, timed and dated image software, all-weather enclosure, power supply and control panel.

| | Craft@Hrs | Unit | Material | Labor | Equipment | Total |
|---|---|---|---|---|---|---|
|    Lamp-pole mount of camera | B5@4.00 | Ea | 630.00 | 142.00 | — | 772.00 |
|    Power supply positioning | CE@2.00 | Ea | 525.00 | 117.00 | — | 642.00 |
|    Wiring of controls and digital links | CE@2.00 | Ea | 420.00 | 117.00 | — | 537.00 |
|    Startup and adjustment | CE@1.00 | Ea | — | 58.40 | — | 58.40 |

**Subway platform egress barrier** Pedestrian safety barrier, grade 304L stainless steel, with bolted mounting plates, electrical gate closure, master control switch relay and manual override. To American Railroad Association standards. Add the cost of removing any existing concrete surface, electrical service and foundation.

| | Craft@Hrs | Unit | Material | Labor | Equipment | Total |
|---|---|---|---|---|---|---|
|    Gate positioning and bolting | B5@2.00 | Ea | 3,780.00 | 70.90 | — | 3,850.90 |
|    Wiring of controls | CE@2.00 | Ea | 263.00 | 117.00 | — | 380.00 |
|    Startup and adjustment | CE@1.00 | Ea | — | 58.40 | — | 58.40 |

**Pedestrian walkway bridge** Over-street fully screened pedestrian walkway bridge. Pratt, Howe or Warren type double diagonal truss bridge span element, with open grid weathering steel metal walkway, 60-foot span with twin level staircases on both ends of bridge and galvanized mesh sidewall traffic protection screening. Complies with Federal Highway Association and National Steel Bridge Association design standards. Equipment includes a backhoe and a stinger truck crane. Add the foundation cost.

| | Craft@Hrs | Unit | Material | Labor | Equipment | Total |
|---|---|---|---|---|---|---|
|    Foundation plate/upright crane-in | B5@24.0 | Ea | 137,000.00 | 851.00 | 500.00 | 138,351.00 |
|    Craning in, bolt-up and weld of span | B5@16.0 | Ea | 446,000.00 | 567.00 | 750.00 | 447,317.00 |
|    Tensioning and adjustment of | | | | | | |
|      stressed diagonal brace elements | B5@24.0 | Ea | 1,580.00 | 851.00 | — | 2,431.00 |

# 13 Special Construction

| | Craft@Hrs | Unit | Material | Labor | Equipment | Total |
|---|---|---|---|---|---|---|
| Sandblast and paint structure | PT@16.0 | Ea | 15,000.00 | 596.00 | 600.00 | 16,196.00 |
| Mounting of protective screening | B5@24.0 | Ea | 105,000.00 | 851.00 | — | 105,851.00 |
| Inspection and test | MW@16.0 | Ea | — | 842.00 | — | 842.00 |

| | Craft@Hrs | Unit | Material | Labor | Total |
|---|---|---|---|---|---|
| **Decorative and ornamental lighting** for convention centers, religious, educational and entertainment venues. Costs include vendor-supplied computer simulated lighting studies, concrete anchoring, trenching of electrical cable, main breaker, adjustable lighting array switch with digital timing actuation, calibration and aiming of installed fixtures. Add the cost of electrical service and the foundation. | | | | | |
| Bolting of fixtures | E4@0.50 | Ea | 315.00 | 24.60 | 339.60 |
| Wiring of main breaker and timer | CE@4.00 | Ea | 210.00 | 233.00 | 443.00 |
| Light aiming and calibration | E4@0.50 | Ea | — | 24.60 | 24.60 |
| **Subway and metro traffic control egress portals** Includes swipe card reader, three-prong turnstile, twin 304 stainless steel stall median with maintenance doors, wiring to central controller module and lockset. Add the cost of electrical service and the foundation. | | | | | |
| Positioning of equipment | CE@2.00 | Ea | 4,730.00 | 117.00 | 4,847.00 |
| Wiring of controls | CE@4.00 | Ea | 105.00 | 233.00 | 338.00 |
| Startup and adjustment | CE@1.00 | Ea | — | 58.40 | 58.40 |
| **Magnetic metal detector** for arch-type pedestrian traffic portals for schools, courthouses, airports and public safety facilities. Includes typical power supply wiring, permanent grouted and leveled cement foundation, trenched wiring, calibration, test, and operating personnel training and licensing. Installed to Department of Homeland Security and International Air Traffic Association specifications and standards. No electrical service or foundation costs included. | | | | | |
| Trenching of prewired conduit | E4@5.00 | CLF | 50.00 | 246.00 | 296.00 |
| Pouring and grouting of foundation | CM@4.00 | Ea | 200.00 | 201.00 | 401.00 |
| Equipment and positioning | CE@2.00 | Ea | 14,300.00 | 117.00 | 14,417.00 |
| Wiring of controls | CE@4.00 | Ea | 105.00 | 233.00 | 338.00 |
| Startup and adjustment | CE@1.00 | Ea | — | 58.40 | 58.40 |
| **MRI parcel inspection station** Magnetic resonance interference baggage inspection station for airports and public buildings. Includes foot-pedal-controlled baggage conveyor, inspection device with monitor screen, alarm matrix with emergency intercom, permanent grouted and leveled cement foundation, calibration, test and training of operation personnel. Installed to Department of Homeland Security and International Air Traffic Association specifications and standards. Add the cost of electrical service and the foundation. | | | | | |
| Equipment and positioning | CE@2.00 | Ea | 137,000.00 | 117.00 | 137,117.00 |
| Wiring of controls | CE@2.00 | Ea | 263.00 | 117.00 | 380.00 |
| Startup and adjustment | CE@1.00 | Ea | — | 58.40 | 58.40 |
| **Commercial high-definition digital cable-based FCC-licensed studio** Facilities for production, post-production (editing) and broadcast. Includes architectural modifications to wall, floor and ceiling elements of a pre-existing office or warehouse structure to address standards-controlled sound mitigation and electronic spurious emissions limitation requirements. Add the cost of industry-specific television production equipment such as television cameras, mounts, lighting, microphones, recording equipment and esthetic trim such as paint and wall coverings in administration areas. Design parameters reflect compliance with standards specified by the Audio Engineering Society, Institute of Broadcast Sound, American Federation of TV and Recording Artists and the Society of Motion Picture and Television Engineers. | | | | | |
| Wall modifications, soundproofing | | | | | |
| Staggered studs | C8@.021 | SF | 0.65 | .96 | 1.61 |
| Insulation | C8@.006 | SF | 0.53 | .27 | .80 |
| Specialty perfboard | C8@.007 | SF | 0.79 | .32 | 1.11 |

# 13 Special Construction

| | Craft@Hrs | Unit | Material | Labor | Total |
|---|---|---|---|---|---|
| Ceiling modifications, soundproofing | | | | | |
| Staggered studs | C8@.022 | SF | 0.65 | 1.01 | 1.66 |
| Insulation | C8@.008 | SF | 0.53 | .37 | .90 |
| Specialty perfboard | C8@.009 | SF | 0.79 | .41 | 1.20 |
| Floor modifications (over existing slab) | | | | | |
| Prefab soundproof floor panels | FL@.020 | SF | 2.10 | 1.01 | 3.11 |
| Acoustical mahogany parquet | FL@.020 | SF | 3.94 | 1.01 | 4.95 |
| Prefab control room modules | | | | | |
| Leitch Technology Corporation or equal | CC@.030 | SF | 7.88 | 1.55 | 9.43 |
| Scrim and lighting tracks, less scrims and lighting | E4@1.50 | Ea | 788.00 | 73.70 | 861.70 |
| Electrical service | | | | | |
| Sunken outlet trays, 480 volt, three phase, 60 Hz | E4@1.00 | Ea | 315.00 | 49.10 | 364.10 |
| Main electrical switchgear panel, 500 amp | E4@23.2 | Ea | 4,600.00 | 1,140.00 | 5,740.00 |
| Ground fault and lightning protection | E4@16.0 | Ea | 1,260.00 | 786.00 | 2,046.00 |
| Electronic control panel support framing and mounts | | | | | |
| Wood control panel supports, custom | C8@.256 | LF | 78.10 | 11.70 | 89.80 |
| Prefab, custom, 20 gauge galvanized steel | D4@.785 | LF | 143.00 | 36.30 | 179.30 |
| Control room soundproof glazing | G1@.086 | SF | 24.10 | 3.96 | 28.06 |
| Uninterruptible power source | | | | | |
| Engine generator, 500 KW | E4@60.0 | Ea | 131,000.00 | 2,950.00 | 133,950.00 |
| Radio frequency emissions screening | | | | | |
| Copper grounded | E4@1.16 | Ea | 32.20 | 57.00 | 89.20 |
| Electrostatic discharge protective grounding | E4@1.13 | Ea | 29.50 | 55.50 | 85.00 |
| Digital cable enclosure tray | | | | | |
| and connector panel | BE@0.50 | Ea | 117.00 | 19.80 | 136.80 |
| Animation studio | | | | | |
| Multiple workstation cubicle gallery | C8@.028 | SF | 11.90 | 1.28 | 13.18 |

# 14 Conveying Equipment

**Dumbwaiters** These costs do not include allowance for constructing the hoistway walls or electrical work.
Manual, 2 stop, 25 feet per minute, no doors, by rated capacity, 24" x 24" x 36" high car

| | Craft@Hrs | Unit | Material | Labor | Total |
|---|---|---|---|---|---|
| 25 pounds to 50 pounds | CV@39.8 | Ea | 2,010.00 | 2,130.00 | 4,140.00 |
| 75 pounds to 200 pounds | CV@42.7 | Ea | 2,720.00 | 2,290.00 | 5,010.00 |
| Add for each additional stop | — | Ea | — | — | 1,380.00 |
| Electric, with machinery mounted above, floor loading, no security gates included | | | | | |
| 50 lbs, 25 FPM, 2 stop, no doors | CV@42.2 | Ea | 2,380.00 | 2,260.00 | 4,640.00 |
| 50 lbs, 25 FPM, 2 stop, manual doors | CV@106. | Ea | 5,900.00 | 5,680.00 | 11,580.00 |
| 50 lbs, 50 FPM, 2 stop, manual doors | CV@107. | Ea | 5,940.00 | 5,730.00 | 11,670.00 |
| 75 lbs, 25 FPM, 2 stop, no doors | CV@42.2 | Ea | 2,380.00 | 2,260.00 | 4,640.00 |
| 75 lbs, 25 FPM, 2 stop, manual doors | CV@107. | Ea | 7,140.00 | 5,730.00 | 12,870.00 |
| 75 lbs, 50 FPM, 2 stop, manual doors | CV@111. | Ea | 7,160.00 | 5,940.00 | 13,100.00 |
| 100 lbs, 25 FPM, 2 stop, no doors | CV@50.6 | Ea | 3,470.00 | 2,710.00 | 6,180.00 |
| 100 lbs, 25 FPM, 2 stop, manual doors | CV@107. | Ea | 7,340.00 | 5,730.00 | 13,070.00 |
| 100 lbs, 50 FPM, 2 stop, manual doors | CV@111. | Ea | 7,500.00 | 5,940.00 | 13,440.00 |
| 100 lbs, 100 FPM, 5 stop, manual doors | CV@162. | Ea | 7,870.00 | 8,680.00 | 16,550.00 |
| 200 lbs, 25 FPM, 2 stop, no doors | CV@52.3 | Ea | 3,560.00 | 2,800.00 | 6,360.00 |
| 200 lbs, 25 FPM, 2 stop, manual doors | CV@111. | Ea | 7,500.00 | 5,940.00 | 13,440.00 |
| 200 lbs, 100 FPM, 5 stop, manual doors | CV@162. | Ea | 9,180.00 | 8,680.00 | 17,860.00 |

| | Craft@Hrs | Unit | Material | Labor | Total |
|---|---|---|---|---|---|
| 300 lbs, 50 FPM, 2 stop, manual doors | CV@113. | Ea | 15,600.00 | 6,050.00 | 21,650.00 |
| 300 lbs, 100 FPM, 5 stop, manual doors | CV@171. | Ea | 20,800.00 | 9,160.00 | 29,960.00 |
| 400 lbs, 50 FPM, 2 stop, manual doors | CV@114. | Ea | 15,800.00 | 6,100.00 | 21,900.00 |
| 400 lbs, 100 FPM, 5 stop, manual doors | CV@174. | Ea | 23,000.00 | 9,320.00 | 32,320.00 |
| 500 lbs, 50 FPM, 2 stop, manual doors | CV@117. | Ea | 16,400.00 | 6,270.00 | 22,670.00 |
| 500 lbs, 100 FPM, 5 stop, manual doors | CV@178. | Ea | 24,200.00 | 9,530.00 | 33,730.00 |

**Elevators, Passenger** Typical subcontract costs excluding the shaftwall.

Hydraulic, office type

100 FPM, 10 to 13 passenger, automatic exit door, 7' x 5' cab, 2,500 lb capacity

| | | | | | |
|---|---|---|---|---|---|
| 3 stop | — | LS | — | — | 60,700.00 |
| 4 stop | — | LS | — | — | 65,600.00 |
| Each additional stop over 4 | — | Ea | — | — | 5,870.00 |

150 FPM, 10 to 13 passenger, center opening automatic door, 7' x 5' cab, 2,500 lb capacity

| | | | | | |
|---|---|---|---|---|---|
| 3 stop | — | LS | — | — | 64,400.00 |
| 4 stop | — | LS | — | — | 68,200.00 |
| Each additional stop over 4 | — | Ea | — | — | 8,380.00 |

Geared, 350 FPM, 3,500 lb automatic, selective collective, 5' x 8' cab

| | | | | | |
|---|---|---|---|---|---|
| 5 stop | — | LS | — | — | 108,000.00 |
| 6 stop | — | LS | — | — | 117,000.00 |
| 10 stop | — | LS | — | — | 140,000.00 |
| 15 stop | — | LS | — | — | 188,000.00 |
| Each additional stop | — | Ea | — | — | 7,170.00 |
| Add for 5,000 lb capacity | — | LS | — | — | 3,770.00 |

Gearless, 500 FPM, 3,500 lb, 6' x 9' cab

| | | | | | |
|---|---|---|---|---|---|
| 10 stop | — | LS | — | — | 223,000.00 |
| 15 stop | — | LS | — | — | 262,000.00 |
| Add for each additional stop | — | Ea | — | — | 6,310.00 |

Gearless, 700 FPM, 4,500 lb, 6' x 9' cab, passenger

| | | | | | |
|---|---|---|---|---|---|
| 10 stop | — | LS | — | — | 242,000.00 |
| 15 stop | — | LS | — | — | 274,000.00 |
| 20 stop | — | LS | — | — | 303,000.00 |
| Add for each additional stop | — | Ea | — | — | 7,540.00 |

Gearless, 1,000 FPM, 4,500 lb, 6' x 9' cab, passenger

| | | | | | |
|---|---|---|---|---|---|
| 10 stop | — | LS | — | — | 258,000.00 |
| Add for each additional stop | — | Ea | — | — | 9,930.00 |
| Sidewalk elevators, 2 stops, 2,500 pounds | — | LS | — | — | 45,400.00 |

**Elevators, Freight** Typical subcontract costs excluding the shaftwall.

Hydraulic, 2,500 pound capacity, 100 FPM, single entry cab

| | | | | | |
|---|---|---|---|---|---|
| Manual vertical doors, 2 stops | — | LS | — | — | 49,500.00 |

Hydraulic, 6,000 pound capacity, 100 FPM, single entry cab

| | | | | | |
|---|---|---|---|---|---|
| Manual vertical doors, 5 stops | — | LS | — | — | 72,700.00 |

Hydraulic, 6,000 pound capacity, 100 FPM, single entry cab

| | | | | | |
|---|---|---|---|---|---|
| Powered vertical doors, 5 stops | — | LS | — | — | 85,500.00 |

Hydraulic, 10,000 pound capacity, 100 FPM, single entry cab

| | | | | | |
|---|---|---|---|---|---|
| Manual vertical doors, 5 stops | — | LS | — | — | 81,300.00 |
| Powered vertical doors, 5 stops | — | LS | — | — | 91,900.00 |

Geared, 2,500 pound capacity, 100 FPM, single entry cab

| | | | | | |
|---|---|---|---|---|---|
| Manual doors, 2 stops | — | LS | — | — | 78,900.00 |

# 14 Conveying Equipment

| | Craft@Hrs | Unit | Material | Labor | Total |
|---|---|---|---|---|---|
| **Moving Stairs and Walks** Typical subcontract prices. | | | | | |
| Escalators, 90 FPM, steel trim, 32" step width, opaque balustrade | | | | | |
| 13' rise | — | Ea | — | — | 110,000.00 |
| 15' rise | — | Ea | — | — | 116,000.00 |
| 17' rise | — | Ea | — | — | 122,000.00 |
| 19' rise | — | Ea | — | — | 126,000.00 |
| 21' rise | — | Ea | — | — | 148,000.00 |
| Add for 48" width | — | % | — | — | 15.0 |
| **Hoists and Cranes** These costs do not include electrical work. | | | | | |
| Electric hoists, manual trolley, 20 FPM lift. Add cost of monorail below | | | | | |
| 1/2 ton, swivel mount, 25' lift | D4@9.34 | Ea | 4,860.00 | 432.00 | 5,292.00 |
| 1/2 ton, swivel mount, 45' lift | D4@13.7 | Ea | 5,150.00 | 634.00 | 5,784.00 |
| 1/2 ton, swivel mount, 75' lift | D4@18.0 | Ea | 5,740.00 | 833.00 | 6,573.00 |
| 1 ton, geared trolley, 20' lift | D4@11.9 | Ea | 5,810.00 | 550.00 | 6,360.00 |
| 1 ton, geared trolley, 30' lift | D4@15.8 | Ea | 6,140.00 | 731.00 | 6,871.00 |
| 1 ton, geared trolley, 55' lift | D4@19.2 | Ea | 6,750.00 | 888.00 | 7,638.00 |
| 2 ton, geared trolley, 15' lift | D4@15.8 | Ea | 7,110.00 | 731.00 | 7,841.00 |
| 2 ton, geared trolley, 35' lift | D4@15.8 | Ea | 7,550.00 | 731.00 | 8,281.00 |
| 2 ton, geared trolley, 55' lift | D4@15.8 | Ea | 8,240.00 | 731.00 | 8,971.00 |
| 2 ton, geared trolley, 70' lift | D4@15.8 | Ea | 8,630.00 | 731.00 | 9,361.00 |
| Electric hoists, power trolley, 20 FPM trolley speed. Add cost of monorail from below | | | | | |
| 1/2 ton, 50 FPM lift to 25' | D4@8.94 | Ea | 5,740.00 | 413.00 | 6,153.00 |
| 1/2 ton, 100 FPM lift to 25' | D4@15.8 | Ea | 8,420.00 | 731.00 | 9,151.00 |
| 1/2 ton, 50 FPM lift to 45' | D4@8.94 | Ea | 6,050.00 | 413.00 | 6,463.00 |
| 1/2 ton, 100 FPM lift to 45' | D4@15.8 | Ea | 8,840.00 | 731.00 | 9,571.00 |
| 1/2 ton, 50 FPM lift to 75' | D4@11.9 | Ea | 6,610.00 | 550.00 | 7,160.00 |
| 1/2 ton, 100 FPM lift to 75' | D4@18.0 | Ea | 9,280.00 | 833.00 | 10,113.00 |
| 1 ton, 50 FPM lift to 20' | D4@15.8 | Ea | 5,850.00 | 731.00 | 6,581.00 |
| 1 ton, 50 FPM lift to 30' | D4@15.8 | Ea | 6,200.00 | 731.00 | 6,931.00 |
| 1 ton, 50 FPM lift to 55' | D4@18.0 | Ea | 6,880.00 | 833.00 | 7,713.00 |
| 2 ton, 50 FPM lift to 15' | D4@17.9 | Ea | 7,030.00 | 828.00 | 7,858.00 |
| 2 ton, 50 FPM lift to 35' | D4@19.9 | Ea | 7,730.00 | 920.00 | 8,650.00 |
| 2 ton, 50 FPM lift to 55' | D4@19.9 | Ea | 8,280.00 | 920.00 | 9,200.00 |
| 2 ton, 50 FPM lift to 70' | D4@19.9 | Ea | 8,380.00 | 920.00 | 9,300.00 |
| Monorail for electric hoists, channel type | | | | | |
| 100 pounds per LF | D4@.443 | LF | 11.10 | 20.50 | 31.60 |
| 200 pounds per LF | D4@.555 | LF | 17.60 | 25.70 | 43.30 |
| 300 pounds per LF | D4@.761 | LF | 26.50 | 35.20 | 61.70 |
| Jib cranes, self-supporting, swinging 8' boom, 220 degree rotation | | | | | |
| 1,000 pounds | D4@7.90 | Ea | 1,740.00 | 365.00 | 2,105.00 |
| 2,000 pounds | D4@11.9 | Ea | 1,860.00 | 550.00 | 2,410.00 |
| 3,000 pounds | D4@13.5 | Ea | 2,090.00 | 624.00 | 2,714.00 |
| 4,000 pounds | D4@13.5 | Ea | 2,470.00 | 624.00 | 3,094.00 |
| 6,000 pounds | D4@15.8 | Ea | 2,640.00 | 731.00 | 3,371.00 |
| 10,000 pounds | D4@19.7 | Ea | 3,580.00 | 911.00 | 4,491.00 |
| Jib cranes, wall mounted, swinging 8' boom, 180 degree rotation | | | | | |
| 1,000 pounds | D4@8.19 | Ea | 939.00 | 379.00 | 1,318.00 |
| 2,000 pounds | D4@13.5 | Ea | 1,010.00 | 624.00 | 1,634.00 |
| 4,000 pounds | D4@13.5 | Ea | 1,540.00 | 624.00 | 2,164.00 |
| 6,000 pounds | D4@15.8 | Ea | 1,740.00 | 731.00 | 2,471.00 |
| 10,000 pounds | D4@19.7 | Ea | 3,180.00 | 911.00 | 4,091.00 |

# 14 Conveying Equipment

| | Craft@Hrs | Unit | Material | Labor | Total |
|---|---|---|---|---|---|

## Material Handling Systems

Conveyors, typical subcontract price. Foundations, support structures or electrical work not included

| | Craft@Hrs | Unit | Material | Labor | Total |
|---|---|---|---|---|---|
| Belt type, 24" width | — | LF | — | — | 222.00 |
| Mail conveyors, automatic, electronic | | | | | |
|   Horizontal | — | LF | — | — | 1,920.00 |
|   Vertical, per 12' floor | — | Ea | — | — | 24,400.00 |

Chutes, linen or solid waste handling, prefabricated unit including roof vent, 1-1/2 hour "B" rated doors, discharge and sprinkler system, gravity feed. Costs shown are for each floor based on 8' to 10' floor to floor height

| | Craft@Hrs | Unit | Material | Labor | Total |
|---|---|---|---|---|---|
| Light duty, aluminum, 20" diameter | D4@5.85 | Ea | 910.00 | 271.00 | 1,181.00 |
| Standard, 18 gauge steel, 24" diameter | D4@5.85 | Ea | 1,010.00 | 271.00 | 1,281.00 |
| Standard, 18 gauge steel, 30" diameter | D4@5.85 | Ea | 1,220.00 | 271.00 | 1,491.00 |
| Heavy duty, 18 gauge stainless steel | | | | | |
|   24" diameter | D4@5.85 | Ea | 1,520.00 | 271.00 | 1,791.00 |
|   30" diameter | D4@5.85 | Ea | 1,620.00 | 271.00 | 1,891.00 |
| Manual door with stainless steel rim | D4@2.17 | Ea | 435.00 | 100.00 | 535.00 |
| Disinfecting and sanitizing unit | D4@2.17 | Ea | 204.00 | 100.00 | 304.00 |
| Discharge storage unit | | | | | |
|   Aluminum | D4@1.00 | Ea | 764.00 | 46.30 | 810.30 |
|   Stainless steel | D4@1.00 | Ea | 1,400.00 | 46.30 | 1,446.30 |

**Turntables** These costs do not include electrical work.

Baggage handling carousels

| | Craft@Hrs | Unit | Material | Labor | Total |
|---|---|---|---|---|---|
| Round, 20' diameter | D4@208. | Ea | 37,100.00 | 9,620.00 | 46,720.00 |
| Round, 25' diameter | D4@271. | Ea | 48,800.00 | 12,500.00 | 61,300.00 |

**Pneumatic Tube Systems** Typical subcontract prices.

| | Craft@Hrs | Unit | Material | Labor | Total |
|---|---|---|---|---|---|
| 3" diameter, two station, single pipe, 100' between stations | — | LS | — | — | 6,360.00 |
| 3" diameter, two station, twin pipe, 100' between stations | — | LS | — | — | 8,150.00 |

# 21 Fire Suppression

**Sprinkler systems** Typical subcontract prices including the subcontractor's overhead and profit. These prices include design drawings, valves and connection to piping to within 5'0" outside the building. Costs will be higher where room sizes are smaller or where coverage per head averages less than 110 SF. Make additional allowances if a booster pump is required.

| | Craft@Hrs | Unit | Material | Labor | Total |
|---|---|---|---|---|---|
| Exposed systems, wet, complete, cost per SF of floor protected | | | | | |
|   5,000 SF | — | SF | — | — | 4.74 |
|   Over 5,000 to 15,000 SF | — | SF | — | — | 4.00 |
|   15,000 SF or more | — | SF | — | — | 3.55 |
| Concealed systems, wet, complete, cost per SF of floor protected | | | | | |
|   5,000 SF | — | SF | — | — | 4.60 |
|   Over 5,000 to 15,000 SF | — | SF | — | — | 3.70 |
|   15,000 SF or more | — | SF | — | — | 2.89 |
|   Add for dry systems | — | % | — | — | 20.0 |

| | Craft@Hrs | Unit | Material | Labor | Total |
|---|---|---|---|---|---|

**Fire sprinkler system components** Make additional allowances to shut down and drain the system when required

Sprinkler heads only (remove and replace)

| | Craft@Hrs | Unit | Material | Labor | Total |
|---|---|---|---|---|---|
| Brass pendent or upright head, 155-200 degree | SP@.350 | Ea | 7.30 | 21.60 | 28.90 |
| Brass pendent or upright head, 286 degree | SP@.350 | Ea | 7.30 | 21.60 | 28.90 |
| Brass pendent or upright head, 360 degree | SP@.350 | Ea | 4.92 | 21.60 | 26.52 |
| Brass pendent or upright head, 400-500 degree | SP@.350 | Ea | 21.80 | 21.60 | 43.40 |
| Chrome pendent or upright head, 155-200 degree | SP@.350 | Ea | 7.82 | 21.60 | 29.42 |
| Chrome pendent or upright head, 286 degree | SP@.350 | Ea | 7.82 | 21.60 | 29.42 |
| Dry pendent or upright head | SP@.350 | Ea | 18.40 | 21.60 | 40.00 |
| Relocate wet sprinkler head and branch drop | SP@1.50 | Ea | 35.00 | 92.70 | 127.70 |
| Add for heads more than 14' above floor | — | % | — | 20.0 | — |

**Wet system components** (where freezing is not a hazard)

Zone valves

| | Craft@Hrs | Unit | Material | Labor | Total |
|---|---|---|---|---|---|
| 2" OS&Y gate valve, flanged or grooved | SP@1.75 | Ea | 363.00 | 108.00 | 471.00 |
| 3" OS&Y gate valve, flanged or grooved | SP@2.00 | Ea | 384.00 | 124.00 | 508.00 |
| 4" OS&Y gate valve, flanged or grooved | SP@2.25 | Ea | 442.00 | 139.00 | 581.00 |
| 6" OS&Y gate valve, flanged or grooved | SP@3.00 | Ea | 681.00 | 185.00 | 866.00 |
| 8" OS&Y gate valve, flanged or grooved | SP@3.50 | Ea | 1,080.00 | 216.00 | 1,296.00 |
| 4" alarm valve, flanged or grooved | SP@2.65 | Ea | 510.00 | 164.00 | 674.00 |
| 6" alarm valve, flanged or grooved | SP@3.50 | Ea | 627.00 | 216.00 | 843.00 |
| 8" alarm valve, flanged or grooved | SP@4.25 | Ea | 914.00 | 263.00 | 1,177.00 |
| Alarm valve trim only (retard chamber and gauges) | SP@1.25 | Ea | 343.00 | 77.20 | 420.20 |
| 4" alarm valve package, complete | SP@4.00 | Ea | 1,610.00 | 247.00 | 1,857.00 |
| 6" alarm valve package, complete | SP@4.50 | Ea | 1,770.00 | 278.00 | 2,048.00 |
| 8" alarm valve package, complete | SP@5.25 | Ea | 2,170.00 | 324.00 | 2,494.00 |
| Retard pressure switch | SP@1.75 | Ea | 289.00 | 108.00 | 397.00 |

Check valves

| | Craft@Hrs | Unit | Material | Labor | Total |
|---|---|---|---|---|---|
| 3" swing check valve | SP@2.00 | Ea | 202.00 | 124.00 | 326.00 |
| 4" swing check valve | SP@2.25 | Ea | 183.00 | 139.00 | 322.00 |
| 6" swing check valve | SP@3.00 | Ea | 359.00 | 185.00 | 544.00 |
| 3" wafer check valve | SP@2.00 | Ea | 239.00 | 124.00 | 363.00 |
| 4" wafer check valve | SP@2.25 | Ea | 257.00 | 139.00 | 396.00 |
| 6" wafer check valve | SP@3.00 | Ea | 422.00 | 185.00 | 607.00 |
| 8" wafer check valve | SP@3.50 | Ea | 600.00 | 216.00 | 816.00 |
| 10" wafer check valve | SP@4.25 | Ea | 1,000.00 | 263.00 | 1,263.00 |
| 3" double check detector assembly | SP@3.75 | Ea | 2,560.00 | 232.00 | 2,792.00 |
| 4" double check detector assembly | SP@4.50 | Ea | 2,790.00 | 278.00 | 3,068.00 |
| 6" double check detector assembly | SP@5.75 | Ea | 4,380.00 | 355.00 | 4,735.00 |
| 8" double check detector assembly | SP@6.25 | Ea | 7,910.00 | 386.00 | 8,296.00 |

**Dry system components** (distribution piping holds no water)

Dry valves (deluge)

| | Craft@Hrs | Unit | Material | Labor | Total |
|---|---|---|---|---|---|
| 3" dry pipe valve | SP@2.25 | Ea | 908.00 | 139.00 | 1,047.00 |
| 4" dry pipe valve | SP@2.65 | Ea | 1,030.00 | 164.00 | 1,194.00 |
| 6" dry pipe valve | SP@3.50 | Ea | 1,310.00 | 216.00 | 1,526.00 |
| Dry valve trim and gauges | SP@1.00 | Ea | 428.00 | 61.80 | 489.80 |

**Wall hydrants** for sprinkler systems, brass

| | Craft@Hrs | Unit | Material | Labor | Total |
|---|---|---|---|---|---|
| Single outlet, 2-1/2" x 2-1/2" | SP@2.00 | Ea | 328.00 | 124.00 | 452.00 |
| 2-way outlet, 2-1/2" x 2-1/2" x 4" | SP@2.50 | Ea | 527.00 | 154.00 | 681.00 |
| 3-way outlet, 2-1/2" x 2-1/2" x 3-1/3" x 4" | SP@2.75 | Ea | 956.00 | 170.00 | 1,126.00 |
| Fire dept. connection, 4" (Siamese) | SP@2.50 | Ea | 527.00 | 154.00 | 681.00 |
| Pumper connection, 6" | SP@3.00 | Ea | 1,040.00 | 185.00 | 1,225.00 |
| Roof manifold, with valves | SP@3.90 | Ea | 615.00 | 241.00 | 856.00 |

# 21 Fire Suppression

|  | Craft@Hrs | Unit | Material | Labor | Total |
|---|---|---|---|---|---|
| **Miscellaneous sprinkler system components** | | | | | |
| Air maintenance device | SP@.781 | Ea | 191.00 | 48.30 | 239.30 |
| Low air supervisory unit | SP@1.11 | Ea | 590.00 | 68.60 | 658.60 |
| Low air pressure switch | SP@.784 | Ea | 191.00 | 48.40 | 239.40 |
| Pressure switch (double circuit, open and close) | SP@1.92 | Ea | 185.00 | 119.00 | 304.00 |
| 1/2" ball drip at check valve and fire dept. connection | SP@.193 | Ea | 21.30 | 11.90 | 33.20 |
| Water powered gong (local alarm) | SP@1.94 | Ea | 231.00 | 120.00 | 351.00 |
| Cabinet with 6 spare heads and wrench | SP@.291 | LS | 88.60 | 18.00 | 106.60 |
| Inspector's test connection | SP@.973 | Ea | 73.80 | 60.10 | 133.90 |
| Escutcheon for sprinkler heads, chrome | SP@.071 | Ea | 1.25 | 4.39 | 5.64 |
| Field testing and flushing, subcontract | — | LS | — | — | 193.00 |
| Disinfection of distribution system, subcontract | — | LS | — | — | 193.00 |
| Hydraulic design fee, subcontract | | | | | |
|    Sprinklers in a high density or high risk area, typical | — | LS | — | — | 2,660.00 |
| Underground valve vault (if required), subcontract | — | LS | — | — | 3,120.00 |
| **Fire hose cabinets,** recessed, 24" x 30" x 5-1/2" with rack, full glass door, 1-1/2" hose, valve and nozzle. Standard. | | | | | |
| Painted steel,   75' hose, 24" x 30" | SP@1.95 | Ea | 505.00 | 120.00 | 625.00 |
| Painted steel, 100' hose, 24" x 36" | SP@1.95 | Ea | 503.00 | 120.00 | 623.00 |
| Aluminum, 75' hose | SP@1.95 | Ea | 695.00 | 120.00 | 815.00 |
| Stainless steel, 75' hose | SP@1.95 | Ea | 902.00 | 120.00 | 1,022.00 |
| **Fire extinguishers**  Factory charged, complete with hose and horn and wall mounting bracket | | | | | |
| 5 lb. carbon dioxide | C8@.488 | Ea | 159.00 | 22.40 | 181.40 |
| 10 lb. carbon dioxide | C8@.488 | Ea | 198.00 | 22.40 | 220.40 |
| 20 lb. carbon dioxide | C8@.488 | Ea | 285.00 | 22.40 | 307.40 |
| 5 lb. Halotron | C8@.488 | Ea | 189.00 | 22.40 | 211.40 |
| 11 lb. Halotron | C8@.488 | Ea | 375.00 | 22.40 | 397.40 |
| 15.5 lb. Halotron | C8@.488 | Ea | 468.00 | 22.40 | 490.40 |
| **Extinguisher cabinets**  No extinguishers included, painted steel, recessed | | | | | |
| 9" x 24" x 5" full glass door | C8@1.67 | Ea | 135.00 | 76.50 | 211.50 |
| 9" x 24" x 5" break glass door | C8@1.67 | Ea | 162.00 | 76.50 | 238.50 |
| 12" x 27" x 8" full glass door | C8@1.67 | Ea | 157.00 | 76.50 | 233.50 |
| 12" x 27" x 8" break glass door | C8@1.67 | Ea | 186.00 | 76.50 | 262.50 |
| Add for semi-recessed or surface mount | — | % | 10.0 | — | — |
| Add for chrome cabinets | — | % | 15.0 | — | — |
| Add for stainless steel cabinets | — | Ea | 202.00 | — | 202.00 |

# 22 Plumbing

**Carbon Steel Pipe and Fittings (Black Steel)**  Pipe is ERW – A53 (electric resistance welded) with a factory-welded seam. Installation heights indicate the working height above the building floor. Based on orders aggregating 1,000 foot quantities. Assembly costs include hangers as required for the pipe diameter, a reducing tee and an elbow each 16 feet for 1/2" pipe, to each 50 feet for 12" pipe. Expect to pay as much as double these prices for smaller quantities purchased at retail. For more detailed coverage, see *National Plumbing & HVAC Estimator*, http://CraftsmanSiteLicense.com.

**1/2" carbon steel (A53) threaded pipe and fittings**
Pipe, no fittings or supports included, 1/2", threaded ends

|  | Craft@Hrs | Unit | Material | Labor | Total |
|---|---|---|---|---|---|
| Schedule 40 to 10' high | M5@.060 | LF | .88 | 2.98 | 3.86 |
| Schedule 40 over 10' to 20' | M5@.070 | LF | .88 | 3.48 | 4.36 |
| Schedule 80 to 10' high | M5@.070 | LF | 1.48 | 3.48 | 4.96 |
| Schedule 80 over 10' to 20' | M5@.080 | LF | 1.48 | 3.98 | 5.46 |
| Pipe assembly, Schedule 40, fittings, | | | | | |
|    hangers and supports | M5@.130 | LF | 1.34 | 6.46 | 7.80 |
| Add for seamless pipe (A106) | — | % | 50.0 | — | — |

# 22 Plumbing

| | Craft@Hrs | Unit | Material | Labor | Total |
|---|---|---|---|---|---|
| **90-degree ells, 1/2", threaded** | | | | | |
| 150 lb, malleable iron | M5@.120 | Ea | .93 | 5.97 | 6.90 |
| 300 lb, malleable iron | M5@.132 | Ea | 5.21 | 6.56 | 11.77 |
| **45-degree ells, 1/2", threaded** | | | | | |
| 150 lb, malleable iron | M5@.120 | Ea | 1.52 | 5.97 | 7.49 |
| 300 lb, malleable iron | M5@.132 | Ea | 6.77 | 6.56 | 13.33 |
| **Tees, 1/2", threaded** | | | | | |
| 150 lb, malleable iron | M5@.180 | Ea | 1.26 | 8.95 | 10.21 |
| 300 lb, malleable iron | M5@.198 | Ea | 7.33 | 9.84 | 17.17 |
| 150 lb, reducing, malleable iron | M5@.170 | Ea | 2.69 | 8.45 | 11.14 |
| 300 lb, reducing, malleable iron | M5@.180 | Ea | 10.10 | 8.95 | 19.05 |
| **Caps, 1/2" threaded** | | | | | |
| 150 lb, malleable iron | M5@.090 | Ea | .95 | 4.47 | 5.42 |
| 300 lb, malleable iron | M5@.100 | Ea | 3.52 | 4.97 | 8.49 |
| **Couplings, 1/2" threaded** | | | | | |
| 150 lb, malleable iron | M5@.120 | Ea | 1.27 | 5.97 | 7.24 |
| 300 lb, malleable iron | M5@.132 | Ea | 3.92 | 6.56 | 10.48 |
| **Flanges, forged steel, 1/2"** | | | | | |
| 150 lb, threaded flange | M5@.489 | Ea | 10.60 | 24.30 | 34.90 |
| 300 lb, threaded flange | M5@.781 | Ea | 13.70 | 38.80 | 52.50 |
| 600 lb, threaded flange | M5@.859 | Ea | 47.70 | 42.70 | 90.40 |
| **Unions, 1/2"** | | | | | |
| 150 lb, malleable iron | M5@.140 | Ea | 4.16 | 6.96 | 11.12 |
| 300 lb, malleable iron | M5@.150 | Ea | 6.45 | 7.46 | 13.91 |
| Add for galvanized 1/2" pipe | — | % | 32.0 | — | — |
| Add for galvanized 1/2" 150 lb. malleable fittings | — | % | 15.0 | — | — |
| Add for galvanized 1/2" 300 lb. malleable fittings | — | % | 55.0 | — | — |
| Add for galvanized 1/2" pipe assembly, Schedule 40 | — | % | 25.0 | — | — |

## 3/4" carbon steel (A53) threaded pipe and fittings

| | Craft@Hrs | Unit | Material | Labor | Total |
|---|---|---|---|---|---|
| **Pipe, no fittings or supports included, 3/4", threaded** | | | | | |
| Schedule 40 to 10' high | M5@.070 | LF | .98 | 3.48 | 4.46 |
| Schedule 40 over 10' to 20' | M5@.080 | LF | .98 | 3.98 | 4.96 |
| Schedule 80 to 10' high | M5@.080 | LF | 1.85 | 3.98 | 5.83 |
| Schedule 80 over 10' to 20' | M5@.095 | LF | 1.85 | 4.72 | 6.57 |
| **Pipe assembly, Schedule 40, fittings,** | | | | | |
| hangers and supports | M5@.140 | LF | 1.58 | 6.96 | 8.54 |
| Add for seamless pipe (A106) | — | % | 55.0 | — | — |
| **90-degree ells, 3/4" threaded** | | | | | |
| 150 lb, malleable iron | M5@.130 | Ea | 1.16 | 6.46 | 7.62 |
| 300 lb, malleable iron | M5@.143 | Ea | 6.02 | 7.11 | 13.13 |
| **45-degree ells, 3/4" threaded** | | | | | |
| 150 lb, malleable iron | M5@.130 | Ea | 1.87 | 6.46 | 8.33 |
| 300 lb, malleable iron | M5@.143 | Ea | 7.52 | 7.11 | 14.63 |
| **Tees, 3/4" threaded** | | | | | |
| 150 lb, malleable iron | M5@.190 | Ea | 1.82 | 9.44 | 11.26 |
| 300 lb, malleable iron | M5@.209 | Ea | 8.05 | 10.40 | 18.45 |
| 150 lb, reducing, malleable iron | M5@.180 | Ea | 3.42 | 8.95 | 12.37 |
| 300 lb, reducing, malleable iron | M5@.205 | Ea | 12.70 | 10.20 | 22.90 |
| **Caps, 3/4" threaded** | | | | | |
| 150 lb, malleable iron | M5@.100 | Ea | 1.27 | 4.97 | 6.24 |
| 300 lb, malleable iron | M5@.110 | Ea | 4.69 | 5.47 | 10.16 |

| | Craft@Hrs | Unit | Material | Labor | Total |
|---|---|---|---|---|---|
| Couplings, 3/4" threaded | | | | | |
| 150 lb, malleable iron | M5@.130 | Ea | 1.50 | 6.46 | 7.96 |
| 300 lb, malleable iron | M5@.143 | Ea | 4.58 | 7.11 | 11.69 |
| Unions, 3/4" | | | | | |
| 150 lb, malleable iron | M5@.150 | Ea | 4.77 | 7.46 | 12.23 |
| 300 lb, malleable iron | M5@.165 | Ea | 7.27 | 8.20 | 15.47 |
| Flanges, forged steel, 3/4" | | | | | |
| 150 lb, threaded flange | M5@.585 | Ea | 11.20 | 29.10 | 40.30 |
| 300 lb, threaded flange | M5@.781 | Ea | 14.80 | 38.80 | 53.60 |
| 600 lb, threaded flange | M5@.859 | Ea | 47.70 | 42.70 | 90.40 |
| Add for galvanized 3/4" pipe | — | % | 32.0 | — | — |
| Add for galvanized 3/4" 150 lb. malleable fittings | — | % | 25.0 | — | — |
| Add for galvanized 3/4" 300 lb. malleable fittings | — | % | 55.0 | — | — |
| Add for galvanized 3/4" pipe assembly, Schedule 40 | — | % | 30.0 | — | — |

## 1" carbon steel (A53) threaded pipe and fittings

| | Craft@Hrs | Unit | Material | Labor | Total |
|---|---|---|---|---|---|
| Pipe, no fittings or supports included, 1", threaded | | | | | |
| Schedule 40 to 10' high | M5@.090 | LF | 1.50 | 4.47 | 5.97 |
| Schedule 40 over 10' to 20' | M5@.110 | LF | 1.50 | 5.47 | 6.97 |
| Schedule 80 to 10' high | M5@.100 | LF | 2.52 | 4.97 | 7.49 |
| Schedule 80 over 10' to 20' | M5@.120 | LF | 2.52 | 5.97 | 8.49 |
| Pipe assembly, Schedule 40, fittings, hangers and supports | M5@.160 | LF | 2.16 | 7.95 | 10.11 |
| Add for seamless pipe (A106) | — | % | 50.0 | — | — |
| 90-degree ells, 1" threaded | | | | | |
| 150 lb, malleable iron | M5@.180 | Ea | 1.98 | 8.95 | 10.93 |
| 300 lb, malleable iron | M5@.198 | Ea | 7.68 | 9.84 | 17.52 |
| 45-degree ells, 1" threaded | | | | | |
| 150 lb, malleable iron | M5@.180 | Ea | 2.38 | 8.95 | 11.33 |
| 300 lb, malleable iron | M5@.198 | Ea | 8.35 | 9.84 | 18.19 |
| Tees, 1" threaded | | | | | |
| 150 lb, malleable iron | M5@.230 | Ea | 3.09 | 11.40 | 14.49 |
| 300 lb, malleable iron | M5@.253 | Ea | 9.67 | 12.60 | 22.27 |
| 150 lb, reducing, malleable iron | M5@.205 | Ea | 5.91 | 10.20 | 16.11 |
| 300 lb, reducing, malleable iron | M5@.220 | Ea | 16.40 | 10.90 | 27.30 |
| Caps, 1" threaded | | | | | |
| 150 lb, malleable iron | M5@.140 | Ea | 1.58 | 6.96 | 8.54 |
| 300 lb, malleable iron | M5@.155 | Ea | 6.04 | 7.71 | 13.75 |
| Couplings, 1" threaded | | | | | |
| 150 lb, malleable iron | M5@.180 | Ea | 2.28 | 8.95 | 11.23 |
| 300 lb, malleable iron | M5@.198 | Ea | 5.40 | 9.84 | 15.24 |
| Flanges, forged steel, 1" | | | | | |
| 150 lb, threaded flange | M5@.731 | Ea | 9.82 | 36.30 | 46.12 |
| 300 lb, threaded flange | M5@1.05 | Ea | 16.40 | 52.20 | 68.60 |
| 600 lb, threaded flange | M5@1.15 | Ea | 47.70 | 57.20 | 104.90 |
| Unions, 1" | | | | | |
| 150 lb, malleable iron | M5@.210 | Ea | 3.35 | 10.40 | 13.75 |
| 300 lb, malleable iron | M5@.230 | Ea | 8.32 | 11.40 | 19.72 |
| Add for galvanized 1" pipe | — | % | 32.0 | — | — |
| Add for galvanized 1" 150 lb. malleable fittings | — | % | 20.0 | — | — |
| Add for galvanized 1" 300 lb. malleable fittings | — | % | 60.0 | — | — |
| Add for galvanized 1" pipe assembly, Schedule 40 | — | % | 30.0 | — | — |

| | Craft@Hrs | Unit | Material | Labor | Total |
|---|---|---|---|---|---|
| **1-1/4" carbon steel (A53) threaded pipe and fittings** | | | | | |
| Pipe, no fittings or supports included, 1-1/4", threaded | | | | | |
| Schedule 40 to 10' high | M5@.100 | LF | 1.94 | 4.97 | 6.91 |
| Schedule 40 over 10' to 20' | M5@.120 | LF | 1.94 | 5.97 | 7.91 |
| Schedule 80 to 10' high | M5@.110 | LF | 3.30 | 5.47 | 8.77 |
| Schedule 80 over 10' to 20' | M5@.130 | LF | 3.30 | 6.46 | 9.76 |
| Pipe assembly, Schedule 40, fittings, hangers and supports | M5@.200 | LF | 2.82 | 9.94 | 12.76 |
| Add for seamless pipe (A106) | — | % | 45.0 | — | — |
| 90-degree ells, 1-1/4" threaded | | | | | |
| 150 lb, malleable iron | M5@.240 | Ea | 3.24 | 11.90 | 15.14 |
| 300 lb, malleable iron | M5@.264 | Ea | 10.70 | 13.10 | 23.80 |
| 45-degree ells, 1-1/4" threaded | | | | | |
| 150 lb, malleable iron | M5@.240 | Ea | 4.20 | 11.90 | 16.10 |
| 300 lb, malleable iron | M5@.264 | Ea | 12.90 | 13.10 | 26.00 |
| Tees, 1-1/4" threaded | | | | | |
| 150 lb, malleable iron | M5@.310 | Ea | 5.07 | 15.40 | 20.47 |
| 300 lb, malleable iron | M5@.341 | Ea | 12.90 | 17.00 | 29.90 |
| 150 lb, reducing, malleable iron | M5@.290 | Ea | 8.51 | 14.40 | 22.91 |
| 300 lb, reducing, malleable iron | M5@.310 | Ea | 20.20 | 15.40 | 35.60 |
| Caps, 1-1/4" threaded | | | | | |
| 150 lb, malleable iron | M5@.180 | Ea | 2.03 | 8.95 | 10.98 |
| 300 lb, malleable iron | M5@.200 | Ea | 8.21 | 9.94 | 18.15 |
| Couplings, 1-1/4" threaded | | | | | |
| 150 lb, malleable iron | M5@.240 | Ea | 2.92 | 11.90 | 14.82 |
| 300 lb, malleable iron | M5@.264 | Ea | 6.45 | 13.10 | 19.55 |
| Unions, 1-1/4" | | | | | |
| 150 lb, malleable iron | M5@.280 | Ea | 8.91 | 13.90 | 22.81 |
| 300 lb, malleable iron | M5@.310 | Ea | 14.60 | 15.40 | 30.00 |
| Flanges, forged steel, 1-1/4" | | | | | |
| 150 lb, threaded flange | M5@.788 | Ea | 15.40 | 39.20 | 54.60 |
| 300 lb, threaded flange | M5@1.04 | Ea | 22.70 | 51.70 | 74.40 |
| 600 lb, threaded flange | M5@1.14 | Ea | 50.10 | 56.70 | 106.80 |
| Add for galvanized 1-1/4" pipe | — | % | 30.0 | — | — |
| Add for galvanized 1-1/4" 150 lb. malleable fittings | — | % | 20.0 | — | — |
| Add for galvanized 1-1/4" 300 lb. malleable fittings | — | % | 65.0 | — | — |
| Add for galvanized 1-1/4" pipe assembly, Schedule 40 | — | % | 30.0 | — | — |
| **1-1/2" carbon steel (A53) threaded pipe and fittings** | | | | | |
| Pipe, no fittings or supports included, 1-1/2", threaded | | | | | |
| Schedule 40 to 10' high | M5@.110 | LF | 2.30 | 5.47 | 7.77 |
| Schedule 40 over 10' to 20' | M5@.130 | LF | 2.30 | 6.46 | 8.76 |
| Schedule 80 to 10' high | M5@.120 | LF | 3.91 | 5.97 | 9.88 |
| Schedule 80 over 10' to 20' | M5@.140 | LF | 3.91 | 6.96 | 10.87 |
| Pipe assembly, Schedule 40, fittings, hangers and supports | M5@.230 | LF | 4.98 | 11.40 | 16.38 |
| Add for seamless pipe (A106) | — | % | 45.0 | — | — |
| 90-degree ells, 1-1/2" threaded | | | | | |
| 150 lb, malleable iron | M5@.300 | Ea | 4.28 | 14.90 | 19.18 |
| 300 lb, malleable iron | M5@.333 | Ea | 12.90 | 16.60 | 29.50 |

| | Craft@Hrs | Unit | Material | Labor | Total |
|---|---|---|---|---|---|
| **45-degree ells, 1-1/2" threaded** | | | | | |
| 150 lb, malleable iron | M5@.390 | Ea | 5.18 | 19.40 | 24.58 |
| 300 lb, malleable iron | M5@.429 | Ea | 16.80 | 21.30 | 38.10 |
| **Tees, 1-1/2" threaded** | | | | | |
| 150 lb, malleable iron | M5@.390 | Ea | 6.22 | 19.40 | 25.62 |
| 300 lb, malleable iron | M5@.429 | Ea | 16.00 | 21.30 | 37.30 |
| 150 lb, reducing, malleable iron | M5@.360 | Ea | 11.70 | 17.90 | 29.60 |
| 300 lb, reducing, malleable iron | M5@.390 | Ea | 30.00 | 19.40 | 49.40 |
| **Caps, 1-1/2" threaded** | | | | | |
| 150 lb, malleable iron | M5@.230 | Ea | 2.76 | 11.40 | 14.16 |
| 300 lb, malleable iron | M5@.250 | Ea | 9.72 | 12.40 | 22.12 |
| **Couplings, 1-1/2" threaded** | | | | | |
| 150 lb, malleable iron | M5@.300 | Ea | 3.91 | 14.90 | 18.81 |
| 300 lb, malleable iron | M5@.330 | Ea | 9.67 | 16.40 | 26.07 |
| **Unions, 1-1/2"** | | | | | |
| 150 lb, malleable iron | M5@.360 | Ea | 11.00 | 17.90 | 28.90 |
| 300 lb, malleable iron | M5@.400 | Ea | 15.80 | 19.90 | 35.70 |
| **Flanges, forged steel, 1-1/2"** | | | | | |
| 150 lb, threaded flange | M5@.840 | Ea | 15.40 | 41.80 | 57.20 |
| 300 lb, threaded flange | M5@1.15 | Ea | 23.20 | 57.20 | 80.40 |
| 600 lb, threaded flange | M5@1.26 | Ea | 50.10 | 62.60 | 112.70 |
| Add for galvanized 1-1/2" pipe | — | % | 30.0 | — | — |
| Add for galvanized 1-1/2" 150 lb. malleable fittings | — | % | 20.0 | — | — |
| Add for galvanized 1-1/2" 300 lb. malleable fittings | — | % | 55.0 | — | — |
| Add for galvanized 1-1/2" pipe assembly, Schedule 40 | — | % | 30.0 | — | — |

## 2" carbon steel (A53) threaded pipe and fittings

| | Craft@Hrs | Unit | Material | Labor | Total |
|---|---|---|---|---|---|
| **Pipe, no fittings or supports included, 2", threaded** | | | | | |
| Schedule 40 to 10' high | M5@.120 | LF | 3.41 | 5.97 | 9.38 |
| Schedule 40 over 10' to 20' | M5@.140 | LF | 3.41 | 6.96 | 10.37 |
| Schedule 80 to 10' high | M5@.130 | LF | 5.34 | 6.46 | 11.80 |
| Schedule 80 over 10' to 20' | M5@.150 | LF | 5.34 | 7.46 | 12.80 |
| Pipe assembly, Schedule 40, fittings, hangers and supports | M5@.250 | LF | 4.75 | 12.40 | 17.15 |
| Add for seamless pipe (A106) | — | % | 40.0 | — | — |
| **90-degree ells, 2" threaded** | | | | | |
| 150 lb, malleable iron | M5@.380 | Ea | 7.30 | 18.90 | 26.20 |
| 300 lb, malleable iron | M5@.418 | Ea | 18.00 | 20.80 | 38.80 |
| **45-degree ells, 2" threaded** | | | | | |
| 150 lb, malleable iron | M5@.380 | Ea | 7.75 | 18.90 | 26.65 |
| 300 lb, malleable iron | M5@.418 | Ea | 25.10 | 20.80 | 45.90 |
| **Tees, 2" threaded** | | | | | |
| 150 lb, malleable iron | M5@.490 | Ea | 10.50 | 24.40 | 34.90 |
| 300 lb, malleable iron | M5@.539 | Ea | 23.30 | 26.80 | 50.10 |
| 150 lb, reducing, malleable iron | M5@.460 | Ea | 10.80 | 22.90 | 33.70 |
| 300 lb, reducing, malleable iron | M5@.500 | Ea | 30.00 | 24.90 | 54.90 |
| **Caps, 2" threaded** | | | | | |
| 150 lb, malleable iron | M5@.290 | Ea | 4.07 | 14.40 | 18.47 |
| 300 lb, malleable iron | M5@.320 | Ea | 9.72 | 15.90 | 25.62 |

| | Craft@Hrs | Unit | Material | Labor | Total |
|---|---|---|---|---|---|
| Couplings, 2" threaded | | | | | |
|    150 lb, malleable iron | M5@.380 | Ea | 5.69 | 18.90 | 24.59 |
|    300 lb, malleable iron | M5@.418 | Ea | 13.50 | 20.80 | 34.30 |
| Flanges, forged steel, 2" | | | | | |
|    150 lb, threaded flange | M5@.290 | Ea | 16.70 | 14.40 | 31.10 |
|    300 lb, threaded flange | M5@.310 | Ea | 26.60 | 15.40 | 42.00 |
|    600 lb, threaded flange | M5@.320 | Ea | 68.50 | 15.90 | 84.40 |
| Unions, 2" | | | | | |
|    150 lb, malleable iron | M5@.450 | Ea | 6.01 | 22.40 | 28.41 |
|    300 lb, malleable iron | M5@.500 | Ea | 16.70 | 24.90 | 41.60 |
| Add for galvanized 2" pipe, add | — | % | 30.0 | — | — |
| Add for galvanized 2" 50 lb. malleable fittings | — | % | 20.0 | — | — |
| Add for galvanized 2" 300 lb. malleable fittings | — | % | 55.0 | — | — |
| Add for galvanized 2" pipe assembly, Schedule 40 | — | % | 30.0 | — | — |

## 3" carbon steel pipe (A53) and welded fittings

| | Craft@Hrs | Unit | Material | Labor | Total |
|---|---|---|---|---|---|
| Pipe, no fittings or supports included, 3", plain end | | | | | |
|    Schedule 40 to 10' high | Ml@.190 | LF | 6.16 | 10.10 | 16.26 |
|    Schedule 40 over 10' to 20' high | Ml@.230 | LF | 6.16 | 12.20 | 18.36 |
|    Schedule 80 to 10' high | Ml@.280 | LF | 13.10 | 14.80 | 27.90 |
|    Schedule 80 over 10' to 20' high | Ml@.335 | LF | 13.10 | 17.80 | 30.90 |
|    Pipe assembly, Schedule 40, fittings, hangers and supports | Ml@.380 | LF | 8.68 | 20.10 | 28.78 |
|    Add for seamless pipe (A106) | — | % | 40.0 | — | — |
|    Add for galvanized pipe assembly, Schedule 40 | — | % | 30.0 | — | — |
| Butt welded joints, 3" pipe | | | | | |
|    Schedule 40 pipe | Ml@.930 | Ea | — | 49.30 | 49.30 |
|    Schedule 80 pipe | Ml@1.11 | Ea | — | 58.80 | 58.80 |
| 90-degree ells, 3" | | | | | |
|    Schedule 40 carbon steel, butt weld, long turn | Ml@1.33 | Ea | 10.10 | 70.50 | 80.60 |
|    Schedule 80 carbon steel, butt weld, long turn | Ml@1.77 | Ea | 14.40 | 93.80 | 108.20 |
| 45-degree ells, 3" | | | | | |
|    Schedule 40, carbon steel, butt weld | Ml@1.33 | Ea | 7.92 | 70.50 | 78.42 |
|    Schedule 80, carbon steel, butt weld | Ml@1.77 | Ea | 11.00 | 93.80 | 104.80 |
| Tees, 3" | | | | | |
|    Schedule 40, carbon steel, butt weld | Ml@2.00 | Ea | 25.60 | 106.00 | 131.60 |
|    Schedule 80, carbon steel, butt weld | Ml@2.66 | Ea | 51.50 | 141.00 | 192.50 |
|    Reducing, Schedule 40, carbon steel, butt weld | Ml@1.89 | Ea | 33.20 | 100.00 | 133.20 |
| Caps, 3" | | | | | |
|    Schedule 40, carbon steel, butt weld | Ml@.930 | Ea | 12.20 | 49.30 | 61.50 |
|    Schedule 80, carbon steel, butt weld | Ml@1.11 | Ea | 8.63 | 58.80 | 67.43 |
| Reducing couplings | | | | | |
|    Schedule 40, carbon steel, butt weld | Ml@1.20 | Ea | 14.30 | 63.60 | 77.90 |
| Weld-o-lets, 3" | | | | | |
|    Schedule 40, carbon steel | Ml@2.00 | Ea | 43.80 | 106.00 | 149.80 |
| Welded pipe flanges, forged steel, 3" butt weld | | | | | |
|    150 lb, slip-on flange | Ml@.730 | Ea | 19.20 | 38.70 | 57.90 |
|    300 lb, slip-on flange | Ml@.980 | Ea | 26.30 | 51.90 | 78.20 |
|    600 lb, slip-on flange | Ml@1.03 | Ea | 52.60 | 54.60 | 107.20 |
|    150 lb, weld neck flange | Ml@.730 | Ea | 24.00 | 38.70 | 62.70 |
|    300 lb, weld neck flange | Ml@.980 | Ea | 29.80 | 51.90 | 81.70 |

| | Craft@Hrs | Unit | Material | Labor | Total |
|---|---|---|---|---|---|
| 600 lb, weld neck flange | Ml@1.03 | Ea | 50.90 | 54.60 | 105.50 |
| 150 lb, threaded flange | Ml@.460 | Ea | 34.80 | 24.40 | 59.20 |
| 300 lb, threaded flange | Ml@.500 | Ea | 42.50 | 26.50 | 69.00 |
| 600 lb, threaded flange | Ml@.510 | Ea | 69.60 | 27.00 | 96.60 |
| **Flange bolt & gasket kits, 3"** | | | | | |
| Full face, red rubber gasket, 150 lb | Ml@.750 | Ea | 8.03 | 39.70 | 47.73 |
| Full face, red rubber gasket, 300 lb | Ml@1.00 | Ea | 8.03 | 53.00 | 61.03 |

## 4" carbon steel pipe (A53) and welded fittings

| | Craft@Hrs | Unit | Material | Labor | Total |
|---|---|---|---|---|---|
| Pipe, no fittings or supports included, 4", plain end | | | | | |
| Schedule 40 to 10' high | Ml@.260 | LF | 8.77 | 13.80 | 22.57 |
| Schedule 40 over 10' to 20' high | Ml@.360 | LF | 8.77 | 19.10 | 27.87 |
| Schedule 80 to 10' high | Ml@.340 | LF | 19.20 | 18.00 | 37.20 |
| Schedule 80 over 10' to 20' high | Ml@.400 | LF | 19.20 | 21.20 | 40.40 |
| Pipe assembly, Schedule 40, fittings, hangers and supports | Ml@.430 | LF | 12.00 | 22.80 | 34.80 |
| Add for seamless pipe (A106) | — | % | 35.0 | — | — |
| Add for galvanized pipe assembly, Schedule 40 | — | % | 30.0 | — | — |
| Butt welded joints, 4" pipe | | | | | |
| Schedule 40 pipe | Ml@1.25 | Ea | — | 66.20 | 66.20 |
| Schedule 80 pipe | Ml@1.50 | Ea | — | 79.50 | 79.50 |
| 90-degree ells, 4" | | | | | |
| Schedule 40 carbon steel, butt weld, long turn | Ml@1.33 | Ea | 16.20 | 70.50 | 86.70 |
| Schedule 80 carbon steel, butt weld, long turn | Ml@2.37 | Ea | 34.40 | 126.00 | 160.40 |
| 45-degree ells, 4" | | | | | |
| Schedule 40, carbon steel, butt weld | Ml@1.33 | Ea | 11.70 | 70.50 | 82.20 |
| Schedule 80, carbon steel, butt weld | Ml@2.37 | Ea | 16.60 | 126.00 | 142.60 |
| Tees, 4" | | | | | |
| Schedule 40, carbon steel, butt weld | Ml@2.67 | Ea | 34.90 | 141.00 | 175.90 |
| Schedule 80, carbon steel, butt weld | Ml@3.55 | Ea | 54.20 | 188.00 | 242.20 |
| Reducing, Schedule 40 carbon steel, butt weld | Ml@2.44 | Ea | 35.60 | 129.00 | 164.60 |
| Caps, 4" | | | | | |
| Schedule 40, carbon steel, butt weld | Ml@1.25 | Ea | 15.80 | 66.20 | 82.00 |
| Schedule 80, carbon steel, butt weld | Ml@1.50 | Ea | 11.50 | 79.50 | 91.00 |
| Reducing couplings, 4" | | | | | |
| Schedule 40, carbon steel, butt weld | Ml@1.60 | Ea | 15.00 | 84.80 | 99.80 |
| Weld-o-lets, 4" | | | | | |
| Schedule 40, carbon steel | Ml@2.67 | Ea | 56.90 | 141.00 | 197.90 |
| Welded flanges, forged steel, 4" | | | | | |
| 150 lb, slip-on flange | Ml@.980 | Ea | 25.80 | 51.90 | 77.70 |
| 300 lb, slip-on flange | Ml@1.30 | Ea | 41.60 | 68.90 | 110.50 |
| 600 lb, slip-on flange | Ml@1.80 | Ea | 102.00 | 95.40 | 197.40 |
| 150 lb, weld neck flange | Ml@.980 | Ea | 32.00 | 51.90 | 83.90 |
| 300 lb, weld neck flange | Ml@1.30 | Ea | 46.00 | 68.90 | 114.90 |
| 600 lb, weld neck flange | Ml@1.80 | Ea | 91.00 | 95.40 | 186.40 |
| 150 lb, threaded flange | Ml@.600 | Ea | 37.00 | 31.80 | 68.80 |
| 300 lb, threaded flange | Ml@.640 | Ea | 47.00 | 33.90 | 80.90 |
| 600 lb, threaded flange | Ml@.650 | Ea | 118.00 | 34.40 | 152.40 |
| Flange bolt & gasket kits, 4" | | | | | |
| Full face, red rubber gasket, 150 lb | Ml@1.00 | Ea | 9.68 | 53.00 | 62.68 |
| Full face, red rubber gasket, 300 lb | Ml@1.20 | Ea | 14.30 | 63.60 | 77.90 |

# 22 Plumbing

| | Craft@Hrs | Unit | Material | Labor | Total |
|---|---|---|---|---|---|
| **6" carbon steel pipe (A53) and welded fittings** | | | | | |
| Pipe, no fittings or supports included, 6", plain end | | | | | |
|     Schedule 40 to 10' high | M8@.420 | LF | 15.10 | 22.90 | 38.00 |
|     Schedule 40 over 10' to 20' | M8@.500 | LF | 15.10 | 27.30 | 42.40 |
|     Schedule 80 to 10' high | M8@.460 | LF | 28.80 | 25.10 | 53.90 |
|     Schedule 80 over 10' to 20' | M8@.550 | LF | 28.80 | 30.00 | 58.80 |
|     Pipe assembly, Schedule 40, fittings, hangers and supports | M8@.800 | LF | 19.30 | 43.70 | 63.00 |
|     Add for seamless pipe (A106) | — | % | 40.0 | — | — |
|     Add for galvanized pipe assembly, Schedule 40 | — | % | 30.0 | — | — |
| Butt welded joints, 6" pipe | | | | | |
|     Schedule 40 pipe | M8@1.87 | Ea | — | 102.00 | 102.00 |
|     Schedule 80 pipe | M8@2.24 | Ea | — | 122.00 | 122.00 |
| 90-degree ells, 6" | | | | | |
|     Schedule 40 carbon steel, butt weld, long turn | M8@2.67 | Ea | 45.70 | 146.00 | 191.70 |
|     Schedule 80 carbon steel, butt weld, long turn | M8@3.55 | Ea | 65.50 | 194.00 | 259.50 |
| 45-degree ells, 6" | | | | | |
|     Schedule 40, carbon steel, butt weld | M8@2.67 | Ea | 30.00 | 146.00 | 176.00 |
|     Schedule 80, carbon steel, butt weld | M8@3.55 | Ea | 46.70 | 194.00 | 240.70 |
| Tees, 6" | | | | | |
|     Schedule 40, carbon steel, butt weld | M8@4.00 | Ea | 64.00 | 219.00 | 283.00 |
|     Schedule 80, carbon steel, butt weld | M8@5.32 | Ea | 87.00 | 291.00 | 378.00 |
|     Reducing, Schedule 40 carbon steel, butt weld | M8@3.78 | Ea | 80.40 | 207.00 | 287.40 |
| Caps, 6" | | | | | |
|     Schedule 40, carbon steel, butt weld | M8@1.87 | Ea | 18.40 | 102.00 | 120.40 |
|     Schedule 80, carbon steel, butt weld | M8@2.24 | Ea | 21.50 | 122.00 | 143.50 |
| Reducing couplings, 6" | | | | | |
|     Schedule 40, carbon steel, butt weld | M8@2.30 | Ea | 27.80 | 126.00 | 153.80 |
| Weld-o-lets, 6" | | | | | |
|     Schedule 40, carbon steel | M8@3.65 | Ea | 233.00 | 199.00 | 432.00 |
| Welded flanges, forged steel, 6" | | | | | |
|     150 lb, slip-on flange | M8@1.47 | Ea | 40.90 | 80.30 | 121.20 |
|     300 lb, slip-on flange | M8@1.95 | Ea | 77.60 | 107.00 | 184.60 |
|     150 lb, weld neck flange | M8@1.47 | Ea | 53.60 | 80.30 | 133.90 |
|     300 lb, weld neck flange | M8@1.95 | Ea | 82.90 | 107.00 | 189.90 |
| Flange bolt & gasket kits, 6" | | | | | |
|     Full face, red rubber gasket, 150 lb | M8@1.20 | Ea | 16.40 | 65.60 | 82.00 |
|     Full face, red rubber gasket, 300 lb | M8@1.65 | Ea | 26.30 | 90.10 | 116.40 |
| **8" carbon steel pipe (A53) and welded fittings** | | | | | |
| Pipe, no fittings or supports included, 8" plain end | | | | | |
|     Schedule 40 to 10' high | M8@.510 | LF | 19.70 | 27.90 | 47.60 |
|     Schedule 40 over 10' to 20' | M8@.600 | LF | 19.70 | 32.80 | 52.50 |
|     Schedule 80 to 10' high | M8@.560 | LF | 38.10 | 30.60 | 68.70 |
|     Schedule 80 over 10' to 20' high | M8@.670 | LF | 38.10 | 36.60 | 74.70 |
|     Add for seamless pipe (A106) | — | % | 40.0 | — | — |
| Butt welded joints, 8" pipe | | | | | |
|     Schedule 40 pipe | M8@2.24 | Ea | — | 122.00 | 122.00 |
|     Schedule 80 pipe | M8@2.68 | Ea | — | 146.00 | 146.00 |
| 90-degree ells, 8" | | | | | |
|     Schedule 40 carbon steel, butt weld, long turn | M8@3.20 | Ea | 91.60 | 175.00 | 266.60 |
|     Schedule 80 carbon steel, butt weld, long turn | M8@4.26 | Ea | 105.00 | 233.00 | 338.00 |

| | Craft@Hrs | Unit | Material | Labor | Total |
|---|---|---|---|---|---|
| 45-degree ells, 8" | | | | | |
|    Schedule 40, carbon steel, butt weld | M8@3.20 | Ea | 62.20 | 175.00 | 237.20 |
|    Schedule 80, carbon steel, butt weld | M8@4.26 | Ea | 66.20 | 233.00 | 299.20 |
| Tees, 8" | | | | | |
|    Schedule 40, carbon steel, butt weld | M8@4.80 | Ea | 130.00 | 262.00 | 392.00 |
|    Schedule 80, carbon steel, butt weld | M8@6.38 | Ea | 153.00 | 349.00 | 502.00 |
|    Reducing, Schedule 40, carbon steel, butt weld | M8@4.40 | Ea | 173.00 | 240.00 | 413.00 |
| Caps, 8" | | | | | |
|    Schedule 40, carbon steel, butt weld | M8@2.24 | Ea | 28.90 | 122.00 | 150.90 |
|    Schedule 80, carbon steel, butt weld | M8@2.68 | Ea | 29.10 | 146.00 | 175.10 |
| Reducing couplings, 8" | | | | | |
|    Schedule 40, carbon steel, butt weld | M8@2.88 | Ea | 64.90 | 157.00 | 221.90 |
| Weld-o-lets, 8" | | | | | |
|    Schedule 40, carbon steel | M8@4.76 | Ea | 386.00 | 260.00 | 646.00 |
| Welded flanges, forged steel, 8" | | | | | |
|    150 lb, slip-on flange | M8@1.77 | Ea | 56.00 | 96.70 | 152.70 |
|    300 lb, slip-on flange | M8@2.35 | Ea | 107.00 | 128.00 | 235.00 |
|    150 lb, weld neck flange | M8@1.77 | Ea | 78.00 | 96.70 | 174.70 |
|    300 lb, weld neck flange | M8@2.35 | Ea | 113.00 | 128.00 | 241.00 |
| Flange bolt & gasket kits, 8" | | | | | |
|    Full face, red rubber gasket, 150 lb | M8@1.40 | Ea | 15.80 | 76.50 | 92.30 |
|    Full face, red rubber gasket, 300 lb | M8@1.85 | Ea | 24.80 | 101.00 | 125.80 |

## 10" carbon steel pipe (A53) and welded fittings

| | Craft@Hrs | Unit | Material | Labor | Total |
|---|---|---|---|---|---|
| Pipe, no fittings or supports included, 10", plain end | | | | | |
|    Schedule 40 to 20' high | M8@.720 | LF | 28.10 | 39.30 | 67.40 |
|    Schedule 80 to 20' high | M8@.830 | LF | 66.00 | 45.30 | 111.30 |
|    Add for seamless pipe (A106) | — | % | 40.0 | — | — |
| Butt welded joints, 10" pipe | | | | | |
|    Schedule 40 pipe | M8@2.70 | Ea | — | 148.00 | 148.00 |
|    Schedule 80 pipe | M8@3.22 | Ea | — | 176.00 | 176.00 |
| 90-degree ells, 10" | | | | | |
|    Schedule 40 carbon steel, butt weld, long turn | M8@4.00 | Ea | 164.00 | 219.00 | 383.00 |
|    Schedule 80 carbon steel, butt weld, long turn | M8@5.75 | Ea | 191.00 | 314.00 | 505.00 |
| 45-degree ells, 10" | | | | | |
|    Schedule 40, carbon steel, butt weld | M8@4.00 | Ea | 95.30 | 219.00 | 314.30 |
|    Schedule 80, carbon steel, butt weld | M8@5.75 | Ea | 106.00 | 314.00 | 420.00 |
| Tees, 10" | | | | | |
|    Schedule 40, carbon steel, butt weld | M8@6.00 | Ea | 293.00 | 328.00 | 621.00 |
|    Schedule 80, carbon steel, butt weld | M8@8.61 | Ea | 247.00 | 470.00 | 717.00 |
|    Reducing, Schedule 40, carbon steel, butt weld | M8@5.60 | Ea | 293.00 | 306.00 | 599.00 |
| Caps, 10" | | | | | |
|    Schedule 40, carbon steel, butt weld | M8@2.80 | Ea | 48.70 | 153.00 | 201.70 |
|    Schedule 80, carbon steel, butt weld | M8@3.62 | Ea | 43.10 | 198.00 | 241.10 |
| Reducing couplings, 10" | | | | | |
|    Schedule 40, carbon steel, butt weld | M8@3.60 | Ea | 57.40 | 197.00 | 254.40 |
| Weld-o-lets, 10" | | | | | |
|    Schedule 40, carbon steel | M8@6.35 | Ea | 529.00 | 347.00 | 876.00 |
| Welded pipe flanges, forged steel, 10" | | | | | |
|    150 lb, slip-on flange | M8@2.20 | Ea | 92.40 | 120.00 | 212.40 |
|    150 lb, weld neck flange | M8@2.20 | Ea | 127.00 | 120.00 | 247.00 |
|    300 lb, weld neck flange | M8@2.90 | Ea | 257.00 | 158.00 | 415.00 |

| | Craft@Hrs | Unit | Material | Labor | Total |
|---|---|---|---|---|---|
| Flange bolt & gasket kits, 10" | | | | | |
| Full face, red rubber gasket, 150 lb | M8@1.80 | Ea | 31.20 | 98.30 | 129.50 |
| Full face, red rubber gasket, 300 lb | M8@2.20 | Ea | 39.60 | 120.00 | 159.60 |
| **12" carbon steel pipe (A53) and welded fittings** | | | | | |
| Pipe, no fittings or supports included, 12", plain end | | | | | |
| Schedule 40 to 20' high | M8@.910 | LF | 33.90 | 49.70 | 83.60 |
| Schedule 80 to 20' high | M8@1.05 | LF | 75.90 | 57.40 | 133.30 |
| Add for seamless pipe (A106) | — | % | 40.0 | — | — |
| Butt welded joints, 12" pipe | | | | | |
| Schedule 40 pipe | M8@3.25 | Ea | — | 178.00 | 178.00 |
| Schedule 80 pipe | M8@3.88 | Ea | — | 212.00 | 212.00 |
| 90-degree ells, 12" | | | | | |
| Schedule 40 carbon steel, butt weld, long turn | M8@4.80 | Ea | 244.00 | 262.00 | 506.00 |
| Schedule 80 carbon steel, butt weld, long turn | M8@7.76 | Ea | 269.00 | 424.00 | 693.00 |
| 45-degree ells, 12" | | | | | |
| Schedule 40, carbon steel, butt weld | M8@4.80 | Ea | 139.00 | 262.00 | 401.00 |
| Schedule 80, carbon steel, butt weld | M8@7.76 | Ea | 159.00 | 424.00 | 583.00 |
| Tees, 12" | | | | | |
| Schedule 40, carbon steel, butt weld | M8@7.20 | Ea | 336.00 | 393.00 | 729.00 |
| Schedule 80, carbon steel, butt weld | M8@11.6 | Ea | 357.00 | 634.00 | 991.00 |
| Reducing, Schedule 40, carbon steel, butt weld | M8@6.80 | Ea | 437.00 | 371.00 | 808.00 |
| Caps, 12" | | | | | |
| Schedule 40, carbon steel, butt weld | M8@3.36 | Ea | 61.00 | 184.00 | 245.00 |
| Schedule 80, carbon steel, butt weld | M8@4.88 | Ea | 57.10 | 267.00 | 324.10 |
| Reducing couplings, 12" | | | | | |
| Schedule 40, carbon steel, butt weld | M8@4.30 | Ea | 169.00 | 235.00 | 404.00 |
| Weld-o-lets, 12" | | | | | |
| Schedule 40, carbon steel | M8@8.48 | Ea | 649.00 | 463.00 | 1,112.00 |
| Welded flanges, forged steel, 12" | | | | | |
| 150 lb, slip-on flange | M8@2.64 | Ea | 188.00 | 144.00 | 332.00 |
| 300 lb, slip-on flange | M8@3.50 | Ea | 316.00 | 191.00 | 507.00 |
| 150 lb, weld neck flange | M8@2.64 | Ea | 240.00 | 144.00 | 384.00 |
| 300 lb, weld neck flange | M8@3.50 | Ea | 420.00 | 191.00 | 611.00 |
| Flange bolt & gasket kits, 12" | | | | | |
| Full face, red rubber gasket, 150 lb | M8@2.20 | Ea | 41.10 | 120.00 | 161.10 |
| Full face, red rubber gasket, 300 lb | M8@2.60 | Ea | 53.10 | 142.00 | 195.10 |

**Copper Pressure Pipe and Soldered Copper Fittings** Note that copper pipe and fitting prices can change very quickly as the price of copper changes. No hangers included. See pipe hangers in the sections that follow. Labor cost assumes pipe is installed either horizontally or vertically in a building and up to 10' above floor level. Add 25% to the labor cost for heights over 10' above floor level. For more detailed coverage of copper pressure pipe, see *National Plumbing & HVAC Estimator*, http://CraftsmanSiteLicense.com.

**Type M hard copper pipe** Used for water supply pressure piping above ground inside residential buildings. Smaller quantities purchased from building material retailers may cost more. Larger quantities purchased from specialty suppliers will cost less. Add the cost of soldered wrought copper fittings and supports in the sections that follow.

| | Craft@Hrs | Unit | Material | Labor | Total |
|---|---|---|---|---|---|
| 1/2" pipe | P6@.032 | LF | .99 | 1.61 | 2.60 |
| 3/4" pipe | P6@.035 | LF | 1.63 | 1.76 | 3.39 |
| 1" pipe | P6@.038 | LF | 2.89 | 1.91 | 4.80 |
| 1-1/4" pipe | P6@.042 | LF | 4.37 | 2.11 | 6.48 |

| | Craft@Hrs | Unit | Material | Labor | Total |
|---|---|---|---|---|---|
| 1-1/2" pipe | P6@.046 | LF | 5.44 | 2.31 | 7.75 |
| 2" pipe | P6@.053 | LF | 9.53 | 2.66 | 12.19 |
| 2-1/2" pipe | P6@.060 | LF | 12.10 | 3.01 | 15.11 |
| 3" pipe | P6@.066 | LF | 15.40 | 3.31 | 18.71 |
| 4" pipe | P6@.080 | LF | 26.70 | 4.02 | 30.72 |

**Type L hard copper pipe** is commonly used for water supply pressure piping in commercial buildings, either above or below ground. Larger quantities will cost less. Smaller quantities purchased from building material retailers will cost more. Add the cost of soldered wrought copper fittings and supports in the sections that follow.

| | Craft@Hrs | Unit | Material | Labor | Total |
|---|---|---|---|---|---|
| 1/2" pipe | P6@.032 | LF | 1.42 | 1.61 | 3.03 |
| 3/4" pipe | P6@.035 | LF | 2.25 | 1.76 | 4.01 |
| 1" pipe | P6@.038 | LF | 3.63 | 1.91 | 5.54 |
| 1-1/4" pipe | P6@.042 | LF | 5.09 | 2.11 | 7.20 |
| 1-1/2" pipe | P6@.046 | LF | 6.45 | 2.31 | 8.76 |
| 2" pipe | P6@.053 | LF | 9.40 | 2.66 | 12.06 |
| 2-1/2" pipe | P6@.060 | LF | 16.90 | 3.01 | 19.91 |
| 3" pipe | P6@.066 | LF | 22.80 | 3.31 | 26.11 |
| 4" pipe | P6@.080 | LF | 36.30 | 4.02 | 40.32 |

**Type L soft copper tube** Type L tube is commonly used for water supply pressure piping in commercial buildings, either above or below ground. Commercial quantities will cost less. Smaller quantities purchased from building material retailers will cost more. Add the cost of copper fittings and supports in the sections that follow.

| | Craft@Hrs | Unit | Material | Labor | Total |
|---|---|---|---|---|---|
| 1/4", 60' coil | P6@.020 | LF | .83 | 1.00 | 1.83 |
| 3/8", per foot | P6@.020 | LF | 1.48 | 1.00 | 2.48 |
| 1/2", per foot | P6@.025 | LF | 2.73 | 1.26 | 3.99 |
| 1/2", 10' coil | P6@.025 | LF | 2.36 | 1.26 | 3.62 |
| 1/2", 20' coil | P6@.025 | LF | 2.24 | 1.26 | 3.50 |
| 1/2", 60' coil | P6@.025 | LF | 2.42 | 1.26 | 3.68 |
| 1/2", 100' coil | P6@.025 | LF | 2.25 | 1.26 | 3.51 |
| 5/8", 20' coil | P6@.030 | LF | 2.27 | 1.51 | 3.78 |
| 3/4", 60' coil | P6@.030 | LF | 2.70 | 1.51 | 4.21 |
| 3/4", 100' coil | P6@.030 | LF | 3.59 | 1.51 | 5.10 |
| 1", per foot | P6@.035 | LF | 4.56 | 1.76 | 6.32 |
| 1", 100' coil | P6@.035 | LF | 4.42 | 1.76 | 6.18 |
| 1-1/4", 60' coil | P6@.042 | LF | 6.32 | 2.11 | 8.43 |
| 1-1/2", 60' coil | P6@.046 | LF | 8.08 | 2.31 | 10.39 |

**Type K hard copper pipe** Installed in an open trench to 5' deep. No excavation or backfill included. Type K is acceptable for applications above and below ground, inside and outside of buildings. Use corporation (compression) fittings rather than soldered joints in underground applications. Based on commercial quantity purchases. Smaller quantities will cost up to 35% more. Add the cost of fittings and hangers when required.

| | Craft@Hrs | Unit | Material | Labor | Total |
|---|---|---|---|---|---|
| 1/2" pipe | P6@.032 | LF | 2.00 | 1.61 | 3.61 |
| 3/4" pipe | P6@.035 | LF | 3.72 | 1.76 | 5.48 |

**Type K soft copper tube** Type K is acceptable for applications above and below ground, inside and outside of buildings. Use corporation (compression) fittings rather than soldered joints in underground applications. Commercial quantities will cost less. Smaller quantities purchased from building material retailers will cost more. Add the cost of copper fittings and supports in the sections that follow.

| | Craft@Hrs | Unit | Material | Labor | Total |
|---|---|---|---|---|---|
| 1/2", 60' coil | P6@.025 | LF | 2.70 | 1.26 | 3.96 |
| 3/4", 60' coil | P6@.030 | LF | 5.44 | 1.51 | 6.95 |
| 3/4", 100' coil | P6@.030 | LF | 5.53 | 1.51 | 7.04 |
| 1", 100' coil | P6@.035 | LF | 6.92 | 1.76 | 8.68 |

| | Craft@Hrs | Unit | Material | Labor | Total |
|---|---|---|---|---|---|

**Copper pressure fittings** Soldered wrought copper fittings. Solder is 95% tin and 5% antimony (lead free). Installed either horizontally or vertically in a building up to 10' above floor level.

1/2" copper pressure fittings

| | | | | | |
|---|---|---|---|---|---|
| 1/2" 90-degree ell | P6@.107 | Ea | .50 | 5.37 | 5.87 |
| 1/2" 45-degree el | P6@.107 | Ea | 1.05 | 5.37 | 6.42 |
| 1/2" tee | P6@.129 | Ea | .71 | 6.48 | 7.19 |
| 1/2" x 1/2" x 3/8" reducing tee | P6@.121 | Ea | 6.28 | 6.08 | 12.36 |
| 1/2" cap | P6@.069 | Ea | .60 | 3.47 | 4.07 |
| 1/2" coupling | P6@.107 | Ea | .36 | 5.37 | 5.73 |
| 1/2" copper x male pipe thread adapter | P6@.086 | Ea | .84 | 4.32 | 5.16 |

3/4" copper pressure fittings

| | | | | | |
|---|---|---|---|---|---|
| 3/4" 90-degree ell | P6@.150 | Ea | 1.03 | 7.53 | 8.56 |
| 3/4" 45-degree ell | P6@.150 | Ea | 1.89 | 7.53 | 9.42 |
| 3/4" tee | P6@.181 | Ea | 1.77 | 9.09 | 10.86 |
| 3/4" x 3/4" x 1/2" reducing tee | P6@.170 | Ea | 2.85 | 8.54 | 11.39 |
| 3/4" cap | P6@.096 | Ea | .89 | 4.82 | 5.71 |
| 3/4" coupling | P6@.150 | Ea | .70 | 7.53 | 8.23 |
| 3/4" copper x male pipe thread adapter | P6@.105 | Ea | 1.84 | 5.27 | 7.11 |

1" copper pressure fittings

| | | | | | |
|---|---|---|---|---|---|
| 1" 90-degree ell | P6@.193 | Ea | 3.52 | 9.69 | 13.21 |
| 1" 45-degree ell | P6@.193 | Ea | 5.85 | 9.69 | 15.54 |
| 1" tee | P6@.233 | Ea | 9.07 | 11.70 | 20.77 |
| 1" x 1" x 3/4" reducing tee | P6@.219 | Ea | 9.38 | 11.00 | 20.38 |
| 1" cap | P6@.124 | Ea | 2.22 | 6.23 | 8.45 |
| 1" coupling | P6@.193 | Ea | 2.22 | 9.69 | 11.91 |
| 1" copper x male pipe thread adapter | P6@.134 | Ea | 6.53 | 6.73 | 13.26 |

1-1/4" copper pressure fittings

| | | | | | |
|---|---|---|---|---|---|
| 1-1/4" 90-degree ell | P6@.236 | Ea | 5.72 | 11.90 | 17.62 |
| 1-1/4" 45-degree ell | P6@.236 | Ea | 8.04 | 11.90 | 19.94 |
| 1-1/4" tee | P6@.285 | Ea | 13.50 | 14.30 | 27.80 |
| 1-1/4" x 1-1/4" x 1" reducing tee | P6@.268 | Ea | 15.40 | 13.50 | 28.90 |
| 1-1/4" cap | P6@.151 | Ea | 3.11 | 7.58 | 10.69 |
| 1-1/4" coupling | P6@.236 | Ea | 3.85 | 11.90 | 15.75 |
| 1-1/4" copper x male pipe thread adapter | P6@.163 | Ea | 9.82 | 8.19 | 18.01 |

1-1/2" copper pressure fittings

| | | | | | |
|---|---|---|---|---|---|
| 1-1/2" 90-degree ell | P6@.278 | Ea | 9.33 | 14.00 | 23.33 |
| 1-1/2" 45-degree ell | P6@.278 | Ea | 9.73 | 14.00 | 23.73 |
| 1-1/2" tee | P6@.337 | Ea | 19.80 | 16.90 | 36.70 |
| 1-1/2" x 1-1/2" x 1-1/4" reducing tee | P6@.302 | Ea | 17.90 | 15.20 | 33.10 |
| 1-1/2" cap | P6@.178 | Ea | 4.56 | 8.94 | 13.50 |
| 1-1/2" coupling | P6@.278 | Ea | 6.09 | 14.00 | 20.09 |
| 1-1/2" copper x male pipe thread adapter | P6@.194 | Ea | 12.70 | 9.74 | 22.44 |

2" copper pressure fittings

| | | | | | |
|---|---|---|---|---|---|
| 2" 90-degree ell | P6@.371 | Ea | 15.70 | 18.60 | 34.30 |
| 2" 45-degree ell | P6@.371 | Ea | 14.50 | 18.60 | 33.10 |
| 2" tee | P6@.449 | Ea | 26.80 | 22.50 | 49.30 |
| 2" x 2" x 1-1/2" reducing tee | P6@.422 | Ea | 24.20 | 21.20 | 45.40 |
| 2" cap | P6@.238 | Ea | 8.33 | 12.00 | 20.33 |
| 2" coupling | P6@.371 | Ea | 10.70 | 18.60 | 29.30 |
| 2" copper x male pipe thread adapter | P6@.259 | Ea | 18.10 | 13.00 | 31.10 |

| | Craft@Hrs | Unit | Material | Labor | Total |
|---|---|---|---|---|---|
| 2-1/2" copper pressure fittings | | | | | |
| 2-1/2" 90-degree ell | P6@.457 | Ea | 32.00 | 23.00 | 55.00 |
| 2-1/2" 45-degree ell | P6@.457 | Ea | 38.90 | 23.00 | 61.90 |
| 2-1/2" tee | P6@.552 | Ea | 61.90 | 27.70 | 89.60 |
| 2-1/2" x 2 1/2" x 2" reducing tee | P6@.519 | Ea | 72.90 | 26.10 | 99.00 |
| 2-1/2" cap | P6@.292 | Ea | 27.50 | 14.70 | 42.20 |
| 2-1/2" coupling | P6@.457 | Ea | 18.90 | 23.00 | 41.90 |
| 2-1/2" copper x male pipe thread adapter | P6@.319 | Ea | 58.00 | 16.00 | 74.00 |
| 3" copper pressure fittings | | | | | |
| 3" 90-degree ell | P6@.543 | Ea | 40.20 | 27.30 | 67.50 |
| 3" 45-degree ell | P6@.543 | Ea | 50.60 | 27.30 | 77.90 |
| 3" tee | P6@.656 | Ea | 94.90 | 32.90 | 127.80 |
| 3" x 3" x 2" reducing tee | P6@.585 | Ea | 82.90 | 29.40 | 112.30 |
| 3" cap | P6@.347 | Ea | 33.30 | 17.40 | 50.70 |
| 3" coupling | P6@.543 | Ea | 28.80 | 27.30 | 56.10 |
| 3" copper x male pipe thread adapter | P6@.375 | Ea | 97.40 | 18.80 | 116.20 |
| 4" copper pressure fittings | | | | | |
| 4" 90-degree ell | P6@.714 | Ea | 97.40 | 35.90 | 133.30 |
| 4" 45-degree ell | P6@.714 | Ea | 106.00 | 35.90 | 141.90 |
| 4" tee | P6@.863 | Ea | 207.00 | 43.30 | 250.30 |
| 4" x 4" x 3" reducing tee | P6@.811 | Ea | 156.00 | 40.70 | 196.70 |
| 4" cap | P6@.457 | Ea | 70.10 | 23.00 | 93.10 |
| 4" coupling | P6@.714 | Ea | 67.90 | 35.90 | 103.80 |
| 4" copper x male pipe thread adapter | P6@.485 | Ea | 208.00 | 24.40 | 232.40 |

**Flexible PVC-Jacketed Copper Tubing** For fuel oil, gas and propane transfer lines. Flame retardant and dielectric. Fungus, acid and moisture corrosion resistant. Acceptable for direct burial underground or under concrete. Installed in an open trench.

| | Craft@Hrs | Unit | Material | Labor | Total |
|---|---|---|---|---|---|
| 3/8" OD | PM@.010 | LF | 1.13 | .42 | 1.55 |
| 1/2" OD | PM@.010 | LF | 1.35 | .42 | 1.77 |
| 5/8" OD | PM@.010 | LF | 1.88 | .42 | 2.30 |

Shrinkage kit for flexible PVC-jacketed copper tubing. Add for shrinkage tubing kit to allow for the coupling of sections with standard flare fittings while maintaining the integrity of the connection.

| | | Unit | Material | Labor | Total |
|---|---|---|---|---|---|
| 3/8" kit | — | Ea | 10.50 | — | 10.50 |
| 1/2" kit | — | Ea | 11.00 | — | 11.00 |
| 5/8" kit | — | Ea | 12.80 | — | 12.80 |

**Schedule 40 PVC Pressure Pipe and Socket-weld Fittings** Installed in a building vertically or horizontally up to 10' above floor level.

| | Craft@Hrs | Unit | Material | Labor | Total |
|---|---|---|---|---|---|
| 1/2" Schedule 40 PVC pressure pipe and fittings | | | | | |
| 1/2" pipe | P6@.020 | LF | .13 | 1.00 | 1.13 |
| 1/2" 90-degree ell | P6@.100 | Ea | .23 | 5.02 | 5.25 |
| 1/2" tee | P6@.130 | Ea | .36 | 6.53 | 6.89 |
| 3/4" Schedule 40 PVC pressure pipe and fittings | | | | | |
| 3/4" pipe | P6@.025 | LF | .18 | 1.26 | 1.44 |
| 3/4" 90-degree ell | P6@.115 | Ea | .32 | 5.78 | 6.10 |
| 3/4" tee | P6@.140 | Ea | .35 | 7.03 | 7.38 |
| 1" Schedule 40 PVC pressure pipe and fittings | | | | | |
| 1" pipe | P6@.030 | LF | .26 | 1.51 | 1.77 |
| 1" 90-degree ell | P6@.120 | Ea | .60 | 6.03 | 6.63 |
| 1" tee | P6@.170 | Ea | .60 | 8.54 | 9.14 |

# 22 Plumbing

|  | Craft@Hrs | Unit | Material | Labor | Total |
|---|---|---|---|---|---|
| **1-1/4" Schedule 40 PVC pressure pipe and fittings** | | | | | |
| 1-1/4" pipe | P6@.035 | LF | .36 | 1.76 | 2.12 |
| 1-1/4" 90-degree ell | P6@.160 | Ea | 1.28 | 8.04 | 9.32 |
| 1-1/4" tee | P6@.220 | Ea | 1.28 | 11.00 | 12.28 |
| **1-1/2" Schedule 40 PVC pressure pipe and fittings** | | | | | |
| 1-1/2" pipe | P6@.040 | LF | .41 | 2.01 | 2.42 |
| 1-1/2" 90-degree ell | P6@.180 | Ea | 1.30 | 9.04 | 10.34 |
| 1-1/2" tee | P6@.250 | Ea | 1.65 | 12.60 | 14.25 |
| **2" Schedule 40 PVC pressure pipe and fittings** | | | | | |
| 2" pipe | P6@.046 | LF | .56 | 2.31 | 2.87 |
| 2" 90-degree ell | P6@.200 | Ea | 2.08 | 10.00 | 12.08 |
| 2" tee | P6@.280 | Ea | 2.33 | 14.10 | 16.43 |
| **2-1/2" Schedule 40 PVC pressure pipe and fittings** | | | | | |
| 2-1/2" pipe | P6@.053 | LF | .82 | 2.66 | 3.48 |
| 2-1/2" 90-degree ell | P6@.250 | Ea | 6.08 | 12.60 | 18.68 |
| 2-1/2" tee | P6@.350 | Ea | 7.89 | 17.60 | 25.49 |
| **3" Schedule 40 PVC pressure pipe and fittings** | | | | | |
| 3" pipe | P6@.055 | LF | 1.04 | 2.76 | 3.80 |
| 3" 90-degree ell | P6@.300 | Ea | 7.37 | 15.10 | 22.47 |
| 3" tee | P6@.420 | Ea | 9.02 | 21.10 | 30.12 |
| **4" Schedule 40 PVC pressure pipe and fittings** | | | | | |
| 4" pipe | P6@.070 | LF | 1.53 | 3.52 | 5.05 |
| 4" 90-degree ell | P6@.500 | Ea | 13.10 | 25.10 | 38.20 |
| 4" tee | P6@.560 | Ea | 18.70 | 28.10 | 46.80 |

**Schedule 80 PVC Pressure Pipe and Socket-Weld Fittings** Installed in a building vertically or horizontally up to 10' above floor level.

|  | Craft@Hrs | Unit | Material | Labor | Total |
|---|---|---|---|---|---|
| **1/2" Schedule 80 PVC pressure pipe and fittings** | | | | | |
| 1/2" pipe | P6@.021 | LF | .36 | 1.05 | 1.41 |
| 1/2" 90-degree ell | P6@.105 | Ea | .79 | 5.27 | 6.06 |
| 1/2" tee | P6@.135 | Ea | 2.05 | 6.78 | 8.83 |
| **3/4" Schedule 80 PVC pressure pipe and fittings** | | | | | |
| 3/4" pipe | P6@.026 | LF | .40 | 1.31 | 1.71 |
| 3/4" 90-degree ell | P6@.120 | Ea | 1.03 | 6.03 | 7.06 |
| 3/4" tee | P6@.145 | Ea | 2.12 | 7.28 | 9.40 |
| **1" Schedule 80 PVC pressure pipe and fittings** | | | | | |
| 1" pipe | P6@.032 | LF | .63 | 1.61 | 2.24 |
| 1" 90-degree ell | P6@.125 | Ea | 1.62 | 6.28 | 7.90 |
| 1" tee | P6@.180 | Ea | 2.66 | 9.04 | 11.70 |
| **1-1/4" Schedule 80 PVC pressure pipe and fittings** | | | | | |
| 1-1/4" pipe | P6@.037 | LF | .72 | 1.86 | 2.58 |
| 1-1/4" 90-degree ell | P6@.170 | Ea | 2.15 | 8.54 | 10.69 |
| 1-1/4" tee | P6@.230 | Ea | 7.33 | 11.60 | 18.93 |
| **1-1/2" Schedule 80 PVC pressure pipe and fittings** | | | | | |
| 1-1/2" pipe | P6@.042 | LF | .97 | 2.11 | 3.08 |
| 1-1/2" 90-degree ell | P6@.190 | Ea | 2.30 | 9.54 | 11.84 |
| 1-1/2" tee | P6@.265 | Ea | 7.33 | 13.30 | 20.63 |
| **2" Schedule 80 PVC pressure pipe and fittings** | | | | | |
| 2" pipe | P6@.047 | LF | 1.32 | 2.36 | 3.68 |
| 2" 90-degree ell | P6@.210 | Ea | 2.81 | 10.50 | 13.31 |
| 2" tee | P6@.295 | Ea | 9.16 | 14.80 | 23.96 |

| | Craft@Hrs | Unit | Material | Labor | Total |
|---|---|---|---|---|---|
| **2-1/2" Schedule 80 PVC pressure pipe and fittings** | | | | | |
| 2-1/2" pipe | P6@.052 | LF | 2.12 | 2.61 | 4.73 |
| 2-1/2" 90-degree ell | P6@.262 | Ea | 6.55 | 13.20 | 19.75 |
| 2-1/2" tee | P6@.368 | Ea | 9.96 | 18.50 | 28.46 |
| **3" Schedule 80 PVC pressure pipe and fittings** | | | | | |
| 3" pipe | P6@.057 | LF | 2.84 | 2.86 | 5.70 |
| 3" 90-degree ell | P6@.315 | Ea | 7.37 | 15.80 | 23.17 |
| 3" tee | P6@.440 | Ea | 12.40 | 22.10 | 34.50 |
| **4" Schedule 80 PVC pressure pipe and fittings** | | | | | |
| 4" pipe | P6@.074 | LF | 3.81 | 3.72 | 7.53 |
| 4" 90-degree ell | P6@.520 | Ea | 10.20 | 26.10 | 36.30 |
| 4" tee | P6@.585 | Ea | 13.10 | 29.40 | 42.50 |

**CPVC Chlorinated Pressure Pipe and Socket-Weld Fittings** Installed in a building vertically or horizontally up to 10' above floor level. CPVC pipe is the plastic alternative to traditional copper pipe. For hot and cold water distribution to 210 degrees F. Tan in color. Complies with ASTM D 2846 and has the same outside diameter as copper tubing. SDR 11 pressure pipe has the same pressure rating for all pipe diameters. FlowGuard Gold® pipe has a gold stripe.

| | Craft@Hrs | Unit | Material | Labor | Total |
|---|---|---|---|---|---|
| **1/2" CPVC chlorinated pressure pipe and fittings** | | | | | |
| #4120 SDR 11 CPVC pipe | P6@.020 | LF | .31 | 1.00 | 1.31 |
| FlowGuard Gold pipe | P6@.020 | LF | .39 | 1.00 | 1.39 |
| 90-degree elbow | P6@.100 | Ea | .27 | 5.02 | 5.29 |
| 90-degree drop-ear elbow | P6@.100 | Ea | .98 | 5.02 | 6.00 |
| FlowGuard drop-ear reducing elbow | P6@.100 | Ea | .59 | 5.02 | 5.61 |
| 45-degree elbow | P6@.100 | Ea | .36 | 5.02 | 5.38 |
| Coupling | P6@.100 | Ea | .25 | 5.02 | 5.27 |
| 90-degree street elbow | P6@.100 | Ea | .28 | 5.02 | 5.30 |
| Tee, 1/2" x 1/2" x 1/2" | P6@.130 | Ea | .31 | 6.53 | 6.84 |
| Tee, 1/2" x 3/4" x 3/4" | P6@.150 | Ea | .61 | 7.53 | 8.14 |
| Union | P6@.200 | Ea | 2.96 | 10.00 | 12.96 |
| Compression coupling | P6@.200 | Ea | 3.01 | 10.00 | 13.01 |
| Bushing, 1/2" x 1/8" bushing | P6@.080 | Ea | .63 | 4.02 | 4.65 |
| Cap | P6@.060 | Ea | .27 | 3.01 | 3.28 |
| **3/4" CPVC chlorinated pressure pipe and fittings** | | | | | |
| #4120 SDR 11 CPVC pipe | P6@.025 | LF | .50 | 1.26 | 1.76 |
| FlowGuard Gold pipe | P6@.025 | LF | .65 | 1.26 | 1.91 |
| 90-degree elbow | P6@.115 | Ea | .42 | 5.78 | 6.20 |
| 45-degree elbow | P6@.115 | Ea | .43 | 5.78 | 6.21 |
| Coupling | P6@.115 | Ea | .28 | 5.78 | 6.06 |
| 90-degree street elbow | P6@.115 | Ea | .47 | 5.78 | 6.25 |
| 90-degree reducing elbow, 3/4" x 1/2" | P6@.115 | Ea | .47 | 5.78 | 6.25 |
| Tee, 3/4" x 3/4" x 3/4" | P6@.140 | Ea | .59 | 7.03 | 7.62 |
| Tee, 3/4" x 1/2" x 1/2" | P6@.140 | Ea | .69 | 7.03 | 7.72 |
| Tee, 3/4" x 1/2" x 3/4" | P6@.150 | Ea | .61 | 7.53 | 8.14 |
| Tee, 3/4" x 3/4" x 1/2" | P6@.150 | Ea | .68 | 7.53 | 8.21 |
| Tee, 3/4" x 1/2" x 1/2" | P6@.150 | Ea | .69 | 7.53 | 8.22 |
| Cap | P6@.070 | Ea | .33 | 3.52 | 3.85 |
| Bushing, 1" x 1/2" | P6@.080 | Ea | .99 | 4.02 | 5.01 |
| Bushing, 3/4" x 1/2" | P6@.080 | Ea | .35 | 4.02 | 4.37 |
| Union | P6@.250 | Ea | 4.45 | 12.60 | 17.05 |
| Reducing coupling, 3/4" x 1/2" | P6@.115 | Ea | .33 | 5.78 | 6.11 |
| Compression coupling | P6@.200 | Ea | 3.57 | 10.00 | 13.57 |

| | Craft@Hrs | Unit | Material | Labor | Total |
|---|---|---|---|---|---|
| **1" CPVC chlorinated pressure pipe and fittings** | | | | | |
| FlowGuard Gold pipe | P6@.030 | Ea | 1.44 | 1.51 | 2.95 |
| 45-degree elbow | P6@.120 | Ea | 1.05 | 6.03 | 7.08 |
| Coupling | P6@.120 | Ea | 1.09 | 6.03 | 7.12 |
| Tee, 1" x 1" x 1" | P6@.170 | Ea | 2.69 | 8.54 | 11.23 |
| Tee, 1" x 1" x 3/4" | P6@.170 | Ea | 2.78 | 8.54 | 11.32 |
| Cap | P6@.070 | Ea | 0.78 | 3.52 | 4.30 |
| Bushing, 1" x 3/4" | P6@.080 | Ea | 1.09 | 4.02 | 5.11 |
| **CPVC and brass transition fittings. FIPT means female iron pipe thread. MIPT means male iron pipe thread.** | | | | | |
| 1/2" union, slip x FIPT | P6@.150 | Ea | 3.92 | 7.53 | 11.45 |
| 1/2" drop ear 90 degree elbow, slip x FIPT | P6@.090 | Ea | 4.56 | 4.52 | 9.08 |
| 1/2" union, slip x MIPT | P6@.250 | Ea | 3.35 | 12.60 | 15.95 |
| 1/2" x 3/4" union, slip x MIPT | P6@.250 | Ea | 5.47 | 12.60 | 18.07 |
| 3/4" union, slip x FIPT | P6@.170 | Ea | 4.77 | 8.54 | 13.31 |
| 3/4" union, slip x MIPT | P6@.250 | Ea | 5.58 | 12.60 | 18.18 |
| **CPVC to copper pipe transition fittings. One slip joint and one threaded joint. FIPT means female iron pipe thread. MIPT means male iron pipe thread.** | | | | | |
| 1/2" slip x FIPT | P6@.080 | Ea | 3.45 | 4.02 | 7.47 |
| 3/4" slip x FIPT | P6@.085 | Ea | 4.55 | 4.27 | 8.82 |
| 1/2" slip x MIPT | P6@.080 | Ea | 3.45 | 4.02 | 7.47 |
| 3/4" slip x MIPT | P6@.085 | Ea | 4.47 | 4.27 | 8.74 |
| **CPVC adapter fittings. One slip joint and one threaded joint. Joins CVPV pipe to threaded fittings or valves. FIPT means female iron pipe thread. MIPT means male iron pipe thread.** | | | | | |
| 1/2" slip x FIPT | P6@.090 | Ea | .93 | 4.52 | 5.45 |
| 1/2" slip x MIPT | P6@.090 | Ea | .29 | 4.52 | 4.81 |
| 1/2" SPIG x compression | P6@.090 | Ea | .63 | 4.52 | 5.15 |
| 3/4" slip x MIPT | P6@.100 | Ea | .43 | 5.02 | 5.45 |
| 3/4" slip x FIPT | P6@.100 | Ea | 1.07 | 5.02 | 6.09 |
| 1" slip x FIPT | P6@.120 | Ea | 1.31 | 6.03 | 7.34 |
| 1" slip x MIPT | P6@.120 | Ea | 1.81 | 6.03 | 7.84 |
| **CPVC snap-on tubing strap. Secures CPVC pipe in a horizontal or vertical position. Polypropylene.** | | | | | |
| 1/2" pipe | P6@.110 | Ea | .09 | 5.52 | 5.61 |
| 3/4" pipe | P6@.115 | Ea | .12 | 5.78 | 5.90 |

**Schedule 80 CPVC Chlorinated Pressure Pipe and Socket-weld Fittings** Installed in a building vertically or horizontally up to 10' above floor level. Typical use is in fire protection systems.

| | Craft@Hrs | Unit | Material | Labor | Total |
|---|---|---|---|---|---|
| **3/4" Schedule 80 CPVC chlorinated pressure pipe and fittings** | | | | | |
| 3/4" pipe | SP@.055 | LF | 1.93 | 3.40 | 5.33 |
| 3/4" 90-degree ell | SP@.100 | Ea | 1.81 | 6.18 | 7.99 |
| 3/4" tee | SP@.150 | Ea | 3.37 | 9.27 | 12.64 |
| **1" Schedule 80 CPVC chlorinated pressure pipe and fittings** | | | | | |
| 1" pipe | SP@.055 | LF | 1.89 | 3.40 | 5.29 |
| 1" 90-degree ell | SP@.120 | Ea | 2.86 | 7.41 | 10.27 |
| 1" tee | SP@.170 | Ea | 4.13 | 10.50 | 14.63 |
| **1-1/4" Schedule 80 CPVC chlorinated pressure pipe and fittings** | | | | | |
| 1-1/4" pipe | SP@.065 | LF | 2.98 | 4.02 | 7.00 |
| 1-1/4" 90-degree ell | SP@.170 | Ea | 6.27 | 10.50 | 16.77 |
| 1-1/4" tee | SP@.230 | Ea | 9.71 | 14.20 | 23.91 |
| **1-1/2" Schedule 80 CPVC chlorinated pressure pipe and fittings** | | | | | |
| 1-1/2" pipe | SP@.075 | LF | 3.12 | 4.63 | 7.75 |
| 1-1/2" 90-degree ell | SP@.180 | Ea | 6.90 | 11.10 | 18.00 |
| 1-1/2" tee | SP@.250 | Ea | 9.96 | 15.40 | 25.36 |

| | Craft@Hrs | Unit | Material | Labor | Total |
|---|---|---|---|---|---|
| 2" Schedule 80 CPVC chlorinated pressure pipe and fittings | | | | | |
| 2" pipe | SP@.085 | LF | 4.31 | 5.25 | 9.56 |
| 2" 90-degree ell | SP@.200 | Ea | 8.34 | 12.40 | 20.74 |
| 2" tee | SP@.280 | Ea | 11.00 | 17.30 | 28.30 |

**PEX Tube and Fittings** Cross-linked polyethylene tube installed in a building vertically or horizontally up to 10' above floor level. Using crimped connections. By copper tube size. Based on 100' rolls of color coded (red, blue or white) tube. For domestic water systems and in-floor hydronics heating systems.

| | Craft@Hrs | Unit | Material | Labor | Total |
|---|---|---|---|---|---|
| 3/8" PEX tube | | | | | |
| 3/8" AQUAPEX | P6@.019 | LF | .14 | .95 | 1.09 |
| 3/8" Oxygen Barrier PEX | P6@.019 | LF | .33 | .95 | 1.28 |
| 3/8" ThermaPEX | P6@.019 | LF | .49 | .95 | 1.44 |
| 3/8" hePEX | P6@.019 | LF | .67 | .95 | 1.62 |
| 3/8" PEX fittings | | | | | |
| 3/8" PEX brass elbow | P6@.160 | Ea | .69 | 8.04 | 8.73 |
| 3/8" PEX brass coupler | P6@.160 | Ea | .51 | 8.04 | 8.55 |
| 3/8" PEX crimp ring | P6@.008 | Ea | .16 | .40 | .56 |
| 3/8" x 3/8" x 3/8" PEX brass tee | P6@.240 | Ea | 1.21 | 12.10 | 13.31 |
| 3/8" PEX economy clamp | P6@.020 | Ea | .08 | 1.00 | 1.08 |
| 1/2" PEX tube | | | | | |
| 1/2" AQUAPEX, white | P6@.020 | LF | .38 | 1.00 | 1.38 |
| 1/2" AQUAPEX, red or blue | P6@.020 | LF | .40 | 1.00 | 1.40 |
| 1/2" PEX, red or blue | P6@.020 | LF | .33 | 1.00 | 1.33 |
| 1/2" ViegaPEX, red or blue | P6@.020 | LF | .39 | 1.00 | 1.39 |
| 1/2" Pre-Insulated AQUAPEX | P6@.020 | LF | 1.86 | 1.00 | 2.86 |
| 1/2" Oxygen Barrier PEX | P6@.020 | LF | .41 | 1.00 | 1.41 |
| 1/2" ThermaPEX | P6@.020 | LF | .51 | 1.00 | 1.51 |
| 1/2" hePEX | P6@.020 | LF | .69 | 1.00 | 1.69 |
| 1/2" PEX fittings | | | | | |
| 1/2" PEX brass elbow | P6@.160 | Ea | .30 | 8.04 | 8.34 |
| 1/2" PEX brass coupler | P6@.160 | Ea | .39 | 8.04 | 8.43 |
| 1/2" PEX ball valve | P6@.200 | Ea | 3.62 | 10.00 | 13.62 |
| 1/2" PEX crimp ring | P6@.008 | Ea | .16 | .40 | .56 |
| 1/2" x 1/2" x 1/2" PEX brass tee | P6@.240 | Ea | .75 | 12.10 | 12.85 |
| 1/2" PEX x 1/2" copper pipe adapter | P6@.160 | Ea | .64 | 8.04 | 8.68 |
| 1/2" PEX economy clamp | P6@.020 | Ea | .13 | 1.00 | 1.13 |
| 1/2" PEX suspension clip | P6@.020 | Ea | .20 | 1.00 | 1.20 |
| 5/8" PEX tube | | | | | |
| 5/8" Oxygen Barrier PEX | P6@.021 | LF | .53 | 1.05 | 1.58 |
| 5/8" ThermaPEX | P6@.021 | LF | .64 | 1.05 | 1.69 |
| 5/8" hePEX | P6@.021 | LF | .83 | 1.05 | 1.88 |
| 5/8" PEX fittings | | | | | |
| 5/8" PEX brass elbow | P6@.160 | Ea | 1.20 | 8.04 | 9.24 |
| 5/8" PEX brass coupler | P6@.160 | Ea | .80 | 8.04 | 8.84 |
| 5/8" PEX crimp ring | P6@.009 | Ea | .21 | .45 | .66 |
| 5/8" x 5/8" x 5/8" PEX brass tee | P6@.240 | Ea | 1.61 | 12.10 | 13.71 |
| 5/8" PEX x 1/2" copper pipe adapter | P6@.160 | Ea | 1.35 | 8.04 | 9.39 |
| 5/8" PEX x 3/4" copper pipe adapter | P6@.160 | Ea | 1.52 | 8.04 | 9.56 |

|  | Craft@Hrs | Unit | Material | Labor | Total |
|---|---|---|---|---|---|
| **3/4" PEX tube** | | | | | |
| 3/4" AQUAPEX, white | P6@.021 | LF | .69 | 1.05 | 1.74 |
| 3/4" AQUAPEX, red or blue | P6@.021 | LF | .65 | 1.05 | 1.70 |
| 3/4" PEX, red or blue | P6@.021 | LF | .58 | 1.05 | 1.63 |
| 3/4" ViegaPEX, red or blue | P6@.021 | LF | .63 | 1.05 | 1.68 |
| 3/4" Pre-Insulated AQUAPEX | P6@.021 | LF | 2.15 | 1.05 | 3.20 |
| 3/4" Oxygen Barrier PEX | P6@.021 | LF | .67 | 1.05 | 1.72 |
| 3/4" ThermaPEX | P6@.021 | LF | .90 | 1.05 | 1.95 |
| 3/4" hePEX | P6@.021 | LF | 1.04 | 1.05 | 2.09 |
| **3/4" PEX fittings** | | | | | |
| 3/4" PEX brass elbow | P6@.160 | Ea | .85 | 8.04 | 8.89 |
| 3/4" PEX brass coupler | P6@.160 | Ea | .54 | 8.04 | 8.58 |
| 3/4" PEX ball valve | P6@.200 | Ea | 5.20 | 10.00 | 15.20 |
| 3/4" PEX crimp ring | P6@.009 | Ea | .21 | .45 | .66 |
| 3/4" x 3/4" x 3/4" PEX brass tee | P6@.240 | Ea | 1.17 | 12.10 | 13.27 |
| 3/4" PEX x 3/4" copper pipe adapter | P6@.160 | Ea | .81 | 8.04 | 8.85 |
| 3/4" PEX economy clamp | P6@.160 | Ea | .19 | 8.04 | 8.23 |
| 3/4" PEX suspension clip | P6@.020 | Ea | .25 | 1.00 | 1.25 |
| **1" PEX tube** | | | | | |
| 1" AQUAPEX, white | P6@.025 | LF | 1.42 | 1.26 | 2.68 |
| 1" AQUAPEX, red or blue | P6@.025 | LF | 1.38 | 1.26 | 2.64 |
| 1" PEX, red or blue | P6@.025 | LF | .90 | 1.26 | 2.16 |
| 1" ViegaPEX, red or blue | P6@.025 | LF | 1.09 | 1.26 | 2.35 |
| 1" Pre-Insulated AQUAPEX | P6@.025 | LF | 2.71 | 1.26 | 3.97 |
| 1" Oxygen Barrier PEX | P6@.025 | LF | .99 | 1.26 | 2.25 |
| 1" ThermaPEX | P6@.025 | LF | 1.50 | 1.26 | 2.76 |
| 1" hePEX | P6@.025 | LF | 1.73 | 1.26 | 2.99 |
| **1" PEX fittings** | | | | | |
| 1" PEX brass elbow | P6@.200 | Ea | 2.36 | 10.00 | 12.36 |
| 1" PEX brass coupler | P6@.200 | Ea | 1.10 | 10.00 | 11.10 |
| 1" PEX x 1" copper pipe adapter | P6@.165 | Ea | 2.35 | 8.29 | 10.64 |
| 1" PEX ball valve | P6@.220 | Ea | 7.51 | 11.00 | 18.51 |
| 1" PEX crimp ring | P6@.010 | Ea | .22 | .50 | .72 |
| 1" x 1" x 1" PEX brass tee | P6@.240 | Ea | 2.30 | 12.10 | 14.40 |
| **PEX manifolds** | | | | | |
| 1/2" 14 port manifold (6 hot, 8 cold) | P6@1.25 | Ea | 121.00 | 62.80 | 183.80 |
| 1/2" 18 port manifold (8 hot, 10 cold) | P6@1.50 | Ea | 147.00 | 75.30 | 222.30 |
| 1/2" 24 port manifold (9 hot, 15 cold) | P6@1.95 | Ea | 187.00 | 97.90 | 284.90 |
| 1/2" 36 port manifold (14 hot, 22 cold) | P6@2.90 | Ea | 261.00 | 146.00 | 407.00 |
| 1" copper manifold w/ 6 ports | P6@.600 | Ea | 19.40 | 30.10 | 49.50 |
| 1" copper manifold w/ 12 ports | P6@1.20 | Ea | 35.30 | 60.30 | 95.60 |
| 2-Loop radiant heat manifold | P6@1.00 | Ea | 210.00 | 50.20 | 260.20 |
| 4-Loop radiant heat manifold | P6@1.50 | Ea | 304.00 | 75.30 | 379.30 |
| 8-Loop radiant heat manifold | P6@1.00 | Ea | 491.00 | 50.20 | 541.20 |

**PEX-AL Tube and Fittings** Cross-linked polyethylene tube with aluminum middle layer, installed in a building vertically or horizontally up to 10' above floor level. Using crimped connections. By copper tube size. Based on 100' rolls of tube. For domestic water systems and in-floor hydronics heating systems.

| 1/2" PEX-AL tube and fittings | | | | | |
|---|---|---|---|---|---|
| 1/2" PEX-AL tubing | P6@.020 | LF | .53 | 1.00 | 1.53 |
| 1/2" PEX-AL press coupling | P6@.160 | Ea | 6.23 | 8.04 | 14.27 |

| | Craft@Hrs | Unit | Material | Labor | Total |
|---|---|---|---|---|---|
| 1/2" PEX-AL x 3/4" male adapter | P6@.160 | Ea | 11.00 | 8.04 | 19.04 |
| 1/2" PEX-AL x 3/4" copper pipe adapter | P6@.180 | Ea | 6.53 | 9.04 | 15.57 |
| 1/2" PEX crimp ring | P6@.008 | Ea | .16 | .40 | .56 |
| 1/2" x 1/2" x 1/2" PEX brass tee | P6@.240 | Ea | .75 | 12.10 | 12.85 |
| 1/2" PEX economy clamp | P6@.020 | Ea | .13 | 1.00 | 1.13 |
| 1/2" PEX suspension clip | P6@.020 | Ea | .20 | 1.00 | 1.20 |
| **5/8" PEX-AL tube and fittings** | | | | | |
| 5/8" PEX-AL tubing | P6@.021 | LF | .73 | 1.05 | 1.78 |
| 5/8" PEX-AL press coupling | P6@.160 | Ea | 10.40 | 8.04 | 18.44 |
| 5/8" PEX-AL press elbow | P6@.160 | Ea | 6.76 | 8.04 | 14.80 |
| 5/8" PEX-AL x 3/4" male adapter | P6@.160 | Ea | 12.10 | 8.04 | 20.14 |
| 5/8" PEX-AL x 3/4" copper pipe adapter | P6@.180 | Ea | 11.50 | 9.04 | 20.54 |
| 5/8" PEX crimp ring | P6@.008 | Ea | .16 | .40 | .56 |
| 5/8" x 5/8" x 5/8" PEX brass tee | P6@.240 | Ea | .75 | 12.10 | 12.85 |
| 5/8" PEX economy clamp | P6@.020 | Ea | .13 | 1.00 | 1.13 |
| 5/8" PEX suspension clip | P6@.020 | Ea | .20 | 1.00 | 1.20 |
| **3/4" PEX-AL tube and fittings** | | | | | |
| 3/4" PEX-AL tubing | P6@.021 | LF | 1.25 | 1.05 | 2.30 |
| 3/4" PEX-AL press coupling | P6@.160 | Ea | 10.70 | 8.04 | 18.74 |
| 3/4" PEX-AL press elbow | P6@.160 | Ea | 9.07 | 8.04 | 17.11 |
| 3/4" PEX-AL press tee | P6@.240 | Ea | 17.20 | 12.10 | 29.30 |
| 3/4" PEX-AL x 3/4" male adapter | P6@.160 | Ea | 13.30 | 8.04 | 21.34 |
| 3/4" PEX-AL x 3/4" copper pipe adapter | P6@.180 | Ea | 9.53 | 9.04 | 18.57 |
| 3/4" PEX crimp ring | P6@.008 | Ea | .16 | .40 | .56 |
| 3/4" x 5/8" x 5/8" PEX brass tee | P6@.240 | Ea | .75 | 12.10 | 12.85 |
| 3/4" PEX economy clamp | P6@.020 | Ea | .13 | 1.00 | 1.13 |
| 3/4" PEX suspension clip | P6@.020 | Ea | .20 | 1.00 | 1.20 |
| **1" PEX-AL tube and fittings** | | | | | |
| 1" PEX-AL tubing | P6@.025 | LF | 1.67 | 1.26 | 2.93 |
| 1" PEX-AL press coupling | P6@.160 | Ea | 13.90 | 8.04 | 21.94 |
| 1" PEX-AL press elbow | P6@.160 | Ea | 13.60 | 8.04 | 21.64 |
| 1" PEX-AL press tee | P6@.240 | Ea | 20.50 | 12.10 | 32.60 |
| 1" PEX-AL x 3/4" male adapter | P6@.160 | Ea | 18.50 | 8.04 | 26.54 |
| 1" PEX-AL x 3/4" copper pipe adapter | P6@.180 | Ea | 12.20 | 9.04 | 21.24 |
| 1" PEX crimp ring | P6@.008 | Ea | .16 | .40 | .56 |
| 1" x 5/8" x 5/8" PEX brass tee | P6@.240 | Ea | .75 | 12.10 | 12.85 |
| 1" PEX economy clamp | P6@.020 | Ea | .13 | 1.00 | 1.13 |
| 1" PEX suspension clip | P6@.020 | Ea | .20 | 1.00 | 1.20 |

**Copper DWV Pipe and Fittings** Straight rigid lengths. For commercial drain, waste and vent (DWV) systems. Labor assumes soldered joints with pipe installed vertically in a building up to 10' above floor level. Add the cost of supports.

| 1-1/4" copper DWV pipe and fittings | Craft@Hrs | Unit | Material | Labor | Total |
|---|---|---|---|---|---|
| Cut pipe, per foot | P6@.042 | LF | 11.90 | 2.11 | 14.01 |
| 10' lengths | P6@.042 | LF | 4.26 | 2.11 | 6.37 |
| Coupling | P6@.200 | Ea | 3.14 | 10.00 | 13.14 |
| 90-degree elbow | P6@.200 | Ea | 6.91 | 10.00 | 16.91 |
| 45-degree elbow | P6@.200 | Ea | 5.13 | 10.00 | 15.13 |
| Tee | P6@.240 | Ea | 13.60 | 12.10 | 25.70 |
| Trap adapter | P6@.145 | Ea | 11.40 | 7.28 | 18.68 |

# 22 Plumbing

|  | Craft@Hrs | Unit | Material | Labor | Total |
|---|---|---|---|---|---|
| 1-1/2" copper DWV pipe and fittings | | | | | |
| Cut pipe, per foot | P6@.045 | LF | 8.89 | 2.26 | 11.15 |
| 10' lengths | P6@.045 | LF | 4.93 | 2.26 | 7.19 |
| Coupling | P6@.200 | Ea | 3.98 | 10.00 | 13.98 |
| 90-degree elbow | P6@.350 | Ea | 9.22 | 17.60 | 26.82 |
| 45-degree elbow | P6@.200 | Ea | 7.24 | 10.00 | 17.24 |
| Wye, cast brass | P6@.240 | Ea | 41.60 | 12.10 | 53.70 |
| Tee | P6@.240 | Ea | 16.60 | 12.10 | 28.70 |
| Test cap | P6@.075 | Ea | 1.19 | 3.77 | 4.96 |
| Male adapter, C x MIP | P6@.150 | Ea | 10.90 | 7.53 | 18.43 |
| Female adapter, FTG x C | P6@.200 | Ea | 8.28 | 10.00 | 18.28 |
| Trap adapter | P6@.155 | Ea | 14.10 | 7.78 | 21.88 |
| Trap adapter, with cleanout | P6@.235 | Ea | 45.80 | 11.80 | 57.60 |
| 2" copper DWV pipe and fittings | | | | | |
| Cut pipe, per foot | P6@.056 | LF | 11.80 | 2.81 | 14.61 |
| 10' lengths | P6@.056 | LF | 6.40 | 2.81 | 9.21 |
| Coupling | P6@.270 | Ea | 5.57 | 13.60 | 19.17 |
| 90-degree elbow | P6@.270 | Ea | 18.50 | 13.60 | 32.10 |
| 45-degree elbow | P6@.270 | Ea | 12.00 | 13.60 | 25.60 |
| Reducing bushing, 2" to 1-1/2" | P6@.140 | Ea | 12.20 | 7.03 | 19.23 |
| Wye, cast brass | P6@.320 | Ea | 36.30 | 16.10 | 52.40 |
| Tee, 2" x 2" x 2" | P6@.320 | Ea | 19.90 | 16.10 | 36.00 |
| Tee, 2" x 2" x 1-1/2" | P6@.240 | Ea | 20.20 | 12.10 | 32.30 |
| Male adapter C x MIP | P6@.185 | Ea | 14.00 | 9.29 | 23.29 |
| Female adapter C x FIP | P6@.259 | Ea | 16.10 | 13.00 | 29.10 |
| Trap adapter, with cleanout | P6@.325 | Ea | 76.30 | 16.30 | 92.60 |

**PVC SDR 35 DWV Drainage Pipe** Solvent weld, installed in an open trench inside a building (main building drain), including typical fittings (one wye and 45-degree elbow every 20'). SDR 35 is thin wall pipe suitable for below grade applications only. Add for excavation and backfill. Use these figures for preliminary estimates.

|  | Craft@Hrs | Unit | Material | Labor | Total |
|---|---|---|---|---|---|
| 3" pipe assembly below grade | P6@.160 | LF | 3.02 | 8.04 | 11.06 |
| 4" pipe assembly below grade | P6@.175 | LF | 4.61 | 8.79 | 13.40 |
| 6" pipe assembly below grade | P6@.220 | LF | 9.68 | 11.00 | 20.68 |
| Add for interior excavation and backfill (shallow) | CL@.180 | LF | — | 7.18 | 7.18 |

**ABS DWV Pipe and Fittings** Acrylonitrile Butadiene Styrene (ASTM D 2661-73) pipe is made from rubber-based resins and used primarily for drain, waste and vent systems. The Uniform Plumbing Code limits ABS to residential buildings and below grade applications in commercial and public buildings. Solvent-weld ABS is easier to install than cast iron or copper drainage pipe. ABS does not rot, rust, corrode or collect waste and resists mechanical damage, even at low temperatures.

|  | Craft@Hrs | Unit | Material | Labor | Total |
|---|---|---|---|---|---|
| 1-1/2" pipe, 10' lengths | P6@.020 | LF | .63 | 1.00 | 1.63 |
| 2" pipe, 10' lengths | P6@.025 | LF | .87 | 1.26 | 2.13 |
| 3" pipe, 10' lengths | P6@.030 | LF | 1.75 | 1.51 | 3.26 |
| 4" pipe, 10' lengths | P6@.040 | LF | 2.19 | 2.01 | 4.20 |
| Deduct for below grade applications | — | % | — | -15.0 | — |
| ABS 90-degree elbow (1/4 bend), hub and hub except as noted. Street elbows are hub and spigot. | | | | | |
| 1-1/2" ell | P6@.060 | Ea | 1.10 | 3.01 | 4.11 |
| 1-1/2" long sweep | P6@.200 | Ea | 2.32 | 10.00 | 12.32 |
| 1-1/2" street ell | P6@.120 | Ea | 2.03 | 6.03 | 8.06 |
| 1-1/2" venting | P6@.150 | Ea | 1.33 | 7.53 | 8.86 |
| 2" ell | P6@.125 | Ea | 1.49 | 6.28 | 7.77 |
| 2", long sweep | P6@.250 | Ea | 2.85 | 12.60 | 15.45 |
| 2", street ell | P6@.125 | Ea | 2.50 | 6.28 | 8.78 |

# 22 Plumbing

| | Craft@Hrs | Unit | Material | Labor | Total |
|---|---|---|---|---|---|
| 2", venting | P6@.150 | Ea | 2.34 | 7.53 | 9.87 |
| 3" ell | P6@.175 | Ea | 4.45 | 8.79 | 13.24 |
| 3", long sweep | P6@.500 | Ea | 6.28 | 25.10 | 31.38 |
| 3", street ell | P6@.150 | Ea | 5.11 | 7.53 | 12.64 |
| 3", venting | P6@.150 | Ea | 5.54 | 7.53 | 13.07 |
| 3" x 3" x 1-1/2", hub x hub x hub, side inlet | P6@.500 | Ea | 11.60 | 25.10 | 36.70 |
| 3" x 3" x 2", hub x hub x hub, side inlet | P6@.500 | Ea | 10.80 | 25.10 | 35.90 |
| 3" x 3" x 2" low heel elbow | P6@.500 | Ea | 7.20 | 25.10 | 32.30 |
| 3" x 3" x 1-1/2" low heel elbow | P6@.175 | Ea | 10.80 | 8.79 | 19.59 |
| 4" ell | P6@.175 | Ea | 8.31 | 8.79 | 17.10 |
| 4", long sweep | P6@.500 | Ea | 12.60 | 25.10 | 37.70 |
| 4", street ell | P6@.175 | Ea | 11.00 | 8.79 | 19.79 |
| 4" x 3" closet elbow, spigot x hub | P6@.150 | Ea | 9.69 | 7.53 | 17.22 |
| 4" x 3" closet elbow, hub x hub | P6@.260 | Ea | 5.88 | 13.10 | 18.98 |
| ABS 60-degree elbow (1/6 bend), hub and hub | | | | | |
| 1-1/2" bend | P6@.200 | Ea | 1.80 | 10.00 | 11.80 |
| 2" bend | P6@.250 | Ea | 2.33 | 12.60 | 14.93 |
| 3" bend | P6@.500 | Ea | 9.27 | 25.10 | 34.37 |
| 4" bend | P6@.500 | Ea | 13.90 | 25.10 | 39.00 |
| ABS 45-degree elbow (1/8 bend). Regular bends are hub x hub. Street bends are hub and spigot. | | | | | |
| 1-1/2" bend | P6@.120 | Ea | 1.34 | 6.03 | 7.37 |
| 1-1/2" street bend | P6@.120 | Ea | 1.43 | 6.03 | 7.46 |
| 2" bend | P6@.125 | Ea | 2.46 | 6.28 | 8.74 |
| 2" street bend | P6@.125 | Ea | 2.32 | 6.28 | 8.60 |
| 3" bend | P6@.175 | Ea | 4.99 | 8.79 | 13.78 |
| 3" street bend | P6@.150 | Ea | 4.56 | 7.53 | 12.09 |
| 4" bend | P6@.175 | Ea | 7.33 | 8.79 | 16.12 |
| 4" street bend | P6@.175 | Ea | 9.41 | 8.79 | 18.20 |
| ABS 22-1/2 degree elbow (1/16 bend). Regular bends are hub x hub. Street bends are hub and spigot. | | | | | |
| 1-1/2" bend | P6@.120 | Ea | 1.66 | 6.03 | 7.69 |
| 1-1/2", street bend | P6@.120 | Ea | 4.55 | 6.03 | 10.58 |
| 2" bend | P6@.125 | Ea | 2.14 | 6.28 | 8.42 |
| 2", street bend | P6@.125 | Ea | 3.73 | 6.28 | 10.01 |
| 3" bend | P6@.170 | Ea | 8.00 | 8.54 | 16.54 |
| 3" street bend | P6@.150 | Ea | 8.37 | 7.53 | 15.90 |
| 4" bend | P6@.175 | Ea | 7.62 | 8.79 | 16.41 |
| 4" street bend | P6@.175 | Ea | 11.50 | 8.79 | 20.29 |
| ABS coupling, hub x hub | | | | | |
| 1-1/2" | P6@.120 | Ea | .66 | 6.03 | 6.69 |
| 2" | P6@.125 | Ea | .86 | 6.28 | 7.14 |
| 2" x 1-1/2", reducing | P6@.125 | Ea | 2.13 | 6.28 | 8.41 |
| 3" | P6@.175 | Ea | 1.80 | 8.79 | 10.59 |
| 3" x 2", reducing | P6@.175 | Ea | 5.49 | 8.79 | 14.28 |
| 4" | P6@.250 | Ea | 3.37 | 12.60 | 15.97 |
| 4" x 3", reducing | P6@.250 | Ea | 7.63 | 12.60 | 20.23 |
| ABS wye, connects two branch lines to a main horizontal drain line. Hub x hub x hub. | | | | | |
| 1-1/2" | P6@.150 | Ea | 2.86 | 7.53 | 10.39 |
| 2" | P6@.190 | Ea | 3.83 | 9.54 | 13.37 |
| 2" x 1-1/2" | P6@.190 | Ea | 3.79 | 9.54 | 13.33 |
| 3" | P6@.260 | Ea | 7.15 | 13.10 | 20.25 |
| 3" x 1-1/2" | P6@.240 | Ea | 6.24 | 12.10 | 18.34 |
| 3" x 2" | P6@.260 | Ea | 4.69 | 13.10 | 17.79 |
| 4" | P6@.350 | Ea | 13.60 | 17.60 | 31.20 |
| 4" x 2" | P6@.350 | Ea | 11.00 | 17.60 | 28.60 |
| 4" x 3" | P6@.360 | Ea | 10.80 | 18.10 | 28.90 |

| | Craft@Hrs | Unit | Material | Labor | Total |
|---|---|---|---|---|---|
| **ABS double wye**, joins three ABS branch lines to a main horizontal drain line. Hub x hub x hub x hub. | | | | | |
| 1-1/2" | P6@.260 | Ea | 4.91 | 13.10 | 18.01 |
| 2" | P6@.290 | Ea | 6.24 | 14.60 | 20.84 |
| 3" | P6@.325 | Ea | 11.80 | 16.30 | 28.10 |
| 4" x 3" | P6@.350 | Ea | 21.30 | 17.60 | 38.90 |
| **ABS sanitary tee (TY)**, connects a branch line to a vertical ABS drain line. Regular sanitary tees are hub x hub x hub. Street sanitary tees are spigot x hub x hub. | | | | | |
| 1-1/2" | P6@.150 | Ea | 1.66 | 7.53 | 9.19 |
| 1-1/2", long sweep | P6@.150 | Ea | 4.74 | 7.53 | 12.27 |
| 1-1/2", street tee | P6@.150 | Ea | 4.60 | 7.53 | 12.13 |
| 2" x 1-1/2" | P6@.160 | Ea | 1.61 | 8.04 | 9.65 |
| 2" x 1-1/2" x 1-1/2" | P6@.160 | Ea | 1.88 | 8.04 | 9.92 |
| 2" x 1-1/2" x 2" | P6@.160 | Ea | 2.43 | 8.04 | 10.47 |
| 2" | P6@.190 | Ea | 2.55 | 9.54 | 12.09 |
| 2", long sweep | P6@.190 | Ea | 5.12 | 9.54 | 14.66 |
| 2" x 1-1/2", long sweep | P6@.190 | Ea | 5.47 | 9.54 | 15.01 |
| 2" street tee | P6@.190 | Ea | 5.23 | 9.54 | 14.77 |
| 2" x 1-1/2" x 1-1/2" street tee | P6@.190 | Ea | 5.61 | 9.54 | 15.15 |
| 3" | P6@.260 | Ea | 5.91 | 13.10 | 19.01 |
| 3", long sweep | P6@.260 | Ea | 10.70 | 13.10 | 23.80 |
| 3" x 1-1/2" | P6@.210 | Ea | 5.58 | 10.50 | 16.08 |
| 3" x 1-1/2", long sweep | P6@.200 | Ea | 7.16 | 10.00 | 17.16 |
| 3" x 2" | P6@.235 | Ea | 5.47 | 11.80 | 17.27 |
| 3" x 2", long sweep | P6@.260 | Ea | 5.58 | 13.10 | 18.68 |
| 4" | P6@.350 | Ea | 10.50 | 17.60 | 28.10 |
| 4", long sweep | P6@.350 | Ea | 11.50 | 17.60 | 29.10 |
| 4" x 2" | P6@.275 | Ea | 12.00 | 13.80 | 25.80 |
| 4" x 2", long sweep | P6@.350 | Ea | 15.50 | 17.60 | 33.10 |
| 4" x 3" | P6@.325 | Ea | 10.10 | 16.30 | 26.40 |
| 4" x 3", long sweep | P6@.360 | Ea | 17.80 | 18.10 | 35.90 |
| **ABS bungalow fitting (Wisconsin tee)**. Connects vent (3" top inlet) and horizontal water closet discharge (3" branch) and bathtub discharge (1-1/2" branch ) to a vertical soil stack. One fitting connects discharge and venting for a set of washroom fixtures. | | | | | |
| 3", 1-1/2" left inlet | P6@.325 | Ea | 14.00 | 16.30 | 30.30 |
| 3", 2" right inlet | P6@.325 | Ea | 12.00 | 16.30 | 28.30 |
| **ABS double sanitary tee (double TY/cross)**, brings two ABS-DWV branch lines into a vertical waste or soil stack. All hub except as noted. | | | | | |
| 1-1/2" | P6@.250 | Ea | 3.69 | 12.60 | 16.29 |
| 2" x 1-1/2" | P6@.260 | Ea | 4.53 | 13.10 | 17.63 |
| 2" | P6@.290 | Ea | 7.28 | 14.60 | 21.88 |
| 2" double fixture tee | P6@.290 | Ea | 8.38 | 14.60 | 22.98 |
| 2" x 1-1/2" x 1-1/2" x 1-1/2" | P6@.260 | Ea | 8.12 | 13.10 | 21.22 |
| 3" | P6@.325 | Ea | 13.60 | 16.30 | 29.90 |
| 3" double fixture tee, spigot x hub | P6@.450 | Ea | 17.50 | 22.60 | 40.10 |
| 3" x 1-1/2" | P6@.310 | Ea | 7.06 | 15.60 | 22.66 |
| 3" x 2" | P6@.335 | Ea | 6.55 | 16.80 | 23.35 |
| 4" | P6@.350 | Ea | 26.20 | 17.60 | 43.80 |
| **ABS cleanout tee, two-way** | | | | | |
| 3", hub x hub x hub | P6@.175 | Ea | 14.10 | 8.79 | 22.89 |
| 4", hub x hub x hub | P6@.250 | Ea | 28.90 | 12.60 | 41.50 |

# 22 Plumbing

| | Craft@Hrs | Unit | Material | Labor | Total |
|---|---|---|---|---|---|
| ABS test tee, provides a cleanout opening flush with a finished wall. Includes a cleanout plug. FIPT means female iron pipe thread. | | | | | |
| 1-1/2", hub x hub x cleanout | P6@.130 | Ea | 5.09 | 6.53 | 11.62 |
| 2", hub x hub x cleanout | P6@.170 | Ea | 7.72 | 8.54 | 16.26 |
| 3", hub x hub x cleanout | P6@.220 | Ea | 14.00 | 11.00 | 25.00 |
| 4", hub x hub x cleanout | P6@.240 | Ea | 22.30 | 12.10 | 34.40 |
| 1-1/2", hub x hub x FIPT | P6@.150 | Ea | 4.11 | 7.53 | 11.64 |
| 2", hub x hub x FIPT | P6@.190 | Ea | 5.47 | 9.54 | 15.01 |
| 3", hub x hub x FIPT | P6@.260 | Ea | 9.25 | 13.10 | 22.35 |
| 4", hub x hub x FIPT | P6@.350 | Ea | 13.80 | 17.60 | 31.40 |
| Inset test cap with knockout | | | | | |
| 1-1/2" | P6@.100 | Ea | .21 | 5.02 | 5.23 |
| 2" | P6@.100 | Ea | .26 | 5.02 | 5.28 |
| 3" | P6@.100 | Ea | .32 | 5.02 | 5.34 |
| 4" | P6@.100 | Ea | .38 | 5.02 | 5.40 |
| Mechanical plastic test plug with galvanized wing nut, natural rubber O-ring. Max air pressure of 5 psi. | | | | | |
| 1-1/2" | P6@.030 | Ea | 3.42 | 1.51 | 4.93 |
| 2" | P6@.030 | Ea | 4.18 | 1.51 | 5.69 |
| 3" | P6@.030 | Ea | 4.83 | 1.51 | 6.34 |
| 4" | P6@.030 | Ea | 6.25 | 1.51 | 7.76 |
| ABS closet flange. Closet flanges are used for mounting water closets. | | | | | |
| 4" x 3", with knock out | P6@.225 | Ea | 3.41 | 11.30 | 14.71 |
| 4" x 3", flush fit | P6@.225 | Ea | 5.43 | 11.30 | 16.73 |
| 4" x 3", flange x hub | P6@.225 | Ea | 6.77 | 11.30 | 18.07 |
| 4" x 3", flange x spigot | P6@.225 | Ea | 11.20 | 11.30 | 22.50 |
| 4" x 3", offset, adjustable | P6@.225 | Ea | 11.00 | 11.30 | 22.30 |
| 4" closet flange riser | P6@.180 | Ea | 3.44 | 9.04 | 12.48 |
| ABS female adapter, joins ABS pipe to standard male threaded pipe. FIPT means female iron pipe thread. | | | | | |
| 1-1/2", hub x FIPT | P6@.080 | Ea | 1.49 | 4.02 | 5.51 |
| 1-1/2", spigot x FIPT | P6@.080 | Ea | 1.62 | 4.02 | 5.64 |
| 2", hub x FIPT | P6@.085 | Ea | 2.80 | 4.27 | 7.07 |
| 2", spigot x FIPT | P6@.085 | Ea | 2.08 | 4.27 | 6.35 |
| 3", hub x FIPT | P6@.155 | Ea | 4.46 | 7.78 | 12.24 |
| 3", spigot x FIPT | P6@.155 | Ea | 4.42 | 7.78 | 12.20 |
| 4", hub x FIPT | P6@.220 | Ea | 7.57 | 11.00 | 18.57 |
| 4", spigot x FIPT | P6@.220 | Ea | 7.60 | 11.00 | 18.60 |
| ABS male adapter, joins ABS pipe to standard female threaded pipe. MIPT means male iron pipe thread. | | | | | |
| 1-1/2", hub x MIPT | P6@.080 | Ea | 1.17 | 4.02 | 5.19 |
| 2", hub x MIPT | P6@.085 | Ea | 2.03 | 4.27 | 6.30 |
| 3", hub x MIPT | P6@.155 | Ea | 4.56 | 7.78 | 12.34 |
| 4", hub x MIPT | P6@.220 | Ea | 10.70 | 11.00 | 21.70 |
| ABS trap adapter, joins ABS pipe to sink or bath waste outlets | | | | | |
| 1-1/2", hub x slip jam nut | P6@.200 | Ea | 1.86 | 10.00 | 11.86 |
| 1-1/2", spigot x slip jam nut | P6@.200 | Ea | 2.14 | 10.00 | 12.14 |
| 1-1/2" x 1-1/4", spigot x slip jam nut | P6@.200 | Ea | 1.45 | 10.00 | 11.45 |
| 1-1/2" x 1-1/4", hub x slip jam nut | P6@.200 | Ea | 2.07 | 10.00 | 12.07 |
| ABS cleanout adapter, fits inside pipe, flush finish | | | | | |
| 2", insert | P6@.100 | Ea | 3.89 | 5.02 | 8.91 |
| 3", insert | P6@.150 | Ea | 4.40 | 7.53 | 11.93 |
| 4", insert | P6@.320 | Ea | 6.06 | 16.10 | 22.16 |

# 22 Plumbing

| | Craft@Hrs | Unit | Material | Labor | Total |
|---|---|---|---|---|---|
| **ABS-DWV cleanout plug, male pipe threads** | | | | | |
| 1-1/2" | P6@.090 | Ea | 1.09 | 4.52 | 5.61 |
| 2" | P6@.120 | Ea | 1.44 | 6.03 | 7.47 |
| 3" | P6@.240 | Ea | 2.18 | 12.10 | 14.28 |
| 4" | P6@.320 | Ea | 3.85 | 16.10 | 19.95 |
| 2", countersunk | P6@.120 | Ea | 2.37 | 6.03 | 8.40 |
| 3", countersunk | P6@.240 | Ea | 2.93 | 12.10 | 15.03 |
| **ABS soil pipe adapter, joins ABS DWV to cast iron pipe soil pipe** | | | | | |
| 3" ABS hub x cast iron spigot | P6@.155 | Ea | 5.61 | 7.78 | 13.39 |
| 4" ABS hub to cast iron spigot | P6@.250 | Ea | 8.05 | 12.60 | 20.65 |
| 4" cast iron spigot to 3" DWV hub | P6@.220 | Ea | 6.00 | 11.00 | 17.00 |
| **ABS "P"-trap, used at fixture waste outlets to prevent sewer gas from entering living space via the fixture drain** | | | | | |
| 1-1/2", all hub | P6@.220 | Ea | 3.11 | 11.00 | 14.11 |
| 1-1/2", hub x hub, with union | P6@.280 | Ea | 5.61 | 14.10 | 19.71 |
| 1-1/2", hub, with cleanout | P6@.155 | Ea | 9.66 | 7.78 | 17.44 |
| 1-1/2", hub x spigot, with union | P6@.220 | Ea | 5.62 | 11.00 | 16.62 |
| 2", all hub | P6@.280 | Ea | 5.47 | 14.10 | 19.57 |
| 2", hub x hub, with union | P6@.280 | Ea | 8.87 | 14.10 | 22.97 |
| 2", hub, with cleanout | P6@.195 | Ea | 12.80 | 9.79 | 22.59 |
| **ABS return bend, bottom portion of "P"-trap, no cleanout** | | | | | |
| 1-1/2", all hub | P6@.120 | Ea | 2.48 | 6.03 | 8.51 |
| 2", all hub | P6@.125 | Ea | 4.11 | 6.28 | 10.39 |
| **ABS cap, permanently caps ABS-DWV piping** | | | | | |
| 1-1/2" | P6@.030 | Ea | 2.58 | 1.51 | 4.09 |
| 2" | P6@.040 | Ea | 4.11 | 2.01 | 6.12 |
| 3" | P6@.050 | Ea | 6.26 | 2.51 | 8.77 |
| 4" | P6@.090 | Ea | 8.38 | 4.52 | 12.90 |
| **ABS flush bushing, inserts into ABS fitting hub to reduce to a smaller pipe size** | | | | | |
| 2" x 1-1/2" | P6@.062 | Ea | 1.08 | 3.11 | 4.19 |
| 3" x 1-1/2" | P6@.088 | Ea | 3.90 | 4.42 | 8.32 |
| 3" x 2" | P6@.088 | Ea | 2.75 | 4.42 | 7.17 |
| 4" x 2" | P6@.125 | Ea | 7.93 | 6.28 | 14.21 |
| 4" x 3" | P6@.125 | Ea | 7.01 | 6.28 | 13.29 |
| **ABS heavy-duty floor drain** | | | | | |
| 3", less sediment bucket | P6@.280 | Ea | 39.60 | 14.10 | 53.70 |
| 4", less sediment bucket | P6@.300 | Ea | 40.10 | 15.10 | 55.20 |

## Hangers and Supports

| | Craft@Hrs | Unit | Material | Labor | Total |
|---|---|---|---|---|---|
| **Angle bracket hangers, steel, by rod size** | | | | | |
| 3/8" angle bracket hanger | M5@.070 | Ea | 1.97 | 3.48 | 5.45 |
| 1/2" angle bracket hanger | M5@.070 | Ea | 2.20 | 3.48 | 5.68 |
| 5/8" angle bracket hanger | M5@.070 | Ea | 4.85 | 3.48 | 8.33 |
| 3/4" angle bracket hanger | M5@.070 | Ea | 7.49 | 3.48 | 10.97 |
| 7/8" angle bracket hanger | M5@.070 | Ea | 13.30 | 3.48 | 16.78 |
| **Angle supports, wall mounted, welded steel** | | | | | |
| 12" x 18", medium weight | M5@.950 | Ea | 83.90 | 47.20 | 131.10 |
| 18" x 24", medium weight | M5@1.10 | Ea | 105.00 | 54.70 | 159.70 |
| 24" x 30", medium weight | M5@1.60 | Ea | 132.00 | 79.50 | 211.50 |
| 12" x 30", heavy weight | M5@1.10 | Ea | 119.00 | 54.70 | 173.70 |
| 18" x 24", heavy weight | M5@1.10 | Ea | 168.00 | 54.70 | 222.70 |
| 24" x 30", heavy weight | M5@1.60 | Ea | 189.00 | 79.50 | 268.50 |
| **"C"-clamps with lock nut, steel, by rod size** | | | | | |
| 3/8" "C"-clamp | M5@.070 | Ea | 2.02 | 3.48 | 5.50 |
| 1/2" "C"-clamp | M5@.070 | Ea | 2.45 | 3.48 | 5.93 |

# 22 Plumbing

|  | Craft@Hrs | Unit | Material | Labor | Total |
|---|---|---|---|---|---|
| 5/8" "C"-clamp | M5@.070 | Ea | 4.96 | 3.48 | 8.44 |
| 3/4" "C"-clamp | M5@.070 | Ea | 7.13 | 3.48 | 10.61 |
| 7/8" "C"-clamp | M5@.070 | Ea | 19.20 | 3.48 | 22.68 |
| Top beam clamps, steel, by rod size | | | | | |
| 3/8" top beam clamp | M5@.070 | Ea | 2.38 | 3.48 | 5.86 |
| 1/2" top beam clamp | M5@.070 | Ea | 3.85 | 3.48 | 7.33 |
| 5/8" top beam clamp | M5@.070 | Ea | 4.72 | 3.48 | 8.20 |
| 3/4" top beam clamp | M5@.070 | Ea | 6.89 | 3.48 | 10.37 |
| 7/8" top beam clamp | M5@.070 | Ea | 9.03 | 3.48 | 12.51 |
| Clevis hangers, standard duty, by intended pipe size, no threaded rod or beam clamp included | | | | | |
| 1/2" clevis hanger | M5@.156 | Ea | 1.87 | 7.75 | 9.62 |
| 3/4" clevis hanger | M5@.156 | Ea | 1.87 | 7.75 | 9.62 |
| 1" clevis hanger | M5@.156 | Ea | 1.98 | 7.75 | 9.73 |
| 1-1/4" clevis hanger | M5@.173 | Ea | 2.26 | 8.60 | 10.86 |
| 1-1/2" clevis hanger | M5@.173 | Ea | 2.47 | 8.60 | 11.07 |
| 2" clevis hanger | M5@.173 | Ea | 2.75 | 8.60 | 11.35 |
| 2-1/2" clevis hanger | M5@.189 | Ea | 3.08 | 9.40 | 12.48 |
| 3" clevis hanger | M5@.189 | Ea | 3.85 | 9.40 | 13.25 |
| 4" clevis hanger | M5@.189 | Ea | 4.76 | 9.40 | 14.16 |
| 6" clevis hanger | M5@.202 | Ea | 8.03 | 10.00 | 18.03 |
| 8" clevis hanger | M5@.202 | Ea | 11.70 | 10.00 | 21.70 |
| Add for epoxy coated | — | % | 25.0 | — | — |
| Add for copper plated | — | % | 35.0 | — | — |
| Deduct for light duty | — | % | -20.0 | — | — |
| Swivel ring hangers, light duty, by intended pipe size, band hanger, galvanized steel, no hanger rod, beam clamp or insert included | | | | | |
| 1/2" swivel ring hanger | M5@.125 | Ea | .47 | 6.21 | 6.68 |
| 3/4" swivel ring hanger | M5@.125 | Ea | .47 | 6.21 | 6.68 |
| 1" swivel ring hanger | M5@.125 | Ea | .47 | 6.21 | 6.68 |
| 1-1/4" swivel ring hanger | M5@.125 | Ea | .54 | 6.21 | 6.75 |
| 1-1/2" swivel ring hanger | M5@.125 | Ea | .63 | 6.21 | 6.84 |
| 2" swivel ring hanger | M5@.125 | Ea | .75 | 6.21 | 6.96 |
| 2-1/2" swivel ring hanger | M5@.135 | Ea | 1.03 | 6.71 | 7.74 |
| 3" swivel ring hanger | M5@.135 | Ea | 1.22 | 6.71 | 7.93 |
| 4" swivel ring hanger | M5@.135 | Ea | 1.69 | 6.71 | 8.40 |
| 6" swivel ring hanger | M5@.145 | Ea | 3.18 | 7.21 | 10.39 |
| 8" swivel ring hanger | M5@.145 | Ea | 3.81 | 7.21 | 11.02 |
| Add for epoxy coating | — | % | 45.0 | — | — |
| Add for copper plated | — | % | 55.0 | — | — |
| Riser clamps, carbon steel, by intended pipe size | | | | | |
| 1/2" riser clamp | M5@.100 | Ea | 1.60 | 4.97 | 6.57 |
| 3/4" riser clamp | M5@.100 | Ea | 2.35 | 4.97 | 7.32 |
| 1" riser clamp | M5@.100 | Ea | 2.41 | 4.97 | 7.38 |
| 1-1/4" riser clamp | M5@.115 | Ea | 2.86 | 5.72 | 8.58 |
| 1-1/2" riser clamp | M5@.115 | Ea | 3.04 | 5.72 | 8.76 |
| 2" riser clamp | M5@.125 | Ea | 3.18 | 6.21 | 9.39 |
| 2-1/2" riser clamp | M5@.135 | Ea | 3.35 | 6.71 | 10.06 |
| 3" riser clamp | M5@.140 | Ea | 3.69 | 6.96 | 10.65 |
| 4" riser clamp | M5@.145 | Ea | 4.63 | 7.21 | 11.84 |
| 6" riser clamp | M5@.155 | Ea | 8.03 | 7.71 | 15.74 |
| 8" riser clamp | M5@.165 | Ea | 13.00 | 8.20 | 21.20 |
| Add for epoxy coating | — | % | 25.0 | — | — |
| Add for copper plated | — | % | 35.0 | — | — |

# 22 Plumbing

| | Craft@Hrs | Unit | Material | Labor | Total |
|---|---|---|---|---|---|
| Pipe clamps, medium duty, by intended pipe size | | | | | |
| 3" pipe clamp | M5@.189 | Ea | 4.84 | 9.40 | 14.24 |
| 4" pipe clamp | M5@.189 | Ea | 6.89 | 9.40 | 16.29 |
| 6" pipe clamp | M5@.202 | Ea | 16.80 | 10.00 | 26.80 |
| Pipe rolls | | | | | |
| 3" pipe roll, complete | M5@.403 | Ea | 15.00 | 20.00 | 35.00 |
| 4" pipe roll, complete | M5@.518 | Ea | 15.80 | 25.70 | 41.50 |
| 6" pipe roll, complete | M5@.518 | Ea | 15.80 | 25.70 | 41.50 |
| 8" pipe roll, complete | M5@.578 | Ea | 27.80 | 28.70 | 56.50 |
| 3" pipe roll stand, complete | M5@.323 | Ea | 40.10 | 16.10 | 56.20 |
| 4" pipe roll stand, complete | M5@.604 | Ea | 60.00 | 30.00 | 90.00 |
| 6" pipe roll stand, complete | M5@.775 | Ea | 60.00 | 38.50 | 98.50 |
| 8" pipe roll stand, complete | M5@.775 | Ea | 105.00 | 38.50 | 143.50 |
| Pipe straps, galvanized, two hole, lightweight | | | | | |
| 2-1/2" pipe strap | M5@.119 | Ea | 1.92 | 5.92 | 7.84 |
| 3" pipe strap | M5@.128 | Ea | 2.47 | 6.36 | 8.83 |
| 3-1/2" pipe strap | M5@.135 | Ea | 3.14 | 6.71 | 9.85 |
| 4" pipe strap | M5@.144 | Ea | 3.50 | 7.16 | 10.66 |
| Plumber's tape, galvanized, 3/4" x 10' roll | | | | | |
| 26 gauge | — | LF | 1.69 | — | 1.69 |
| 22 gauge | — | LF | 2.23 | — | 2.23 |
| Sliding pipe guides, welded | | | | | |
| 3" pipe guide | M5@.354 | Ea | 17.20 | 17.60 | 34.80 |
| 4" pipe guide | M5@.354 | Ea | 23.20 | 17.60 | 40.80 |
| 6" pipe guide | M5@.354 | Ea | 33.50 | 17.60 | 51.10 |
| 8" pipe guide | M5@.471 | Ea | 50.20 | 23.40 | 73.60 |
| Threaded hanger rod, by rod size, in 6' to 10' lengths, per LF | | | | | |
| 3/8" rod | — | LF | .40 | — | .40 |
| 1/2" rod | — | LF | .61 | — | .61 |
| 5/8" rod | — | LF | .84 | — | .84 |
| 3/4" rod | — | LF | 1.47 | — | 1.47 |
| "U"-bolts, with hex nuts, light duty | | | | | |
| 4" "U"-bolt | M5@.089 | Ea | 5.41 | 4.42 | 9.83 |
| 6" "U"-bolt | M5@.135 | Ea | 9.95 | 6.71 | 16.66 |
| Wall sleeves, cast iron | | | | | |
| 4" sleeve | M5@.241 | Ea | 7.45 | 12.00 | 19.45 |
| 6" sleeve | M5@.284 | Ea | 9.23 | 14.10 | 23.33 |
| 8" sleeve | M5@.377 | Ea | 12.10 | 18.70 | 30.80 |

**Gate Valves,** for water, gas and steam applications; complies with American Society of Mechanical Engineers Standard B16, Manufacturers Standardization Society, Instrumentation Society of America standards.

| | Craft@Hrs | Unit | Material | Labor | Total |
|---|---|---|---|---|---|
| 1/2" bronze gate valves | | | | | |
| 125 lb, threaded | P6@.210 | Ea | 10.50 | 10.50 | 21.00 |
| 125 lb, soldered | P6@.200 | Ea | 10.10 | 10.00 | 20.10 |
| 150 lb, threaded | P6@.210 | Ea | 18.10 | 10.50 | 28.60 |
| 200 lb, threaded | P6@.210 | Ea | 23.50 | 10.50 | 34.00 |
| 3/4" bronze gate valves | | | | | |
| 125 lb, threaded | P6@.250 | Ea | 12.50 | 12.60 | 25.10 |
| 125 lb, soldered | P6@.240 | Ea | 11.90 | 12.10 | 24.00 |
| 150 lb, threaded | P6@.250 | Ea | 24.60 | 12.60 | 37.20 |
| 200 lb, threaded | P6@.250 | Ea | 30.50 | 12.60 | 43.10 |

| | Craft@Hrs | Unit | Material | Labor | Total |
|---|---|---|---|---|---|
| 1" bronze gate valves | | | | | |
|    125 lb, threaded | P6@.300 | Ea | 18.00 | 15.10 | 33.10 |
|    125 lb, soldered | P6@.290 | Ea | 17.60 | 14.60 | 32.20 |
|    150 lb, threaded | P6@.300 | Ea | 29.00 | 15.10 | 44.10 |
|    200 lb, threaded | P6@.300 | Ea | 37.60 | 15.10 | 52.70 |
| 1-1/2" bronze gate valves | | | | | |
|    125 lb, threaded | P6@.450 | Ea | 29.70 | 22.60 | 52.30 |
|    125 lb, soldered | P6@.440 | Ea | 28.70 | 22.10 | 50.80 |
|    150 lb, threaded | P6@.450 | Ea | 52.20 | 22.60 | 74.80 |
|    200 lb, threaded | P6@.450 | Ea | 63.60 | 22.60 | 86.20 |
| 2" bronze gate valves | | | | | |
|    125 lb, threaded | P6@.500 | Ea | 49.70 | 25.10 | 74.80 |
|    125 lb, soldered | P6@.480 | Ea | 48.00 | 24.10 | 72.10 |
|    150 lb, threaded | P6@.500 | Ea | 82.00 | 25.10 | 107.10 |
|    200 lb, threaded | P6@.500 | Ea | 93.30 | 25.10 | 118.40 |
| Cast iron, bronze trim, non-rising stem | | | | | |
|    2-1/2" 125 lb, cast iron, flanged | MI@.600 | Ea | 208.00 | 31.80 | 239.80 |
|    2-1/2" 250 lb, cast iron, flanged | MI@.600 | Ea | 409.00 | 31.80 | 440.80 |
|    3" 125 lb, cast iron, flanged | MI@.750 | Ea | 225.00 | 39.70 | 264.70 |
|    4" 125 lb, cast iron, flanged | MI@1.35 | Ea | 332.00 | 71.50 | 403.50 |
|    4" 250 lb, cast iron, flanged | MI@1.35 | Ea | 849.00 | 71.50 | 920.50 |

**Steel Gate, Globe and Check Valves**

| | Craft@Hrs | Unit | Material | Labor | Total |
|---|---|---|---|---|---|
| Cast steel Class 150 gate valves, OS&Y, flanged ends | | | | | |
|    2-1/2" valve | MI@.600 | Ea | 853.00 | 31.80 | 884.80 |
|    3" valve | MI@.750 | Ea | 853.00 | 39.70 | 892.70 |
|    4" valve | MI@1.35 | Ea | 1,050.00 | 71.50 | 1,121.50 |
| Cast steel Class 300 gate valves, OS&Y, flanged ends | | | | | |
|    2-1/2" valve | MI@.600 | Ea | 1,120.00 | 31.80 | 1,151.80 |
|    3" valve | MI@.750 | Ea | 1,120.00 | 39.70 | 1,159.70 |
|    4" valve | MI@1.35 | Ea | 1,550.00 | 71.50 | 1,621.50 |

**OS&Y Gate Valves** Forged steel class 800 bolted bonnet outside stem and yoke valves for water, gas and steam applications, complies with American Society of Mechanical Engineers Standard B16, Manufacturers Standardization Society, Instrumentation Society of America standards.

| | Craft@Hrs | Unit | Material | Labor | Total |
|---|---|---|---|---|---|
| Threaded ends OS&Y gate valves | | | | | |
|    1/2" valve | P6@.210 | Ea | 87.90 | 10.50 | 98.40 |
|    3/4" valve | P6@.250 | Ea | 120.00 | 12.60 | 132.60 |
|    1" valve | P6@.300 | Ea | 217.00 | 15.10 | 232.10 |
|    1-1/2" valve | P6@.450 | Ea | 273.00 | 22.60 | 295.60 |
|    2" valve | P6@.500 | Ea | 417.00 | 25.10 | 442.10 |
| 2-1/2" flanged iron OS&Y gate valves | | | | | |
|    125 lb | MI@.600 | Ea | 289.00 | 31.80 | 320.80 |
|    250 lb | MI@.600 | Ea | 652.00 | 31.80 | 683.80 |
| 3" flanged iron OS&Y gate valves | | | | | |
|    125 lb | MI@.750 | Ea | 307.00 | 39.70 | 346.70 |
|    250 lb | MI@.750 | Ea | 748.00 | 39.70 | 787.70 |
| 4" flanged iron OS&Y gate valves | | | | | |
|    125 lb | MI@1.35 | Ea | 427.00 | 71.50 | 498.50 |
|    250 lb | MI@1.35 | Ea | 1,100.00 | 71.50 | 1,171.50 |
| 6" flanged iron OS&Y gate valves | | | | | |
|    125 lb | MI@2.50 | Ea | 726.00 | 132.00 | 858.00 |
|    250 lb | MI@2.50 | Ea | 1,820.00 | 132.00 | 1,952.00 |

# 22 Plumbing

|  | Craft@Hrs | Unit | Material | Labor | Total |
|---|---|---|---|---|---|

**Butterfly Valves, Ductile Iron**
Wafer body, flanged, with lever lock, 150 lb valves have aluminum bronze disk, stainless steel stem and EPDM seals. 200 lb nickel plated ductile iron disk and Buna-N seals.

2-1/2" butterfly valve

| | | | | | |
|---|---|---|---|---|---|
| 150 lb | MI@.300 | Ea | 91.70 | 15.90 | 107.60 |
| 200 lb | MI@.400 | Ea | 130.00 | 21.20 | 151.20 |

3" butterfly valve

| | | | | | |
|---|---|---|---|---|---|
| 150 lb | MI@.400 | Ea | 101.00 | 21.20 | 122.20 |
| 200 lb | MI@.450 | Ea | 138.00 | 23.80 | 161.80 |

4" butterfly valve

| | | | | | |
|---|---|---|---|---|---|
| 150 lb | MI@.500 | Ea | 125.00 | 26.50 | 151.50 |
| 200 lb | MI@.550 | Ea | 172.00 | 29.10 | 201.10 |

6" butterfly valve

| | | | | | |
|---|---|---|---|---|---|
| 150 lb | MI@.750 | Ea | 196.00 | 39.70 | 235.70 |
| 200 lb | MI@.750 | Ea | 276.00 | 39.70 | 315.70 |

8" butterfly valve, with gear operator

| | | | | | |
|---|---|---|---|---|---|
| 150 lb | MI@.950 | Ea | 277.00 | 50.30 | 327.30 |
| 200 lb | MI@.950 | Ea | 508.00 | 50.30 | 558.30 |

10" butterfly valve, with gear operator

| | | | | | |
|---|---|---|---|---|---|
| 150 lb | MI@1.40 | Ea | 389.00 | 74.20 | 463.20 |
| 200 lb | MI@1.40 | Ea | 650.00 | 74.20 | 724.20 |

**Bronze Globe Valves**  125 lb and 150 lb ratings are saturated steam pressure. 200 lb rating is working steam pressure, union bonnet. 125 lb valves have bronze disks. 150 lb valves have composition disks. 300 lb valves have regrinding disks. For gas and steam applications, complies with American Society of Mechanical Engineers Standard B16, Manufacturers Standardization Society, Instrumentation Society of America and American Petroleum Institute standards.

1/2" threaded bronze globe valves

| | | | | | |
|---|---|---|---|---|---|
| 125 lb | P6@.210 | Ea | 13.70 | 10.50 | 24.20 |
| 150 lb | P6@.210 | Ea | 24.30 | 10.50 | 34.80 |
| 300 lb | P6@.210 | Ea | 36.60 | 10.50 | 47.10 |

3/4" threaded bronze globe valves

| | | | | | |
|---|---|---|---|---|---|
| 125 lb | P6@.250 | Ea | 18.60 | 12.60 | 31.20 |
| 150 lb | P6@.250 | Ea | 33.10 | 12.60 | 45.70 |
| 300 lb | P6@.250 | Ea | 46.60 | 12.60 | 59.20 |

1" threaded bronze globe valves

| | | | | | |
|---|---|---|---|---|---|
| 125 lb | P6@.300 | Ea | 26.50 | 15.10 | 41.60 |
| 150 lb | P6@.300 | Ea | 51.50 | 15.10 | 66.60 |
| 300 lb | P6@.300 | Ea | 67.00 | 15.10 | 82.10 |

1-1/2" threaded bronze globe valves

| | | | | | |
|---|---|---|---|---|---|
| 125 lb | P6@.450 | Ea | 50.00 | 22.60 | 72.60 |
| 150 lb | P6@.450 | Ea | 97.20 | 22.60 | 119.80 |
| 300 lb | P6@.450 | Ea | 128.00 | 22.60 | 150.60 |

2" threaded bronze globe valves

| | | | | | |
|---|---|---|---|---|---|
| 125 lb | P6@.500 | Ea | 81.20 | 25.10 | 106.30 |
| 150 lb | P6@.500 | Ea | 147.00 | 25.10 | 172.10 |
| 300 lb | P6@.500 | Ea | 178.00 | 25.10 | 203.10 |

**Iron Body Globe Valves**  Flanged, iron body, OS&Y (outside stem & yoke) 125 psi steam pressure and 200 psi cold water pressure rated.

| | | | | | |
|---|---|---|---|---|---|
| 3" flanged OS&Y, globe valve | MI@.750 | Ea | 378.00 | 39.70 | 417.70 |
| 3" 250 lb | MI@.750 | Ea | 735.00 | 39.70 | 774.70 |
| 4" flanged OS&Y, globe valve | MI@1.35 | Ea | 537.00 | 71.50 | 608.50 |

|  | Craft@Hrs | Unit | Material | Labor | Total |
|---|---|---|---|---|---|
| 4" 250 lb | MI@1.35 | Ea | 694.00 | 71.50 | 765.50 |
| 6" flanged OS&Y, globe valve | MI@2.50 | Ea | 997.00 | 132.00 | 1,129.00 |
| 6" 250 lb | MI@2.50 | Ea | 1,820.00 | 132.00 | 1,952.00 |
| 8" flanged OS&Y, globe valve | MI@3.00 | Ea | 1,760.00 | 159.00 | 1,919.00 |
| 8" 250 lb | MI@3.00 | Ea | 3,100.00 | 159.00 | 3,259.00 |

**Swing Check Valves** for water and steam applications, complies with American Society of Mechanical Engineers Standard B16, Manufacturers Standardization Society, Instrumentation Society of America standards.

| 1/2" bronze check valves | Craft@Hrs | Unit | Material | Labor | Total |
|---|---|---|---|---|---|
| 125 lb, threaded, swing | P6@.210 | Ea | 11.60 | 10.50 | 22.10 |
| 125 lb, soldered, swing | P6@.200 | Ea | 11.30 | 10.00 | 21.30 |
| 125 lb, threaded, vertical lift check | P6@.210 | Ea | 57.90 | 10.50 | 68.40 |
| 150 lb, threaded, swing, bronze disk | P6@.210 | Ea | 38.90 | 10.50 | 49.40 |
| 200 lb, threaded, swing, bronze disk | P6@.210 | Ea | 76.40 | 10.50 | 86.90 |
| 3/4" bronze check valves |  |  |  |  |  |
| 125 lb, threaded, swing | P6@.250 | Ea | 16.70 | 12.60 | 29.30 |
| 125 lb, soldered, swing | P6@.240 | Ea | 16.00 | 12.10 | 28.10 |
| 125 lb, threaded, swing, vertical check | P6@.250 | Ea | 78.90 | 12.60 | 91.50 |
| 150 lb, threaded, swing, bronze disk | P6@.250 | Ea | 48.50 | 12.60 | 61.10 |
| 200 lb, threaded, swing, bronze disk | P6@.250 | Ea | 90.40 | 12.60 | 103.00 |
| 1" bronze check valves |  |  |  |  |  |
| 125 lb, threaded, swing | P6@.300 | Ea | 22.10 | 15.10 | 37.20 |
| 125 lb, soldered, swing | P6@.290 | Ea | 21.20 | 14.60 | 35.80 |
| 125 lb, threaded, swing, vertical check | P6@.300 | Ea | 98.90 | 15.10 | 114.00 |
| 150 lb, threaded, swing, bronze disk | P6@.300 | Ea | 65.50 | 15.10 | 80.60 |
| 200 lb, threaded, swing, bronze disk | P6@.300 | Ea | 140.00 | 15.10 | 155.10 |
| 1-1/2" bronze check valves |  |  |  |  |  |
| 125 lb, threaded, swing | P6@.450 | Ea | 43.40 | 22.60 | 66.00 |
| 125 lb, soldered, swing | P6@.440 | Ea | 42.50 | 22.10 | 64.60 |
| 125 lb, threaded, swing, vertical check | P6@.450 | Ea | 167.00 | 22.60 | 189.60 |
| 150 lb, threaded, swing, swing, bronze disk | P6@.450 | Ea | 113.00 | 22.60 | 135.60 |
| 200 lb, threaded, swing, bronze disk | P6@.450 | Ea | 226.00 | 22.60 | 248.60 |
| 2" bronze check valves |  |  |  |  |  |
| 125 lb, threaded, swing | P6@.500 | Ea | 72.80 | 25.10 | 97.90 |
| 125 lb, soldered, swing | P6@.480 | Ea | 71.00 | 24.10 | 95.10 |
| 125 lb, threaded, vertical check | P6@.500 | Ea | 249.00 | 25.10 | 274.10 |
| 150 lb, threaded, swing, bronze disk | P6@.500 | Ea | 166.00 | 25.10 | 191.10 |
| 200 lb, threaded, swing, bronze disk | P6@.500 | Ea | 318.00 | 25.10 | 343.10 |
| 2-1/2" flanged iron body check valve |  |  |  |  |  |
| Swing check, bronze trim, 125 psi steam | MI@.600 | Ea | 144.00 | 31.80 | 175.80 |
| Swing check, iron trim, 125 psi steam | MI@.600 | Ea | 203.00 | 31.80 | 234.80 |
| Swing check, bronze trim, 250 psi steam | MI@.750 | Ea | 440.00 | 39.70 | 479.70 |
| 3" flanged iron body check valve |  |  |  |  |  |
| Swing check, bronze trim, 125 psi steam | MI@.750 | Ea | 174.00 | 39.70 | 213.70 |
| Swing check, iron trim, 125 psi steam | MI@.750 | Ea | 278.00 | 39.70 | 317.70 |
| Swing check, bronze trim, 250 psi steam | MI@1.25 | Ea | 546.00 | 66.20 | 612.20 |
| 4" flanged iron body check valve |  |  |  |  |  |
| Swing check, bronze trim, 125 psi steam | MI@1.35 | Ea | 282.00 | 71.50 | 353.50 |
| Swing check, iron trim, 125 psi steam | MI@1.35 | Ea | 458.00 | 71.50 | 529.50 |
| Swing check, bronze trim, 250 psi steam | MI@1.85 | Ea | 693.00 | 98.00 | 791.00 |
| 6" flanged iron body check valve |  |  |  |  |  |
| Swing check, bronze trim, 125 psi steam | MI@2.50 | Ea | 457.00 | 132.00 | 589.00 |
| Swing check, iron trim, 125 psi steam | MI@2.50 | Ea | 730.00 | 132.00 | 862.00 |
| Swing check, bronze trim, 250 psi steam | MI@3.00 | Ea | 1,300.00 | 159.00 | 1,459.00 |

|  | Craft@Hrs | Unit | Material | Labor | Total |
|---|---|---|---|---|---|
| 8" flanged iron body check valve | | | | | |
|   Swing check, bronze trim, 125 psi steam | Ml@3.00 | Ea | 844.00 | 159.00 | 1,003.00 |
|   Swing check, iron trim, 125 psi steam | Ml@3.00 | Ea | 1,340.00 | 159.00 | 1,499.00 |
| 10" flanged iron body check valve | | | | | |
|   Swing check, bronze trim, 125 psi steam | Ml@4.00 | Ea | 1,720.00 | 212.00 | 1,932.00 |
|   Swing check, iron trim, 125 psi steam | Ml@4.00 | Ea | 2,760.00 | 212.00 | 2,972.00 |
| 12" flanged iron body check valve | | | | | |
|   Swing check, bronze trim, 125 psi steam | Ml@4.50 | Ea | 2,270.00 | 238.00 | 2,508.00 |
|   Swing check, iron trim, 125 psi steam | Ml@4.50 | Ea | 3,600.00 | 238.00 | 3,838.00 |

**Miscellaneous Valves and Regulators** For water and steam applications, complies with American Society of Mechanicals Engineers Standard B16, Manufacturers Standardization Society, Instrumentation Society of America standards.

|  | Craft@Hrs | Unit | Material | Labor | Total |
|---|---|---|---|---|---|
| Angle valves, bronze, 150 lb, threaded, Teflon disk | | | | | |
|   1/2" valve | P6@.210 | Ea | 55.10 | 10.50 | 65.60 |
|   3/4" valve | P6@.250 | Ea | 75.20 | 12.60 | 87.80 |
|   1" valve | P6@.300 | Ea | 108.00 | 15.10 | 123.10 |
|   1-1/4" valve | P6@.400 | Ea | 141.00 | 20.10 | 161.10 |
|   1-1/2" valve | P6@.450 | Ea | 273.00 | 22.60 | 295.60 |
| Lavatory supply valve with hand wheel shutoff, flexible compression connection tube with length as 1/2" angle or straight stops | | | | | |
|   12" tube | P6@.294 | Ea | 20.30 | 14.80 | 35.10 |
|   15" tube | P6@.294 | Ea | 21.70 | 14.80 | 36.50 |
|   20" tube | P6@.473 | Ea | 24.80 | 23.80 | 48.60 |
| Gas stops, lever handle | | | | | |
|   1/2" stop | P6@.210 | Ea | 9.78 | 10.50 | 20.28 |
|   3/4" stop | P6@.250 | Ea | 13.00 | 12.60 | 25.60 |
|   1" stop | P6@.300 | Ea | 20.10 | 15.10 | 35.20 |
|   1-1/4" stop | P6@.450 | Ea | 30.00 | 22.60 | 52.60 |
|   1-1/2" stop | P6@.500 | Ea | 52.30 | 25.10 | 77.40 |
|   2" stop | P6@.750 | Ea | 80.90 | 37.70 | 118.60 |
| Hose bibbs, threaded | | | | | |
|   3/4", brass | P6@.350 | Ea | 4.99 | 17.60 | 22.59 |
| Backflow preventers reduced pressure | | | | | |
|   3/4", threaded, bronze, with 2 gate valves | P6@1.00 | Ea | 209.00 | 50.20 | 259.20 |
|   1", threaded, bronze, with 2 gate valves | P6@1.25 | Ea | 253.00 | 62.80 | 315.80 |
|   2", threaded, with 2 gate valves | P6@2.13 | Ea | 373.00 | 107.00 | 480.00 |
|   2-1/2", with 2 gate valves, flanged, iron body | Ml@2.00 | Ea | 1,910.00 | 106.00 | 2,016.00 |
|   3", with 2 gate valves, flanged, iron body | Ml@3.00 | Ea | 2,400.00 | 159.00 | 2,559.00 |
|   4", with 2 gate valves, flanged, iron body | Ml@4.50 | Ea | 2,860.00 | 238.00 | 3,098.00 |
|   6", without gate valves, flanged, iron body | Ml@8.00 | Ea | 4,700.00 | 424.00 | 5,124.00 |
|   6", with 2 gate valves, flanged, iron body | Ml@9.00 | Ea | 5,440.00 | 477.00 | 5,917.00 |
|   8", without gate valves, flanged, iron body | Ml@11.5 | Ea | 8,240.00 | 609.00 | 8,849.00 |
|   8", with 2 gate valves, flanged, iron body | Ml@12.5 | Ea | 11,900.00 | 662.00 | 12,562.00 |
| Ball valves, bronze, 150 lb, threaded, Teflon seat | | | | | |
|   1/2" valve | P6@.210 | Ea | 6.61 | 10.50 | 17.11 |
|   3/4" valve | P6@.250 | Ea | 9.77 | 12.60 | 22.37 |
|   1" valve | P6@.300 | Ea | 13.90 | 15.10 | 29.00 |
|   1-1/4" valve | P6@.400 | Ea | 24.00 | 20.10 | 44.10 |
|   1-1/2" valve | P6@.450 | Ea | 29.30 | 22.60 | 51.90 |
|   2" valve | P6@.500 | Ea | 39.40 | 25.10 | 64.50 |
|   2-1/2" valve | Ml@.750 | Ea | 75.00 | 39.70 | 114.70 |
|   3" valve | Ml@.950 | Ea | 102.00 | 50.30 | 152.30 |

# 22 Plumbing

| | Craft@Hrs | Unit | Material | Labor | Total |
|---|---|---|---|---|---|
| **Hose gate valves, with cap, 200 lb** | | | | | |
| 2-1/2", non-rising stem | P6@1.00 | Ea | 81.80 | 50.20 | 132.00 |
| **Pressure regulator valves, bronze, 300 lb threaded, 25 to 75 PSI, with "Y" strainer** | | | | | |
| 3/4" valve | P6@.250 | Ea | 117.00 | 12.60 | 129.60 |
| 1" valve | P6@.300 | Ea | 155.00 | 15.10 | 170.10 |
| 1-1/4" valve | P6@.400 | Ea | 241.00 | 20.10 | 261.10 |
| 1-1/2" valve | P6@.450 | Ea | 371.00 | 22.60 | 393.60 |
| 2" valve | P6@.500 | Ea | 468.00 | 25.10 | 493.10 |
| 2-1/2" valve | P6@.750 | Ea | 919.00 | 37.70 | 956.70 |
| 3" valve | P6@.950 | Ea | 1,060.00 | 47.70 | 1,107.70 |
| **Water control valves, threaded, 3-way with actuator** | | | | | |
| 2-1/2", CV 54 | MI@.750 | Ea | 1,720.00 | 39.70 | 1,759.70 |
| 3", CV 80 | MI@.950 | Ea | 2,060.00 | 50.30 | 2,110.30 |
| 4", CV 157 | MI@1.30 | Ea | 3,930.00 | 68.90 | 3,998.90 |

## Piping Specialties

Stainless steel bellows type expansion joints, 150 lb. Flanged units include bolt and gasket sets.

| | Craft@Hrs | Unit | Material | Labor | Total |
|---|---|---|---|---|---|
| 3" pipe, 4" travel, welded | MI@1.33 | Ea | 979.00 | 70.50 | 1,049.50 |
| 3" pipe, 6" travel, welded | MI@1.33 | Ea | 1,070.00 | 70.50 | 1,140.50 |
| 4" pipe, 3" travel, welded | MI@1.78 | Ea | 1,020.00 | 94.30 | 1,114.30 |
| 4" pipe, 6" travel, welded | MI@1.78 | Ea | 1,270.00 | 94.30 | 1,364.30 |
| 5" pipe, 4" travel, flanged | MI@2.20 | Ea | 2,710.00 | 117.00 | 2,827.00 |
| 5" pipe, 7" travel, flanged | MI@2.20 | Ea | 2,800.00 | 117.00 | 2,917.00 |
| 6" pipe, 4" travel, flanged | MI@2.67 | Ea | 2,890.00 | 141.00 | 3,031.00 |
| 8" pipe, 8" travel, flanged | MI@3.20 | Ea | 4,280.00 | 170.00 | 4,450.00 |
| **Ball joints, welded ends, 150 lb** | | | | | |
| 3" ball joint | MI@2.40 | Ea | 454.00 | 127.00 | 581.00 |
| 4" ball joint | MI@3.35 | Ea | 669.00 | 178.00 | 847.00 |
| 5" ball joint | MI@4.09 | Ea | 1,130.00 | 217.00 | 1,347.00 |
| 6" ball joint | MI@4.24 | Ea | 1,210.00 | 225.00 | 1,435.00 |
| 8" ball joint | MI@4.79 | Ea | 2,120.00 | 254.00 | 2,374.00 |
| 10" ball joint | MI@5.81 | Ea | 2,600.00 | 308.00 | 2,908.00 |
| 14" ball joint | MI@7.74 | Ea | 3,650.00 | 410.00 | 4,060.00 |

## In-Line Circulating Pumps
Flanged, for hot or cold water circulation. By flange size and pump horsepower rating. See flange costs below. No electrical work included.

| | Craft@Hrs | Unit | Material | Labor | Total |
|---|---|---|---|---|---|
| **Horizontal drive, iron body** | | | | | |
| 3/4" to 1-1/2", 1/12 HP | MI@1.25 | Ea | 208.00 | 66.20 | 274.20 |
| 2", 1/4 HP | MI@1.40 | Ea | 362.00 | 74.20 | 436.20 |
| 2-1/2" or 3", 1/4 HP | MI@1.40 | Ea | 519.00 | 74.20 | 593.20 |
| 3/4" to 1-1/2", 1/6 HP | MI@1.25 | Ea | 307.00 | 66.20 | 373.20 |
| 3/4" to 1-1/2", 1/4 HP | MI@1.40 | Ea | 412.00 | 74.20 | 486.20 |
| 3", 1/3 HP | MI@1.50 | Ea | 704.00 | 79.50 | 783.50 |
| 3", 1/2 HP | MI@1.70 | Ea | 737.00 | 90.10 | 827.10 |
| 3", 3/4 HP | MI@1.80 | Ea | 850.00 | 95.40 | 945.40 |
| **Horizontal drive, all bronze** | | | | | |
| 3/4" to 1-1/2", 1/12 HP | MI@1.25 | Ea | 318.00 | 66.20 | 384.20 |
| 3/4" to 1-1/2", 1/6 HP | MI@1.25 | Ea | 478.00 | 66.20 | 544.20 |
| 3/4" to 2", 1/4 HP | MI@1.40 | Ea | 567.00 | 74.20 | 641.20 |
| 2-1/2" or 3", 1/4 HP | MI@1.50 | Ea | 940.00 | 79.50 | 1,019.50 |
| 3", 1/3 HP | MI@1.50 | Ea | 1,250.00 | 79.50 | 1,329.50 |
| 3", 1/2 HP | MI@1.70 | Ea | 1,280.00 | 90.10 | 1,370.10 |
| 3", 3/4 HP | MI@1.80 | Ea | 1,330.00 | 95.40 | 1,425.40 |

# 22 Plumbing

| | Craft@Hrs | Unit | Material | Labor | Total |
|---|---|---|---|---|---|
| **Vertical drive, iron body** | | | | | |
| 3/4" to 1/2", 1/6 HP | MI@1.25 | Ea | 423.00 | 66.20 | 489.20 |
| 2", 1/4 HP | MI@1.40 | Ea | 499.00 | 74.20 | 573.20 |
| 2-1/2", 1/4 HP | MI@1.40 | Ea | 591.00 | 74.20 | 665.20 |
| 3", 1/4 HP | MI@1.40 | Ea | 591.00 | 74.20 | 665.20 |
| | | | | | |
| **Pump connectors, vibration isolators** In line, stainless steel braided connections, threaded | | | | | |
| 1/2" x 9" | M5@.210 | Ea | 6.00 | 10.40 | 16.40 |
| 1/2" x 12" | M5@.300 | Ea | 6.39 | 14.90 | 21.29 |
| 1/2" x 16" | M5@.400 | Ea | 6.90 | 19.90 | 26.80 |
| 1/2" x 20" | M5@.450 | Ea | 7.03 | 22.40 | 29.43 |
| 1/2" x 30" | MI@.500 | Ea | 8.55 | 26.50 | 35.05 |
| 1/2" x 36" | MI@.750 | Ea | 10.50 | 39.70 | 50.20 |
| | | | | | |
| **Flanges for circulating pumps** Per set of two, including bolts and gaskets | | | | | |
| 3/4" to 1-1/2", iron | M5@.290 | Ea | 8.74 | 14.40 | 23.14 |
| 2", iron | M5@.290 | Ea | 12.40 | 14.40 | 26.80 |
| 2-1/2" to 3", iron | M5@.420 | Ea | 39.70 | 20.90 | 60.60 |
| 3/4" to 1", bronze | M5@.290 | Ea | 19.50 | 14.40 | 33.90 |
| 2", bronze | M5@.290 | Ea | 78.20 | 14.40 | 92.60 |
| 2-1/2" to 3", iron | M5@.420 | Ea | 40.50 | 20.90 | 61.40 |
| 3/4" copper sweat flanges | M5@.290 | Ea | 12.20 | 14.40 | 26.60 |
| 1" copper sweat flanges | M5@.290 | Ea | 18.10 | 14.40 | 32.50 |
| 1-1/2" copper sweat flanges | M5@.290 | Ea | 20.80 | 14.40 | 35.20 |
| | | | | | |
| **Base-Mounted Centrifugal Pumps** Single stage cast iron pumps with flanges but no electrical connection. | | | | | |
| 50 GPM at 100' head | MI@3.15 | Ea | 2,670.00 | 167.00 | 2,837.00 |
| 50 GPM at 200' head | MI@4.15 | Ea | 3,010.00 | 220.00 | 3,230.00 |
| 50 GPM at 300' head | MI@5.83 | Ea | 3,390.00 | 309.00 | 3,699.00 |
| 200 GPM at 100' head | MI@6.70 | Ea | 3,430.00 | 355.00 | 3,785.00 |
| 200 GPM at 200' head | MI@7.15 | Ea | 3,540.00 | 379.00 | 3,919.00 |
| 200 GPM at 300' head | MI@10.2 | Ea | 3,840.00 | 540.00 | 4,380.00 |
| 240 GPM, 7-1/2 HP | MI@6.20 | Ea | 2,240.00 | 329.00 | 2,569.00 |
| 300 GPM, 5 HP | MI@3.91 | Ea | 2,560.00 | 207.00 | 2,767.00 |
| 340 GPM, 7-1/2 HP | MI@6.72 | Ea | 2,920.00 | 356.00 | 3,276.00 |
| 370 GPM, 5 HP | MI@4.02 | Ea | 2,620.00 | 213.00 | 2,833.00 |
| 375 GPM, 7-1/2 HP | MI@7.14 | Ea | 2,910.00 | 378.00 | 3,288.00 |
| 465 GPM, 10 HP | MI@7.63 | Ea | 3,030.00 | 404.00 | 3,434.00 |

Additional costs for base-mounted centrifugal pumps. Add the cost of a concrete pad, control wiring, an appliance dolly ($14 per day), come-a-long ($16 per day) and a 1/2-ton chain hoist ($21 per day) if required.

| | Craft@Hrs | Unit | Material | Labor | Total |
|---|---|---|---|---|---|
| Place 2' x 2' x 1/2" vibration pads | CF@.750 | Ea | 38.80 | 26.30 | 65.10 |
| Bore a feedwater hole through concrete | P1@.250 | Ea | — | 8.94 | 8.94 |

**Vertical Turbine Pumps 3,550 RPM** Add the cost of electrical work.
Cast iron single-stage vertical turbine pumps

| | Craft@Hrs | Unit | Material | Labor | Total |
|---|---|---|---|---|---|
| 50 GPM at 50' head | MI@9.9 | Ea | 4,650.00 | 525.00 | 5,175.00 |
| 50 GPM at 100' head | MI@9.9 | Ea | 5,370.00 | 525.00 | 5,895.00 |
| 100 GPM at 300' head | MI@9.9 | Ea | 6,930.00 | 525.00 | 7,455.00 |
| 200 GPM at 50' head | MI@9.9 | Ea | 5,800.00 | 525.00 | 6,325.00 |

# 22 Plumbing

| | Craft@Hrs | Unit | Material | Labor | Total |
|---|---|---|---|---|---|
| Cast iron multi-stage vertical turbine pumps | | | | | |
| 50 GPM at 100' head | D9@15.6 | Ea | 5,290.00 | 755.00 | 6,045.00 |
| 50 GPM at 200' head | D9@15.6 | Ea | 6,080.00 | 755.00 | 6,835.00 |
| 50 GPM at 300' head | D9@15.6 | Ea | 6,930.00 | 755.00 | 7,685.00 |
| 100 GPM at 100' head | D9@15.6 | Ea | 5,690.00 | 755.00 | 6,445.00 |
| 100 GPM at 200' head | D9@15.6 | Ea | 6,410.00 | 755.00 | 7,165.00 |
| 100 GPM at 300' head | D9@15.6 | Ea | 7,370.00 | 755.00 | 8,125.00 |
| 200 GPM at 100' head | D9@15.6 | Ea | 6,230.00 | 755.00 | 6,985.00 |
| 200 GPM at 200' head | D9@20.5 | Ea | 7,370.00 | 992.00 | 8,362.00 |
| 200 GPM at 300' head | D9@20.5 | Ea | 7,870.00 | 992.00 | 8,862.00 |
| Bronze single-stage vertical turbine pumps | | | | | |
| 50 GPM at 50' head | D9@15.6 | Ea | 4,630.00 | 755.00 | 5,385.00 |
| 100 GPM at 50' head | D9@15.6 | Ea | 4,950.00 | 755.00 | 5,705.00 |
| 200 GPM at 50' head | D9@15.6 | Ea | 5,370.00 | 755.00 | 6,125.00 |
| Bronze multi-stage vertical turbine pumps | | | | | |
| 50 GPM at 50' head | D9@15.6 | Ea | 5,290.00 | 755.00 | 6,045.00 |
| 50 GPM at 100' head | D9@15.6 | Ea | 5,690.00 | 755.00 | 6,445.00 |
| 50 GPM at 150' head | D9@15.6 | Ea | 5,810.00 | 755.00 | 6,565.00 |
| 100 GPM at 50' head | D9@15.6 | Ea | 5,290.00 | 755.00 | 6,045.00 |
| 100 GPM at 100' head | D9@15.6 | Ea | 5,690.00 | 755.00 | 6,445.00 |
| 100 GPM at 150' head | D9@15.6 | Ea | 5,990.00 | 755.00 | 6,745.00 |
| 200 GPM at 50' head | D9@15.6 | Ea | 5,690.00 | 755.00 | 6,445.00 |
| 200 GPM at 100' head | D9@15.6 | Ea | 6,230.00 | 755.00 | 6,985.00 |
| 200 GPM at 150' head | D9@15.6 | Ea | 6,870.00 | 755.00 | 7,625.00 |
| **Sump Pumps** Bronze 1,750 RPM sump pumps. These costs do not include electrical work. | | | | | |
| 25 GPM at 25' head | MI@17.1 | Ea | 3,730.00 | 906.00 | 4,636.00 |
| 25 GPM at 50' head | MI@17.1 | Ea | 3,880.00 | 906.00 | 4,786.00 |
| 25 GPM at 100' head | MI@17.1 | Ea | 4,070.00 | 906.00 | 4,976.00 |
| 25 GPM at 150' head | MI@17.1 | Ea | 5,420.00 | 906.00 | 6,326.00 |
| 50 GPM at 25' head | MI@20.1 | Ea | 3,610.00 | 1,070.00 | 4,680.00 |
| 50 GPM at 50' head | MI@20.1 | Ea | 3,760.00 | 1,070.00 | 4,830.00 |
| 50 GPM at 100' head | MI@20.1 | Ea | 3,870.00 | 1,070.00 | 4,940.00 |
| 50 GPM at 150' head | MI@20.1 | Ea | 5,560.00 | 1,070.00 | 6,630.00 |
| 100 GPM at 25' head | MI@22.8 | Ea | 3,830.00 | 1,210.00 | 5,040.00 |
| 100 GPM at 50' head | MI@22.8 | Ea | 3,770.00 | 1,210.00 | 4,980.00 |
| 100 GPM at 100' head | MI@22.8 | Ea | 4,870.00 | 1,210.00 | 6,080.00 |
| 100 GPM at 150' head | MI@22.8 | Ea | 5,530.00 | 1,210.00 | 6,740.00 |
| **Miscellaneous Pumps** These costs do not include electrical work. | | | | | |
| Salt water pump, 135 GPM at 300' head | D9@19.9 | Ea | 15,400.00 | 963.00 | 16,363.00 |
| Sewage pumps, vertical mount, centrifugal | | | | | |
| 1/2 HP, 25 GPM at 25' head | P6@8.00 | Ea | 1,180.00 | 402.00 | 1,582.00 |
| 1-1/2 HP, 100 GPM at 70' head | P6@14.6 | Ea | 5,190.00 | 733.00 | 5,923.00 |
| 10 HP, 300 GPM at 70' head | P6@18.3 | Ea | 8,440.00 | 919.00 | 9,359.00 |
| Medical vacuum pump, duplex, tank mount with 30 gallon tank, | | | | | |
| 1/2 HP, 20" vacuum | D4@20.6 | Ea | 8,900.00 | 953.00 | 9,853.00 |
| Oil pump, base mounted, 1-1/2 GPM at 15' head, | | | | | |
| 1/3 HP, 120 volt | D4@1.50 | Ea | 259.00 | 69.40 | 328.40 |

| | Craft@Hrs | Unit | Material | Labor | Total |
|---|---|---|---|---|---|

**Closed Cell Elastomeric Pipe Insulation** Semi split, non-self sealing, thickness as shown, by nominal pipe diameter. R factor equals 3.58 at 220 degrees F. Make additional allowances for scaffolding if required. Also see fittings, flanges and valves at the end of this section. For more detailed coverage of pipe insulation, see *National Plumbing & HVAC Estimator*, http://CraftsmanSiteLicense.com

| | Craft@Hrs | Unit | Material | Labor | Total |
|---|---|---|---|---|---|
| 1/4" pipe | P6@.039 | LF | .23 | 1.96 | 2.19 |
| 3/8" pipe | P6@.039 | LF | .24 | 1.96 | 2.20 |
| 1/2 " pipe | P6@.039 | LF | .17 | 1.96 | 2.13 |
| 3/4" pipe | P6@.039 | LF | .21 | 1.96 | 2.17 |
| 1" pipe | P6@.042 | LF | .34 | 2.11 | 2.45 |
| 1-1/4" pipe | P6@.042 | LF | .40 | 2.11 | 2.51 |
| 1-1/2" pipe | P6@.042 | LF | .60 | 2.11 | 2.71 |
| 2" pipe | P6@.046 | LF | .84 | 2.31 | 3.15 |
| 2-1/2", 3/8" thick | P6@.056 | LF | 1.57 | 2.81 | 4.38 |
| 2-1/2", 1/2" thick | P6@.056 | LF | 2.12 | 2.81 | 4.93 |
| 2-1/2", 3/4" thick | P6@.056 | LF | 2.74 | 2.81 | 5.55 |
| 3", 3/8" thick | P6@.062 | LF | 1.68 | 3.11 | 4.79 |
| 3", 1/2" thick | P6@.062 | LF | 2.39 | 3.11 | 5.50 |
| 3", 3/4" thick | P6@.062 | LF | 3.10 | 3.11 | 6.21 |
| 4", 3/8" thick | P6@.068 | LF | 2.10 | 3.41 | 5.51 |
| 4", 1/2" thick | P6@.068 | LF | 2.80 | 3.41 | 6.21 |
| 4", 3/4" thick | P6@.068 | LF | 3.64 | 3.41 | 7.05 |
| Add for self sealing | — | % | 75.0 | — | — |

Additional cost for insulating pipe fittings and flanges:
    For each fitting or flange, add the cost of insulating 3 LF of pipe of the same size.
Additional cost for insulating valves: Valve body only. Add the cost of insulating 5 LF of pipe of the same size.
    Body and bonnet or yoke valves. Add the cost of insulating 10 LF of pipe of the same size.

**Fiberglass pipe insulation with AP-T Plus (all purpose self-sealing) jacket** By nominal pipe diameter. R factor equals 2.56 at 300 degrees F. Add the cost of scaffolding, if required. Also see fittings, flanges and valves at the end of this section. Available in 3 foot lengths, price is per linear foot.

| | Craft@Hrs | Unit | Material | Labor | Total |
|---|---|---|---|---|---|
| 1/2" diameter pipe | | | | | |
| 1/2" thick insulation | P6@.038 | LF | 2.09 | 1.91 | 4.00 |
| 1" thick insulation | P6@.038 | LF | 2.49 | 1.91 | 4.40 |
| 1-1/2" thick insulation | P6@.040 | LF | 4.96 | 2.01 | 6.97 |
| 3/4" diameter pipe | | | | | |
| 1/2" thick insulation | P6@.038 | LF | 2.30 | 1.91 | 4.21 |
| 1" thick insulation | P6@.038 | LF | 2.84 | 1.91 | 4.75 |
| 1-1/2" thick insulation | P6@.040 | LF | 5.23 | 2.01 | 7.24 |
| 1" diameter pipe | | | | | |
| 1/2" thick insulation | P6@.040 | LF | 2.42 | 2.01 | 4.43 |
| 1" thick insulation | P6@.040 | LF | 2.49 | 2.01 | 4.50 |
| 1-1/2" thick insulation | P6@.042 | LF | 5.49 | 2.11 | 7.60 |
| 1-1/4" diameter pipe | | | | | |
| 1/2" thick insulation | P6@.040 | LF | 3.28 | 2.01 | 5.29 |
| 1" thick insulation | P6@.040 | LF | 3.28 | 2.01 | 5.29 |
| 1-1/2" thick insulation | P6@.042 | LF | 5.90 | 2.11 | 8.01 |
| 1-1/2" diameter pipe | | | | | |
| 1/2" thick insulation | P6@.042 | LF | 2.87 | 2.11 | 4.98 |
| 1" thick insulation | P6@.042 | LF | 3.72 | 2.11 | 5.83 |
| 1-1/2" thick insulation | P6@.044 | LF | 6.36 | 2.21 | 8.57 |
| 2" diameter pipe | | | | | |
| 1/2" thick insulation | P6@.044 | LF | 3.09 | 2.21 | 5.30 |
| 1" thick insulation | P6@.044 | LF | 4.24 | 2.21 | 6.45 |
| 1-1/2" thick insulation | P6@.046 | LF | 6.83 | 2.31 | 9.14 |

| | Craft@Hrs | Unit | Material | Labor | Total |
|---|---|---|---|---|---|
| **Meters, Gauges and Indicators** | | | | | |
| Water meters | | | | | |
| Disc type, 1-1/2" | P6@1.00 | Ea | 178.00 | 50.20 | 228.20 |
| Turbine type, 1-1/2" | P6@1.00 | Ea | 270.00 | 50.20 | 320.20 |
| Displacement type, AWWA C7000, 1" | P6@.941 | Ea | 275.00 | 47.30 | 322.30 |
| Displacement type, AWWA C7000, 1-1/2" | P6@1.23 | Ea | 950.00 | 61.80 | 1,011.80 |
| Displacement type, AWWA C7000, 2" | P6@1.57 | Ea | 1,430.00 | 78.80 | 1,508.80 |
| Curb stop and waste valve, 1" | P6@1.00 | Ea | 100.00 | 50.20 | 150.20 |
| Thermoflow indicators, soldered | | | | | |
| 1-1/4" indicator | M5@2.01 | Ea | 390.00 | 99.90 | 489.90 |
| 2" indicator | M5@2.01 | Ea | 445.00 | 99.90 | 544.90 |
| 2-1/2" indicator | MI@1.91 | Ea | 652.00 | 101.00 | 753.00 |
| Cast steel steam meters, in-line, flanged, 300 PSI | | | | | |
| 1" pipe, threaded | M5@1.20 | Ea | 3,710.00 | 59.70 | 3,769.70 |
| 2" pipe | MI@1.14 | Ea | 4,220.00 | 60.40 | 4,280.40 |
| 3" pipe | MI@2.31 | Ea | 4,700.00 | 122.00 | 4,822.00 |
| 4" pipe | MI@2.71 | Ea | 5,220.00 | 144.00 | 5,364.00 |
| Cast steel by-pass steam meters, 2", by line pipe size | | | | | |
| 6" line | MI@16.7 | Ea | 9,050.00 | 885.00 | 9,935.00 |
| 8" line | MI@17.7 | Ea | 9,300.00 | 938.00 | 10,238.00 |
| 10" line | MI@18.5 | Ea | 9,680.00 | 980.00 | 10,660.00 |
| 12" line | MI@19.7 | Ea | 10,000.00 | 1,040.00 | 11,040.00 |
| 14" line | MI@21.4 | Ea | 10,600.00 | 1,130.00 | 11,730.00 |
| 16" line | MI@22.2 | Ea | 11,100.00 | 1,180.00 | 12,280.00 |
| Add for steam meter pressure-compensated counter | — | Ea | 1,540.00 | — | 1,540.00 |
| Add for contactor used with dial counter | — | Ea | 711.00 | — | 711.00 |
| Add for wall or panel-mounted remote totalizer with contactor | — | Ea | 1,180.00 | — | 1,180.00 |
| Add for direct reading pressure gauge | M5@.250 | Ea | 86.50 | 12.40 | 98.90 |
| Add for direct reading thermometer, stainless steel case, with trim, 2% accuracy | | | | | |
| 3" dial | M5@.250 | Ea | 66.50 | 12.40 | 78.90 |
| **Plumbing Specialties** | | | | | |
| Access doors for plumbing, primer coated steel 14 gauge door, 18 gauge frame, with cam lock, for masonry, drywall, tile, wall or ceiling. L&L Louvers | | | | | |
| Standard, wall or ceiling | | | | | |
| 8" x 8" | SM@.500 | Ea | 26.40 | 29.20 | 55.60 |
| 12" x 12" | SM@.500 | Ea | 28.40 | 29.20 | 57.60 |
| 14" x 14" | SM@.500 | Ea | 33.50 | 29.20 | 62.70 |
| 16" x 16" | SM@.500 | Ea | 36.50 | 29.20 | 65.70 |
| 18" x 18" | SM@.500 | Ea | 41.60 | 29.20 | 70.80 |
| 22" x 22" | SM@.500 | Ea | 55.80 | 29.20 | 85.00 |
| 22" x 30" | SM@.500 | Ea | 71.10 | 29.20 | 100.30 |
| 24" x 24" | SM@.500 | Ea | 58.90 | 29.20 | 88.10 |
| 24" x 36" | SM@.500 | Ea | 85.30 | 29.20 | 114.50 |
| 30" x 30" | SM@.500 | Ea | 116.00 | 29.20 | 145.20 |
| Add for cylinder lock | — | Ea | 15.20 | — | 15.20 |
| Add for louvers | — | % | 150.0 | — | — |
| Add for fire rating, non-insulated | — | % | 400.0 | — | — |
| Add for fire rating and insulated | — | % | 500.0 | — | — |
| Add for stainless steel, 304-2B | — | % | 300.0 | — | — |

| | Craft@Hrs | Unit | Material | Labor | Total |
|---|---|---|---|---|---|

**Cleanouts** For drainage piping. A cleanout may be required: at the end of each waste line, or where a main building drain exits the building, or at regular 100' intervals for straight runs longer than 100', or at the base of soil or waste stacks, or where a 90-degree change in direction of a main building drain takes place. Consult your plumbing code for the regulations that apply to your area.

Finished grade end cleanouts (Malcolm) with Neo-Loc & Dura-Coated cast iron body

| | Craft@Hrs | Unit | Material | Labor | Total |
|---|---|---|---|---|---|
| 2" pipe, 4-1/8" round top | P6@.500 | Ea | 97.00 | 25.10 | 122.10 |
| 3" pipe, 5-1/8" round top | P6@.500 | Ea | 130.00 | 25.10 | 155.10 |
| 4" pipe, 6-1/8" round top | P6@.500 | Ea | 130.00 | 25.10 | 155.10 |

Cast iron in-line cleanouts (Barrett). Make additional allowances for access doors

| | Craft@Hrs | Unit | Material | Labor | Total |
|---|---|---|---|---|---|
| 2" pipe | P6@.350 | Ea | 36.00 | 17.60 | 53.60 |
| 4" pipe | P6@.400 | Ea | 67.40 | 20.10 | 87.50 |
| 6" pipe | P6@.400 | Ea | 238.00 | 20.10 | 258.10 |

Cleanout access cover accessories

| | Craft@Hrs | Unit | Material | Labor | Total |
|---|---|---|---|---|---|
| Round polished bronze cover | — | Ea | 21.50 | — | 21.50 |
| Square chrome plated cover | — | Ea | 196.00 | — | 196.00 |
| Square bronze hinged cover box | — | Ea | 140.00 | — | 140.00 |

**Floor drains,** body and grate only, add rough-in

| | Craft@Hrs | Unit | Material | Labor | Total |
|---|---|---|---|---|---|
| 3" x 2" plastic grate and body | P6@.25 | Ea | 13.10 | 12.60 | 25.70 |
| 5" x 3" plastic grate and body | P6@.25 | Ea | 18.40 | 12.60 | 31.00 |
| 3" x 2" cast iron body, nickel-bronze grate | P6@.50 | Ea | 38.00 | 25.10 | 63.10 |
| 5" x 3" cast iron body, nickel-bronze grate | P6@.50 | Ea | 43.50 | 25.10 | 68.60 |
| 6" x 4" cast iron body, nickel-bronze grate | P6@.50 | Ea | 50.60 | 25.10 | 75.70 |
| Add for a sediment bucket | — | Ea | 8.05 | — | 8.05 |
| Add for plastic backwater valve | — | Ea | 31.90 | — | 31.90 |
| Add for cast iron backwater valve | — | Ea | 235.00 | — | 235.00 |
| Add for trap seal primer fitting | — | Ea | 6.20 | — | 6.20 |

Drain, general use, medium duty, PVC plastic, 3" or 4" pipe

| | Craft@Hrs | Unit | Material | Labor | Total |
|---|---|---|---|---|---|
| 3" x 4", stainless steel strainer | P6@1.16 | Ea | 9.72 | 58.30 | 68.02 |
| 3" x 4", brass strainer | P6@1.32 | Ea | 24.00 | 66.30 | 90.30 |
| 4" with 5" brass strainer & spigot outlet | P6@1.55 | Ea | 44.70 | 77.80 | 122.50 |

Roof drains, cast iron mushroom strainer, bottom outlet

| | Craft@Hrs | Unit | Material | Labor | Total |
|---|---|---|---|---|---|
| 2" to 4" drain | P6@1.75 | Ea | 180.00 | 87.90 | 267.90 |
| 5" or 6" drain | P6@1.98 | Ea | 253.00 | 99.40 | 352.40 |
| 8" drain | P6@2.36 | Ea | 331.00 | 119.00 | 450.00 |

Roof (jack) flashing, round

| | Craft@Hrs | Unit | Material | Labor | Total |
|---|---|---|---|---|---|
| 4" jack | SM@.311 | Ea | 8.23 | 18.20 | 26.43 |
| 6" jack | SM@.416 | Ea | 8.94 | 24.30 | 33.24 |
| 8" jack | SM@.516 | Ea | 12.30 | 30.20 | 42.50 |

**Plumbing Equipment** Valves, supports, vents, gas or electric hookup and related equipment are not included except as noted.

**Water heaters, commercial, gas fired** 3 year warranty, Glass-lined. Set in place only. Make additional allowances for pipe, circulating pump, gas and flue connections.

Standard efficiency

| | Craft@Hrs | Unit | Material | Labor | Total |
|---|---|---|---|---|---|
| 50 gallons, 98 MBtu, 95 GPH recovery | P6@2.00 | Ea | 820.00 | 100.00 | 920.00 |
| 60 gallons, 55 MBtu, 50 GPH recovery | P6@2.00 | Ea | 1,470.00 | 100.00 | 1,570.00 |
| 67 gallons, 114 MBtu, 108 GPH recovery | P6@2.00 | Ea | 1,730.00 | 100.00 | 1,830.00 |
| 76 gallons, 180 MBtu, 175 GPH recovery | P6@2.25 | Ea | 2,350.00 | 113.00 | 2,463.00 |
| 82 gallons, 156 MBtu, 151 GPH recovery | P6@2.25 | Ea | 2,230.00 | 113.00 | 2,343.00 |
| 91 gallons, 300 MBtu, 291 GPH recovery | P6@2.50 | Ea | 2,760.00 | 126.00 | 2,886.00 |

| | Craft@Hrs | Unit | Material | Labor | Total |
|---|---|---|---|---|---|
| Energy Miser. Meets ASHRAE 90.1b-1992 standards. | | | | | |
| 50 gallons, 98 MBtu, 95 GPH recovery | P6@2.00 | Ea | 2,130.00 | 100.00 | 2,230.00 |
| 65 gallons, 360 MBtu, 350 GPH recovery | P6@2.00 | Ea | 3,110.00 | 100.00 | 3,210.00 |
| 67 gallons, 114 MBtu, 108 GPH recovery | P6@2.00 | Ea | 2,180.00 | 100.00 | 2,280.00 |
| 72 gallons, 250 MBtu, 242 GPH recovery | P6@2.25 | Ea | 2,910.00 | 113.00 | 3,023.00 |
| 76 gallons, 180 MBtu, 175 GPH recovery | P6@2.25 | Ea | 2,700.00 | 113.00 | 2,813.00 |
| 82 gallons, 156 MBtu, 151 GPH recovery | P6@2.25 | Ea | 2,640.00 | 113.00 | 2,753.00 |
| 85 gallons, 400 MBtu, 388 GPH recovery | P6@2.25 | Ea | 3,270.00 | 113.00 | 3,383.00 |
| 91 gallons, 200 MBtu, 194 GPH recovery | P6@2.50 | Ea | 2,910.00 | 126.00 | 3,036.00 |
| 91 gallons, 300 MBtu, 291 GPH recovery | P6@2.50 | Ea | 2,940.00 | 126.00 | 3,066.00 |
| 100 gallons, 199 MBtu, 200 GPH recovery | P6@2.50 | Ea | 3,590.00 | 126.00 | 3,716.00 |
| Power vent kit | P6@2.00 | Ea | 688.00 | 100.00 | 788.00 |
| Equal flow manifold kit, duplex | — | Ea | 473.00 | — | 473.00 |
| Equal flow manifold kit, triplex | — | Ea | 942.00 | — | 942.00 |
| Add for propane fired units, any of above | — | Ea | 122.00 | — | 122.00 |

**Water heaters, commercial, electric** 208/240 volts, with surface mounted thermostat, Rheem Glass-lined Energy Miser. Set in place only. Make additional allowances for pipe, circulating pump, and electrical connections.

| | Craft@Hrs | Unit | Material | Labor | Total |
|---|---|---|---|---|---|
| 50 gallons, 9 kW | P6@1.75 | Ea | 3,050.00 | 87.90 | 3,137.90 |
| 50 gallons, 27 kW | P6@1.75 | Ea | 3,790.00 | 87.90 | 3,877.90 |
| 85 gallons, 18 kW | P6@2.00 | Ea | 3,620.00 | 100.00 | 3,720.00 |
| 120 gallons, 45 kW | P6@3.00 | Ea | 5,480.00 | 151.00 | 5,631.00 |

**Water heater stands** Installed at same time as water heater. Raises flame elements 18" off floor. Secures to wall with seismic clips.

| | Craft@Hrs | Unit | Material | Labor | Total |
|---|---|---|---|---|---|
| 80 gallon capacity | — | Ea | 40.80 | — | 40.80 |
| 100 gallon capacity | — | Ea | 46.60 | — | 46.60 |
| Fully enclosed, 80 gallon capacity | — | Ea | 125.00 | — | 125.00 |

**Water heater safety pans**
1" side outlet drain, 16 gauge steel, installed at same time as water heater

| | Craft@Hrs | Unit | Material | Labor | Total |
|---|---|---|---|---|---|
| 14" diameter | — | Ea | 15.50 | — | 15.50 |
| 16" diameter | — | Ea | 17.20 | — | 17.20 |
| 19" diameter | — | Ea | 21.20 | — | 21.20 |
| 22" diameter | — | Ea | 23.60 | — | 23.60 |
| 24" diameter | — | Ea | 26.10 | — | 26.10 |
| 26-1/2" diameter | — | Ea | 28.70 | — | 28.70 |

1" bottom outlet drain, 16 gauge steel, installed at same time as water heater

| | Craft@Hrs | Unit | Material | Labor | Total |
|---|---|---|---|---|---|
| 16" diameter | — | Ea | 15.10 | — | 15.10 |
| 19" diameter | — | Ea | 19.40 | — | 19.40 |
| 22" diameter | — | Ea | 23.10 | — | 23.10 |
| 24" diameter | — | Ea | 25.50 | — | 25.50 |
| 26-1/2" diameter | — | Ea | 28.00 | — | 28.00 |

**Water heater earthquake restraints** State approved, installed at the same time as water heater

| | Craft@Hrs | Unit | Material | Labor | Total |
|---|---|---|---|---|---|
| Up to 52 gal heater, double strap | — | Ea | 20.70 | — | 20.70 |
| Up to 100 gal heater, double strap | — | Ea | 24.40 | — | 24.40 |
| Free standing restraint system – 75 gal capacity | — | Ea | 171.00 | — | 171.00 |
| Free standing restraint system – 100 gal capacity | — | Ea | 185.00 | — | 185.00 |
| Wall mount platform with restraints – 75 gal capacity | — | Ea | 163.00 | — | 163.00 |
| Wall mount platform with restraints – 100 gal capacity | — | Ea | 182.00 | — | 182.00 |

| | Craft@Hrs | Unit | Material | Labor | Total |
|---|---|---|---|---|---|

**Water heater connection assembly**  Includes typical pipe, fittings, valves, gages, and hangers

| | Craft@Hrs | Unit | Material | Labor | Total |
|---|---|---|---|---|---|
| Copper pipe connection, 3/4" supply | P6@2.25 | Ea | 158.00 | 113.00 | 271.00 |
| Copper pipe connection, 1" supply | P6@2.75 | Ea | 279.00 | 138.00 | 417.00 |
| Copper pipe connection, 1-1/4" supply | P6@3.50 | Ea | 344.00 | 176.00 | 520.00 |
| Copper pipe connection, 1-1/2" supply | P6@3.75 | Ea | 355.00 | 188.00 | 543.00 |
| Copper pipe connection, 2" supply | P6@4.50 | Ea | 380.00 | 226.00 | 606.00 |
| Copper pipe connection, 2-1/2" supply | P6@5.75 | Ea | 792.00 | 289.00 | 1,081.00 |
| Copper pipe connection, 3" supply | P6@6.50 | Ea | 1,220.00 | 326.00 | 1,546.00 |
| Carbon steel gas pipe connection, 3/4" | P6@1.40 | Ea | 92.50 | 70.30 | 162.80 |
| Carbon steel gas pipe connection, 1" | P6@1.50 | Ea | 120.00 | 75.30 | 195.30 |
| Carbon steel gas pipe connection, 1-1/2" | P6@2.00 | Ea | 177.00 | 100.00 | 277.00 |
| Double wall vent pipe connection, 4" | P6@1.40 | Ea | 104.00 | 70.30 | 174.30 |

**Instantaneous water heaters, electric**  Maximum operating pressure of 150 PSI. Includes plumbing hookup only. Add for electrical work. 1/2" compression fittings used except where noted. www.eemaxinc.com
120/208/240 volt single point for point-of-use handwashing. 33-degree F temperature rise at 1/2 GPM. Includes 3/8" compression fittings.

| | Craft@Hrs | Unit | Material | Labor | Total |
|---|---|---|---|---|---|
| 2.4/3.0 kW unit | P6@1.00 | Ea | 431.00 | 50.20 | 481.20 |

208/240 volt flow controlled for kitchen, bar sink and dual lavatory sinks. 38-degree F temperature rise at 1.0 GPM.

| | Craft@Hrs | Unit | Material | Labor | Total |
|---|---|---|---|---|---|
| 8.3 kW unit | P6@1.00 | Ea | 298.00 | 50.20 | 348.20 |

240/277 volt thermostatic for applications where accurate temperature control is required. May be used in combination with other water heating units. 41-degree F temperature rise at 1.0 GPM.

| | Craft@Hrs | Unit | Material | Labor | Total |
|---|---|---|---|---|---|
| 6.0 kW unit | P6@1.00 | Ea | 383.00 | 50.20 | 433.20 |

240 volt twin heating module for residential showers, mop sinks, industrial processes, campgrounds and resort cabins. May be used in combination with other water heating units. 51-degree F temperature rise at 2.0 GPM.

| | Craft@Hrs | Unit | Material | Labor | Total |
|---|---|---|---|---|---|
| 15.0 kW unit | P6@1.00 | Ea | 557.00 | 50.20 | 607.20 |

277 volt high capacity instantaneous water heater. May be used in combination with other water heating units. 35-degree F temperature rise at 3.5 GPM.

| | Craft@Hrs | Unit | Material | Labor | Total |
|---|---|---|---|---|---|
| 18.0 kW unit | P6@1.00 | Ea | 1,230.00 | 50.20 | 1,280.20 |

240 volt high purity for microchip manufacturing, pharmaceutical production, ultrasonic cleaning and batch chemical mixing. Maintains water purity up to 18 MEG OHM quality. 38-degree F temperature rise at 1.0 GPM.

| | Craft@Hrs | Unit | Material | Labor | Total |
|---|---|---|---|---|---|
| 5.5 kW unit | P6@1.00 | Ea | 97.80 | 50.20 | 148.00 |

## Solar Water Heating Systems

Solar collector panels, no piping included

| | Craft@Hrs | Unit | Material | Labor | Total |
|---|---|---|---|---|---|
| Typical 24 SF panel | P6@.747 | Ea | 299.00 | 37.50 | 336.50 |
| Supports for solar panels, per panel | P6@2.25 | Ea | 32.00 | 113.00 | 145.00 |
| Control panel for solar heating system | P6@1.56 | Ea | 110.00 | 78.30 | 188.30 |

Solar collector system pumps, no wiring or piping included

| | Craft@Hrs | Unit | Material | Labor | Total |
|---|---|---|---|---|---|
| 5.4 GPM, 7.9' head | P6@1.89 | Ea | 218.00 | 94.90 | 312.90 |
| 10.5 GPM, 11.6' head or 21.6 GPM, 8.5' head | P6@2.25 | Ea | 395.00 | 113.00 | 508.00 |
| 12.6 GPM, 20.5' head or 22.5 GPM, 13.3' head | P6@3.26 | Ea | 607.00 | 164.00 | 771.00 |
| 46.5 GPM, 13.7' head | P6@4.59 | Ea | 907.00 | 231.00 | 1,138.00 |
| 60.9 GPM, 21.0' head | P6@5.25 | Ea | 1,040.00 | 264.00 | 1,304.00 |

Solar hot water storage tanks, no piping and no pad included

| | Craft@Hrs | Unit | Material | Labor | Total |
|---|---|---|---|---|---|
| 40 gallon, horizontal | P6@1.12 | Ea | 260.00 | 56.20 | 316.20 |
| 110 gallon, horizontal | P6@3.54 | Ea | 1,060.00 | 178.00 | 1,238.00 |
| 180 gallon, vertical | P6@5.43 | Ea | 1,720.00 | 273.00 | 1,993.00 |
| 650 gallon, horizontal | P6@5.84 | Ea | 2,530.00 | 293.00 | 2,823.00 |
| 1,250 gallon, horizontal | P6@5.84 | Ea | 3,730.00 | 293.00 | 4,023.00 |
| 1,500 gallon, vertical | P6@11.4 | Ea | 5,920.00 | 573.00 | 6,493.00 |

# 22 Plumbing

| | Craft@Hrs | Unit | Material | Labor | Total |
|---|---|---|---|---|---|
| 2,200 gallon, horizontal | P6@17.1 | Ea | 5,380.00 | 859.00 | 6,239.00 |
| 2,700 gallon, vertical | P6@14.9 | Ea | 6,120.00 | 748.00 | 6,868.00 |

**Grease interceptors**

| | | | | | |
|---|---|---|---|---|---|
| 2" cast iron, 4 GPM, 8 lb capacity | P6@4.00 | Ea | 598.00 | 201.00 | 799.00 |
| 2" cast iron, 10 GPM, 20 lb capacity | P6@5.00 | Ea | 972.00 | 251.00 | 1,223.00 |
| 2" steel, 15 GPM, 30 lb capacity | P6@7.00 | Ea | 1,440.00 | 352.00 | 1,792.00 |
| 3" steel, 20 GPM, 40 lb capacity | P6@8.00 | Ea | 1,780.00 | 402.00 | 2,182.00 |

**Hair interceptors**

| | | | | | |
|---|---|---|---|---|---|
| 1-1/2" cast iron | P6@.650 | Ea | 202.00 | 32.60 | 234.60 |
| 1-1/2" nickel bronze | P6@.750 | Ea | 284.00 | 37.70 | 321.70 |

**Sewage ejector packaged systems** with vertical pumps, float switch controls, poly tank, cover and fittings, no electrical work or pipe connections included.
Simplex sewage ejector systems (single pump)

| | | | | | |
|---|---|---|---|---|---|
| 1/2 HP, 2" line, 20" x 30" tank | P6@4.75 | Ea | 561.00 | 239.00 | 800.00 |
| 3/4 HP, 2" line, 24" x 36" tank | P6@4.75 | Ea | 1,420.00 | 239.00 | 1,659.00 |
| 1 HP, 3" line, 24" x 36" tank | P6@6.00 | Ea | 1,650.00 | 301.00 | 1,951.00 |

Duplex sewage ejector systems (two pumps), alternator not included

| | | | | | |
|---|---|---|---|---|---|
| 1/2 HP, 2" line, 30" x 36" tank | P6@6.50 | Ea | 2,170.00 | 326.00 | 2,496.00 |
| 3/4 HP, 2" line, 30" x 36" tank | P6@6.50 | Ea | 2,810.00 | 326.00 | 3,136.00 |
| 1 HP, 3" line, 30" x 36" tank | P6@8.00 | Ea | 3,160.00 | 402.00 | 3,562.00 |
| Duplex pump controls (alternator) | — | Ea | 1,160.00 | — | 1,160.00 |

**Sump pumps** With float switch, no electrical work included
Submersible type

| | | | | | |
|---|---|---|---|---|---|
| 1/4 HP, 1-1/2" outlet, ABS plastic | P6@2.00 | Ea | 118.00 | 100.00 | 218.00 |
| 1/3 HP, 1-1/4" outlet, cast iron | P6@2.00 | Ea | 174.00 | 100.00 | 274.00 |
| 1/3 HP, 1-1/4" outlet, ABS plastic | P6@2.00 | Ea | 143.00 | 100.00 | 243.00 |

Upright type

| | | | | | |
|---|---|---|---|---|---|
| 1/3 HP, 1-1/4" outlet, cast iron | P6@2.00 | Ea | 139.00 | 100.00 | 239.00 |
| 1/3 HP, 1-1/2" outlet, ABS plastic | P6@2.00 | Ea | 102.00 | 100.00 | 202.00 |

**Hot water storage tanks** Insulated, floor mounted, includes pipe connection

| | | | | | |
|---|---|---|---|---|---|
| 80 gallons | P6@4.00 | Ea | 573.00 | 201.00 | 774.00 |
| 120 gallons | P6@4.50 | Ea | 878.00 | 226.00 | 1,104.00 |
| 200 gallons | P6@5.25 | Ea | 2,360.00 | 264.00 | 2,624.00 |
| 980 gallons | P6@20.0 | Ea | 8,100.00 | 1,000.00 | 9,100.00 |
| 1,230 gallons | P6@22.0 | Ea | 9,640.00 | 1,100.00 | 10,740.00 |
| 1,615 gallons | P6@26.0 | Ea | 13,300.00 | 1,310.00 | 14,610.00 |

**Septic tanks** Includes typical excavation and piping

| | | | | | |
|---|---|---|---|---|---|
| 500 gallons, polyethylene | U2@15.2 | Ea | 576.00 | 711.00 | 1,287.00 |
| 1,000 gallons, fiberglass | U2@16.1 | Ea | 817.00 | 753.00 | 1,570.00 |
| 2,000 gallons, concrete | U2@17.0 | Ea | 1,240.00 | 795.00 | 2,035.00 |
| 5,000 gallons, concrete | U2@22.3 | Ea | 4,610.00 | 1,040.00 | 5,650.00 |

**Water softeners** Automatic two tank unit, 3/4" valve

| | | | | | |
|---|---|---|---|---|---|
| 1.0 CF, 30,000 grain capacity | P6@3.59 | Ea | 424.00 | 180.00 | 604.00 |
| 2.0 CF, 60,000 grain capacity | P6@3.59 | Ea | 577.00 | 180.00 | 757.00 |

| | Craft@Hrs | Unit | Material | Labor | Total |
|---|---|---|---|---|---|

**Water filters**  In-line, filters for taste, odor, chlorine and sediment to 20 microns, with by-pass assembly. Includes pipe connection.

| | Craft@Hrs | Unit | Material | Labor | Total |
|---|---|---|---|---|---|
| Housing, by-pass and carbon filter, complete | P6@2.50 | Ea | 165.00 | 126.00 | 291.00 |
| Water filter housing only | P6@1.00 | Ea | 41.30 | 50.20 | 91.50 |
| By-pass assembly only, 3/4" copper | P6@1.25 | Ea | 89.50 | 62.80 | 152.30 |
| Carbon core cartridge, 20 microns | P6@.250 | Ea | 20.60 | 12.60 | 33.20 |
| Ceramic disinfection and carbon cartridge | P6@.250 | Ea | 55.00 | 12.60 | 67.60 |
| Lead, chlorine, taste and odor cartridge | P6@.250 | Ea | 41.30 | 12.60 | 53.90 |

**Water sterilizers**  Ultraviolet, point-of-use water. Add the cost of electrical and pipe connections. In gallons per minute (GPM) of rated capacity. 2 GPM and larger units include light-emitting diode (LED).

| | Craft@Hrs | Unit | Material | Labor | Total |
|---|---|---|---|---|---|
| 1/2 GPM point of use filter, 1/2" connection | P6@2.00 | Ea | 275.00 | 100.00 | 375.00 |
| 3/4 GPM point of use filter, 1/2" connection | P6@2.00 | Ea | 294.00 | 100.00 | 394.00 |
| 2 GPM, 1/2" connection | P6@2.50 | Ea | 325.00 | 126.00 | 451.00 |
| 2 GPM, with alarm, 1/2" connection | P6@2.50 | Ea | 423.00 | 126.00 | 549.00 |
| 2 GPM, with alarm and monitor, 1/2" connection | P6@3.00 | Ea | 627.00 | 151.00 | 778.00 |
| 5 GPM, 3/4" connection | P6@2.50 | Ea | 596.00 | 126.00 | 722.00 |
| 5 GPM, with alarm, 3/4" connection | P6@2.50 | Ea | 678.00 | 126.00 | 804.00 |
| 5 GPM, with alarm and monitor, 3/4" connection | P6@3.00 | Ea | 773.00 | 151.00 | 924.00 |
| 8 GPM, with alarm, 3/4" connection | P6@2.50 | Ea | 627.00 | 126.00 | 753.00 |
| 8 GPM, with alarm and monitor, 3/4" connection | P6@3.00 | Ea | 839.00 | 151.00 | 990.00 |
| 12 GPM, with alarm, 3/4" connection | P6@2.50 | Ea | 800.00 | 126.00 | 926.00 |
| 12 GPM, with alarm and monitor, 3/4" connection | P6@3.00 | Ea | 978.00 | 151.00 | 1,129.00 |

**Reverse osmosis water filters**  Economy system. Add the cost of electrical and pipe connections. In gallons per day of rated capacity.

| | Craft@Hrs | Unit | Material | Labor | Total |
|---|---|---|---|---|---|
| 12 gallons per day | P6@3.50 | Ea | 305.00 | 176.00 | 481.00 |
| 16 gallons per day | P6@3.50 | Ea | 325.00 | 176.00 | 501.00 |
| 30 gallons per day | P6@3.50 | Ea | 459.00 | 176.00 | 635.00 |
| 60 gallons per day | P6@3.50 | Ea | 508.00 | 176.00 | 684.00 |
| 100 gallons per day | P6@3.50 | Ea | 774.00 | 176.00 | 950.00 |

**Rough-in Plumbing Assemblies**  Includes rough-ins for commercial, industrial and residential buildings. This section shows costs for typical rough-ins of drain, waste and vent (DWV) and water supply piping using either plastic or metal pipe and fittings. Plastic ABS or PVC DWV pipe can be combined with plastic "poly b" supply pipe for a complete residential rough-in assembly. Cast iron or copper drainage DWV pipe can be combined with copper supply pipe for a complete commercial rough-in assembly. (Check local plumbing and building codes to determine acceptable materials for above and below grade installations in your area.)  Detailed estimates of the piping requirements for each type plumbing fixture rough-in were made using the Uniform Plumbing Code as a guideline. For scheduling purposes, estimate that a plumber and an apprentice can install rough-in piping for 3 to 4 plumbing fixtures in an 8-hour day. The same plumber and apprentice will install, connect and test 3 to 4 fixtures a day. When roughing-in less than 4 fixtures in a building, increase the rough-in cost 25% for each fixture less than 4. Add the cost of the fixture and final connection assembly (as shown in Fixtures and Final Connection Assemblies) to the rough-in assembly cost. Rule of Thumb: Use plastic drainage and supply pipe in economy residential applications. Use plastic drainage and copper supply in quality residential applications. Use cast iron or copper DWV drainage and copper supply pipe in quality commercial applications. In some jurisdictions, plastic drainage pipe is allowed below grade for commercial applications only. In residential applications, plastic drainage pipe may be allowed in above and below grade applications. The difference is the liability of the owner to the patrons of a public building vs. the liability of the owner of a private residence in case of fire. ABS pipe releases toxic gases when it burns.

| | Craft@Hrs | Unit | Material | Labor | Total |
|---|---|---|---|---|---|
| **Water closet rough-in** | | | | | |
| Floor mounted, tank type, rough-in includes 7' of 3", 8' of 1-1/2" DWV pipe and 8' of 1/2" supply pipe with typical fittings, hangers, and strapping | | | | | |
| Plastic DWV and PEX supply rough-in | P6@1.90 | Ea | 22.50 | 95.40 | 117.90 |
| Plastic DWV and copper supply rough-in | P6@2.00 | Ea | 27.70 | 100.00 | 127.70 |
| Cast iron and copper supply rough-in | P6@2.25 | Ea | 104.00 | 113.00 | 217.00 |
| Copper DWV and copper supply rough-in | P6@2.25 | Ea | 182.00 | 113.00 | 296.00 |
| Floor mounted flush valve type, rough-in includes 7' of 3", 8' of 1-1/2" DWV pipe and 8' of 1" supply pipe with typical fittings, hangers, and strapping | | | | | |
| Plastic DWV and PVC schedule 40 supply rough-in | P6@1.90 | Ea | 21.30 | 95.40 | 116.70 |
| Plastic DWV and copper supply rough-in | P6@2.00 | Ea | 44.40 | 100.00 | 144.40 |
| Cast iron and copper supply rough-in | P6@2.25 | Ea | 121.00 | 113.00 | 234.00 |
| Copper DWV and copper supply rough-in | P6@2.25 | Ea | 182.00 | 113.00 | 295.00 |
| Wall hung, tank type, rough-in includes 5' of 4", 8' of 1-1/2" DWV pipe and 8' of 1/2" supply pipe with typical fittings, hangers, and strapping | | | | | |
| Plastic DWV and PEX supply rough-in | P6@1.80 | Ea | 21.10 | 90.40 | 111.50 |
| Plastic DWV and copper supply rough-in | P6@1.90 | Ea | 26.20 | 95.40 | 121.60 |
| Cast iron and copper supply rough-in | P6@2.35 | Ea | 101.00 | 118.00 | 219.00 |
| Copper DWV and copper supply rough-in | P6@2.35 | Ea | 207.00 | 118.00 | 325.00 |
| Add for fixture carrier | P6@.750 | Ea | 244.00 | 37.70 | 281.70 |
| Wall hung flush valve type, rough-in includes 5' of 4", 6' of 1-1/2" DWV pipe and 8' of 1" supply pipe with typical fittings, hangers, and strapping | | | | | |
| Plastic DWV and PVC schedule 40 supply rough-in | P6@1.70 | Ea | 18.40 | 85.40 | 103.80 |
| Plastic DWV and copper supply rough-in | P6@1.00 | Ea | 41.60 | 50.20 | 91.80 |
| Cast iron and copper supply rough-in | P6@1.95 | Ea | 106.00 | 97.90 | 203.90 |
| Copper DWV and copper supply rough-in | P6@1.95 | Ea | 193.00 | 97.90 | 290.90 |
| Add for fixture carrier | P6@.750 | Ea | 244.00 | 37.70 | 281.70 |
| **Urinal rough-in** | | | | | |
| Wall hung urinal, rough-in includes 15' of DWV pipe and 8' of supply pipe with typical fittings, hangers and strapping | | | | | |
| Plastic DWV and supply rough-in | P6@2.00 | Ea | 11.90 | 100.00 | 111.90 |
| Plastic DWV and copper supply rough-in | P6@2.10 | Ea | 24.70 | 105.00 | 129.70 |
| Copper DWV and copper supply rough-in | P6@2.35 | Ea | 125.00 | 118.00 | 243.00 |
| Add for fixture carrier | P6@.850 | Ea | 186.00 | 42.70 | 228.70 |
| Floor mounted urinal, rough-in includes 15' of DWV pipe and 8' of supply pipe with typical fittings, hangers and strapping | | | | | |
| Plastic DWV and supply rough-in | P6@2.15 | Ea | 11.90 | 108.00 | 119.90 |
| Plastic DWV and copper supply rough-in | P6@2.25 | Ea | 24.70 | 113.00 | 137.70 |
| Copper DWV and copper supply rough-in | P6@2.50 | Ea | 125.00 | 126.00 | 251.00 |
| **Lavatory rough-in** | | | | | |
| Lavatory, rough-in includes 15' of 1-1/2" DWV pipe and 15' of 1/2" supply pipe with typical fittings, hangers and strapping | | | | | |
| Plastic DWV and PEX supply rough-in | P6@1.70 | Ea | 17.00 | 85.40 | 102.40 |
| Plastic DWV and copper supply rough-in | P6@1.75 | Ea | 26.60 | 87.90 | 114.50 |
| Copper DWV and copper supply rough-in | P6@1.90 | Ea | 127.00 | 95.40 | 222.40 |
| Add for carrier | P6@.750 | Ea | 186.00 | 37.70 | 223.70 |
| **Bathtub and shower**  Rough-in includes 15' of 1-1/2" DWV pipe and 15' of 1/2" supply pipe with typical fittings, hangers, and strapping | | | | | |
| Plastic DWV and PEX supply rough-in | P6@1.95 | Ea | 17.00 | 97.90 | 114.90 |
| Plastic DWV and copper supply rough-in | P6@2.10 | Ea | 26.60 | 105.00 | 131.60 |
| Copper DWV and copper supply rough-in | P6@2.35 | Ea | 127.00 | 118.00 | 245.00 |

| | Craft@Hrs | Unit | Material | Labor | Total |
|---|---|---|---|---|---|

**Kitchen sink**  Rough-in includes 15' of 1-1/2" DWV pipe and 15' of 1/2" supply pipe with typical, fittings, hangers and strapping

| | Craft@Hrs | Unit | Material | Labor | Total |
|---|---|---|---|---|---|
| Plastic DWV and PEX supply rough-in | P6@1.70 | Ea | 17.00 | 85.40 | 102.40 |
| Plastic DWV and copper supply rough-in | P6@1.75 | Ea | 26.60 | 87.90 | 114.50 |
| Copper DWV and copper supply rough-in | P6@1.90 | Ea | 127.00 | 95.40 | 222.40 |

**Drinking fountain and water cooler**  Rough-in includes 15' of 1-1/2" DWV pipe and 10' of 1/2" supply pipe with typical, fittings, strapping and hangers

| | Craft@Hrs | Unit | Material | Labor | Total |
|---|---|---|---|---|---|
| Plastic DWV and PEX supply rough-in | P6@1.80 | Ea | 14.80 | 90.40 | 105.20 |
| Plastic DWV and copper supply rough-in | P6@2.00 | Ea | 21.20 | 100.00 | 121.20 |
| Copper DWV and copper supply rough-in | P6@2.20 | Ea | 122.00 | 110.00 | 232.00 |

**Wash fountain**  Rough-in includes 20' of 1-1/2" DWV pipe and 20' of 1/2" supply pipe with typical fittings, hangers and strapping

| | Craft@Hrs | Unit | Material | Labor | Total |
|---|---|---|---|---|---|
| Plastic DWV and supply rough-in | P6@1.90 | Ea | 22.60 | 95.40 | 118.00 |
| Plastic DWV and copper supply rough-in | P6@2.00 | Ea | 35.50 | 100.00 | 135.50 |
| Copper DWV and copper supply rough-in | P6@2.10 | Ea | 169.00 | 105.00 | 274.00 |

**Service sink**  Rough-in includes 15' of DWV pipe and 15' of supply pipe with typical fittings, hangers, and strapping
Wall-hung slop sink

| | Craft@Hrs | Unit | Material | Labor | Total |
|---|---|---|---|---|---|
| Plastic DWV and supply rough-in | P6@2.25 | Ea | 17.00 | 113.00 | 130.00 |
| Plastic DWV and copper supply rough-in | P6@2.35 | Ea | 26.60 | 118.00 | 144.60 |
| Cast iron DWV and copper supply rough-in | P6@2.45 | Ea | 127.00 | 123.00 | 250.00 |

Floor-mounted mop sink

| | Craft@Hrs | Unit | Material | Labor | Total |
|---|---|---|---|---|---|
| Plastic DWV and supply rough-in | P6@2.35 | Ea | 17.00 | 118.00 | 135.00 |
| Plastic DWV and copper supply rough-in | P6@2.45 | Ea | 26.60 | 123.00 | 149.60 |
| Cast iron DWV and copper supply rough-in | P6@2.60 | Ea | 127.00 | 131.00 | 258.00 |

**Emergency drench shower and eye wash station**  Rough-in includes a valved water supply with 15' of supply pipe, typical fittings and hangers. Add the cost of a floor drain if required. See section below.

| | Craft@Hrs | Unit | Material | Labor | Total |
|---|---|---|---|---|---|
| 3/4" PVC schedule 40 | P6@1.25 | Ea | 12.70 | 62.80 | 75.50 |
| 3/4" Copper pipe | P6@1.50 | Ea | 36.70 | 75.30 | 112.00 |

**Surgeons' scrub sink**  Rough-in includes 7' of 2" and 8' of 1-1/2" DWV pipe and 15' of 3/4" supply pipe with typical fittings and hangers

| | Craft@Hrs | Unit | Material | Labor | Total |
|---|---|---|---|---|---|
| Plastic DWV and supply rough-in | P6@2.25 | Ea | 15.10 | 113.00 | 128.10 |
| Plastic DWV and copper supply rough-in | P6@2.35 | Ea | 39.10 | 118.00 | 157.10 |
| Cast iron DWV and copper supply rough-in | P6@2.75 | Ea | 115.00 | 138.00 | 253.00 |

**Floor sink, floor drain**  Rough-in includes 15' of drain and 15' of vent pipe, "P"-trap, typical fittings and hangers. Does not include the actual floor drain unit.

| | Craft@Hrs | Unit | Material | Labor | Total |
|---|---|---|---|---|---|
| 2" plastic DWV rough-in | P6@1.50 | Ea | 37.40 | 75.30 | 112.70 |
| 3" plastic DWV rough-in | P6@1.75 | Ea | 78.10 | 87.90 | 166.00 |
| 2" cast iron DWV rough-in | P6@2.00 | Ea | 188.00 | 100.00 | 288.00 |
| 3" cast iron DWV rough-in | P6@2.50 | Ea | 248.00 | 126.00 | 374.00 |
| 4" cast iron DWV rough-in | P6@2.50 | Ea | 320.00 | 126.00 | 446.00 |

**Fixtures and final connection assemblies**

Fixture costs are based on Good to Better quality white fixtures of standard dimensions and include all trim and valves. Add 30% for colored fixtures. Detailed estimates of the piping requirements for each type of plumbing fixture were made using the *Uniform Plumbing Code* as a guideline.

# 22 Plumbing

|  | Craft@Hrs | Unit | Material | Labor | Total |
|---|---|---|---|---|---|
| **Water closets** | | | | | |
| Floor mounted, tank type, elongated vitreous china bowl | | | | | |
|     Water closet, trim and valves (add rough-in) | P6@2.10 | Ea | 232.00 | 105.00 | 337.00 |
|     Deduct for round bowl | — | Ea | -39.86 | — | -39.86 |
| Floor mounted flush valve type, elongated vitreous china bowl | | | | | |
|     Water closet, trim and valves (add rough-in) | P6@2.60 | Ea | 300.00 | 131.00 | 431.00 |
| Floor mounted flush valve type, elongated vitreous china bowl, | | | | | |
| meets ADA requirements (18" high) | | | | | |
|     Water closet, trim and valves (add rough-in) | P6@2.60 | Ea | 355.00 | 131.00 | 486.00 |
| Wall hung, tank type, elongated vitreous china bowl, cast iron carrier | | | | | |
|     Water closet, trim and valves (add rough-in) | P6@3.00 | Ea | 678.00 | 151.00 | 829.00 |
|     Deduct for round bowl | — | Ea | -78.38 | — | -78.38 |
| Wall hung flush valve type elongated vitreous china bowl, cast iron carrier | | | | | |
|     Water closet, trim and valves (add rough-in) | P6@3.55 | Ea | 479.00 | 178.00 | 657.00 |
|     Add for bed pan cleanser (flush valve and rim lugs) | P6@1.50 | Ea | 208.00 | 75.30 | 283.30 |
| **Urinals** | | | | | |
| Wall hung urinal, vitreous china, washout flush action, 3/4" top spud with hand operated flush valve/vacuum breaker | | | | | |
|     Urinal, trim and valves (add rough-in) | P6@2.35 | Ea | 389.00 | 118.00 | 507.00 |
|     Add for fixture carrier | P6@1.25 | Ea | 113.00 | 62.80 | 175.80 |
| Floor mounted urinal, 18" wide, sloping front vitreous china, washout flush action, 3/4" top spud with hand operated flush valve/vacuum breaker | | | | | |
|     Urinal, trim and valves (add rough-in) | P6@5.25 | Ea | 940.00 | 264.00 | 1,204.00 |
| **Water conserving urinals, faucets and flush valves** | | | | | |
| Wall mounted urinal, ABS-structural plastic. Fit between studs framed at 16" on center. | | | | | |
|     Add the cost of wall finishing and rough-in plumbing | | | | | |
|         10 oz water per flush | P6@1.50 | Ea | 298.00 | 75.30 | 373.30 |
|     Battery-powered sensor faucet, 6 volt DC. Continuous discharge up to 10 seconds. Less than one quart per cycle. www.totousa.com | | | | | |
|         Standard spout | P6@1.50 | Ea | 384.00 | 75.30 | 459.30 |
|         Gooseneck spout | P6@1.50 | Ea | 423.00 | 75.30 | 498.30 |
|     AC-powered sensor faucet, 12 volt. Continuous discharge up to 10 seconds. Less than one quart per cycle | | | | | |
|         Standard spout | P6@1.50 | Ea | 384.00 | 75.30 | 459.30 |
|         Gooseneck spout | P6@1.50 | Ea | 423.00 | 75.30 | 498.30 |
|     Battery-powered sensor flush valves, 6 volt DC, exposed. Light flushing after brief use and full flushing after extended use. Includes rough fittings | | | | | |
|         Urinal flush valve | P6@2.0 | Ea | 346.00 | 100.00 | 446.00 |
|         Toilet flush valve, 1.6 or 3.5 gallons | P6@2.0 | Ea | 378.00 | 100.00 | 478.00 |
|     AC-powered sensor flush valves, 24 volt. Light flushing after brief use and full flushing after extended use. Includes rough fittings | | | | | |
|         Urinal flush valve, exposed | P6@2.0 | Ea | 369.00 | 100.00 | 469.00 |
|         Urinal flush valve, concealed | P6@2.0 | Ea | 377.00 | 100.00 | 477.00 |
|         Toilet flush valve | P6@2.0 | Ea | 400.00 | 100.00 | 500.00 |
| **Lavatories** | | | | | |
| Wall-hung vitreous china lavatory with single handle faucet and pop-up waste | | | | | |
|     Lavatory, trim and valves (add rough-in) | P6@2.50 | Ea | 397.00 | 126.00 | 523.00 |
|     Add for wheelchair type (ADA standards) | — | Ea | 183.00 | — | 183.00 |
| Countertop mounted vitreous china lavatory with single handle water and pop-up waste | | | | | |
|     Lavatory, trim and valves (add rough-in) | P6@2.00 | Ea | 295.00 | 100.00 | 395.00 |
| Countertop mounted enameled steel lavatory with single handle water faucet and pop-up waste | | | | | |
|     Lavatory, trim and valves (add rough-in) | P6@2.00 | Ea | 249.00 | 100.00 | 349.00 |

# 22 Plumbing

| | Craft@Hrs | Unit | Material | Labor | Total |
|---|---|---|---|---|---|
| **Bathtubs and Showers** | | | | | |
| Recessed enameled steel tub with shower | | | | | |
| Tub, trim and valves (add rough-in) | P6@2.50 | Ea | 417.00 | 126.00 | 543.00 |
| Recessed enameled cast iron tub with shower | | | | | |
| Tub, trim and valves (add rough-in) | P6@4.50 | Ea | 815.00 | 226.00 | 1,041.00 |
| Acrylic one-piece tub and shower enclosure | | | | | |
| Enclosure, trim and valves (add rough-in) | P6@4.75 | Ea | 765.00 | 239.00 | 1,004.00 |
| Acrylic three-piece tub and shower enclosure (renovation) | | | | | |
| Enclosure, trim and valves (add rough-in) | P6@5.75 | Ea | 1,030.00 | 289.00 | 1,319.00 |
| Fiberglass one-piece tub and shower enclosure | | | | | |
| Enclosure, trim and valves (add rough-in) | P6@4.75 | Ea | 565.00 | 239.00 | 804.00 |
| Fiberglass three-piece tub and shower enclosure | | | | | |
| Enclosure, trim and valves (add rough-in) | P6@5.75 | Ea | 852.00 | 289.00 | 1,141.00 |
| Acrylic one-piece shower stall | | | | | |
| Shower stall, trim and valves (add rough-in) | P6@3.50 | Ea | 727.00 | 176.00 | 903.00 |
| Add for larger stall | — | Ea | 126.00 | — | 126.00 |
| Acrylic three-piece shower stall (renovation) | | | | | |
| Shower stall, trim and valves (add rough-in) | P6@4.25 | Ea | 953.00 | 213.00 | 1,166.00 |
| Add for larger stall | — | Ea | 149.00 | — | 149.00 |
| Fiberglass one-piece shower stall | | | | | |
| Shower stall, trim and valves (add rough-in) | P6@3.50 | Ea | 490.00 | 176.00 | 666.00 |
| Add for larger stall | — | Ea | 100.00 | — | 100.00 |
| Add for corner entry | — | Ea | 81.60 | — | 81.60 |
| Fiberglass three-piece shower stall (renovation) | | | | | |
| Shower stall, trim and valves (add rough-in) | P6@4.25 | Ea | 628.00 | 213.00 | 841.00 |
| Add for larger stall | — | Ea | 139.00 | — | 139.00 |
| Add for corner entry | — | Ea | 113.00 | — | 113.00 |
| Acrylic shower base | | | | | |
| Base, trim and valves (add rough-in) | P6@2.50 | Ea | 307.00 | 126.00 | 433.00 |
| Add for larger base | — | Ea | 12.60 | — | 12.60 |
| Add for corner entry | — | Ea | 12.60 | — | 12.60 |
| Fiberglass shower base | | | | | |
| Base, trim and valves (add rough-in) | P6@2.50 | Ea | 346.00 | 126.00 | 472.00 |
| Add for larger base | — | Ea | 12.60 | — | 12.60 |
| Add for corner entry | — | Ea | 12.60 | — | 12.60 |
| Tub and shower doors | | | | | |
| Three panel sliding tub doors | P6@.500 | Ea | 314.00 | 25.10 | 339.10 |
| Three panel sliding shower doors | P6@.500 | Ea | 275.00 | 25.10 | 300.10 |
| Two panel sliding shower doors | P6@.500 | Ea | 232.00 | 25.10 | 257.10 |
| One panel swinging shower door | P6@.500 | Ea | 113.00 | 25.10 | 138.10 |
| **Kitchen sinks** | | | | | |
| Single compartment kitchen sink, 20" long, 21" wide, 8" deep, self rimming, countertop mounted, stainless steel with single handle faucet and deck spray | | | | | |
| Sink, trim and valves (add rough-in) | P6@2.00 | Ea | 247.00 | 100.00 | 347.00 |
| Double compartment kitchen sink, 32" long, 21" wide, 8" deep, self rimming, countertop mounted, stainless steel with single handle faucet and deck spray | | | | | |
| Sink, trim and valves (add rough-in) | P6@2.15 | Ea | 294.00 | 108.00 | 402.00 |

# 22 Plumbing

|  | Craft@Hrs | Unit | Material | Labor | Total |
|---|---|---|---|---|---|
| **Drinking fountain and water cooler** | | | | | |
| Wall hung drinking fountains, 10" wide, 10" high, non-electric, white vitreous china | | | | | |
| Wall hung fountain, and trim (add rough-in) | P6@2.00 | Ea | 406.00 | 100.00 | 506.00 |
| Recessed drinking fountains, 14" wide, 26" high, non-electric, white vitreous china | | | | | |
| Recessed fountain, and trim (add rough-in) | P6@2.50 | Ea | 719.00 | 126.00 | 845.00 |
| Semi-recessed water cooler, vitreous china, refrigerated (no electrical work included) | | | | | |
| Semi-recessed cooler, and trim (add rough-in) | P6@2.50 | Ea | 125.00 | 126.00 | 251.00 |
| Add for fully recessed | P6@.750 | Ea | 933.00 | 37.70 | 970.70 |
| Free-standing water cooler, refrigerated (no electrical work included) | | | | | |
| Wall hung water cooler, and trim (add rough-in) | P6@2.50 | Ea | 823.00 | 126.00 | 949.00 |
| **Wash fountains** | | | | | |
| 54" circular wash fountain, precast stone (terrazzo), with foot operated water spray control | | | | | |
| Wash fountain, trim and valves (add rough-in) | P6@3.65 | Ea | 2,700.00 | 183.00 | 2,883.00 |
| Add for stainless steel in lieu of precast stone | — | Ea | 622.00 | — | 622.00 |
| 54" semi-circular wash fountain, precast stone (terrazzo), with foot operated water spray control | | | | | |
| Wash fountain, trim and valves (add rough-in) | P6@3.65 | Ea | 2,430.00 | 183.00 | 2,613.00 |
| Add for stainless steel in lieu of precast stone | — | Ea | 436.00 | — | 436.00 |
| 36" circular wash fountain, precast stone (terrazzo), with foot operated water spray control | | | | | |
| Wash fountain, trim and valves (add rough-in) | P6@3.50 | Ea | 2,160.00 | 176.00 | 2,336.00 |
| Add for stainless steel in lieu of precast stone | — | Ea | 373.00 | — | 373.00 |
| 36" semi-circular wash fountain, precast stone, (terrazzo), with foot operated water spray control | | | | | |
| Wash fountain, trim and valves (add rough-in) | P6@3.50 | Ea | 1,980.00 | 176.00 | 2,156.00 |
| Add for stainless steel in lieu of precast stone | — | Ea | 246.00 | — | 246.00 |
| **Service sinks** | | | | | |
| Wall hung enameled cast iron slop sink, with wall-mounted faucet, hose and "P"-trap standard | | | | | |
| Slop sink, trim and valves (add rough-in) | P6@3.30 | Ea | 763.00 | 166.00 | 929.00 |
| Floor mounted mop sink, precast molded stone (terrazzo), with wall mounted faucet and hose | | | | | |
| Mop sink, 24" x 24", 6" curbs (add rough-in) | P6@2.60 | Ea | 425.00 | 131.00 | 556.00 |
| Mop sink, 32" x 32", 6" curbs (add rough-in) | P6@2.60 | Ea | 457.00 | 131.00 | 588.00 |
| Mop sink, 36" x 36", 6" curbs (add rough-in) | P6@2.60 | Ea | 499.00 | 131.00 | 630.00 |
| Add for 12" curbs, 6" drop front & stainless steel caps | — | Ea | 242.00 | — | 242.00 |
| **Drench shower and emergency eye wash stations** | | | | | |
| Drench shower, plastic pipe and head, wall mounted | P6@2.00 | Ea | 173.00 | 100.00 | 273.00 |
| Drench shower galvanized pipe, free standing | P6@2.50 | Ea | 372.00 | 126.00 | 498.00 |
| Drench shower, emergency eyewash combination | P6@2.75 | Ea | 1,010.00 | 138.00 | 1,148.00 |
| Emergency eyewash station | P6@2.00 | Ea | 403.00 | 100.00 | 503.00 |
| Walk-thru station, 18 heads | P6@12.5 | Ea | 2,600.00 | 628.00 | 3,228.00 |
| **Surgeons' scrub sink** | | | | | |
| Wall-hung wash sink, white enameled cast iron, 48" long, 18" deep, trough-type with 2 twin-handled faucets and two soap dishes | | | | | |
| Wash sink, trim and valves (add rough-in) | P6@2.50 | Ea | 1,090.00 | 126.00 | 1,216.00 |
| Add for foot or knee pedal operation | P6@.500 | Ea | 126.00 | 25.10 | 151.10 |
| Add for electronic infrared hands free operation | P6@.500 | Ea | 157.00 | 25.10 | 182.10 |
| Add for stainless steel in lieu of enameled cast iron | — | Ea | 190.00 | — | 190.00 |
| **Floor sinks** EMCO Supply | | | | | |
| 12" square sink, cast iron top and body (add rough-in) | P6@1.00 | Ea | 193.00 | 50.20 | 243.20 |
| Add for an acid-resistant finish to the sink | — | Ea | 32.10 | — | 32.10 |

# 22 Plumbing

| | Craft@Hrs | Unit | Material | Labor | Total |
|---|---|---|---|---|---|

**Sauna Room** Commercial dry type, with insulated prefabricated 6' wide by 12' long by 8' high enclosure, with premounted cedar inner walls and ceiling and three-tier benches, roof vent, electric stone floor-mounted heater module with thermostat regulator, grouted stone floor, sanitary floor drain with cleanout, adjoining three-stall fiberglass shower stall with fixtures and drain.

| | Craft@Hrs | Unit | Material | Labor | Total |
|---|---|---|---|---|---|
| Commercial sauna enclosure module | — | Ea | 6,500.00 | — | 6,500.00 |
| Prep & sealing of floor | B9@8.0 | Ea | 400.00 | 261.00 | 661.00 |
| Pouring of floor | B9@8.0 | Ea | — | 261.00 | 261.00 |
| Tiling of floor | B1@8.0 | Ea | 800.00 | 266.00 | 1,066.00 |
| Ceiling fabrication | B1@8.0 | Ea | — | 266.00 | 266.00 |
| Running of vents | SW@8.0 | Ea | 200.00 | 328.00 | 528.00 |
| Running of drain | P6@4.0 | Ea | 100.00 | 201.00 | 301.00 |
| Electrical for heat and lighting | CE@4.0 | Ea | 150.00 | 233.00 | 383.00 |
| Mounting of benches | B1@4.0 | Ea | — | 133.00 | 133.00 |
| Installation of sealed portal | B1@4.0 | Ea | 250.00 | 133.00 | 383.00 |
| Running of CW/HW piping and valves | P6@8.0 | Ea | 325.00 | 402.00 | 727.00 |
| 3-stall shower module installation and piping | P6@16.0 | Ea | 750.00 | 804.00 | 1,554.00 |
| Subtotal | — | Ea | 2,975.00 | 3,288.00 | 6,263.00 |
| Grand Total (with Sauna Module) | — | Ea | 9,475.00 | 3,288.00 | 12,763.00 |

# 23 HVAC

**Gas-fired heating equipment** No gas pipe or electric runs included
Unit heaters, gas fired, ceiling suspended, with flue and gas valve, propeller fan with aluminized heat exchanger, Modine PAE series

| | Craft@Hrs | Unit | Material | Labor | Total |
|---|---|---|---|---|---|
| 50 MBtu input | M5@4.75 | Ea | 833.00 | 236.00 | 1,069.00 |
| 75 MBtu input | M5@4.75 | Ea | 910.00 | 236.00 | 1,146.00 |
| 125 MBtu input | M5@5.25 | Ea | 1,140.00 | 261.00 | 1,401.00 |
| 175 MBtu input | M5@5.50 | Ea | 1,280.00 | 273.00 | 1,553.00 |
| 225 MBtu input | MI@6.75 | Ea | 1,450.00 | 358.00 | 1,808.00 |
| 300 MBtu input | MI@7.50 | Ea | 1,810.00 | 397.00 | 2,207.00 |
| 400 MBtu input | MI@8.75 | Ea | 2,390.00 | 464.00 | 2,854.00 |
| Add for single stage intermittent ignition pilot | — | LS | 188.00 | — | 188.00 |
| Add for high efficiency units (PV series) | — | % | 35.0 | — | — |
| Add for stainless steel burner | — | % | 20.0 | — | — |
| Add for air blower style fan | — | % | 10.0 | — | — |
| Add for switch and thermostat | CE@1.00 | Ea | 51.30 | 58.40 | 109.70 |
| Add for gas supply pipe connection | M5@1.75 | Ea | 83.30 | 87.00 | 170.30 |
| Add for typical electric run, BX | E4@2.50 | Ea | 103.00 | 123.00 | 226.00 |

Furnaces, wall type, upflow, gas fired with gas valve, flue vent and electronic ignition

| | Craft@Hrs | Unit | Material | Labor | Total |
|---|---|---|---|---|---|
| 35 MBtu input | M5@4.25 | Ea | 872.00 | 211.00 | 1,083.00 |
| 50 MBtu input | M5@4.25 | Ea | 983.00 | 211.00 | 1,194.00 |
| 65 MBtu input | M5@4.25 | Ea | 993.00 | 211.00 | 1,204.00 |
| Add for gas supply pipe connection | M5@1.75 | Ea | 83.00 | 87.00 | 170.00 |
| Add for power vent kit (direct sidewall vent) | M5@2.50 | Ea | 113.00 | 124.00 | 237.00 |
| Add for thermostat | CE@1.00 | Ea | 45.40 | 58.40 | 103.80 |

Furnaces, non-condensing gas fired, multi-positional, mid-efficiency with gas valve, flue vent and electronic ignition

| | Craft@Hrs | Unit | Material | Labor | Total |
|---|---|---|---|---|---|
| 40 MBtu input, 80% AFUE | M5@4.25 | Ea | 1,000.00 | 211.00 | 1,211.00 |
| 60 MBtu input, 80% AFUE | M5@4.25 | Ea | 1,100.00 | 211.00 | 1,311.00 |
| 80 MBtu input, 80% AFUE | M5@4.25 | Ea | 1,130.00 | 211.00 | 1,341.00 |

| | Craft@Hrs | Unit | Material | Labor | Total |
|---|---|---|---|---|---|
| 100 MBtu input, 80% AFUE | M5@4.50 | Ea | 1,160.00 | 224.00 | 1,384.00 |
| 120 MBtu input, 80% AFUE | M5@4.50 | Ea | 1,250.00 | 224.00 | 1,474.00 |
| 140 MBtu input, 80% AFUE | M5@4.50 | Ea | 1,390.00 | 224.00 | 1,614.00 |
| Add for gas supply pipe connection | M5@1.75 | Ea | 83.00 | 87.00 | 170.00 |
| Add for power vent kit (direct sidewall vent) | M5@2.50 | Ea | 108.00 | 124.00 | 232.00 |
| Add for thermostat | CE@1.00 | Ea | 45.40 | 58.40 | 103.80 |

Furnaces, condensing gas fired, upflow, high-efficiency with gas valve, flue vent and electronic ignition

| | Craft@Hrs | Unit | Material | Labor | Total |
|---|---|---|---|---|---|
| 50 MBtu input, 90% AFUE | M5@4.25 | Ea | 1,780.00 | 211.00 | 1,991.00 |
| 70 MBtu input, 90% AFUE | M5@4.25 | Ea | 1,990.00 | 211.00 | 2,201.00 |
| 90 MBtu input, 90% AFUE | M5@4.25 | Ea | 2,000.00 | 211.00 | 2,211.00 |
| 110 MBtu input, 90% AFUE | M5@4.50 | Ea | 2,160.00 | 224.00 | 2,384.00 |
| 130 MBtu input, 90% AFUE | M5@4.50 | Ea | 2,170.00 | 224.00 | 2,394.00 |
| Add for gas supply pipe connection | M5@1.75 | Ea | 83.00 | 87.00 | 170.00 |
| Add for power vent kit (direct sidewall vent) | M5@2.50 | Ea | 113.00 | 124.00 | 237.00 |

Air curtain, gas fired, ceiling suspended, includes flue, vent, and electrical ignition

| | Craft@Hrs | Unit | Material | Labor | Total |
|---|---|---|---|---|---|
| 256 MBtu, 2,840 CFM | MI@7.50 | Ea | 1,980.00 | 397.00 | 2,377.00 |
| 290 MBtu, 4,740 CFM | MI@8.25 | Ea | 2,160.00 | 437.00 | 2,597.00 |
| 400 MBtu, 5,660 CFM | MI@9.75 | Ea | 2,390.00 | 517.00 | 2,907.00 |
| 525 MBtu, 7,580 CFM | MI@11.0 | Ea | 2,510.00 | 583.00 | 3,093.00 |
| 630 MBtu, 9,480 CFM | MI@12.5 | Ea | 2,780.00 | 662.00 | 3,442.00 |
| 850 MBtu, 11,400 CFM | MI@16.0 | Ea | 3,970.00 | 848.00 | 4,818.00 |
| Add for gas supply pipe connection | M5@1.75 | Ea | 83.00 | 87.00 | 170.00 |

## Electric Heat

Air curtain electric heaters, wall mounted, includes electrical connection only

| | Craft@Hrs | Unit | Material | Labor | Total |
|---|---|---|---|---|---|
| 37" long,  9.5 kW | H9@6.33 | Ea | 834.00 | 370.00 | 1,204.00 |
| 49" long, 12.5 kW | H9@8.05 | Ea | 1,180.00 | 470.00 | 1,650.00 |
| 61" long, 16.0 kW | H9@11.2 | Ea | 1,420.00 | 654.00 | 2,074.00 |

Baseboard electric heaters, with remote thermostat

| | Craft@Hrs | Unit | Material | Labor | Total |
|---|---|---|---|---|---|
| 185 watts per LF | CE@.175 | LF | 28.00 | 10.20 | 38.20 |

**Hydronic hot water generators.** U.S. Green Council and ASME "H" stamp certified to a rating of 95% AFUE (annual fuel usage efficiency) with EPA Tier II Nox emissions compliant burner, Kunkel safety valve, low-water cutoff, digital panel controls and automatic level controls, Parker Boiler 210-L series or equal. For commercial and light industrial applications. Equipment cost is for a 2-ton capacity forklift.

| | Craft@Hrs | Unit | Material | Labor | Equipment | Total |
|---|---|---|---|---|---|---|
| 4 BHP,    133 MBtu/Hr | M5@4.00 | Ea | 2,600.00 | 199.00 | 63.70 | 2,862.70 |
| 8 BHP,    263 MBtu/Hr | M5@4.00 | Ea | 3,460.00 | 199.00 | 63.70 | 3,722.70 |
| 12 BHP,    404 MBtu/Hr | M5@4.00 | Ea | 4,490.00 | 199.00 | 63.70 | 4,752.70 |
| 15 BHP,    514 MBtu/Hr | M5@5.00 | Ea | 6,780.00 | 249.00 | 79.60 | 7,108.60 |
| 19 BHP,    624 MBtu/Hr | M5@5.00 | Ea | 8,990.00 | 249.00 | 79.60 | 9,318.60 |
| 22 BHP,    724 MBtu/Hr | M5@6.00 | Ea | 9,350.00 | 298.00 | 95.60 | 9,743.60 |
| 29 BHP,    962 MBtu/Hr | M5@6.00 | Ea | 11,200.00 | 298.00 | 95.60 | 11,593.60 |
| 34 BHP, 1,125 MBtu/Hr | MI@8.00 | Ea | 12,500.00 | 424.00 | 84.90 | 13,008.90 |
| 40 BHP, 1,336 MBtu/Hr | MI@8.00 | Ea | 14,400.00 | 424.00 | 84.90 | 14,908.90 |
| 47 BHP, 1,571 MBtu/Hr | MI@9.00 | Ea | 14,900.00 | 477.00 | 95.60 | 15,472.60 |
| 53 BHP, 1,758 MBtu/Hr | MI@10.0 | Ea | 16,100.00 | 530.00 | 106.00 | 16,736.00 |
| 63 BHP, 2,100 MBtu/Hr | MI@12.0 | Ea | 16,700.00 | 636.00 | 127.00 | 17,463.00 |
| 89 BHP, 3,001 MBtu/Hr | MI@18.0 | Ea | 21,500.00 | 954.00 | 191.00 | 22,645.00 |
| 118 BHP, 4,001 MBtu/Hr | MI@22.0 | Ea | 22,200.00 | 1,170.00 | 234.00 | 23,604.00 |

# 23 HVAC

|  | Craft@Hrs | Unit | Material | Labor | Equipment | Total |
|---|---|---|---|---|---|---|

LEED certification (Leadership in Energy and Environmental Design) requires 88% boiler efficiency, recording controls, compliance with SCAQMD (S. Calif. Air Quality Management District) 1146.2 emission standards and zone thermostats.

|  | Craft@Hrs | Unit | Material | Labor | Equipment | Total |
|---|---|---|---|---|---|---|
| Add for LEED-certified boiler with USGBC rating | — | Ea | — | 10% | — | — |
| Add for LEED registration and inspection | — | Ea | — | — | — | 2,000.00 |
| Add for LEED central performance monitor/recorder | — | Ea | 2,440.00 | — | — | 2,440.00 |
| Add for LEED-certified zone recording thermostats | — | Ea | 364.00 | — | — | 364.00 |

**Installation of packaged hydronic boilers.** Add the cost of control wiring, supply lines (electric, feedwater and gas), drain line, circulating pump, expansion tank, vent stack, permits, final inspection and rental of an appliance dolly ($14 per day), come-a-long ($16 per day) and a 1/2-ton chain hoist ($21 per day) if required.

|  | Craft@Hrs | Unit | Material | Labor | Equipment | Total |
|---|---|---|---|---|---|---|
| Place 4' x 4' vibration pads | CF@.750 | Ea | 38.80 | 26.30 | — | 65.10 |
| Connect gas and feedwater lines | P1@2.50 | Ea | — | 89.40 | — | 89.40 |
| Mount interior boiler drain | P1@.500 | Ea | 7.76 | 17.90 | — | 25.66 |
| Bore pipe hole through basement wall | P1@.250 | Ea | — | 8.94 | — | 8.94 |
| Bore stack vent through exterior wall | P1@.250 | Ea | — | 8.94 | — | 8.94 |
| Mount and edge-seal stack | P1@.500 | Ea | 40.00 | 17.90 | — | 57.90 |
| Mount piping for circulating pump | P1@.450 | Ea | — | 16.10 | — | 16.10 |
| Install expansion tank |  |  |  |  |  |  |
| 2.1 gallon | P1@.300 | Ea | 32.59 | 10.70 | — | 43.29 |
| 4.5 gallon | P1@.300 | Ea | 54.32 | 10.70 | — | 65.02 |

## Hydronic Heating

Hot water baseboard fin tube radiation, copper tube and aluminum fin element, commercial quality, 18 gauge enclosure. Add for control wiring.

|  | Craft@Hrs | Unit | Material | Labor | Total |
|---|---|---|---|---|---|
| 9-7/8" enclosure and 1-1/4" element | P6@.280 | LF | 28.10 | 14.10 | 42.20 |
| 13-7/8" enclosure and 1-1/4" element, 2 rows | P6@.320 | LF | 48.60 | 16.10 | 64.70 |
| 9-7/8" enclosure only | P6@.130 | LF | 8.80 | 6.53 | 15.33 |
| 13-7/8" enclosure only | P6@.150 | LF | 13.50 | 7.53 | 21.03 |
| 1-1/4" element only | P6@.150 | LF | 20.00 | 7.53 | 27.53 |
| 1" element only | P6@.150 | LF | 10.00 | 7.53 | 17.53 |
| 3/4" element only | P6@.140 | LF | 8.04 | 7.03 | 15.07 |
| Add for pipe connection and control valve | P6@2.25 | Ea | 184.00 | 113.00 | 297.00 |
| Add for corners, fillers and caps, average per foot | — | % | 10.0 | — | — |
| End cap, 4" | P6@.090 | Ea | 6.07 | 4.52 | 10.59 |
| End cap, 2" | P6@.090 | Ea | 3.05 | 4.52 | 7.57 |
| Valve access door | P6@.250 | Ea | 15.20 | 12.60 | 27.80 |
| Inside corner | P6@.120 | Ea | 6.04 | 6.03 | 12.07 |
| Outside corner | P6@.120 | Ea | 8.58 | 6.03 | 14.61 |
| Filler sleeve | P6@.090 | Ea | 3.78 | 4.52 | 8.30 |

Hot water baseboard fin tube radiation, steel element, commercial quality, 16 gauge, sloping top cover. Make additional allowances for control wiring.

|  | Craft@Hrs | Unit | Material | Labor | Total |
|---|---|---|---|---|---|
| 9-7/8" enclosure and 1-1/4" steel element | P6@.330 | LF | 33.60 | 16.60 | 50.20 |
| 9-7/8" enclosure only | P6@.130 | LF | 10.10 | 6.53 | 16.63 |
| 1-1/4" steel element only | P6@.160 | LF | 23.30 | 8.04 | 31.34 |
| Add for pipe connection and control valve | P6@2.25 | Ea | 190.00 | 113.00 | 303.00 |
| Add for corners, fillers and caps, average per foot | — | % | 10.0 | — | — |
| End cap, 4" | P6@.090 | Ea | 12.00 | 4.52 | 16.52 |
| End cap, 2" | P6@.090 | Ea | 5.75 | 4.52 | 10.27 |
| Valve access door | P6@.250 | Ea | 15.80 | 12.60 | 28.40 |
| Inside corner | P6@.120 | Ea | 10.60 | 6.03 | 16.63 |
| Outside corner | P6@.120 | Ea | 14.10 | 6.03 | 20.13 |
| Filler sleeve | P6@.090 | Ea | 7.65 | 4.52 | 12.17 |

# 23 HVAC

| | Craft@Hrs | Unit | Material | Labor | Total |
|---|---|---|---|---|---|
| **Unit heaters, hot water, horizontal, fan driven** | | | | | |
| 12.5 MBtu,    200 CFM | M5@2.00 | Ea | 412.00 | 99.40 | 511.40 |
| 17    MBtu,    300 CFM | M5@2.00 | Ea | 455.00 | 99.40 | 554.40 |
| 25    MBtu,    500 CFM | M5@2.50 | Ea | 520.00 | 124.00 | 644.00 |
| 30    MBtu,    700 CFM | M5@3.00 | Ea | 547.00 | 149.00 | 696.00 |
| 50    MBtu, 1,000 CFM | M5@3.50 | Ea | 636.00 | 174.00 | 810.00 |
| 60    MBtu, 1,300 CFM | M5@4.00 | Ea | 756.00 | 199.00 | 955.00 |
| **Unit heaters, hot water, vertical, fan driven** | | | | | |
| 12.5 MBtu,    200 CFM | M5@2.00 | Ea | 463.00 | 99.40 | 562.40 |
| 17    MBtu,    300 CFM | M5@2.00 | Ea | 471.00 | 99.40 | 570.40 |
| 25    MBtu,    500 CFM | M5@2.50 | Ea | 611.00 | 124.00 | 735.00 |
| 30    MBtu,    700 CFM | M5@3.00 | Ea | 637.00 | 149.00 | 786.00 |
| 50    MBtu, 1,000 CFM | M5@3.50 | Ea | 768.00 | 174.00 | 942.00 |
| 60    MBtu, 1,300 CFM | M5@4.00 | Ea | 844.00 | 199.00 | 1,043.00 |

## Steam Heating

| | Craft@Hrs | Unit | Material | Labor | Total |
|---|---|---|---|---|---|
| **Steam baseboard radiation, per linear foot, wall mounted** | | | | | |
| 1/2" tube and cover | P6@.456 | LF | 10.50 | 22.90 | 33.40 |
| 3/4" tube and cover | P6@.493 | LF | 11.30 | 24.80 | 36.10 |
| 3/4" tube and cover, high capacity | P6@.493 | LF | 13.20 | 24.80 | 38.00 |
| 1/2" tube alone | P6@.350 | LF | 5.71 | 17.60 | 23.31 |
| 3/4" tube alone | P6@.377 | LF | 6.50 | 18.90 | 25.40 |
| Cover only | P6@.137 | LF | 6.84 | 6.88 | 13.72 |
| **Unit heater, steam, horizontal, fan driven** | | | | | |
| 18    MBtu,    300 CFM | M5@2.00 | Ea | 384.00 | 99.40 | 483.40 |
| 45    MBtu,    500 CFM | M5@2.50 | Ea | 637.00 | 124.00 | 761.00 |
| 60    MBtu,    700 CFM | M5@3.00 | Ea | 719.00 | 149.00 | 868.00 |
| 85    MBtu, 1,000 CFM | M5@3.50 | Ea | 758.00 | 174.00 | 932.00 |
| **Unit heater, steam, vertical, fan driven** | | | | | |
| 12.5 MBtu,    200 CFM | M5@2.00 | Ea | 349.00 | 99.40 | 448.40 |
| 17    MBtu,    300 CFM | M5@2.00 | Ea | 355.00 | 99.40 | 454.40 |
| 40    MBtu,    500 CFM | M5@2.50 | Ea | 689.00 | 124.00 | 813.00 |
| 60    MBtu,    700 CFM | M5@3.00 | Ea | 798.00 | 149.00 | 947.00 |
| 70    MBtu, 1,000 CFM | M5@3.50 | Ea | 1,010.00 | 174.00 | 1,184.00 |

## Hydronic Heating Specialties

| | Craft@Hrs | Unit | Material | Labor | Total |
|---|---|---|---|---|---|
| **Air eliminator-purger, screwed connections** | | | | | |
| 3/4", cast iron, steam & water, 150 PSI | PF@.250 | Ea | 487.00 | 14.90 | 501.90 |
| 3/4", cast iron, 150 PSI water | PF@.250 | Ea | 157.00 | 14.90 | 171.90 |
| 1", cast iron, 150 PSI water | PF@.250 | Ea | 178.00 | 14.90 | 192.90 |
| 3/8", brass, 125 PSI steam | PF@.250 | Ea | 128.00 | 14.90 | 142.90 |
| 1/2", cast iron, 250 PSI steam | PF@.250 | Ea | 265.00 | 14.90 | 279.90 |
| 3/4", cast iron, 250 PSI steam | PF@.250 | Ea | 332.00 | 14.90 | 346.90 |
| Airtrol fitting, 3/4" | PF@.300 | Ea | 56.30 | 17.90 | 74.20 |
| Air eliminator vents, 1/4" | P1@.150 | Ea | 18.90 | 5.36 | 24.26 |
| **Atmospheric vacuum breakers, screwed** | | | | | |
| 1/2" vacuum breaker | PF@.210 | Ea | 27.30 | 12.50 | 39.80 |
| 3/4" vacuum breaker | PF@.250 | Ea | 29.10 | 14.90 | 44.00 |
| 1" vacuum breaker | PF@.300 | Ea | 46.20 | 17.90 | 64.10 |
| 1-1/4" vacuum breaker | PF@.400 | Ea | 77.00 | 23.80 | 100.80 |
| 1-1/2" vacuum breaker | PF@.450 | Ea | 90.40 | 26.80 | 117.20 |
| 2" vacuum breaker | PF@.500 | Ea | 135.00 | 29.80 | 164.80 |

| | Craft@Hrs | Unit | Material | Labor | Total |
|---|---|---|---|---|---|

Circuit balancing valves, for hot water applications, complies with American Society of Mechanical Engineers Standard B16, Manufacturers Standardization Society, Instrumentation Society of America standards.

| | Craft@Hrs | Unit | Material | Labor | Total |
|---|---|---|---|---|---|
| 1/2" circuit balancing valve | PF@.210 | Ea | 34.50 | 12.50 | 47.00 |
| 3/4" circuit balancing valve | PF@.250 | Ea | 44.30 | 14.90 | 59.20 |
| 1" circuit balancing valve | PF@.300 | Ea | 56.70 | 17.90 | 74.60 |
| 1-1/2" circuit balancing valve | PF@.450 | Ea | 74.30 | 26.80 | 101.10 |

Flow check valves, brass, threaded, horizontal type

| | Craft@Hrs | Unit | Material | Labor | Total |
|---|---|---|---|---|---|
| 3/4" check valve | PF@.250 | Ea | 28.40 | 14.90 | 43.30 |
| 1" check valve | PF@.300 | Ea | 40.80 | 17.90 | 58.70 |
| 1-1/2" valve | PF@.350 | Ea | 90.20 | 20.80 | 111.00 |
| 2" valve | PF@.400 | Ea | 90.40 | 23.80 | 114.20 |

Pressure reducing valves, bronze body, for water, gas and steam applications, complies with American Society of Mechanical Engineers Standard B16, Manufacturers Standardization Society, Instrumentation Society of America standards.

| | Craft@Hrs | Unit | Material | Labor | Total |
|---|---|---|---|---|---|
| 1/2", 25-75 PSI | PF@.210 | Ea | 116.00 | 12.50 | 128.50 |
| 1/2", 10-35 PSI | PF@.210 | Ea | 134.00 | 12.50 | 146.50 |
| 3/4", 25-75 PSI | PF@.250 | Ea | 138.00 | 14.90 | 152.90 |
| 3/4", 10-35 PSI | PF@.250 | Ea | 155.00 | 14.90 | 169.90 |
| 1", 25-75 PSI | PF@.300 | Ea | 217.00 | 17.90 | 234.90 |
| 1", 10-35 PSI | PF@.300 | Ea | 237.00 | 17.90 | 254.90 |
| 1-1/4", 25-75 PSI | PF@.350 | Ea | 436.00 | 20.80 | 456.80 |
| 1-1/4", 10-35 PSI | PF@.350 | Ea | 474.00 | 20.80 | 494.80 |
| 1-1/2", 25-75 PSI | PF@.250 | Ea | 469.00 | 14.90 | 483.90 |
| 1-1/2", 10-35 PSI | PF@.250 | Ea | 535.00 | 14.90 | 549.90 |
| 2", 25-75 PSI | PF@.250 | Ea | 700.00 | 14.90 | 714.90 |
| 2", 10-35 PSI | PF@.250 | Ea | 737.00 | 14.90 | 751.90 |

Pressure regulators, feed water 25-75 PSI, cast iron, screwed connections

| | Craft@Hrs | Unit | Material | Labor | Total |
|---|---|---|---|---|---|
| 3/4" regulator | PF@.250 | Ea | 84.20 | 14.90 | 99.10 |
| 1" regulator | PF@.300 | Ea | 130.00 | 17.90 | 147.90 |
| 1-1/4" regulator | PF@.400 | Ea | 176.00 | 23.80 | 199.80 |
| 1-1/2" regulator | PF@.450 | Ea | 253.00 | 26.80 | 279.80 |
| 2" regulator | PF@.600 | Ea | 406.00 | 35.70 | 441.70 |
| 2-1/2" regulator | PF@.700 | Ea | 609.00 | 41.70 | 650.70 |

Combination pressure relief and reducing valves

| | Craft@Hrs | Unit | Material | Labor | Total |
|---|---|---|---|---|---|
| 1/2", brass body | PF@.210 | Ea | 104.00 | 12.50 | 116.50 |
| 3/4", cast iron body | PF@.250 | Ea | 162.00 | 14.90 | 176.90 |

Wye pattern bronze strainers, for water, gas and steam applications, complies with American Society of Mechanical Engineers Standard B16, Manufacturers Standardization Society, Instrumentation Society of America standards.

Wye pattern strainers, screwed connections, 250 PSI, cast iron body

| | Craft@Hrs | Unit | Material | Labor | Total |
|---|---|---|---|---|---|
| 3/4" strainer | PF@.260 | Ea | 12.40 | 15.50 | 27.90 |
| 1" strainer | PF@.330 | Ea | 15.70 | 19.60 | 35.30 |
| 1-1/4" strainer | PF@.440 | Ea | 22.70 | 26.20 | 48.90 |
| 1-1/2" strainer | PF@.495 | Ea | 26.40 | 29.50 | 55.90 |
| 2" strainer | PF@.550 | Ea | 41.50 | 32.70 | 74.20 |

Wye pattern strainers, screwed connections, 250 PSI, cast bronze body

| | Craft@Hrs | Unit | Material | Labor | Total |
|---|---|---|---|---|---|
| 1/2" strainer | PF@.210 | Ea | 27.90 | 12.50 | 40.40 |
| 3/4" strainer | PF@.250 | Ea | 37.80 | 14.90 | 52.70 |
| 1" strainer | PF@.300 | Ea | 48.20 | 17.90 | 66.10 |
| 1-1/4" strainer | PF@.400 | Ea | 68.10 | 23.80 | 91.90 |
| 1-1/2" strainer | PF@.450 | Ea | 88.30 | 26.80 | 115.10 |
| 2" strainer | PF@.500 | Ea | 152.00 | 29.80 | 181.80 |

# 23 HVAC

| | Craft@Hrs | Unit | Material | Labor | Total |
|---|---|---|---|---|---|
| **Wye pattern strainers, screwed connections, 600 PSI, stainless steel body** | | | | | |
| 3/4" strainer | PF@.260 | Ea | 298.00 | 15.50 | 313.50 |
| 1" strainer | PF@.330 | Ea | 408.00 | 19.60 | 427.60 |
| 1-1/4" strainer | PF@.440 | Ea | 496.00 | 26.20 | 522.20 |
| 1-1/2" strainer | PF@.495 | Ea | 682.00 | 29.50 | 711.50 |
| 2" strainer | PF@.550 | Ea | 947.00 | 32.70 | 979.70 |
| | | | | | |
| **Wye pattern strainers, flanged connections, 125 PSI, cast iron body** | | | | | |
| 2" strainer | PF@.500 | Ea | 114.00 | 29.80 | 143.80 |
| 2-1/2" strainer | PF@.600 | Ea | 128.00 | 35.70 | 163.70 |
| 3" strainer | PF@.750 | Ea | 147.00 | 44.70 | 191.70 |
| 4" strainer | PF@1.35 | Ea | 253.00 | 80.40 | 333.40 |
| 6" strainer | PF@2.50 | Ea | 510.00 | 149.00 | 659.00 |
| 8" strainer | PF@3.00 | Ea | 866.00 | 179.00 | 1,045.00 |
| | | | | | |
| **Tempering valves, screwed connections, high temperature for hot water applications, complies with American Society of Mechanical Engineers Standard B16, Manufacturers Standardization Society, Instrumentation Society of America standards.** | | | | | |
| 3/4", screwed | PF@.250 | Ea | 460.00 | 14.90 | 474.90 |
| 1", screwed | PF@.300 | Ea | 659.00 | 17.90 | 676.90 |
| 1-1/4", screwed | PF@.400 | Ea | 954.00 | 23.80 | 977.80 |
| 1-1/2", screwed | PF@.450 | Ea | 1,140.00 | 26.80 | 1,166.80 |
| 2", screwed | PF@.500 | Ea | 1,440.00 | 29.80 | 1,469.80 |
| | | | | | |
| **Thermostatic mixing valves, 110 to 150 degrees F** | | | | | |
| 1/2", soldered | PF@.240 | Ea | 58.90 | 14.30 | 73.20 |
| 3/4", threaded | PF@.250 | Ea | 76.90 | 14.90 | 91.80 |
| 3/4", soldered | PF@.300 | Ea | 64.70 | 17.90 | 82.60 |
| 1", threaded | PF@.300 | Ea | 295.00 | 17.90 | 312.90 |
| | | | | | |
| **Float and thermostatic traps, cast iron body, parallel connection** | | | | | |
| 3/4" (15 to 75 PSI) | PF@.250 | Ea | 157.00 | 14.90 | 171.90 |
| 1" (15 to 30 PSI) | PF@.300 | Ea | 186.00 | 17.90 | 203.90 |
| 3/4"-1" (125 PSI) | PF@.250 | Ea | 242.00 | 14.90 | 256.90 |
| 3/4"-1" (175 PSI) | PF@.300 | Ea | 530.00 | 17.90 | 547.90 |
| 1-1/4" (15 to 30 PSI) | PF@.450 | Ea | 263.00 | 26.80 | 289.80 |
| 1-1/4"-1-1/2" (75 PSI) | PF@.450 | Ea | 444.00 | 26.80 | 470.80 |
| 1-1/4"-1-1/2" (125 PSI) | PF@.500 | Ea | 446.00 | 29.80 | 475.80 |
| 1-1/4"-1-1/2" (175 PSI) | PF@.500 | Ea | 641.00 | 29.80 | 670.80 |
| 1-1/2" (15 to 30 PSI) | PF@.450 | Ea | 394.00 | 26.80 | 420.80 |
| 2" (15 to 30 PSI) | PF@.450 | Ea | 1,340.00 | 26.80 | 1,366.80 |
| 2" (75 PSI) | PF@.550 | Ea | 1,350.00 | 32.70 | 1,382.70 |
| 2" (125 PSI) | PF@.600 | Ea | 1,360.00 | 35.70 | 1,395.70 |
| 2-1/2" (15 to 250 PSI) | PF@.800 | Ea | 3,450.00 | 47.60 | 3,497.60 |
| | | | | | |
| **Liquid level gauges** | | | | | |
| 3/4", aluminum | PF@.250 | Ea | 270.00 | 14.90 | 284.90 |
| 3/4", 125 PSI PVC | PF@.250 | Ea | 282.00 | 14.90 | 296.90 |
| 1/2", 175 PSI bronze | PF@.210 | Ea | 53.00 | 12.50 | 65.50 |
| 3/4", 150 PSI stainless steel | PF@.250 | Ea | 261.00 | 14.90 | 275.90 |
| 1", 150 PSI stainless steel | PF@.300 | Ea | 293.00 | 17.90 | 310.90 |

# 23 HVAC

| | Craft@Hrs | Unit | Material | Labor | Total |
|---|---|---|---|---|---|
| **Tanks** Includes fittings but no excavation, dewatering, shoring, supports, piping, backfill, paving, or concrete | | | | | |
| Underground steel oil storage tanks | | | | | |
| 550 gallon | MI@4.86 | Ea | 1,600.00 | 258.00 | 1,858.00 |
| 2,500 gallon | M8@18.5 | Ea | 3,710.00 | 1,010.00 | 4,720.00 |
| 5,000 gallon | M8@24.0 | Ea | 6,640.00 | 1,310.00 | 7,950.00 |
| 10,000 gallon | M8@37.3 | Ea | 11,100.00 | 2,040.00 | 13,140.00 |
| Above-ground steel oil storage tanks | | | | | |
| 275 gallon | MI@2.98 | Ea | 555.00 | 158.00 | 713.00 |
| 500 gallon | MI@3.80 | Ea | 1,190.00 | 201.00 | 1,391.00 |
| 1,000 gallon | M8@7.64 | Ea | 1,990.00 | 417.00 | 2,407.00 |
| 1,500 gallon | M8@8.29 | Ea | 2,200.00 | 453.00 | 2,653.00 |
| 2,000 gallon | M8@10.5 | Ea | 2,830.00 | 574.00 | 3,404.00 |
| 5,000 gallon | M8@29.6 | Ea | 7,800.00 | 1,620.00 | 9,420.00 |
| Add for vent caps | M8@.155 | Ea | 1.77 | 8.47 | 10.24 |
| Add for filler cap | M8@.155 | Ea | 39.50 | 8.47 | 47.97 |
| Underground fiberglass oil storage tanks | | | | | |
| 550 gallon | MI@8.36 | Ea | 3,870.00 | 443.00 | 4,313.00 |
| 1,000 gallon | M8@16.8 | Ea | 7,000.00 | 918.00 | 7,918.00 |
| 2,000 gallon | M8@16.8 | Ea | 7,910.00 | 918.00 | 8,828.00 |
| 4,000 gallon | M8@25.9 | Ea | 9,620.00 | 1,410.00 | 11,030.00 |
| 6,000 gallon | M8@33.5 | Ea | 10,800.00 | 1,830.00 | 12,630.00 |
| 8,000 gallon | M8@33.5 | Ea | 12,100.00 | 1,830.00 | 13,930.00 |
| 10,000 gallon | M8@33.5 | Ea | 13,100.00 | 1,830.00 | 14,930.00 |
| 12,000 gallon | M8@84.4 | Ea | 16,300.00 | 4,610.00 | 20,910.00 |
| 15,000 gallon | M8@89.0 | Ea | 19,100.00 | 4,860.00 | 23,960.00 |
| 20,000 gallon | M8@89.0 | Ea | 23,700.00 | 4,860.00 | 28,560.00 |
| Propane tanks with valves, underground | | | | | |
| 1,000 gallon | M8@10.0 | Ea | 2,760.00 | 546.00 | 3,306.00 |

# 26 Electrical

**Typical In-Place Costs** Preliminary estimates per square foot of floor area for all electrical work, including service entrance, distribution and lighting fixtures, by building type. These costs include the subcontractor's overhead and profit.

| | Craft@Hrs | Unit | Material | Labor | Total |
|---|---|---|---|---|---|
| Commercial stores | — | SF | — | — | 9.11 |
| Market buildings | — | SF | — | — | 10.20 |
| Recreation facilities | — | SF | — | — | 11.00 |
| Schools | — | SF | — | — | 12.70 |
| Colleges | — | SF | — | — | 17.20 |
| Hospitals | — | SF | — | — | 29.70 |
| Office buildings | — | SF | — | — | 11.44 |
| Warehouses | — | SF | — | — | 5.22 |
| Parking lots | — | SF | — | — | 1.92 |
| Garages | — | SF | — | — | 8.51 |

**Underground Power and Communication Duct** Type BE LTMC8 (encased burial) PVC duct in 20' lengths. Add for excavation, shoring, dewatering, spacers, concrete envelope around duct, backfill, compaction, wire, pavement removal and re-paving as required. Direct burial (DB) power and communication duct will cost about 25% more.

| | Craft@Hrs | Unit | Material | Labor | Total |
|---|---|---|---|---|---|
| Single duct run | | | | | |
| 2-1/2" duct | E1@.053 | LF | .52 | 2.67 | 3.19 |
| 3" duct | E1@.055 | LF | 1.26 | 2.77 | 4.03 |
| 4" duct | E1@.070 | LF | 1.89 | 3.52 | 5.41 |

# 26 Electrical

| | Craft@Hrs | Unit | Material | Labor | Total |
|---|---|---|---|---|---|

**Concrete Envelope for Duct**  No wire, duct, excavation, pavement removal or replacement included. Based on ready-mix concrete. Includes allowance for waste.

| | Craft@Hrs | Unit | Material | Labor | Total |
|---|---|---|---|---|---|
| 9" x  9" (.56 CF per linear foot) | E2@.020 | LF | 1.83 | .98 | 2.81 |
| 12" x 12" (1.00 CF per linear foot) | E2@.036 | LF | 3.30 | 1.77 | 5.07 |
| 12" x 24" (2.00 CF per linear foot) | E2@.072 | LF | 6.59 | 3.54 | 10.13 |
| Per CY | E2@1.21 | CY | 100.00 | 59.40 | 159.40 |

**Precast Concrete Pull and Junction Box**  Reinforced concrete with appropriate inserts, cast iron frame, cover and pulling hooks. Overall dimensions as shown.

Precast handholes, 4' deep

| | Craft@Hrs | Unit | Material | Labor | Total |
|---|---|---|---|---|---|
| 2' wide, 3' long | E1@4.24 | Ea | 996.00 | 213.00 | 1,209.00 |
| 3' wide, 3' long | E1@5.17 | Ea | 1,500.00 | 260.00 | 1,760.00 |
| 4' wide, 4' long | E1@13.2 | Ea | 2,080.00 | 664.00 | 2,744.00 |

Precast power manholes, 7' deep

| | Craft@Hrs | Unit | Material | Labor | Total |
|---|---|---|---|---|---|
| 4' wide,  6' long | E1@24.6 | Ea | 3,070.00 | 1,240.00 | 4,310.00 |
| 6' wide,  8' long | E1@26.6 | Ea | 3,400.00 | 1,340.00 | 4,740.00 |
| 8' wide, 10' long | E1@26.6 | Ea | 3,580.00 | 1,340.00 | 4,920.00 |

**Light and Power Circuits**  Costs per circuit including stranded THHN copper wire (except where noted), conduit, handy box, cover, straps, fasteners and connectors, but without fixture, receptacle or switch, except as noted. Circuits for an exit light, fire alarm bell or suspended ceiling lighting system, 20 amp, with 4" junction box and 30' of 1/2" EMT conduit and copper THHN wire

| | Craft@Hrs | Unit | Material | Labor | Total |
|---|---|---|---|---|---|
| 3 #12 solid wire | E4@2.68 | Ea | 20.70 | 132.00 | 152.70 |
| 4 #12 solid wire | E4@2.90 | Ea | 24.00 | 142.00 | 166.00 |
| 5 #12 solid wire | E4@3.14 | Ea | 27.30 | 154.00 | 181.30 |

Under-slab circuit, with 4" square junction box, 10' of 1/2" RSC and 20' of 1/2" PVC conduit

| | Craft@Hrs | Unit | Material | Labor | Total |
|---|---|---|---|---|---|
| 3 #12 wire, 20 amp | E4@1.75 | Ea | 27.30 | 86.00 | 113.30 |
| 4 #12 wire, 20 amp | E4@1.97 | Ea | 30.60 | 96.80 | 127.40 |
| 5 #12 wire, 20 amp | E4@2.21 | Ea | 33.90 | 109.00 | 142.90 |
| 3 #10 wire, 30 amp | E4@2.00 | Ea | 33.60 | 98.30 | 131.90 |
| 4 #10 wire, 30 amp | E4@2.31 | Ea | 39.00 | 113.00 | 152.00 |

Under-slab circuit, with 4" square junction box, 10' of 1/2" RSC conduit, 20' of 1/2" PVC conduit and 6' of liquid-tight flexible conduit

| | Craft@Hrs | Unit | Material | Labor | Total |
|---|---|---|---|---|---|
| 3 #12 wire, 20 amp | E4@2.41 | Ea | 34.80 | 118.00 | 152.80 |
| 4 #12 wire, 20 amp | E4@2.69 | Ea | 38.10 | 132.00 | 170.10 |
| 3 #10 wire, 30 amp | E4@2.75 | Ea | 41.10 | 135.00 | 176.10 |
| 4 #10 wire, 30 amp | E4@3.81 | Ea | 46.50 | 187.00 | 233.50 |

20 amp switch circuit, 277 volt, with 30' of 1/2" EMT conduit, switch, cover and 2 indenter connectors

One gang box

| | Craft@Hrs | Unit | Material | Labor | Total |
|---|---|---|---|---|---|
| 2 #12 wire, Single pole switch | E4@2.82 | Ea | 21.90 | 139.00 | 160.90 |
| 3 #12 wire, Three-way switch | E4@2.93 | Ea | 27.50 | 144.00 | 171.50 |
| 3 #12 wire, Four-way switch | E4@3.14 | Ea | 37.30 | 154.00 | 191.30 |

Two gang box

| | Craft@Hrs | Unit | Material | Labor | Total |
|---|---|---|---|---|---|
| 2 #12 wire, Single pole switch | E4@2.95 | Ea | 29.60 | 145.00 | 174.60 |
| 3 #12 wire, Three-way switch | E4@3.08 | Ea | 37.50 | 151.00 | 188.50 |
| 3 #12 wire, Four-way switch | E4@3.24 | Ea | 57.00 | 159.00 | 216.00 |

| | Craft@Hrs | Unit | Material | Labor | Total |
|---|---|---|---|---|---|
| 15 amp switch circuit, 125 volt, with 30' of #14 two conductor Romex non-metallic cable (NM) with ground, switch, cover and switch connection | | | | | |
| One gang box | | | | | |
|   Single pole switch | E4@1.05 | Ea | 28.80 | 51.60 | 80.40 |
|   Three-way switch | E4@1.18 | Ea | 31.10 | 58.00 | 89.10 |
| Two gang box | | | | | |
|   Single pole switch | E4@1.17 | Ea | 36.50 | 57.50 | 94.00 |
|   Single pole switch & receptacle | E4@1.36 | Ea | 42.70 | 66.80 | 109.50 |
|   Single pole & 3-way switch | E4@1.43 | Ea | 41.00 | 70.30 | 111.30 |
|   One two pole & 3-way switch | E4@1.66 | Ea | 45.50 | 81.60 | 127.10 |
| 20 amp 2 pole duplex receptacle outlet circuit, 125 volt, 30' of 1/2" EMT conduit, wire and specification grade receptacle, connectors and box | | | | | |
| 3 #12 wire | E4@2.80 | Ea | 25.90 | 138.00 | 163.90 |
| 4 #12 wire | E4@3.06 | Ea | 29.20 | 150.00 | 179.20 |
| 5 #12 wire | E4@3.26 | Ea | 32.50 | 160.00 | 192.50 |
| Duplex receptacle outlet circuit with 30' of Romex non-metallic cable (NM) with ground, receptacle, cover and connection | | | | | |
| 15 amp, 125 volt, #14/2 wire | E4@1.11 | Ea | 35.90 | 54.50 | 90.40 |
| 15 amp, 125 volt, #14/2 wire with ground fault circuit interrupter (GFCI) | E4@1.30 | Ea | 57.50 | 63.90 | 121.40 |
| 20 amp, 125 volt, #12/2 wire | E4@1.36 | Ea | 35.80 | 66.80 | 102.60 |
| 30 amp, 250 volt, #10/3 wire, dryer circuit | E4@1.77 | Ea | 65.10 | 87.00 | 152.10 |
| 50 amp, 250 volt, #8/3 wire, range circuit | E4@2.06 | Ea | 41.10 | 101.00 | 142.10 |

**Rigid Galvanized Steel Conduit,** standard wall. Installed exposed in a building either vertically or horizontally up to 10' above floor level. No wire, fittings or supports included. Each 10' length comes with one coupling. Add the cost of fittings, boxes, supports and wire.

| | Craft@Hrs | Unit | Material | Labor | Total |
|---|---|---|---|---|---|
| 1/2" conduit | E4@.043 | LF | 1.32 | 2.11 | 3.43 |
| 3/4" conduit | E4@.054 | LF | 1.51 | 2.65 | 4.16 |
| 1" conduit | E4@.066 | LF | 2.43 | 3.24 | 5.67 |
| 1-1/4" conduit | E4@.077 | LF | 3.14 | 3.78 | 6.92 |
| 1-1/2" conduit | E4@.100 | LF | 3.47 | 4.91 | 8.38 |
| 2" conduit | E4@.116 | LF | 4.42 | 5.70 | 10.12 |
| 2-1/2" conduit | E4@.143 | LF | 6.15 | 7.03 | 13.18 |
| 3" conduit | E4@.164 | LF | 9.95 | 8.06 | 18.01 |
| 4" conduit | E4@.221 | LF | 13.70 | 10.90 | 24.60 |
| Add for red plastic coating | — | % | 80.0 | — | — |
| Deduct for rigid steel conduit installed in a concrete slab or open trench | — | % | — | -40.0 | — |

90- or 45- degree ells, standard weight rigid galvanized steel, threaded both ends

| | Craft@Hrs | Unit | Material | Labor | Total |
|---|---|---|---|---|---|
| 1/2" elbow | E4@.170 | Ea | 3.64 | 8.35 | 11.99 |
| 3/4" elbow | E4@.190 | Ea | 3.98 | 9.33 | 13.31 |
| 1" elbow | E4@.210 | Ea | 4.45 | 10.30 | 14.75 |
| 1-1/4" elbow | E4@.311 | Ea | 5.49 | 15.30 | 20.79 |
| 1-1/2" elbow | E4@.378 | Ea | 6.30 | 18.60 | 24.90 |
| 2" elbow | E4@.448 | Ea | 10.50 | 22.00 | 32.50 |
| 2-1/2" elbow | E4@.756 | Ea | 19.80 | 37.10 | 56.90 |
| 3" elbow | E4@1.10 | Ea | 27.40 | 54.00 | 81.40 |
| 4" elbow | E4@1.73 | Ea | 44.60 | 85.00 | 129.60 |
| Add for PVC coated ells | — | % | 40.0 | — | — |

| | Craft@Hrs | Unit | Material | Labor | Total |
|---|---|---|---|---|---|
| **Rigid steel threaded couplings** | | | | | |
| 1/2" | E4@.062 | Ea | 1.28 | 3.05 | 4.33 |
| 3/4" | E4@.080 | Ea | 1.48 | 3.93 | 5.41 |
| 1" | E4@.092 | Ea | 3.03 | 4.52 | 7.55 |
| 1-1/4" | E4@.102 | Ea | 3.33 | 5.01 | 8.34 |
| 1-1/2" | E4@.140 | Ea | 3.81 | 6.88 | 10.69 |
| 2" | E4@.170 | Ea | 4.49 | 8.35 | 12.84 |
| 2-1/2" | E4@.202 | Ea | 9.00 | 9.92 | 18.92 |
| 3" | E4@.230 | Ea | 15.70 | 11.30 | 27.00 |
| 4" | E4@.425 | Ea | 26.10 | 20.90 | 47.00 |
| **Three-piece "Erickson" unions, by pipe size** | | | | | |
| 1" | E4@.394 | Ea | 5.35 | 19.40 | 24.75 |
| **Insulated grounding bushing** | | | | | |
| 1/2" | E4@.062 | Ea | 2.42 | 3.05 | 5.47 |
| 3/4" | E4@.080 | Ea | 2.81 | 3.93 | 6.74 |
| 1" | E4@.092 | Ea | 3.88 | 4.52 | 8.40 |
| 1-1/4" | E4@.102 | Ea | 4.25 | 5.01 | 9.26 |
| 1-1/2" | E4@.140 | Ea | 5.03 | 6.88 | 11.91 |
| 2" | E4@.170 | Ea | 5.72 | 8.35 | 14.07 |
| 3" | E4@.230 | Ea | 13.50 | 11.30 | 24.80 |
| 4" | E4@.425 | Ea | 20.40 | 20.90 | 41.30 |
| **Insulated thermoplastic bushing** | | | | | |
| 1/2" | E4@.020 | Ea | .15 | .98 | 1.13 |
| 3/4" | E4@.025 | Ea | .20 | 1.23 | 1.43 |
| 1" | E4@.030 | Ea | .58 | 1.47 | 2.05 |
| 1-1/4" | E4@.040 | Ea | 1.33 | 1.97 | 3.30 |
| 1-1/2" | E4@.045 | Ea | .76 | 2.21 | 2.97 |
| 2" | E4@.050 | Ea | .80 | 2.46 | 3.26 |
| 2-1/2" | E4@.050 | Ea | 1.53 | 2.46 | 3.99 |
| 3" | E4@.070 | Ea | 1.67 | 3.44 | 5.11 |
| **Type A, C, LB, E, LL or LR threaded conduit body with cover and gasket** | | | | | |
| 1/2" | E4@.191 | Ea | 7.84 | 9.38 | 17.22 |
| 3/4" | E4@.238 | Ea | 9.28 | 11.70 | 20.98 |
| 1" | E4@.285 | Ea | 12.80 | 14.00 | 26.80 |
| 1-1/4" | E4@.358 | Ea | 17.70 | 17.60 | 35.30 |
| 1-1/2" | E4@.456 | Ea | 23.40 | 22.40 | 45.80 |
| 2" | E4@.598 | Ea | 33.30 | 29.40 | 62.70 |
| 2-1/2" | E4@1.43 | Ea | 76.60 | 70.30 | 146.90 |
| 3" | E4@1.77 | Ea | 118.00 | 87.00 | 205.00 |
| 4" | E4@2.80 | Ea | 208.00 | 138.00 | 346.00 |
| **Type T threaded conduit body, with cover and gasket** | | | | | |
| 1/2" | E4@.285 | Ea | 9.77 | 14.00 | 23.77 |
| 3/4" | E4@.358 | Ea | 12.00 | 17.60 | 29.60 |
| 1" | E4@.433 | Ea | 15.10 | 21.30 | 36.40 |
| 1-1/4" | E4@.534 | Ea | 22.10 | 26.20 | 48.30 |
| 1-1/2" | E4@.679 | Ea | 36.30 | 33.40 | 69.70 |
| 2" | E4@.889 | Ea | 42.60 | 43.70 | 86.30 |
| 2-1/2" | E4@2.15 | Ea | 99.20 | 106.00 | 205.20 |
| 3" | E4@2.72 | Ea | 142.00 | 134.00 | 276.00 |
| 4" | E4@4.22 | Ea | 299.00 | 207.00 | 506.00 |

| | Craft@Hrs | Unit | Material | Labor | Total |
|---|---|---|---|---|---|
| Service entrance cap, threaded | | | | | |
| 1/2" | E4@.319 | Ea | 4.89 | 15.70 | 20.59 |
| 3/4" | E4@.373 | Ea | 5.25 | 18.30 | 23.55 |
| 1" | E4@.376 | Ea | 5.52 | 18.50 | 24.02 |
| 1-1/4" | E4@.482 | Ea | 6.61 | 23.70 | 30.31 |
| 1-1/2" | E4@.536 | Ea | 10.90 | 26.30 | 37.20 |
| 2" | E4@.591 | Ea | 22.30 | 29.00 | 51.30 |
| 2-1/2" | E4@.671 | Ea | 33.40 | 33.00 | 66.40 |
| 3" | E4@.749 | Ea | 73.60 | 36.80 | 110.40 |
| 4" | E4@.912 | Ea | 93.40 | 44.80 | 138.20 |
| Add for heights over 10' to 20' | — | % | — | 20.0 | — |

**Electric Metallic Tubing (EMT)** Installed exposed in a building either vertically or horizontally up to 10' above floor level. No wire, fittings or supports included except as noted.

| | Craft@Hrs | Unit | Material | Labor | Total |
|---|---|---|---|---|---|
| 1/2" conduit | E4@.036 | LF | .29 | 1.77 | 2.06 |
| 3/4" conduit | E4@.043 | LF | .49 | 2.11 | 2.60 |
| 1" conduit | E4@.055 | LF | .85 | 2.70 | 3.55 |
| 1-1/4" conduit | E4@.067 | LF | 1.21 | 3.29 | 4.50 |
| 1-1/2" conduit | E4@.070 | LF | 1.56 | 3.44 | 5.00 |
| 2" conduit | E4@.086 | LF | 1.92 | 4.23 | 6.15 |
| 2-1/2" conduit | E4@.106 | LF | 3.90 | 5.21 | 9.11 |
| 3" conduit | E4@.127 | LF | 4.67 | 6.24 | 10.91 |
| 4" conduit | E4@.170 | LF | 6.72 | 8.35 | 15.07 |
| EMT 90-degree elbow | | | | | |
| 1/2" pull elbow | E4@.143 | Ea | 3.53 | 7.03 | 10.56 |
| 3/4" pull elbow | E4@.170 | Ea | 5.11 | 8.35 | 13.46 |
| 1" combination elbow | E4@.190 | Ea | 7.79 | 9.33 | 17.12 |
| 1-1/4" combination elbow | E4@.216 | Ea | 10.40 | 10.60 | 21.00 |
| 1-1/2" pull elbow | E4@.288 | Ea | 4.04 | 14.10 | 18.14 |
| 2" service entrance elbow | E4@.358 | Ea | 32.30 | 17.60 | 49.90 |
| 2-1/2" pull elbow | E4@.648 | Ea | 22.20 | 31.80 | 54.00 |
| 3" pull elbow | E4@.858 | Ea | 36.30 | 42.20 | 78.50 |
| 4" pull elbow | E4@1.33 | Ea | 52.60 | 65.30 | 117.90 |
| EMT set screw connector | | | | | |
| 1/2" EMT connector | E4@.085 | Ea | .59 | 4.18 | 4.77 |
| 3/4" EMT connector | E4@.096 | Ea | .72 | 4.72 | 5.44 |
| 1" EMT connector | E4@.124 | Ea | 1.18 | 6.09 | 7.27 |
| 1-1/4" EMT connector | E4@.143 | Ea | 1.54 | 7.03 | 8.57 |
| 1-1/2" EMT connector | E4@.162 | Ea | 2.06 | 7.96 | 10.02 |
| 2" EMT connector | E4@.237 | Ea | 3.03 | 11.60 | 14.63 |
| 2-1/2" EMT connector | E4@.363 | Ea | 8.03 | 17.80 | 25.83 |
| 3" EMT connector | E4@.479 | Ea | 9.70 | 23.50 | 33.20 |
| 4" EMT connector | E4@.655 | Ea | 14.50 | 32.20 | 46.70 |
| Add for insulated set-screw connector | — | % | 25.0 | — | — |
| EMT compression connector | | | | | |
| 1/2" EMT connector | E4@.085 | Ea | .67 | 4.18 | 4.85 |
| 3/4" EMT connector | E4@.096 | Ea | 1.06 | 4.72 | 5.78 |
| 1" EMT connector | E4@.124 | Ea | 1.57 | 6.09 | 7.66 |
| 1-1/4" EMT connector | E4@.143 | Ea | 2.55 | 7.03 | 9.58 |
| 1-1/2" EMT connector | E4@.162 | Ea | 3.16 | 7.96 | 11.12 |
| 2" EMT connector | E4@.237 | Ea | 4.49 | 11.60 | 16.09 |

| | Craft@Hrs | Unit | Material | Labor | Total |
|---|---|---|---|---|---|
| 2-1/2" EMT connector | E4@.363 | Ea | 14.50 | 17.80 | 32.30 |
| 3" EMT connector | E4@.479 | Ea | 19.40 | 23.50 | 42.90 |
| 4" EMT connector | E4@.655 | Ea | 30.40 | 32.20 | 62.60 |
| Add for insulated compression connector | — | % | 25.0 | — | — |
| **EMT set screw coupling** | | | | | |
| 1/2" EMT coupling | E4@.085 | Ea | .59 | 4.18 | 4.77 |
| 3/4" EMT coupling | E4@.096 | Ea | .84 | 4.72 | 5.56 |
| 1" EMT coupling | E4@.124 | Ea | 1.33 | 6.09 | 7.42 |
| 1-1/4" EMT coupling | E4@.143 | Ea | 1.70 | 7.03 | 8.73 |
| 1-1/2" EMT coupling | E4@.162 | Ea | 2.79 | 7.96 | 10.75 |
| 2" EMT coupling | E4@.237 | Ea | 3.28 | 11.60 | 14.88 |
| 2-1/2" EMT coupling | E4@.363 | Ea | 6.33 | 17.80 | 24.13 |
| 3" EMT coupling | E4@.479 | Ea | 7.66 | 23.50 | 31.16 |
| 4" EMT coupling | E4@.655 | Ea | 12.20 | 32.20 | 44.40 |
| **EMT compression coupling** | | | | | |
| 1/2" EMT coupling | E4@.085 | Ea | .82 | 4.18 | 5.00 |
| 3/4" EMT coupling | E4@.096 | Ea | 1.06 | 4.72 | 5.78 |
| 1" EMT coupling | E4@.124 | Ea | 1.57 | 6.09 | 7.66 |
| 1-1/4" EMT coupling | E4@.143 | Ea | 2.80 | 7.03 | 9.83 |
| 1-1/2" EMT coupling | E4@.162 | Ea | 4.01 | 7.96 | 11.97 |
| 2" EMT coupling | E4@.237 | Ea | 4.82 | 11.60 | 16.42 |
| 2-1/2" EMT coupling | E4@.363 | Ea | 18.20 | 17.80 | 36.00 |
| 3" EMT coupling | E4@.479 | Ea | 23.10 | 23.50 | 46.60 |
| 4" EMT coupling | E4@.655 | Ea | 36.60 | 32.20 | 68.80 |
| Add for heights over 10' to 20' | — | % | — | 20.0 | — |

**EMT Conduit with Wire** THHN wire pulled in EMT conduit bent each 10', with one tap-on coupling each 10', one set screw connector each 50' and a strap and lag screw each 6'. Installed exposed on walls or ceilings. Add 15% to the labor cost for heights over 10' to 20'. Use these figures for preliminary estimates.

| | Craft@Hrs | Unit | Material | Labor | Total |
|---|---|---|---|---|---|
| **1-1/2" EMT conduit and wire as shown below** | | | | | |
| 2 #1 aluminum, 1 #8 copper | E4@.140 | LF | 2.73 | 6.88 | 9.61 |
| 3 #1 aluminum, 1 #8 copper | E4@.156 | LF | 3.18 | 7.66 | 10.84 |
| 4 #1 aluminum, 1 #8 copper | E4@.173 | LF | 3.63 | 8.50 | 12.13 |
| 4 #4 copper, 1 #8 copper | E4@.162 | LF | 4.68 | 7.96 | 12.64 |
| 2 #3 copper, 1 #8 copper | E4@.140 | LF | 3.64 | 6.88 | 10.52 |
| 3 #3 copper, 1 #8 copper | E4@.155 | LF | 4.54 | 7.62 | 12.16 |
| 4 #3 copper, 1 #8 copper | E4@.170 | LF | 5.44 | 8.35 | 13.79 |
| 2 #1 copper, 1 #8 copper | E4@.146 | LF | 4.84 | 7.17 | 12.01 |
| 3 #1 copper, 1 #8 copper | E4@.164 | LF | 6.34 | 8.06 | 14.40 |
| 4 #1 copper, 1 #8 copper | E4@.183 | LF | 7.84 | 8.99 | 16.83 |
| **2" EMT conduit and wire as shown below** | | | | | |
| 2 #2/0 aluminum, 1 #6 copper | E4@.170 | LF | 3.52 | 8.35 | 11.87 |
| 3 #2/0 aluminum, 1 #6 copper | E4@.190 | LF | 4.11 | 9.33 | 13.44 |
| 4 #2/0 aluminum, 1 #6 copper | E4@.210 | LF | 4.70 | 10.30 | 15.00 |
| 2 #3/0 aluminum, 1 #6 copper | E4@.174 | LF | 3.66 | 8.55 | 12.21 |
| 3 #3/0 aluminum, 1 #6 copper | E4@.195 | LF | 4.32 | 9.58 | 13.90 |
| 4 #3/0 aluminum, 1 #6 copper | E4@.218 | LF | 4.98 | 10.70 | 15.68 |
| 2 #1/0 copper, 1 #6 copper | E4@.171 | LF | 5.99 | 8.40 | 14.39 |
| 3 #1/0 copper, 1 #6 copper | E4@.193 | LF | 7.81 | 9.48 | 17.29 |
| 4 #1/0 copper, 1 #6 copper | E4@.213 | LF | 9.63 | 10.50 | 20.13 |
| 2 #3/0 copper, 1 #4 copper | E4@.175 | LF | 8.07 | 8.60 | 16.67 |
| 3 #3/0 copper, 1 #4 copper | E4@.197 | LF | 10.80 | 9.68 | 20.48 |

# 26 Electrical

| | Craft@Hrs | Unit | Material | Labor | Total |
|---|---|---|---|---|---|
| **2-1/2" EMT conduit and wire as shown below** | | | | | |
| 2 250 MCM alum, 1 #4 copper | E4@.210 | LF | 6.71 | 10.30 | 17.01 |
| 3 250 MCM alum, 1 #4 copper | E4@.237 | LF | 7.76 | 11.60 | 19.36 |
| 4 #3 copper, 1 #4 copper | E4@.216 | LF | 8.21 | 10.60 | 18.81 |
| 2 #4 copper, 1 #2 copper | E4@.186 | LF | 6.50 | 9.14 | 15.64 |
| 3 #4 copper, 1 #2 copper | E4@.198 | LF | 7.21 | 9.73 | 16.94 |
| 4 #4 copper, 1 #2 copper | E4@.210 | LF | 7.92 | 10.30 | 18.22 |
| 3 250 MCM copper, 1 #2 copper | E4@.253 | LF | 10.70 | 12.40 | 23.10 |
| **3" EMT conduit and wire as shown below** | | | | | |
| 4 250 MCM alum, 1 #4 copper | E4@.288 | LF | 9.59 | 14.10 | 23.69 |
| 4 250 MCM copper, 1 #2 copper | E4@.290 | LF | 13.30 | 14.20 | 27.50 |
| 5 250 MCM copper, 1 #2 copper | E4@.319 | LF | 15.20 | 15.70 | 30.90 |

**EMT Conduit Circuits** Cost per circuit based on 30' run from the panel. Includes THHN copper wire pulled in conduit, 2 compression connectors, 2 couplings, conduit bending, straps, bolts and washers. Three-wire circuits are 120, 240 or 277 volt 1 phase, with 2 conductors and ground. Four-wire circuits are 120/208 or 480 volt 3 phase, with 3 conductors and ground. Five-wire circuits are 120/208 or 480 volt 3 phase, with 3 conductors, neutral and ground. Installed exposed in vertical or horizontal runs to 10' above floor level in a building. Add 15% to the manhours and labor costs for heights over 10' to 20'. Use these figures for preliminary estimates. Based on commercial quantity discounts.

| | Craft@Hrs | Unit | Material | Labor | Total |
|---|---|---|---|---|---|
| **15 amp circuits** | | | | | |
| 3 #14 wire, 1/2" conduit | E4@2.39 | Ea | 14.40 | 117.00 | 131.40 |
| 4 #14 wire, 1/2" conduit | E4@2.58 | Ea | 16.30 | 127.00 | 143.30 |
| **20 amp circuits** | | | | | |
| 3 #12 wire, 1/2" conduit | E4@2.49 | Ea | 16.80 | 122.00 | 138.80 |
| 4 #12 wire, 1/2" conduit | E4@2.72 | Ea | 19.50 | 134.00 | 153.50 |
| 5 #12 wire, 1/2" conduit | E4@2.95 | Ea | 22.30 | 145.00 | 167.30 |
| **30 amp circuits** | | | | | |
| 3 #10 wire, 1/2" conduit | E4@2.75 | Ea | 25.59 | 135.00 | 160.59 |
| 4 #10 wire, 1/2" conduit | E4@3.08 | Ea | 31.30 | 151.00 | 182.30 |
| 5 #10 wire, 3/4" conduit | E4@3.58 | Ea | 43.10 | 176.00 | 219.10 |
| **50 amp circuits** | | | | | |
| 3 #8 wire, 3/4" conduit | E4@2.98 | Ea | 39.80 | 146.00 | 185.80 |
| 4 #8 wire, 3/4" conduit | E4@3.32 | Ea | 48.20 | 163.00 | 211.20 |
| 5 #8 wire, 1" conduit | E4@4.02 | Ea | 67.40 | 198.00 | 265.40 |
| **65 amp circuits** | | | | | |
| 2 #6, 1 #8 wire, 1" conduit | E4@3.42 | Ea | 59.60 | 168.00 | 227.60 |
| 3 #6, 1 #8 wire, 1" conduit | E4@3.76 | Ea | 72.50 | 185.00 | 257.50 |
| 4 #6, 1 #8 wire, 1-1/4" conduit | E4@4.48 | Ea | 96.10 | 220.00 | 316.10 |
| **85 amp circuits in 1-1/4" conduit** | | | | | |
| 2 #4, 1 #8 wire | E4@3.89 | Ea | 87.10 | 191.00 | 278.10 |
| 3 #4, 1 #8 wire | E4@4.25 | Ea | 108.00 | 209.00 | 317.00 |
| 4 #4, 1 #8 wire | E4@4.48 | Ea | 130.00 | 220.00 | 350.00 |
| **100 amp circuits in 1-1/2" conduit** | | | | | |
| 2 #3, 1 #8 wire | E4@4.20 | Ea | 109.00 | 206.00 | 315.00 |
| 3 #3, 1 #8 wire | E4@4.66 | Ea | 136.00 | 229.00 | 365.00 |
| 4 #3, 1 #8 wire | E4@5.10 | Ea | 163.00 | 251.00 | 414.00 |
| **125 amp circuits in 1-1/2" conduit** | | | | | |
| 2 #1, 1 #8 wire | E4@4.40 | Ea | 145.00 | 216.00 | 361.00 |
| 3 #1, 1 #8 wire | E4@4.92 | Ea | 190.00 | 242.00 | 432.00 |
| 4 #1, 1 #8 wire | E4@5.47 | Ea | 235.00 | 269.00 | 504.00 |

| | Craft@Hrs | Unit | Material | Labor | Total |
|---|---|---|---|---|---|
| **150 amp circuits in 2" conduit** | | | | | |
| 2 #1/0, 1 #6 wire | E4@5.16 | Ea | 180.00 | 254.00 | 434.00 |
| 3 #1/0, 1 #6 wire | E4@5.78 | Ea | 234.00 | 284.00 | 518.00 |
| 4 #1/0, 1 #6 wire | E4@6.43 | Ea | 289.00 | 316.00 | 605.00 |
| **200 amp circuits in 2" conduit** | | | | | |
| 2 #3/0, 1 #4 wire | E4@5.49 | Ea | 242.00 | 270.00 | 512.00 |
| 3 #3/0, 1 #4 wire | E4@6.27 | Ea | 324.00 | 308.00 | 632.00 |
| 4 #3/0, 1 #4 wire | E4@7.80 | Ea | 405.00 | 383.00 | 788.00 |
| **225 amp circuits in 2-1/2" conduit** | | | | | |
| 2 #4/0, 1 #2 wire | E4@6.48 | Ea | 372.00 | 318.00 | 690.00 |
| 3 #4/0, 1 #2 wire | E4@7.33 | Ea | 482.00 | 360.00 | 842.00 |
| 4 #4/0, 1 #2 wire | E4@8.19 | Ea | 592.00 | 402.00 | 994.00 |
| **250 amp circuits in 2-1/2" conduit** | | | | | |
| 2 250 MCM, 1 #2 wire | E4@6.63 | Ea | 264.00 | 326.00 | 590.00 |
| 3 250 MCM, 1 #2 wire | E4@8.24 | Ea | 320.00 | 405.00 | 725.00 |
| 4 250 MCM, 1 #2 wire | E4@9.56 | Ea | 376.00 | 470.00 | 846.00 |

**Intermediate Metal Conduit (IMC)**  Installed exposed in a building either vertically or horizontally up to 10' above floor level. No wire, fittings or supports included.

| | Craft@Hrs | Unit | Material | Labor | Total |
|---|---|---|---|---|---|
| 1/2" conduit | E4@.038 | LF | 1.05 | 1.87 | 2.92 |
| 3/4" conduit | E4@.048 | LF | 1.16 | 2.36 | 3.52 |
| 1" conduit | E4@.056 | LF | 1.77 | 2.75 | 4.52 |
| 1-1/4" conduit | E4@.073 | LF | 2.62 | 3.59 | 6.21 |
| 1-1/2" conduit | E4@.086 | LF | 2.59 | 4.23 | 6.82 |
| 2" conduit | E4@.101 | LF | 2.66 | 4.96 | 7.62 |
| 2-1/2" conduit | E4@.127 | LF | 7.64 | 6.24 | 13.88 |
| 3" conduit | E4@.146 | LF | 8.87 | 7.17 | 16.04 |

**PVC Schedule 40 Conduit**  Installed in or under a building slab. No wire, fittings or supports included.

| | Craft@Hrs | Unit | Material | Labor | Total |
|---|---|---|---|---|---|
| 1/2" | E4@.030 | LF | .16 | 1.47 | 1.63 |
| 3/4" | E4@.036 | LF | .18 | 1.77 | 1.95 |
| 1" | E4@.046 | LF | .30 | 2.26 | 2.56 |
| 1-1/4" | E4@.056 | LF | .43 | 2.75 | 3.18 |
| 1-1/2" | E4@.061 | LF | .49 | 3.00 | 3.49 |
| 2" | E4@.072 | LF | .54 | 3.54 | 4.08 |
| 2-1/2" | E4@.089 | LF | 1.24 | 4.37 | 5.61 |
| 3" | E4@.105 | LF | 1.53 | 5.16 | 6.69 |
| 3-12" | E4@.130 | LF | 1.67 | 6.39 | 8.06 |
| 4" | E4@.151 | LF | 2.39 | 7.42 | 9.81 |
| 5" | E4@.176 | LF | 3.21 | 8.65 | 11.86 |

**PVC Schedule 80 Conduit**  Installed in or under a building slab. No wire, fittings or supports included.

| | Craft@Hrs | Unit | Material | Labor | Total |
|---|---|---|---|---|---|
| 1/2" | E4@.030 | LF | .16 | 1.47 | 1.63 |
| 3/4" | E4@.036 | LF | .22 | 1.77 | 1.99 |
| 1" | E4@.046 | LF | .37 | 2.26 | 2.63 |
| 1-1/4" | E4@.056 | LF | .45 | 2.75 | 3.20 |
| 2-1/2" | E4@.089 | LF | 1.31 | 4.37 | 5.68 |
| 3" | E4@.105 | LF | 1.56 | 5.16 | 6.72 |

| | Craft@Hrs | Unit | Material | Labor | Total |
|---|---|---|---|---|---|
| **Non-metallic conduit coupling** | | | | | |
| 1/2" coupling | E4@.020 | Ea | .27 | .98 | 1.25 |
| 3/4" coupling | E4@.051 | Ea | .33 | 2.51 | 2.84 |
| 1" coupling | E4@.062 | Ea | .50 | 3.05 | 3.55 |
| 1-1/4" coupling | E4@.078 | Ea | .70 | 3.83 | 4.53 |
| 2" coupling | E4@.102 | Ea | 1.05 | 5.01 | 6.06 |
| 2-1/2" coupling | E4@.113 | Ea | 1.58 | 5.55 | 7.13 |
| 3" coupling | E4@.129 | Ea | 2.02 | 6.34 | 8.36 |
| 4" coupling | E4@.156 | Ea | 4.75 | 7.66 | 12.41 |
| 2-1/2" long line coupling | E4@.113 | Ea | 3.84 | 5.55 | 9.39 |
| 4" long line coupling | E4@.156 | Ea | 6.42 | 7.66 | 14.08 |
| **90-degree Schedule 40 non-metallic elbow** | | | | | |
| 1/2" ell | E4@.082 | Ea | .91 | 4.03 | 4.94 |
| 3/4" ell | E4@.102 | Ea | 1.04 | 5.01 | 6.05 |
| 1" ell | E4@.124 | Ea | 1.57 | 6.09 | 7.66 |
| 1-1/4" ell | E4@.154 | Ea | 2.53 | 7.57 | 10.10 |
| 1-1/2" ell | E4@.195 | Ea | 3.16 | 9.58 | 12.74 |
| 2" ell | E4@.205 | Ea | 6.50 | 10.10 | 16.60 |
| 2-1/2" ell | E4@.226 | Ea | 7.34 | 11.10 | 18.44 |
| 3" ell | E4@.259 | Ea | 9.93 | 12.70 | 22.63 |
| 4" ell | E4@.313 | Ea | 22.10 | 15.40 | 37.50 |
| 1/2" access pull ell | E4@.082 | Ea | 3.09 | 4.03 | 7.12 |
| 3/4" access pull ell | E4@.102 | Ea | 4.05 | 5.01 | 9.06 |
| 1-1/4" ell, 24" radius | E4@.154 | Ea | 8.48 | 7.57 | 16.05 |
| 1-1/2" ell, 24" radius | E4@.195 | Ea | 10.20 | 9.58 | 19.78 |
| 2" ell, 24" radius | E4@.205 | Ea | 12.10 | 10.10 | 22.20 |
| 2-1/2" ell, 24" radius | E4@.226 | Ea | 13.60 | 11.10 | 24.70 |
| 3" ell, 24" radius | E4@.259 | Ea | 14.50 | 12.70 | 27.20 |
| 2" ell, 36" radius | E4@.205 | Ea | 18.50 | 10.10 | 28.60 |
| 2-1/2" ell, 36" radius | E4@.226 | Ea | 20.10 | 11.10 | 31.20 |
| Deduct for 45 or 30 degree ells | — | % | -10.0 | — | — |
| **90-degree Schedule 80 non-metallic elbow** | | | | | |
| 1" ell | E4@.124 | Ea | 2.47 | 6.09 | 8.56 |
| 1-1/4" ell | E4@.154 | Ea | 4.21 | 7.57 | 11.78 |
| 1-1/2" ell | E4@.195 | Ea | 4.55 | 9.58 | 14.13 |
| 2" ell | E4@.205 | Ea | 6.13 | 10.10 | 16.23 |
| 2-1/2" ell | E4@.226 | Ea | 13.10 | 11.10 | 24.20 |
| 1" ell, 24" radius | E4@.124 | Ea | 9.88 | 6.09 | 15.97 |
| 2" ell, 24" radius | E4@.205 | Ea | 11.70 | 10.10 | 21.80 |
| 2-1/2" ell, 24" radius | E4@.226 | Ea | 18.80 | 11.10 | 29.90 |
| 2" ell, 36" radius | E4@.205 | Ea | 25.80 | 10.10 | 35.90 |
| **PVC conduit end bell** | | | | | |
| 1" bell end | E4@.105 | Ea | 1.75 | 5.16 | 6.91 |
| 1-1/4" bell end | E4@.115 | Ea | 1.99 | 5.65 | 7.64 |
| 1-1/2" bell end | E4@.124 | Ea | 2.18 | 6.09 | 8.27 |
| 2" bell end | E4@.143 | Ea | 2.72 | 7.03 | 9.75 |
| 2-1/2" bell end | E4@.221 | Ea | 3.55 | 10.90 | 14.45 |
| 3" bell end | E4@.285 | Ea | 4.21 | 14.00 | 18.21 |
| **PVC bell end reducer** | | | | | |
| 3" to 2-1/2" | E4@.200 | Ea | 11.70 | 9.83 | 21.53 |
| 4" to 3" | E4@.250 | Ea | 14.80 | 12.30 | 27.10 |

|  | Craft@Hrs | Unit | Material | Labor | Total |
|---|---|---|---|---|---|
| Thermoplastic insulating bushing | | | | | |
| 1/2" bushing | E4@.025 | Ea | .15 | 1.23 | 1.38 |
| 3/4" bushing | E4@.025 | Ea | .19 | 1.23 | 1.42 |
| 1" bushing | E4@.030 | Ea | .56 | 1.47 | 2.03 |
| 1-1/4" bushing | E4@.040 | Ea | 1.34 | 1.97 | 3.31 |
| 1-1/2" bushing | E4@.040 | Ea | .62 | 1.97 | 2.59 |
| 2" bushing | E4@.050 | Ea | 1.39 | 2.46 | 3.85 |
| 2-1/2" bushing | E4@.050 | Ea | 1.54 | 2.46 | 4.00 |
| 3" bushing | E4@.070 | Ea | 1.68 | 3.44 | 5.12 |
| PVC reducer bushing, Schedule 40, socket ends | | | | | |
| 3/4" x 1/2" bushing | E4@.030 | Ea | .93 | 1.47 | 2.40 |
| 1" x 1/2" bushing | E4@.030 | Ea | 1.76 | 1.47 | 3.23 |
| 1" x 3/4" bushing | E4@.040 | Ea | 2.00 | 1.97 | 3.97 |
| 1-1/4" x 1" bushing | E4@.050 | Ea | 2.25 | 2.46 | 4.71 |
| 1-1/2" x 1" bushing | E4@.050 | Ea | 2.15 | 2.46 | 4.61 |
| 1-1/2" x 1-1/4" bushing | E4@.050 | Ea | 2.76 | 2.46 | 5.22 |
| 2" x 1-1/4" bushing | E4@.050 | Ea | 3.08 | 2.46 | 5.54 |
| 2" x 1-1/2" bushing | E4@.070 | Ea | 3.16 | 3.44 | 6.60 |
| 3" x 2" bushing | E4@.100 | Ea | 6.58 | 4.91 | 11.49 |
| PVC conduit box adapter | | | | | |
| 1/2" adapter | E4@.050 | Ea | .54 | 2.46 | 3.00 |
| 3/4" adapter | E4@.060 | Ea | .58 | 2.95 | 3.53 |
| 1" adapter | E4@.080 | Ea | .63 | 3.93 | 4.56 |
| 1-1/4" adapter | E4@.100 | Ea | .87 | 4.91 | 5.78 |
| 1-1/2" adapter | E4@.100 | Ea | .90 | 4.91 | 5.81 |
| 2" adapter | E4@.150 | Ea | 1.20 | 7.37 | 8.57 |
| PVC female conduit adapter, for joining conduit to threaded fittings | | | | | |
| 1/2" adapter | E4@.050 | Ea | .54 | 2.46 | 3.00 |
| 3/4" adapter | E4@.060 | Ea | .58 | 2.95 | 3.53 |
| 1" adapter | E4@.080 | Ea | .63 | 3.93 | 4.56 |
| 1-1/4" adapter | E4@.100 | Ea | .87 | 4.91 | 5.78 |
| 1-1/2" adapter | E4@.100 | Ea | .90 | 4.91 | 5.81 |
| 2" adapter | E4@.150 | Ea | 1.20 | 7.37 | 8.57 |
| 2-1/2" adapter | E4@.178 | Ea | 1.44 | 8.75 | 10.19 |
| 4" adapter | E4@.244 | Ea | 4.46 | 12.00 | 16.46 |
| PVC male conduit adapter, for joining conduit to threaded fittings | | | | | |
| 1/2" adapter | E4@.050 | Ea | .29 | 2.46 | 2.75 |
| 3/4" adapter | E4@.060 | Ea | .45 | 2.95 | 3.40 |
| 1" adapter | E4@.080 | Ea | .64 | 3.93 | 4.57 |
| 1-1/4" adapter | E4@.100 | Ea | .81 | 4.91 | 5.72 |
| 1-1/2" adapter | E4@.100 | Ea | .86 | 4.91 | 5.77 |
| 2" adapter | E4@.150 | Ea | 1.19 | 7.37 | 8.56 |
| 2-1/2" adapter | E4@.178 | Ea | 1.64 | 8.75 | 10.39 |
| 3" adapter | E4@.244 | Ea | 3.06 | 12.00 | 15.06 |
| Type T PVC conduit body (access fitting) | | | | | |
| 1/2" conduit body | E4@.150 | Ea | 3.00 | 7.37 | 10.37 |
| 3/4" conduit body | E4@.184 | Ea | 3.51 | 9.04 | 12.55 |
| 1" conduit body | E4@.225 | Ea | 4.13 | 11.10 | 15.23 |
| 1-1/4" conduit body | E4@.300 | Ea | 5.23 | 14.70 | 19.93 |
| 1-1/2" conduit body | E4@.350 | Ea | 7.88 | 17.20 | 25.08 |
| 2" conduit body | E4@.400 | Ea | 12.50 | 19.70 | 32.20 |

# 26 Electrical

|  | Craft@Hrs | Unit | Material | Labor | Total |
|---|---|---|---|---|---|
| **Type C PVC conduit body (access fitting)** | | | | | |
| 1/2" conduit body | E4@.100 | Ea | 3.12 | 4.91 | 8.03 |
| 3/4" conduit body | E4@.132 | Ea | 3.89 | 6.49 | 10.38 |
| 1" conduit body | E4@.164 | Ea | 5.23 | 8.06 | 13.29 |
| 1-1/4" conduit body | E4@.197 | Ea | 6.76 | 9.68 | 16.44 |
| 1-1/2" conduit body | E4@.234 | Ea | 7.65 | 11.50 | 19.15 |
| 2" conduit body | E4@.263 | Ea | 12.70 | 12.90 | 25.60 |
| **Type LB PVC conduit body (access fitting)** | | | | | |
| 1/2" conduit body | E4@.100 | Ea | 2.64 | 4.91 | 7.55 |
| 3/4" conduit body | E4@.132 | Ea | 3.36 | 6.49 | 9.85 |
| 1" conduit body | E4@.164 | Ea | 3.84 | 8.06 | 11.90 |
| 1-1/4" conduit body | E4@.197 | Ea | 5.41 | 9.68 | 15.09 |
| 1-1/2" conduit body | E4@.234 | Ea | 6.90 | 11.50 | 18.40 |
| 2" conduit body | E4@.263 | Ea | 9.69 | 12.90 | 22.59 |
| 3" conduit body | E4@.300 | Ea | 29.30 | 14.70 | 44.00 |
| 4" conduit body | E4@.350 | Ea | 47.20 | 17.20 | 64.40 |
| **Type LL PVC conduit body (access fitting)** | | | | | |
| 1/2" conduit body | E4@.100 | Ea | 2.62 | 4.91 | 7.53 |
| 3/4" conduit body | E4@.132 | Ea | 3.36 | 6.49 | 9.85 |
| 1" conduit body | E4@.164 | Ea | 3.80 | 8.06 | 11.86 |
| 1-1/4" conduit body | E4@.197 | Ea | 4.82 | 9.68 | 14.50 |
| 1-1/2" conduit body | E4@.234 | Ea | 6.33 | 11.50 | 17.83 |
| 2" conduit body | E4@.263 | Ea | 10.60 | 12.90 | 23.50 |
| **Type LR PVC conduit body (access fitting)** | | | | | |
| 1/2" conduit body | E4@.100 | Ea | 2.25 | 4.91 | 7.16 |
| 3/4" conduit body | E4@.132 | Ea | 3.37 | 6.49 | 9.86 |
| 1" conduit body | E4@.164 | Ea | 3.33 | 8.06 | 11.39 |
| 1-1/4" conduit body | E4@.197 | Ea | 5.67 | 9.68 | 15.35 |
| 1-1/2" conduit body | E4@.234 | Ea | 7.10 | 11.50 | 18.60 |
| 2" conduit body | E4@.263 | Ea | 11.00 | 12.90 | 23.90 |
| **Two-hole PVC clamp** | | | | | |
| 1/2" clamp | E4@.100 | Ea | .19 | 4.91 | 5.10 |
| 3/4" clamp | E4@.100 | Ea | .25 | 4.91 | 5.16 |
| 1" clamp | E4@.125 | Ea | .31 | 6.14 | 6.45 |
| 1-1/4" clamp | E4@.150 | Ea | .42 | 7.37 | 7.79 |
| 1-1/2" clamp | E4@.150 | Ea | .44 | 7.37 | 7.81 |
| 2" clamp | E4@.175 | Ea | .49 | 8.60 | 9.09 |
| 3" clamp | E4@.175 | Ea | 1.54 | 8.60 | 10.14 |

**PVC Conduit with Wire** Includes THHN wire pulled in Type 40 rigid PVC heavy wall conduit placed in a slab. Use these figures for preliminary estimates. Includes one rigid coupling every 10'. Add for fittings as required. Based on commercial quantity discounts.

|  | Craft@Hrs | Unit | Material | Labor | Total |
|---|---|---|---|---|---|
| **1/2" PVC conduit with wire** | | | | | |
| 3 #12 solid copper | E4@.054 | LF | .49 | 2.65 | 3.14 |
| 4 #12 solid copper | E4@.061 | LF | .60 | 3.00 | 3.60 |
| 5 #12 solid copper | E4@.069 | LF | .71 | 3.39 | 4.10 |
| 3 #10 solid copper | E4@.062 | LF | .70 | 3.05 | 3.75 |
| 4 #10 solid copper | E4@.073 | LF | .88 | 3.59 | 4.47 |
| **3/4" PVC conduit with wire** | | | | | |
| 5 #10 solid copper | E4@.089 | LF | 1.09 | 4.37 | 5.46 |
| 3 #8 copper | E4@.069 | LF | 1.03 | 3.39 | 4.42 |
| 4 #8 copper | E4@.080 | LF | 1.31 | 3.93 | 5.24 |

# 26 Electrical

|  | Craft@Hrs | Unit | Material | Labor | Total |
|---|---|---|---|---|---|
| **1" PVC conduit with wire** | | | | | |
| 2 #6 copper, 1 #8 copper | E4@.081 | LF | 1.44 | 3.98 | 5.42 |
| 3 #6 copper, 1 #8 copper | E4@.092 | LF | 1.87 | 4.52 | 6.39 |
| 5 #8 copper | E4@.100 | LF | 1.70 | 4.91 | 6.61 |
| **1-1/2" PVC conduit with wire** | | | | | |
| 4 #6 copper, 1 #8 copper | E4@.117 | LF | 2.48 | 5.75 | 8.23 |
| 2 #4 copper, 1 #8 copper | E4@.097 | LF | 2.18 | 4.77 | 6.95 |
| 2 #1 alum, 1 #8 copper | E4@.102 | LF | 1.66 | 5.01 | 6.67 |
| 3 #1 alum, 1 #8 copper | E4@.119 | LF | 2.11 | 5.85 | 7.96 |
| 4 #4 copper, 1 #8 copper | E4@.124 | LF | 3.60 | 6.09 | 9.69 |
| 3 #3 copper, 1 #8 copper | E4@.116 | LF | 3.46 | 5.70 | 9.16 |
| 4 #3 copper, 1 #8 copper | E4@.132 | LF | 4.36 | 6.49 | 10.85 |
| 2 #1 copper, 1 #8 copper | E4@.108 | LF | 3.76 | 5.31 | 9.07 |
| 3 #1 copper, 1 #8 copper | E4@.125 | LF | 5.26 | 6.14 | 11.40 |
| 4 #1 copper, 1 #8 copper | E4@.143 | LF | 6.76 | 7.03 | 13.79 |
| **2" PVC conduit with wire** | | | | | |
| 4 #1 alum, 1 #8 copper | E4@.154 | LF | 2.62 | 7.57 | 10.19 |
| 2 2/0 alum, 1 #6 copper | E4@.123 | LF | 2.15 | 6.04 | 8.19 |
| 3 2/0 alum, 1 #6 copper | E4@.143 | LF | 2.73 | 7.03 | 9.76 |
| 4 2/0 alum, 1 #6 copper | E4@.163 | LF | 3.32 | 8.01 | 11.33 |
| 2 3/0 alum, 1 #6 copper | E4@.127 | LF | 2.29 | 6.24 | 8.53 |
| 3 3/0 alum, 1 #6 copper | E4@.148 | LF | 2.94 | 7.27 | 10.21 |
| 4 3/0 alum, 1 #6 copper | E4@.171 | LF | 3.60 | 8.40 | 12.00 |
| 2 1/0 & 1 #6 copper | E4@.124 | LF | 2.01 | 6.09 | 8.10 |
| 3 1/0 & 1 #6 copper | E4@.146 | LF | 2.53 | 7.17 | 9.70 |
| 4 1/0 & 1 #6 copper | E4@.167 | LF | 3.04 | 8.20 | 11.24 |
| 2 3/0 & 1 #4 copper | E4@.136 | LF | 2.57 | 6.68 | 9.25 |
| 3 3/0 & 1 #4 copper | E4@.162 | LF | 3.22 | 7.96 | 11.18 |
| **2-1/2" PVC conduit with wire** | | | | | |
| 2 250 MCM alum, 1 #4 copper | E4@.156 | LF | 4.05 | 7.66 | 11.71 |
| 3 250 MCM alum, 1 #4 copper | E4@.183 | LF | 5.10 | 8.99 | 14.09 |
| 4 3/0 alum, 1 #4 copper | E4@.206 | LF | 3.92 | 10.10 | 14.02 |
| 2 4/0 alum, 1 #2 copper | E4@.164 | LF | 3.90 | 8.06 | 11.96 |
| 3 4/0 alum, 1 #2 copper | E4@.193 | LF | 4.64 | 9.48 | 14.12 |
| 4 4/0 alum, 1 #2 copper | E4@.221 | LF | 5.38 | 10.90 | 16.28 |
| 2 250 MCM copper, 1 #2 copper | E4@.170 | LF | 6.14 | 8.35 | 14.49 |
| **3" PVC conduit with wire** | | | | | |
| 4 250 MCM alum, 1 #4 copper | E4@.226 | LF | 6.44 | 11.10 | 17.54 |
| 2 350 MCM alum, 1 #2 copper | E4@.186 | LF | 5.45 | 9.14 | 14.59 |
| 3 350 MCM alum, 1 #2 copper | E4@.217 | LF | 6.81 | 10.70 | 17.51 |
| 3 250 MCM alum, 1 #2 copper | E4@.217 | LF | 5.86 | 10.70 | 16.56 |
| 4 250 MCM alum, 1 #2 copper | E4@.248 | LF | 6.91 | 12.20 | 19.11 |
| **3-1/2" PVC conduit with wire** | | | | | |
| 4 350 MCM alum, 1 #2 copper | E4@.264 | LF | 8.32 | 13.00 | 21.32 |
| **4" PVC conduit with wire** | | | | | |
| 6 350 MCM alum, 1 #2 copper | E4@.293 | LF | 11.80 | 14.40 | 26.20 |
| 7 350 MCM alum, 1 #2 copper | E4@.391 | LF | 13.10 | 19.20 | 32.30 |
| 3 500 MCM alum, 1 1/0 copper | E4@.311 | LF | 9.83 | 15.30 | 25.13 |
| 4 500 MCM alum, 1 1/0 copper | E4@.355 | LF | 11.70 | 17.40 | 29.10 |

## 26 Electrical

|  | Craft@Hrs | Unit | Material | Labor | Total |
|---|---|---|---|---|---|

**Flexible Aluminum Conduit Type RWA** installed in a building to 10' above floor level. No wire, fittings or supports included.

| | Craft@Hrs | Unit | Material | Labor | Total |
|---|---|---|---|---|---|
| 3/8" conduit, 25' coil | E4@.025 | LF | .36 | 1.23 | 1.59 |
| 1/2" conduit, cut length | E4@.028 | LF | .45 | 1.38 | 1.83 |
| 1/2" conduit, 50' coil | E4@.028 | LF | .39 | 1.38 | 1.77 |
| 1/2" conduit, 500' reel | E4@.028 | LF | .36 | 1.38 | 1.74 |
| 3/4" conduit, cut length | E4@.030 | LF | .83 | 1.47 | 2.30 |
| 3/4" conduit, 25' coil | E4@.030 | LF | .49 | 1.47 | 1.96 |
| 1" conduit, cut length | E4@.033 | LF | 1.11 | 1.62 | 2.73 |
| 1" conduit, 50' coil | E4@.033 | LF | .89 | 1.62 | 2.51 |
| 1-1/4" conduit, cut length | E4@.035 | LF | 1.72 | 1.72 | 3.44 |
| 1-1/4" conduit, 50' coil | E4@.035 | LF | 1.39 | 1.72 | 3.11 |
| 1-1/2" conduit, 25' coil | E4@.038 | LF | 2.65 | 1.87 | 4.52 |
| 2" conduit, cut length | E4@.040 | LF | 3.71 | 1.97 | 5.68 |
| 2" conduit, 25' coil | E4@.040 | LF | 2.95 | 1.97 | 4.92 |

**Flexible Steel Conduit** installed in a building to 10' above floor level. No wire, fittings or supports included.

| | Craft@Hrs | Unit | Material | Labor | Total |
|---|---|---|---|---|---|
| 3/8" conduit, cut length | E4@.025 | LF | .66 | 1.23 | 1.89 |
| 3/8" conduit, 25' coil | E4@.025 | LF | .45 | 1.23 | 1.68 |
| 3/8" conduit, 100' coil | E4@.025 | LF | .60 | 1.23 | 1.83 |
| 3/8" conduit, 250' coil | E4@.025 | LF | .35 | 1.23 | 1.58 |
| 1/2" conduit, cut length | E4@.028 | LF | .75 | 1.38 | 2.13 |
| 1/2" conduit, 25' coil | E4@.028 | LF | .49 | 1.38 | 1.87 |
| 1/2" conduit, 100' coil | E4@.028 | LF | .42 | 1.38 | 1.80 |
| 3/4" conduit, cut length | E4@.030 | LF | 1.09 | 1.47 | 2.56 |
| 3/4" conduit, 25' coil | E4@.030 | LF | .68 | 1.47 | 2.15 |
| 3/4" conduit, 100' coil | E4@.030 | LF | .63 | 1.47 | 2.10 |
| 3/4" conduit, 500' reel | E4@.030 | LF | .53 | 1.47 | 2.00 |
| 1" conduit, cut length | E4@.033 | LF | 1.46 | 1.62 | 3.08 |
| 1" conduit, 50' coil | E4@.033 | LF | 1.22 | 1.62 | 2.84 |
| 1-1/4" conduit, cut length | E4@.035 | LF | 1.69 | 1.72 | 3.41 |
| 1-1/4" conduit, 50' coil | E4@.035 | LF | 1.57 | 1.72 | 3.29 |
| 1-1/2" conduit, cut length | E4@.038 | LF | 3.46 | 1.87 | 5.33 |
| 1-1/2" conduit, 25' coil | E4@.038 | LF | 2.45 | 1.87 | 4.32 |
| 2" conduit, cut length | E4@.040 | LF | 3.89 | 1.97 | 5.86 |
| 2" conduit, 25' coil | E4@.040 | LF | 3.26 | 1.97 | 5.23 |
| 90-degree flex screw-in connector | | | | | |
| 3/8" connector | E4@.070 | Ea | 1.01 | 3.44 | 4.45 |
| 1/2" connector | E4@.086 | Ea | .96 | 4.23 | 5.19 |
| 3/4" connector | E4@.100 | Ea | 2.48 | 4.91 | 7.39 |
| 1" connector | E4@.113 | Ea | 7.17 | 5.55 | 12.72 |
| 1-1/4" connector | E4@.148 | Ea | 7.41 | 7.27 | 14.68 |
| 1-1/2" connector | E4@.191 | Ea | 15.60 | 9.38 | 24.98 |
| 2" connector | E4@.260 | Ea | 14.80 | 12.80 | 27.60 |
| Flex screw-in coupling | | | | | |
| 1/2" coupling | E4@.086 | Ea | .72 | 4.23 | 4.95 |
| 3/4" coupling | E4@.100 | Ea | 2.10 | 4.91 | 7.01 |
| 1" coupling | E4@.113 | Ea | 2.15 | 5.55 | 7.70 |
| 1-1/4" coupling | E4@.148 | Ea | 4.21 | 7.27 | 11.48 |
| 1-1/2" coupling | E4@.191 | Ea | 5.30 | 9.38 | 14.68 |
| 2" coupling | E4@.260 | Ea | 9.75 | 12.80 | 22.55 |
| Add for height over 10' to 20' | — | % | — | 20.0 | — |

| | Craft@Hrs | Unit | Material | Labor | Total |
|---|---|---|---|---|---|
| **Flexible Non-metallic Conduit,** (ENT) blue conduit installed in a building to 10' above floor level. NER-290 adopted by BOCA, ICBO and SBCC codes. No wire, fittings or supports included. | | | | | |
| 1/2" conduit, cut length | E4@.021 | LF | .31 | 1.03 | 1.34 |
| 1/2" conduit, 10' coil | E4@.021 | LF | .31 | 1.03 | 1.34 |
| 1/2" conduit, 200' coil | E4@.021 | LF | .25 | 1.03 | 1.28 |
| 3/4" conduit, cut length | E4@.022 | LF | .50 | 1.08 | 1.58 |
| 3/4" conduit, 10' coil | E4@.022 | LF | .44 | 1.08 | 1.52 |
| 3/4" conduit, 100' coil | E4@.022 | LF | .40 | 1.08 | 1.48 |
| 1" conduit, 10' coil | E4@.030 | LF | .36 | 1.47 | 1.83 |
| **Liquid-Tight Flexible Metal Conduit** installed exposed in a building either vertically or horizontally up to 10' above floor level. No wire, fittings or supports included. | | | | | |
| 1/2" conduit, cut length | E4@.045 | LF | 1.30 | 2.21 | 3.51 |
| 3/4" conduit, 25' coil | E4@.047 | LF | 1.18 | 2.31 | 3.49 |
| 1" conduit, cut length | E4@.050 | LF | 3.40 | 2.46 | 5.86 |
| 1-1/4" conduit, cut length | E4@.055 | LF | 3.11 | 2.70 | 5.81 |
| 1-1/2" conduit, cut length | E4@.065 | LF | 3.84 | 3.19 | 7.03 |
| 2" conduit, cut length | E4@.070 | LF | 4.29 | 3.44 | 7.73 |
| Liquid-tight straight flex connector | | | | | |
| 1/2" connector | E4@.129 | Ea | 3.26 | 6.34 | 9.60 |
| 3/4" connector | E4@.140 | Ea | 4.56 | 6.88 | 11.44 |
| 1" connector | E4@.221 | Ea | 7.43 | 10.90 | 18.33 |
| 1-1/4" connector | E4@.244 | Ea | 8.20 | 12.00 | 20.20 |
| 1-1/2" connector | E4@.293 | Ea | 12.30 | 14.40 | 26.70 |
| 2" connector | E4@.492 | Ea | 20.60 | 24.20 | 44.80 |
| 90-degree liquid-tight flex connector | | | | | |
| 1/2" connector | E4@.129 | Ea | 5.56 | 6.34 | 11.90 |
| 3/4" connector | E4@.140 | Ea | 8.46 | 6.88 | 15.34 |
| 1" connector | E4@.221 | Ea | 15.60 | 10.90 | 26.50 |
| 1-1/4" connector | E4@.244 | Ea | 17.70 | 12.00 | 29.70 |
| 1-1/2" connector | E4@.293 | Ea | 21.00 | 14.40 | 35.40 |
| 2" connector | E4@.492 | Ea | 32.40 | 24.20 | 56.60 |
| **Non-Metallic Liquid-Tight Flexible Conduit** installed exposed in a building either vertically or horizontally up to 10' above floor level. No wire, fittings or supports included. | | | | | |
| 3/8" conduit, 100-foot rolls | E4@.051 | LF | .45 | 2.51 | 2.96 |
| 1/2" conduit, 100-foot rolls | E4@.051 | LF | .46 | 2.51 | 2.97 |
| 3/4" conduit, 100-foot rolls | E4@.067 | LF | .60 | 3.29 | 3.89 |
| 1" conduit, 100-foot rolls | E4@.067 | LF | 1.65 | 3.29 | 4.94 |
| 1-1/4" conduit, 100-foot rolls | E4@.085 | LF | 2.20 | 4.18 | 6.38 |
| 1-1/2" conduit, 50-foot rolls | E4@.085 | LF | 2.95 | 4.18 | 7.13 |
| 2" conduit, 50-foot rolls | E4@.105 | LF | 3.80 | 5.16 | 8.96 |
| Non-metallic liquid-tight flex connector | | | | | |
| 1/2" zero to 90 degree connector | E4@.060 | Ea | 2.33 | 2.95 | 5.28 |
| 3/4" zero to 90 degree connector | E4@.070 | Ea | 3.68 | 3.44 | 7.12 |
| 1" zero to 90 degree connector | E4@.080 | Ea | 2.71 | 3.93 | 6.64 |
| 1-1/4" zero to 90 degree connector | E4@.100 | Ea | 5.80 | 4.91 | 10.71 |
| 1-1/2" zero to 90 degree connector | E4@.100 | Ea | 7.69 | 4.91 | 12.60 |
| 2" zero to 90 degree connector | E4@.140 | Ea | 13.20 | 6.88 | 20.08 |
| 1/2" 90 degree connector | E4@.060 | Ea | 2.49 | 2.95 | 5.44 |
| 3/4" 90 degree connector | E4@.070 | Ea | 3.89 | 3.44 | 7.33 |
| 1" 90 degree connector | E4@.080 | Ea | 3.33 | 3.93 | 7.26 |
| 1-1/4" 90 degree connector | E4@.100 | Ea | 10.30 | 4.91 | 15.21 |
| 1-1/2" 90 degree connector | E4@.100 | Ea | 13.20 | 4.91 | 18.11 |
| 2" 90 degree connector | E4@.140 | Ea | 19.00 | 6.88 | 25.88 |
| Add for height over 10' to 20' | — | % | — | 20.0 | — |

|  | Craft@Hrs | Unit | Material | Labor | Total |
|---|---|---|---|---|---|

**Conduit Supports and Hangers** By conduit size. Installed exposed in a building to 10' above floor level. Hangers and supports installed at heights over 10' will add about 20% to the labor cost.

Right angle conduit supports

| | Craft@Hrs | Unit | Material | Labor | Total |
|---|---|---|---|---|---|
| 3/8" support | E4@.030 | Ea | .37 | 1.47 | 1.84 |
| 1/2" support or 3/4" support | E4@.040 | Ea | .37 | 1.97 | 2.34 |
| 1" support | E4@.050 | Ea | .56 | 2.46 | 3.02 |
| 1-1/4" support | E4@.060 | Ea | .62 | 2.95 | 3.57 |
| 1-1/2" support | E4@.060 | Ea | .67 | 2.95 | 3.62 |
| 2" support | E4@.100 | Ea | .90 | 4.91 | 5.81 |
| 2-1/2" support | E4@.100 | Ea | 1.07 | 4.91 | 5.98 |
| 3" support | E4@.150 | Ea | 1.33 | 7.37 | 8.70 |
| 3-1/2" support | E4@.150 | Ea | 1.77 | 7.37 | 9.14 |
| 4" support | E4@.150 | Ea | 3.70 | 7.37 | 11.07 |
| Add for parallel conduit supports | — | % | 50.0 | — | — |
| Add for edge conduit supports | — | % | 120.0 | — | — |

Conduit hangers, rigid with bolt, labor includes the cost of cutting threaded rod to length and attaching the rod. Add the material cost of the threaded rod from below.

| | Craft@Hrs | Unit | Material | Labor | Total |
|---|---|---|---|---|---|
| 1/2" hanger | E4@.097 | Ea | .64 | 4.77 | 5.41 |
| 3/4" hanger | E4@.097 | Ea | .66 | 4.77 | 5.43 |
| 1" hanger | E4@.108 | Ea | 1.08 | 5.31 | 6.39 |
| 1-1/4" hanger | E4@.108 | Ea | 1.50 | 5.31 | 6.81 |
| 1-1/2" hanger | E4@.108 | Ea | 1.82 | 5.31 | 7.13 |
| 2" hanger | E4@.119 | Ea | 1.99 | 5.85 | 7.84 |
| 2-1/2" hanger | E4@.119 | Ea | 2.56 | 5.85 | 8.41 |
| 3" hanger | E4@.119 | Ea | 3.06 | 5.85 | 8.91 |
| 3-1/2" hanger | E4@.129 | Ea | 4.36 | 6.34 | 10.70 |
| 4" hanger | E4@.129 | Ea | 10.30 | 6.34 | 16.64 |

All threaded rod, plated steel, per linear foot of rod

| | Craft@Hrs | Unit | Material | Labor | Total |
|---|---|---|---|---|---|
| 1/4", 20 thread | — | LF | .48 | — | .48 |
| 5/16", 18 thread | — | LF | .59 | — | .59 |
| 3/8", 16 thread | — | LF | .64 | — | .64 |
| 1/2", 13 thread | — | LF | 1.01 | — | 1.01 |
| 5/8", 11 thread | — | LF | 1.40 | — | 1.40 |
| Deduct for plain steel rod | — | % | -10.0 | — | — |

Rod beam clamps, for 1/4" or 3/8" threaded rod

| | Craft@Hrs | Unit | Material | Labor | Total |
|---|---|---|---|---|---|
| 1" x 1" base, 15/16" jaw, 450 lb rating | E4@.200 | Ea | 2.09 | 9.83 | 11.92 |
| 2" x 2" base, 1" jaw, 1,300 lb rating | E4@.200 | Ea | 4.01 | 9.83 | 13.84 |
| 2-5/8" x 2-1/2" base, 1" jaw, 1,300 lb rating | E4@.234 | Ea | 5.85 | 11.50 | 17.35 |

Conduit clamps for EMT conduit

| | Craft@Hrs | Unit | Material | Labor | Total |
|---|---|---|---|---|---|
| 1/2" or 3/4" clamp | E4@.062 | Ea | .20 | 3.05 | 3.25 |
| 1" clamp | E4@.069 | Ea | .27 | 3.39 | 3.66 |
| 1-1/4" clamp | E4@.069 | Ea | .40 | 3.39 | 3.79 |
| 1-1/2" clamp | E4@.069 | Ea | .42 | 3.39 | 3.81 |
| 2" clamp | E4@.075 | Ea | .48 | 3.68 | 4.16 |

One hole heavy-duty stamped steel conduit straps

| | Craft@Hrs | Unit | Material | Labor | Total |
|---|---|---|---|---|---|
| 1/2" strap | E4@.039 | Ea | .14 | 1.92 | 2.06 |
| 3/4" strap | E4@.039 | Ea | .18 | 1.92 | 2.10 |
| 1" strap | E4@.039 | Ea | .32 | 1.92 | 2.24 |
| 1-1/4" strap | E4@.039 | Ea | .48 | 1.92 | 2.40 |
| 1-1/2" strap | E4@.039 | Ea | .64 | 1.92 | 2.56 |
| 2" strap | E4@.039 | Ea | .83 | 1.92 | 2.75 |
| 2-1/2" strap | E4@.053 | Ea | 2.86 | 2.60 | 5.46 |

| | Craft@Hrs | Unit | Material | Labor | Total |
|---|---|---|---|---|---|
| 3" strap | E4@.053 | Ea | 3.99 | 2.60 | 6.59 |
| 3-1/2" strap | E4@.059 | Ea | 5.99 | 2.90 | 8.89 |
| 4" strap | E4@.059 | Ea | 6.20 | 2.90 | 9.10 |
| Add for malleable iron one hole straps | — | % | 30.0 | — | — |
| Deduct for nail drive straps | — | % | -20.0 | -30.0 | — |
| Deduct for two hole stamped steel EMT straps | — | % | -70.0 | — | — |

Hanger channel, 1-5/8" x 1-5/8", based on 12" length
No holes, solid back,

| | Craft@Hrs | Unit | Material | Labor | Total |
|---|---|---|---|---|---|
| 12 gauge steel | E4@.119 | Ea | 3.48 | 5.85 | 9.33 |

Holes 1-1/2" on center,

| | Craft@Hrs | Unit | Material | Labor | Total |
|---|---|---|---|---|---|
| 12 gauge aluminum | E4@.119 | Ea | 3.99 | 5.85 | 9.84 |

Channel strap for rigid steel conduit or EMT conduit, by conduit size, with bolts

| | Craft@Hrs | Unit | Material | Labor | Total |
|---|---|---|---|---|---|
| 1/2" strap | E4@.020 | Ea | 1.00 | .98 | 1.98 |
| 3/4" strap | E4@.020 | Ea | 1.12 | .98 | 2.10 |
| 1" strap | E4@.023 | Ea | 1.20 | 1.13 | 2.33 |
| 1-1/4" strap | E4@.023 | Ea | 1.40 | 1.13 | 2.53 |
| 1-1/2" strap | E4@.023 | Ea | 1.58 | 1.13 | 2.71 |
| 2" strap | E4@.023 | Ea | 1.79 | 1.13 | 2.92 |
| 2-1/2" strap | E4@.039 | Ea | 1.96 | 1.92 | 3.88 |
| 3" strap | E4@.039 | Ea | 2.15 | 1.92 | 4.07 |
| 3-1/2" strap | E4@.039 | Ea | 2.63 | 1.92 | 4.55 |
| 4" strap | E4@.062 | Ea | 2.92 | 3.05 | 5.97 |
| 5" strap | E4@.062 | Ea | 3.60 | 3.05 | 6.65 |
| 6" strap | E4@.062 | Ea | 4.84 | 3.05 | 7.89 |

Lag screws, flattened end with stove bolt

| | Craft@Hrs | Unit | Material | Labor | Total |
|---|---|---|---|---|---|
| 1/4" x 3" | E4@.092 | Ea | .33 | 4.52 | 4.85 |
| 1/4" x 4" | E4@.092 | Ea | .47 | 4.52 | 4.99 |
| 5/16" x 3-1/2" | E4@.092 | Ea | .54 | 4.52 | 5.06 |
| 3/8" x 4" | E4@.119 | Ea | .64 | 5.85 | 6.49 |
| 1/2" x 5" | E4@.129 | Ea | 1.26 | 6.34 | 7.60 |

Lag screw short expansion shields, without screws. Labor is for drilling hole

| | Craft@Hrs | Unit | Material | Labor | Total |
|---|---|---|---|---|---|
| 1/4" or 5/16" x 1" shield | E4@.175 | Ea | .21 | 8.60 | 8.81 |
| 3/8" x 1-3/4" shield | E4@.272 | Ea | .15 | 13.40 | 13.55 |
| 1/2" x 2" shield | E4@.321 | Ea | .50 | 15.80 | 16.30 |
| 5/8" x 2" shield | E4@.321 | Ea | 1.34 | 15.80 | 17.14 |
| 5/8" x 3-1/2" shield | E4@.345 | Ea | 2.02 | 16.90 | 18.92 |

Self drilling masonry anchors

| | Craft@Hrs | Unit | Material | Labor | Total |
|---|---|---|---|---|---|
| 1/4" anchor | E4@.127 | Ea | .55 | 6.24 | 6.79 |
| 5/16" anchor | E4@.127 | Ea | .75 | 6.24 | 6.99 |
| 3/8" anchor | E4@.191 | Ea | .86 | 9.38 | 10.24 |
| 1/2" anchor | E4@.191 | Ea | 1.35 | 9.38 | 10.73 |
| 5/8" anchor | E4@.259 | Ea | 2.25 | 12.70 | 14.95 |
| 3/4" anchor | E4@.259 | Ea | 3.71 | 12.70 | 16.41 |

**Wiremold Raceway** Steel, one-piece, installed on a finished wall in a building. Raceway including one coupling each 10 feet

| | Craft@Hrs | Unit | Material | Labor | Total |
|---|---|---|---|---|---|
| V500 2 wire surface mounted raceway | E4@.059 | LF | .92 | 2.90 | 3.82 |
| V700 2 wire surface mounted raceway | E4@.067 | LF | 1.16 | 3.29 | 4.45 |

# 26 Electrical

| | Craft@Hrs | Unit | Material | Labor | Total |
|---|---|---|---|---|---|
| **Wiremold fittings** | | | | | |
| V700 90-degree flat elbow | E4@.050 | Ea | 4.93 | 2.46 | 7.39 |
| V700 90-degree twist elbow | E4@.050 | Ea | 8.53 | 2.46 | 10.99 |
| V700 90-degree inside elbow | E4@.050 | Ea | 4.98 | 2.46 | 7.44 |
| V700 90-degree outside elbow | E4@.050 | Ea | 4.93 | 2.46 | 7.39 |
| V700 90-degree internal pull elbow | E4@.050 | Ea | 13.50 | 2.46 | 15.96 |
| V700 90-degree internal twist elbow | E4@.150 | Ea | 4.80 | 7.37 | 12.17 |
| V500 45-degree flat elbow | E4@.150 | Ea | 4.93 | 7.37 | 12.30 |
| V500 90-degree flat elbow | E4@.150 | Ea | 2.47 | 7.37 | 9.84 |
| V500/700 adjustable offset connector | E4@.150 | Ea | 9.89 | 7.37 | 17.26 |
| V500 conduit connector | E4@.150 | Ea | 4.32 | 7.37 | 11.69 |
| V700 connection cover | E4@.200 | Ea | .49 | 9.83 | 10.32 |
| V700 3-way branch "T"-fitting | E4@.100 | Ea | 6.24 | 4.91 | 11.15 |
| V700 to 1/2" conduit transition fitting | E4@.050 | Ea | 6.16 | 2.46 | 8.62 |
| V500/700 flexible section, 18" | E4@.350 | Ea | 19.30 | 17.20 | 36.50 |
| V500/700 galvanized box connector | E4@.200 | Ea | 6.72 | 9.83 | 16.55 |
| V500/700 coupling | E4@.200 | Ea | .42 | 9.83 | 10.25 |
| V500/700 support clip | — | Ea | .47 | — | .47 |
| **Wiremold device box, V500/700 surface raceway** | | | | | |
| 1 gang, 1-3/8" diameter, 1 KO | E4@.250 | Ea | 6.87 | 12.30 | 19.17 |
| 1 gang, 2-1/4" diameter, 1 KO | E4@.250 | Ea | 14.70 | 12.30 | 27.00 |
| 1 gang, 4-5/8" x 2-7/8" | E4@.200 | Ea | 6.32 | 9.83 | 16.15 |
| 2 gang, 4-3/4" x 4-3/4" | E4@.300 | Ea | 11.40 | 14.70 | 26.10 |
| 2 gang, 15/16" diameter, 1 KO | E4@.300 | Ea | 5.22 | 14.70 | 19.92 |
| 2 gang, 1-3/4" diameter, 2 KO | E4@.300 | Ea | 12.30 | 14.70 | 27.00 |
| 2 gang, 2-1/4" diameter, 1 KO | E4@.300 | Ea | 17.10 | 14.70 | 31.80 |
| Corner box, 2-1/2" diameter | E4@.250 | Ea | 13.60 | 12.30 | 25.90 |
| Duplex receptacle box, 4-1/8" x 2" x 1-3/8" | E4@.360 | Ea | 12.80 | 17.70 | 30.50 |
| Fixture box, 4-3/4" diameter | E4@.250 | Ea | 10.20 | 12.30 | 22.50 |
| Fixture box, 6-3/8" diameter | E4@.250 | Ea | 12.00 | 12.30 | 24.30 |
| Fixture box, 5-1/2" diameter | E4@.250 | Ea | 10.30 | 12.30 | 22.60 |
| Junction and pull box, 1" x 5-1/2" diameter | E4@.250 | Ea | 9.31 | 12.30 | 21.61 |
| Extension box, 4-3/4" diameter | E4@.250 | Ea | 10.30 | 12.30 | 22.60 |
| **Wiremold outlet box, V700 surface raceway** | | | | | |
| Regular outlet box | E4@.250 | Ea | 6.64 | 12.30 | 18.94 |
| Deep outlet box | E4@.250 | Ea | 8.00 | 12.30 | 20.30 |
| Extra deep outlet box | E4@.250 | Ea | 7.22 | 12.30 | 19.52 |
| Fan box | E4@.250 | Ea | 11.70 | 12.30 | 24.00 |
| Starter box | E4@.250 | Ea | 6.77 | 12.30 | 19.07 |
| Utility box, 2" x 4-1/8" x 1-3/8" | E4@.360 | Ea | 8.01 | 17.70 | 25.71 |
| **Plugmould® V2000 multi-outlet raceway** | | | | | |
| Entrance end fitting | E4@.100 | Ea | 6.01 | 4.91 | 10.92 |
| 6 outlets, 36" long | E4@.450 | Ea | 34.60 | 22.10 | 56.70 |
| 6 outlets, 40" long | E4@.450 | Ea | 40.30 | 22.10 | 62.40 |
| 8 outlets, 52" long | E4@.550 | Ea | 51.70 | 27.00 | 78.70 |
| 10 outlets, 60" long | E4@.650 | Ea | 40.20 | 31.90 | 72.10 |
| **Cable Tray and Duct** | | | | | |
| Cable tray, steel, ladder type | | | | | |
| 4" wide | E4@.144 | LF | 17.50 | 7.07 | 24.57 |
| 6" wide | E4@.166 | LF | 33.70 | 8.16 | 41.86 |
| 8" wide | E4@.187 | LF | 52.90 | 9.19 | 62.09 |
| 10" wide | E4@.211 | LF | 97.40 | 10.40 | 107.80 |
| 12" wide | E4@.238 | LF | 74.40 | 11.70 | 86.10 |

| | Craft@Hrs | Unit | Material | Labor | Total |
|---|---|---|---|---|---|

Steel underfloor duct, including typical supports, fittings and accessories

| | Craft@Hrs | Unit | Material | Labor | Total |
|---|---|---|---|---|---|
| 3-1/4" wide, 1 cell | E4@.097 | LF | 7.75 | 4.77 | 12.52 |
| 3-1/4" wide, 2 cell | E4@.104 | LF | 15.70 | 5.11 | 20.81 |
| 7-1/4" wide, 1 cell | E4@.148 | LF | 18.30 | 7.27 | 25.57 |
| 7-1/4" wide, 2 cell | E4@.148 | LF | 27.90 | 7.27 | 35.17 |

## Wire and Cable

Copper wire and cable prices can change very quickly. Quarterly price updates for *National Estimator* are free and automatic during the year of issue. You'll be prompted when it's time to collect the next update. A Web connection is required.

**Service Entrance Cable** 600 volt stranded service entrance wire pulled in conduit except as noted. No excavation or conduit included. By American Wire Gauge.

SEU service entrance cable. Type XHHW wire wound in a concentrically applied neutral and jacketed with gray sunlight-resistant PVC.

| | Craft@Hrs | Unit | Material | Labor | Total |
|---|---|---|---|---|---|
| 8 gauge, 3 conductor, copper | E4@.012 | LF | .44 | .59 | 1.03 |
| 6 gauge, 3 conductor, copper | E4@.012 | LF | 2.21 | .59 | 2.80 |
| 6, 6, and 8 gauge conductors, copper | E4@.012 | LF | 1.38 | .59 | 1.97 |
| 6 gauge, 3 conductor, aluminum | E4@.012 | LF | .99 | .59 | 1.58 |
| 4, 4, 4, and 6 gauge conductor, copper | E4@.014 | LF | 3.51 | .69 | 4.20 |
| 4-4 gauge, 4 conductor, copper | E4@.014 | LF | 3.05 | .69 | 3.74 |
| 4-4 gauge, 4 conductor, aluminum | E4@.014 | LF | 1.46 | .69 | 2.15 |
| 3-3-3 gauge, 5 conductor, copper | E4@.015 | LF | 5.31 | .74 | 6.05 |
| 2-2-2 gauge, 4 conductor, copper | E4@.016 | LF | 4.66 | .79 | 5.45 |
| 2-2-2 gauge, 4 conductor, aluminum | E4@.016 | LF | 1.83 | .79 | 2.62 |
| 2-2 gauge, 2 conductor, copper | E4@.016 | LF | 5.69 | .79 | 6.48 |
| 1-1-1 gauge, 3 conductor, aluminum | E4@.018 | LF | 3.16 | .88 | 4.04 |
| 4/0, 4/0, 4/0, 2/0 conductor, aluminum | E4@.026 | LF | 4.74 | 1.28 | 6.02 |

USE underground service entrance wire. Stranded conductor insulated with abrasion, heat, moisture, and sunlight-resistant cross-linked polyethylene (XLP).

| | Craft@Hrs | Unit | Material | Labor | Total |
|---|---|---|---|---|---|
| 8 gauge, copper | E4@.009 | LF | .75 | .44 | 1.19 |
| 6 gauge, copper | E4@.010 | LF | .70 | .49 | 1.19 |
| 2 gauge, copper | E4@.016 | LF | 1.23 | .79 | 2.02 |
| 1/0 gauge, copper | E4@.017 | LF | 2.37 | .84 | 3.21 |
| 2/0 gauge, copper | E4@.018 | LF | 2.62 | .88 | 3.50 |
| 4/0 gauge, copper | E4@.018 | LF | 3.51 | .88 | 4.39 |
| 250 MCM, copper | E4@.020 | LF | 2.84 | .98 | 3.82 |
| 500 MCM, copper | E4@.022 | LF | 5.57 | 1.08 | 6.65 |
| Deduct for aluminum USE service entrance wire | — | % | -40.0 | — | — |

URD triplex direct-burial aluminum service entrance cable

| | Craft@Hrs | Unit | Material | Labor | Total |
|---|---|---|---|---|---|
| 6 gauge, Erskine | E4@.011 | LF | .99 | .54 | 1.53 |
| 4 gauge, Vassar | E4@.012 | LF | 1.11 | .59 | 1.70 |
| 2 gauge, Ramapo | E4@.013 | LF | 1.53 | .64 | 2.17 |
| 2/0 gauge, Converse | E4@.017 | LF | 2.16 | .84 | 3.00 |

**Bare Copper Ground Wire** Pulled in conduit with conductor wires. No excavation or duct included. By American Wire Gauge. Cost per linear foot based on typical discounts for the quantities listed.

| | Craft@Hrs | Unit | Material | Labor | Total |
|---|---|---|---|---|---|
| 12 gauge, solid, cut length | E4@.004 | LF | .53 | .20 | .73 |
| 12 gauge, solid, 1,250' reel | E4@.004 | LF | .18 | .20 | .38 |
| 10 gauge, solid, cut length | E4@.005 | LF | .59 | .25 | .84 |
| 10 gauge, solid, 800' reel | E4@.005 | LF | .29 | .25 | .54 |
| 8 gauge, cut length | E4@.005 | LF | .87 | .25 | 1.12 |
| 8 gauge, solid, 500' coil | E4@.005 | LF | .44 | .25 | .69 |

| | Craft@Hrs | Unit | Material | Labor | Total |
|---|---|---|---|---|---|
| 6 gauge, solid, cut length | E4@.005 | LF | 1.16 | .25 | 1.41 |
| 6 gauge, solid, 315' coil | E4@.005 | LF | .70 | .25 | .95 |
| 4 gauge, solid, cut length | E4@.006 | LF | 1.59 | .29 | 1.88 |
| 4 gauge, solid, 198' coil | E4@.006 | LF | 1.11 | .29 | 1.40 |
| 4 gauge, stranded, cut length | E4@.006 | LF | 1.72 | .29 | 2.01 |
| 4 gauge, stranded, 198' coil | E4@.006 | LF | 1.17 | .29 | 1.46 |
| 2 gauge, cut length | E4@.008 | LF | 2.07 | .39 | 2.46 |
| 2 gauge, stranded, cut length | E4@.008 | LF | 2.21 | .39 | 2.60 |
| 2 gauge, solid, 125' reel | E4@.008 | LF | 1.74 | .39 | 2.13 |
| 2 gauge, stranded, 125' reel | E4@.008 | LF | 1.83 | .39 | 2.22 |

**THHN/THWN-2 Copper Building Wire** Pulled in conduit for feeder and branch circuits. No conduit included. THHN has a heat-resistant thermoplastic cover approved for working temperatures up to 90 degrees Celsius in dry locations. THWN resists moisture, oil and gasoline and is approved for temperatures to 75 degrees Celsius in dry, wet or oil locations. THHN/THWN is made from soft annealed copper and insulated with heat- and moisture-resistant PVC and covered with nylon (polyamide). Available in colors to simplify identification of conductors after wire is pulled in conduit. By American Wire Gauge (AWG). Stranded wire has 19 strands. Cost per linear foot based on typical discounts for the quantities listed. Copper wire prices are highly volatile.

| | Craft@Hrs | Unit | Material | Labor | Total |
|---|---|---|---|---|---|
| 14 gauge, solid, cut length | E4@.006 | LF | .28 | .29 | .57 |
| 14 gauge, solid, 50' coil | E4@.006 | LF | .24 | .29 | .53 |
| 14 gauge, solid, 100' coil | E4@.006 | LF | .19 | .29 | .48 |
| 14 gauge, solid, 500' coil | E4@.006 | LF | .07 | .29 | .36 |
| 14 gauge, stranded, cut length | E4@.006 | LF | .25 | .29 | .54 |
| 14 gauge, stranded, 500' coil | E4@.006 | LF | .07 | .29 | .36 |
| 14 gauge, stranded, 2,500' reel | E4@.006 | LF | .08 | .29 | .37 |
| 12 gauge, solid, cut length | E4@.007 | LF | .36 | .34 | .70 |
| 12 gauge, solid, 50' coil | E4@.007 | LF | .24 | .34 | .58 |
| 12 gauge, solid, 100' coil | E4@.007 | LF | .26 | .34 | .60 |
| 12 gauge, solid, 500' coil | E4@.007 | LF | .11 | .34 | .45 |
| 12 gauge, solid, 2,500' reel | E4@.007 | LF | .11 | .34 | .45 |
| 12 gauge, stranded, cut length | E4@.007 | LF | .33 | .34 | .67 |
| 12 gauge, stranded, 50' coil | E4@.007 | LF | .26 | .34 | .60 |
| 12 gauge, stranded, 100' coil | E4@.007 | LF | .28 | .34 | .62 |
| 12 gauge, stranded, 500' coil | E4@.007 | LF | .09 | .34 | .43 |
| 12 gauge, stranded, 2,500' reel | E4@.007 | LF | .09 | .34 | .43 |
| 10 gauge, solid, cut length | E4@.008 | LF | .49 | .39 | .88 |
| 10 gauge, solid, 500' coil | E4@.008 | LF | .18 | .39 | .57 |
| 10 gauge, stranded, cut length | E4@.008 | LF | .43 | .39 | .82 |
| 10 gauge, stranded, 50' coil | E4@.008 | LF | .28 | .39 | .67 |
| 10 gauge, stranded, 100' coil | E4@.008 | LF | .26 | .39 | .65 |
| 10 gauge, stranded, 500' coil | E4@.008 | LF | .19 | .39 | .58 |
| 10 gauge, stranded, 2,500' reel | E4@.008 | LF | .20 | .39 | .59 |
| 8 gauge, solid, cut length | E4@.011 | LF | 1.00 | .54 | 1.54 |
| 8 gauge, stranded, cut length | E4@.009 | LF | .54 | .44 | .98 |
| 8 gauge, stranded, 500' coil | E4@.009 | LF | .28 | .44 | .72 |
| 8 gauge, stranded, 1,000' coil | E4@.009 | LF | .28 | .44 | .72 |
| 6 gauge, stranded, cut length | E4@.010 | LF | .61 | .49 | 1.10 |
| 6 gauge, stranded, 500' coil | E4@.010 | LF | .43 | .49 | .92 |
| 6 gauge, stranded, 1,000' reel | E4@.010 | LF | .43 | .49 | .92 |
| 4 gauge, stranded, cut length | E4@.012 | LF | .74 | .59 | 1.33 |
| 4 gauge, stranded, 500' reel | E4@.012 | LF | .71 | .59 | 1.30 |
| 4 gauge, stranded, 1,000' reel | E4@.012 | LF | .70 | .59 | 1.29 |
| 4 gauge, stranded, 2,500' reel | E4@.012 | LF | .70 | .59 | 1.29 |
| 3 gauge, stranded, cut length | E4@.013 | LF | 1.03 | .64 | 1.67 |

| | Craft@Hrs | Unit | Material | Labor | Total |
|---|---|---|---|---|---|
| 3 gauge, stranded, 500' reel | E4@.013 | LF | .90 | .64 | 1.54 |
| 2 gauge, stranded, cut length | E4@.013 | LF | 1.26 | .64 | 1.90 |
| 2 gauge, stranded, 500' reel | E4@.013 | LF | 1.18 | .64 | 1.82 |
| 2 gauge, stranded, 1,000' reel | E4@.013 | LF | 1.17 | .64 | 1.81 |
| 2 gauge, stranded, 2,500' reel | E4@.013 | LF | 1.17 | .64 | 1.81 |
| 1 gauge, stranded, cut length | E4@.014 | LF | 1.76 | .69 | 2.45 |
| 1 gauge, stranded, 500' reel | E4@.014 | LF | 1.50 | .69 | 2.19 |
| 1/0 gauge, stranded, cut length | E4@.015 | LF | 1.80 | .74 | 2.54 |
| 1/0 gauge, stranded, 500' reel | E4@.015 | LF | 1.82 | .74 | 2.56 |
| 1/0 gauge, stranded, 1,000' reel | E4@.015 | LF | 1.81 | .74 | 2.55 |
| 2/0 gauge, stranded, cut length | E4@.016 | LF | 2.45 | .79 | 3.24 |
| 2/0 gauge, stranded, 500' reel | E4@.016 | LF | 2.22 | .79 | 3.01 |
| 2/0 gauge, stranded, 1,000' reel | E4@.016 | LF | 2.18 | .79 | 2.97 |
| 3/0 gauge, stranded, cut length | E4@.017 | LF | 2.72 | .84 | 3.56 |
| 3/0 gauge, stranded, 500' reel | E4@.017 | LF | 2.72 | .84 | 3.56 |
| 3/0 gauge, stranded, 1,000' reel | E4@.017 | LF | 2.71 | .84 | 3.55 |
| 4/0 gauge, stranded, cut length | E4@.018 | LF | 3.54 | .88 | 4.42 |
| 4/0 gauge, stranded, 500' reel | E4@.018 | LF | 3.66 | .88 | 4.54 |
| 4/0 gauge, stranded, 1,000' reel | E4@.018 | LF | 3.66 | .88 | 4.54 |
| 250 MCM, stranded, cut length | E4@.020 | LF | 1.91 | .98 | 2.89 |
| 250 MCM, stranded, 500' reel | E4@.020 | LF | 1.86 | .98 | 2.84 |
| 250 MCM, stranded, 1,000' reel | E4@.020 | LF | 2.74 | .98 | 3.72 |
| 350 MCM, stranded, cut length | E4@.038 | LF | 4.34 | 1.87 | 6.21 |
| 350 MCM, stranded, 1,000' reel | E4@.038 | LF | 2.29 | 1.87 | 4.16 |
| 500 MCM, stranded, cut length | E4@.046 | LF | 5.84 | 2.26 | 8.10 |
| 500 MCM, stranded, 1,000' reel | E4@.046 | LF | 5.49 | 2.26 | 7.75 |

**XHHW Copper Building Wire** Pulled in conduit for feeder and branch circuits. No conduit included. Abrasion-, moisture- and heat-resistant thermoplastic intended for harsh environments. Maximum operating temperature is 90 degrees Celsius in dry locations and 75 degrees Celsius in wet locations. Flame-retardant cross-linked synthetic polymer insulation. XHHW has no outer jacket or covering. Available in colors to simplify identification of conductors after wire is pulled in conduit. By American Wire Gauge (AWG). Stranded wire has 19 strands. Cost per linear foot based on typical discounts for the quantities listed. Copper wire prices are highly volatile.

| | Craft@Hrs | Unit | Material | Labor | Total |
|---|---|---|---|---|---|
| 14 gauge stranded, cut length | E4@.006 | LF | .49 | .29 | .78 |
| 14 gauge stranded, 500' coil | E4@.006 | LF | .19 | .29 | .48 |
| 12 gauge solid, cut length | E4@.007 | LF | .25 | .34 | .59 |
| 12 gauge stranded, cut length | E4@.007 | LF | .48 | .34 | .82 |
| 12 gauge, solid, 500' coil | E4@.007 | LF | .23 | .34 | .57 |
| 12 gauge, stranded, 500' coil | E4@.007 | LF | .26 | .34 | .60 |
| 12 gauge, solid, 2,500' reel | E4@.007 | LF | .23 | .34 | .57 |
| 10 gauge, solid, cut length | E4@.008 | LF | .57 | .39 | .96 |
| 10 gauge, solid, 500' coil | E4@.008 | LF | .25 | .39 | .64 |
| 4 gauge, stranded, cut length | E4@.012 | LF | 1.81 | .59 | 2.40 |
| 4 gauge, stranded, 500' reel | E4@.012 | LF | 2.07 | .59 | 2.66 |
| 2 gauge, stranded, cut length | E4@.014 | LF | 2.70 | .69 | 3.39 |
| 2 gauge, stranded, 500' reel | E4@.014 | LF | 2.68 | .69 | 3.37 |
| 1/0 gauge, stranded, cut length | E4@.015 | LF | 3.55 | .74 | 4.29 |
| 1/0 gauge, stranded, 500' reel | E4@.015 | LF | 2.68 | .74 | 3.42 |
| 2/0 gauge, stranded, cut length | E4@.016 | LF | 3.92 | .79 | 4.71 |
| 2/0 gauge, stranded, 500' reel | E4@.016 | LF | 3.65 | .79 | 4.44 |
| 3/0 gauge, stranded, cut length | E4@.017 | LF | 4.74 | .84 | 5.58 |
| 4/0 gauge, stranded, cut length | E4@.018 | LF | 6.50 | .88 | 7.38 |
| 4/0 gauge, stranded, 500' reel | E4@.018 | LF | 7.83 | .88 | 8.71 |

| | Craft@Hrs | Unit | Material | Labor | Total |
|---|---|---|---|---|---|

**Type UF-B Direct Burial Sheathed Cable** Installed in an open trench. No excavation or backfill included. UF-B is designed for burial in a trench but may also be used to wire interior branch circuits in residential and agricultural buildings. The cable jacket is sunlight-, fungus- and moisture-resistant gray PVC. Full-size ground wire. Copper wire prices are highly volatile.

| | Craft@Hrs | Unit | Material | Labor | Total |
|---|---|---|---|---|---|
| 14 gauge, 2 conductor, cut length | E4@.005 | LF | .64 | .25 | .89 |
| 14 gauge, 2 conductor, 25' coil | E4@.005 | LF | .62 | .25 | .87 |
| 14 gauge, 2 conductor, 50' coil | E4@.005 | LF | .60 | .25 | .85 |
| 14 gauge, 2 conductor, 100' coil | E4@.005 | LF | .54 | .25 | .79 |
| 14 gauge, 2 conductor, 250' coil | E4@.005 | LF | .30 | .25 | .55 |
| 14 gauge, 2 conductor, 1,000' coil | E4@.005 | LF | .25 | .25 | .50 |
| 14 gauge, 3 conductor, cut length | E4@.005 | LF | .89 | .25 | 1.14 |
| 14 gauge, 3 conductor, 250' coil | E4@.005 | LF | .48 | .25 | .73 |
| 14 gauge, 3 conductor, 1,000' reel | E4@.005 | LF | .47 | .25 | .72 |
| 12 gauge, 2 conductor, cut length | E4@.005 | LF | .85 | .25 | 1.10 |
| 12 gauge, 2 conductor, 25' coil | E4@.005 | LF | .81 | .25 | 1.06 |
| 12 gauge, 2 conductor, 50' coil | E4@.005 | LF | .74 | .25 | .99 |
| 12 gauge, 2 conductor, 100' coil | E4@.005 | LF | .67 | .25 | .92 |
| 12 gauge, 2 conductor, 250' coil | E4@.005 | LF | .43 | .25 | .68 |
| 12 gauge, 2 conductor, 1,000' reel | E4@.005 | LF | .39 | .25 | .64 |
| 12 gauge, 3 conductor, cut length | E4@.005 | LF | 1.46 | .25 | 1.71 |
| 12 gauge, 3 conductor, 100' coil | E4@.005 | LF | .77 | .25 | 1.02 |
| 12 gauge, 3 conductor, 250' coil | E4@.005 | LF | .52 | .25 | .77 |
| 12 gauge, 3 conductor, 1,000' coil | E4@.005 | LF | .59 | .25 | .84 |
| 10 gauge, 2 conductor, cut length | E4@.005 | LF | 1.46 | .25 | 1.71 |
| 10 gauge, 2 conductor, 100' coil | E4@.005 | LF | .84 | .25 | 1.09 |
| 10 gauge, 2 conductor, 250' coil | E4@.005 | LF | .59 | .25 | .84 |
| 10 gauge, 2 conductor, 1,000' reel | E4@.005 | LF | .66 | .25 | .91 |
| 10 gauge, 3 conductor, cut length | E4@.006 | LF | 2.09 | .29 | 2.38 |
| 10 gauge, 3 conductor, 100' coil | E4@.006 | LF | 1.08 | .29 | 1.37 |
| 10 gauge, 3 conductor, 250' coil | E4@.006 | LF | .78 | .29 | 1.07 |
| 10 gauge, 3 conductor, 1,000' reel | E4@.006 | LF | .97 | .29 | 1.26 |
| 8 gauge, 2 conductor, cut length | E4@.007 | LF | 2.19 | .34 | 2.53 |
| 8 gauge, 2 conductor, 500' reel | E4@.007 | LF | 1.47 | .34 | 1.81 |
| 8 gauge, 3 conductor, cut length | E4@.007 | LF | 2.62 | .34 | 2.96 |
| 8 gauge, 3 conductor, 125' reel | E4@.007 | LF | 1.78 | .34 | 2.12 |
| 8 gauge, 3 conductor, 500' reel | E4@.007 | LF | 1.76 | .34 | 2.10 |
| 6 gauge, 2 conductor, cut length | E4@.008 | LF | 2.78 | .39 | 3.17 |
| 6 gauge, 2 conductor, 500' reel | E4@.008 | LF | 1.92 | .39 | 2.31 |
| 6 gauge, 3 conductor, 125' reel | E4@.008 | LF | 2.42 | .39 | 2.81 |
| 6 gauge, 3 conductor, 500' reel | E4@.008 | LF | 2.40 | .39 | 2.79 |

**Type NM-B (Romex) Sheathed Cable** Installed in wood frame walls for outlets and switched fixtures, including boring out and pulling cable. NM-B cable can also be run through voids in masonry block or tile walls if little moisture is expected. Conductor insulation is color-coded PVC. 14 AWG (American Wire Gauge) is white, 12 AWG is yellow, 10 AWG is orange, and 8 and 6 AWG are black. Rated at 15 amps. With full-size ground wire. Cost per linear foot based on typical discounts for the quantities listed.

| | Craft@Hrs | Unit | Material | Labor | Total |
|---|---|---|---|---|---|
| 14 gauge, 2 conductor, cut length | E4@.027 | LF | .82 | 1.33 | 2.15 |
| 14 gauge, 2 conductor, 25' coil | E4@.027 | LF | .45 | 1.33 | 1.78 |
| 14 gauge, 2 conductor, 50' coil | E4@.027 | LF | .40 | 1.33 | 1.73 |
| 14 gauge, 2 conductor, 100' coil | E4@.027 | LF | .31 | 1.33 | 1.64 |
| 14 gauge, 2 conductor, 1,000' reel | E4@.027 | LF | .17 | 1.33 | 1.50 |
| 14 gauge, 3 conductor, cut length | E4@.030 | LF | 1.12 | 1.47 | 2.59 |
| 14 gauge, 3 conductor, 25' coil | E4@.030 | LF | .65 | 1.47 | 2.12 |

| | Craft@Hrs | Unit | Material | Labor | Total |
|---|---|---|---|---|---|
| 14 gauge, 3 conductor, 50' coil | E4@.030 | LF | .59 | 1.47 | 2.06 |
| 14 gauge, 3 conductor, 100' coil | E4@.030 | LF | .52 | 1.47 | 1.99 |
| 14 gauge, 3 conductor, 1,000' coil | E4@.030 | LF | .25 | 1.47 | 1.72 |
| 12 gauge, 2 conductor, cut length | E4@.030 | LF | .74 | 1.47 | 2.21 |
| 12 gauge, 2 conductor, 25' coil | E4@.030 | LF | .68 | 1.47 | 2.15 |
| 12 gauge, 2 conductor, 50' coil | E4@.030 | LF | .62 | 1.47 | 2.09 |
| 12 gauge, 2 conductor, 100' coil | E4@.030 | LF | .55 | 1.47 | 2.02 |
| 12 gauge, 3 conductor, cut length | E4@.031 | LF | 1.53 | 1.52 | 3.05 |
| 12 gauge, 3 conductor, 25' coil | E4@.031 | LF | 1.10 | 1.52 | 2.62 |
| 12 gauge, 3 conductor, 50' coil | E4@.031 | LF | .99 | 1.52 | 2.51 |
| 12 gauge, 3 conductor, 100' coil | E4@.031 | LF | .77 | 1.52 | 2.29 |
| 12 gauge, 3 conductor, 250' coil | E4@.031 | LF | .42 | 1.52 | 1.94 |
| 12 gauge, 3 conductor, 1,000' coil | E4@.031 | LF | .42 | 1.52 | 1.94 |
| 10 gauge, 2 conductor, cut length | E4@.031 | LF | 1.66 | 1.52 | 3.18 |
| 10 gauge, 2 conductor, 25' coil | E4@.031 | LF | 1.06 | 1.52 | 2.58 |
| 10 gauge, 2 conductor, 50' coil | E4@.031 | LF | .99 | 1.52 | 2.51 |
| 10 gauge, 2 conductor, 100' coil | E4@.031 | LF | .81 | 1.52 | 2.33 |
| 10 gauge, 2 conductor, 250' coil | E4@.031 | LF | .46 | 1.52 | 1.98 |
| 10 gauge, 2 conductor, 1,000' coil | E4@.031 | LF | .46 | 1.52 | 1.98 |
| 10 gauge, 3 conductor, cut length | E4@.035 | LF | 2.18 | 1.72 | 3.90 |
| 10 gauge, 3 conductor, 25' coil | E4@.035 | LF | 1.41 | 1.72 | 3.13 |
| 10 gauge, 3 conductor, 50' coil | E4@.035 | LF | 1.39 | 1.72 | 3.11 |
| 10 gauge, 3 conductor, 100' coil | E4@.035 | LF | 1.15 | 1.72 | 2.87 |
| 10 gauge, 3 conductor, 1,000' reel | E4@.035 | LF | .69 | 1.72 | 2.41 |
| 8 gauge, 2 conductor, cut length | E4@.040 | LF | 1.81 | 1.97 | 3.78 |
| 8 gauge, 2 conductor, 125' coil | E4@.040 | LF | .81 | 1.97 | 2.78 |
| 8 gauge, 2 conductor, 500' reel | E4@.040 | LF | .81 | 1.97 | 2.78 |
| 8 gauge, 3 conductor, 125' coil | E4@.043 | LF | 1.07 | 2.11 | 3.18 |
| 8 gauge, 3 conductor, 500' reel | E4@.043 | LF | 1.07 | 2.11 | 3.18 |
| 6 gauge, 2 conductor, cut length | E4@.043 | LF | 2.51 | 2.11 | 4.62 |
| 6 gauge, 2 conductor, 125' coil | E4@.043 | LF | 1.16 | 2.11 | 3.27 |
| 6 gauge, 2 conductor, 500' reel | E4@.043 | LF | 1.15 | 2.11 | 3.26 |
| 6 gauge, 3 conductor, cut length | E4@.047 | LF | 3.64 | 2.31 | 5.95 |
| 6 gauge, 3 conductor, 125' reel | E4@.047 | LF | 1.54 | 2.31 | 3.85 |
| 6 gauge, 3 conductor, 500' reel | E4@.047 | LF | 1.53 | 2.31 | 3.84 |
| 4 gauge, 3 conductor, cut length | E4@.051 | LF | 6.10 | 2.51 | 8.61 |
| 2 gauge, 3 conductor, cut length | E4@.056 | LF | 8.65 | 2.75 | 11.40 |
| 2 gauge, 3 conductor, 500' reel | E4@.056 | LF | 5.62 | 2.75 | 8.37 |
| Deduct for no ground wire | | | | | |
| #14 and #12 cable | — | % | -5.0 | -5.0 | — |
| #10, #8 and #6 wire | — | % | -15.0 | -5.0 | — |

**MC Aluminum-Clad Cable** For use in dry sites. Exposed, concealed, or in approved raceway. Open or messenger-supported aerial runs. For powering lighting, control and signal circuits. Includes neutral, ground and one or two phase conductors. Installed with clamps or staples in a frame building or embedded in masonry, including approved bushings at terminations. Note that wire prices can change very quickly.

| | Craft@Hrs | Unit | Material | Labor | Total |
|---|---|---|---|---|---|
| 14 gauge, 2 conductor, cut length | E4@.009 | LF | .99 | .44 | 1.43 |
| 14 gauge, 2 conductor, 25' coil | E4@.009 | LF | .58 | .44 | 1.02 |
| 14 gauge, 2 conductor, 50' coil | E4@.009 | LF | .51 | .44 | .95 |
| 14 gauge, 2 conductor, 100' coil | E4@.009 | LF | .46 | .44 | .90 |
| 14 gauge, 2 conductor, 250' coil | E4@.009 | LF | .41 | .44 | .85 |
| 14 gauge, 2 conductor, 500' reel | E4@.009 | LF | .41 | .44 | .85 |
| 14 gauge, 3 conductor, cut length | E4@.010 | LF | 1.18 | .49 | 1.67 |

| | Craft@Hrs | Unit | Material | Labor | Total |
|---|---|---|---|---|---|
| 14 gauge, 3 conductor, 25' coil | E4@.010 | LF | .84 | .49 | 1.33 |
| 14 gauge, 3 conductor, 100' coil | E4@.010 | LF | .76 | .49 | 1.25 |
| 14 gauge, 3 conductor, 250' coil | E4@.010 | LF | .69 | .49 | 1.18 |
| 14 gauge, 3 conductor, 500' reel | E4@.010 | LF | .65 | .49 | 1.14 |
| 12 gauge, 2 conductor, cut length | E4@.010 | LF | 1.03 | .49 | 1.52 |
| 12 gauge, 2 conductor, 25' coil | E4@.010 | LF | .60 | .49 | 1.09 |
| 12 gauge, 2 conductor, 50' coil | E4@.010 | LF | .54 | .49 | 1.03 |
| 12 gauge, 2 conductor, 100' coil | E4@.010 | LF | .48 | .49 | .97 |
| 12 gauge, 2 conductor, 250' coil | E4@.010 | LF | .43 | .49 | .92 |
| 12 gauge, 2 conductor, 500' reel | E4@.010 | LF | .43 | .49 | .92 |
| 12 gauge, 2 conductor, 1,000' reel | E4@.010 | LF | .44 | .49 | .93 |
| 12 gauge, 3 conductor, cut length | E4@.010 | LF | 1.34 | .49 | 1.83 |
| 12 gauge, 3 conductor, 25' coil | E4@.010 | LF | .86 | .49 | 1.35 |
| 12 gauge, 3 conductor, 100' coil | E4@.010 | LF | .79 | .49 | 1.28 |
| 12 gauge, 3 conductor, 250' coil | E4@.010 | LF | .71 | .49 | 1.20 |
| 12 gauge, 3 conductor, 500' reel | E4@.010 | LF | .71 | .49 | 1.20 |
| 10 gauge, 2 conductor, cut length | E4@.011 | LF | 1.39 | .54 | 1.93 |
| 10 gauge, 2 conductor, 250' reel | E4@.011 | LF | 1.01 | .54 | 1.55 |
| 10 gauge, 3 conductor, cut length | E4@.011 | LF | 2.04 | .54 | 2.58 |
| 10 gauge, 3 conductor 250' reel | E4@.011 | LF | 1.38 | .54 | 1.92 |

**Flexible Armored AC Cable,** galvanized steel, 16 gauge integral bond wire to armor grounding path. Paper wrap conductor insulation covering. By American Wire Gauge and number of conductors. Installed exposed in a building either vertically or horizontally up to 10' above floor level. Sometimes referred to as "BX" cable.

| | Craft@Hrs | Unit | Material | Labor | Total |
|---|---|---|---|---|---|
| #14 wire, 2 conductor | E4@.030 | LF | .83 | 1.47 | 2.30 |
| #14 wire, 3 conductor | E4@.030 | LF | 1.33 | 1.47 | 2.80 |
| #12 wire, 2 conductor | E4@.030 | LF | .99 | 1.47 | 2.46 |
| #12 wire, 3 conductor | E4@.035 | LF | 1.50 | 1.72 | 3.22 |
| #12 wire, 4 conductor | E4@.038 | LF | 1.71 | 1.87 | 3.58 |
| #10 wire, 2 conductor | E4@.031 | LF | 1.64 | 1.52 | 3.16 |
| #10 wire, 3 conductor | E4@.039 | LF | 1.98 | 1.92 | 3.90 |
| # 6 wire, 3 conductor | E4@.047 | LF | 3.14 | 2.31 | 5.45 |
| # 4 wire, 3 conductor | E4@.051 | LF | 5.30 | 2.51 | 7.81 |
| # 2 wire, 3 conductor | E4@.056 | LF | 7.54 | 2.75 | 10.29 |

**Power Cable** Single conductor medium voltage ozone-resistant stranded copper cable pulled in conduit. No conduit included. Labor cost assumes three bundled conductors are pulled at one time on runs up to 100 feet. Labor cost will be higher on longer cable pulls and about 50% lower when cable is laid in an open trench. No splicing included. Listed by American Wire Gauge. Note that wire prices can change very quickly. Check with supplier.

5,000 volt, tape shielded, crosslinked polyethylene (XLP) insulated, with PVC jacket

| | Craft@Hrs | Unit | Material | Labor | Total |
|---|---|---|---|---|---|
| #6 gauge | E4@.014 | LF | 6.00 | .69 | 6.69 |
| #4 gauge | E4@.016 | LF | 5.79 | .79 | 6.58 |
| #2 gauge | E4@.021 | LF | 8.36 | 1.03 | 9.39 |
| #1/0 gauge | E4@.026 | LF | 10.70 | 1.28 | 11.98 |
| #2/0 gauge | E4@.028 | LF | 13.10 | 1.38 | 14.48 |
| #4/0 gauge | E4@.038 | LF | 17.50 | 1.87 | 19.37 |
| 350 MCM | E4@.047 | LF | 26.30 | 2.31 | 28.61 |
| 500 MCM | E4@.053 | LF | 33.50 | 2.60 | 36.10 |

15,000 volt, tape shielded, ethylene propylene rubber (EPR) insulated, with PVC jacket

| | Craft@Hrs | Unit | Material | Labor | Total |
|---|---|---|---|---|---|
| #2 gauge | E4@.024 | LF | 11.10 | 1.18 | 12.28 |
| #1/0 | E4@.030 | LF | 13.80 | 1.47 | 15.27 |
| #2/0 | E4@.032 | LF | 15.60 | 1.57 | 17.17 |
| 350 MCM | E4@.055 | LF | 29.00 | 2.70 | 31.70 |
| 500 MCM | E4@.062 | LF | 36.00 | 3.05 | 39.05 |

| | Craft@Hrs | Unit | Material | Labor | Total |
|---|---|---|---|---|---|

**Snow Melting Cable** For concrete pavement. Self-regulating heater cable 208-277 volt AC, encased in concrete pavement for walkways, steps, loading ramps or parking garages. Based on Raychem ElectroMelt™ System for concrete pavement. Cable is cut to length at the site and laid in a serpentine pattern using 12-inch center-to-center spacing. Use nylon cable ties to fasten cable to the top of reinforcing before concrete is placed. For scheduling purposes estimate that a crew of one can lay out, cut to length, install and tie 600 LF of cable in an 8-hour day. Based on 1.1 LF of cable per SF of pavement, including ties, "return bends" and waste, this yields 545 SF of pavement. These costs do not include reinforcing, concrete or concrete placing. A rule of thumb for sizing circuit breakers: For startup at 0 degrees F with 220 volts AC, allow 0.20 amps per LF of cable.

| | Craft@Hrs | Unit | Material | Labor | Total |
|---|---|---|---|---|---|
| Heater cable (1.1 LF per SF of pavement) | CE@.015 | SF | 12.00 | .88 | 12.88 |
| Power connection kit, including end seals | CE@.500 | Ea | 36.00 | 29.20 | 65.20 |
| Cable splice kit | CE@.250 | Ea | 16.80 | 14.60 | 31.40 |
| Cable expansion joint kit | CE@.250 | Ea | 25.10 | 14.60 | 39.70 |
| ElectroMelt™ junction box | CE@.500 | Ea | 136.00 | 29.20 | 165.20 |
| System controller, automatic | CE@1.50 | Ea | 2,880.00 | 87.60 | 2,967.60 |
| Thermostat, line sensing | CE@.500 | Ea | 294.00 | 29.20 | 323.20 |
| Ground fault protection device | CE@.969 | Ea | 434.00 | 56.60 | 490.60 |

**Ice Melting Cable** For roofs, gutters and downspouts. Self-regulating heater cable 120 or 208-277 volts AC, run exposed on the surface of the roof or within gutters. Based on Raychem IceStop™ System for roofs and gutters. Cable is cut to length on site and laid in a serpentine pattern using 24-inch center-to-center spacing and 32-inch loop height. Fasten to the roof with Raychem clips. For scheduling purposes estimate that a crew of one can cut to length and install 400 LF of cable in an 8-hour day. Based on 1.2 LF of cable per SF of protected roof area, including clips, "return bends" and waste, this yields 330 SF of protected area. Figure each LF of gutter as equal to one SF of roof. Figure each downspout as equal to 15 SF of roof. These costs do not include the roofing, gutters, or downspouts. A rule of thumb for sizing circuit breakers: For startup at 0 degrees F with 120 volts AC, allow 0.17 amps per LF of cable. For startup at 0 degrees F with 208-277 volts AC, allow 0.11 amps per LF of cable.

| | Craft@Hrs | Unit | Material | Labor | Total |
|---|---|---|---|---|---|
| Heater cable (1.2 LF per SF of protected area) | CE@.033 | LF | 7.20 | 1.93 | 9.13 |
| Power connection kit | CE@.500 | Ea | 36.00 | 29.20 | 65.20 |
| Cable splice kit | CE@.374 | Ea | 16.80 | 21.80 | 38.60 |
| End seal kit | CE@.374 | Ea | 15.60 | 21.80 | 37.40 |
| Downspout hanger | CE@.374 | Ea | 30.00 | 21.80 | 51.80 |
| Metal roof mounting kit | CE@.500 | Ea | 30.00 | 29.20 | 59.20 |

**Electrical Outlet Boxes** Steel boxes installed on an exposed wall or ceiling. The material cost for phenolic, PVC and fiberglass boxes will be 50% to 70% less. These costs include fasteners but no switch or receptacle.

Square outlet boxes

| | Craft@Hrs | Unit | Material | Labor | Total |
|---|---|---|---|---|---|
| 4" x 4" x 1-1/2" deep with | | | | | |
| 1/2" and 3/4" knockouts | E4@.187 | Ea | 1.39 | 9.19 | 10.58 |
| Side mounting bracket | E4@.187 | Ea | 3.39 | 9.19 | 12.58 |
| 4" x 4" x 2-1/8" deep with | | | | | |
| 1/2" and 3/4" knockouts | E4@.245 | Ea | 3.39 | 12.00 | 15.39 |
| Side mounting bracket | E4@.245 | Ea | 3.25 | 12.00 | 15.25 |
| Add for | | | | | |
| 1-1/2" deep extension rings | E4@.120 | Ea | 3.09 | 5.90 | 8.99 |
| Romex or BX clamps | — | Ea | .48 | — | .48 |
| Steel flush cover blanks | E4@.039 | Ea | 1.48 | 1.92 | 3.40 |
| Plaster rings to 3/4" deep | E4@.080 | Ea | 2.12 | 3.93 | 6.05 |

Octagon outlet boxes

| | Craft@Hrs | Unit | Material | Labor | Total |
|---|---|---|---|---|---|
| 4" x 1-1/2" deep with | | | | | |
| 1/2" knockouts | E4@.187 | Ea | 1.37 | 9.19 | 10.56 |
| Mounting bracket, horizontal | E4@.187 | Ea | 4.22 | 9.19 | 13.41 |
| 4" x 2-1/8" deep with | | | | | |
| 1/2" and 3/4" knockouts | E4@.245 | Ea | 5.73 | 12.00 | 17.73 |

| | Craft@Hrs | Unit | Material | Labor | Total |
|---|---|---|---|---|---|
| Add for | | | | | |
| 1-1/2" deep extension rings | E4@.120 | Ea | 2.76 | 5.90 | 8.66 |
| 1-1/2" with 18.5" adjustable bar set | E4@.282 | Ea | 3.37 | 13.90 | 17.27 |
| 3" deep concrete ring | E4@.516 | Ea | 2.45 | 25.40 | 27.85 |
| 4" steel flush cover blanks | E4@.039 | Ea | 1.03 | 1.92 | 2.95 |
| Plaster ring to 3/4" deep | E4@.080 | Ea | 2.20 | 3.93 | 6.13 |
| Handy boxes, 4" x 2-1/8" | | | | | |
| 1-1/2" deep, 1/2" knockouts | E4@.187 | Ea | 2.49 | 9.19 | 11.68 |
| 1-7/8" deep, 3/4" knockouts | E4@.187 | Ea | 1.08 | 9.19 | 10.27 |
| 2-1/2" deep, 3/4" knockouts | E4@.245 | Ea | 3.39 | 12.00 | 15.39 |
| 2-1/8" deep, with side mounting bracket | E4@.245 | Ea | 3.25 | 12.00 | 15.25 |
| 1-1/2" deep extension rings | E4@.120 | Ea | 4.17 | 5.90 | 10.07 |
| Blank or switch cover | E4@.039 | Ea | .98 | 1.92 | 2.90 |
| Weatherproof box and cover | E4@.059 | Ea | 4.78 | 2.90 | 7.68 |
| Switch boxes, 3" x 2", square corner gangable boxes with mounting ears | | | | | |
| 2" deep, 1/2" knockouts | E4@.245 | Ea | 2.66 | 12.00 | 14.66 |
| 2-1/2" deep, 1/2" knockouts | E4@.245 | Ea | 7.67 | 12.00 | 19.67 |
| 3-1/2" deep, 3/4" knockouts | E4@.324 | Ea | 5.90 | 15.90 | 21.80 |
| Gang switch boxes, 2" x 3" x 1-5/8" | | | | | |
| 2 gang | E4@.242 | Ea | 9.98 | 11.90 | 21.88 |
| 3 gang | E4@.342 | Ea | 11.80 | 16.80 | 28.60 |
| 4 gang | E4@.342 | Ea | 16.50 | 16.80 | 33.30 |
| 5 gang | E4@.482 | Ea | 22.90 | 23.70 | 46.60 |
| 6 gang | E4@.482 | Ea | 41.90 | 23.70 | 65.60 |
| Floor boxes, watertight, cast iron, round | | | | | |
| 4-3/16" x 3-3/4" deep, non-adjustable | E4@1.54 | Ea | 71.10 | 75.70 | 146.80 |
| 3-3/4" x 2" deep, semi-adjustable | E4@1.54 | Ea | 82.90 | 75.70 | 158.60 |
| 4-3/16" x 3-3/4" deep, adjustable | E4@1.54 | Ea | 98.60 | 75.70 | 174.30 |
| 2-1/8" single, round floor cover plates | E4@.156 | Ea | 46.40 | 7.66 | 54.06 |
| Galvanized or gray enamel NEMA Class 1 (indoor) pull boxes, with screw cover, 4" deep | | | | | |
| 4" x  4" | E4@.373 | Ea | 8.33 | 18.30 | 26.63 |
| 4" x  6" | E4@.373 | Ea | 9.95 | 18.30 | 28.25 |
| 6" x  6" | E4@.373 | Ea | 12.50 | 18.30 | 30.80 |
| 6" x  8" | E4@.391 | Ea | 13.70 | 19.20 | 32.90 |
| 8" x  8" | E4@.391 | Ea | 15.00 | 19.20 | 34.20 |
| 8" x 10" | E4@.444 | Ea | 18.40 | 21.80 | 40.20 |
| 8" x 12" | E4@.483 | Ea | 17.80 | 23.70 | 41.50 |
| 10" x 10" | E4@.492 | Ea | 16.80 | 24.20 | 41.00 |
| 10" x 12" | E4@.539 | Ea | 31.40 | 26.50 | 57.90 |
| 12" x 12" | E4@.593 | Ea | 35.40 | 29.10 | 64.50 |
| 12" x 16" | E4@.663 | Ea | 49.60 | 32.60 | 82.20 |
| 12" x 18" | E4@.785 | Ea | 51.60 | 38.60 | 90.20 |
| Galvanized or gray enamel NEMA Class 1 (indoor) pull boxes, with screw cover, 6" deep | | | | | |
| 6" x  6" | E4@.373 | Ea | 12.70 | 18.30 | 31.00 |
| 8" x  8" | E4@.391 | Ea | 17.60 | 19.20 | 36.80 |
| 8" x 10" | E4@.444 | Ea | 19.70 | 21.80 | 41.50 |
| 10" x 10" | E4@.492 | Ea | 23.30 | 24.20 | 47.50 |
| 10" x 12" | E4@.539 | Ea | 26.70 | 26.50 | 53.20 |
| 12" x 12" | E4@.593 | Ea | 33.40 | 29.10 | 62.50 |
| 12" x 16" | E4@.663 | Ea | 39.50 | 32.60 | 72.10 |
| 12" x 18" | E4@.785 | Ea | 42.70 | 38.60 | 81.30 |
| 16" x 16" | E4@.929 | Ea | 48.30 | 45.60 | 93.90 |
| 18" x 18" | E4@1.10 | Ea | 52.40 | 54.00 | 106.40 |
| 18" x 24" | E4@1.30 | Ea | 95.70 | 63.90 | 159.60 |
| 24" x 24" | E4@1.54 | Ea | 120.00 | 75.70 | 195.70 |

| | Craft@Hrs | Unit | Material | Labor | Total |
|---|---|---|---|---|---|
| **Cast aluminum NEMA Class 3R (weatherproof) screw cover junction boxes, recessed, flush mounted, with cover** | | | | | |
| 6" x 6" x 4" deep | E4@.930 | Ea | 220.00 | 45.70 | 265.70 |
| 6" x 8" x 4" deep | E4@.930 | Ea | 271.00 | 45.70 | 316.70 |
| 6" x 6" x 6" deep | E4@.930 | Ea | 271.00 | 45.70 | 316.70 |
| 6" x 12" x 6" deep | E4@1.42 | Ea | 403.00 | 69.80 | 472.80 |
| 8" x 8" x 4" deep | E4@.982 | Ea | 309.00 | 48.20 | 357.20 |
| 8" x 8" x 6" deep | E4@1.14 | Ea | 332.00 | 56.00 | 388.00 |
| 8" x 12" x 6" deep | E4@1.30 | Ea | 502.00 | 63.90 | 565.90 |
| 10" x 10" x 6" deep | E4@1.33 | Ea | 284.00 | 65.30 | 349.30 |
| 12" x 12" x 6" deep | E4@1.35 | Ea | 596.00 | 66.30 | 662.30 |
| 12" x 12" x 8" deep | E4@1.73 | Ea | 701.00 | 85.00 | 786.00 |
| 12" x 24" x 6" deep | E4@1.97 | Ea | 1,330.00 | 96.80 | 1,426.80 |
| 18" x 36" x 8" deep | E4@2.56 | Ea | 3,080.00 | 126.00 | 3,206.00 |
| **Hinged cover panel enclosures, NEMA 1, enamel, with cover** | | | | | |
| 16" x 12" x 7" deep | E4@1.30 | Ea | 119.00 | 63.90 | 182.90 |
| 20" x 20" x 7" deep | E4@1.30 | Ea | 163.00 | 63.90 | 226.90 |
| 30" x 20" x 7" deep | E4@1.77 | Ea | 173.00 | 87.00 | 260.00 |
| 24" x 20" x 9" deep | E4@1.30 | Ea | 202.00 | 63.90 | 265.90 |
| 30" x 24" x 9" deep | E4@1.77 | Ea | 230.00 | 87.00 | 317.00 |
| 36" x 30" x 9" deep | E4@1.77 | Ea | 287.00 | 87.00 | 374.00 |

## Electrical Receptacles

Standard commercial grade except where noted. Ivory or brown receptacles. White or gray receptacles will cost about 10% more. Labor includes connecting wire and securing the device in the box. Add the cost of the box and the cover plate.

| | Craft@Hrs | Unit | Material | Labor | Total |
|---|---|---|---|---|---|
| **15 amp, 125 volt self-grounding, 2 pole, 3 wire duplex receptacles** | | | | | |
| Residential grade, grounded duplex | CE@.162 | Ea | .51 | 9.46 | 9.97 |
| Screwless, not self-grounding | CE@.162 | Ea | 2.19 | 9.46 | 11.65 |
| Side terminals | CE@.162 | Ea | 2.19 | 9.46 | 11.65 |
| Back and side terminals | CE@.162 | Ea | 2.18 | 9.46 | 11.64 |
| Feed thru wiring, back and side terminals | CE@.162 | Ea | 5.04 | 9.46 | 14.50 |
| Safety ground, side terminals | CE@.162 | Ea | 8.06 | 9.46 | 17.52 |
| Ground fault circuit interrupter receptacle | CE@.324 | Ea | 10.80 | 18.90 | 29.70 |
| Add for NEMA 5 single receptacles | — | % | 15.0 | — | — |
| **20 amp, 125 volt self-grounding, 2 pole, 3 wire duplex receptacles** | | | | | |
| Side terminals | CE@.162 | Ea | 9.54 | 9.46 | 19.00 |
| Back and side terminals | CE@.162 | Ea | 11.70 | 9.46 | 21.16 |
| Feed thru wiring, back and side terminals | CE@.162 | Ea | 12.20 | 9.46 | 21.66 |
| Hospital grade | CE@.162 | Ea | 12.70 | 9.46 | 22.16 |
| Ground fault circuit interrupter receptacle | CE@.324 | Ea | 18.10 | 18.90 | 37.00 |
| Add for NEMA 5 single receptacles | — | % | 15.0 | — | — |
| **250 volt receptacles, self-grounding, 2 pole, 3 wire, back & side terminals** | | | | | |
| 15 amp, duplex | CE@.187 | Ea | 11.50 | 10.90 | 22.40 |
| 15 amp, single | CE@.187 | Ea | 7.95 | 10.90 | 18.85 |
| 20 amp, duplex | CE@.187 | Ea | 8.94 | 10.90 | 19.84 |
| 20 amp, duplex, feed thru wiring | CE@.187 | Ea | 14.10 | 10.90 | 25.00 |
| 20 amp, single | CE@.187 | Ea | 9.85 | 10.90 | 20.75 |
| Clock receptacle, 2 pole, 15 amp, 125 volt | CE@.329 | Ea | 7.11 | 19.20 | 26.31 |
| 120/208 volt 20 amp 4 pole, duplex receptacle | CE@.795 | Ea | 31.10 | 46.40 | 77.50 |
| **125/250 volt 3 pole receptacles, flush mount** | | | | | |
| 15 amp/10 amp | CE@.714 | Ea | 9.37 | 41.70 | 51.07 |
| 20 amp | CE@.820 | Ea | 12.00 | 47.90 | 59.90 |
| 277 volt 30 amp 2 pole receptacle | CE@.268 | Ea | 49.40 | 15.60 | 65.00 |
| Dryer receptacles, 250 volt, 30/50 amp, 3 wire | CE@.536 | Ea | 24.80 | 31.30 | 56.10 |

| | Craft@Hrs | Unit | Material | Labor | Total |
|---|---|---|---|---|---|
| Accessories for 50 amp dryer receptacles | | | | | |
| Flush mounted plug | CE@.288 | Ea | 43.50 | 16.80 | 60.30 |
| Surface mounted plug | CE@.288 | Ea | 29.80 | 16.80 | 46.60 |
| Cord sets for dryer receptacles | | | | | |
| 36 inch, three #10 wires | CE@.288 | Ea | 18.20 | 16.80 | 35.00 |
| 48 inch, three #10 wires | CE@.288 | Ea | 20.60 | 16.80 | 37.40 |
| 60 inch, three #10 wires | CE@.288 | Ea | 23.90 | 16.80 | 40.70 |

**Electrical Switches**  Commercial grade, 120 to 277 volt rating, except where noted. Ivory or brown. Add 10% for white or gray. Labor includes connecting wire to the switch and securing the switch in the box. Add the cost of the box and cover plate.

| | Craft@Hrs | Unit | Material | Labor | Total |
|---|---|---|---|---|---|
| 15 amp switches, back and side wired | | | | | |
| One pole quiet switch, residential (side wired only) | CE@.112 | Ea | 4.65 | 6.54 | 11.19 |
| One pole switch, commercial, minimum quality | CE@.112 | Ea | 5.25 | 6.54 | 11.79 |
| Two pole switch | CE@.309 | Ea | 8.94 | 18.00 | 26.94 |
| Three-way switch | CE@.227 | Ea | 4.68 | 13.30 | 17.98 |
| Four-way switch | CE@.309 | Ea | 12.60 | 18.00 | 30.60 |
| 20 amp switches, back and side wired | | | | | |
| One pole switch | CE@.187 | Ea | 4.64 | 10.90 | 15.54 |
| Two pole switch | CE@.433 | Ea | 9.33 | 25.30 | 34.63 |
| Three-way switch | CE@.291 | Ea | 5.83 | 17.00 | 22.83 |
| Four-way switch | CE@.435 | Ea | 14.70 | 25.40 | 40.10 |
| 30 amp switches, side wired | | | | | |
| One pole switch | CE@.246 | Ea | 16.30 | 14.40 | 30.70 |
| Two pole switch | CE@.475 | Ea | 24.20 | 27.70 | 51.90 |
| Three-way switch | CE@.372 | Ea | 23.70 | 21.70 | 45.40 |
| Four-way switch | CE@.475 | Ea | 35.20 | 27.70 | 62.90 |
| 20 amp weatherproof switches, lever handle, with cover | | | | | |
| One pole switch | CE@.187 | Ea | 11.80 | 10.90 | 22.70 |
| Two pole switch | CE@.433 | Ea | 22.70 | 25.30 | 48.00 |
| Three-way switch | CE@.291 | Ea | 15.60 | 17.00 | 32.60 |
| Single pole, two gang, 10 amp | CE@.358 | Ea | 26.70 | 20.90 | 47.60 |
| Dimmer switches, push for off | | | | | |
| 600 watt, one pole | CE@.417 | Ea | 16.60 | 24.30 | 40.90 |
| 600 watt, three-way | CE@.626 | Ea | 16.80 | 36.50 | 53.30 |
| Fluorescent dimmer, 10 lamp load | CE@.626 | Ea | 91.90 | 36.50 | 128.40 |
| Astro dial time switch, 40 amp, with box | CE@3.00 | Ea | 177.00 | 175.00 | 352.00 |
| 15 minute timer switch, wall box mounted | CE@.426 | Ea | 27.60 | 24.90 | 52.50 |
| Single pole, 1 throw time switch, 277 volt | CE@.890 | Ea | 73.60 | 51.90 | 125.50 |
| Float switches for sump pumps, automatic 10A, 125/250/480 VAC | CE@1.03 | Ea | 40.30 | 60.10 | 100.40 |
| One-way 15 amp toggle switch with neon pilot | CE@.327 | Ea | 8.49 | 19.10 | 27.59 |
| Three-way 15 amp toggle switch with neon pilot | CE@.372 | Ea | 12.70 | 21.70 | 34.40 |
| Photoelectric switches, flush, with wall plate | | | | | |
| 120 volt, 1,000 watt | CE@.426 | Ea | 9.96 | 24.90 | 34.86 |
| 208 volt, 1,800 watt | CE@.626 | Ea | 12.10 | 36.50 | 48.60 |
| 480 volt, 3,000 watt | CE@.795 | Ea | 16.40 | 46.40 | 62.80 |

| | Craft@Hrs | Unit | Material | Labor | Total |
|---|---|---|---|---|---|
| Button control stations, surface mounted, NEMA Class 1, standard duty, 120/240 volt, with enclosure | | | | | |
| Start-stop switch, 2 button | CE@.624 | Ea | 31.00 | 36.40 | 67.40 |
| Start-stop switch with lockout, 2 button | CE@.624 | Ea | 46.60 | 36.40 | 83.00 |
| Forward, reverse and stop buttons, 3 button | CE@.624 | Ea | 62.90 | 36.40 | 99.30 |
| Forward, reverse, stop and lockout, 3 button | CE@.624 | Ea | 63.30 | 36.40 | 99.70 |
| Manual toggle starter switches, surface mounted, NEMA Class 1, 120/240 volt, non-reversing, with enclosure | | | | | |
| Size 0 motors, 2 pole | CE@.613 | Ea | 126.00 | 35.80 | 161.80 |
| Size 1 motors, 3 pole | CE@.698 | Ea | 163.00 | 40.70 | 203.70 |
| Size 1-1/2 motors, 2 pole | CE@.901 | Ea | 187.00 | 52.60 | 239.60 |
| Manual button starter switches, surface mounted, NEMA Class 1, 110 to 240 volt, 2 pole, 1 phase, with enclosure, start, stop, reset, with relay | | | | | |
| Size 00 motors | CE@.712 | Ea | 168.00 | 41.60 | 209.60 |
| Size 0 motors | CE@.712 | Ea | 182.00 | 41.60 | 223.60 |
| Size 1 motors | CE@.820 | Ea | 204.00 | 47.90 | 251.90 |
| Size 1-1/2 motors | CE@.975 | Ea | 245.00 | 56.90 | 301.90 |

**Electrical Box Cover Plates**  Includes the cover plate and screws. Add the cost of the box and switch or receptacle. Labor includes attaching the cover plate to the box. Almond or ivory plastic except as noted.

| | Craft@Hrs | Unit | Material | Labor | Total |
|---|---|---|---|---|---|
| Residential grade electrical box cover plates | | | | | |
| 1 gang plate, duplex receptacle | CE@.050 | Ea | .25 | 2.92 | 3.17 |
| 1 gang plate, toggle switch | CE@.050 | Ea | .25 | 2.92 | 3.17 |
| 2 gang plate, 2 duplex receptacles | CE@.100 | Ea | 1.08 | 5.84 | 6.92 |
| 2 gang plate, two toggle switches | CE@.100 | Ea | 1.08 | 5.84 | 6.92 |
| 2 gang plate, oversize, toggle or receptacle | CE@.100 | Ea | 2.11 | 5.84 | 7.95 |
| 3 gang plate, toggle or receptacle | CE@.150 | Ea | 2.18 | 8.76 | 10.94 |
| Decora multi-gang sectional plate, end or center | CE@.050 | Ea | 2.19 | 2.92 | 5.11 |
| Stainless steel electrical box cover plates | | | | | |
| 1 gang plate, blank, duplex or toggle | CE@.050 | Ea | 1.60 | 2.92 | 4.52 |
| 1 gang plate, power outlet | CE@.050 | Ea | 2.14 | 2.92 | 5.06 |
| 2 gang plate, blank, duplex or toggle | CE@.100 | Ea | 3.18 | 5.84 | 9.02 |
| 3 gang plate, blank, duplex or toggle | CE@.150 | Ea | 3.48 | 8.76 | 12.24 |
| Specification grade electrical box cover plates | | | | | |
| 1 gang plate, single receptacle or toggle switch | CE@.050 | Ea | 5.18 | 2.92 | 8.10 |
| 1 gang plate, duplex receptacle or toggle switch | CE@.050 | Ea | 7.71 | 2.92 | 10.63 |
| 1 gang plate, GFIC and rocker switch | CE@.050 | Ea | 7.71 | 2.92 | 10.63 |
| 2 gang plate, double duplex or toggle switches | CE@.100 | Ea | 6.61 | 5.84 | 12.45 |
| 2 gang plate, double GFIC and rocker switches | CE@.100 | Ea | 6.61 | 5.84 | 12.45 |
| 3 gang plate, duplex receptacles or switches | CE@.150 | Ea | 10.20 | 8.76 | 18.96 |

**Grounding Devices**  No excavation or concrete included.

| | Craft@Hrs | Unit | Material | Labor | Total |
|---|---|---|---|---|---|
| Copper-clad grounding rods, driven | | | | | |
| 5/8" x 8' | E4@.365 | Ea | 11.20 | 17.90 | 29.10 |
| 5/8" x 10' | E4@.420 | Ea | 15.00 | 20.60 | 35.60 |
| 3/4" x 8' | E4@.365 | Ea | 22.40 | 17.90 | 40.30 |
| 3/4" x 10' | E4@.420 | Ea | 23.50 | 20.60 | 44.10 |
| Grounding locknuts | | | | | |
| 1/2" locknuts | E4@.050 | Ea | 1.36 | 2.46 | 3.82 |
| 3/4" locknuts | E4@.060 | Ea | 1.73 | 2.95 | 4.68 |
| 1" locknuts | E4@.080 | Ea | 2.34 | 3.93 | 6.27 |
| 1-1/4" locknuts | E4@.100 | Ea | 3.13 | 4.91 | 8.04 |
| 1-1/2" locknuts | E4@.100 | Ea | 4.22 | 4.91 | 9.13 |
| 2" locknuts | E4@.150 | Ea | 5.74 | 7.37 | 13.11 |
| 2-1/2" locknuts | E4@.200 | Ea | 11.60 | 9.83 | 21.43 |
| 3" locknuts | E4@.200 | Ea | 14.60 | 9.83 | 24.43 |
| 4" locknuts | E4@.300 | Ea | 30.50 | 14.70 | 45.20 |

| | Craft@Hrs | Unit | Material | Labor | Total |
|---|---|---|---|---|---|
| Ground rod clamps | | | | | |
| 5/8" clamp | E4@.249 | Ea | 1.99 | 12.20 | 14.19 |
| 3/4" clamp | E4@.249 | Ea | 2.84 | 12.20 | 15.04 |
| Coupling for threaded ground rod | | | | | |
| 5/8" clamp | E4@.249 | Ea | 11.20 | 12.20 | 23.40 |
| 3/4" clamp | E4@.249 | Ea | 15.60 | 12.20 | 27.80 |
| Copper bonding connector straps, 3/4" x .050" | | | | | |
| 1" strap | E4@.286 | Ea | 11.80 | 14.10 | 25.90 |
| 2" strap | E4@.300 | Ea | 15.20 | 14.70 | 29.90 |
| 3" strap | E4@.315 | Ea | 20.40 | 15.50 | 35.90 |
| 4" strap | E4@.330 | Ea | 24.20 | 16.20 | 40.40 |
| Brazed connections for wire | | | | | |
| #6 wire | E4@.194 | Ea | 1.14 | 9.53 | 10.67 |
| #2 wire | E4@.194 | Ea | 1.34 | 9.53 | 10.87 |
| #1/0 wire | E4@.295 | Ea | 3.61 | 14.50 | 18.11 |
| #4/0 wire | E4@.391 | Ea | 8.90 | 19.20 | 28.10 |
| Grounding using a 5/8" x 8' copper clad ground rod, includes 10' grounding conductor, clamps, connectors, locknut, and rod. | | | | | |
| 100 amp service,#8 copper wire | E4@1.13 | Ea | 34.80 | 55.50 | 90.30 |
| 150 amp service,#6 copper wire | E4@1.16 | Ea | 37.90 | 57.00 | 94.90 |

**Electric Motors**  General purpose, 3 phase, open drip-proof, industrial duty, 1,700-1,780 RPM, 208-230/460 volt, rigid welded or solid base, furnished and placed, no hookup included. Based on Dayton motors.

| | Craft@Hrs | Unit | Material | Labor | Total |
|---|---|---|---|---|---|
| 1/2 HP | E4@1.99 | Ea | 181.00 | 97.80 | 278.80 |
| 3/4 HP | E4@1.99 | Ea | 202.00 | 97.80 | 299.80 |
| 1 HP | E4@1.99 | Ea | 214.00 | 97.80 | 311.80 |
| 1-1/2 HP | E4@1.99 | Ea | 223.00 | 97.80 | 320.80 |
| 2 HP | E4@1.99 | Ea | 229.00 | 97.80 | 326.80 |
| 3 HP | E4@1.99 | Ea | 247.00 | 97.80 | 344.80 |
| 5 HP | E4@2.22 | Ea | 289.00 | 109.00 | 398.00 |
| 7-1/2 HP | E4@2.40 | Ea | 426.00 | 118.00 | 544.00 |
| 10 HP | E4@2.51 | Ea | 494.00 | 123.00 | 617.00 |

**Electric Motor Connection**  Includes a NEMA-1 fusible heavy duty safety switch, connectors, fittings, ells, adapters, supports, anchors and conduit with wire appropriate for the load connected but no starter or motor.

| | Craft@Hrs | Unit | Material | Labor | Total |
|---|---|---|---|---|---|
| Single phase, using 10' RSC, 20' PVC and 6' flex conduit, NEMA Class 1 (indoor) general duty switch, 240 V | | | | | |
| For motor to 2 HP, 20 amps | E4@6.35 | Ea | 145.00 | 312.00 | 457.00 |
| For 2.5 to 3 HP motor, 30 amps | E4@6.63 | Ea | 157.00 | 326.00 | 483.00 |
| For 3 HP motor, 45 amps | E4@6.68 | Ea | 327.00 | 328.00 | 655.00 |
| For 8 to 15 HP motor, 90 amps | E4@13.4 | Ea | 477.00 | 658.00 | 1,135.00 |
| Three phase wiring using 10' RSC, 20' PVC and 6' flex conduit, NEMA Class 1 (indoor) switch, 600 V | | | | | |
| For motor to 2 HP, 20 amps | E4@7.18 | Ea | 299.00 | 353.00 | 652.00 |
| For 3 to 5 HP motor, 30 amps | E4@7.90 | Ea | 311.00 | 388.00 | 699.00 |
| For 7 to 10 HP motor, 45 amps | E4@8.30 | Ea | 373.00 | 408.00 | 781.00 |
| For 25 to 50 HP motor, 90 amps | E4@9.30 | Ea | 774.00 | 457.00 | 1,231.00 |
| For 50 to 100 HP motor, 135 amps | E4@14.1 | Ea | 1,190.00 | 693.00 | 1,883.00 |
| For 100 to 200 HP motor, 270 amps | E4@21.9 | Ea | 2,840.00 | 1,080.00 | 3,920.00 |
| Single phase, using 10' RSC, 20' PVC and 6' flex conduit, NEMA Class 3R (weatherproof) switch, 240 V | | | | | |
| For motor to 2 HP, 20 amps | E4@6.63 | Ea | 186.00 | 326.00 | 512.00 |
| For 2 to 5 HP motor, 30 amps | E4@6.97 | Ea | 198.00 | 342.00 | 540.00 |
| For 5 to 7.5 HP motor, 45 amps | E4@6.94 | Ea | 274.00 | 341.00 | 615.00 |

# 26 Electrical

| | Craft@Hrs | Unit | Material | Labor | Total |
|---|---|---|---|---|---|
| **Three phase using 10' RSC, 20' PVC and 6' flex conduit, NEMA Class 3R (weatherproof) switch, 600 V** | | | | | |
| For motor to 5 HP motor, 20 amps | E4@7.10 | Ea | 466.00 | 349.00 | 815.00 |
| For 1 to 10 HP motor, 30 amps | E4@7.25 | Ea | 477.00 | 356.00 | 833.00 |
| For 10 to 25 HP motor, 45 amps | E4@9.51 | Ea | 563.00 | 467.00 | 1,030.00 |
| For 25 to 50 HP motor, 90 amps | E4@9.51 | Ea | 980.00 | 467.00 | 1,447.00 |
| For 60 to 100 HP motor, 135 amps | E4@12.1 | Ea | 1,440.00 | 594.00 | 2,034.00 |
| For 125 to 200 HP motor, 270 amps | E4@21.6 | Ea | 3,410.00 | 1,060.00 | 4,470.00 |

**Electric Motor Connection**  Complete installation, including magnetic motor starter, overload relay, all wiring, a heavy duty fusible safety switch, junction box and cover, connectors, fittings, adapters, supports, anchors and conduit appropriate for the load connected. No motor included.

| | Craft@Hrs | Unit | Material | Labor | Total |
|---|---|---|---|---|---|
| **Single phase, using 10' RSC, 20' PVC and 6' flex conduit, NEMA Class 1 (indoor) starter, 230 V** | | | | | |
| For motor to 2 HP, 20 amps | E4@8.11 | Ea | 305.00 | 398.00 | 703.00 |
| For 2 to 3 HP motor, 30 amps | E4@8.32 | Ea | 353.00 | 409.00 | 762.00 |
| For 3 to 8 HP motor, 45 amps | E4@8.45 | Ea | 506.00 | 415.00 | 921.00 |
| For 8 to 15 HP motor, 90 amps | E4@15.5 | Ea | 870.00 | 762.00 | 1,632.00 |
| **Three phase wiring using 10' RSC, 20' PVC and 6' flex conduit, NEMA Class 1 (indoor) starter, 480 V** | | | | | |
| For motor to 2 HP, 20 amps | E4@9.90 | Ea | 831.00 | 486.00 | 1,317.00 |
| For 2 to 5 HP motor, 30 amps | E4@10.6 | Ea | 885.00 | 521.00 | 1,406.00 |
| For 7 to 25 HP motor, 45 amps | E4@20.5 | Ea | 1,360.00 | 1,010.00 | 2,370.00 |
| For 25 to 50 HP motor, 90 amps | E4@23.1 | Ea | 2,350.00 | 1,130.00 | 3,480.00 |
| **Single phase, using 10' RSC, 20' PVC and 6' flex conduit, NEMA Type 3R (weatherproof) starter, 230 V** | | | | | |
| For motor to 2 HP, 20 amps | E4@8.91 | Ea | 424.00 | 438.00 | 862.00 |
| For 2 to 5 HP motor, 30 amps | E4@9.15 | Ea | 491.00 | 450.00 | 941.00 |
| For 5 to 7.5 HP motor, 45 amps | E4@9.25 | Ea | 694.00 | 454.00 | 1,148.00 |
| **Three phase using 10' RSC, 20' PVC and 6' flex conduit, NEMA Type 3R (weatherproof) starter, 480 V** | | | | | |
| For motor to 5 HP, 20 amps | E4@10.9 | Ea | 1,210.00 | 536.00 | 1,746.00 |
| For 5 to 10 HP motor, 30 amps | E4@11.7 | Ea | 1,280.00 | 575.00 | 1,855.00 |
| For 10 to 25 HP motor, 45 amps | E4@22.5 | Ea | 1,980.00 | 1,110.00 | 3,090.00 |
| For 25 to 50 HP motor, 90 amps | E4@25.4 | Ea | 3,380.00 | 1,250.00 | 4,630.00 |

**Motor Starters**  Magnetic operated full voltage non-reversing motor controllers with thermal overload relays and enclosure.

| | Craft@Hrs | Unit | Material | Labor | Total |
|---|---|---|---|---|---|
| **Two pole, 1 phase contactors, NEMA Class 1 (indoor), electrically held with holding interlock, one reset** | | | | | |
| Size 00, 1/3 HP, 9 amp | E4@1.47 | Ea | 249.00 | 72.20 | 321.20 |
| Size 0,  1 HP, 18 amp | E4@1.62 | Ea | 288.00 | 79.60 | 367.60 |
| Size 1,  2 HP, 27 amp | E4@1.73 | Ea | 327.00 | 85.00 | 412.00 |
| Size 2,  3 HP, 45 amp | E4@1.73 | Ea | 419.00 | 85.00 | 504.00 |
| Size 3,  7-1/2 HP, 90 amp | E4@2.02 | Ea | 597.00 | 99.20 | 696.20 |
| **Three pole polyphase NEMA Class 1 (indoor) AC magnetic combination starters with fusible disconnect and overload relays but no heaters, 208 to 240 volts** | | | | | |
| Size 0,  3 HP, 30 amps | E4@2.75 | Ea | 787.00 | 135.00 | 922.00 |
| Size 1,  7.5 HP, 30 amps | E4@3.89 | Ea | 828.00 | 191.00 | 1,019.00 |
| Size 1, 10 HP, 60 amps | E4@3.89 | Ea | 828.00 | 191.00 | 1,019.00 |
| Size 2, 10 HP, 60 amps | E4@5.34 | Ea | 1,310.00 | 262.00 | 1,572.00 |
| Size 2, 25 HP, 100 amps | E4@5.34 | Ea | 1,310.00 | 262.00 | 1,572.00 |
| Size 3, 25 HP, 100 amps | E4@6.76 | Ea | 2,170.00 | 332.00 | 2,502.00 |
| Size 3, 50 HP, 200 amps | E4@6.76 | Ea | 2,170.00 | 332.00 | 2,502.00 |
| Size 4, 40 HP, 200 amps | E4@12.0 | Ea | 4,190.00 | 590.00 | 4,780.00 |
| Size 4, 75 HP, 400 amps | E4@12.0 | Ea | 4,190.00 | 590.00 | 4,780.00 |

# 26 Electrical

|  | Craft@Hrs | Unit | Material | Labor | Total |
|---|---|---|---|---|---|
| On and off switches for starter enclosures, any motor starter above | | | | | |
| Switch kit without pilot light | E4@.744 | Ea | 65.60 | 36.60 | 102.20 |
| Switch kit with pilot light | E4@.744 | Ea | 99.00 | 36.60 | 135.60 |
| Accessories for any motor starter | | | | | |
| Add for 440 to 600 volt starters | — | % | 2.0 | — | — |
| Add for NEMA Type 3R (rainproof) enclosure | — | % | 40.0 | 10.0 | — |
| Add for NEMA Type 4 (waterproof) enclosure | — | % | 40.0 | 20.0 | — |
| Add for NEMA Type 12 (dust-tight) enclosure | — | % | 15.0 | 15.0 | — |
| Add for starters in an oversize enclosure | — | % | 21.0 | — | — |
| Deduct for starters with circuit breakers | — | % | -4.0 | — | — |

**Safety Switches** Wall mounted switches with enclosures as noted. No fuses or hubs included.

Heavy-duty (NEMA-1) 600 volt 2, 3 or 4 pole fusible safety switches

| | Craft@Hrs | Unit | Material | Labor | Total |
|---|---|---|---|---|---|
| 30 amp | E4@3.06 | Ea | 291.00 | 150.00 | 441.00 |
| 60 amp | E4@3.89 | Ea | 337.00 | 191.00 | 528.00 |
| 100 amp | E4@4.20 | Ea | 629.00 | 206.00 | 835.00 |
| 200 amp | E4@6.92 | Ea | 908.00 | 340.00 | 1,248.00 |
| 400 amp | E4@11.4 | Ea | 2,230.00 | 560.00 | 2,790.00 |
| 600 amp | E4@14.3 | Ea | 3,320.00 | 703.00 | 4,023.00 |
| 800 amp | E4@19.6 | Ea | 4,630.00 | 963.00 | 5,593.00 |
| 1,200 amp | E4@22.7 | Ea | 7,800.00 | 1,120.00 | 8,920.00 |

Heavy-duty rainproof (NEMA-3R) 600 volt 2, 3 or 4 pole fusible safety switches

| | Craft@Hrs | Unit | Material | Labor | Total |
|---|---|---|---|---|---|
| 30 amp | E4@3.32 | Ea | 471.00 | 163.00 | 634.00 |
| 60 amp | E4@4.61 | Ea | 559.00 | 226.00 | 785.00 |
| 100 amp | E4@4.77 | Ea | 868.00 | 234.00 | 1,102.00 |
| 200 amp | E4@7.51 | Ea | 1,200.00 | 369.00 | 1,569.00 |
| 400 amp | E4@12.4 | Ea | 2,880.00 | 609.00 | 3,489.00 |
| 600 amp | E4@15.6 | Ea | 5,780.00 | 766.00 | 6,546.00 |
| 800 amp | E4@21.4 | Ea | 8,670.00 | 1,050.00 | 9,720.00 |
| 1,200 amp | E4@25.8 | Ea | 9,510.00 | 1,270.00 | 10,780.00 |

Heavy-duty watertight (NEMA-4) 600 volt 3 pole, 4 wire fusible safety switches

| | Craft@Hrs | Unit | Material | Labor | Total |
|---|---|---|---|---|---|
| 30 amp | E4@3.76 | Ea | 1,080.00 | 185.00 | 1,265.00 |
| 60 amp | E4@5.21 | Ea | 1,750.00 | 256.00 | 2,006.00 |
| 100 amp | E4@5.34 | Ea | 2,750.00 | 262.00 | 3,012.00 |
| 200 amp | E4@8.52 | Ea | 3,870.00 | 419.00 | 4,289.00 |
| 400 amp | E4@13.7 | Ea | 7,660.00 | 673.00 | 8,333.00 |
| 600 amp | E4@17.0 | Ea | 10,900.00 | 835.00 | 11,735.00 |

Heavy-duty dust-tight (NEMA-12) 600 volt 3 pole or 4 pole solid neutral fusible safety switches

| | Craft@Hrs | Unit | Material | Labor | Total |
|---|---|---|---|---|---|
| 30 amp | E4@3.32 | Ea | 493.00 | 163.00 | 656.00 |
| 60 amp | E4@4.20 | Ea | 483.00 | 206.00 | 689.00 |
| 100 amp | E4@4.48 | Ea | 773.00 | 220.00 | 993.00 |
| 200 amp | E4@7.51 | Ea | 1,210.00 | 369.00 | 1,579.00 |
| 400 amp | E4@12.7 | Ea | 2,750.00 | 624.00 | 3,374.00 |
| 600 amp | E4@15.9 | Ea | 4,640.00 | 781.00 | 5,421.00 |

General duty (NEMA-1) 240 volt 3 pole, 4 wire non-fusible safety switches

| | Craft@Hrs | Unit | Material | Labor | Total |
|---|---|---|---|---|---|
| 30 amp | E4@2.58 | Ea | 94.60 | 127.00 | 221.60 |
| 100 amp | E4@3.60 | Ea | 285.00 | 177.00 | 462.00 |
| 200 amp | E4@6.50 | Ea | 602.00 | 319.00 | 921.00 |
| 400 amp | E4@10.8 | Ea | 1,570.00 | 531.00 | 2,101.00 |
| 600 amp | E4@12.7 | Ea | 2,930.00 | 624.00 | 3,554.00 |

| | Craft@Hrs | Unit | Material | Labor | Total |
|---|---|---|---|---|---|
| **General duty rainproof (NEMA-3R) 240 volt 3 pole, 4 wire non-fusible safety switches** | | | | | |
| 30 amp | E4@2.88 | Ea | 148.00 | 141.00 | 289.00 |
| 60 amp | E4@3.76 | Ea | 223.00 | 185.00 | 408.00 |
| 100 amp | E4@3.89 | Ea | 411.00 | 191.00 | 602.00 |
| 200 amp | E4@7.23 | Ea | 736.00 | 355.00 | 1,091.00 |
| 400 amp | E4@11.8 | Ea | 1,840.00 | 580.00 | 2,420.00 |
| 600 amp | E4@14.0 | Ea | 3,970.00 | 688.00 | 4,658.00 |
| **Heavy duty watertight (NEMA-4) 240 volt 3 pole, 4 wire fusible safety switches** | | | | | |
| 30 amp | E4@3.19 | Ea | 1,070.00 | 157.00 | 1,227.00 |
| 60 amp | E4@3.76 | Ea | 1,260.00 | 185.00 | 1,445.00 |
| 100 amp | E4@4.33 | Ea | 2,580.00 | 213.00 | 2,793.00 |
| 200 amp | E4@7.80 | Ea | 3,500.00 | 383.00 | 3,883.00 |
| 400 amp | E4@13.0 | Ea | 7,690.00 | 639.00 | 8,329.00 |
| 600 amp | E4@15.4 | Ea | 10,800.00 | 757.00 | 11,557.00 |
| **Heavy duty dust-tight (NEMA-12) 240 volt 3 pole, 4 wire fusible safety switches** | | | | | |
| 30 amp | E4@2.88 | Ea | 320.00 | 141.00 | 461.00 |
| 60 amp | E4@3.76 | Ea | 411.00 | 185.00 | 596.00 |
| 100 amp | E4@4.04 | Ea | 591.00 | 198.00 | 789.00 |
| 200 amp | E4@7.23 | Ea | 791.00 | 355.00 | 1,146.00 |
| 400 amp | E4@11.8 | Ea | 1,990.00 | 580.00 | 2,570.00 |
| 600 amp | E4@14.0 | Ea | 3,320.00 | 688.00 | 4,008.00 |
| **Add for conduit hubs** | | | | | |
| 3/4" to 1-1/2", to 100 amp | — | Ea | 9.74 | — | 9.74 |
| 2", 200 amp | — | Ea | 17.00 | — | 17.00 |
| 2-1/2", 200 amp | — | Ea | 28.70 | — | 28.70 |

**Service Entrance and Distribution Switchboards** MA Class 1 indoor, 600 volt, 3 phase, 4 wire, for 240/480 volt insulated case main breakers. Basic structure is 90" high by 21" deep. Width varies with equipment capacity. Labor cost includes setting and leveling on a prepared pad but excludes the pad cost. Add breaker, instrumentation and accessory costs for a complete installation. Based on Westinghouse Pow-R-Gear. These switchboards are custom designed for each installation. Costs can vary widely. Multiple units ordered at the same time may reduce costs per unit by 25% or more.

| | Craft@Hrs | Unit | Material | Labor | Total |
|---|---|---|---|---|---|
| 600 amp bus | E4@22.9 | Ea | 4,280.00 | 1,130.00 | 5,410.00 |
| 1,000 amp bus | E4@22.9 | Ea | 4,420.00 | 1,130.00 | 5,550.00 |
| 1,200 amp bus | E4@22.9 | Ea | 5,360.00 | 1,130.00 | 6,490.00 |
| 1,600 amp bus | E4@22.9 | Ea | 6,190.00 | 1,130.00 | 7,320.00 |
| 2,000 amp bus | E4@22.9 | Ea | 6,770.00 | 1,130.00 | 7,900.00 |
| 2,500 amp bus | E4@22.9 | Ea | 8,250.00 | 1,130.00 | 9,380.00 |
| 3,000 amp bus | E4@22.9 | Ea | 9,470.00 | 1,130.00 | 10,600.00 |
| 4,000 amp bus | E4@22.9 | Ea | 12,200.00 | 1,130.00 | 13,330.00 |

240/480 volt draw-out main breakers, 100K amp interrupt capacity. Includes connecting and testing breakers only. Based on Pow-R-Trip

| | Craft@Hrs | Unit | Material | Labor | Total |
|---|---|---|---|---|---|
| 100 to 800 amp, manual operation | E4@10.2 | Ea | 12,500.00 | 501.00 | 13,001.00 |
| 1,000 to 1,600 amp, manual operation | E4@17.5 | Ea | 26,100.00 | 860.00 | 26,960.00 |
| 2,000 amp, manual operation | E4@23.2 | Ea | 31,900.00 | 1,140.00 | 33,040.00 |
| 2,500 amp, manual operation | E4@23.2 | Ea | 54,400.00 | 1,140.00 | 55,540.00 |
| 3,000 amp, manual operation | E4@23.2 | Ea | 60,600.00 | 1,140.00 | 61,740.00 |
| 4,000 amp, manual operation | E4@23.2 | Ea | 73,000.00 | 1,140.00 | 74,140.00 |
| 100 to 800 amp electric operation | E4@10.2 | Ea | 16,800.00 | 501.00 | 17,301.00 |
| 1,000 to 1,600 amp electric operation | E4@17.5 | Ea | 28,600.00 | 860.00 | 29,460.00 |
| 2,000 amp electric operation | E4@23.2 | Ea | 37,900.00 | 1,140.00 | 39,040.00 |
| 2,500 amp electric operation | E4@23.2 | Ea | 64,900.00 | 1,140.00 | 66,040.00 |
| 3,000 amp electric operation | E4@23.2 | Ea | 70,700.00 | 1,140.00 | 71,840.00 |

# 26 Electrical

| | Craft@Hrs | Unit | Material | Labor | Total |
|---|---|---|---|---|---|
| Additional space for future draw-out breakers | | | | | |
| 100 to 800 amp | — | Ea | 3,360.00 | — | 3,360.00 |
| 1,000 to 1,600 amp | — | Ea | 4,910.00 | — | 4,910.00 |
| 2,000 amp | — | Ea | 5,540.00 | — | 5,540.00 |
| 2,500 amp | — | Ea | 5,910.00 | — | 5,910.00 |
| 3,000 amp | — | Ea | 10,800.00 | — | 10,800.00 |

**Service Entrance and Distribution Switchboards, Weatherproof**

600 volt, 3 phase, 4 wire, for 480 volt insulated case main breakers. Basic structure is 90" high by 21" deep. Width varies with equipment capacity. Labor cost includes setting and leveling on a prepared pad but excludes the pad cost. Add breaker, instrumentation and accessory costs from below for a complete installation. These switchboards are custom designed for each installation. Costs can vary widely. Multiple units ordered at the same time can reduce costs per unit by 25% or more.

| | Craft@Hrs | Unit | Material | Labor | Total |
|---|---|---|---|---|---|
| 600 amp bus | E4@23.2 | Ea | 5,050.00 | 1,140.00 | 6,190.00 |
| 1,000 amp bus | E4@23.2 | Ea | 5,920.00 | 1,140.00 | 7,060.00 |
| 1,200 amp bus | E4@23.2 | Ea | 7,600.00 | 1,140.00 | 8,740.00 |
| 1,600 amp bus | E4@23.2 | Ea | 7,300.00 | 1,140.00 | 8,440.00 |
| 2,000 amp bus | E4@23.2 | Ea | 8,060.00 | 1,140.00 | 9,200.00 |
| 2,500 amp bus | E4@23.2 | Ea | 9,700.00 | 1,140.00 | 10,840.00 |
| 3,000 amp bus | E4@23.2 | Ea | 9,880.00 | 1,140.00 | 11,020.00 |
| 4,000 amp bus | E4@23.2 | Ea | 14,400.00 | 1,140.00 | 15,540.00 |

480 volt manual operated draw-out breakers. By amp interrupt capacity (AIC) as noted. Includes connecting and testing factory installed breakers only.

| | | | | | |
|---|---|---|---|---|---|
| 50 to 800 amp, 30K AIC | E4@10.2 | Ea | 17,100.00 | 501.00 | 17,601.00 |
| 50 to 1,600 amp, 50K AIC | E4@17.5 | Ea | 35,400.00 | 860.00 | 36,260.00 |
| 2,000 amp | E4@23.2 | Ea | 45,600.00 | 1,140.00 | 46,740.00 |
| 1,200 to 3,200 amp | E4@23.2 | Ea | 80,200.00 | 1,140.00 | 81,340.00 |
| 4,000 amp | E4@23.2 | Ea | 128,000.00 | 1,140.00 | 129,140.00 |

480 volt manual operated draw-out breakers. 200K amp interrupt capacity (AIC) as noted. Includes connecting and testing factory installed breakers only.

| | | | | | |
|---|---|---|---|---|---|
| 50 to 800 amp | E4@10.2 | Ea | 26,200.00 | 501.00 | 26,701.00 |
| 50 to 1,600 amp | E4@17.5 | Ea | 46,700.00 | 860.00 | 47,560.00 |
| 1,200 to 3,200 amp | E4@23.2 | Ea | 119,000.00 | 1,140.00 | 120,140.00 |
| 4,000 amp | E4@23.2 | Ea | 194,000.00 | 1,140.00 | 195,140.00 |

480 volt electrically operated draw-out breakers. By amp interrupt capacity (AIC) as noted. Includes connecting and testing factory installed breakers only.

| | | | | | |
|---|---|---|---|---|---|
| 50 to 800 amp, 30K AIC | E4@10.2 | Ea | 23,200.00 | 501.00 | 23,701.00 |
| 50 to 1,600 amp, 50K AIC | E4@17.5 | Ea | 48,100.00 | 860.00 | 48,960.00 |
| 2,000 amp | E4@23.2 | Ea | 61,400.00 | 1,140.00 | 62,540.00 |
| 1,200 to 3,200 amp | E4@23.2 | Ea | 92,300.00 | 1,140.00 | 93,440.00 |
| 4,000 amp | E4@23.2 | Ea | 143,000.00 | 1,140.00 | 144,140.00 |

480 volt electrically operated draw-out breakers. 200K amp interrupt capacity (AIC) as noted. Includes connecting and testing factory installed breakers only.

| | | | | | |
|---|---|---|---|---|---|
| 50 to 800 amp | E4@10.2 | Ea | 31,900.00 | 501.00 | 32,401.00 |
| 50 to 1,600 amp | E4@17.5 | Ea | 64,300.00 | 860.00 | 65,160.00 |
| 1,200 to 3,200 amp | E4@23.2 | Ea | 132,000.00 | 1,140.00 | 133,140.00 |
| 4,000 amp | E4@23.2 | Ea | 194,000.00 | 1,140.00 | 195,140.00 |
| Additional space for future draw-out breakers | | | | | |
| 50 to 800 amp | — | Ea | 3,960.00 | — | 3,960.00 |
| 50 to 1,600 amp | — | Ea | 6,420.00 | — | 6,420.00 |
| 2,000 amp | — | Ea | 8,770.00 | — | 8,770.00 |
| 1,200 to 3,200 amp | — | Ea | 13,600.00 | — | 13,600.00 |
| 4,000 amp | — | Ea | 25,100.00 | — | 25,100.00 |

# 26 Electrical

| | Craft@Hrs | Unit | Material | Labor | Total |
|---|---|---|---|---|---|

**Switchboard Instrumentation and Accessories** Add these costs to the cost of the basic enclosure and breakers to find the complete switchboard cost. These figures are for factory-installed instrumentation and accessories and include the cost of connecting and testing only.

Bus duct connection, 3 phase, 4 wire

| | Craft@Hrs | Unit | Material | Labor | Total |
|---|---|---|---|---|---|
| 225 amp | CE@.516 | Ea | 764.00 | 30.10 | 794.10 |
| 400 amp | CE@.516 | Ea | 890.00 | 30.10 | 920.10 |
| 600 amp | CE@.516 | Ea | 1,080.00 | 30.10 | 1,110.10 |
| 800 amp | CE@.516 | Ea | 1,710.00 | 30.10 | 1,740.10 |
| 1,000 amp | CE@.516 | Ea | 1,970.00 | 30.10 | 2,000.10 |
| 1,350 amp | CE@1.01 | Ea | 2,550.00 | 59.00 | 2,609.00 |
| 1,600 amp | CE@1.01 | Ea | 2,960.00 | 59.00 | 3,019.00 |
| 2,000 amp | CE@1.01 | Ea | 3,610.00 | 59.00 | 3,669.00 |
| 2,500 amp | CE@1.01 | Ea | 4,300.00 | 59.00 | 4,359.00 |
| 3,000 amp | CE@1.01 | Ea | 4,900.00 | 59.00 | 4,959.00 |
| 4,000 amp | CE@1.01 | Ea | 5,420.00 | 59.00 | 5,479.00 |
| Recording meters | | | | | |
| Voltmeter | CE@1.01 | Ea | 10,900.00 | 59.00 | 10,959.00 |
| Ammeter | CE@1.01 | Ea | 11,100.00 | 59.00 | 11,159.00 |
| Wattmeter | CE@1.01 | Ea | 13,800.00 | 59.00 | 13,859.00 |
| Varmeter | CE@1.01 | Ea | 14,700.00 | 59.00 | 14,759.00 |
| Power factor meter | CE@1.01 | Ea | 14,500.00 | 59.00 | 14,559.00 |
| Frequency meter | CE@1.01 | Ea | 14,500.00 | 59.00 | 14,559.00 |
| Watt-hour meter, 3 element | CE@1.01 | Ea | 18,800.00 | 59.00 | 18,859.00 |
| Non-recording meters | | | | | |
| Voltmeter | CE@.516 | Ea | 1,830.00 | 30.10 | 1,860.10 |
| Ammeter for incoming line | CE@.516 | Ea | 1,880.00 | 30.10 | 1,910.10 |
| Ammeter for feeder circuits | CE@.516 | Ea | 781.00 | 30.10 | 811.10 |
| Wattmeter | CE@1.01 | Ea | 6,070.00 | 59.00 | 6,129.00 |
| Varmeter | CE@1.01 | Ea | 5,490.00 | 59.00 | 5,549.00 |
| Power factor meter | CE@1.01 | Ea | 4,430.00 | 59.00 | 4,489.00 |
| Frequency meter | CE@1.01 | Ea | 5,240.00 | 59.00 | 5,299.00 |
| Watt-hour meter, 3 element | CE@1.01 | Ea | 5,490.00 | 59.00 | 5,549.00 |
| Instrument phase select switch | CE@.516 | Ea | 947.00 | 30.10 | 977.10 |
| Adjustable short time pickup and delay | CE@.516 | Ea | 1,430.00 | 30.10 | 1,460.10 |
| Ground fault trip pickup and delay, 4 wire | CE@.525 | Ea | 1,520.00 | 30.60 | 1,550.60 |
| Ground fault test panel | CE@.516 | Ea | 1,640.00 | 30.10 | 1,670.10 |
| Shunt trip for breakers | CE@.516 | Ea | 1,100.00 | 30.10 | 1,130.10 |
| Key interlock for breakers | CE@.516 | Ea | 1,100.00 | 30.10 | 1,130.10 |
| Breaker lifting and transport truck | — | Ea | 7,220.00 | — | 7,220.00 |
| Breaker lifting device | — | Ea | 6,290.00 | — | 6,290.00 |
| Current transformers, by primary capacity | | | | | |
| 800 amps and less | CE@1.01 | Ea | 1,360.00 | 59.00 | 1,419.00 |
| 1,000 to 1,500 amps | CE@1.01 | Ea | 2,010.00 | 59.00 | 2,069.00 |
| 2,000 to 6,000 amps | CE@1.01 | Ea | 2,380.00 | 59.00 | 2,439.00 |
| Current transformer mount | — | Ea | 644.00 | — | 644.00 |
| Potential transformer | CE@1.01 | Ea | 1,770.00 | 59.00 | 1,829.00 |
| Potential transformer mount | — | Ea | 405.00 | — | 405.00 |

| | Craft@Hrs | Unit | Material | Labor | Total |
|---|---|---|---|---|---|

## Feeder Section Breaker Panels

Flush or surface mounted 277/480 volt, 3 phase, 4 wire, FA frame panels.

| | Craft@Hrs | Unit | Material | Labor | Total |
|---|---|---|---|---|---|
| 225 amp, with 22K amp interrupt capacity breaker | E4@6.76 | Ea | 1,810.00 | 332.00 | 2,142.00 |
| 225 amp, with 25K amp interrupt capacity breaker | E4@6.76 | Ea | 3,360.00 | 332.00 | 3,692.00 |
| 400 amp, with 30K amp interrupt capacity breaker | E4@10.8 | Ea | 4,040.00 | 531.00 | 4,571.00 |
| 400 amp, with 35K amp interrupt capacity breaker | E4@10.8 | Ea | 4,840.00 | 531.00 | 5,371.00 |
| 225 amp, with main lugs only | E4@6.76 | Ea | 612.00 | 332.00 | 944.00 |
| 400 amp, with main lugs only | E4@10.8 | Ea | 861.00 | 531.00 | 1,392.00 |
| 600 amp, with main lugs only | E4@19.3 | Ea | 1,150.00 | 948.00 | 2,098.00 |

## Transformers  Indoor dry-type transformer, floor mounted, including connections.

Single phase light and power circuit transformer, 240/480 volt primary, 120/240 secondary, no taps for voltage change

| | Craft@Hrs | Unit | Material | Labor | Total |
|---|---|---|---|---|---|
| 1 KVA | E4@3.76 | Ea | 290.00 | 185.00 | 475.00 |
| 2 KVA | E4@4.20 | Ea | 432.00 | 206.00 | 638.00 |
| 3 KVA | E4@4.33 | Ea | 551.00 | 213.00 | 764.00 |
| 5 KVA | E4@5.47 | Ea | 716.00 | 269.00 | 985.00 |
| 7.5 KVA | E4@8.34 | Ea | 888.00 | 410.00 | 1,298.00 |
| 10 KVA | E4@9.38 | Ea | 1,100.00 | 461.00 | 1,561.00 |
| 15 KVA | E4@12.6 | Ea | 1,280.00 | 619.00 | 1,899.00 |
| 25 KVA | E4@12.6 | Ea | 1,720.00 | 619.00 | 2,339.00 |

Three phase light and power circuit transformers, 480 volt primary, 120/208 volt secondary, with 6 taps for voltage change

| | Craft@Hrs | Unit | Material | Labor | Total |
|---|---|---|---|---|---|
| 30 KVA | E4@10.5 | Ea | 3,280.00 | 516.00 | 3,796.00 |
| 45 KVA | E4@12.6 | Ea | 3,940.00 | 619.00 | 4,559.00 |
| 50 KVA | E4@12.6 | Ea | 5,230.00 | 619.00 | 5,849.00 |
| 75 KVA | E4@14.8 | Ea | 5,940.00 | 727.00 | 6,667.00 |
| 112.5 KVA | E4@15.7 | Ea | 7,890.00 | 771.00 | 8,661.00 |
| 150 KVA | E4@16.7 | Ea | 10,300.00 | 820.00 | 11,120.00 |
| 225 KVA | E4@18.9 | Ea | 13,900.00 | 929.00 | 14,829.00 |
| 300 KVA | E4@21.2 | Ea | 17,800.00 | 1,040.00 | 18,840.00 |

Three phase 5,000 volt load center transformers, 4,160 volt primary, 480/277 volt secondary, with voltage change taps, insulation class H, 80 degree system

| | Craft@Hrs | Unit | Material | Labor | Total |
|---|---|---|---|---|---|
| 225 KVA | E4@29.5 | Ea | 16,700.00 | 1,450.00 | 18,150.00 |
| 300 KVA | E4@33.7 | Ea | 19,900.00 | 1,660.00 | 21,560.00 |
| 500 KVA | E4@42.5 | Ea | 30,100.00 | 2,090.00 | 32,190.00 |
| 750 KVA | E4@53.1 | Ea | 39,400.00 | 2,610.00 | 42,010.00 |
| 1,000 KVA | E4@63.2 | Ea | 50,400.00 | 3,110.00 | 53,510.00 |

## Panelboards  Surface mounted lighting and appliance distribution panelboards for bolt-on circuit breakers. No breakers included.

3 wire 120/240 volt, 250 amp, single phase AC panelboards with main lugs only for bolt-on breakers

| | Craft@Hrs | Unit | Material | Labor | Total |
|---|---|---|---|---|---|
| 18 circuits | E4@10.7 | Ea | 282.00 | 526.00 | 808.00 |
| 30 circuits | E4@15.3 | Ea | 287.00 | 752.00 | 1,039.00 |
| 42 circuits | E4@20.9 | Ea | 379.00 | 1,030.00 | 1,409.00 |
| Deduct for 3 wire 120/240 panelboards | | | | | |
| With main lugs only | — | % | -10.0 | -15.0 | — |

4 wire 120/208 volt, 250 amp, 3 phase AC panelboards for bolt-on breakers, with main circuit breaker only

| | Craft@Hrs | Unit | Material | Labor | Total |
|---|---|---|---|---|---|
| 18 circuits | E4@12.9 | Ea | 229.00 | 634.00 | 863.00 |
| 30 circuits | E4@21.5 | Ea | 325.00 | 1,060.00 | 1,385.00 |
| 42 circuits | E4@30.1 | Ea | 422.00 | 1,480.00 | 1,902.00 |
| Deduct for 3 wire panelboards | | | | | |
| With circuit breaker mains only | — | % | -10.0 | -15.0 | — |

| | Craft@Hrs | Unit | Material | Labor | Total |
|---|---|---|---|---|---|

**Combination Service Entrance Device**  Meter socket, service disconnect and loadcenter.
Surface mount NEMA Type 3R enclosure with ring type utility meter socket, service disconnect, and integral HOM loadcenter. Single phase 3-wire. 120/240 Volt AC, 22,000 amp short circuit current rating. Overhead or underground service feed. Add the cost of distribution breakers and conduit hubs.

| | Craft@Hrs | Unit | Material | Labor | Total |
|---|---|---|---|---|---|
| 100 amp, 16 spaces, 24 circuits | E4@1.20 | Ea | 54.70 | 59.00 | 113.70 |
| 125 amp, 16 spaces, 24 circuits | E4@1.20 | Ea | 98.60 | 59.00 | 157.60 |
| 200 amp, 20 spaces, 40 circuits | E4@1.20 | Ea | 135.00 | 59.00 | 194.00 |
| 200 amp, 20 spaces, 40 circuits | E4@2.60 | Ea | 124.00 | 128.00 | 252.00 |
| 200 amp, 30 spaces, 40 circuits | E4@2.60 | Ea | 121.00 | 128.00 | 249.00 |
| 200 amp, 40 spaces, 40 circuits | E4@2.60 | Ea | 140.00 | 128.00 | 268.00 |
| 320 amp, 30 spaces, 40 circuits, no bypass | E4@3.10 | Ea | 567.00 | 152.00 | 719.00 |
| 320 amp, 30 spaces, 40 circuits, test block bypass | E4@3.10 | Ea | 616.00 | 152.00 | **768.00** |

Surface mount NEMA Type 3R class 4120 enclosure with ring type utility meter socket, dual main breakers and HO load center. Single phase, 3-wire, 120/240 Volts AC, 25,000 amp short circuit current rating. Underground service feed. With 100 percent branch neutrals, horn by-pass and 5th jaw factory installed. Add the cost of distribution breakers and conduit hubs.

| | Craft@Hrs | Unit | Material | Labor | Total |
|---|---|---|---|---|---|
| 400 amp, 30 spaces, 40 circuits, test block bypass | E4@3.50 | Ea | 670.00 | 172.00 | 842.00 |
| 400 amp, 30 spaces, 40 circuits, no test block | E4@3.50 | Ea | 680.00 | 172.00 | 852.00 |

Surface or semi-flush mount NEMA Type 1 (indoor) enclosure with multiple meter positions, main breaker and loadcenter. Line and load at top or bottom. 1 phase, 3-wire 120/240 Volt AC. 65,000 amp short circuit rating. Add the cost of distribution breakers and conduit hubs.

| | Craft@Hrs | Unit | Material | Labor | Total |
|---|---|---|---|---|---|
| 125 amp, 200 amp bus, 2 meter positions | E4@1.20 | Ea | 191.00 | 59.00 | 250.00 |
| 125 amp, 300 amp bus, 3 meter positions | E4@1.20 | Ea | 311.00 | 59.00 | 370.00 |
| 125 amp, 400 amp bus, 4 meter position | E4@1.20 | Ea | 432.00 | 59.00 | 491.00 |
| 125 amp, 400 amp bus, 6 meter positions | E4@1.20 | Ea | 608.00 | 59.00 | 667.00 |

**Loadcenters**  For plug-in branch breakers. Surface or recess mount. Add the cost of distribution breakers and conduit hubs.
Indoor (NEMA 1) single phase 3-wire 120/240 volt loadcenter. Class 1170. 22,000 RMS amps short circuit rating. With factory installed main breaker only.

| | Craft@Hrs | Unit | Material | Labor | Total |
|---|---|---|---|---|---|
| 100 amps,  8 spaces, 16 circuits | E4@1.20 | Ea | 54.30 | 59.00 | 113.30 |
| 100 amps, 12 spaces, 12 circuits | E4@1.20 | Ea | 61.50 | 59.00 | 120.50 |
| 100 amps, 20 spaces, 20 circuits | E4@1.20 | Ea | 54.00 | 59.00 | 113.00 |
| 100 amps, 24 spaces, 24 circuits | E4@1.20 | Ea | 98.30 | 59.00 | 157.30 |
| 100 amps, 30 spaces, 30 circuits | E4@1.20 | Ea | 100.00 | 59.00 | 159.00 |
| 125 amps, 12 spaces, 24 circuits | E4@1.52 | Ea | 74.00 | 74.70 | 148.70 |
| 125 amps, 24 spaces, 24 circuits | E4@1.52 | Ea | 102.00 | 74.70 | 176.70 |
| 150 amps, 30 spaces, 30 circuits | E4@2.30 | Ea | 112.00 | 113.00 | 225.00 |
| 200 amps, 20 spaces, 40 circuits | E4@2.60 | Ea | 81.20 | 128.00 | 209.20 |
| 200 amps, 30 spaces, 30 circuits | E4@2.60 | Ea | 107.00 | 128.00 | 235.00 |
| 200 amps, 40 spaces, 40 circuits | E4@2.60 | Ea | 140.00 | 128.00 | 268.00 |

Outdoor (NEMA 4) single phase 3-wire 120/240 volt loadcenter, Class 1170, 22,000 amps short circuit rating. With factory installed main breaker only.

| | Craft@Hrs | Unit | Material | Labor | Total |
|---|---|---|---|---|---|
| 100 amps,  8 spaces, 16 circuits | E4@1.20 | Ea | 79.90 | 59.00 | 138.90 |
| 100 amps, 12 spaces, 12 circuits | E4@1.20 | Ea | 72.30 | 59.00 | 131.30 |
| 100 amps, 20 spaces, 20 circuits | E4@1.20 | Ea | 98.30 | 59.00 | 157.30 |
| 125 amps, 24 spaces, 24 circuits | E4@1.52 | Ea | 153.00 | 74.70 | 227.70 |
| 150 amps, 30 spaces, 30 circuits | E4@2.30 | Ea | 180.00 | 113.00 | 293.00 |
| 200 amps, 20 spaces, 40 circuits | E4@2.60 | Ea | 157.00 | 128.00 | 285.00 |
| 200 amps, 30 spaces, 30 circuits | E4@2.60 | Ea | 186.00 | 128.00 | 314.00 |
| 200 amps, 40 spaces, 40 circuits | E4@2.60 | Ea | 230.00 | 128.00 | 358.00 |

# 26 Electrical

| | Craft@Hrs | Unit | Material | Labor | Total |
|---|---|---|---|---|---|
| **Indoor (NEMA 1) three phase 4 wire 120/208 volt loadcenter with factory-installed main circuit breaker only** | | | | | |
| 100 amps, 12 circuits | E4@1.21 | Ea | 148.00 | 59.40 | 207.40 |
| 100 amps, 18 circuits | E4@1.60 | Ea | 200.00 | 78.60 | 278.60 |
| 125 amps, 30 circuits | E4@1.91 | Ea | 397.00 | 93.80 | 490.80 |
| 150 amps, 24 circuits | E4@2.18 | Ea | 383.00 | 107.00 | 490.00 |
| 150 amps, 42 circuits | E4@2.82 | Ea | 456.00 | 139.00 | 595.00 |
| 200 amps, 42 circuits | E4@3.63 | Ea | 456.00 | 178.00 | 634.00 |
| 225 amps, 42 circuits | E4@3.78 | Ea | 479.00 | 186.00 | 665.00 |
| 400 amps, 24 circuits | E4@3.91 | Ea | 965.00 | 192.00 | 1,157.00 |
| 400 amps, 42 circuits | E4@4.51 | Ea | 1,110.00 | 222.00 | 1,332.00 |
| **Three phase 4 wire 120/208 volt loadcenters with main lugs only.** | | | | | |
| 125 amps, 12 circuits | E4@1.19 | Ea | 88.00 | 58.50 | 146.50 |
| 150 amps, 18 circuits | E4@1.47 | Ea | 132.00 | 72.20 | 204.20 |
| 200 amps, 30 circuits | E4@1.90 | Ea | 178.00 | 93.30 | 271.30 |
| 200 amps, 40 circuits | E4@2.18 | Ea | 238.00 | 107.00 | 345.00 |
| 225 amps, 42 circuits | E4@2.75 | Ea | 251.00 | 135.00 | 386.00 |
| 400 amps, 24 circuits | E4@3.63 | Ea | 458.00 | 178.00 | 636.00 |
| 400 amps, 42 circuits | E4@3.78 | Ea | 509.00 | 186.00 | 695.00 |
| 600 amps, 24 circuits | E4@3.91 | Ea | 557.00 | 192.00 | 749.00 |
| 600 amps, 42 circuits | E4@6.53 | Ea | 683.00 | 321.00 | 1,004.00 |
| **Circuit Breakers** No enclosures included. 10,000 amp interrupt capacity except as noted. | | | | | |
| Plug-in molded case 120/240 volt, 100 amp frame circuit breakers, 1" or 1/2" module | | | | | |
| 15 to 50 amps, single pole | CE@.270 | Ea | 8.42 | 15.80 | 24.22 |
| 15 to 60 amps, two pole | CE@.374 | Ea | 8.71 | 21.80 | 30.51 |
| 70 amps, two pole | CE@.563 | Ea | 25.80 | 32.90 | 58.70 |
| 80 to 100 amps, two pole | CE@.680 | Ea | 34.20 | 39.70 | 73.90 |
| Plug-in molded case 240 volt, 100 amp frame circuit breakers | | | | | |
| 15 thru 60 amps, three pole | CE@.863 | Ea | 29.80 | 50.40 | 80.20 |
| 60 thru 100 amps, two pole | CE@.577 | Ea | 25.80 | 33.70 | 59.50 |
| 60 thru 100 amps, three pole | CE@.897 | Ea | 55.00 | 52.40 | 107.40 |
| Plug-in molded case 480 volt, 100 amp frame circuit breakers | | | | | |
| 70 thru 100 amps, two pole | CE@1.22 | Ea | 332.00 | 71.20 | 403.20 |
| 70 thru 100 amps, three pole | CE@1.71 | Ea | 418.00 | 99.80 | 517.80 |
| Plug-in molded case 600 volt, 100 amp frame circuit breakers | | | | | |
| 15 thru 60 amps, two pole | CE@.577 | Ea | 278.00 | 33.70 | 311.70 |
| 15 thru 60 amps, three pole | CE@.897 | Ea | 357.00 | 52.40 | 409.40 |
| 70 thru 100 amps, two pole | CE@1.22 | Ea | 372.00 | 71.20 | 443.20 |
| 70 thru 100 amps, three pole | CE@1.71 | Ea | 468.00 | 99.80 | 567.80 |
| Plug-in molded case 250/600 volt, 225 amp frame circuit breakers | | | | | |
| 125 thru 225 amps, two pole | CE@2.06 | Ea | 868.00 | 120.00 | 988.00 |
| 125 thru 225 amps, three pole | CE@2.57 | Ea | 1,100.00 | 150.00 | 1,250.00 |
| Plug-in molded case 250/600 volt, 400 amp frame circuit breakers | | | | | |
| 250 thru 400 amps, two pole | CE@2.88 | Ea | 1,470.00 | 168.00 | 1,638.00 |
| 250 thru 400 amps, three pole | CE@4.33 | Ea | 1,720.00 | 253.00 | 1,973.00 |
| Plug-in molded case 250/600 volt, 800 amp frame circuit breakers | | | | | |
| 400 thru 600 amps, three pole | CE@6.04 | Ea | 2,180.00 | 353.00 | 2,533.00 |
| 700 thru 800 amps, two pole | CE@4.73 | Ea | 3,030.00 | 276.00 | 3,306.00 |
| 700 thru 800 amps, three pole | CE@6.71 | Ea | 3,910.00 | 392.00 | 4,302.00 |
| Plug-in molded case 600 volt, 1,200 amp frame circuit breakers | | | | | |
| 600 thru 1,200 amps, two pole | CE@9.60 | Ea | 5,010.00 | 560.00 | 5,570.00 |
| 600 thru 1,200 amps, three pole | CE@11.9 | Ea | 6,420.00 | 695.00 | 7,115.00 |

# 26 Electrical

| | Craft@Hrs | Unit | Material | Labor | Total |
|---|---|---|---|---|---|
| Space only for factory assembled panels | | | | | |
| Single pole space | — | Ea | 22.60 | — | 22.60 |
| Two pole space | — | Ea | 20.70 | — | 20.70 |
| Three pole space | — | Ea | 27.90 | — | 27.90 |
| Bolt-on molded case 120/240 volt circuit breakers | | | | | |
| 15 thru 60 amps, one pole | CE@.320 | Ea | 10.60 | 18.70 | 29.30 |
| 15 thru 60 amps, two pole | CE@.484 | Ea | 22.70 | 28.30 | 51.00 |
| 15 thru 60 amps, three pole | CE@.638 | Ea | 120.00 | 37.20 | 157.20 |
| 70 thru 100 amps, two pole | CE@.712 | Ea | 126.00 | 41.60 | 167.60 |
| 70 thru 100 amps, three pole | CE@.946 | Ea | 147.00 | 55.20 | 202.20 |
| Bolt-on molded case 480 volt circuit breakers, 65,000 amp interrupt capacity | | | | | |
| 15 thru 60 amps, two pole | CE@.484 | Ea | 129.00 | 28.30 | 157.30 |
| 15 thru 60 amps, three pole | CE@.712 | Ea | 225.00 | 41.60 | 266.60 |
| 70 thru 100 amps, three pole | CE@.804 | Ea | 342.00 | 46.90 | 388.90 |

**Circuit Breaker Panels** 4 wire 3 phase 240 volts, flush-mounted, with main breaker, assembled

| | Craft@Hrs | Unit | Material | Labor | Total |
|---|---|---|---|---|---|
| 100 amp service, 12 circuits | E4@4.60 | Ea | 502.00 | 226.00 | 728.00 |
| 100 amp service, 18 circuits | E4@6.90 | Ea | 533.00 | 339.00 | 872.00 |
| 100 amp service, 24 circuits | E4@8.10 | Ea | 577.00 | 398.00 | 975.00 |
| 100 amp service, 30 circuits | E4@8.20 | Ea | 621.00 | 403.00 | 1,024.00 |
| 100 amp service, 36 circuits | E4@8.30 | Ea | 663.00 | 408.00 | 1,071.00 |
| 100 amp service, 42 circuits | E4@8.40 | Ea | 708.00 | 413.00 | 1,121.00 |
| 225 amp service, 12 circuits | E4@8.50 | Ea | 1,100.00 | 418.00 | 1,518.00 |
| 225 amp service, 18 circuits | E4@8.60 | Ea | 1,120.00 | 423.00 | 1,543.00 |
| 225 amp service, 24 circuits | E4@8.70 | Ea | 1,160.00 | 427.00 | 1,587.00 |
| 225 amp service, 30 circuits | E4@10.0 | Ea | 1,210.00 | 491.00 | 1,701.00 |
| 225 amp service, 36 circuits | E4@12.8 | Ea | 1,250.00 | 629.00 | 1,879.00 |
| 225 amp service, 42 circuits | E4@15.1 | Ea | 1,290.00 | 742.00 | 2,032.00 |

**Fuses** No enclosures included.

| | Craft@Hrs | Unit | Material | Labor | Total |
|---|---|---|---|---|---|
| 250 volt cartridge fuses, one time non-renewable | | | | | |
| To 30 amps | CE@.077 | Ea | 1.11 | 4.49 | 5.60 |
| 35 to 60 amps | CE@.077 | Ea | 2.58 | 4.49 | 7.07 |
| 70 to 100 amps | CE@.077 | Ea | 7.78 | 4.49 | 12.27 |
| 110 to 200 amps | CE@.077 | Ea | 13.80 | 4.49 | 18.29 |
| 225 to 400 amps | CE@.077 | Ea | 29.10 | 4.49 | 33.59 |
| 450 to 600 amps | CE@.077 | Ea | 66.50 | 4.49 | 70.99 |
| 600 volt cartridge fuses, one time non-renewable | | | | | |
| To 30 amps | CE@.077 | Ea | 10.70 | 4.49 | 15.19 |
| 35 to 60 amps | CE@.077 | Ea | 18.30 | 4.49 | 22.79 |
| 70 to 100 amps | CE@.077 | Ea | 48.60 | 4.49 | 53.09 |
| 110 to 200 amps | CE@.077 | Ea | 72.20 | 4.49 | 76.69 |
| 225 to 400 amps | CE@.087 | Ea | 145.00 | 5.08 | 150.08 |
| 450 to 600 amps | CE@.124 | Ea | 312.00 | 7.24 | 319.24 |
| 250 volt Class H renewable fuses | | | | | |
| To 10 amps | CE@.077 | Ea | 11.60 | 4.49 | 16.09 |
| 12 to 30 amps | CE@.077 | Ea | 12.50 | 4.49 | 16.99 |
| 35 to 60 amps | CE@.077 | Ea | 13.20 | 4.49 | 17.69 |
| 70 to 100 amps | CE@.077 | Ea | 46.10 | 4.49 | 50.59 |
| 110 to 200 amps | CE@.077 | Ea | 106.00 | 4.49 | 110.49 |
| 225 to 400 amps | CE@.080 | Ea | 216.00 | 4.67 | 220.67 |
| 450 to 600 amps | CE@.122 | Ea | 311.00 | 7.12 | 318.12 |

|  | Craft@Hrs | Unit | Material | Labor | Total |
|---|---|---|---|---|---|
| **600 volt Class H renewable fuses** | | | | | |
| To 10 amps | CE@.077 | Ea | 30.90 | 4.49 | 35.39 |
| 12 to 30 amps | CE@.077 | Ea | 24.60 | 4.49 | 29.09 |
| 35 to 60 amps | CE@.077 | Ea | 33.10 | 4.49 | 37.59 |
| 70 to 100 amps | CE@.077 | Ea | 92.10 | 4.49 | 96.59 |
| 110 to 200 amps | CE@.077 | Ea | 181.00 | 4.49 | 185.49 |
| 225 to 400 amps | CE@.153 | Ea | 417.00 | 8.93 | 425.93 |
| 450 to 600 amps | CE@.223 | Ea | 501.00 | 13.00 | 514.00 |
| **250 volt Class RK5 non-renewable cartridge fuses** | | | | | |
| 10 to 30 amps | CE@.077 | Ea | 2.42 | 4.49 | 6.91 |
| 35 to 60 amps | CE@.077 | Ea | 6.07 | 4.49 | 10.56 |
| 70 to 100 amps | CE@.077 | Ea | 8.46 | 4.49 | 12.95 |
| 110 to 200 amps | CE@.077 | Ea | 15.40 | 4.49 | 19.89 |
| 225 to 400 amps | CE@.230 | Ea | 61.10 | 13.40 | 74.50 |
| 450 to 600 amps | CE@.297 | Ea | 86.10 | 17.30 | 103.40 |
| 650 to 1,200 amps | CE@.372 | Ea | 107.00 | 21.70 | 128.70 |
| 1,400 to 1,600 amps | CE@.804 | Ea | 144.00 | 46.90 | 190.90 |
| 1,800 to 2,000 amps | CE@1.00 | Ea | 184.00 | 58.40 | 242.40 |
| **600 volt Class RK5 non-renewable cartridge fuses** | | | | | |
| 10 to 30 amps | CE@.077 | Ea | 6.31 | 4.49 | 10.80 |
| 35 to 60 amps | CE@.077 | Ea | 10.90 | 4.49 | 15.39 |
| 70 to 100 amps | CE@.077 | Ea | 22.30 | 4.49 | 26.79 |
| 110 to 200 amps | CE@.129 | Ea | 44.50 | 7.53 | 52.03 |
| 225 to 400 amps | CE@.257 | Ea | 88.70 | 15.00 | 103.70 |
| 450 to 600 amps | CE@.354 | Ea | 128.00 | 20.70 | 148.70 |
| 650 to 1,200 amps | CE@.372 | Ea | 203.00 | 21.70 | 224.70 |
| **600 volt Class L fuses, bolt-on** | | | | | |
| 600 to 1,200 amps | CE@.333 | Ea | 311.00 | 19.40 | 330.40 |
| 1,500 to 1,600 amps | CE@.615 | Ea | 349.00 | 35.90 | 384.90 |
| 1,400 to 2,000 amps | CE@.809 | Ea | 388.00 | 47.20 | 435.20 |
| 2,500 amps | CE@1.10 | Ea | 622.00 | 64.20 | 686.20 |
| 3,000 amps | CE@1.25 | Ea | 631.00 | 73.00 | 704.00 |
| 3,500 to 4,000 amps | CE@1.74 | Ea | 1,050.00 | 102.00 | 1,152.00 |
| 5,000 amps | CE@2.33 | Ea | 1,350.00 | 136.00 | 1,486.00 |

|  | Craft@Hrs | Unit | Material | Labor | Equipment | Total |
|---|---|---|---|---|---|---|
| **Overhead Electrical Distribution** | | | | | | |
| Poles, pressure-treated wood, Class 4, Type C, set with crane or boom, including augured holes. Equipment is a flatbed truck with a boom and an auger. | | | | | | |
| 25' high | E1@7.62 | Ea | 328.00 | 383.00 | 96.80 | 807.80 |
| 30' high | E1@7.62 | Ea | 422.00 | 383.00 | 96.80 | 901.80 |
| 35' high | E1@9.72 | Ea | 549.00 | 489.00 | 124.00 | 1,162.00 |
| 40' high | E1@11.5 | Ea | 651.00 | 578.00 | 146.00 | 1,375.00 |
| **Cross arms, wood, with typical hardware** | | | | | | |
| 4' high | E1@2.86 | Ea | 78.10 | 144.00 | 36.30 | 258.40 |
| 5' high | E1@2.86 | Ea | 97.70 | 144.00 | 36.30 | 278.00 |
| 6' high | E1@4.19 | Ea | 141.00 | 211.00 | 53.20 | 405.20 |
| 8' high | E1@4.91 | Ea | 187.00 | 247.00 | 62.40 | 496.40 |
| **Transformers, pole mounted, single phase 7620/13200 Y primary, 120/240 secondary, with two 2-1/2% taps above and below** | | | | | | |
| 10 KVA | E1@4.14 | Ea | 1,280.00 | 208.00 | 52.60 | 1,540.60 |
| 15 KVA | E1@4.14 | Ea | 1,480.00 | 208.00 | 52.60 | 1,740.60 |
| 25 KVA | E1@4.14 | Ea | 1,800.00 | 208.00 | 52.60 | 2,060.60 |

| | Craft@Hrs | Unit | Material | Labor | Equipment | Total |
|---|---|---|---|---|---|---|
| 36.5 KVA | E1@4.14 | Ea | 2,420.00 | 208.00 | 52.60 | 2,680.60 |
| 50 KVA | E1@4.14 | Ea | 2,760.00 | 208.00 | 52.60 | 3,020.60 |
| 75 KVA | E1@6.16 | Ea | 4,090.00 | 310.00 | 78.30 | 4,478.30 |
| 100 KVA | E1@6.16 | Ea | 4,670.00 | 310.00 | 78.30 | 5,058.30 |
| 167 KVA | E1@6.16 | Ea | 7,160.00 | 310.00 | 78.30 | 7,548.30 |
| 250 KVA | E1@6.16 | Ea | 7,320.00 | 310.00 | 78.30 | 7,708.30 |
| 333 KVA | E1@6.16 | Ea | 7,920.00 | 310.00 | 78.30 | 8,308.30 |
| 500 KVA | E1@6.16 | Ea | 11,000.00 | 310.00 | 78.30 | 11,388.30 |
| Fused cutouts, pole mounted, 5 KV | | | | | | |
| 50 amp | E1@2.56 | Ea | 133.00 | 129.00 | 32.50 | 294.50 |
| 100 amp | E1@2.56 | Ea | 136.00 | 129.00 | 32.50 | 297.50 |
| 250 amp | E1@2.56 | Ea | 153.00 | 129.00 | 32.50 | 314.50 |
| Switches, disconnect, pole mounted, pole arm throw, 5 KV, set of 3 with throw and lock | | | | | | |
| 400 amp | E1@45.8 | Ea | 2,450.00 | 2,300.00 | 582.00 | 5,332.00 |
| 600 amp | E1@45.8 | Ea | 4,640.00 | 2,300.00 | 582.00 | 7,522.00 |
| 1,200 amp | E1@51.4 | Ea | 9,000.00 | 2,590.00 | 582.00 | 12,172.00 |

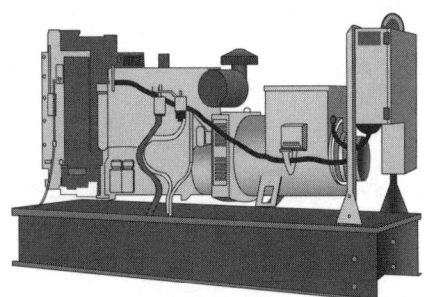 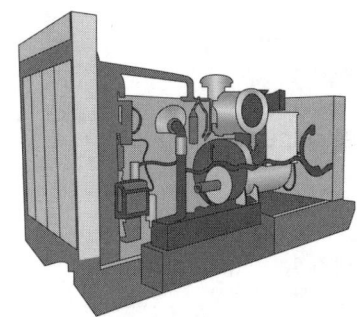

**Motor generator sets** Dual fuel (oil or gas) four-stroke, turbocharged eight-cylinder, 60 Hz, 480 volts, 3 phase, 1,800 RPM, UL 2200 listed, emissions certified to EPA Tier 2 rating, with National Electrical Code compliant 3- or 4-pole generator and UL 1446 insulation. Control package includes digital engine and generator monitoring, metering, ground fault interruption and lightning shielding, R448 voltage regulator, RFI filter. Includes paralleling and synchronizing automatic transfer switch or (for standby emergencies) automatic dual lockout automatic transfer switch, fused main circuit breaker, electronic load-sharing module for generator ganging, electric starter motor, battery, and trickle charger, lube oil cooling system with radiator and fan, oil pump and sump, water cooling system with water pump, radiator and fan, mounting skid, sound- and weather-proof enclosure, stainless steel exhaust silencer and stack, natural gas fuel train with diaphragm regulator valve, modulating metering valve, pressure gauge and flowmeter, diesel fuel train with fuel pump, fuel pre-heater, day tank with level gauge, main dual-fuel engine inlet regulating valve, remote annunciation panel with alarms and warning lights, and NFPA-rated $CO_2$ fire extinguishing system. Add the cost of connecting electrical loads and electrical controls, a concrete pad for mounting on the ground or structural steel supports for mounting on a roof, approval of prints by an inspection bureau and post-construction inspection by insurance and fire inspectors. These figures assume exhaust heat is released into the atmosphere. Suitable for standby or primary power generation in commercial, light industrial, government and institutional buildings, schools and hospitals. Equipment cost assumes use of a 2-ton capacity crane.

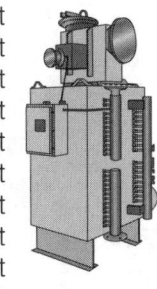

| | Craft@Hrs | Unit | Material | Labor | Equipment | Total |
|---|---|---|---|---|---|---|
| 13 KW net electrical output | D6@24.0 | Ea | 6,100.00 | 1,280.00 | 1,040.00 | 8,420.00 |
| 17 KW net electrical output | D6@24.0 | Ea | 8,090.00 | 1,280.00 | 1,040.00 | 10,410.00 |
| 30 KW net electrical output | D6@24.0 | Ea | 14,100.00 | 1,280.00 | 1,040.00 | 16,420.00 |
| 50 KW net electrical output | D6@32.0 | Ea | 23,400.00 | 1,710.00 | 1,050.00 | 26,160.00 |
| 80 KW net electrical output | D6@32.0 | Ea | 37,700.00 | 1,710.00 | 1,050.00 | 40,460.00 |
| 100 KW net electrical output | D6@32.0 | Ea | 47,000.00 | 1,710.00 | 1,050.00 | 49,760.00 |
| 150 KW net electrical output | D6@40.0 | Ea | 70,600.00 | 2,140.00 | 1,530.00 | 74,270.00 |
| 200 KW net electrical output | D6@40.0 | Ea | 94,200.00 | 2,140.00 | 1,530.00 | 97,870.00 |
| 300 KW net electrical output | D6@40.0 | Ea | 141,000.00 | 2,140.00 | 1,530.00 | 144,670.00 |
| 400 KW net electrical output | D6@80.0 | Ea | 188,000.00 | 4,270.00 | 1,710.00 | 193,980.00 |
| 545 KW net electrical output | D6@80.0 | Ea | 257,000.00 | 4,270.00 | 1,710.00 | 262,980.00 |

| | Craft@Hrs | Unit | Material | Labor | Equipment | Total |
|---|---|---|---|---|---|---|

Add-on boiler modules for motor generators. Add these costs to the appropriate costs above when exhaust heat from the motor generator will be captured to power an ebullient hot water or steam boiler which is part of the building HVAC system. Based on 60 PSI hot water or 15 PSI steam. Includes boiler, temperature, pressure and level gauges, boiler feed pump, condensate return or circulating pump loop, main valve, pressure regulating valve with vent, drain line and enclosure extension. In pounds for per hour (PPH) rating, hot water or steam.

| | Craft@Hrs | Unit | Material | Labor | Equipment | Total |
|---|---|---|---|---|---|---|
| 3,500 PPH (200 KW generator) | D5@40.0 | Ea | 29,600.00 | 2,010.00 | — | 31,610.00 |
| 5,000 PPH (300 KW generator) | D5@40.0 | Ea | 44,400.00 | 2,010.00 | — | 46,410.00 |
| 6,750 PPH (400 KW generator) | D5@40.0 | Ea | 59,300.00 | 2,010.00 | — | 61,310.00 |
| 8,000 PPH (545 KW generator) | D5@40.0 | Ea | 80,400.00 | 2,010.00 | — | 82,410.00 |

**Pollution control modules for motor generator sets.** N0x, S0x and VM (volatile materials) emissions reduction wet free-standing module for motor driven prime power and standby dual-fuel (natural gas or oil) electrical generator sets. Conforms with 2010 California Code, EPA Tier IV and European ISO standards. Verantis, Croll-Reynolds or equal. Either vertical counter-flow or packed bed wet type. For commercial and light industrial applications, schools, hospitals and governmental/institutional applications. Oil-fired turbine generators will add about 15% to the total cost, including a second stage, a second stage pump, venturi particulate separator and sludge tank. Add the cost of a permit. To calculate the annual lease rate for a complete installation, divide the total installed cost by five. Then add 2% per year to cover vendor-supplied maintenance.

| | Craft@Hrs | Unit | Material | Labor | Equipment | Total |
|---|---|---|---|---|---|---|
| 13 KW electrical output, net | — | Ea | 5,500.00 | — | — | 5,500.00 |
| 17 KW electrical output, net | — | Ea | 7,200.00 | — | — | 7,200.00 |
| 30 KW electrical output, net | — | Ea | 12,750.00 | — | — | 12,750.00 |
| 50 KW electrical output, net | — | Ea | 21,250.00 | — | — | 21,250.00 |
| 80 KW electrical output, net | — | Ea | 34,000.00 | — | — | 34,000.00 |
| 100 KW electrical output, net | — | Ea | 42,500.00 | — | — | 42,500.00 |
| 150 KW electrical output, net | — | Ea | 63,750.00 | — | — | 63,750.00 |
| 200 KW electrical output, net | — | Ea | 40,000.00 | — | — | 40,000.00 |
| 300 KW electrical output, net | — | Ea | 42,000.00 | — | — | 42,000.00 |
| 400 KW electrical output, net | — | Ea | 50,000.00 | — | — | 50,000.00 |
| 545 KW electrical output, net | — | Ea | 75,000.00 | — | — | 75,000.00 |

Installation costs for pollution control modules for gas turbine generators.

| | Craft@Hrs | Unit | Material | Labor | Equipment | Total |
|---|---|---|---|---|---|---|
| Form and pour foundation (typical) | U1@.040 | SF | 2.05 | 1.95 | — | 4.00 |
| Position housing, fans and pumps | U1@4.00 | Ea | — | 195.00 | — | 195.00 |
| Tie-down and leveling | U1@4.00 | Ea | 500.00 | 195.00 | — | 695.00 |
| Mount piping system supports | U1@6.00 | Ea | 1.50 | 293.00 | 10.83 | 305.33 |
| Rig and mount plenum and ducts | SM@5.00 | Ea | — | 292.00 | | 292.00 |
| Rig and mount tank for emission reduction fluid and housing inlet metering valve | U1@4.00 | Ea | — | 195.00 | — | 195.00 |
| Weld system piping | U1@8.00 | Ea | 2.10 | 391.00 | 8.44 | 401.54 |
| Wire electrical system | BE@8.00 | Ea | 1.80 | 317.00 | 9.46 | 328.26 |
| Wire control system | BE@16.0 | Ea | 1.90 | 633.00 | — | 634.90 |
| Level and balance rotating equipment | P1@8.00 | Ea | 200.00 | 286.00 | — | 486.00 |
| Calibration and test | U1@2.00 | Ea | — | 97.70 | — | 97.70 |
| Witnessed run trials | P1@4.00 | Ea | — | 143.00 | 140.28 | 283.28 |

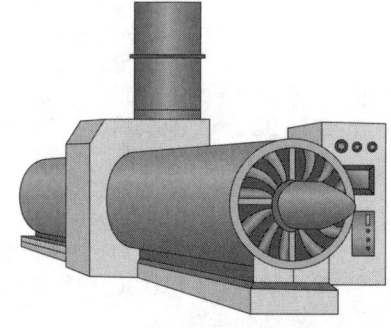

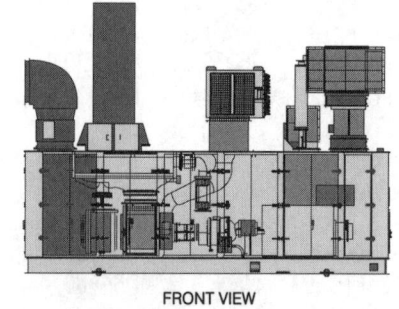

FRONT VIEW

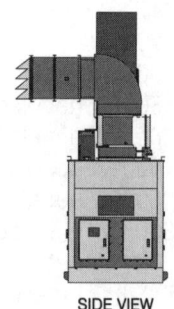

SIDE VIEW

|  | Craft@Hrs | Unit | Material | Labor | Equipment | Total |
|---|---|---|---|---|---|---|

**Gas turbine generator sets** Dual fuel (oil or gas) single shaft turbine with 11 stage compression, 2 stage hot section and single stage power turbine, 60 Hz, 480 volts, 3 phase, 1,800 RPM, UL 2200 listed, emissions compliance certified to California B.A.C.T. rating. Includes a *National Electrical Code* compliant 3- or 4-pole generator with UL 1446 insulation, digital engine and generator monitoring, metering and protection with ground fault interruption and lightning shielding, R448 voltage regulator, RFI filter, Woodward API 611 governor and overspeed valve relay, paralleling and synchronizing automatic transfer switch or (for standby emergencies) automatic dual lockout automatic transfer switch, fused main circuit breaker, electronic load-sharing module for generator ganging, electric starter motor, battery, and trickle charger, lube oil cooling system with radiator and fan, oil pump and sump, water cooling system with water pump, radiator and fan, mounting skid, sound- and weather-proof enclosure, turbine inlet air filter, stainless steel exhaust silencer and stack, natural gas fuel train with diaphragm regulator valve, modulating metering valve, pressure gauge and flowmeter, liquid fuel train with fuel pump, fuel preheater, day tank with level gauge, main dual-fuel engine inlet regulating valve, remote annunciation panel with alarms and warning lights, and NFPA-rated $CO_2$ fire extinguishing system. Add the cost of approval of prints by an inspection bureau, connecting electrical loads and electrical controls, a concrete pad for mounting on the ground or structural steel supports for mounting on a roof, grouting, precision leveling, startup, commissioning, and post-construction inspection by insurance and fire inspectors. These figures assume turbine exhaust heat is released into the atmosphere. Suitable for standby or primary power generation for commercial and light to heavy industrial applications, schools, hospitals, government and institutional use and utility peak-shaving service. Equipment cost is for a 2-ton capacity crane.

|  | Craft@Hrs | Unit | Material | Labor | Equipment | Total |
|---|---|---|---|---|---|---|
| 1,000 KW electrical output | D6@80.0 | Ea | 996,000.00 | 4,270.00 | 2,120.00 | 1,002,390.00 |
| 3,500 KW electrical output | D6@80.0 | Ea | 3,470,000.00 | 4,270.00 | 2,120.00 | 3,476,390.00 |
| 5,000 KW electrical output | D6@120 | Ea | 4,930,000.00 | 6,410.00 | 3,240.00 | 4,939,650.00 |
| 6,000 KW electrical output | D6@120 | Ea | 5,940,000.00 | 6,410.00 | 3,240.00 | 5,949,650.00 |
| 7,500 KW electrical output | D6@160 | Ea | 9,400,000.00 | 8,550.00 | 4,770.00 | 9,413,320.00 |

**Add-on heat recovery system for gas turbine generators** Add these costs to the appropriate gas turbine generator costs above when exhaust heat will be used to drive a 60 PSI hot water to 600 PSI steam boiler which is part of the building HVAC system. Includes a D- or A-Type watertube bundle, superheater and mud drums, primary stack, secondary stack with service exhaust gas sealed damper bypass plenum, temperature, pressure and level gauges, duplex boiler feed pump, auxiliary dual-fuel continuously-modulating burner, digital burner control and ignition annunciator panel with alarms, condensate return or circulating pump loop, main valve, twin pressure relief valves with vent, drain line, thermally and acoustically insulated enclosure extension, chemical feedwater treatment tank, pump with bladed mixer and relay valve. By pounds for per hour (PPH) rating, hot water or steam. Equipment cost is for a 2-ton capacity crane.

|  | Craft@Hrs | Unit | Material | Labor | Equipment | Total |
|---|---|---|---|---|---|---|
| 3,000 PPH (1,000 KW turbine generator) | D6@80.0 | Ea | 200,000.00 | 4,270.00 | 2,120.00 | 206,390.00 |
| 10,500 PPH (3,500 KW turbine generator) | D6@80.0 | Ea | 684,000.00 | 4,270.00 | 2,120.00 | 690,390.00 |
| 15,000 PPH (5,000 KW turbine generator) | D6@120 | Ea | 986,000.00 | 6,410.00 | 3,240.00 | 995,650.00 |
| 18,000 PPH (6,000 KW turbine generator) | D6@120 | Ea | 1,180,000.00 | 6,410.00 | 3,240.00 | 1,189,650.00 |
| 22,500 PPH (7,500 KW turbine generator) | D6@160 | Ea | 1,850,000.00 | 8,550.00 | 4,770.00 | 1,863,320.00 |
| 34,500 PPH (11,500 KW turbine generator) | D6@160 | Ea | 2,630,000.00 | 8,550.00 | 4,770.00 | 2,643,320.00 |

| | Craft@Hrs | Unit | Material | Labor | Equipment | Total |
|---|---|---|---|---|---|---|

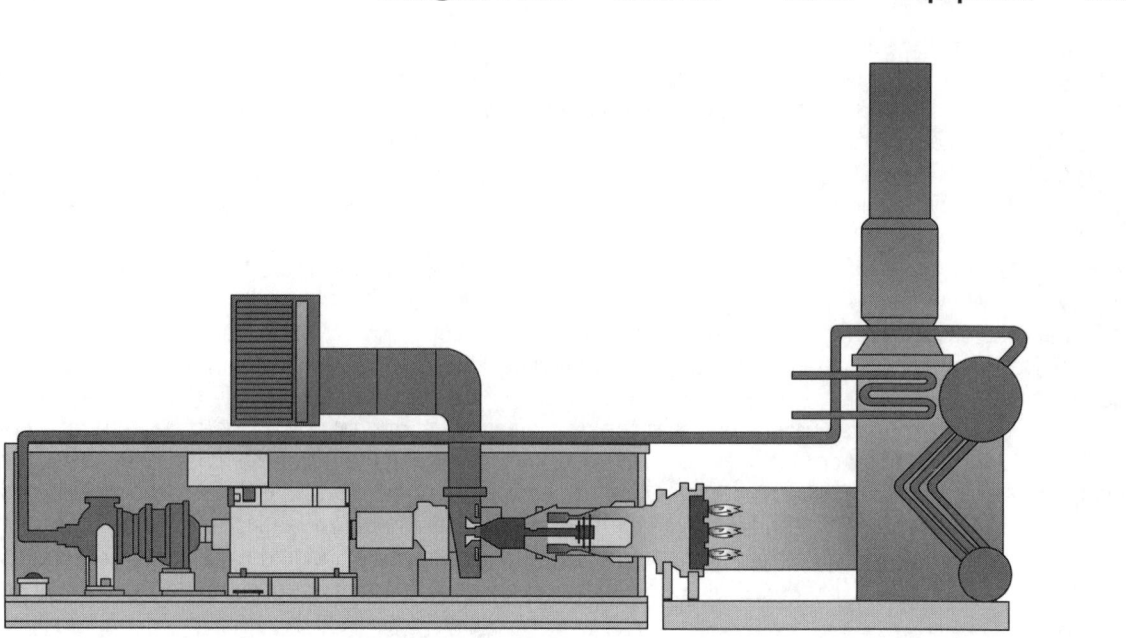

COMBINED CYCLE GAS TURBINE

Add for combined cycle gas turbine heat recovery generator sets. These costs assume exhaust heat drives a turbogenerator for electric power cogeneration. Combined cycle operations increase generating capacity by approximately 20% over simple cycle gas turbine generators. Includes the steam turbogenerator, backpressure single stage with Woodward API 611 governor and overspeed cutoff valve, main inlet throttling valve, *National Electrical Code* compliant 4-pole 480 volt, 60 HZ, 3-phase electrical generator with paralleling and synchronizing automatic transfer switch, main breaker with ground fault interruption and lightning protection, ambient forced and air-cooled fintube coil and fan condenser array with condensate pump and return line, bypass drain, emergency vent to atmosphere, digital and thermostatically controlled annunciator panel with pressure/temp/flow rate metering, alarms and emergency cutoff capabilities in a NEMA 10 (wet environment) enclosure. Add the cost of electrical connection and control tie-in, grouting, precision equipment leveling, concrete pad or structural steel roof mount support, vibration isolation mounts, permitting, approvals or inspection. For a combined cycle gas turbine heat recovery generator, add the following percentages to the combined cost of the appropriate gas turbine generator set and matching heat recovery system.

| | | | | | | |
|---|---|---|---|---|---|---|
| Add for power generation driven by exhaust heat | — | % | 30.0 | 50.0 | 100.0 | — |

**Power generator exhaust stack**  Plate steel exhaust stack for high mass flow rate power generators. Mill-rolled, seam-welded and shop fabricated for field erection. Stack has ionically-bonded internal silicon lining with external aluminized coating, external spiral windbreaker banding, two gasketed and bolted flanges per stack section, buttressed base support sleeve, inlet and service bypass. T-plenum includes a soundproofing enclosure and structural steel support frame. Mounting includes a vibration-isolating spring-loaded circular base plus concrete-anchored guy wiring, eyelets and adjustable guy wire extension tensioners. Add the cost of a reinforced 3000 P.S.I. laser-leveled concrete pad. Equipment cost is for rental of a 3-ton 200-foot extensible truck mounted cable-drive spider gantry crane.  Cost per stack section. Each exhaust assembly includes one T-plenum and enough stack sections to meet code requirements and manufacturer specifications. Typical stack height is 75' to 200'. Labor costs assume half of the stack is erected in place upright. The balance is assembled on the ground and lifted into place with a spider crane.

| | Craft@Hrs | Unit | Material | Labor | Equipment | Total |
|---|---|---|---|---|---|---|
| 10" diameter stack T-plenum with frame | D5@36.0 | Ea | 17,400.00 | 1,810.00 | — | 19,210.00 |
| 16" diameter stack T-plenum with frame | D5@36.0 | Ea | 23,200.00 | 1,810.00 | — | 25,010.00 |
| 20" diameter stack T-plenum with frame | D5@40.0 | Ea | 30,200.00 | 2,010.00 | — | 32,210.00 |
| 24" diameter stack T-plenum with frame | D5@48.0 | Ea | 36,000.00 | 2,410.00 | — | 38,410.00 |
| 30" diameter stack T-plenum with frame | D5@60.0 | Ea | 40,700.00 | 3,010.00 | — | 43,710.00 |

| | Craft@Hrs | Unit | Material | Labor | Equipment | Total |
|---|---|---|---|---|---|---|
| 48" diameter stack T-plenum with frame | D5@72.0 | Ea | 48,800.00 | 3,610.00 | — | 52,410.00 |
| 60" diameter stack T-plenum with frame | D5@96.0 | Ea | 58,100.00 | 4,820.00 | — | 62,920.00 |
| 72" diameter stack T-plenum with frame | D5@120 | Ea | 90,600.00 | 6,020.00 | — | 96,620.00 |
| 96" diameter stack T-plenum with frame | D5@160 | Ea | 127,000.00 | 8,030.00 | — | 135,030.00 |
| 132" diameter stack T-plenum with frame | D5@180 | Ea | 256,000.00 | 9,030.00 | — | 265,030.00 |
| 10" diameter stack, 10' high section | D6@24.0 | Ea | 872.00 | 1,280.00 | 4,320.00 | 6,472.00 |
| 16" diameter stack, 10' high section | D6@24.0 | Ea | 1,040.00 | 1,280.00 | 4,320.00 | 6,640.00 |
| 20" diameter stack, 10' high section | D6@24.0 | Ea | 1,400.00 | 1,280.00 | 4,320.00 | 7,000.00 |
| 24" diameter stack, 10' high section | D6@32.0 | Ea | 1,850.00 | 1,710.00 | 4,320.00 | 7,880.00 |
| 30" diameter stack, 10' high section | D6@32.0 | Ea | 2,220.00 | 1,710.00 | 4,320.00 | 8,250.00 |
| 48" diameter stack, 10' high section | D6@32.0 | Ea | 2,560.00 | 1,710.00 | 4,320.00 | 8,590.00 |
| 60" diameter stack, 20' high section | D6@40.0 | Ea | 3,490.00 | 2,140.00 | 6,480.00 | 12,110.00 |
| 72" diameter stack, 20' high section | D6@40.0 | Ea | 4,180.00 | 2,140.00 | 6,480.00 | 12,800.00 |
| 96" diameter stack, 20' high section | D6@40.0 | Ea | 4,420.00 | 2,140.00 | 6,480.00 | 13,040.00 |
| 132" diameter stack, 20' high section | D6@80.0 | Ea | 5,230.00 | 4,270.00 | 6,480.00 | 15,980.00 |

**Pollution control modules for gas turbine generators.** N0x, S0x, VM (volatile materials) and particle emissions reduction wet module, free-standing, simple cycle. For natural gas fired prime (baseload) power and standby emergency electrical generators. Verantis, Croll-Reynolds, or equal. Complies with 2010 California Code, Tier IV, EPA and European ISO standards. Either vertical counter-flow or packed-bed wet type. For commercial and light to heavy industrial applications, schools, hospitals, and governmental/institutional applications and utility peak-shaving about 15% to the total cost, including a second stage, a second stage pump, venturi particulate separator and sludge tank. To calculate the annual lease rate for a complete installation, divide the total installed cost by five. Then add 2% per year to cover vendor-supplied maintenance.

| | | Unit | Material | Labor | Equipment | Total |
|---|---|---|---|---|---|---|
| 1,000 KW electrical | — | Ea | 75,000.00 | — | — | 75,000.00 |
| 3,500 KW electrical | — | Ea | 125,000.00 | — | — | 125,000.00 |
| 5,000 KW electrical | — | Ea | 150,000.00 | — | — | 150,000.00 |
| 6,000 KW electrical | — | Ea | 160,000.00 | — | — | 160,000.00 |
| 7,500 KW electrical | — | Ea | 175,000.00 | — | — | 175,000.00 |
| 11,500 KW electrical | — | Ea | 250,000.00 | — | — | 250,000.00 |

Installation costs for pollution control modules for gas turbine generators.

| | | Unit | Material | Labor | Equipment | Total |
|---|---|---|---|---|---|---|
| Position main housing, fans and pumps | U1@4.00 | Ea | — | 195.00 | — | 195.00 |
| Tie-down and leveling | U1@4.00 | Ea | 500.00 | 195.00 | — | 695.00 |
| Mount piping system frame supports | U1@6.00 | LF | 1.50 | 293.00 | — | 294.50 |
| Rig and mount plenum and ducts | SM@5.00 | Ea | — | 292.00 | — | 292.00 |
| Rig and mount tank for emission reduction fluid | | | | | | |
| and housing inlet metering valve | U1@4.00 | Ea | — | 195.00 | — | 195.00 |
| Weld systems piping | U1@8.00 | LF | 2.10 | 391.00 | — | 393.10 |
| Wire electrical system | BE@8.00 | LF | 1.80 | 317.00 | — | 318.80 |
| Wire controls system | BE@16.0 | LF | 1.90 | 633.00 | — | 634.90 |
| Rotating equipment system level & balance | P1@8.00 | Ea | 200.00 | 286.00 | — | 486.00 |
| Calibration & test | U1@2.00 | Ea | — | 97.70 | — | 97.70 |
| Witnessed run trials (add the permit fee) | P1@4.00 | Ea | — | 143.00 | — | 143.00 |

| | Craft@Hrs | Unit | Material | Labor | Total |
|---|---|---|---|---|---|

**Integrated Photovoltaic Systems** For light commercial, government and institutional applications. In arrays of from 50 to 200 KW-hour capacity. Façade, skylight or vertically mounted (integrated into glazing). Includes photovoltaic modules (thin-film or polycrystalline/crystalline ribbon, or amorphous silicon), a charge controller to regulate the power into and out of the battery storage bank (in stand-alone systems), a power storage system or utility grid integration, an inverter to convert DC output to AC, appropriate support and mounting hardware, wiring, and safety disconnects. Thin film single crystal cells are laser-etched and vapor deposited into 2-pane 5/8" to 1" glazing. Polycrystalline or crystalline ribbon solar cells and amorphous silicon solar cells are opaque and are used most commonly for retrofit roof, pergola and canopy mounting. A 6 SF solar unit has 36 cells in a 18" x 4' panel with a 4" x 4" junction box. Add the cost of the permit, roof load resistance and façade deflection reinforcement and standby power generator, if required. Complies with Federal Spec 48 14 00.

| | Craft@Hrs | Unit | Material | Labor | Total |
|---|---|---|---|---|---|
| Vertical framing mullions, 11 ga. x 1" x 3", aluminum | IW@.009 | Lb | 1.24 | .57 | 1.81 |
| Horizontal framing mullions, 11 ga. x 2" x 3", aluminum with galvanized steel internal bracing | IW@.019 | Lb | 2.07 | 1.21 | 3.28 |
| Thin film single crystal solar cells | G1@.045 | SF | 78.00 | 2.07 | 80.07 |
| Polycrystalline or crystalline ribbon solar cells | G1@.035 | SF | 71.00 | 1.61 | 72.61 |
| Amorphous solar cells | G1@.035 | SF | 28.00 | 1.61 | 29.61 |
| Add for thermally reflective or retentive coatings | — | % | 15.0 | — | — |
| Charge controller with fusing and emergency interrupts, 2.25 watt-hour | E4@4.0 | Ea | 181.12 | 197.00 | 378.12 |
| Power storage system, utility grid paralleling switchgear and service bypass, 2.20 watt-hour | E4@8.0 | Ea | 362.24 | 393.00 | 755.24 |
| Battery pack and fault protection, 6.00 watt-hour | E4@2.0 | Ea | 300.00 | 98.30 | 398.30 |
| DC to AC inverter with power quality system filters, 1.85 watt-hour | E4@16.0 | Ea | 724.48 | 786.00 | 1,510.48 |
| Electrical wiring for integrated PV system | E4@.007 | LF | .30 | .34 | .64 |
| Inspection and test of integrated PV system | E4@8.0 | Ea | — | 393.00 | 393.00 |
| Typical installed cost per SF of cell (1.5 watts), including poly cells, supports, charge controller, power storage, grid interconnect, inverter, wiring, inspection and test, 50 to 200 KW array | — | LS | — | — | 105.30 |
| Add for architectural services and approvals | — | % | — | — | 10.0 |

**Fluorescent Lighting** Commercial, industrial and architectural grade. Rapid start. With ballasts but no lamps, wire or conduit. Includes setting and connecting prewired fixtures only.

Surface mounted fluorescent fixtures, better quality, baked enamel finish, with hinged wraparound acrylic prismatic lens suitable for office or classroom use, 4' long, Lithonia LB series

| | Craft@Hrs | Unit | Material | Labor | Total |
|---|---|---|---|---|---|
| Two 40 watt lamp fixture (#240-A) | CE@.773 | Ea | 68.40 | 45.10 | 113.50 |
| Four 40 watt lamp fixture (#440-A) | CE@.773 | Ea | 129.00 | 45.10 | 174.10 |
| Eight 40 watt lamp fixture, 8' long (#8T440-A) | CE@.960 | Ea | 196.00 | 56.00 | 252.00 |

Surface mounted fluorescent fixtures, baked enamel finish, with wraparound acrylic prismatic lens suitable for industrial use, 4' long, Lithonia series

| | Craft@Hrs | Unit | Material | Labor | Total |
|---|---|---|---|---|---|
| One 40 watt lamp fixture (CB140A) | CE@.806 | Ea | 61.00 | 47.00 | 108.00 |
| Two 40 watt lamp fixture (LB240A) | CE@.858 | Ea | 52.20 | 50.10 | 102.30 |
| Four 40 watt lamp fixture (LB440A) | CE@1.07 | Ea | 89.60 | 62.50 | 152.10 |
| Add for 40 watt lamps | — | Ea | 3.49 | — | 3.49 |

Surface mounted steel sided fluorescent fixtures with acrylic prismatic diffuser, 7-1/16" x 4-1/2"

| | Craft@Hrs | Unit | Material | Labor | Total |
|---|---|---|---|---|---|
| Two 40 watt lamps | CE@.955 | Ea | 74.70 | 55.70 | 130.40 |
| Four 40 watt lamps | CE@1.05 | Ea | 104.00 | 61.30 | 165.30 |
| Add for 40 watt straight lamps | — | Ea | 3.49 | — | 3.49 |

| | Craft@Hrs | Unit | Material | Labor | Total |
|---|---|---|---|---|---|
| **Wall surface mounted fluorescent fixtures, baked enamel finish, 2 lamps, 8" wide 5" high, hinged acrylic prismatic lens, Lightolier** | | | | | |
| 26" long fixture | CE@.737 | Ea | 156.00 | 43.00 | 199.00 |
| 38" long fixture | CE@.795 | Ea | 181.00 | 46.40 | 227.40 |
| 50" long fixture | CE@.858 | Ea | 175.00 | 50.10 | 225.10 |
| Add for 24" 20 watt lamps | — | Ea | 3.19 | — | 3.19 |
| Add for 36" 30 watt lamps | — | Ea | 4.94 | — | 4.94 |
| Add for 48" 40 watt lamps | — | Ea | 2.26 | — | 2.26 |
| **Industrial pendent or surface mounted open striplights, baked enamel finish, Lithonia C series** | | | | | |
| One or two 40 watt lamps, 4' long (C240) | CE@.865 | Ea | 35.40 | 50.50 | 85.90 |
| One or two 75 watt lamps, 8' long (C296) | CE@1.02 | Ea | 52.20 | 59.50 | 111.70 |
| One or two 55 watt lamps, 6' long (C272) | CE@1.02 | Ea | 52.20 | 59.50 | 111.70 |
| Add for 40 watt lamps | — | Ea | 2.43 | — | 2.43 |
| Add for 75 or 55 watt lamps | — | Ea | 5.77 | — | 5.77 |
| Add for 24" stem and canopy set, per fixture | — | Ea | 5.77 | — | 5.77 |
| **Industrial pendent or surface mounted fluorescent fixtures, 12" wide baked enamel, steel reflector-shield, Lithonia L/LA series** | | | | | |
| Two 40 watt lamps, 4' long (L-240-120) | CE@.903 | Ea | 38.00 | 52.70 | 90.70 |
| Two 75 watt lamps, 8' long (L-296-120) | CE@1.21 | Ea | 74.70 | 70.60 | 145.30 |
| Add for 40 watt lamps | — | Ea | 2.43 | — | 2.43 |
| Add for 75 watt lamps | — | Ea | 5.81 | — | 5.81 |
| **Ceiling grid mounted fluorescent troffer fixtures, baked enamel finish, with hinged acrylic prismatic lens with energy saving ballast, Day-Brite** | | | | | |
| Two 40 watt "U" lamp fixture, 2' x 2' | CE@.908 | Ea | 55.20 | 53.00 | 108.20 |
| Two 40 watt lamp fixture, 2' x 4' | CE@.998 | Ea | 62.60 | 58.30 | 120.90 |
| Four 40 watt lamp fixture, 2' x 4' | CE@1.05 | Ea | 67.60 | 61.30 | 128.90 |
| Add for plaster frames, per fixture | — | Ea | 8.06 | — | 8.06 |
| **Air-handling grid mounted floating door, heat transfer troffer fixtures with acrylic lens, no ductwork included** | | | | | |
| 1' x 4', two 40 watt lamps | CE@.908 | Ea | 104.00 | 53.00 | 157.00 |
| 2' x 2', one U-shape 40 watt lamp | CE@.809 | Ea | 110.00 | 47.20 | 157.20 |
| 2' x 4', two 40 watt lamps | CE@1.00 | Ea | 138.00 | 58.40 | 196.40 |
| 2' x 4', three 40 watt lamps | CE@1.05 | Ea | 172.00 | 61.30 | 233.30 |
| 2' x 4', four 40 watt lamps | CE@1.05 | Ea | 172.00 | 61.30 | 233.30 |
| Add for 40 watt 4' lamps | — | Ea | 2.43 | — | 2.43 |
| Add for 40 watt "U" lamps | — | Ea | 10.10 | — | 10.10 |
| **Surface mounted vandal-resistant luminaire with tamper-resistant screw-type latching, Sylvania** | | | | | |
| 1 lamp | CE@.761 | Ea | 115.00 | 44.40 | 159.40 |
| 2 lamps | CE@.761 | Ea | 117.00 | 44.40 | 161.40 |
| 4 lamps | CE@.761 | Ea | 198.00 | 44.40 | 242.40 |
| **Parabolic fluorescent luminaire, 6" x 24" x 24" with semi-specular anodized aluminum reflector and parabolic louvers, I.C.E. AR series** | | | | | |
| Two 40 watt lamp, bracket or pendent mount | CE@.955 | Ea | 72.50 | 55.70 | 128.20 |
| Four 40 watt lamps, bracket or pendent mount | CE@.955 | Ea | 116.00 | 55.70 | 171.70 |
| **Round surface mounted fluorescent fixtures, aluminum housing, matte black finish, polycarbonate or acrylic opal globe, Circleline** | | | | | |
| 22 to 32 watt, 14" diameter | CE@.478 | Ea | 45.60 | 27.90 | 73.50 |
| 32 to 40 watt, 18" diameter | CE@.478 | Ea | 63.10 | 27.90 | 91.00 |
| Add for circline tubes | — | Ea | 7.22 | — | 7.22 |

| | Craft@Hrs | Unit | Material | Labor | Total |
|---|---|---|---|---|---|
| Outdoor sign fluorescent fixtures, anodized aluminum housing with parabolic specular reflector and acrylic lens, with high output lamps and remote ballast | | | | | |
| One 60 watt lamp. 48" long fixture | CE@2.39 | Ea | 220.00 | 140.00 | 360.00 |
| One 85 watt lamp, 72" long fixture | CE@2.39 | Ea | 231.00 | 140.00 | 371.00 |
| One 110 watt lamp, 96" long fixture | CE@2.39 | Ea | 255.00 | 140.00 | 395.00 |
| Two 85 watt lamps, 144" long fixture | CE@2.39 | Ea | 346.00 | 140.00 | 486.00 |
| Add for high output T12 800 MA lamps, sign white | | | | | |
| 60 watt | — | Ea | 6.62 | — | 6.62 |
| 85 watt | — | Ea | 6.62 | — | 6.62 |
| 115 watt | — | Ea | 18.50 | — | 18.50 |
| Cost per linear foot of fixture | CE@.179 | LF | 59.00 | 10.40 | 69.40 |
| Add for 90-degree elbows | CE@.475 | Ea | 153.00 | 27.70 | 180.70 |
| Add for in-line connectors | CE@.475 | Ea | 76.50 | 27.70 | 104.20 |
| Wet or damp location surface mounted fluorescent fixtures, hinge mounted acrylic diffuse lens, fiberglass or plastic housing with gaskets, Hubbell | | | | | |
| Two 40 watt lamps, 4' long fixture | CE@1.21 | Ea | 131.00 | 70.60 | 201.60 |
| Two 40 watt lamps, 8' long fixture | CE@1.86 | Ea | 285.00 | 109.00 | 394.00 |
| Add for 40 watt lamps | — | Ea | 2.43 | — | 2.43 |

## High Intensity Discharge Lighting

Commercial and architectural grade fixtures. With ballasts but no lamps, wire, conduit or concrete except as noted. See lamp costs below. Labor includes setting and connecting prewired fixtures only.

| | Craft@Hrs | Unit | Material | Labor | Total |
|---|---|---|---|---|---|
| Recess mounted commercial luminaire, 2' x 2' x 13", aluminum reflector glass lens, Wide-Lite series | | | | | |
| 175 watt, metal halide | CE@1.15 | Ea | 620.00 | 67.10 | 687.10 |
| 250 watt, high pressure sodium | CE@1.15 | Ea | 710.00 | 67.10 | 777.10 |
| 400 watt, metal halide | CE@1.15 | Ea | 626.00 | 67.10 | 693.10 |
| 400 watt, high pressure sodium | CE@1.15 | Ea | 779.00 | 67.10 | 846.10 |
| Surface mounted commercial luminaire, 2' x 2' x 15", baked enamel finish with tempered glass lens, Holophane series | | | | | |
| 70 watt, high pressure sodium | CE@1.42 | Ea | 374.00 | 82.90 | 456.90 |
| 100 watt, high pressure sodium | CE@1.42 | Ea | 411.00 | 82.90 | 493.90 |
| 150 watt, high pressure sodium | CE@1.42 | Ea | 411.00 | 82.90 | 493.90 |
| 175 watt, metal halide | CE@1.42 | Ea | 394.00 | 82.90 | 476.90 |
| 400 watt, metal halide | CE@1.42 | Ea | 423.00 | 82.90 | 505.90 |
| Recess mounted indirect metal halide luminaire, aluminum reflector with glass lens | | | | | |
| 175 watt, square | CE@1.71 | Ea | 620.00 | 99.80 | 719.80 |
| 250 watt, square | CE@1.71 | Ea | 620.00 | 99.80 | 719.80 |
| 400 watt, square | CE@1.71 | Ea | 626.00 | 99.80 | 725.80 |
| 175 watt, round | CE@1.71 | Ea | 595.00 | 99.80 | 694.80 |
| 250 watt, round | CE@1.71 | Ea | 595.00 | 99.80 | 694.80 |
| 400 watt, round | CE@1.71 | Ea | 601.00 | 99.80 | 700.80 |
| Surface mounted indirect high pressure sodium luminaire, aluminum reflector with glass lens | | | | | |
| 250 watt, square | CE@1.71 | Ea | 684.00 | 99.80 | 783.80 |
| 400 watt, square | CE@1.71 | Ea | 750.00 | 99.80 | 849.80 |
| 250 watt, round | CE@1.71 | Ea | 658.00 | 99.80 | 757.80 |
| 400 watt, round | CE@1.71 | Ea | 724.00 | 99.80 | 823.80 |
| Outdoor wall pack, aluminum housing with hinged polycarbonate access door and lens, Sylvania | | | | | |
| 100 watt, high pressure sodium | CE@.955 | Ea | 221.00 | 55.70 | 276.70 |
| 150 watt, high pressure sodium | CE@.955 | Ea | 195.00 | 55.70 | 250.70 |
| 175 watt, mercury vapor | CE@.955 | Ea | 190.00 | 55.70 | 245.70 |
| 175 watt, metal halide | CE@.955 | Ea | 242.00 | 55.70 | 297.70 |

| | Craft@Hrs | Unit | Material | Labor | Total |
|---|---|---|---|---|---|
| **Utility wall pack, aluminum housing with hinged polycarbonate or tempered glass lens, Holophane Wallpack 2** | | | | | |
| 70 watt, high pressure sodium | CE@.955 | Ea | 374.00 | 55.70 | 429.70 |
| 100 watt, high pressure sodium | CE@.955 | Ea | 402.00 | 55.70 | 457.70 |
| 150 watt, high pressure sodium | CE@.955 | Ea | 411.00 | 55.70 | 466.70 |
| 175 watt, metal halide | CE@.955 | Ea | 394.00 | 55.70 | 449.70 |
| 250 watt, mercury vapor | CE@.955 | Ea | 362.00 | 55.70 | 417.70 |
| Add for photo cell automatic control, factory installed | — | Ea | 23.30 | — | 23.30 |
| **High bay metal halide industrial fixture, cast aluminum housing, spun aluminum reflector, lens is tempered glass** | | | | | |
| 400 watt | CE@2.61 | Ea | 222.00 | 152.00 | 374.00 |
| 1,000 watt | CE@2.61 | Ea | 378.00 | 152.00 | 530.00 |
| Add for power hook and cord | CE@.250 | Ea | 65.40 | 14.60 | 80.00 |
| Add for thru-wire power hook receptacle | CE@.199 | Ea | 55.60 | 11.60 | 67.20 |
| Add for wire guard on lens | CE@.199 | Ea | 65.40 | 11.60 | 77.00 |
| **Low bay high pressure sodium industrial fixture, cast aluminum with epoxy finish, acrylic lens** | | | | | |
| 100 watt fixture | CE@1.42 | Ea | 328.00 | 82.90 | 410.90 |
| 150 watt fixture | CE@1.39 | Ea | 340.00 | 81.10 | 421.10 |
| 250 watt fixture | CE@1.39 | Ea | 484.00 | 81.10 | 565.10 |
| Add for power hook and cord | CE@.250 | Ea | 60.40 | 14.60 | 75.00 |
| Add for thru-wire power hook receptacle | CE@.199 | Ea | 53.10 | 11.60 | 64.70 |
| Add for wire guard on lens | CE@.199 | Ea | 71.80 | 11.60 | 83.40 |
| **Low bay metal halide industrial downlight, cast aluminum with epoxy finish, shock resistant glass lens** | | | | | |
| 100 watt fixture | CE@1.07 | Ea | 408.00 | 62.50 | 470.50 |
| 400 watt fixture | CE@1.39 | Ea | 511.00 | 81.10 | 592.10 |
| Add for power hook and cord | CE@.250 | Ea | 60.40 | 14.60 | 75.00 |
| Add for thru-wire power hook receptacle | CE@.199 | Ea | 53.10 | 11.60 | 64.70 |
| Add for wire guard on lens | CE@.199 | Ea | 73.40 | 11.60 | 85.00 |
| **Lamps for high intensity discharge luminaires** | | | | | |
| 400 watt, mercury vapor | — | Ea | 23.10 | — | 23.10 |
| 1,000 watt, mercury vapor | — | Ea | 97.00 | — | 97.00 |
| 35 watt, high pressure sodium | — | Ea | 28.30 | — | 28.30 |
| 70 watt, high pressure sodium | — | Ea | 28.30 | — | 28.30 |
| 100 watt, high pressure sodium | — | Ea | 28.30 | — | 28.30 |
| 150 watt, high pressure sodium | — | Ea | 39.50 | — | 39.50 |
| 250 watt, high pressure sodium | — | Ea | 66.50 | — | 66.50 |
| 400 watt, high pressure sodium | — | Ea | 79.00 | — | 79.00 |
| 1,000 watt, high pressure sodium | — | Ea | 95.90 | — | 95.90 |
| 175 watt, metal halide | — | Ea | 36.10 | — | 36.10 |
| 250 watt, metal halide | — | Ea | 36.10 | — | 36.10 |
| 400 watt, metal halide | — | Ea | 62.10 | — | 62.10 |
| 1,000 watt, metal halide | — | Ea | 114.00 | — | 114.00 |
| Add for diffuse coated lamps | — | Ea | 6.29 | — | 6.29 |

**Incandescent Lighting** Commercial and architectural grade. No wire, conduit or concrete included. Includes setting and connecting prewired fixtures only.

| | Craft@Hrs | Unit | Material | Labor | Total |
|---|---|---|---|---|---|
| **Step illumination light fixtures, 16 gauge steel housing with white enamel finish, 5" high, 4" deep, 11" long** | | | | | |
| Polycarbonate or tempered glass lens fixture | CE@1.07 | Ea | 180.00 | 62.50 | 242.50 |
| Louvered sheet metal front fixture | CE@1.07 | Ea | 154.00 | 62.50 | 216.50 |
| Add for brushed aluminum trim, 2 lamps | — | Ea | 42.70 | — | 42.70 |
| **Wall surface mounted fluorescent fixtures, baked enamel finish, 2 lamps, 8" wide 5" high, hinged acrylic prismatic lens, Lightolier** | | | | | |
| 26" long fixture | CE@.737 | Ea | 156.00 | 43.00 | 199.00 |
| 38" long fixture | CE@.795 | Ea | 181.00 | 46.40 | 227.40 |
| 50" long fixture | CE@.858 | Ea | 175.00 | 50.10 | 225.10 |

| | Craft@Hrs | Unit | Material | Labor | Total |
|---|---|---|---|---|---|
| Add for 24" 20 watt lamps | — | Ea | 3.19 | — | 3.19 |
| Add for 36" 30 watt lamps | — | Ea | 4.94 | — | 4.94 |
| Add for 48" 40 watt lamps | — | Ea | 2.26 | — | 2.26 |

Recessed round eyelid wall wash, 6" milligroove baffle, Prescolite

| | Craft@Hrs | Unit | Material | Labor | Total |
|---|---|---|---|---|---|
| 150 watt fixture | CE@1.07 | Ea | 152.00 | 62.50 | 214.50 |

Exterior wall fixtures, cast anodized aluminum with white polycarbonate globe, gaskets between globe and housing and between housing and base, Prescolite

| | Craft@Hrs | Unit | Material | Labor | Total |
|---|---|---|---|---|---|
| Cylinder globe fixture | CE@.320 | Ea | 80.70 | 18.70 | 99.40 |
| Round globe fixture | CE@.430 | Ea | 80.70 | 25.10 | 105.80 |
| Add for 100 watt lamp | — | Ea | 4.46 | — | 4.46 |
| Add for 150 watt lamp | — | Ea | 4.55 | — | 4.55 |

Obstruction light, meets Federal Aviation Administration Specification L-810, cast aluminum housing, two one-piece red, heat-resistant fresnel globes, with photoelectric control, mounted on 1" rigid steel conduit, including junction box and mounting plate

| | Craft@Hrs | Unit | Material | Labor | Total |
|---|---|---|---|---|---|
| Two 100 watt lamp fixture | CE@4.91 | Ea | 133.00 | 287.00 | 420.00 |

**Vandal Resistant Lighting** Commercial and architectural grade. Includes setting and connecting prewired fixtures but no wire, conduit or lamps.

Ceiling mounted 12" x 12" x 6" fixture, polycarbonate prismatic lens

| | Craft@Hrs | Unit | Material | Labor | Total |
|---|---|---|---|---|---|
| Two 100 watt incandescent lamps | CE@.764 | Ea | 87.20 | 44.60 | 131.80 |
| 35 watt, high pressure sodium lamp | CE@.764 | Ea | 254.00 | 44.60 | 298.60 |
| 50 watt, high pressure sodium lamp | CE@.764 | Ea | 259.00 | 44.60 | 303.60 |
| 70 watt, high pressure sodium lamp | CE@.764 | Ea | 274.00 | 44.60 | 318.60 |
| 22 watt fluorescent circline lamp | CE@.955 | Ea | 61.20 | 55.70 | 116.90 |

Wall mounted 6" wide, 9" high, 7" deep fixture, polycarbonate prismatic diffuser

| | Craft@Hrs | Unit | Material | Labor | Total |
|---|---|---|---|---|---|
| One 100 watt incandescent lamp | CE@.572 | Ea | 62.80 | 33.40 | 96.20 |
| 13 watt fluorescent lamp | CE@.764 | Ea | 98.80 | 44.60 | 143.40 |
| 35 watt, high pressure sodium lamp | CE@.475 | Ea | 266.00 | 27.70 | 293.70 |
| 50 watt, high pressure sodium lamp | CE@.475 | Ea | 267.00 | 27.70 | 294.70 |
| 70 watt, high pressure sodium lamp | CE@.475 | Ea | 254.00 | 27.70 | 281.70 |

Lamps for vandal-resistant fixtures

| | Craft@Hrs | Unit | Material | Labor | Total |
|---|---|---|---|---|---|
| 100 watt incandescent lamp | — | Ea | 4.47 | — | 4.47 |
| 13 watt fluorescent lamp | — | Ea | 9.30 | — | 9.30 |
| 35 to 70 watt high pressure sodium lamp | — | Ea | 66.50 | — | 66.50 |

**Lighted Exit Signs** Includes setting and connecting prewired fixtures but no wire or conduit.

Universal mount (wall or ceiling), Thinline series, 6" letters on one side, stencil face

| | Craft@Hrs | Unit | Material | Labor | Total |
|---|---|---|---|---|---|
| Single | CE@.764 | Ea | 27.70 | 44.60 | 72.30 |
| Twin | CE@.764 | Ea | 27.70 | 44.60 | 72.30 |

Universal mount, 6" letters on two sides of polycarbonate

| | Craft@Hrs | Unit | Material | Labor | Total |
|---|---|---|---|---|---|
| housing, two 15 watt lamps | CE@.764 | Ea | 43.40 | 44.60 | 88.00 |
| Add for two circuit supply | — | Ea | 9.08 | — | 9.08 |
| Add for 24" stem hanger | — | Ea | 12.40 | — | 12.40 |
| Add for battery backup for exit signs | CE@.381 | Ea | 21.70 | 22.20 | 43.90 |

**Explosion Proof Lighting**

No wire, conduit or concrete included. Includes setting and connecting prewired fixtures only.

Explosion proof fixtures, cast aluminum housing, heat-resistant prestressed globe, with fiberglass reinforced polyester reflector

| | Craft@Hrs | Unit | Material | Labor | Total |
|---|---|---|---|---|---|
| 60-200 watt incandescent fixture | CE@1.74 | Ea | 77.10 | 102.00 | 179.10 |
| 70 watt HID fixture | CE@1.74 | Ea | 298.00 | 102.00 | 400.00 |
| 150 watt HID fixture | CE@1.74 | Ea | 422.00 | 102.00 | 524.00 |

| | Craft@Hrs | Unit | Material | Labor | Total |
|---|---|---|---|---|---|

Explosion proof (Class 1 Division 1) fixtures, cast aluminum housing, heat-resistant prestressed globe, with fiberglass reinforced polyester reflector

| | Craft@Hrs | Unit | Material | Labor | Total |
|---|---|---|---|---|---|
| 250 watt, high pressure sodium fixture | CE@2.57 | Ea | 830.00 | 150.00 | 980.00 |
| 175 watt, metal halide fixture | CE@2.57 | Ea | 637.00 | 150.00 | 787.00 |
| 400 watt, metal halide fixture | CE@2.57 | Ea | 798.00 | 150.00 | 948.00 |

Explosion proof (Class 1 Division 2) fixtures, cast aluminum housing, heat-resistant prestressed globe, with fiberglass reinforced polyester reflector

| | Craft@Hrs | Unit | Material | Labor | Total |
|---|---|---|---|---|---|
| 70 watt, high pressure sodium fixture | CE@2.57 | Ea | 428.00 | 150.00 | 578.00 |
| 250 watt, high pressure sodium fixture | CE@2.57 | Ea | 586.00 | 150.00 | 736.00 |
| 175 watt, metal halide fixture | CE@2.57 | Ea | 398.00 | 150.00 | 548.00 |
| 400 watt, metal halide fixture | CE@2.57 | Ea | 564.00 | 150.00 | 714.00 |

**Emergency Lighting** Includes setting and connecting prewired fixtures but no wire or conduit.

Shelf or wall mounted solid state battery pack with charger, lead-acid battery, baked enamel finish, with 2 halogen lamps, 90 minute operation, indoor use only

| | Craft@Hrs | Unit | Material | Labor | Total |
|---|---|---|---|---|---|
| 6 volt, 15 watt | CE@1.74 | Ea | 101.00 | 102.00 | 203.00 |
| 6 volt, 30 watt | CE@1.74 | Ea | 155.00 | 102.00 | 257.00 |
| 6 volt, 50 watt | CE@1.74 | Ea | 183.00 | 102.00 | 285.00 |
| 6 volt, 100 watt | CE@1.74 | Ea | 246.00 | 102.00 | 348.00 |
| 12 volt, 100 watt | CE@1.74 | Ea | 248.00 | 102.00 | 350.00 |

Self diagnostic emergency lighting fixtures, microprocessor based, continuously monitors battery/lamps

| | Craft@Hrs | Unit | Material | Labor | Total |
|---|---|---|---|---|---|
| Emergency fixture, two lamps, wall mount | CE@.825 | Ea | 188.00 | 48.20 | 236.20 |
| LED exit sign with emergency fixture, wall mount | CE@.825 | Ea | 331.00 | 48.20 | 379.20 |
| LED exit sign only, wall mount | CE@.825 | Ea | 282.00 | 48.20 | 330.20 |

| | Craft@Hrs | Unit | Material | Labor | Equipment | Total |
|---|---|---|---|---|---|---|

**Yard and Street Lighting** Includes setting and connecting prewired fixtures with ballast but no pole, concrete work, excavation, wire, lamps or conduit. See pole costs below. Equipment is a flatbed truck with a boom & auger.

Mast arm mounted rectangular high pressure sodium flood, aluminum housing with baked enamel finish, anodized aluminum reflector, tempered glass lens, slip-fitting for mast mount, photoelectric cell

| | Craft@Hrs | Unit | Material | Labor | Equipment | Total |
|---|---|---|---|---|---|---|
| 70 watt | E1@1.55 | Ea | 441.00 | 78.00 | 19.70 | 538.70 |
| 100 watt | E1@1.55 | Ea | 456.00 | 78.00 | 19.70 | 553.70 |
| 150 watt | E1@1.55 | Ea | 457.00 | 78.00 | 19.70 | 554.70 |
| 250 watt | E1@1.55 | Ea | 487.00 | 78.00 | 19.70 | 584.70 |
| 400 watt | E1@1.55 | Ea | 581.00 | 78.00 | 19.70 | 678.70 |
| Add for 6' x 2" mounting arm | E1@.547 | Ea | 68.00 | 27.50 | 19.70 | 115.20 |

Yoke mounted high intensity discharge flood, die-cast anodized aluminum housing with tempered glass lens

| | Craft@Hrs | Unit | Material | Labor | Equipment | Total |
|---|---|---|---|---|---|---|
| 250 watt, high pressure sodium | E1@2.71 | Ea | 384.00 | 136.00 | 34.40 | 554.40 |
| 250 watt, metal halide | E1@2.71 | Ea | 336.00 | 136.00 | 34.40 | 506.40 |
| 400 watt, high pressure sodium | E1@2.71 | Ea | 395.00 | 136.00 | 34.40 | 565.40 |
| 400 watt, metal halide | E1@2.71 | Ea | 343.00 | 136.00 | 34.40 | 513.40 |
| 1,000 watt, high pressure sodium | E1@2.71 | Ea | 644.00 | 136.00 | 34.40 | 814.40 |
| 1,000 watt, metal halide | E1@2.71 | Ea | 516.00 | 136.00 | 34.40 | 686.40 |
| 1,000 watt, mercury vapor | E1@2.71 | Ea | 84.10 | 136.00 | 34.40 | 254.50 |

Round flood, aluminum reflector, impact-resistant glass lens

| | Craft@Hrs | Unit | Material | Labor | Equipment | Total |
|---|---|---|---|---|---|---|
| 400 watt, high pressure sodium | E1@2.71 | Ea | 447.00 | 136.00 | 34.40 | 617.40 |
| 1,000 watt, high pressure sodium | E1@2.71 | Ea | 714.00 | 136.00 | 34.40 | 884.40 |
| 400 watt, metal halide | E1@2.71 | Ea | 410.00 | 136.00 | 34.40 | 580.40 |
| 1,000 watt, metal halide | E1@2.71 | Ea | 562.00 | 136.00 | 34.40 | 732.40 |

Cobra head high intensity flood, anodized aluminum reflector, glass lens, die cast aluminum housing, with photoelectric control but no bulb or mounting arm

| | Craft@Hrs | Unit | Material | Labor | Equipment | Total |
|---|---|---|---|---|---|---|
| 1,000 watt, high pressure sodium | E1@2.71 | Ea | 650.00 | 136.00 | 34.40 | 820.40 |
| 1,000 watt, mercury vapor | E1@2.71 | Ea | 477.00 | 136.00 | 34.40 | 647.40 |
| 1,000 watt, metal halide | E1@2.71 | Ea | 515.00 | 136.00 | 34.40 | 685.40 |

| | Craft@Hrs | Unit | Material | Labor | Equipment | Total |
|---|---|---|---|---|---|---|
| Square metal halide flood, anodized aluminum housing, tempered glass door, trunnion or yoke mount | | | | | | |
| 400 watt | E1@2.71 | Ea | 552.00 | 136.00 | 34.40 | 722.40 |
| 1,000 watt | E1@2.71 | Ea | 653.00 | 136.00 | 34.40 | 823.40 |

## Yard and Street Lighting Poles

Hole for a pole foundation dug with a truck-mounted auger, no soil disposal included

| | Craft@Hrs | Unit | Material | Labor | Equipment | Total |
|---|---|---|---|---|---|---|
| Per CF of undisturbed soil | E1@.064 | CF | — | 3.22 | .81 | 4.03 |

Concrete pole foundations, formed, poured and finished, with anchor bolts. Material costs shown include concrete, forms and anchor bolts

| | Craft@Hrs | Unit | Material | Labor | Equipment | Total |
|---|---|---|---|---|---|---|
| 12" diameter, 36" deep, for 12' pole | E1@.746 | Ea | 23.90 | 37.50 | 9.48 | 70.88 |
| 24" diameter, 72" deep, for 30' pole | E1@1.14 | Ea | 95.40 | 57.30 | 14.50 | 167.20 |
| 30" diameter, 76" deep, for 40' pole | E1@1.54 | Ea | 125.00 | 77.50 | 19.60 | 222.10 |

**Yard and street light poles,** including base plate but no top brackets. See labor costs below.

| | Unit | Round steel | Square steel | Alumi-num |
|---|---|---|---|---|
| 12' high | Ea | 729.00 | 472.00 | 293.00 |
| 16' high | Ea | 850.00 | 553.00 | 363.00 |
| 20' high, 6" wide | Ea | 1,160.00 | 696.00 | 516.00 |
| 25' high, 6" wide | Ea | 1,390.00 | 850.00 | 1,150.00 |
| 30' high, 8" wide | Ea | 1,590.00 | 1,290.00 | 1,280.00 |
| 35' high, 8" wide | Ea | 2,060.00 | 1,610.00 | 1,990.00 |
| 40' high, 9" wide | Ea | 2,420.00 | 1,830.00 | 2,280.00 |
| 50' high, 10" wide | Ea | 2,700.00 | 2,650.00 | — |

POLE BASE

| | Craft@Hrs | Unit | Material | Labor | Equipment | Total |
|---|---|---|---|---|---|---|
| **Labor setting metal light poles** Equipment is a flatbed truck with a boom & auger | | | | | | |
| To 20' high | E1@2.89 | Ea | — | 145.00 | 36.70 | 181.70 |
| 25' to 30' high | E1@4.35 | Ea | — | 219.00 | 55.30 | 274.30 |
| 35' to 40' high | E1@5.73 | Ea | — | 288.00 | 72.80 | 360.80 |
| 50' high | E1@8.57 | Ea | — | 431.00 | 109.00 | 540.00 |
| Deduct for aluminum poles | — | % | — | -20.0 | -20.0 | — |
| Top brackets for steel and aluminum light poles, installed prior to setting pole | | | | | | |
| Single arm | CE@.500 | Ea | 96.30 | 29.20 | 6.35 | 131.85 |
| Two arm | CE@.500 | Ea | 113.00 | 29.20 | 6.35 | 148.55 |
| Three arm | CE@.991 | Ea | 147.00 | 57.80 | 12.60 | 217.40 |
| Four arm | CE@.991 | Ea | 183.00 | 57.80 | 12.60 | 253.40 |

## Area Lighting, Installed on Poles

These costs include excavation for pole, pole foundation (2' x 2' x 3' deep), 150' of 2" PVC conduit run with 40' of RSC, 600' of #8 copper wire and terminations, 35' x 8" square steel pole and one arm and 480 volt mercury vapor 1,000 watt luminaire with lamp for each fixture. Equipment is a flatbed truck with a boom & auger.

Cost per pole including fixtures as shown

CONCRETE POLE

FIXTURE

TOP VIEW

| | Craft@Hrs | Unit | Material | Labor | Equipment | Total |
|---|---|---|---|---|---|---|
| 1 fixture per pole | E1@30.4 | Ea | 3,590.00 | 1,530.00 | 386.00 | 5,506.00 |
| 2 fixtures per pole | E1@32.5 | Ea | 4,920.00 | 1,630.00 | 413.00 | 6,963.00 |
| 3 fixtures per pole | E1@34.8 | Ea | 6,260.00 | 1,750.00 | 442.00 | 8,452.00 |
| 4 fixtures per pole | E1@36.8 | Ea | 7,590.00 | 1,850.00 | 468.00 | 9,908.00 |

# 27 Communications

| | Craft@Hrs | Unit | Material | Labor | Total |
|---|---|---|---|---|---|

**Security and Alarm Systems** Sensor costs include connection but no wire runs.

Monitor panels, including accessory section and connection to signal and power supply wiring, cost per panel

1 zone wall-mounted cabinet with 1 monitor panel, tone, standard line
| supervision, and 115 volt power supply | E4@2.02 | Ea | 984.00 | 99.20 | 1,083.20 |

5 zone wall-mounted cabinet with 5 monitor panels, 10 zone monitor rack, 5 monitor panel blanks, tone,
| standard line supervision and 115 volt power supply | E4@8.21 | Ea | 5,580.00 | 403.00 | 5,983.00 |

10 zone wall-mounted cabinet with 10 monitor panels, 10 zone monitor rack, tone, standard line
| supervision and 115 volt power supply | E4@12.1 | Ea | 6,410.00 | 594.00 | 7,004.00 |

10 zone monitor rack and 10 monitor panels with tone and standard line supervision
| but no power supply and no cabinet | E4@11.1 | Ea | 5,300.00 | 545.00 | 5,845.00 |

1 zone wall-mounted cabinet with 1 monitor panel, tone, high security line supervision
| and 115 volt power supply | E4@2.02 | Ea | 1,310.00 | 99.20 | 1,409.20 |
| high security line supervision but no power supply | E4@7.82 | Ea | 6,960.00 | 384.00 | 7,344.00 |

10 zone wall-mounted cabinet with 10 monitor panels, 10 zone monitor rack, tone, high security
| line supervision and 115 volt power supply | E4@12.0 | Ea | 8,150.00 | 590.00 | 8,740.00 |
| Emergency power indicator for monitor panel | E4@1.14 | Ea | 261.00 | 56.00 | 317.00 |

Monitor panels

Panel with accessory section, tone and standard
| line supervision | E4@.990 | Ea | 430.00 | 48.60 | 478.60 |

Panel with accessory section, tone and high security
| line supervision | E4@.904 | Ea | 647.00 | 44.40 | 691.40 |

Monitor racks
| 1 zone with 115 volt power supply | E4@.904 | Ea | 354.00 | 44.40 | 398.40 |
| 10 zone with 115 volt power supply | E4@.904 | Ea | 1,820.00 | 44.40 | 1,864.40 |
| 10 zone with no power supply | E4@.904 | Ea | 1,410.00 | 44.40 | 1,454.40 |

Monitor cabinets
| 1 zone monitor for wall mounted cabinet | E4@.990 | Ea | 555.00 | 48.60 | 603.60 |
| 5 zone monitor for wall mounted cabinet | E4@.990 | Ea | 712.00 | 48.60 | 760.60 |
| 10 zone monitor for wall mounted cabinet | E4@.990 | Ea | 868.00 | 48.60 | 916.60 |
| 20 zone monitor for wall mounted cabinet | E4@.990 | Ea | 1,170.00 | 48.60 | 1,218.60 |
| 50 zone monitor for wall mounted cabinet | E4@.990 | Ea | 2,630.00 | 48.60 | 2,678.60 |
| 50 zone monitor for floor mounted cabinet | E4@1.04 | Ea | 2,860.00 | 51.10 | 2,911.10 |

Balanced magnetic door switches
| Surface mounted | E4@2.02 | Ea | 108.00 | 99.20 | 207.20 |
| Surface mounted with remote test | E4@2.02 | Ea | 152.00 | 99.20 | 251.20 |
| Flush mounted | E4@2.02 | Ea | 104.00 | 99.20 | 203.20 |
| Mounted bracket or spacer | E4@.249 | Ea | 6.28 | 12.20 | 18.48 |

Photoelectric sensors
| 500' range, 12 volt DC | E4@2.02 | Ea | 297.00 | 99.20 | 396.20 |
| 800' range, 12 volt DC | E4@2.02 | Ea | 334.00 | 99.20 | 433.20 |
| Fence type, 6 beam, without wire or trench | E4@3.89 | Ea | 10,800.00 | 191.00 | 10,991.00 |
| Fence type, 9 beam, without wire or trench | E4@3.89 | Ea | 13,500.00 | 191.00 | 13,691.00 |

Capacitance wire grid systems
| Surface type | E4@.990 | Ea | 76.10 | 48.60 | 124.70 |
| Duct type | E4@.990 | Ea | 62.00 | 48.60 | 110.60 |
| Tube grid kit | E4@.990 | Ea | 112.00 | 48.60 | 160.60 |
| Vibration sensors, to 30 per zone | E4@2.02 | Ea | 127.00 | 99.20 | 226.20 |
| Audio sensors, to 30 per zone | E4@2.02 | Ea | 157.00 | 99.20 | 256.20 |
| Inertia sensors, outdoor, without trenching | E4@2.02 | Ea | 104.00 | 99.20 | 203.20 |
| Inertia sensors, indoor | E4@2.02 | Ea | 69.90 | 99.20 | 169.10 |
| Electric sensor cable, 300 meters, no trenching | E4@3.89 | Ea | 1,580.00 | 191.00 | 1,771.00 |

| | Craft@Hrs | Unit | Material | Labor | Total |
|---|---|---|---|---|---|
| Ultrasonic transmitters, to 20 per zone | | | | | |
|   Omni-directional | E4@2.02 | Ea | 68.00 | 99.20 | 167.20 |
|   Directional | E4@2.02 | Ea | 76.10 | 99.20 | 175.30 |
| Ultrasonic transceivers, to 20 per zone | | | | | |
|   Omni-directional | E4@2.02 | Ea | 76.10 | 99.20 | 175.30 |
|   Directional | E4@2.02 | Ea | 81.50 | 99.20 | 180.70 |
|   High security | E4@2.02 | Ea | 277.00 | 99.20 | 376.20 |
|   Standard security | E4@2.02 | Ea | 112.00 | 99.20 | 211.20 |
| Microwave perimeter sensor, to 4 per zone | E4@3.89 | Ea | 4,450.00 | 191.00 | 4,641.00 |
| Passive interior infrared sensors, to 20 per zone | | | | | |
|   Wide pattern | E4@2.02 | Ea | 616.00 | 99.20 | 715.20 |
|   Narrow pattern | E4@2.02 | Ea | 584.00 | 99.20 | 683.20 |
| Access and secure control units | | | | | |
|   For balanced magnetic door switches | E4@3.06 | Ea | 337.00 | 150.00 | 487.00 |
|   For photoelectric sensors | E4@3.06 | Ea | 616.00 | 150.00 | 766.00 |
|   For capacitance sensors | E4@3.06 | Ea | 625.00 | 150.00 | 775.00 |
|   For audio and vibration sensors | E4@3.06 | Ea | 721.00 | 150.00 | 871.00 |
|   For inertia sensors | E4@3.06 | Ea | 647.00 | 150.00 | 797.00 |
|   For nimmer (m.b.) detector | E4@3.06 | Ea | 538.00 | 150.00 | 688.00 |
|   For electric cable sensors | E4@3.06 | Ea | 3,250.00 | 150.00 | 3,400.00 |
|   For ultrasonic sensors | E4@3.06 | Ea | 990.00 | 150.00 | 1,140.00 |
|   For microwave sensors | E4@3.06 | Ea | 609.00 | 150.00 | 759.00 |
| Accessories | | | | | |
|   Tamper assembly for monitor cabinet | E4@.990 | Ea | 68.00 | 48.60 | 116.60 |
|   Monitor panel blank | E4@.249 | Ea | 7.59 | 12.20 | 19.79 |
|   Audible alarm | E4@.990 | Ea | 79.50 | 48.60 | 128.10 |
|   Audible alarm control | E4@.990 | Ea | 315.00 | 48.60 | 363.60 |
|   Termination screw terminal cabinet, 25 pair | E4@.990 | Ea | 229.00 | 48.60 | 277.60 |
|   Termination screw terminal cabinet, 50 pair | E4@1.36 | Ea | 371.00 | 66.80 | 437.80 |
|   Termination screw terminal cabinet, 150 pair | E4@3.45 | Ea | 612.00 | 169.00 | 781.00 |
|   Universal termination with remote test | | | | | |
|     for cabinets and control panels | E4@1.47 | Ea | 55.90 | 72.20 | 128.10 |
|   Universal termination without remote test | E4@.990 | Ea | 32.70 | 48.60 | 81.30 |
|   High security line supervision termination | | | | | |
|     With remote test and tone | E4@1.47 | Ea | 283.00 | 72.20 | 355.20 |
|     With tone only | E4@1.47 | Ea | 261.00 | 72.20 | 333.20 |
|   12" door cord for capacitance sensor | E4@.249 | Ea | 8.72 | 12.20 | 20.92 |
|   Insulation block kit for capacitance sensor | E4@.990 | Ea | 43.40 | 48.60 | 92.00 |
|   Termination block for capacitance sensor | E4@.249 | Ea | 9.23 | 12.20 | 21.43 |
|   200 zone graphic display | E4@3.06 | Ea | 4,340.00 | 150.00 | 4,490.00 |
|   210 zone multiplexer event recorder unit | E4@3.89 | Ea | 4,140.00 | 191.00 | 4,331.00 |
|   Guard alert display | E4@3.06 | Ea | 974.00 | 150.00 | 1,124.00 |
|   Uninterruptable power supply (battery) | E4@2.02 | Ea | 1,990.00 | 99.20 | 2,089.20 |
|   12 volt, 40VA plug-in transformer | E4@.495 | Ea | 37.40 | 24.30 | 61.70 |
|   18 volt, 40VA plug-in transformer | E4@.495 | Ea | 25.60 | 24.30 | 49.90 |
|   24 volt, 40VA plug-in transformer | E4@.495 | Ea | 18.60 | 24.30 | 42.90 |
|   Test relay | E4@.332 | Ea | 57.60 | 16.30 | 73.90 |
|   Sensor test control (average for sensors) | E4@.990 | Ea | 75.50 | 48.60 | 124.10 |
|   Cable, 2 pair 22 gauge, twisted, 1 pair shielded, run in exposed walls or | | | | | |
|     per LF per pull, pulled in raceway and connected | E4@.008 | LF | 1.55 | .39 | 1.94 |

Security access digital outdoor keypad/reader, modular single-unit type. Up to sixty users. Accepts cards, keyed-in passwords or both, non-volatile memory. Accepts any magnetic strip card format. Add the cost of the door actuator mechanism and the door. Cost includes control wiring.

| | Craft@Hrs | Unit | Material | Labor | Total |
|---|---|---|---|---|---|
|   Digital access outdoor keypad/reader | CE@4.0 | Ea | 314.00 | 233.00 | 547.00 |

# 27 Communications

| | Craft@Hrs | Unit | Material | Labor | Total |
|---|---|---|---|---|---|
| **Fire Alarm and Detection Systems** No wiring included except as noted. | | | | | |
| Pedestal mounted master fire alarm box | E4@3.89 | Ea | 1,340.00 | 191.00 | 1,531.00 |
| Automatic coded signal transmitter | E4@3.89 | Ea | 707.00 | 191.00 | 898.00 |
| Fire alarm control panel, simplex | E4@3.89 | Ea | 662.00 | 191.00 | 853.00 |
| Fire detection annunciator panel | | | | | |
|     8 zone drop | E4@3.89 | Ea | 438.00 | 191.00 | 629.00 |
|     12 zone drop | E4@5.03 | Ea | 576.00 | 247.00 | 823.00 |
|     16 zone drop | E4@5.93 | Ea | 895.00 | 291.00 | 1,186.00 |
|     8 zone lamp panel only | E4@4.33 | Ea | 145.00 | 213.00 | 358.00 |
|     12 zone lamp panel only | E4@5.93 | Ea | 214.00 | 291.00 | 505.00 |
|     16 zone lamp panel only | E4@7.23 | Ea | 284.00 | 355.00 | 639.00 |
| Battery charger and cabinet, simplex | E4@6.19 | Ea | 381.00 | 304.00 | 685.00 |
|     Add for nickel cadmium batteries | E4@3.89 | Ea | 414.00 | 191.00 | 605.00 |
| Fire alarm pull stations, manual operation | E4@.497 | Ea | 21.40 | 24.40 | 45.80 |
| Fire alarm 10" bell with outlet box | E4@.497 | Ea | 69.60 | 24.40 | 94.00 |
| Magnetic door holder | E4@1.33 | Ea | 65.20 | 65.30 | 130.50 |
| Combination door holder and closer | E4@2.48 | Ea | 264.00 | 122.00 | 386.00 |
| Detex door lock | E4@2.72 | Ea | 31.30 | 134.00 | 165.30 |
| Thermodetector | E4@.497 | Ea | 9.43 | 24.40 | 33.83 |
| Ionization AC smoke detector, with wiring | E4@.710 | Ea | 161.00 | 34.90 | 195.90 |
| Fixed temp. and rate of rise smoke detector | E4@.746 | Ea | 14.50 | 36.70 | 51.20 |
| Carbon dioxide pressure switch | E4@.746 | Ea | 4.99 | 36.70 | 41.69 |
| | | | | | |
| **Telephone Wiring** | | | | | |
| Plywood backboard mount for circuit wiring, set on masonry wall | | | | | |
|     4' x 4' x 3/4" | E4@.938 | Ea | 11.90 | 46.10 | 58.00 |
|     4' x 8' x 3/4" | E4@1.10 | Ea | 23.90 | 54.00 | 77.90 |
| Cable tap in a manhole or junction box | | | | | |
|     25 to 50 pair cable | E4@7.80 | Ea | 115.00 | 383.00 | 498.00 |
|     100 to 200 pair cable | E4@10.1 | Ea | 132.00 | 496.00 | 628.00 |
|     300 pair cable | E4@12.0 | Ea | 134.00 | 590.00 | 724.00 |
|     400 pair cable | E4@13.0 | Ea | 145.00 | 639.00 | 784.00 |
| Cable terminations in a manhole or junction box | | | | | |
|     25 pair cable | E4@1.36 | Ea | 10.40 | 66.80 | 77.20 |
|     50 pair cable | E4@1.36 | Ea | 11.90 | 66.80 | 78.70 |
|     100 pair cable | E4@2.72 | Ea | 23.80 | 134.00 | 157.80 |
|     150 pair cable | E4@3.45 | Ea | 36.00 | 169.00 | 205.00 |
|     200 pair cable | E4@4.61 | Ea | 53.80 | 226.00 | 279.80 |
|     300 pair cable | E4@5.03 | Ea | 71.50 | 247.00 | 318.50 |
|     400 pair cable | E4@5.34 | Ea | 94.90 | 262.00 | 356.90 |
| Communications cable, pulled in conduit or walls and connected | | | | | |
|     2 pair cable | E4@.008 | LF | .12 | .39 | .51 |
|     25 pair cable | E4@.030 | LF | 1.43 | 1.47 | 2.90 |
|     50 pair cable | E4@.043 | LF | 3.02 | 2.11 | 5.13 |
|     75 pair cable | E4@.043 | LF | 3.77 | 2.11 | 5.88 |
|     100 pair cable | E4@.043 | LF | 4.39 | 2.11 | 6.50 |
|     150 pair cable | E4@.045 | LF | 6.12 | 2.21 | 8.33 |
|     200 pair cable | E4@.051 | LF | 7.97 | 2.51 | 10.48 |
|     300 pair cable | E4@.051 | LF | 11.80 | 2.51 | 14.31 |
| Telephone outlets (no junction box included) | | | | | |
|     Wall outlet | E4@.396 | Ea | 6.24 | 19.50 | 25.74 |
|     Floor outlet, flush mounted | E4@.495 | Ea | 7.75 | 24.30 | 32.05 |

| | Craft@Hrs | Unit | Material | Labor | Total |
|---|---|---|---|---|---|

## Closed Circuit Television Systems

Outdoor television camera in weatherproof housing, cable preinstalled.

| | Craft@Hrs | Unit | Material | Labor | Total |
|---|---|---|---|---|---|
| Wall or roof mount, professional grade | E4@.500 | Ea | 1,620.00 | 24.60 | 1,644.60 |
| Wall or roof mount, economy | E4@.500 | Ea | 225.00 | 24.60 | 249.60 |
| Self-terminating outlet box, cover plate | E4@.245 | Ea | 5.06 | 12.00 | 17.06 |
| Coaxial cable (RG59/J), 75 ohms, no raceway included | E4@.006 | LF | .34 | .29 | .63 |
| In-line cable taps (PTU) for 36" TV system | E4@.689 | Ea | 8.00 | 33.90 | 41.90 |
| Cable blocks for in-line taps | — | Ea | 4.16 | — | 4.16 |

Desk-type TV monitor, 19" diagonal screen, 75 ohms, labor includes hookup only

| | Craft@Hrs | Unit | Material | Labor | Total |
|---|---|---|---|---|---|
| 4 camera inputs | E4@.500 | Ea | 343.00 | 24.60 | 367.60 |

Security CCTV digital video recorder central control station, modular single-unit type. Allows up to 16 discrete channels and simultaneous real-time monitoring of all 16 channels. Digital video recorder saves pictures every 3 seconds per channel. Password control access to DVD data only. Cost includes monitor, PC, software, cable connectors for sixteen stations, keyboard, mouse, software training and installation. Add the cost of wiring to digital CCTV's.

| | Craft@Hrs | Unit | Material | Labor | Total |
|---|---|---|---|---|---|
| Security CCTV recorder control station | CE@4.0 | Ea | 5,620.00 | 233.00 | 5,853.00 |

## Public Address Paging Systems

Signal cable, #22 solid copper polyethylene insulated, PVC jacketed wire

| | Craft@Hrs | Unit | Material | Labor | Total |
|---|---|---|---|---|---|
| Pulled in conduit, no conduit included | E4@.008 | LF | .15 | .39 | .54 |

Microphone, desk type, push to talk, with

| | Craft@Hrs | Unit | Material | Labor | Total |
|---|---|---|---|---|---|
| 7' of three conductor cable | E4@.250 | Ea | 105.00 | 12.30 | 117.30 |

Wall mounted enameled steel amplifier cabinet (18" high, 24" wide, 6" deep) with electrical receptacle

| | Craft@Hrs | Unit | Material | Labor | Total |
|---|---|---|---|---|---|
| and lockable door | E4@1.67 | Ea | 308.00 | 82.00 | 390.00 |

Amplifier, 100 watts, 4 inputs, installation includes hookup only, no wire pull

| | Craft@Hrs | Unit | Material | Labor | Total |
|---|---|---|---|---|---|
| Complete unit | E4@.500 | Ea | 355.00 | 24.60 | 379.60 |

Wall mounted paging horns, transformer type, 250 to 13,000 Hertz, 8 ohms, 30 Watts, 10" diameter, 12" deep, gray baked enamel, labor includes hookup only

| | Craft@Hrs | Unit | Material | Labor | Total |
|---|---|---|---|---|---|
| Interior/exterior 15/30 Watts | E4@.250 | Ea | 69.00 | 12.30 | 81.30 |

Ceiling mounted 8" speaker with 12" grill, labor includes installation and hookup only, no wire pull

| | Craft@Hrs | Unit | Material | Labor | Total |
|---|---|---|---|---|---|
| 8" ceiling speaker | E4@.330 | Ea | 26.70 | 16.20 | 42.90 |
| 8" ceiling speaker with volume | E4@.330 | Ea | 28.80 | 16.20 | 45.00 |

Outdoor weather resistant speakers

| | Craft@Hrs | Unit | Material | Labor | Total |
|---|---|---|---|---|---|
| Landscape weather resistant | E4@.330 | Ea | 210.00 | 16.20 | 226.20 |
| Rock landscape (high end) | E4@.330 | Ea | 415.00 | 16.20 | 431.20 |

**Computer Network Cable** Pulled in conduit or walls. Add for connectors and cable testing. Prices are for cable purchased in bulk on 1,000' spools.

Coaxial cable

Thin Ethernet, 50 ohm, 10BASE2 (RG58). Used in modern networks for Wireless Network high gain antennae

| | Craft@Hrs | Unit | Material | Labor | Total |
|---|---|---|---|---|---|
| PVC cable | E4@.008 | LF | .13 | .39 | .52 |
| Plenum cable | E4@.008 | LF | .33 | .39 | .72 |

Fiber optic cable

Singlemode fiber optic cable, duplex is 4 strand, simplex is 1 strand

| | Craft@Hrs | Unit | Material | Labor | Total |
|---|---|---|---|---|---|
| Singlemode duplex fiber optic cables (8.3/125) | E4@.012 | LF | .19 | .59 | .78 |
| Singlemode simplex fiber optic cables (8.3/125) | E4@.012 | LF | .18 | .59 | .77 |

Multimode fiber optic cable, duplex is a zip type

| | Craft@Hrs | Unit | Material | Labor | Total |
|---|---|---|---|---|---|
| Multimode duplex fiber optic cables (50/125) | E4@.012 | LF | .22 | .59 | .81 |
| Multimode duplex fiber optic cables (62.5/125) | E4@.012 | LF | .22 | .59 | .81 |
| Multimode simplex fiber optic cables (50/125) | E4@.012 | LF | .22 | .59 | .81 |
| Multimode simplex fiber optic cables (62.5/125) | E4@.012 | LF | .22 | .59 | .81 |

|  | Craft@Hrs | Unit | Material | Labor | Total |
|---|---|---|---|---|---|
| Twisted pair Category 5, 100 Mbps, material price based on purchase of 1000' rolls, use solid for horizontal runs. STP is shielded | | | | | |
| 4 pair UTP PVC stranded cable | E4@.008 | LF | .12 | .39 | .51 |
| 4 pair UTP PVC solid cable | E4@.008 | LF | .12 | .39 | .51 |
| 4 pair UTP plenum solid cable | E4@.008 | LF | .21 | .39 | .60 |
| 4 pair STP PVC stranded cable | E4@.008 | LF | .18 | .39 | .57 |
| 4 pair STP PVC solid cable | E4@.008 | LF | .18 | .39 | .57 |
| 4 pair STP plenum solid cable | E4@.008 | LF | .21 | .39 | .60 |
| Twisted pair Category 5e, 350 Mbps, material price based on purchase of 1,000' rolls, use solid for horizontal runs. STP is shielded. | | | | | |
| 4 pair UTP PVC solid | E4@.008 | LF | .17 | .39 | .56 |
| 4 pair UTP PVC stranded | E4@.008 | LF | .17 | .39 | .56 |
| 4 pair UTP plenum solid | E4@.008 | LF | .25 | .39 | .64 |
| 4 pair STP PVC stranded | E4@.008 | LF | .18 | .39 | .57 |
| 4 pair STP PVC solid | E4@.008 | LF | .18 | .39 | .57 |
| 4 pair STP plenum solid | E4@.008 | LF | .26 | .39 | .65 |
| Twisted pair Category 6, 1 Gbps, material price based on purchase of 1,000' rolls, use solid for horizontal runs | | | | | |
| 4 pair UTP PVC stranded | E4@.008 | LF | .18 | .39 | .57 |
| 4 pair UTP PVC solid | E4@.008 | LF | .19 | .39 | .58 |
| 4 pair UTP plenum solid | E4@.008 | LF | .31 | .39 | .70 |
| Add for multiple cable pulls of thin Ethernet or unshielded twisted pair cables when cables are pulled simultaneously in plenum or PVC, based on quantity of cables pulled | | | | | |
| Add for each cable pulled simultaneously, over one | E4@.250 | Ea | — | 12.30 | 12.30 |

## Punch-down blocks

| Plywood backboard mount for patch panel or punch-down blocks, set on a masonry wall | | | | | |
|---|---|---|---|---|---|
| 4' x 4' x 3/4" | E4@.938 | Ea | 11.90 | 46.10 | 58.00 |
| 4' x 8' x 3/4" | E4@1.10 | Ea | 23.90 | 54.00 | 77.90 |
| Distribution frame for punch-down blocks | | | | | |
| Holds 5 punch-down blocks | E4@.250 | Ea | 33.50 | 12.30 | 45.80 |
| Holds 10 punch-down blocks | E4@.250 | Ea | 39.80 | 12.30 | 52.10 |
| Punch-down blocks, labor is to install block in distribution frame only, add labor to punch-down wire pairs as shown below | | | | | |
| 50 pair | E4@.250 | Ea | 18.30 | 12.30 | 30.60 |
| 100 pair | E4@.250 | Ea | 26.60 | 12.30 | 38.90 |
| Punch-down wire pairs, labor only, per pair | E4@.020 | Ea | — | .98 | .98 |

## Computer network cable connectors

| Coax | | | | | |
|---|---|---|---|---|---|
| Thin Ethernet | | | | | |
| BNC type, crimp on, female | E4@.100 | Ea | 3.74 | 4.91 | 8.65 |
| Fiber optic cable, terminated, 10' length | | | | | |
| ST-ST singlemode duplex riser | E4@.100 | Ea | 57.00 | 4.91 | 61.91 |
| ST-SC singlemode duplex riser | E4@.100 | Ea | 57.00 | 4.91 | 61.91 |
| SC-SC singlemode duplex riser | E4@.100 | Ea | 57.00 | 4.91 | 61.91 |
| FC-SC singlemode duplex riser | E4@.100 | Ea | 57.00 | 4.91 | 61.91 |
| ST-ST singlemode simplex riser | E4@.100 | Ea | 31.20 | 4.91 | 36.11 |
| ST-SC singlemode simplex riser | E4@.100 | Ea | 31.20 | 4.91 | 36.11 |
| SC-SC singlemode simplex riser | E4@.100 | Ea | 31.20 | 4.91 | 36.11 |
| FC-SC singlemode simplex riser | E4@.100 | Ea | 31.20 | 4.91 | 36.11 |
| ST-ST multimode duplex riser | E4@.100 | Ea | 26.00 | 4.91 | 30.91 |

# 27 Communications

| | Craft@Hrs | Unit | Material | Labor | Total |
|---|---|---|---|---|---|
| ST-SC multimode duplex riser | E4@.100 | Ea | 26.00 | 4.91 | 30.91 |
| SC-SC multimode duplex riser | E4@.100 | Ea | 26.00 | 4.91 | 30.91 |
| FC-SC multimode duplex riser | E4@.100 | Ea | 77.90 | 4.91 | 82.81 |
| FDDI-FDDI multimode duplex riser | E4@.100 | Ea | 98.60 | 4.91 | 103.51 |
| FDDI OFNP plenum | E4@.100 | Ea | 93.40 | 4.91 | 98.31 |
| Fiber optic cable connectors, special tools required | | | | | |
| SC multimode connector | E4@.150 | Ea | 8.30 | 7.37 | 15.67 |
| SC singlemode connector | E4@.150 | Ea | 9.34 | 7.37 | 16.71 |
| FC multimode connector | E4@.150 | Ea | 9.34 | 7.37 | 16.71 |
| FC singlemode connector | E4@.150 | Ea | 9.34 | 7.37 | 16.71 |
| ST multimode connector | E4@.150 | Ea | 8.30 | 7.37 | 15.67 |
| ST singlemode connector | E4@.150 | Ea | 9.34 | 7.37 | 16.71 |
| FDDI for round cable | E4@.150 | Ea | 15.70 | 7.37 | 23.07 |
| FDDI for Zip cable | E4@.150 | Ea | 15.70 | 7.37 | 23.07 |
| Shielded twisted pair (STP) | | | | | |
| IBM Type 2, 6 pair data connector | E4@.100 | Ea | 2.18 | 4.91 | 7.09 |
| IBM Type 6, 2 pair data connector | E4@.150 | Ea | 3.06 | 7.37 | 10.43 |
| IBM Type 9, 2 pair data connector | E4@.150 | Ea | 3.06 | 7.37 | 10.43 |
| Twisted pair connectors (UTP) and (STP), modular hand tool required, material prices based on package of 50 | | | | | |
| RJ11 | E4@.100 | Ea | .54 | 4.91 | 5.45 |
| Cat5-rated RJ45 connectors for solid wire | E4@.100 | Ea | .75 | 4.91 | 5.66 |
| Cat5-rated RJ45 connectors for stranded wire | E4@.100 | Ea | .75 | 4.91 | 5.66 |
| Cat6-rated RJ45 connectors for stranded wire | E4@.100 | Ea | .75 | 4.91 | 5.66 |
| Shielded RJ45 connectors for solid or stranded wire | E4@.100 | Ea | 2.13 | 4.91 | 7.04 |
| **Cable testing** Testing, with dedicated cable tester | | | | | |
| Coax cable, per conductor | E4@.125 | Ea | — | 6.14 | 6.14 |
| Fiber optic cable, per segment | E4@.160 | Ea | — | 7.86 | 7.86 |
| Shielded twisted pair (STP), per pair | E4@.125 | Ea | — | 6.14 | 6.14 |
| Unshielded twisted pair (UTP), per pair | E4@.125 | Ea | — | 6.14 | 6.14 |
| **Wall plates** | | | | | |
| Coax wall plates with jacks (no junction box included) | | | | | |
| One or two jacks, including termination | E4@.396 | Ea | .80 | 19.50 | 20.30 |
| Modular wall plates with jacks for UTP (no junction box included), including termination | | | | | |
| One 4 or 6 wire jack (RJ11) | E4@.396 | Ea | 2.55 | 19.50 | 22.05 |
| One 8 wire jack (RJ45) | E4@.396 | Ea | 2.55 | 19.50 | 22.05 |
| Two 8 wire jacks (RJ45) | E4@.450 | Ea | 3.57 | 22.10 | 25.67 |

# 31 Earthwork

| | Craft@Hrs | Unit | Material | Labor | Equipment | Total |
|---|---|---|---|---|---|---|

**Site Clearing** Using a dozer as described below, a 3/4 CY wheel loader and two 10 CY dump trucks. Clearing costs will be higher on smaller jobs and where the terrain limits production. These costs include hauling 6 miles to a legal dump. Dump fees are not included. See Waste Disposal. Figures in parentheses show the approximate "loose" volume of the material to be loaded and hauled to the dump. For estimating purposes, assume that the cost of moving equipment on and off the job site will be equal to the cost of clearing one acre. For more detailed coverage, see *National Earthwork and Heavy Equipment Estimator,* http://CraftsmanSiteLicense.com.

| | Craft@Hrs | Unit | Material | Labor | Equipment | Total |
|---|---|---|---|---|---|---|
| Clear light brush and grub roots | | | | | | |
| Using a 105 HP D-5 dozer (350 CY) | C2@34.8 | Acre | — | 1,660.00 | 1,210.00 | 2,870.00 |

# 31 Earthwork

| | Craft@Hrs | Unit | Material | Labor | Equipment | Total |
|---|---|---|---|---|---|---|
| **Clear medium brush and small trees, grub roots** | | | | | | |
| Using a 130 HP D-6 dozer (420 CY) | C2@41.8 | Acre | — | 2,000.00 | 1,560.00 | 3,560.00 |
| **Clear brush and trees to 6" trunk diameter** | | | | | | |
| Using a 335 HP D-8 dozer (455 CY) | C2@45.2 | Acre | — | 2,160.00 | 1,780.00 | 3,940.00 |
| **Clear wooded area, pull stumps** | | | | | | |
| Using a 460 HP D-9 dozer (490 CY) | C2@48.7 | Acre | — | 2,330.00 | 3,240.00 | 5,570.00 |

**Rock Excavation** Rock drilling costs. Based on using a pneumatic truck-mounted wagon drill. Equipment includes one wheel-mounted air compressor and one 4-1/2" pneumatic wagon drill with hydraulic swing boom and drill bits. Costs shown are per linear foot of hole drilled. Add blasting costs below. Use $2,500 as a minimum job charge. LF per hour shown is for a 2-man crew.

| | Craft@Hrs | Unit | Material | Labor | Equipment | Total |
|---|---|---|---|---|---|---|
| Easy work, 38 LF per hour | S1@.052 | LF | — | 2.47 | 2.61 | 5.08 |
| Moderate work, 30 LF per hour | S1@.067 | LF | — | 3.18 | 3.36 | 6.54 |
| Most work, 25 LF per hour | S1@.080 | LF | — | 3.80 | 4.01 | 7.81 |
| Hard work, 20 LF per hour | S1@.100 | LF | — | 4.74 | 5.01 | 9.75 |
| Dense rock, 15 LF per hour | S1@.132 | LF | — | 6.26 | 6.62 | 12.88 |
| Drilling 2-1/2" hole for | | | | | | |
| rock bolts, 24 LF/hour | S1@.086 | LF | — | 4.08 | 4.31 | 8.39 |

**Blasting costs** Based on two cycles per hour and 20 linear foot lifts. These costs assume 75% fill of holes with explosives. Equipment includes one flatbed truck. Costs shown are per cubic yard of area blasted. Loading or hauling of blasted material not included. Use $3,500 as a minimum job charge. Load explosives in holes and detonate, by pattern spacing. CY per hour shown is for a 5-man crew.

| | Craft@Hrs | Unit | Material | Labor | Equipment | Total |
|---|---|---|---|---|---|---|
| Using explosives, per pound | — | Lb | 4.36 | — | — | 4.36 |
| 6' x 6' pattern (26 CY per hour) | C1@.192 | CY | 13.10 | 7.84 | .76 | 21.70 |
| 7' x 7' pattern (35 CY per hour) | C1@.144 | CY | 9.78 | 5.88 | .57 | 16.23 |
| 7' x 8' pattern (47 CY per hour) | C1@.106 | CY | 7.63 | 4.33 | .42 | 12.38 |
| 9' x 9' pattern (60 CY per hour) | C1@.084 | CY | 6.54 | 3.43 | .33 | 10.30 |
| 10' x 10' pattern (72 CY per hour) | C1@.069 | CY | 5.45 | 2.82 | .27 | 8.54 |
| 12' x 12' pattern (108 CY per hour) | C1@.046 | CY | 3.70 | 1.88 | .18 | 5.76 |
| 14' x 14' pattern (147 CY per hour) | C1@.034 | CY | 2.61 | 1.39 | .13 | 4.13 |

**Site Grading** Layout, staking, flagmen, lights, watering, ripping, rock breaking, loading or hauling not included. Site preparation grading based on using one crawler tractor. Use $3,000.00 as a minimum job charge. (One acre is 43,560 SF or 4,840 SY.)

| | Craft@Hrs | Unit | Material | Labor | Equipment | Total |
|---|---|---|---|---|---|---|
| **General area rough grading with 100 HP D-4 tractor** | | | | | | |
| (.5 acres per hour) | S1@4.00 | Acre | — | 190.00 | 97.80 | 287.80 |
| **General area rough grading with 335 HP D-8 tractor** | | | | | | |
| (1 acre per hour) | S1@2.00 | Acre | — | 94.90 | 148.00 | 242.90 |
| **Site grading based on using one 100 HP (10,000 lb) motor grader** | | | | | | |
| General area grading | | | | | | |
| (.7 acres per hour) | S1@2.86 | Acre | — | 136.00 | 68.20 | 204.20 |
| **Fine grading subgrade to 1/10'** | | | | | | |
| (.45 acres per hour) | S1@4.44 | Acre | — | 211.00 | 106.00 | 317.00 |
| Cut slope or shape embankment to 2' high | | | | | | |
| (.25 acres per hour) | S1@8.00 | Acre | — | 380.00 | 191.00 | 571.00 |
| Rough grade sub-base course on roadway | | | | | | |
| (.2 acres per hour) | S1@10.0 | Acre | — | 474.00 | 238.00 | 712.00 |
| Finish grade base or leveling course on roadway | | | | | | |
| (.17 acres per hour) | S1@11.8 | Acre | — | 560.00 | 281.00 | 841.00 |
| Finish grading for building slab, windrow excess | | | | | | |
| (1,000 SF per hour) | S1@.002 | SF | — | .09 | .05 | .14 |

| | Craft@Hrs | Unit | Material | Labor | Equipment | Total |
|---|---|---|---|---|---|---|

**Embankment Grading** Earth embankment spreading, shaping, compacting and watering, and finish shaping. Use $15,000 as a minimum job charge, loading or hauling of earth fill not included.

Spreading and shaping. Spread and shape earth from loose piles, based on using a D-8 tractor.

| | Craft@Hrs | Unit | Material | Labor | Equipment | Total |
|---|---|---|---|---|---|---|
| 6" to 10" lifts (164 CY per hour) | T0@.006 | CY | — | .33 | .89 | 1.22 |

Compacting and watering. (Add for cost of water from below.) CY per hour shown is for a 3-man crew. Based on using a self-propelled 100 HP vibrating roller and a 3,500 gallon truck. Productivity assumes 3 passes at 7' wide.

| | Craft@Hrs | Unit | Material | Labor | Equipment | Total |
|---|---|---|---|---|---|---|
| 6" lifts (500 CY per hour) | C3@.006 | CY | — | .28 | .22 | .50 |

Compacting with a sheepsfoot roller towed behind a D-7 tractor and a 3,500 gallon truck. Productivity assumes 3 passes at 5' wide

| | Craft@Hrs | Unit | Material | Labor | Equipment | Total |
|---|---|---|---|---|---|---|
| 6" lifts (185 CY per hour) | C3@.016 | CY | — | .74 | .96 | 1.70 |
| 8" lifts (245 CY per hour) | C3@.012 | CY | — | .56 | .72 | 1.28 |

Cost of water for compacted earth embankments. Based on water at $2.50 per 1,000 gallons and 66 gallons per cubic yard of compacted material. Assumes optimum moisture at 10%, natural moisture of 2% and evaporation of 2%. Placed in conjunction with compaction shown above.

| | Craft@Hrs | Unit | Material | Labor | Equipment | Total |
|---|---|---|---|---|---|---|
| Cost per CY of compacted embankment | — | CY | .25 | — | — | .25 |

Finish shaping of embankments. Earth embankment slopes and swales up to 1 in 4 incline. Based on using a 150 HP grader and a 15-ton self-propelled rubber tired roller. SY per hour shown is for a 3-man crew.

| | Craft@Hrs | Unit | Material | Labor | Equipment | Total |
|---|---|---|---|---|---|---|
| (200 SY per hour) | SS@.016 | SY | — | .80 | .76 | 1.56 |

Finish shaping of embankments with a D-8 tractor

| | Craft@Hrs | Unit | Material | Labor | Equipment | Total |
|---|---|---|---|---|---|---|
| (150 SY per hour) | T0@.007 | SY | — | .38 | 1.04 | 1.42 |

Finish shaping of embankment slopes and swales by hand. SY per hour shown is for 1 man.

| | Craft@Hrs | Unit | Material | Labor | Equipment | Total |
|---|---|---|---|---|---|---|
| Slopes up to 1 in 4 (16 SY per hour) | CL@.063 | SY | — | 2.51 | — | 2.51 |
| Slopes over 1 in 4 (12.5 SY per hour) | CL@.080 | SY | — | 3.19 | — | 3.19 |

**Ripping rock** Based on using a tractor-mounted ripper. Equipment includes a D-8 dozer and a D-9 dozer. Costs shown are per CY of area ripped. Loading or hauling of material not included. Use $2,000 as a minimum job charge.

| | Craft@Hrs | Unit | Material | Labor | Equipment | Total |
|---|---|---|---|---|---|---|
| Clay or glacial tills, D-9 cat tractor with 2-shank ripper (500 CY per hour) | T0@.002 | CY | — | .11 | .42 | .53 |
| Clay or glacial tills, D-8 cat tractor with 1-shank ripper (335 CY per hour) | T0@.003 | CY | — | .16 | .44 | .60 |
| Shale, sandstone, or limestone, D-9 cat tractor with 2-shank ripper (125 CY per hour) | T0@.008 | CY | — | .44 | 1.66 | 2.10 |
| Shale, sandstone, or limestone, D-8 cat tractor with 1-shank ripper (95 CY per hour) | T0@.011 | CY | — | .60 | 1.63 | 2.23 |
| Slate, metamorphic rock, D-9 cat tractor with 2-shank ripper (95 CY per hour) | T0@.011 | CY | — | .60 | 2.29 | 2.89 |
| Slate, metamorphic rock, D-8 cat tractor with 1-shank ripper (63 CY per hour) | T0@.016 | CY | — | .88 | 2.37 | 3.25 |
| Granite or basalt, D-9 cat tractor with 2-shank ripper (78 CY per hour) | T0@.013 | CY | — | .71 | 2.70 | 3.41 |
| Granite or basalt, D-8 cat tractor with 1-shank ripper (38 CY per hour) | T0@.027 | CY | — | 1.48 | 3.99 | 5.47 |

**Rock loosening** Based on using pneumatic jackhammers. Equipment includes one compressor and two jackhammers with chisel points. Costs shown are per CY of area loosened. Loading and hauling not included. CY per hour shown is for a 2-man crew. Use $800 as a minimum job charge.

| | Craft@Hrs | Unit | Material | Labor | Equipment | Total |
|---|---|---|---|---|---|---|
| Igneous or dense rock (1.4 CY per hour) | CL@1.43 | CY | — | 57.00 | 29.10 | 86.10 |
| Most weathered rock (2.4 CY per hour) | CL@.835 | CY | — | 33.30 | 17.00 | 50.30 |
| Soft sedimentary rock (4 CY per hour) | CL@.500 | CY | — | 19.90 | 10.20 | 30.10 |

# 31 Earthwork

| | Craft@Hrs | Unit | Material | Labor | Equipment | Total |
|---|---|---|---|---|---|---|

**Trench Excavation and Backfill** These costs and productivity are based on utility line trenches and continuous footings where the spoil is piled adjacent to the trench. Linear feet (LF), cubic yards (CY), square feet (SF), or square yards (SY) per hour shown are based on a 2-man crew. Increase costs by 10% to 25% when spoil is loaded in trucks. Hauling, shoring, dewatering or unusual conditions are not included.

12" wide bucket, for 12" wide trench. Depths 3' to 5'. Equipment is a wheel loader with 1 CY bucket and integral backhoe. Use $400 as a minimum job charge.

| | | | | | | |
|---|---|---|---|---|---|---|
| Light soil (60 LF per hour) | S1@.033 | LF | — | 1.57 | .58 | 2.15 |
| Medium soil (55 LF per hour) | S1@.036 | LF | — | 1.71 | .63 | 2.34 |
| Heavy or wet soil (35 LF per hour) | S1@.057 | LF | — | 2.70 | 1.00 | 3.70 |

18" wide bucket, for 18" wide trench. Depths 3' to 5'. Equipment is a wheel loader with 1 CY bucket and integral backhoe. Use $400 as a minimum job charge.

| | | | | | | |
|---|---|---|---|---|---|---|
| Light soil (55 LF per hour) | S1@.036 | LF | — | 1.71 | .63 | 2.34 |
| Medium soil (50 LF per hour) | S1@.040 | LF | — | 1.90 | .70 | 2.60 |
| Heavy or wet soil (30 LF per hour) | S1@.067 | LF | — | 3.18 | 1.18 | 4.36 |

24" wide bucket, for 24" wide trench. Depths 3' to 5'. Equipment is a wheel loader with 1 CY bucket and integral backhoe. Use $400 as a minimum job charge.

| | | | | | | |
|---|---|---|---|---|---|---|
| Light soil (50 LF per hour) | S1@.040 | LF | — | 1.90 | .70 | 2.60 |
| Medium soil (45 LF per hour) | S1@.044 | LF | — | 2.09 | .77 | 2.86 |
| Heavy or wet soil (25 LF per hour) | S1@.080 | LF | — | 3.80 | 1.41 | 5.21 |

Truck-mounted Gradall with 1 CY bucket. Use $700 as a minimum job charge.

| | | | | | | |
|---|---|---|---|---|---|---|
| Light soil (40 CY per hour) | S1@.050 | CY | — | 2.37 | 3.00 | 5.37 |
| Medium soil (34 CY per hour) | S1@.059 | CY | — | 2.80 | 3.54 | 6.34 |
| Heavy or wet soil (27 CY per hour) | S1@.074 | CY | — | 3.51 | 4.43 | 7.94 |
| Loose rock (23 CY per hour) | S1@.087 | CY | — | 4.13 | 5.21 | 9.34 |

Crawler-mounted hydraulic backhoe with 3/4 CY bucket. Use $500 as a minimum job charge.

| | | | | | | |
|---|---|---|---|---|---|---|
| Light soil (33 CY per hour) | S1@.061 | CY | — | 2.89 | 2.90 | 5.79 |
| Medium soil (27 CY per hour) | S1@.074 | CY | — | 3.51 | 3.52 | 7.03 |
| Heavy or wet soil (22 CY per hour) | S1@.091 | CY | — | 4.32 | 4.33 | 8.65 |
| Loose rock (18 CY per hour) | S1@.112 | CY | — | 5.31 | 5.33 | 10.64 |

Crawler-mounted hydraulic backhoe with 1 CY bucket. Use $500 as a minimum job charge.

| | | | | | | |
|---|---|---|---|---|---|---|
| Light soil (65 CY per hour) | S1@.031 | CY | — | 1.47 | .91 | 2.38 |
| Medium soil (53 CY per hour) | S1@.038 | CY | — | 1.80 | 1.11 | 2.91 |
| Heavy or wet soil (43 CY per hour) | S1@.047 | CY | — | 2.23 | 1.38 | 3.61 |
| Loose rock (37 CY per hour) | S1@.055 | CY | — | 2.61 | 1.61 | 4.22 |
| Blasted rock (34 CY per hour) | S1@.058 | CY | — | 2.75 | 1.70 | 4.45 |

Crawler-mounted hydraulic backhoe with 1-1/2 CY bucket. Use $500 as a minimum job charge.

| | | | | | | |
|---|---|---|---|---|---|---|
| Light soil (83 CY per hour) | S1@.024 | CY | — | 1.14 | .87 | 2.01 |
| Medium soil (70 CY per hour) | S1@.029 | CY | — | 1.38 | 1.05 | 2.43 |
| Heavy or wet soil (57 CY per hour) | S1@.035 | CY | — | 1.66 | 1.26 | 2.92 |
| Loose rock (48 CY per hour) | S1@.042 | CY | — | 1.99 | 1.52 | 3.51 |
| Blasted rock (43 CY per hour) | S1@.047 | CY | — | 2.23 | 1.70 | 3.93 |

Crawler-mounted hydraulic backhoe with 2 CY bucket. Use $700 as a minimum job charge.

| | | | | | | |
|---|---|---|---|---|---|---|
| Light soil (97 CY per hour) | S1@.021 | CY | — | 1.00 | .90 | 1.90 |
| Medium soil (80 CY per hour) | S1@.025 | CY | — | 1.19 | 1.07 | 2.26 |
| Heavy or wet soil (65 CY per hour) | S1@.031 | CY | — | 1.47 | 1.32 | 2.79 |
| Loose rock (55 CY per hour) | S1@.036 | CY | — | 1.71 | 1.54 | 3.25 |
| Blasted rock (50 CY per hour) | S1@.040 | CY | — | 1.90 | 1.71 | 3.61 |

Crawler-mounted hydraulic backhoe with 2-1/2 CY bucket. Use $750 as a minimum job charge.

| | | | | | | |
|---|---|---|---|---|---|---|
| Light soil (122 CY per hour) | S1@.016 | CY | — | .76 | 1.04 | 1.80 |
| Medium soil (100 CY per hour) | S1@.020 | CY | — | .95 | 1.29 | 2.24 |
| Heavy or wet soil (82 CY per hour) | S1@.024 | CY | — | 1.14 | 1.55 | 2.69 |
| Loose rock (68 CY per hour) | S1@.029 | CY | — | 1.38 | 1.88 | 3.26 |
| Blasted rock (62 CY per hour) | S1@.032 | CY | — | 1.52 | 2.07 | 3.59 |

| | Craft@Hrs | Unit | Material | Labor | Equipment | Total |
|---|---|---|---|---|---|---|
| Chain-boom Ditch Witch digging trench to 12" wide and 5' deep. Use $400 as a minimum job charge. | | | | | | |
| Light soil (10 CY per hour) | S1@.200 | CY | — | 9.49 | 2.46 | 11.95 |
| Most soils (8.5 CY per hour) | S1@.235 | CY | — | 11.10 | 2.89 | 13.99 |
| Heavy soil (7 CY per hour) | S1@.286 | CY | — | 13.60 | 3.52 | 17.12 |
| Trenchers, chain boom, 55 HP, digging trench to 18" wide and 8' deep. Use $500 as a minimum job charge. | | | | | | |
| Light soil (62 CY per hour) | S1@.032 | CY | — | 1.52 | .73 | 2.25 |
| Most soils (49 CY per hour) | S1@.041 | CY | — | 1.95 | .94 | 2.89 |
| Heavy soil (40 CY per hour) | S1@.050 | CY | — | 2.37 | 1.15 | 3.52 |
| Trenchers, chain boom, 100 HP, digging trench to 24" wide and 8' deep. Use $500 as a minimum job charge. | | | | | | |
| Light soil (155 CY per hour) | S1@.013 | CY | — | .62 | .49 | 1.11 |
| Most soils (125 CY per hour) | S1@.016 | CY | — | .76 | .60 | 1.36 |
| Heavy soil (100 CY per hour) | S1@.020 | CY | — | .95 | .75 | 1.70 |
| Trim trench bottom to 1/10' | | | | | | |
| By hand (200 SF per hour) | CL@.010 | SF | — | .40 | — | .40 |

**Backfill trenches** from loose material piled adjacent to a trench. No compaction included. Soil previously excavated.

| | Craft@Hrs | Unit | Material | Labor | Equipment | Total |
|---|---|---|---|---|---|---|
| Backfill trenches by hand | | | | | | |
| (8 CY per hour) | CL@.250 | CY | — | 9.97 | — | 9.97 |
| Wheel loader 55 HP. Use $400 as a minimum job charge | | | | | | |
| Typical soils (50 CY per hour) | S1@.040 | CY | — | 1.90 | .61 | 2.51 |
| D-3 crawler dozer. Use $475 as a minimum job charge | | | | | | |
| Typical soils (25 CY per hour) | S1@.080 | CY | — | 3.80 | 1.67 | 5.47 |
| 3/4 CY crawler loader. Use $450 as a minimum job charge | | | | | | |
| Typical soils (33 CY per hr) | S1@.061 | CY | — | 2.89 | 1.24 | 4.13 |
| D-7 crawler dozer. Use $700 as a minimum job charge | | | | | | |
| Typical soils (130 CY per hour) | S1@.015 | CY | — | .71 | .77 | 1.48 |
| Wheel loader 55 HP. Sand or gravel bedding. Use $400 as a minimum job charge | | | | | | |
| Typical soils (80 CY per hour) | S1@.025 | CY | — | 1.19 | .38 | 1.57 |
| Fine grade bedding by hand | | | | | | |
| (22 SY per hr) | CL@.090 | SY | — | 3.59 | — | 3.59 |
| Compaction of soil in trenches in 8" layers. Use $200 as a minimum job charge | | | | | | |
| Pneumatic tampers | | | | | | |
| (20 CY per hour) | CL@.050 | CY | — | 1.99 | 1.48 | 3.47 |
| Vibrating rammers | | | | | | |
| (10 CY per hour) | CL@.100 | CY | — | 3.99 | .93 | 4.92 |

**Dragline Excavation** For mass excavation, footings or foundations. Costs shown include casting the excavated soil adjacent to the excavation or loading it into trucks. Hauling costs are not included. Cubic yards (CY) per hour are as measured in an undisturbed condition (bank measure) and are based on a 2-man crew. Use $3,000 as a minimum job charge.

| | Craft@Hrs | Unit | Material | Labor | Equipment | Total |
|---|---|---|---|---|---|---|
| 1-1/2 CY dragline with a 45-ton crawler crane | | | | | | |
| Loam or light clay (73 CY per hour) | H2@.027 | CY | — | 1.35 | 1.13 | 2.48 |
| Sand or gravel (67 CY per hour) | H2@.030 | CY | — | 1.50 | 1.26 | 2.76 |
| Heavy clay (37 CY per hour) | H2@.054 | CY | — | 2.69 | 2.26 | 4.95 |
| Unclassified soil (28 CY per hour) | H2@.071 | CY | — | 3.54 | 2.97 | 6.51 |

| | Craft@Hrs | Unit | Material | Labor | Equipment | Total |
|---|---|---|---|---|---|---|
| **2 CY dragline with a 60-ton crawler crane** | | | | | | |
| Loam or light clay (88 CY per hour) | H2@.023 | CY | — | 1.15 | 1.36 | 2.51 |
| Sand or gravel (85 CY per hour) | H2@.024 | CY | — | 1.20 | 1.42 | 2.62 |
| Heavy clay (48 CY per hour) | H2@.042 | CY | — | 2.10 | 2.48 | 4.58 |
| Unclassified soil (36 CY per hour) | H2@.056 | CY | — | 2.79 | 3.31 | 6.10 |
| **2-1/2 CY dragline with a 100-ton crawler crane** | | | | | | |
| Loam or light clay (102 CY per hour) | H2@.019 | CY | — | .95 | 1.51 | 2.46 |
| Sand or gravel (98 CY per hour) | H2@.021 | CY | — | 1.05 | 1.67 | 2.72 |
| Heavy clay (58 CY per hour) | H2@.035 | CY | — | 1.75 | 2.78 | 4.53 |
| Unclassified soil (44 CY per hour) | H2@.045 | CY | — | 2.25 | 3.58 | 5.83 |
| **3 CY dragline with a 150-ton crawler crane** | | | | | | |
| Loam or light clay (116 CY per hour) | H2@.017 | CY | — | .85 | 1.56 | 2.41 |
| Sand or gravel (113 CY per hour) | H2@.018 | CY | — | .90 | 1.66 | 2.56 |
| Heavy clay (70 CY per hour) | H2@.029 | CY | — | 1.45 | 2.67 | 4.12 |
| Unclassified soil (53 CY per hour) | H2@.038 | CY | — | 1.90 | 3.50 | 5.40 |

**Backhoe Excavation** For mass excavation, footing or foundation costs shown include casting the excavated soil adjacent to the excavation or loading it into trucks. Hauling costs are not included. Cubic yards (CY) per hour are as measured in an undisturbed condition (bank measure) and are based on a 2-man crew. Equipment costs are based on using a crawler-mounted hydraulic backhoe. Use $2,500 as a minimum job charge

| | Craft@Hrs | Unit | Material | Labor | Equipment | Total |
|---|---|---|---|---|---|---|
| **1 CY backhoe** | | | | | | |
| Light soil (69 CY per hour) | S1@.029 | CY | — | 1.38 | .57 | 1.95 |
| Most soils (57 CY per hour) | S1@.035 | CY | — | 1.66 | .68 | 2.34 |
| Wet soil, loose rock (46 CY per hour) | S1@.043 | CY | — | 2.04 | .84 | 2.88 |
| **1-1/2 CY backhoe** | | | | | | |
| Light soil (90 CY per hour) | S1@.022 | CY | — | 1.04 | .44 | 1.48 |
| Most soils (77 CY per hour) | S1@.026 | CY | — | 1.23 | .52 | 1.75 |
| Wet soil, loose rock (60 CY per hour) | S1@.033 | CY | — | 1.57 | .66 | 2.23 |
| Blasted rock (54 CY per hour) | S1@.044 | CY | — | 2.09 | .88 | 2.97 |
| **2 CY backhoe** | | | | | | |
| Light soil (100 CY per hour) | S1@.020 | CY | — | .95 | .81 | 1.76 |
| Most soils (85 CY per hour) | S1@.023 | CY | — | 1.09 | 1.93 | 3.02 |
| Wet soil, loose rock (70 CY per hour) | S1@.029 | CY | — | 1.38 | 1.18 | 2.56 |

**Moving and Loading Excavated Materials** Costs shown include moving the material 50' and dumping it into piles or loading it into trucks. Add 25% for each 50' of travel beyond the first 50'. Hauling costs are not included. Cubic yards (CY) per hour are for material in a loose condition (previously excavated) and are based on a 2-man crew performing the work. Equipment costs are based on using a wheel-mounted front end loader. Use $2,000 as a minimum job charge.

| | Craft@Hrs | Unit | Material | Labor | Equipment | Total |
|---|---|---|---|---|---|---|
| **3/4 CY loader** | | | | | | |
| (32 CY per hour) | S1@.062 | CY | — | 2.94 | .96 | 3.90 |
| **1 CY loader** | | | | | | |
| (55 CY per hour) | S1@.036 | CY | — | 1.71 | .65 | 2.36 |
| **2 CY loader** | | | | | | |
| (90 CY per hour) | S1@.022 | CY | — | 1.04 | .46 | 1.50 |
| **3-1/4 CY loader** | | | | | | |
| (125 CY per hour) | S1@.016 | CY | — | .76 | .57 | 1.33 |
| **5 CY loader** | | | | | | |
| (218 CY per hour) | S1@.009 | CY | — | .43 | .43 | .86 |

| | Craft@Hrs | Unit | Material | Labor | Equipment | Total |
|---|---|---|---|---|---|---|

**Hauling excavated material** Using trucks. Costs shown include 4 minutes for loading, 3 minutes for dumping and travel time based on the one-way distance, speed and cycles per hour as noted. Costs for equipment to excavate and load the material are not included. Truck capacity shown is based on loose cubic yards of material. Allow the following percentage amounts when estimating quantities based on undisturbed (bank measure) materials for swell when exporting from the excavation site or shrinkage when importing from borrow site: clay (33%), common earth (25%), granite (65%), mud (21%), sand or gravel (12%).
3 CY dump truck. Use $550 as a minimum job charge.

| | Craft@Hrs | Unit | Material | Labor | Equipment | Total |
|---|---|---|---|---|---|---|
| 1 mile haul at 20 MPH | | | | | | |
| (4.2 cycles and 12.6 CY per hour) | TD@.029 | CY | — | 1.29 | .80 | 2.09 |
| 3 mile haul at 30 MPH | | | | | | |
| (2.9 cycles and 8.7 CY per hour) | TD@.044 | CY | — | 1.96 | 1.22 | 3.18 |
| 6 mile haul at 40 MPH | | | | | | |
| (2.1 cycles and 6.3 CY per hour) | TD@.058 | CY | — | 2.59 | 1.61 | 4.20 |

5 CY dump truck. Use $600 as a minimum job charge.

| | Craft@Hrs | Unit | Material | Labor | Equipment | Total |
|---|---|---|---|---|---|---|
| 1 mile haul at 20 MPH | | | | | | |
| (4.2 cycles and 21 CY per hour) | TD@.020 | CY | — | .89 | .70 | 1.59 |
| 3 mile haul at 30 MPH | | | | | | |
| (2.9 cycles and 14.5 CY per hour) | TD@.028 | CY | — | 1.25 | .98 | 2.23 |
| 6 mile haul at 40 MPH | | | | | | |
| (2.1 cycles and 10.5 CY per hour) | TD@.041 | CY | — | 1.83 | 1.44 | 3.27 |

10 CY dump truck. Use $800 as a minimum job charge.

| | Craft@Hrs | Unit | Material | Labor | Equipment | Total |
|---|---|---|---|---|---|---|
| 1 mile haul at 20 MPH | | | | | | |
| (4.2 cycles and 42 CY per hour) | TD@.015 | CY | — | .67 | .63 | 1.30 |
| 3 mile haul at 30 MPH | | | | | | |
| (2.9 cycles and 29 CY per hour) | TD@.022 | CY | — | .98 | .92 | 1.90 |
| 6 mile haul at 40 MPH | | | | | | |
| (2.1 cycles and 21 CY per hour) | TD@.029 | CY | — | 1.29 | 1.22 | 2.51 |

25 CY off-highway dump truck. Use $2,000 as a minimum job charge.

| | Craft@Hrs | Unit | Material | Labor | Equipment | Total |
|---|---|---|---|---|---|---|
| 1 mile haul at 20 MPH | | | | | | |
| (4.2 cycles and 105 CY per hour) | TD@.009 | CY | — | .40 | 1.05 | 1.45 |
| 3 mile haul at 30 MPH | | | | | | |
| (2.9 cycles and 72.5 CY per hour) | TD@.014 | CY | — | .62 | 1.63 | 2.25 |
| 6 mile haul at 40 MPH | | | | | | |
| (2.1 cycles and 52.5CY per hour) | TD@.019 | CY | — | .85 | 2.21 | 3.06 |

**Dozer Excavation** Mass excavation using a crawler tractor with a dozing blade attached. Costs shown include excavation and pushing soil 150' to a stockpile. Based on good access and a maximum of 5% grade. Increase or reduce costs 20% for each 50' of push more or less than 150'. Cubic yards (CY) per hour shown are undisturbed bank measure and are based on a 1-man crew. Use the following as minimum job charges: $850 if using a D-4 or a D-6 dozer; $1,350 if using a D-7 or D-8 dozer; or $2,000 if using a D-9 dozer.
Gravel or loose sand

| | Craft@Hrs | Unit | Material | Labor | Equipment | Total |
|---|---|---|---|---|---|---|
| 100 HP D-4 dozer with "S"-blade | | | | | | |
| (29 CY per hour) | TO@.034 | CY | — | 1.87 | 1.66 | 3.53 |
| 140 HP D-6 dozer with "S"-blade | | | | | | |
| (56 CY per hour) | TO@.018 | CY | — | .99 | 1.29 | 2.28 |
| 200 HP D-7 dozer with "S"-blade | | | | | | |
| (78 CY per hour) | TO@.013 | CY | — | .71 | 1.34 | 2.05 |
| 335 HP D-8 dozer with "U"-blade | | | | | | |
| (140 CY per hour) | TO@.007 | CY | — | .38 | 1.04 | 1.42 |
| 460 HP D-9 dozer with "U"-blade | | | | | | |
| (205 CY per hour) | TO@.005 | CY | — | .27 | 1.04 | 1.31 |

| | Craft@Hrs | Unit | Material | Labor | Equipment | Total |
|---|---|---|---|---|---|---|
| Loam or soft clay | | | | | | |
| 100 HP D-4 dozer with "S"-blade | | | | | | |
| (26 CY per hour) | TO@.038 | CY | — | 2.09 | 1.86 | 3.95 |
| 140 HP D-6 dozer with "S"-blade | | | | | | |
| (50 CY per hour) | TO@.020 | CY | — | 1.10 | 1.43 | 2.53 |
| 200 HP D-7 dozer with "S"-blade | | | | | | |
| (70 CY per hour) | TO@.014 | CY | — | .77 | 1.44 | 2.21 |
| 335 HP D-8 dozer with "U"-blade | | | | | | |
| (125 CY per hour) | TO@.008 | CY | — | .44 | 1.18 | 1.62 |
| 460 HP D-9 dozer with "U" blade | | | | | | |
| (185 CY per hour) | TO@.005 | CY | — | .27 | 1.04 | 1.31 |
| Shale, sandstone or blasted rock | | | | | | |
| 100 HP D-4 dozer with "S"-blade | | | | | | |
| (20 CY per hour) | TO@.051 | CY | — | 2.80 | 2.49 | 5.29 |
| 140 HP D-6 dozer with "S"-blade | | | | | | |
| (38 CY per hour) | TO@.026 | CY | — | 1.43 | 1.86 | 3.29 |
| 200 HP D-7 dozer with "S"-blade | | | | | | |
| (53 CY per hour) | TO@.019 | CY | — | 1.04 | 1.95 | 2.99 |
| 335 HP D-8 dozer with "U"- blade | | | | | | |
| (95 CY per hour) | TO@.011 | CY | — | .60 | 1.63 | 2.23 |
| 460 HP D-9 dozer with "U"- blade | | | | | | |
| (140 CY per hour) | TO@.007 | CY | — | .38 | 1.45 | 1.83 |

**Scraper-Hauler Excavation** Mass excavation using a self-propelled scraper-hauler. Equipment costs include the scraper-hauler and a crawler tractor with a dozer blade attached pushing it 10 minutes each hour. Cubic yards (CY) per hour shown are undisturbed bank measure. Work done in clay, shale or soft rock will cost 10% to 25% more.
15 CY self-propelled 200 HP scraper-hauler and a D-8 335 HP tractor pushing. Use $6,000 as a minimum job charge.

| | Craft@Hrs | Unit | Material | Labor | Equipment | Total |
|---|---|---|---|---|---|---|
| 1,000' haul | | | | | | |
| (9 cycles and 135 CY per hr) | SS@.016 | CY | — | .80 | 1.26 | 2.06 |
| 2,500' haul | | | | | | |
| (6 cycles and 90 CY per hr) | SS@.024 | CY | — | 1.20 | 1.88 | 3.08 |
| 4,000' haul | | | | | | |
| (4.5 cycles and 68 CY per hr) | SS@.032 | CY | — | 1.60 | 2.51 | 4.11 |

24 CY self-propelled 350 HP scraper-hauler and a D-9 460 HP tractor pushing. Use $8,000 as a minimum job charge.

| | Craft@Hrs | Unit | Material | Labor | Equipment | Total |
|---|---|---|---|---|---|---|
| 1,000' haul | | | | | | |
| (9 cycles and 225 CY per hr) | SS@.010 | CY | — | .50 | 1.07 | 1.57 |
| 2,500' haul | | | | | | |
| (6 cycles and 150 CY per hr) | SS@.014 | CY | — | .70 | 1.49 | 2.19 |
| 4,000' haul | | | | | | |
| (4.5 cycles and 113 CY per hr) | SS@.019 | CY | — | .95 | 2.03 | 2.98 |

32 CY self-propelled 550 HP scraper-hauler and a D-9 460 HP tractor pushing. Use $10,000 as a minimum job charge.

| | Craft@Hrs | Unit | Material | Labor | Equipment | Total |
|---|---|---|---|---|---|---|
| 1,000' haul | | | | | | |
| (9 cycles and 315 CY per hr) | SS@.007 | CY | — | .35 | .81 | 1.16 |
| 2,500' haul | | | | | | |
| (6 cycles and 219 CY per hr) | SS@.010 | CY | — | .50 | 1.16 | 1.66 |
| 4,000' haul | | | | | | |
| (4.5 cycles and 158 CY per hr) | SS@.014 | CY | — | .70 | 1.62 | 2.32 |

**Manual Excavation** Based on one throw (except where noted) of shoveled earth from trench or pit. Shoring or dewatering not included.

| | Craft@Hrs | Unit | Material | Labor | Equipment | Total |
|---|---|---|---|---|---|---|
| Trench or pit in light soil (silty sand or loess, etc.) | | | | | | |
| Up to 4' deep (.82 CY per hour) | CL@1.22 | CY | — | 48.70 | — | 48.70 |
| Over 4' to 6' deep (.68 CY per hour) | CL@1.46 | CY | — | 58.20 | — | 58.20 |
| Over 6' deep, two throws | | | | | | |
| (.32 CY per hour) | CL@3.16 | CY | — | 126.00 | — | 126.00 |

# 31 Earthwork

| | Craft@Hrs | Unit | Material | Labor | Equipment | Total |
|---|---|---|---|---|---|---|
| Trench or pit in medium soil (sandy clay, for example) | | | | | | |
| Up to 4' deep (.68 CY per hour) | CL@1.46 | CY | — | 58.20 | — | 58.20 |
| Over 4' to 6' deep (.54 CY per hour) | CL@1.86 | CY | — | 74.20 | — | 74.20 |
| Over 6' deep, two throws | | | | | | |
| (.27 CY per hour) | CL@3.66 | CY | — | 146.00 | — | 146.00 |
| Trench or pit in heavy soil (clayey materials, shales, caliche, etc.) | | | | | | |
| Up to 4' deep (.54 CY per hour) | CL@1.86 | CY | — | 74.20 | — | 74.20 |
| Over 4' to 6' deep (.46 CY per hour) | CL@2.16 | CY | — | 86.10 | — | 86.10 |
| Over 6' deep, 2 throws | | | | | | |
| (.25 CY per hour) | CL@3.90 | CY | — | 156.00 | — | 156.00 |
| Wheelbarrow (5 CF) on firm ground, load by shovel from pile, wheel 300 feet and dump | | | | | | |
| Light soil (.74 CY per hour) | CL@1.35 | CY | — | 53.80 | — | 53.80 |
| Medium soil (.65 CY per hour) | CL@1.54 | CY | — | 61.40 | — | 61.40 |
| Heavy soil (.56 CY per hour) | CL@1.79 | CY | — | 71.40 | — | 71.40 |
| Loose rock (.37 CY per hour) | CL@2.70 | CY | — | 108.00 | — | 108.00 |
| Hand trim and shape | | | | | | |
| Around utility lines (15 CF hour) | CL@.065 | CF | — | 2.59 | — | 2.59 |
| For slab on grade (63 SY per hour) | CL@.016 | SY | — | .64 | — | .64 |
| Trench bottom (100 SF per hour) | CL@.010 | SF | — | .40 | — | .40 |

**Wellpoint Dewatering** Based on header pipe connecting 2" diameter jetted wellpoints 5' on center. Includes header pipe, wellpoints, filter sand, pumps and swing joints. The header pipe length is usually equal to the perimeter of the area excavated.

Typical cost for installation and removal of wellpoint system with wellpoints 14' deep and placed 5' on center along 6" header pipe. Based on equipment rented for 1 month. Use 100 LF as a minimum job charge and add cost for operator as required. Also add (if required) for additional months of rental, fuel for the pumps, a standby pump, water truck for jetting, outflow pipe, permits and consultant costs.

| 6" header pipe and accessories, system per LF installed, rented 1 month and removed | | | | | | |
|---|---|---|---|---|---|---|
| 100 LF system | C5@.535 | LF | 1.40 | 24.00 | 52.80 | 78.20 |
| 200 LF system | C5@.535 | LF | 1.40 | 24.00 | 31.50 | 56.90 |
| 500 LF system | C5@.535 | LF | 1.40 | 24.00 | 27.30 | 52.70 |
| 1,000 LF system | C5@.535 | LF | 1.40 | 24.00 | 23.00 | 48.40 |
| Add for 8" header pipe | C5@.010 | LF | .05 | .45 | 1.78 | 2.28 |
| Add for 18' wellpoint depth | C5@.096 | LF | .11 | 4.30 | 5.78 | 10.19 |
| Add for second month | — | LF | — | — | 17.00 | 17.00 |
| Add for each additional month | — | LF | — | — | 10.10 | 10.10 |
| Add for operator, per hour | T0@1.00 | Hr | — | 55.00 | — | 55.00 |

**Roadway and Embankment Earthwork Cut & Fill** Mass excavating, hauling, placing and compacting (cut & fill) using a 200 HP, 15 CY capacity self-propelled scraper-hauler with attachments. Productivity is based on a 1,500' haul with 6" lifts compacted to 95% per AASHO requirements. Cubic yards (CY) or square yards (SY) per hour shown are bank measurement. Use $3,000 as a minimum job charge.

| Cut, fill and compact, clearing or finishing not included | | | | | | |
|---|---|---|---|---|---|---|
| Most soil types (45 CY per hour) | T0@.022 | CY | — | 1.21 | 2.97 | 4.18 |
| Rock and earth mixed (40 CY per hour) | T0@.025 | CY | — | 1.37 | 3.37 | 4.74 |
| Rippable rock (25 CY per hour) | T0@.040 | CY | — | 2.20 | 5.40 | 7.60 |
| Earth banks and levees, clearing or finishing not included | | | | | | |
| Cut, fill and compact (85 CY per hour) | T0@.012 | CY | — | .66 | 1.62 | 2.28 |
| Channelize compacted fill | | | | | | |
| (55 CY per hour) | T0@.018 | CY | — | .99 | 2.43 | 3.42 |

| | Craft@Hrs | Unit | Material | Labor | Equipment | Total |
|---|---|---|---|---|---|---|
| Roadway subgrade preparation | | | | | | |
| Scarify and compact (91 CY per hour) | T0@.011 | CY | — | .60 | 1.48 | 2.08 |
| Roll and compact base course | | | | | | |
| (83 CY per hour) | T0@.012 | CY | — | .66 | 1.62 | 2.28 |
| Excavation, open ditches (38 CY per hour) | T0@.026 | CY | — | 1.43 | 3.51 | 4.94 |
| Trimming and finishing | | | | | | |
| Banks, swales or ditches (140 CY per hour) | T0@.007 | CY | — | .38 | .94 | 1.32 |
| Scarify asphalt pavement (48 SY per hour) | T0@.021 | SY | — | 1.15 | 2.83 | 3.98 |

**Soil Covers**

| | Craft@Hrs | Unit | Material | Labor | Equipment | Total |
|---|---|---|---|---|---|---|
| Using staked burlap and tar over straw | CL@.019 | SY | .85 | .76 | — | 1.61 |
| Using copolymer-base sprayed-on liquid soil sealant | | | | | | |
| On slopes for erosion control | CL@.012 | SY | 1.77 | .48 | — | 2.25 |
| On level ground for dust abatement | CL@.012 | SY | 3.54 | .48 | — | 4.02 |

**Soil Stabilization** See also Slope Protection. Lime slurry injection treatment. Equipment is a 100 HP grader, a 1-ton self-propelled steel roller and a 10-ton 100 HP vibratory steel roller. Use $3,000 as a minimum job charge. Cost per CY of soil treated.

| | Craft@Hrs | Unit | Material | Labor | Equipment | Total |
|---|---|---|---|---|---|---|
| (25 CY per hour) | S7@.240 | CY | 12.90 | 11.40 | 6.76 | 31.06 |

**Vibroflotation treatment** Equipment includes a 50-ton hydraulic crane, a 1 CY wheel loader for 15 minutes of each hour, and a vibroflotation machine with pumps. Use $3,500 as a minimum job charge. Typical cost per CY of soil treated.

| | Craft@Hrs | Unit | Material | Labor | Equipment | Total |
|---|---|---|---|---|---|---|
| Low cost (92 CY per hour) | S7@.065 | CY | — | 3.08 | 2.98 | 6.06 |
| High cost (46 CY per hour) | S7@.130 | CY | — | 6.17 | 5.97 | 12.14 |

**Soil cement treatment** Includes materials and placement for 7% cement mix. Equipment is a cross shaft mixer, a 2-ton 3- wheel steel roller and a vibratory roller. Costs shown are per CY of soil treated. Use $3,000 as a minimum job charge.

| | Craft@Hrs | Unit | Material | Labor | Equipment | Total |
|---|---|---|---|---|---|---|
| On level ground | | | | | | |
| (5 CY per hour) | S7@1.20 | CY | 11.80 | 56.90 | 30.70 | 99.40 |
| On slopes under 3 in 1 | | | | | | |
| (4.3 CY per hour) | S7@1.40 | CY | 11.80 | 66.40 | 35.80 | 114.00 |
| On slopes over 3 in 1 | | | | | | |
| (4 CY per hour) | S7@1.50 | CY | 11.80 | 71.20 | 38.30 | 121.30 |

**Construction fabrics (geotextiles)** Mirafi products. Costs shown include 10% for lapover. Drainage fabric (Mirafi 140NC), used to line trenches, allows water passage but prevents soil migration into drains.

| | Craft@Hrs | Unit | Material | Labor | Equipment | Total |
|---|---|---|---|---|---|---|
| Per SY of fabric placed by hand | CL@.016 | SY | .96 | .64 | — | 1.60 |
| Stabilization fabric (Mirafi 500X), used to prevent mixing of unstable soils with road base aggregate | | | | | | |
| Per SY of fabric placed by hand | CL@.008 | SY | .80 | .32 | — | 1.12 |
| Prefabricated drainage fabric (Miradrain), used to drain subsurface water from structural walls | | | | | | |
| Per SY of fabric placed by hand | CL@.006 | SY | .96 | .24 | — | 1.20 |
| Erosion control mat (Miramat TM8), used to reduce surface soil erosion on slopes or in ditches while promoting seed growth | | | | | | |
| Per SY of mat placed by hand | CL@.020 | SY | .58 | .80 | — | 1.38 |

**Slope Protection** See also Soil Stabilization. Gabions as manufactured by Maccaferri Inc. Typical prices for gabions as described, placed by hand. Prices can be expected to vary based on quantity. Cost for stone is not included. Add for stone as required. Costs shown are per each gabion.

| | Craft@Hrs | Unit | Material | Labor | Equipment | Total |
|---|---|---|---|---|---|---|
| Galvanized mesh | | | | | | |
| 6' x 3' x 3' | CL@2.70 | Ea | 146.00 | 108.00 | — | 254.00 |
| 9' x 3' x 3' | CL@2.90 | Ea | 85.10 | 116.00 | — | 201.10 |
| 12' x 3' x 3' | CL@3.10 | Ea | 181.00 | 124.00 | — | 305.00 |
| Erosion and drainage control polypropylene geotextile fabric (Mirafi FW700) | | | | | | |
| For use under riprap, hand placed | CL@.011 | SY | 2.34 | .44 | — | 2.78 |

# 31 Earthwork

| | Craft@Hrs | Unit | Material | Labor | Equipment | Total |
|---|---|---|---|---|---|---|

**Riprap** Dumped from trucks and placed using a 15-ton hydraulic crane and a 10 CY dump truck. Riprap costs will vary widely. Add $1.00 per CY per mile for trucking to the site. Cubic yards (CY) per hour shown in parentheses are based on a 7-man crew. Use $2,500 as a minimum job charge.

| | Craft@Hrs | Unit | Material | Labor | Equipment | Total |
|---|---|---|---|---|---|---|
| Typical riprap cost, per CY | — | CY | 67.10 | — | — | 67.10 |
| 5 to 7 CF pieces (9.4 CY per hour) | S5@.745 | CY | 67.10 | 31.80 | 10.40 | 109.30 |
| 5 to 7 CF pieces, sacked and placed, excluding bag cost (3 CY per hour) | S5@2.33 | CY | 67.10 | 99.60 | 32.40 | 199.10 |
| Loose small riprap stone, under 30 lbs each, hand placed (2 CY per hour) | CL@.500 | CY | 67.10 | 19.90 | — | 87.00 |

**Ornamental large rock** Rock prices include local delivery of 10 CY minimum. Cubic yards (CY) per hour shown in parentheses are based on a 7-man crew. Equipment is a 15-ton hydraulic crane and a 10 CY dump truck. Use $2,500 as a minimum job charge. Expect wide variation in costs of ornamental rock.

| | Craft@Hrs | Unit | Material | Labor | Equipment | Total |
|---|---|---|---|---|---|---|
| Volcanic cinder, 1,200 lbs per CY (8 CY per hour) | S5@.875 | CY | 314.00 | 37.40 | 12.20 | 363.60 |
| Featherock, 1,000 lbs per CY (12 CY per hour) | S5@.583 | CY | 345.00 | 24.90 | 8.10 | 378.00 |

**Rock fill** Dumped from trucks and placed with a 100 HP D-4 tractor. Rock prices include local delivery of 10 CY minimum. Cubic yards (CY) per hour shown in parentheses are based on a 3-man crew. Use $1,350 as a minimum job charge.

| | Craft@Hrs | Unit | Material | Labor | Equipment | Total |
|---|---|---|---|---|---|---|
| Drain rock, 3/4" to 1-1/2" (12 CY per hour) | S6@.250 | CY | 24.80 | 11.20 | 4.08 | 40.08 |
| Bank run gravel (12 CY per hour) | S6@.250 | CY | 27.60 | 11.20 | 4.08 | 42.88 |
| Pea gravel (12 CY per hour) | S6@.250 | CY | 27.60 | 11.20 | 4.08 | 42.88 |
| Straw bales secured to ground with reinforcing bars, bale is 3'6" long, 2' wide, 1'6" high. | | | | | | |
| Per bale of straw, placed by hand | CL@.166 | Ea | 5.00 | 6.62 | — | 11.62 |
| Sedimentation control fence, installed vertically at bottom of slope, with stakes attached, | | | | | | |
| 36" high woven fabric (Mirafi 100X) | CL@.003 | LF | .28 | .12 | — | .40 |

**Shoring, Bulkheads and Underpinning** Hot-rolled steel sheet piling, driven and pulled, based on good driving conditions and using sheets 30' to 50' in length with 32 lbs of 38.5 KSI steel per SF. Material cost assumes the piling is salvaged. Use $16,000 as a minimum job charge.

| | Craft@Hrs | Unit | Material | Labor | Equipment | Total |
|---|---|---|---|---|---|---|
| Using steel at, per ton | — | Ton | 1,150.00 | — | — | 1,150.00 |
| Assumed steel salvage value, per ton | — | Ton | 859.00 | — | — | 859.00 |
| Typical driving and pulling cost, per ton | — | Ton | — | — | 221.00 | 221.00 |
| Up to 20' deep | S8@.096 | SF | 4.63 | 4.70 | 4.31 | 13.64 |
| Over 20' to 35' deep | S8@.088 | SF | 4.63 | 4.30 | 3.97 | 12.90 |
| Over 35' to 50' deep | S8@.087 | SF | 4.63 | 4.26 | 3.90 | 12.79 |
| Over 50' deep | S8@.085 | SF | 4.63 | 4.16 | 3.83 | 12.62 |

Extra costs for sheet steel piling
Add for:

| | Craft@Hrs | Unit | Material | Labor | Equipment | Total |
|---|---|---|---|---|---|---|
| 50' to 65' lengths | — | SF | .11 | — | — | .11 |
| 65' to 100' lengths | — | SF | .27 | — | — | .27 |
| Epoxy coal tar 16 mil factory finish | — | SF | 1.80 | — | — | 1.80 |
| 32 lb per SF piling material left in place | — | SF | 4.95 | — | — | 4.95 |
| 38 lbs per SF (MZ 38) steel | — | SF | 1.04 | — | — | 1.04 |
| 50 KSI high-strength steel | — | SF | .61 | — | — | .61 |
| Marine grade 50 KSI high strength steel | — | SF | 1.12 | — | — | 1.12 |

| | Craft@Hrs | Unit | Material | Labor | Equipment | Total |
|---|---|---|---|---|---|---|
| Deductive costs for sheet steel piling | | | | | | |
| Deduct for: | | | | | | |
| 28 lbs per SF (PS 28) steel | — | SF | -.80 | — | — | -.80 |
| 23 lbs per SF (PS 23) steel | — | SF | -1.12 | — | — | -1.12 |
| Deduct labor & equipment cost | | | | | | |
| when piling is left in place | — | SF | — | -1.50 | -.53 | -2.03 |

Timber trench sheeting and bracing per square foot of trench wall measured on one side of trench. Costs include pulling and salvage, depths to 12'. Equipment includes a 10-ton hydraulic crane and a 1-ton flat-bed truck. Use $4,000 as a minimum job charge.

| | Craft@Hrs | Unit | Material | Labor | Equipment | Total |
|---|---|---|---|---|---|---|
| Less than 6' wide, open bracing | S5@.053 | SF | 1.74 | 2.27 | .54 | 4.55 |
| 6' to 10' wide, open bracing | S5@.064 | SF | 1.83 | 2.74 | .65 | 5.22 |
| 11' to 16' wide, open bracing | S5@.079 | SF | 2.15 | 3.38 | .80 | 6.33 |
| Over 16' wide, open bracing | S5@.091 | SF | 2.35 | 3.89 | .92 | 7.16 |
| Less than 6' wide, closed sheeting | S5@.114 | SF | 1.99 | 4.87 | 1.15 | 8.01 |
| 6' to 10' wide, closed sheeting | S5@.143 | SF | 1.78 | 6.11 | 1.45 | 9.34 |
| 11' to 16' wide, closed sheeting | S5@.163 | SF | 2.12 | 6.97 | 1.65 | 10.74 |
| Over 16' wide, closed sheeting | S5@.199 | SF | 2.67 | 8.51 | 2.01 | 13.19 |

Additional costs for bracing on trenches over 12' deep

| | Craft@Hrs | Unit | Material | Labor | Equipment | Total |
|---|---|---|---|---|---|---|
| Normal bracing, to 15' deep | S5@.050 | SF | .33 | 2.14 | .51 | 2.98 |
| One line of bracing, over 15' to 22' deep | S5@.053 | SF | .40 | 2.27 | .54 | 3.21 |
| Two lines of bracing, over 22' to 35' deep | S5@.061 | SF | .68 | 2.61 | .62 | 3.91 |
| Three lines of bracing, over 35' to 45' deep | S5@.064 | SF | .76 | 2.74 | .65 | 4.15 |

**Pile Foundations** These costs assume solid ground for support of the pile driving equipment with all site work performed by others at no cost prior to pile driving being performed. Costs shown per linear foot (LF) are per vertical foot of pile depth. Standby or idle time, access roads, rig mats, test piles or special engineering required by unusual conditions are not included.

**Pile testing** Testing costs will vary with the soil type, method of testing, and seismic requirements for any given job site. Use the following as guideline costs for estimating purposes. Typical costs including cost for moving test equipment on and off job site. Add cost below for various types of piling.

| | Craft@Hrs | Unit | Material | Labor | Equipment | Total |
|---|---|---|---|---|---|---|
| 50- to 100-ton range | S8@224. | Ea | — | 11,000.00 | 8,620.00 | 19,620.00 |
| Over 100-ton to 200-ton range | S8@280. | Ea | — | 13,700.00 | 10,800.00 | 24,500.00 |
| Over 200-ton to 300-ton range | S8@336. | Ea | — | 16,400.00 | 12,900.00 | 29,300.00 |

**Prestressed concrete piles** Equipment includes a 60-ton crawler mounted crane equipped with pile driving attachments, a 4-ton forklift, a truck with hand tools, a portable air compressor complete with hoses and pneumatic operated tools, a welding machine and an oxygen-acetylene cutting torch. Use $28,000 as minimum job charge, including 1,000 LF of piling and the cost for moving the pile driving equipment on and off the job site.

| | Craft@Hrs | Unit | Material | Labor | Equipment | Total |
|---|---|---|---|---|---|---|
| Minimum job charge | — | LS | — | — | — | 30,000.00 |

For jobs requiring over 1,000 LF of piling, add to minimum charge as follows. (Figures in parentheses show approximate production per day based on a 7-man crew.)

12" square

| | Craft@Hrs | Unit | Material | Labor | Equipment | Total |
|---|---|---|---|---|---|---|
| 30' or 40' long piles (550 LF per day) | S8@.102 | LF | 16.20 | 4.99 | 4.30 | 25.49 |
| 50' or 60' long piles (690 LF per day) | S8@.082 | LF | 16.20 | 4.01 | 3.46 | 23.67 |

14" square

| | Craft@Hrs | Unit | Material | Labor | Equipment | Total |
|---|---|---|---|---|---|---|
| 30' or 40' long piles (500 LF per day) | S8@.113 | LF | 19.80 | 5.53 | 4.76 | 30.09 |
| 50' or 60' long piles (645 LF per day) | S8@.087 | LF | 19.80 | 4.26 | 3.67 | 27.73 |

| | Craft@Hrs | Unit | Material | Labor | Equipment | Total |
|---|---|---|---|---|---|---|
| 16" square | | | | | | |
| 30' or 40' long piles (420 LF per day) | S8@.134 | LF | 28.20 | 6.55 | 5.65 | 40.40 |
| 50' or 60' long piles (550 LF per day) | S8@.102 | LF | 28.20 | 4.99 | 4.30 | 37.49 |
| 18" square | | | | | | |
| 30' or 40' long piles (400 LF per day) | S8@.143 | LF | 35.20 | 6.99 | 6.03 | 48.22 |
| 50' or 60' long piles (525 LF per day) | S8@.107 | LF | 35.20 | 5.23 | 4.51 | 44.94 |

**Additional costs for prestressed concrete piles** Pre-drilling holes for piles. Equipment includes a flatbed truck with a boom and a power-operated auger. Use $2,200 as a minimum job charge, including drilling 100 LF of holes and the cost to move the drilling equipment on and off the job site.

| | Craft@Hrs | Unit | Material | Labor | Equipment | Total |
|---|---|---|---|---|---|---|
| Minimum job charge | — | LS | — | — | — | 2,200.00 |

For jobs requiring more than 100 LF of holes drilled, add to minimum cost as shown below

| | Craft@Hrs | Unit | Material | Labor | Equipment | Total |
|---|---|---|---|---|---|---|
| Pre-drilling first 10 LF, per hole | C9@1.50 | LS | — | 71.30 | 53.20 | 124.50 |
| Add per LF of drilling, after first 10 LF | C9@.177 | LF | — | 8.42 | 6.28 | 14.70 |

**Steel "HP" shape piles** Equipment includes a 60-ton crawler-mounted crane equipped with pile driving attachments, a 4-ton forklift, a truck with hand tools, a portable air compressor complete with hoses and pneumatic operated tools, a welding machine and an oxygen-acetylene cutting torch. Use $25,000 as minimum job charge, including 1,000 LF of piling and cost for moving pile driving equipment on and off the job site.

| | Craft@Hrs | Unit | Material | Labor | Equipment | Total |
|---|---|---|---|---|---|---|
| Minimum job charge | — | LS | — | — | — | 20,000.00 |

For jobs requiring over 1,000 LF of piling add to minimum charge as follows. (Figures in parentheses show approximate production per day based on a 7-man crew.)

| | Craft@Hrs | Unit | Material | Labor | Equipment | Total |
|---|---|---|---|---|---|---|
| HP8  8" x  8", 36 lbs per LF | | | | | | |
| 30' or 40' long piles (800 LF per day) | S8@.070 | LF | 13.10 | 3.42 | 2.95 | 19.47 |
| 50' or 60' long piles (880 LF per day) | S8@.064 | LF | 13.10 | 3.13 | 2.70 | 18.93 |
| HP10  10" x 10", 42 lbs per LF | | | | | | |
| 30' or 40' long piles (760 LF per day) | S8@.074 | LF | 14.80 | 3.62 | 3.12 | 21.54 |
| 50' or 60' long piles (800 LF per day) | S8@.070 | LF | 14.80 | 3.42 | 2.95 | 21.17 |
| HP10  10" x 10", 48 lbs per LF | | | | | | |
| 30' or 40' long piles (760 LF per day) | S8@.074 | LF | 17.10 | 3.62 | 3.12 | 23.84 |
| 50' or 60' long piles (800 LF per day) | S8@.070 | LF | 17.10 | 3.42 | 2.95 | 23.47 |
| HP12  12" x 12", 53 lbs per LF | | | | | | |
| 30' or 40' long piles (730 LF per day) | S8@.077 | LF | 18.20 | 3.77 | 3.24 | 25.21 |
| 50' or 60' long piles (760 LF per day) | S8@.074 | LF | 18.20 | 3.62 | 3.12 | 24.94 |
| HP12  12" x 12", 74 lbs per LF | | | | | | |
| 30' or 40' long piles (730 LF per day) | S8@.077 | LF | 25.40 | 3.77 | 3.24 | 32.41 |
| 50' or 60' long piles (760 LF per day) | S8@.074 | LF | 25.40 | 3.62 | 3.12 | 32.14 |
| HP14  14" x 14", 73 lbs per LF | | | | | | |
| 30' or 40' long piles (690 LF per day) | S8@.081 | LF | 24.40 | 3.96 | 3.41 | 31.77 |
| 50' or 60' long piles (730 LF per day) | S8@.077 | LF | 24.40 | 3.77 | 3.24 | 31.41 |
| HP14  14" x 14", 89 lbs per LF | | | | | | |
| 30' or 40' long piles (630 LF per day) | S8@.089 | LF | 29.80 | 4.35 | 3.75 | 37.90 |
| 50' or 60' long piles (690 LF per day) | S8@.081 | LF | 29.80 | 3.96 | 3.41 | 37.17 |
| HP14  14" x 14", 102 lbs per LF | | | | | | |
| 30' or 40' long piles (630 LF per day) | S8@.089 | LF | 34.10 | 4.35 | 3.75 | 42.20 |
| 50' or 60' long piles (690 LF per day) | S8@.081 | LF | 34.10 | 3.96 | 3.41 | 41.47 |
| HP14  14" x 14", 117 lbs per LF | | | | | | |
| 30' or 40' long piles (630 LF per day) | S8@.089 | LF | 39.10 | 4.35 | 3.75 | 47.20 |
| 50' or 60' long piles (690 LF per day) | S8@.081 | LF | 39.10 | 3.96 | 3.41 | 46.47 |

# 31 Earthwork

|  | Craft@Hrs | Unit | Material | Labor | Equipment | Total |
|---|---|---|---|---|---|---|
| Additional costs related to any of above | | | | | | |
| Standard driving points or splices | | | | | | |
| HP8 | S8@1.50 | Ea | 41.80 | 73.40 | 22.10 | 137.30 |
| HP10 | S8@1.75 | Ea | 45.40 | 85.60 | 25.80 | 156.80 |
| HP12 | S8@2.00 | Ea | 57.10 | 97.80 | 29.50 | 184.40 |
| HP14 | S8@2.50 | Ea | 67.00 | 122.00 | 36.90 | 225.90 |
| Cutting pile off to required elevation | | | | | | |
| HP8 or HP10 | S8@1.00 | Ea | 4.55 | 48.90 | 14.70 | 68.15 |
| HP12 or HP14 | S8@1.50 | Ea | 6.86 | 73.40 | 22.10 | 102.36 |

**Steel pipe piles** Equipment includes a 45-ton crawler-mounted crane equipped with pile driving attachments, a 2-ton forklift, a truck with hand tools, a portable air compressor complete with hoses and pneumatic operated tools, a welding machine and an oxygen-acetylene cutting torch. Use $20,000 as minimum job charge, including 500 LF of piling and cost for moving pile driving equipment on and off the job site.

| Minimum job charge | — | LS | — | — | — | 20,000.00 |
|---|---|---|---|---|---|---|

For jobs requiring over 500 LF of piling add to minimum charge as follows. (Figures in parentheses show approximate production per day based on a 7-man crew.)

|  | Craft@Hrs | Unit | Material | Labor | Equipment | Total |
|---|---|---|---|---|---|---|
| Steel pipe piles, non-filled | | | | | | |
| 8" (660 LF per day) | S8@.085 | LF | 11.20 | 4.16 | 3.07 | 18.43 |
| 10" (635 LF per day) | S8@.088 | LF | 15.90 | 4.30 | 3.17 | 23.37 |
| 12" (580 LF per day) | S8@.096 | LF | 19.60 | 4.70 | 3.46 | 27.76 |
| 14" (500 LF per day) | S8@.112 | LF | 21.80 | 5.48 | 4.04 | 31.32 |
| 16" (465 LF per day) | S8@.120 | LF | 25.10 | 5.87 | 4.33 | 35.30 |
| 18" (440 LF per day) | S8@.128 | LF | 28.60 | 6.26 | 4.62 | 39.48 |
| Steel pipe piles, concrete filled | | | | | | |
| 8" (560 LF per day) | S8@.100 | LF | 13.30 | 4.89 | 3.61 | 21.80 |
| 10" (550 LF per day) | S8@.101 | LF | 16.60 | 4.94 | 3.64 | 25.18 |
| 12" (500 LF per day) | S8@.112 | LF | 21.90 | 5.48 | 4.04 | 31.42 |
| 14" (425 LF per day) | S8@.131 | LF | 23.60 | 6.41 | 4.72 | 34.73 |
| 16" (400 LF per day) | S8@.139 | LF | 28.00 | 6.80 | 5.01 | 39.81 |
| 18" (375 LF per day) | S8@.149 | LF | 32.20 | 7.29 | 5.37 | 44.86 |
| Splices for steel pipe piles | | | | | | |
| 8" pile | S8@1.12 | Ea | 52.10 | 54.80 | 5.20 | 112.10 |
| 10" pile | S8@1.55 | Ea | 58.60 | 75.80 | 21.00 | 155.40 |
| 12" pile | S8@2.13 | Ea | 63.10 | 104.00 | 28.90 | 196.00 |
| 14" pile | S8@2.25 | Ea | 83.40 | 110.00 | 30.50 | 223.90 |
| 16" pile | S8@2.39 | Ea | 133.00 | 117.00 | 32.40 | 282.40 |
| 18" pile | S8@2.52 | Ea | 159.00 | 123.00 | 34.20 | 316.20 |
| Standard points for steel pipe piles | | | | | | |
| 8" point | S8@1.49 | Ea | 79.90 | 72.90 | 20.20 | 173.00 |
| 10" point | S8@1.68 | Ea | 108.00 | 82.20 | 22.80 | 213.00 |
| 12" point | S8@1.84 | Ea | 145.00 | 90.00 | 25.00 | 260.00 |
| 14" point | S8@2.03 | Ea | 182.00 | 99.30 | 27.50 | 308.80 |
| 16" point | S8@2.23 | Ea | 268.00 | 109.00 | 30.30 | 407.30 |
| 18" point | S8@2.52 | Ea | 354.00 | 123.00 | 34.20 | 511.20 |

**Wood piles** End bearing Douglas fir, treated with 12 lb creosote. Piles up to 40' long are 8" minimum tip diameter. Over 45' long, 7" is the minimum tip diameter. Equipment includes a 45-ton crawler-mounted crane equipped with pile driving attachments, a 2-ton forklift, a truck with hand tools, a portable air compressor complete with hoses and pneumatic operated tools. Use $20,000 as minimum job charge, including 1,000 LF of piling and cost for moving pile driving equipment on and off the job site.

| Minimum job charge | — | LS | — | — | — | 24,000.00 |
|---|---|---|---|---|---|---|

# 31 Earthwork

|  | Craft@Hrs | Unit | Material | Labor | Equipment | Total |
|---|---|---|---|---|---|---|

For jobs requiring over 1,000 LF of piling, add to minimum charge as follows. (Figures in parentheses show approximate production per day based on a 7-man crew.)

| | Craft@Hrs | Unit | Material | Labor | Equipment | Total |
|---|---|---|---|---|---|---|
| To 30' long (725 LF per day) | S8@.077 | LF | 5.29 | 3.77 | 2.59 | 11.65 |
| Over 30' to 40' (800 LF per day) | S8@.069 | LF | 5.38 | 3.37 | 2.32 | 11.07 |
| Over 40' to 50' (840 LF per day) | S8@.067 | LF | 5.92 | 3.28 | 2.25 | 11.45 |
| Over 50' to 60' (920 LF per day) | S8@.061 | LF | 6.33 | 2.98 | 2.05 | 11.36 |
| Over 60' to 80' (950 LF per day) | S8@.059 | LF | 8.09 | 2.89 | 1.99 | 12.97 |
| Add for drive shoe, per pile | — | Ea | 24.50 | — | — | 24.50 |

**Caisson Foundations** Drilled in stable soil, no shoring required, filled with reinforced concrete. Add for belled bottoms, if required, from the following section. Equipment includes a truck-mounted A-frame hoist and a power operated auger and a concrete pump. Concrete is based on using 3,000 PSI pump mix delivered to the job site in ready-mix trucks and includes a 5% allowance for waste. Spiral caisson reinforcing is based on using 3/8" diameter hot rolled steel spirals with main vertical bars as shown, shop fabricated, delivered ready to install, with typical engineering and drawings. Costs for waste disposal are not included. Refer to Construction Materials Disposal section and add for same. Use $15,000 as a minimum charge, including 200 LF of reinforced concrete filled caissons and cost for moving the equipment on and off the job.

| | Craft@Hrs | Unit | Material | Labor | Equipment | Total |
|---|---|---|---|---|---|---|
| Minimum job charge | — | LS | — | — | — | 15,000.00 |

For jobs requiring over 200 LF of reinforced concrete-filled caissons add to minimum charge as follows. (Figures in parentheses show approximate production per day based on a 3-man crew)

16" diameter, drilled caisson filled with reinforced concrete

| | Craft@Hrs | Unit | Material | Labor | Equipment | Total |
|---|---|---|---|---|---|---|
| Drilling (680 VLF per day) | S6@.035 | VLF | — | 1.57 | 2.56 | 4.13 |
| Spiral reinforcing plus 6 #6 bars (590 VLF per day) | S6@.041 | VLF | 25.90 | 1.84 | 2.99 | 30.73 |
| Concrete filling (615 VLF per day) | S6@.039 | VLF | 5.60 | 1.75 | 2.86 | 10.21 |
| **Total for 16" caisson** | **S6@.115** | **VLF** | **31.50** | **5.16** | **8.41** | **45.07** |

24" diameter, drilled caisson filled with reinforced concrete

| | Craft@Hrs | Unit | Material | Labor | Equipment | Total |
|---|---|---|---|---|---|---|
| Drilling (570 VLF per day) | S6@.042 | VLF | — | 1.89 | 3.08 | 4.97 |
| Spiral reinforcing plus 6 #6 bars (280 VLF per day) | S6@.085 | VLF | 44.30 | 3.82 | 6.22 | 54.34 |
| Concrete filling (280 VLF per day) | S6@.086 | VLF | 12.50 | 3.86 | 6.30 | 22.66 |
| **Total for 24" caisson** | **S6@.213** | **VLF** | **56.80** | **9.57** | **15.60** | **81.97** |

36" diameter, drilled caisson filled with reinforced concrete

| | Craft@Hrs | Unit | Material | Labor | Equipment | Total |
|---|---|---|---|---|---|---|
| Drilling (440 LF per day) | S6@.055 | VLF | — | 2.47 | 4.03 | 6.50 |
| Spiral reinforcing plus 8 #10 bars (135 VLF per day) | S6@.176 | VLF | 80.20 | 7.91 | 12.90 | 101.01 |
| Concrete filling (125 VLF per day) | S6@.194 | VLF | 28.20 | 8.71 | 14.20 | 51.11 |
| **Total for 36" caisson** | **S6@.425** | **VLF** | **108.40** | **19.09** | **31.13** | **158.62** |

Bell-bottom footings, concrete filled, for pre-drilled caissons. Add to the cost of any of the caissons above. (Figures in parentheses show approximate production per day based on a 3-man crew.)

| | Craft@Hrs | Unit | Material | Labor | Equipment | Total |
|---|---|---|---|---|---|---|
| 6' diameter bell bottom concrete filled (8 per day) | S6@3.00 | Ea | 332.00 | 135.00 | 220.00 | 687.00 |
| 8' diameter bell bottom concrete filled (4 per day | S6@6.00 | Ea | 788.00 | 270.00 | 439.00 | 1,497.00 |

|  | Craft@Hrs | Unit | Material | Labor | Equipment | Total |
|---|---|---|---|---|---|---|

**Asphalt Paving** To arrive at the total cost for new asphalt paving jobs, add the cost of the subbase preparation and aggregate to the costs for the asphalt binder course, the rubberized interliner and the wear course. Cost for site preparation is not included. These costs are intended to be used for on-site roadways and parking lots. Use $7,500 as a minimum charge. For more detailed coverage, see *National Earthwork and Heavy Equipment Estimator*, http://CraftsmanSiteLicense.com.

**Subbase and aggregate** Equipment includes a pneumatic-tired articulated 10,000 pound motor grader, a 4-ton two axle steel drum roller and a 3,500 gallon water truck.
Fine grading and compacting existing subbase

| | Craft@Hrs | Unit | Material | Labor | Equipment | Total |
|---|---|---|---|---|---|---|
| To within plus or minus 1/10' | P5@.066 | SY | .01 | 2.89 | 1.95 | 4.85 |

Stone aggregate base, Class 2. Includes grading and compacting of base material.

| | Craft@Hrs | Unit | Material | Labor | Equipment | Total |
|---|---|---|---|---|---|---|
| Class 2 stone aggregate base, per CY | — | CY | 30.10 | — | — | 30.10 |
| Class 2 stone aggregate base, per ton | — | Ton | 21.50 | — | — | 21.50 |
| 1" thick base (36 SY per CY) | P5@.008 | SY | .84 | .35 | .24 | 1.43 |
| 4" thick base (9 SY per CY) | P5@.020 | SY | 3.34 | .88 | .59 | 4.81 |
| 6" thick base (6 SY per CY) | P5@.033 | SY | 5.02 | 1.45 | .97 | 7.44 |
| 8" thick base (4.5 SY per CY) | P5@.039 | SY | 6.69 | 1.71 | 1.15 | 9.55 |
| 10" thick base (3.6 SY per CY) | P5@.042 | SY | 8.36 | 1.84 | 1.24 | 11.44 |
| 12" thick base (3 SY per CY) | P5@.048 | SY | 10.00 | 2.10 | 1.42 | 13.52 |

**Asphalt materials** Equipment includes a 10' wide self-propelled paving machine and a 4-ton two axle steel drum roller.
Binder course, spread and rolled

| | Craft@Hrs | Unit | Material | Labor | Equipment | Total |
|---|---|---|---|---|---|---|
| Binder course, per ton | — | Ton | 91.90 | — | — | 91.90 |
| 1" thick (18 SY per ton) | P5@.028 | SY | 5.06 | 1.23 | 1.02 | 7.31 |
| 1-1/2" thick (12 SY per ton) | P5@.036 | SY | 8.45 | 1.58 | 1.47 | 11.50 |
| 2" thick (9 SY per ton) | P5@.039 | SY | 11.30 | 1.71 | 1.59 | 14.60 |
| 3" thick (6 SY per ton) | P5@.050 | SY | 16.90 | 2.19 | 2.04 | 21.13 |

Rubberized asphalt interlayer with tack coat. Based on using 10' wide rolls 3/16" thick x 500' long and one coat of mopped-on bituminous tack coat, MC70 (0.15 gallon per SY) before allowance for waste.

| | Craft@Hrs | Unit | Material | Labor | Equipment | Total |
|---|---|---|---|---|---|---|
| Rubberized asphalt interlayer, per 5,000 SF roll | — | Roll | 275.00 | — | — | 275.00 |
| Bituminous tack coat, per gallon | — | Gal | 3.01 | — | — | 3.01 |
| Laid, cut, mopped and rolled | P5@.005 | SY | 1.28 | .22 | .20 | 1.70 |

Wear course

| | Craft@Hrs | Unit | Material | Labor | Equipment | Total |
|---|---|---|---|---|---|---|
| Bituminous wear course, per ton | — | Ton | 106.00 | — | — | 106.00 |
| 1" thick (18 SY per ton) | P5@.028 | SY | 5.84 | 1.23 | 1.14 | 8.21 |
| 1-1/2" thick (12 SY per ton) | P5@.036 | SY | 8.77 | 1.58 | 1.47 | 11.82 |
| 2" thick (9 SY per ton) | P5@.039 | SY | 11.80 | 1.71 | 1.59 | 15.10 |
| 3" thick (6 SY per ton) | P5@.050 | SY | 17.50 | 2.19 | 2.04 | 21.73 |

Add to total installed cost of paving as described, if required

| | Craft@Hrs | Unit | Material | Labor | Equipment | Total |
|---|---|---|---|---|---|---|
| Bituminous prime coat | P5@.001 | SY | .93 | .04 | .04 | 1.01 |
| Tar emulsion protective seal coat | P5@.034 | SY | 3.25 | 1.49 | 1.39 | 6.13 |
| Asphalt slurry seal | P5@.043 | SY | 4.12 | 1.89 | 1.75 | 7.76 |

Asphaltic concrete curb, straight, placed during paving. Equipment includes an asphalt curbing machine

| | Craft@Hrs | Unit | Material | Labor | Equipment | Total |
|---|---|---|---|---|---|---|
| 8" x 6", placed by hand | P5@.141 | LF | 2.18 | 6.18 | — | 8.36 |
| 8" x 6", machine extruded | P5@.067 | LF | 1.98 | 2.94 | .60 | 5.52 |
| 8" x 8", placed by hand | P5@.141 | LF | 3.49 | 6.18 | — | 9.67 |
| 8" x 8", machine extruded | P5@.068 | LF | 2.64 | 2.98 | .29 | 5.91 |
| Add for curved sections | P5@.034 | LF | — | 1.49 | .15 | 1.64 |

Asphaltic concrete speed bumps, 3" high, 24" wide (135 LF and 2 tons per CY), placed on parking lot surface and rolled

| | Craft@Hrs | Unit | Material | Labor | Equipment | Total |
|---|---|---|---|---|---|---|
| 100 LF job | P5@.198 | LF | 1.40 | 8.68 | 2.25 | 12.33 |
| Painting speed bumps, 24" wide, white | P5@.022 | LF | .50 | .96 | — | 1.46 |

See also the section on Speed Bumps and Parking Blocks following Pavement Striping and Marking.

| | Craft@Hrs | Unit | Material | Labor | Equipment | Total |
|---|---|---|---|---|---|---|

**Asphalt Paving for Roadways, Assembly Costs**  On-site roadway, with 4" thick aggregate base, 2" thick asphalt binder course, rubberized asphalt interlayer, and 2" thick wear course. This is often called 4" paving over 4" aggregate base.

| | Craft@Hrs | Unit | Material | Labor | Equipment | Total |
|---|---|---|---|---|---|---|
| Fine grading and compact existing subbase | P5@.066 | SY | .01 | 2.89 | 2.69 | 5.59 |
| Stone aggregate base, Class 2, 4" thick | P5@.020 | SY | 3.34 | .88 | .82 | 5.04 |
| Binder course 2" thick (9 SY per ton) | P5@.039 | SY | 11.30 | 1.71 | 1.59 | 14.60 |
| Rubberized asphalt interlayer | P5@.005 | SY | 1.28 | .22 | .20 | 1.70 |
| Wear course 2" thick (9 SY per ton) | P5@.039 | SY | 11.80 | 1.71 | 1.59 | 15.10 |
| Tar emulsion protective seal coat | P5@.034 | SY | 3.51 | 1.49 | 1.39 | 6.39 |

Total estimated cost for typical 10' wide roadway, 4" paving over 4" aggregate base

| | Craft@Hrs | Unit | Material | Labor | Equipment | Total |
|---|---|---|---|---|---|---|
| Cost per SY of roadway | P5@.203 | SY | 31.24 | 8.90 | 8.28 | 48.42 |
| Cost per SF roadway | P5@.023 | SF | 3.47 | 1.01 | .92 | 5.40 |
| Cost per LF, 10' wide roadway | P5@.230 | LF | 34.71 | 10.08 | 9.20 | 53.99 |

**Asphalt Paving for Parking Areas, Assembly Costs**  Equipment includes a 10' wide self-propelled paving machine and a 4-ton two axle steel drum roller. On-site parking lot with 6" thick aggregate base, 3" thick asphalt binder course, rubberized asphalt interliner, and 3" thick wear course. This is often called 6" paving over 6" aggregate base.

| | Craft@Hrs | Unit | Material | Labor | Equipment | Total |
|---|---|---|---|---|---|---|
| Fine grading and compact existing subbase | P5@.066 | SY | .01 | 2.89 | 2.69 | 5.59 |
| Stone aggregate base, Class 2, 6" thick | P5@.033 | SY | 5.02 | 1.45 | 1.35 | 7.82 |
| Binder course 3" thick | P5@.050 | SY | 16.90 | 2.19 | 2.04 | 21.13 |
| Rubberized asphalt interlayer | P5@.005 | SY | 1.28 | .22 | .20 | 1.70 |
| Wear course 3" thick | P5@.050 | SY | 17.50 | 2.19 | 2.04 | 21.73 |
| Tar emulsion protective seal coat | P5@.034 | SY | 3.51 | 1.49 | 1.39 | 6.39 |

Total estimated cost of typical 1,000 SY parking lot, 6" paving over 6" aggregate base

| | Craft@Hrs | Unit | Material | Labor | Equipment | Total |
|---|---|---|---|---|---|---|
| Cost per SY of parking lot | P5@.238 | SY | 44.22 | 10.43 | 9.71 | 64.36 |
| Cost per SF of parking lot | P5@.026 | SF | 4.91 | 1.14 | 1.08 | 7.13 |

**Cold Milling of Asphalt Surface to 1" Depth**  Equipment includes a motorized street broom, a 20" pavement miller, two 5 CY dump trucks and a pickup truck. Hauling and disposal not included. Use $3,000 as a minimum charge.

| | Craft@Hrs | Unit | Material | Labor | Equipment | Total |
|---|---|---|---|---|---|---|
| Milling asphalt to 1" depth | C6@.017 | SY | — | .75 | .58 | 1.33 |

**Asphalt Pavement Repairs** to conform to existing street, strips of 100 to 500 SF. Equipment includes one wheel-mounted 150 CFM air compressor, paving breaker, jackhammer bits, one 55 HP wheel loader with integral backhoe, one medium weight pneumatic tamper, one 1-ton two axle steel roller and one 5 CY dump truck These costs include the cost of loading and hauling to a legal dump within 6 miles. Dump fees are not included. See Waste Disposal.

Asphalt pavement repairs, assembly costs. Use $3,600 as a minimum job charge.

| | Craft@Hrs | Unit | Material | Labor | Equipment | Total |
|---|---|---|---|---|---|---|
| Cut and remove 2" asphalt and 36" subbase | P5@.250 | SY | — | 11.00 | 6.11 | 17.11 |

Place 30" fill compacted to 90%, 6" aggregate base, 2" thick asphalt and tack coat

| | Craft@Hrs | Unit | Material | Labor | Equipment | Total |
|---|---|---|---|---|---|---|
| 30" fill (30 SY per CY at $11.10 CY) | P5@.099 | SY | .37 | 4.34 | 2.42 | 7.13 |
| 6" thick base (6 SY per CY) | P5@.033 | SY | 5.09 | 1.45 | .81 | 7.35 |
| 2" thick asphalt (9 SY per ton) | P5@.039 | SY | 11.30 | 1.71 | .95 | 13.96 |
| Tar emulsion protective seal coat | P5@.034 | SY | 3.51 | 1.49 | .83 | 5.83 |
| Total per SY | P5@.455 | SY | 20.27 | 19.95 | 11.12 | 51.34 |

| | Craft@Hrs | Unit | Material | Labor | Equipment | Total |
|---|---|---|---|---|---|---|

**Asphalt overlay** on existing asphalt pavement. Equipment includes a 10' wide self-propelled paving machine and a 4-ton two axle steel drum roller. Costs shown assume unobstructed access to area receiving overlay. Use $6,000 as a minimum job charge.

Rubberized asphalt interlayer with tack coat. Based on using 10' wide rolls 3/16" thick x 500' long and one coat of mopped-on bituminous tack coat, MC70 (with .15 gallon per SY) before allowance for waste.

| | Craft@Hrs | Unit | Material | Labor | Equipment | Total |
|---|---|---|---|---|---|---|
| Rubberized asphalt inner layer, per 5,000 SF roll | — | Roll | 293.00 | — | — | 293.00 |
| Bituminous tack coat, per gallon | — | Gal | 4.30 | — | — | 4.30 |
| Laid, cut, mopped and rolled | P5@.005 | SY | 1.26 | .22 | .20 | 1.68 |
| Wear course | | | | | | |
| Bituminous wear course, per ton | — | Ton | 103.00 | — | — | 103.00 |
| 1" thick (18 SY per ton) | P5@.028 | SY | 5.69 | 1.23 | 1.14 | 8.06 |
| 1-1/2" thick (12 SY per ton) | P5@.036 | SY | 8.55 | 1.58 | 1.47 | 11.60 |
| 2" thick (9 SY per ton) | P5@.039 | SY | 11.40 | 1.71 | 1.59 | 14.70 |
| Add to overlaid paving as described above, if required | | | | | | |
| Bituminous prime coat, MC70 | P5@.001 | SY | 1.22 | .04 | .04 | 1.30 |
| Tar emulsion protective seal coat | P5@.034 | SY | 3.69 | 1.49 | 1.39 | 6.57 |
| Asphalt slurry seal | P5@.043 | SY | 4.82 | 1.89 | 1.75 | 8.46 |

**Concrete Paving** No grading or base preparation included.

| | Craft@Hrs | Unit | Material | Labor | Equipment | Total |
|---|---|---|---|---|---|---|
| Steel edge forms to 15", rented | C8@.028 | LF | — | 1.28 | .82 | 2.10 |

2,000 PSI concrete placed directly from the chute of a ready-mix truck, vibrated and broom finished. Costs include 4% allowance for waste but no forms, reinforcing, or joints.

| | Craft@Hrs | Unit | Material | Labor | Equipment | Total |
|---|---|---|---|---|---|---|
| 4" thick (80 SF per CY) | P8@.013 | SF | 1.23 | .59 | .02 | 1.84 |
| 6" thick (54 SF per CY) | P8@.016 | SF | 1.84 | .72 | .02 | 2.58 |
| 8" thick (40 SF per CY) | P8@.019 | SF | 2.46 | .86 | .02 | 3.34 |
| 10" thick (32 SF per CY) | P8@.021 | SF | 3.08 | .95 | .03 | 4.06 |
| 12" thick (27 SF per CY) | P8@.024 | SF | 3.70 | 1.08 | .03 | 4.81 |
| 15" thick (21 SF per CY) | P8@.028 | SF | 4.69 | 1.26 | .04 | 5.99 |

**Clean and Sweep Pavement** Equipment cost is $53.00 per hour for a self-propelled vacuum street sweeper, 7 CY maximum capacity. Use $300 as a minimum charge for sweeping using this equipment.

| | Craft@Hrs | Unit | Material | Labor | Equipment | Total |
|---|---|---|---|---|---|---|
| Pavement sweeping, using equipment | T0@.200 | MSF | — | 11.00 | 20.70 | 31.70 |
| Clean and sweep pavement, by hand | CL@1.44 | MSF | — | 57.40 | — | 57.40 |

**Joint Treatment for Concrete Pavement** Seal joint in concrete pavement with ASTM D-3569 PVC coal tar including 5% waste). Cost per linear foot (LF), by joint size.

| | Craft@Hrs | Unit | Material | Labor | Equipment | Total |
|---|---|---|---|---|---|---|
| PVC coal tar, per gallon | — | Gal | 52.20 | — | — | 52.20 |
| 3/8" x 1/2"   (128 LF per gallon) | PA@.002 | LF | .44 | .10 | — | .54 |
| 1/2" x 1/2"   (96 LF per gallon) | PA@.002 | LF | .58 | .10 | — | .68 |
| 1/2" x 3/4"   (68 LF per gallon) | PA@.003 | LF | .78 | .15 | — | .93 |
| 1/2" x 1"   (50 LF per gallon) | PA@.003 | LF | 1.09 | .15 | — | 1.24 |
| 1/2" x 1-1/4"   (40 LF per gallon) | PA@.004 | LF | 1.40 | .21 | — | 1.61 |
| 1/2" x 1-1/2"   (33 LF per gallon) | PA@.006 | LF | 1.66 | .31 | — | 1.97 |
| 3/4" x 3/4"   (46 LF per gallon) | PA@.004 | LF | 1.17 | .21 | — | 1.38 |
| 3/4" x 1-1/4"   (33 LF per gallon) | PA@.006 | LF | 1.66 | .31 | — | 1.97 |
| 3/4" x 1-1/2"   (27 LF per gallon) | PA@.007 | LF | 2.04 | .36 | — | 2.40 |
| 3/4" x 2"   (22 LF per gallon) | PA@.008 | LF | 2.50 | .41 | — | 2.91 |
| 1" x 1"   (25 LF per gallon) | PA@.008 | LF | 2.19 | .41 | — | 2.60 |

| | Craft@Hrs | Unit | Material | Labor | Equipment | Total |
|---|---|---|---|---|---|---|
| **Backer rod for joints in concrete pavement** | | | | | | |
| 1/2" backer rod | PA@.002 | LF | .06 | .10 | — | .16 |
| 3/4" or 1" backer rod | PA@.002 | LF | .09 | .10 | — | .19 |

Clean out old joint sealer in concrete pavement by sandblasting and reseal with PVC coal tar sealer, including new backer rod. Use $400 as a minimum job charge.

| | Craft@Hrs | Unit | Material | Labor | Equipment | Total |
|---|---|---|---|---|---|---|
| 1/2" x 1/2" | PA@.009 | LF | .54 | .46 | .27 | 1.27 |
| 1/2" x 3/4" | PA@.013 | LF | .78 | .67 | .40 | 1.85 |
| 1/2" x 1" | PA@.013 | LF | 1.04 | .67 | .40 | 2.11 |
| 1/2" x 1-1/4" | PA@.018 | LF | 1.27 | .93 | .55 | 2.75 |
| 1/2" x 1-1/2" | PA@.022 | LF | 1.54 | 1.13 | .67 | 3.34 |

| | Craft@Hrs | Unit | Material | Labor | Total |
|---|---|---|---|---|---|

**Brick Paving** Extruded hard red brick. These costs do not include site preparation. Material costs shown include 2% for waste. (Figures in parentheses show quantity of brick before allowance for waste.) For expanded coverage of brick paving, see the *National Concrete & Masonry Estimator*, http://CraftsmanSiteLicense.com

| | Craft@Hrs | Unit | Material | Labor | Total |
|---|---|---|---|---|---|
| **Brick laid flat without mortar** | | | | | |
| Using brick, per thousand | — | M | 364.00 | — | 364.00 |
| **Running bond** | | | | | |
| 4" x 8" x 2-1/4" (4.5 per SF) | M1@.114 | SF | 1.64 | 5.31 | 6.95 |
| 4" x 8" x 1-5/8" (4.5 per SF) | M1@.112 | SF | 1.61 | 5.22 | 6.83 |
| 3-5/8" x 7-5/8" x 2-1/4" (5.2 per SF) | M1@.130 | SF | 1.86 | 6.06 | 7.92 |
| 3-5/8" x 7-5/8" x 1-5/8" (5.2 per SF) | M1@.127 | SF | 1.86 | 5.92 | 7.78 |
| Add for herringbone pattern | M1@.033 | SF | — | 1.54 | 1.54 |
| **Basketweave** | | | | | |
| 4" x 8" x 2-1/4" (4.5 per SF) | M1@.133 | SF | 1.67 | 6.20 | 7.87 |
| 4" x 8" x 1-5/8" (4.5 per SF) | M1@.130 | SF | 1.61 | 6.06 | 7.67 |
| **Brick laid on edge without mortar** | | | | | |
| **Running bond** | | | | | |
| 4" x 8" x 2-1/4" (8 per SF) | M1@.251 | SF | 2.96 | 11.70 | 14.66 |
| 3-5/8" x 7-5/8" x 2-1/4" (8.4 per SF) | M1@.265 | SF | 3.01 | 12.40 | 15.41 |

Brick laid flat with 3/8" mortar joints and mortar setting bed. Mortar included in material cost.

| | Craft@Hrs | Unit | Material | Labor | Total |
|---|---|---|---|---|---|
| **Running bond** | | | | | |
| 4" x 8" x 2-1/4" (3.9 per SF) | M1@.127 | SF | 2.29 | 5.92 | 8.21 |
| 4" x 8" x 1-5/8" (3.9 per SF) | M1@.124 | SF | 2.25 | 5.78 | 8.03 |
| 3-5/8" x 7-5/8" x 2-1/4" (4.5 per SF) | M1@.144 | SF | 2.52 | 6.71 | 9.23 |
| 3-5/8" x 7-5/8" x 1-5/8" (4.5 per SF) | M1@.140 | SF | 2.46 | 6.53 | 8.99 |
| Add for herringbone pattern | M1@.033 | SF | — | 1.54 | 1.54 |
| **Basketweave** | | | | | |
| 4" x 8" x 2-1/4" (3.9 per SF) | M1@.147 | SF | 1.44 | 6.85 | 8.29 |
| 4" x 8" x 1-5/8" (3.9 per SF) | M1@.144 | SF | 1.40 | 6.71 | 8.11 |
| 3-5/8" x 7-5/8" x 2-1/4" (4.5 per SF) | M1@.147 | SF | 1.67 | 6.85 | 8.52 |
| 3-5/8" x 7-5/8" x 1-5/8" (4.5 per SF) | M1@.144 | SF | 1.61 | 6.71 | 8.32 |

| | Craft@Hrs | Unit | Material | Labor | Total |
|---|---|---|---|---|---|
| **Brick laid on edge with 3/8" mortar joints and setting bed, running bond, including mortar** | | | | | |
| 4" x 8" x 2-1/4" (6.5 per SF) | M1@.280 | SF | 3.26 | 13.10 | 16.36 |
| 3-5/8" x 7-5/8" x 2-1/4" (6.9 per SF) | M1@.314 | SF | 3.40 | 14.60 | 18.00 |
| **Mortar color for brick laid in mortar, add per SF of paved area** | | | | | |
| Red or yellow | — | SF | .10 | — | .10 |
| Black or brown | — | SF | .15 | — | .15 |
| Green | — | SF | .38 | — | .38 |

# 32 Exterior Improvements

| | Craft@Hrs | Unit | Material | Labor | Total |
|---|---|---|---|---|---|
| **Base underlayment for paving brick** | | | | | |
| Add for 2" sand base (124 SF per ton) | M1@.009 | SF | .06 | .42 | .48 |
| Add for 15 lb felt underlayment | M1@.002 | SF | .06 | .09 | .15 |
| Add for asphalt tack coat (80 SF per gallon) | M1@.005 | SF | .06 | .23 | .29 |
| Add for 2% neoprene tack coat (40 SF per gal.) | M1@.005 | SF | .17 | .23 | .40 |
| **Brick edging, 8" deep, set in concrete with dry joints** | | | | | |
| Using brick, per thousand | — | M | 363.00 | — | 363.00 |
| Headers 4" x 8" x 2-1/4" (3 per LF) | M1@.147 | LF | 1.61 | 6.85 | 8.46 |
| Headers 3-5/8" x 7-5/8" x 2-1/4" (3.3 per LF) | M1@.159 | LF | 1.72 | 7.41 | 9.13 |
| Rowlocks 3-5/8" x 7-5/8" x 2-1/4" (5.3 per LF) | M1@.251 | LF | 2.46 | 11.70 | 14.16 |
| Rowlocks 4" x 8" x 2-1/4" (5.3 per LF) | M1@.265 | LF | 2.46 | 12.40 | 14.86 |
| **Brick edging, 8" deep, set in concrete with 3/8" mortar joints** | | | | | |
| Using brick, per thousand | — | M | 370.00 | — | 370.00 |
| Headers 4" x 8" x 2-1/4" (2.75 per LF) | M1@.159 | LF | 1.01 | 7.41 | 8.42 |
| Headers 3-5/8" x 7-5/8" x 2-1/4" (3 per LF) | M1@.173 | LF | 1.11 | 8.07 | 9.18 |
| Rowlocks 3-5/8" x 7-5/8" x 2-1/4" (4.6 per LF) | M1@.298 | LF | 1.70 | 13.90 | 15.60 |
| Rowlocks 4" x 8" x 2-1/4" (4.6 per LF) | M1@.314 | LF | 1.70 | 14.60 | 16.30 |

**Masonry, Stone and Slate Paving** These costs do not include site preparation. Costs for stone and slate vary widely. (Quantities shown in parentheses are before waste allowance.) Use $500 as a minimum job charge.

| | Craft@Hrs | Unit | Material | Labor | Total |
|---|---|---|---|---|---|
| **Interlocking 9" x 4-1/2" concrete pavers, dry joints, natural gray** | | | | | |
| 2-3/8" thick (3.5 per SF) | M1@.058 | SF | 2.68 | 2.70 | 5.38 |
| 3-1/8" thick (3.5 per SF) | M1@.063 | SF | 2.90 | 2.94 | 5.84 |
| Add for standard colors (3.5 per SF) | — | SF | .21 | — | .21 |
| Add for custom colors (3.5 per SF) | — | SF | .42 | — | .42 |
| **Patio blocks, 8" x 16" x 2", dry joints** | | | | | |
| Natural gray (1.125 per SF) | M1@.052 | SF | 1.40 | 2.42 | 3.82 |
| Standard colors (1.125 per SF) | M1@.052 | SF | 1.56 | 2.42 | 3.98 |
| **Flagstone pavers in sand bed with dry joints** | | | | | |
| Random ashlar, 1-1/4" | M1@.193 | SF | 3.02 | 9.00 | 12.02 |
| Random ashlar, 2-1/2" | M1@.228 | SF | 4.57 | 10.60 | 15.17 |
| Irregular fitted, 1-1/4" | M1@.208 | SF | 2.54 | 9.70 | 12.24 |
| Irregular fitted, 2-1/2" | M1@.251 | SF | 3.88 | 11.70 | 15.58 |
| **Flagstone pavers in mortar bed with mortar joints. Add the cost of mortar below** | | | | | |
| Random ashlar, 1-1/4" | M1@.208 | SF | 3.10 | 9.70 | 12.80 |
| Random ashlar, 2-1/2" | M1@.251 | SF | 4.70 | 11.70 | 16.40 |
| Irregular fitted, 1-1/4" | M1@.228 | SF | 2.62 | 10.60 | 13.22 |
| Irregular fitted, 2-1/2" | M1@.277 | SF | 3.97 | 12.90 | 16.87 |
| **Granite pavers, gray, sawn, thermal finish, 3/4" joint. Add bedding costs from the section that follows** | | | | | |
| 4" x 4" x 2" thick | M1@.184 | SF | 23.70 | 8.58 | 32.28 |
| 4" x 4" x 3" thick | M1@.184 | SF | 28.50 | 8.58 | 37.08 |
| 4" x 4" x 4" thick | M1@.184 | SF | 32.30 | 8.58 | 40.88 |
| 4" x 8" x 2" thick | M1@.171 | SF | 20.50 | 7.97 | 28.47 |
| 4" x 8" x 3" thick | M1@.184 | SF | 23.30 | 8.58 | 31.88 |
| 4" x 8" x 4" thick | M1@.184 | SF | 25.70 | 8.58 | 34.28 |
| 8" x 8" x 2" thick | M1@.171 | SF | 16.90 | 7.97 | 24.87 |
| 8" x 8" x 3" thick | M1@.171 | SF | 18.20 | 7.97 | 26.17 |
| 8" x 8" x 4" thick | M1@.184 | SF | 19.40 | 8.58 | 27.98 |
| 12" x 12" x 2" thick | M1@.171 | SF | 8.28 | 7.97 | 16.25 |
| 12" x 12" x 3" thick | M1@.171 | SF | 9.83 | 7.97 | 17.80 |
| 12" x 12" x 4" thick | M1@.171 | SF | 10.40 | 7.97 | 18.37 |

# 32 Exterior Improvements

| | Craft@Hrs | Unit | Material | Labor | Total |
|---|---|---|---|---|---|
| Mortar bed, 1/2" thick, 24 SF per CF of mortar | | | | | |
| SF of bed | M1@.039 | SF | .25 | 1.82 | 2.07 |
| Sand bed, 2" thick, 150 SF per CY of sand | | | | | |
| Per SF of bed | M1@.024 | SF | .06 | 1.12 | 1.18 |
| Limestone flag paving | | | | | |
| Sand bed, dry joints | | | | | |
| Random ashlar, 1-1/4" | M1@.193 | SF | 3.30 | 9.00 | 12.30 |
| Random ashlar, 2-1/2" | M1@.228 | SF | 6.29 | 10.60 | 16.89 |
| Irregular fitted, 1-1/2" | M1@.208 | SF | 2.86 | 9.70 | 12.56 |
| Irregular fitted, 2-1/2" | M1@.251 | SF | 4.86 | 11.70 | 16.56 |
| Mortar bed and joints | | | | | |
| Random ashlar, 1" | M1@.208 | SF | 2.79 | 9.70 | 12.49 |
| Random ashlar, 2-1/2" | M1@.251 | SF | 6.11 | 11.70 | 17.81 |
| Irregular fitted, 1" | M1@.228 | SF | 2.51 | 10.60 | 13.11 |
| Irregular fitted, 2-1/2" | M1@.277 | SF | 4.67 | 12.90 | 17.57 |
| Slate flag paving, natural cleft | | | | | |
| Sand bed, dry joints | | | | | |
| Random ashlar, 1-1/4" | M1@.173 | SF | 10.00 | 8.07 | 18.07 |
| Irregular fitted, 1-1/4" | M1@.193 | SF | 9.27 | 9.00 | 18.27 |
| Mortar bed and joints | | | | | |
| Random ashlar, 3/4" | M1@.173 | SF | 7.78 | 8.07 | 15.85 |
| Random ashlar, 1" | M1@.208 | SF | 9.14 | 9.70 | 18.84 |
| Irregular fitted, 3/4" | M1@.193 | SF | 7.25 | 9.00 | 16.25 |
| Irregular fitted, 1" | M1@.228 | SF | 8.46 | 10.60 | 19.06 |
| Add for sand rubbed slate | — | SF | 1.86 | — | 1.86 |

| | Craft@Hrs | Unit | Material | Labor | Equipment | Total |
|---|---|---|---|---|---|---|

**Curbs, Gutters and Driveway Aprons** Costs assume 3 uses of the forms and 2,000 PSI concrete placed directly from the chute of a ready-mix truck. Concrete cast-in-place curb. Excavation or backfill not included. Use $750 as a minimum job charge.

**Curbs**

| | Craft@Hrs | Unit | Material | Labor | Equipment | Total |
|---|---|---|---|---|---|---|
| Using concrete, per cubic yard | — | CY | 98.60 | — | — | 98.60 |
| Vertical curb | | | | | | |
| 6" x 12" straight curb | P9@.109 | LF | 3.54 | 5.15 | 1.46 | 10.15 |
| 6" x 12" curved curb | P9@.150 | LF | 3.60 | 7.09 | 2.01 | 12.70 |
| 6" x 18" straight curb | P9@.119 | LF | 5.25 | 5.63 | 1.59 | 12.47 |
| 6" x 18" curved curb | P9@.166 | LF | 5.34 | 7.85 | 2.22 | 15.41 |
| 6" x 24" straight curb | P9@.123 | LF | 6.97 | 5.81 | 1.64 | 14.42 |
| 6" x 24" curved curb | P9@.179 | LF | 7.05 | 8.46 | 2.39 | 17.90 |
| Rolled curb and gutter, 6" roll | | | | | | |
| 6" x 18" base, straight curb | P9@.143 | LF | 6.97 | 6.76 | 1.91 | 15.64 |
| 6" x 18" base, curved curb | P9@.199 | LF | 7.05 | 9.41 | 2.66 | 19.12 |
| 6" x 24" base, straight curb | P9@.168 | LF | 8.76 | 7.94 | 2.25 | 18.95 |
| 6" x 24" base, curved curb | P9@.237 | LF | 8.87 | 11.20 | 3.17 | 23.24 |

**Driveway aprons.** Concrete cast-in-place driveway aprons, including 4% waste, wire mesh, forms and finishing. Per square foot of apron, no site preparation included. Use $200 as a minimum job charge.

| | Craft@Hrs | Unit | Material | Labor | Equipment | Total |
|---|---|---|---|---|---|---|
| 4" thick (80 SF per CY) | P9@.024 | SF | 2.22 | 1.13 | .20 | 3.55 |
| 6" thick (54 SF per CY) | P9@.032 | SF | 2.72 | 1.51 | .27 | 4.50 |

# 32 Exterior Improvements

| | Craft@Hrs | Unit | Material | Labor | Equipment | Total |
|---|---|---|---|---|---|---|

**Walkways** Concrete, 2,000 PSI placed directly from the chute of a ready-mix truck, vibrated, plain scored, and broom finished. Costs include 2% allowance for waste but no forms and reinforcing.

| | Craft@Hrs | Unit | Material | Labor | Equipment | Total |
|---|---|---|---|---|---|---|
| 4" thick (80 SF per CY) | P8@.013 | SF | 1.21 | .59 | .02 | 1.82 |
| Add for integral colors, most pastels | — | SF | 1.25 | — | — | 1.25 |
| Add for 1/2" color top course | P8@.012 | SF | 1.40 | .54 | .02 | 1.96 |
| Add for steel trowel finish | CM@.011 | SF | — | .55 | .03 | .58 |
| Add for seeded aggregate finish | CM@.011 | SF | .12 | .55 | .03 | .70 |
| Add for exposed aggregate wash process | CM@.006 | SF | .06 | .30 | .02 | .38 |

Asphalt walkway, temporary, including installation and breakout, but no hauling. Use $500 as a minimum job charge.

| | Craft@Hrs | Unit | Material | Labor | Equipment | Total |
|---|---|---|---|---|---|---|
| 2" to 2-1/2" thick | P5@.015 | SF | 1.74 | .66 | .53 | 2.93 |

## Headers and Dividers

| | Craft@Hrs | Unit | Material | Labor | Equipment | Total |
|---|---|---|---|---|---|---|
| 2" x 4", treated pine | C8@.027 | LF | .56 | 1.24 | — | 1.80 |
| 2" x 6", treated pine | C8@.028 | LF | .82 | 1.28 | — | 2.10 |
| 2" x 6", redwood | C8@.028 | LF | 1.23 | 1.28 | — | 2.51 |
| 2" x 4", patio type dividers, untreated | C8@.030 | LF | .59 | 1.37 | — | 1.96 |

**Porous Paving Systems** Thin-walled HDPE plastic rings connected by an interlocking geogrid structure, installed on a porous base course for commercial applications. Rings transfer loads from the surface to the grid structure to an engineered course base. Equipment includes a 1/2 CY backhoe. Installed on an existing prepared base.
Porous paving system, grass. Rolled out over base course and seeded with grass mixture. Includes fertilizer and soil polymer mix for spreading over base.

| | Craft@Hrs | Unit | Material | Labor | Equipment | Total |
|---|---|---|---|---|---|---|
| 6.6' wide by 164' long rolls | S6@.100 | CSF | 174.00 | 4.49 | 1.45 | 179.94 |
| Add for 1" sand fill | S6@.030 | CSF | 6.13 | 1.35 | — | 7.48 |

Porous paving system, gravel. Rolled out over base course and covered with decorative fill gravel.

| | Craft@Hrs | Unit | Material | Labor | Equipment | Total |
|---|---|---|---|---|---|---|
| 6.6' wide by 164' long rolls | S6@.110 | CSF | 174.00 | 4.94 | 1.60 | 180.54 |
| Add for 1" gravel fil | S6@.030 | CSF | 7.73 | 1.35 | — | 9.08 |

## Guardrails and Bumpers
Equipment is a 4,000 lb forklift. Use $4,000 as a minimum job charge

| | Craft@Hrs | Unit | Material | Labor | Equipment | Total |
|---|---|---|---|---|---|---|
| Steel guardrail, standard beam | H1@.250 | LF | 21.70 | 13.10 | 1.99 | 36.79 |
| Posts, wood, treated, 4" x 4" x 36" high. | H1@1.00 | Ea | 4.19 | 52.60 | 7.96 | 64.75 |

Guardrail two sides,

| | Craft@Hrs | Unit | Material | Labor | Equipment | Total |
|---|---|---|---|---|---|---|
| Channel rail and side blocks | H1@1.00 | LF | 37.00 | 52.60 | 7.96 | 97.56 |
| Sight screen | H1@.025 | LF | 1.90 | 1.31 | .20 | 3.41 |

Highway median barricade, precast concrete

| | Craft@Hrs | Unit | Material | Labor | Equipment | Total |
|---|---|---|---|---|---|---|
| 2'8" high x 10' long | H1@.500 | Ea | 340.00 | 26.30 | 3.98 | 370.28 |
| Add for reflectors, 2 per 10' section | — | LS | 13.70 | — | — | 13.70 |

Parking bumper, precast concrete, including dowels

| | Craft@Hrs | Unit | Material | Labor | Equipment | Total |
|---|---|---|---|---|---|---|
| 3' wide | CL@.381 | Ea | 27.20 | 15.20 | 6.07 | 48.47 |
| 6' wide | CL@.454 | Ea | 47.90 | 18.10 | 7.23 | 73.23 |

**Bollards** Posts mounted in walkways to limit vehicle entry. Equipment is a 4,000 lb capacity forklift. Add for concrete footings, if required, from costs shown at the end of this section. Use $1,000 as a minimum job charge.
Cast iron bollards, ornamental, surface mounted

| | Craft@Hrs | Unit | Material | Labor | Equipment | Total |
|---|---|---|---|---|---|---|
| 12" diameter base, 42" high | S1@1.00 | Ea | 961.00 | 47.40 | 15.90 | 1,024.30 |
| 17" diameter base, 42" high | S1@1.00 | Ea | 1,620.00 | 47.40 | 15.90 | 1,683.30 |

Lighted, electrical work not included,

| | Craft@Hrs | Unit | Material | Labor | Equipment | Total |
|---|---|---|---|---|---|---|
| 17" diameter base, 42" high | S1@1.00 | Ea | 1,810.00 | 47.40 | 15.90 | 1,873.30 |

Concrete bollards, precast, exposed aggregate, surface mounted

| | Craft@Hrs | Unit | Material | Labor | Equipment | Total |
|---|---|---|---|---|---|---|
| Square, 12" x 12" x 30" high | S1@1.00 | Ea | 379.00 | 47.40 | 15.90 | 442.30 |
| Round, 12" diameter x 30" high | S1@1.00 | Ea | 387.00 | 47.40 | 15.90 | 450.30 |

# 32 Exterior Improvements

| | Craft@Hrs | Unit | Material | Labor | Equipment | Total |
|---|---|---|---|---|---|---|
| Granite bollards, doweled to concrete foundation or slab (foundation not included.) Expect these costs to vary widely. | | | | | | |
| Square, smooth matte finish, flat top | | | | | | |
| 16" x 16" x 30", 756 lbs | S1@1.00 | Ea | 949.00 | 47.40 | 15.90 | 1,012.30 |
| 16" x 16" x 54", 1,320 lbs | S1@1.50 | Ea | 1,200.00 | 71.20 | 23.90 | 1,295.10 |
| Square, rough finish, pyramid top | | | | | | |
| 12" x 12" x 24", 330 lbs | S1@1.00 | Ea | 424.00 | 47.40 | 15.90 | 487.30 |
| Round, smooth matte finish, flat top | | | | | | |
| 12" diameter, 18" high | S1@1.00 | Ea | 917.00 | 47.40 | 15.90 | 980.30 |
| Octagonal, smooth matte finish, flat top, | | | | | | |
| 24" x 24" x 20", 740 lbs | S1@1.00 | Ea | 1,380.00 | 47.40 | 15.90 | 1,443.30 |
| Pipe bollards, concrete filled steel pipe, painted yellow, 8' long, set 4' in concrete, includes digging hole | | | | | | |
| 6" diameter pipe | S1@.750 | Ea | 119.00 | 35.60 | 11.90 | 166.50 |
| 8" diameter pipe | S1@.750 | Ea | 197.00 | 35.60 | 11.90 | 244.50 |
| 12" diameter pipe | S1@.750 | Ea | 320.00 | 35.60 | 11.90 | 367.50 |
| Wood bollards, pressure treated timber, includes hand-digging and backfill of hole but no disposal of excess soil. Length shown is portion exposed, costs include 3' bury depth. | | | | | | |
| 6" x  6" x 36" high | S1@.750 | Ea | 28.40 | 35.60 | 11.90 | 75.90 |
| 8" x  8" x 30" high | S1@.750 | Ea | 46.50 | 35.60 | 11.90 | 94.00 |
| 8" x  8" x 36" high | S1@.750 | Ea | 50.70 | 35.60 | 11.90 | 98.20 |
| 8" x  8" x 42" high | S1@.750 | Ea | 55.00 | 35.60 | 11.90 | 102.50 |
| 12" x 12" x 24" high | S1@.750 | Ea | 94.70 | 35.60 | 11.90 | 142.20 |
| 12" x 12" x 36" high | S1@.750 | Ea | 114.00 | 35.60 | 11.90 | 161.50 |
| 12" x 12" x 42" high | S1@.750 | Ea | 124.00 | 35.60 | 11.90 | 171.50 |
| Footings for bollards, add if required | | | | | | |
| Permanent concrete footing | C8@.500 | Ea | 35.70 | 22.90 | — | 58.60 |
| Footing for removable bollards | C8@.500 | Ea | 161.00 | 22.90 | — | 183.90 |

**Pavement Striping and Marking**  Use $150 as a minimum job charge. Pavement line markings.

| | Craft@Hrs | Unit | Material | Labor | Equipment | Total |
|---|---|---|---|---|---|---|
| Single line striping, 4" wide solid | PA@.005 | LF | .15 | .26 | .04 | .45 |
| Single line striping, 4" wide skip | PA@.005 | LF | .07 | .26 | .04 | .37 |
| Red curb painting | PA@.006 | LF | .61 | .31 | .05 | .97 |
| Parking lot spaces. For estimating purposes figure one space per 300 SF of parking pavement area. | | | | | | |
| Single line striping, | | | | | | |
| approx 25 LF per space | PA@.115 | Ea | 3.73 | 5.91 | 1.01 | 10.65 |
| Dual line striping, | | | | | | |
| approx 45 LF per space | PA@.175 | Ea | 6.70 | 9.00 | 1.54 | 17.24 |
| Traffic symbols | | | | | | |
| Arrows | PA@.250 | Ea | 13.90 | 12.90 | 2.19 | 28.99 |
| Lettering, 2' x 8' template | PA@.500 | Ea | 89.30 | 25.70 | 4.39 | 119.39 |
| Handicapped symbol | | | | | | |
| One color | PA@.250 | Ea | 11.30 | 12.90 | 2.19 | 26.39 |
| Two color | PA@.500 | Ea | 14.60 | 25.70 | 4.39 | 44.69 |

| | Craft@Hrs | Unit | Material | Labor | Total |
|---|---|---|---|---|---|
| **Speed Bumps and Parking Blocks**  Prefabricated solid plastic, pre-drilled. Based on Super Saver Bumps or Super Saver Blocks as manufactured by The Traffic Safety Store (www. trafficsafetystore.com) Costs shown include installation hardware. Labor includes drilling holes in substrate for anchors. Available in yellow, gray, blue, white. | | | | | |
| Speed bumps, based on quantity of 16 bumps | | | | | |
| 6' x 2" x 10" (25 lbs each) | CL@.750 | Ea | 80.00 | 29.90 | 109.90 |
| 9' x 2" x 10" (45 lbs each) | CL@1.00 | Ea | 130.00 | 39.90 | 169.90 |
| Parking blocks, based on quantity of 25 blocks | | | | | |
| 6' x 4" x 6" (16 lbs each) | CL@.750 | Ea | 37.00 | 29.90 | 66.90 |

# 32 Exterior Improvements

| | Craft@Hrs | Unit | Material | Labor | Total |
|---|---|---|---|---|---|

**Traffic Signs**  Reflectorized steel, with a high strength "U"-channel galvanized steel pipe post 10' long set 2' into the ground. Includes digging of hole with a manual auger and backfill.

| | Craft@Hrs | Unit | Material | Labor | Total |
|---|---|---|---|---|---|
| Stop, 24" x 24" | CL@1.25 | Ea | 200.00 | 49.90 | 249.90 |
| Speed limit, Exit, etc., 18" x 24" | CL@1.25 | Ea | 183.00 | 49.90 | 232.90 |
| Warning, 24" x 24" | CL@1.25 | Ea | 195.00 | 49.90 | 244.90 |
| Yield, 30" triangle | CL@1.25 | Ea | 193.00 | 49.90 | 242.90 |
| Handicapped parking, 12" x 18" | CL@1.25 | Ea | 141.00 | 49.90 | 190.90 |
| Add for breakaway galvanized square sign post | — | Ea | 65.70 | — | 65.70 |

| | Craft@Hrs | Unit | Material | Labor | Equipment | Total |
|---|---|---|---|---|---|---|

**Post Holes**  Drilling 2' deep by 10" to 12" diameter fence post holes. Equipment is a gasoline powered 12" auger. Costs will be higher on slopes or where access is limited. These costs do not include layout, setting of fence posts or disposal of excavated material.

| | Craft@Hrs | Unit | Material | Labor | Equipment | Total |
|---|---|---|---|---|---|---|
| Light to medium soil, laborer working with a hand auger | | | | | | |
| 4 holes per hour | CL@.250 | Ea | — | 9.97 | 2.88 | 12.85 |
| Medium to heavy soil, laborer working with a breaking bar and hand auger | | | | | | |
| 2 holes per hour | CL@.500 | Ea | — | 19.90 | 5.76 | 25.66 |
| Medium to heavy soil, 2 laborers working with a gas-powered auger | | | | | | |
| 15 holes per hour | CL@.133 | Ea | — | 5.30 | 1.53 | 6.83 |
| Broken rock, laborer and equipment operator working with a compressor and jackhammer | | | | | | |
| 2 holes per hour | S1@1.00 | Ea | — | 47.40 | 7.52 | 54.92 |
| Rock, laborer and an equipment operator working with a truck-mounted drill | | | | | | |
| 12 holes per hour | S1@.166 | Ea | — | 7.88 | 4.09 | 11.97 |

**Fencing, Chain Link**  Fence, industrial grade 9 gauge 2" x 2" galvanized steel chain link fabric and framework, including 2" or 2-3/8" line posts, as indicated, set at 10' intervals, set in concrete in augured post holes and 1-5/8" top rail, tension panels at corners and abrupt grade changes. Add for gates, gate posts and corner posts as required. Equipment is a gasoline-powered 12" auger. Use $600 as a minimum job charge. For scheduling purposes, estimate that a 2-man crew will install 125 to 130 LF of fence per 8-hour day.

| | Craft@Hrs | Unit | Material | Labor | Equipment | Total |
|---|---|---|---|---|---|---|
| 4' high galvanized fence, 2" line posts | C4@.073 | LF | 18.00 | 3.03 | .28 | 21.31 |
| 5' high galvanized fence, 2" line posts | C4@.091 | LF | 20.10 | 3.77 | .35 | 24.22 |
| 6' high galvanized fence, 2" line posts | C4@.106 | LF | 22.00 | 4.39 | .41 | 26.80 |
| 6' high galvanized fence, 2-3/8" line posts | C4@.109 | LF | 24.70 | 4.52 | .42 | 29.64 |
| 7' high galvanized fence, 2-3/8" line posts | C4@.127 | LF | 24.90 | 5.26 | .49 | 30.65 |
| 8' high galvanized fence, 2-3/8" line posts | C4@.146 | LF | 29.30 | 6.05 | .56 | 35.91 |
| 10' high galvanized fence, 2-3/8" line posts | C4@.182 | LF | 33.90 | 7.54 | .70 | 42.14 |
| 12' high galvanized fence, 2-3/8" line posts | C4@.218 | LF | 39.50 | 9.04 | .84 | 49.38 |
| Add for vertical aluminum privacy slats | C4@.016 | SF | 2.05 | .66 | — | 2.71 |
| Add for vinyl coated fabric and posts | — | % | 50.0 | — | — | — |
| Add for under 200 LF quantities | C4@.006 | LF | .85 | .25 | .02 | 1.12 |
| Deduct for highway quantities | — | LF | -1.67 | — | — | -1.67 |
| Deduct for 11 gauge chain link fabric | — | SF | -1.67 | — | — | -1.67 |
| Deduct for "C"-section line posts | — | LF | -2.45 | — | — | -2.45 |

# 32 Exterior Improvements

|  | Craft@Hrs | Unit | Material | Labor | Equipment | Total |
|---|---|---|---|---|---|---|

**Gates** Driveway or walkway. 9 gauge 2" x 2" galvanized steel chain link with frame and tension bars. Costs shown are per square foot of frame area. Add for hardware and gate posts, from below.

| | | | | | | |
|---|---|---|---|---|---|---|
| With 1-3/8" diameter pipe frame | C4@.025 | SF | 9.91 | 1.04 | — | 10.95 |
| With 1-5/8" diameter pipe frame | C4@.025 | SF | 11.40 | 1.04 | — | 12.44 |

Gate hardware, per gate

| | | | | | | |
|---|---|---|---|---|---|---|
| Walkway gate | C4@.250 | Ea | 46.10 | 10.40 | — | 56.50 |
| Driveway gate | C4@.250 | Ea | 112.00 | 10.40 | — | 122.40 |
| Add for vinyl-coated fabric and frame | — | SF | .81 | — | — | .81 |

Corner, end or gate posts, complete, heavyweight, 2-1/2" outside diameter galvanized, with fittings and brace. Costs include 2/3 CF sack of concrete per post.

| | | | | | | |
|---|---|---|---|---|---|---|
| 4' high | C4@.612 | Ea | 61.10 | 25.40 | 2.35 | 88.85 |
| 5' high | C4@.668 | Ea | 69.30 | 27.70 | 2.57 | 99.57 |
| 6' high | C4@.735 | Ea | 68.20 | 30.50 | 2.82 | 101.52 |
| 7' high | C4@.800 | Ea | 70.30 | 33.20 | 3.07 | 106.57 |
| 8' high | C4@.882 | Ea | 92.60 | 36.60 | 3.39 | 132.59 |
| 10' high | C4@.962 | Ea | 105.00 | 39.90 | 3.70 | 148.60 |
| 12' high, with middle rail | C4@1.00 | Ea | 116.00 | 41.50 | 3.84 | 161.34 |
| Add for vinyl-coated posts | | | | | | |
| Per vertical foot | — | VLF | .85 | — | — | .85 |
| Deduct for 2" outside diameter | — | VLF | -1.16 | — | — | -1.16 |

Barbed wire topper for galvanized chain link fence

| | | | | | | |
|---|---|---|---|---|---|---|
| Single strand barbed wire | C4@.004 | LF | .35 | .17 | — | .52 |
| 3 strands on one side, with support arm | C4@.008 | LF | 2.16 | .33 | — | 2.49 |
| 3 strand on each side, with support arm | C4@.013 | LF | 4.41 | .54 | — | 4.95 |
| Double coil of 24" to 30" barbed tape | C4@.038 | LF | 15.90 | 1.58 | — | 17.48 |

**Baseball Backstops** Galvanized 9 gauge mesh chain link backstop with 2-5/8" OD galvanized vertical end posts, 2-3/8" OD galvanized vertical back posts and 1-5/8" OD galvanized framing. Includes 4' deep overhang, 20 LF behind home plate, 16 LF on both first and third base lines, rear and wing planks 3' high and two 12' long benches. Equipment is a 4,000 lb forklift.

| | | | | | | |
|---|---|---|---|---|---|---|
| Traditional backstop, 20' H x 58' W x 14' D | S6@24.0 | Ea | 7,210.00 | 1,080.00 | 255.00 | 8,545.00 |
| Traditional backstop, 30' H x 68' W x 14' D | S6@24.0 | Ea | 10,500.00 | 1,080.00 | 255.00 | 11,835.00 |
| Arched backstop, 20' H x 62' W x 20' D | S6@44.0 | Ea | 8,140.00 | 1,980.00 | 255.00 | 10,375.00 |
| Arched backstop, 18' H x 62' W x 20' D | S6@44.0 | Ea | 7,570.00 | 1,980.00 | 255.00 | 9,805.00 |

Add for concrete footings, 12" diameter x 24" deep, includes digging by hand and spreading excavated material adjacent to hole, using fence post mix bagged concrete.

| | | | | | | |
|---|---|---|---|---|---|---|
| Per footing | S6@4.00 | Ea | 6.05 | 180.00 | — | 186.05 |

**Sprinkler Irrigation Systems** Typical complete system costs, including PVC pipe, heads, valves, fittings, trenching and backfill. Per SF of area watered. Add 10% for irrigation systems installed in areas subject to freezing hazard. Equipment is a 55 HP pneumatic tired riding type trencher.

Large areas using pop-up impact heads, 200 to 300 SF per head, manual system

| | | | | | | |
|---|---|---|---|---|---|---|
| To 5,000 SF job | S4@.005 | SF | .17 | .20 | .09 | .46 |
| Over 5,000 SF to 10,000 SF | S4@.004 | SF | .15 | .16 | .07 | .38 |
| Over 10,000 SF | S4@.003 | SF | .13 | .12 | .06 | .31 |
| Add for automatic controls | S4@.001 | SF | .04 | .04 | .02 | .10 |

# 32 Exterior Improvements

|  | Craft@Hrs | Unit | Material | Labor | Equipment | Total |
|---|---|---|---|---|---|---|
| **Small areas, under 5,000 SF, spray heads** | | | | | | |
| Strip, automatic, shrub type | S4@.008 | SF | .26 | .32 | .15 | .73 |
| Commercial type, manual | S4@.007 | SF | .25 | .28 | .13 | .66 |
| Residential type, manual | S4@.007 | SF | .24 | .28 | .13 | .65 |
| Add for automatic controls | S4@.001 | SF | .04 | .04 | .02 | .10 |
| **Trenching for sprinkler systems, in normal non-compacted topsoil** | | | | | | |
| **Main lines with a 55 HP riding type trencher, 18" wide** | | | | | | |
| 12" deep, 143 LF per hour | S1@.007 | LF | — | .33 | .19 | .52 |
| 18" deep, 125 LF per hour | S1@.008 | LF | — | .38 | .22 | .60 |
| 24" deep, 100 LF per hour | S1@.010 | LF | — | .47 | .28 | .75 |
| **Lateral lines with a 20 HP riding type trencher, 12" wide** | | | | | | |
| 8" deep, 200 LF per hour | S1@.005 | LF | — | .24 | .09 | .33 |
| 12" deep, 125 LF per hour | S1@.008 | LF | — | .38 | .14 | .52 |
| **Add for hard soil** | | | | | | |
| (shelf, rock field, hardpan) | S1@.006 | LF | — | .28 | .11 | .39 |
| **Boring under walkways and pavement** | | | | | | |
| To 8' wide, by hand, 8 LF per hour | CL@.125 | LF | — | 4.99 | — | 4.99 |
| **Backfill and compact trench, by hand** | | | | | | |
| Mains to 24" deep, 125 LF per hour | CL@.008 | LF | — | .32 | — | .32 |
| Laterals to 12" deep, 190 LF per hour | CL@.005 | LF | — | .20 | — | .20 |

|  | Craft@Hrs | Unit | Material | Labor | Total |
|---|---|---|---|---|---|
| **Restore turf above pipe after sprinkler installation** | | | | | |
| Replace and roll sod | CL@.019 | SY | 4.46 | .76 | 5.22 |
| Reseed by hand, cover, water and mulch | CL@.010 | SY | .24 | .40 | .64 |
| **Connection to existing water line** | | | | | |
| Residential or small commercial tap | CL@.939 | Ea | 21.50 | 37.40 | 58.90 |
| Residential or small commercial stub | CL@.683 | Ea | 16.70 | 27.20 | 43.90 |
| Medium commercial stub | CL@1.25 | Ea | 45.20 | 49.90 | 95.10 |
| **Atmospheric vacuum breaker** | | | | | |
| 3/4" | CL@.470 | Ea | 12.80 | 18.70 | 31.50 |
| 1" | CL@.500 | Ea | 17.10 | 19.90 | 37.00 |
| 1-1/2" | CL@.575 | Ea | 43.20 | 22.90 | 66.10 |
| 2" | CL@.625 | Ea | 75.20 | 24.90 | 100.10 |
| **Pressure type vacuum breaker, with threaded ball valves** | | | | | |
| 3/4" | CL@.700 | Ea | 112.00 | 27.90 | 139.90 |
| 1" | CL@.750 | Ea | 127.00 | 29.90 | 156.90 |
| 1-1/2" | CL@.850 | Ea | 306.00 | 33.90 | 339.90 |
| 2" | CL@.939 | Ea | 349.00 | 37.40 | 386.40 |
| **Double check valve assembly with two ball valves** | | | | | |
| 1-1/2" valve | CL@.750 | Ea | 310.00 | 29.90 | 339.90 |
| 2" valve | CL@.939 | Ea | 371.00 | 37.40 | 408.40 |
| 3" valve | CL@1.25 | Ea | 1,280.00 | 49.90 | 1,329.90 |
| 4" valve | CL@1.50 | Ea | 1,990.00 | 59.80 | 2,049.80 |
| **Reduced pressure backflow preventer with gate valves** | | | | | |
| 1" backflow preventer | CL@.750 | Ea | 162.00 | 29.90 | 191.90 |
| 1-1/2" backflow preventer | CL@.939 | Ea | 303.00 | 37.40 | 340.40 |
| 2" backflow preventer | CL@.939 | Ea | 340.00 | 37.40 | 377.40 |
| 3" backflow preventer | CL@1.25 | Ea | 1,260.00 | 49.90 | 1,309.90 |
| 4" backflow preventer | CL@1.50 | Ea | 1,950.00 | 59.80 | 2,009.80 |
| 6" backflow preventer | CL@1.88 | Ea | 2,530.00 | 75.00 | 2,605.00 |
| **Valve boxes, plastic box with plastic lid** | | | | | |
| 7" round | CL@.314 | Ea | 5.32 | 12.50 | 17.82 |
| 10" round | CL@.375 | Ea | 11.60 | 15.00 | 26.60 |

| | Craft@Hrs | Unit | Material | Labor | Total |
|---|---|---|---|---|---|

Sprinkler controllers, solid state electronics, 115-volt/24-volt electric timer, add electrical connection and feeder wire to valves

| | Craft@Hrs | Unit | Material | Labor | Total |
|---|---|---|---|---|---|
| Indoor mount, 4 station | CL@.960 | Ea | 21.60 | 38.30 | 59.90 |
| Indoor mount, 6 station | CL@.960 | Ea | 54.50 | 38.30 | 92.80 |
| Indoor mount, 8 station | CL@1.10 | Ea | 58.80 | 43.90 | 102.70 |
| Outdoor mount, 6-8 station, PC programmable | CL@1.10 | Ea | 105.00 | 43.90 | 148.90 |
| Outdoor mount, 12 station | CL@1.10 | Ea | 80.20 | 43.90 | 124.10 |
| Pedestal for outdoor units, set in concrete | CL@1.00 | Ea | 53.40 | 39.90 | 93.30 |

Underground control wire for automatic valves, includes waterproof connectors, wire laid in an open trench, two conductors per valve, costs are per pair of conductors

| | Craft@Hrs | Unit | Material | Labor | Total |
|---|---|---|---|---|---|
| 4 conductors | CL@.001 | LF | .21 | .04 | .25 |
| 5 conductors | CL@.001 | LF | .22 | .04 | .26 |
| 7 conductors | CL@.002 | LF | .30 | .08 | .38 |

Anti-siphon solenoid flow control valves, with atmospheric vacuum breaker. UV-resistant PVC, 24 volt. Add feeder wire as required.

| | Craft@Hrs | Unit | Material | Labor | Total |
|---|---|---|---|---|---|
| 3/4", residential | CL@.341 | Ea | 16.50 | 13.60 | 30.10 |
| 1" residential | CL@.341 | Ea | 17.00 | 13.60 | 30.60 |
| 3/4", commercial | CL@.341 | Ea | 27.80 | 13.60 | 41.40 |
| 1" commercial | CL@.418 | Ea | 28.90 | 16.70 | 45.60 |
| 1-1/2" commercial | CL@.796 | Ea | 38.40 | 31.70 | 70.10 |
| 2" commercial | CL@.866 | Ea | 41.40 | 34.50 | 75.90 |

Manual sprinkler control valves, anti-siphon, brass, with union

| | Craft@Hrs | Unit | Material | Labor | Total |
|---|---|---|---|---|---|
| 3/4" valve | CL@.518 | Ea | 42.80 | 20.70 | 63.50 |
| 1" valve | CL@.610 | Ea | 55.30 | 24.30 | 79.60 |
| 1-1/4" valve | CL@.677 | Ea | 87.40 | 27.00 | 114.40 |
| 1-1/2" valve | CL@.796 | Ea | 96.00 | 31.70 | 127.70 |
| 2" valve | CL@.866 | Ea | 139.00 | 34.50 | 173.50 |

Angle valves, brass, with union

| | Craft@Hrs | Unit | Material | Labor | Total |
|---|---|---|---|---|---|
| 3/4" valve | CL@.518 | Ea | 31.20 | 20.70 | 51.90 |
| 1" valve | CL@.610 | Ea | 39.30 | 24.30 | 63.60 |
| 1-1/4" valve | CL@.677 | Ea | 86.00 | 27.00 | 113.00 |
| 1-1/2" valve | CL@.796 | Ea | 92.50 | 31.70 | 124.20 |
| 2" valve | CL@.866 | Ea | 129.00 | 34.50 | 163.50 |

**PVC Schedule 40 pipe and fittings,** except as noted. Solvent welded. Plain end. Schedule 40 denotes wall thickness. Class 315, 200 and 160 identify the normal working pressure, 315, 200 or 160 PSI. Add the cost of excavation and backfill.

| | Craft@Hrs | Unit | Material | Labor | Total |
|---|---|---|---|---|---|
| 1/2" pipe | CL@.005 | LF | .16 | .20 | .36 |
| 1/2" pipe, Class 315 | CL@.005 | LF | .11 | .20 | .31 |
| 1/2" ell | CL@.120 | Ea | .30 | 4.79 | 5.09 |
| 1/2" tee | CL@.140 | Ea | .30 | 5.58 | 5.88 |
| 3/4" pipe | CL@.005 | LF | .20 | .20 | .40 |
| 3/4" pipe, Class 200 | CL@.005 | LF | .14 | .20 | .34 |
| 3/4" ell | CL@.160 | Ea | .36 | 6.38 | 6.74 |
| 3/4" tee | CL@.180 | Ea | .35 | 7.18 | 7.53 |
| 1" pipe | CL@.006 | LF | .30 | .24 | .54 |
| 1" pipe, Class 200 | CL@.006 | LF | .17 | .24 | .41 |
| 1" ell | CL@.180 | Ea | .51 | 7.18 | 7.69 |
| 1" tee | CL@.220 | Ea | .68 | 8.77 | 9.45 |
| 1-1/4" pipe | CL@.006 | LF | .42 | .24 | .66 |
| 1-1/4" pipe, Class 160 | CL@.006 | LF | .22 | .24 | .46 |
| 1-1/4" ell | CL@.180 | Ea | 1.05 | 7.18 | 8.23 |
| 1-1/4" tee | CL@.250 | Ea | 1.36 | 9.97 | 11.33 |

| | Craft@Hrs | Unit | Material | Labor | Total |
|---|---|---|---|---|---|
| 1-1/2" pipe | CL@.007 | LF | .45 | .28 | .73 |
| 1-1/2" pipe, Class 160 | CL@.007 | LF | .29 | .28 | .57 |
| 1-1/2" ell | CL@.180 | Ea | 1.34 | 7.18 | 8.52 |
| 1-1/2" tee | CL@.250 | Ea | 1.71 | 9.97 | 11.68 |
| 2" pipe | CL@.008 | LF | .60 | .32 | .92 |
| 2" ell | CL@.200 | Ea | 1.93 | 7.98 | 9.91 |
| 2" tee | CL@.300 | Ea | 2.61 | 12.00 | 14.61 |
| 3" pipe | CL@.008 | LF | 1.24 | .32 | 1.56 |
| 3" ell | CL@.240 | Ea | 7.53 | 9.57 | 17.10 |
| 3" tee | CL@.360 | Ea | 9.97 | 14.40 | 24.37 |
| Polyethylene pipe, 80 PSI, black, flexible 100' rolls | | | | | |
| 3/4" diameter | CL@.006 | LF | .15 | .24 | .39 |
| 1" diameter | CL@.006 | LF | .22 | .24 | .46 |
| 1-1/4" diameter | CL@.007 | LF | .34 | .28 | .62 |
| 1-1/2" diameter | CL@.008 | LF | .54 | .32 | .86 |
| Polyethylene pipe, 100 PSI, black, flexible 100' rolls | | | | | |
| 1-1/2" diameter | CL@.008 | LF | .49 | .32 | .81 |
| 2" diameter | CL@.009 | LF | .67 | .36 | 1.03 |
| Add for 125 PSI polyethylene pipe | — | % | 20.0 | — | — |
| Polyethylene pipe to male iron pipe thread adapter, brass | | | | | |
| 3/4" male adapter | CL@.112 | Ea | 3.65 | 4.47 | 8.12 |
| 1" male adapter | CL@.172 | Ea | 4.63 | 6.86 | 11.49 |
| 1-1/4" male adapter | CL@.200 | Ea | 5.14 | 7.98 | 13.12 |

## Shrub and Lawn Sprinkler Heads and Risers

| | Craft@Hrs | Unit | Material | Labor | Total |
|---|---|---|---|---|---|
| Shrub bubbler, plastic nozzle, brass adjustment screw | | | | | |
| 1/4 circle | P1@.120 | Ea | 0.82 | 4.29 | 5.11 |
| 1/2 circle | P1@.120 | Ea | 0.82 | 4.29 | 5.11 |
| Full circle | P1@.120 | Ea | 0.82 | 4.29 | 5.11 |
| Mushroom spray | P1@.120 | Ea | 1.06 | 4.29 | 5.35 |
| Flush head plastic lawn sprinkler, brass insert | | | | | |
| 1/4 circle | P1@.120 | Ea | 1.13 | 4.29 | 5.42 |
| 1/2 circle | P1@.120 | Ea | 1.13 | 4.29 | 5.42 |
| Full circle | P1@.120 | Ea | 1.13 | 4.29 | 5.42 |
| Strip sprinkler | P1@.120 | Ea | 2.54 | 4.29 | 6.83 |
| Brass 2" pop-up lawn sprinkler | | | | | |
| 1/4 circle | P1@.120 | Ea | 3.27 | 4.29 | 7.56 |
| 1/2 circle | P1@.120 | Ea | 3.27 | 4.29 | 7.56 |
| Full circle | P1@.120 | Ea | 3.27 | 4.29 | 7.56 |
| Spring-loaded plastic 4" pop-up lawn sprinkler | | | | | |
| 1/4 circle | P1@.120 | Ea | 2.11 | 4.29 | 6.40 |
| 1/2 circle | P1@.120 | Ea | 2.11 | 4.29 | 6.40 |
| Full circle | P1@.120 | Ea | 1.97 | 4.29 | 6.26 |
| Adjustable circle | P1@.120 | Ea | 2.26 | 4.29 | 6.55 |
| Gear-driven rotating lawn sprinkler, 40 to 360 degrees | | | | | |
| 15' to 30' circle | P1@.140 | Ea | 9.57 | 5.01 | 14.58 |
| Pop-up impact sprinkler, adjustable, 40 to 360 degrees | | | | | |
| To 80' radius | P1@.140 | Ea | 13.90 | 5.01 | 18.91 |
| Sprinkler riser nipples | | | | | |
| 1/2" x close | P1@.030 | Ea | .15 | 1.07 | 1.22 |
| 1/2" x 2" | P1@.030 | Ea | .16 | 1.07 | 1.23 |
| 1/2" x 6" cut-off | P1@.030 | Ea | .16 | 1.07 | 1.23 |
| 1/2" x 6" flexible cut-off | P1@.030 | Ea | .16 | 1.07 | 1.23 |
| 1/2" x 6" | P1@.030 | Ea | .30 | 1.07 | 1.37 |

| | Craft@Hrs | Unit | Material | Labor | Total |
|---|---|---|---|---|---|
| 1/2" x 12" | P1@.030 | Ea | .44 | 1.07 | 1.51 |
| 1/2" threaded ell | P1@.030 | Ea | .55 | 1.07 | 1.62 |
| 1/2" and 3/4" x 6" cut off | P1@.030 | Ea | .72 | 1.07 | 1.79 |
| 1/2" and 3/4" x 6" flexible cut-off | P1@.030 | Ea | .72 | 1.07 | 1.79 |
| 3/4" x close | P1@.030 | Ea | .24 | 1.07 | 1.31 |
| 3/4" x 2" | P1@.030 | Ea | .25 | 1.07 | 1.32 |
| 3/4" x 6" cut off | P1@.030 | Ea | .72 | 1.07 | 1.79 |
| 3/4" x 6" flexible cut-off | P1@.030 | Ea | .72 | 1.07 | 1.79 |
| 3/4" threaded ell | P1@.030 | Ea | .56 | 1.07 | 1.63 |

Quick coupling valves, add swing joint costs below

| | Craft@Hrs | Unit | Material | Labor | Total |
|---|---|---|---|---|---|
| 3/4" regular | CL@.089 | Ea | 27.30 | 3.55 | 30.85 |
| 3/4" 2-piece | CL@.089 | Ea | 32.00 | 3.55 | 35.55 |
| 1" regular | CL@.094 | Ea | 38.80 | 3.75 | 42.55 |
| 1" 2-piece | CL@.094 | Ea | 46.40 | 3.75 | 50.15 |
| 1-1/4" regular | CL@.108 | Ea | 48.80 | 4.31 | 53.11 |
| 1-1/2" regular | CL@.124 | Ea | 58.70 | 4.95 | 63.65 |
| Add for locking vinyl cover | — | Ea | 12.60 | — | 12.60 |

Double swing sprinkler riser. Two close nipples, two PVC ells, one cut-off riser.

| | Craft@Hrs | Unit | Material | Labor | Total |
|---|---|---|---|---|---|
| 1/2" pipe | P1@.150 | Ea | 1.55 | 5.36 | 6.91 |
| 3/4" pipe | P1@.150 | Ea | 2.30 | 5.36 | 7.66 |

Triple swing sprinkler riser. Three close nipples, three PVC ells, one cut-off riser.

| | Craft@Hrs | Unit | Material | Labor | Total |
|---|---|---|---|---|---|
| 1/2" pipe | P1@.150 | Ea | 2.25 | 5.36 | 7.61 |
| 3/4" pipe | P1@.150 | Ea | 3.09 | 5.36 | 8.45 |

Hose bibb

| | Craft@Hrs | Unit | Material | Labor | Total |
|---|---|---|---|---|---|
| 1/2" with 12" galvanized riser | CL@.041 | Ea | 9.83 | 1.64 | 11.47 |
| 3/4" with 12" galvanized riser | CL@.041 | Ea | 11.20 | 1.64 | 12.84 |
| 1/2" x 8" frost-proof with vacuum breaker | CL@.041 | Ea | 24.10 | 1.64 | 25.74 |
| 3/4" x 8" frost-proof with vacuum breaker | CL@.041 | Ea | 26.30 | 1.64 | 27.94 |

## Planting Bed Preparation

Spreading topsoil from piles on site, based on topsoil delivered to the site. Topsoil prices can be expected to vary widely.

| | Craft@Hrs | Unit | Material | Labor | Equipment | Total |
|---|---|---|---|---|---|---|
| Hand work, 10 CY job, level site, 10' throw | CL@.500 | CY | 33.50 | 19.90 | — | 53.40 |
| Move 25' in wheelbarrow | CL@1.00 | CY | 33.50 | 39.90 | — | 73.40 |
| With 3/4 CY loader 100 CY job, level site | TO@.080 | CY | 33.50 | 4.40 | 2.48 | 40.38 |

Spreading granular or powdered soil conditioners, based on fertilizer at $.26 per pound. Spread 20 pounds per 1,000 square feet (MSF). Add cost for mixing soil and fertilizer from below.

| | Craft@Hrs | Unit | Material | Labor | Equipment | Total |
|---|---|---|---|---|---|---|
| By hand, 2,250 SF per hour | CL@.445 | MSF | 5.81 | 17.70 | — | 23.51 |
| Hand broadcast spreader, 9,000 SF per hour | CL@.111 | MSF | 5.81 | 4.43 | .29 | 10.53 |
| Push gravity spreader, 3,000 SF per hour | CL@.333 | MSF | 5.81 | 13.30 | 1.46 | 20.57 |
| Push broadcast spreader, 20,000 SF per hour | CL@.050 | MSF | 5.81 | 1.99 | .22 | 8.02 |
| Tractor drawn broadcast spreader, 87 MSF per hour | CL@.011 | MSF | 5.81 | .44 | .40 | 6.65 |

Spreading organic soil conditioners, based on peat humus at $.19 per pound. Spread 200 pounds per 1,000 square feet (MSF). Add mixing cost from below

| | Craft@Hrs | Unit | Material | Labor | Equipment | Total |
|---|---|---|---|---|---|---|
| By hand, 1,000 SF per hour | CL@1.00 | MSF | 20.50 | 39.90 | — | 60.40 |

| | Craft@Hrs | Unit | Material | Labor | Equipment | Total |
|---|---|---|---|---|---|---|
| Push gravity spreader, | | | | | | |
| 2,750 SF per hour | CL@.363 | MSF | 43.50 | 14.50 | 1.59 | 59.59 |
| Manure spreader, | | | | | | |
| 34,000 SF per hour | CL@.029 | MSF | 43.50 | 1.16 | .12 | 44.78 |

Fertilizer and soil conditioners

| | Craft@Hrs | Unit | Material | Labor | Equipment | Total |
|---|---|---|---|---|---|---|
| Most nitrate fertilizers, | | | | | | |
| ammonium sulfate | — | Lb | .18 | — | — | .18 |
| Lawn and garden fertilizer, | | | | | | |
| calcium nitrate | — | Lb | .19 | — | — | .19 |
| Hydrated lime sodium nitrate | — | Lb | .16 | — | — | .16 |
| Ground limestone | — | Ton | 149.00 | — | — | 149.00 |
| Ground dolomitic limestone | — | Ton | 154.00 | — | — | 154.00 |
| Composted manure | — | Lb | .03 | — | — | .03 |
| Peat humus | — | Lb | .03 | — | — | .03 |
| Vermiculite, perlite | — | CF | 2.71 | — | — | 2.71 |

Spreading lime at 70 pounds per 1,000 SF. Based on an 8 acre job using a tractor-drawn spreader at $25.00 per hour and 1 acre per hour. Add mixing cost from below. Note: 1 acre equals 43,560 square feet.

| | Craft@Hrs | Unit | Material | Labor | Equipment | Total |
|---|---|---|---|---|---|---|
| Lime | TO@1.00 | Acre | 339.00 | 55.00 | 34.10 | 428.10 |
| Lime | TO@.656 | Ton | 193.00 | 36.10 | 22.40 | 251.50 |

Mixing soil and fertilizer in place. Medium soil. Light soil (sand or soft loam) will decrease costs shown by 20% to 30%. Heavy soil (clay, wet or rocky soil) will increase costs shown by 20% to 40%. Add soil conditioner cost from sections preceding this section.

| | Craft@Hrs | Unit | Material | Labor | Equipment | Total |
|---|---|---|---|---|---|---|
| Mixing by hand | | | | | | |
| 2" deep, 9 SY per hour | CL@.111 | SY | — | 4.43 | — | 4.43 |
| 4" deep, 7.5 SY per hour | CL@.133 | SY | — | 5.30 | — | 5.30 |
| 6" deep, 5.5 SY per hour | CL@.181 | SY | — | 7.22 | — | 7.22 |
| Mixing with a 11 HP rear-tined hydraulic tiller | | | | | | |
| 2" deep, 110 SY per hour | CL@.009 | SY | — | .36 | .17 | .53 |
| 4" deep, 90 SY per hour | CL@.011 | SY | — | .44 | .20 | .64 |
| 6" deep, 65 SY per hour | CL@.015 | SY | — | .60 | .28 | .88 |
| 8" deep, 45 SY per hour | CL@.022 | SY | — | .88 | .40 | 1.28 |
| Mixing with a 3' wide tractor-driven tiller 20 HP tractor | | | | | | |
| 2" deep, 50,000 SF per hour | TO@.020 | MSF | — | 1.10 | .58 | 1.68 |
| 4" deep, 40,000 SF per hour | TO@.025 | MSF | — | 1.37 | .72 | 2.09 |
| 6" deep, 30,000 SF per hour | TO@.033 | MSF | — | 1.81 | .95 | 2.76 |
| 8" deep, 20,000 SF per hour | TO@.050 | MSF | — | 2.75 | 1.44 | 4.19 |
| Mixing with a 6' wide tractor-driven tiller and 40 HP tractor | | | | | | |
| 2" deep, 90,000 SF per hour | TO@.011 | MSF | — | .60 | .35 | .95 |
| 4" deep, 70,000 SF per hour | TO@.014 | MSF | — | .77 | .45 | 1.22 |
| 6" deep, 53,000 SF per hour | TO@.019 | MSF | — | 1.04 | .61 | 1.65 |
| 8" deep, 35,000 SF per hour | TO@.028 | MSF | — | 1.54 | .90 | 2.44 |
| Leveling the surface for lawn or planting bed | | | | | | |
| By hand, 110 SY per hour | CL@.009 | SY | — | .36 | — | .36 |
| With 6' drag harrow, | | | | | | |
| at 60,000 SF per hour | TO@.017 | MSF | — | .93 | .65 | 1.58 |
| With 12' drag harrow, | | | | | | |
| at 115,000 SF per hour | TO@.009 | MSF | — | .49 | .37 | .86 |
| Plant bed preparation | | | | | | |
| With a 3/4 CY wheel loader, | | | | | | |
| at 12.5 CY per hour | TO@.080 | CY | — | 4.40 | 3.26 | 7.66 |
| By hand to 18" deep, 8.5 SY/hr | CL@.118 | SY | — | 4.71 | — | 4.71 |

# 32 Exterior Improvements

| | Craft@Hrs | Unit | Material | Labor | Equipment | Total |
|---|---|---|---|---|---|---|
| Mixing planting soil | | | | | | |
| By hand, .9 CY per hour | CL@1.11 | CY | — | 44.30 | — | 44.30 |
| With 3/4 CY wheel loader, 7 CY per hour | T0@.143 | CY | — | 7.86 | 5.83 | 13.69 |
| With a soil shredder, 11 CY per hour | T0@.090 | CY | — | 4.95 | 3.67 | 8.62 |

## Seeding and Planting

Seeding with a hand broadcast spreader, 10 pounds per 1,000 SF. The cost of grass seed will vary greatly depending on variety — see below.

| | Craft@Hrs | Unit | Material | Labor | Equipment | Total |
|---|---|---|---|---|---|---|
| 4,000 SF per hour, hand seeding | CL@.250 | MSF | 25.10 | 9.97 | .66 | 35.73 |
| 10,000 SF per hour, | | | | | | |
| Push spreader seeding | CL@.100 | MSF | 12.50 | 3.99 | .44 | 16.93 |
| Mechanical seeding | T0@13.1 | Acre | 438.00 | 720.00 | 481.00 | 1,639.00 |
| Mechanical seeding | T0@.300 | MSF | 10.10 | 16.50 | 11.00 | 37.60 |

Seeding with push broadcast spreader, 5 pounds per 1,000 SF

Seeding with a mechanical seeder at $28 per hour and 175 pounds per acre, 5 acre job

Hydroseeding subcontract (spray application of seed, binder and fertilizer slurry). Costs will vary based upon site

Sealing soil, application by a hydroseeding unit. Use $750 as a minimum job charge.

Grass seed

Sodding, placed on level ground at 25 SY per hour. Add delivery cost below. Use 50 SY as a minimum job charge.

Work grass seed into soil, no seed included

conditions, quality/type of seed. Use $1,500 as a minimum job charge.

| | Craft@Hrs | Unit | Material | Labor | Equipment | Total |
|---|---|---|---|---|---|---|
| 1,000 to 2,000 SF job | — | MSF | — | — | — | 180.00 |
| 2,001 to 4,000 SF job | — | MSF | — | — | — | 150.00 |
| 4,001 to 6,000 SF job | — | MSF | — | — | — | 120.00 |
| 6,001 to 10,000 SF job | — | MSF | — | — | — | 110.00 |
| 10,001 to 20,000 SF job | — | MSF | — | — | — | 100.00 |
| 20,001 to 50,000 SF job | — | MSF | — | — | — | 90.00 |
| 50,001 to 100,000 SF job | — | MSF | — | — | — | 80.00 |
| Over 100,000 SF job | — | MSF | — | — | — | 70.00 |
| Copolymer based liquid | | | | | | |
| for erosion control | C4@.020 | SY | 1.77 | .83 | .19 | 2.79 |
| Annual ryegrass | — | Lb | 1.58 | — | — | 1.58 |
| Most fescue | — | Lb | 1.68 | — | — | 1.68 |
| Kentucky bluegrass | — | Lb | 4.54 | — | — | 4.54 |
| Bermuda | — | Lb | 3.25 | — | — | 3.25 |
| Heat-and drought-resistant | | | | | | |
| premium blend | — | Lb | 5.12 | — | — | 5.12 |
| Dichondra | — | Lb | 17.50 | — | — | 17.50 |
| Centipede | — | Lb | 37.80 | — | — | 37.80 |
| Northeast, Midwest, bluegrass blend | CL@.040 | SY | 3.33 | 1.60 | — | 4.93 |
| Great Plains, South, bluegrass | CL@.040 | SY | 3.53 | 1.60 | — | 5.13 |
| West, bluegrass, tall fescue | CL@.040 | SY | 3.11 | 1.60 | — | 4.71 |
| South, Bermuda grass, | | | | | | |
| Centipede grass | CL@.040 | SY | 2.41 | 1.60 | — | 4.01 |
| South, Zoysia grass, St. Augustine | CL@.040 | SY | 6.84 | 1.60 | — | 8.44 |
| Typical delivery cost, to 90 miles | — | SY | .60 | — | — | .60 |
| Add for staking sod on slopes | CL@.010 | SY | .10 | .40 | — | .50 |
| By hand, 220 SY per hour | CL@.005 | SY | — | .20 | — | .20 |
| 6' harrow and 40 HP tractor | | | | | | |
| 60 MSF per hour | T0@.017 | MSF | — | .93 | .71 | 1.64 |
| 12' harrow and 60 HP tractor | | | | | | |
| 100 MSF per hour | T0@.010 | MSF | — | .55 | .45 | 1.00 |

# 32 Exterior Improvements

| | Craft@Hrs | Unit | Material | Labor | Equipment | Total |
|---|---|---|---|---|---|---|
| **Apply top dressing over seed or stolons** | | | | | | |
| 300 pounds per 1,000 SF (2.7 pounds per SY) | | | | | | |
| By hand, 65 SY per hour | CL@.015 | SY | .59 | .60 | — | 1.19 |
| With manure spreader, | | | | | | |
| 10,000 SF per hour | CL@.125 | MSF | 59.00 | 4.99 | — | 63.99 |
| **Roll sod or soil surface with a roller** | | | | | | |
| With a push roller, 400 SY per hour | CL@.002 | SY | — | .08 | — | .08 |
| With a 20 HP tractor, | | | | | | |
| 25,000 SF per hour | TO@.040 | MSF | — | 2.20 | .96 | 3.16 |
| **Sprigging, by hand** | | | | | | |
| 6" spacing, 50 SY per hour | CL@.020 | SY | .45 | .80 | — | 1.25 |
| 9" spacing, 100 SY per hour | CL@.010 | SY | .43 | .40 | — | .83 |
| 12" spacing, 150 SY per hour | CL@.007 | SY | .41 | .28 | — | .69 |
| Hybrid Bermuda grass stolons, | | | | | | |
| per bushel | — | Ea | 4.65 | — | — | 4.65 |
| **Ground cover, large areas, no surface preparation** | | | | | | |
| Ice plant (80 SF per flat) | CL@.008 | SF | .27 | .32 | — | .59 |
| Strawberry (75 SF per flat) | CL@.008 | SF | .30 | .32 | — | .62 |
| Ivy and similar (70 SF per flat) | CL@.008 | SF | .32 | .32 | — | .64 |
| **Gravel bed, pea gravel (1.4 tons per CY)** | | | | | | |
| Spread by hand | CL@.555 | CY | 42.60 | 22.10 | — | 64.70 |
| **Trees and shrubs. See also Landscaping in the Residential Division** | | | | | | |
| Shrubs, most varieties | | | | | | |
| 1 gallon, 100 units | CL@.085 | Ea | 11.20 | 3.39 | — | 14.59 |
| 1 gallon, over 100 units | CL@.076 | Ea | 10.10 | 3.03 | — | 13.13 |
| 5 gallons, 100 units | CL@.443 | Ea | 29.80 | 17.70 | — | 47.50 |
| Trees, most varieties, complete, staked, typical costs | | | | | | |
| 5 gallon, 4' to 6' high | CL@.716 | Ea | 44.40 | 28.60 | — | 73.00 |
| 15 gallon, 8' to 10' high | CL@1.37 | Ea | 115.00 | 54.60 | — | 169.60 |
| 20" x 24" box, 10' to 12' high | CL@4.54 | Ea | 580.00 | 181.00 | — | 761.00 |
| Specimen size 36" box, | | | | | | |
| 14' to 20' high | CL@11.6 | Ea | 1,250.00 | 463.00 | — | 1,713.00 |
| **Guaranteed establishment of plants** | | | | | | |
| Add, as a % of total contract price | — | % | — | — | — | 12.0 |
| **Edging** | | | | | | |
| Plastic benderboard, | | | | | | |
| 4" x 3/8", staked | CL@.013 | LF | .88 | .52 | — | 1.40 |
| Redwood benderboard, | | | | | | |
| 4" x 5/16", | CL@.013 | LF | .44 | .52 | — | .96 |
| Redwood headerboard, 2" x 4", staked | | | | | | |
| Construction Common grade | CL@.127 | LF | 1.16 | 5.06 | — | 6.22 |
| Decomposed granite edging | | | | | | |
| Saturated and rolled, 3" deep | CL@.011 | SF | .42 | .44 | — | .86 |
| **Landscape stepping stones (concrete pavers)** | | | | | | |
| Laid on level ground, 14" x 14", | | | | | | |
| 30 lb. each | CL@.024 | SF | 2.16 | .96 | — | 3.12 |
| **Pressure-treated piles. in 2' to 5' sections,** | | | | | | |
| 9" to 12" diameter | CL@.045 | LF | 2.82 | 1.79 | — | 4.61 |

# 32 Exterior Improvements

| | Craft@Hrs | Unit | Material | Labor | | Total |
|---|---|---|---|---|---|---|

**Tree Grates** Two piece concrete grates surrounding a tree trunk, set in concrete. Add for delivery.

| | Craft@Hrs | Unit | Material | Labor | Total |
|---|---|---|---|---|---|
| 36" x 36" x 2-3/4" square (270 lbs.) | C5@4.00 | Ea | 268.00 | 179.00 | 447.00 |
| 48" x 48" x 2-3/4" square (512 lbs.) | C5@5.00 | Ea | 298.00 | 224.00 | 522.00 |
| 60" x 60" x 2-3/4" square (750 lbs.) | C5@6.50 | Ea | 335.00 | 291.00 | 626.00 |
| 36" diameter x 2-3/4" round (230 lbs.) | C5@4.00 | Ea | 274.00 | 179.00 | 453.00 |
| 48" diameter x 2-3/4" round (480 lbs.) | C5@5.00 | Ea | 316.00 | 224.00 | 540.00 |
| Add for cast steel frame | — | % | 45.0 | 20.0 | — |

# 33 Utilities

| | Craft@Hrs | Unit | Material | Labor | Equipment | Total |
|---|---|---|---|---|---|---|

**Cast Iron Flanged Pipe** AWWA C151 pipe with cast iron flanges on both ends. Pipe is laid and connected in an open trench. Equipment includes a wheel-mounted 1 CY backhoe for lifting and placing the pipe and fittings. Trench excavation, dewatering, backfill and compaction are not included. Use $3,000 as a minimum job charge. See next section for fittings, valves and accessories.

| | Craft@Hrs | Unit | Material | Labor | Equipment | Total |
|---|---|---|---|---|---|---|
| 4" pipe, 260 pounds per 18' section | U1@.057 | LF | 12.60 | 2.78 | .56 | 15.94 |
| 6" pipe, 405 pounds per 18' section | U1@.075 | LF | 14.70 | 3.66 | .73 | 19.09 |
| 8" pipe, 570 pounds per 18' section | U1@.075 | LF | 21.00 | 3.66 | .73 | 25.39 |
| 10" pipe, 740 pounds per 18' section | U1@.094 | LF | 28.60 | 4.59 | .92 | 34.11 |
| 12" pipe, 930 pounds per 18' section | U1@.094 | LF | 38.80 | 4.59 | .92 | 44.31 |

**Cast Iron Fittings, Valves and Accessories** Based on type ASA A21, 10-A21-11-6 and 4C110-64, C111-64 fittings. Installed in an open trench and connected. Equipment includes a wheel-mounted 1 CY backhoe for lifting and placing. Trench excavation, dewatering, backfill and compaction are not included. These items are also suitable for use with the Ductile Iron Pipe in the section that follows. Costs estimated using this section may be combined with the costs estimated from the Cast Iron Flanged Pipe section or the Ductile Iron Pipe section for arriving at the minimum job charge shown in those sections.

Cast iron mechanical joint 90 degree ells

| | Craft@Hrs | Unit | Material | Labor | Equipment | Total |
|---|---|---|---|---|---|---|
| 4" ell | U1@.808 | Ea | 94.00 | 39.50 | 7.88 | 141.38 |
| 6" ell | U1@1.06 | Ea | 142.00 | 51.80 | 10.30 | 204.10 |
| 8" ell | U1@1.06 | Ea | 207.00 | 51.80 | 10.30 | 269.10 |
| 10" ell | U1@1.32 | Ea | 311.00 | 64.50 | 12.90 | 388.40 |
| 12" ell | U1@1.32 | Ea | 389.00 | 64.50 | 12.90 | 466.40 |
| 14" ell | U1@1.32 | Ea | 891.00 | 64.50 | 12.90 | 968.40 |
| 16" ell | U1@1.65 | Ea | 1,020.00 | 80.60 | 16.10 | 1,116.70 |
| Deduct for 1/8 or 1/16 bends | — | % | -20.0 | — | — | — |

Cast iron mechanical joint tees

| | Craft@Hrs | Unit | Material | Labor | Equipment | Total |
|---|---|---|---|---|---|---|
| 4" x 4" | U1@1.22 | Ea | 146.00 | 59.60 | 11.90 | 217.50 |
| 6" x 4" | U1@1.59 | Ea | 185.00 | 77.60 | 15.50 | 278.10 |
| 6" x 6" | U1@1.59 | Ea | 203.00 | 77.60 | 15.50 | 296.10 |
| 8" x 8" | U1@1.59 | Ea | 292.00 | 77.60 | 15.50 | 385.10 |
| 10" x 10" | U1@1.59 | Ea | 423.00 | 77.60 | 15.50 | 516.10 |
| 12" x 12" | U1@2.48 | Ea | 549.00 | 121.00 | 24.20 | 694.20 |
| Add for wyes | — | % | 40.0 | — | — | — |
| Add for crosses | — | % | 30.0 | 33.0 | 33.0 | — |

Cast iron mechanical joint reducers

| | Craft@Hrs | Unit | Material | Labor | Equipment | Total |
|---|---|---|---|---|---|---|
| 6" x 4" | U1@1.06 | Ea | 107.00 | 51.80 | 10.30 | 169.10 |
| 8" x 6" | U1@1.06 | Ea | 159.00 | 51.80 | 10.30 | 221.10 |
| 10" x 8" | U1@1.32 | Ea | 208.00 | 64.50 | 12.90 | 285.40 |
| 12" x 10" | U1@1.32 | Ea | 279.00 | 64.50 | 12.90 | 356.40 |

| | Craft@Hrs | Unit | Material | Labor | Equipment | Total |
|---|---|---|---|---|---|---|
| **AWWA mechanical joint gate valves** | | | | | | |
| 3" valve | U1@3.10 | Ea | 357.00 | 151.00 | 30.20 | 538.20 |
| 4" valve | U1@4.04 | Ea | 505.00 | 197.00 | 39.40 | 741.40 |
| 6" valve | U1@4.80 | Ea | 802.00 | 234.00 | 46.80 | 1,082.80 |
| 8" valve | U1@5.14 | Ea | 1,380.00 | 251.00 | 50.10 | 1,681.10 |
| 10" valve | U1@5.67 | Ea | 2,150.00 | 277.00 | 55.30 | 2,482.30 |
| 12" valve | U1@6.19 | Ea | 3,640.00 | 302.00 | 60.40 | 4,002.40 |
| **Indicator post valve** | | | | | | |
| 6" upright, adjustable | U1@6.50 | Ea | 726.00 | 317.00 | 63.40 | 1,106.40 |
| **Fire hydrant with spool** | | | | | | |
| 6" | U1@5.00 | Ea | 1,250.00 | 244.00 | 48.80 | 1,542.80 |
| **Backflow preventers, installed at ground level** | | | | | | |
| 3" backflow preventer | U1@3.50 | Ea | 1,850.00 | 171.00 | 34.10 | 2,055.10 |
| 4" backflow preventer | U1@4.00 | Ea | 2,490.00 | 195.00 | 39.00 | 2,724.00 |
| 6" backflow preventer | U1@5.00 | Ea | 2,840.00 | 244.00 | 48.80 | 3,132.80 |
| 8" backflow preventer | U1@6.00 | Ea | 6,290.00 | 293.00 | 58.50 | 6,641.50 |
| **Block structure for backflow preventer** | | | | | | |
| 5' x 7' x 3'4" | M1@32.0 | Ea | 931.00 | 1,490.00 | — | 2,421.00 |
| **Shear gate** | | | | | | |
| 8", with adjustable release handle | U1@2.44 | Ea | 1,630.00 | 119.00 | 23.80 | 1,772.80 |
| **Adjustable length valve boxes, for up to 20" valves** | | | | | | |
| 5' depth | U1@1.75 | Ea | 99.50 | 85.50 | 17.10 | 202.10 |
| **Concrete thrust blocks, minimum formwork** | | | | | | |
| 1/4 CY thrust block | C8@1.20 | Ea | 67.20 | 55.00 | — | 122.20 |
| 1/2 CY thrust block | C8@2.47 | Ea | 135.00 | 113.00 | — | 248.00 |
| 3/4 CY thrust block | C8@3.80 | Ea | 201.00 | 174.00 | — | 375.00 |
| 1 CY thrust block | C8@4.86 | Ea | 268.00 | 223.00 | — | 491.00 |
| **Tapping saddles, double strap, iron body, tap size to 2"** | | | | | | |
| Pipe to 4" | U1@2.38 | Ea | 45.20 | 116.00 | 23.20 | 184.40 |
| 8" pipe | U1@2.89 | Ea | 59.30 | 141.00 | 28.20 | 228.50 |
| 10" pipe | U1@3.15 | Ea | 60.50 | 154.00 | 30.70 | 245.20 |
| 12" pipe | U1@3.52 | Ea | 79.90 | 172.00 | 34.30 | 286.20 |
| 14" pipe, bronze body | U1@3.78 | Ea | 80.00 | 185.00 | 36.90 | 301.90 |
| **Tapping valves, mechanical joint** | | | | | | |
| 4" valve | U1@6.93 | Ea | 673.00 | 338.00 | 67.60 | 1,078.60 |
| 6" valve | U1@8.66 | Ea | 905.00 | 423.00 | 84.40 | 1,412.40 |
| 8" valve | U1@9.32 | Ea | 1,330.00 | 455.00 | 90.10 | 1,875.10 |
| 10" valve | U1@10.4 | Ea | 2,030.00 | 508.00 | 101.00 | 2,639.00 |
| 12" valve | U1@11.5 | Ea | 3,090.00 | 562.00 | 112.00 | 3,764.00 |
| **Labor and equipment tapping pipe, by hole size** | | | | | | |
| 4" pipe | U1@5.90 | Ea | — | 288.00 | 57.50 | 345.50 |
| 6" pipe | U1@9.32 | Ea | — | 455.00 | 90.90 | 545.90 |
| 8" pipe | U1@11.7 | Ea | — | 571.00 | 114.00 | 685.00 |
| 10" pipe | U1@15.5 | Ea | — | 757.00 | 151.00 | 908.00 |
| 12" pipe | U1@23.4 | Ea | — | 1,140.00 | 228.00 | 1,368.00 |

| | Craft@Hrs | Unit | Material | Labor | Equipment | Total |
|---|---|---|---|---|---|---|

**Ductile Iron Pressure Pipe** Mechanical joint 150 lb Class 52 ductile iron pipe. Installed in an open trench to 5' deep. Equipment includes a wheel-mounted 1 CY backhoe for lifting and placing the pipe. Trench excavation, dewatering, backfill and compaction are not included. Cast iron fittings and valves appear in the preceding sections. Use $4,000 as a minimum job charge.

| | Craft@Hrs | Unit | Material | Labor | Equipment | Total |
|---|---|---|---|---|---|---|
| 4" pipe | U1@.140 | LF | 15.30 | 6.84 | 1.37 | 23.51 |
| 6" pipe | U1@.180 | LF | 22.00 | 8.79 | 1.76 | 32.55 |
| 8" pipe | U1@.200 | LF | 30.50 | 9.77 | 1.95 | 42.22 |
| 10" pipe | U1@.230 | LF | 45.00 | 11.20 | 2.24 | 58.44 |
| 12" pipe | U1@.260 | LF | 61.60 | 12.70 | 2.54 | 76.84 |
| 14" pipe | U1@.300 | LF | 94.40 | 14.60 | 2.93 | 111.93 |
| 16" pipe | U1@.340 | LF | 121.00 | 16.60 | 3.32 | 140.92 |

**PVC Water Pipe** Installed in an open trench. Equipment includes a wheel-mounted 1 CY backhoe for lifting and placing the pipe. Trench excavation, bedding material and backfill are not included. Use $2,500 as a minimum job charge.

Class 200 PVC cold water pressure pipe

| | Craft@Hrs | Unit | Material | Labor | Equipment | Total |
|---|---|---|---|---|---|---|
| 1-1/2" pipe | U1@.029 | LF | .47 | 1.42 | .28 | 2.17 |
| 2" pipe | U1@.036 | LF | .69 | 1.76 | .35 | 2.80 |
| 2-1/2" pipe | U1@.037 | LF | 1.03 | 1.81 | .36 | 3.20 |
| 3" pipe | U1@.065 | LF | 1.48 | 3.17 | .63 | 5.28 |
| 4" pipe | U1@.068 | LF | 2.47 | 3.32 | .66 | 6.45 |
| 6" pipe | U1@.082 | LF | 5.40 | 4.00 | .80 | 10.20 |

90-degree ells Schedule 40

| | Craft@Hrs | Unit | Material | Labor | Equipment | Total |
|---|---|---|---|---|---|---|
| 1-1/2" ells | P6@.385 | Ea | 1.51 | 19.30 | — | 20.81 |
| 2" ells | P6@.489 | Ea | 2.45 | 24.60 | — | 27.05 |
| 2-1/2" ells | P6@.724 | Ea | 7.40 | 36.40 | — | 43.80 |
| 3" ells | P6@1.32 | Ea | 8.76 | 66.30 | — | 75.06 |
| 4" ells | P6@1.57 | Ea | 15.60 | 78.80 | — | 94.40 |
| 6" ells | P6@1.79 | Ea | 50.40 | 89.90 | — | 140.30 |
| Deduct for 45-degree ells | — | % | -5.0 | — | — | — |

Tees

| | Craft@Hrs | Unit | Material | Labor | Equipment | Total |
|---|---|---|---|---|---|---|
| 1-1/2" tees | P6@.581 | Ea | 2.01 | 29.20 | — | 31.21 |
| 2" tees | P6@.724 | Ea | 3.03 | 36.40 | — | 39.43 |
| 2-1/2" tees | P6@.951 | Ea | 9.66 | 47.80 | — | 57.46 |
| 3" tees | P6@1.57 | Ea | 12.80 | 78.80 | — | 91.60 |
| 4" tees | P6@2.09 | Ea | 23.20 | 105.00 | — | 128.20 |
| 6" tees | P6@2.70 | Ea | 77.20 | 136.00 | — | 213.20 |

**Polyethylene Pipe** Belled ends with rubber ring joints. Installed in an open trench. Equipment includes a wheel-mounted 1/2 CY backhoe for lifting and placing the pipe. No excavation, bedding material, shoring, backfill or dewatering included. Use $2,000 as a minimum job charge.

160 PSI polyethylene pipe, Standard Dimension Ratio (SDR) 11 to 1

| | Craft@Hrs | Unit | Material | Labor | Equipment | Total |
|---|---|---|---|---|---|---|
| 3" pipe | U1@.108 | LF | 8.34 | 5.27 | .96 | 14.57 |
| 4" pipe | U1@.108 | LF | 13.80 | 5.27 | .96 | 20.03 |
| 5" pipe | U1@.108 | LF | 21.60 | 5.27 | .96 | 27.83 |
| 6" pipe | U1@.108 | LF | 30.00 | 5.27 | .96 | 36.23 |
| 8" pipe | U1@.145 | LF | 50.70 | 7.08 | 1.29 | 59.07 |
| 10" pipe | U1@.145 | LF | 79.80 | 7.08 | 1.29 | 88.17 |
| 12" pipe | U1@.145 | LF | 112.00 | 7.08 | 1.29 | 120.37 |

# 33 Utilities

| | Craft@Hrs | Unit | Material | Labor | Equipment | Total |
|---|---|---|---|---|---|---|
| 100 PSI polyethylene pipe, Standard Dimension Ratio (SDR) 17 to 1 | | | | | | |
| 4" pipe | U1@.108 | LF | 9.41 | 5.27 | .96 | 15.64 |
| 6" pipe | U1@.108 | LF | 20.20 | 5.27 | .96 | 26.43 |
| 8" pipe | U1@.145 | LF | 34.10 | 7.08 | 1.29 | 42.47 |
| 10" pipe | U1@.145 | LF | 52.30 | 7.08 | 1.29 | 60.67 |
| 12" pipe | U1@.145 | LF | 74.10 | 7.08 | 1.29 | 82.47 |
| Low pressure polyethylene pipe, Standard Dimension Ratio (SDR) 21 to 1 | | | | | | |
| 4" pipe | U1@.108 | LF | 7.78 | 5.27 | .96 | 14.01 |
| 5" pipe | U1@.108 | LF | 11.60 | 5.27 | .96 | 17.83 |
| 6" pipe | U1@.108 | LF | 17.00 | 5.27 | .96 | 23.23 |
| 7" pipe | U1@.145 | LF | 19.20 | 7.08 | 1.29 | 27.57 |
| 8" pipe | U1@.145 | LF | 28.20 | 7.08 | 1.29 | 36.57 |
| 10" pipe | U1@.145 | LF | 42.40 | 7.08 | 1.29 | 50.77 |
| 12" pipe | U1@.145 | LF | 61.30 | 7.08 | 1.29 | 69.67 |
| 14" pipe | U1@.169 | LF | 74.10 | 8.25 | 1.50 | 83.85 |
| 16" pipe | U1@.169 | LF | 96.70 | 8.25 | 1.50 | 106.45 |
| 18" pipe | U1@.169 | LF | 121.00 | 8.25 | 1.50 | 130.75 |
| 20" pipe | U1@.212 | LF | 150.00 | 10.40 | 1.88 | 162.28 |
| 22" pipe | U1@.212 | LF | 182.00 | 10.40 | 1.88 | 194.28 |
| 24" pipe | U1@.212 | LF | 215.00 | 10.40 | 1.88 | 227.28 |
| Fittings, molded, bell ends | | | | | | |
| 4" 90-degree ell, 160 PSI | U1@2.00 | Ea | 191.00 | 97.70 | 17.80 | 306.50 |
| 4" 45-degree ell, 160 PSI | U1@2.00 | Ea | 191.00 | 97.70 | 17.80 | 306.50 |
| 6" 90-degree ell, 160 PSI | U1@2.00 | Ea | 390.00 | 97.70 | 17.80 | 505.50 |
| 6" 45-degree ell, 160 PSI | U1@2.00 | Ea | 390.00 | 97.70 | 17.80 | 505.50 |
| 4" to 3" reducer, 160 PSI | U1@2.00 | Ea | 114.00 | 97.70 | 17.80 | 229.50 |
| 6" to 4" reducer, 160 PSI | U1@2.00 | Ea | 272.00 | 97.70 | 17.80 | 387.50 |
| 3" transition section, 160 PSI | U1@2.00 | Ea | 239.00 | 97.70 | 17.80 | 354.50 |
| 6" transition section, 160 PSI | U1@2.00 | Ea | 1,060.00 | 97.70 | 17.80 | 1,175.50 |
| 10" by 4" branch saddle, 160 PSI | U1@3.00 | Ea | 346.00 | 146.00 | 26.70 | 518.70 |
| 3" flanged adapter, low pressure | U1@1.00 | Ea | 180.00 | 48.80 | 8.89 | 237.69 |
| 4" flanged adapter, low pressure | U1@1.00 | Ea | 257.00 | 48.80 | 8.89 | 314.69 |
| 6" flanged adapter, low pressure | U1@1.00 | Ea | 338.00 | 48.80 | 8.89 | 395.69 |
| 8" flanged adapter, low pressure | U1@1.50 | Ea | 488.00 | 73.20 | 13.30 | 574.50 |
| 10" flanged adapter, low pressure | U1@1.50 | Ea | 688.00 | 73.20 | 13.30 | 774.50 |
| 4" flanged adapter, 160 PSI | U1@1.00 | Ea | 280.00 | 48.80 | 8.89 | 337.69 |
| 6" flanged adapter, 160 PSI | U1@1.00 | Ea | 394.00 | 48.80 | 8.89 | 451.69 |
| Fittings, fabricated, belled ends | | | | | | |
| 10" 90-degree ell, 100 PSI | U1@3.00 | Ea | 1,980.00 | 146.00 | 26.70 | 2,152.70 |
| 10" 45-degree ell, 100 PSI | U1@3.00 | Ea | 1,100.00 | 146.00 | 26.70 | 1,272.70 |
| 4" 30-degree ell, 100 PSI | U1@2.00 | Ea | 320.00 | 97.70 | 17.80 | 435.50 |
| 8" to 6" reducer, 100 PSI | U1@3.00 | Ea | 584.00 | 146.00 | 26.70 | 756.70 |
| 10" to 8" reducer, 100 PSI | U1@3.00 | Ea | 688.00 | 146.00 | 26.70 | 860.70 |
| 4" 45-degree wye, 100 PSI | U1@2.00 | Ea | 850.00 | 97.70 | 17.80 | 965.50 |

| | Craft@Hrs | Unit | Material | Labor | Equipment | Total |
|---|---|---|---|---|---|---|

**Water Main Sterilization** Gas chlorination method, typical costs for sterilizing new water mains prior to inspection, includes operator and equipment. Add $200 to total costs for setting up and removing equipment. Minimum cost will be $500. These costs include the subcontractor's overhead and profit. Typical cost per 1,000 LF of main with diameter as shown.

| | Craft@Hrs | Unit | Material | Labor | Equipment | Total |
|---|---|---|---|---|---|---|
| 3" or 4" diameter | — | MLF | — | — | — | 55.60 |
| 6" diameter | — | MLF | — | — | — | 97.90 |
| 8" diameter | — | MLF | — | — | — | 130.00 |
| 10" diameter | — | MLF | — | — | — | 149.00 |
| 12" diameter | — | MLF | — | — | — | 185.00 |
| 14" diameter | — | MLF | — | — | — | 220.00 |
| 16" diameter | — | MLF | — | — | — | 234.00 |
| 18" diameter | — | MLF | — | — | — | 289.00 |
| 24" diameter | — | MLF | — | — | — | 382.00 |
| 36" diameter | — | MLF | — | — | — | 549.00 |
| 48" diameter | — | MLF | — | — | — | 745.00 |

**Water Tanks, Elevated** Typical subcontract prices for 100' high steel tanks, including design, fabrication and erection. Complete tank includes 36" pipe riser with access manhole, 8" overflow to ground, ladder with protective cage, vent, balcony catwalk, concrete foundation, painting and test. No well, pump, fencing, drainage or water distribution piping included. Costs are per tank with gallon capacity as shown. Tanks built in areas without earthquake or high wind risk may cost 15% to 25% less.

| | Craft@Hrs | Unit | Material | Labor | Total |
|---|---|---|---|---|---|
| 75,000 gallon | — | Ea | — | — | 360,000.00 |
| 100,000 gallon | — | Ea | — | — | 386,000.00 |
| 150,000 gallon | — | Ea | — | — | 460,000.00 |
| 200,000 gallon | — | Ea | — | — | 602,000.00 |
| 300,000 gallon | — | Ea | — | — | 746,000.00 |
| 400,000 gallon | — | Ea | — | — | 864,000.00 |
| Add or subtract for each foot of height more or less than 100, per 100 gallons of capacity | — | Ea | — | — | .86 |
| Cathodic protection system, per tank | — | LS | — | — | 18,300.00 |

Electric service, including float switches, obstruction marker lighting system, and hookup of cathodic protection system

| | Craft@Hrs | Unit | Material | Labor | Total |
|---|---|---|---|---|---|
| Per tank | — | LS | — | — | 13,900.00 |

| | Craft@Hrs | Unit | Material | Labor | Equipment | Total |
|---|---|---|---|---|---|---|

**PVC Rigid Sewer and Drain Pipe** Rigid polyvinyl chloride. Solvent weld. ASTM D-3034, plain end. Installed in an open trench. No excavation, shoring, backfill or dewatering included.

| | Craft@Hrs | Unit | Material | Labor | Equipment | Total |
|---|---|---|---|---|---|---|
| 4" pipe, 10' lengths | CL@.073 | LF | .90 | 2.91 | — | 3.81 |
| 4" pipe, 13' lengths | CL@.073 | LF | 1.66 | 2.91 | — | 4.57 |
| 6" pipe, 13' lengths | CL@.079 | LF | 3.60 | 3.15 | — | 6.75 |
| Elbow fittings (1/4 bend) | | | | | | |
| 4" elbow | CL@.250 | Ea | 5.06 | 9.97 | — | 15.03 |
| 6" elbow | CL@.332 | Ea | 20.80 | 13.20 | — | 34.00 |
| Tee fittings | | | | | | |
| 4" elbow | CL@.375 | Ea | 4.84 | 15.00 | — | 19.84 |
| 6" elbow | CL@.500 | Ea | 21.00 | 19.90 | — | 40.90 |

| | Craft@Hrs | Unit | Material | Labor | Equipment | Total |
|---|---|---|---|---|---|---|

**PVC Sewer Pipe**  13' lengths, bell and spigot ends with rubber ring gasket in bell. Installed in open trench. Equipment includes a wheel-mounted 1 CY backhoe for lifting and placing the pipe. Trench excavation, dewatering, backfill and compaction are not included. Use $2,500 as a minimum job charge.

| | Craft@Hrs | Unit | Material | Labor | Equipment | Total |
|---|---|---|---|---|---|---|
| 4" pipe | C5@.047 | LF | 1.66 | 2.11 | .46 | 4.23 |
| 6" pipe | C5@.056 | LF | 3.60 | 2.51 | .56 | 6.67 |
| 8" pipe | C5@.061 | LF | 6.42 | 2.73 | .59 | 9.74 |
| 10" pipe | C5@.079 | LF | 10.30 | 3.54 | .77 | 14.61 |
| 12" pipe | C5@.114 | LF | 14.30 | 5.11 | 1.11 | 20.52 |

**Sewer Main Cleaning**  Ball method, typical costs for cleaning new sewer mains prior to inspection, includes operator and equipment. Add $200 to total costs for setting up and removing equipment. Minimum cost will be $500. These costs include the subcontractor's overhead and profit. Typical costs per 1,000 LF of main diameter listed.

| | Craft@Hrs | Unit | Material | Labor | Equipment | Total |
|---|---|---|---|---|---|---|
| 6" diameter | — | MLF | — | — | — | 96.80 |
| 8" diameter | — | MLF | — | — | — | 124.00 |
| 10" diameter | — | MLF | — | — | — | 149.00 |
| 12" diameter | — | MLF | — | — | — | 185.00 |
| 14" diameter | — | MLF | — | — | — | 221.00 |
| 16" diameter | — | MLF | — | — | — | 249.00 |
| 18" diameter | — | MLF | — | — | — | 289.00 |
| 24" diameter | — | MLF | — | — | — | 382.00 |
| 36" diameter | — | MLF | — | — | — | 575.00 |
| 48" diameter | — | MLF | — | — | — | 746.00 |

**Polyethylene Flexible Drainage Tubing**  Corrugated drainage tubing, plain or perforated and snap-on ABS fittings. Installed in an open trench. Excavation, bedding material and backfill not included.

| | Craft@Hrs | Unit | Material | Labor | Equipment | Total |
|---|---|---|---|---|---|---|
| 3" (10' length) | CL@.009 | LF | .72 | .36 | — | 1.08 |
| 3" (100' coil) | CL@.009 | LF | .55 | .36 | — | .91 |
| 4" (10' length) | CL@.011 | LF | .61 | .44 | — | 1.05 |
| 4" (100' coil) | CL@.011 | LF | .55 | .44 | — | .99 |
| 6" (100' coil) | CL@.014 | LF | 1.37 | .56 | — | 1.93 |
| 8" (20' or 40' length) | CL@.017 | LF | 3.71 | .68 | — | 4.39 |
| 10" (20' length) | CL@.021 | LF | 8.14 | .84 | — | 8.98 |
| 12" (20' length) | CL@.022 | LF | 11.70 | .88 | — | 12.58 |

Polyethylene tube snap fittings, elbows or tees

| | Craft@Hrs | Unit | Material | Labor | Equipment | Total |
|---|---|---|---|---|---|---|
| 3" fitting | CL@.025 | Ea | 7.08 | 1.00 | — | 8.08 |
| 4" fitting | CL@.025 | Ea | 8.45 | 1.00 | — | 9.45 |
| 6" fitting | CL@.025 | Ea | 14.10 | 1.00 | — | 15.10 |

Filter sock sieve. Keeps soil out of perforated pipe. Reduces need for aggregate filters. By pipe size. Per linear foot of sock.

| | Craft@Hrs | Unit | Material | Labor | Equipment | Total |
|---|---|---|---|---|---|---|
| 3", 10' length | CL@.006 | LF | .69 | .24 | — | .93 |
| 4", 10' length | CL@.007 | LF | .67 | .28 | — | .95 |
| 4", 100' length | CL@.007 | LF | .26 | .28 | — | .54 |

**Hi-Q drain pipe**  Corrugated high-density polyethylene drainage tubing and culvert pipe installed in an open trench. Excavation, bedding material and backfill not included. 20' lengths.

| | Craft@Hrs | Unit | Material | Labor | Equipment | Total |
|---|---|---|---|---|---|---|
| 4" | CL@.009 | LF | 1.33 | .36 | — | 1.69 |
| 6" | CL@.014 | LF | 3.16 | .56 | — | 3.72 |
| 12" culvert | CL@.022 | LF | 8.04 | .88 | — | 8.92 |
| 15" culvert | CL@.026 | LF | 16.10 | 1.04 | — | 17.14 |
| 18" culvert | CL@.045 | LF | 29.40 | 1.79 | — | 31.19 |
| 24" culvert | CL@.090 | LF | 51.90 | 3.59 | — | 55.49 |

| | Craft@Hrs | Unit | Material | Labor | Equipment | Total |
|---|---|---|---|---|---|---|
| Corrugated drain pipe fittings | | | | | | |
| 4" snap adapter | CL@.025 | Ea | 2.56 | 1.00 | — | 3.56 |
| 6" snap adapter | CL@.025 | Ea | 5.14 | 1.00 | — | 6.14 |
| 6" x 4" reducer | CL@.025 | Ea | 5.16 | 1.00 | — | 6.16 |
| 6" snap coupling | CL@.025 | Ea | 4.46 | 1.00 | — | 5.46 |
| 6" wye | CL@.037 | Ea | 12.90 | 1.48 | — | 14.38 |
| 6" tee | CL@.037 | Ea | 11.50 | 1.48 | — | 12.98 |
| 6" blind tee | CL@.037 | Ea | 9.63 | 1.48 | — | 11.11 |
| 6" split end cap | CL@.025 | Ea | 3.86 | 1.00 | — | 4.86 |
| 12" split coupling | CL@.050 | Ea | 12.90 | 1.99 | — | 14.89 |
| 15" split coupling | CL@.090 | Ea | 16.80 | 3.59 | — | 20.39 |

**Non-Reinforced Concrete Pipe** Smooth wall, standard strength, 3' lengths with tongue-and-groove mortar joint ends. Installed in an open trench. Equipment includes a wheel-mounted 1 CY backhoe for lifting and placing the pipe. Trench excavation, bedding material and backfill are not included. Use $3,500 as a minimum job charge.

| | Craft@Hrs | Unit | Material | Labor | Equipment | Total |
|---|---|---|---|---|---|---|
| 6" pipe | C5@.076 | LF | 3.30 | 3.41 | .74 | 7.45 |
| 8" pipe | C5@.088 | LF | 5.09 | 3.95 | .86 | 9.90 |
| 10" pipe | C5@.088 | LF | 6.08 | 3.95 | .86 | 10.89 |
| 12" pipe | C5@.093 | LF | 8.48 | 4.17 | .91 | 13.56 |
| 15" pipe | C5@.105 | LF | 10.50 | 4.71 | 1.02 | 16.23 |
| 18" pipe | C5@.117 | LF | 13.30 | 5.25 | 1.14 | 19.69 |
| 21" pipe | C5@.117 | LF | 16.90 | 5.25 | 1.14 | 23.29 |
| 24" pipe | C5@.130 | LF | 24.80 | 5.83 | 1.27 | 31.90 |
| 30" pipe | C5@.143 | LF | 43.40 | 6.41 | 1.39 | 51.20 |
| 36" pipe | C5@.158 | LF | 47.00 | 7.08 | 1.54 | 55.62 |

**Non-Reinforced Perforated Concrete Underdrain Pipe** Perforated smooth wall, standard strength, 3' lengths with tongue and groove mortar joint ends. Installed in an open trench. Equipment includes a wheel-mounted 1 CY backhoe for lifting and placing the pipe. Trench excavation, bedding material and backfill are not included. Use $2,200 as a minimum job charge.

| | Craft@Hrs | Unit | Material | Labor | Equipment | Total |
|---|---|---|---|---|---|---|
| 6" pipe | C5@.076 | LF | 3.91 | 3.41 | .74 | 8.06 |
| 8" pipe | C5@.104 | LF | 5.93 | 4.66 | 1.01 | 11.60 |

**Reinforced Concrete Pipe** Class III, 1350 "D"-load, ASTM C-76, 8' lengths. Tongue-and-groove mortar joint ends. Installed in an open trench. Equipment cost shown is for a wheel-mounted 1 CY backhoe for lifting and placing the pipe. Trench excavation, dewatering, backfill and compaction are not included. Use $3,000 as a minimum job charge.

| | Craft@Hrs | Unit | Material | Labor | Equipment | Total |
|---|---|---|---|---|---|---|
| 18" pipe | C5@.272 | LF | 16.90 | 12.20 | 2.65 | 31.75 |
| 24" pipe | C5@.365 | LF | 20.90 | 16.40 | 3.56 | 40.86 |
| 30" pipe | C5@.426 | LF | 25.30 | 19.10 | 4.15 | 48.55 |
| 36" pipe | C5@.511 | LF | 40.30 | 22.90 | 4.98 | 68.18 |
| 42" pipe | C5@.567 | LF | 49.50 | 25.40 | 5.53 | 80.43 |
| 48" pipe | C5@.623 | LF | 59.70 | 27.90 | 6.07 | 93.67 |
| 54" pipe | C5@.707 | LF | 71.60 | 31.70 | 6.89 | 110.19 |
| 60" pipe | C5@.785 | LF | 87.10 | 35.20 | 7.65 | 129.95 |
| 66" pipe | C5@.914 | LF | 107.00 | 41.00 | 8.91 | 156.91 |
| 72" pipe | C5@.914 | LF | 125.00 | 41.00 | 8.91 | 174.91 |
| 78" pipe | C5@.914 | LF | 142.00 | 41.00 | 8.91 | 191.91 |
| 84" pipe | C5@1.04 | LF | 166.00 | 46.60 | 10.10 | 222.70 |
| 90" pipe | C5@1.04 | LF | 177.00 | 46.60 | 10.10 | 233.70 |
| 96" pipe | C5@1.18 | LF | 199.00 | 52.90 | 11.50 | 263.40 |

| | Craft@Hrs | Unit | Material | Labor | Equipment | Total |
|---|---|---|---|---|---|---|

**Reinforced Elliptical Concrete Pipe** Class III, 1350 "D", C-502-72, 8' lengths, tongue-and-groove mortar joint ends. Installed in an open trench. Equipment includes a wheel-mounted 1 CY backhoe for lifting and placing the pipe. Trench excavation, dewatering, backfill and compaction are not included. Use $3,500 as a minimum job charge.

| | Craft@Hrs | Unit | Material | Labor | Equipment | Total |
|---|---|---|---|---|---|---|
| 19" x 30" pipe, 24" pipe equivalent | C5@.356 | LF | 32.10 | 16.00 | 3.47 | 51.57 |
| 24" x 38" pipe, 30" pipe equivalent | C5@.424 | LF | 43.40 | 19.00 | 4.13 | 66.53 |
| 29" x 45" pipe, 36" pipe equivalent | C5@.516 | LF | 69.90 | 23.10 | 5.03 | 98.03 |
| 34" x 53" pipe, 42" pipe equivalent | C5@.581 | LF | 86.70 | 26.00 | 5.66 | 118.36 |
| 38" x 60" pipe, 48" pipe equivalent | C5@.637 | LF | 104.00 | 28.60 | 6.21 | 138.81 |
| 48" x 76" pipe, 60" pipe equivalent | C5@.797 | LF | 172.00 | 35.70 | 7.77 | 215.47 |
| 53" x 83" pipe, 66" pipe equivalent | C5@.923 | LF | 211.00 | 41.40 | 9.00 | 261.40 |
| 58" x 91" pipe, 72" pipe equivalent | C5@.923 | LF | 249.00 | 41.40 | 9.00 | 299.40 |

**Flared Concrete End Sections** For round concrete drain pipe, precast. Equipment includes a wheel-mounted 1 CY backhoe for lifting and placing the items. Trench excavation, dewatering, backfill and compaction are not included. Use $2,500 as a minimum job charge.

| | Craft@Hrs | Unit | Material | Labor | Equipment | Total |
|---|---|---|---|---|---|---|
| 12" opening, 530 lbs | C5@2.41 | Ea | 326.00 | 108.00 | 23.50 | 457.50 |
| 15" opening, 740 lbs | C5@2.57 | Ea | 384.00 | 115.00 | 25.10 | 524.10 |
| 18" opening, 990 lbs | C5@3.45 | Ea | 429.00 | 155.00 | 33.60 | 617.60 |
| 24" opening, 1,520 lbs | C5@3.45 | Ea | 525.00 | 155.00 | 33.60 | 713.60 |
| 30" opening, 2,190 lbs | C5@3.96 | Ea | 719.00 | 178.00 | 38.60 | 935.60 |
| 36" opening, 4,100 lbs | C5@3.96 | Ea | 901.00 | 178.00 | 38.60 | 1,117.60 |
| 42" opening, 5,380 lbs | C5@4.80 | Ea | 1,280.00 | 215.00 | 46.80 | 1,541.80 |
| 48" opening, 6,550 lbs | C5@5.75 | Ea | 1,520.00 | 258.00 | 56.10 | 1,834.10 |
| 54" opening, 8,000 lbs | C5@7.10 | Ea | 1,940.00 | 318.00 | 69.20 | 2,327.20 |

**Precast Reinforced Concrete Box Culvert Pipe** ASTM C-850 with tongue-and-groove mortar joint ends. Installed in an open trench. Equipment includes a wheel-mounted 1 CY backhoe for lifting and placing the pipe. Trench excavation, dewatering, backfill and compaction are not included. Laying length is 8' for smaller cross sections and 6' or 4' for larger cross sections. Use $8,500 as a minimum job charge.

| | Craft@Hrs | Unit | Material | Labor | Equipment | Total |
|---|---|---|---|---|---|---|
| 4' x 3' | C5@.273 | LF | 127.00 | 12.20 | 2.66 | 141.86 |
| 5' x 5' | C5@.336 | LF | 188.00 | 15.10 | 3.28 | 206.38 |
| 6' x 4' | C5@.365 | LF | 219.00 | 16.40 | 3.56 | 238.96 |
| 6' x 6' | C5@.425 | LF | 255.00 | 19.10 | 4.14 | 278.24 |
| 7' x 4' | C5@.442 | LF | 264.00 | 19.80 | 4.31 | 288.11 |
| 7' x 7' | C5@.547 | LF | 326.00 | 24.50 | 5.33 | 355.83 |
| 8' x 4' | C5@.579 | LF | 283.00 | 26.00 | 5.65 | 314.65 |
| 8' x 8' | C5@.613 | LF | 368.00 | 27.50 | 5.98 | 401.48 |
| 9' x 6' | C5@.658 | LF | 394.00 | 29.50 | 6.42 | 429.92 |
| 9' x 9' | C5@.773 | LF | 467.00 | 34.70 | 7.54 | 509.24 |
| 10' x 6' | C5@.787 | LF | 470.00 | 35.30 | 7.67 | 512.97 |
| 10' x 10' | C5@.961 | LF | 576.00 | 43.10 | 9.37 | 628.47 |

**Corrugated Metal Pipe, Galvanized** 20' lengths, installed in an open trench. Costs include couplers. Equipment includes a wheel-mounted 1 CY backhoe for lifting and placing the pipe. Trench excavation, dewatering, backfill and compaction are not included. Use $3,500 as a minimum job charge.
Round pipe

| | Craft@Hrs | Unit | Material | Labor | Equipment | Total |
|---|---|---|---|---|---|---|
| 8", 16 gauge (.064) | C5@.102 | LF | 11.00 | 4.57 | .99 | 16.56 |
| 10", 16 gauge (.064) | C5@.102 | LF | 11.40 | 4.57 | .99 | 16.96 |
| 12", 16 gauge (.064) | C5@.120 | LF | 12.00 | 5.38 | 1.17 | 18.55 |
| 15", 16 gauge (.064) | C5@.120 | LF | 14.70 | 5.38 | 1.17 | 21.25 |

| | Craft@Hrs | Unit | Material | Labor | Equipment | Total |
|---|---|---|---|---|---|---|
| 18", 16 gauge (.064) | C5@.152 | LF | 18.00 | 6.81 | 1.48 | 26.29 |
| 24", 14 gauge (.079) | C5@.196 | LF | 27.80 | 8.79 | 1.91 | 38.50 |
| 30", 14 gauge (.079) | C5@.196 | LF | 34.30 | 8.79 | 1.91 | 45.00 |
| 36", 14 gauge (.079) | C5@.231 | LF | 41.70 | 10.40 | 2.25 | 54.35 |
| 42", 14 gauge (.079) | C5@.231 | LF | 48.50 | 10.40 | 2.25 | 61.15 |
| 48", 14 gauge (.079) | C5@.272 | LF | 55.00 | 12.20 | 2.65 | 69.85 |
| 60", 12 gauge (.109) | C5@.311 | LF | 93.50 | 13.90 | 3.03 | 110.43 |
| Oval pipe | | | | | | |
| 18" x 11", 16 gauge (.064) | C5@.133 | LF | 17.60 | 5.96 | 1.30 | 24.86 |
| 22" x 13", 16 gauge (.064) | C5@.167 | LF | 21.00 | 7.49 | 1.63 | 30.12 |
| 29" x 18", 14 gauge (.079) | C5@.222 | LF | 32.80 | 9.95 | 2.16 | 44.91 |
| 36" x 22", 14 gauge (.079) | C5@.228 | LF | 40.70 | 10.20 | 2.22 | 53.12 |
| 43" x 27", 14 gauge (.079) | C5@.284 | LF | 43.90 | 12.70 | 2.77 | 59.37 |
| Flared end sections for oval pipe | | | | | | |
| 18" x 11" | C5@1.59 | Ea | 95.20 | 71.30 | 15.50 | 182.00 |
| 22" x 13" | C5@1.75 | Ea | 113.00 | 78.50 | 17.06 | 208.56 |
| 29" x 18" | C5@1.85 | Ea | 172.00 | 82.90 | 18.00 | 272.90 |
| 36" x 22" | C5@2.38 | Ea | 275.00 | 107.00 | 23.20 | 405.20 |
| 43" x 27" | C5@2.85 | Ea | 455.00 | 128.00 | 27.80 | 610.80 |
| Add for bituminous coating, with paved invert | | | | | | |
| On round pipe | — | % | 20.0 | — | — | — |
| On oval pipe or flared ends | — | % | 24.0 | — | — | — |
| Deduct for aluminum pipe, | | | | | | |
| round, oval or flared ends | — | % | -12.0 | — | — | — |

**Corrugated Metal Nestable Pipe**  Split-in-half, Type I, 16 gauge. Trench excavation, backfill, compaction and grading are not included. Equipment includes a wheel-mounted 1 CY backhoe for lifting and placing the pipe. Use $5,500 as a minimum job charge.

| | Craft@Hrs | Unit | Material | Labor | Equipment | Total |
|---|---|---|---|---|---|---|
| 12" diameter, galvanized | C5@.056 | LF | 20.00 | 2.51 | .55 | 23.06 |
| 18" diameter, galvanized | C5@.064 | LF | 25.90 | 2.87 | .62 | 29.39 |
| Add for bituminous coating, with paved invert | — | % | 20.0 | — | — | — |

**Drainage Tile**  Vitrified clay drainage tile pipe. Standard weight, 6' joint length with 1 coupler per joint. Installed in an open trench. Couplers consist of a rubber sleeve with stainless steel compression bands. Equipment includes a wheel-mounted 1/2 CY backhoe for lifting and placing the pipe. Excavation, bedding material or backfill are not included. Use $2,000 as a minimum job charge.

| | Craft@Hrs | Unit | Material | Labor | Equipment | Total |
|---|---|---|---|---|---|---|
| 4" tile | S6@.095 | LF | 3.30 | 4.27 | 1.13 | 8.70 |
| 6" tile | S6@.119 | LF | 7.11 | 5.35 | 1.41 | 13.87 |
| 8" tile | S6@.143 | LF | 10.20 | 6.42 | 1.70 | 18.32 |
| 10" tile | S6@.165 | LF | 16.20 | 7.41 | 1.96 | 25.57 |
| Add for covering 4" to 10" tile, 30 lb felt building paper | CL@.048 | LF | .38 | 1.91 | — | 2.29 |
| Elbows (1/4 bends), costs include one coupler | | | | | | |
| 4" elbow | S6@.286 | Ea | 21.80 | 12.80 | 3.39 | 37.99 |
| 6" elbow | S6@.360 | Ea | 38.50 | 16.20 | 4.27 | 58.97 |
| 8" elbow | S6@.430 | Ea | 65.30 | 19.30 | 5.10 | 89.70 |
| 10" elbow | S6@.494 | Ea | 113.00 | 22.20 | 5.86 | 141.06 |
| Tees, costs include two couplers | | | | | | |
| 4" tee | S6@.430 | Ea | 36.50 | 19.30 | 5.10 | 60.90 |
| 6" tee | S6@.540 | Ea | 61.80 | 24.30 | 6.40 | 92.50 |
| 8" tee | S6@.643 | Ea | 91.00 | 28.90 | 7.62 | 127.52 |
| 10" tee | S6@.744 | Ea | 198.00 | 33.40 | 8.82 | 240.22 |

| | Craft@Hrs | Unit | Material | Labor | Total |
|---|---|---|---|---|---|

**Fiberglass Trench Drain System** System includes 16 sloped channels, each 6 feet long, 16 channel frames, 64 iron grates, each 18 inches long, 128 grate locking bolts, outlet and end caps. Neutral channels extend the system run beyond 96 feet. Corrosion resistant polyester or vinylester. Channel has 8" internal width and built-in slope of 1%. Installed in an open trench.

| | Craft@Hrs | Unit | Material | Labor | Total |
|---|---|---|---|---|---|
| Typical 96' fiberglass drainage system | P6@10.0 | Ea | 4,950.00 | 502.00 | 5,452.00 |
| System components | | | | | |
| 6' sloped channel | — | Ea | 154.00 | — | 154.00 |
| 6' channel frame | — | Ea | 61.60 | — | 61.60 |
| 18" cast iron grate | — | Ea | 28.80 | — | 28.80 |
| 6" grate locking bolt | — | Ea | .79 | — | .79 |
| 6" outlet cap | — | Ea | 22.10 | — | 22.10 |
| 6" closed end cap | — | Ea | 22.30 | — | 22.30 |
| Add for 6' neutral channels | — | Ea | 155.00 | — | 155.00 |

**Precast Trench Drain System** System includes 30 framed, interlocking tongue-and-groove connection channels made of precast polyester or vinylester, one meter long each, with integral cast-in metal rail edge, 30 snap-in galvanized steel slotted grates, 24" x 24" catch basin, 60 grate locking bolts, outlet and end caps. Channel has 4" internal width and built-in slope of .6%. All channels have preformed round and oval drillouts for vertical outlet connection with underground piping. Neutral channels extend the system run beyond 30 meters. Channel and grating combination certified to Load Class D (90,000 lbs). Installed in an open trench.

| | Craft@Hrs | Unit | Material | Labor | Total |
|---|---|---|---|---|---|
| Typical 30 meter galvanized steel drainage system | P6@8.0 | Ea | 2,550.00 | 402.00 | 2,952.00 |
| System components | | | | | |
| 1 meter sloped channel with frame | — | Ea | 60.90 | — | 60.90 |
| 1 meter galvanized steel grate | — | Ea | 13.90 | — | 13.90 |
| 24" x 24" catch basin | — | Ea | 98.70 | — | 98.70 |
| Grate locking bolt | — | Ea | 3.25 | — | 3.25 |
| 4" outlet cap | — | Ea | 7.28 | — | 7.28 |
| 4" closed end cap | — | Ea | 7.31 | — | 7.31 |
| Add for 1 meter neutral channels | — | Ea | 60.90 | — | 60.90 |

| | Craft@Hrs | Unit | Material | Labor | Equipment | Total |
|---|---|---|---|---|---|---|

**Corrugated Polyethylene Culvert Pipe** Heavy-duty, 20' lengths. Installed in an open trench. Equipment includes a wheel-mounted 1 CY backhoe for lifting and placing the pipe. Trench excavation, dewatering, backfill and compaction are not included. Based on 500' minimum job. Various sizes may be combined for total footage. Couplings are split type that snap onto the pipe. Use $2,500 as a minimum job charge.

| | Craft@Hrs | Unit | Material | Labor | Equipment | Total |
|---|---|---|---|---|---|---|
| 8" diameter pipe | C5@.035 | LF | 2.89 | 1.57 | .34 | 4.80 |
| 10" diameter pipe | C5@.038 | LF | 6.39 | 1.70 | .37 | 8.46 |
| 12" diameter pipe | C5@.053 | LF | 9.07 | 2.38 | .52 | 11.97 |
| 18" diameter pipe | C5@.111 | LF | 18.30 | 4.98 | 1.08 | 24.36 |
| 8" coupling | C5@.105 | Ea | 7.74 | 4.71 | 1.02 | 13.47 |
| 10" coupling | C5@.114 | Ea | 7.50 | 5.11 | 1.11 | 13.72 |
| 12" coupling | C5@.158 | Ea | 11.70 | 7.08 | 1.54 | 20.32 |
| 18" coupling | C5@.331 | Ea | 24.20 | 14.80 | 3.23 | 42.23 |

**Vitrified Clay Pipe** C-200, extra strength, installed in open trench. 4" through 12" are 6'0" joint length, over 12" are either 7' or 7'6" joint length. Costs shown include one coupler per joint. Couplers consist of a rubber sleeve with stainless steel compression bands. Equipment includes a wheel-mounted 1 CY backhoe for lifting and placing the pipe. Trench excavation, dewatering, backfill and compaction are not included. Use $3,500 as a minimum job charge.

| | Craft@Hrs | Unit | Material | Labor | Equipment | Total |
|---|---|---|---|---|---|---|
| 4" pipe | C5@.081 | LF | 4.73 | 3.63 | .79 | 9.15 |
| 6" pipe | C5@.175 | LF | 10.20 | 7.85 | 1.71 | 19.76 |
| 8" pipe | C5@.183 | LF | 14.60 | 8.20 | 1.78 | 24.58 |

| | Craft@Hrs | Unit | Material | Labor | Equipment | Total |
|---|---|---|---|---|---|---|
| 10" pipe | C5@.203 | LF | 23.30 | 9.10 | 1.96 | 34.36 |
| 12" pipe | C5@.225 | LF | 33.00 | 10.10 | 2.19 | 45.29 |
| 15" pipe | C5@.233 | LF | 31.30 | 10.40 | 2.27 | 43.97 |
| 18" pipe | C5@.254 | LF | 45.10 | 11.40 | 2.48 | 58.98 |
| 21" pipe | C5@.283 | LF | 61.90 | 12.70 | 2.76 | 77.36 |
| 24" pipe | C5@.334 | LF | 82.10 | 15.00 | 3.26 | 100.36 |
| 27" pipe | C5@.396 | LF | 93.70 | 17.80 | 3.86 | 115.36 |
| 30" pipe | C5@.452 | LF | 114.00 | 20.30 | 4.41 | 138.71 |

1/4 bends including couplers

| | Craft@Hrs | Unit | Material | Labor | Equipment | Total |
|---|---|---|---|---|---|---|
| 4" bend | C5@.494 | Ea | 31.30 | 22.10 | 4.82 | 58.22 |
| 6" bend | C5@.524 | Ea | 54.40 | 23.50 | 5.11 | 83.01 |
| 8" bend | C5@.549 | Ea | 72.70 | 24.60 | 5.35 | 102.65 |
| 10" bend | C5@.610 | Ea | 125.00 | 27.30 | 5.95 | 158.25 |
| 12" bend | C5@.677 | Ea | 196.00 | 30.30 | 6.60 | 232.90 |
| 15" bend | C5@.715 | Ea | 422.00 | 32.10 | 6.97 | 461.07 |
| 18" bend | C5@.844 | Ea | 579.00 | 37.80 | 8.23 | 625.03 |
| 21" bend | C5@.993 | Ea | 789.00 | 44.50 | 9.66 | 843.16 |
| 24" bend | C5@1.17 | Ea | 1,030.00 | 52.50 | 11.40 | 1,093.90 |
| 30" bend | C5@1.37 | Ea | 1,470.00 | 61.40 | 13.40 | 1,544.80 |

Wyes and tees including couplers

| | Craft@Hrs | Unit | Material | Labor | Equipment | Total |
|---|---|---|---|---|---|---|
| 4" wye or tee | C5@.744 | Ea | 40.30 | 33.40 | 7.25 | 80.95 |
| 6" wye or tee | C5@.787 | Ea | 68.90 | 35.30 | 7.67 | 111.87 |
| 8" wye or tee | C5@.823 | Ea | 102.00 | 36.90 | 8.02 | 146.92 |
| 10" wye or tee | C5@.915 | Ea | 219.00 | 41.00 | 8.92 | 268.92 |
| 12" wye or tee | C5@1.02 | Ea | 288.00 | 45.70 | 9.95 | 343.65 |
| 15" wye or tee | C5@1.10 | Ea | 473.00 | 49.30 | 10.70 | 533.00 |
| 18" wye or tee | C5@1.30 | Ea | 640.00 | 58.30 | 12.70 | 711.00 |
| 21" wye or tee | C5@1.53 | Ea | 840.00 | 68.60 | 14.90 | 923.50 |
| 24" wye or tee | C5@1.80 | Ea | 100.00 | 80.70 | 17.60 | 198.30 |
| 27" wye or tee | C5@2.13 | Ea | 1,210.00 | 95.50 | 20.80 | 1,326.30 |
| 30" wye or tee | C5@2.50 | Ea | 1,190.00 | 112.00 | 24.40 | 1,326.40 |

**Catch Basins** Catch basin including 6" concrete top and base. Equipment includes a wheel-mounted 1 CY backhoe for lifting and placing the items. Trench excavation, bedding material and backfill are not included. Use $3,000 as a minimum job charge.

4' diameter precast concrete basin

| | Craft@Hrs | Unit | Material | Labor | Equipment | Total |
|---|---|---|---|---|---|---|
| 4' deep | S6@19.0 | Ea | 1,600.00 | 853.00 | 247.00 | 2,700.00 |
| 6' deep | S6@23.7 | Ea | 1,990.00 | 1,060.00 | 308.00 | 3,358.00 |
| 8' deep | S6@28.9 | Ea | 2,440.00 | 1,300.00 | 376.00 | 4,116.00 |

Light-duty grate, gray iron, asphalt coated
Grate on pipe bell

| | Craft@Hrs | Unit | Material | Labor | Equipment | Total |
|---|---|---|---|---|---|---|
| 6" diameter, 13 lb | S6@.205 | Ea | 15.20 | 9.21 | 2.67 | 27.08 |
| 8" diameter, 25 lb | S6@.247 | Ea | 41.30 | 11.10 | 3.21 | 55.61 |

Frame and grate

| | Craft@Hrs | Unit | Material | Labor | Equipment | Total |
|---|---|---|---|---|---|---|
| 8" diameter, 55 lb | S6@.424 | Ea | 88.20 | 19.00 | 5.51 | 112.71 |
| 17" diameter, 135 lb | S6@.823 | Ea | 214.00 | 37.00 | 10.70 | 261.70 |

Medium-duty, frame and grate

| | Craft@Hrs | Unit | Material | Labor | Equipment | Total |
|---|---|---|---|---|---|---|
| 11" diameter, 70 lb | S6@.441 | Ea | 106.00 | 19.80 | 5.73 | 131.53 |
| 15" diameter, 120 lb | S6@.770 | Ea | 191.00 | 34.60 | 10.00 | 235.60 |
| Radial grate, 20" diameter, 140 lb | S6@1.68 | Ea | 219.00 | 75.50 | 21.80 | 316.30 |

# 33 Utilities

| | Craft@Hrs | Unit | Material | Labor | Equipment | Total |
|---|---|---|---|---|---|---|
| **Heavy-duty frame and grate** | | | | | | |
| Flat grate | | | | | | |
| 11-1/2" diameter, 85 lb | S6@.537 | Ea | 138.00 | 24.10 | 6.98 | 169.08 |
| 20" diameter, 235 lb | S6@2.01 | Ea | 375.00 | 90.30 | 26.10 | 491.40 |
| 21" diameter, 315 lb | S6@2.51 | Ea | 458.00 | 113.00 | 32.60 | 603.60 |
| 24" diameter, 350 lb | S6@2.51 | Ea | 507.00 | 113.00 | 32.60 | 652.60 |
| 30" diameter, 555 lb | S6@4.02 | Ea | 745.00 | 181.00 | 52.30 | 978.30 |
| Convex or concave grate | | | | | | |
| 20" diameter, 200 lb | S6@1.83 | Ea | 266.00 | 82.20 | 23.80 | 372.00 |
| 20" diameter, 325 lb | S6@2.51 | Ea | 434.00 | 113.00 | 32.60 | 579.60 |
| **Beehive grate and frame** | | | | | | |
| 11" diameter, 80 lb | S6@.495 | Ea | 191.00 | 22.20 | 6.44 | 219.64 |
| 15" diameter, 120 lb | S6@1.83 | Ea | 182.00 | 82.20 | 23.80 | 288.00 |
| 21" diameter, 285 lb | S6@2.51 | Ea | 383.00 | 113.00 | 32.60 | 528.60 |
| 24" diameter, 375 lb | S6@2.87 | Ea | 530.00 | 129.00 | 37.30 | 696.30 |

**Manholes** Equipment includes a wheel-mounted 1 CY backhoe for lifting and placing the items. Trench excavation, bedding material and backfill are not included. Use $3,000 as a minimum job charge.

Precast concrete manholes, including concrete top and base. No excavation or backfill included. Add for steps, frames and lids from the costs listed below

| | Craft@Hrs | Unit | Material | Labor | Equipment | Total |
|---|---|---|---|---|---|---|
| 3' x 6' to 8' deep | S6@17.5 | Ea | 1,860.00 | 786.00 | 228.00 | 2,874.00 |
| 3' x 9' to 12' deep | S6@19.4 | Ea | 2,180.00 | 871.00 | 252.00 | 3,303.00 |
| 3' x 13' to 16' deep | S6@23.1 | Ea | 2,810.00 | 1,040.00 | 300.00 | 4,150.00 |
| 3' diameter, 4' deep | S6@12.1 | Ea | 1,100.00 | 544.00 | 257.00 | 1,901.00 |
| 4' diameter, 5' deep | S6@22.8 | Ea | 1,330.00 | 1,020.00 | 296.00 | 2,646.00 |
| 4' diameter, 6' deep | S6@26.3 | Ea | 1,500.00 | 1,180.00 | 342.00 | 3,022.00 |
| 4' diameter, 7' deep | S6@30.3 | Ea | 1,640.00 | 1,360.00 | 394.00 | 3,394.00 |
| 4' diameter, 8' deep | S6@33.7 | Ea | 1,830.00 | 1,510.00 | 438.00 | 3,778.00 |
| 4' diameter, 9' deep | S6@38.2 | Ea | 1,950.00 | 1,720.00 | 497.00 | 4,167.00 |
| 4' diameter, 10' deep | S6@43.3 | Ea | 2,120.00 | 1,950.00 | 563.00 | 4,633.00 |
| **Concrete block radial manholes, 4' inside diameter, no excavation or backfill included** | | | | | | |
| 4' deep | M1@8.01 | Ea | 444.00 | 373.00 | — | 817.00 |
| 6' deep | M1@13.4 | Ea | 653.00 | 625.00 | — | 1,278.00 |
| 8' deep | M1@20.0 | Ea | 939.00 | 932.00 | — | 1,871.00 |
| 10' deep | M1@26.7 | Ea | 966.00 | 1,240.00 | — | 2,206.00 |
| Depth over 10', add per LF | M1@2.29 | LF | 123.00 | 107.00 | — | 230.00 |
| 2' depth cone block for 30" grate | M1@5.34 | LF | 185.00 | 249.00 | — | 434.00 |
| 2'6" depth cone block for 24" grate | M1@6.43 | LF | 241.00 | 300.00 | — | 541.00 |
| **Manhole steps, cast iron, heavy type, asphalt coated** | | | | | | |
| 10" x 14-1/2" in job-built manhole | M1@.254 | Ea | 21.90 | 11.80 | — | 33.70 |
| 12" x 10" in precast manhole | — | Ea | 20.10 | — | — | 20.10 |
| **Gray iron manhole frames, asphalt coated, standard sizes** | | | | | | |
| Light-duty frame and lid | | | | | | |
| 15" diameter, 65 lb | S6@.660 | Ea | 114.00 | 29.60 | 8.58 | 152.18 |
| 22" diameter, 140 lb | S6@1.42 | Ea | 250.00 | 63.80 | 18.50 | 332.30 |
| Medium-duty frame and lid | | | | | | |
| 11" diameter, 75 lb | S6@1.51 | Ea | 133.00 | 67.80 | 19.60 | 220.40 |
| 20" diameter, 185 lb | S6@3.01 | Ea | 306.00 | 135.00 | 39.10 | 480.10 |
| Heavy-duty frame and lid | | | | | | |
| 17" diameter, 135 lb | S6@2.51 | Ea | 233.00 | 113.00 | 32.60 | 378.60 |
| 21" diameter, 315 lb | S6@3.77 | Ea | 518.00 | 169.00 | 49.00 | 736.00 |
| 24" diameter, 375 lb | S6@4.30 | Ea | 599.00 | 193.00 | 55.90 | 847.90 |

| | Craft@Hrs | Unit | Material | Labor | Equipment | Total |
|---|---|---|---|---|---|---|
| **Connect new drain line to existing manhole, no excavation or backfill included. Typical cost.** | | | | | | |
| Per connection | S6@6.83 | Ea | 181.00 | 307.00 | 88.80 | 576.80 |
| **Connect existing drain line to new manhole, no excavation or backfill included. Typical cost.** | | | | | | |
| Per connection | S6@3.54 | Ea | 110.00 | 159.00 | 46.00 | 315.00 |
| **Manhole repairs and alterations, typical costs** | | | | | | |
| Repair manhole leak with grout | S6@15.8 | Ea | 524.00 | 710.00 | 205.00 | 1,439.00 |
| Repair inlet leak with grout | S6@15.8 | Ea | 637.00 | 710.00 | 205.00 | 1,552.00 |
| Grout under manhole frame | S6@1.29 | Ea | 10.80 | 57.90 | 16.80 | 85.50 |
| Drill and grout pump setup cost | S6@17.4 | LS | 637.00 | 782.00 | 223.00 | 1,642.00 |
| Grout for pressure grouting | — | Gal | 10.70 | — | — | 10.70 |
| Replace brick in manhole wall | M1@8.79 | SF | 15.50 | 410.00 | — | 425.50 |
| Replace brick under manhole frame | M1@8.79 | LS | 51.80 | 410.00 | — | 461.80 |
| Raise existing frame and cover 2" | CL@4.76 | LS | 287.00 | 190.00 | — | 477.00 |
| Raise existing frame and cover more than 2", per each 1" added | CL@.250 | LS | 99.10 | 9.97 | — | 109.07 |
| **TV inspection of pipe interior, subcontract** | — | LF | — | — | — | 4.40 |
| **Grouting concrete pipe joints** | | | | | | |
| 6" to 30" diameter | S6@.315 | Ea | 26.10 | 14.10 | 4.10 | 44.30 |
| 33" to 60" diameter | S6@1.12 | Ea | 51.50 | 50.30 | 14.60 | 116.40 |
| 66" to 72" diameter | S6@2.05 | Ea | 104.00 | 92.10 | 26.70 | 222.80 |

**Accessories for Site Utilities** Equipment includes a wheel-mounted 1 CY backhoe for lifting and placing the items. Trench excavation, bedding material and backfill are not included. Costs estimated using this section may be combined with costs estimated from other site work utility sections for arriving at a minimum job charge.

| | Craft@Hrs | Unit | Material | Labor | Equipment | Total |
|---|---|---|---|---|---|---|
| **Curb inlets, gray iron, asphalt coated, heavy-duty frame, grate and curb box** | | | | | | |
| 20" x 11", 260 lb | S6@2.75 | Ea | 413.00 | 124.00 | 35.80 | 572.80 |
| 20" x 16.5", 300 lb | S6@3.04 | Ea | 449.00 | 137.00 | 39.50 | 625.50 |
| 20" x 17", 400 lb | S6@4.30 | Ea | 569.00 | 193.00 | 55.90 | 817.90 |
| 19" x 18", 500 lb | S6@5.06 | Ea | 681.00 | 227.00 | 65.80 | 973.80 |
| 30" x 17", 600 lb | S6@6.04 | Ea | 737.00 | 271.00 | 78.50 | 1,086.50 |
| **Gutter inlets, gray iron, asphalt coated, heavy-duty frame and grate** | | | | | | |
| 8" x 11.5", 85 lb | S6@.444 | Ea | 138.00 | 19.90 | 5.77 | 163.67 |
| 22" x 17", 260 lb, concave | S6@1.84 | Ea | 395.00 | 82.70 | 23.90 | 501.60 |
| 22.3" x 22.3", 475 lb | S6@2.87 | Ea | 428.00 | 129.00 | 37.30 | 594.30 |
| 29.8" x 17.8", 750 lb | S6@3.37 | Ea | 657.00 | 151.00 | 43.80 | 851.80 |
| **Trench inlets, ductile iron, light-duty frame and grate for pedestrian traffic** | | | | | | |
| 53.5" x 8.3", 100 lb | S6@.495 | Ea | 134.00 | 22.20 | 6.44 | 162.64 |
| **Trench inlets, gray iron, asphalt coated frame and grate or solid cover** | | | | | | |
| Light-duty, 1-1/4" | | | | | | |
| 8" wide grate | S6@.247 | LF | 56.90 | 11.10 | 3.21 | 71.21 |
| 12" wide grate | S6@.275 | LF | 82.20 | 12.40 | 3.58 | 98.18 |
| 8" wide solid cover | S6@.247 | LF | 63.10 | 11.10 | 3.21 | 77.41 |
| 12" wide solid cover | S6@.275 | LF | 90.10 | 12.40 | 3.58 | 106.08 |
| Heavy-duty, 1-3/4" | | | | | | |
| 8" wide grate | S6@.275 | LF | 61.80 | 12.40 | 3.58 | 77.78 |
| 12" wide grate | S6@.309 | LF | 91.90 | 13.90 | 4.02 | 109.82 |
| 8" wide solid cover | S6@.275 | LF | 71.20 | 12.40 | 3.58 | 87.18 |
| 12" wide solid cover | S6@.309 | LF | 98.30 | 13.90 | 4.02 | 116.22 |

|  | Craft@Hrs | Unit | Material | Labor | Equipment | Total |
|---|---|---|---|---|---|---|

**Gas Distribution Lines** Installed in an open trench. Equipment includes a wheel-mounted 1 CY backhoe for lifting and placing the pipe. No excavation, bedding material or backfill included. Use $2,000 as a minimum job charge.

Polyethylene pipe, 60 PSI

|  | Craft@Hrs | Unit | Material | Labor | Equipment | Total |
|---|---|---|---|---|---|---|
| 1-1/4" diameter, coils | U1@.064 | LF | .93 | 3.13 | .62 | 4.68 |
| 1-1/2" diameter, coils | U1@.064 | LF | 1.30 | 3.13 | .62 | 5.05 |
| 2" diameter, coils | U1@.071 | LF | 1.67 | 3.47 | .69 | 5.83 |
| 3" diameter, coils | U1@.086 | LF | 3.30 | 4.20 | .84 | 8.34 |
| 3" diameter, 38' long with couplings | U1@.133 | LF | 2.74 | 6.49 | 1.30 | 10.53 |
| 4" diameter, 38' long with couplings | U1@.168 | LF | 5.51 | 8.20 | 1.64 | 15.35 |
| 6" diameter, 38' long with couplings | U1@.182 | LF | 11.80 | 8.89 | 1.77 | 22.46 |
| 8" diameter, 38' long with couplings | U1@.218 | LF | 21.30 | 10.60 | 2.13 | 34.03 |

**Meters and pressure regulators** Natural gas diaphragm meters, direct digital reading, not including temperature and pressure compensation. For use with pressure regulators below.

|  | Craft@Hrs | Unit | Material | Labor | Equipment | Total |
|---|---|---|---|---|---|---|
| 425 CFH at 10 PSI | P6@4.38 | Ea | 291.00 | 220.00 | — | 511.00 |
| 425 CFH at 25 PSI | P6@4.38 | Ea | 385.00 | 220.00 | — | 605.00 |

Natural gas pressure regulators with screwed connections

|  | Craft@Hrs | Unit | Material | Labor | Equipment | Total |
|---|---|---|---|---|---|---|
| 3/4" or 1" connection | P6@.772 | Ea | 29.80 | 38.80 | — | 68.60 |
| 1-1/4" or 1-1/2" connection | P6@1.04 | Ea | 157.00 | 52.20 | — | 209.20 |
| 2" connection | P6@1.55 | Ea | 1,290.00 | 77.80 | — | 1,367.80 |

**Pipe Jacking** Typical costs for jacking .50" thick wall pipe casing under an existing roadway. Costs include leaving casing in place. Add 15% when ground water is present. Add 100% for light rock conditions. Includes jacking pits on both sides. Equipment includes a 1 CY wheel-mounted backhoe plus a 2-ton truck equipped for this type of work. Size shown is casing diameter. Use $2,200 as a minimum job charge.

|  | Craft@Hrs | Unit | Material | Labor | Equipment | Total |
|---|---|---|---|---|---|---|
| 2" casing | C5@.289 | LF | 7.52 | 13.00 | 4.66 | 25.18 |
| 3" casing | C5@.351 | LF | 8.89 | 15.70 | 5.66 | 30.25 |
| 4" casing | C5@.477 | LF | 11.70 | 21.40 | 7.69 | 40.79 |
| 6" casing | C5@.623 | LF | 15.00 | 27.90 | 10.10 | 53.00 |
| 8" casing | C5@.847 | LF | 19.80 | 38.00 | 13.70 | 71.50 |
| 10" casing | C5@1.24 | LF | 29.30 | 55.60 | 20.00 | 104.90 |
| 12" casing | C5@1.49 | LF | 35.60 | 66.80 | 24.00 | 126.40 |
| 16" casing | C5@1.78 | LF | 42.30 | 79.80 | 28.70 | 150.80 |
| 17" casing | C5@2.09 | LF | 48.50 | 93.70 | 33.70 | 175.90 |
| 24" casing | C5@7.27 | LF | 171.00 | 326.00 | 117.00 | 614.00 |
| 30" casing | C5@7.88 | LF | 186.00 | 353.00 | 127.00 | 666.00 |
| 36" casing | C5@8.61 | LF | 205.00 | 386.00 | 139.00 | 730.00 |
| 42" casing | C5@9.29 | LF | 218.00 | 416.00 | 150.00 | 784.00 |
| 48" casing | C5@10.2 | LF | 240.00 | 457.00 | 165.00 | 862.00 |

# Index

**647**

**654**

**664**

**665**

# Practical References for Builders

## Steel-Frame House Construction

Framing with steel has obvious advantages over wood, yet building with steel requires new skills that can present challenges to the wood builder. This book explains the secrets of steel framing techniques for building homes, whether pre-engineered or built stick by stick. It shows you the techniques, the tools, the materials, and how you can make it happen. Includes hundreds of photos and illustrations, plus a FREE download with steel framing details and a database of steel materials and manhours, with an estimating program. **320 pages, 8½ x 11, $39.75**

## National Painting Cost Estimator

A complete guide to estimating painting costs for just about any type of residential, commercial, or industrial painting, whether by brush, spray, or roller. Shows typical costs and bid prices for fast, medium, and slow work, including material costs per gallon; square feet covered per gallon; square feet covered per manhour; labor, material, overhead, and taxes per 100 square feet; and how much to add for profit. Includes a CD-ROM with an electronic version of the book with *National Estimator*, a stand-alone *Windows*™ estimating program, plus an interactive multimedia video that shows how to use the disk to compile construction cost estimates.
**448 pages, 8½ x 11, $63.00. Revised annually**

## Basic Lumber Engineering for Builders

Beam and lumber requirements for many jobs aren't always clear, especially with changing building codes and lumber products. Most of the time you rely on your own "rules of thumb" when figuring spans or lumber engineering. This book can help you fill the gap between what you can find in the building code span tables and what you need to pay a certified engineer to do. With its large, clear illustrations and examples, this book shows you how to figure stresses for pre-engineered wood or wood structural members, how to calculate loads, and how to design your own girders, joists and beams. Included FREE with the book — an easy-to-use limited version of NorthBridge Software's *Wood Beam Sizing* program.
**272 pages, 8½ x 11, $38.00**

## Contractor's Guide to *QuickBooks Pro* 2010

This user-friendly manual walks you through *QuickBooks Pro*'s detailed setup procedure and explains step-by-step how to create a first-rate accounting system. You'll learn in days, rather than weeks, how to use *QuickBooks Pro* to get your contracting business organized, with simple, fast accounting procedures. On the CD included with the book you'll find a *QuickBooks Pro* file for a construction company. Open it, enter your own company's data, and add info on your suppliers and subs. You also get a complete estimating program, including a database, and a job costing program that lets you export your estimates to *QuickBooks Pro*. It even includes many useful construction forms to use in your business.
**344 pages, 8½ x 11, $57.00** *(See Checklist for additional versions.)*

## Plumber's Handbook Revised

This new edition shows what will and won't pass inspection in drainage, vent, and waste piping, septic tanks, water supply, graywater recycling systems, pools and spas, fire protection, and gas piping systems. All tables, standards, and specifications are completely up-to-date with recent plumbing code changes. Covers common layouts for residential work, how to size piping, select and hang fixtures, practical recommendations, and trade tips. It's the approved reference for the plumbing contractor's exam in many states. Includes an extensive set of multiple-choice questions after each chapter, with answers and explanations in the back of the book, along with a complete sample plumber's exam. **352 pages, 8½ x 11, $41.50**

## Finish Carpentry: Efficient Techniques for Custom Interiors

Professional finish carpentry demands expert skills, precise tools, and a solid understanding of how to do the work. This book explains how to install moldings, paneled walls and ceilings, and just about every aspect of interior trim — including doors and windows. Covers built-in bookshelves, coffered ceilings, and skylight wells and soffits, including paneled ceilings with decorative beams. **276 pages, 8½ x 11, $34.95**

## Construction Estimating Reference Data

Provides the 300 most useful manhour tables for practically every item of construction. Labor requirements are listed for sitework, concrete work, masonry, steel, carpentry, thermal and moisture protection, doors and windows, finishes, mechanical and electrical. Each section details the work being estimated and gives appropriate crew size and equipment needed. Includes a CD-ROM with an electronic version of the book with *National Estimator*, a stand-alone *Windows*™ estimating program, plus an inter-active multi-media video that shows how to use the disk to compile construction cost estimates. **432 pages, 11 x 8½, $39.50**

## CD Estimator

If your computer has *Windows*™ and a CD-ROM drive, CD Estimator puts at your fingertips over 150,000 construction costs for new construction, remodeling, renovation & insurance repair, home improvement, framing & finish carpentry, electrical, concrete & masonry, painting, earthwork & heavy equipment and plumbing & HVAC. Quarterly cost updates are available at no charge on the Internet. You'll also have the *National Estimator* program — a stand-alone estimating program for *Windows*™ that *Remodeling* magazine called a "computer wiz," and *Job Cost Wizard*, a program that lets you export your estimates to *QuickBooks Pro* for actual job costing. A 60-minute interactive video teaches you how to use this CD-ROM to estimate construction costs. And to top it off, to help you create professional-looking estimates, the disk includes over 40 construction estimating and bidding forms in a format that's perfect for nearly any *Windows*™ word processing or spreadsheet program.
**CD Estimator is $108.50**

## Construction Forms for Contractors

This practical guide contains 78 practical forms, letters and checklists, guaranteed to help you streamline your office, organize your jobsites, gather and organize records and documents, keep a handle on your subs, reduce estimating errors, administer change orders and lien issues, monitor crew productivity, track your equipment use, and more. Includes accounting forms, change order forms, forms for customers, estimating forms, field work forms, HR forms, lien forms, office forms, bids and proposals, subcontracts, and more. All are also on the CD-ROM included, in *Excel* spreadsheets, as formatted Rich Text that you can fill out on your computer, and as PDFs.
**360 pages, 8½ x 11, $48.50**

## Excavation & Grading Handbook Revised

The foreman's, superintendent's and operator's guide to highway, subdivision and pipeline jobs: how to read plans and survey stake markings, set grade, excavate, compact, pave and lay pipe on nearly any job. Includes hundreds of informative, on-the-job photos and diagrams that even experienced pros will find invaluable. This new edition has been completely revised to be current with state-of-the-art equipment usage and the most efficient excavating and grading techniques. You'll learn how to read topo maps, use a laser level, set crows feet, cut drainage channels, lay or remove asphaltic concrete, and use GPS and sonar for absolute precision. For those in training, each chapter has a set of self-test questions, and a Study Center CD-ROM included has all 250 questions in a simple interactive format to make learning easy and fun. **512 pages, 8½ x 11, $42.00**

## Construction Contract Writer

Relying on a "one-size-fits-all" boilerplate construction contract to fit your jobs can be dangerous — almost as dangerous as a handshake agreement. *Construction Contract Writer* lets you draft a contract in minutes that precisely fits your needs and the particular job, and meets both state and federal requirements. You just answer a series of questions — like an interview — to construct a legal contract for each project you take on. Anticipate where disputes could arise and settle them in the contract before they happen. Include the warranty protection you intend, the payment schedule, and create subcontracts from the prime contract by just clicking a box. Includes a feedback button to an attorney on the Craftsman staff to help should you get stumped — *No extra charge.*
**$99.95**. Download the *Construction Contract Writer* at: http://www.constructioncontractwriter.com

## Builder's Guide to Drainage & Retaining Walls

Creating an adequate drainage system is the critical first step in most construction projects. Retaining walls are integral to the design plan, when used to achieve level grade. These two topics are interrelated, and this book provides a complete introduction to doing both right. Here you'll find the data you need on average rainfall and runoff requirements so you can figure gutters, downspouts and roof drainage, and install appropriate waterproofing and dampproofing in basements and crawlspaces. The book gives details for determining slope drainage, doing the grading, complying with street drainage requirements, and dealing with lots with septic systems. Many types of retaining wall are covered, including modular unit walls, bin and crib walls, and more. Also provides tables for estimating concrete quantities for both poured-in-place and masonry and concrete walls. Includes detailed recommendations for retaining walls in most of the situations you're likely to encounter. Includes a CD-ROM that brings you the entire book in an Adobe PDF file, for jobsite reference and quick word search. (This file has no print capability.) A second PDF file includes 100 sample details you can print and carry along on jobs. Published by Builder's Book Inc. **294 pages, 8½ x 11, $59.95**

## Standard Estimating Practice

Estimating isn't always an easy job. Sometimes snap decisions can produce negative long-term effects. This book was designed by the American Society of Professional Estimators as a set of standards to guide professional estimators. It's intended to help every estimator develop estimates that are uniform and verifiable. Every step that should be included in the estimate is listed, as well as aspects in the plans to consider when you're estimating a job, and what you should look for that may not be included. The result should help you produce more consistently accurate estimates. **506 pages, 8½ x 11, $89.00**

## Construction Estimating

This unusually well-organized book shows the best and easiest way to estimate materials for room additions or residential structures. It gives estimating tables and procedures needed to make a fast, accurate, and complete material list of the structural members found in wood- and steel-framed buildings. This book is divided into 72 units, each of them covering a separate element in the estimating procedure. Covers estimating foundations, floor framing, wall framing, ceiling framing, roof framing, roofing materials, exterior and interior finish materials, hardware, steel joist floor framing, steel stud framing, and steel ceiling joist and rafter framing. **496 pages, 8½ x 11, $49.50**

## Roofing Construction & Estimating

Installation, repair and estimating for nearly every type of roof covering available today in residential and commercial structures: asphalt shingles, roll roofing, wood shingles and shakes, clay tile, slate, metal, built-up, and elastomeric. Covers sheathing and underlayment techniques, as well as secrets for installing leakproof valleys. Many estimating tips help you minimize waste, as well as insure a profit on every job. Troubleshooting techniques help you identify the true source of most leaks. Over 300 large, clear illustrations help you find the answer to just about all your roofing questions. **432 pages, 8½ x 11, $38.00**

## Craftsman's Construction Installation Encyclopedia

Step-by-step installation instructions for just about any residential construction, remodeling or repair task, arranged alphabetically, from *Acoustic tile* to *Wood flooring*. Includes hundreds of illustrations that show how to build, install, or remodel each part of the job, as well as manhour tables for each work item so you can estimate and bid with confidence. Also includes a CD-ROM with all the material in the book, handy look-up features, and the ability to capture and print out for your crew the instructions and diagrams for any job. **792 pages, 8½ x 11, $65.00**

## Estimating With Microsoft *Excel* (3rd Edition)

Step-by-step instructions show you how to create your own customized automated spreadsheet estimating program for use with *Excel* 2007. You'll learn how to use the magic of *Excel* to create all the forms you need; detail sheets, cost breakdown summaries, and more. With *Excel* as your tool, you can easily estimate costs for all phases of the job, from pulling permits, to concrete, rebar, and roofing. You'll see how to create your own formulas and macros and apply them in your everyday projects. If you've wanted to use *Excel*, but were unsure of how to make use of all its features, let this new book show you how. Includes a CD-ROM that illustrates examples in the book and provides you with templates you can use to set up your own estimating system. **158 pages, 7 x 9½, $44.95**

## Estimating Excavation

How to calculate the amount of dirt you'll have to move and the cost of owning and operating the machines you'll do it with. Detailed, step-by-step instructions on how to assign bid prices to each part of the job, including labor and equipment costs. Also, the best ways to set up an organized and logical estimating system, take off from contour maps, estimate quantities in irregular areas, and figure your overhead. **448 pages, 8½ x 11, $39.50**

## Concrete Construction

Just when you think you know all there is about concrete, many new innovations create faster, more efficient ways to do the work. This comprehensive concrete manual has both the tried-and-tested methods and materials, and more recent innovations. It covers everything you need to know about concrete, along with Styrofoam forming systems, fiber reinforcing adjuncts, and some architectural innovations, like architectural foam elements, that can help you offer more in the jobs you bid on. Every chapter provides detailed, step-by-step instructions for each task, with hundreds of photographs and drawings that show exactly how the work is done. To keep your jobs organized, there are checklists for each stage of the concrete work, from planning, to finishing and protecting your pours. Whether you're doing residential or commercial work, this manual has the instructions, illustrations, charts, estimating data, rules of thumb and examples every contractor can apply on their concrete jobs. **288 pages, 8½ x 11, $28.75**

## National Building Cost Manual

Square-foot costs for residential, commercial, industrial, military, schools, greenhouses, churches and farm buildings. Includes important variables that can make any building unique from a cost standpoint. Quickly work up a reliable budget estimate based on actual materials and design features, area, shape, wall height, number of floors, and support requirements. Now includes easy-to-use software that calculates total in-place cost estimates. Use the regional cost adjustment factors provided to tailor the estimate to any jobsite in the U.S. Then view, print, email or save the detailed PDF report as needed. **264 pages, 8½ x 11, $53.00. Revised annually**

## National Electrical Estimator

This year's prices for installation of all common electrical work: conduit, wire, boxes, fixtures, switches, outlets, loadcenters, panelboards, raceway, duct, signal systems, and more. Provides material costs, manhours per unit, and total installed cost. Explains what you should know to estimate each part of an electrical system. Includes a CD-ROM with an electronic version of the book with *National Estimator*, a stand-alone *Windows*™ estimating program, plus an interactive multimedia video that shows how to use the disk to compile construction cost estimates. **552 pages, 8½ x 11, $62.75. Revised annually**

## Estimating Home Building Costs, Revised

Estimate every phase of residential construction from site costs to the profit margin you include in your bid. Shows how to keep track of manhours and make accurate labor cost estimates for site clearing and excavation, footings, foundations, framing and sheathing finishes, electrical, plumbing, and more. Provides and explains sample cost estimate worksheets with complete instructions for each job phase. This practical guide to estimating home construction costs has been updated with digital *Excel* estimating forms and worksheets that ensure accurate and complete estimates for your residential projects. Enter your project information on the worksheets and *Excel* automatically totals each material and labor cost from every stage of construction to a final cost estimate worksheet. Load the enclosed CD-ROM into your computer and create your own estimate as you follow along with the step-by-step techniques in this book. **336 pages, 8½ x 11, $38.00**

## Estimating & Bidding for Builders & Remodelers

This 4th edition has all the information you need for estimating and bidding new construction and home improvement projects. It shows how to select jobs that will be profitable, do a labor and materials take-off from the plans, calculate overhead and figure your markup, and schedule the work. Includes a CD with an easy-to-use construction estimating program and a database of 50,000 current labor and material cost estimates for new construction and home improvement work, with area modifiers for every zip code. Price updates on the Web are free and automatic. **272 pages, 8½ x 11, $89.50**

## Residential Construction Performance Guidelines, 4th Ed.

Created and reviewed by more than 300 builders and remodelers, this guide gives cut-and-dried construction standards that should apply to new construction and remodeling. It defines corrective action necessary to bring all construction up to standards. Standards are listed for sitework, foundations, interior concrete slabs, basement and crawl spaces for block walls and poured walls, wood-floor framing, beams, columns and posts, plywood and joists, walls, wall insulation, windows, doors, exterior finishes and trim, roofs, roof sheathing, roof installation and leaks, plumbing, sanitary and sewer systems, electrical, interior climate control, HVAC systems, cabinets and countertops, floor finishes and more.
**120 pages, 6½ x 8½, $44.95. Published by NAHB Remodelers Council**

## Markup & Profit: A Contractor's Guide

In order to succeed in a construction business, you have to be able to price your jobs to cover all labor, material and overhead expenses, *and* make a decent profit. The problem is knowing what markup to use. You don't want to lose jobs because you charged too much, and you don't want to work for free because you charged too little. If you know how to calculate markup, you can apply it to your job costs to find the right sales price for your work. This book gives you tried and tested formulas, with step-by-step instructions and easy-to-follow examples, so you can easily figure the markup that's right for *your* business. Includes a CD-ROM with forms and checklists for your use. **320 pages, 8½ x 11, $32.50**

## National Home Improvement Estimator

Current labor and material prices for home improvement projects. Provides manhours for each job, recommended crew size, and the labor cost for removal and installation work. Material prices are current, with location adjustment factors and free monthly updates on the Web. Gives step-by-step instructions for the work, with helpful diagrams, and home improvement shortcuts and tips from experts. Includes a CD-ROM with an electronic version of the book, and *National Estimator*, a stand-alone *Windows*™ estimating program, plus an interactive multimedia tutorial that shows how to use the disk to compile home improvement cost estimates. **520 pages, 8½ x 11, $63.75. Revised annually**

## Building Code Compliance for Contractors & Inspectors

An answer book for both contractors and building inspectors, this manual explains what it takes to pass inspections under the 2009 *International Residential Code*. It includes a code checklist for every trade, covering some of the most common reasons why inspectors reject residential work – footings, foundations, slabs, framing, sheathing, plumbing, electrical, HVAC, energy conservation and final inspection. The requirement for each item on the checklist is explained, and the code section cited so you can look it up or show it to the inspector. Knowing in advance what the inspector wants to see gives you an (almost unfair) advantage. To pass inspection, do your own pre-inspection before the inspector arrives. If your work requires getting permits and passing inspections, put this manual to work on your next job. If you're considering a career in code enforcement, this can be your guidebook. **8½ x 11, 232 pages, $32.50**

## Residential Wiring to the 2008 *NEC*

This completely revised manual explains in simple terms how to install rough and finish wiring in new construction, alterations, and additions. It takes you from basic electrical theory to current wiring methods that comply with the 2008 *National Electrical Code*. You'll find complete instructions on troubleshooting and repairs of existing wiring, and how to extend service into additions and remodels. Hundreds of drawings and photos show you the tools and gauges you need, and how to plan and install the wiring. Includes demand factors, circuit loads, the formulas you need, and over 20 pages of the most-needed 2008 *NEC* tables to help your wiring pass inspection the first time. Includes a CD-ROM with an Interactive Study Center that helps you retain what you've learned, and study for the electrician's exam. Also on the CD is the entire book in PDF format, with easy search features so you can quickly find answers to your residential wiring questions. **304 pages, 8½ x 11, $42.00**

---

**Download all of Craftsman's most popular costbooks for one low price with the Craftsman Site License:**
**http://www.craftsmansitelicense.com**

---